Intermediate Algebra

6TH EDITION

Intermediate Algebra

6TH EDITION

Mervin L. Keedy

Purdue University

Marvin L. Bittinger

Indiana University—Purdue University at Indianapolis

ADDISON-WESLEY PUBLISHING COMPANY

Reading, Massachusetts • Menlo Park, California • New York
Don Mills, Ontario • Wokingham, England • Amsterdam • Bonn
Sydney • Singapore • Tokyo • Madrid • San Juan

Sponsoring Editor	Elizabeth Burr
Managing Editor	Karen Guardino
Production Supervisor	Jack Casteel
Design, Editorial, and Production Services	Quadrata, Inc.
Illustrator	ST Associates, Inc., and Scientific Illustrators
Art Consultant	Loretta Bailey
Manufacturing Supervisor	Roy Logan
Cover Design	Marshall Henrichs
Cover Photograph	Eric Fordham

PHOTO CREDITS

1, UPI/Bettmann Newsphotos **61,** © Tim Davis 1982, Photo Researchers, Inc. **127,** Eva Demjen/Stock, Boston **175,** Coco McCoy, Rainbow Pictures **235,** *PSSC Physics*, 2nd ed., 1965; D.C. Heath and Company with Education Development Center, Inc., Newton MA. **309,** Delta Queen **373,** 1988 Mejrejen Ferguson, PhotoEdit **437,** Barbara Rios, Photo Researchers, Inc. **511,** Cary Wolinsky/Stock, Boston **541,** unknown **575,** NASA **626 (left),** AP/Wide World Photos **626 (right),** Rick Haston

Library of Congress Cataloging-in-Publication Data

Keedy, Mervin Laverne.
Intermediate algebra/Mervin L. Keedy, Marvin L. Bittinger.—6th ed.
p. cm.
ISBN 0-201-19695-6
1. Algebra. I. Bittinger, Marvin L. II. Title.
QA154.2.K434 1991
512.9—dc20
90-30808
CIP

Reprinted with corrections, May 1991

2 3 4 5 6 7 8 9 10 - DO - 95 94 93 92 91

Contents

3 Graphs of Equations and Inequalities 127

4 Systems of Equations and Inequalities 175

5 Polynomials 235

6 Rational Expressions and Equations 309

7 Radical Expressions and Equations 373

8 Quadratic Equations 437

9 Conic Sections, Relations, and Functions 511

10 Exponential and Logarithmic Functions 575

Preface

Intended for students who have a firm background in introductory algebra, this text is appropriate for a one-term course in intermediate algebra. It is the third in a series of texts that includes the following:

Keedy/Bittinger: *Basic Mathematics*, Sixth Edition,
Keedy/Bittinger: *Introductory Algebra*, Sixth Edition,
Keedy/Bittinger: *Intermediate Algebra*, Sixth Edition.

Intermediate Algebra, Sixth Edition, provides the necessary preparation for any introductory college-level mathematics course, including courses in college algebra, precalculus, finite mathematics, or brief calculus.

Intermediate Algebra, Sixth Edition, is a significant revision of the Fifth Edition, with respect to content, pedagogy, and an expanded supplements package. Its unique approach, which has been developed over many years, is designed to help today's students both learn *and* retain mathematical concepts. The Sixth Edition is accompanied by a comprehensive supplements package that has been integrated with the text to provide maximum support for both instructor and student.

Following are some distinctive features of the approach and pedagogy that we feel will help meet some of the challenges all instructors face teaching developmental mathematics.

APPROACH

CAREFUL DEVELOPMENT OF CONCEPTS We have divided each section into discrete and manageable learning objectives. Within the presentation of each objective, there is a careful buildup of difficulty through a series of developmental and followup examples. These enable students to thoroughly understand the mathematical concepts involved at each step. Each objective is constructed in a similar way, which gives students a high level of comfort with both the text and their learning process.

FOCUS ON "WHY" Throughout the text, we present the appropriate mathematical rationale for a topic, rather than mathematical "shortcuts." For example, when manipulating rational expressions, we remove factors of 1 rather than cancel, although cancellation is mentioned with appropriate cautions. This helps prevent student errors in cancellation and other incorrectly remembered shortcuts in later courses.

PROBLEM SOLVING We include real-life applications and problem-solving techniques throughout the text to motivate students and encourage them to think about how mathematics can be used. We also introduce a five-step problem-solving process early in the text and use the basic steps of this process (Familiarize, Translate, Solve, Check, and State the Answer) whenever a problem is solved.

GRAPHING Although introduced in Chapter 2, graphing is integrated throughout the remainder of the text to provide a visual interpretation of different types of equations and their solutions. This gives students a better intuitive understanding of the material. In addition, familiarity and practice with graphing techniques make students more comfortable with this essential tool when they move on to later courses.

PEDAGOGY

INTERACTIVE WORKTEXT APPROACH The pedagogy of this text is designed to provide students with a clear set of learning objectives, and involve them with the development of the material, providing immediate and continual reinforcement.

Section objectives are keyed to appropriate sections of the text, exercises, and answers, so that students can easily find appropriate review material if they are unable to do an exercise.

Numerous *margin exercises* throughout the text provide immediate reinforcement of concepts covered in each section.

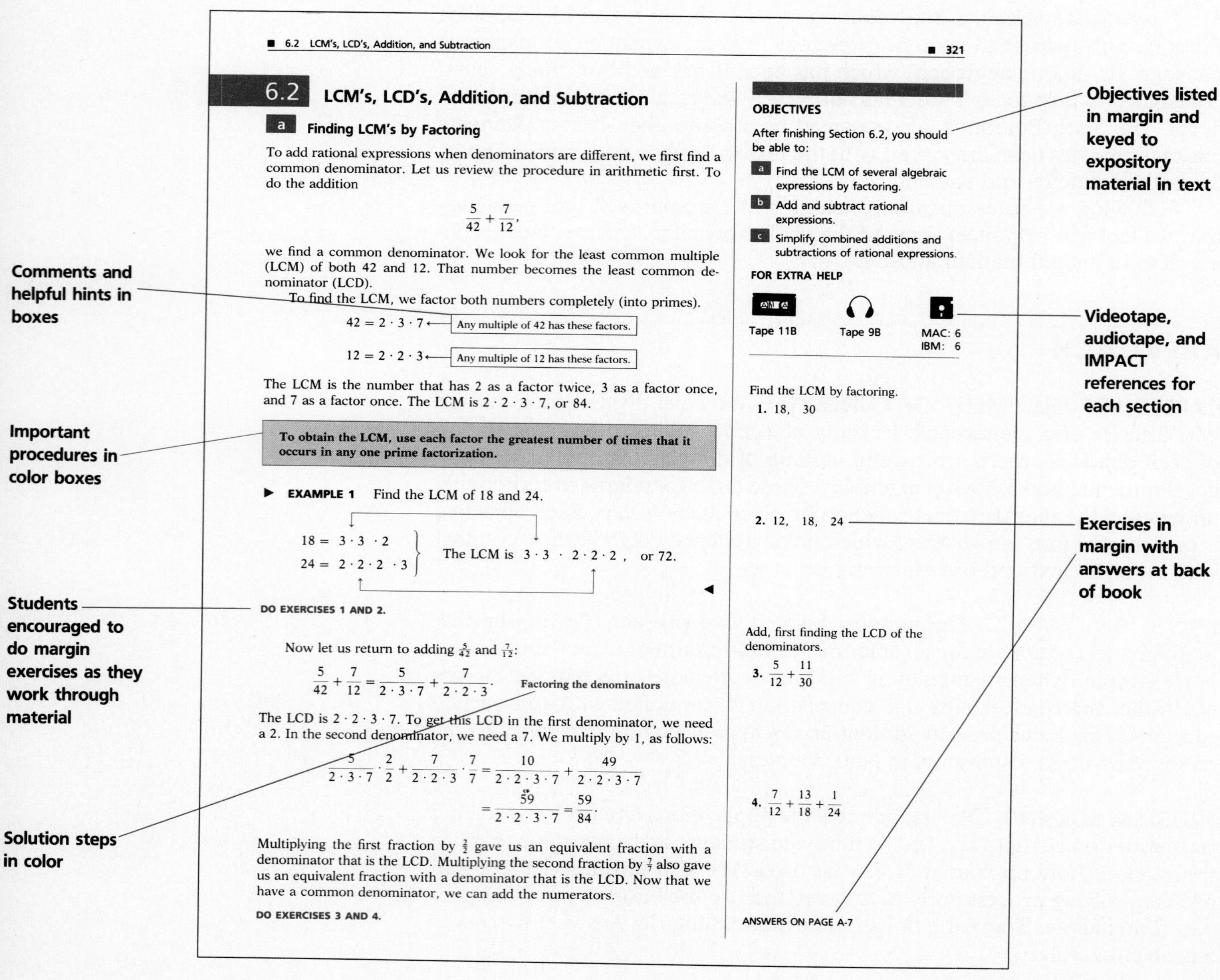

■ 6.2 LCM's, LCD's, Addition, and Subtraction ■ 321

6.2 LCM's, LCD's, Addition, and Subtraction

a Finding LCM's by Factoring

To add rational expressions when denominators are different, we first find a common denominator. Let us review the procedure in arithmetic first. To do the addition

$$\frac{5}{42}+\frac{7}{12},$$

we find a common denominator. We look for the least common multiple (LCM) of both 42 and 12. That number becomes the least common denominator (LCD).

To find the LCM, we factor both numbers completely (into primes).

$42 = 2 \cdot 3 \cdot 7$ ← Any multiple of 42 has these factors.

$12 = 2 \cdot 2 \cdot 3$ ← Any multiple of 12 has these factors.

The LCM is the number that has 2 as a factor twice, 3 as a factor once, and 7 as a factor once. The LCM is $2 \cdot 2 \cdot 3 \cdot 7$, or 84.

To obtain the LCM, use each factor the greatest number of times that it occurs in any one prime factorization.

► **EXAMPLE 1** Find the LCM of 18 and 24.

$18 = 3 \cdot 3 \cdot 2$
$24 = 2 \cdot 2 \cdot 2 \cdot 3$ } The LCM is $3 \cdot 3 \cdot 2 \cdot 2 \cdot 2$, or 72. ◄

DO EXERCISES 1 AND 2.

Now let us return to adding $\frac{5}{42}$ and $\frac{7}{12}$:

$$\frac{5}{42}+\frac{7}{12}=\frac{5}{2\cdot3\cdot7}+\frac{7}{2\cdot2\cdot3}.$$ Factoring the denominators

The LCD is $2 \cdot 2 \cdot 3 \cdot 7$. To get this LCD in the first denominator, we need a 2. In the second denominator, we need a 7. We multiply by 1, as follows:

$$\frac{5}{2\cdot3\cdot7}\cdot\frac{2}{2}+\frac{7}{2\cdot2\cdot3}\cdot\frac{7}{7}=\frac{10}{2\cdot2\cdot3\cdot7}+\frac{49}{2\cdot2\cdot3\cdot7}$$
$$=\frac{59}{2\cdot2\cdot3\cdot7}=\frac{59}{84}.$$

Multiplying the first fraction by $\frac{2}{2}$ gave us an equivalent fraction with a denominator that is the LCD. Multiplying the second fraction by $\frac{7}{7}$ also gave us an equivalent fraction with a denominator that is the LCD. Now that we have a common denominator, we can add the numerators.

DO EXERCISES 3 AND 4.

OBJECTIVES

After finishing Section 6.2, you should be able to:

a Find the LCM of several algebraic expressions by factoring.
b Add and subtract rational expressions.
c Simplify combined additions and subtractions of rational expressions.

FOR EXTRA HELP

Tape 11B | Tape 9B | MAC: 6 IBM: 6

Find the LCM by factoring.

1. 18, 30

2. 12, 18, 24

Add, first finding the LCD of the denominators.

3. $\frac{5}{12}+\frac{11}{30}$

4. $\frac{7}{12}+\frac{13}{18}+\frac{1}{24}$

ANSWERS ON PAGE A-7

Exercise sets on tearout sheets for each section

Exercises keyed to objectives and material in text

Exercise Set 6.2 ■ 327

NAME SECTION DATE

EXERCISE SET 6.2

a Find the LCM by factoring.

1. 12, 18 **2.** 15, 20 **3.** 18, 48 **4.** 45, 54

5. 24, 36 **6.** 30, 75 **7.** 9, 15, 5 **8.** 27, 35, 63

ANSWERS

1.

2.

3.

4.

5.

Answer space provided for quick and easy grading

STUDY AID REFERENCES Many valuable study aids accompany this text. Each section is referenced to appropriate videotape, audiotape, and software diskette numbers to make it easy for students to find and use the correct support materials.

VERBALIZATION SKILLS AND "THINKING IT THROUGH" Students' perception that mathematics is a foreign language is a significant barrier to their ability to think mathematically and is a major cause of math anxiety. In the Sixth Edition we have encouraged students to think through mathematical situations, synthesize concepts, and verbalize mathematics whenever possible.

"*Thinking it Through*" exercises at the end of each chapter encourage students to both think and write about key mathematical ideas that they have encountered in the chapter.

"*Synthesis Exercises*" at the end of most exercise sets require students to synthesize several learning objectives or to think through and provide insight into the present material.

In addition, many important definitions, such as the laws of exponents, are presented verbally as well as symbolically, to help students learn to read mathematical notation.

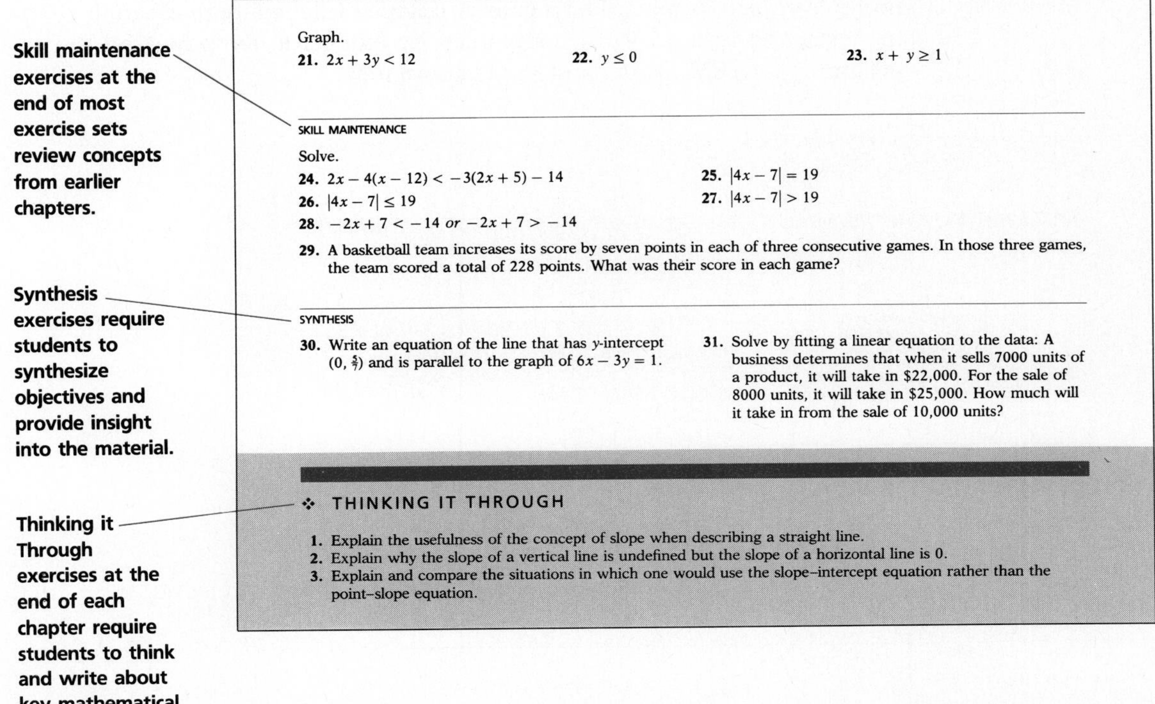

Skill maintenance exercises at the end of most exercise sets review concepts from earlier chapters.

Synthesis exercises require students to synthesize objectives and provide insight into the material.

Thinking it Through exercises at the end of each chapter require students to think and write about key mathematical ideas.

Graph.

21. $2x + 3y < 12$ **22.** $y \leq 0$ **23.** $x + y \geq 1$

SKILL MAINTENANCE

Solve.

24. $2x - 4(x - 12) < -3(2x + 5) - 14$

25. $|4x - 7| = 19$

26. $|4x - 7| \leq 19$

27. $|4x - 7| > 19$

28. $-2x + 7 < -14$ *or* $-2x + 7 > -14$

29. A basketball team increases its score by seven points in each of three consecutive games. In those three games, the team scored a total of 228 points. What was their score in each game?

SYNTHESIS

30. Write an equation of the line that has y-intercept $(0, \frac{5}{7})$ and is parallel to the graph of $6x - 3y = 1$.

31. Solve by fitting a linear equation to the data: A business determines that when it sells 7000 units of a product, it will take in $22,000. For the sale of 8000 units, it will take in $25,000. How much will it take in from the sale of 10,000 units?

THINKING IT THROUGH

1. Explain the usefulness of the concept of slope when describing a straight line.
2. Explain why the slope of a vertical line is undefined but the slope of a horizontal line is 0.
3. Explain and compare the situations in which one would use the slope–intercept equation rather than the point–slope equation.

SKILL MAINTENANCE Because retention of skills is critical to students' future success, skill maintenance is a major emphasis of the Sixth Edition.

Each chapter begins with a "*Points to Remember*" box, which highlights key formulas and definitions from previous chapters.

In addition, we include *Skill Maintenance Exercises* at the end of most exercise sets. These review skills and concepts from earlier sections of the text.

At the end of each chapter, our *Summary and Review* summarizes important properties and formulas and includes extensive review exercises.

Each *Chapter Test* tests four review objectives from preceding chapters as well as the chapter objectives.

We also include a *Cumulative Review* at the end of each chapter; this reviews material from all preceding chapters.

At the back of the text are answers to all review exercises, together with section and objective references, so that students know exactly what material to restudy if they miss a review exercise.

TESTING AND SKILL ASSESSMENT Accurate assessment of student comprehension is an important factor in a student's long-term success. In the Sixth Edition, we have provided many assessment opportunities.

A *Diagnostic Pretest* at the beginning of the text can place students in the appropriate chapter for their skill level, and identifies both familiar material and specific trouble areas later in the text.

Chapter Pretests diagnose student skills and place the students appropriately within each chapter, allowing them to concentrate on topics with which they have particular difficulty.

Chapter Tests at the end of each chapter allow students to review and test comprehension of chapter skills.

Answers to each question on all tests are included at the back of the text.

For additional testing options, we have developed a printed test bank with many alternative forms of each chapter test in both open-ended and multiple-choice formats. For a greater degree of flexibility in creating chapter tests, the text is also accompanied by extensive computerized testing programs for IBM, MAC, and Apple computers.

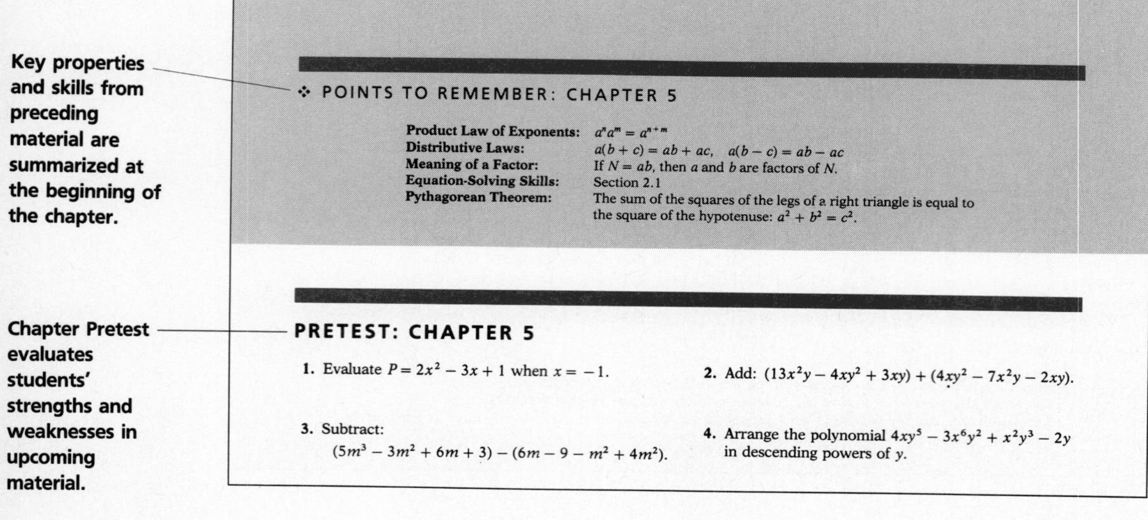

❖ POINTS TO REMEMBER: CHAPTER 5

Product Law of Exponents: $a^n a^m = a^{n+m}$
Distributive Laws: $a(b + c) = ab + ac$, $a(b - c) = ab - ac$
Meaning of a Factor: If $N = ab$, then a and b are factors of N.
Equation-Solving Skills: Section 2.1
Pythagorean Theorem: The sum of the squares of the legs of a right triangle is equal to the square of the hypotenuse: $a^2 + b^2 = c^2$.

PRETEST: CHAPTER 5

1. Evaluate $P = 2x^2 - 3x + 1$ when $x = -1$.
2. Add: $(13x^2y - 4xy^2 + 3xy) + (4xy^2 - 7x^2y - 2xy)$.
3. Subtract: $(5m^3 - 3m^2 + 6m + 3) - (6m - 9 - m^2 + 4m^2)$.
4. Arrange the polynomial $4xy^5 - 3x^6y^2 + x^2y^3 - 2y$ in descending powers of y.

FLEXIBILITY OF TEACHING MODES

The flexible worktext format of *Intermediate Algebra* allows the book to be used in many ways.

- **In a standard lecture.** To use the book in a lecture format, the instructor lectures in a conventional manner and encourages students to do the margin exercises while studying on their own. This greatly enhances the readability of the text.
- **For a modified lecture.** To bring student-centered activity into the class, the instructor stops lecturing and has the students do margin exercises.
- **For a no-lecture class.** The instructor makes assignments that students do on their own, including working the margin exercises. During the class period following the assignment, the instructor answers questions, and students have an extra day or two to polish their work before handing it in. In the meantime, they are working on the next assignment. This method provides individualization while keeping a class together. It also minimizes the number of instructor hours required and has been found to work well with large classes.
- **In a learning laboratory.** Because this text is highly readable and easy to understand, it can be used in a learning laboratory or any other self-study situation.

KEY CONTENT CHANGES

In response to both extensive user comments and reviewer feedback, there have been many organizational changes and revisions to the Sixth Edition. Detailed information about the changes made in this material is available in the form of a Conversion Guide. Please ask your local Addison-Wesley sales representative for more information. Following is a list of the major organizational changes for this revision:

- Where possible, short sections have been combined to streamline the presentation and reduce the overall number of sections.
- More exercises have been added throughout the text, increasing the overall number by approximately 15%.
- A five-step problem-solving process is now introduced early in the text, and these steps are used throughout the text whenever a problem is solved.
- Chapter 1 has been revised to provide a more streamlined review of the operations and properties of real numbers. The arithmetic review material has been omitted, and the introduction to the real-number system now uses set terminology.
- Chapter 2 has been split into two chapters. The first contains material on solving equations, and the second contains the introduction to graphing linear equations.
- The graphing material has been revised to provide a more logical presentation of concepts involving slope.
- The material on inequalities from Chapter 4 has been moved earlier and integrated into appropriate equation-solving material in Chapters 2, 3, and 4.

- The introduction to functions from Chapter 5 has been moved to a later point in the text, where it is covered at the end of the chapter on conic sections. This material now includes new material on composition of functions and inverse functions.

SUPPLEMENTS

This text is accompanied by a comprehensive supplements package. Below is a brief list of these supplements, followed by a detailed description of each one.

For the Instructor	For the Student
Teacher's Edition	Student's Solutions Manual
Instructor's Solutions Manual	Videotapes
Instructor's Resource Guide	Audiotapes
Printed Test Bank	The Math Hotline
Lab Resource Manual	Comprehensive Tutorial Software
Answer Book	Drill and Practice Software
Computerized Testing	

SUPPLEMENTS FOR THE INSTRUCTOR

All supplements for the instructor are free upon adoption of this text.

Teacher's Edition

This is a specially bound version of the student text with exercise answers printed in a third color. It also includes additional information on the skill maintenance exercises, suggested syllabi for different length courses, and some information about the teaching aids that accompany the text.

Instructor's Solutions Manual

This manual by Judith A. Penna contains worked-out solutions to all even-numbered exercises and discussions of the "Thinking It Through" sections.

Instructor's Resource Guide

This guide contains the following:

- Additional "Thinking It Through" exercises.
- Extra practice problems for some of the most challenging topics in the text.
- Teaching essays on math anxiety and study skills.
- Indexes to the videotapes, the audiotapes, and the software that accompany the text.
- Number lines and grids for test preparation.
- Conversion guide that cross-references the Fifth Edition to the Sixth Edition.
- Black-line transparency masters including a selection of key definitions, procedures, problem-solving strategies, graphs, and figures to use in class.

Printed Test Bank

This is an extensive collection of alternative chapter test forms, including the following:

- 5 alternative test forms for each chapter with questions in the same topic order as the objectives presented in the chapter.
- 5 alternative forms for each chapter with the questions in a different order on each test form.
- 3 multiple-choice test forms for each chapter.
- 2 cumulative review tests for each chapter.
- 9 alternative forms of the final examination, 3 with questions organized by chapter, 3 with questions scrambled, and 3 with multiple-choice questions.

Lab Resource Manual

This manual contains a selection of essays on setting up learning labs, including information on running large testing centers and setting up mastery learning programs. It also includes a directory of learning lab coordinators who are available to answer questions.

Answer Book

The Answer Book will contain answers to all the exercises in the text for you to make available to your students.

Computerized Testing

OmniTest (IBM PC), AWTest (Apple II series)

This text is accompanied by algorithm-driven testing systems for both IBM and Apple. With both machine versions, it is easy to create up to 99 variations of a customized test with just a few keystrokes, choosing from over 300 open-ended and multiple-choice test items. Instructors can also print out tests in chapter-test format.

The IBM testing program, OmniTest, also allows users to enter their own test items and edit existing items in an easy-to-use WYSIWYG format.

LXR·Test™ (Macintosh)

This is a versatile and flexible test-item bank of more than 1200 multiple-choice and open-ended test items with complete math graphics and full editing capabilities. Tests can be created by selecting specific test items or by requesting the computer to select items randomly from designated objectives. LXR·TEST can create multiple test versions by scrambling the order of multiple-choice distractors or the order of the questions themselves.

SUPPLEMENTS FOR THE STUDENT

Student's Solutions Manual

This manual by Judith A. Penna contains completely worked-out solutions with step-by-step annotations for all the odd-numbered exercises in the text. It is free to adopting instructors and may be purchased by your students from Addison-Wesley Publishing Company.

Videotapes

Using the chalkboard and manipulative aids, Donna DeSpain lectures in detail, works out exercises, and solves problems from most sections in the text on 19 70-minute videotapes. These tapes are ideal for students who have missed a lecture or who need extra help. Each section in the text is referenced to the appropriate tape number and section, underneath the icon ▬. A complete set of videotapes is free to qualifying adopters.

Audiotapes

The audiotapes are designed to lead students through the material in each text section. Bill Saler explains solution steps to examples, cautions students about common errors, and instructs them to stop the tape and do exercises in the margin. He then reviews the margin-exercise solutions, pointing out potential errors. Each section in the text is referenced to the appropriate tape number and section, underneath the icon. Audiotapes are free to qualifying adopters.

The Math Hotline

This telephone hotline is open 24 hours a day for students to receive detailed hints for exercises that have been developed by Larry Bittinger. Exercises covered include all the odd-numbered exercises in the exercise sets with the exception of the skill maintenance and synthesis exercises.

Tutorial Software

A variety of tutorial software packages is available to accompany this text. Please contact your Addison-Wesley representative for a software sampler that contains demonstration disks for these packages and a summary of our distribution policy.

Comprehensive Tutorials

IMPACT: An Interactive Mathematics Tutorial **by Wayne Mackey and Doug Proffer, Collin County Community College (IBM PC or MACINTOSH).**

This software was developed exclusively for Addison-Wesley and is keyed section by section to this text. Icons at the beginning of each section reference the appropriate disk number. The disk menus correspond to the text's section numbers.

IMPACT is designed to generate practice exercises based on the exercise sets in this book. If students are having trouble with a particular exercise, they can ask to see an example or a step-by-step solution to the problem they are working on. Each step of the step-by-step solutions is treated interactively to keep students involved in the solution of the problem and help them identify precisely where they are having trouble. *IMPACT* also keeps detailed records of students' scores.

Instructional Software for Algebra **(Apple II series).**

This software covers selected algebra topics. It also gives students brief explanations and examples, followed by practice exercises with interactive feedback for student error.

Drill and Practice Packages

The Math Lab **by Chris Avery and Chris Barker, DeAnza College (Apple II series, IBM PC, or Macintosh).**

Students choose the topic, level of difficulty, and number of exercises. If they get a wrong answer, *The Math Lab* will prompt them with the first step of the solution. This software also keeps detailed records of student scores.

Professor Weissman's Software by Martin Weissman, Essex County College (IBM PC or compatible).

Professor Weissman's Software generates exercises based on the student's selection of topic and level of difficulty. If they get a wrong answer, the software gives them a step-by-step solution. The level of difficulty increases if students are successful.

In the back of this text is a coupon for *Professor Weissman's Software* that allows students to buy the software directly from Martin Weissman at a discount.

The Algebra Problem Solver by Michael Hoban and Kathirgama Nathan, La Guardia Community College (IBM PC).

After selecting the topic and exercise type, students can enter their own exercises or request an exercise from the computer. In each case, *The Algebra Problem Solver* will give the student detailed, annotated, step-by-step solutions.

ACKNOWLEDGMENTS

Many of you who teach developmental mathematics have helped to shape the Sixth Edition of this text by reviewing, answering surveys, participating in focus groups, filling out questionnaires and spending time with us on your campuses. Our heartfelt thanks to all of you, and many apologies to anyone we have missed on the following list.

TEXTBOOK REVIEWERS

John E. Alberghini, *Manchester Community College;* Mary Jean Brod, *University of Montana;* Dr. Louis F. Bush, *San Diego City College;* Linda Cook, *Jefferson College;* Karen J. Emerson, *St. Petersburg Junior College;* Katherine J. Huppler, *Saint Cloud State University;* Phyllis Jore, *Valencia Community College;* Norman Mittman, *Northeastern Illinois University;* Marilyn P. Persson, *University of Kansas;* David Price, *Tarrant County Junior College;* Barbara Sallach, *New Mexico State University;* Mark Serebransky, *Camden County College;* Ara B. Sullenberger, *Tarrant County Junior College;* Eunice Waddington, *Tri County Technical College*

FORMAL AND INFORMAL FOCUS GROUP PARTICIPANTS

Geoff Akst, *Borough of Manhattan Community College;* Betty Jo Baker, *Lansing Community College;* Gene Beuthin, *Saginaw Valley State University;* Rheta Beaver, *Valencia Community College;* Roy Boersema, *Front Range Community College;* Dale Boye, *Schoolcraft College;* Jim Brenner, *Black Hawk College;* Ben Cheatham, *Valencia Community College;* Karen Clark, *Tacoma Community College;* Tom Clark, *Lane Community College;* Sally Copeland, *Johnson County Community College;* Ernie Danforth, *Corning Community College;* Sarah Evangelista, *Temple University;* Bill Freed, *Concordia College;* Sally Glover-Richard, *Pierce Community College;* Valerie Hayward, *Orange Coast College;* Eric Heinz, *Catonsville Community College;* Bruce Hoelter, *Raritan Valley Community College;* Lou Hoezle, *Bucks County Community College;* Linda Horner, *Broward Community College;* Mary Indelicato,

Normandale Community College; Tom Jebson, *Pierce Community College;* Jeff Jones, *County College of Morris;* Judith Jones, *Valencia Community College;* Virginia Keen, *West Michigan University;* Roxanne King, *Prince Georges Community College;* Lee Marva Lacy, *Glendale Community College;* Ginny Licata, *Camden County College;* Randy Liefson, *Pierce Community College;* Charlie Luttrell, *Frederick Community College;* Marilyn MacDonald, *Red Deer College;* Sharon MacKendrick, *New Mexico State University;* Annette Magyar, *Southwestern Michigan College;* Bob Malena, *Community College of Allegheny County;* Marilyn Masterson, *Lansing Community College;* Don McNair, *Lane Community College;* John Pazdar, *Greater Hartford Community College;* Donald Perry, *Lee College;* Jeanne Romeo, *Delta College;* Jack Rotman, *Lansing Community College;* Winona Sathre, *Valencia Community College;* Billie Stacey, *Sinclair Community College;* John Steele, *Lane Community College;* Dave Steinfort, *Grand Rapids Junior College;* Betty Swift, *Cerritos College;* Bill Wittinfeld, *Tacoma Community College; Faculty of St. Petersburg Junior College*

QUESTIONNAIRE RESPONDEES

Tony Abruzzo, *University of New Mexico;* Boyd Benson, *Rio Hondo College;* Murray Butler, *Patrick Henry State Junior College;* Debra Caplinger, *Patrick Henry State Junior College;* Max Cisneros, *Albuquerque Technical Vocational;* Michelle Fleck, *College of Eastern Utah;* Linda Long, *Ricks College;* George Pimmata, *Suffolk County Community College*

We also wish to thank the many people without whose committed efforts our work could not have been completed. In particular, we would like to thank Judy Beecher, Barbara Johnson, and Judy Penna for their work on proofreading the manuscript and overseeing the production process. We would also like to thank Pat Pasternak, who did a marvelous job typing the text manuscript and answer section, and Bill Saler, Martha Cox, Nancy Woods, and Lauren Page, who did a thorough and conscientious job of checking the manuscript.

M.L.K.
M.L.B.

To The Student

This text has many features that can help you succeed in intermediate algebra. To familiarize yourself with these, you might read the preface that starts on page ix and study the annotated pages that are included. Following are a few suggestions on how to use these features to enhance your learning process.

BEFORE YOU START THE TEXT

If you are in a classroom setting, your instructor might ask you to take the diagnostic pretest at the beginning of the text, checking your answers at the back of the text, to find out what material you already know and what material you need to spend time on. You can also use this pretest to skip material that you already know from an independent learning situation.

BEFORE YOU START A CHAPTER

The chapter opening page gives you an idea of the material that you are about to study and how it can be used. The chapter opening introduction also tells you what sections you will need to review in order to do the skill maintenance exercises on the chapter test. It's a good idea to restudy these sections to keep the material fresh in your mind for the midterm or final examination.

The first page of each chapter lists "Points to Remember" that will be needed to work certain examples and exercises in the chapter. You should try to review any skills listed here before beginning the chapter and learn any formulas or definitions.

This same page also includes a chapter pretest. You can work through this and check your answers at the back of the text to identify sections that you might skip or sections that give you particular difficulty and need extra concentration.

WORKING THROUGH A SECTION

First you should read the learning objectives for the section. The symbol next to an objective (a, b, c) appears next to the text, exercises, and answers that correspond to that objective, so you can always refer back to the appropriate material when you need to review a topic.

You will also notice that there are references to the audiotapes, videotapes, and software that are available for extra help for the section underneath the objective listing. The software referenced is a program called *IMPACT: An Interactive Mathematics Tutorial.*

As you work through a section, you will see an instruction to "Do Exercises x–xx." This refers to the exercises in the margin of the page. You should always stop and do these to practice what you have just studied because they greatly enhance the readability of the text. Answers to the margin exercises are at the back of the text.

After you have completed a section, you should do the assigned exercises in the exercise set. The exercises are keyed to the section objectives, so that if you get an incorrect answer, you know that you should restudy the text section that corresponds to the symbol.

Answers to all the odd-numbered exercises are at the back of the text. A solutions manual with complete worked-out solutions to all the odd-numbered exercises is available from Addison-Wesley Publishing Company.

PREPARING FOR A CHAPTER TEST

To prepare for a chapter test, you can review your homework and restudy sections that were particularly difficult. You should also learn the "Important Properties and Formulas" that begin the chapter's summary and review and study the review sections that are listed at the beginning of the review exercises.

After studying, you might set aside a block of time to work through the summary and review as if it were a test. You can check your answers at the back of the text after you are done. The answers are coded to sections and objectives, so you can restudy any areas in which you are having trouble. You can also take the chapter test as practice, again checking your answers at the back of the text.

If you are still having difficulties with a topic, you might try either going to see your instructor or working with the videotapes, audiotapes, or tutorial software that are referenced at the beginning of the text sections. Be sure to start studying in time to get extra help before you must take the test.

PREPARING FOR A MIDTERM OR FINAL EXAMINATION

To keep material fresh in your mind for a midterm or final examination, you can work through the cumulative reviews at the end of each chapter. You can also use these as practice midterms or finals. In addition, there is a final examination at the end of the text. The answers to all exercises in the cumulative reviews and the final examination are at the back of the text.

OTHER STUDY TIPS

There is a saying in the real-estate business: "The three most important things to consider when buying a house are *location, location, location.*" When trying to learn mathematics, the three most important things are *time, time, time.* Try to carefully analyze your situation. Be sure to allow yourself *time* to do the lesson. Are you taking too many courses? Are you working so much that you do not have *time* to study? Are you taking *time* to maintain daily preparation? Other study tips are provided on pages marked "Sidelights" in the text.

Intermediate Algebra

6TH EDITION

DIAGNOSTIC PRETEST

Chapter 1

1. Subtract: $1.45 - (-2.12)$.

2. Multiply: $-\frac{5}{6}\left(\frac{2}{15}\right)$.

3. Simplify: $2x - 2[x - (4 + 3x)]$.

4. Simplify: $\left[\frac{-3x^2y^{-3}}{2x^{-1}y^4}\right]^{-2}$.

Chapter 2

Solve.

5. $3(x + 1) = 2 - (x - 2)$

6. $2 \le 4 - x \le 7$

7. $|2x + 3| > 1$

8. Three less than eight times a number is two more than six times the number. Find the number.

Chapter 3

Graph on a plane.

9. $3x \le 6 - y$

10. $x = -1$

11. Find an equation of the line containing the pair of points $(1, 3)$ and $(-1, 5)$.

12. Find an equation of the line containing the point $(0, 3)$ and perpendicular to the line $3x - y = 7$.

Chapter 4

Solve.

13. $3x - y = 5,$
$x + 2y = 3$

14. $x - 4y + 2z = -1,$
$2x + y - z = 8,$
$-x - 3y + z = -5$

15. A motorboat took 6 hr to make a downstream trip with a 3-mph current. The return trip against the same current took 8 hr. Find the speed of the boat in still water.

16. Graph. Find the coordinates of any vertices formed.
$x \ge 3,$
$x \le 6 - 3y,$
$x - 2y \le 6$

Chapter 5

17. Multiply: $(3x - 5y)^2$.

18. Factor: $x^4 - 1$.

19. Solve: $x^2 - 18 = 7x$.

20. Three times the square of a number is two more than five times the number. Find the number.

Chapter 6

21. Divide and simplify:
$$\frac{x^2 - 9}{x^2 + 3x + 2} \div \frac{x^2 - 6x + 9}{2x + 4}.$$

22. Simplify:
$$\frac{x - \frac{1}{x}}{1 + \frac{1}{x}}.$$

Solve.

23. $\frac{3}{x} + \frac{2}{x-2} = \frac{1}{x}$

24. One folding machine in a print shop can fold an order of pamphlets in 4 hr. Another machine can do the same job in 3 hr. How long will it take if both folding machines are used?

Chapter 7

For Exercises 25–27, assume that all expressions under radicals represent nonnegative numbers.

25. Multiply and simplify:

$$(\sqrt{6} + \sqrt{8x^3})(\sqrt{6} - 2\sqrt{2y^2}).$$

26. Add and simplify:

$$\sqrt{75} + \sqrt{300} + 3\sqrt{27}.$$

27. Rationalize the denominator:

$$\frac{2\sqrt{x} - \sqrt{y}}{\sqrt{x} - \sqrt{y}}.$$

28. Solve:

$$\sqrt[3]{x-5} = -2.$$

Chapter 8

Solve.

29. $x^2 + 2x + 5 = 0$

30. $(x-3)^2 + (x-3) - 12 = 0$

31. $\frac{x}{2} = \frac{x+1}{x+3}$

32. Graph $y = 2x^2 + 5x + 3$. Label the vertex and the line of symmetry.

Chapter 9

Graph.

33. $\frac{x^2}{25} + \frac{y^2}{9} = 1$

34. $x^2 + y^2 - 6x + 4y + 9 = 0$

35. Solve:

$$x^2 - y^2 = 6,$$
$$xy = 4.$$

36. Find a formula for the inverse of $f(x) = 2x + 5$.

Chapter 10

37. Graph: $y = \log_2 x$.

38. Express as a single logarithm:

$$\frac{1}{2}\log x - \log y.$$

Solve.

39. $\log_4 1 = x$

40. $\log_2 (x-1) = 1 - \log_2 (x+1)$

INTRODUCTION Why do we study mathematics? One reason is to be able to solve problems. Before we can do so, there are certain algebraic manipulations that we must learn. These manipulations, such as simplifying expressions, are based on the properties of numbers.

This chapter is a review for those who have recently studied introductory algebra. For those who have not, this chapter will bring mathematical backgrounds together before solving equations and problems in Chapter 2. ❖

1 Algebra and Real Numbers

AN APPLICATION

The earned-run average of a pitcher is $9R/I$, where R stands for the number of earned runs allowed and I stands for the number of innings pitched. Dwight Gooden of the New York Mets allowed 41 earned runs in 211 innings. What was his earned-run average?

THE MATHEMATICS

We substitute 41 for R and 211 for I in the expression for earned-run average:

$$\frac{9 \cdot R}{I} = \frac{9 \cdot 41}{211} \approx 1.75.$$

This is an *algebraic expression.*

❖ POINTS TO REMEMBER: CHAPTER 1

Area of a rectangle:	$A = l \cdot w$
Area of a square:	$A = s^2$
Area of a triangle:	$A = \frac{1}{2} \cdot b \cdot h$
Area of a parallelogram:	$A = b \cdot h$
Simple-interest formula:	$I = P \cdot r \cdot t$

PRETEST: CHAPTER 1

1. Evaluate $x - 3y$ when $x = 5$ and $y = -4$.

2. Translate to an algebraic expression: Forty-seven percent of some number.

3. True or false: $-8 \leq -4$.

4. Simplify: $|-3.6|$.

5. Add: $-8 + 24$.

6. Subtract: $-3.4 - 8.2$.

7. Multiply: $-\frac{3}{5} \cdot \frac{10}{7}$.

8. Divide: $\frac{-200}{-25}$.

9. Multiply: $-4(x - 3y)$.

10. Factor: $6x - 18xy + 24$.

11. Collect like terms: $5x - 8y + 23 - 6x + 14y - 12$.

12. Simplify: $8(x + 2) - 6(2x + 12)$.

13. Find an equivalent expression: $-(-2x + 5y - 24)$.

14. Simplify: $(5x^3)(-4x^{-6})$.

15. Simplify: $\frac{36x^3y^{-10}}{-9x^5y^{-8}}$.

16. Simplify: $(-5x^{-6}y^5)^{-2}$.

Convert to scientific notation.

17. 0.0000000786

18. 457,890,000,000

Convert to decimal notation.

19. 7.89×10^{13}

20. 7.89×10^{-6}

1.1 Introduction to Algebra and Expressions

Our goal in this section is to introduce you to intermediate algebra. We will study evaluating expressions and translating to expressions of the type used in algebra.

You have probably already taken a course in introductory algebra. In your study of intermediate algebra, you will not only review and strengthen your skills of introductory algebra, but also learn many new skills.

a Algebraic Expressions

Basic to a study of algebra is the use of algebraic expressions. In arithmetic, you worked with expressions such as

$$31 + 76, \quad 14 \times 35, \quad 26 - 17, \quad \frac{7}{8}, \quad \text{and} \quad 3^2.$$

In algebra, we use certain letters, or **variables,** for numbers and work with *algebraic expressions* such as

$$31 + x, \quad 14 \times t, \quad 26 - y, \quad \frac{a}{8}, \quad \text{and} \quad x^2.$$

How do these expressions arise? Most often, they arise in problem solving. For example, consider the chart shown here, one that you might see in a magazine. Suppose we want to know how much greater the life expectancy of women is in the United States than in the Soviet Union.

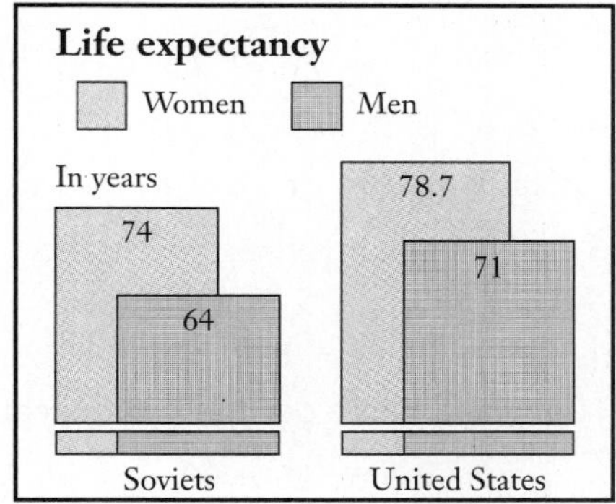

In algebra, we attempt to translate this problem to an equation. It might be done as follows:

Life exp. W in USSR	plus	how much	is	life exp. W in US
↓	↓	↓	↓	↓
74	$+$	x	$=$	78.7

Note that we have an algebraic expression on the left. To find the number x, we subtract 74 on both sides of the equation:

$$x = 78.7 - 74.$$

Then we carry out the subtraction and obtain the answer, 4.7 years.

In arithmetic, you probably would do this subtraction right away without considering an equation. In algebra, you will find most problems extremely difficult to solve without first translating to an equation.

DO EXERCISE 1 (IN THE MARGIN AT THE RIGHT).

OBJECTIVES

After finishing Section 1.1, you should be able to:

a Evaluate an algebraic expression by substitution.

b Translate a phrase to an algebraic expression.

FOR EXTRA HELP

Tape 1A | Tape 1A | MAC: 1 IBM: 1

1. Translate this problem to an equation, using the chart at the left: How much greater is the life expectancy of men in the United States than in the Soviet Union?

ANSWER ON PAGE A-1

An **algebraic expression** consists of variables, numerals, and operation signs. When we replace a variable by a number, we say that we are **substituting** for the variable. When we calculate the results, we get a number. This process is called **evaluating the expression.**

EXAMPLE 1 Evaluate $x - y$ for $x = 83$ and $y = 49$.

We substitute 83 for x and 49 for y and carry out the subtraction:

$$x - y = 83 - 49 = 34.$$

The number 34 is called the *value* of the expression. ◀

EXAMPLE 2 Evaluate a/b for $a = 63$ and $b = 7$.

We substitute 63 for a and 7 for b and carry out the division:

$$\frac{a}{b} = \frac{63}{7} = 9.$$ ◀

DO EXERCISES 2–4 (IN THE MARGIN).

In arithmetic, we use × or a dot · to mean multiplication. Thus,

$7 \cdot 9$ and 7×9 have the same meaning.

In algebra, when two letters, or a number and a letter, are written together, that also means that they are to be multiplied. Numbers to be multiplied are called **factors.** We usually write a factor that is a number before any factor named by a letter. For example,

$4y$ means $4 \cdot y$ or $4 \times y$ or $4(y)$

and

ab means $a \cdot b$ or $a \times b$ or $a(b)$.

Now we evaluate expressions involving products.

EXAMPLE 3 Evaluate $6y$ for $y = 15$.

We substitute 15 for y and carry out the multiplication:

$$6y = 6 \cdot 15 = 90.$$ ◀

EXAMPLE 4 The area A of a triangle with base b and height h is given by $A = \frac{1}{2}bh$. Find the area when b is 24.5 in. and h is 16 in.

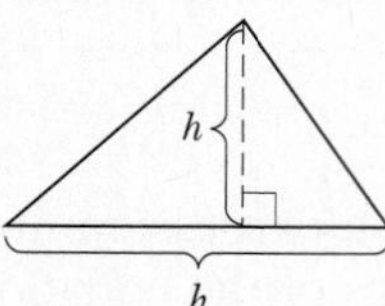

We substitute 24.5 for b and 16 for h and carry out the multiplication:

$$A = \tfrac{1}{2}bh$$
$$= \tfrac{1}{2} \times 24.5 \times 16 = 196 \text{ sq in.}$$ ◀

DO EXERCISES 5 AND 6.

2. Evaluate $a + b$ for $a = 48$ and $b = 36$.

3. Evaluate $x - y$ for $x = 97$ and $y = 29$.

4. Evaluate a/b for $a = 400$ and $b = 8$.

5. Evaluate $8t$ for $t = 15$.

6. Find the area of a triangle when h is 24 ft and b is 8 ft.

ANSWERS ON PAGE A-1

▶ **EXAMPLE 5** Evaluate $14m/n$ for $m = 8$ and $n = 16$.

We substitute 8 for m and 16 for n and carry out the calculation:

$$\frac{14m}{n} = \frac{14 \cdot 8}{16} = \frac{112}{16} = 7.$$ ◀

DO EXERCISE 7.

▶ **EXAMPLE 6** *Earned-run average.* The earned-run average of a pitcher is $9R/I$, where R stands for the number of earned runs allowed and I stands for the number of innings pitched. It is assumed that the number of innings in the baseball games being played is 9, even though a pitcher may pitch a different number of innings. Dwight Gooden of the New York Mets recently allowed 41 earned runs in 211 innings. What was his earned-run average?

The number of earned runs is 41, so we substitute 41 for R. The number of innings pitched is 211, so we substitute 211 for I. Then we multiply 41 by 9, and divide the result by 211:

$$\frac{9R}{I} = \frac{9 \cdot 41}{211} = \frac{369}{211} \approx 1.75.$$ **"≈" means "approximately equal to"** ◀

DO EXERCISE 8.

b Translating to Algebraic Expressions

In algebra, we translate problems to equations. The parts of equations are translations of phrases to algebraic expressions. To help you become familiar with algebra and to ease such translations when we do them later, we now practice translating.

▶ **EXAMPLE 7** Translate to an algebraic expression: Eight less than some number.

We let t represent the number. We can use any variable we wish, such as x, y, m, n, and so on. If we knew the number to be 23, then the translation would be $23 - 8$. If we knew the number to be 345, then the translation would be $345 - 8$. Since we are using a variable for the number, the translation is $t - 8$. ◀

DO EXERCISE 9.

▶ **EXAMPLE 8** Translate to an algebraic expression: Twenty-two more than some number.

This time we let y represent the number. If we knew the number to be 47, then the translation would be $47 + 22$, or $22 + 47$. If we knew the number to be 17.95, then the translation would be $17.95 + 22$, or $22 + 17.95$. Since we are using a variable, the translation is $y + 22$, or $22 + y$. ◀

7. Evaluate $\frac{10p}{q}$ when $p = 60$ and $q = 25$.

8. Bret Saberhagen of the Kansas City Royals recently allowed 54 earned runs in 173 innings. What was his earned-run average?

9. Translate to an algebraic expression: Forty-seven more than some number.

Translate to an algebraic expression.

10. 16 less than some number

11. 16 less some number

12. One-fourth of some number

13. Six more than eight times some number

ANSWERS ON PAGE A-1

14. The difference between two numbers m and n

15. Sixty-nine percent of some number

16. Three hundred less than the product of two numbers

17. The sum of two numbers

ANSWERS ON PAGE A-1

▶ **EXAMPLE 9** Translate each of the following to an algebraic expression.

Phrase	Algebraic expression
Nine more than some number	$m + 9$, or $9 + m$
Nine less than some number	$a - 9$
Nine less some number	$9 - q$
The difference between two numbers a and b	$a - b$
Half of a number	$\frac{1}{2}x$, or $x/2$, or $\frac{x}{2}$
Seven more than five times a number	$7 + 5t$, or $5t + 7$
Three less than the product of two numbers	$pq - 3$
Seventy-three percent of some number	$73\%z$, or $0.73z$

◀

DO EXERCISES 10–17. (EXERCISES 10–13 ARE ON THE PRECEDING PAGE.)

From time to time you will find a "*Sidelights*" like the one below. These are optional, but you may find them helpful and of interest. They will include such topics as study tips, career opportunities involving mathematics, applications, computer–calculator exercises, or other mathematical topics.

❖ SIDELIGHTS

Study Tips

Many students begin the study of a text by opening to the first section assigned by an instructor. There are many ways in which you can enhance your use of this book, and they have been outlined carefully in a page in the preface titled *To the student.* If you have not read that page, do so now before you start the exercise set on the next page.

There are some points on that page that bear repeating here.

- *Be sure to note the special symbols* **a**, **b**, **c**, *and so on, that correspond to the objectives you are to learn.* They appear many places throughout the text. The first time you will see them is in the headings for the section. The second time you will see them is in the exercise set, as follows. You will also see them in the answers to the Review Exercises, the Chapter Tests, and the Cumulative Reviews. These allow you to reference back when you need to review a topic.
- *Be sure to note also the symbols in the margin under the list of objectives at the beginning of the section.* These refer to the many distinctive study aids that accompany the book.
- *Be sure to stop and do the margin exercises as you study a section.* When our students come to us troubled about how they are doing in the course, the first question we ask them is "Are you doing the margin exercises when directed to do so?" This is one of the most effective ways to enhance your ability to learn mathematics from this text. Don't deprive yourself of its benefits!
- *When you study the book, don't mark points you think are important, but mark the points you do not understand!* The book is written with all kinds of processes that highlight important points. Use your efforts to mark where you are having trouble. Then when you go to class or a math lab or a tutoring session, you are prepared to ask questions that close in on your difficulties.
- *Try to keep one section ahead of your syllabus.* We have tried to write a book that is readable for students. If you study ahead of your lectures, you can concentrate on just the lectures, rather than trying to write everything down. You can then take notes only of special points or of questions related to what is happening in class.

NAME SECTION DATE

EXERCISE SET 1.1

ANSWERS

a Substitute to find values of the expressions.

1. The length of a rectangle is 8 yd longer than the width. Suppose the variable w stands for the width. Then $w + 8$ stands for the length. What is the length when the width is 5 yd? 14 yd? 52 yd?

2. Employee A takes five times as long to do a job as employee B. Suppose t stands for the time it takes B to do the job. Then $5t$ stands for the time it takes A. How long did it take A if B took 16 sec? 90 sec? 7 min?

3. The area A of a rectangle with length l and width w is given by $A = lw$. Find the area when $l = 16$ cm (centimeters) and $w = 9$ cm.

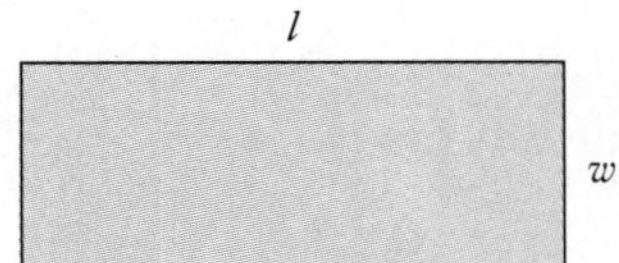

4. The area A of a parallelogram with base b and height h is given by $A = bh$. Find the area of a parallelogram with a height of 15.4 cm and a base of 6.5 cm.

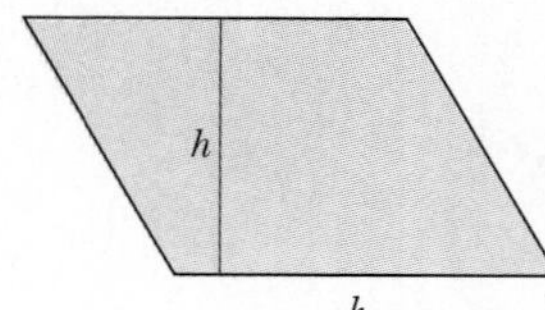

Evaluate.

5. $97y$ for $y = 3$

6. $16x$ for $x = 8$

7. $\frac{x}{y}$ for $x = 15$ and $y = 3$

8. $\frac{m}{n}$ for $m = 24$ and $n = 2$

9. $\frac{m - n}{8}$ for $m = 36$ and $n = 4$

10. $\frac{p + q}{5}$ for $p = 20$ and $q = 20$

11. $\frac{5z}{y}$ for $z = 9$ and $y = 2$

12. $\frac{18m}{n}$ for $m = 7$ and $n = 18$

Simple interest. The *simple interest I* on a principal of P dollars at interest rate r for time t, in years, is given by $I = Prt$.

13. Find the simple interest on a principal of \$2500 at 8% for 2 years. (*Hint:* 8% = 0.08.)

14. Find the simple interest on a principal of \$6875 at 7.5% for 1 year.

1. ______

2. ______

3. ______

4. ______

5. ______

6. ______

7. ______

8. ______

9. ______

10. ______

11. ______

12. ______

13. ______

14. ______

ANSWERS

15. ______
16. ______
17. ______
18. ______
19. ______
20. ______
21. ______
22. ______
23. ______
24. ______
25. ______
26. ______
27. ______
28. ______
29. ______
30. ______
31. ______
32. ______
33. ______
34. ______
35. ______
36. ______
37. ______
38. ______
39. ______
40. ______
41. ______
42. ______
43. ______
44. ______
45. ______
46. ______

b Translate to an algebraic expression.

15. 7 more than m

16. 9 more than t

17. 11 less than c

18. 47 less than d

19. 26 greater than q

20. 11 greater than z

21. b more than a

22. c more than d

23. x less than y

24. c less than b

25. 28% of x

26. Sixty-seven percent of m

27. The sum of a and b

28. The sum of m and n

29. Twice x

30. Four times p

31. Seven times t

32. Nine times d

33. The difference between 17 and b

34. The difference between p and q

35. 8 more than some number

36. One more than some number

37. 54 less than some number

38. 47 less than some number

39. 54 less some number

40. 47 less some number

41. A number x plus three times y

42. A number a minus 2 times b

SYNTHESIS

This heading refers to the fact that the exercises that follow are more challenging, requiring you to put together objectives of this section or preceding sections of the book.

Translate to an equation.

43. The distance d that an object travels at speed r in time t is speed times time. Write an equation for d.

44. You invest P dollars at 11% simple interest. Write an equation for the number of dollars N in the account a year from now.

Evaluate.

45. $\frac{256y}{32x}$ for $y = 3$ and $x = 4$

46. $\frac{y + x}{2} + \frac{3y}{x}$ for $x = 2$ and $y = 4$

1.2 The Real-Number System

Sets that are parts of other sets are called **subsets.** In this section, we become acquainted with the set of *real numbers* and its various subsets.

a Set Notation and the Set of Real Numbers

Set Notation

The numbers used in algebra are the so-called real numbers. There is a real number for every point on a number line.

The set containing the numbers -5, 0, and 3 can be named $\{-5, 0, 3\}$. This method of naming sets is called **roster notation.** Three important subsets of the real numbers are listed below using roster notation.

> ***Natural numbers*** **= {1, 2, 3, . . .}. These are the numbers used for counting.**
>
> ***Whole numbers*** **= {0, 1, 2, 3, . . .}. The set of natural numbers with 0 included.**
>
> ***Integers*** **= {. . . , −4, −3, −2, −1, 0, 1, 2, 3, 4, . . .}. The set of whole numbers and their opposites.**

The **integers** consist of the whole numbers and their opposites. We illustrate them on a number line as follows. The **opposite** of a number is found by reflecting it across the number 0. Thus the opposite of 3 is -3. The opposite of -4 is 4. The opposite of 0 is 0.

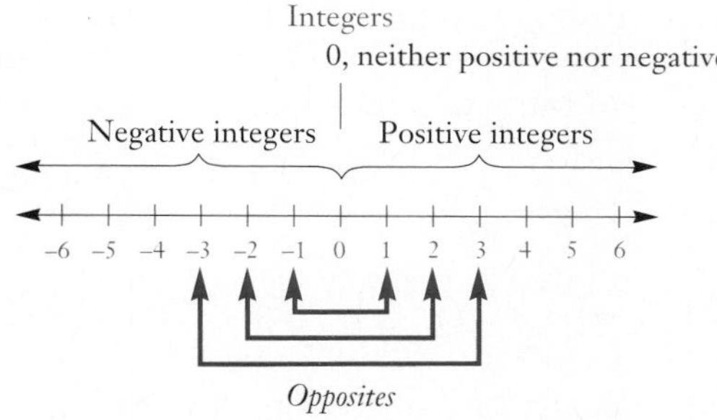

The natural numbers are called **positive integers.** The opposites of the natural numbers are called **negative integers.** Zero is neither positive nor negative. This gives us the set of integers, which extends infinitely to the left and right of 0 on a number line.

We read the symbol -3 as either "the opposite of 3" or "negative 3."

DO EXERCISES 1–3.

To describe other subsets of real numbers, we sometimes need another kind of set notation called **set-builder notation.** To use set-builder notation, we specify conditions by which we know whether a number, or other object, is in the set. For example, the set of all odd natural numbers less than 11 can

OBJECTIVES

After finishing Section 1.2, you should be able to:

- **a** Use roster and set-builder notation to name sets, and distinguish between various kinds of real numbers.
- **b** Determine which of two real numbers is greater and indicate which, using < and >; given an inequality like $a < b$, write another inequality with the same meaning; and determine whether an inequality like $-2 \le 3$ is true.
- **c** Graph inequalities on a number line.
- **d** Find the absolute value of a real number.

FOR EXTRA HELP

Tape 1B | Tape 1A | MAC: 1 IBM: 1

Find the opposite of the number.

1. 9

2. -6

3. 0

ANSWERS ON PAGE A-1

4. Name the set consisting of the first seven odd whole numbers using both roster notation and set-builder notation.

ANSWER ON PAGE A-1

be described as follows:

$$\{x \mid x \text{ is an odd natural number less than } 11\}.$$

The set of ←
all numbers x ←
such that ←
x is an odd natural number less than 11 ←

In this case, we can easily write another name for the preceding set using roster notation as follows:

$$\{1, 3, 5, 7, 9\}.$$

▶ **EXAMPLE 1** Name the set consisting of the first six even whole numbers using both roster notation and set-builder notation.

Roster notation: $\{0, 2, 4, 6, 8, 10\}$.
Set-builder notation: $\{x \mid x \text{ is one of the first six even whole numbers}\}$ ◀

DO EXERCISE 4.

The advantage of set-builder notation is that we can use it to describe very large sets that may be difficult to describe using roster notation. Such is the case when we try to describe or name the set of *rational numbers.*

> ***Rational numbers*** $= \left\{\frac{a}{b} \,\middle|\, a \text{ and } b \text{ are integers and } b \neq 0\right\}.$
>
> **This is read "the set of numbers $\frac{a}{b}$, where a and b are integers and $b \neq 0$."**

We can also describe the set of **rational numbers** as the set of all quotients of integers with nonzero divisors. Every rational number is a real number. Thus the rational numbers can be named by fractional notation. The following are examples of rational numbers:

$$\frac{4}{5}, \quad -\frac{4}{5}, \quad \frac{9}{1}, \quad 6, \quad -4, \quad 0, \quad \frac{68}{-7}, \quad 1.5, \quad -0.12$$

$$\left(-\frac{4}{5} \text{ can be named } \frac{-4}{5} \text{ or } \frac{4}{-5}\right), \qquad \left(1.5 \text{ can be named } \frac{3}{2}\right).$$

Rational numbers can also be named using decimal, or percent notation. For example,

$$\frac{5}{8} = 0.625 = 62.5\% \quad \text{and} \quad \frac{6}{11} = 0.545454\ldots = 0.\overline{54} = 54.\overline{54}\%.$$

The bar in $0.\overline{54}$ indicates the repeating part of decimal notation.

Note that this new set of numbers, the rational numbers, contains the whole numbers and the integers.

The Real Numbers

The number line has a point for every rational number. Is there a rational number for every point of the line? The answer is *no.* There are some points of the line for which there is no rational number. These points correspond to what are called **irrational numbers.**

What kinds of numbers correspond to points that are *not* rational numbers, but irrational numbers? One example is the number π, which you are familiar with using to find the area and circumference of a circle: $A = \pi r^2$ and $C = 2\pi r$. Another is the square root of 2, named $\sqrt{2}$.

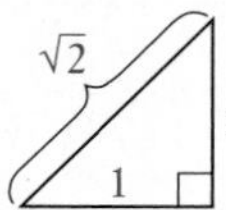

It can be thought of as the length of the diagonal of a square with sides of length 1. The irrational number $\sqrt{2}$ is the number that when multiplied by itself gives 2. There is no rational number that can be multiplied by itself to get 2. But the following are rational *approximations:*

1.4 is an approximation of $\sqrt{2}$ because $(1.4)^2 = 1.96$.

1.41 is a better approximation because $(1.41)^2 = 1.9881$.

1.4142 is an even better approximation because $(1.4142)^2 = 1.99996164$.

We can find rational approximations for square roots using a calculator.

Decimal notation for rational numbers either terminates or repeats. Decimal notation for irrational numbers neither terminates nor repeats. Some other examples of irrational numbers are $\sqrt{3}$, $-\sqrt{8}$, $\sqrt{13}$, and 0.454554555455554 . . . . Whenever we take the square root of a number that is not a perfect square, we will get an irrational number.

The number system in which all points on the number line correspond to numbers is called the system of **real numbers.**

***Real numbers* = {*x* | *x* corresponds to a point on a number line}.**

The real numbers consist of the rational numbers and the irrational numbers. Thus we can also describe the real numbers as follows.

***Real numbers* = {*x* | *x* is rational or *x* is irrational}.**

The following figure shows the relationships among various kinds of numbers.

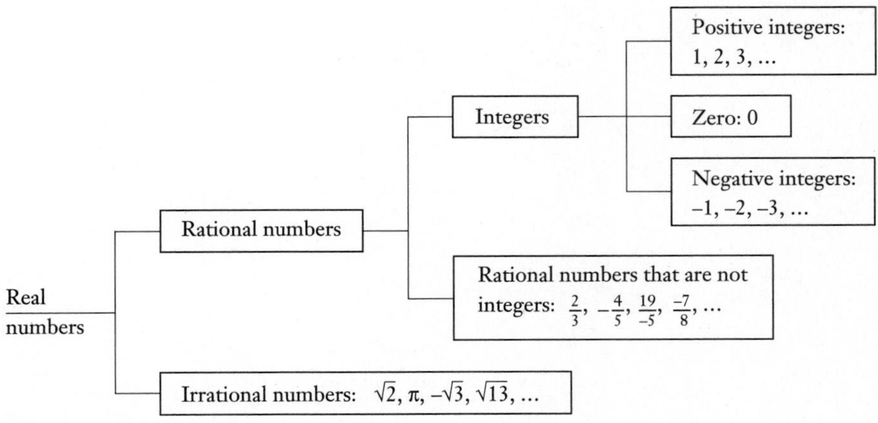

DO EXERCISE 5.

5. Given the numbers

$20,\ -10,\ -5.34,\ 18.999,\ \frac{11}{45},\ \sqrt{7},\ -\sqrt{2},\ 0,\ -\frac{2}{3},$ 9.34334333433334 . . . :

a) Name the natural numbers.

b) Name the whole numbers.

c) Name the integers.

d) Name the irrational numbers.

e) Name the rational numbers.

f) Name the real numbers.

ANSWERS ON PAGE A-1

Insert < or > for ▩ to write a true sentence.

6. -5 ▩ -4

7. $-\frac{1}{4}$ ▩ $-\frac{1}{2}$

8. 87 ▩ 67

9. -9.8 ▩ -4.5

10. 6.78 ▩ -6.77

11. $-\frac{4}{5}$ ▩ -0.86

12. $\frac{14}{29}$ ▩ $\frac{17}{32}$

13. $-\frac{12}{13}$ ▩ $-\frac{14}{15}$

14. 1.8 ▩ 1.08

ANSWERS ON PAGE A-1

b Order for the Real Numbers

Real numbers are named in order on the number line, with larger numbers named further to the right. For any two numbers on the line, the one to the left is less than the one to the right.

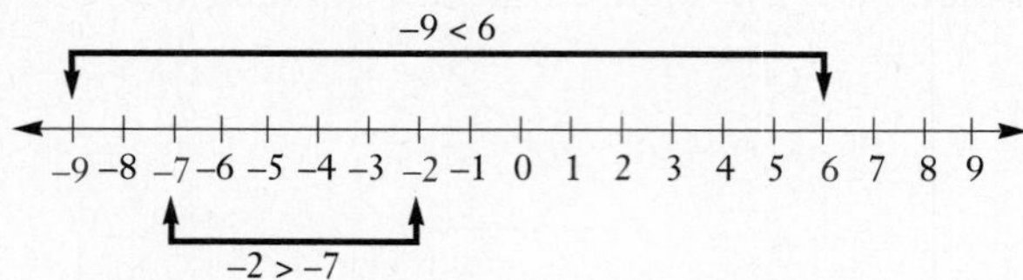

We use the symbol **<** to mean "**is less than.**" The sentence $-9 < 6$ means "-9 is less than 6." The symbol **>** means "**is greater than.**" The sentence $-2 > -7$ means "-2 is greater than -7." A handy mental device is to think of > or < as an arrowhead that points to the smaller number.

► **EXAMPLES** Use either < or > for ▩ to write a true sentence.

2. 4 ▩ 9 — Since 4 is to the left of 9, 4 is less than 9, so $4 < 9$.

3. -8 ▩ 3 — Since -8 is to the left of 3, we have $-8 < 3$.

4. 7 ▩ -12 — Since 7 is to the right of -12, then $7 > -12$.

5. -21 ▩ -5 — Since -21 is to the left of -5, we have $-21 < -5$.

6. -2.7 ▩ $-\frac{3}{2}$ — The answer is $-2.7 < -\frac{3}{2}$.

7. 1.5 ▩ -2.7 — The answer is $1.5 > -2.7$.

8. 4.79 ▩ 4.97 — The answer is $4.79 < 4.97$.

9. -8.45 ▩ 1.32 — The answer is $-8.45 < 1.32$.

10. $\frac{5}{8}$ ▩ $\frac{7}{11}$ — We convert to decimal notation: $\frac{5}{8} = 0.625$ and $\frac{7}{11} = 0.6363\ldots$. Thus $\frac{5}{8} < \frac{7}{11}$. ◄

DO EXERCISES 6–14.

Note that $-8 < 5$ and $5 > -8$ are both true. These are **inequalities.** Every true inequality yields another true inequality if we interchange the numbers or variables and reverse the direction of the inequality sign.

$a < b$ also has the meaning $b > a$.

▶ **EXAMPLES** Write another inequality with the same meaning.

11. $a < -5$ The inequality $-5 > a$ has the same meaning.

12. $-3 > -8$ The inequality $-8 < -3$ has the same meaning. ◀

DO EXERCISES 15 AND 16.

Note that all positive real numbers are greater than zero and all negative real numbers are less than zero.

If x is a positive real number, then $x > 0$.
If x is a negative real number, then $x < 0$.

Expressions like $a \leq b$ and $b \geq a$ are also **inequalities.** We read $a \leq b$ as "**a is less than or equal to b.**" We read $a \geq b$ as "**a is greater than or equal to b.**"

▶ **EXAMPLES** Write true or false.

13. $-8 \leq 5$ True since $-8 < 5$ is true.

14. $-8 \leq -8$ True since $-8 = -8$ is true.

15. $-7 \geq 4$ False since neither $-7 > 4$ nor $-7 = 4$ is true. ◀

DO EXERCISES 17–19.

C Graphing Inequalities on a Number Line

Some replacements for the variable in an inequality make it true and some make it false. A replacement that makes it true is called a **solution.** The set of all solutions is called the **solution set.** A **graph** of an inequality is a drawing that represents its solution set.

▶ **EXAMPLE 16** Graph the inequality $x > -3$ on a number line.

The solutions consist of all real numbers greater than -3, so we shade all numbers greater than -3. Note that -3 is not a solution. We indicate this by using an open circle at -3.

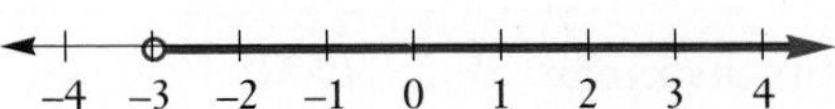

The graph represents the solution set $\{x | x > -3\}$. ◀

Write another inequality with the same meaning.

15. $x > 6$

16. $-4 \leq 7$

Write true or false.

17. $6 \geq -9$

18. $-18 \leq -18$

19. $-7 \leq -10$

ANSWERS ON PAGE A-1

Graph the inequality.

20. $x > -1$

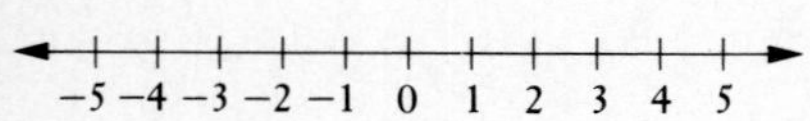

21. $x \leq 5$

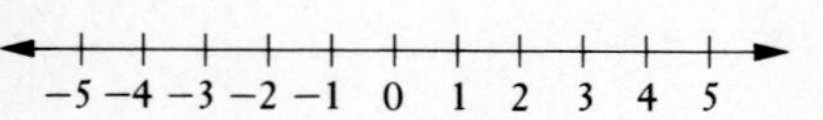

Find the absolute value.

22. $\left|-\frac{1}{4}\right|$

23. $|2|$

24. $\left|\frac{3}{2}\right|$

25. $|-2.3|$

ANSWERS ON PAGE A-1

▶ **EXAMPLE 17** Graph the inequality $x \leq 2$ on a number line.

We make a drawing that represents the solution set $\{x \mid x \leq 2\}$. This time the graph consists of 2 as well as the numbers less than 2. We shade all numbers to the left of 2 and use a solid circle at 2 to indicate that it is also a solution.

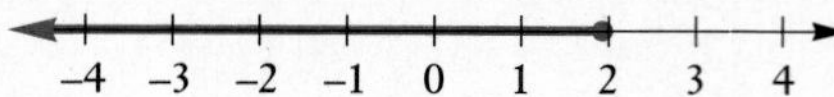

◀

DO EXERCISES 20 AND 21.

d Absolute Value

From the number line, we see that numbers like 6 and -6 are the same distance from zero. We call the distance from zero the **absolute value** of the number.

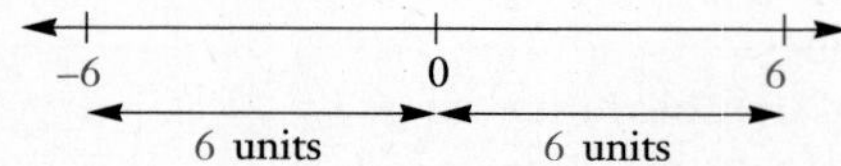

The *absolute value* of a number is its distance from zero on a number line. We use the symbol $|x|$ to represent the absolute value of a number x.

To find absolute value:

1. If a number is negative, make it positive.

2. If a number is positive or zero, leave it alone.

▶ **EXAMPLES** Find the absolute value.

18. $|-7|$ The distance of -7 from 0 is 7, so $|-7|$ is 7.

19. $|12|$ The distance of 12 from 0 is 12, so $|12|$ is 12.

20. $|0|$ The distance of 0 from 0 is 0, so $|0|$ is 0.

21. $\left|\frac{4}{5}\right| = \frac{4}{5}$

22. $|-3.86| = 3.86$ ◀

DO EXERCISES 22–25.

To the student and the instructor: The *Skill Maintenance Exercises,* which occur at the end of most exercise sets, review any skill that has been previously studied in the text.

NAME SECTION DATE

EXERCISE SET 1.2

a Given the numbers -5, 0, 2, $-\frac{1}{2}$, -3, $\frac{7}{8}$, 14, $-\frac{8}{3}$, 2.43, $7\frac{1}{2}$, $\sqrt{3}$, $\sqrt{16}$, $-\frac{15}{3}$, $0.121221222122221\ldots$:

1. Name the natural numbers.

2. Name the whole numbers.

3. Name the rational numbers.

4. Name the integers.

5. Name the real numbers.

6. Name the irrational numbers.

Given the numbers $-\sqrt{5}$, -3.43, -11, 12, 0, $\frac{11}{34}$, $-\frac{7}{13}$, π, $-3.565665666566665\ldots$:

7. Name the whole numbers.

8. Name the natural numbers.

9. Name the integers.

10. Name the rational numbers.

11. Name the irrational numbers.

12. Name the real numbers.

Use roster notation to name the set.

13. The set of all letters in the word "math"

14. The set of all letters in the word "solve"

15. The set of all even whole numbers less than 15

16. The set of all odd whole numbers less than 15

17. The set of all even natural numbers

18. The set of all odd natural numbers

Use set-builder notation to name the set.

19. $\{0, 1, 2, 3, 4, 5\}$

20. $\{4, 5, 6, 7, 8, 9, 10\}$

21. The set of all rational numbers

22. The set of all real numbers

23. The set of real numbers greater than -3

24. The set of all real numbers less than or equal to 21

b Use either $<$ or $>$ for ■ to write a true sentence.

25. 7 ■ 0

26. 8 ■ 0

27. -10 ■ 5

28. 9 ■ -9

29. -8 ■ 8

30. 0 ■ -11

31. -8 ■ -3

32. -6 ■ -3

33. -5 ■ -13

34. -3 ■ -6

35. -8.8 ■ -3.3

36. -16.5 ■ 0.022

37. 37.2 ■ -1.67

38. -13.99 ■ -8.45

39. $\frac{6}{13}$ ■ $\frac{13}{25}$

40. $-\frac{14}{15}$ ■ $-\frac{27}{53}$

ANSWERS

1. ____
2. ____
3. ____
4. ____
5. ____
6. ____
7. ____
8. ____
9. ____
10. ____
11. ____
12. ____
13. ____
14. ____
15. ____
16. ____
17. ____
18. ____
19. ____
20. ____
21. ____
22. ____
23. ____
24. ____
25. ____
26. ____
27. ____
28. ____
29. ____
30. ____
31. ____
32. ____
33. ____
34. ____
35. ____
36. ____
37. ____
38. ____
39. ____
40. ____

ANSWERS

41.
42.
43.
44.
45.
46.
47.
48.
49. See graph.
50. See graph.
51. See graph.
52. See graph.
53. See graph.
54. See graph.
55. See graph.
56. See graph.
57.
58.
59.
60.
61.
62.
63.
64.
65.
66.
67.
68.
69.
70.
71.
72.
73.
74.
75.

Write an inequality with the same meaning.

41. $-8 > x$ **42.** $x < 7$ **43.** $-12 \leq y$ **44.** $10 \geq t$

Write true or false.

45. $6 \leq -6$ **46.** $-7 \leq -7$ **47.** $5 \geq -8$ **48.** $-11 \geq -13$

c Graph the inequality.

49. $x < -2$

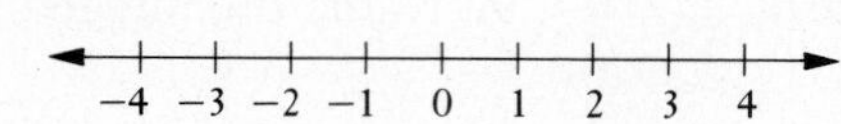

50. $x < -1$

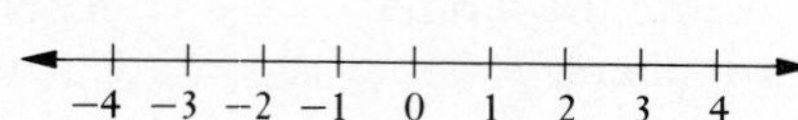

51. $x \leq -2$

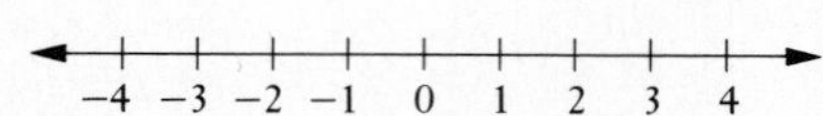

52. $x \geq -1$

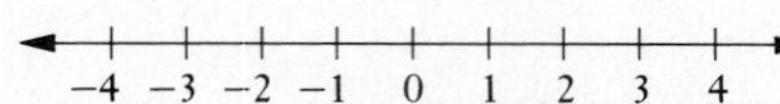

53. $x > -4$

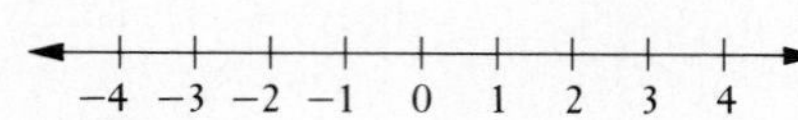

54. $x < 0$

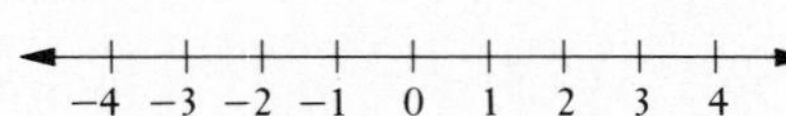

55. $x \geq 2$

$-4 \quad -3 \quad -2 \quad -1 \quad 0 \quad 1 \quad 2 \quad 3 \quad 4$

56. $x \leq 0$

$-4 \quad -3 \quad -2 \quad -1 \quad 0 \quad 1 \quad 2 \quad 3 \quad 4$

d Find the absolute value.

57. $|-5|$ **58.** $|-6|$ **59.** $|18|$ **60.** $|12|$ **61.** $|-46|$

62. $|465|$ **63.** $\left|-\frac{2}{3}\right|$ **64.** $\left|-\frac{13}{8}\right|$ **65.** $\left|\frac{0}{7}\right|$ **66.** $|16.4|$

SKILL MAINTENANCE

67. Evaluate $8xy$ when $x = 8$ and $y = 20$.

68. Translate to an algebraic expression: Seventy-eight percent of y.

SYNTHESIS

69. List ten examples of rational numbers.

70. List ten examples of rational numbers that are *not* integers.

Use either $\leq$ or $\geq$ for ▒ to write true sentences.

71. $|-3|$ ▒ 5 **72.** $|-5|$ ▒ $|-2|$

73. $|4|$ ▒ $|-7|$ **74.** $|-8|$ ▒ $|8|$

75. List the following numbers in order from least to greatest.

$$1.1\%,\ \frac{1}{11},\ \frac{2}{7},\ 0.3\%,\ 0.11,\ \frac{1}{8}\%,\ 0.009,\ \frac{99}{1000},\ 0.286,\ \frac{1}{8},\ \frac{9}{100},\ 1\%$$

1.3 Operations on Real Numbers

We now review the addition, subtraction, multiplication, and division of real numbers.

a Addition

To explain addition of real numbers, we can use a number line. To perform the addition $a + b$, we start at a and then move according to b. If b is positive, we move to the right. If b is negative, we move to the left.

▶ **EXAMPLES**

1. $6 + (-4) = 2$

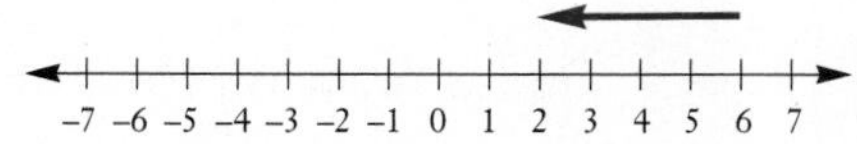

Start at 6. Move 4 units to the left. The answer is 2.

2. $-3 + 2 = -1$

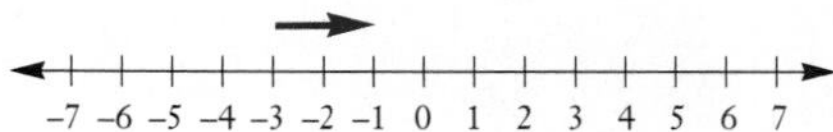

Start at -3. Move 2 units to the right. The answer is -1.

3. $-1 + (-5) = -6$

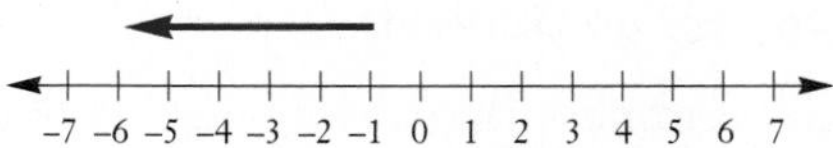

Start at -1. Move 5 units to the left. The answer is -6. ◀

DO EXERCISES 1–4.

You may have noticed some patterns in the preceding examples. These lead us to rules for adding without using a number line.

> Rules for Addition of Real Numbers
>
> 1. ***Positive numbers:*** **Add the numbers. The result is positive.**
> 2. ***Negative numbers:*** **Add absolute values. Make the answer negative.**
> 3. ***A positive and a negative number:*** **Subtract the smaller absolute value from the larger. Then:**
>   a) **If the positive number has the greater absolute value, make the answer positive.**
>   b) **If the negative number has the greater absolute value, make the answer negative.**
>   c) **If the numbers have the same absolute value, make the answer 0.**
> 4. ***One number is zero:*** **The sum is the other number.**

Rule 4 is known as the **Identity Property of Zero.** It says that for any real number a, $a + 0 = a$.

OBJECTIVES

After finishing Section 1.3, you should be able to:

- **a** Add real numbers.
- **b** Find the opposite, or additive inverse, of a number.
- **c** Subtract real numbers.
- **d** Multiply real numbers.
- **e** Divide real numbers.

FOR EXTRA HELP

Tape 1C

Tape 1B

MAC: 1
IBM: 1

Add using a number line.

1. $-3 + 5$

2. $9 + (-5)$

3. $6 + (-10)$

4. $-5 + 5$

ANSWERS ON PAGE A-1

Add.

5. $-7+(-11)$

6. $-8.9+(-9.7)$

7. $-\frac{6}{5}+\left(-\frac{23}{5}\right)$

8. $-\frac{3}{10}+\left(-\frac{2}{5}\right)$

9. $-7+7$

10. $-7.4+0$

11. $4+(-7)$

12. $-7.8+4.5$

13. $\frac{3}{8}+\left(-\frac{5}{8}\right)$

14. $-\frac{3}{5}+\frac{7}{10}$

Find the additive inverse, or opposite, of the number.

15. -14

16. $\frac{2}{3}$

17. 0

ANSWERS ON PAGE A-1

► **EXAMPLES** Add without using a number line.

4. $-13+(-8)=-21$ Two negatives. Think: Add the absolute values, 13 and 8, getting 21. Make the answer *negative*, -21.

5. $-2.1+8.5=6.4$ The absolute values are 2.1 and 8.5. Subtract 2.1 from 8.5. The positive number has the larger absolute value, so the answer is *positive*, 6.4.

6. $-48+31=-17$ The absolute values are 48 and 31. Subtract 31 from 48. The negative number has the larger absolute value, so the answer is *negative*, -17.

7. $2.6+(-2.6)=0$ The numbers have the same absolute value. The sum is 0.

8. $-\frac{5}{9}+0=-\frac{5}{9}$ One number is zero. The sum is $-\frac{5}{9}$.

9. $-9.4+3.8=-5.6$

10. $-\frac{3}{4}+\frac{9}{4}=\frac{6}{4}=\frac{3}{2}$

11. $-\frac{2}{3}+\frac{5}{8}=-\frac{16}{24}+\frac{15}{24}=-\frac{1}{24}$ ◄

DO EXERCISES 5–14.

b Additive Inverses, or Opposites

Suppose we add two numbers that are opposites, such as 4 and -4. The result is 0. When opposites are added, the result is always 0. Such numbers are also called **additive inverses.** Every real number has an additive inverse.

> **Two numbers whose sum is 0 are called *additive inverses,* or *opposites* of each other.**

► **EXAMPLES** Find the additive inverse, or opposite, of the number.

12. 8.6 The additive inverse of 8.6 is -8.6 because $8.6+(-8.6)=0$.

13. 0 The additive inverse of 0 is 0 because $0+0=0$.

14. $-\frac{7}{9}$ The additive inverse of $-\frac{7}{9}$ is $\frac{7}{9}$ because $-\frac{7}{9}+\frac{7}{9}=0$. ◄

To name the additive inverse, or opposite, we use the symbol $-$, and read the symbolism $-a$ as "the opposite of a" or "the additive inverse of a."

DO EXERCISES 15–17.

► **EXAMPLE 15** Evaluate $-x$ and $-(-x)$ (a) when $x=23$ and (b) when $x=-5$.

a) If $x=23$, then $-x=-23$. **The additive inverse, or opposite, of 23 is -23.**
If $x=23$, then $-(-x)=-(-23)=23$. **The opposite of the opposite of 23 is 23.**

b) If $x=-5$, then $-x=-(-5)=5$.
If $x=-5$, then $-(-x)=-(-(-5))=-5$. ◄

Note in Example 15(b) that an extra set of parentheses is used to show that we are substituting the negative number -5 for x. Symbolism like $--x$ is not considered meaningful.

DO EXERCISES 18–21.

A symbol such as -8 is usually read "negative 8." It could be read "the additive inverse of 8," because the additive inverse of 8 is negative 8. It could also be read "the opposite of 8," because the opposite of 8 is -8. Thus a symbol like $-x$ can be read "the additive inverse of x" or "the opposite of x" but *not* "negative x," because we do not know whether it represents a positive number, a negative number, or 0. Check this out by referring to the preceding examples.

We can use the symbolism $-a$ for the additive inverse of a to restate the definition of additive inverse.

> **For any real number a, the additive inverse of a, $-a$, is such that**
> $$a + (-a) = (-a) + a = 0.$$

Signs of Numbers

A negative number is sometimes said to have a "negative sign." A positive number is said to have a "positive sign." When we replace a number by its additive inverse, or opposite, we can say that we have "changed its sign."

► **EXAMPLES** Change the sign. (Find the additive inverse, or opposite.)

16. -3 $\quad -(-3) = 3$

17. -10 $\quad -(-10) = 10$

18. 0 $\quad -0 = 0$

19. 14 $\quad -(14) = -14$ ◄

DO EXERCISE 22.

We can now use the concept of opposite to give a more formal definition of absolute value.

> **For any real number a, the *absolute value* of a, denoted $|a|$, is given by**
> $$|a| = \begin{cases} a, & \text{if } a \geq 0 \quad \text{For example, } |8| = 8 \text{ and } |0| = 0. \\ -a, & \text{if } a < 0. \quad \text{For example, } |-5| = -(-5) = 5. \end{cases}$$
> **(The absolute value of a is a if a is nonnegative. The absolute value of a is the opposite of a if a is negative.)**

C Subtraction

Subtraction is defined as follows.

> **The difference $a - b$ is the number that when added to b gives a. That is, $a - b = c$, if c is a number such that $a = c + b$.**

To see that the definition is right, think of $3 - 5$. When you subtract, you get -2. To check, you add -2 to 5, to see if -2 is the number that when added to 5 gives 3. This definition is not generally used to carry out subtraction of real numbers. From the definition, the following can be proven.

> **For any real numbers a and b, $a - b = a + (-b)$. (We can subtract by adding the opposite (additive inverse) of the number being subtracted.)**

18. Evaluate $-a$ when $a = 9$.

19. Evaluate $-a$ when $a = -\frac{3}{5}$.

20. Evaluate $-(-a)$ when $a = -5.9$.

21. Evaluate $-(-a)$ when $a = \frac{2}{3}$.

22. Change the sign.

a) 11

b) -17

c) 0

d) x

e) $-x$

ANSWERS ON PAGE A-1

Subtract.

23. $8 - (-9)$

24. $-10 - 6$

25. $5 - 8$

26. $-23.7 - 5.9$

27. $-2 - (-5)$

28. $-\frac{11}{12} - \left(-\frac{23}{12}\right)$

29. $\frac{2}{3} - \left(-\frac{5}{6}\right)$

30. a) $17 - 23$

b) $-17 - 23$

c) $-17 - (-23)$

31. Look for a pattern and complete.

$4 \cdot 5 = 20$	$-2 \cdot 5 =$
$3 \cdot 5 = 15$	$-3 \cdot 5 =$
$2 \cdot 5 =$	$-4 \cdot 5 =$
$1 \cdot 5 =$	$-5 \cdot 5 =$
$0 \cdot 5 =$	$-6 \cdot 5 =$
$-1 \cdot 5 =$	

Multiply.

32. $-4 \cdot 6$

33. $(3.5)(-8.1)$

34. $-\frac{4}{5} \cdot 10$

35. Look for a pattern and complete.

$4(-5) = -20$	$-1(-5) =$
$3(-5) = -15$	$-2(-5) =$
$2(-5) =$	$-3(-5) =$
$1(-5) =$	$-4(-5) =$
$0(-5) =$	$-5(-5) =$

Multiply.

36. $-8(-9)$

37. $\left(-\frac{4}{5}\right) \cdot \left(-\frac{2}{3}\right)$

38. $(-4.7)(-9.1)$

ANSWERS ON PAGE A-1

▶ **EXAMPLES** Subtract.

20. $3 - 5 = 3 + (-5) = -2$ **Changing the sign of 5 and adding**

21. $7 - (-3) = 7 + (3) = 10$ **Changing the sign of -3 and adding**

22. $-19.4 - 5.6 = -19.4 + (-5.6) = -25$

23. $-\frac{4}{3} - \left(-\frac{2}{5}\right) = -\frac{4}{3} + \frac{2}{5} = -\frac{20}{15} + \frac{6}{15} = -\frac{14}{15}$ ◀

DO EXERCISES 23–30.

d Multiplication

We know how to multiply positive numbers. What happens when we multiply a positive number and a negative number?

DO EXERCISE 31.

> **To multiply a positive number and a negative number, we multiply their absolute values. Then we make the answer negative.**

▶ **EXAMPLES**

24. $-3 \cdot 5 = -15$

25. $6 \cdot (-7) = -42$

26. $(-1.2)(4.5) = -5.4$

27. $3 \cdot \left(-\frac{1}{2}\right) = \frac{3}{1} \cdot \left(-\frac{1}{2}\right) = -\frac{3}{2}$ ◀

DO EXERCISES 32–34.

What happens when we multiply two negative numbers?

DO EXERCISE 35.

> **To multiply two negative numbers, we multiply their absolute values. The answer is positive.**

▶ **EXAMPLES** Multiply.

28. $-3 \cdot (-5) = 15$

29. $-5.2(-10) = 52$

30. $(-8.8)(-3.5) = 30.8$

31. $\left(-\frac{3}{4}\right) \cdot \left(-\frac{5}{2}\right) = \frac{15}{8}$ ◀

DO EXERCISES 36–38.

e Division

> **The quotient a/b is defined to be the number c (if it exists) that when multiplied by b gives a. That is,**
>
> $$\frac{a}{b} = c,$$
>
> **if c is a number such that $a = cb$.**

The definition of division parallels the one for subtraction. Using this definition and the rules for multiplying, we can see how to handle signs when dividing.

▶ **EXAMPLES** Divide.

32. $\frac{10}{-2} = -5$, because $-5 \cdot (-2) = 10$

33. $\frac{-32}{4} = -8$, because $-8 \cdot (4) = -32$

34. $\frac{-25}{-5} = 5$, because $5 \cdot (-5) = -25$ ◀

In division, if both numbers are negative, the answer is positive. If one is positive and one is negative, the answer is negative.

▶ **EXAMPLES** Divide.

35. $\frac{40}{-4} = -10$

36. $-10 \div 5 = -2$

37. $\frac{-10}{-40} = \frac{1}{4}$, or 0.25

38. $\frac{-10}{-3} = \frac{10}{3}$, or $3.33\overline{3}$ ◀

We can restate the rules for division and multiplication as follows.

To multiply or divide two real numbers:

1. **Multiply or divide the absolute values.**
2. **If the signs are the same, the answer is positive.**
3. **If the signs are different, the answer is negative.**

DO EXERCISES 39–42.

Division by Zero

We do not divide a nonzero number n by zero. Let's see why. By the definition of division, $n/0$ would be some number that when multiplied by 0 gives n. But when any number is multiplied by 0, the result is 0. The only possibility for n would be 0.

Consider 0/0. We might say that it is 5 because $5 \cdot 0 = 0$. We might also say that it is -8 because $-8 \cdot 0 = 0$. In fact, 0/0 could be any number at all. So we agree not to divide by 0. Division by 0 is undefined and not possible.

▶ **EXAMPLES** Divide, if possible.

39. $\frac{7}{0}$ Undefined: Division by 0.

40. $\frac{0}{7} = 0$ The quotient is 0 because $0 \cdot 7 = 0$.

41. $\frac{4}{x - x}$ Undefined: $x - x = 0$ for any x. ◀

DO EXERCISES 43–46.

Divide.

39. $\frac{-28}{-14}$

40. $125 \quad (-5)$

41. $\frac{-75}{25}$

42. $-4.2 \div (-21)$

Divide, if possible.

43. $\frac{8}{0}$

44. $\frac{0}{9}$

45. $\frac{17}{2x - 2x}$

46. $\frac{3x - 3x}{x - x}$

ANSWERS ON PAGE A-1

Find the reciprocal of the number.

47. $\frac{3}{8}$

48. $-\frac{4}{5}$

49. 18

50. -4.3

51. 0.5

52. Complete this table.

	Opposite (additive inverse)	Reciprocal (multiplicative inverse)
$\frac{2}{3}$	$-\frac{2}{3}$	$\frac{3}{2}$
$\frac{4}{5}$		
$-\frac{3}{4}$		
0.25		
8		
-5		
0		

Divide by multiplying by the reciprocal of the divisor.

53. $-\frac{3}{4} \div \frac{7}{8}$

54. $-\frac{12}{5} \div \left(-\frac{7}{15}\right)$

ANSWERS ON PAGE A-2

Division and Reciprocals

Two numbers whose product is 1 are called **reciprocals** (or **multiplicative inverses**) of each other.

Every nonzero real number a has a reciprocal (multiplicative inverse) $1/a$. The reciprocal of a positive number is positive. The reciprocal of a negative number is negative.

▶ **EXAMPLES** Find the reciprocal of the number.

42. $\frac{4}{5}$ The reciprocal is $\frac{5}{4}$, because $\frac{4}{5} \cdot \frac{5}{4} = 1$.

43. 8 The reciprocal is $\frac{1}{8}$, because $8 \cdot \frac{1}{8} = 1$.

44. $-\frac{2}{3}$ The reciprocal is $-\frac{3}{2}$, because $-\frac{2}{3} \cdot \left(-\frac{3}{2}\right) = 1$.

45. 0.25 The reciprocal is $\frac{1}{0.25}$ or 4, because $0.25 \cdot 4 = 1$.

Remember that a number and its reciprocal (multiplicative inverse) have the same sign. Do *not* change sign when taking the reciprocal of a number. When finding an opposite (additive inverse), change the sign. ◀

DO EXERCISES 47–52.

We know that we can subtract by adding an opposite, or additive inverse. Similarly, we can divide by multiplying by a reciprocal.

For any real numbers a and b, $b \neq 0$,

$$a \div b = \frac{a}{b} = a \cdot \frac{1}{b}.$$

(To divide, we can multiply by the reciprocal of the divisor.)

We sometimes say that we "invert the divisor and multiply."

▶ **EXAMPLES** Divide by multiplying by the reciprocal of the divisor.

46. $\frac{1}{4} \div \frac{3}{5} = \frac{1}{4} \cdot \frac{5}{3} = \frac{5}{12}$ **"Inverting" the divisor, $\frac{3}{5}$, and multiplying**

47. $\frac{2}{3} \div \left(-\frac{4}{9}\right) = \frac{2}{3} \cdot \left(-\frac{9}{4}\right) = -\frac{18}{12}$, or $-\frac{3}{2}$ ◀

DO EXERCISES 53 AND 54.

The following properties can be used to make sign changes in fractional notation.

For any numbers a and b, $b \neq 0$,

$$\frac{-a}{b} = \frac{a}{-b} = -\frac{a}{b} \quad \text{and} \quad \frac{-a}{-b} = \frac{a}{b}.$$

NAME SECTION DATE

EXERCISE SET 1.3

a Add.

1. $-12+(-16)$ **2.** $-11+(-18)$ **3.** $8+(-3)$ **4.** $9+(-4)$

5. $-8+(-8)$ **6.** $-6+(-6)$ **7.** $7+(-11)$ **8.** $9+(-12)$

9. $-16+9$ **10.** $-23+8$ **11.** $-24+0$ **12.** $0+(-34)$

13. $-8.4+9.6$ **14.** $-6.3+8.2$ **15.** $-2.62+(-6.24)$

16. $-5.83+(-7.43)$ **17.** $-\frac{2}{7}+\frac{3}{7}$ **18.** $-\frac{5}{6}+\frac{1}{6}$

19. $-\frac{11}{12}+\left(-\frac{5}{12}\right)$ **20.** $-\frac{3}{8}+\left(-\frac{7}{8}\right)$ **21.** $\frac{2}{5}+\left(-\frac{3}{10}\right)$

22. $-\frac{3}{4}+\frac{1}{8}$ **23.** $-\frac{2}{5}+\frac{3}{4}$ **24.** $-\frac{5}{6}+\left(-\frac{7}{8}\right)$

b Evaluate $-a$ when a is each of the following.

25. 7 **26.** -8 **27.** -2.8 **28.** 0

Find the opposite (additive inverse).

29. 10 **30.** -8 **31.** 0 **32.** $-2x$

c Subtract.

33. $5-7$ **34.** $9-12$ **35.** $-5-7$ **36.** $-9-12$

ANSWERS

1. ____
2. ____
3. ____
4. ____
5. ____
6. ____
7. ____
8. ____
9. ____
10. ____
11. ____
12. ____
13. ____
14. ____
15. ____
16. ____
17. ____
18. ____
19. ____
20. ____
21. ____
22. ____
23. ____
24. ____
25. ____
26. ____
27. ____
28. ____
29. ____
30. ____
31. ____
32. ____
33. ____
34. ____
35. ____
36. ____

ANSWERS

37. ______
38. ______
39. ______
40. ______
41. ______
42. ______
43. ______
44. ______
45. ______
46. ______
47. ______
48. ______
49. ______
50. ______
51. ______
52. ______
53. ______
54. ______
55. ______
56. ______
57. ______
58. ______
59. ______
60. ______
61. ______
62. ______
63. ______
64. ______
65. ______
66. ______
67. ______
68. ______
69. ______
70. ______
71. ______
72. ______

37. $23 - 23$ **38.** $23 - (-23)$ **39.** $-23 - 23$ **40.** $-23 - (-23)$

41. $-6 - (-11)$ **42.** $-7 - (-12)$ **43.** $10 - (-5)$ **44.** $28 - (-16)$

45. $15.8 - 27.4$ **46.** $17.2 - 34.9$ **47.** $-18.01 - 11.24$

48. $-19.04 - 15.76$ **49.** $-\frac{21}{4} - \left(-\frac{7}{4}\right)$ **50.** $-\frac{16}{5} - \left(-\frac{3}{5}\right)$

51. $-\frac{1}{2} - \left(-\frac{1}{12}\right)$ **52.** $-\frac{3}{4} - \left(-\frac{3}{2}\right)$ **53.** $-\frac{7}{8} - \frac{5}{6}$

54. $-\frac{2}{3} - \frac{4}{5}$ **55.** $\frac{1}{3} - \frac{4}{5}$ **56.** $-\frac{4}{7} - \left(-\frac{5}{9}\right)$

d Multiply.

57. $3(-7)$ **58.** $5(-8)$ **59.** $-2 \cdot 4$ **60.** $-5 \cdot 9$

61. $-8(-2)$ **62.** $-7(-3)$ **63.** $-9 \cdot 14$ **64.** $-8 \cdot 17$

65. $-6(-5.7)$ **66.** $-7(-6.1)$ **67.** $-\frac{3}{5} \cdot \frac{4}{7}$ **68.** $-\frac{5}{4} \cdot \frac{11}{3}$

69. $-3\left(-\frac{2}{3}\right)$ **70.** $-5\left(-\frac{3}{5}\right)$ **71.** $-3(-4)(5)$ **72.** $-6(-8)(9)$

73. $(-4.2)(-6.3)$

74. $(-7.4)(-9.6)$

75. $-\frac{9}{11} \cdot \left(-\frac{11}{9}\right)$

76. $-\frac{13}{7} \cdot \left(-\frac{5}{2}\right)$

77. $-\frac{2}{3} \cdot \left(-\frac{2}{3}\right) \cdot \left(-\frac{2}{3}\right)$

78. $-\frac{4}{5} \cdot \left(-\frac{4}{5}\right) \cdot \left(-\frac{4}{5}\right)$

e Divide, if possible.

79. $\frac{-8}{4}$

80. $\frac{-16}{2}$

81. $\frac{56}{-8}$

82. $\frac{63}{-7}$

83. $-77 \quad (-11)$

84. $-48 \div (-6)$

85. $\frac{-5.4}{-18}$

86. $\frac{-8.4}{-12}$

87. $\frac{9}{0}$

88. $\frac{84}{0}$

89. $\frac{0}{16}$

90. $\frac{0}{28}$

91. $\frac{9}{y-y}$

92. $\frac{2x-2x}{2x-2x}$

Find the reciprocal of the number.

93. $\frac{3}{4}$

94. $\frac{9}{10}$

95. $-\frac{7}{8}$

96. $-\frac{5}{6}$

97. 15

98. -75

99. 4.4

100. -5.5

Divide.

101. $\frac{2}{7} \div \left(-\frac{11}{3}\right)$

102. $\frac{3}{5} \div \left(-\frac{6}{7}\right)$

103. $-\frac{10}{3} \div \left(-\frac{2}{15}\right)$

104. $-\frac{12}{5} \div \left(-\frac{3}{10}\right)$

105. $18.6 \div (-3.1)$

106. $39.9 \div (-13.3)$

ANSWERS

73. ____
74. ____
75. ____
76. ____
77. ____
78. ____
79. ____
80. ____
81. ____
82. ____
83. ____
84. ____
85. ____
86. ____
87. ____
88. ____
89. ____
90. ____
91. ____
92. ____
93. ____
94. ____
95. ____
96. ____
97. ____
98. ____
99. ____
100. ____
101. ____
102. ____
103. ____
104. ____
105. ____
106. ____

ANSWERS

107. ______
108. ______
109. ______
110. ______
111. ______
112. ______
113. ______
114. ______
115. ______
116. ______
117. ______
118. ______
119. ______
120. ______
121. ______
122. ______
123. ______
124. ______
125. ______
126. ______
127. ______
128. ______
129. ______
130. ______
131. ______

107. $(-75.5) \div (-15.1)$ **108.** $(-12.1) \div (-0.11)$ **109.** $-48 \div 0.4$

110. $520 \div (-0.13)$ **111.** $\frac{3}{4} \div \left(-\frac{2}{3}\right)$ **112.** $\frac{5}{8} \div \left(-\frac{1}{2}\right)$

113. $-\frac{5}{4} \div \left(-\frac{3}{4}\right)$ **114.** $-\frac{5}{9} \div \left(-\frac{5}{6}\right)$ **115.** $-\frac{2}{3} \div \left(-\frac{4}{9}\right)$

116. $-\frac{3}{5} \div \left(-\frac{5}{8}\right)$ **117.** $-\frac{3}{8} \div \left(-\frac{8}{3}\right)$ **118.** $-\frac{5}{8} \div \left(-\frac{5}{6}\right)$

119. $-6.6 \div 3.3$ **120.** $-44.1 \div (-6.3)$ **121.** $\frac{-12}{-13}$

122. $\frac{-1.9}{20}$ **123.** $\frac{48.6}{-30}$ **124.** $\frac{-17.8}{3.2}$ **125.** $\frac{-9}{17-17}$ **126.** $\frac{-8}{-6+6}$

127. Complete the following table.

Number	Opposite (additive inverse)	Reciprocal (multiplicative inverse)
$\frac{2}{3}$		
$-\frac{5}{4}$		
0		
1		
-4.5		

SKILL MAINTENANCE

128. Translate to an algebraic expression: The sum of p and q.

129. Evaluate $\frac{y}{x}$ when $y = 72$ and $x = 9$.

SYNTHESIS

130. What number can be added to 11.7 to obtain $-7\frac{3}{4}$?

131. The reciprocal of an electric resistance is called **conductance.** When two resistors are connected in parallel, the conductance is the sum of the conductances,

$$\frac{1}{r_1} + \frac{1}{r_2}.$$

Find the conductance of two resistors of 10 ohms and 5 ohms when connected in parallel.

1.4 Properties of Real Numbers

a Equivalent Expressions

In solving equations and performing other operations in algebra, we manipulate expressions in various ways. For example, instead of

$$x + 2x$$

we might write

$$3x,$$

knowing that the two expressions represent the same number for any replacement of x that is meaningful. In that sense, the expressions $x + 2x$ and $3x$ are **equivalent**.

> **Two expressions that have the same value for all meaningful replacements are called *equivalent*.**

The expressions $x - x$ and $5x$ are *not* equivalent.

DO EXERCISES 1 AND 2.

In this section, we will consider several laws of real numbers that will allow us to find equivalent expressions. The first two laws are the **identity properties of 0 and 1.**

> The Identity Property of 0
>
> **For any real number a, $a + 0 = 0 + a = a$. (The number 0 is the *additive identity*.)**
>
> The Identity Property of 1
>
> **For any real number a, $a \cdot 1 = 1 \cdot a = a$. (The number 1 is the *multiplicative identity*.)**

We often refer to the use of the identity property of 1 as "multiplying by 1."

► **EXAMPLE 1** Use multiplying by 1 to find an expression equivalent to $\frac{3}{5}$ with a denominator of $5x$.

We multiply by 1, using x/x as a name for 1:

$$\frac{3}{5} = \frac{3}{5} \cdot 1 = \frac{3}{5} \cdot \frac{x}{x} = \frac{3x}{5x}.$$ ◄

Note that the expressions $3/5$ and $3x/5x$ are equivalent. They have the same value for any meaningful replacement. Note too that 0 is not a meaningful replacement in $3x/5x$, but for all nonzero real numbers the expressions $3/5$ and $3x/5x$ have the same value.

DO EXERCISE 3.

In algebra, we consider an expression like $3/5$ to be a "simplified" form of $3x/5x$. To find such simplified expressions, we factor numerator and denominator in order to remove a factor of 1.

OBJECTIVES

After finishing Section 1.4, you should be able to:

- **a** Find equivalent fractional expressions by multiplying by 1 and simplify fractional expressions.
- **b** Use the commutative and associative laws to find equivalent expressions.
- **c** Use the distributive laws to multiply expressions like 8 and $x - y$ and to factor expressions like $4x - 12$.
- **d** Collect like terms.
- **e** Simplify expressions by removing parentheses and collecting like terms.

FOR EXTRA HELP

Tape 1D

Tape 1B

MAC: 1
IBM: 1

1. Complete the following table by evaluating each expression for the given values.

	$x + 2x$	$3x$
$x = 5$		
$x = -6$		
$x = 4.8$		

2. Complete the following table by evaluating each expression for the given values.

	$6x - x$	$5x$
$x = 3$		
$x = -6$		
$x = 4.8$		

3. Use multiplying by 1 to find an expression equivalent to $\frac{2}{3}$ with a denominator of $3y$.

ANSWERS ON PAGE A-2

Simplify.

4. $\dfrac{2y}{3y}$

5. $-\dfrac{20m}{12m}$

6. Evaluate $x + y$ and $y + x$ when $x = -3$ and $y = 5$.

7. Evaluate xy and yx when $x = -2$ and $y = 7$.

ANSWERS ON PAGE A-2

▶ **EXAMPLE 2** Simplify: $-\dfrac{24y}{16y}$.

$$-\frac{24y}{16y} = -\frac{3 \cdot 8y}{2 \cdot 8y}$$ **We look for the largest common factor of the numerator and the denominator and factor each.**

$$= -\frac{3}{2} \cdot \frac{8y}{8y}$$ **Factoring the expression**

$$= -\frac{3}{2} \cdot 1$$ $\dfrac{8y}{8y} = 1$

$$= -\frac{3}{2}$$ **Removing a factor of 1 using the identity property of 1 in reverse** ◀

DO EXERCISES 4 AND 5.

b The Commutative and the Associative Laws

Let us examine the expressions $x + y$ and $y + x$, as well as xy and yx.

▶ **EXAMPLE 3** Evaluate $x + y$ and $y + x$ when $x = 5$ and $y = 8$.

We substitute 5 for x and 8 for y in both expressions:

$$x + y = 5 + 8 = 13; \qquad y + x = 8 + 5 = 13.$$ ◀

▶ **EXAMPLE 4** Evaluate xy and yx when $x = 26$ and $y = 13$.

We substitute 26 for x and 13 for y in both expressions:

$$xy = 26 \cdot 13 = 338; \qquad yx = 13 \cdot 26 = 338.$$ ◀

DO EXERCISES 6 AND 7.

Note that the expressions $x + y$ and $y + x$ have the same values no matter what the variables stand for. Thus they are equivalent. They assert that when we add two numbers, the order in which we add does not matter. Similarly, when we multiply two numbers, the order in which we multiply does not matter. Thus the expressions xy and yx are equivalent. They have the same values no matter what the variables stand for. These are examples of general patterns or laws.

> The Commutative Laws
>
> ***Addition.*** **For any numbers *a* and *b*,**
>
> $$a + b = b + a.$$
>
> **(We can change the order when adding without affecting the answer.)**
>
> ***Multiplication.*** **For any numbers *a* and *b*,**
>
> $$ab = ba.$$
>
> **(We can change the order when multiplying without affecting the answer.)**

Using a commutative law, we know that $x + 4$ and $4 + x$ are equivalent. Similarly, $5x$ and $x(5)$ are equivalent. Thus, in an algebraic expression, we can replace one by the other and the result would be equivalent to the original expression.

Now let us examine the expressions $a + (b + c)$ and $(a + b) + c$. Note that these expressions use parentheses as grouping symbols, and they also involve three numbers. Calculations within parentheses are to be done first.

▶ **EXAMPLE 5** Calculate and compare: $4 + (8 + 5)$ and $(4 + 8) + 5$.

$4 + (8 + 5) = 4 + 13$ **Calculating within parentheses first; adding the 8 and the 5**

$= 17$

$(4 + 8) + 5 = 12 + 5$ **Calculating within parentheses first; adding the 4 and the 8**

$= 17$

The two expressions are equivalent. Moving the parentheses to group the additions differently did not affect the value of the expression. ◀

▶ **EXAMPLE 6** Calculate and compare: $7 \cdot (4 \cdot 2)$ and $(7 \cdot 4) \cdot 2$.

$7 \cdot (4 \cdot 2) = 7 \cdot 8 = 56$ $\qquad$ $(7 \cdot 4) \cdot 2 = 28 \cdot 2 = 56$ ◀

DO EXERCISES 8 AND 9.

When only addition is involved, parentheses can be placed any way we please without affecting the answer. When only multiplication is involved, parentheses can be placed any way we please without affecting the answer.

The Associative Laws

Addition. **For any numbers *a*, *b*, and *c*,**

$$a + (b + c) = (a + b) + c.$$

(Numbers can be grouped in any manner for addition.)

Multiplication. **For any numbers *a*, *b*, and *c*,**

$$a \cdot (b \cdot c) = (a \cdot b) \cdot c.$$

(Numbers can be grouped in any manner for multiplication.

When only additions or only multiplications are involved, parentheses may be placed any way we please. Thus we often omit them. For example,

$x + (y + 3)$ means $x + y + 3$, and $l(wh)$ means lwh.

▶ **EXAMPLE 7** Use the commutative and the associative laws to write at least three expressions equivalent to $(x + 8) + y$.

a) $(x + 8) + y = x + (8 + y)$ **Using the associative law first and then the commutative law**

$= x + (y + 8)$

b) $(x + 8) + y = y + (x + 8)$ **Using the commutative law and then the commutative law again**

$= y + (8 + x)$

c) $(x + 8) + y = 8 + (x + y)$ **Using the commutative law first and then the associative law** ◀

DO EXERCISES 10 AND 11.

8. Calculate and compare: $10 + (9 + 2)$ and $(10 + 9) + 2$.

9. Calculate and compare: $11 \cdot (5 \cdot 8)$ and $(11 \cdot 5) \cdot 8$.

10. Use the commutative laws to write an expression equivalent to each of $y + 5$, ab, and $8 + mn$.

11. Use the commutative and the associative laws to write at least three expressions equivalent to $(2 \cdot x) \cdot y$.

ANSWERS ON PAGE A-2

Evaluate each expression when $a = 5$, $b = -2$, $c = 4$, and $d = 10$.

12. $ab + ad$ and $a(b + d)$

13. $cb - cd$ and $c(b - d)$

14. Evaluate each expression when $x = -2$ and determine whether the expressions in (a) and (b) are equivalent.

a) $3x + 1$

b) $3(x + 1)$

Simplify and compare.

15. a) $5(6 + 4)$

b) $5 \cdot 6 + 5 \cdot 4$

16. a) $59(100 - 1)$

b) $59 \cdot 100 - 59 \cdot 1$

Multiply.

17. $8(y - 10)$

18. $a(x + y - z)$

19. $10\left(4x - 6y + \frac{1}{2}z\right)$

ANSWERS ON PAGE A-2

C The Distributive Laws

We agree that in an expression like $(4 \cdot 10) + (9 \cdot 2)$, we can omit the parentheses. Thus, $4 \cdot 10 + 9 \cdot 2$ means $(4 \cdot 10) + (9 \cdot 2)$. In other words, we do the multiplications first. We make a similar agreement for subtraction. That is, $4 \cdot 10 - 9 \cdot 2$ means $(4 \cdot 10) - (9 \cdot 2)$. The multiplications are to be done first.

▶ **EXAMPLE 8** Evaluate the expressions $8x + 8y$ and $8(x + y)$ when $x = 4$ and $y = 5$. Then simplify.

a) $8x + 8y = 8 \cdot 4 + 8 \cdot 5$ **Substituting**
$= 32 + 40$ **Multiplying**
$= 72$ **Adding**

b) $8(x + y) = 8(4 + 5)$ **Substituting**
$= 8(9)$ **Adding**
$= 72$ **Multiplying** ◀

DO EXERCISES 12–14.

The expressions $8x + 8y$ and $8(x + y)$ are **equivalent.** This fact is the result of another law about real numbers. We can either add and then multiply, or multiply and then add.

> The Distributive Law of Multiplication Over Addition
>
> **For any numbers a, b, and c, $a(b + c) = ab + ac$.**

> Note that we cannot omit the parentheses in the expression $a(b + c)$. If we did, we would have $ab + c$, which by our agreement means $(ab) + c$.

There is another distributive law.

> The Distributive Law of Multiplication Over Subtraction
>
> **For any numbers a, b, and c, $a(b - c) = ab - ac$.**

We can subtract and then multiply, or we can multiply and then subtract. We often refer to "*the* distributive law" when we mean *either* of these laws.

DO EXERCISES 15 AND 16.

The distributive laws are the basis of multiplying in algebra as well as in arithmetic. In the following examples, note that we multiply each number or letter inside the parentheses by the factor outside.

▶ **EXAMPLES** Multiply.

9. $4(x - 2) = 4 \cdot x - 4 \cdot 2 = 4x - 8$

10. $b(s - t + f) = bs - bt + bf$

11. $-3(y + 4) = -3 \cdot y + (-3) \cdot 4 = -3y - 12$

12. $-2x(y - 1) = -2x \cdot y - (-2x) \cdot 1 = -2xy + 2x$ ◀

DO EXERCISES 17–19.

What do we mean by the **terms** of the expression? When there are only addition signs, it is easy to tell, but if there are some subtraction signs, we might not be as sure. If there are subtraction signs, we can always find an equivalent expression using addition signs.

▶ **EXAMPLE 13** What are the terms of $3x - 4y + 2z$?

We first find an equivalent expression that uses addition signs:

$3x - 4y + 2z = 3x + (-4y) + 2z$. **Using the property $a - b = a + (-b)$**

Thus the terms are $3x$, $-4y$, and $2z$. ◀

DO EXERCISE 20.

The reverse of multiplying is called **factoring.**

> **To *factor* an expression is to find an equivalent expression that is a product. If $N = ab$, then a and b are factors of N.**

▶ **EXAMPLES** Factor.

14. $8x + 8y = 8(x + y)$ **8 and $x + y$ are factors.**

15. $cx - cy = c(x - y)$ **c and $x - y$ are factors.** ◀

Whenever the terms of an expression have a factor in common, we can "remove" that factor using the distributive laws. We proceed as in Examples 14 and 15, but we may have to factor some of the terms first in order to display the common factor.

Generally, we try to factor out the largest factor common to all the terms. In the following example, we might factor out 3, but there is a larger factor common to the terms, 9. So we factor out the 9.

▶ **EXAMPLE 16** Factor: $9x + 27y$.

$$9x + 27y = 9x + 9 \cdot (3y) = 9(x + 3y)$$ ◀

We often have to supply a factor of 1 when factoring out a common factor, as in the next example, which is from a formula about simple interest.

▶ **EXAMPLE 17** Factor: $P + Prt$.

$P + Prt = P \cdot 1 + Prt$ **Writing P as a product of P and 1**

$= P(1 + rt)$ **Using the distributive law**

It is a common error to omit this 1. If you do that, when you factor out P, you might leave out an entire term. ◀

DO EXERCISES 21–25.

20. What are the terms of $-5x - 7y + 67t - \frac{4}{5}$?

Factor.

21. $9x + 9y$

22. $ac - ay$

23. $6x - 12$

24. $35x - 25y + 15w + 5$

25. $bs + bt - bw$

ANSWERS ON PAGE A-2

Collect like terms.

26. $9x + 11x$

27. $5x - 12x$

28. $5x + x$

29. $x - 7x$

30. $22x - 2.5y + 1.4x + 6.4y$

31. $\frac{2}{3}x - \frac{3}{4}y + \frac{4}{5}x - \frac{5}{6}y + 23$

Multiply.

32. $-1 \cdot 24$

33. $-1 \cdot 0$

34. $-1 \cdot (-10)$

ANSWERS ON PAGE A-2

d Collecting Like Terms

If two terms have the same letter, or letters, we say that they are **like terms,** or **similar terms.** If two terms have no letters at all but are just numbers, they are similar terms. We can simplify by **collecting** or **combining like terms,** using the distributive laws, which we can apply on the right of the equals sign, because of the commutative law of multiplication.

▶ **EXAMPLES** Collect like terms.

18. $3x + 5x = (3 + 5)x = 8x$

19. $x - 3x = 1 \cdot x - 3 \cdot x = (1 - 3)x = -2x$

20. $2x + 3y - 5x - 2y = 2x + 3y + (-5x) + (-2y)$ **Subtracting by adding an opposite**

$= 2x + (-5x) + 3y + (-2y)$ **Using the commutative law**

$= (2 - 5)x + (3 - 2)y$ **Using a distributive law**

$= -3x + y$ **Adding**

21. $3x + 2x + 5 + 7 = (3 + 2)x + (5 + 7) = 5x + 12$

22. $4.2x - 6.7y - 5.8x + 23y = (4.2 - 5.8)x + (-6.7 + 23)y$
$= -1.6x + 16.3y$

23. $-\frac{1}{4}a + \frac{1}{2}b - \frac{3}{5}a - \frac{2}{5}b = \left(-\frac{1}{4} - \frac{3}{5}\right)a + \left(\frac{1}{2} - \frac{2}{5}\right)b$
$= \left(-\frac{5}{20} - \frac{12}{20}\right)a + \left(\frac{5}{10} - \frac{4}{10}\right)b$
$= -\frac{17}{20}a + \frac{1}{10}b$ ◀

You need not write the intermediate steps when you can do the computations mentally.

DO EXERCISES 26–31.

e Multiplying by −1 and Removing Parentheses

What happens when we multiply a number by -1?

▶ **EXAMPLES**

24. $-1 \cdot 9 = -9$ **25.** $-1 \cdot \left(-\frac{3}{5}\right) = \frac{3}{5}$ **26.** $-1 \cdot 0 = 0$ ◀

DO EXERCISES 32–34.

> The Property of −1
>
> **For any number a, $-1 \cdot a = -a$. (Negative 1 times a is the opposite of a; in other words, changing the sign is the same as multiplying by -1.)**

From this fact, we know that we can replace $-$ by -1 or the reverse, in any expression. In that way, we can find an equivalent expression for an opposite.

▶ **EXAMPLES** Find an equivalent expression.

27. $-(3x) = -1(3x)$ **Replacing − by −1 using the property of −1**
$= (-1 \cdot 3)x$ **Using an associative law**
$= -3x$ **Multiplying**

28. $-(-9y) = -1(-9y)$ **Replacing − by −1**
$= [-1(-9)]y$ **Using an associative law**
$= 9y$ **Multiplying** ◀

DO EXERCISES 35 AND 36.

▶ **EXAMPLES** For each opposite, find an equivalent expression that does not have parentheses.

29. $-(4 + x) = -1(4 + x)$ **Replacing − by −1**
$= -1 \cdot 4 + (-1) \cdot x$ **Multiplying using a distributive law**
$= -4 + (-x)$ **Replacing $-1 \cdot x$ by $-x$**
$= -4 - x$ **Adding an opposite is the same as subtracting.**

30. $-(3x - 2y + 4) = -1(3x - 2y + 4)$
$= -1 \cdot 3x - (-1)2y + (-1)4$ **Using a distributive law**
$= -3x - (-2y) + (-4)$ **Multiplying**
$= -3x + [-(-2y)] + (-4)$ **Meaning of subtraction**
$= -3x + 2y - 4$

31. $-(a - b) = -1(a - b)$
$= -1 \cdot a - (-1) \cdot b$
$= -a + [-(-1)b]$
$= -a + b = b - a$

> Example 31 illustrates something that you should remember, because it is a shortcut. The opposite of an expression $a - b$ is $b - a$; that is, $-(a - b) = b - a$.

◀

DO EXERCISES 37–41.

The above examples show that we can find an equivalent expression for an opposite by multiplying every term by -1. We could also say that we change the sign of every term inside the parentheses. Thus we can skip some steps.

▶ **EXAMPLE 32** Find an equivalent expression that does not have parentheses:

$$-\left(-9t + 7z - \frac{1}{4}w\right).$$

We have

$$-\left(-9t + 7z - \frac{1}{4}w\right) = 9t - 7z + \frac{1}{4}w.$$ **Changing the sign of every term** ◀

DO EXERCISES 42–44.

Find an equivalent expression.

35. $-(9x)$

36. $-(-24t)$

For each opposite, find an equivalent expression that does not have parentheses.

37. $-(7 - y)$

38. $-(x - y)$

39. $-(9x + 6y + 11)$

40. $-(23x - 7y - 2)$

41. $-(-3x - 2y - 1)$

Find an equivalent expression that does not have parentheses.

42. $-(-2x - 5z + 24)$

43. $-(3x - 2y)$

44. $-\left(\frac{1}{4}t + 41w - 5d - 23\right)$

ANSWERS ON PAGE A-2

In some expressions commonly encountered in algebra, there are parentheses preceded by subtraction signs. These parentheses can be removed by changing the sign of *every* term inside. In this way, we simplify by finding a less complicated equivalent expression.

▶ **EXAMPLES** Remove parentheses and simplify.

33. $6x - (4x + 2) = 6x + [-(4x + 2)]$ **Subtracting by adding the opposite**
$= 6x - 4x - 2$ **Changing the sign of every term inside**
$= 2x - 2$ **Collecting like terms**

34. $3y - 4 - (9y - 7) = 3y - 4 - 9y + 7$
$= -6y + 3$, or $3 - 6y$

If parentheses are preceded by an addition sign, *no* signs are changed when they are removed.

35. $3y + (3x - 8) - (5 - 12y) = 3y + 3x - 8 - 5 + 12y$
$= 15y + 3x - 13$

36. $\frac{1}{3}(15x - 4) - (5x + 2y) + 1 = \frac{1}{3} \cdot 15x - \frac{1}{3} \cdot 4 - 5x - 2y + 1$
$= 5x - \frac{4}{3} - 5x - 2y + 1$
$= -2y - \frac{1}{3}$ ◀

DO EXERCISES 45–48.

We now consider subtracting an expression consisting of several terms preceded by a number other than -1.

▶ **EXAMPLES** Remove parentheses and simplify.

37. $x - 3(x + y) = x + [-3(x + y)]$ **Subtracting by adding the opposite**
$= x - 3x - 3y$ **Removing parentheses by multiplying $x + y$ by -3**
$= -2x - 3y$ **Collecting like terms**

38. $3y - 2(4y - 5) = 3y + [-2(4y - 5)]$ **Subtracting by adding the opposite**
$= 3y - 8y + 10$ **Removing parentheses by multiplying $4y - 5$ by -2**
$= -5y + 10$ **Collecting like terms**

A common error is to forget to change this sign. *Remember:* When multiplying like this, change the sign of *every* term inside the parentheses. ◀

DO EXERCISES 49–51.

Remove parentheses and simplify.

45. $6x - (3x + 8)$

46. $6y - 4 - (2y - 5)$

47. $6x - (9y - 4) - (8x + 10)$

48. $7x - (-9y - 4) + (8x - 10)$

Remove parentheses and simplify.

49. $x - 2(y + x)$

50. $3x - 5(2y - 4x)$

51. $(4a - 3b) - \frac{1}{4}(4a - 3) + 5$

ANSWERS ON PAGE A-2

NAME SECTION DATE

EXERCISE SET 1.4

a Use multiplying by 1 to find an expression equivalent to the given one with the indicated denominator.

1. $\frac{3x}{5}$; 10 **2.** $\frac{4y}{7}$; 14 **3.** $\frac{3}{4}$; $4x$ **4.** $\frac{2}{3}$; $3y$

Simplify.

5. $\frac{25x}{15x}$ **6.** $\frac{36y}{18y}$ **7.** $\frac{100}{25x}$ **8.** $\frac{625}{15y}$

b Use a commutative law to find an equivalent expression.

9. $m + 9$ **10.** $t + 7$ **11.** pq **12.** ab

13. $8 + ab$ **14.** $mn + 12$ **15.** $xy + z$ **16.** $r + pq$

Use an associative law to find an expression equivalent to each of the following.

17. $m + (n + 2)$ **18.** $5 \cdot (p \cdot q)$ **19.** $(7 \cdot x) \cdot y$ **20.** $(7 + p) + q$

Use the commutative and the associative laws to find three equivalent expressions.

21. $(a + b) + 8$ **22.** $(4 + x) + y$ **23.** $7 \cdot (a \cdot b)$ **24.** $(8 \cdot m) \cdot n$

c Evaluate the expression when $x = -2$, $y = 3$, and $z = -4$.

25. $xy + x$ **26.** $x(y + 1)$ **27.** $(x + y)z$ **28.** $xz + yz$

29. $xy - xz$ **30.** $x(y - z)$ **31.** $3x + 7$ **32.** $3(x + 7)$

ANSWERS

1. ______
2. ______
3. ______
4. ______
5. ______
6. ______
7. ______
8. ______
9. ______
10. ______
11. ______
12. ______
13. ______
14. ______
15. ______
16. ______
17. ______
18. ______
19. ______
20. ______
21. ______
22. ______
23. ______
24. ______
25. ______
26. ______
27. ______
28. ______
29. ______
30. ______
31. ______
32. ______

ANSWERS

33. ______
34. ______
35. ______
36. ______
37. ______
38. ______
39. ______
40. ______
41. ______
42. ______
43. ______
44. ______
45. ______
46. ______
47. ______
48. ______
49. ______
50. ______
51. ______
52. ______
53. ______
54. ______
55. ______
56. ______
57. ______
58. ______
59. ______
60. ______
61. ______
62. ______
63. ______
64. ______
65. ______
66. ______

The expression $P(1 + rt)$ gives the value of an account of P dollars, invested at a rate r (in percent notation) for a time t (in years). Find the value of an account under the following conditions.

33. $P = \$120, \quad r = 6\%, \quad t = 1$ yr

34. $P = \$500, \quad r = 8\%, \quad t = \frac{1}{2}$ yr

Multiply.

35. $3(a + 1)$

36. $8(x + 1)$

37. $4(x - y)$

38. $9(a - b)$

39. $-5(2a + 3b)$

40. $-2(3c + 5d)$

41. $2a(b - c + d)$

42. $5x(y - z + w)$

43. $2\pi r(h + 1)$

44. $P(1 + rt)$

45. $\frac{1}{2}h(a + b)$

46. $\frac{1}{4}\pi r(1 + s)$

What are the terms of the following?

47. $4a - 5b + 6$

48. $5x - 9y + 12$

49. $2x - 3y - 2z$

50. $5a - 7b - 9c$

Factor.

51. $18x + 18y$

52. $7a + 7b$

53. $9p - 9$

54. $12x - 12$

55. $7x - 21$

56. $6y - 36$

57. $xy + x$

58. $ab + a$

59. $2x - 2y + 2z$

60. $3x + 3y - 3z$

61. $3x + 6y - 3$

62. $4a + 8b - 4$

63. $ab + ac - ad$

64. $xy - xz + xw$

65. $\frac{1}{4}\pi rr + \frac{1}{4}\pi rs$

66. $\frac{1}{2}ah + \frac{1}{2}bh$

d Collect like terms.

67. $4a + 5a$

68. $9x + 3x$

69. $8b - 11b$

70. $9c - 12c$

71. $14y + y$

72. $13x + x$

73. $12a - a$

74. $15x - x$

75. $t - 9t$

76. $x - 6x$

77. $5x - 3x + 8x$

78. $3x - 11x + 2x$

79. $5x - 8y + 3x$

80. $9a - 10b + 4a$

81. $7c + 8d - 5c + 2d$

82. $12a + 3b - 5a + 6b$

83. $4x - 7 + 18x + 25$

84. $13p + 5 - 4p + 7$

85. $1.3x + 1.4y - 0.11x - 0.47y$

86. $0.17a + 1.7b - 12a - 38b$

87. $\frac{2}{3}a + \frac{5}{6}b - 27 - \frac{4}{5}a - \frac{7}{6}b$

88. $-\frac{1}{4}x - \frac{1}{2}x + \frac{1}{4}y + \frac{1}{2}y - 34$

The *perimeter* of a rectangle is the distance around it. The perimeter P is given by $P = 2l + 2w$.

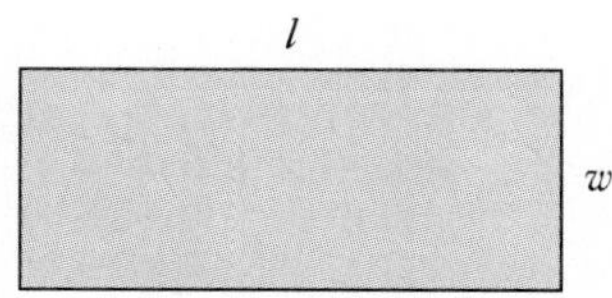

89. Find an equivalent expression for perimeter by factoring.

90. The standard football field has $l = 360$ ft and $w = 160$ ft. Evaluate both expressions to find the perimeter.

e For each opposite, find an equivalent expression that does not have parentheses.

91. $-(-4b)$

92. $-(-5x)$

93. $-(a + 2)$

94. $-(b + 9)$

95. $-(b - 3)$

96. $-(x - 8)$

97. $-(t - y)$

98. $-(r - s)$

ANSWERS

67. ______
68. ______
69. ______
70. ______
71. ______
72. ______
73. ______
74. ______
75. ______
76. ______
77. ______
78. ______
79. ______
80. ______
81. ______
82. ______
83. ______
84. ______
85. ______
86. ______
87. ______
88. ______
89. ______
90. ______
91. ______
92. ______
93. ______
94. ______
95. ______
96. ______
97. ______
98. ______

ANSWERS

99.

100.

101.

102.

103.

104.

105.

106.

107.

108.

109.

110.

111.

112.

113.

114.

115.

116.

117.

118.

119.

120.

121.

122.

123.

124.

125.

126.

127.

128.

129.

130.

131.

99. $-(a + b + c)$

100. $-(x + y + z)$

101. $-(8x - 6y + 13)$

102. $-(9a - 7b + 24)$

103. $-(-2c + 5d - 3e + 4f)$

104. $-(-4x + 8y - 5w + 9z)$

105. $-\left(-1.2x + 56.7y - 34z - \frac{1}{4}\right)$

106. $-\left(-x + 2y - \frac{2}{3}z - 56.3w\right)$

Simplify by removing parentheses and collecting like terms.

107. $a + (2a + 5)$

108. $x + (5x + 9)$

109. $4m - (3m - 1)$

110. $5a - (4a - 3)$

111. $3d - 7 - (5 - 2d)$

112. $8x - 9 - (7 - 5x)$

113. $-2(x + 3) - 5(x - 4)$

114. $-9(y + 7) - 6(y - 3)$

115. $5x - 7(2x - 3) - 4$

116. $8y - 4(5y - 6) + 9$

117. $8x - (-3y + 7) + (9x - 11)$

118. $-5t + (4t - 12) - 2(3t + 7)$

119. $\frac{1}{4}(24x - 8) - \frac{1}{2}(-8x + 6) - 14$

120. $-\frac{1}{2}(10t - w) + \frac{1}{4}(-28t + 4) + 1$

SKILL MAINTENANCE

Subtract.

121. $-12 - (-19)$

122. $-34.2 - 67.8$

123. $-\frac{11}{5} - \left(-\frac{17}{10}\right)$

Multiply.

124. $-\frac{1}{4}\left(-\frac{1}{2}\right)$

125. $-45(20)$

126. $0.23(-200)$

127. $-45(-90)$

SYNTHESIS

Make substitutions to determine whether the expressions seem to be equivalent.

128. $5x + 2;\quad 5(x + 2)$

129. $\frac{1}{2}x;\quad \frac{x}{2}$

130. $\frac{2x}{3};\quad \frac{2}{3}x$

131. $\frac{5x}{9};\quad \frac{5}{9x}$

1.5 Exponential Notation and Order of Operations

a Exponential Notation

Exponential notation is a shorthand device. For $3 \cdot 3 \cdot 3 \cdot 3$, we write 3^4. The latter is called **exponential notation.** In the expression 3^4, the number 3 is called the **base** and the 4 is called the **exponent.**

> **Exponential notation a^n, where n is an integer greater than 1, means**
>
> $$\underbrace{a \cdot a \cdot a \cdots a \cdot a}_{n \text{ factors.}}$$
>
> **We read "a^n" as "a to the nth power," or simply "a to the nth."**
> **We read "a^2" as "a-squared" and "a^3" as "a-cubed."**

> CAUTION! a^n does *not* mean to multiply n times a. For example, 3^2 means $3 \cdot 3$, or 9, not $3 \cdot 2$, or 6.

▶ **EXAMPLES** Write exponential notation.

1. $7 \cdot 7 \cdot 7 = 7^3$
2. $xxx = x^3$
3. $2x \cdot 2x \cdot 2x \cdot 2x = (2x)^4$ ◀

DO EXERCISES 1–3.

▶ **EXAMPLES** Rewrite without an exponent.

4. $5^2 = 5 \cdot 5$, or 25
5. $\left(\frac{1}{2}\right)^3 = \frac{1}{2} \cdot \frac{1}{2} \cdot \frac{1}{2}$, or $\frac{1}{8}$
6. $(4y)^2 = 4y \cdot 4y$, or $16yy$
7. $(0.1)^4 = (0.1)(0.1)(0.1)(0.1)$, or 0.0001 ◀

DO EXERCISES 4–8.

▶ **EXAMPLES** Simplify.

8. $(7x)^2 = 7x \cdot 7x = 49x^2$
9. $(-3y)^3 = (-3y)(-3y)(-3y) = -27y^3$ ◀

DO EXERCISES 9 AND 10.

In general, an exponent tells how many times the base occurs as a factor. What happens when the exponent is 1 or 0? We cannot have the base occurring as a factor 1 time or 0 times because there are no products. Look for a pattern below. Think of dividing by 10 on the right.

$$10^4 = 10 \cdot 10 \cdot 10 \cdot 10 = 10{,}000$$
$$10^3 = 10 \cdot 10 \cdot 10 = 1000$$
$$10^2 = 10 \cdot 10 = 100$$
$$10^1 = ?$$
$$10^0 = ?$$

In order for the pattern to continue, 10^1 would have to be 10 and 10^0 would have to be 1. We will *agree* that exponents of 1 and 0 have that meaning.

OBJECTIVES

After finishing Section 1.5, you should be able to:

a Rewrite expressions with or without whole-number exponents.

b Rewrite expressions with or without negative integers as exponents.

c Simplify expressions using the rules for order of operations, and simplify expressions containing parentheses within parentheses.

FOR EXTRA HELP

Tape 2A | Tape 2A | MAC: 1 IBM: 1

Write exponential notation.

1. $8 \cdot 8 \cdot 8 \cdot 8$
2. mmm
3. $4y \cdot 4y \cdot 4y \cdot 4y \cdot 4y$

Rewrite without exponents.

4. 3^4
5. $\left(\frac{1}{4}\right)^2$
6. y^2
7. $(0.2)^3$
8. $(5x)^4$

Simplify.

9. $(5y)^2$
10. $(-2)^3$

ANSWERS ON PAGE A-2

Rewrite without exponents.

11. 8^1

12. $(-31)^1$

13. 3^0

14. $(-7)^0$

15. y^0, where $y \neq 0$

Rewrite using a positive exponent.

16. m^{-4}

17. $(-4)^{-3}$

18. $\dfrac{1}{x^{-3}}$

Rewrite using a negative exponent.

19. $\dfrac{1}{q^3}$

20. $\dfrac{1}{(-5)^4}$

ANSWERS ON PAGE A-2

For any number a, we agree that a^1 means a.
For any nonzero number a, we agree that a^0 means 1.

CAUTION! a^0 does *not* mean a times 0! It means 1. We will see in Section 1.6 why we do not allow the base a to be zero.

▶ **EXAMPLES** Rewrite without an exponent.

10. $4^1 = 4$

11. $(-9y)^1 = -9y$

12. $6^0 = 1$

13. $(-37.4)^0 = 1$ ◀

DO EXERCISES 11–15.

b Negative Integers as Exponents

How shall we define negative integers as exponents? Look for a pattern below. Again, think of dividing by 10 on the right.

$$10^2 = 100$$
$$10^1 = 10$$
$$10^0 = 1$$
$$10^{-1} = ?$$
$$10^{-2} = ?$$

In order for the pattern to continue, 10^{-1} would have to be $\frac{1}{10}$ and 10^{-2} would have to be $\frac{1}{100}$. This leads to the following agreement.

If n is any integer, a^{-n} is given the meaning $1/a^n$. In other words, a^n and a^{-n} are reciprocals.

▶ **EXAMPLES** Rewrite using a positive exponent.

14. $y^{-5} = \dfrac{1}{y^5}$

15. $(-2)^{-3} = \dfrac{1}{(-2)^3}$

16. $\dfrac{1}{t^{-4}} = t^4$ (t^{-4} and t^4 are reciprocals) ◀

CAUTION! A negative exponent does *not* necessarily indicate that an answer is negative! For example, 3^{-2} means $1/3^2$, which is $1/9$.

DO EXERCISES 16–18.

▶ **EXAMPLES** Rewrite using a negative exponent.

17. $\dfrac{1}{x^2} = x^{-2}$

18. $\dfrac{1}{(-7)^4} = (-7)^{-4}$ ◀

DO EXERCISES 19 AND 20.

c Order of Operations

What does $8 + 2 \cdot 5^3$ mean? If we add 8 and 2 and multiply by 5^3, or 125, we get 1250. If we multiply 2 times 125 and add 8, we get 258. Both results cannot be correct. To avoid such difficulties, we make agreements about which operations should be done first.

> Rules for Order of Operations
>
> 1. **Do all the calculations within parentheses before operations outside.**
> 2. **Evaluate all exponential expressions.**
> 3. **Do all multiplications and divisions in order from left to right.**
> 4. **Do all additions and subtractions in order from left to right.**

Computers are programmed using these rules.

▶ **EXAMPLE 19** Simplify: $-43 \cdot 56 - 17$.

There are no parentheses or powers so we start with the third step.

$-43 \cdot 56 - 17 = -2408 - 17$ **Carrying out all multiplications and divisions in order from left to right**

$= -2425$ **Carrying out all additions and subtractions in order from left to right** ◀

▶ **EXAMPLE 20** Simplify and compare: $(8 - 10)^2$ and $8^2 - 10^2$.

$$(8 - 10)^2 = (-2)^2 = 4;$$
$$8^2 - 10^2 = 64 - 100 = -36$$

We see that $(8 - 10)^2$ and $8^2 - 10^2$ are *not* the same. ◀

▶ **EXAMPLE 21** Simplify: $3^4 + 62 \cdot 8 - 2(29 + 33 \cdot 4)$.

$3^4 + 62 \cdot 8 - 2(29 + 33 \cdot 4)$

$= 3^4 + 62 \cdot 8 - 2(29 + 132)$ **Carrying out operations inside parentheses first**

$= 3^4 + 62 \cdot 8 - 2(161)$ **Completing the addition inside parentheses**

$= 81 + 62 \cdot 8 - 2(161)$ **Evaluating exponential expressions**

$= 81 + 496 - 322$ **Doing all multiplications**

$= 577 - 322$ **Doing all additions and subtractions in order from left to right**

$= 255$

DO EXERCISES 21–23. ◀

When parentheses occur within parentheses, we can make them different shapes, such as [ ] (also called "brackets") and { } (usually called "braces"). All of these have the same meaning. When parentheses occur within parentheses, computations in the *innermost* ones are to be done first.

▶ **EXAMPLE 22** Simplify: $5 - \{6 - [3 - (7 + 3)]\}$.

$5 - \{6 - [3 - (7 + 3)]\} = 5 - \{6 - [3 - 10]\}$ **Adding 7 + 3**

$= 5 - \{6 - (-7)\}$ **Subtracting 3 − 10**

$= 5 - 13$ **Subtracting 6 −(−7)**

$= -8$ ◀

Simplify.

21. $43 - 52 \cdot 80$

22. $62 \cdot 8 + 4^3 - (5^2 - 64 \div 4)$

23. Simplify and compare: $(7 - 4)^2$ and $7^2 - 4^2$.

ANSWERS ON PAGE A-2

Simplify.

24. $6 - \{5 - [2 - (8 + 20)]\}$

25. $5 + \{6 - [2 + (5 - 2)]\}$

Simplify.

26. $\dfrac{8 \cdot 7 - |6 - 8|}{5^2 + 6^3}$

27. $\dfrac{(8 - 3)^2 + (7 - 10)^2}{3^2 - 2^3}$

Simplify.

28. $15x - \{2[2(x - 5) - 6(x + 3)] + 4\}$

29. $9a + \{3a - 2[(a - 4) - (a + 2)]\}$

ANSWERS ON PAGE A-2

▶ **EXAMPLE 23** Simplify: $7 - [3(2 - 5) - 4(2 + 3)]$.

$7 - [3(2 - 5) - 4(2 + 3)] = 7 - [3(-3) - 4(5)]$ **Doing the calculations in the innermost parentheses**

$= 7 - [-9 - 20]$

$= 7 - [-29]$

$= 36$ ◀

DO EXERCISES 24 AND 25.

In addition to the usual grouping symbols—parentheses, brackets, and braces—a fraction bar and absolute-value signs can act as grouping symbols.

▶ **EXAMPLE 24** Calculate: $\dfrac{12|7 - 9| + 8 \cdot 5}{3^2 + 2^3}$.

An equivalent expression with brackets as grouping symbols is

$$[12|7 - 9| + 8 \cdot 5] \div [3^2 + 2^3].$$

What this shows, in effect, is that we do the calculations in the numerator and in the denominator, and then divide the results:

$$\frac{12|7 - 9| + 8 \cdot 5}{3^2 + 2^3} = \frac{12|-2| + 8 \cdot 5}{9 + 8}$$

$$= \frac{12(2) + 8 \cdot 5}{17}$$

$$= \frac{24 + 40}{17}$$

$$= \frac{64}{17}.$$ ◀

DO EXERCISES 26 AND 27.

When expressions with parentheses contain variables, we still work from the inside out when simplifying.

▶ **EXAMPLE 25** Simplify: $6y - \{4[3(y - 2) - 4(y + 2)] - 3\}$.

$6y - \{4[3(y - 2) - 4(y + 2)] - 3\}$

$= 6y - \{4[3y - 6 - 4y - 8] - 3\}$ **Multiplying to remove the innermost parentheses using a distributive law**

$= 6y - \{4[-y - 14] - 3\}$ **Collecting like terms in the inner parentheses**

$= 6y - \{-4y - 56 - 3\}$ **Multiplying to remove the inner parentheses using a distributive law**

$= 6y - \{-4y - 59\}$ **Collecting like terms**

$= 6y + 4y + 59$ **Removing parentheses**

$= 10y + 59$ **Collecting like terms** ◀

DO EXERCISES 28 AND 29.

NAME SECTION DATE

EXERCISE SET 1.5

a Write exponential notation.

1. $4 \cdot 4 \cdot 4 \cdot 4 \cdot 4$

2. $6 \cdot 6 \cdot 6$

3. $5 \cdot 5 \cdot 5 \cdot 5 \cdot 5 \cdot 5$

4. $x \cdot x \cdot x \cdot x$

5. $mmmm$

6. $ttttt$

7. $3a \cdot 3a \cdot 3a \cdot 3a$

8. $5x \cdot 5x \cdot 5x \cdot 5x \cdot 5x$

9. $5 \cdot 5 \cdot c \cdot c \cdot c \cdot d \cdot d \cdot d \cdot d$

10. $2 \cdot 2 \cdot 2 \cdot r \cdot r \cdot r \cdot r \cdot t \cdot t$

Rewrite without an exponent.

11. 2^5

12. 7^3

13. $(-3)^4$

14. $(-8)^2$

15. x^4

16. y^6

17. $(-4b)^3$

18. $(-3x)^4$

19. $(ab)^4$

20. $(xyz)^3$

21. 5^1

22. $(\sqrt{6})^1$

23. $(3z)^0$

24. $\left(\frac{9}{7}\right)^1$

25. $(\sqrt{8})^0$

26. $(-4)^0$

27. $\left(\frac{7}{8}\right)^1$

28. $(5xy)^0$

b Rewrite using a positive exponent.

29. x^{-3}

30. y^{-4}

31. $\frac{1}{a^{-2}}$

32. $\frac{1}{y^{-7}}$

33. $(-11)^{-1}$

34. $(-4)^{-3}$

ANSWERS

1. ______
2. ______
3. ______
4. ______
5. ______
6. ______
7. ______
8. ______
9. ______
10. ______
11. ______
12. ______
13. ______
14. ______
15. ______
16. ______
17. ______
18. ______
19. ______
20. ______
21. ______
22. ______
23. ______
24. ______
25. ______
26. ______
27. ______
28. ______
29. ______
30. ______
31. ______
32. ______
33. ______
34. ______

ANSWERS

35. ______
36. ______
37. ______
38. ______
39. ______
40. ______
41. ______
42. ______
43. ______
44. ______
45. ______
46. ______
47. ______
48. ______
49. ______
50. ______
51. ______
52. ______
53. ______
54. ______
55. ______
56. ______
57. ______
58. ______
59. ______
60. ______
61. ______
62. ______
63. ______
64. ______
65. ______
66. ______
67. ______
68. ______

Rewrite using a negative exponent.

35. $\frac{1}{3^4}$ **36.** $\frac{1}{9^2}$ **37.** $\frac{1}{b^3}$

38. $\frac{1}{n^5}$ **39.** $\frac{1}{(-16)^2}$ **40.** $\frac{1}{(-8)^6}$

C Simplify.

41. $[10 - 3(6 - 1)]$ **42.** $[8 - 6(9 - 5)]$

43. $9[8 - 7(5 - 2)]$ **44.** $10[7 - 4(8 - 5)]$

45. $[5(8 - 6) + 12] - [24 - (8 - 4)]$ **46.** $[9(7 - 4) + 19] - [25 - (7 + 3)]$

47. $[64 \div (-4)] \div (-2)$ **48.** $[48 \div (-3)] \div \left(-\frac{1}{4}\right)$

49. $17(-24) + 50$ **50.** $20 \cdot 10 - 16 \cdot 25$

51. $(5 + 7)^2$; $5^2 + 7^2$ **52.** $(9 - 12)^2$; $9^2 - 12^2$

53. $2^3 + 2^4 - 20 \cdot 30$ **54.** $7 \cdot 8 - 3^2 - 2^3$

55. $5^3 + 36 \cdot 72 - (18 + 25 \cdot 4)$ **56.** $4^3 + 20 \cdot 10 + 7^2 - 23$

57. $(13 \cdot 2 - 8 \cdot 4)^2$ **58.** $(9 \cdot 8 + 3 \cdot 3)^2$

59. 🖩 $4000 \cdot (1 + 0.12)^3$ **60.** 🖩 $5000 \cdot (4 + 1.16)^2$

61. 🖩 $(20 \cdot 4 + 13 \cdot 8)^2 - (39 \cdot 59)^3$ **62.** 🖩 $(43 \cdot 6 - 14 \cdot 7)^3 + (33 \cdot 34)^2$

63. $18 - 2 \cdot 3 - 9$ **64.** $18 - (2 \cdot 3 - 9)$

65. $(18 - 2 \cdot 3) - 9$ **66.** $(18 - 2)(3 - 9)$

67. $[24 \div (-3)] \div \left(-\frac{1}{2}\right)$ **68.** $[(-32) \div (-2)] \div (-2)$

69. $15 \cdot (-24) + 50$

70. $30 \cdot 20 - 15 \cdot 24$

71. $4 \div (8 - 10)^2 + 1$

72. $16 \div (19 - 15)^2 - 7$

73. $6^3 + 25 \cdot 71 - (16 + 25 \cdot 4)$

74. $5^3 + 20 \cdot 40 + 8^2 - 29$

75. ▦ $5000 \cdot (1 + 0.16)^3$

76. ▦ $4000 \cdot (3 + 1.14)^2$

77. $4 \cdot 5 - 2 \cdot 6 + 4$

78. $8(7 - 3)/4$

79. $4 \cdot (6 + 8)/(4 + 3)$

80. $4^3/8$

81. $[2 \cdot (5 - 3)]^2$

82. $5^3 - 7^2$

83. $8(-7) + 6(-5)$

84. $10(-5) + 1(-1)$

85. $19 - 5(-3) + 3$

86. $14 - 2(-6) + 7$

87. $9 \div (-3) + 16 \div 8$

88. $-32 - 8 \div 4 - (-2)$

89. $7 + 10 - (-10 \div 2)$

90. $(3 - 8)^2$

91. $3^2 - 8^2$

92. $12 - 20^3$

93. $20 + 4^3 \div (-8)$

94. $2 \times 10^3 - 5000$

95. $-7(3^4) + 18$

96. $6[9 - (3 - 4)]$

97. $8[(6 - 13) - 11]$

98. $1000 \div (-100) \div 10$

99. $256 \div (-32) \div (-4)$

100. $\dfrac{20 - 6^2}{9^2 + 3^2}$

101. $\dfrac{5^2 - |4^3 - 8|}{9^2 - 2^2 - 1^5}$

102. $\dfrac{4|6 - 7| - 5 \cdot 4}{6 \cdot 7 - 8|4 - 1|}$

103. $\dfrac{30(8 - 3) - 4(10 - 3)}{10|2 - 6| - 2(5 + 2)}$

104. $\dfrac{5^3 - 3^2 + 12 \cdot 5}{-32 \div (-16) \div (-4)}$

ANSWERS

69. ________
70. ________
71. ________
72. ________
73. ________
74. ________
75. ________
76. ________
77. ________
78. ________
79. ________
80. ________
81. ________
82. ________
83. ________
84. ________
85. ________
86. ________
87. ________
88. ________
89. ________
90. ________
91. ________
92. ________
93. ________
94. ________
95. ________
96. ________
97. ________
98. ________
99. ________
100. ________
101. ________
102. ________
103. ________
104. ________

ANSWERS

105. ______
106. ______
107. ______
108. ______
109. ______
110. ______
111. ______
112. ______
113. ______
114. ______
115. ______
116. ______
117. ______
118. ______
119. ______
120. ______
121. ______
122. ______
123. ______
124. ______
125. ______
126. ______
127. ______
128. ______
129. ______

Simplify.

105. $9a - [7 - 5(7a - 3)]$

106. $12b - [9 - 7(5b - 6)]$

107. $5\{-2 + 3[4 - 2(3 + 5)]\}$

108. $7\{-7 + 8[5 - 3(4 + 6)]\}$

109. $[10(x + 3) - 4] + [2(x - 1) + 6]$

110. $[9(x + 5) - 7] + [4(x - 12) + 9]$

111. $[7(x + 5) - 19] - [4(x - 6) + 10]$

112. $[6(x + 4) - 12] - [5(x - 8) + 11]$

113. $3\{[7(x - 2) + 4] - [2(2x - 5) + 6]\}$

114. $4\{[8(x - 3) + 9] - [4(3x - 7) + 2]\}$

115. $4\{[5(x - 3) + 2^2] - 3[2(x + 5) - 9^2]\}$

116. $3\{[6(x - 4) + 5^2] - 2[5(x + 8) - 10^2]\}$

117. $2y + \{8[3(2y - 5) - (8y + 9)] + 6\}$

118. $7b - \{5[4(3b - 8) - (9b + 10)] + 14\}$

119. $[8(x - 2) + 9x] - \{7[3(2y - 5) - (8y + 7)] + 9\}$

120. $[11(a - 3) + 12a] - \{6[4(3b - 7) - (9b + 10)] + 11\}$

121. $-3[9(x - 4) + 5x] - 8\{3[5(3y + 4)] - 12\}$

122. $-6[8(y - 7) + 9y] - 7\{5[7(4z + 3)] - 14\}$

SKILL MAINTENANCE

123. The standard tennis court has length 78 ft and width 36 ft. Find the perimeter in two ways using two equivalent expressions for perimeter. Find the area.

124. Write an algebraic expression: Two times y plus three times x.

SYNTHESIS

125. Determine whether it is true that $(-x)^2 = x^2$, for any real number x. Explain why or why not.

126. Determine whether it is true that for any real numbers a and b, $ab = (-a)(-b)$. Explain why or why not.

Simplify.

127. $z - \{2z + [3z - (4z + 5x) - 6z] + 7z\} - 8z$

128. $\{x + [f - (f + x)] + [x - f]\} + 3x$

129. $x - \{x + 1 - [x + 2 - (x - 3 - \{x + 4 - [x - 5 + (x - 6)]\})]\}$

1.6 Properties of Exponents and Scientific Notation

We often need to find ways to determine equivalent exponential expressions. We do this with several rules or properties regarding exponents.

a Multiplication and Division

To see how to multiply, or simplify, in an expression such as $a^3 \cdot a^2$, we use the definition of exponential notation:

$$a^3 \cdot a^2 = \underbrace{a \cdot a \cdot a}_{\text{3 factors}} \cdot \underbrace{a \cdot a}_{\text{2 factors}} = a^5$$

The exponent in a^5 is the *sum* of those in $a^3 \cdot a^2$. In general, the exponents are added when we multiply, but note that the base must be the same in all factors. This is true for any integer exponents, even those that may be negative or zero.

> The Product Rule
>
> **For any number a and any integers m and n,**
>
> $$a^m \cdot a^n = a^{m+n}.$$
>
> **(When multiplying with exponential notation, we can add the exponents if the bases are the same.)**

► **EXAMPLES** Multiply and simplify.

1. $x^4 \cdot x^3 = x^{4+3} = x^7$

2. $4^5 \cdot 4^{-3} = 4^{5+(-3)} = 4^2 = 16$

3. $(-2)^{-3}(-2)^7 = (-2)^{-3+7} = (-2)^4 = 16$

4. $(8x^n)(6x^{2n}) = 8 \cdot 6x^{n+2n} = 48x^{3n}$

5. $(8x^4y^{-2})(-3x^{-3}y) = 8 \cdot (-3) \cdot x^4 \cdot x^{-3} \cdot y^{-2} \cdot y^1$ Using the associative and commutative laws

$= -24x^{4-3}y^{-2+1}$ Using the product rule

$= -24xy^{-1} = -\dfrac{24x}{y}$

Note that we give answers using positive exponents. In some situations, this may not be appropriate, but we do so here. ◄

DO EXERCISES 1–7.

We consider a division:

$$\frac{8^5}{8^3} = \frac{8 \cdot 8 \cdot 8 \cdot 8 \cdot 8}{8 \cdot 8 \cdot 8} = \frac{8 \cdot 8 \cdot 8}{8 \cdot 8 \cdot 8} \cdot 8 \cdot 8 = 8 \cdot 8 = 8^2.$$

We can obtain the result by subtracting exponents. This is always the case, even if exponents are negative or zero.

OBJECTIVES

After finishing Section 1.6, you should be able to:

a Use exponential notation in multiplication and division.

b Use exponential notation in raising a power to a power, and in raising a product or a quotient to a power.

c Convert between decimal and scientific notation and use scientific notation with multiplication and division.

FOR EXTRA HELP

Tape 2B Tape 2A MAC: 1 IBM: 1

Multiply and simplify.

1. $8^{-3}8^7$

2. y^7y^{-2}

3. $(9x^{-4})(-2x^7)$

4. $(-3x^{-4})(25x^{-10})$

5. $(-7x^{3n})(6x^{5n})$

6. $(5x^{-3}y^4)(-2x^{-9}y^{-2})$

7. $(4x^{-2}y^4)(15x^2y^{-3})$

ANSWERS ON PAGE A-2

Divide and simplify.

8. $\dfrac{4^8}{4^5}$

9. $\dfrac{5^4}{5^{-2}}$

10. $\dfrac{10^{-8}}{10^{-2}}$

11. $\dfrac{45x^{5n}}{-9x^{3n}}$

12. $\dfrac{42y^7x^6}{-21y^{-3}x^{10}}$

13. $\dfrac{33a^5b^{-2}}{22a^2b^{-4}}$

ANSWERS ON PAGE A-2

The Quotient Rule

For any nonzero number a and any integers m and n,

$$\frac{a^m}{a^n} = a^{m-n}.$$

(When dividing with exponential notation, we can subtract the exponent of the denominator from the exponent of the numerator, if the bases are the same.)

► **EXAMPLES** Divide and simplify.

6. $\dfrac{5^7}{5^3} = 5^{7-3} = 5^4$ **Subtracting exponents using the quotient rule**

7. $\dfrac{5^7}{5^{-3}} = 5^{7-(-3)} = 5^{7+3} = 5^{10}$ **Subtracting exponents (adding an opposite)**

8. $\dfrac{9^{-2}}{9^5} = 9^{-2-5} = 9^{-7} = \dfrac{1}{9^7}$

9. $\dfrac{7^{-4}}{7^{-5}} = 7^{-4-(-5)} = 7^{-4+5} = 7^1 = 7$

10. $\dfrac{16x^4y^7}{-8x^3y^9} = \dfrac{16}{-8} \cdot \dfrac{x^4}{x^3} \cdot \dfrac{y^7}{y^9} = -2xy^{-2} = -\dfrac{2x}{y^2}$

The answers $\dfrac{-2x}{y^2}$ or $\dfrac{2x}{-y^2}$ would also be correct here.

11. $\dfrac{40x^{-2n}}{4x^{5n}} = \dfrac{40}{4}x^{-2n-5n} = 10x^{-7n} = \dfrac{10}{x^{7n}}$

12. $\dfrac{14x^7y^{-3}}{4x^5y^{-5}} = \dfrac{14}{4} \cdot \dfrac{x^7}{x^5} \cdot \dfrac{y^{-3}}{y^{-5}} = \dfrac{7}{2}x^2y^2$ ◄

In exercises such as Examples 6–12 above, it may help to think as follows: After writing the base, write the top exponent. Then write a subtraction sign. Then write the bottom exponent. Then do the subtraction. For example,

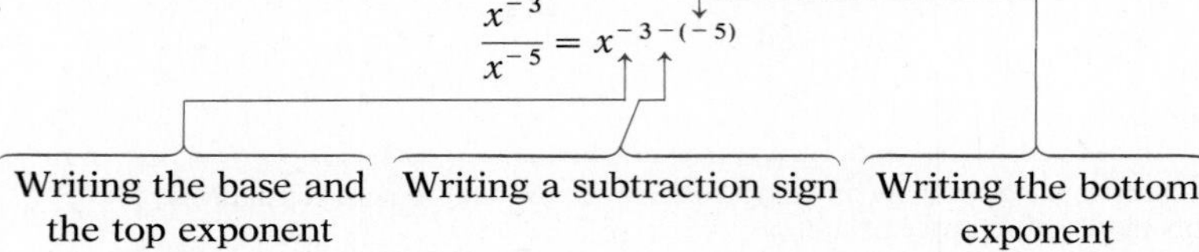

DO EXERCISES 8–13.

Earlier, we stated that we do not define 0^0. Now we can see why. Note the following. We know that 0^0 is equal to 0^{1-1}. But 0^{1-1} is also equal to 0/0. We have already seen that we must leave 0/0 undefined, so we also leave 0^0 undefined.

b Raising Powers to Powers and Products and Quotients to Powers

Next, consider an expression like $(5^2)^4$. In this case, we are raising 5^2 to the fourth power:

$$\begin{aligned}(5^2)^4 &= (5^2)(5^2)(5^2)(5^2)\\ &= (5 \cdot 5)(5 \cdot 5)(5 \cdot 5)(5 \cdot 5)\\ &= 5 \cdot 5 \cdot 5 \cdot 5 \cdot 5 \cdot 5 \cdot 5 \cdot 5 \qquad \text{Using an associative law}\\ &= 5^8.\end{aligned}$$

Note that here we could have multiplied the exponents:

$$(5^2)^4 = 5^{2 \cdot 4} = 5^8$$

Likewise, $(y^8)^3 = (y^8)(y^8)(y^8) = y^{24}$. Once again, we get the same result if we multiply the exponents:

$$(y^8)^3 = y^{8 \cdot 3} = y^{24}$$

> The Power Rule
>
> **For any real number a and any integers m and n,**
>
> $$(a^m)^n = a^{mn}.$$
>
> **(To raise a power to a power, multiply the exponents.)**

▶ **EXAMPLES** Simplify

13. $(x^5)^7 = 3^{5 \cdot 7} = 3^{35}$ **Multiply exponents.**

14. $(y^{-2})^{-2} = y^{(-2)(-2)} = y^4$

15. $(x^{-5})^4 = x^{-5 \cdot 4} = x^{-20} = \dfrac{1}{x^{20}}$

16. $(x^4)^{-2t} = x^{4(-2t)} = x^{-8t} = \dfrac{1}{x^{8t}}$ ◀

DO EXERCISES 14–16.

When an expression inside parentheses is raised to a power, the inside expression is the base. Let us compare $2a^3$ and $(2a)^3$:

$$2a^3 = 2 \cdot a \cdot a \cdot a; \qquad \text{The base is } a.$$

$$\begin{aligned}(2a)^3 &= (2a)(2a)(2a) \qquad \text{The base is } 2a.\\ &= (2 \cdot 2 \cdot 2)(a \cdot a \cdot a) \qquad \text{Using the associative law of multiplication}\\ &= 2^3a^3 = 8a^3.\end{aligned}$$

We see that $2a^3$ and $(2a)^3$ are *not* equivalent. We also see that we can evaluate the power $(2a)^3$ by raising each factor to the power 3. This leads us to the following rule for raising a product to a power.

> Raising a Product to a Power
>
> **For any real numbers a and b and any integer n,**
>
> $$(ab)^n = a^nb^n.$$
>
> **(To raise a product to the nth power, raise each factor to the nth power.)**

Simplify.

14. $(3^7)^6$

15. $(z^{-4})^{-5}$

16. $(t^2)^{-7m}$

ANSWERS ON PAGE A-2

Simplify.

17. $(2xy)^3$

18. $(4x^{-2}y^7)^2$

19. $(-2x^4y^2)^5$

20. $(10x^{-4}y^7z^{-2})^3$

Simplify.

21. $\left(\frac{x^{-3}}{y^4}\right)^{-3}$

22. $\left(\frac{3x^2y^{-3}}{y^5}\right)^2$

23. $\left[\frac{-3a^{-5}b^3}{2a^{-2}b^{-4}}\right]^{-3}$

ANSWERS ON PAGE A-2

► **EXAMPLES** Simplify.

17. $(3x^2y^{-2})^3 = 3^3(x^2)^3(y^{-2})^3 = 3^3x^6y^{-6} = 27x^6y^{-6} = \frac{27x^6}{y^6}$

18. $(5x^3y^{-5}z^2)^4 = 5^4(x^3)^4(y^{-5})^4(z^2)^4 = 625x^{12}y^{-20}z^8 = \frac{625x^{12}z^8}{y^{20}}$ ◄

DO EXERCISES 17–20.

There is a similar rule for raising a quotient to a power.

> Raising a Quotient to a Power
>
> **For any real numbers a and b, $b \neq 0$, and any integer n,**
>
> $$\left(\frac{a}{b}\right)^n = \frac{a^n}{b^n}.$$
>
> **(To raise a quotient to the nth power, raise the numerator to the nth power and divide by the denominator to the nth power.)**

► **EXAMPLES** Simplify. Write the answer using positive exponents.

19. $\left(\frac{x^2}{y^{-3}}\right)^{-5} = \frac{x^{2\cdot(-5)}}{y^{-3\cdot(-5)}} = \frac{x^{-10}}{y^{15}} = \frac{1}{x^{10}y^{15}}$

20. $\left(\frac{2x^3y^{-2}}{3y^4}\right)^5 = \left(\frac{2}{3}x^3y^{-2-4}\right)^5 = \left(\frac{2}{3}x^3y^{-6}\right)^5 = \left(\frac{2}{3}\right)^5(x^3)^5(y^{-6})^5$

$$= \frac{2^5}{3^5}x^{15}y^{-30} = \frac{32x^{15}}{243y^{30}}$$

21. $\left[\frac{-3a^{-5}b^3}{2a^{-2}b^{-4}}\right]^{-2} = \left[\frac{-3}{2}a^{-5-(-2)}b^{3-(-4)}\right]^{-2} = \left(\frac{-3}{2}a^{-3}b^7\right)^{-2}$

$$= \left(\frac{-3}{2}\right)^{-2}(a^{-3})^{-2}(b^7)^{-2} = \frac{1}{(\frac{-3}{2})^2}a^6b^{-14} = \frac{1}{\frac{9}{4}}a^6b^{-14}$$

$$= \frac{4}{9}a^6b^{-14} = \frac{4a^6}{9b^{14}}$$ ◄

Another way to carry out Example 21 is to first write expressions with positive exponents, as follows:

$$\left[\frac{-3a^{-5}b^3}{2a^{-2}b^{-4}}\right]^{-2} = \left[\frac{2a^{-2}b^{-4}}{-3a^{-5}b^3}\right]^2 = \left[\frac{2a^5}{-3a^2b^3b^4}\right]^2 = \left[\frac{2a^{5-2}}{-3b^{3+4}}\right]^2$$

$$= \left[\frac{2a^3}{-3b^7}\right]^2 = \frac{2^2(a^3)^2}{(-3)^2(b^7)^2} = \frac{4a^6}{9b^{14}}.$$

DO EXERCISES 21–23.

The following is a summary of the laws of exponents introduced in Sections 1.5 and 1.6.

> $a^1 = a$, $a^0 = 1\ (a \neq 0)$, $a^{-n} = \frac{1}{a^n}$, $a^m a^n = a^{m+n}$,
>
> $(a^m)^n = a^{mn}$, $\frac{a^m}{a^n} = a^{m-n}$, $(ab)^n = a^n b^n$, $\left(\frac{a}{b}\right)^n = \frac{a^n}{b^n}$

C Scientific Notation

There are many kinds of symbolism, or *notation*, for numbers. You are already familiar with fractional notation, decimal notation, and percent notation. Now we study another notation, **scientific notation,** which is especially useful when calculations involve very large or very small numbers and when estimating. The following are examples of scientific notation:

The mass of the earth:

$$5.98 \times 10^{24} \text{ kg} = 5{,}980{,}000{,}000{,}000{,}000{,}000{,}000{,}000 \text{ kg};$$

The mass of a hydrogen atom:

$$1.7 \times 10^{-24} \text{ gm} = 0.0000000000000000000000017 \text{ gm}.$$

> ***Scientific notation* for a number is an expression of the type**
> $$N \times 10^n,$$
> **where 1 is less than or equal to N and N is less than 10 ($1 \le N < 10$) and N is expressed in decimal notation. The expression 10^n is also considered to be scientific notation when $N = 1$.**

You should try to make conversions to scientific notation mentally as much as possible. Here is a handy mental device.

> **A positive exponent indicates a large number and a negative exponent indicates a small number.**

► **EXAMPLES** Convert mentally to scientific notation.

22. Light travels 9,460,000,000,000 km in one year.

$9{,}460{,}000{,}000{,}000 = 9.46 \times 10^{12}$ 9.460,000,000,000. (12 places)

Large number, so the exponent is positive.

23. $0.0000018 = 1.8 \times 10^{-6}$ 0.000001.8 (6 places)

Small number, so the exponent is negative. ◄

► **EXAMPLES** Convert mentally to decimal notation.

24. $4.893 \times 10^5 = 489{,}300$ 4.89300. (5 places)

Positive exponent, so the answer is a large number.

25. $8.7 \times 10^{-8} = 0.000000087$ 0.00000008.7 (8 places)

Negative exponent, so the answer is a small number. ◄

When using a calculator, we might express a number such as 560,000,000 using scientific notation in a form like

5.6 E 8, or 5.6 8.

Similarly, a number like 0.00000028 might be expressed like

2.8 E −7 or 2.8 −7.

Each of the following is *not* scientific notation.

13.95×10^{13}, — This number is greater than 10.

0.468×10^{-8} — This number is less than 1.

DO EXERCISES 24–27.

Convert to scientific notation.

24. Light travels 5,880,000,000,000 mi in one year.

25. 0.000000000257

Convert to decimal notation.

26. 4.567×10^{-13}

27. The distance from the earth to the sun is 9.3×10^7 mi.

ANSWERS ON PAGE A-2

Multiply and write scientific notation for the answer.

28. $(9.1 \times 10^{-17})(8.2 \times 10^{3})$

29. $(1.12 \times 10^{-8})(5 \times 10^{-7})$

Divide and write scientific notation for the answer.

30. $\dfrac{4.2 \times 10^{5}}{2.1 \times 10^{2}}$

31. $\dfrac{1.1 \times 10^{-4}}{2.0 \times 10^{-7}}$

32. There are 2864 members of the Professional Bowlers Association. There are 234 million people in the United States. What part of the population are members of the Professional Bowlers Association? Write decimal notation, percent notation, and scientific notation for the answer.

33. The mass of the planet Jupiter is about 318 times the mass of the earth. Write scientific notation for the mass of Jupiter. See Example 29.

ANSWERS ON PAGE A-2

Multiplying and dividing in scientific notation is easy because we can use the properties of exponents.

► **EXAMPLE 26** Multiply and write scientific notation for the answer: $(3.1 \times 10^{5})(4.5 \times 10^{-3})$.

We apply the commutative and the associative laws to get

$$(3.1 \times 10^{5})(4.5 \times 10^{-3}) = (3.1 \times 4.5)(10^{5} \times 10^{-3}) = 13.95 \times 10^{2}.$$

To find scientific notation for the result, we convert 13.95 to scientific notation and then simplify:

$$13.95 \times 10^{2} = (1.395 \times 10^{1}) \times 10^{2} = 1.395 \times 10^{3}.$$ ◄

DO EXERCISES 28 AND 29.

► **EXAMPLE 27** Divide and write scientific notation for the answer:

$$\frac{6.4 \times 10^{-7}}{8.0 \times 10^{6}}.$$

$$\frac{6.4 \times 10^{-7}}{8.0 \times 10^{6}} = \frac{6.4}{8.0} \times \frac{10^{-7}}{10^{6}}$$ Factoring shows two divisions.

$$= 0.8 \times 10^{-13}$$ **Doing the divisions separately**

$$= (8.0 \times 10^{-1}) \times 10^{-13}$$ **Converting 0.8 to scientific notation**

$$= 8.0 \times 10^{-14}$$ The answer 0.8×10^{-13} is not incorrect, except that we have not converted it to scientific notation. ◄

DO EXERCISES 30 AND 31.

► **EXAMPLE 28** There are 300,000 words in the English language. The average person knows about 10,000 of them. What part of the total number of words does the average person know? Write decimal notation rounded to three decimal places. Then write both percent notation and scientific notation for the answer.

The part of the total number of words that the average person knows is

$$\frac{10{,}000}{300{,}000} \approx 0.03333;$$ **This is decimal notation.**

$$= 3.333\%;$$ **This is percent notation.**

$$= 3.333 \times 10^{-2}.$$ **This is scientific notation.** ◄

► **EXAMPLE 29** The mass of the earth is about 5.98×10^{24} kg. The mass of the sun is about 333,000 times the mass of the earth. Write scientific notation for the mass of the sun.

The mass of the sun is 333,000 times the mass of the earth. We convert each number to scientific notation and multiply.

$$\begin{aligned}(333{,}000)(5.98 \times 10^{24}) &= (3.33 \times 10^{5})(5.98 \times 10^{24})\\ &= (3.33 \times 5.98)(10^{5} \times 10^{24})\\ &= 19.9134 \times 10^{29}\\ &= (1.99134 \times 10^{1}) \times 10^{29}\\ &= 1.99134 \times 10^{30}\end{aligned}$$ ◄

DO EXERCISES 32 AND 33.

NAME SECTION DATE

EXERCISE SET 1.6

a Multiply and simplify.

1. $5^6 \cdot 5^3$
2. $6^2 \cdot 6^6$
3. $8^{-6} \cdot 8^2$
4. $9^{-5} \cdot 9^3$
5. $8^{-2} \cdot 8^{-4}$
6. $9^{-1} \cdot 9^{-6}$
7. $b^2 \cdot b^{-5}$
8. $a^4 \cdot a^{-3}$
9. $a^{-3} \cdot a^4 \cdot a^2$
10. $x^{-8} \cdot x^5 \cdot x^3$
11. $(2x)^3 \cdot (3x)^2$
12. $(9y)^2 \cdot (2y)^3$
13. $(14m^2n^3)(-2m^3n^2)$
14. $(6x^5y^{-2})(-3x^2y^3)$
15. $(-2x^{-3})(7x^{-8})$
16. $(6x^{-4}y^3)(-4x^{-8}y^{-2})$
17. $(15x^{4t})(7x^{-6t})$
18. $(9x^{-4n})(-4x^{-8n})$

Divide and simplify.

19. $\frac{6^8}{6^3}$
20. $\frac{7^9}{7^4}$
21. $\frac{4^3}{4^{-2}}$
22. $\frac{5^8}{5^{-3}}$
23. $\frac{10^{-3}}{10^6}$
24. $\frac{12^{-4}}{12^8}$
25. $\frac{9^{-4}}{9^{-6}}$
26. $\frac{2^{-7}}{2^{-5}}$
27. $\frac{x^{-4n}}{x^{6n}}$
28. $\frac{y^{-3t}}{y^{8t}}$
29. $\frac{w^{-11q}}{w^{-6q}}$
30. $\frac{m^{-7t}}{m^{-5t}}$
31. $\frac{a^3}{a^{-2}}$
32. $\frac{y^4}{y^{-5}}$
33. $\frac{9a^2}{(-3a)^2}$
34. $\frac{24a^5b^3}{-8a^4b}$
35. $\frac{-24x^6y^7}{18x^{-3}y^9}$
36. $\frac{14a^4b^{-3}}{-8a^8b^{-5}}$
37. $\frac{-18x^{-2}y^3}{-12x^{-5}y^5}$
38. $\frac{-14a^{14}b^{-5}}{-18a^{-2}b^{-10}}$

ANSWERS

1. ____
2. ____
3. ____
4. ____
5. ____
6. ____
7. ____
8. ____
9. ____
10. ____
11. ____
12. ____
13. ____
14. ____
15. ____
16. ____
17. ____
18. ____
19. ____
20. ____
21. ____
22. ____
23. ____
24. ____
25. ____
26. ____
27. ____
28. ____
29. ____
30. ____
31. ____
32. ____
33. ____
34. ____
35. ____
36. ____
37. ____
38. ____

ANSWERS

39. ______
40. ______
41. ______
42. ______
43. ______
44. ______
45. ______
46. ______
47. ______
48. ______
49. ______
50. ______
51. ______
52. ______
53. ______
54. ______
55. ______
56. ______
57. ______
58. ______
59. ______
60. ______
61. ______
62. ______
63. ______
64. ______
65. ______
66. ______
67. ______
68. ______
69. ______
70. ______
71. ______
72. ______
73. ______
74. ______
75. ______
76. ______

b Simplify.

39. $(4^3)^2$

40. $(5^4)^5$

41. $(8^4)^{-3}$

42. $(9^3)^{-4}$

43. $(6^{-4})^{-3}$

44. $(7^{-8})^{-5}$

45. $(3x^2y^2)^3$

46. $(2a^3b^4)^5$

47. $(-2x^3y^{-4})^{-2}$

48. $(-3a^2b^{-5})^{-3}$

49. $(-6a^{-2}b^3c)^{-2}$

50. $(-8x^{-4}y^5z^2)^{-4}$

51. $\left(\dfrac{4^{-3}}{3^4}\right)^3$

52. $\left(\dfrac{5^2}{4^{-3}}\right)^{-3}$

53. $\left(\dfrac{2x^3y^{-2}}{3y^{-3}}\right)^3$

54. $\left(\dfrac{-4x^4y^{-2}}{5x^{-1}y^4}\right)^{-4}$

55. $\left(\dfrac{125a^2b^{-3}}{5a^4b^{-2}}\right)^{-5}$

56. $\left(\dfrac{-200x^3y^{-5}}{8x^5y^{-7}}\right)^{-4}$

57. $\left(\dfrac{-6^5y^4z^{-5}}{2^{-2}y^{-2}z^3}\right)^6$

58. $\left(\dfrac{9^{-2}x^{-4}y}{3^{-3}x^{-3}y^2}\right)^8$

59. $[(-2x^{-4}y^{-2})^{-3}]^{-2}$

60. $[(-4a^{-4}b^{-5})^{-3}]^4$

61. $\left(\dfrac{3a^{-2}b}{5a^{-7}b^5}\right)^{-7}$

62. $\left(\dfrac{2x^2y^{-2}}{3x^8y^7}\right)^9$

63. $\dfrac{10^{2a+1}}{10^{a+1}}$

64. $\dfrac{11^{b+2}}{11^{3b-3}}$

65. $\dfrac{9a^{x-2}}{3a^{2x+2}}$

66. $\dfrac{-12x^{a+1}}{4x^{2-a}}$

67. $\dfrac{45x^{2a+4}y^{b+1}}{-9x^{a+3}y^{2+b}}$

68. $\dfrac{-28x^{b+5}y^{4+c}}{7x^{b-5}y^{c-4}}$

69. $(8^x)^{4y}$

70. $(7^{2p})^{3q}$

71. $(12^{3-a})^{2b}$

72. $(x^{a-1})^{3b}$

73. $(5x^{a-1}y^{b+1})^{2c}$

74. $(4x^{3a}y^{2b})^{5c}$

75. $\dfrac{4x^{2a+3}y^{2b-1}}{2x^{a+1}y^{b+1}}$

76. $\dfrac{25x^{a+b}y^{b-a}}{-5x^{a-b}y^{b+a}}$

C Convert the number to scientific notation.

77. 47,000,000,000

78. 2,600,000,000,000

79. The gross national product (GNP) one year was \$932,000,000,000.

80. Each year there are 1,095,000 new single-family homes built in the United States.

81. 0.000000016

82. 0.000000263

83. 0.00000000007

84. 0.00000000000009

Convert the number to decimal notation.

85. 6.73×10^{8}

86. 9.24×10^{7}

87. The wavelength of a certain red light is 6.6×10^{-5} cm.

88. The mass of an electron is 9.11×10^{-28} g.

89. An electron has a charge of 4.8×10^{-11} electrostatic units.

90. The population of the United States is 2.44×10^{8}.

91. 8.923×10^{-10}

92. 7.034×10^{-2}

Multiply and write the answer in scientific notation.

93. $(2.3 \times 10^{6})(4.2 \times 10^{-11})$

94. $(6.5 \times 10^{3})(5.2 \times 10^{-8})$

95. $(2.34 \times 10^{-8})(5.7 \times 10^{-4})$

96. $(3.26 \times 10^{-6})(8.2 \times 10^{9})$

Divide and write the answer in scientific notation.

97. $\dfrac{8.5 \times 10^{8}}{3.4 \times 10^{5}}$

98. $\dfrac{5.1 \times 10^{6}}{3.4 \times 10^{3}}$

99. $\dfrac{4.0 \times 10^{-6}}{8.0 \times 10^{-3}}$

100. $\dfrac{7.5 \times 10^{-9}}{2.5 \times 10^{-4}}$

ANSWERS

77. ________

78. ________

79. ________

80. ________

81. ________

82. ________

83. ________

84. ________

85. ________

86. ________

87. ________

88. ________

89. ________

90. ________

91. ________

92. ________

93. ________

94. ________

95. ________

96. ________

97. ________

98. ________

99. ________

100. ________

ANSWERS

101. ______________

102. ______________

103. ______________

104. ______________

105. ______________

106. ______________

107. ______________

108. ______________

109. ______________

110. ______________

111. ______________

112. ______________

113. ______________

114. ______________

115. ______________

116. ______________

Write the answers to Exercises 101–108 in scientific notation.

101. Each day we purchase 25,000 new automobiles. How many automobiles are purchased in one year? There are 244 million people in the country. How many new automobiles are purchased per person in one year?

102. Each day we purchase 574,000 record albums. How many record albums are purchased in one year? There are 244 million people in the country. How many record albums are purchased per person in one year?

103. Light traveling 300,000 kilometers per second (km/s) takes 4,500,000 seconds to reach the earth from the sun. Find the distance from the sun to the earth. (*Hint:* Distance = Speed × Time.)

104. The distance light travels in 100 yr is approximately 5.87×10^{14} mi. How far does light travel in 13 weeks?

105. Americans eat 6.5 million gal of popcorn each day. How much popcorn do they eat in one year?

106. Americans drink 3 million gal of orange juice in one day. How much orange juice is consumed in this country in one year?

107. The average discharge at the mouth of the Amazon River is 4,200,000 cubic feet per second. How much water is discharged from the Amazon River in one hour? one year?

108. There are 300,000 words in the English language. The exceptional person knows about 20,000 of them. What part of the total number of words does the exceptional person know?

SKILL MAINTENANCE

Simplify.

109. $9x - (-4y + 8) + (10x - 12)$

110. $-6t - (5t - 13) + 2(4 - 6t)$

SYNTHESIS

Simplify.

111. $\dfrac{(2^{-2})^{-4} \times (2^3)^{-2}}{(2^{-2})^2 \cdot (2^5)^{-3}}$

112. $\left[\dfrac{(-3x^{-2}y^5)^{-3}}{(2x^4y^{-8})^{-2}}\right]^2$

113. $\left[\left(\dfrac{a^{-2}}{b^7}\right)^{-3} \cdot \left(\dfrac{a^4}{b^{-3}}\right)^2\right]^{-1}$

Simplify. Assume that variables in exponents represent integers.

114. $(m^{x-b}n^{x+b})^x(m^bn^{-b})^x$

115. $\left[\dfrac{(2x^ay^b)^3}{(-2x^ay^b)^2}\right]^2$

116. $(x^by^a \cdot x^ay^b)^c$

SUMMARY AND REVIEW: CHAPTER 1

IMPORTANT PROPERTIES AND FORMULAS

Properties of Real Numbers

Commutative Laws: $a + b = b + a,\ ab = ba$
Associative Laws: $a + (b + c) = (a + b) + c,\ a(bc) = (ab)c$
Distributive Laws: $a(b + c) = ab + ac,\ a(b - c) = ab - ac$
Identities: $a + 0 = a,\ a \cdot 1 = a$
Inverses: $a + (-a) = 0,\ a \cdot \frac{1}{a} = 1$

Properties of Exponents: $a^1 = a$ $\quad a^0 = 1$ $\quad a^{-n} = \frac{1}{a^n}$

Product Rule: $a^m \cdot a^n = a^{m+n}$
Quotient Rule: $\frac{a^m}{a^n} = a^{m-n}$
Power Rule: $(a^m)^n = a^{mn}$
Raising a Product to a Power: $(ab)^n = a^n b^n$
Raising a Quotient to a Power: $\left(\frac{a}{b}\right)^n = \frac{a^n}{b^n}$
Scientific Notation: $N \times 10^n$, or 10^n, where N is such that $1 \leq N < 10$.

REVIEW EXERCISES

The review exercises that follow are for practice. Answers are at the back of the book. If you miss an exercise, restudy the section and objective indicated alongside the answer.

1. The area A of a circle is given by $A = \pi r^2$, where r is the radius. Find the area when $r = 10$ cm. Use 3.14 for π.

Evaluate.

2. $5x - 7$ when $x = 2$
3. $\frac{x - y}{2}$ when $x = 20$ and $y = 4$

Translate to an algebraic expression.

4. Five times x
5. 9 less than t
6. Twenty-eight percent of y
7. Use $<$ or $>$ for ▒ to write a true sentence: -3.9 ▒ 2.9.
8. Write another inequality with the same meaning as $18 > x$.
9. True or false: $-13 \geq 5$
10. True or false: $7 \leq 7$

Graph on a number line.

11. $x > -4$
12. $x \leq 1$

Simplify.

13. $|-7.23|$
14. $|9 - 9|$
15. $\left|\frac{21}{8}\right|$

Add, subtract, multiply, or divide.

16. $6 + (-8)$
17. $-3.8 + (-4.1)$
18. $\frac{3}{4} + \left(-\frac{13}{4}\right)$
19. $-8 - (-3)$
20. $-17.3 - 9.3$
21. $\frac{3}{2} - \left(-\frac{13}{4}\right)$
22. $(-3.8)(-2.7)$
23. $-\frac{2}{3}\left(\frac{9}{14}\right)$

24. $-6(-7)(4)$

25. $-12 \div 3$

26. $\frac{-84}{-4}$

27. $\frac{49}{-7}$

28. $\frac{5}{6} \div (-\frac{10}{7})$

29. $-\frac{7}{3} \div \frac{5}{11}$

30. $-\frac{5}{2} \div (-\frac{15}{16})$

31. $-108 \div 4.5$

Use a commutative law to write an equivalent expression.

32. $11 + a$

33. $8y$

Use an associative law to write an equivalent expression.

34. $(9 + a) + b$

35. $8(xy)$

Find $-a$ when a is as given. (In other words, find the opposite of the number.)

36. $a = -7$

37. $a = 2.3$

38. $a = 0$

Multiply.

39. $-3(2x - y)$

40. $4ab(2c + 1)$

Factor.

41. $5x + 10y - 5z$

42. $ptr + pts$

Collect like terms.

43. $2x + 6y - 5x - y$

44. $7c - 6 + 9c + 2$

45. Find an equivalent expression: $-(-9c + 4d - 3)$.

Simplify.

46. $4(x - 3) - 3(x - 5)$

47. $12x - 3(2x - 5)$

48. $7x - [4 - 5(3x - 2)]$

49. $4m-3[3(4m-2)-(5m+2)]+12$

50. $2^3 - 3^4 + (13 \cdot 5 + 67)$

51. $64 \div (-4) + (-5)(20)$

Multiply or divide and simplify.

52. $(2x^4y^{-3})(-5x^3y^{-2})$

53. $\dfrac{-15x^2y^{-5}}{10x^6y^{-8}}$

Simplify.

54. $(-3a^{-4}bc^3)^{-2}$

55. $\left[\dfrac{-2x^4y^{-4}}{3x^{-2}y^6}\right]^{-4}$

56. Divide and write scientific notation for the answer.

$$\frac{2.2 \times 10^7}{3.2 \times 10^{-3}}$$

57. Multiply and write scientific notation for the answer.

$$(3.2 \times 10^4)(4.1 \times 10^{-6})$$

58. Which of the following are rational numbers?

$$\sqrt{4}, \quad \sqrt{3}, \quad -\frac{2}{3}, \quad 0.45\overline{45}, \quad -23.788$$

SYNTHESIS

59. Which of the following are equivalent?

a) $3x - 3y$ b) $3x - y$ c) $x^{-2}x^5$ d) x^{-10} e) x^{-3} f) $(x^{-2})^5$ g) $x(yz)$ h) $x(y + z)$ i) $3(x - y)$ j) $xy + xz$

60. Simplify: $(x^y \cdot x^{3y})^3$.

61. If $a = 2^x$ and $b = 2^{x+5}$, find $a^{-1}b$.

Simplify.

62. $[-(7a - b) - (a + 5b)] - [2(a + \frac{1}{2}b) + 3(7a - \frac{5}{3}b)]$

63. $0.01\{0.1(x - 2y) - [0.001(3x + y) - (0.2x - 0.1y)]\} - (x - y)$

❖ THINKING IT THROUGH

1. Give five examples of rational numbers that are not integers.
2. Explain and compare the commutative, associative, and distributive laws.
3. Explain the meaning of a negative exponent in as many different ways as you can.
4. Give two expressions that are equivalent. Give two that are not equivalent.

NAME SECTION DATE

TEST: CHAPTER 1

1. Evaluate $3x - 3y$ when $x = 2$ and $y = -4$.

2. Translate to an algebraic expression: Thirty-six percent of m.

Graph on a number line.

3. $x \leq -1$

−4 −3 −2 −1 0 1 2 3 4

4. $x > -2$

−4 −3 −2 −1 0 1 2 3 4

True or false.

5. $-8 \leq -6$

6. $-6 \geq -6$

7. Write another inequality with the same meaning as $5 > a$.

8. Use $<$ or $>$ for ■ to write a true sentence: -4.5 ■ -8.7.

9. Convert to scientific notation: 0.0000437.

Simplify.

10. $\left|\frac{7}{8}\right|$

11. $|-13.4|$

12. $|0|$

Find $-a$ when a is as given. (In other words, find the opposite, or additive inverse of the number.)

13. $a = 8$

14. $a = -13$

15. $a = \frac{1}{4}$

Add.

16. $7 + (-9)$

17. $-5.3 + (-7.8)$

18. $-\frac{5}{2} + \left(-\frac{7}{4}\right)$

Subtract.

19. $-6 - (-5)$

20. $-18.2 - 11.5$

21. $\frac{19}{4} - \left(-\frac{3}{2}\right)$

Multiply.

22. $(-4.1)(8.2)$

23. $-\frac{4}{5}\left(-\frac{15}{16}\right)$

24. $-6(-4)(-11)2$

Divide.

25. $\frac{-10}{2}$

26. $-75 \div (-5)$

27. $\frac{7}{-5}$

28. $-\frac{4}{3} \div \frac{5}{7}$

29. $-\frac{11}{3} \div \left(-\frac{7}{6}\right)$

30. $-459.2 \div 5.6$

ANSWERS

1. ______
2. ______
3. ______
4. ______
5. ______
6. ______
7. ______
8. ______
9. ______
10. ______
11. ______
12. ______
13. ______
14. ______
15. ______
16. ______
17. ______
18. ______
19. ______
20. ______
21. ______
22. ______
23. ______
24. ______
25. ______
26. ______
27. ______
28. ______
29. ______
30. ______

ANSWERS

31. ______
32. ______
33. ______
34. ______
35. ______
36. ______
37. ______
38. ______
39. ______
40. ______
41. ______
42. ______
43. ______
44. ______
45. ______
46. ______
47. ______
48. ______
49. ______
50. ______
51. ______
52. ______
53. ______
54. ______
55. ______
56. ______
57. ______
58. ______
59. ______
60. ______

Use a commutative law to write an equivalent expression.

31. pq

32. $t + 4$

Use an associative law to write an equivalent expression.

33. $3 + (t + w)$

34. $(4a)b$

Multiply.

35. $-2(3a - 4b)$

36. $3\pi r(s + 1)$

Factor.

37. $ab - ac + 2ad$

38. $2ah + h$

Collect like terms.

39. $6y - 8x + 4y + 3x$

40. $4a - 7 + 17a + 21$

41. Find an equivalent expression: $-(-9x + 7y - 22)$.

Simplify.

42. $-3(x + 2) - 4(x - 5)$

43. $9y - 5(4y - 6)$

44. $4x - [6 - 3(2x - 5)]$

45. $3a - 2[5(2a - 5) - (8a + 4)] + 10$

46. $45 \cdot 20 - (5 + 7)^2$

47. $(8 - 13)^2$; $8^2 - 13^2$

Multiply or divide and simplify.

48. $(3a^4b^{-2})(-2a^5b^{-3})$

49. $\dfrac{-12x^3y^{-4}}{8x^7y^{-6}}$

50. $(5a^{4n})(-10a^{5n})$

51. $\dfrac{-60x^{3t}}{12x^{7t}}$

Simplify.

52. $(-3a^{-3}b^2c)^{-4}$

53. $\left[\dfrac{-3x^3y^{-3}}{5x^{-2}y^5}\right]^{-3}$

54. $(10a^{-5}5b^{-3}4c)^{-6}$

55. $\left[\dfrac{-5a^{-2}b^8}{10a^{10}b^{-4}}\right]^{-4}$

56. The mass of the earth is 5.98×10^{24} kg. The mass of the planet Pluto is about 0.002 times the mass of the earth. Write scientific notation for the mass of Pluto.

57. Which of the following are irrational numbers?

-45, $\sqrt{7}$, $\sqrt{9}$, $\dfrac{-4}{5}$, $2.3\overline{76}$, π

SYNTHESIS

58. Which of the following are equivalent?

a) $x^{-3}x^{-4}$ b) x^{12} c) x^{-12} d) $5x + 5$ e) $(x^{-3})^{-4}$
f) $5(x + 1)$ g) $5x$ h) $5 + 5x$ i) $5(xy)$ j) $(5x)y$

59. Simplify: $(m^{a-b}m^{3b-a})^4$.

60. Determine whether $-x^2$ and $(-x)^2$ are equivalent. Think of $-x^2$ as $-1 \cdot x^2$.

INTRODUCTION In this chapter, the manipulations presented in Chapter 1 are applied to the solving of equations and inequalities, which are in turn used to solve problems. We also learn to solve formulas for letters. This is an important skill in many applications of mathematics.

The review sections to be tested in addition to the material in this chapter are 1.3, 1.4, and 1.6. ❖

Solving Equations and Inequalities

2

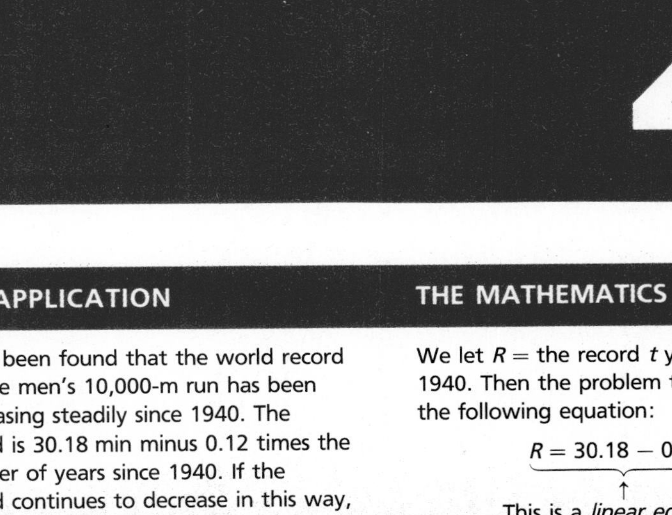

AN APPLICATION

It has been found that the world record for the men's 10,000-m run has been decreasing steadily since 1940. The record is 30.18 min minus 0.12 times the number of years since 1940. If the record continues to decrease in this way, what will it be in 1998?

THE MATHEMATICS

We let $R =$ the record t years from 1940. Then the problem translates to the following equation:

$$\underbrace{R = 30.18 - 0.12t}.$$

↑

This is a *linear equation*.

❖ POINTS TO REMEMBER: CHAPTER 2

Simple-Interest Formula: $I = Prt$
Sum of the Angles of a Triangle = 180°
Perimeter of a Rectangle: $P = 2l + 2w$
Consecutive Integers: $x,\quad x+1,\quad x+2,\quad x+3,$ etc.
Consecutive Even Integers: $x,\quad x+2,\quad x+4,\quad x+6,$ etc.
Consecutive Odd Integers: $x,\quad x+2,\quad x+4,\quad x+6,$ etc.

PRETEST: CHAPTER 2

Solve.

1. $-6x = 42$

2. $-3.2 + y = 18.1$

3. $9 - 2x = 11$

4. $4 - 3(y + 1) = 2(y - 3) - 7$

5. $-7x < 21$

6. $2x - 1 \le 5x + 8$

7. $-1 \le 3x + 2 \le 4$

8. $|2y - 7| = 9$

9. $4a + 5 < -3$ or $4a + 5 > 3$

10. $|3a + 5| > 1$

11. $|3x + 4| = |x - 7|$

12. $-8(x - 7) \ge 10(2x + 3) - 12$

13. Solve for r: $P = 3rq$.

14. Find the distance between 8.2 and 10.6.

Solve.

15. The perimeter of a rectangular field is 170 m. The length is 5 m more than the width. Find the dimensions of the field.

16. Money is borrowed at 11% simple interest. After one year, $832.50 pays off the loan. How much was originally borrowed?

17. You can rent a car for $45 per day with unlimited mileage, or for $25 per day with an extra charge of 18¢ per mile. For what numbers of miles traveled would the unlimited mileage plan save you money?

18. Find the intersection:

$$\{-4, -3, 1, 4, 9\} \cap \{-3, -2, 2, 4, 5\}.$$

19. Find the union: $\{-1, 1\} \cup \{0\}$.

20. Simplify: $\left|\dfrac{-5y}{2x}\right|$.

2.1 Solving Equations

a Equations

In order to solve problems, we must be able to solve *equations.* What is the meaning of an equation?

An *equation* is a number sentence that says that the expressions on either side of the equals sign, =, represent the same number.

▶ **EXAMPLES**

1. $1 + 10 = 11$ is an equation. It is true.

2. $7 - 8 = 9 - 13$ is an equation. It is false.

3. $x - 9 = 3$ is an equation. It is neither true nor false until we determine what x represents. ◀

DO EXERCISES 1–3.

▶ **EXAMPLE 4** Tell the meaning of the equation $7 - 8 = 9 - 13$.

$7 - 8 = 9 - 13$ says that $7 - 8$ and $9 - 13$ represent the same number. In this case, the equation is false. ◀

DO EXERCISES 4 AND 5.

Solutions of Equations

If an equation contains a variable, then the equation may be neither true nor false. Some replacements may make it true and some may make it false.

The replacements making an equation true are called its *solutions.* The set of all solutions is called the *solution set.* When we find all solutions, no matter how, we say that we have *solved* the equation. There are various procedures for solving equations.

▶ **EXAMPLE 5** Find the replacement that makes $2x + 3 = 13$ true.

The number 5 makes the equation $2x + 3 = 13$ true, because if we substitute 5 into the left side, we get $2 \cdot 5 + 3$, or 13. ◀

DO EXERCISES 6 AND 7 ON THE FOLLOWING PAGE.

OBJECTIVES

After finishing Section 2.1, you should be able to:

- **a** Tell what an equation means and what is meant by a solution of an equation.
- **b** Solve equations using the addition principle.
- **c** Solve equations using the multiplication principle.
- **d** Solve equations using the addition and multiplication principles together, removing parentheses where appropriate.

FOR EXTRA HELP

Tape 3A　Tape 2B　MAC: 2 IBM: 2

Consider the following equations.

a) $3 + 4 = 7$
b) $5 - 1 = 2$
c) $21 + 2 = 24$
d) $x - 5 = 12$
e) $9 - x = x$
f) $13 + 2 = 15$

1. Which equations are true?

2. Which equations are false?

3. Which equations are neither true nor false?

Tell what the equation means.

4. $89 + 2 = 3 + 4$

5. $6 - 7 = 4 - 5$

ANSWERS ON PAGE A-2

6. Consider the replacements 5, 8, 4, 0, and 2. Which of them make the equation $2x + 3 = 11$ false?

7. Find the replacement that makes the equation $3x + 2 = 11$ true.

8. a) Determine whether

$3x + 2 = 11$ and $x = 3$

are equivalent.

b) Determine whether

$2x + 3 = 11$ and $x = -6$

are equivalent.

ANSWERS ON PAGE A-2

Equivalent Equations

Consider the equation

$$x = 5.$$

The solution to this equation is easily "seen" to be 5. If we replace x by 5, we get

$$5 = 5, \quad \text{which is true.}$$

Now consider the equation

$$2x + 3 = 13.$$

In Example 5, we discovered that the solution of this equation is also 5, but the fact that 5 is the solution is not as obvious. We now consider principles that allow us to start with an equation, and end up with an equation like $x = 5$, in which the variable is alone on one side, and for which the solution is easy to find. The equations $2x + 3 = 13$ and $x = 5$ are **equivalent.**

Equations with the same solutions are called *equivalent equations.*

DO EXERCISE 8.

b The Addition Principle

One of the principles we use in solving equations concerns adding. Consider the equation $a = b$. It says that a and b represent the same number. Suppose this is true and then add a number c to a. We will get the same answer if we add c to b, because a and b are the same number.

The Addition Principle

If $a = b$ is true, then $a + c = b + c$ is true for any number c.

When we use the addition principle, we sometimes say that we "add the same number on both sides of an equation." This is also true for subtraction, because we can express every subtraction as an addition; that is,

$$a - c = b - c \qquad \text{means} \qquad a + (-c) = b + (-c).$$

The addition principle also tells us that we can "subtract the same number on both sides of an equation."

► **EXAMPLE 6** Solve: $x + 6 = -15$.

$x + 6 = -15$

$x + 6 - 6 = -15 - 6$ **Using the addition principle: adding -6 on both sides or subtracting 6 on both sides. (Note that 6 and -6 are opposites.)**

$x + 0 = -21$ **Simplifying**

$x = -21$ **Using the identity property of 0: $x + 0 = x$**

Check:

$x + 6 = -15$	
$-21 + 6$	-15
-15	TRUE

The solution is -21. ◄

In Example 6, we wanted to get x alone so that we could easily see the solution, so we added the opposite of 6. This eliminated the 6 on the left, giving us the *additive identity* 0, which when added to x is x. We started with $x + 6 = -15$; using the addition principle, we derived a simpler equation, $x = -21$, from which it was easy to *see* the solution. The equations $x + 6 = -15$ and $x = -21$ are *equivalent.*

▶ **EXAMPLE 7** Solve: $y - 4.7 = 13.9$.

$y - 4.7 = 13.9$

$y - 4.7 + 4.7 = 13.9 + 4.7$ **Using the addition principle: adding 4.7 on both sides. Note that −4.7 and 4.7 are opposites.**

$y + 0 = 13.9 + 4.7$

$y = 13.9 + 4.7$ **Using the identity property of 0: $y + 0 = y$**

$y = 18.6$ **Doing the calculation 13.9 + 4.7**

Check:

$y - 4.7 = 13.9$	
$18.6 - 4.7$	13.9
13.9	TRUE

The solution is 18.6. ◀

▶ **EXAMPLE 8** Solve: $-\frac{3}{8} + x = -\frac{5}{7}$.

$-\frac{3}{8} + x = -\frac{5}{7}$

$\frac{3}{8} + (-\frac{3}{8}) + x = \frac{3}{8} - \frac{5}{7}$ **Using the addition principle: adding $\frac{3}{8}$**

$0 + x = \frac{3}{8} \cdot \frac{7}{7} - \frac{5}{7} \cdot \frac{8}{8}$ **Multiplying by 1 to obtain the least common denominator**

$x = \frac{21}{56} - \frac{40}{56}$

$x = -\frac{19}{56}$

Check:

$-\frac{3}{8} + x = -\frac{5}{7}$	
$-\frac{3}{8} + (-\frac{19}{56})$	$-\frac{5}{7}$
$-\frac{3}{8} \cdot \frac{7}{7} + (-\frac{19}{56})$	
$-\frac{21}{56} + (-\frac{19}{56})$	
$-\frac{40}{56}$	
$-\frac{5}{7}$	TRUE

The solution is $-\frac{19}{56}$. ◀

DO EXERCISES 9–12.

c The Multiplication Principle

Suppose that an equation $a = b$ is true and we multiply a by a number c. We will get the same answer if we multiply b by c. This is because a and b are the same number.

> The Multiplication Principle
>
> **If an equation $a = b$ is true, then an equation $a \cdot c = b \cdot c$ is true for any number c.**

Solve using the addition principle.

9. $x + 9 = 2$

10. $x + \frac{1}{4} = -\frac{3}{5}$

11. $13 = -25 + y$

12. $y - 61.4 = 78.9$

ANSWERS ON PAGE A-2

Solve using the multiplication principle.

13. $8x = 10$

14. $-\frac{3}{7}y = 21$

15. $-4x = -\frac{6}{7}$

16. $-12.6 = 4.2y$

ANSWERS ON PAGE A-2

▶ **EXAMPLE 9** Solve: $22 = -\frac{4}{5}x$.

$$22 = -\tfrac{4}{5}x$$
$$-\tfrac{5}{4} \cdot 22 = -\tfrac{5}{4} \cdot (-\tfrac{4}{5})x \qquad \textbf{Multiplying by } -\tfrac{5}{4}\textbf{, the reciprocal of } -\tfrac{4}{5}$$
$$-\tfrac{55}{2} = 1 \cdot x$$
$$-\tfrac{55}{2} = x \qquad \textbf{Using the identity property of 1: } 1 \cdot x = x$$

Check:

$22 = -\frac{4}{5}x$	
22	$-\frac{4}{5}\ (-\frac{55}{2})$
	22 TRUE

The solution is $-\frac{55}{2}$. ◀

In Example 9, in order to get x alone, we multiplied by the *multiplicative inverse*, or *reciprocal* of $-\frac{4}{5}$. When we multiplied, we got the *multiplicative identity* 1 times x, or $1 \cdot x$, which simplified to x. This enabled us to eliminate the $-\frac{4}{5}$ on the right.

This multiplication principle also tells us that we can "divide on both sides by a nonzero number." This is because division is the same as multiplying by a reciprocal. That is,

$$\frac{a}{c} = \frac{b}{c} \quad \text{means} \quad a \cdot \frac{1}{c} = b \cdot \frac{1}{c}, \quad \text{when } c \neq 0.$$

In practice it is usually more convenient to "divide" both sides of the equation if the number in front of the variable is in decimal notation or is an integer. When the number is in fractional notation, it is more convenient to "multiply" by a reciprocal.

▶ **EXAMPLE 10** Solve: $4x = 9$.

$$4x = 9$$
$$\frac{4x}{4} = \frac{9}{4} \qquad \textbf{Using the multiplication principle: multiplying by } \tfrac{1}{4} \textbf{ on both sides or dividing on both sides by 4}$$
$$1 \cdot x = \tfrac{9}{4} \qquad \textbf{Simplifying}$$
$$x = \tfrac{9}{4} \qquad \textbf{Using the identity property of 1: } 1 \cdot x = x$$

Check:

$4x = 9$	
$4 \cdot \frac{9}{4}$	9
9	TRUE

The solution is $\frac{9}{4}$. ◀

▶ **EXAMPLE 11** Solve: $-0.05y = -5.5$.

$$-0.05y = -5.5$$
$$\frac{-0.05}{-0.05}y = \frac{-5.5}{-0.05} \qquad \textbf{Dividing by } -0.05 \textbf{ on both sides}$$
$$1 \cdot y = \frac{5.5}{0.05}$$
$$y = 110 \qquad \textbf{Carrying out the division}$$

The check is left to the student. The solution is 110. ◀

DO EXERCISES 13–16.

d Using the Principles Together

Let's see how we can use the addition and multiplication principles together.

► **EXAMPLE 12** Solve: $3x - 4 = 13$.

$$3x - 4 = 13$$

$$3x - 4 + 4 = 13 + 4 \quad \textbf{Using the addition principle: adding 4}$$

$$3x + 0 = 17 \quad \textbf{Adding}$$

$$3x = 17$$

$$\frac{3x}{3} = \frac{17}{3} \quad \textbf{Dividing by 3 on both sides}$$

$$x = \frac{17}{3}$$

Check:

$3x - 4$	$= 13$
$3 \cdot \frac{17}{3} - 4$	13
$17 - 4$	
13	TRUE

The solution is $\frac{17}{3}$, or $5\frac{2}{3}$.

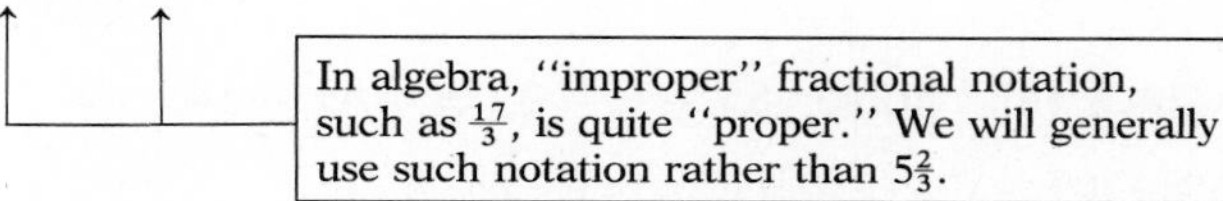

◄

DO EXERCISE 17.

In a situation such as Example 12, it is easier to first use the addition principle. In a situation in which fractions are involved, it may be easier to use the multiplication principle first.

► **EXAMPLE 13** Clear of fractions and solve: $\frac{3}{16}x + \frac{1}{8} = \frac{11}{8}$.

We multiply on both sides by the least common denominator—in this case, 16.

$$\frac{3}{16}x + \frac{1}{8} = \frac{11}{8}$$

$$16\left(\frac{3}{16}x + \frac{1}{8}\right) = 16\left(\frac{11}{8}\right) \quad \textbf{Multiplying by 16}$$

$$16 \cdot \frac{3}{16}x + 16 \cdot \frac{1}{8} = 22$$

Carrying out the multiplication. We use a distributive law on the left, being careful to multiply *both* terms by 16.

$$3x + 2 = 22$$

$$3x = 20$$

$$x = \frac{20}{3}$$

The number $\frac{20}{3}$ checks and is the solution. ◄

17. Solve: $-4 + 9x = 8$.

ANSWER ON PAGE A-2

18. Clear of fractions and solve:

$$\frac{2}{3} - \frac{5}{6}y = \frac{1}{3}.$$

19. Clear of decimals and solve:

$$6.3x - 9.1 = 3.$$

Solve.

20. $\frac{5}{2}x + \frac{9}{2}x = 21$

21. $1.4x - 0.9x + 0.7 = -2.2$

22. $-4x + 2 + 5x = 3x - 15$

ANSWERS ON PAGE A-2

▶ **EXAMPLE 14** Clear of decimals and solve: $12.4 - 5.12x = 3.14$.

We multiply on both sides by a power of ten—in this case, 10^2 or 100—to clear the equation of decimals:

$$12.4 - 5.12x = 3.14$$

$$100(12.4 - 5.12x) = 100(3.14) \quad \textbf{Multiplying by 100}$$

$$100(12.4) - 100(5.12x) = 314 \quad \textbf{Carrying out the multiplication. We use a distributive law on the left, being careful to multiply } \textbf{\textit{both}} \textbf{ terms by 100.}$$

$$1240 - 512x = 314 \quad \textbf{Simplifying}$$

$$-512x = -1240 + 314 \quad \textbf{Adding } -1240$$

$$-512x = -926$$

$$x = \frac{-926}{-512} \quad \textbf{Dividing by } -512$$

$$x = \frac{463}{256}. \quad \textbf{Simplifying}$$ ◀

DO EXERCISES 18 AND 19.

When there are like terms in an equation, we combine them on each side. Then if there are still like terms on opposite sides, we get them on the same side using the addition principle.

▶ **EXAMPLE 15** Solve: $8x + 6 - 2x = -4x - 14$.

$$8x + 6 - 2x = -4x - 14$$

$$6x + 6 = -4x - 14 \quad \textbf{Combining like terms on the left}$$

$$6x + 4x + 6 = -14 \quad \textbf{Adding } 4x \textbf{ on both sides}$$

$$10x + 6 = -14 \quad \textbf{Combining like terms on the left}$$

$$10x = -20 \quad \textbf{Subtracting 6 on both sides}$$

$$x = \frac{-20}{10} \quad \textbf{Dividing by 10}$$

$$x = -2 \quad \textbf{Simplifying}$$

Check:

$8x + 6 - 2x$	$= -4x - 14$	
$8(-2) + 6 - 2(-2)$	$-4(-2) - 14$	
$-16 + 6 + 4$	$8 - 14$	
-6	-6	TRUE

The solution is -2. ◀

Here is a summary of the method of solving equations, as used in the preceding examples.

1. **Clear the equation of fractions or decimals if that is needed. (This is optional, but can ease computation.)**
2. **Collect like terms on each side of the equation, if possible.**
3. **Use the addition principle to get all like terms with letters on one side and all other terms on the other side.**
4. **Collect like terms on each side again, if possible.**
5. **Use the multiplication principle to solve for the variable.**

DO EXERCISES 20–22.

Special Cases

There are equations with no solution.

▶ **EXAMPLE 16** Solve: $-8x + 5 = 14 - 8x$.

$$-8x + 5 = 14 - 8x$$

$$5 = 14 \quad \textbf{Adding } 8x\textbf{, we get a false equation.}$$

No matter what number we try for x, we get a false sentence. Thus the equation has no solution. ◀

There are equations for which any real number is a solution.

▶ **EXAMPLE 17** Solve: $-8x + 5 = 5 - 8x$.

$$-8x + 5 = 5 - 8x$$

$$5 = 5 \quad \textbf{Adding } 8x\textbf{, we get a true equation.}$$

Replacing x by any real number gives a true sentence. Thus any real number is a solution. ◀

DO EXERCISES 23 AND 24.

Equations containing parentheses can often be solved by first multiplying to remove parentheses and then proceeding as before.

▶ **EXAMPLE 18** Solve: $3(7 - 2x) = 14 - 8(x - 1)$.

$3(7 - 2x) = 14 - 8(x - 1)$

$21 - 6x = 14 - 8x + 8$ **Multiplying, using the distributive law, to remove parentheses**

$21 - 6x = 22 - 8x$ **Combining like terms**

$8x - 6x = 22 + (-21)$ **Adding -21 and also $8x$**

$2x = 1$ **Combining like terms**

$x = \frac{1}{2}$ **Multiplying by $\frac{1}{2}$**

Check:

$3(7 - 2x)$	$= 14 - 8(x - 1)$	
$3(7 - 2 \cdot \frac{1}{2})$	$14 - 8(\frac{1}{2} - 1)$	
$3(7 - 1)$	$14 - 8(-\frac{1}{2})$	
$3 \cdot 6$	$14 + 4$	
18	18	TRUE

The solution is $\frac{1}{2}$.

On the right side of the equation, note that $14 - 8(x - 1) = 14 - 8x + 8$. ◀

DO EXERCISES 25 AND 26.

Solve.

23. $4 + 7x = 7x + 9$

24. $3 + 9x = 9x + 3$

Solve.

25. $30 + 7(x - 1) = 3(2x + 7)$

26. $3(y - 1) - 1 = 2 - 5(y + 5)$

ANSWERS ON PAGE A-2

❖ SIDELIGHTS

Repeating Decimals and Fractional Notation

Given a repeating decimal, how can we find fractional notation?

EXAMPLE 1 Find fractional notation for $0.\overline{6}$.

We first let n represent the number. Then we multiply by 10, moving the decimal point one place to the right. Now

$$\begin{aligned} 10n &= 6.666\overline{6} \\ n &= 0.666\overline{6} \\ \hline 9n &= 6 \end{aligned}$$

This is the original number. We subtract n from $10n$ on the left, to get $9n$. On the right, when we subtract, we get 6.

Next, we divide by 9:

$$n = \tfrac{6}{9}, \quad \text{or } \tfrac{2}{3}.$$

▦ We check the answer by dividing: $2 \div 3 = 0.\overline{6}$. ◀

EXAMPLE 2 Find fractional notation for $0.8\overline{3}$.

First we multiply by 10. Letting n represent the number, we get

$$10n = 8.3333\overline{3}.$$

Now the repeating part starts at the decimal point, as in Example 1. We multiply by 10 again, to get

$$100n = 83.3333\overline{3}.$$

Next, we subtract $10n$ from $100n$:

$$\begin{aligned} 100n &= 83.333333\overline{3} \\ 10n &= 8.333333\overline{3} \\ \hline 90n &= 75 \end{aligned}$$

Now we divide by 90:

$$n = \tfrac{75}{90}, \quad \text{or } \tfrac{5}{6}.$$

▦ *Check:* Divide 5 by 6: $5 \div 6 = 0.8\overline{3}$. ◀

EXERCISES

Find fractional notation.

1. $0.\overline{4}$ **2.** $0.1\overline{6}$ **3.** $8.\overline{3}$ **4.** $0.\overline{12}$

5. $0.\overline{123}$ **6.** $0.12\overline{5}$ **7.** $0.42\overline{3}$ **8.** $6.1\overline{82}$

NAME SECTION DATE

EXERCISE SET 2.1

Remember to review the objectives before doing the exercises.

b Solve using the addition principle. Don't forget to check.

1. $x + 5 = 14$

2. $y + 7 = 19$

3. $-22 = x - 18$

4. $-26 = y - 19$

5. $-8 + y = 15$

6. $-9 + t = 17$

7. $-12 + z = -51$

8. $-37 + x = -89$

9. $p - 2.96 = 83.9$

10. $z - 14.9 = -5.73$

11. $-\frac{3}{8} + x = -\frac{5}{24}$

12. $x + \frac{1}{12} = -\frac{5}{6}$

c Solve using the multiplication principle. Don't forget to check.

13. $5x = 20$

14. $3x = 21$

15. $-4x = 88$

16. $-11y = 121$

17. $4 = 24t$

18. $8 = 40y$

19. $-3z = -96$

20. $-8y = -120$

21. $4.8y = -28.8$

22. $0.39t = -2.73$

23. $\frac{3}{2}t = -\frac{1}{4}$

24. $-\frac{5}{6}y = -\frac{7}{8}$

d Solve using the principles together. Don't forget to check.

25. $4x - 12 = 60$

26. $4x - 6 = 70$

27. $5x - 10 = 45$

28. $6z - 7 = 11$

29. $9t + 4 = -104$

30. $5x + 7 = -108$

31. $-\frac{7}{3}x + \frac{2}{3} = -18$

32. $-\frac{9}{2}y + 4 = -\frac{91}{2}$

33. $\frac{6}{5}x + \frac{4}{10}x = \frac{32}{10}$

34. $\frac{9}{5}y + \frac{4}{10}y = \frac{66}{10}$

35. $0.9y - 0.7y = 4.2$

36. $0.8t - 0.3t = 6.5$

ANSWERS

1. ______
2. ______
3. ______
4. ______
5. ______
6. ______
7. ______
8. ______
9. ______
10. ______
11. ______
12. ______
13. ______
14. ______
15. ______
16. ______
17. ______
18. ______
19. ______
20. ______
21. ______
22. ______
23. ______
24. ______
25. ______
26. ______
27. ______
28. ______
29. ______
30. ______
31. ______
32. ______
33. ______
34. ______
35. ______
36. ______

ANSWERS

37.
38.
39.
40.
41.
42.
43.
44.
45.
46.
47.
48.
49.
50.
51.
52.
53.
54.
55.
56.
57.
58.
59.
60.
61.
62.
63.
64.
65.
66.
67.
68.
69.
70.
71.
72.
73.
74.
75.

37. $8x + 48 = 3x - 12$ **38.** $15x + 20 = 8x - 22$ **39.** $7y - 1 = 23 - 5y$

40. $3x - 15 = 15 - 3x$ **41.** $4x - 3 = 5 + 12x$ **42.** $9t - 4 = 14 + 15t$

43. $5 - 4a = a - 13$ **44.** $8 - 5x = x - 16$ **45.** $3m - 7 = -7 - 4m - m$

46. $5x - 8 = -8 + 3x - x$ **47.** $5x + 3 = 11 - 4x + x$ **48.** $6y + 20 = 10 + 3y + y$

49. $-7 + 9x = 9x - 7$ **50.** $-3t + 4 = 5 - 3t$ **51.** $6y - 8 = 9 + 6y$

52. $5 - 3x = -3x + 5$ **53.** $2(x + 6) = 8x$ **54.** $3(y + 5) = 8y$

55. $80 = 10(3t + 2)$ **56.** $27 = 9(5y - 2)$ **57.** $180(n - 2) = 900$

58. $210(x - 3) = 840$ **59.** $5y - (2y - 10) = 25$ **60.** $8x - (3x - 5) = 40$

61. $7(3x + 6) = 11 - (x + 2)$ **62.** $9(2x + 8) = 20 - (x + 5)$

63. $\frac{1}{8}(16y + 8) - 17 = -\frac{1}{4}(8y - 16)$ **64.** $\frac{1}{6}(12t + 48) - 20 = -\frac{1}{8}(24t - 144)$

65. $3[5 - 3(4 - t)] - 2 = 5[3(5t - 4) + 8] - 26$

66. $6[4(8 - y) - 5(9 + 3y)] - 21 = -7[3(7 + 4y) - 4]$

67. $\frac{2}{3}\left(\frac{7}{8} + 4x\right) - \frac{5}{8} = \frac{3}{8}$ **68.** $\frac{3}{4}\left(3x - \frac{1}{2}\right) + \frac{2}{3} = \frac{1}{3}$

SKILL MAINTENANCE

Multiply or divide and simplify.

69. $(6x^5y^{-4})(-3x^{-3}y^{-7})$ **70.** $\dfrac{6x^5y^{-4}}{-3x^{-3}y^{-7}}$

71. Multiply: $-4(3x - 2y + z)$. **72.** Factor: $4x - 10y + 2$.

SYNTHESIS

Solve.

73. $\frac{3x}{2} + \frac{5x}{3} - \frac{13x}{6} - \frac{2}{3} = \frac{5}{6}$ **74.** $\frac{2x - 5}{6} + \frac{4 - 7x}{8} = \frac{10 + 6x}{3}$

75. $2x - 4 - (x + 1) - 3(x - 2) = 6(2x - 3) - 3(6x - 1) - 8$

2.2 Solving Problems

a Five Steps for Problem Solving

Since you have already studied some algebra, you have probably had some experience with problem solving. The following five-step strategy can be very helpful.

Five Steps for Problem Solving

1. ***Familiarize*** **yourself with the problem situation.**
2. ***Translate*** **the problem to an equation.**
3. ***Solve*** **the equation.**
4. ***Check*** **the answer in the original problem.**
5. ***State*** **the answer to the problem clearly.**

Of the five steps, probably the most important is the first one: becoming familiar with the problem situation. Here are some hints for familiarization.

To *familiarize* yourself with the problem situation:

1. **If a problem is given in words, read it carefully.**
2. **Reread the problem, perhaps aloud. Try to verbalize the problem to yourself.**
3. **List the information given and the question to be answered. Choose a variable (or variables) to represent the unknown(s) and clearly state what the variable represents. Be descriptive! For example, let L = length (in meters), d = distance (in miles), and so on.**
4. **Find further information if necessary. Look up a formula on the inside front cover of this book or in a reference book. Talk to a reference librarian or an expert in the field.**
5. **Make a table of the information given and the information you have collected. Look for patterns that may help in the translation to an equation.**
6. **Make a drawing and label it with known information. Also, indicate unknown information, using specific units if given.**
7. **Guess or estimate the answer.**

▶ **EXAMPLE 1** A 28-ft rope is cut into two pieces. One piece is 3 ft longer than the other. How long are the pieces?

1. *Familiarize.* We let x = the length of one piece (in feet) and $x + 3$ = the length of the other piece (in feet). Then we draw a picture.

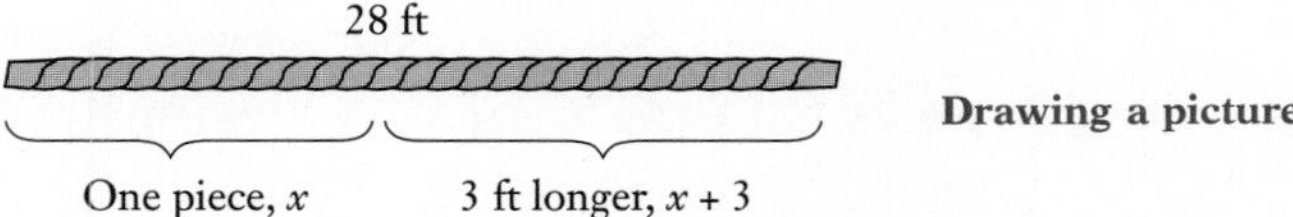

Drawing a picture

2. *Translate.* The picture can help in translating. Here is one way to do it. Note that "is" translates to "=."

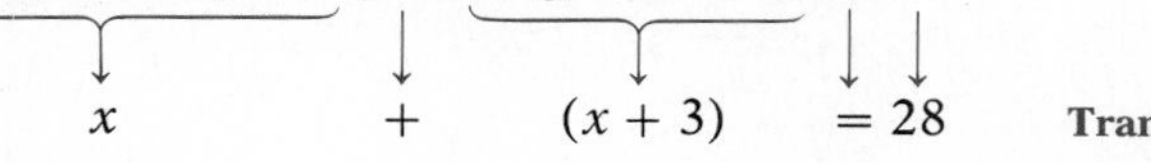

Translation

OBJECTIVE

After finishing Section 2.2, you should be able to:

a Solve problems by translating to equations.

FOR EXTRA HELP

Tape 4A

Tape 2B

MAC: 2
IBM: 2

1. A 32-ft rope is cut into two pieces, one three times as long as the other. How long are the pieces?

2. If seven times a certain number is subtracted from 6, the result is five times the number. What is the number? (*Note:* Seven times a certain number subtracted from 6 translates to $6 - 7x$.)

3. What percent of 84 is 11.76?

ANSWERS ON PAGE A-2

3. *Solve.*

$$x + (x + 3) = 28$$

$$2x + 3 = 28 \quad \text{Removing parentheses and collecting like terms}$$

$$2x = 25$$

$$x = \tfrac{25}{2}, \text{ or } 12\tfrac{1}{2}.$$

4. *Check.* Do we have an answer to the *problem*? If one piece is $12\frac{1}{2}$ ft long, the one that is 3 ft longer must be $15\frac{1}{2}$ ft long. The lengths of the pieces add up to 28 ft, so this checks.

5. *State.* One piece is $12\frac{1}{2}$ ft long and the other is $15\frac{1}{2}$ ft long. ◀

DO EXERCISE 1.

▶ **EXAMPLE 2** Five plus twice a number is seven times the number. What is the number?

1. *Familiarize.* This time it does not make sense to draw a picture. We merely let $x =$ the number.

2. *Translate.*

5 plus twice a number is seven times the number.

$$5 + 2x = 7 \cdot x \quad \text{Translation}$$

3. *Solve.*

$$5 + 2x = 7x$$

$$5 = 7x - 2x$$

$$5 = 5x$$

$$1 = x.$$

4. *Check.* Twice 1 is 2. If we add 5, we get 7. This is $7 \cdot 1$. This checks.

5. *State.* The answer to the *problem* is 1. ◀

DO EXERCISE 2.

▶ **EXAMPLE 3** What percent of 76 is 13.68?

1. *Familiarize.* Again, a picture is not needed. We let $y =$ the number.

2. *Translate.*

What percent of 76 is 13.68? Note that "of" translates to a multiplication sign.

$$y \quad \% \quad \times 76 = 13.68$$

3. *Solve.*

$$y \times (0.01) \times 76 = 13.68 \quad \text{Substituting "} \times 0.01 \text{" for "\%"}$$

$$0.76y = 13.68$$

$$y = \frac{13.68}{0.76}, \text{ or } 18 \quad \text{Solving}$$

4. *Check.* $18\% \cdot 76 = 0.18 \times 76 = 13.68$. The answer checks.

5. *State.* The solution of the problem is 18%. ◀

DO EXERCISE 3.

► **EXAMPLE 4** One angle of a triangle is five times as large as the first angle. The measure of the third angle is 2° less than that of the first angle. How large are the angles?

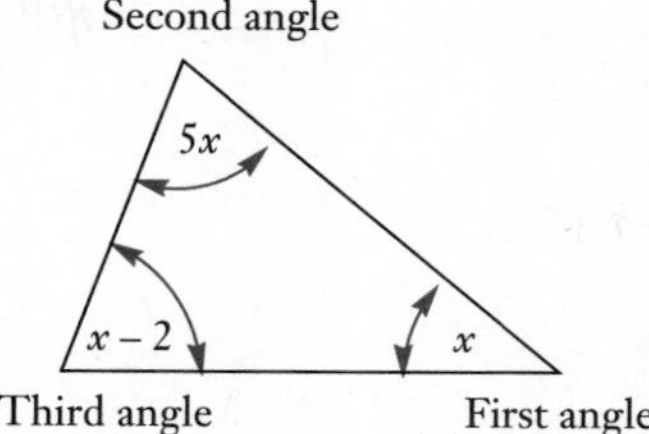

1. *Familiarize.* We draw a picture. We let $x =$ the measure of the first angle, $5x =$ the measure of the second angle, and $x - 2 =$ the measure of the third angle.
2. *Translate.* To translate, we need to recall a geometric fact—that is, the measures of the angles of any triangle add up to 180°.

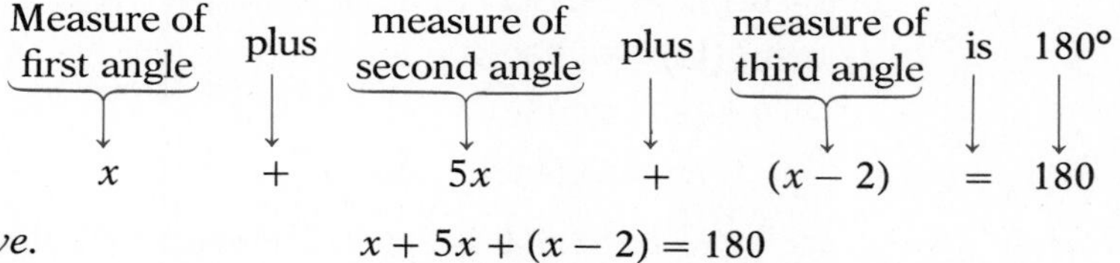

3. *Solve.*

$$x + 5x + (x - 2) = 180$$
$$7x - 2 = 180$$
$$7x = 182$$
$$x = 26$$

The solution of the equation is 26.

4. *Check.* The angles will then have measures as follows:

First angle $x = 26°$,

Second angle $5x = 130°$,

Third angle $x - 2 = 24°$.

These add up to 180° so they give an answer to the *problem.*

5. *State.* The first angle has measure 26°, the second 130°, and the third 24°. ◄

DO EXERCISE 4.

► **EXAMPLE 5** It has been found that the world record for the men's 10,000-m run has been decreasing steadily since 1940. The record is 30.18 min minus 0.12 times the number of years since 1940. If the record continues to decrease in this way, what will it be in 1998?

1. *Familiarize.* We let $t =$ time, in years, since 1940, and $R =$ the running record, in minutes.
2. *Translate.*

Record is 30.18 min minus 0.12 times number of years since 1940

$$R \quad = 30.18 \quad - \quad 0.12 \quad \cdot \quad t$$

$$R = 30.18 - 0.12t$$

3. *Solve.* In 1998, t will be 58, so we have

$$R = 30.18 - 0.12 \cdot 58.$$

We solve by carrying out the calculation to get 23.22.

4. *Check.* The record 23.22 checks by nature of the calculation.
5. *State.* The *predicted* record in 1998 is 23.22 min. ◄

DO EXERCISE 5.

4. The first angle of a triangle is twice as large as the second. The measure of the third angle is 20° greater than that of the second angle. How large are the angles?

5. ▦ It has been found that the world record for the women's 100-m dash has also been decreasing steadily. The record is 11.01 sec minus 0.03125 times the number of years since 1976. If the record continues to decrease in this way, what will it be in 1996?

ANSWERS ON PAGE A-2

6. A clothing store drops the price of suits 25% to a sale price of \$93. What was the former price?

ANSWER ON PAGE A-2

▶ **EXAMPLE 6** Once the pilots of Pan American Airlines shocked the business world by taking a pay cut of 11% to a new salary of \$48,950 per year. What was their former salary?

1. *Familiarize.* We let $x =$ the former salary.
2. *Translate.*

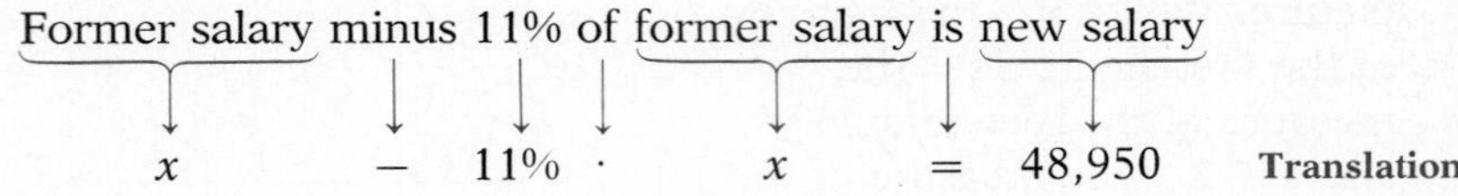

3. *Solve.*

$$x - 11\% \cdot x = 48{,}950$$
$$1x - 0.11x = 48{,}950 \quad \text{Replacing 11\% by 0.11}$$
$$(1 - 0.11)x = 48{,}950$$
$$0.89x = 48{,}950 \quad \text{Collecting like terms}$$
$$x = 55{,}000 \quad \text{Dividing by 0.89}$$

4. *Check.* Note that 11% of \$55,000 is \$6050. Subtracting this decrease from \$55,000 (the former salary), we get \$49,950 (the new salary). This checks.
5. *State.* The former salary was \$55,000. ◀

Example 6 and Margin Exercise 6 are both concerned with a percent of decrease.

DO EXERCISE 6.

The following are examples of **consecutive integers:** 19, 20, 21, 22; as well as −34, −33, −32, −31. Note that consecutive integers can be represented in the form x, $x + 1$, $x + 2$, and so on.

The following are examples of **consecutive even integers:** 20, 22, 24, 26; as well as −34, −32, −30, −28. Note that consecutive even integers can be represented in the form x, $x + 2$, $x + 4$, $x + 6$, and so on.

The following are examples of **consecutive odd integers:** 17, 19, 21, 23; as well as −77, −75, −73, −71. Note that consecutive odd integers can be represented in the form x, $x + 2$, $x + 4$, $x + 6$, and so on.

▶ **EXAMPLE 7** The sum of two consecutive integers is 35. What are the integers?

1. *Familiarize.* Since the integers are consecutive, we know that one of them is one greater than the other. Thus we let $x =$ the first integer and $x + 1 =$ the next consecutive integer.
2. *Translate.* We translate as follows:

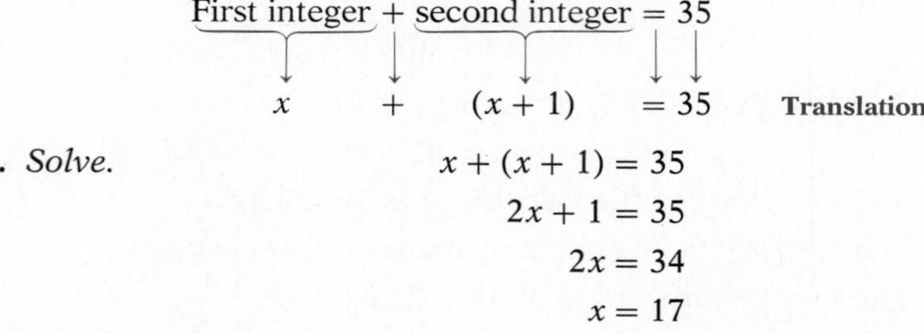

3. *Solve.*

$$x + (x + 1) = 35$$
$$2x + 1 = 35$$
$$2x = 34$$
$$x = 17$$

4. *Check.* The answers are 17 and 18. These are both integers and consecutive. Their sum is 35, so the answers check in the problem.
5. *State.* The consecutive integers are 17 and 18. ◀

DO EXERCISE 7.

7. The sum of two consecutive odd integers is 36. What are the integers? (*Note:* Consecutive odd integers are next to each other, such as -7 and -5. If x is an odd integer, the next consecutive *odd* integer is $x + 2$.)

▶ **EXAMPLE 8** The perimeter of a rectangle (distance around it) is 322 meters (m). The length is 25 m greater than the width. Find the dimensions.

1. *Familiarize.* We first draw a picture.

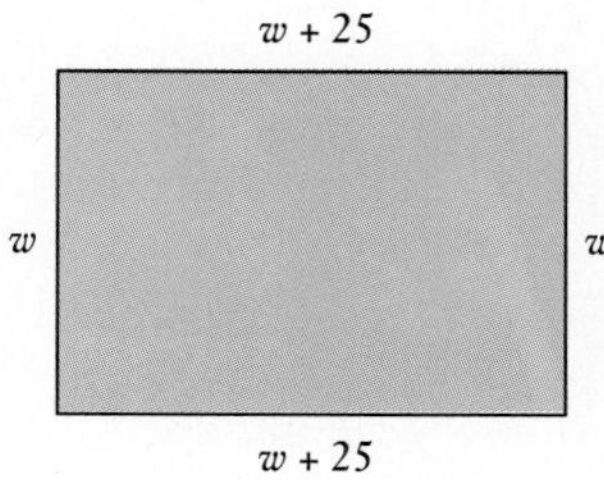

We have let $w =$ the width (in meters). Then $w + 25 =$ the length (in meters).

2. *Translate.* We first recall the formula for perimeter: $P = 2l + 2w$. (If you did not remember, you might need to look it up.) To translate, we substitute $w + 25$ for l and 322 for P, as follows.

$$2l + 2w = P$$
$$2(w + 25) + 2w = 322 \qquad \text{Substituting } w + 25 \text{ for } l \text{ and 322 for } P$$

3. *Solve.*

$$2(w + 25) + 2w = 322$$
$$2w + 50 + 2w = 322 \qquad \text{Removing parentheses using a distributive law}$$
$$4w + 50 = 322 \qquad \text{Collecting like terms}$$
$$4w = 272 \qquad \text{Subtracting 50 on both sides}$$
$$w = 68 \qquad \text{Dividing by 4 on both sides}$$

Then $w = 68$ (width) and $w + 25 = 93$ (length).

4. *Check.* The length is 25 more than the width. The perimeter is $2 \cdot 93 + 2 \cdot 68$, which is 322.
5. *State.* The dimensions of the rectangle are 68 m and 93 m. ◀

DO EXERCISE 8.

8. The length of a rectangle is twice the width. The perimeter is 39 m. Find the dimensions. (*Hint:* The perimeter is the total distance around the rectangle, not just halfway around.)

ANSWERS ON PAGE A-2

❖ SIDELIGHTS

Study Tips: Extra Tips on Problem Solving

We will often present some tips and guidelines to enhance your learning abilities. The following tips are focused on problem solving. They summarize some points already considered and propose some new tips.

- *The following are the five steps for problem solving:*

1. ***Familiarize* yourself with the problem situation.**
2. ***Translate* the problem to an equation.** As you study more mathematics, you will find that the translation may be to some other kind of mathematical language, such as an inequality.
3. ***Solve* the equation.** If the translation is to some other kind of mathematical language, you would carry out some other kind of mathematical manipulation.
4. ***Check* the answer in the original equation.** This does not mean to check in the translated equation. It means to go back to the original worded problem.
5. ***State* the answer to the problem clearly.**

For Step 4 on checking, some further comment is appropriate. *You may be able to translate to an equation and to solve the equation, but none of the solutions of the equation are solutions of the original problem.* To see how this can happen, consider the following problem:

EXAMPLE The sum of two consecutive even integers is 537. Find the integers.

1. *Familiarize.* Suppose we let $x =$ the first number. Then $x + 2 =$ the second number.
2. *Translate.* The problem can be translated to the following equation:

$$x + (x + 2) = 537.$$

3. *Solve.* We solve the equation as follows:

$$2x + 2 = 537$$
$$2x = 535$$
$$x = \frac{535}{2}, \quad \text{or } 267.5.$$

4. *Check.* Then $x + 2 = 269.5$. However, the numbers 267.5 and 269.5 are not only not even, but they are not integers.
5. *State.* The problem has no solution. ◀

The following are some other tips for problem solving.

- *To be good at problem solving, do lots of problems.* The situation is similar to what happens to some people learning a skill such as playing golf. At first, they are not successful, but the more they practice and work at improving their skills, the more successful they become. For problem solving, do more than just two or three odd-numbered problems assigned. Do them all and if you have time, do the even-numbered problems. Then find another book on the same subject and do problems in that book.
- *Look for patterns when solving problems.* By using the preceding tip and doing lots of problems, you will eventually see patterns in similar kinds of problems. For example, there is a pattern in the way that you solve problems involving consecutive integers.
- *When translating to an equation, or some other mathematical language, consider the dimensions of the variables and the constants in the equation.* The variables that represent length should all be in the same unit, those that represent money should be all in dollars or all in cents, and so on.

NAME SECTION DATE

EXERCISE SET 2.2

ANSWERS

a Solve.

1. A lab technician cuts a 12-in. piece of tubing into two pieces in such a way that one piece is 4 in. longer than the other. How long are the pieces?

2. An electrician cuts a 30-ft piece of wire into two pieces. One piece is 2 ft longer than the other. How long are the pieces?

3. A plumber wants to cut a piece of pipe 4 m long into two pieces so that one piece is $\frac{2}{3}$ as long as the other. Find the length of each piece.

4. A sailor wants to cut a piece of rope 5 m long into two pieces so that one piece is $\frac{3}{5}$ as long as the other. Find the length of each piece.

5. In a recent year, the average sales per year at a McDonalds restaurant was $475,000 more than at a Burger King. Total sales at the two stores was $2,525,000. What was the average sales at each store?

6. In a recent year, the total number of airplanes made by the Boeing Aircraft Company and McDonnell Douglas Company was 1500. Boeing had 700 more orders than McDonnell Douglas. How many orders did each have?

7. An F-14 Tomcat fighter plane, used by the Navy, eventually has to be renovated at a cost of $480,000. This cost is 2% of the original cost of the airplane. What was its original cost?

8. The U.S. Postal Service reported recently that if a company sent out 18,000 pieces of junk mail, then 14,040 of those pieces would actually be opened and read. What percent is this of the number sent out?

9. A car rental agency charges a daily rate of $39.95 plus $0.29 per mile. A businesswoman spends $102.30 for such a car on a one-day trip. How many miles did she drive? (*Hint:* This problem translates in a manner similar to Example 5.)

10. A worker on a production line is paid a base salary of $190 per week plus $0.85 for each unit produced. One week the worker earned $551.25. How many units were produced? (*Hint:* This problem translates in a manner similar to Example 5.)

1. ____________
2. ____________
3. ____________
4. ____________
5. ____________
6. ____________
7. ____________
8. ____________
9. ____________
10. ____________

ANSWERS

11. ______

12. ______

13. ______

14. ______

15. ______

16. ______

17. ______

18. ______

19. ______

20. ______

21. ______

22. ______

23. ______

24. ______

11. AT&T recently charged a daily rate of 30¢ for the first minute plus 26¢ for each additional minute on a call within a mileage band of 926 to 1910 miles. A person makes a call under these constraints at a cost of $11.74. How long was the call?

12. For the same kind of call as in Exercise 11, an evening call would cost 19.5¢ plus 16.9¢ for each additional minute. A person makes an evening call under these conditions at a cost of $6.279. How long was the call?

13. Eighteen plus five times a number is seven times the number. What is the number?

14. Sixteen plus three times a number is four times the number. What is the number?

15. Fifteen more than three times a number is the same as ten less than six times the number. What is the number?

16. Sixty more than nine times a number is the same as two less than ten times the number. What is the number?

17. A pro shop in a bowling alley decreases the price of a urethane bowling ball 24% to a sale price of $78.66. What was the former price?

18. An appliance store decreases the price of a certain type of television 18% to a sale price of $410. What was the former price?

19. Money is borrowed at 9% simple interest. After one year, $708.50 pays off the loan. How much was originally borrowed? (*Hint:* See Exercises 13 and 14 of Exercise Set 1.1.)

20. Money is borrowed at 7% simple interest. After one year, $856 pays off the loan. How much was originally borrowed? (*Hint:* See Exercises 13 and 14 of Exercise Set 1.1.)

21. The second angle of a triangle is three times the first, and the third is 12° less than twice the first. Find the measures of the angles.

22. The second angle of a triangle is four times the first, and the third is 5° more than twice the first. Find the measures of the angles.

23. The perimeter of a college basketball court is 96 m and the length is 14 m more than the width. What are the dimensions?

24. The perimeter of a certain soccer field is 310 m. The length is 65 m more than the width. What are the dimensions?

25. Find three consecutive integers such that the sum of the first, twice the second, and three times the third is 80.

26. Find two consecutive even integers such that two times the first plus three times the second is 76.

27. After a person gets a 20% raise in salary, the new salary is $9600. What was the old salary? (*Hint:* What number plus 20% of that number is 9600?)

28. A person gets a 17% raise, bringing the salary to $21,645. What was the salary before the raise?

29. The total cost for tuition plus room and board at State University is $2584. Tuition costs $704 more than room and board. What is the tuition fee?

30. The cost of a private pilot course is $1275. The flight portion costs $625 more than the ground school portion. What is the cost of each?

31. Eleven less than seven times a number is five more than six times the number. Find the number.

32. Fourteen less than eight times a number is three more than four times the number.

33. The sum of two consecutive odd integers is 137. Find the integers.

34. The sum of two consecutive even integers is 659. Find the integers.

35. A student's scores on five tests are 93%, 89%, 72%, 80%, and 96%. What must the student score on the sixth test so that the average will be 88%?

36. Three numbers are such that the second is six less than three times the first, and the third is two more than two thirds of the second. The sum of the three numbers is 172. Find the largest number.

37. Find three consecutive odd integers such that the sum of the first, two times the second, and three times the third is 70.

38. The first angle of a triangle is four times as large as a second angle. The third angle measures 15° less than twice the first angle. Find the measures of the angles.

ANSWERS

25. ____________

26. ____________

27. ____________

28. ____________

29. ____________

30. ____________

31. ____________

32. ____________

33. ____________

34. ____________

35. ____________

36. ____________

37. ____________

38. ____________

ANSWERS

39. ____________

40. ____________

41. ____________

42. ____________

43. ____________

44. ____________

45. ____________

46. ____________

47. ____________

48. ____________

SYNTHESIS

39. A piece of wire 100 cm long is to be cut into two pieces and those pieces are each to be bent to make a square. The length of a side of one square is to be 2 cm greater than the length of a side of the other. How should the wire be cut?

40. A piece of wire 100 cm long is to be cut into two pieces and those pieces are each to be bent to make a square. The length of a side of one square is to be $\frac{2}{3}$ of the length of a side of the other square. How should the wire be cut?

41. The National Basketball Association operates what is called a 24-second clock during its games. This means that the offensive team has 24 seconds in which to attempt a field goal. If it does not make such an attempt, it loses the ball. The NBA arrived at the 24 seconds by dividing the total number of seconds in a game by the average number of points scored per game. There are 48 minutes in a game. What is the average number of points scored in a game?

42. The yearly changes in the population census of a city for three consecutive years are, respectively, 20% increase, 30% increase, and 20% decrease. What is the total percent change from the beginning to the end of the third year, to the nearest percent?

43. The height and sides of a triangle are represented by four consecutive integers. The height is the first integer and the base is the third integer. The perimeter of the triangle is 42 in. Find the area of the triangle.

44. The salary of an employee was reduced $n\%$ during a time when a company was having financial difficulty. By what percent would the company then have to raise the salary in order to bring it back to where it was before the reduction?

45. A literature professor has a book collection containing 400 novels. The number of horror novels is 46 percent of the number of science fiction novels; the number of science fiction novels is 65% of the number of romance novels; and the number of mystery novels is 17% of the number of horror novels. How many science fiction novels does the professor have? Round to the nearest one.

46. Suppose your watch loses one and one-half seconds every hour. You have a friend whose watch gains one second every hour. The watches show the same time now. After how many more seconds will the watches show the same time again?

47. Suppose the figure $ABCD$ below is a square. Point A is folded onto the midpoint of $\overline{AB}$ and point D is folded onto the midpoint of $\overline{DC}$. The perimeter of the smaller figure formed is 25 in. Find the area of the square $ABCD$.

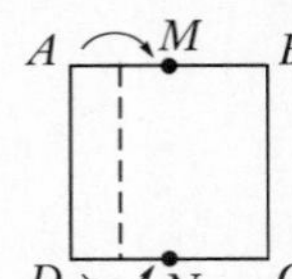

48. Consider the geometric figure below. Suppose that $L \parallel M$, $m\angle 8 = 5x + 25$, and $m\angle 4 = 8x + 4$. Find $m\angle 2$ and $m\angle 1$.

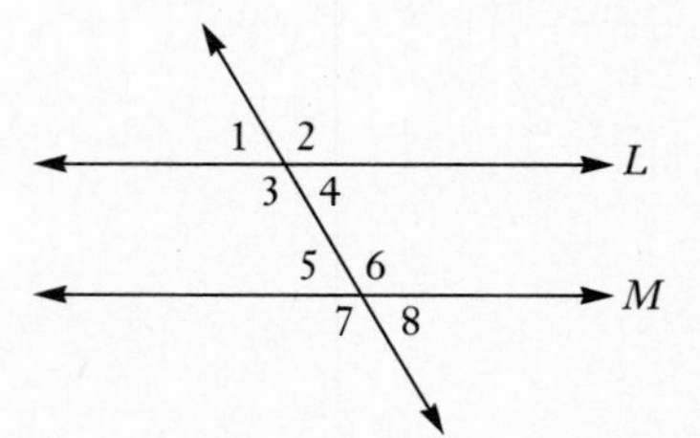

2.3 Formulas

a Solving Formulas

A **formula** is a kind of recipe, or rule, for doing a certain kind of calculation. Formulas are often given by equations. Here is a formula for finding the area of a trapezoid:

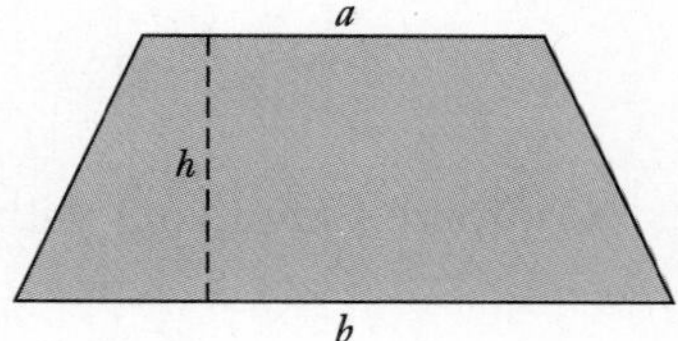

$$A = \frac{1}{2} h(a + b).$$

The formula says that to find the area of a trapezoid, take half the product of the height, h, and the sum of the lengths of the parallel sides, a and b. Suppose that we know the area A and the lengths of the parallel sides a and b, and we want to find the height h. To do this, we can get h alone on one side of the equation. We "solve" for h.

► **EXAMPLE 1** Solve the trapezoid formula $A = \frac{1}{2}h(a + b)$ for h.

$A = \frac{1}{2}h(a + b)$ **We want this letter alone.**

$2A = h(a + b)$ **Multiplying by 2 to clear of the fraction**

$\frac{2A}{a + b} = h$ **Dividing on both sides by $a + b$** ◄

When we solve a formula for a given letter, we use the same principles that we would use to solve any equation. The following summarizes steps to follow in solving formulas for letters.

To solve a formula for a given letter, identify the letter, and:

1. **Multiply on both sides to clear of fractions or decimals, if necessary.**
2. **Collect like terms on each side, if necessary. This may require factoring if a variable is in more than one term.**
3. **Using the addition principle, get all terms with the letter to be solved for on one side of the equation and all other terms on the other side.**
4. **Collect like terms again, if necessary.**
5. **Solve for the letter in question using the multiplication principle.**

DO EXERCISE 1.

► **EXAMPLE 2** Solve for b: $A = \frac{5}{2}(b - 20)$.

$A = \frac{5}{2}(b - 20)$ **We want this letter alone.**

$2A = 5(b - 20)$ **Multiplying by 2**

$2A = 5b - 100$ **Removing parentheses**

$2A + 100 = 5b$

$\frac{2A + 100}{5} = b$, or $b = \frac{2A}{5} + 20$. ◄

DO EXERCISE 2.

OBJECTIVE

After finishing Section 2.3, you should be able to:

a Solve formulas for a specified letter.

FOR EXTRA HELP

Tape 4B | Tape 3A | MAC: 2 IBM: 2

1. A formula for the area of a triangle is $A = \frac{1}{2}bh$. Solve for b.

2. Solve for c: $P = \frac{3}{5}(c + 10)$.

ANSWERS ON PAGE A-3

3. Solve for m: $H = 2r + 3m$.

4. Solve for Q: $T = Q + Qiy$.

5. Solve for b: $A = \frac{1}{2}h(a + b)$.

ANSWERS ON PAGE A-3

▶ **EXAMPLE 3** Solve for r: $H = 2r + 3m$.

$$H = 2r + 3m \quad \textbf{We want this letter alone.}$$
$$H - 3m = 2r \quad \textbf{Subtracting } 3m$$
$$\frac{H - 3m}{2} = r \quad \textbf{Dividing by 2}$$ ◀

Compare Example 3 with an ordinary equation with numbers.

▶ **EXAMPLE 3A** Solve for x: $11 = 2x + 3 \cdot 5$.

$$11 = 2x + 3 \cdot 5$$
$$11 - 3 \cdot 5 = 2x \quad \textbf{Subtracting } 3 \cdot 5$$
$$\frac{11 - 3 \cdot 5}{2} = x \quad \textbf{Dividing by 2}$$ ◀

In this case we could simplify further; in Example 3, we could not.

DO EXERCISE 3.

▶ **EXAMPLE 4** The formula $A = P + Prt$ gives the amount to which a principal P, in dollars, will grow at simple interest rate r in t years. Solve the formula for P.

$$A = P + Prt \quad \textbf{We want this letter alone.}$$
$$A = P(1 + rt) \quad \textbf{Factoring (or collecting like terms)}$$
$$\frac{A}{1 + rt} = P \quad \textbf{Dividing on both sides by } 1 + rt$$ ◀

DO EXERCISE 4.

▶ **EXAMPLE 5** Solve the trapezoid formula $A = \frac{1}{2}h(a + b)$ for a.

$$A = \tfrac{1}{2}h(a + b) \quad \textbf{We want this letter alone.}$$
$$2A = h(a + b) \quad \textbf{Multiplying by 2 to clear of the fraction}$$
$$2A = ha + hb \quad \textbf{Using a distributive law}$$
$$2A - hb = ha \quad \textbf{Adding } -hb$$
$$\frac{2A - hb}{h} = a \quad \textbf{Dividing by } h$$ ◀

There are other correct answers to this problem. One of them is $(2A/h) - b = a$. It can be found by multiplying by $2/h$ at the outset.

DO EXERCISE 5.

NAME SECTION DATE

EXERCISE SET 2.3

a Solve.

1. $A = lw$, for l
(area of a rectangle)

2. $A = lw$, for w

3. $W = EI$, for I
(an electricity formula)

4. $W = EI$, for E

5. $d = rt$, for t
(a motion formula)

6. $d = rt$, for r

7. $I = Prt$, for t
(a simple-interest formula)

8. $I = Prt$, for P

9. $E = mc^2$, for m
(a relativity formula)

10. $E = mc^2$, for c^2

11. $P = 2l + 2w$, for l
(a perimeter of a rectangle)

12. $P = 2l + 2w$, for w

13. $c^2 = a^2 + b^2$, for a^2
(the Pythagorean theorem for right triangles)

14. $c^2 = a^2 + b^2$, for b^2

15. $Ax + By = C$, for x
(equation of a line)

16. $Ax + By = C$, for y

ANSWERS

1. ______
2. ______
3. ______
4. ______
5. ______
6. ______
7. ______
8. ______
9. ______
10. ______
11. ______
12. ______
13. ______
14. ______
15. ______
16. ______

ANSWERS

17. ______

18. ______

19. ______

20. ______

21. ______

22. ______

23. ______

24. ______

25. ______

26. ______

27. ______

28. ______

29. ______

30. ______

31. ______

32. ______

33. ______

34. ______

35. ______

36. ______

37. ______

38. ______

17. $A = \pi r^2$, for r^2 (area of a circle)

18. $A = \pi r^2$, for π

19. $W = \frac{11}{2}(h - 40)$, for h

20. $C = \frac{5}{9}(F - 32)$, for F (a temperature formula)

21. $V = \frac{4}{3}\pi r^3$, for r^3 (volume of a sphere)

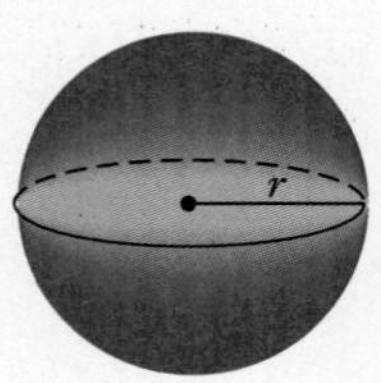

22. $V = \frac{4}{3}\pi r^3$, for π

23. $A = \frac{1}{2}h(c - d)$, for c

24. $A = \frac{1}{2}h(c - d)$, for d

25. $F = \dfrac{mv^2}{r}$, for m (a physics formula)

26. $F = \dfrac{mv^2}{r}$, for v^2

27. $P = 5a - 3ab$, for a; for b

28. $Q = 4p + 7pq$, for p; for q

SKILL MAINTENANCE

Divide.

29. $\dfrac{80}{-16}$

30. $-2000 \div (-8)$

31. $-\dfrac{1}{2} \div \dfrac{1}{4}$

32. $120 \div (-4.8)$

SYNTHESIS

Solve.

33. $s = v_1 t + \frac{1}{2}at^2$, for a; for v_1

34. $A = \pi rs + \pi r^2$, for s

35. $\dfrac{P_1 V_1}{T_1} = \dfrac{P_2 V_2}{T_2}$, for V_1; for P_2

36. $\dfrac{P_1 V_1}{T_1} = \dfrac{P_2 V_2}{T_2}$, for T_2; for P_1

37. The area of the shaded triangle ABE is 20 cm². Find the area of the trapezoid.

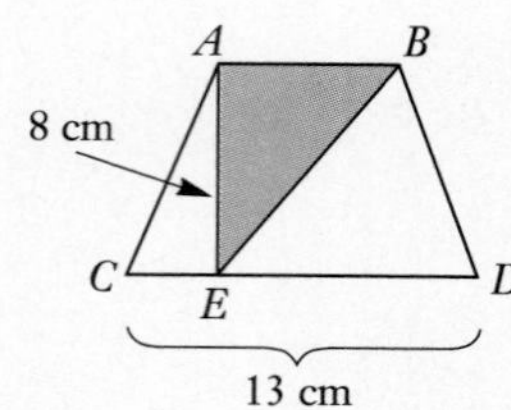

38. In Exercise 7, you solved the formula $I = Prt$ for t. Now use it to find how long it will take a deposit of \$75 to earn \$3 interest when invested at 5% simple interest.

2.4 Inequalities and Problem Solving

a Solving Inequalities

We can extend our equation-solving skills to the solving of inequalities. An **inequality** is any sentence having one of the verbs $<$, $>$, $\leq$, or $\geq$ (see Section 1.2)—for example,

$$-2 < a, \qquad x > 4, \qquad x + 3 \leq 6, \quad \text{and} \quad 16 - 7y \geq 10y - 4.$$

Some replacements for the variable in an inequality make it true, and some make it false. A replacement that makes it true is called a **solution.** The set of all solutions is called the **solution set.** When we have found the set of all solutions of an inequality, we say that we have **solved** the inequality.

▶ **EXAMPLES** Determine whether the given number is a solution of the inequality.

1. $x + 3 < 6$; 5

We substitute and get $5 + 3 < 6$, or $8 < 6$, a false sentence. Therefore, 5 is not a solution.

2. $2x - 3 > -3$; 1

We substitute and get $2(1) - 3 > -3$, or $-1 > -3$, a true sentence. Therefore, 1 is a solution.

3. $4x - 1 \leq 3x + 2$; 3

We substitute and get $4(3) - 1 \leq 3(3) + 2$, or $11 \leq 11$, a true sentence. Therefore, 3 is a solution. ◀

DO EXERCISES 1–3.

The Addition Principle

There is an addition principle for inequalities that is similar to the one for solving equations.

> The Addition Principle for Inequalities
>
> **If the same number is added (or subtracted) on both sides of an inequality, we get another true inequality.**
>
> **If $a < b$ is true, then $a + c < b + c$ is true.**
> **If $a > b$ is true, then $a + c > b + c$ is true.**
> **If $a \leq b$ is true, then $a + c \leq b + c$ is true.**
> **If $a \geq b$ is true, then $a + c \geq b + c$ is true.**

To solve inequalities, we try to get the variable alone on one side in order to get an inequality for which we can determine the solutions easily. This would also allow us to easily draw a graph of the inequality as in Section 1.2.

OBJECTIVES

After finishing Section 2.4, you should be able to:

a Determine whether a given number is a solution of an inequality, solve an inequality using the addition principle, and then graph the inequality.

b Solve an inequality using the multiplication principle and then graph the inequality.

c Solve an inequality using both principles and then graph the inequality.

d Translate sentences to inequalities and solve problems by translating to inequalities.

FOR EXTRA HELP

Tape 4C | Tape 3A | MAC: 2 IBM: 2

Determine whether the given number is a solution of the inequality.

1. $3 - x < 2$;

2. $3x + 2 > -1$;

3. $3x + 2 \leq 4x - 3$;

ANSWERS ON PAGE A-3

Solve. Then graph.

4. $x + 6 > 9$

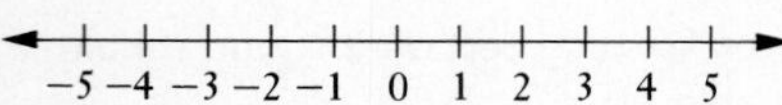

5. $x + 4 \leq 7$

ANSWERS ON PAGE A-3

▶ **EXAMPLE 4** Solve: $x + 5 > 3$. Then graph.

We have

$$x + 5 > 3$$

$$x + 5 - 5 > 3 - 5 \quad \text{Using the addition principle: adding } -5 \text{ or subtracting } 5$$

$$x > -2.$$

Any real number greater than -2 is a solution. The following are some of the solutions:

$$-1, \quad 0, \quad 7.3, \quad \pi, \quad \tfrac{17}{5}, \quad 2345, \quad 78.56.$$

To describe all the solutions, we could try to write them in a set as follows: $\{-1, 0, 7.3, \pi, \frac{17}{5}, 2345, 76.56, \ldots\}$. However, we can never list all the solutions this way; and seeing this set, without knowing what the inequality is, makes it difficult to know what real numbers we are considering. Instead, we use set-builder notation as considered in Section 1.2. The notation is

$$\{x \mid x > -2\},$$

which is read:

"The set of all x such that x is greater than -2."

Henceforth, you should use this notation when naming solution sets of inequalities. The graph of the inequality is as follows:

−3 −2 −1 0 1 2 3 4 5 6 7

We cannot check all the solutions of an inequality by substitution, as we can check the solutions of an equation. There are too many of them. In this case, we could do a partial check by substituting a number greater than -2, say 8, into the original inequality:

$x + 5 > 3$	
$8 + 5$	3
13	TRUE

Since $13 > 3$ is true, 8 is a solution. Any number greater than -2 is a solution. ◀

When two inequalities have the same solutions, we say that they are *equivalent*.

Whenever we use the addition principle, adding a specific number on both sides, the first and last inequalities will be equivalent.

DO EXERCISES 4 AND 5.

▶ **EXAMPLE 5** Solve: $4x - 1 \geq 5x - 2$. Then graph.

We have

$$4x - 1 \geq 5x - 2$$
$$4x - 1 + 2 \geq 5x - 2 + 2 \quad \textbf{Adding 2}$$
$$4x + 1 \geq 5x \quad \textbf{Simplifying}$$
$$4x + 1 - 4x \geq 5x - 4x \quad \textbf{Subtracting } 4x$$
$$1 \geq x. \quad \textbf{Simplifying}$$

We know that $1 \geq x$ has the same meaning as $x \leq 1$. Thus any number less than or equal to 1 is a solution. We can express the solution set as $\{x | 1 \geq x\}$ or as $\{x | x \leq 1\}$. The latter is used most often. The graph is as follows:

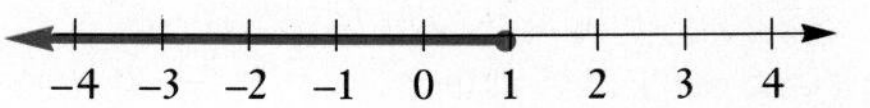

◀

DO EXERCISE 6.

6. Solve: $2x - 3 \geq 3x - 1$. Then graph.

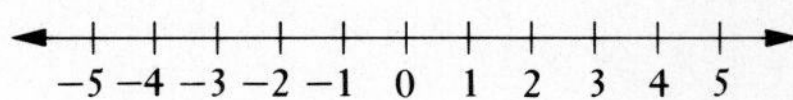

▶ **EXAMPLE 6** Solve: $y + \frac{4}{5} < \frac{2}{3}$.

We have

$$y + \tfrac{4}{5} < \tfrac{2}{3}$$
$$y + \tfrac{4}{5} - \tfrac{4}{5} < \tfrac{2}{3} - \tfrac{4}{5} \quad \textbf{Subtracting } \tfrac{4}{5}$$
$$y < \tfrac{2}{3} \cdot \tfrac{5}{5} - \tfrac{4}{5} \cdot \tfrac{3}{3} \quad \textbf{Multiplying by 1 to find a common denominator}$$
$$y < \tfrac{10}{15} - \tfrac{12}{15}$$
$$y < -\tfrac{2}{15}.$$

Any number less than $-\frac{2}{15}$ is a solution. The solution set is $\{y | y < -\frac{2}{15}\}$. ◀

DO EXERCISES 7 AND 8.

Solve.

7. $x - \dfrac{1}{3} \geq \dfrac{5}{4}$

8. $7y - 1 < -3 + 6y$

b The Multiplication Principle

The multiplication principle for inequalities is somewhat different from the multiplication principle for equations. Consider this true inequality:

$$-4 < 9. \quad \textbf{True}$$

If we multiply both numbers by 2, we get another true inequality:

$$-4(2) < 9(2), \quad \text{or} \quad -8 < 18. \quad \textbf{True}$$

If we multiply both numbers by -3, we get a false inequality:

$$-4(-3) < 9(-3), \quad \text{or} \quad 12 < -27. \quad \textbf{False}$$

However, if we now reverse the inequality symbol above, we get a true inequality:

$$12 > -27. \quad \textbf{True}$$

↑ The < symbol has been reversed!

ANSWERS ON PAGE A-3

> **The Multiplication Principle for Inequalities**
>
> **If we multiply (or divide) on both sides of a true inequality by a positive number, we get another true inequality. If we multiply (or divide) by a negative number and the inequality symbol is reversed, we get another true inequality.**
>
> **For a positive real number c:**
>
> **If $a < b$ is true, then $ac < bc$ is true.**
> **If $a > b$ is true, then $ac > bc$ is true.**
> **If $a \leq b$ is true, then $ac \leq bc$ is true.**
> **If $a \geq b$ is true, then $ac \geq bc$ is true.**
>
> **For a negative real number c:**
>
> **If $a < b$ is true, then $ac > bc$ is true.**
> **If $a > b$ is true, then $ac < bc$ is true.**
> **If $a \leq b$ is true, then $ac \geq bc$ is true.**
> **If $a \geq b$ is true, then $ac \leq bc$ is true.**

The important thing to remember is that if we multiply or divide by a negative number, we must reverse the inequality symbol.

When we solve an inequality using the multiplication principle, we can multiply or divide by any number except zero.

► **EXAMPLE 7** Solve: $3y < \frac{3}{4}$. Then graph.

We have

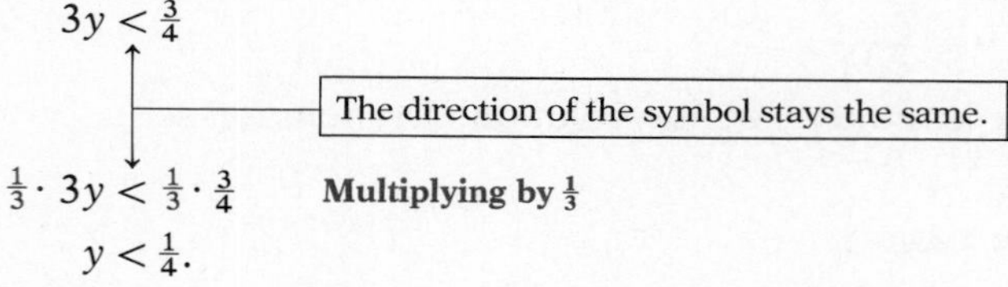

Any number less than $\frac{1}{4}$ is a solution. The solution set is $\{y \mid y < \frac{1}{4}\}$. The graph is as follows:

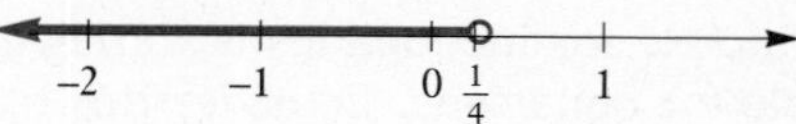

◄

► **EXAMPLE 8** Solve: $-4x < \frac{4}{5}$. Then graph.

We have

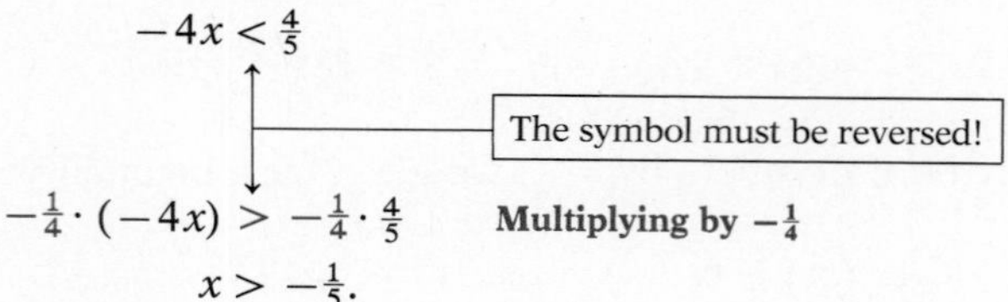

Any number greater than $-\frac{1}{5}$ is a solution. The solution set is $\{x \mid x > -\frac{1}{5}\}$. The graph is as follows:

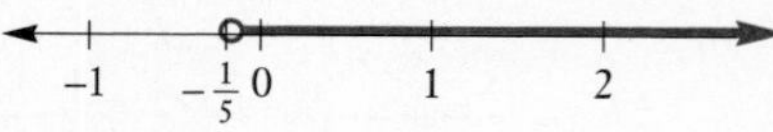

◄

▶ **EXAMPLE 9** Solve: $-5x \geq -80$. Then graph.

We have

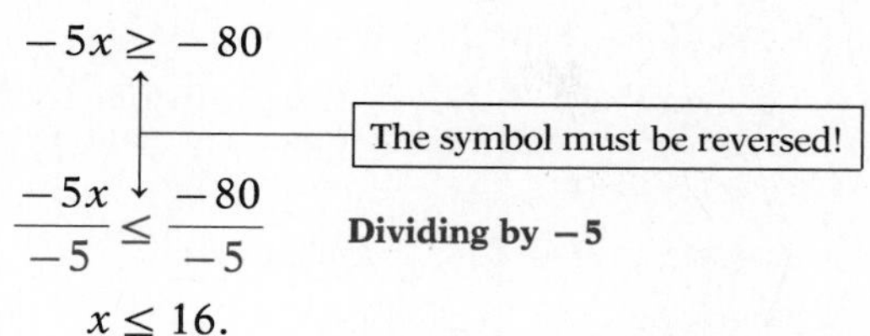

$$-5x \geq -80$$

$$\frac{-5x}{-5} \leq \frac{-80}{-5} \quad \textbf{Dividing by } -5$$

$$x \leq 16.$$

The solution set is $\{x|x \leq 16\}$. The graph is as follows:

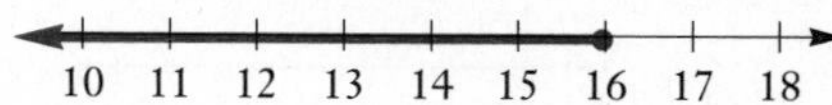

◀

DO EXERCISES 9–11.

C Using the Principles Together

We use the addition and multiplication principles together in solving inequalities in much the same way that we did for equations.

▶ **EXAMPLE 10** Solve: $16 - 7y \geq 10y - 4$.

We have

$$16 - 7y \geq 10y - 4$$

$$-16 + 16 - 7y \geq -16 + 10y - 4 \quad \textbf{Adding } -16$$

$$-7y \geq 10y - 20$$

$$-10y - 7y \geq -10y + 10y - 20 \quad \textbf{Adding } -10y$$

$$-17y \geq -20$$

The symbol must be reversed!

$$\frac{-17y}{-17} \leq \frac{-20}{-17} \quad \textbf{Dividing by } -17$$

$$y \leq \frac{20}{17}.$$

The solution set is $\{y|y \leq \frac{20}{17}\}$. ◀

Solve. Then graph.

9. $5y \leq \frac{3}{2}$

−5 −4 −3 −2 −1 0 1 2 3 4 5

10. $-2y > \frac{5}{6}$

−5 −4 −3 −2 −1 0 1 2 3 4 5

11. $-\frac{1}{3}x \leq -4$

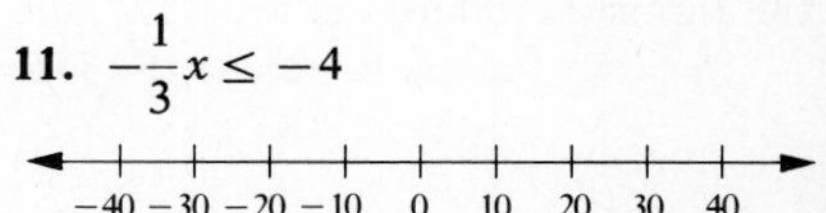

ANSWERS ON PAGE A-3

Solve.

12. $6 - 5y \geq 7$

13. $3x + 5x < 4$

14. $17 - 5(y - 2) \leq$ $45y + 8(2y - 3) - 39y$

Translate to an inequality.

15. A number is less than -5.

16. A number is greater than 4.85.

17. That car is worth at most \$6700 in trade-in allowance.

18. I must be offered at least \$27,500 in order for me to accept the job.

19. Thirteen more than twice a number is less than 78.2.

20. Four times the sum of two numbers is greater than $6\frac{1}{2}$.

ANSWERS ON PAGE A-3

▶ **EXAMPLE 11** Solve: $-3(x + 8) - 5x < 4(x - 9) + 27$.

We have

$$-3(x + 8) - 5x < 4(x - 9) + 27$$

$$-3x - 24 - 5x < 4x - 36 + 27 \quad \text{Using a distributive law to remove parentheses}$$

$$-24 - 8x < 4x - 9$$

$$-24 - 8x + 8x < 4x - 9 + 8x \quad \text{Adding } 8x$$

$$-24 < 12x - 9$$

$$-24 + 9 < 12x - 9 + 9 \quad \text{Adding 9}$$

$$-15 < 12x$$

The symbol stays the same.

$$\frac{-15}{12} < x \quad \text{Dividing by 12}$$

$$-\frac{5}{4} < x.$$

The solution set is $\{x \mid -\frac{5}{4} < x\}$, or $\{x \mid x > -\frac{5}{4}\}$. ◀

DO EXERCISES 12–14.

d Solving Problems Using Inequalities

Certain problems can be translated quite naturally to inequalities rather than equations. In fact, the five-step problem-solving process can still be used.

Let us first practice translating phrases to inequalities.

▶ **EXAMPLES** Translate to an inequality.

12. A number is greater than -7.

$$x > -7$$

13. A number is less than 5.68.

$$y < 5.68$$

14. The number of students at this college is less than or equal to 4500.

$$N \leq 4500$$

15. The cost of producing that machine is at least \$85,000.

$$C \geq \$85,000$$

16. My grade-point average is at most 3.24.

$$g \leq 3.24$$

17. Eighteen more than three times the sum of two numbers is less than fifty-four.

$$18 + 3(a + b) < 54$$ ◀

DO EXERCISES 15–20.

▶ **EXAMPLE 18** *Records in the women's 100-m dash.* Florence Griffith Joyner set a world record of 10.49 sec in the women's 100-m dash in the 1988 Olympics. The formula

$$R = -0.03125t + 11.01$$

can be used to predict the world record in the women's 100-m dash t years after 1976. For example, to find the record in 1996, we would subtract 1976, to get $t = 20$. Determine (in terms of an inequality) those years for which the world record will be less than 10.4 sec.

1. *Familiarize.* We already have a formula. To become more familiar with it, we might make a substitution for t. Suppose we want to know the record after 50 years, or the year 2026. We substitute 50 for t:

$$R = -0.03125t + 11.01 = -0.03125(50) + 11.01 = 9.4475 \text{ sec.}$$

2. *Translate.* The record R is to be less than 10.4 sec. We have the inequality

$$R < 10.4.$$

To find the times t that satisfy the inequality, we substitute $-0.03125t + 11.01$ for R:

$$-0.03125t + 11.01 < 10.4.$$

3. *Solve.* We solve the inequality:

$$-0.03125t + 11.01 < 10.4$$

$$-0.03125t < -0.61 \quad \textbf{Subtracting 11.01}$$

$$\frac{-0.03125t}{-0.03125} > \frac{-0.61}{-0.03125} \quad \textbf{Dividing by } -0.03125$$

$$t > 19.52. \quad \textbf{Simplifying. You might use a calculator.}$$

4. *Check.* We can check by substituting a value for t greater than 19.52. We did that in the familiarization step.

5. *State.* The record will be less than 10.4 for those races occurring more than 19.52 years after 1976, which will be approximately 1996 or after. ◀

DO EXERCISE 21.

21. *Records in the men's 200-m dash.* The formula

$$R = -0.028t + 20.8$$

can be used to predict the world record in the men's 200-m dash t years after 1920. Determine (in terms of an inequality) those years for which the world record will be less than 19.0 sec. (Check a sports almanac to see if this has happened.)

ANSWER ON PAGE A-3

▶ **EXAMPLE 19** On your new job, you can be paid in one of two ways:

Plan A: A salary of \$600 per month, plus a commission of 4% of sales;

Plan B: A salary of \$800 per month, plus a commission of 6% of sales in excess of \$10,000.

For what amount of sales is plan A better than plan B, if we assume that sales are always more than \$10,000?

1. *Familiarize.* Listing the given information in a table will be helpful.

Plan A: Monthly income	**Plan B: Monthly income**
\$600 salary 4% of sales *Total:* \$600 + 4% of sales	\$800 salary 6% of sales over \$10,000 *Total:* \$800 + 6% of sales over \$10,000

We let S = monthly sales.

22. A painter can be paid in one of two ways:

Plan A: $500 plus $4 per hour;
Plan B: Straight $9 per hour.

Suppose that the job takes n hours. For what values of n is plan A better for the painter?

ANSWER ON PAGE A-3

2. *Translate.* We want to find the sales S for which plan A is better than plan B. Monthly income from plan A is $600 + 4\%S$. Monthly income from plan B is $800 + 6\%(S - 10{,}000)$, where we note that $S - 10{,}000$ represents the amount that exceeds 10,000. Thus we want to find values of S for which

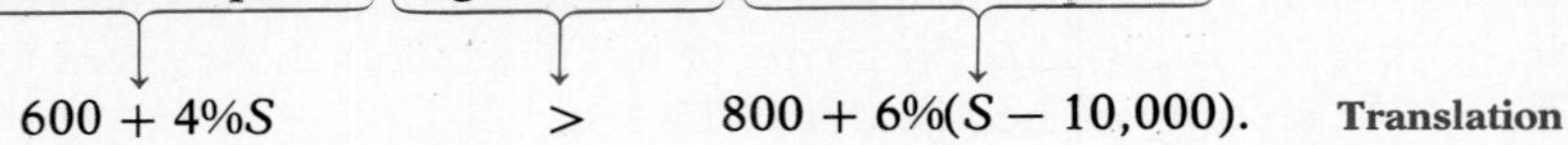

$$600 + 4\%S \quad > \quad 800 + 6\%(S - 10{,}000).$$ **Translation**

3. *Solve.* We solve the inequality:

$$600 + 4\%S > 800 + 6\%S - 6\%(10{,}000)$$
$$600 + 0.04S > 800 + 0.06S - 600$$ **Changing to decimal notation**
$$600 + 0.04S > 200 + 0.06S$$ **Collecting like terms**
$$400 > 0.02S$$ **Subtracting both 200 and 0.04S**
$$20{,}000 > S.$$ **Dividing by 0.02**

4. *Check.* For $S = 20{,}000$, the income from plan A is

$$600 + 4\% \cdot 20{,}000, \quad \text{or} \quad \$1400.$$

The income from plan B is

$$800 + 6\% \cdot 10{,}000, \quad \text{or} \quad \$1400.$$

We calculate for some amount of sales less than $20,000 and for some amount greater than $20,000. Suppose that sales are $15,000. Then

Plan A gives $600 + 4\% \cdot 15{,}000$, or $1200;
Plan B gives $800 + 6\% \cdot 5000$, or $1100.

For sales of $15,000, plan A gives a better income than plan B.

For sales of $25,000,

Plan A gives $600 + 4\% \cdot 25{,}000$, or $1600;
Plan B gives $800 + 6\% \cdot 15{,}000$, or $1700.

Thus for sales of $25,000, plan A gives less income than plan B.

For sales of less than $20,000, plan A is better than plan B, which is what is stated by the inequality $20{,}000 > S$. We cannot check all possible values of S so we will stop here.

5. *State.* For sales of less than $20,000, plan A is better. ◀

DO EXERCISE 22.

NAME SECTION DATE

EXERCISE SET 2.4

a Determine whether the given numbers are solutions of the inequality.

1. $x - 2 \geq 6$; $-4, 0, 4, 8$

2. $3x + 5 \leq -10$; $-5, -10, 0, 27$

3. $t - 8 > 2t - 3$; $0, -8, -9, -3$

4. $5y - 7 < 5 - y$; $2, -3, 0, 3$

Solve. Then graph.

5. $x + 8 > 3$

−6 −5 −4 −3 −2 −1 0 1 2 3 4 5 6

6. $x + 5 > 2$

−6 −5 −4 −3 −2 −1 0 1 2 3 4 5 6

7. $y + 3 < 9$

−6 −5 −4 −3 −2 −1 0 1 2 3 4 5 6

8. $y + 4 < 10$

−6 −5 −4 −3 −2 −1 0 1 2 3 4 5 6

9. $a + 9 \leq -12$

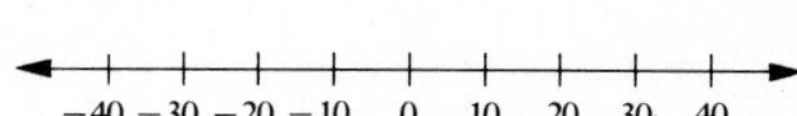

10. $a + 7 \leq -13$

−40 −30 −20 −10 0 10 20 30 40

11. $t + 14 \geq 9$

−6 −5 −4 −3 −2 −1 0 1 2 3 4 5 6

12. $x - 9 \leq 10$

−40 −30 −20 −10 0 10 20 30 40

13. $y - 8 > -14$

−6 −5 −4 −3 −2 −1 0 1 2 3 4 5 6

14. $y - 9 > -18$

−12 −8 −4 0 4 8 12

ANSWERS

1. ______
2. ______
3. ______
4. ______
5. ______
6. ______
7. ______
8. ______
9. ______
10. ______
11. ______
12. ______
13. ______
14. ______

ANSWERS

15. ______
16. ______
17. ______
18. ______
19. ______
20. ______
21. ______
22. ______
23. ______
24. ______
25. ______
26. ______
27. ______
28. ______
29. ______
30. ______
31. ______
32. ______

15. $x - 11 \leq -2$

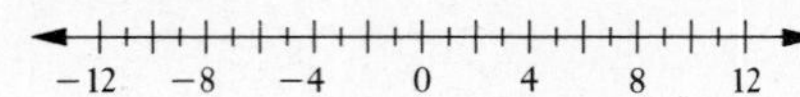

16. $y - 18 \leq -4$

−40 −30 −20 −10 0 10 20 30 40

b Solve. Then graph.

17. $8x \geq 24$

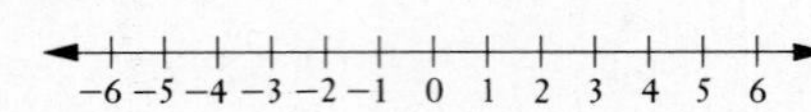

18. $9t < -81$

−12 −8 −4 0 4 8 12

19. $0.3x < -18$

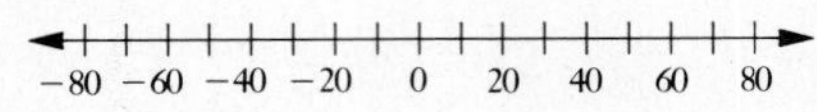

20. $0.5x < 25$

−80 −60 −40 −20 0 20 40 60 80

Solve.

21. $-9x \geq -8.1$

22. $-8y \leq 3.2$

23. $-\frac{3}{4}x \geq -\frac{5}{8}$

24. $-\frac{5}{6}y \leq -\frac{3}{4}$

c Solve. Then graph.

25. $2x + 7 < 19$

−6 −5 −4 −3 −2 −1 0 1 2 3 4 5 6

26. $5y + 13 > 28$

−6 −5 −4 −3 −2 −1 0 1 2 3 4 5 6

27. $5y + 2y \leq -21$

−6 −5 −4 −3 −2 −1 0 1 2 3 4 5 6

28. $-9x + 3x \geq -24$

−6 −5 −4 −3 −2 −1 0 1 2 3 4 5 6

Solve.

29. $2y - 7 < 5y - 9$

30. $8x - 9 < 3x - 11$

31. $0.4x + 5 \leq 1.2x - 4$

32. $0.2y + 1 > 2.4y - 10$

33. $3x - \frac{1}{8} \leq \frac{3}{8} + 2x$

34. $2x - 3 < \frac{13}{4}x + 10 - 1.25x$

35. $4(3y - 2) \geq 9(2y + 5)$

36. $4m + 5 \geq 14(m - 2)$

37. $3(2 - 5x) + 2x < 2(4 + 2x)$

38. $2(0.5 - 3y) + y > (4y - 0.2)8$

39. $5[3m - (m + 4)] > -2(m - 4)$

40. $[8x - 3(3x + 2)] - 5 \geq 3(x + 4) - 2x$

41. $3(r - 6) + 2 > 4(r + 2) - 21$

42. $5(t + 3) + 9 < 3(t - 2) + 6$

43. $19 - (2x + 3) \leq 2(x + 3) + x$

44. $13 - (2c + 2) \geq 2(c + 2) + 3c$

45. $\frac{1}{4}(8y + 4) - 17 < -\frac{1}{2}(4y - 8)$

46. $\frac{1}{3}(6x + 24) - 20 > -\frac{1}{4}(12x - 72)$

47. $2[4 - 2(3 - x)] - 1 \geq 4[2(4x - 3) + 7] - 25$

48. $5[3(7 - t) - 4(8 + 2t)] - 20 \leq -6[2(6 + 3t) - 4]$

49. $\frac{2}{3}(2x - 1) > 10$

50. $\frac{4}{5}(3x + 4) < 20$

51. $\frac{3}{4}(3 + 2x) + 1 \geq 13$

52. $\frac{7}{8}(5 - 4x) - 17 \geq 38$

ANSWERS

33. ______

34. ______

35. ______

36. ______

37. ______

38. ______

39. ______

40. ______

41. ______

42. ______

43. ______

44. ______

45. ______

46. ______

47. ______

48. ______

49. ______

50. ______

51. ______

52. ______

ANSWERS

53. ______
54. ______
55. ______
56. ______
57. ______
58. ______
59. ______
60. ______
61. ______
62. ______
63. ______
64. ______
65. ______
66. ______
67. ______
68. ______
69. ______
70. ______
71. ______
72. ______

53. $\frac{3}{4}(3x - \frac{1}{2}) - \frac{2}{3} < \frac{1}{3}$

54. $\frac{2}{3}(\frac{7}{8} - 4x) - \frac{5}{8} < \frac{3}{8}$

55. $0.7(3x + 6) \geq 1.1 - (x + 2)$

56. $0.9(2x + 8) < 20 - (x + 5)$

57. $a + (a - 3) \leq (a + 2) - (a + 1)$

58. $0.8 - 4(b - 1) > 0.2 + 3(4 - b)$

d Translate to an inequality.

59. A number is less than 8.

60. A number is greater than -1.2.

61. The price of a movie ticket is greater than or equal to $6.

62. The number of people in the flower club is less than or equal to 58.

63. The price of a compact disc is at most $17.95.

64. My salary next year will be at least $35,000.

65. 24 minus three times a number is less than 16 plus the number.

66. -17 more than four times some number is greater than the number minus 23.

67. Fifteen times the sum of two numbers is at least 78.

68. Twice the product of two numbers plus 27 is at most -3.

Solve.

69. A car rents for $30 per day plus 20¢ per mile. You are on a daily budget of $96. What mileages will allow you to stay within the budget?

70. A car can be rented for $35 per day with unlimited mileage, or for $28 per day plus 19¢ per mile. For what daily mileages would the unlimited mileage plan save you money?

71. You are taking a history course in which there will be 4 tests, each worth 100 points. You have scores of 89, 92, and 95 on the first three tests. You must make a total of at least 360 in order to get an A. What scores on the last test will give you an A?

72. You are taking a science course in which there will be 5 tests, each worth 100 points. You have scores of 94, 90, and 89 on the first three tests. You must make a total of at least 450 in order to get an A. What scores on the fourth test will keep you eligible for an A?

73. In planning for a college dance, you find that one band will play for $250 plus 50% of the total ticket sales. Another band will play for a flat fee of $550. In order for the first band to produce more profit for the school than the other band, what is the highest price you can charge per ticket, assuming that 300 people will attend?

74. A bank offers two checking-account plans. Plan A charges a base service charge of $2.00 per month plus 15¢ per check. Plan B charges a base service charge of $4.00 per month plus 9¢ per check. For what numbers of checks per month would plan B be better than plan A?

75. A medical insurance company offers two plans. With plan A, you pay the first $100 of your medical bills and they pay 80% of the rest. With plan B, you pay the first $250 of your medical bills and they pay 90% of the rest. For what total amounts of medical bills will plan B save you money?

76. You can spend $3.50 at the laundromat washing your clothes, or you can have them do it for 40¢ per pound of clothes. For what weights of clothes will it save you money to wash your clothes yourself?

77. On your new job, you can be paid in one of two ways:

Plan A: A salary of $500 per month plus a commission of 4% of gross sales;

Plan B: A salary of $750 per month plus a commission of 5% of gross sales over $8000.

For what amount of gross sales is plan B better than plan A, assuming that gross sales are always more than $8000?

78. On your new job, you can be paid in one of two ways:

Plan A: A salary of $25,000 per year;

Plan B: A salary of $1500 per month plus a commission of 6% of gross sales.

For what amount of gross sales is plan A better than plan B?

79. A mason can be paid in one of two ways:

Plan A: $500 plus $5.00 per hour;

Plan B: Straight $8.00 per hour.

Suppose the job takes n hours to complete. For what values of n is plan A better for the mason than plan B?

80. A mason can be paid in one of two ways:

Plan A: $300 plus $6.00 per hour;

Plan B: Straight $9.50 per hour.

Suppose that the job takes n hours to complete. For what values of n is plan B better for the mason?

ANSWERS

73. ______

74. ______

75. ______

76. ______

77. ______

78. ______

79. ______

80. ______

ANSWERS

81. ______

82. a) ______

b) ______

83. a) ______

b) ______

84. ______

85. ______

86. ______

87. ______

88. ______

89. a) ______

b) ______

90. ______

91. ______

92. ______

93. ______

94. ______

95. ______

96. ______

81. *Converting dress sizes.* The formula

$$I = 2(s + 10)$$

can be used to convert dress sizes s in the United States to dress sizes I in Italy. For what dress sizes in the United States will dress sizes in Italy be larger than 36?

82. *Temperatures of solids.* The formula

$$C = \frac{5}{9}(F - 32)$$

can be used to convert Fahrenheit temperatures F to Celsius temperatures C.

a) Gold is a solid at Celsius temperatures less than 1063°C. Find the Fahrenheit temperatures for which gold is a solid.

b) Silver is a solid at Celsius temperatures less than 960.8°C. Find the Fahrenheit temperatures for which silver is a solid.

83. *Women in the military ranks.* The percentage of women in the total active military duty force has been steadily increasing. The number N of women in the active duty force t years after 1971 is approximated by

$$N = 12{,}197.8t + 44{,}000.$$

a) How many women were in the military in 1971 ($t = 0$)? in 1981 ($t = 10$)? in 1990 ($t = 19$)?

b) For what years will the number of women be at least 250,000?

SKILL MAINTENANCE

84. Multiply: $-6(2a - 5b)$.

85. Factor: $-12a + 30bl$.

86. Collect like terms:

$$-2x + 6y - 8x - 10 + 3y - 40.$$

87. Simplify: $4(a - 2b) - 6(2a - 5b)$.

SYNTHESIS

88. You are going to invest \$25,000, part at 14% and part at 16%. What is the most that can be invested at 14% in order to make at least \$3600 interest per year?

89. *Demand and supply.* The demand D and supply S for a certain product are given by

$$D = 2000 - 60p \quad \text{and} \quad S = 460 + 94p.$$

a) Find those values of p for which demand exceeds supply.

b) Find those values of p for which demand is less than supply.

Determine whether the statement is true or false. If false, give a counterexample.

90. For any real numbers a, b, c, and d, if $a < b$ and $c < d$, then $a + c < b + d$.

91. For any real numbers x and y, if $x < y$, then $x^2 < y^2$.

92. Determine whether the inequalities

$$x < 3 \quad \text{and} \quad 0 \cdot x < 0 \cdot 3$$

are equivalent. Give reasons to support your answer.

Solve.

93. $x + 5 \leq 5 + x$

94. $x + 8 < 3 + x$

95. $x^2 > 0$

96. $x^2 + 1 > 0$

2.5 Sets and Compound Inequalities

Compound inequalities are sentences formed by two or more inequalities that are joined with the word *and* or the word *or*. We solve such sentences; that is, we find the set of all solutions.

a Intersections of Sets and Conjunctions of Sentences

The **intersection** of two sets A and B is the set of all members that are common to A and B. We denote the intersection of sets A and B as

$$A \cap B.$$

The intersection of two sets is often illustrated as follows:

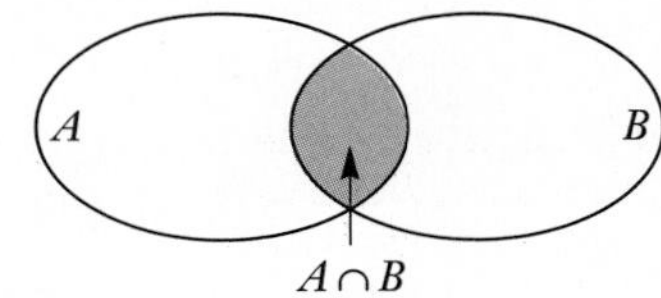

▶ **EXAMPLE 1** Find the intersection:

$$\{1, 2, 3, 4, 5\} \cap \{-2, -1, 0, 1, 2, 3\}.$$

The numbers 1, 2, and 3 are common to the two sets, so the intersection is $\{1, 2, 3\}$.

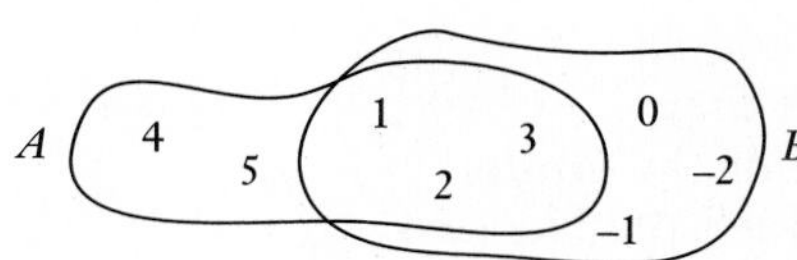

◀

DO EXERCISES 1 AND 2.

When two or more sentences are joined by the word *and* to make a compound sentence, the new sentence is called a **conjunction** of the sentences. The following is an example of a conjunction of inequalities:

$$x > -2 \quad and \quad x < 1.$$

In order for a conjunction to be true, all of the individual sentences must be true. The solution set of a conjunction is the *intersection* of the solution sets of the individual sentences. Let us consider the conjunction

$$x > -2 \quad and \quad x < 1.$$

The graph of each separate sentence is shown below, and the intersection of the solution sets is the last graph.

$\{x \mid x > -2\}$ −7 −6 −5 −4 −3 −2 −1 0 1 2 3 4 5 6 7

$\{x \mid x < 1\}$ −7 −6 −5 −4 −3 −2 −1 0 1 2 3 4 5 6 7

$\{x \mid x > -2\} \cap \{x \mid x < 1\}$
$= \{x \mid x > -2 \text{ and } x < 1\}$ −7 −6 −5 −4 −3 −2 −1 0 1 2 3 4 5 6 7

DO EXERCISE 3.

OBJECTIVES

After finishing Section 2.5, you should be able to:

- **a** Find the intersection of two sets. Solve and graph conjunctions of inequalities.
- **b** Find the union of two sets. Solve and graph disjunctions of inequalities.

FOR EXTRA HELP

Tape 5A   Tape 3B   MAC: 2
IBM: 2

1. Find the intersection:
$\{0, 3, 5, 7\} \cap \{0, 1, 3, 11\}$.

2. Shade the intersection of sets A and B:

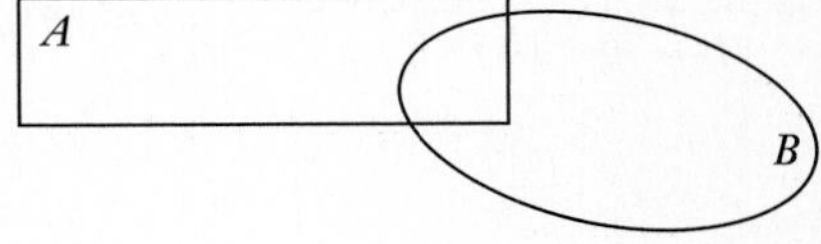

3. Graph: $-1 < x < 4$.

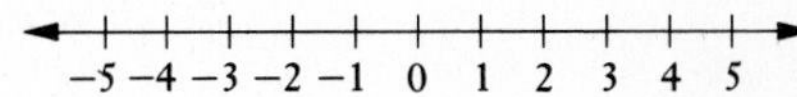

ANSWERS ON PAGE A-3

4. Solve and graph:

$3x + 4 < 10$ *and* $2x - 7 < -13$.

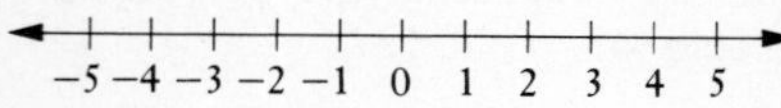

The word "and" corresponds to "intersection" and to the symbol "∩". In order for a number to be a solution of the conjunction, it must be in *both* solution sets.

► **EXAMPLE 2** Solve and graph: $2x - 5 \geq -3$ *and* $5x + 2 \geq 17$.

We solve each inequality separately:

$$\begin{aligned} 2x - 5 &\geq -3 \quad \textit{and} \quad & 5x + 2 &\geq 17 \\ 2x &\geq 2 \quad \textit{and} \quad & 5x &\geq 15 \\ x &\geq 1 \quad \textit{and} \quad & x &\geq 3. \end{aligned}$$

The solution set is the intersection of the individual inequalities.

$\{x \mid x \geq 1\}$

$\{x \mid x \geq 3\}$

$\{x \mid x \geq 1\} \cap \{x \mid x \geq 3\}$
$= \{x \mid x \geq 3\}$

The numbers common to both sets are those that are greater than or equal to 3. Thus the solution set is $\{x \mid x \geq 3\}$. ◄

DO EXERCISE 4.

5. Solve and graph:

$4x + 5 \geq 1$ *and* $2x - 8 < -14$.

If sets have no common members, we say that their intersection is empty, or the **empty set,** which can be named $\varnothing$. The two sets shown here have an empty intersection.

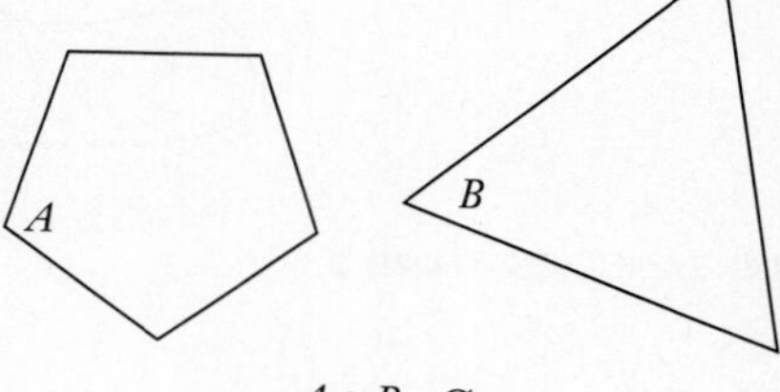

$A \cap B = \varnothing$

► **EXAMPLE 3** Solve and graph: $2x - 3 > 1$ *and* $3x - 1 < 2$.

We solve each inequality separately:

$$\begin{aligned} 2x - 3 &> 1 \quad \textit{and} \quad & 3x - 1 &< 2 \\ 2x &> 4 \quad \textit{and} \quad & 3x &< 3 \\ x &> 2 \quad \textit{and} \quad & x &< 1. \end{aligned}$$

The solution set is the intersection of the individual inequalities.

$\{x \mid x > 2\}$

$\{x \mid x < 1\}$

$\{x \mid x > 2\} \cap \{x \mid x < 1\} = \varnothing$

We see from the individual graphs that there are no real numbers that are solutions of both inequalities. Thus the solution set is the empty set, $\varnothing$. ◄

ANSWERS ON PAGE A-3

DO EXERCISE 5.

The solution set of $x > -2$ *and* $x < 1$ is the intersection

$$\{x|x > -2\} \cap \{x|x < 1\}.$$

The conjunction "$x > -2$ *and* $x < 1$" can be abbreviated

$$-2 < x < 1.$$

This states that x is *between* -2 and 1.

The conjunction $a < x$ *and* $x < b$ can be abbreviated $a < x < b$.
The conjunction $b > x$ *and* $x > a$ can be abbreviated $b > x > a$.

▶ **EXAMPLE 4** Solve and graph: $-1 \le 2x + 5 < 13$.

This inequality is an abbreviation for the following conjunction:

$$-1 \le 2x + 5 \quad \textit{and} \quad 2x + 5 < 13.$$

The word *and* corresponds to set *intersection*. The solution set is thus the intersection of the solution set of $-1 \le 2x + 5$ and the solution set of $2x + 5 < 13$:

$$\{x| -1 \le 2x + 5\} \cap \{x|2x + 5 < 13\}.$$

Method 1. We write the conjunction with the word *and*:

$$\begin{array}{lcr} -1 \le 2x + 5 & \textit{and} & 2x + 5 < 13 \\ -6 \le 2x & \textit{and} & 2x < 8 \\ -3 \le x & \textit{and} & x < 4. \end{array}$$

We now abbreviate the answer:

$$-3 \le x < 4.$$

The solution set is $\{x|-3 \le x < 4\}$.

Method 2. Using Method 1, we did the same thing to each inequality. We can shorten the writing as follows:

$$-1 \le 2x + 5 < 13$$

$$-1 - 5 \le 2x + 5 - 5 < 13 - 5 \qquad \textbf{Subtracting 5}$$

$$-6 \le 2x < 8$$

$$-3 \le x < 4. \qquad \textbf{Dividing by 2}$$

The solution set is $\{x|-3 \le x < 4\}$.

The graph is the intersection of the individual graphs.

$\{x|-3 \le x\}$

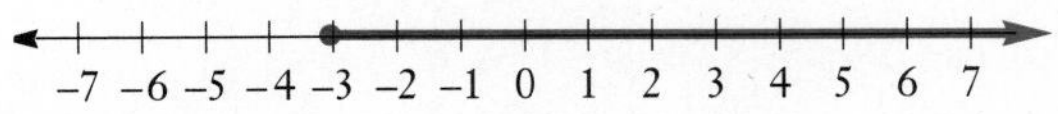

−7 −6 −5 −4 −3 −2 −1 0 1 2 3 4 5 6 7

$\{x|x < 4\}$

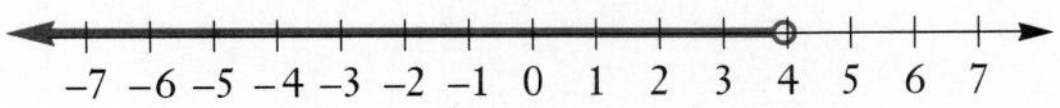

−7 −6 −5 −4 −3 −2 −1 0 1 2 3 4 5 6 7

$\{x|-3 \le x\} \cap \{x|x < 4\}$
$= \{x|-3 \le x < 4\}$

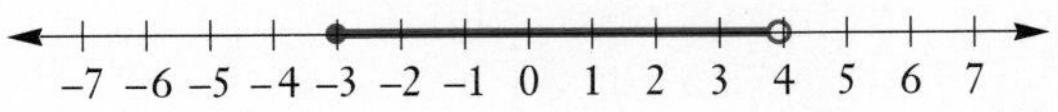

−7 −6 −5 −4 −3 −2 −1 0 1 2 3 4 5 6 7

◀

DO EXERCISE 6.

6. Solve using both methods as in Example 4:

$$-2 < 3x + 4 < 7.$$

ANSWER ON PAGE A-3

7. Solve using both methods as in Example 4:

$$-4 \leq 8 - 2x \leq 4.$$

8. Find the union:

$$\{0, 1, 3, 4\} \cup \{0, 1, 7, 9\}.$$

9. Shade the union of sets A and B:

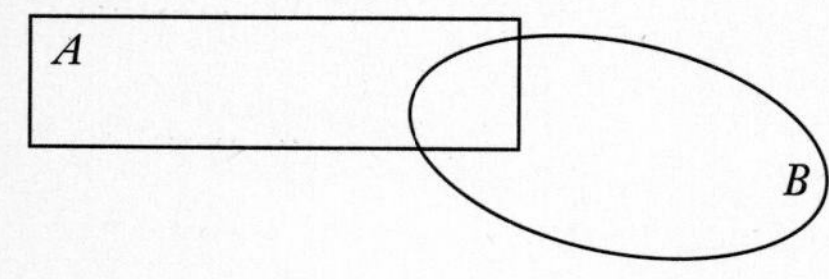

ANSWERS ON PAGE A-3

▶ **EXAMPLE 5** Solve: $3 \leq 5 - 2x < 7$.

We have

$$3 \leq 5 - 2x < 7$$

$$3 - 5 \leq 5 - 2x - 5 < 7 - 5 \quad \textbf{Subtracting 5}$$

$$-2 \leq -2x < 2 \quad \textbf{Simplifying}$$

The symbols must be reversed!

$$\frac{-2}{-2} \geq \frac{-2x}{-2} > \frac{2}{-2} \quad \textbf{Dividing by } -2 \textbf{ and reversing the inequality signs}$$

$$1 \geq x > -1 \quad \textbf{Simplifying}$$

The solution set is $\{x|1 \geq x > -1\}$, or $\{x|-1 < x \leq 1\}$, since the inequalities $1 \geq x > -1$ and $-1 < x \leq 1$ are equivalent. ◀

DO EXERCISE 7.

b Unions of Sets and Disjunctions of Sentences

The **union** of two sets A and B is formed by putting them together. We denote the union of sets A and B as

$$A \cup B.$$

The union of two sets is often illustrated as follows:

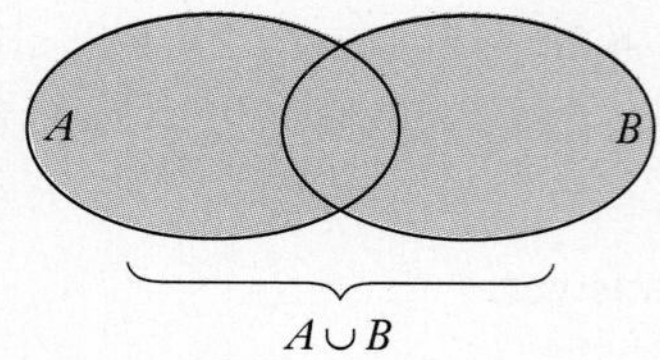

▶ **EXAMPLE 6** Find the union: $\{2, 3, 4\} \cup \{3, 5, 7\}$.

The numbers in either or both sets are 2, 3, 4, 5, and 7, so the union is $\{2, 3, 4, 5, 7\}$.

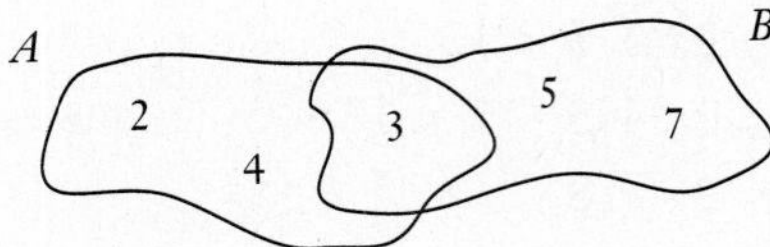

◀

DO EXERCISES 8 AND 9.

When two or more sentences are joined by the word *or* to make a compound sentence, the new sentence is called a **disjunction** of the sentences. The following are some examples:

$x < 0$ *or* $x = 0$ *or* $x > 0$;

y is an odd number *or* y is a prime number;

$x < -3$ *or* $x > 3$.

In order for a disjunction to be true, at least one of the individual sentences must be true. The solution set of a disjunction is the *union* of the individual solution sets. Consider the disjunction

$$x < -3 \quad or \quad x > 3.$$

$\{x|x < -3\}$

$\{x|x > 3\}$

$\{x|x < -3\} \cup \{x|x > 3\}$
$= \{x|x < -3 \text{ or } x > 3\}$

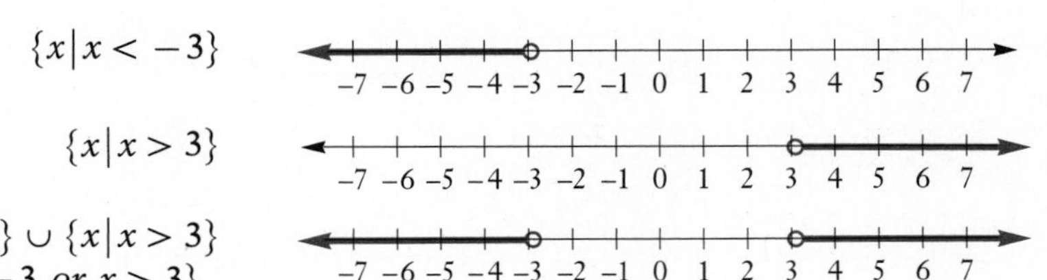

DO EXERCISE 10.

The word "or" corresponds to "union" and the symbol "∪". In order for a number to be in the solution set of a disjunction of two or more sets, it must be in at least one of the sets.

The solution set of $x < -3$ *or* $x > 3$ is the union

$$\{x|x < -3\} \cup \{x|x > 3\} \quad \text{or} \quad \{x|x < -3 \text{ or } x > 3\}.$$

It is tempting to abbreviate a disjunction without writing the word *or* but there is no simple way to do that.

▶ **EXAMPLE 7** Solve and graph: $7 + 2x < 1$ *or* $13 - 5x \leq 3$.

We solve each inequality separately:

$$\begin{aligned} 7 + 2x &< 1 \quad or \quad 13 - 5x \leq 3 \\ 2x &< -6 \quad or \quad -5x \leq -10 \\ x &< -3 \quad or \quad x \geq 2. \end{aligned}$$

To find the solution set of the disjunction, we consider the individual graphs. We graph $x < -3$. We also graph $x \geq 2$. Then we take the union of the two graphs:

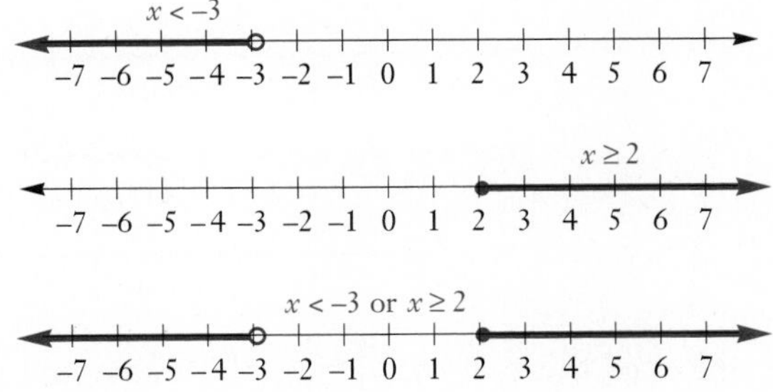

The solution set is $\{x|x < -3 \text{ or } x \geq 2\}$. ◀

10. Graph: $x \leq -2$ *or* $x > 4$.

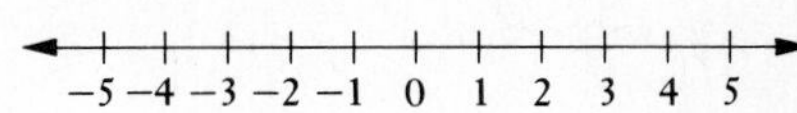

ANSWER ON PAGE A-3

Solve.

11. $x - 4 < -3$ *or* $x - 3 \geq 3$

12. $-2x + 4 \leq -3$ *or* $x + 5 < 3$

13. Solve:

$5x - 7 \leq 13$ *or* $2x - 1 \geq -7$.

14. Solve:

$5x + 4 \leq 14$ *or* $7 - 2x \geq 9$.

ANSWERS ON PAGE A-3

CAUTION! A compound inequality like

$$x < -3 \quad or \quad x \geq 2$$

in Example 7 *cannot* be abbreviated to one like $-3 < x \leq 2$. When the word *or* appears, you must keep that word. There is no short way to write a disjunction.

DO EXERCISES 11 AND 12.

▶ **EXAMPLE 8** Solve: $3x - 11 < 4$ *or* $4x + 9 \geq 1$.

We solve the individual inequalities separately, but we keep writing the word *or*.

$$\begin{aligned} 3x - 11 &< 4 \quad or \quad 4x + 9 \geq 1 \\ 3x &< 15 \quad or \quad 4x \geq -8 \\ x &< 5 \quad or \quad x \geq -2 \end{aligned}$$

Keep the word "or."

To find the solution set, we look at the individual graphs.

$\{x | x < 5\}$

–7 –6 –5 –4 –3 –2 –1 0 1 2 3 4 5 6 7

$\{x | x \geq -2\}$

–7 –6 –5 –4 –3 –2 –1 0 1 2 3 4 5 6 7

$\{x | x < 5\} \cup \{x | x \geq -2\}$
$= \{x | x < 5 \text{ or } x \geq -2\}$
= the set of real numbers

–7 –6 –5 –4 –3 –2 –1 0 1 2 3 4 5 6 7

The union of the two solution sets fills up the entire real-number line. Any real number is in at least one of the individual solution sets. Thus the solution set is the entire set of real numbers. ◀

DO EXERCISE 13.

▶ **EXAMPLE 9** Solve and graph: $4x + 5 < 25$ *or* $2x - 3 < 3$.

We solve the individual inequalities separately:

$$\begin{aligned} 4x + 5 &< 25 \quad or \quad 2x - 3 < 3 \\ 4x &< 20 \quad or \quad 2x < 6 \\ x &< 5 \quad or \quad x < 3. \end{aligned}$$

To find the solution set, we look at the individual graphs.

$\{x | x < 5\}$

–7 –6 –5 –4 –3 –2 –1 0 1 2 3 4 5 6 7

$\{x | x < 3\}$

–7 –6 –5 –4 –3 –2 –1 0 1 2 3 4 5 6 7

$\{x | x < 5\} \cup \{x | x < 3\}$
$= \{x | x < 5\}$

–7 –6 –5 –4 –3 –2 –1 0 1 2 3 4 5 6 7

We see that the union of the two sets is just the individual set $\{x | x < 5\}$. It is the solution set. ◀

DO EXERCISE 14.

NAME SECTION DATE

EXERCISE SET 2.5

a Find the intersection.

1. $\{5, 6, 7, 8\} \cap \{4, 6, 8, 10\}$

2. $\{9, 10, 27\} \cap \{8, 10, 38\}$

3. $\{2, 4, 6, 8\} \cap \{1, 3, 5\}$

4. $\varnothing \cap \varnothing$

5. $\{1, 2, 3, 4\} \cap \{1, 2, 3, 4\}$

6. $\{8, 9, 10\} \cap \varnothing$

Graph.

7. $1 < x < 6$

−6 −5 −4 −3 −2 −1 0 1 2 3 4 5 6

8. $-4 \leq y \leq 3$

−6 −5 −4 −3 −2 −1 0 1 2 3 4 5 6

9. $6 > -x \geq -2$

−6 −5 −4 −3 −2 −1 0 1 2 3 4 5 6

10. $5 > -x \geq -2$

−6 −5 −4 −3 −2 −1 0 1 2 3 4 5 6

Solve.

11. $-10 \leq 3x + 2$ *and* $3x + 2 < 17$

12. $-11 < 4x - 3$ *and* $4x - 3 \leq 13$

13. $3x + 7 \geq 4$ *and* $2x - 5 \geq -1$

14. $4x - 7 < 1$ *and* $7 - 3x > -8$

15. $4 - 3x \geq 10$ *and* $5x - 2 > 13$

16. $5 - 7x > 19$ *and* $2 - 3x < -4$

ANSWERS

1. ________

2. ________

3. ________

4. ________

5. ________

6. ________

7. See graph.

8. See graph.

9. See graph.

10. See graph.

11. ________

12. ________

13. ________

14. ________

15. ________

16. ________

ANSWERS

17. ______

18. ______

19. ______

20. ______

21. ______

22. ______

23. ______

24. ______

25. ______

26. ______

27. ______

28. ______

29. ______

30. ______

31. ______

32. ______

17. $-2 < x + 2 < 8$

18. $-1 < x + 1 \leq 6$

19. $1 < 2y + 5 \leq 9$

20. $3 \leq 5x + 3 \leq 8$

21. $-10 \leq 3x - 5 \leq -1$

22. $-18 \leq -2x - 7 < 0$

23. $2 < x + 3 \leq 9$

24. $-6 \leq x + 1 < 9$

25. $-6 \leq 2x - 3 < 6$

26. $4 > -3m - 7 \geq 2$

27. $-\frac{1}{2} < \frac{1}{4}x - 3 \leq \frac{1}{2}$

28. $-\frac{2}{3} \leq 4 - \frac{1}{4}x < \frac{2}{3}$

29. $-3 < \frac{2x - 5}{4} < 8$

30. $-4 \leq \frac{7 - 3x}{5} \leq 4$

b Find the union.

31. $\{4, 5, 6, 7, 8\} \cup \{1, 4, 6, 11\}$

32. $\{8, 9, 27\} \cup \{2, 8, 27\}$

33. $\{2, 4, 6, 8\} \cup \{1, 3, 5\}$

34. $\{8, 9, 10\} \cup \varnothing$

35. $\{4, 8, 11\} \cup \varnothing$

36. $\varnothing \cup \varnothing$

Graph.

37. $x < -1$ *or* $x > 2$

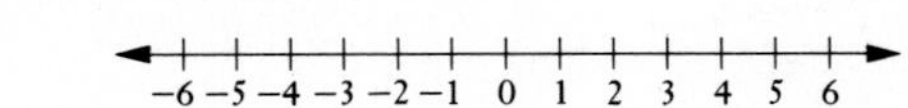

38. $x < -2$ *or* $x > 0$

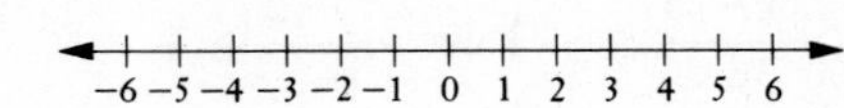

39. $x \leq -3$ *or* $x > 1$

−6 −5 −4 −3 −2 −1 0 1 2 3 4 5 6

40. $x \leq -1$ *or* $x > 3$

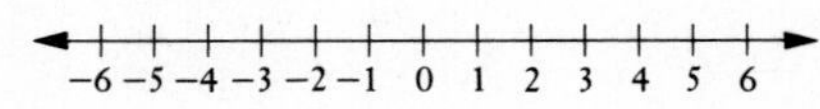

Solve.

41. $x + 7 < -2$ *or* $x + 7 > 2$

42. $x + 9 < -4$ *or* $x + 9 > 4$

43. $2x - 8 \leq -3$ *or* $x - 8 \geq 3$

44. $x - 7 \leq -2$ *or* $3x - 7 \geq 2$

45. $7x + 4 \geq -17$ *or* $6x + 5 \geq -7$

46. $4x - 4 < -8$ *or* $4x - 4 < 12$

47. $7 > -4x + 5$ *or* $10 \leq -4x + 5$

48. $6 > 2x - 1$ *or* $-4 \leq 2x - 1$

ANSWERS

33. ______

34. ______

35. ______

36. ______

37. See graph.

38. See graph.

39. See graph.

40. See graph.

41. ______

42. ______

43. ______

44. ______

45. ______

46. ______

47. ______

48. ______

ANSWERS

49. ______
50. ______
51. ______
52. ______
53. ______
54. ______
55. ______
56. ______
57. ______
58. ______
59. ______
60. ______
61. a) ______
b) ______
62. ______
63. ______
64. ______
65. ______
66. ______
67. ______
68. ______
69. ______
70. ______
71. ______

49. $3x - 7 > -10$ *or* $5x + 2 \leq 22$

50. $3x + 2 < 2$ *or* $4 - 2x < 14$

51. $-2x - 2 < -6$ *or* $-2x - 2 > 6$

52. $-3m - 7 < -5$ *or* $-3m - 7 > 5$

53. $\frac{2}{3}x - 14 < -\frac{5}{6}$ *or* $\frac{2}{3}x - 14 > \frac{5}{6}$

54. $\frac{1}{4} - 3x \leq -3.7$ *or* $\frac{1}{4} - 5x \geq 4.8$

55. $\frac{2x - 5}{6} \leq -3$ *or* $\frac{2x - 5}{6} \geq 4$

56. $\frac{7 - 3x}{5} < -4$ *or* $\frac{7 - 3x}{5} > 4$

SKILL MAINTENANCE

Simplify.

57. $(-2x^{-4}y^6)^5$

58. $(-4a^5b^{-7})(5a^{-12}b^8)$

59. $\frac{-4a^5b^{-7}}{5a^{-12}b^8}$

60. $\left(\frac{56a^5b^{-6}}{28a^7b^{-8}}\right)^{-3}$

SYNTHESIS

61. *Temperatures of liquids.* The formula

$$C = \tfrac{5}{9}(F - 32)$$

can be used to convert Fahrenheit temperatures F to Celsius temperatures C.

a) Gold is a liquid for Celsius temperatures C such that $1063° \leq C < 2660°$. Find such an inequality for the corresponding Fahrenheit temperatures.

b) Silver is a liquid for Celsius temperatures C such that $960.8° \leq C < 2180°$. Find such an inequality for the corresponding Fahrenheit temperatures.

62. *Converting dress sizes.* The formula

$$I = 2(s + 10)$$

can be used to convert dress sizes s in the United States to dress sizes I in Italy. For what dress sizes in the United States will dress sizes in Italy be between 32 and 46?

63. We say that x is *between* a and b if $a < x < b$. Find all the numbers on a number line from which you can subtract 3 and still be between -8 and 8.

Solve.

64. $4m - 8 > 6m + 5$ *or* $5m - 8 < -2$

65. $x - 10 < 5x + 6 \leq x + 10$

66. $2[5(3 - y) - 2(y - 2)] > y + 4$

67. $-\frac{2}{15} \leq \frac{2}{3}x - \frac{2}{5} \leq \frac{2}{15}$

68. $2x - \frac{3}{4} < -\frac{1}{10}$ *or* $2x - \frac{3}{4} > \frac{1}{10}$

69. $3x < 4 - 5x < 5 + 3x$

70. $2x + 3 \leq x - 6$ *or* $3x - 2 \leq 4x + 5$

71. $x + 4 < 2x - 6 \leq x + 12$

2.6 Absolute-Value Equations and Inequalities

a Properties of Absolute Value

We can think of the **absolute value** of a number as its distance from zero on a number line. Recall the formal definition.

> **The absolute value of x, denoted $|x|$, is defined as follows:**
> $$|x| = x \quad \text{if } x \geq 0,$$
> $$|x| = -x \quad \text{if } x < 0.$$

Some simple properties of absolute value allow us to manipulate or simplify algebraic expressions. Those properties follow from the definition.

> Properties of Absolute Value
>
> a) **For any real numbers a and b, $|ab| = |a| \cdot |b|$.**
> **(The absolute value of a product is the product of the absolute values.)**
>
> b) $\left|\frac{a}{b}\right| = \frac{|a|}{|b|}$, **provided that $b \neq 0$.**
> **(The absolute value of a quotient is the quotient of the absolute values.)**
>
> c) $|-a| = |a|$
> **(The absolute value of the opposite of a number is the same as the absolute value of the number.)**

We can use property (a) to prove property (c):

$$|-a| = |-1 \cdot a| = |-1| \cdot |a| = 1 \cdot |a| = |a|.$$

▶ **EXAMPLES** Simplify, leaving as little as possible inside the absolute-value signs.

1. $|5x| = |5| \cdot |x| = 5|x|$
2. $|-3y| = |-3| \cdot |y| = 3|y|$
3. $|7x^2| = |7| \cdot |x^2| = 7|x^2| = 7x^2$ **Since x^2 is never negative for any number x**
4. $\left|\frac{6x}{-3x^2}\right| = \left|\frac{2}{-x}\right| = \frac{|2|}{|-x|} = \frac{2}{|x|}$ ◀

DO EXERCISES 1–5.

b Distance on a Line

The number line below shows that the distance between -3 and 2 is 5.

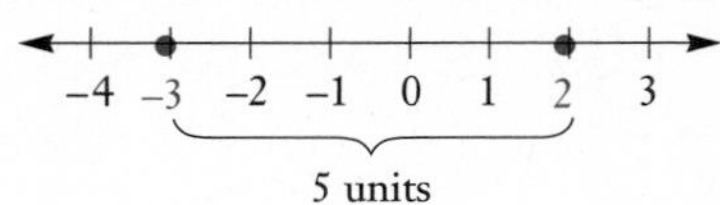

Another way to find the distance between two numbers on a number line is to take the absolute value of the difference, as follows:

$$|-3 - 2| = |-5| = 5, \quad \text{or} \quad |2 - (-3)| = |5| = 5.$$

Note that the order in which we subtract does not matter since we are taking the absolute value after we have subtracted.

OBJECTIVES

After finishing Section 2.6, you should be able to:

a Simplify expressions containing absolute-value symbols.

b Find the distance between two points on a number line.

c Solve and graph equations containing absolute-value symbols.

d Solve equations with two absolute-value expressions.

e Solve and graph inequalities with absolute-value symbols.

FOR EXTRA HELP

Tape 5B | Tape 3B | MAC: 2 IBM: 2

Simplify, leaving as little as possible inside the absolute-value signs.

1. $|7x|$

2. $|x^8|$

3. $|5a^2b|$

4. $\left|\frac{7a}{b^2}\right|$

5. $|-9x|$

ANSWERS ON PAGE A-3

Find the distance between the points.

6. $-6, -35$

7. $19, 14$

8. $0, p$

9. Solve: $|x| = 6$. Then graph using a number line.

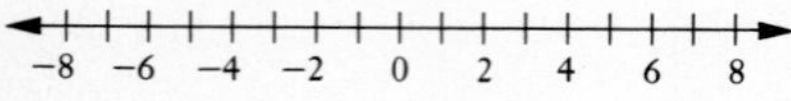

10. Solve: $|x| = -6$.

11. Solve: $|p| = 0$.

ANSWERS ON PAGE A-3

For any real numbers *a* and *b*, the distance between them is $|a - b|$.

We should note that the distance is also $|b - a|$, because $a - b$ and $b - a$ are opposites and hence have the same absolute value.

▶ **EXAMPLE 5** Find the distance between -8 and -92 on a number line.

$$|-8 - (-92)| = |84| = 84, \quad \text{or} \quad |-92 - (-8)| = |-84| = 84$$ ◀

▶ **EXAMPLE 6** Find the distance between x and 0 on a number line.

$$|x - 0| = |x|$$ ◀

DO EXERCISES 6–8.

C Equations with Absolute Value

▶ **EXAMPLE 7** Solve: $|x| = 4$. Then graph using a number line.

Note that $|x| = |x - 0|$, so that $|x - 0|$ is the distance from x to 0. The solutions of the equation are those numbers x whose distance from 0 is 4. Those numbers are -4 and 4. The solution set is $\{-4, 4\}$. The graph consists of just two points, as shown.

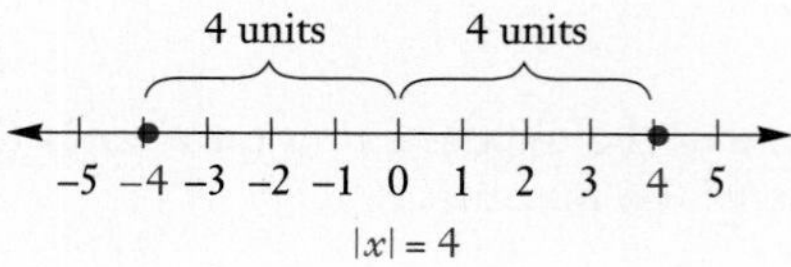

◀

▶ **EXAMPLE 8** Solve: $|x| = 0$.

The only number whose absolute value is 0 is 0 itself. Thus the solution is 0. The solution set is $\{0\}$. ◀

▶ **EXAMPLE 9** Solve: $|x| = -7$.

The absolute value of a number is always nonnegative. There is no number whose absolute value is -7. Thus there is no solution. The solution set is $\varnothing$. ◀

Examples 7–9 lead us to the following principle for solving linear equations with absolute value.

The Absolute-Value Principle

a) For any positive number p, if $|X| = p$, then $X = -p$ or $X = p$. The solutions are $-p$ and p. The solution set is $\{-p, p\}$.

b) The solution of $|X| = 0$ is 0. The solution set is $\{0\}$.

c) If n is negative, $|X| = n$ has no solution. The solution set is $\varnothing$.

DO EXERCISES 9–11.

We can use the absolute-value principle together with the addition and multiplication principles to solve many types of equations with absolute value.

▶ **EXAMPLE 10** Solve: $2|x| + 5 = 9$.

We first use the addition and multiplication principles to get $|x|$ by itself. Then we use the absolute-value principle.

$$2|x| + 5 = 9$$
$$2|x| + 5 - 5 = 9 - 5 \quad \text{Subtracting 5}$$
$$2|x| = 4$$
$$|x| = 2 \quad \text{Dividing by 2}$$
$$x = -2 \quad \text{or} \quad x = 2 \quad \text{Using the absolute-value principle}$$

The solutions are -2 and 2. The solution set is $\{-2, 2\}$. ◀

DO EXERCISES 12–14.

▶ **EXAMPLE 11** Solve: $|x - 2| = 3$.

We can consider solving this equation in two different ways.

Method 1. This method allows us to see the meaning of the solutions graphically. The solution set consists of those numbers that are 3 units from 2 on a number line.

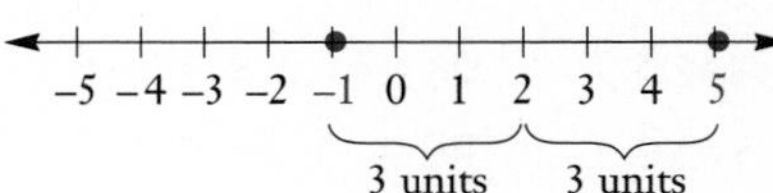

The solutions of $|x - 2| = 3$ are -1 and 5.

Method 2. The method is more efficient. We use the absolute-value principle, replacing X by $x - 2$ and p by 3. Then we solve each equation separately.

$$|X| = a$$
$$|x - 2| = 3$$
$$x - 2 = -3 \quad or \quad x - 2 = 3 \quad \text{Absolute-value principle}$$
$$x = -1 \quad or \quad x = 5$$

The solutions are -1 and 5. The solution set is $\{-1, 5\}$. ◀

DO EXERCISE 15.

▶ **EXAMPLE 12** Solve: $|2x + 5| = 13$.

We use the absolute-value principle, replacing X by $2x + 5$ and p by 13:

$$|X| = p$$
$$|2x + 5| = 13$$
$$2x + 5 = -13 \quad or \quad 2x + 5 = 13 \quad \text{Absolute-value principle}$$
$$2x = -18 \quad or \quad 2x = 8$$
$$x = -9 \quad or \quad x = 4$$

The solutions are -9 and 4. The solution set is $\{-9, 4\}$. ◀

DO EXERCISE 16.

Solve.

12. $|3x| = 6$

13. $4|x| + 10 = 27$

14. $3|x| - 2 = 10$

15. Solve: $|x - 4| = 1$. Use two methods as in Example 11.

16. Solve: $|3x - 4| = 17$.

ANSWERS ON PAGE A-3

17. Solve: $|6 + 2x| = -3$.

▶ **EXAMPLE 13** Solve: $|4 - 7x| = -8$.

Since absolute value is always nonnegative, this equation has no solution. The solution set is $\varnothing$. ◀

DO EXERCISE 17.

d Equations with Two Absolute-Value Expressions

Sometimes equations have two absolute-value expressions. Consider $|a| = |b|$. This means that a and b are the same distance from 0. If a and b are the same distance from 0, then either they are the same number or they are opposites of each other.

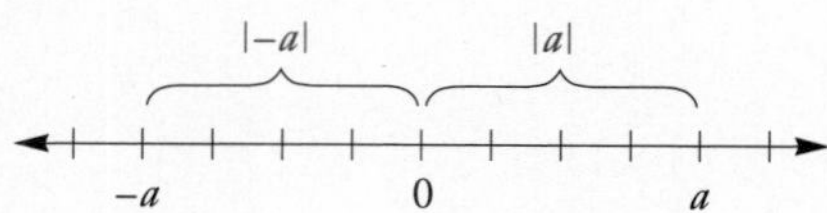

Solve.

18. $|5x - 3| = |x + 4|$

▶ **EXAMPLE 14** Solve: $|2x - 3| = |x + 5|$.

Either $2x - 3 = x + 5$ or $2x - 3 = -(x + 5)$. We solve each equation separately:

$$\begin{aligned} 2x - 3 &= x + 5 &\quad \textit{or}\quad 2x - 3 &= -(x + 5) \\ x - 3 &= 5 &\quad \textit{or}\quad 2x - 3 &= -x - 5 \\ x &= 8 &\quad \textit{or}\quad 3x - 3 &= -5 \\ x &= 8 &\quad \textit{or}\quad 3x &= -2 \\ x &= 8 &\quad \textit{or}\quad x &= -\tfrac{2}{3}. \end{aligned}$$

19. $|x - 3| = |x + 10|$

The solutions are 8 and $-\frac{2}{3}$. The solution set is $\{8, -\frac{2}{3}\}$. ◀

▶ **EXAMPLE 15** Solve: $|x + 8| = |x - 5|$.

We have

$$\begin{aligned} x + 8 &= x - 5 &\quad \textit{or}\quad x + 8 &= -(x - 5) \\ 8 &= -5 &\quad \textit{or}\quad x + 8 &= -x + 5 \\ 8 &= -5 &\quad \textit{or}\quad 2x &= -3 \\ 8 &= -5 &\quad \textit{or}\quad x &= -\tfrac{3}{2}. \end{aligned}$$

The first equation has no solution. The second equation has $-\frac{3}{2}$ as a solution. There is only one solution to the original equation, $-\frac{3}{2}$. The solution set is $\{-\frac{3}{2}\}$. ◀

20. Solve: $|x| = 5$. Then graph using a number line.

−8 −6 −4 −2 0 2 4 6 8

DO EXERCISES 18 AND 19.

e Inequalities with Absolute Value

We can extend our methods for solving equations with absolute value to those for solving inequalities with absolute value.

▶ **EXAMPLE 16** Solve: $|x| = 4$. Then graph using a number line.

From Example 7, we know that the solutions are -4 and 4. The solution set is $\{-4, 4\}$. The graph consists of just two points, as shown here.

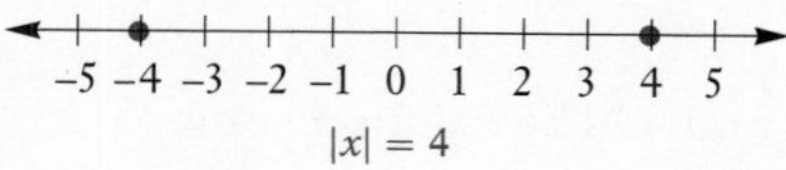

◀

ANSWERS ON PAGE A-3

DO EXERCISE 20.

▶ **EXAMPLE 17** Solve: $|x| < 4$. Then graph.

The solutions of $|x| < 4$ are the solutions of $|x - 0| < 4$ and are those numbers x whose distance from 0 is less than 4. We can check by substituting or by looking at the number line that numbers like -3, -2, -1, $-\frac{1}{2}$, $-\frac{1}{4}$, 0, $\frac{1}{4}$, $\frac{1}{2}$, 1, 2, and 3 are all solutions. In fact, the solutions are all the real numbers x between -4 and 4, such that $-4 < x < 4$. The solution set is $\{x|-4 < x < 4\}$. The graph is as follows.

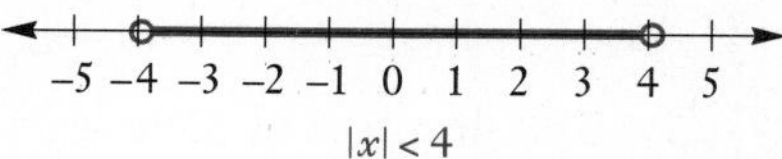

◀

DO EXERCISE 21.

▶ **EXAMPLE 18** Solve: $|x| \geq 4$. Then graph.

The solutions of $|x| \geq 4$ are solutions of $|x - 0| \geq 4$ and are those numbers whose distance from 0 is greater than or equal to 4—in other words, those numbers x such that $x \leq -4$ *or* $x \geq 4$. The solution set is $\{x|x \leq -4$ *or* $x \geq 4\}$. The graph is as follows.

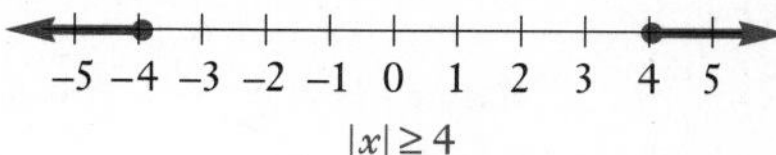

◀

DO EXERCISE 22.

Examples 16–18 illustrate three cases of solving equations and inequalities with absolute value. The expression inside the absolute-value signs can be something besides a single variable. The following is a general principle for solving.

For any positive number p and any expression X:

a) The solutions of $|X| = p$ are those numbers that satisfy $X = -p$ or $X = p$.

b) The solutions of $|X| < p$ are those numbers that satisfy $-p < X < p$.

c) The solutions of $|X| > p$ are those numbers that satisfy $X < -p$ or $X > p$.

a) As an example to illustrate part (a), the solutions of $|5x - 1| = 8$ are those numbers x for which

$$5x - 1 = -8 \quad or \quad 5x - 1 = 8.$$

b) As an example to illustrate part (b), the solutions of $|6x + 7| < 5$ are those numbers x for which

$$-5 < 6x + 7 < 5.$$

c) As an example to illustrate part (c), the solutions of $|2x - 9| > 4$ are those numbers x for which

$$2x - 9 < -4 \quad or \quad 2x - 9 > 4.$$

21. Solve: $|x| < 5$. Then graph.

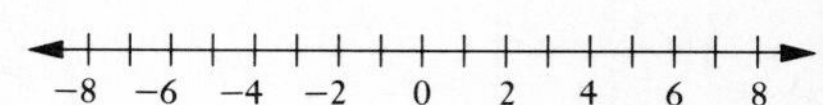

22. Solve: $|x| \geq 5$. Then graph.

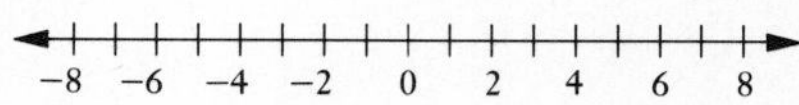

ANSWERS ON PAGE A-3

23. Solve: $|3x + 2| = 8$. Then graph.

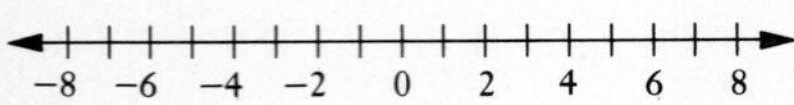

24. Solve: $|2x - 3| < 7$. Then graph.

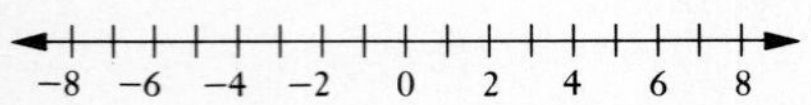

25. Solve: $|7 - 3x| \leq 4$.

26. Solve: $|3x + 2| \geq 5$. Then graph.

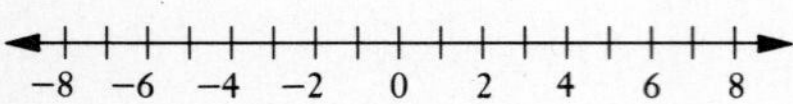

ANSWERS ON PAGE A-3

▶ **EXAMPLE 19** Solve: $|5x - 4| = 11$. Then graph.

We use part (a). In this case, X is $5x - 4$ and p is 11:

$$|X| = p$$
$$|5x - 4| = 11 \qquad \textbf{Replacing } X \textbf{ by } 5x - 4 \textbf{ and } p \textbf{ by } 11$$
$$5x - 4 = -11 \quad or \quad 5x - 4 = 11 \qquad \textbf{Part (a)}$$
$$5x = -7 \quad or \quad 5x = 15 \qquad \textbf{Adding 4}$$
$$x = -\tfrac{7}{5} \quad or \quad x = 3. \qquad \textbf{Dividing by 5}$$

The solution set is $\{-\frac{7}{5}, 3\}$. The graph is as follows.

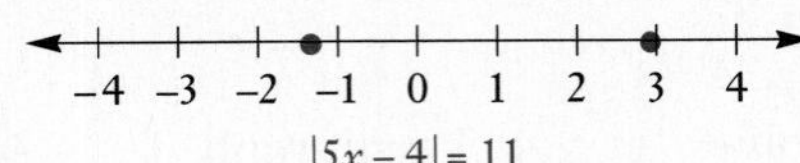

DO EXERCISE 23.

▶ **EXAMPLE 20** Solve: $|3x - 2| < 4$. Then graph.

We use part (b). In this case, X is $3x - 2$ and p is 4:

$$|X| < p$$
$$|3x - 2| < 4 \qquad \textbf{Replacing } X \textbf{ by } 3x - 2 \textbf{ and } p \textbf{ by } 4$$
$$-4 < 3x - 2 < 4 \qquad \textbf{Part (b)}$$
$$-2 < 3x < 6 \qquad \textbf{Adding 2}$$
$$-\tfrac{2}{3} < x < 2. \qquad \textbf{Dividing by 3}$$

The solution set is $\{x | -\frac{2}{3} < x < 2\}$. The graph is as follows.

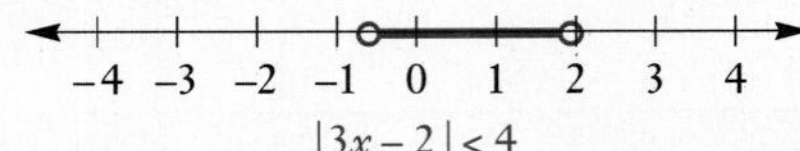

◀

DO EXERCISE 24.

▶ **EXAMPLE 21** Solve: $|8 - 4x| \leq 5$.

We use part (b). In this case, X is $8 - 4x$ and p is 5:

$$|X| \leq p$$
$$|8 - 4x| \leq 5 \qquad \textbf{Replacing } X \textbf{ by } 8 - 4x \textbf{ and } p \textbf{ by } 5$$
$$-5 \leq 8 - 4x \leq 5 \qquad \textbf{Part (b)}$$
$$-13 \leq -4x \leq -3 \qquad \textbf{Subtracting 8}$$
$$\tfrac{13}{4} \geq x \geq \tfrac{3}{4}. \qquad \textbf{Dividing by } -4 \textbf{ and reversing the inequality symbols}$$

The solution set is $\{x | \frac{13}{4} \geq x \geq \frac{3}{4}\}$, or $\{x | \frac{3}{4} \leq x \leq \frac{13}{4}\}$. ◀

DO EXERCISE 25.

▶ **EXAMPLE 22** Solve: $|4x + 2| \geq 6$.

We use part (c). In this case, X is $4x + 2$ and p is 6:

$$|X| \geq p$$
$$|4x + 2| \geq 6 \qquad \textbf{Replacing } X \textbf{ by } 4x + 2 \textbf{ and } p \textbf{ by } 6$$
$$4x + 2 \leq -6 \quad or \quad 4x + 2 \geq 6 \qquad \textbf{Part (c)}$$
$$4x \leq -8 \quad or \quad 4x \geq 4 \qquad \textbf{Subtracting 2}$$
$$x \leq -2 \quad or \quad x \geq 1. \qquad \textbf{Dividing by 4}$$

The solution set is $\{x | x \leq -2 \text{ or } x \geq 1\}$. ◀

DO EXERCISE 26.

NAME SECTION DATE

EXERCISE SET 2.6

a Simplify, leaving as little as possible inside absolute-value signs.

1. $|3x|$ **2.** $|17x|$ **3.** $|9x^2|$ **4.** $|6x^2|$ **5.** $|-4x^2|$

6. $|-10x^2|$ **7.** $|-8y|$ **8.** $|-13y|$ **9.** $\left|\frac{-4}{x}\right|$ **10.** $\left|\frac{y}{7}\right|$

11. $\left|\frac{x^2}{-y}\right|$ **12.** $\left|\frac{x^4}{-y}\right|$ **13.** $\left|\frac{-8x^2}{2x}\right|$ **14.** $\left|\frac{9y}{3y^2}\right|$

b Find the distance between the points on a number line.

15. $-8, -42$ **16.** $-9, -36$ **17.** $26, 15$ **18.** $54, 18$

19. $-3.9, 2.4$ **20.** $-1.8, -3.7$ **21.** $-5, 0$ **22.** $\frac{2}{3}, -\frac{5}{6}$

c Solve.

23. $|x| = 3$ **24.** $|x| = 5$ **25.** $|x| = -3$

26. $|x| = -5$ **27.** $|p| = 0$ **28.** $|y| = 8.6$

29. $|x - 3| = 12$ **30.** $|3x - 2| = 6$ **31.** $|2x - 3| = 4$

ANSWERS

1. ______
2. ______
3. ______
4. ______
5. ______
6. ______
7. ______
8. ______
9. ______
10. ______
11. ______
12. ______
13. ______
14. ______
15. ______
16. ______
17. ______
18. ______
19. ______
20. ______
21. ______
22. ______
23. ______
24. ______
25. ______
26. ______
27. ______
28. ______
29. ______
30. ______
31. ______

ANSWERS

32. ____
33. ____
34. ____
35. ____
36. ____
37. ____
38. ____
39. ____
40. ____
41. ____
42. ____
43. ____
44. ____
45. ____
46. ____
47. ____
48. ____
49. ____
50. ____
51. ____
52. ____
53. ____
54. ____
55. ____
56. ____
57. ____
58. ____
59. ____
60. ____

32. $|5x + 2| = 3$

33. $|4x - 9| = 14$

34. $|9y - 2| = 17$

35. $|x| + 7 = 18$

36. $|x| - 2 = 6.3$

37. $678 = 289 + |t|$

38. $-567 = -1000 + |x|$

39. $|5x| = 40$

40. $|2y| = 18$

41. $|3x| - 4 = 17$

42. $|6x| + 8 = 32$

43. $5|q| - 2 = 9$

44. $7|z| + 2 = 16$

45. $\left|\frac{2x - 1}{3}\right| = 5$

46. $\left|\frac{4 - 5x}{6}\right| = 7$

47. $|m + 5| + 9 = 16$

48. $|t - 7| - 5 = 4$

49. $10 - |2x - 1| = 4$

50. $2|2x - 7| + 11 = 25$

51. $|3x - 4| = -2$

52. $|x - 6| = -8$

53. $|\frac{5}{9} + 3x| = \frac{1}{6}$

54. $|\frac{2}{3} - 4x| = \frac{4}{5}$

d Solve.

55. $|3x + 4| = |x - 7|$

56. $|2x - 8| = |x + 3|$

57. $|x + 5| = |x - 2|$

58. $|x - 7| = |x + 8|$

59. $|2a + 4| = |3a - 1|$

60. $|5p + 7| = |4p + 3|$

61. $|y - 3| = |3 - y|$

62. $|m - 7| = |7 - m|$

63. $|5 - p| = |p + 8|$

64. $|8 - q| = |q + 19|$

65. $\left|\frac{2x - 3}{6}\right| = \left|\frac{4 - 5x}{8}\right|$

66. $\left|\frac{6 - 8x}{5}\right| = \left|\frac{7 + 3x}{2}\right|$

67. $|\frac{1}{2}x - 5| = |\frac{1}{4}x + 3|$

68. $|2 - \frac{2}{3}x| = |4 + \frac{7}{8}x|$

e Solve.

69. $|x| < 3$

70. $|x| \leq 5$

71. $|x| \geq 2$

72. $|y| > 8$

73. $|x - 3| < 1$

74. $|x + 2| \leq 5$

75. $|x + 4| \leq 1$

76. $|x - 2| > 6$

77. $|2x - 3| \leq 4$

78. $|5x + 2| \leq 3$

79. $|2y - 7| > 10$

80. $|3y - 4| > 8$

81. $|4x - 9| \geq 14$

82. $|9y - 2| \geq 17$

83. $|y - 3| < 12$

84. $|p - 2| < 6$

85. $|2x + 3| \leq 4$

86. $|5x + 2| \leq 13$

87. $|4 - 3y| > 8$

88. $|7 - 2y| > 5$

89. $|9 - 4x| \geq 14$

ANSWERS

61. ______

62. ______

63. ______

64. ______

65. ______

66. ______

67. ______

68. ______

69. ______

70. ______

71. ______

72. ______

73. ______

74. ______

75. ______

76. ______

77. ______

78. ______

79. ______

80. ______

81. ______

82. ______

83. ______

84. ______

85. ______

86. ______

87. ______

88. ______

89. ______

ANSWERS

90. ______
91. ______
92. ______
93. ______
94. ______
95. ______
96. ______
97. ______
98. ______
99. ______
100. ______
101. ______
102. ______
103. ______
104. ______
105. ______
106. ______
107. ______
108. ______
109. ______
110. ______
111. ______
112. ______
113. ______
114. ______
115. ______
116. ______
117. ______
118. ______
119. ______
120. ______
121. ______
122. ______
123. ______

90. $|2 - 9p| \geq 17$

91. $|3 - 4x| < 21$

92. $|-5 - 7x| \leq 30$

93. $\left|\frac{1}{2} + 3x\right| \geq 12$

94. $\left|\frac{1}{4}y - 6\right| > 24$

95. $\left|\frac{x - 7}{3}\right| < 4$

96. $\left|\frac{x + 5}{4}\right| \leq 2$

97. $\left|\frac{2 - 5x}{4}\right| \geq \frac{2}{3}$

98. $\left|\frac{1 + 3x}{5}\right| > \frac{7}{8}$

99. $|m + 5| + 9 \leq 16$

100. $|t - 7| + 3 \geq 4$

101. $7 - |3 - 2x| \geq 5$

102. $16 \leq |2x - 3| + 9$

103. $\left|\frac{2x - 1}{0.0059}\right| \leq 1$

104. $\left|\frac{3x - 2}{5}\right| \geq 1$

SKILL MAINTENANCE

105. Add: $-43.5 + (-5.8)$.

106. Subtract: $-43.5 - (-5.8)$.

107. Multiply: $-43.5(-5.8)$.

108. Divide: $-43.5 \div (-5.8)$.

SYNTHESIS

109. From the definition of absolute value, $|x| = x$ only when $x \geq 0$. Thus, $|x + 3| = x + 3$ only when $x + 3 \geq 0$ or $x \geq -3$. Solve $|2x - 5| = 2x - 5$ using this same argument.

Solve.

110. $1 - \left|\frac{1}{4}x + 8\right| = \frac{3}{4}$

111. $|x + 5| = x + 5$

112. $|x - 1| = x - 1$

113. $|7x - 2| = x + 4$

114. $|3x - 4| > -2$

115. $|x - 6| \leq -8$

116. $\left|\frac{5}{9} + 3x\right| < -\frac{1}{6}$

117. $|x + 5| > x$

118. $2 \leq |x - 1| \leq 5$

Find an equivalent inequality with absolute value.

119. $-3 < x < 3$

120. $-5 \leq y \leq 5$

121. $x \leq -6$ *or* $x \geq 6$

122. $-5 < x < 1$

123. $x < -8$ *or* $x > 2$

SUMMARY AND REVIEW: CHAPTER 2

IMPORTANT PROPERTIES AND FORMULAS

Addition Principle for Equations: If $a = b$ is true, then $a + c = b + c$ is true.

Multiplication Principle for Equations: If $a = b$ is true, then $ac = bc$ is true.

Addition Principle for Inequalities: If the same number is added on both sides of an inequality, we get another true inequality.

Multiplication Principle for Inequalities: If we multiply on both sides of a true inequality by a positive number, we get another true inequality. If we multiply by a negative number and the inequality symbol is reversed, we get another true inequality.

Set Intersection: $A \cap B = \{x \mid x \text{ is in } A \text{ and } x \text{ is in } B\}$

Set Union: $A \cup B = \{x \mid x \text{ is in } A \text{ or in } B, \text{ or both}\}$

"$a < x$ *and* $x < b$" is equivalent to "$a < x < b$"

Properties of Absolute Value

$|ab| = |a| \cdot |b|$, $\left|\frac{a}{b}\right| = \frac{|a|}{|b|}$, $|-a| = |a|$, The distance between a and b is $|a - b|$.

Solving Principles for Equations and Inequalities Involving Absolute Value

a) For any positive number p, if $|X| = p$, then $X = -p$ or $X = p$. The solutions are $-p$ and p. The solution set is $\{-p, p\}$.

b) The solution of $|X| = 0$ is 0.

c) If n is negative, then $|X| = n$ has no solution. The solution set is $\varnothing$.

d) The solutions of $|X| < p$ are those numbers that satisfy $-p < X < p$.

e) The solutions of $|X| > p$ are those numbers that satisfy $X < -p$ or $X > p$.

REVIEW EXERCISES

Beginning with this chapter, material from certain sections of preceding chapters will be covered on the chapter tests. Accordingly, the review exercises and the chapter test will contain skill maintenance exercises. Review sections and objectives to be tested in addition to the material in this chapter are [1.3a, b], [1.3c, d], [1.4c, d], and [1.6a, b].

Solve.

1. $-11 + y = -3$

2. $-7x = -3$

3. $-\frac{5}{3}x + \frac{7}{3} = -5$

4. $6(2x - 1) = 3 - (x + 10)$

5. $2.4x + 1.5 = 1.02$

6. $2(3 - x) - 4(x + 1) = 7(1 - x)$

Solve.

7. $C = \frac{4}{11}d + 3$, for d

8. $A = 2a - 3b$, for b

Solve.

9. A piece of rope 27 m long is cut into two pieces so that one piece is four-fifths as long as the other. Find the length of each piece.

10. Find three consecutive integers such that the sum of the first and second is 12 more than the third.

11. Find two consecutive even integers such that three times the first plus two times the second is 74.

12. The total cost for tuition plus room and board at Study University is \$4560. Tuition costs \$640 more than room and board. What is the tuition fee?

Solve.

13. $x \leq -4$

14. $x + 5 > 6$

15. $a + 7 \leq -14$

16. $y - 5 \geq -12$

17. $4y > -15$ **18.** $-0.3y < 9$ **19.** $-6x - 5 < 13$ **20.** $4y + 3 < -6y - 9$

21. $-\frac{1}{2}x - \frac{1}{4} > \frac{1}{2} - \frac{1}{4}x$ **22.** $0.3y - 8 < 2.6y + 15$ **23.** $-2(x - 5) \geq 6(x + 7) - 12$

Solve.

24. You are taking a biology course in which there will be 5 tests. You have scores of 91, 93, 86, and 88. You must make a total of 450 in order to get an A. What scores on the last test will give you an A?

25. You are going to invest \$30,000, part at 13% and part at 15%. What is the most that can be invested at 13% in order to make at least \$4300 interest per year?

Solve.

26. $2x - 5 < -7$ and $3x + 8 \geq 14$ **27.** $-4 < x + 3 \leq 5$ **28.** $-15 < -4x - 5 < 0$

29. $3x < -9$ *or* $-5x < -5$ **30.** $2x + 5 < -17$ *or* $-4x + 10 \leq 34$ **31.** $2x + 7 \leq -5$ *or* $x + 7 \geq 15$

Simplify.

32. $\left|-\frac{3}{x}\right|$ **33.** $\left|\frac{2x}{y^2}\right|$ **34.** $\left|\frac{12y}{-3y^2}\right|$

35. Find the distance between -23 and 39.

Solve.

36. $|x| = 6$ **37.** $|x| < 0$ **38.** $|x| \geq 3.5$ **39.** $|x - 2| = 7$

40. $|2x + 5| < 12$ **41.** $|3x - 4| \geq 15$ **42.** $|2x + 5| = |x - 9|$ **43.** $|5x + 6| = -8$

44. Find the intersection: $\{1, 2, 5, 6, 9\} \cap \{1, 3, 5, 9\}$. **45.** Find the union: $\{1, 2, 5, 6, 9\} \cup \{1, 3, 5, 9\}$.

SKILL MAINTENANCE

46. Add: $-23 + 56$. **47.** Subtract: $-\frac{2}{3} - (-\frac{5}{6})$. **48.** Multiply: $-45(-52.2)$. **49.** Divide: $-\frac{2}{3} \div (-\frac{5}{6})$.

50. Multiply: $10(2x - 3y + 7)$. **51.** Factor: $40x - 8y + 16$.

52. Collect like terms: $-8 + 4x - 2y - 6x + 7y + 20$.

Simplify.

53. $(4a^5b^{-8})(-3a^{-6}b^{11})$ **54.** $(4a^5b^{-8})^{-4}$ **55.** $\frac{-40a^5b^{-11}}{10a^{-2}b^{-14}}$

SYNTHESIS

56. Solve: $|2x + 5| \leq |x + 3|$.

57. Determine whether true or false. If $x < 3$, then $x^2 < 9$. If false, give a counterexample.

❖ THINKING IT THROUGH

1. How does the word "solve" vary in meaning within this chapter?
2. Find the error or errors in the following:

$$\begin{aligned} 7 - 9x + 6x &< -9(x + 2) + 10x \\ 7 - 3x &< -9x - 27 + 10x \\ 7 - 3x &< x + 27 \\ -4x &< 20 \\ x &< -5 \end{aligned}$$

NAME SECTION DATE

TEST: CHAPTER 2

Solve.

1. $-9 + y = 13$
2. $-12x = -8$
3. $0.7x - 0.1 = 2.1 - 0.3x$
4. $5(3x + 6) = 6 - (x + 8)$
5. Solve $h = \frac{2}{11}W + 40$, for W.
6. $|x - 3| = 9$

Solve.

7. Five more than three times a number is five less than four times the number. Find the number.
8. The price of a saw went up 12% to \$14.00. What was the price before the increase?

Solve.

9. $x - 2 < 12$
10. $-0.6y < 30$
11. $-4y - 3 \geq 5$
12. $3a - 5 \leq -2a + 6$
13. $-5y - 1 > -9y + 3$
14. $4(5 - x) < 2x + 5$
15. $-8(2x + 3) + 6(4 - 5x) \geq 2(1 - 7x) - 4(4 + 6x)$
16. You can rent a car for either \$40 per day with unlimited mileage, or \$30 per day with an extra charge of 15¢ a mile. For what numbers of miles traveled would the unlimited mileage plan save you money?
17. A student is taking an intermediate algebra course in which four tests are to be given. To get an A, the student must average at least 90 on the four tests. The student got scores of 89, 91, and 86 on the first three tests. What scores on the last test will allow the student to get an A?

Solve.

18. $5 - 2x \leq 1$ *and* $3x + 2 \geq 14$
19. $-3 < x - 2 < 4$
20. $-11 \leq -5x - 2 < 0$
21. $-3x > 12$ *or* $4x > -10$
22. $x - 7 \leq -5$ *or* $x - 7 \geq -10$
23. $3x - 2 < 7$ *or* $x - 2 > 4$

ANSWERS

1. ______
2. ______
3. ______
4. ______
5. ______
6. ______
7. ______
8. ______
9. ______
10. ______
11. ______
12. ______
13. ______
14. ______
15. ______
16. ______
17. ______
18. ______
19. ______
20. ______
21. ______
22. ______
23. ______

ANSWERS

24. ______

25. ______

26. ______

27. ______

28. ______

29. ______

30. ______

31. ______

32. ______

33. ______

34. ______

35. ______

36. ______

37. ______

38. ______

39. ______

40. ______

41. ______

42. ______

43. ______

44. ______

45. ______

46. ______

Simplify.

24. $\left|\frac{7}{x}\right|$

25. $\left|\frac{-6x^2}{3x}\right|$

26. Find the distance between 4.8 and -3.6.

Solve.

27. $|x| = 9$

28. $|x| > 3$

29. $|4x - 1| < 4.5$

30. $|-5x - 3| \geq 10$

31. $|x + 10| = |x - 12|$

32. $|2 - 5x| = -10$

33. $\left|\frac{6 - x}{7}\right| \leq 15$

34. Find the intersection:
$\{1, 3, 5, 7, 9\} \cap \{3, 5, 11, 13\}$.

35. Find the union:
$\{1, 3, 5, 7, 9\} \cup \{3, 5, 11, 13\}$.

SKILL MAINTENANCE

36. Add: $\frac{3}{4} + (-\frac{5}{8})$.

37. Subtract: $\frac{3}{4} - (-\frac{5}{8})$.

38. Multiply: $-24(-5)$.

39. Divide: $-125 \div 0.5$.

40. Multiply: $-8(2a - 3b)$.

41. Factor: $6a - 10b + 12$.

42. Collect like terms: $2x - 5y + 2.7x - 1.4y$.

Simplify.

43. $(5a^6b^{-12})(-4a^{-5}b^{-10})$

44. $(-5a^6b^{-12})^{-4}$

SYNTHESIS

Solve.

45. $|3x - 4| \leq -3$

46. $7x < 8 - 3x < 6 + 7x$

CUMULATIVE REVIEW: CHAPTERS 1–2

Evaluate.

1. $6x + 11$, for $x = 3$

2. $\frac{4a}{b}$, for $a = 3$ and $b = 6$

Simplify.

3. $-11 + (-4)$

4. $\left|-\frac{4}{7}\right|$

5. $3 - (-14)$

6. $4x - 3(x + 2) + 11$

7. $(5a^2)^0$

8. $(-5)^{-4}$

9. $5 \cdot 7 - 2^2 + 3^2$

10. $[4(6 - 7)]^3$

11. $5x - [11 + 3(2x + 4)]$

12. $a^2 \cdot a^{-5}$

13. $(-3x^5)(-2x^{-3})$

14. $\frac{3^{-4}}{3^{-2}}$

15. $\left(\frac{-3a^2b^{-3}}{2a^{-2}b^4}\right)^2$

16. $\left|\frac{-9x^3}{3x^2}\right|$

Solve.

17. $-6 + x = 12$

18. $\frac{3}{2}x = -\frac{4}{5}$

19. $2x - 3 = 5$

20. $0.5x + 3.2 = 5.2$

21. $9a - 13 + a = 5a + 12$

22. $y = mx + b$, for m

23. $x^2 + y^2 = r^2$, for y^2

24. $y^2 = 4px$, for p

25. $I = \frac{ml^2}{3}$, for m

26. $x - 3 \leq -6$

27. $2x + 5 > 1$

28. $7x - 2 \geq 3x + 1$

29. $\frac{1}{2}(4x - 8) + 3 < -\frac{1}{4}(4x + 8)$

30. $3 < 2x - 1 \leq 13$

31. $2x + 1 \geq -3$ *or* $x + 2 < 4$

32. $|x| = -8$

33. $|x - 2| = 5$

34. $|2x| - 5 = 19$

35. $|6x - 3| \leq 16$

36. $|2x - 5| = |x + 4|$

Multiply.

37. $(-3)(-3.5)$

38. $\left(\frac{5}{8}\right)\left(-\frac{1}{3}\right)$

39. $-4(6a - 2b)$

40. $3x(y - 2z)$

Factor.

41. $5x - 45$

42. $6a - 18b + 54$

Translate to an algebraic expression.

43. 3 more than x

44. The difference between 8 and m

Given the numbers

$$-7.2, \quad -5, \quad -\sqrt{3}, \quad 0, \quad \frac{3}{11}, \quad 2, \quad 8.12, \quad \sqrt{67}.$$

45. Name the irrational numbers.

46. Name the rational numbers.

47. Graph: $x < 1$ *or* $x \geq 2$.

48. Find the distance on a number line between -16 and 10.

49. Use set-builder notation to name the set of all real numbers less than -2.

50. Use $<$ or $>$ for ■ to write a true sentence: -6 ■ -8.

51. Convert to scientific notation: The box office sales for the Disney movie *An American Tail* were $45.8 million.

52. The perimeter of a rectangular field is 172 m. The length is 8 m more than the width. Find the dimensions.

53. Money is borrowed at 8% simple interest. After one year, $820.80 pays off the loan. How much was originally invested?

54. The sum of two consecutive odd integers is 832. Find the integers.

55. A store reduces the price of a coat 22% to a sale price of $109.20. What was the former price?

56. Evaluate $-x$ when $x = -6.4$.

57. Find the intersection: $\{1, 3, 5, 7\} \cap \{1, 4, 7, 10\}$.

58. Find the union: $\{0\} \cup \{1, 2, 3, 4\}$.

SYNTHESIS

59. Evaluate $\dfrac{2x - y}{4} + \dfrac{2x}{3y}$ for $x = 1$ and $y = -2$.

60. Simplify: $7x - \{3[4(x - 2) - 3(x + 1)] - 5(2x - 3)\}$.

61. Simplify: $\left[\dfrac{(-3x^{-3}y^2)^{-2}}{(4x^2y^{-1})^{-3}}\right]^2$.

INTRODUCTION Graphs help us to be able to *see* relationships between quantities. In this chapter, we graph equations whose graphs are straight lines. These are called linear equations. We also consider properties of lines such as slope and intercepts, as well as the graphing of inequalities in two variables.

The review sections to be tested in addition to the material in this chapter are 2.2, 2.4, 2.5, and 2.6. ❖

Graphs of Equations and Inequalities

3

AN APPLICATION

Toll-free 800 telephone listings have been in existence for more than twenty years. The number N of such listings can be estimated by the linear equation

$$N = 68{,}750t + 260{,}000,$$

where $t =$ the number of years since 1983. This means that for 1983, $t = 0$, and for 1994, $t = 11$. Use the equation to predict the number of listings in 1998.

THE MATHEMATICS

To find the number of listings in the year 1998, we subtract 1983 from 1998 to get $t = 15$. Then we substitute as follows:

$$N = 68{,}750(15) + 260{,}000 = 1{,}291{,}250.$$

❖ POINTS TO REMEMBER: CHAPTER 3

Equation-Solving Skills: Section 2.1
Inequality-Solving Skills: Section 2.4
Formula-Solving Skills: Section 2.3

PRETEST: CHAPTER 3

Graph on a plane.

1. $2x - 5y = 20$

2. $x = 4$

3. $y = x - 2$

4. $y = -2$

5. Find the slope and the y-intercept: $y = 5x - 3$.

6. Find the slope, if it exists, of the line containing the points $(7, 4)$ and $(-5, 4)$.

7. Find an equation of the line containing the points $(-3, 7)$ and $(-8, 2)$.

8. Find an equation of the line having the given slope and containing the given point.

$m = 2; \quad (-2, 3)$

9. Find an equation of the line containing the given point and parallel to the given line.

$(2, 5); \quad 2x - 7y = 10$

10. Find an equation of the line containing the given point and perpendicular to the given line.

$(2, 5); \quad 2x - 7y = 10$

Determine whether the graphs of the pair of lines are parallel or perpendicular.

11. $3y - 2x = 21,$
$3x + 2y = 8$

12. $y = 3x + 7,$
$y = 3x - 4$

13. Find the x-intercept of $2x - 3y = 12$.

14. Find the y-intercept of $2x - 3y = 12$.

3.1 Graphs

We see graphs like the following often in newspapers and magazines. Graphs are useful because they allow us to see relationships quickly and clearly.

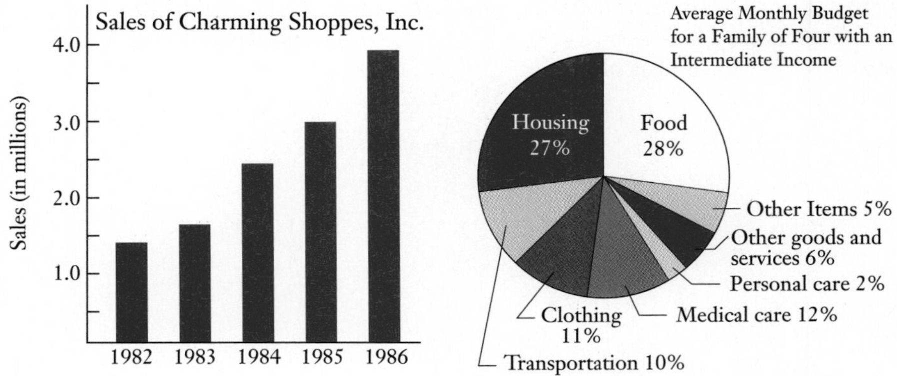

The following graph relates months of 1988 and 1989 to the average number of hours worked per week. *Months* are shown horizontally. With each month there is associated a vertical number, or unit. In this case, the vertical unit is the *average number of hours worked.*

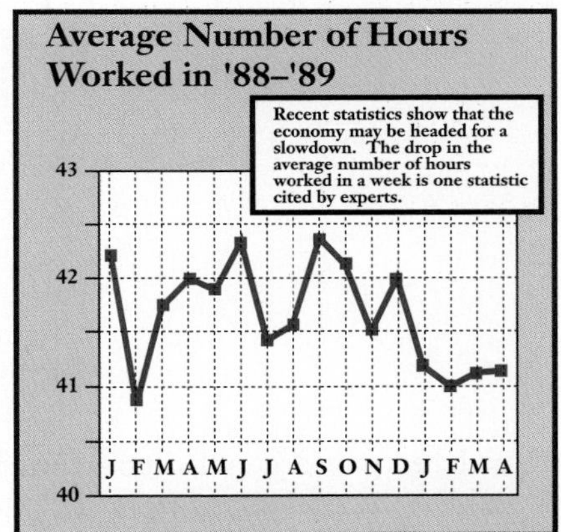

a Plotting Points

We use a number line to graph a single number. We use a plane to graph a pair of numbers. To locate points in a plane, we use two perpendicular number lines, called **axes,** each of which crosses the other at 0. We call this point the **origin.** The horizontal axis is also called the **first axis,** and the vertical axis is also called the **second axis.** Usually the first axis is the x-axis and the second axis is the y-axis. The arrows show the positive directions. We call such a setup a **Cartesian coordinate system** in honor of the great French mathematician and philosopher René Descartes (1597–1650).

OBJECTIVES

After finishing Section 3.1, you should be able to:

- **a** Plot points on a plane, given their coordinates.
- **b** Determine whether an ordered pair of numbers is a solution of an equation with two variables.
- **c** Graph equations of the type $y = mx$ and $y = mx + b$ using tables.

FOR EXTRA HELP

Tape 6A

Tape 4A

MAC: 3
IBM: 3

Plot the points on the graph below.

1. (6, 4)

2. (4, 6)

3. (−3, 5)

4. (−4, −3)

5. (4, −2)

6. (0, 3)

7. (3, 0)

8. (0, −4)

9. (−4, 0)

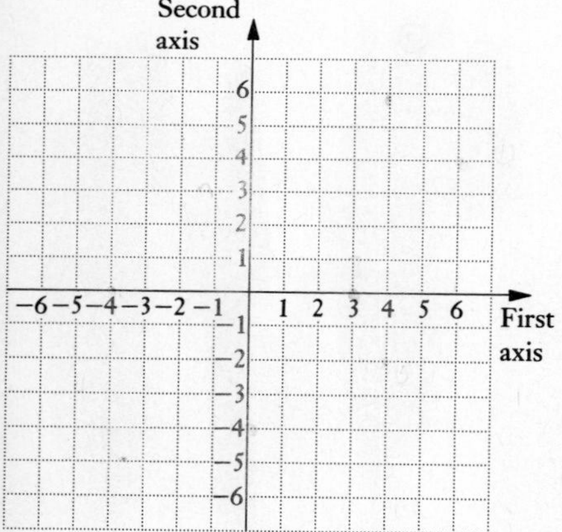

10. What can you say about the coordinates of a point in the third quadrant?

11. What can you say about the coordinates of a point in the fourth quadrant?

ANSWERS ON PAGE A-3

► **EXAMPLE 1** Plot the points (−4, 3), (−5, −3), (0, 4), and (2.5, −2).

To plot (−4, 3), we note that the first number, −4, tells us the distance in the first, or horizontal, direction. We go 4 units *left.* The second number tells us the distance in the second, or vertical, direction. We go 3 units *up.* The point (−4, 3) is then marked, or plotted. We can also plot (−4, 3) by first going *up* 3 units and then going *left* 4 units.

The points (−5, −3), (0, 4), and (2.5, −2) are also plotted at the right.

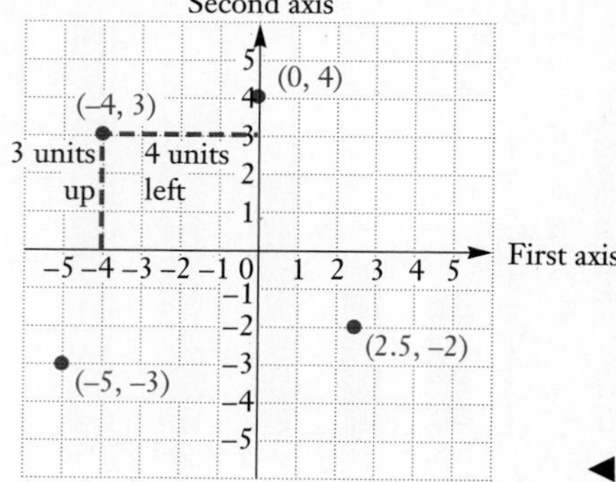

◄

DO EXERCISES 1–9.

Note that (2, 3) and (3, 2) when plotted show different points. These are called **ordered pairs** of numbers because it makes a difference which number comes first.

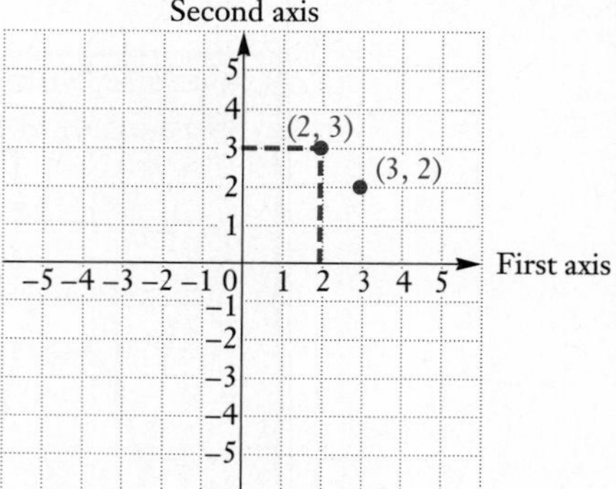

The numbers in an ordered pair are called **coordinates.** In (−4, 3), the *first coordinate* is −4 and the *second coordinate** is 3.

Quadrants

The axes divide the plane into four regions called **quadrants,** as shown here. In region I (the *first* quadrant), both coordinates of a point are positive. In region II (the *second* quadrant), the first coordinate is negative and the second coordinate is positive.

Points with one or more 0's as coordinates, such as (0, −6), (4, 0), and (0, 0), are on axes and *not* in quadrants.

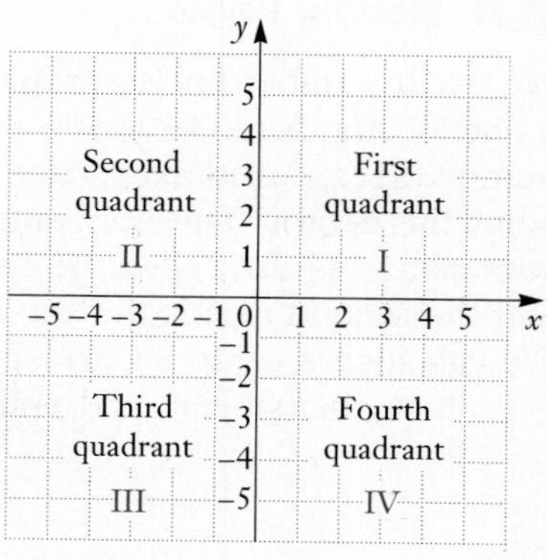

DO EXERCISES 10 AND 11.

* The first coordinate is sometimes called the **abscissa** and the second coordinate is called the **ordinate.**

b Solutions of Equations

If an equation has two variables, its solutions are pairs of numbers. If not directed otherwise, we usually take the variables in alphabetical *order*; hence we get *ordered* pairs of numbers for solutions.

▶ **EXAMPLE 2** Determine whether the following ordered pairs are solutions of the equation $y = 3x - 1$: $(-1, -4)$, $(7, 5)$.

We have

y	$= 3x - 1$
-4	$3(-1) - 1$
-4	$-3 - 1$
	-4

We substitute -1 for x and -4 for y (alphabetical order of variables).

The equation becomes true: $(-1, -4)$ is a solution.

y	$= 3x - 1$
5	$3 \cdot 7 - 1$
5	$21 - 1$
	20

We substitute.

The equation becomes false: $(7, 5)$ is not a solution. ◀

DO EXERCISES 12–14.

12. Determine whether $(9, -2)$ is a solution of the equation $y = 4x + 17$.

13. Determine whether $(0, 23)$ is a solution of $y = 4x + 17$.

14. Determine whether $(2, -2)$ is a solution of $2y + 6x = 8$.

ANSWERS ON PAGE A-3

c Graphing Equations $y = mx$ and $y = mx + b$ Using Tables

The equations considered in Example 2 and in Margin Exercises 12–14 have an infinite number of solutions, which means that we cannot list them all. Because of this, it is convenient to make a drawing that represents the solutions. Such a drawing is called a **graph.**

> **To *graph* an equation means to make a drawing that represents its solutions.**

The graphs of linear equations of the type $y = mx$ and $y = mx + b$ are straight lines. If an equation has a graph that is a line, we can graph it by plotting a few points and then drawing a line through them.

▶ **EXAMPLE 3** Graph: $y = 2x$.

We find some ordered pairs that are solutions, listing the results in a table. We choose *any* number for x and then determine y by substitution. Suppose we choose 3 for x. Then

$$y = 2x = 2 \cdot 3 = 6.$$

We get a solution: the ordered pair (3, 6). Suppose we choose 0 for x. Then

$$y = 2x = 2 \cdot 0 = 0.$$

We get a solution: the ordered pair (0, 0). We make some negative choices for x as well as some positive ones. If a number takes us off the graph paper, we generally do not use it. Continuing in this manner, we get a table like the one shown below. Other values for x could have been chosen. In any case, since $y = 2x$, we get y by doubling x.

15. Graph the equation $y = \frac{1}{2}x$.

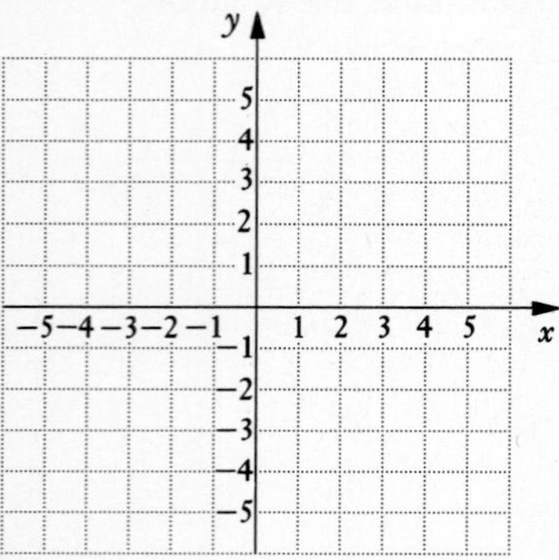

16. Graph the equation $y = 3x$.

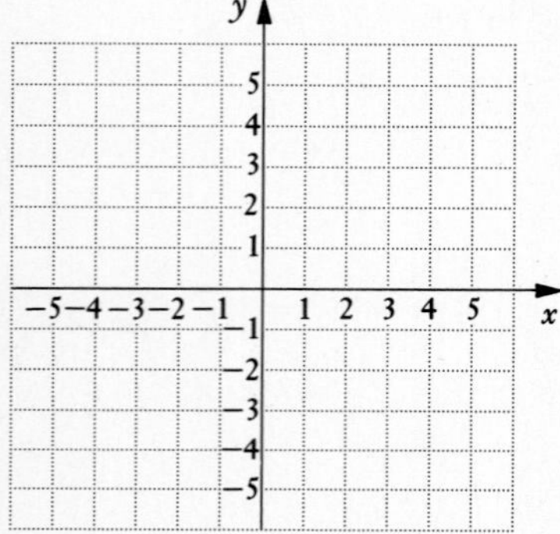

17. Graph the equation $y = -2x$.

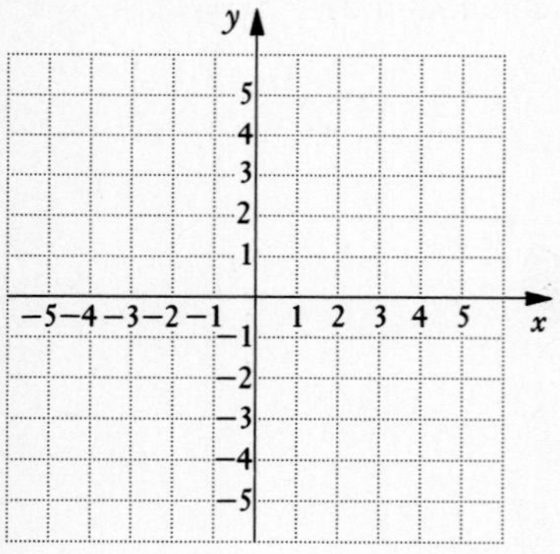

ANSWERS ON PAGE A-3

Now we plot these points. We look for a pattern and see that the graph is a line. We draw the line, or graph, with a ruler and label it $y = 2x$.

x	y $y = 2x$	(x, y)
0	0	(0, 0)
1	2	(1, 2)
3	6	(3, 6)
−2	−4	(−2, −4)
−3	−6	(−3, −6)

(1) Choose *any* x.
(2) Compute y.
(3) Form the pair (x, y).
(4) Plot the points.

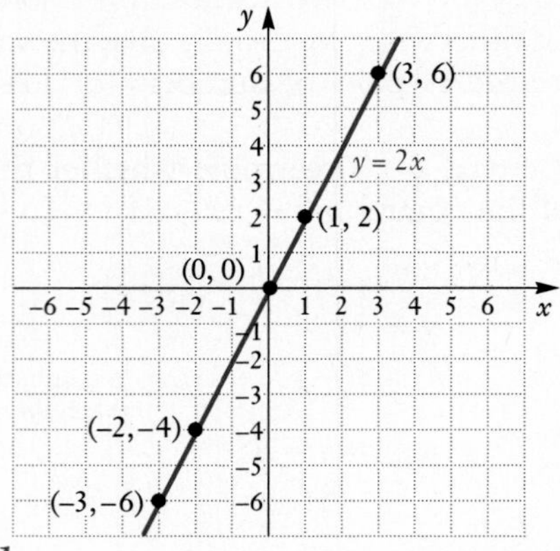

▶ **EXAMPLE 4** Graph: $y = -\frac{1}{5}x$.

To find an ordered pair, we choose any number for x and then determine y. For example,

when $x = -5$, $y = -\frac{1}{5}(-5) = 1$;
when $x = -2$, $y = -\frac{1}{5}(-2) = \frac{2}{5}$.

We find several ordered pairs in this manner, plot them, and draw the line.

x	y $y = -\frac{1}{5}x$	(x, y)
−5	1	(−5, 1)
−2	$\frac{2}{5}$	$(-2, \frac{2}{5})$
0	0	(0, 0)
5	−1	(5, −1)
10	−2	(10, −2)

(1) Choose *any* x.
(2) Compute y.
(3) Form the pair (x, y).
(4) Plot the points.

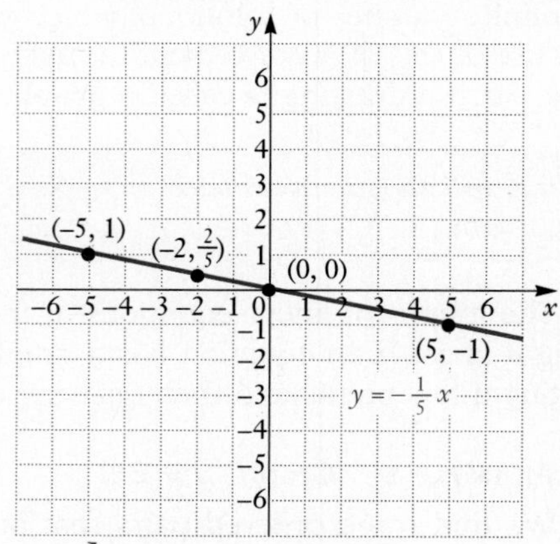

We do not plot (10, −2) because 10 is off the grid that we are using. ◀

DO EXERCISES 15–17.

From your experience with the preceding examples and exercises, you have probably discovered that the graph of any equation $y = mx$ is a straight line through the origin. What will happen if we add a number b on the right side to get an equation $y = mx + b$?

▶ **EXAMPLE 5** Graph $y = 3x - 6$ and compare it with the graph of $y = 3x$.

To find an ordered pair, we choose any number for x and then determine y. For example,

when $x = 0$, $y = 3(0) - 6 = -6$;

when $x = -2$, $y = 3(-2) - 6 = -6 - 6 = -12$.

We find several ordered pairs this way, plot them, and draw the line. It looks just like the graph of $y = 3x$, but it is moved down 6 units.

	y	
x	$y = 3x - 6$	(x, y)
0	-6	$(0, -6)$
-2	-12	$(-2, -12)$
1	-3	$(1, -3)$
3	3	$(3, 3)$

(1) Choose *any* x.
(2) Compute y.
(3) Form the pair (x, y).
(4) Plot the points.

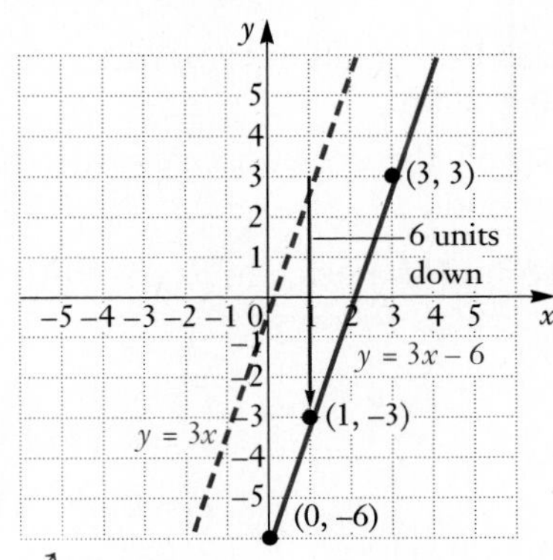

◀

DO EXERCISE 18.

18. Graph the equation $y = x - 2$ and compare it with the graph of $y = x$.

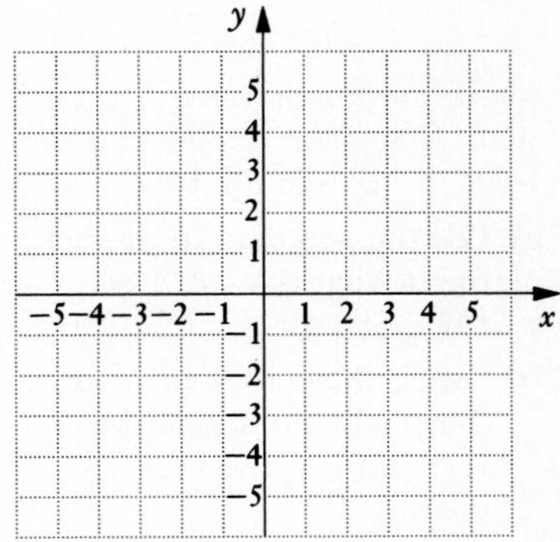

▶ **EXAMPLE 6** Graph $y = -4x + 5$ and compare it with the graph of $y = -4x$.

To find an ordered pair, we choose any number for x and then determine y. For example,

when $x = 0$, $y = -4(0) + 5 = 0 + 5 = 5$;

when $x = -2$, $y = -4(-2) + 5 = 8 + 5 = 13$.

We find several ordered pairs in this manner, plot them, and draw the line.

	y	
x	$y = -4x + 5$	(x, y)
0	5	$(0, 5)$
-2	13	$(-2, 13)$
1	1	$(1, 1)$
2	-3	$(2, -3)$

(1) Choose *any* x.
(2) Compute y.
(3) Form the pair (x, y).
(4) Plot the points.

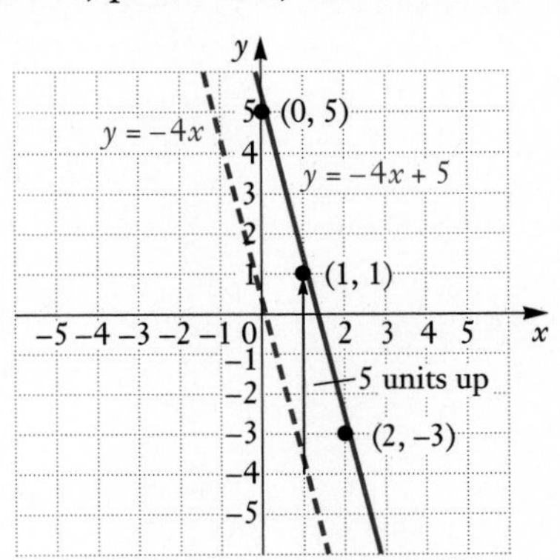

Note that the graph of $y = -4x + 5$ is a line. It looks just like the graph of $y = -4x$, but it is moved up 5 units. ◀

DO EXERCISE 19.

19. Graph the equation $y = -2x + 3$ and compare it with the graph of $y = -2x$.

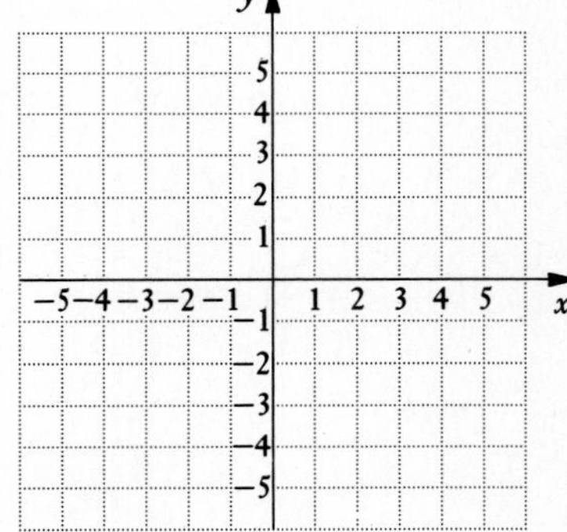

ANSWERS ON PAGE A-3

20. *Driving costs.* Driving costs have been increasing in recent years. The cost per mile C, in cents, is given by

$$C = 0.34t + 21.7,$$

where $t =$ the number of years since 1980. This means that for 1980, $t = 0$, and for 1994, $t = 14$.

a) Use the equation to predict the driving costs in 1991, 1995, and 1998.

b) Graph the equation. Assume that t is the first coordinate and that C is the second coordinate.

ANSWERS ON PAGE A-4

▶ **EXAMPLE 7** *Average price of a movie ticket.* The average price P, in dollars, of a movie ticket can be estimated by the equation

$$P = 0.1522t + 4.29,$$

where $t =$ the number of years since 1990. This means that for 1990, $t = 0$. Suppose we want to know the average price of a movie ticket in 1995. We subtract 1990 from 1995 to get 5, and then let $t = 5$ in the equation. (The price is lower than what might be expected due to senior-citizen discounts, children's prices, and special volume discounts.)

a) Use the equation to predict the average price of a ticket in 1991, 1995, and 2000.

b) Graph the equation. Assume that t is the first coordinate and P is the second.

a) To find the price in years 1991, 1995, and 2000, we substitute and calculate as follows:

$$P = 0.1522(0) + 4.29 = \$4.44,$$
$$P = 0.1522(5) + 4.29 = \$5.05,$$
$$P = 0.1522(10) + 4.29 = \$5.81.$$

b) Using the values computed in (a) and any others we want, we make the following table of values. Note that the number of years t is not negative, since only years from 1990 are considered. Next we plot the points and draw the graph.

Year	Price
1991	\$4.44
1995	\$5.05
2000	\$5.81

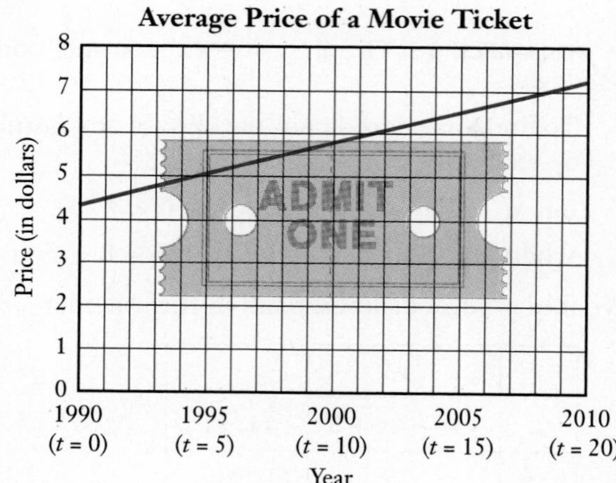

◀

DO EXERCISE 20.

NAME SECTION DATE

EXERCISE SET 3.1

a Plot the following points.

1. $A(5, 3)$, $B(2, 4)$, $C(0, 2)$, $D(0, -6)$, $E(3, 0)$, $F(-2, 0)$, $G(1, -3)$, $H(-5, 3)$, $J(-4, 4)$

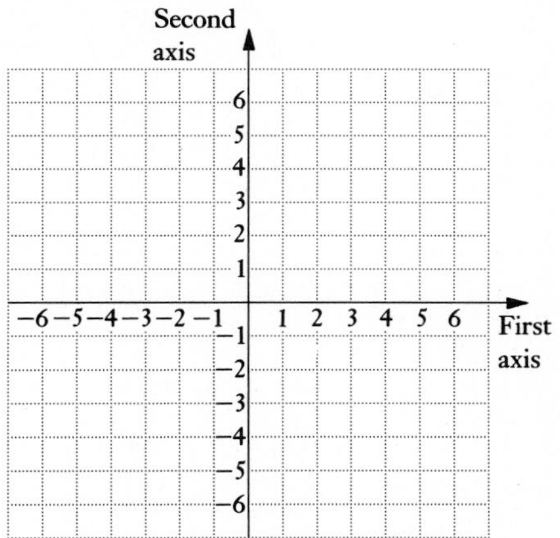

2. $A(3, 5)$, $B(1, 5)$, $C(0, 4)$, $D(0, -4)$, $E(5, 0)$, $F(-5, 0)$, $G(1, -5)$, $H(-7, 4)$, $J(-5, 5)$

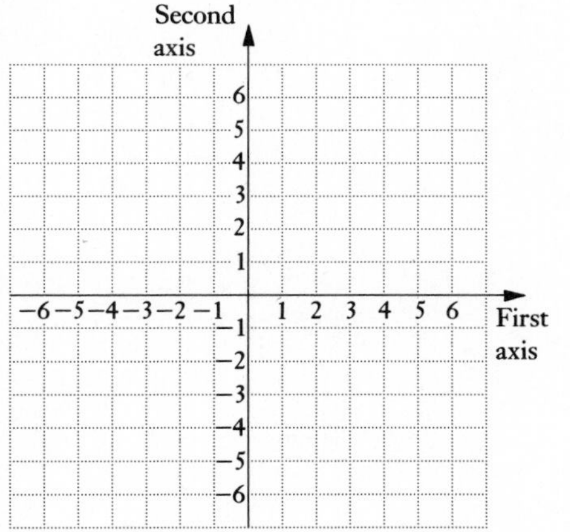

3. Plot the points $M(2, 3)$, $N(5, -3)$, and $P(-2, -3)$. Draw $\overline{MN}$, $\overline{NP}$, and $\overline{MP}$. ($\overline{MN}$ means the line segment from M to N.) What kind of geometric figure is formed? What is its area?

4. Plot the points $Q(-4, 3)$, $R(5, 3)$, $S(2, -1)$, and $T(-7, -1)$. Draw $\overline{QR}$, $\overline{RS}$, $\overline{ST}$, and $\overline{TQ}$. What kind of figure is formed? What is its area?

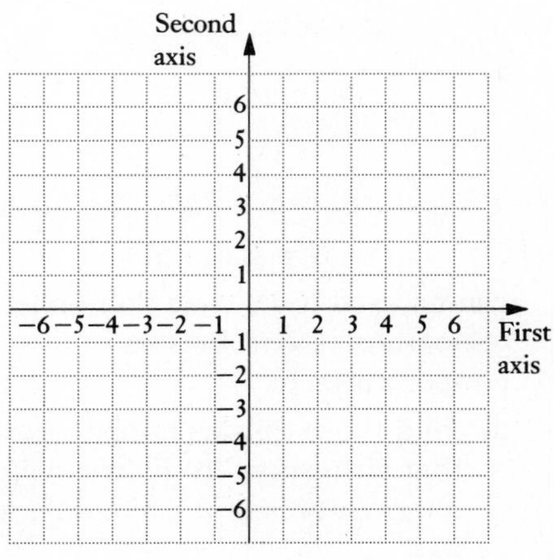

b Determine whether the given ordered pair is a solution of the equation.

5. $(10, 3)$; $3x - 7y = 9$

6. $(-1, 1)$; $5y + 2x = -3$

7. $(4, -2)$; $4y - 5x = 7 + 3x$

8. $(8, -5)$; $3x + 7y = 5 - 2x$

9. $(2, 3)$; $2p + q = 5$

10. $(3, 4)$; $3s + t = 4$

ANSWERS

1. See graph.
2. See graph.
3. See graph.
4. See graph.
5.
6.
7.
8.
9.
10.

C Graph the equation.

11. $y = 5x$

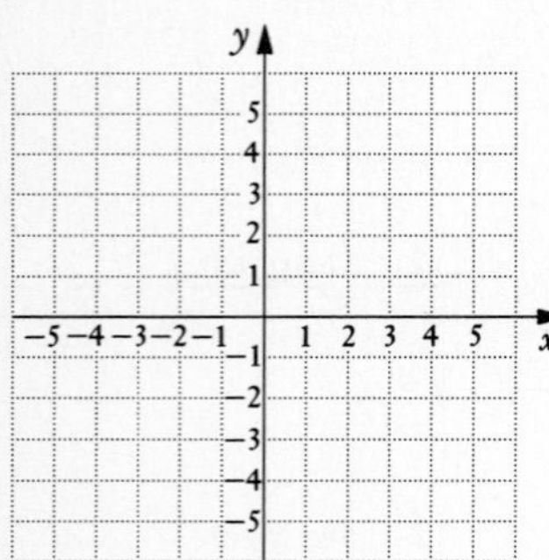

12. $y = -4x$

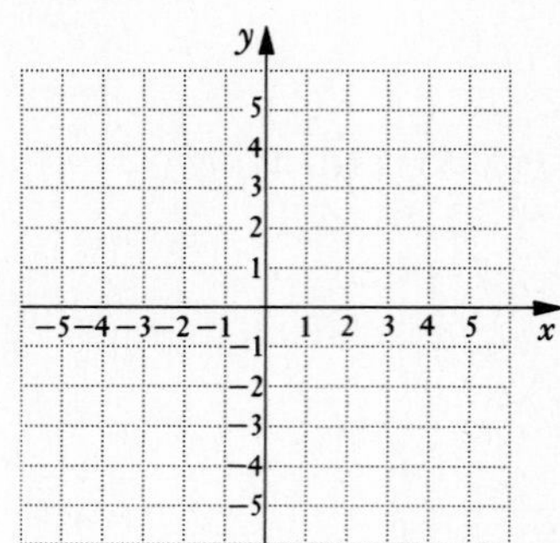

13. $y = -3x$

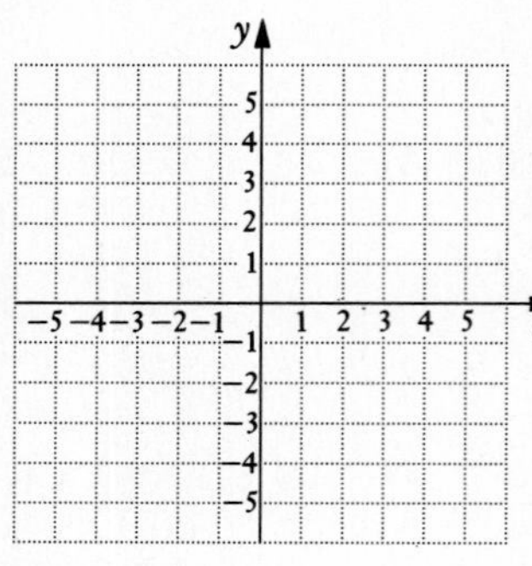

14. $y = \frac{1}{4}x$

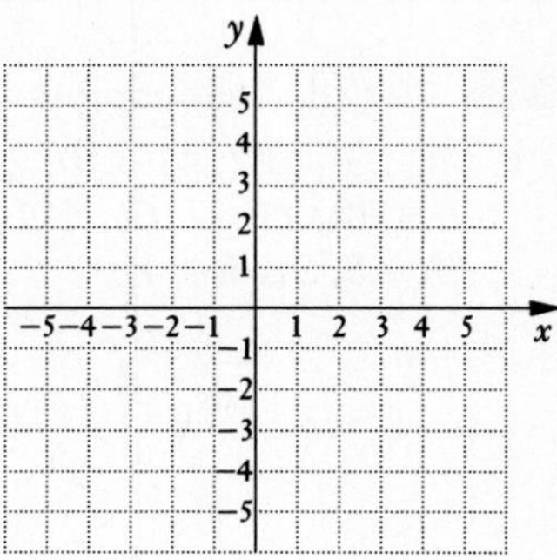

15. $y = x + 3$

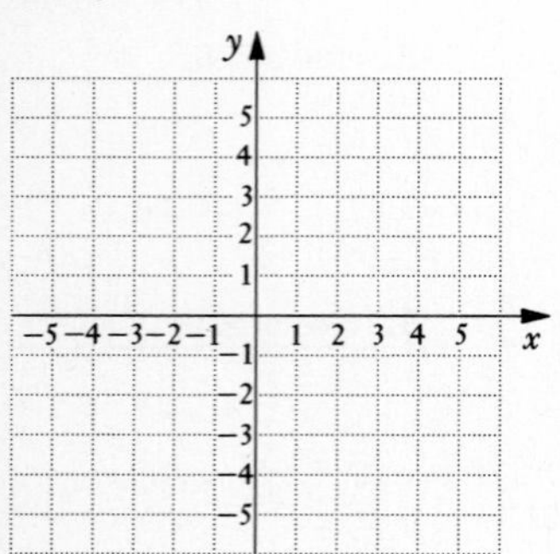

16. $y = 3 - x$

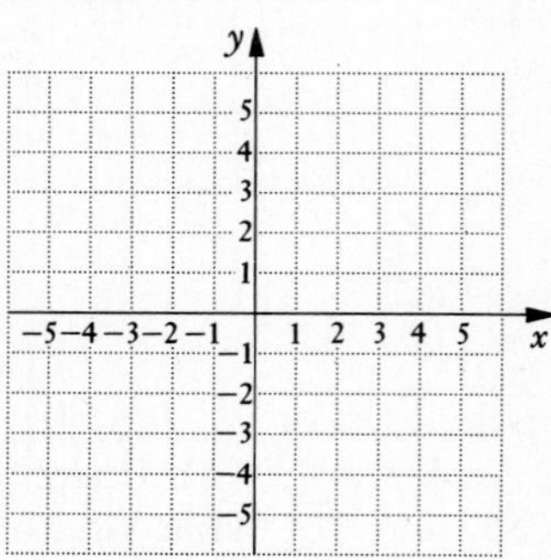

17. $y = \frac{1}{4}x + 2$

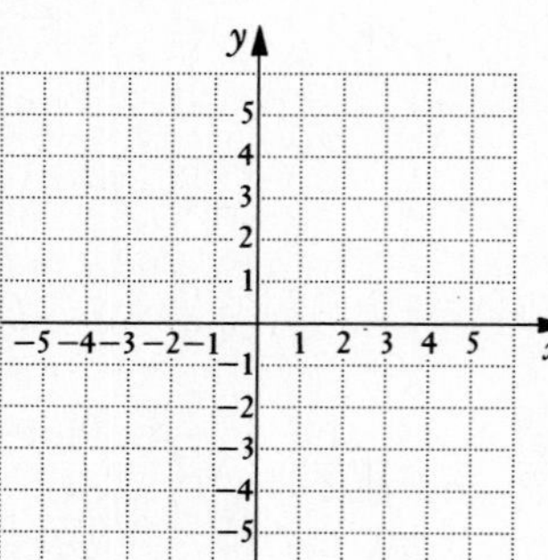

18. $y = 5x - 2$

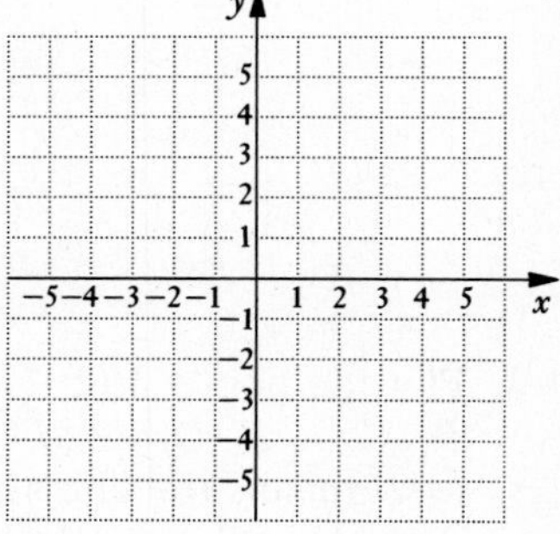

Graph the equation using your own graph paper.

19. $y = -\frac{1}{3}x + 2$

20. $y = -1.5x - 3$

21. $y = 0.3x - 5$

22. $y = \frac{2}{3}x + 1$

ANSWERS

23. a) ______

b) ______

24. a) ______

b) ______

25. ______

26. ______

27. ______

28. ______

23. *Temperatures of solids.* The formula

$$C = \frac{5}{9}(F - 32)$$

can be used to convert Fahrenheit temperatures F to Celsius temperatures C.

a) Find those Celsius temperatures that correspond to Fahrenheit temperatures of 0°, 32°, 41°, and 212°.

b) Graph the equation. Assume that F is the first coordinate and C is the second.

24. *Women in the military ranks.* The percentage of women in the total active military duty force has been steadily increasing. The number N of women in the active duty force t years after 1971 is approximated by

$$N = 12{,}197.8t + 44{,}000.$$

a) How many women were in the military in 1971 ($t = 0$)? in 1981 ($t = 10$)? in 1990 ($t = 19$)?

b) Graph the equation. Assume that t is the first coordinate and N is the second.

SKILL MAINTENANCE

25. Solve: $-6x + 2x - 32 < 64$.

26. Solve: $|3x - 7| \geq 24$.

27. Simplify: $128 \div (-\frac{1}{2}) \div (-64)$.

28. Solve: $-4 < 2x - 5 < 4$.

3.2 More on Graphing Linear Equations

a Graphing Using Intercepts

A **linear equation** is any equation equivalent to one of the type $Ax + By = C$. We call the latter equation the **standard form of a linear equation.** Graphs of linear equations are always straight lines.

The ***x*-intercept** of an equation is a point where the graph crosses the x-axis. The ***y*-intercept** of an equation is a point where the graph crosses the y-axis. We know from geometry that two distinct points determine a straight line. Thus, if we know the intercepts, we can graph the line [assuming both intercepts are not the origin, (0, 0)]. To ensure that a computation error has not been made, it is a good idea to calculate a third point as a check.

Equations of the type $Ax + By = C$, where $A \neq 0$ and $B \neq 0$, can be graphed conveniently using intercepts.

OBJECTIVES

After finishing Section 3.2, you should be able to:

a Graph linear equations using intercepts.

b Graph horizontal and vertical lines of the type $y = b$ and $x = a$.

FOR EXTRA HELP

Tape 6A | Tape 4A | MAC: 3 IBM: 3

▶ **EXAMPLE 1** Graph: $3x + 2y = 12$.

This equation is linear. (We could show that it is by solving for y.) To find the y-intercept, we let $x = 0$ and see what y must be. If we just cover up the $3x$ or ignore it, that amounts to letting x be 0. So we cover up $3x$ and see $2y = 12$. Then y must be 6. The y-intercept is (0, 6).

To find the x-intercept, we can cover up the y-term and look at the rest of the equation. We have

$$3x = 12, \quad \text{or} \quad x = 4.$$

The x-intercept is (4, 0).

We plot these points and draw the line, using a third point as a check. We choose $x = 6$ and solve for y:

$$3(6) + 2y = 12$$
$$18 + 2y = 12$$
$$2y = -6$$
$$y = -3.$$

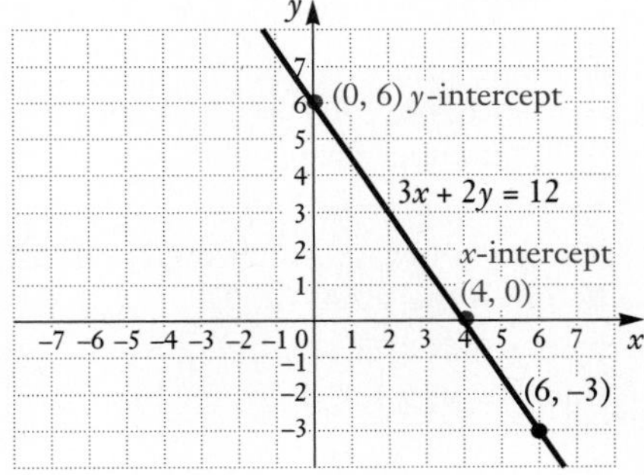

We plot (6, −3) and note that it is on the line. ◀

> **A y-intercept is a point $(0, b)$. To find b, let $x = 0$ in the equation.**
> **An x-intercept is a point $(a, 0)$. To find a, let $y = 0$ in the equation.**

DO EXERCISE 1.

1. Graph using intercepts: $4y - 12 = -6x$.

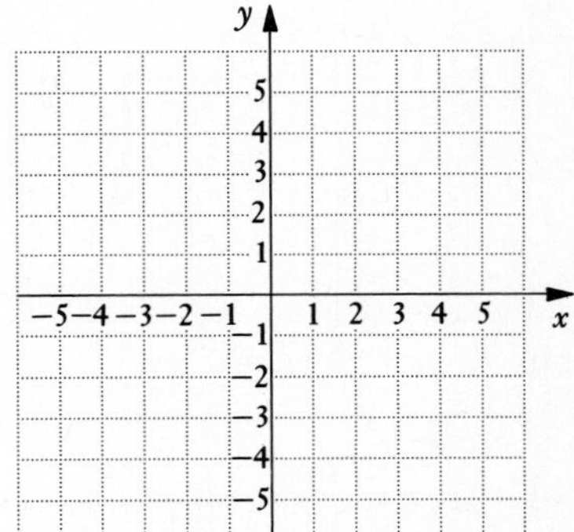

2. Graph: $y = 4$.

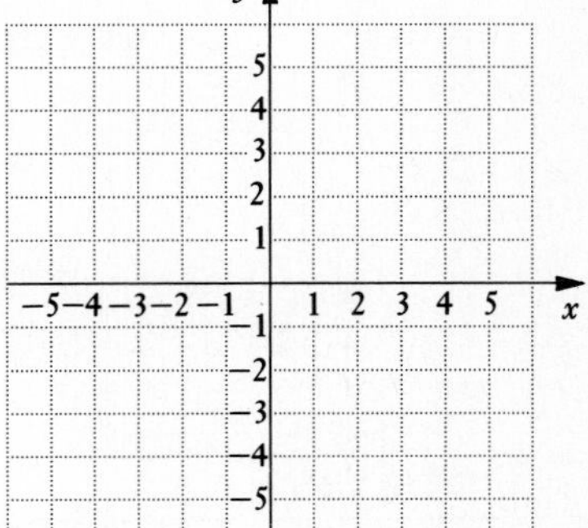

ANSWERS ON PAGE A-4

b Graphing Horizontal and Vertical Lines

Some equations have graphs that are parallel to one of the axes. This happens when either A or B is 0 in $Ax + By = C$. These equations have a missing variable. In the following example, x is missing.

Graph.

3. $x = -5$

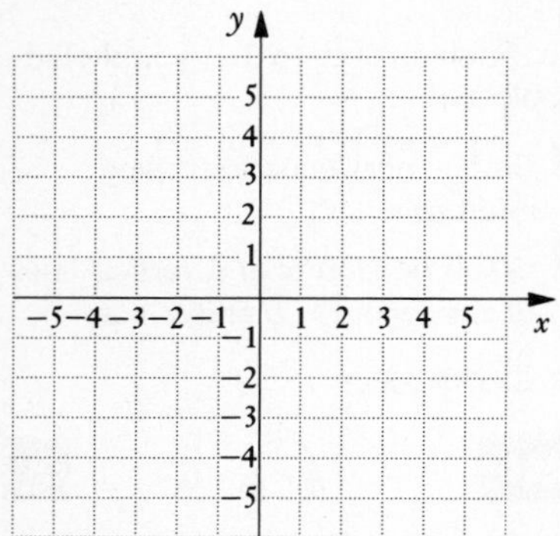

4. $8x - 5 = 19$ (*Hint:* Solve for x.)

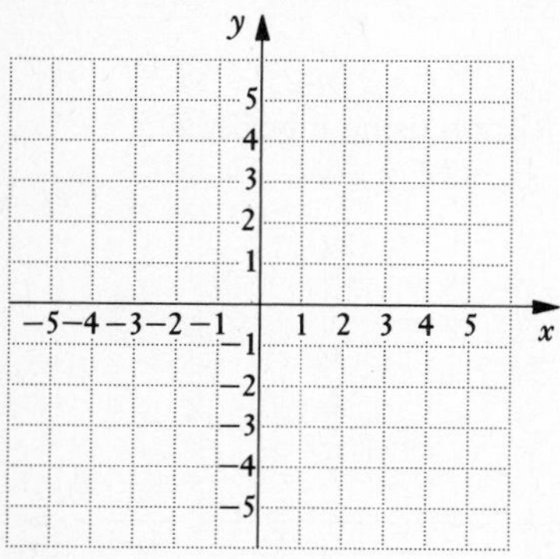

Graph.

5. $2x - 5y = 10$

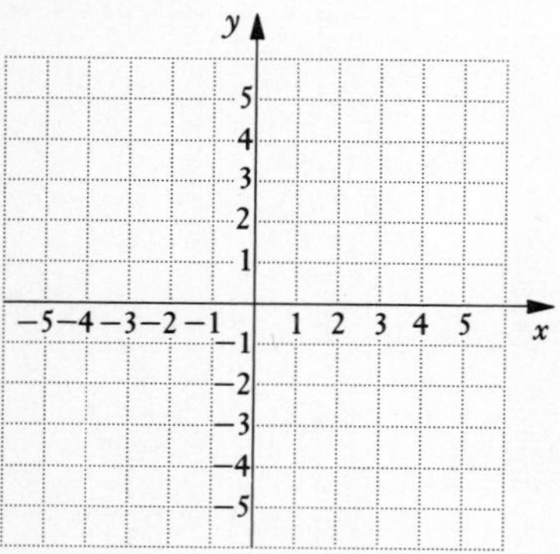

6. $x = -2.4$

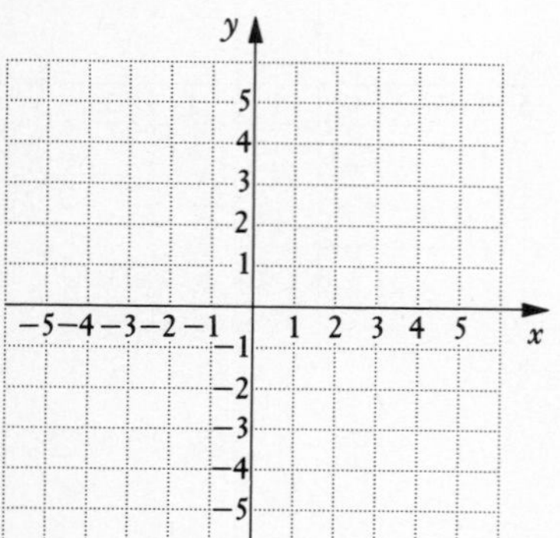

ANSWERS ON PAGE A-4

▶ **EXAMPLE 2** Graph: $y = 3$.

Since x is missing, any number for x will do. Thus all ordered pairs $(x, 3)$ are solutions. The graph is a line parallel to the x-axis.

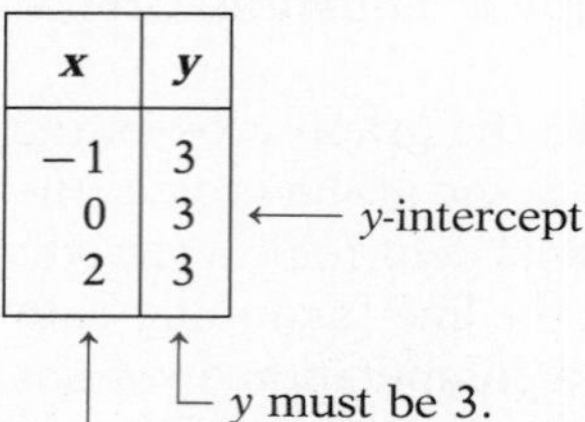

x	y
-1	3
0	3
2	3

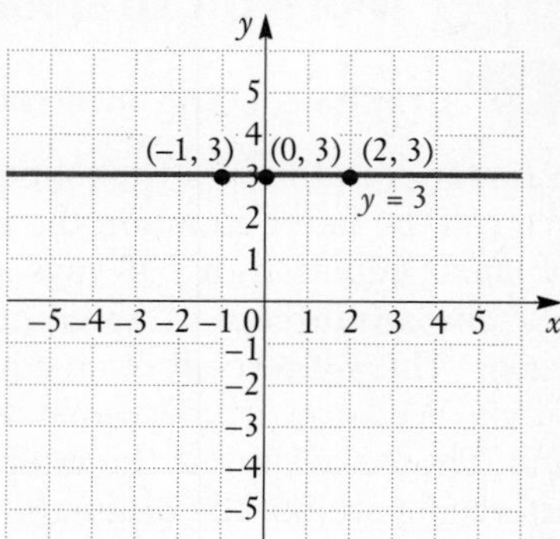

◀

DO EXERCISE 2 ON THE PRECEDING PAGE.

In the following example, y is missing and the graph turns out to be parallel to the y-axis.

▶ **EXAMPLE 3** Graph: $x = -2$.

Since y is missing, any number for y will do. Thus all ordered pairs $(-2, y)$ are solutions. The graph is a line parallel to the y-axis.

x	y
-2	0
-2	3
-2	-4

← x-intercept

x must be -2. Choose *any* number for y.

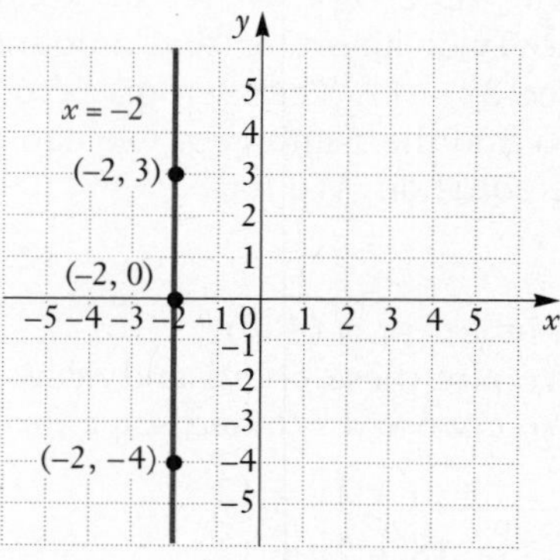

◀

The graph of an equation $x = a$, where a is a constant, is a vertical line through the point $(a, 0)$.

The graph of an equation $y = b$, where b is a constant, is a horizontal line through the point $(0, b)$.

DO EXERCISES 3 AND 4.

We summarize our procedure for graphing linear equations.

To Graph Linear Equations

1. **Is the equation of the type $x = a$ or $y = b$? If so, the graph will be a line parallel to an axis.**
2. **If the line is of the type $y = mx$, both intercepts are the origin $(0, 0)$. Plot $(0, 0)$ and find two other points.**
3. **If the equation is of the type $Ax + By = C$, but not of the type $x = a$, $y = b$, or $y = mx$, graph using intercepts. If the intercepts are too close together, choose another point farther from the origin.**
4. **In any case, use a third point as a check.**

DO EXERCISES 5 AND 6.

NAME SECTION DATE

EXERCISE SET 3.2

a Find the intercepts. Then graph.

1. $x - 3y = 6$

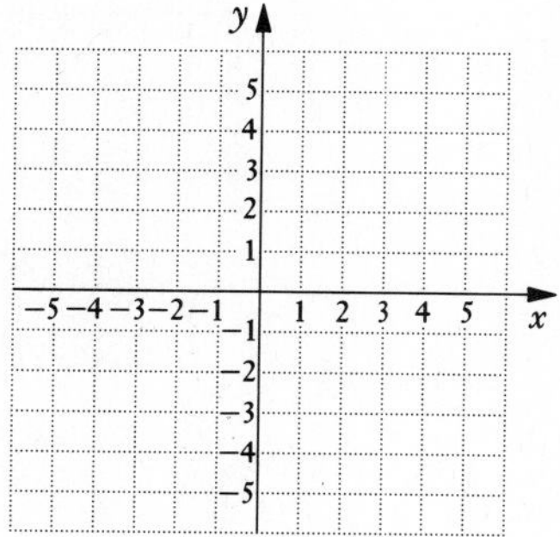

2. $x - 2y = 4$

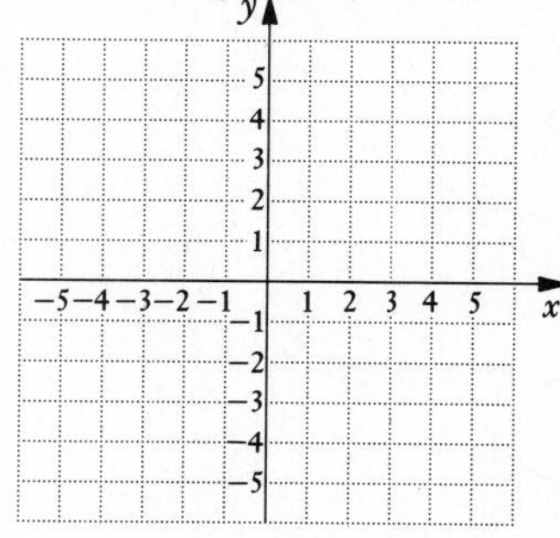

3. $x + 2y = 4$

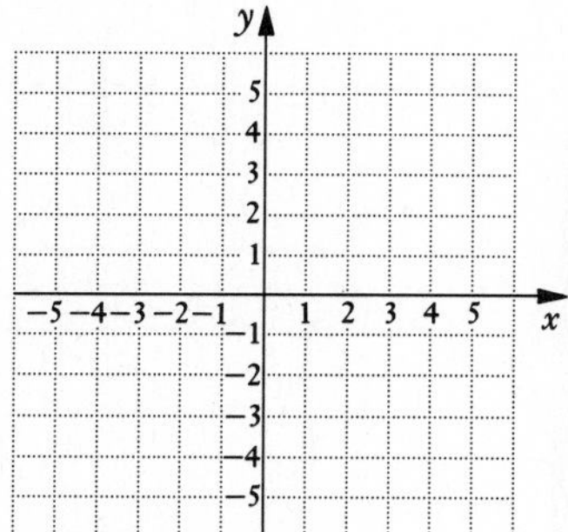

4. $2x + 3y = 6$

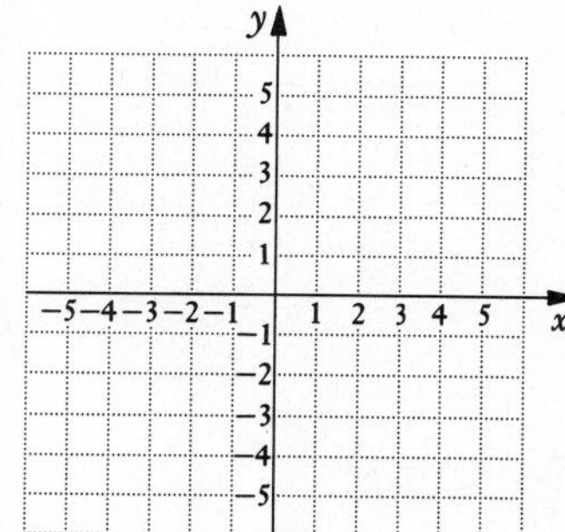

5. $5x - 2y = 10$

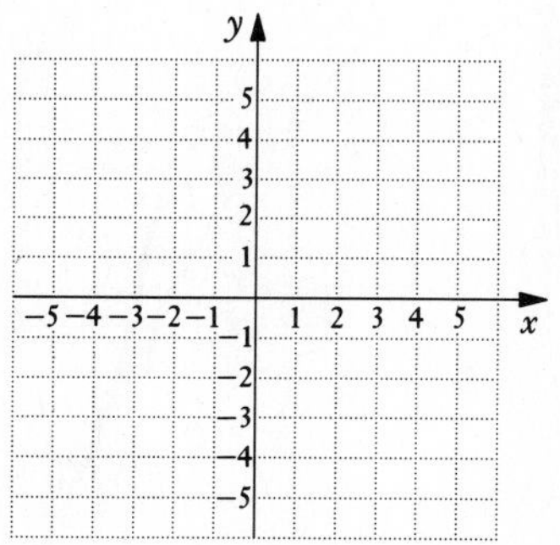

6. $x + 2y = 6$

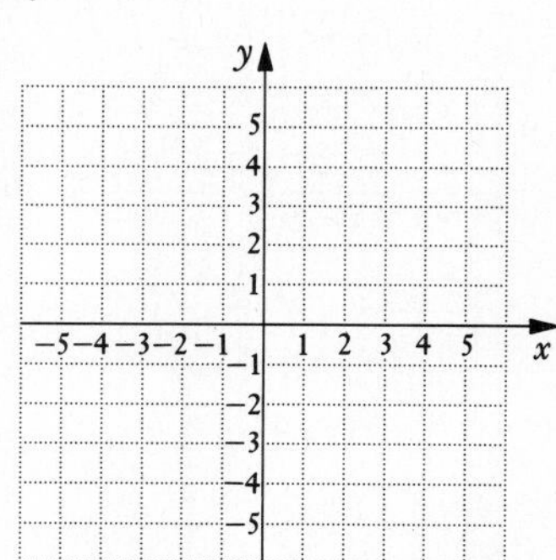

7. $5y = -15 + 3x$

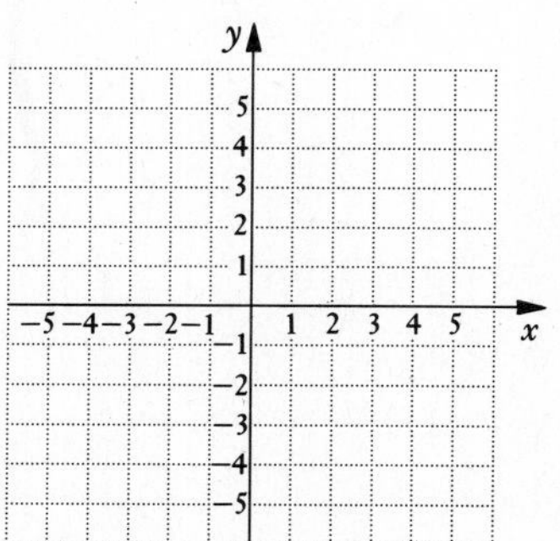

8. $5y = x + 5$

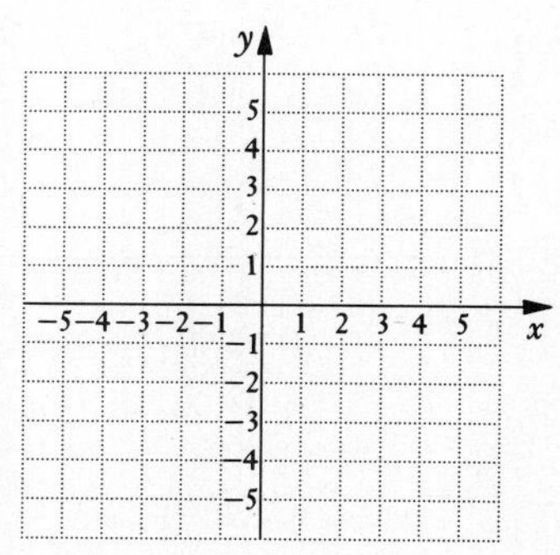

ANSWERS

1. ______

2. ______

3. ______

4. ______

5. ______

6. ______

7. ______

8. ______

ANSWERS

9. ____________

10. ____________

11. ____________

12. ____________

13. ____________

14. ____________

15. ____________

16. ____________

9. $5x - 10 = 5y$

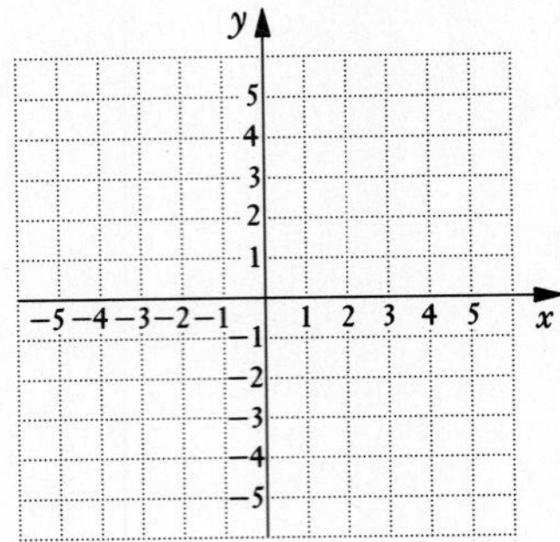

10. $2x - 3y = 6$

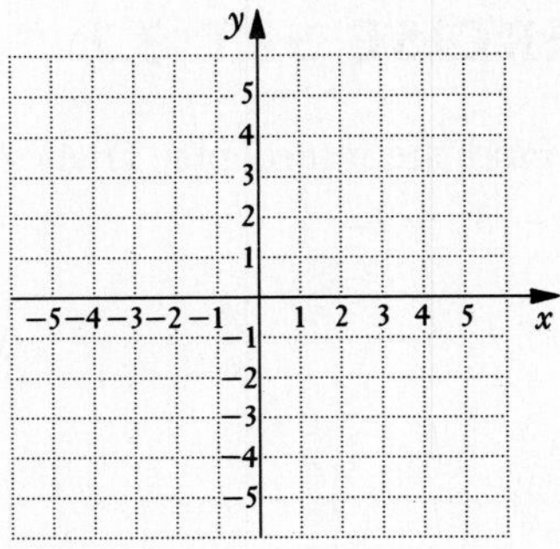

11. $4x + 5y = 20$

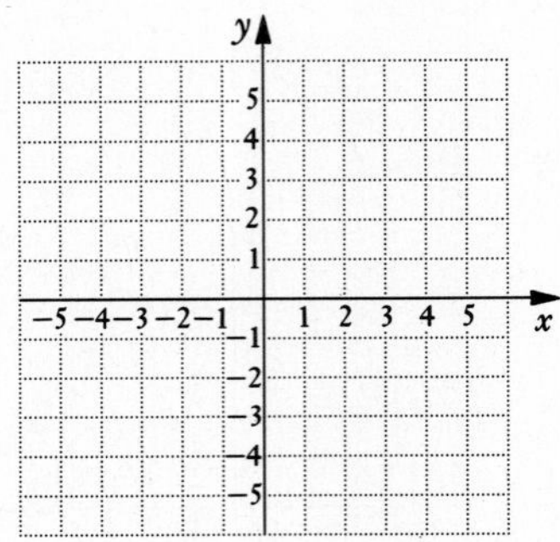

12. $2x + 6y = 12$

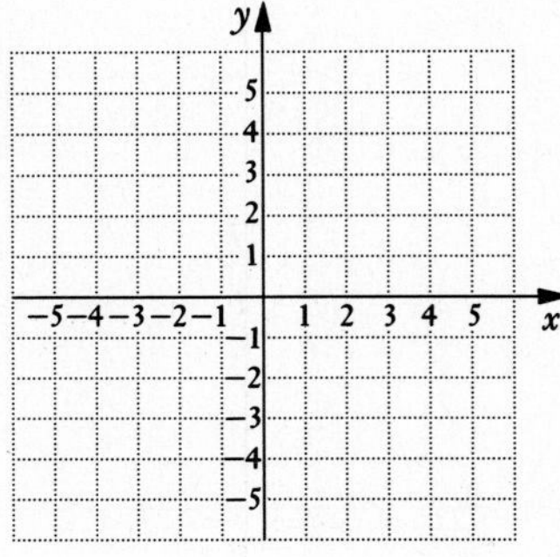

13. $4x - 3y = 12$

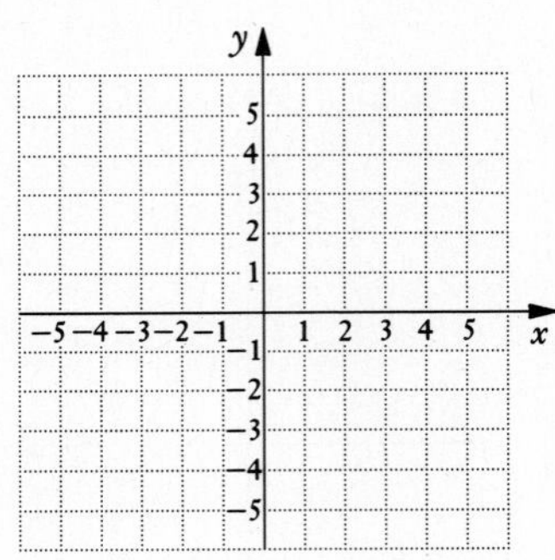

14. $6x - 2y = 6$

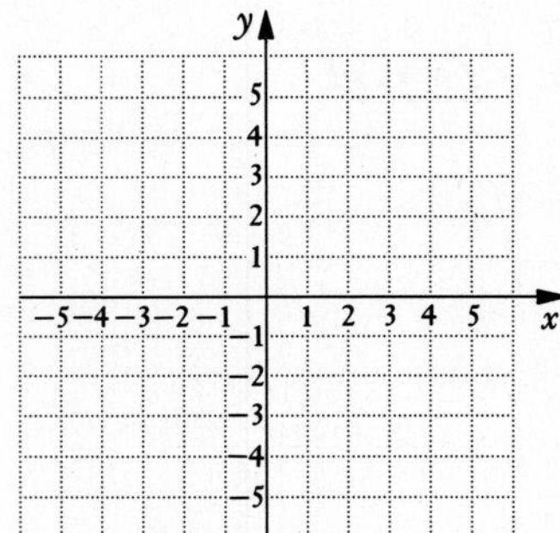

15. $y - 3x = 0$

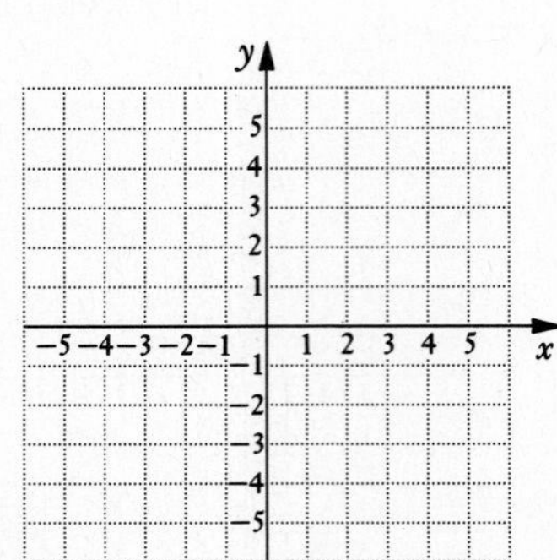

16. $x + 2y = 0$

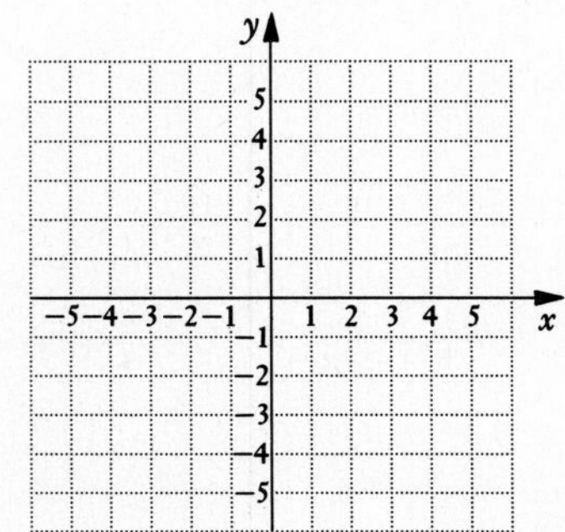

17. $6x - 7 + 3y = 9x - 2y + 8$
(*Hint:* Collect like terms before graphing.)

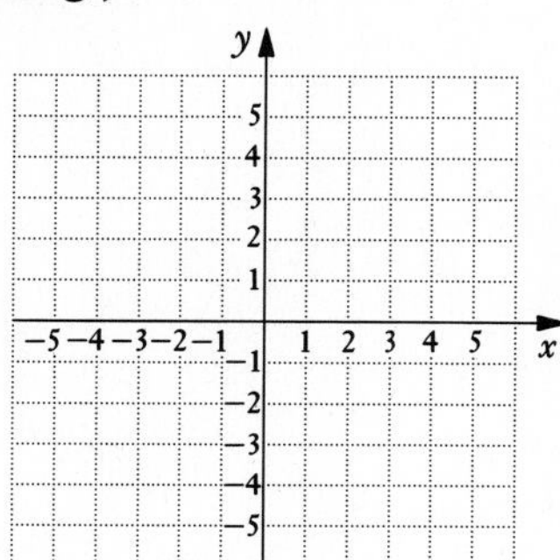

18. $7x - 8 + 4y = 8y + 4x + 4$

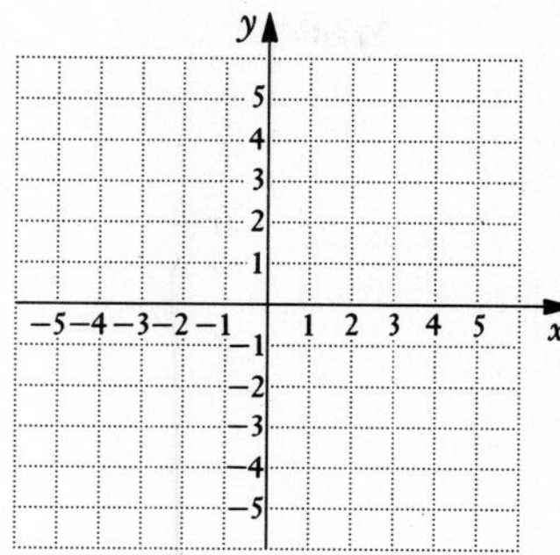

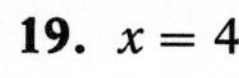
Graph.

19. $x = 4$

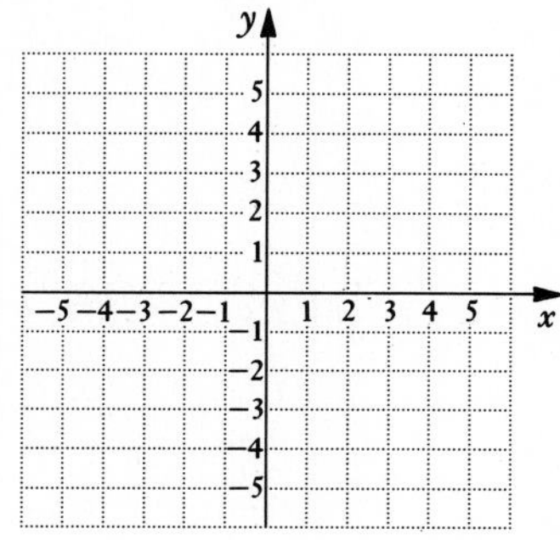

20. $x = -1$

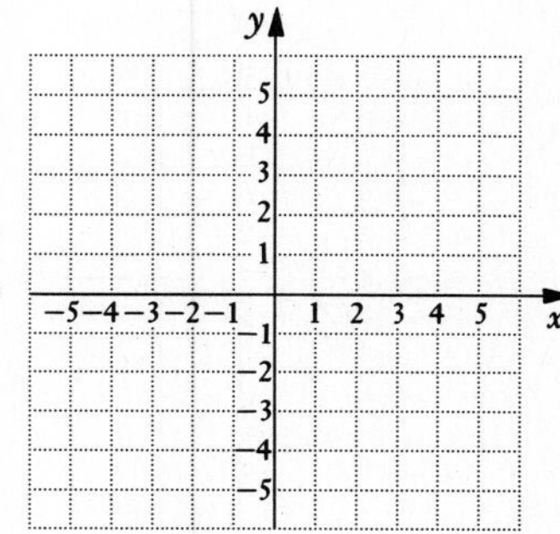

21. $y = -2$

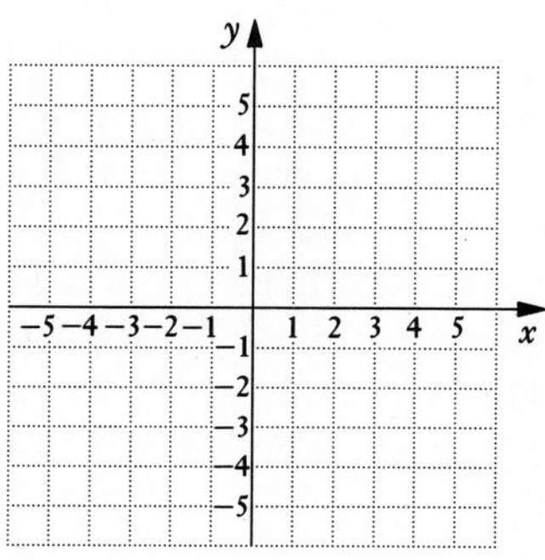

22. $y = \frac{7}{2}$

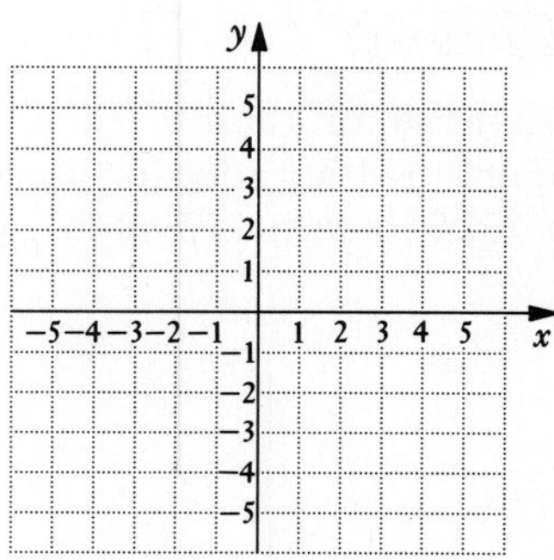

23. $3x + 15 = 0$

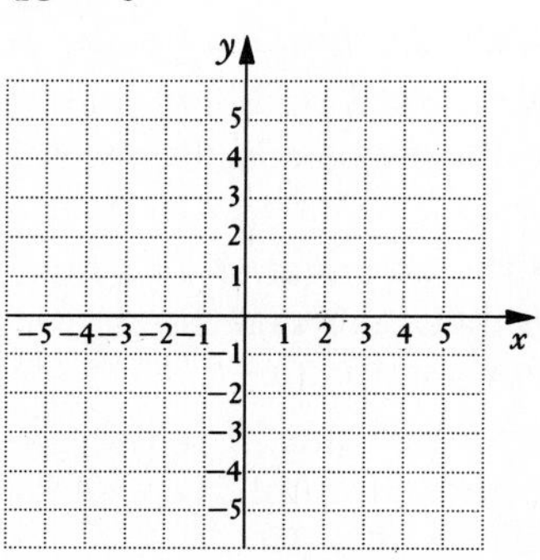

24. $6y - 24 = 0$

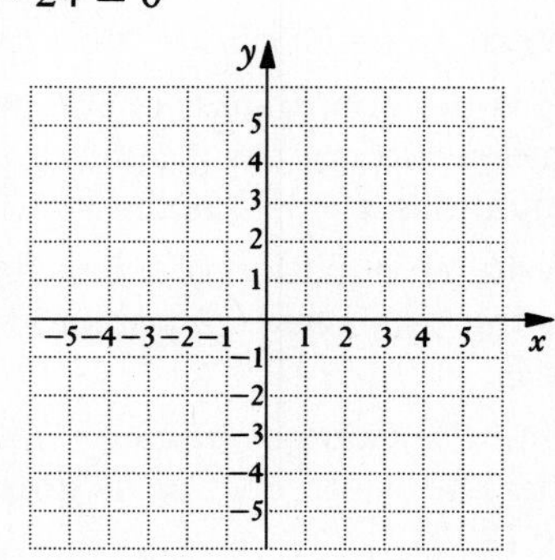

ANSWERS

17. ______

18. ______

19. See graph.

20. See graph.

21. See graph.

22. See graph.

23. See graph.

24. See graph.

ANSWERS

25. See graph.

26. See graph.

27. See graph.

28. See graph.

29.

30.

31.

32.

33.

34.

35.

36.

37.

38.

25. $y = 0$

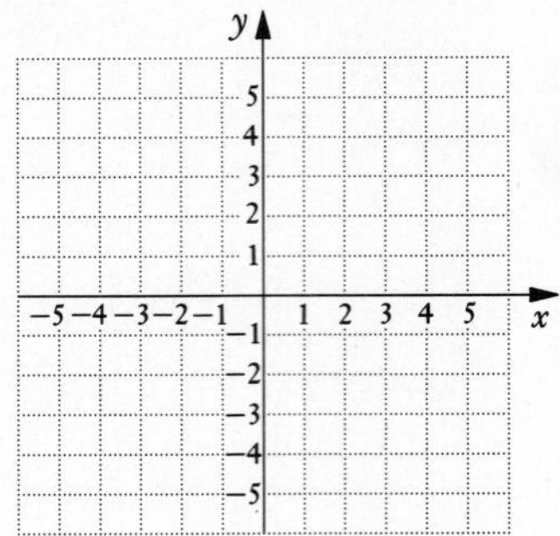

26. $x = 0$

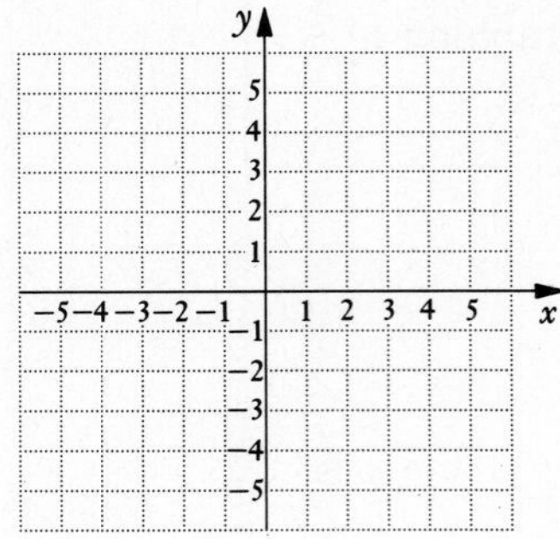

27. $x = \frac{3}{2}$

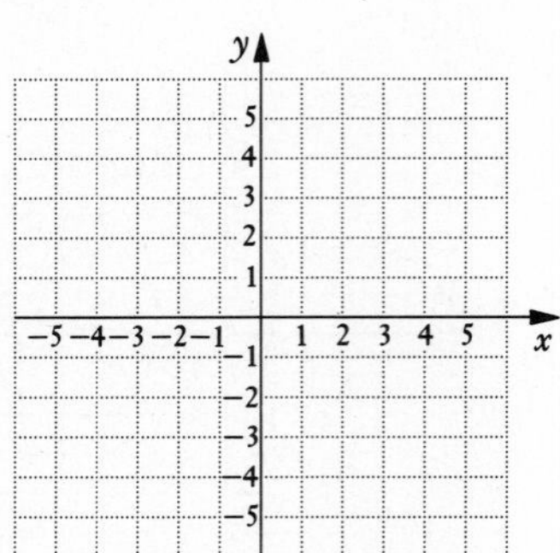

28. $x = -\frac{5}{2}$

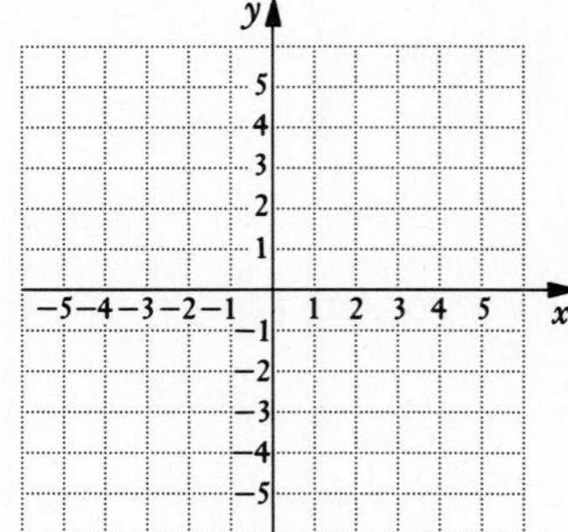

SKILL MAINTENANCE

29. What rate of interest would have to be charged in order for a principal of $320 to earn $17.60 in $\frac{1}{2}$ year?

30. A piece of wire 32.8 ft long is to be cut into two pieces and those pieces are each to be bent to make a square. The length of a side of one square is to be 2.2 ft longer than the length of a side of the other square. How should the wire be cut?

SYNTHESIS

31. Write an equation for the y-axis.

32. Write an equation for the x-axis.

33. Find the coordinates of the point of intersection of the graphs of the equations $x = -4$ and $y = 5$.

34. Write an equation of a line parallel to the x-axis and 6 units below it.

35. Write an equation of a line parallel to the y-axis and 12 units to the right of it.

36. Write an equation of a line parallel to the x-axis and intersecting the y-axis at (0, 3.6).

37. Find the value of m in $y = mx + 3$ so that the x-intercept of its graph will be (4, 0).

38. Find the value of b in $2y = -7x + 3b$ so that the y-intercept of its graph will be $(0, -13)$.

3.3 Graphing Using Slope and y-Intercept

a Slope

Graphs of some linear equations slant upward from left to right. Others slant downward. Some are vertical and some are horizontal. Some slant more steeply than others. We now look for a way to describe these characteristics with numbers.

Consider a line with two points marked P_1 and P_2. As we move from P_1 to P_2, the y-coordinate changes from 1 to 3 and the x-coordinate changes from 2 to 7. The change in y is $3 - 1$, or 2. The change in x is $7 - 2$, or 5.

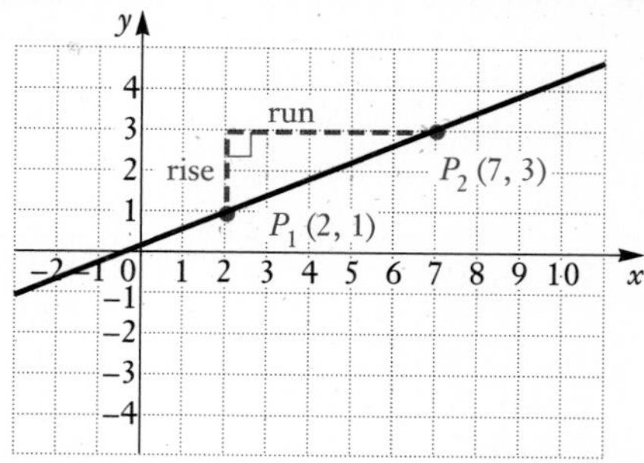

We call the change in y the **rise** and the change in x the **run.** The ratio rise/run is the same for any two points on a line. We call this ratio the **slope.** Slope describes the slant of a line. The slope of the line in the graph above is given by

$$\frac{\text{rise}}{\text{run}}, \quad \text{or} \quad \frac{2}{5}.$$

The *slope* of a line containing points (x_1, y_1) and (x_2, y_2) is given by

$$m = \frac{\text{rise}}{\text{run}}$$

$$= \frac{\text{the change in } y}{\text{the change in } x} = \frac{y_2 - y_1}{x_2 - x_1} = \frac{y_1 - y_2}{x_1 - x_2}.$$

▶ **EXAMPLE 1** Graph the line containing the points $(-4, 3)$ and $(2, -5)$ and find the slope.

The graph is shown below. From $(-4, 3)$ to $(2, -5)$, the change in y, or the rise, is $3 - (-5)$, or 8. The change in x, or the run, is $-4 - 2$, or -6.

$$\text{Slope} = \frac{\text{rise}}{\text{run}} = \frac{\text{change in } y}{\text{change in } x}$$

$$= \frac{3 - (-5)}{-4 - 2}$$

$$= \frac{8}{-6} = -\frac{8}{6}, \quad \text{or} \quad -\frac{4}{3}.$$

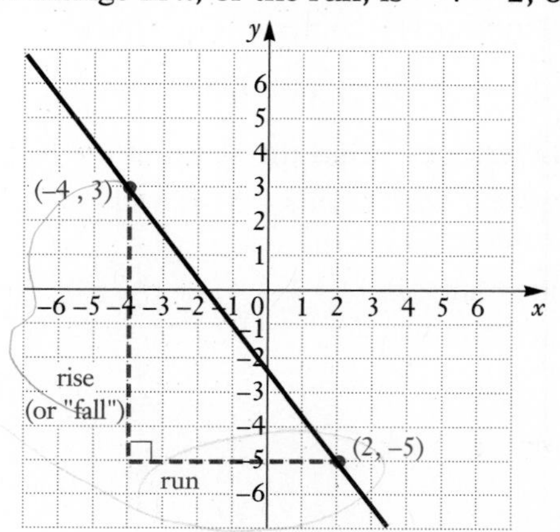

◀

OBJECTIVES

After finishing Section 3.3, you should be able to:

a Given two points of a line, find the slope.

b Find the slope of a line from an equation.

c Given any linear equation, derive the equivalent slope–intercept equation and determine the slope and the y-intercept; and given the slope and the y-intercept of a line, find its slope–intercept equation.

d Given the slope–intercept equation of a line, find the slope and the y-intercept and use them to graph the line.

e Solve application problems involving slope.

FOR EXTRA HELP

Tape 6B | Tape 4B | MAC: 3 IBM: 3

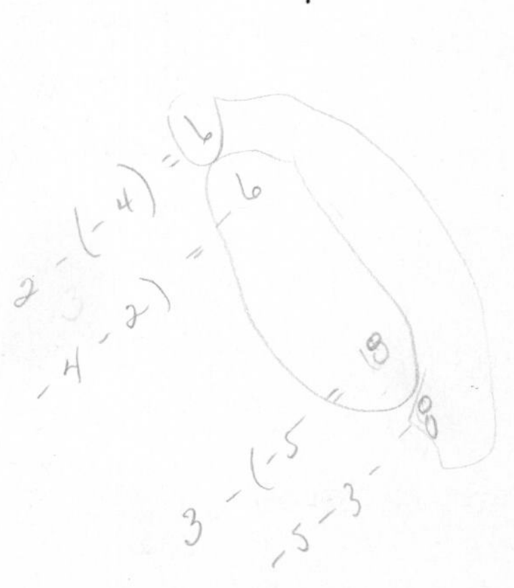

Graph the line through the given points and find its slope.

1. $(-1, -1)$ and $(2, -4)$

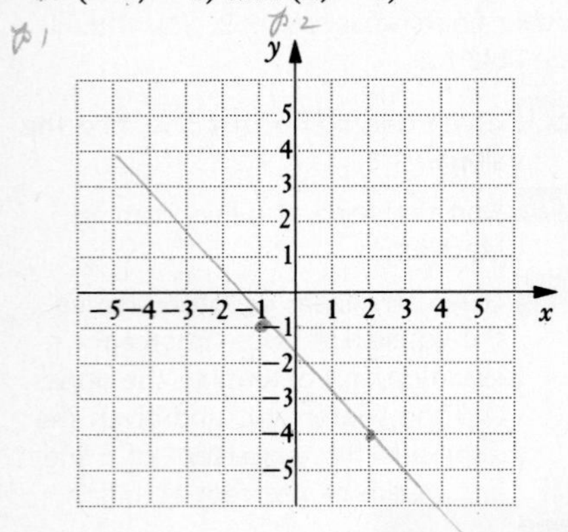

2. $(0, 2)$ and $(3, 1)$

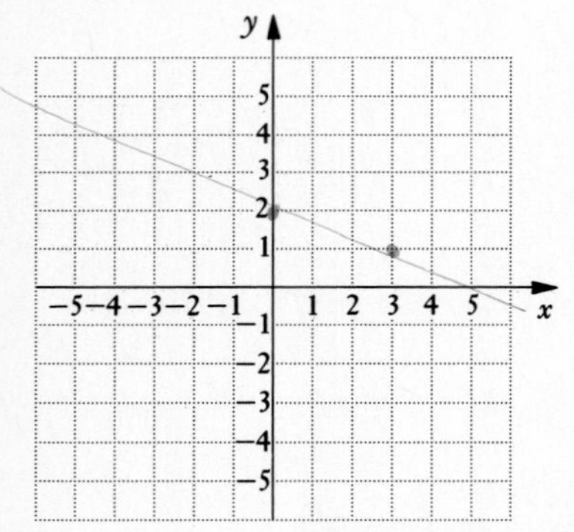

ANSWERS ON PAGE A-4

The formula

$$m = \frac{y_2 - y_1}{x_2 - x_1} = \frac{y_1 - y_2}{x_1 - x_2}$$

tells us that we can subtract in two ways. We must remember, however, to subtract the y-coordinates in the same order that we subtract the x-coordinates. Let's do Example 1 again:

$$\text{Slope} = \frac{\text{change in } y}{\text{change in } x} = \frac{-5 - 3}{2 - (-4)} = \frac{-8}{6} = -\frac{8}{6} = -\frac{4}{3}.$$

We see that both ways give the same slope value.

The slope of a line tells how it slants. A line with positive slope slants up from left to right. The larger the positive slope, the steeper the slant. A line with negative slope slants downward from left to right. The smaller the negative slope, the steeper the line.

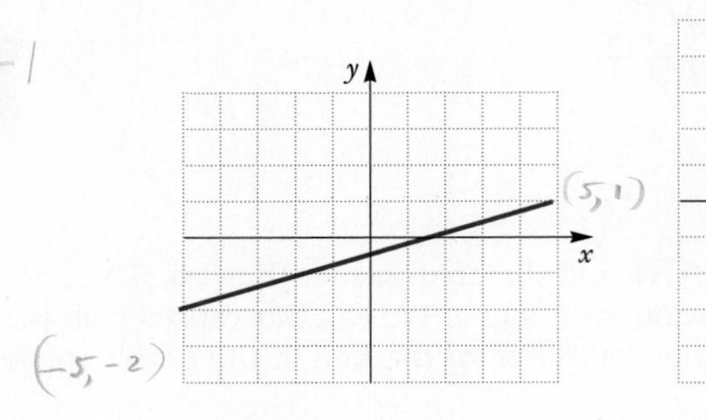

$m = \frac{3}{10}$

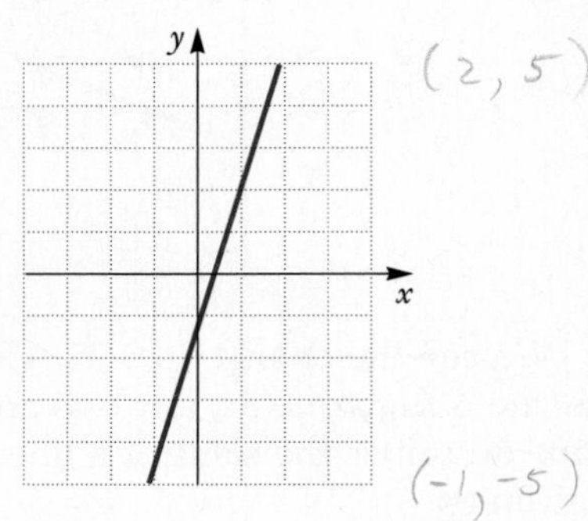

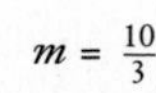

$m = \frac{10}{3}$

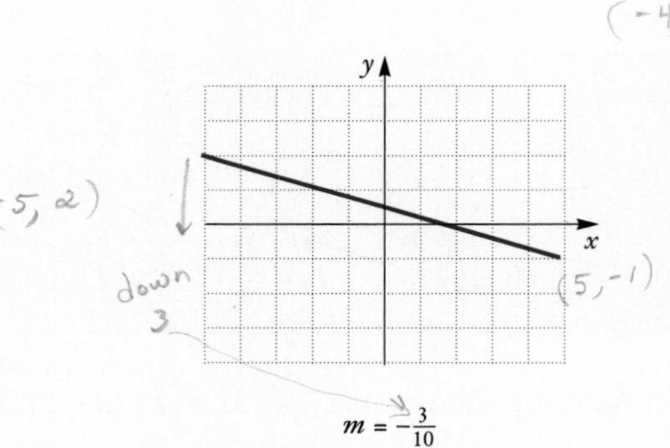

$m = -\frac{3}{10}$

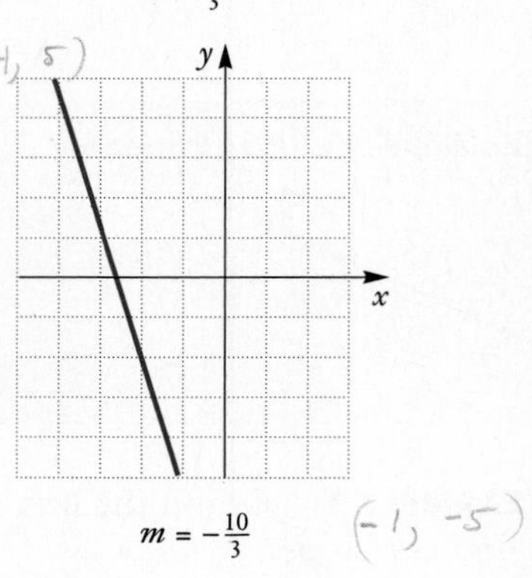

$m = -\frac{10}{3}$

DO EXERCISES 1 AND 2.

b Finding Slopes from Equations

What about the slope of a horizontal or a vertical line?

▶ **EXAMPLE 2** Find the slope of the line $y = 4$.

Consider the points $(-3, 4)$ and $(2, 4)$, which are on the line.

The change in y is $4 - 4$, or 0.

The change in x is $-3 - 2$, or -5.

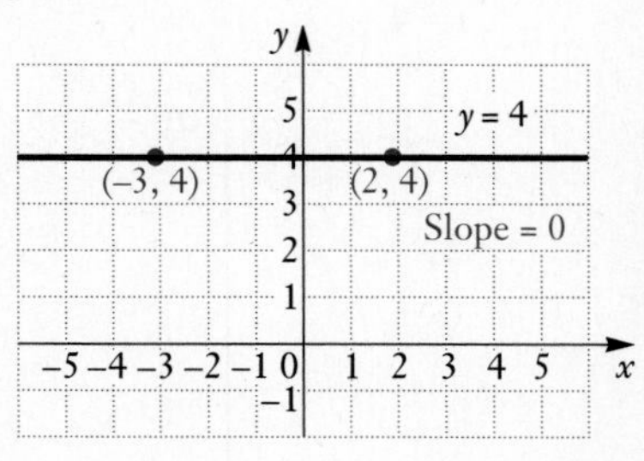

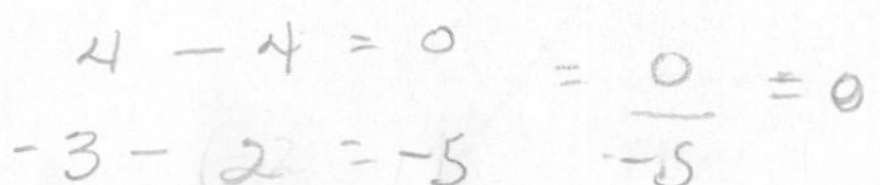

Thus,

$$m = \frac{4 - 4}{-3 - 2} = \frac{0}{-5} = 0.$$

Any two points on a horizontal line have the same y-coordinate. Thus the change in y is always 0, so the slope is 0. ◀

▶ **EXAMPLE 3** Find the slope of the line $x = -3$.

Consider the points $(-3, 3)$ and $(-3, -2)$, which are on the line.

The change in y is $3 - (-2)$, or 5.

The change in x is $-3 - (-3)$, or 0.

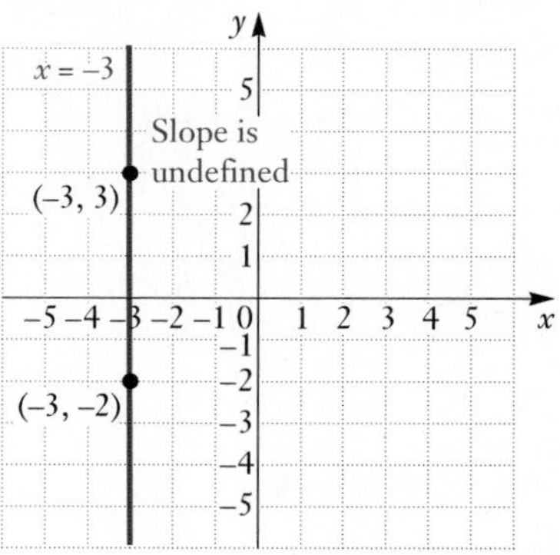

$$m = \frac{3 - (-2)}{-3 - (-3)} = \frac{5}{0} \quad \textbf{Undefined}$$

Since division by 0 is not defined, the slope of this line is not defined. The answer in this example is "The slope of this line is undefined." ◀

A horizontal line has slope 0. The slope of a vertical line is undefined.

DO EXERCISES 3 AND 4.

It is possible to find the slope of a line from its equation. Let us consider the equation

$$y = 2x + 3.$$

We can find two points by choosing convenient values for x, say 0 and 1, and substituting to find the corresponding y-values. We find the two points on the line to be $(0, 3)$ and $(1, 5)$. The slope of the line is found as follows, using the definition of slope:

$$m = \frac{\text{change in } y}{\text{change in } x}$$

$$= \frac{5 - 3}{1 - 0} = \frac{2}{1} = 2.$$

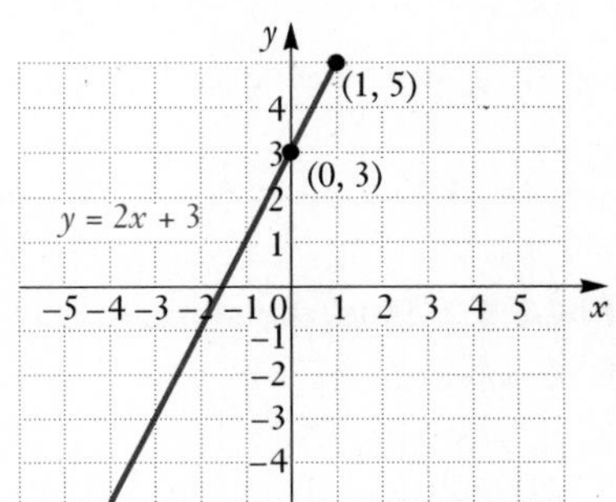

The slope is 2. This is also the coefficient of the x-term in the equation $y = 2x + 3$.

Find the slope, if it exists, of the line.

3. $x = 6$

4. $y = -4.5$

ANSWERS ON PAGE A-4

Find the slope of the line.

5. $y = -2.3x + 7$

6. $6y - 4x = 24$

ANSWERS ON PAGE A-4

> **1. To find the slope of a nonvertical linear equation in x and y, solve the equation for y and get the resulting equation in the form $y = mx + b$. The coefficient of the x-term, m, is the slope of the line.**
>
> **2. The slope of the vertical line $x = a$ is undefined.**

▶ **EXAMPLE 4** Find the slope of the line $2x + 5y = 7$.

We solve for y. We can do so in two ways, first multiplying by $\frac{1}{5}$ or first dividing by 5:

$$\begin{aligned} 2x + 5y &= 7 \\ 5y &= -2x + 7 \\ y &= \frac{1}{5}(-2x + 7) \\ y &= -\frac{2}{5}x + \frac{7}{5}; \end{aligned} \qquad \begin{aligned} 2x + 5y &= 7 \\ 5y &= -2x + 7 \\ y &= \frac{-2x + 7}{5} \\ y &= -\frac{2}{5}x + \frac{7}{5}. \end{aligned}$$

The slope is $-\frac{2}{5}$. ◀

DO EXERCISES 5 AND 6.

C The Slope–Intercept Equation of a Line

In the equation $y = mx + b$, we know that m is the slope. What is the y-intercept? To find out, we let $x = 0$ and solve for y:

$$\begin{aligned} y &= mx + b \\ y &= m(0) + b \\ y &= b. \end{aligned}$$

Thus the y-intercept is $(0, b)$.

> The Slope–Intercept Equation
>
> **The equation $y = mx + b$ is called the *slope–intercept equation*. The slope is m and the y-intercept is $(0, b)$.**

▶ **EXAMPLE 5** Find the slope and the y-intercept of $y = 5x - 4$.

Since the equation is already in the form $y = mx + b$, we simply read the slope and the y-intercept from the equation:

$$y = 5x - 4$$

The slope is 5. The y-intercept is $(0, -4)$. ◀

▶ **EXAMPLE 6** Find the slope and the y-intercept of $2x + 3y = 8$.

We first solve for y:

$$\begin{aligned} 2x + 3y &= 8 \\ 3y &= -2x + 8 && \text{This equation is not yet solved for } y. \\ y &= -\tfrac{2}{3}x + \tfrac{8}{3}. \end{aligned}$$

The slope is $-\frac{2}{3}$, and the y-intercept is $(0, \frac{8}{3})$. ◀

► **EXAMPLE 7** A line has slope -0.7 and y-intercept $(0, 13)$. Find an equation of the line.

We use the slope–intercept equation and substitute -0.7 for m and 13 for b:

$$y = mx + b$$
$$y = -0.7x + 13.$$ ◄

DO EXERCISES 7–9.

Find the slope and the y-intercept.

7. $y = -8x + 23$

8. $5x - 10y = 25$

9. A line has slope 3.4 and y-intercept $(0, -8)$. Find an equation of the line.

d Graphing Using the Slope and the y-Intercept

We can graph lines using the slope–intercept equation.

► **EXAMPLE 8** Graph $y = -\frac{2}{3}x + 1$.

First we plot the y-intercept, $(0, 1)$. We can think of the slope as $\frac{-2}{3}$. Starting at the y-intercept and using the slope, we find another point by moving 2 units down (since the numerator is *negative* and corresponds to the change in y) and 3 units to the right (since the denominator is *positive* and corresponds to the change in x). We get to a new point, $(3, -1)$. In a similar manner, we can move from the point $(3, -1)$ to find another point, $(6, -3)$.

Suppose we think of the slope $-\frac{2}{3}$ as $\frac{2}{-3}$. Then we can start again at $(0, 1)$, but this time we move 2 units up (since the numerator is *positive* and corresponds to the change in y) and 3 units to the left (since the denominator is *negative* and corresponds to the change in x). We get another point on the graph, $(-3, 3)$, and from it we can obtain $(-6, 5)$ and others in a similar manner. We plot the points and draw the line.

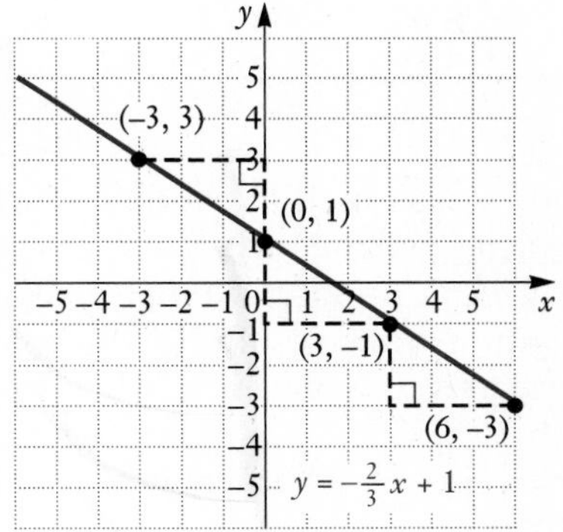

◄

Graph.

10. $y = \frac{3}{4}x - 2$

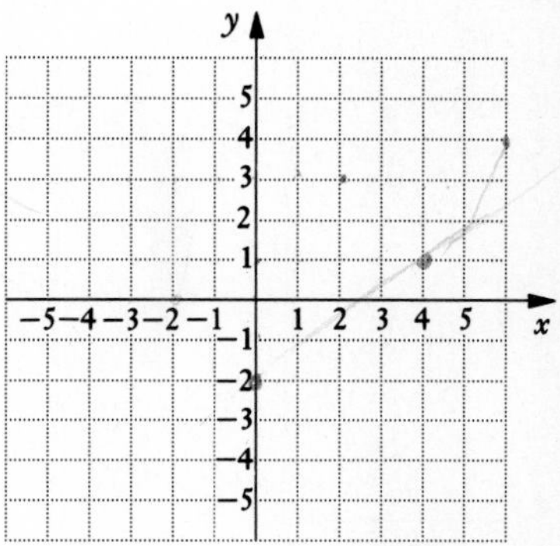

11. $y = \frac{3}{2}x + 1$

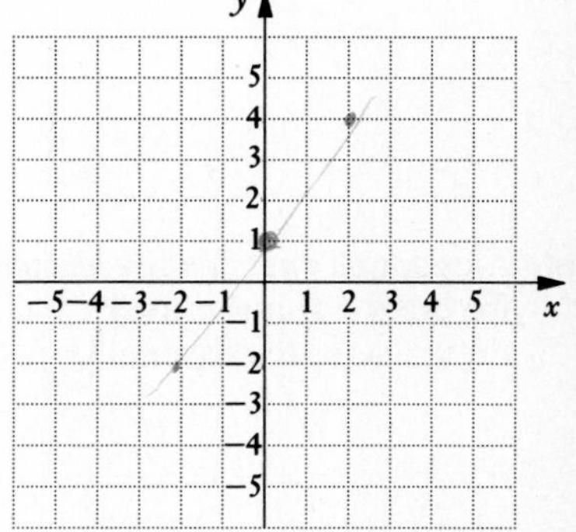

► **EXAMPLE 9** Graph $y = \frac{2}{5}x + 4$.

First we plot the y-intercept, $(0, 4)$. We then consider the slope $\frac{2}{5}$. Starting at the y-intercept and using the slope, we find another point by moving 2 units up (since the numerator is *positive* and corresponds to the change in y) and 5 units to the right (since the denominator is *positive* and corresponds to the change in x). We get to a new point, $(5, 6)$.

ANSWERS ON PAGE A-4

Graph.

12. $y = -\frac{3}{5}x + 5$

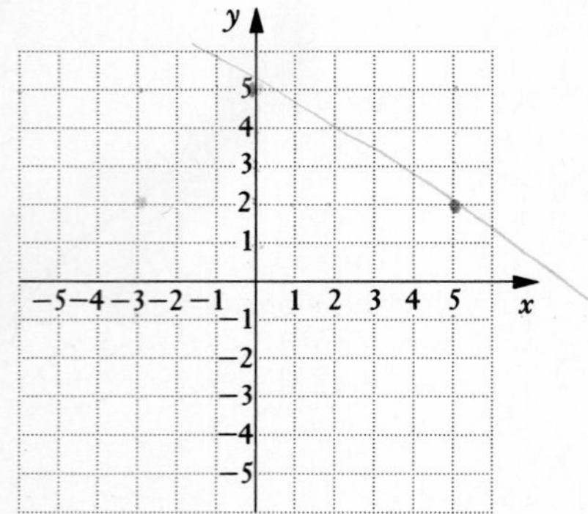

13. $y = -\frac{5}{3}x - 4$

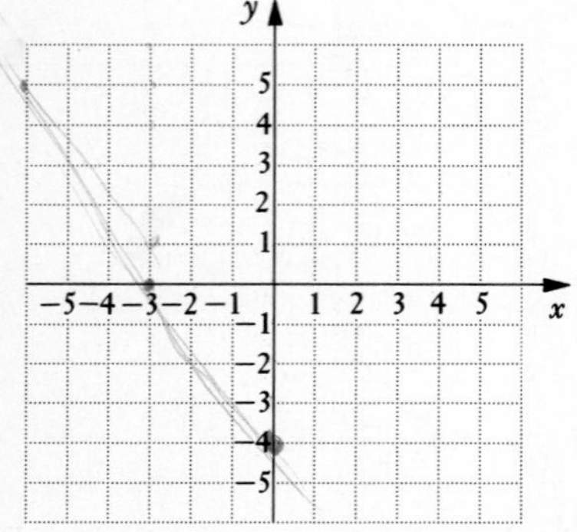

14. A treadmill rises 0.6 ft vertically for every 5 ft horizontally. Find the grade of the treadmill.

ANSWERS ON PAGE A-4

We can also think of the slope $\frac{2}{5}$ as $\frac{-2}{-5}$. We again start at the y-intercept, (0, 4). We move down 2 units (since the numerator is *negative* and corresponds to the change in y) and 5 units to the left (since the denominator is *negative* and corresponds to the change in x). We get to another new point, $(-5, 2)$. We plot the points and draw the line.

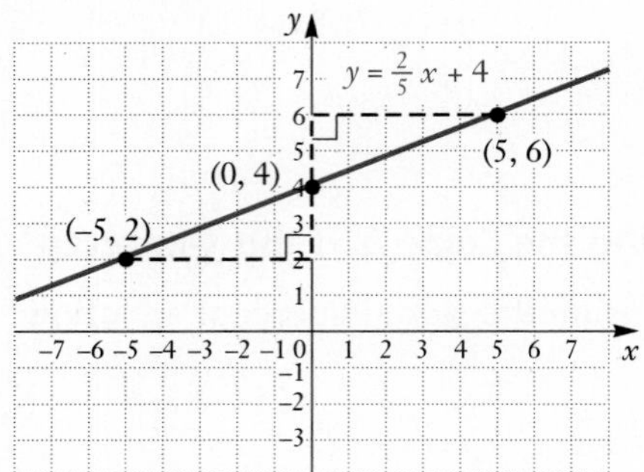

◀

DO EXERCISES 10–13. (EXERCISES 10 AND 11 ARE ON THE PRECEDING PAGE.)

e Applications

Slope has many real-world applications. For example, numbers like 2%, 3%, and 6% are often used to represent the **grade** of a road. Such a number is meant to tell how steep a road is on a hill or mountain. For example, a 3% grade means that for every horizontal distance of 100 ft, the road rises 3 ft.

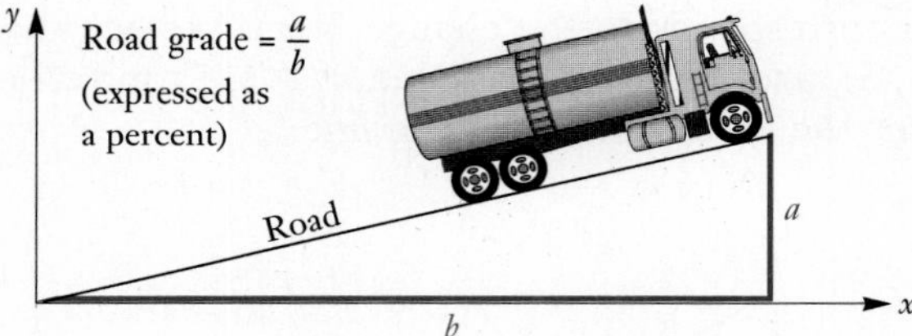

The concept of grade is also used in cardiology when a person runs on a treadmill. A physician may change the slope or grade of a treadmill to measure its effect on heartbeat. Another example occurs in hydrology. When a river flows, the strength or force of the river depends on how much the river falls vertically compared to how much it flows horizontally.

▶ **EXAMPLE 10** A road rises 60 ft vertically for every 1250 ft horizontally. Find the grade of the road.

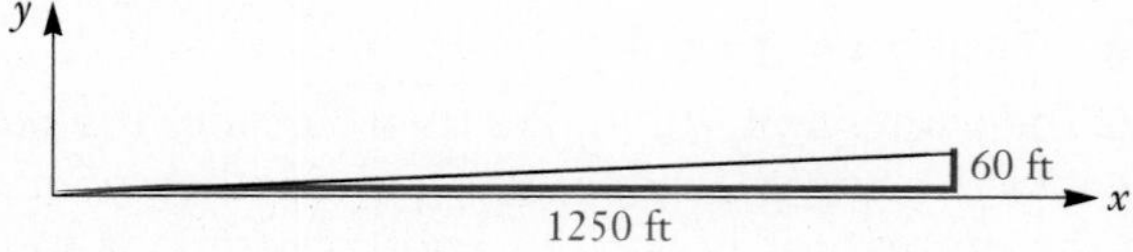

The grade of the road is the slope of the line and is given by

$$m = \frac{60}{1250} = 0.048 = 4.8\%.$$ ◀

DO EXERCISE 14.

NAME SECTION DATE

EXERCISE SET 3.3

a Find the slope, if it exists, of the line containing the pair of points.

1. $(3, 8)$ and $(9, -4)$

2. $(17, -12)$ and $(-9, -15)$

3. $(-8, -7)$ and $(-9, -12)$

4. $(14, 3)$ and $(2, 12)$

5. $(-16.3, 12.4)$ and $(-5.2, 8.7)$

6. $(14.4, -7.8)$ and $(-12.5, -17.6)$

7. $(3.2, -12.8)$ and $(3.2, 2.4)$

8. $(-1.5, 7.6)$ and $(-1.5, 8.8)$

b Find the slope, if it exists, of the line.

9. $3x = 12 + y$

10. $5y - 12 = 3x$

11. $5x - 6 = 15$

12. $-12 = 4x - 7$

13. $5y = 6$

14. $19 = -6y$

15. $y - 6 = 14$

16. $3y - 5 = 8$

17. $12 - 4x = 9 + x$

18. $15 + 7x = 3x - 5$

19. $2y - 4 = 35 + x$

20. $2x - 17 + y = 0$

21. $3y + x = 3y + 2$

22. $x - 4y = 12 - 4y$

23. $3y - 2x = 5 + 9y - 2x$

24. $17y + 4x + 3 = 7 + 4x$

25. $2y - 7x = 10 - 3x$

26. $4 - 5y + 7x = -10$

ANSWERS

1. ______
2. ______
3. ______
4. ______
5. ______
6. ______
7. ______
8. ______
9. ______
10. ______
11. ______
12. ______
13. ______
14. ______
15. ______
16. ______
17. ______
18. ______
19. ______
20. ______
21. ______
22. ______
23. ______
24. ______
25. ______
26. ______

ANSWERS

27. ____
28. ____
29. ____
30. ____
31. ____
32. ____
33. ____
34. ____
35. ____
36. ____
37. ____
38. ____
39. ____
40. ____
41. ____
42. ____
43. ____
44. ____

C Find the slope and the y-intercept of the line.

27. $y = -8x - 9$

28. $y = -3x - 6$

29. $y = 3.8x$

30. $y = -29.4x$

31. $2x + 3y = 8$

32. $5x + 4y = 16$

33. $-8x - 7y = 24$

34. $-2x - 9y = 18$

35. $9x = 3y + 6$

36. $4x = 9y + 36$

37. $-6x = 4y + 3$

38. $y = -19$

Find an equation of the line that has the given characteristics.

39. Slope 5 and y-intercept $(0, 8)$

40. Slope -2 and y-intercept $(0, -4)$

41. Slope 5.8 and y-intercept $(0, -1)$

42. Slope -3.8 and y-intercept $(0, 6)$

43. Slope $-\frac{7}{3}$ and y-intercept $(0, -5)$

44. Slope $\frac{4}{5}$ and y-intercept $(0, 28)$

d Graph the equation using the slope and the y-intercept.

45. $y = \frac{5}{2}x + 1$

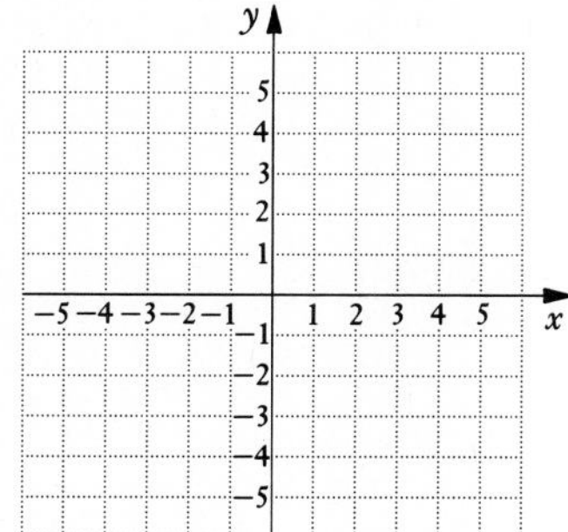

46. $y = \frac{2}{5}x - 4$

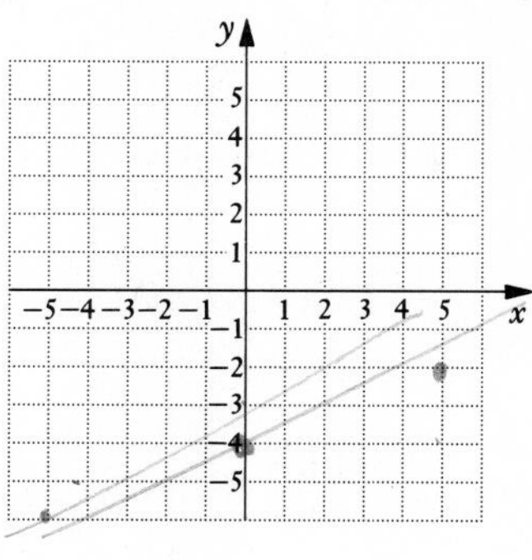

47. $y = -\frac{5}{2}x - 4$

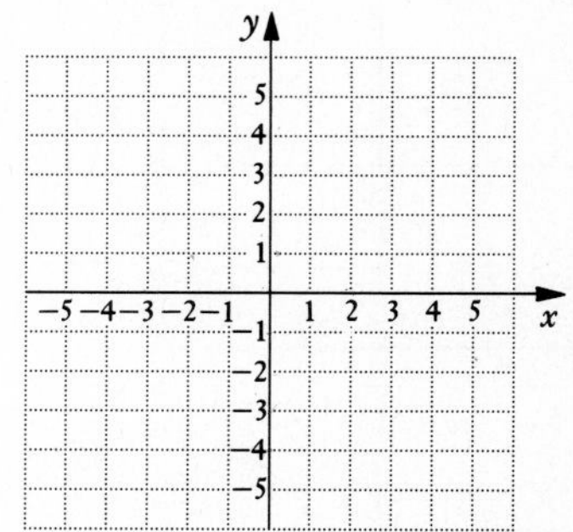

48. $y = \frac{2}{5}x + 3$

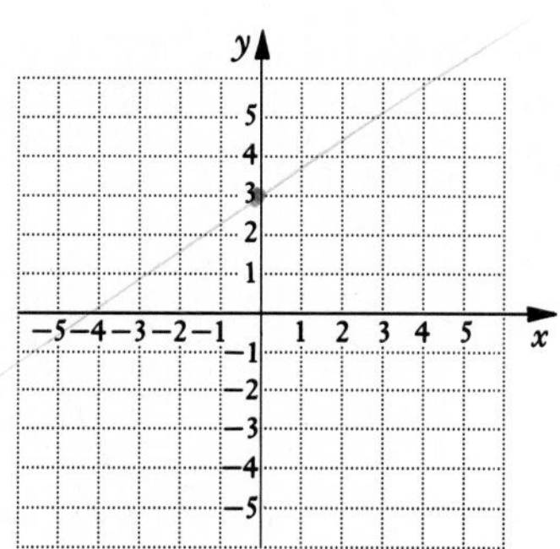

49. $y = 2x - 5$

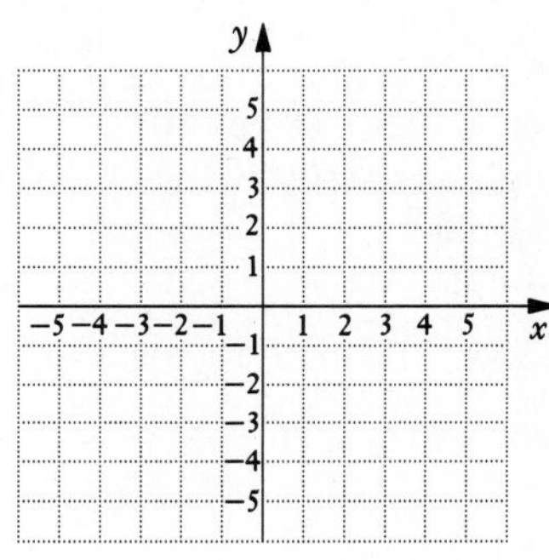

50. $y = -2x + 4$

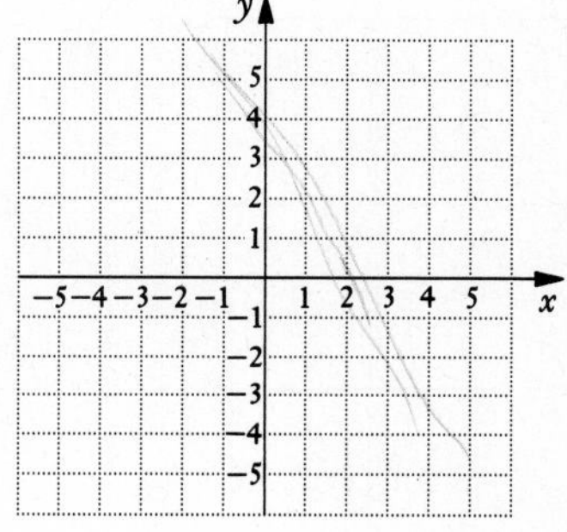

51. $y = \frac{1}{3}x + 6$

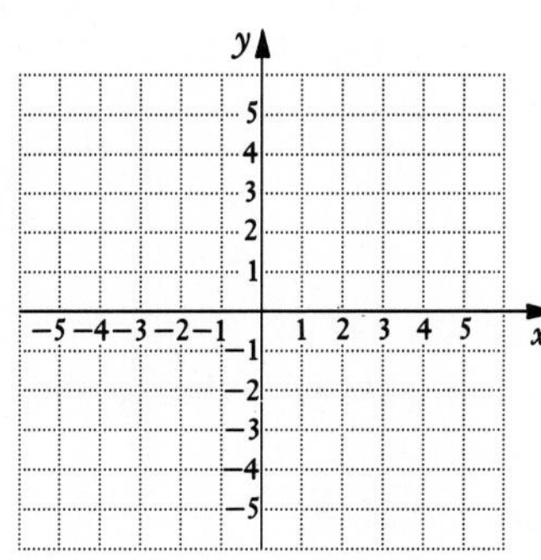

52. $y = -3x + 6$

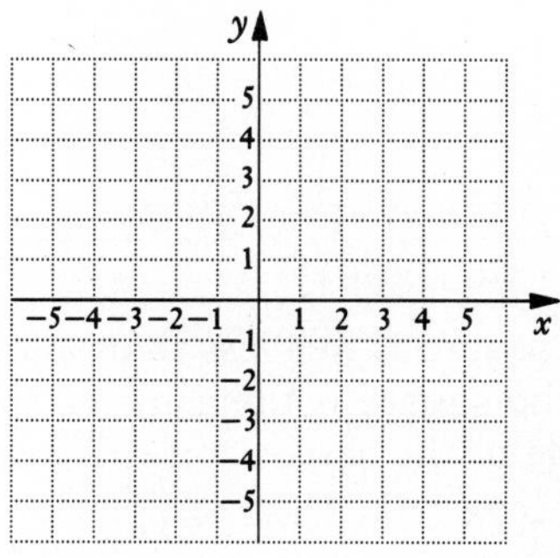

53. $y = -0.25x + 2$

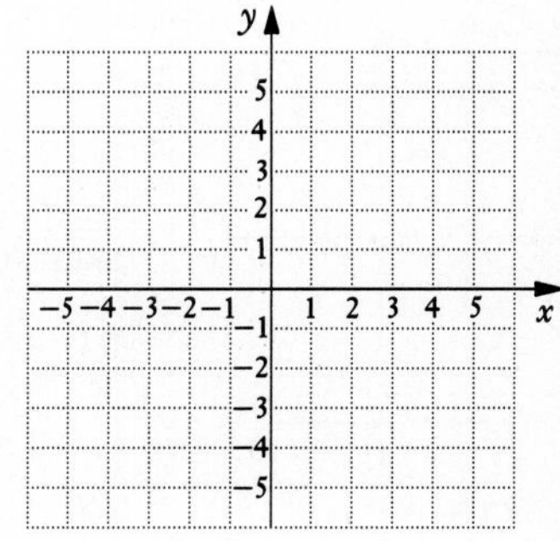

54. $y = 1.5x - 3$

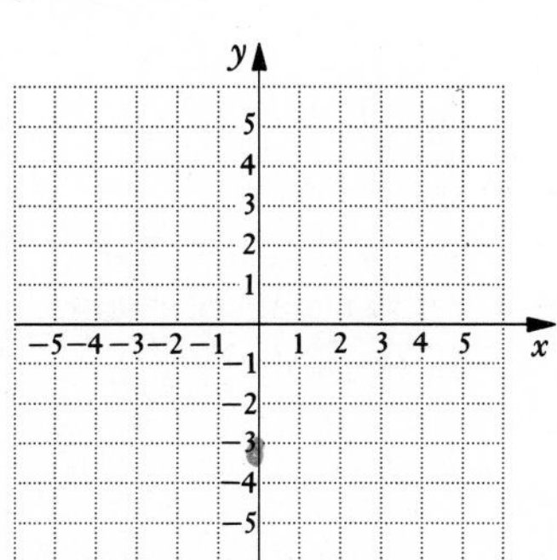

55. $y = -\frac{3}{4}x$

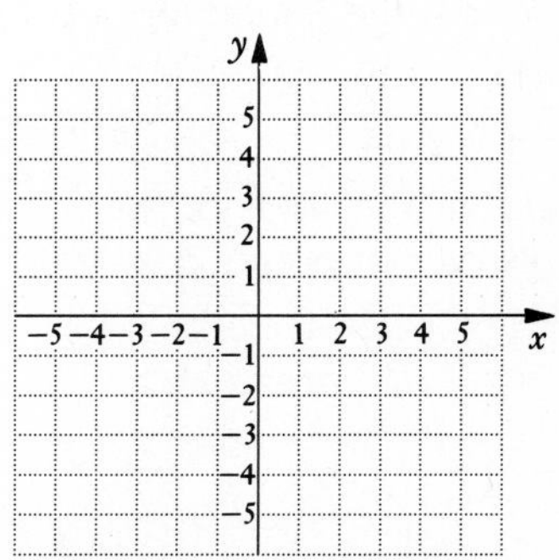

56. $y = \frac{4}{5}x$

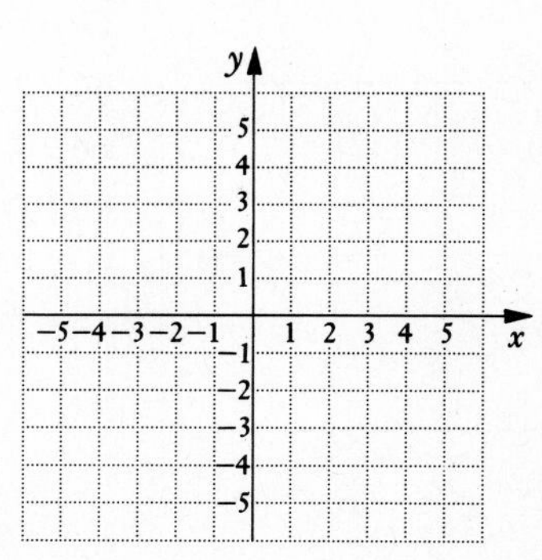

ANSWERS

57. ____
58. ____
59. ____
60. ____
61. ____
62. ____
63. ____
64. ____
65. ____
66. ____
67. ____
68. ____
69. a) ____
b) ____
c) ____

e Solve.

57. A road drops 211.2 ft vertically for every 5280 ft horizontally. What is the grade of the road?

58. A river drops 43.33 ft vertically for every 1238 ft horizontally. What is the slope of the river?

59. A treadmill is 6 ft long and is set at a 13% grade when a heart arrhythmia occurs. How high vertically is the end of the treadmill?

60. A river flows at a slope of 0.08. How many feet does it fall vertically for every 250 ft horizontally?

SKILL MAINTENANCE

61. One side of a square is five less than a side of an equilateral triangle. If the perimeter of the square is the same as the perimeter of the triangle, what is the length of a side of the square? of the triangle?

62. Calculate: $3^2 - 24 \cdot 56 + 144 \div 12$.

63. Solve: $|5x - 8| \geq 32$.

SYNTHESIS

Determine the slope and the y-intercept of the equation.

64. $2x + 5y + 2 = 5x - 10y - 8$

65. $\frac{1}{8}y = -x - \frac{7}{16}$

66. $0.4y - 0.004x = -0.04$

67. $x = -\frac{7}{3}y - \frac{2}{11}$

68. Determine a so that the slope of the line through this pair of points has the given value.

$$(-2, 3a), (4, -a); \quad m = -\tfrac{5}{12}$$

69. Find the slope of the line that contains the given pair of points.

a) $(5b, -6c), (b, -c)$
b) $(b, d), (b, d + e)$
c) $(c + f, a + d), (c - f, -a - d)$

3.4 Other Equations of Lines

a Point–Slope Equations

If we know one point on a line as well as the slope, we can then find an equation of the line. Suppose the known point is (x_1, y_1) and the slope is m. Then consider any other point (x, y) on the line.

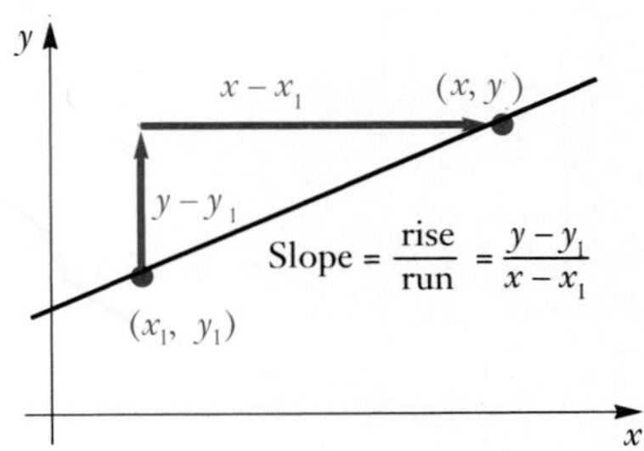

From this drawing, we see that the slope of the line is $\dfrac{y - y_1}{x - x_1}$.

The slope is also m. This gives us the equation

$$\frac{y - y_1}{x - x_1} = m. \quad (1)$$

Multiplying by $x - x_1$ gives us the equation we are looking for.

> **The *point–slope equation* of a line is**
> $$y - y_1 = m(x - x_1).$$

> Equation (1) is easy to remember. You can easily get the point–slope equation from it by multiplying each side by $x - x_1$.

► **EXAMPLE 1** Find an equation of the line containing the point $(\frac{1}{2}, -1)$ with slope 5.

We substitute in the equation $y - y_1 = m(x - x_1)$:

$$y - (-1) = 5(x - \tfrac{1}{2}) \quad \textbf{Substituting}$$
$$y + 1 = 5x - \tfrac{5}{2} \quad \textbf{Simplifying}$$
$$y = 5x - \tfrac{7}{2}.$$

> CAUTION! A common error here is forgetting to multiply *both* terms by 5.

◄

► **EXAMPLE 2** Find an equation of the line with x-intercept $(4, 0)$ and slope 2.

$$y - 0 = 2(x - 4) \quad \textbf{Substituting 2 for } m\textbf{, 4 for } x_1\textbf{, and 0 for } y_1$$
$$y = 2x - 8 \quad \textbf{Simplifying}$$

◄

DO EXERCISES 1 AND 2.

Vertical lines do not have point–slope equations, because their slopes are undefined. All other lines do.

OBJECTIVES

After finishing Section 3.4, you should be able to:

a Given the slope of a line and a point on the line, find an equation of the line.

b Given two points on a line, find an equation of the line.

c Given the equations of two lines, determine whether their graphs are parallel or whether they are perpendicular; and given a line and a point not on the given line, find an equation of the line parallel to the line and containing the point and an equation of the line perpendicular to the line and containing the point.

FOR EXTRA HELP

Tape 6C	Tape 4B	MAC: 3 IBM: 3

1. Find an equation of the line containing the point $(1, -2)$, with slope 3.

2. Find an equation of the line containing the point $(-5, \frac{3}{4})$, with slope $-\frac{4}{3}$.

ANSWERS ON PAGE A-5

3. Find an equation of the line containing the points (4, −3) and (1, 2).

b Finding an Equation Given Two Points

We can use the point–slope equation to find an equation of a line containing two given points.

▶ **EXAMPLE 3** Find an equation of the line containing the points (2, 3) and (1, −4).

We first find the slope of the line using the definition of slope:

$$m = \frac{y_2 - y_1}{x_2 - x_1} = \frac{-4 - 3}{1 - 2} = \frac{-7}{-1} = 7.$$

Since the line passes through (2, 3) and we know the slope to be 7, we can substitute into the point–slope equation:

$$y - 3 = 7(x - 2)$$
$$y - 3 = 7x - 14$$
$$y = 7x - 11.$$

We can also choose to use the point (1, −4). Then substituting, we have $y - (-4) = 7(x - 1)$, which also simplifies to $y = 7x - 11$. ◀

DO EXERCISES 3 AND 4.

4. Find an equation of the line containing the points (−3, −5) and (−4, 12).

c Parallel and Perpendicular Lines

If two lines are vertical, they are parallel. How can we tell whether non-vertical lines are parallel? The answer is simple: We look at their slopes.

Two nonvertical lines are parallel if they have the same slope and different *y*-intercepts.

▶ **EXAMPLE 4** Determine whether the graphs of $y - 3x = 1$ and $-2y = 3x + 2$ are parallel.

We first find the slope–intercept equation of each line. We solve each equation for y:

$$y - 3x = 1 \qquad\qquad -2y = 3x + 2$$
$$y = 3x + 1; \qquad\qquad y = -\tfrac{1}{2}(3x + 2)$$
$$y = (-\tfrac{1}{2})(3x) + (-\tfrac{1}{2})(2)$$
$$y = -\tfrac{3}{2}x - 1.$$

The slopes are 3 and $-\frac{3}{2}$ and are different. Thus the lines are not parallel.

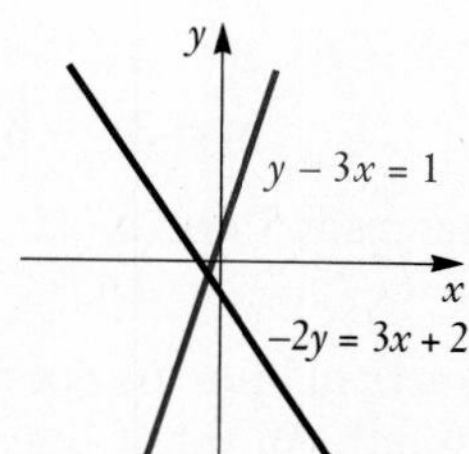

◀

ANSWERS ON PAGE A-5

▶ **EXAMPLE 5** Determine whether the graphs of $3x - y = -5$ and $y - 3x = -2$ are parallel.

We first find the slope–intercept equation of each line. We solve each equation for y:

$$\begin{aligned} 3x - y &= -5 \\ -y &= -3x - 5 \\ y &= 3x + 5; \end{aligned} \qquad \begin{aligned} y - 3x &= -2 \\ y &= 3x - 2. \end{aligned}$$

The slopes are both 3 and are the same. The y-intercepts are different. The lines are parallel.

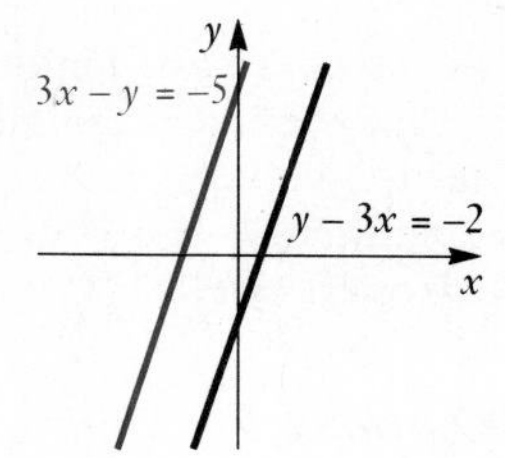

◀

DO EXERCISES 5–7.

Determine whether the graphs of each pair of equations are parallel.

5. $x + 4 = y$,
   $y - x = -3$

6. $y + 4 = 3x$,
   $4x - y = -7$

7. $y = 4x + 5$,
   $2y = 8x + 10$

▶ **EXAMPLE 6** Find an equation of the line containing the point $(-1, 3)$ and parallel to the line $2x + y = 10$.

We first find the slope–intercept equation of the given line by solving for y: $y = -2x + 10$. The line through $(-1, 3)$ must have slope -2. We find an equation of this new line using the point–slope equation:

$$\begin{aligned} y - 3 &= -2[x - (-1)] \\ y - 3 &= -2(x + 1) \\ y - 3 &= -2x - 2 \\ y &= -2x + 1. \end{aligned}$$

The equations $y = -2x + 10$ and $y = -2x + 1$ have the same slope but different y-intercepts. Thus their graphs are parallel.

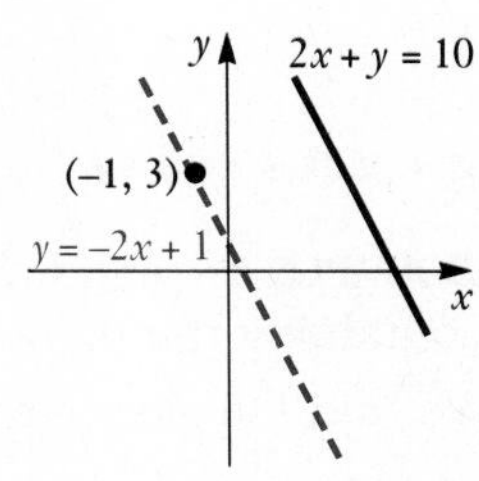

◀

DO EXERCISE 8.

8. Find an equation of the line parallel to the line $8x = 7y - 24$ and containing the point $(2, -1)$.

If one line is vertical and another is horizontal, they are perpendicular. Otherwise, how can we tell whether two lines are perpendicular?

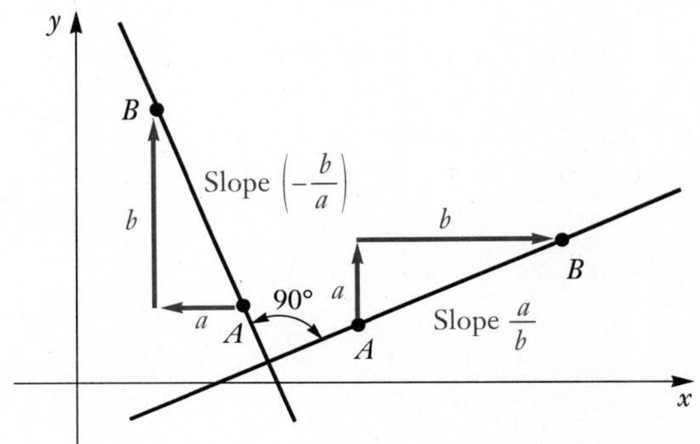

ANSWERS ON PAGE A-5

Determine whether the graphs of each pair of equations are perpendicular.

9. $2y - x = 2,$
$y + 2x = 4$

10. $3y = 2x + 15,$
$2y = 3x + 10$

11. Find an equation of the line perpendicular to the line $2x - 4y = 12$ and containing the point (5, 4).

ANSWERS ON PAGE A-5

Consider a line $\overleftrightarrow{AB}$, as shown above, with slope a/b. Then think of rotating the figure 90° to get a line perpendicular to $\overleftrightarrow{AB}$. For the new line, the rise and run are interchanged, but the run is now negative. Thus the slope of the new line is $-b/a$. Let us multiply the slopes:

$$\frac{a}{b}\left(-\frac{b}{a}\right) = -1.$$

This is the condition under which lines will be perpendicular.

> **Two lines are perpendicular if the product of their slopes is -1. (If one line has slope m, the slope of a line perpendicular to it is $-1/m$. That is, we take the reciprocal and change the sign.) Lines are also perpendicular if one of them is vertical ($x = a$) and one of them is horizontal ($y = b$).**

► **EXAMPLE 7** Determine whether the graphs of $5y = 4x + 10$ and $4y = -5x + 4$ are perpendicular.

We first find the slope–intercept equation of each line.

$$\begin{aligned} 5y &= 4x + 10 \\ y &= \tfrac{1}{5}(4x + 10) \\ y &= \tfrac{1}{5}(4x) + \tfrac{1}{5}(10) \\ y &= \tfrac{4}{5}x + 2. \end{aligned} \qquad \begin{aligned} 4y &= -5x + 4 \\ y &= \tfrac{1}{4}(-5x + 4) \\ y &= \tfrac{1}{4}(-5x) + \tfrac{1}{4}(4) \\ y &= -\tfrac{5}{4}x + 1. \end{aligned}$$

The product of the slopes is -1; that is, $\frac{4}{5} \cdot (-\frac{5}{4}) = -1$. Thus the lines are perpendicular.

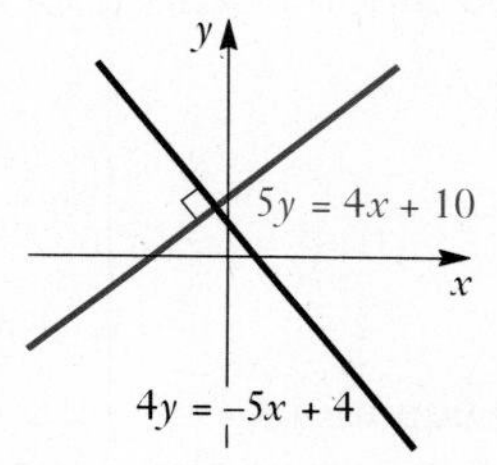

◄

DO EXERCISES 9 AND 10.

► **EXAMPLE 8** Find an equation of the line perpendicular to $4y - x = 20$ and containing the point $(2, -3)$.

We find the slope–intercept equation for the line:

$$\begin{aligned} 4y - x &= 20 \\ 4y &= x + 20 \\ y &= \tfrac{1}{4}(x + 20) \\ y &= \tfrac{1}{4}x + \tfrac{1}{4}(20) \\ y &= \tfrac{1}{4}x + 5. \end{aligned}$$

We know that the slope of the perpendicular line must be -4 because $\frac{1}{4}(-4) = -1$. Using the point–slope equation, we find an equation of this new line having slope -4 and containing the point $(2, -3)$:

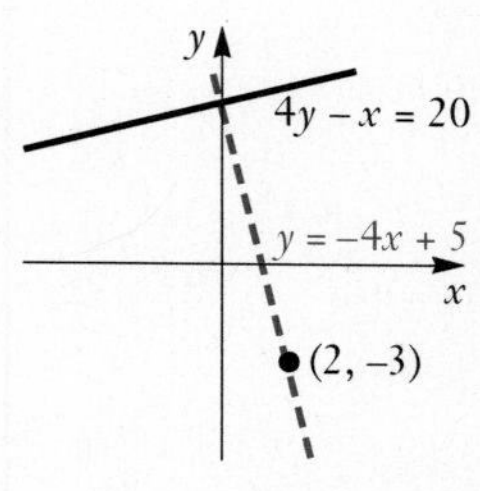

$$\begin{aligned} y - (-3) &= -4(x - 2) \\ y + 3 &= -4x + 8 \\ y &= -4x + 5. \end{aligned}$$

◄

DO EXERCISE 11.

NAME SECTION DATE

EXERCISE SET 3.4

a Find an equation of the line having the given slope and containing the given point. Express the answer in the form $x = a$, $y = b$, or $y = mx + b$.

1. $m = 4$, $(3, 2)$

2. $m = 5$, $(5, 4)$

3. $m = -2$, $(4, 7)$

4. $m = -3$, $(7, 3)$

5. $m = 3$, $(-2, -4)$

6. $m = 1$, $(-5, -7)$

7. $m = -2$, $(8, 0)$

8. $m = -3$, $(-2, 0)$

9. $m = 0$, $(0, -7)$

10. $m = 0$, $(0, 4)$

11. $m = \frac{2}{3}$, $(1, -2)$

12. $m = -\frac{4}{5}$, $(2, 3)$

b Find an equation of the line containing the given pair of points. Express the answer in the form $x = a$, $y = b$, or $y = mx + b$.

13. $(1, 4)$ and $(5, 6)$

14. $(2, 5)$ and $(4, 7)$

15. $(-1, -1)$ and $(2, 2)$

16. $(-5, -5)$ and $(7, 7)$

17. $(-2, 0)$ and $(0, 5)$

18. $(0, -3)$ and $(7, 0)$

19. $(-2, -3)$ and $(-4, -6)$

20. $(-4, -7)$ and $(-2, -1)$

21. $(0, 0)$ and $(5, 2)$

22. $(0, 0)$ and $(-3, 4)$

23. $(\frac{1}{4}, -\frac{1}{2})$ and $(\frac{3}{4}, 6)$

24. $(\frac{2}{3}, \frac{3}{2})$ and $(-3, \frac{5}{6})$

c Determine whether the graphs of each pair of lines are parallel.

25. $x + 6 = y$,
$y - x = -2$

26. $2x - 7 = y$,
$y - 2x = 8$

27. $y + 3 = 5x$,
$3x - y = -2$

28. $y + 8 = -6x$,
$-2x + y = 5$

29. $y = 3x + 9$,
$2y = 6x - 2$

30. $y + 7x = -9$,
$-3y = 21x + 7$

ANSWERS

1. ______
2. ______
3. ______
4. ______
5. ______
6. ______
7. ______
8. ______
9. ______
10. ______
11. ______
12. ______
13. ______
14. ______
15. ______
16. ______
17. ______
18. ______
19. ______
20. ______
21. ______
22. ______
23. ______
24. ______
25. ______
26. ______
27. ______
28. ______
29. ______
30. ______

ANSWERS

31. ______
32. ______
33. ______
34. ______
35. ______
36. ______
37. ______
38. ______
39. ______
40. ______
41. ______
42. ______
43. ______
44. ______
45. ______
46. ______
47. ______
48. ______
49. ______
50. ______
51. ______
52. ______
53. ______
54. ______
55. ______
56. ______
57. ______

Determine whether the graphs of each pair of equations are perpendicular.

31. $y = 4x - 5,$ $4y = 8 - x$

32. $2x - 5y = -3,$ $2x + 5y = 4$

33. $x + 2y = 5,$ $2x + 4y = 8$

34. $y = -x + 7,$ $y = x + 3$

35. $2x - 3y = 7,$ $2y - 3x = 10$

36. $x = y,$ $y = -x$

Write an equation of the line containing the given point and parallel to the given line.

37. $(3, 7)$; $x + 2y = 6$

38. $(0, 3)$; $2x - y = 7$

39. $(2, -1)$; $5x - 7y = 8$

40. $(-4, -5)$; $2x + y = -3$

41. $(-6, 2)$; $3x = 9y + 2$

42. $(-7, 0)$; $2y + 5x = 6$

Write an equation of the line containing the given point and perpendicular to the given line.

43. $(2, 5)$; $2x + y = 3$

44. $(4, 1)$; $x - 3y = 9$

45. $(3, -2)$; $3x + 4y = 5$

46. $(-3, -5)$; $5x - 2y = 4$

47. $(0, 9)$; $2x + 5y = 7$

48. $(-3, -4)$; $6y - 3x = 2$

SKILL MAINTENANCE

Solve.

49. $2x + 3 \leq 5x - 4$

50. $|2x + 3| \leq 13$

51. $|2x + 3| = |x - 4|$

SYNTHESIS

52. Find an equation of the line containing $(4, -2)$ and parallel to the line containing $(-1, 4)$ and $(2, -3)$.

53. Find an equation of the line containing $(-1, 3)$ and perpendicular to the line containing $(3, -5)$ and $(-2, 7)$.

54. Find the value of a so that the graphs of $5y = ax + 5$ and $\frac{1}{4}y = \frac{1}{10}x - 1$ are parallel.

55. Find the value of k so that the graphs of $x + 7y = 70$ and $y + 3 = kx$ are perpendicular.

56. Write an equation of the line that has x-intercept $(-3, 0)$ and y-intercept $(0, \frac{2}{5})$.

57. *Two-point equation.* Prove that the equation of a nonvertical line containing the points (x_1, y_1) and (x_2, y_2) is

$$y - y_1 = \frac{y_2 - y_1}{x_2 - x_1}(x - x_1).$$

3.5 Applications of Linear Equations

a Fitting Equations to Data

There are many situations in the world to which linear equations can be applied. How do we know when we have such a case?

► **EXAMPLE 1** Plot the following data and determine whether a linear equation gives an approximate fit.

The number of times crickets chirp can be used to predict temperature. Here are some measurements made at different temperatures.

Number of chirps per min, N	11	29	47	75	107
Temperature T, in °C	6	8	10	15	20

We make a graph with a T (temperature) axis and an N (number per minute) axis and plot these data. We see that they do lie approximately on a straight line, so we can use a linear equation in this situation.

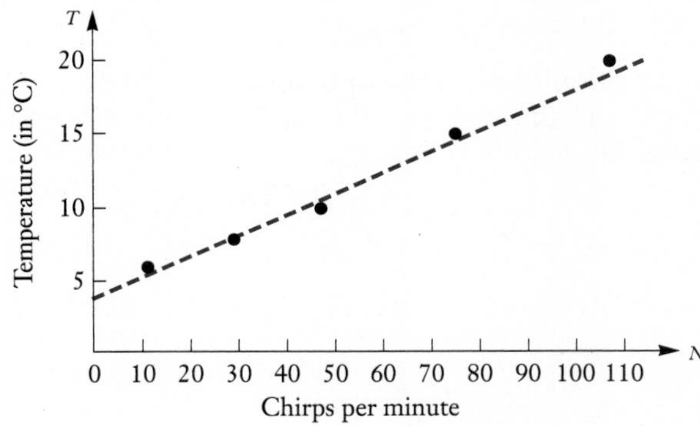

We see that the temperature T depends on the number of chirps N. Thus we call T the *dependent* variable (normally graphed on the vertical or y-axis) and N the *independent* variable (normally graphed on the horizontal or x-axis). ◄

► **EXAMPLE 2** Wind friction, or *resistance*, increases with speed. Here are some measurements made in a wind tunnel. Plot the data and determine whether a linear equation will give an approximate fit.

Velocity, in km/h	10	21	34	40	45	52
Force of resistance, in kg	3	4.2	6.2	7.1	15.1	29.0

OBJECTIVE

After finishing Section 3.5, you should be able to:

a Find a linear equation that fits two data points and use the equation to make predictions.

FOR EXTRA HELP

Tape NC

Tape 5A

MAC: 3
IBM: 3

We make a graph with a V (velocity) horizontal axis and an F (force) vertical axis and plot the data. They do not lie on a straight line, even approximately. Therefore, we cannot use a linear equation in this situation.*

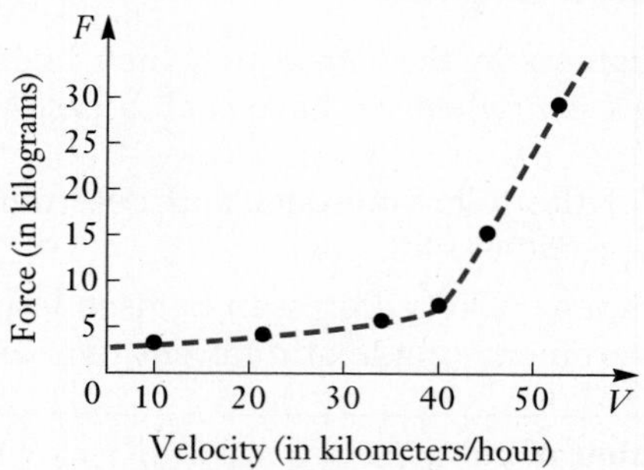

When we have a situation that a linear equation fits, we can find an equation and use it to solve problems and make predictions.

▶ **EXAMPLE 3** *Cricket chirps and temperature.* When crickets chirp 40 times per minute, the temperature is 10°C. When they chirp 112 times per minute, the temperature is 20°C. We know that a linear equation fits this situation. **(a)** Find a linear equation that fits the data. Express T in terms of N. **(b)** From your equation, find the temperature when crickets chirp 76 times per minute; 100 times per minute.

a) To find the equation, we use the two known ordered pairs (40, 10°) and (112, 20°). We call these *data points* (N, T). We first find the slope of the line using the definition of slope:

$$m = \frac{20 - 10}{112 - 40} = \frac{10}{72} = \frac{5}{36}.$$

Then we substitute into the point–slope equation, choosing (40, 10°) for the point:

$$T - 10 = \frac{5}{36}(N - 40)$$

$$T - 10 = \frac{5}{36}N - \frac{5}{36} \cdot 40$$

$$T - 10 = \frac{5}{36}N - \frac{50}{9}$$

$$T = \frac{5}{36}N + \frac{40}{9}, \quad \text{or} \quad \frac{5N + 160}{36}. \qquad \textbf{Solving for } T$$

b) Using this as a formula, we find T when $N = 76$ and when $N = 100$:

$$T = \frac{5 \cdot 76 + 160}{36} = 15°; \qquad T = \frac{5 \cdot 100 + 160}{36} \approx 18.3°.$$ ◀

DO EXERCISE 1.

Equations give good results only within certain limits. A negative number of chirps per minute would be meaningless. Also, imagine what would happen to a cricket at −40°C or at 100°C!

* The situation in Example 2 might fit an equation that is not linear. We will consider such problems later.

1. *Records in the 100-m dash.* It has been found that running records change with time according to linear equations. In 1920, the record for the men's 100-m dash was 10.43 sec. In 1970, it was 9.93 sec.

a) Fit a linear equation to the data points. (Use R for the record and t for the number of years since 1920.) Express R in terms of t.

b) Use your equation to predict the record in 1994; in 1998.

c) In what year will the record be 9.0 sec?

ANSWERS ON PAGE A-5

NAME SECTION DATE

EXERCISE SET 3.5

ANSWERS

a

1. *Life expectancy of females in the United States.* In 1950, the life expectancy of females was 72 yr. In 1970, it was 75 yr. Let E represent the life expectancy and t the number of years since 1950 ($t = 0$ gives 1950 and $t = 10$ gives 1960).
   a) List the data points (t, E).
   b) Fit a linear equation to the data points. Express E in terms of t.
   c) Use the equation of (b) to predict the life expectancy of females in 1996; in 2000.

2. *Life expectancy of males in the United States.* In 1950, the life expectancy of males was 65 yr. In 1970, it was 68 yr. Let E represent life expectancy and t the number of years since 1950.
   a) List the data points (t, E).
   b) Fit a linear equation to the data points. Express E in terms of t.
   c) Use the equation of (b) to predict the life expectancy of males in 1998; in 2001.

3. *Weight vs. height.* It has been shown experimentally that a person's height H is related to that person's weight W by a linear equation. An average-size person who is 70 in. tall weighs 165 lb, and a person 67 in. tall weighs 145 lb.
   a) List the data points (W, H).
   b) Fit a linear equation to the data points. Express H in terms of W.
   c) Use the equation of (b) to estimate the height of an average-size person weighing 130 lb. (This equation is valid only for heights above 40 in.)

4. *Natural gas demand.* In 1950, natural gas demand in the United States was 20 quadrillion joules. In 1960, the demand was 22 quadrillion joules. Let D represent the demand for natural gas t years after 1950.
   a) List the data points (t, D).
   b) Fit a linear equation to the data points. Express D in terms of t.
   c) Use the equation of (b) to predict the natural gas demand in 1998; in 2006.

1. a) ______

b) ______

c) ______

2. a) ______

b) ______

c) ______

3. a) ______

b) ______

c) ______

4. a) ______

b) ______

c) ______

ANSWERS

5. a) ______

b) ______

c) ______

d) ______

6. a) ______

b) ______

c) ______

d) ______

7. a) ______

b) ______

c) ______

8. a) ______

b) ______

c) ______

9. ______

10. ______

11. ______

5. *Records in the 1500-m run.* In 1930, the record for the men's 1500-m run was 3.85 min. In 1950, it was 3.70 min. Let R represent the record in the 1500-m run and t the number of years since 1930.
   a) List the data points (t, R).
   b) Fit a linear equation to the data points. Express R in terms of t.
   c) Use the equation of (b) to predict the record in 1998; in 2002.
   d) When will the record be 3.3 min?

6. *Records in the 400-m run.* In 1930, the record for the men's 400-m run was 46.8 sec. In 1970, it was 43.8 sec. Let R represent the record in the 400-m run and t the number of years since 1930.
   a) List the data points (t, R).
   b) Fit a linear equation to the data points. Express R in terms of t.
   c) Use the equation of (b) to predict the record in 1995; in 1998.
   d) When will the record be 40 sec?

7. *The cost of a taxi ride.* The cost of a taxi ride for 2 mi in Browntown is \$1.75. For 3 mi, the cost is \$2.00.
   a) List the data points (M, C), where $M =$ miles driven and $C =$ cost in dollars.
   b) Fit a linear equation to the data. Express C in terms of M.
   c) Use the equation to find the cost of a 7-mi ride.

8. *The cost of renting a car.* If you rent a car for one day and drive it 100 mi, the cost is \$30. If you drive it 150 mi, the cost is \$37.50.
   a) List the data points (M, C), where $M =$ miles driven and $C =$ cost in dollars.
   b) Fit a linear equation to the data points. Express C in terms of M.
   c) Use the equation to find how much it will cost to rent the car for one day if you drive it 200 mi.

SYNTHESIS

Solve the problem by fitting a linear equation to the data.

9. The value of a copying machine is \$5200 when it is purchased. After 2 years, its value is \$4225. Find its value after 8 years.

10. Water freezes at 32° Fahrenheit and at 0° Celsius. Water boils at 212°F and at 100°C. What Celsius temperature corresponds to a room temperature of 70°F?

11. ▦ A piece of copper wire has a length of 100 cm at 18°C. At 20°C the length of the wire changes to 100.00356 cm. Find the length of the wire at 40°C and at 0°C.

3.6 Graphing Inequalities in Two Variables

A **graph** of an inequality is a drawing that represents its solutions. An inequality in one variable can be graphed on a number line. An inequality in two variables can be graphed on a coordinate plane.

A **linear inequality** is one that we can get from a related linear equation by changing the equals symbol to an inequality symbol. The graph of a linear inequality is a half-plane, sometimes including the graph of the related line along the edge.

a Solutions of Inequalities in Two Variables

The solutions of an inequality in two variables are ordered pairs.

▶ **EXAMPLE 1** Determine whether $(-3, 2)$ is a solution of $5x - 4y > 13$.

We use alphabetical order of variables. We replace x by -3 and y by 2:

$5x - 4y$	> 13	
$5(-3) - 4 \cdot 2$	13	
$-15 - 8$		
-23		FALSE

Since $-23 > 13$ is false, $(-3, 2)$ is not a solution. ◀

▶ **EXAMPLE 2** Determine whether $(6, -7)$ is a solution of $5x - 4y > 13$.

We use alphabetical order of variables. We replace x by 6 and y by -7:

$5x - 4y$	> 13	
$5(6) - 4(-7)$	13	
$30 + 28$		
58		TRUE

Since $58 > 13$ is true, $(6, -7)$ is a solution. ◀

DO EXERCISES 1 AND 2.

b Graphing Inequalities in Two Variables

▶ **EXAMPLE 3** Graph: $y < x$.

We first graph the line $y = x$ for comparison. Every solution of $y = x$ is an ordered pair like (3, 3). The first and second coordinates are the same. The graph of $y = x$ is shown at the left below. We draw it dashed because these points are *not* solutions of $y < x$.

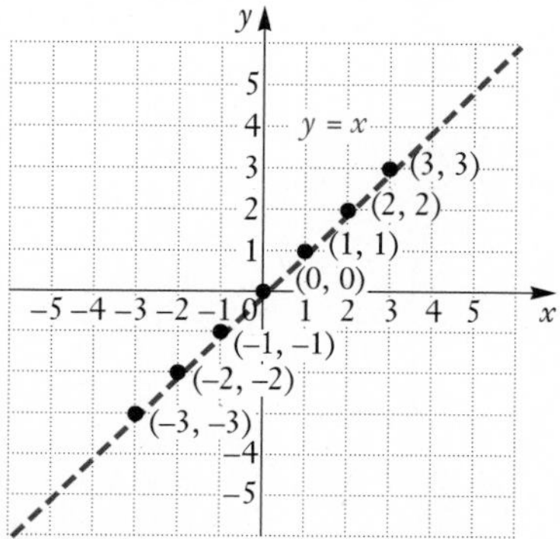

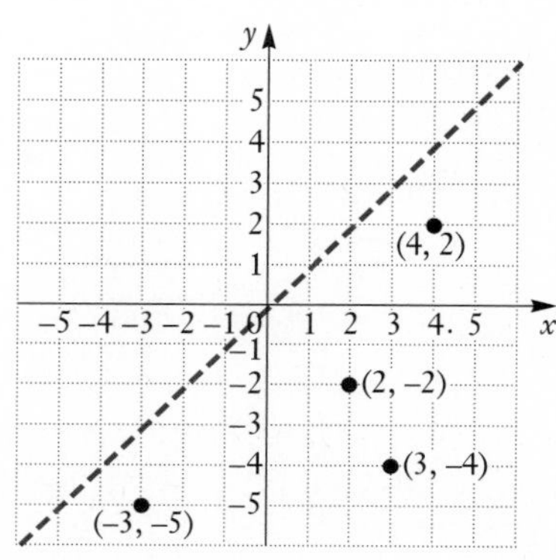

OBJECTIVES

After finishing Section 3.6, you should be able to:

a Determine whether an ordered pair of numbers is a solution of an inequality in two variables.

b Graph linear inequalities in two variables.

FOR EXTRA HELP

Tape 6D

Tape 5A

MAC: 3
IBM: 3

1. Determine whether $(1, -4)$ is a solution of the inequality $4x - 5y < 12$.

2. Determine whether $(4, -3)$ is a solution of the inequality $3y - 2x \leq 6$.

ANSWERS ON PAGE A-5

Now look at the graph on the right on the preceding page. Note that each of the ordered pairs plotted on the half-plane below $y = x$ is a solution of $y < x$. We can check a pair (4, 2) as follows:

$$\begin{array}{c|c} y & < x \\ \hline 2 & 4 \end{array} \quad \text{TRUE}$$

It turns out that any point on the same side of $y = x$ as (4, 2) is also a solution. Thus, if you know that one point in a half-plane is a solution, then all points in that half-plane are solutions. In this text, we will usually indicate this by color shading. We shade the half-plane below $y = x$.

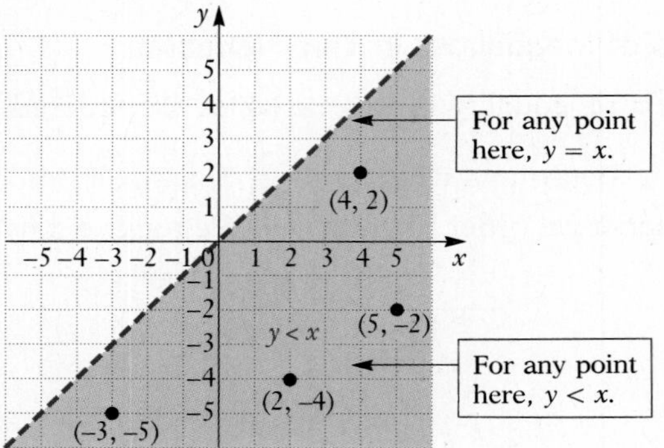

◀

▶ **EXAMPLE 4** Graph: $8x + 3y \geq 24$.

First we sketch the line $8x + 3y = 24$. Points on the line $8x + 3y = 24$ are also in the graph of $8x + 3y \geq 24$, so we draw the line solid. This indicates that all points on the line are solutions. The rest of the solutions are either in the half-plane above the line or the half-plane below the line. To determine which, we select a point that is not on the line and determine whether it is a solution of $8x + 3y \geq 24$. We try $(-3, 4)$ as a test point:

$$\begin{array}{r|l} 8x + 3y & \geq 24 \\ \hline 8(-3) + 3(4) & 24 \\ -24 + 12 & \\ -12 & \end{array} \quad \text{FALSE}$$

We see that $-12 \geq 24$ is *false*. Since $(-3, 4)$ is not a solution, none of the points in the half-plane containing $(-3, 4)$ is a solution. Thus the points in the opposite half-plane are solutions. We shade that half-plane and obtain the graph:

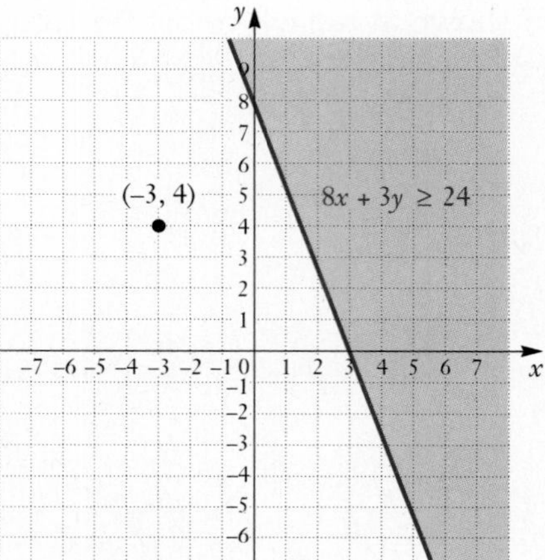

◀

We graph linear inequalities as follows.

To Graph an Inequality in Two Variables

1. **Replace the inequality symbol with an equals sign and graph this related equation.**
2. **If the inequality symbol is $<$ or $>$, draw the line dashed. If the inequality symbol is $\leq$ or $\geq$, draw the line solid.**
3. **The graph consists of a half-plane that is either above or below or left or right of the line and, if the line is solid, the line as well. To determine which half-plane to shade, choose a point not on the line as a test point. Substitute to determine whether that point is a solution. If so, shade the half-plane containing that point. If not, shade the opposite half-plane.**

► **EXAMPLE 5** Graph: $6x - 2y < 12$.

1. We first graph the related equation $6x - 2y = 12$.
2. Since the inequality uses the symbol $<$, points on the line are not solutions of the inequality, so we draw a dashed line.
3. To determine which half-plane to shade, we consider a test point *not* on the line. We try (0, 0) and substitute:

$6x - 2y$	< 12
$6(0) - 2(0)$	12
$0 - 0$	
0	TRUE

Since the inequality $0 < 12$ is *true*, the point (0, 0) is a solution; each point in the half-plane containing (0, 0) is a solution. Thus each point in the opposite half-plane is *not* a solution. The graph is shown below.

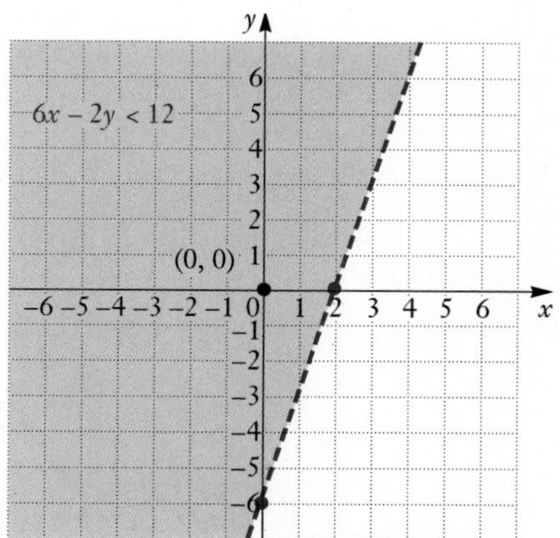

◄

DO EXERCISES 3 AND 4.

► **EXAMPLE 6** Graph $x > -3$ on a plane.

There is a missing variable in this inequality. If we graph the inequality on a line, its graph is as follows:

Graph.

3. $6x - 3y < 18$

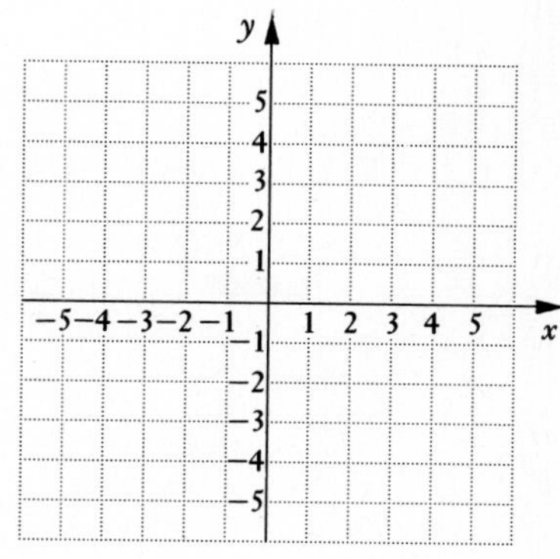

4. $4x + 3y \geq 12$

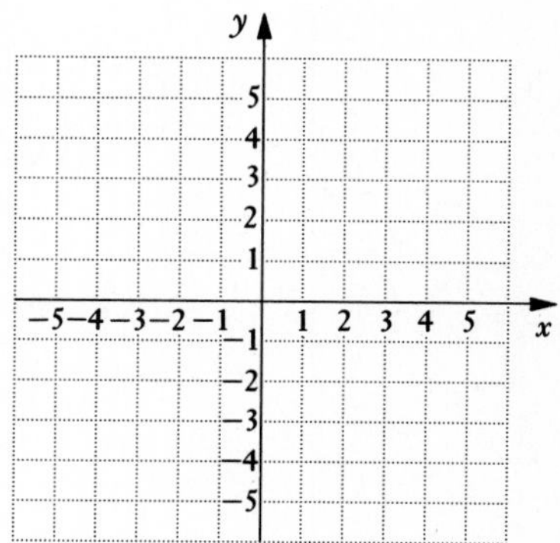

ANSWERS ON PAGE A-5

Graph on a plane.

5. $x < 3$

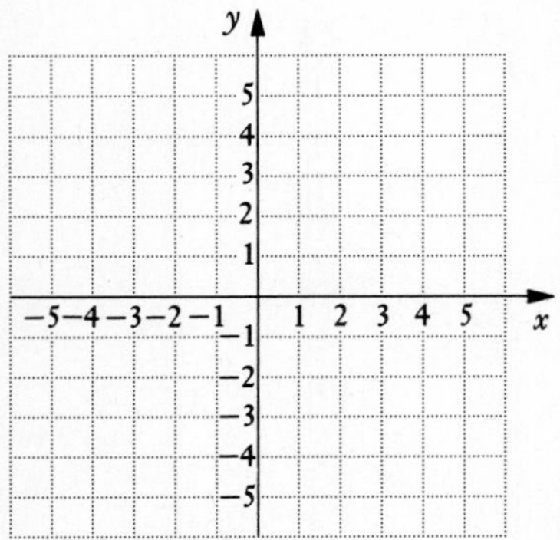

6. $y \geq -4$

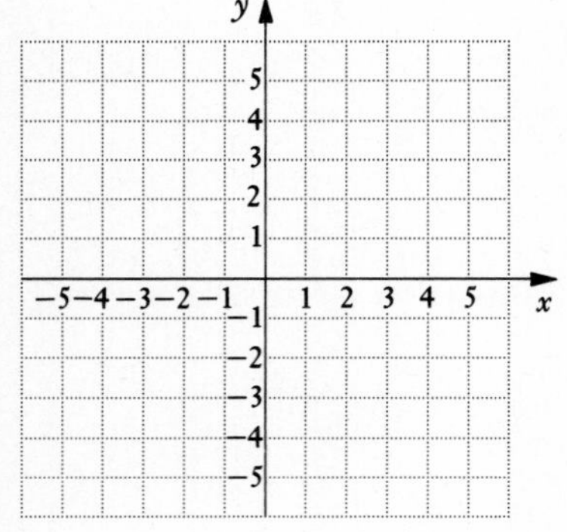

ANSWERS ON PAGE A-5

However, we can also write this inequality as $x + 0y > -3$ and consider graphing it in the plane. We use the same technique that we have used with the other examples. We first graph the related equation $x = -3$ in the plane. We draw the boundary with a dashed line.

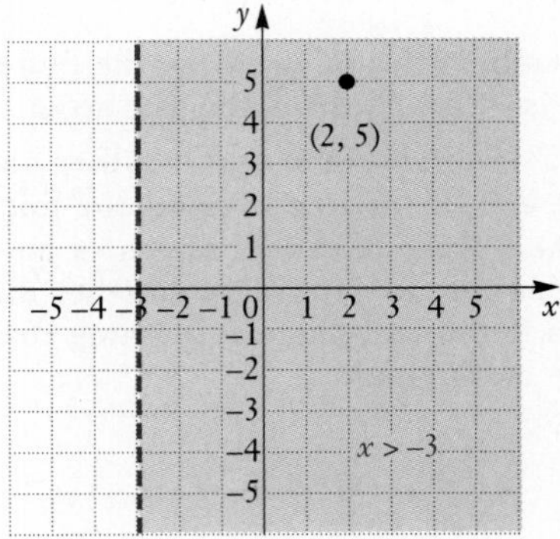

The rest of the graph is a half-plane to the right or left of the line $x = -3$. To determine which, we consider a test point, (2, 5):

$x + 0y > -3$	
$2 + 0(5)$	-3
2	TRUE

Since (2, 5) is a solution, all the pairs in the half-plane containing (2, 5) are solutions. We shade that half-plane.

We see that the solutions of $x > -3$ are all those ordered pairs whose first coordinates are greater than -3. ◀

▶ **EXAMPLE 7** Graph $y \leq 4$ on a plane.

We first graph $y = 4$ using a solid line. We then use $(2, -3)$ as a test point and substitute:

$0x + y \leq 4$	
$0(2) + (-3)$	4
-3	TRUE

We see that $(2, -3)$ is a solution, so all points in the half-plane containing $(2, -3)$ are solutions. Note that this half-plane consists of all ordered pairs whose second coordinate is less than or equal to 4.

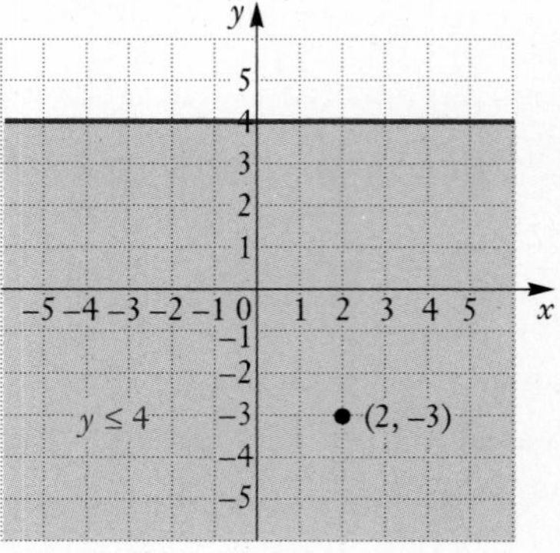

◀

DO EXERCISES 5 AND 6.

NAME SECTION DATE

EXERCISE SET 3.6

ANSWERS

1. ______
2. ______
3. ______
4. ______

a Determine whether the ordered pair is a solution of the inequality.

1. $(-4, 2)$; $2x + y < -5$

2. $(3, -6)$; $4x + 2y \geq 0$

3. $(8, 14)$; $2y - 3x > 5$

4. $(7, 20)$; $3x - y > -1$

b Graph the inequality on a plane.

5. $y > 2x$

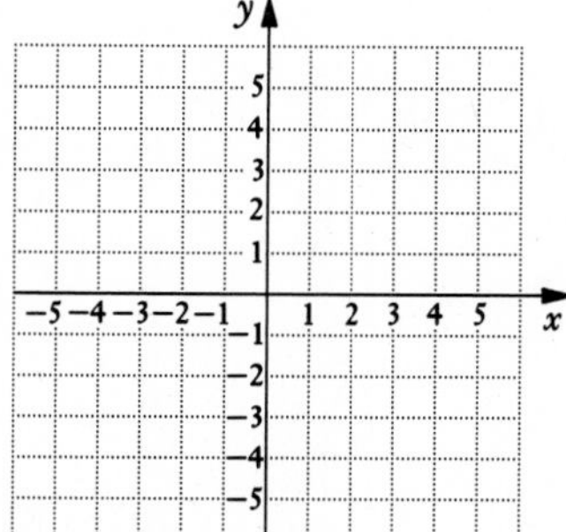

6. $y < 3x$

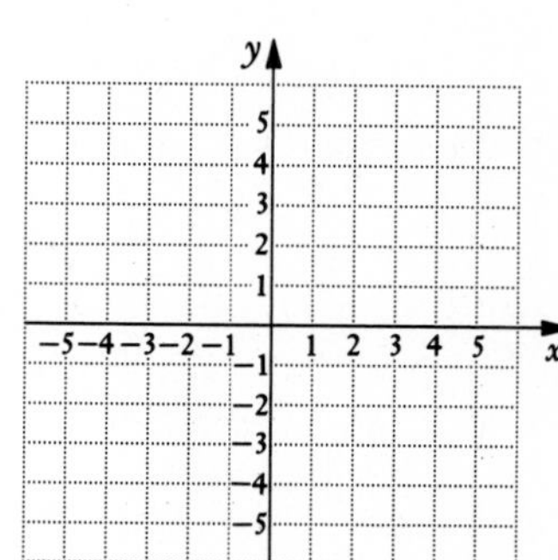

7. $y < x + 1$

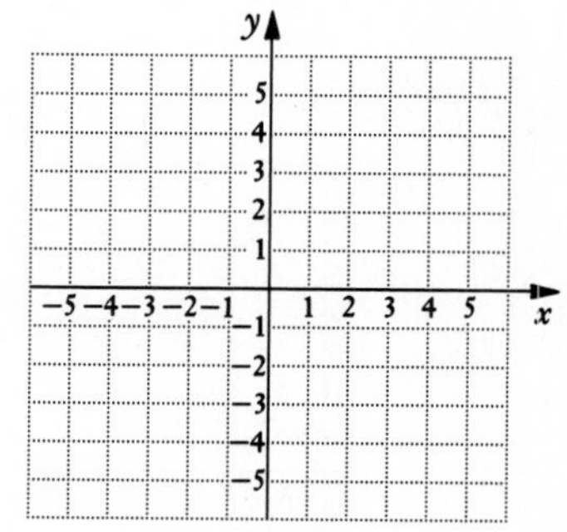

8. $y \leq x - 3$

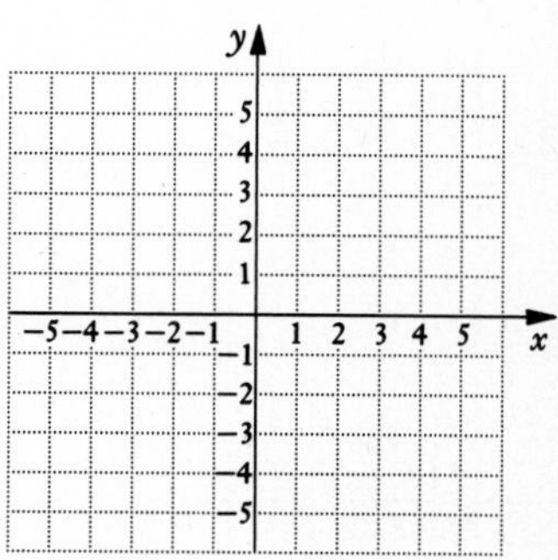

9. $y > x - 2$

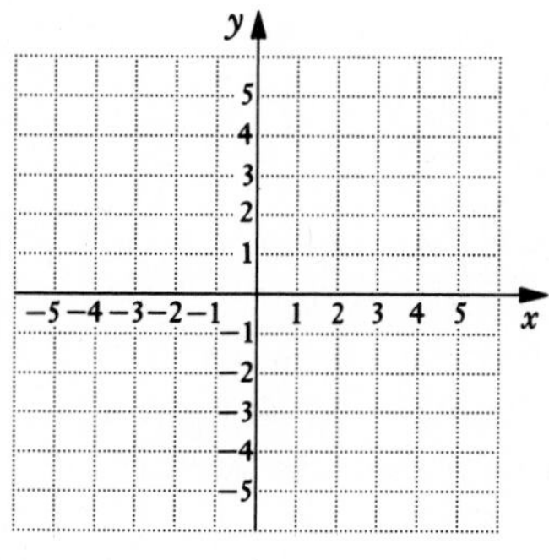

10. $y \geq x + 4$

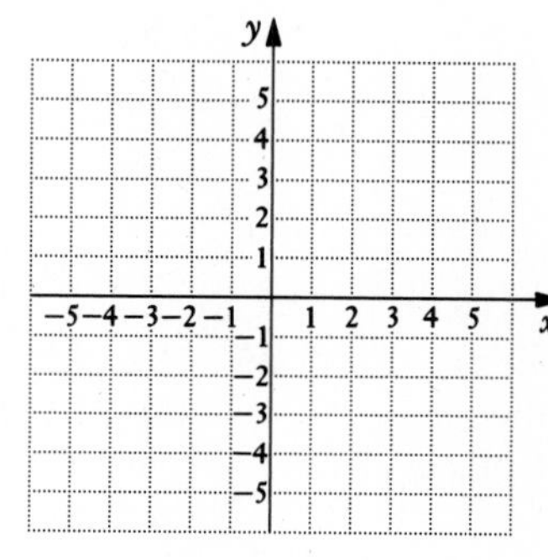

11. $x + y < 4$

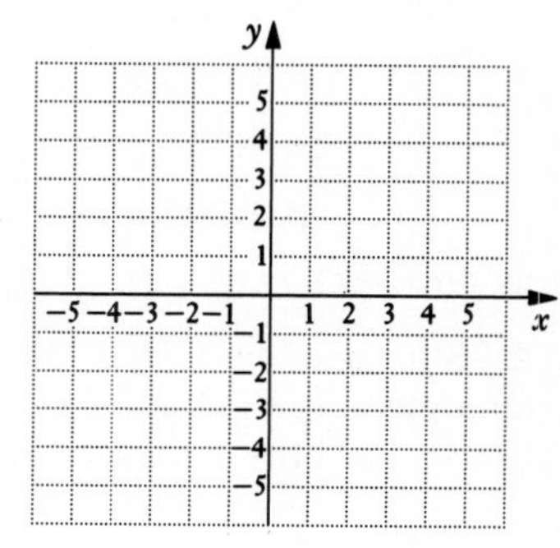

12. $x - y \geq 3$

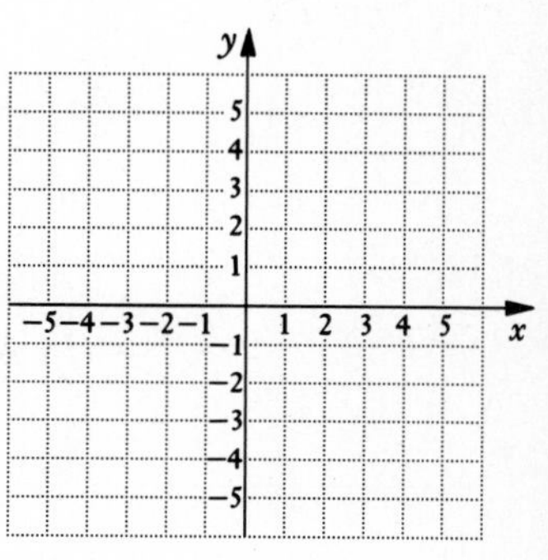

13. $3x + 4y \leq 12$

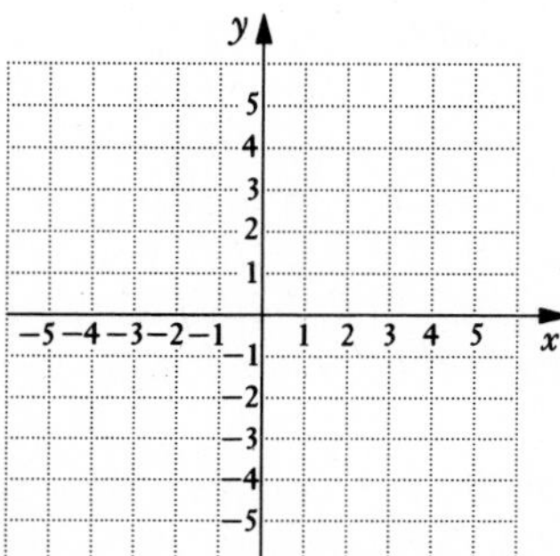

14. $2x + 3y < 6$

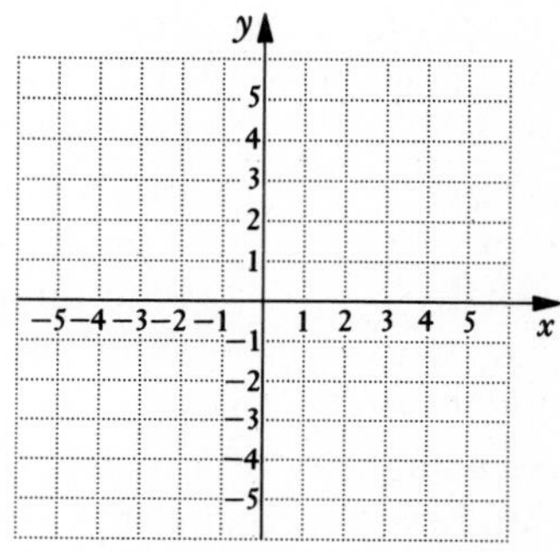

15. $2y - 3x > 6$

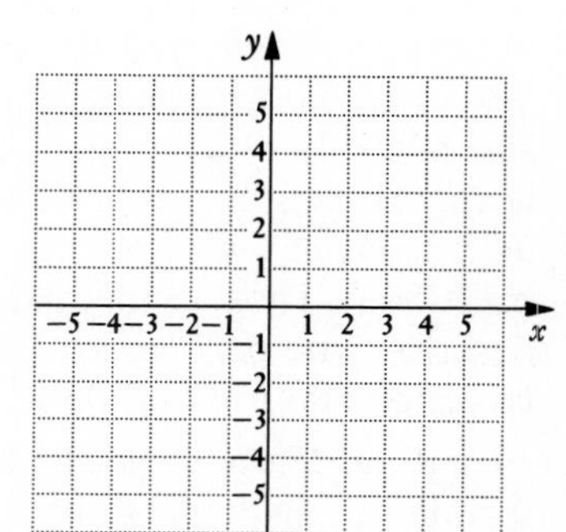

16. $2y - x \leq 4$

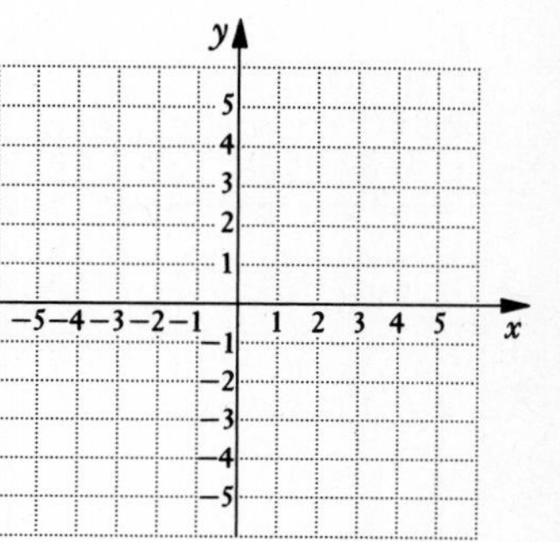

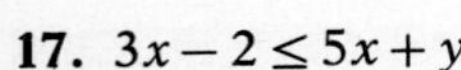

17. $3x - 2 \leq 5x + y$

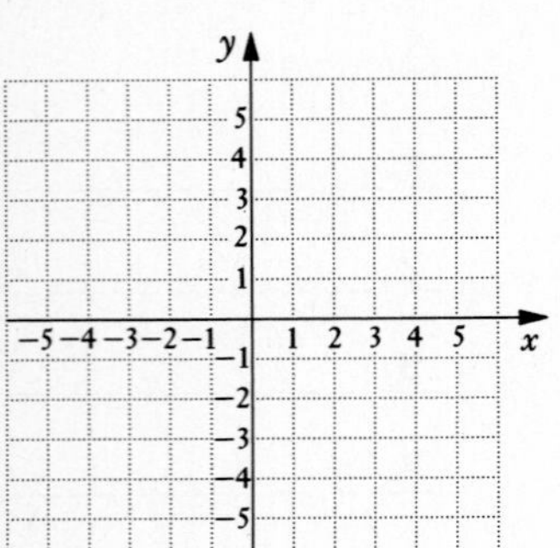

18. $2x - 2y \geq 8 + 2y$

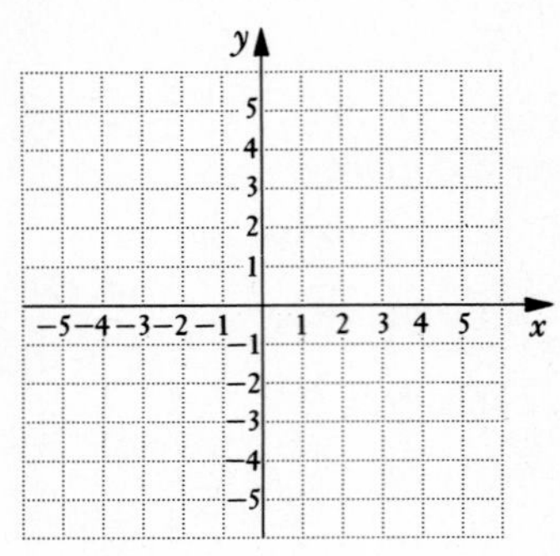

19. $x < -4$

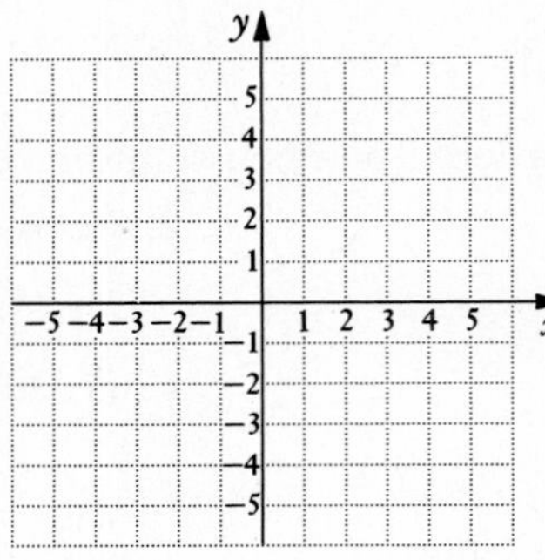

20. $y \geq 2$

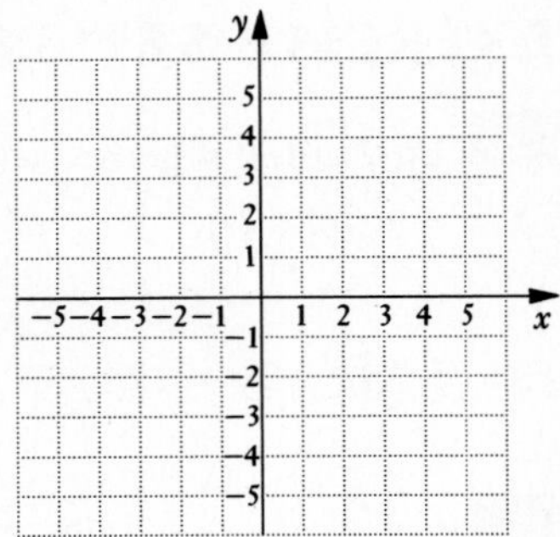

21. $y > -2$

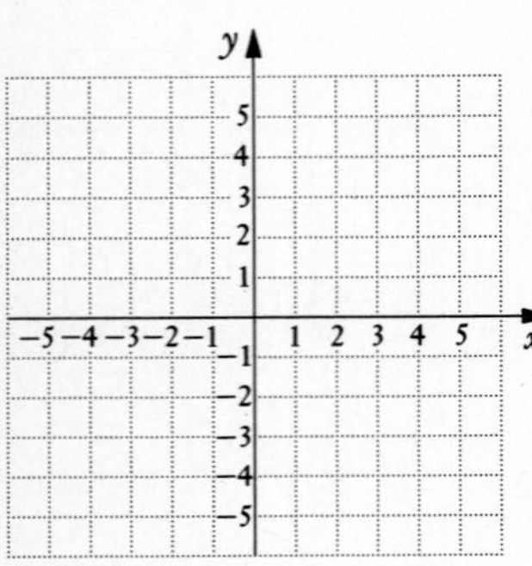

22. $x \leq 5$

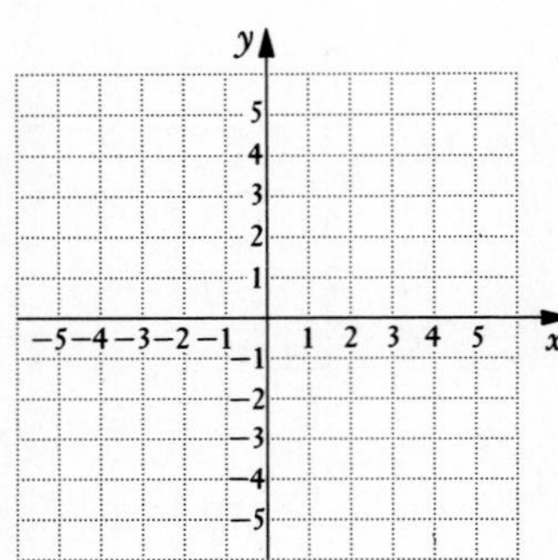

23. $2x + 3y \leq 6$

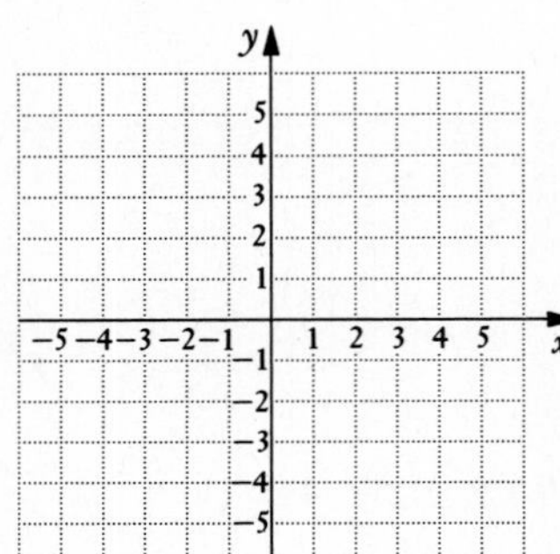

24. $7x + 2y \geq 21$

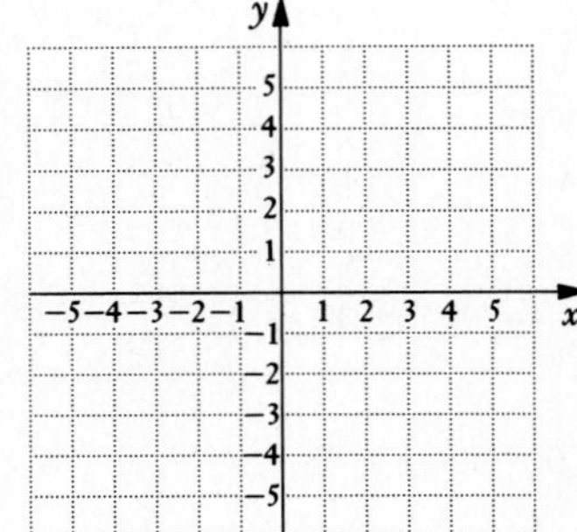

ANSWERS

25. ______

26. ______

27. ______

28. ______

29. ______

SKILL MAINTENANCE

25. A 78-ft rope is cut into two pieces. One piece is 5 ft longer than the other. Find the length of each piece.

26. The perimeter of a rectangle is 456 m. The length is 17 m more than twice the width. Find the area of the rectangle.

SYNTHESIS

27. *Widths of a basketball floor.* Sizes of basketball floors vary due to building size and other constraints such as cost. The length L is to be at most 94 ft and the width W is to be at most 50 ft. Find an inequality that describes the perimeter of a basketball floor. Graph the inequality.

28. *Hockey wins and losses.* A hockey team figures that it needs at least 60 points for the season in order to make the playoffs. A win w is worth 2 points and a tie t is worth 1 point. Find an inequality that describes the situation. Graph the inequality.

29. *Elevators.* Many elevators have a capacity of 1 metric ton (1000 kg). Suppose that c children, each weighing 35 kg, and a adults, each weighing 75 kg, are on an elevator. Find an inequality that asserts that the elevator is overloaded. Graph the inequality.

SUMMARY AND REVIEW: CHAPTER 3

IMPORTANT PROPERTIES AND FORMULAS

$$\text{Slope} = m = \frac{y_2 - y_1}{x_2 - x_1}$$

Equations of Lines

Horizontal Line:	$y = b$, slope 0	*Vertical Line:*	$x = a$, slope undefined
Slope–Intercept Equation:	$y = mx + b$	*Point–Slope Equation:*	$y - y_1 = m(x - x_1)$
Parallel Lines:	$m_1 = m_2$	*Perpendicular Lines:*	$m_1 = -\frac{1}{m_2}$

REVIEW EXERCISES

The review sections and objectives to be tested in addition to the material in this chapter are [2.2a], [2.4c], [2.5a, b], and [2.6c, e].

Graph using the slope and the y-intercept.

1. $y = -3x + 2$

2. $y = \frac{5}{2}x - 3$

Graph.

3. $x = -3$

4. $2y = 8$

Graph using intercepts.

5. $2y = 6 - 3x$

6. $4y + 2x = 8$

7. Find the slope, if it exists, of the line containing the following points.
$(13, 7)$ and $(10, -4)$

8. Find an equation of the line having the given slope and containing the given point.
$m = -3$; $(3, -5)$

9. Find an equation of the line containing the given point and parallel to the given line.
$(-2, 1)$; $4x + 9y = 4$

10. Find an equation of the line containing the given point and perpendicular to the given line.
$(-4, -5)$; $2y + 3x = 8$

11. Find an equation of the line containing the following pair of points.
$(2, -3)$, $(-4, -5)$

12. Find an equation of the line having the given slope and containing the given point.
$m = 2$; $(-2, -3)$

13. Find an equation of the line containing the following pair of points.

$$(-2, 3) \quad \text{and} \quad (-4, 6)$$

14. Find an equation of the line containing the given point and parallel to the given line.

$$(14, -1); \quad 5x + 7y = 8$$

15. Find an equation of the line containing the given point and perpendicular to the given line.

$$(5, 2); \quad 3x + y = 5$$

Determine whether the graphs of each pair of lines are parallel or perpendicular.

16. $y + 5 = -x$, $x - y = 2$

17. $3x - 5 = 7y$, $7y - 3x = 7$

18. $4y + x = 3$, $2x + 8y = 5$

19. $x = 4$, $y = -3$

20. *Taxi costs.* The cost of a taxi ride is \$2.75 for 2.5 mi and \$2.90 for 3.0 mi.

a) List the data points (M, C), where $M =$ the number of miles driven and $C =$ the cost, in dollars.
b) Fit a linear equation to the data points. Express C in terms of M.
c) Use the linear equation to find the cost of a 5-mi ride.

Graph.

21. $2x + 3y < 12$

22. $y \leq 0$

23. $x + y \geq 1$

SKILL MAINTENANCE

Solve.

24. $2x - 4(x - 12) < -3(2x + 5) - 14$

25. $|4x - 7| = 19$

26. $|4x - 7| \leq 19$

27. $|4x - 7| > 19$

28. $-2x + 7 < -14$ *or* $-2x + 7 > -14$

29. A basketball team increases its score by seven points in each of three consecutive games. In those three games, the team scored a total of 228 points. What was their score in each game?

SYNTHESIS

30. Write an equation of the line that has y-intercept $(0, \frac{5}{7})$ and is parallel to the graph of $6x - 3y = 1$.

31. Solve by fitting a linear equation to the data: A business determines that when it sells 7000 units of a product, it will take in \$22,000. For the sale of 8000 units, it will take in \$25,000. How much will it take in from the sale of 10,000 units?

❖ THINKING IT THROUGH

1. Explain the usefulness of the concept of slope when describing a straight line.
2. Explain why the slope of a vertical line is undefined but the slope of a horizontal line is 0.
3. Explain and compare the situations in which one would use the slope–intercept equation rather than the point–slope equation.

NAME SECTION DATE

TEST: CHAPTER 3

Graph using the slope and the y-intercept.

1. $y = -2x - 5$

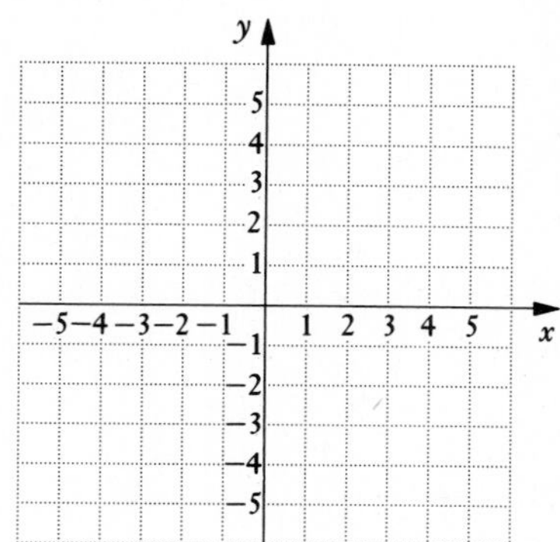

2. $y = -\frac{3}{5}x$

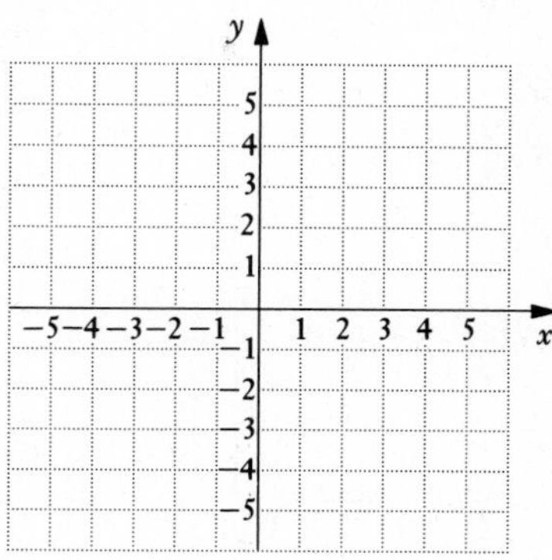

3. Graph using intercepts: $2x + 3y = 6$. Be sure to identify the intercepts.

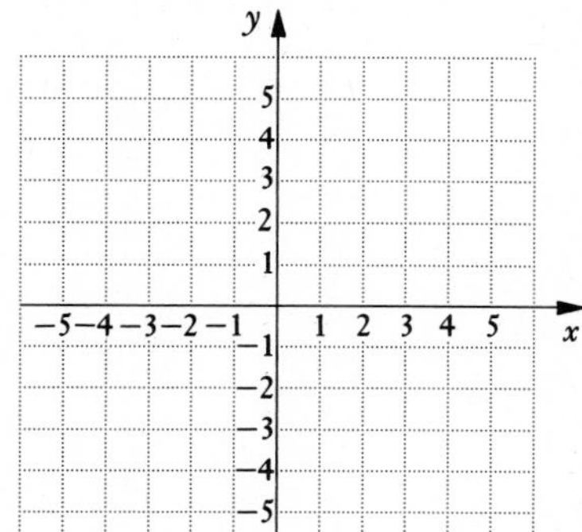

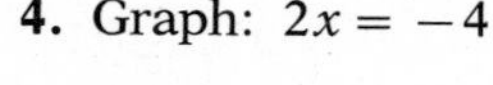

4. Graph: $2x = -4$.

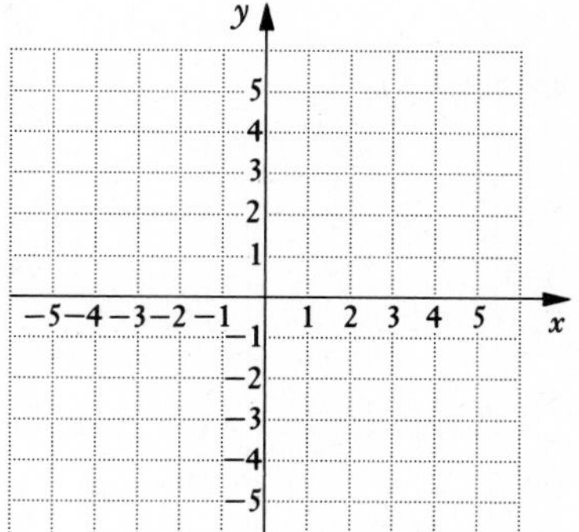

5. Find the slope, if it exists, of the line containing the following points.

$(3, 5)$ and $(10, -4)$

6. Find an equation of the line having the given slope and containing the given point.

$m = -3$; $(1, -2)$

7. Find an equation of the line containing the following pair of points.

$(4, -6)$ and $(-10, 15)$

8. Find an equation of the line containing the given point and parallel to the given line.

$(4, -1)$; $x - 2y = 5$

9. Find an equation of the line containing the given point and perpendicular to the given line.

$(2, 5)$; $x + 3y = 2$

Determine whether the graphs of each pair of lines are parallel or perpendicular.

10. $4y + 2 = 3x$,
$-3x + 4y = -12$

11. $y = -2x + 5$,
$2y - x = 6$

ANSWERS

1. ______

2. ______

3. ______

4. ______

5. ______

6. ______

7. ______

8. ______

9. ______

10. ______

11. ______

ANSWERS

12. a) ______

b) ______

c) ______

13. ______

14. ______

15. ______

16. ______

17. ______

18. ______

19. ______

20. ______

21. ______

22. ______

23. ______

12. *Car rental.* You rent a certain kind of car for a day. If you drive it 250 mi, the cost is \$100. If you drive it 300 mi, the cost is \$110.

a) List the data points (M, C), where $M =$ the number of miles driven and $C =$ the cost, in dollars.

b) Fit a linear equation to the data points. Express C in terms of M.

c) Use the equation to find out how much it will cost to rent the car for a day and drive it 500 mi.

Graph.

13. $x \geq -3$

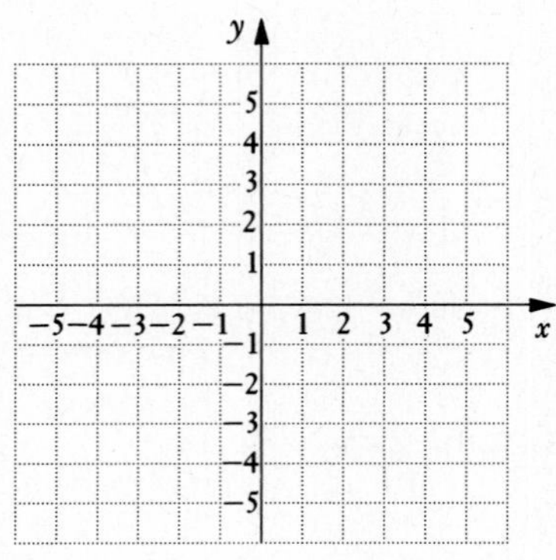

14. $x - 6y \leq -6$

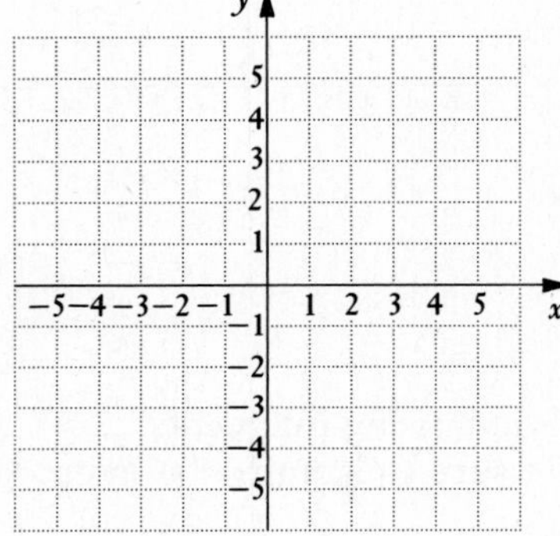

15. $3y > 2x$

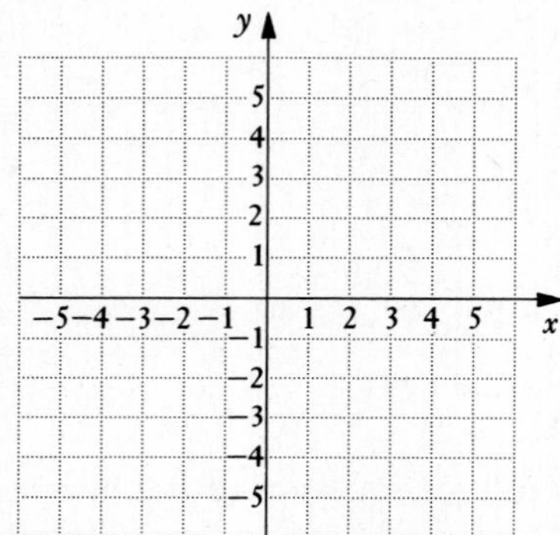

SKILL MAINTENANCE

Solve.

16. $2x - 4(9 - 7x) \geq 3(4x - 6) + 8$

17. $9 - 7x < -23$ *or* $9 - 7x > 23$

18. $|9 - 7x| < 23$

19. $|9 - 7x| = 23$

20. $|9 - 7x| \geq 23$

21. The price of a radio, including 5% sales tax, is \$36.75. Find the price of the radio before the tax was added.

SYNTHESIS

22. Find k such that the line $3x + ky = 17$ is perpendicular to the line $8x - 5y = 26$.

23. Write an equation of the line that has x-intercept $(-1.2, 0)$ and is perpendicular to the graph of $6x - 3y = 1$.

CUMULATIVE REVIEW: CHAPTERS 1–3

1. Evaluate $a^3 + b^0 - c$ when $a = 3$, $b = 7$, and $c = -3$.

Simplify.

2. $|4.3 - 2.1|$

3. $\left|-\frac{2}{3}\right|$

4. $-\frac{1}{3} - \left(-\frac{5}{6}\right)$

5. $-3.2(-11.4)$

6. $2x - 4(3x - 8)$

7. $(-16x^3y^{-4})(-2x^5y^3)$

8. $\frac{27x^0y^3}{-3x^2y^5}$

9. $3(x - 7) - 4[2 - 5(x + 3)]$

10. $-128 \div 16 + 32 \cdot (-10)$

11. $2^3 - (4 \cdot 2 - 3)^2 + 23^0 \cdot 16^1$

Solve.

12. $x + 9.4 = -12.6$

13. $\frac{2}{3}x - \frac{1}{4} = -\frac{4}{5}x$

14. $-2.4t = -48$

15. $4x + 7 = -14$

16. $3n - (4n - 2) = 7$

17. Solve $W = Ax + By$ for x.

18. Solve $M = A + 4AB$ for A.

Solve.

19. $y - 12 \leq -5$

20. $6x - 7 < 2x - 13$

21. $5(1 - 2x) + x < 2(3 + x)$

22. $x + 3 < -1$ *or* $x + 9 \geq 1$

23. $-3 < x + 4 \leq 8$

24. $-8 \leq 2x - 4 \leq -1$

25. $|x| = 8$

26. $|y| > 4$

27. $|4x - 1| \leq 7$

Graph on a plane.

28. $y = -2x + 3$

29. $3x = 2y + 6$

30. $4x + 16 = 0$

31. $-2y = -6$

32. $y \leq 2x + 1$

33. $x - 5y > 5$

34. Find an equation of the line containing the point $(-4, -6)$ and perpendicular to the line whose equation is $4y - x = 3$.

35. Find an equation of the line containing the point $(-4, -6)$ and parallel to the line whose equation is $4y - x = 3$.

36. Find the slope and the y-intercept of $-4y + 9x = 12$.

37. Find the slope, if it exists, of the line containing the points $(2, 7)$ and $(-1, 3)$.

38. Find an equation of the line with slope -3 and containing the point $(2, -11)$.

39. Find an equation of the line containing the points $(-6, 3)$ and $(4, 2)$.

Solve.

40. Nine plus five times a number is 173.4. Find the number.

41. A *mil* is one thousandth of a dollar. The taxation rate in a certain school district is 5 mils for every dollar of assessed valuation. The assessed valuation for the district is 13.4 million dollars. How much tax revenue will be raised?

42. Seventeen more than seven times a number is three less than twelve times the number. What is the number?

43. After a person gets a 20% raise in salary, the new salary is \$10,800. What was the old salary?

44. The perimeter of a lot is 80 m. The length exceeds the width by 6 m. Find the dimensions.

SYNTHESIS

45. An automotive dealer discovers that when \$1000 is spent on radio advertising, weekly sales increase by \$101,000. When \$1250 is spent on radio advertising, weekly sales increase by \$126,000. Assuming that sales increase according to a linear equation, by what would sales increase when \$1500 is spent on radio advertising?

46. Simplify: $(6x^{a+2}y^{b+2})(-2x^{a-2}y^{y+1})$.

47. Solve: $x + 5 < 3x - 7 \leq x + 13$.

48. Which pairs of the following four equations represent perpendicular lines?

(1) $7y - 3x = 21$
(2) $-3x - 7y = 12$
(3) $7y + 3x = 21$
(4) $3y + 7x = 12$

INTRODUCTION As you have probably noticed, the most difficult and time-consuming part of solving a problem is usually the translation of the situation to mathematical language. When you know how to translate using more than one equation, the translation is often much easier. In this chapter, you will learn how to translate to a system of equations. You will also learn to solve systems and then apply that skill to problem solving.

The review sections to be tested in addition to the material in this chapter are 1.5, 2.1, 2.3, and 3.3. ❖

Systems of Equations and Inequalities

AN APPLICATION

One canned juice drink is 15% orange juice; another is 5% orange juice. How many liters of each should be mixed together in order to get 10 liters that is 10% orange juice?

THE MATHEMATICS

Let $x =$ the number of liters of the 15% juice and $y =$ the number of liters of the 5% juice. The problem translates to the following system of equations:

$$x + y = 10,$$

$$0.15x + 0.05y = 1.$$

❖ POINTS TO REMEMBER: CHAPTER 4

Supplementary Angles: Two angles are supplementary if the sum of their measures is 180°.
Complementary Angles: Two angles are complementary if the sum of their measures is 90°.
Perimeter of a Rectangle: $P = 2l + 2w$
Motion Formula: $d = rt$
Sum of the Angle Measures of a Triangle = 180°.
Simple-Interest Formula: $I = Prt$

PRETEST: CHAPTER 4

1. Solve this system by graphing:

$$y = x + 1,$$
$$y + x = 3.$$

2. Solve the system of equations in Exercise 1 using the substitution method.

3. Solve using the elimination method:

$$3x + 5y = 1,$$
$$4x + 3y = -6.$$

4. Classify the system in Exercise 1 as consistent or inconsistent.

5. Classify the system in Exercise 1 as dependent or independent.

6. Solve the system in Exercise 3 using Cramer's rule. Show your work.

7. Solve:

$$\begin{aligned} 3x + 5y - 2z &= 7, \\ 2x + y - 3z &= -5, \\ 4x - 2y + z &= 3. \end{aligned}$$

8. Evaluate:

$$\begin{vmatrix} -3 & 1 \\ 2 & 5 \end{vmatrix}$$

9. Evaluate:

$$\begin{vmatrix} 4 & -1 & 2 \\ 1 & -2 & 0 \\ -1 & 3 & 1 \end{vmatrix}.$$

10. Solve using Cramer's rule. Show your work.

$$\begin{aligned} x + y + z &= 4, \\ x - 2y - z &= 3, \\ 3x - y - 4z &= 2 \end{aligned}$$

11. Two investments are made totaling \$7500. For a certain year, the investments yielded \$690 in simple interest. Part of the \$7500 is invested at 8% and part at 10%. Find the amount invested at each rate.

12. The sum of three numbers is 209. The second is twelve less than twice the first. The third is eighteen more than four times the first. Find the numbers.

13. Mixture A is 32% salt and the rest water. Mixture B is 58% salt and the rest water. How many pounds of each mixture should be combined in order to obtain 120 lb of a mixture that is 44% salt?

14. A motorboat took 4 hr to make a downstream trip with a 6-mph current. The return trip against the same current took 5 hr. Find the speed of the boat in still water.

Graph the system of inequalities. Find the coordinates of any vertices formed.

15. $x + y \leq 4,$
$x - y \leq 4$

16. $x + y \leq 16,$
$3x + 6y \leq 60,$
$x \geq 0,$
$y \geq 0$

4.1 Systems of Equations in Two Variables

a Systems of Equations and Solutions

Many problems can be more easily solved by translating them to two or more equations in two or more variables. See the example on the chapter opening page.

A **solution** of a system of equations in two variables, such as those in Examples 1 and 2, is an ordered pair of numbers that makes *both* equations true.

EXAMPLE 1 Determine whether $(-4, 7)$ is a solution of the system

$$x + y = 3,$$
$$5x - y = -27.$$

In most cases, we use alphabetical order of the variables. Thus we replace x by -4 and y by 7:

$x + y = 3$	
$-4 + 7$	3
3	TRUE

$5x - y = -27$	
$5(-4) - 7$	-27
$-20 - 7$	
-27	TRUE

The pair $(-4, 7)$ makes both equations true, so it is a solution of the *system*. We sometimes describe such a solution by saying that $x = -4$ and $y = 7$.

DO EXERCISES 1 AND 2.

b Solving Systems of Equations by Graphing

Recall that the **graph** of an equation is a drawing that represents its solution set. If the graph of an equation is a line, then every point on that line corresponds to an ordered pair that is a solution of the equation. If we graph a **system** of two linear equations, the point at which the lines intersect will be a solution of *both* equations.

EXAMPLE 2 Solve this system of equations by graphing:

$$y - x = 1,$$
$$y + x = 3.$$

We draw the graph of each equation using any method studied in Chapter 3.

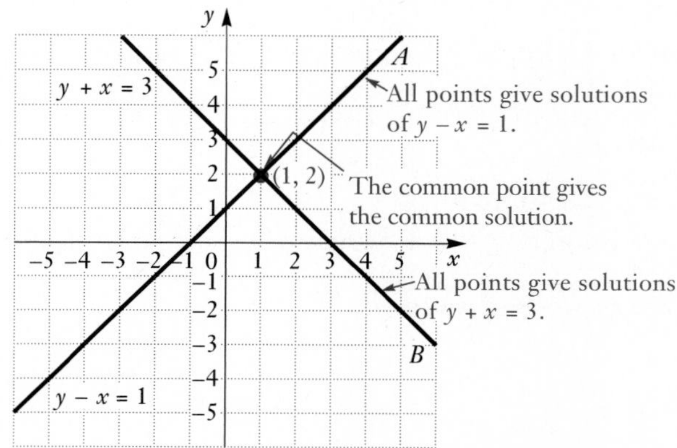

OBJECTIVES

After finishing Section 4.1, you should be able to:

a Determine whether an ordered pair is a solution of a system of two equations.

b Solve a system of two linear equations by graphing and determine whether the system is consistent or inconsistent and whether it is dependent or independent.

FOR EXTRA HELP

Tape 7A

Tape 5B

MAC: 4
IBM: 4

1. Determine whether (20, 40) is a solution of the system

$$2x = y,$$
$$4y + x = 180.$$

2. Determine whether $(-2, 3)$ is a solution of the system

$$2a - 5b = 7,$$
$$5b + 3a = -4.$$

ANSWERS ON PAGE A-5

Solve the system by graphing.

3. $-2x + y = 1,$
$3x + y = 1$

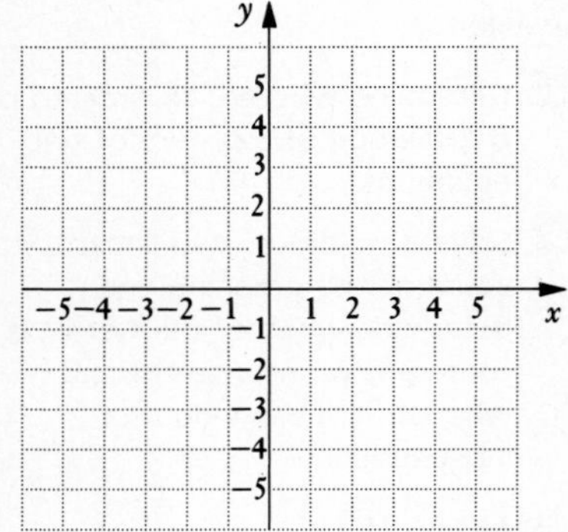

4. $x = 2y,$
$4y + x = 6$

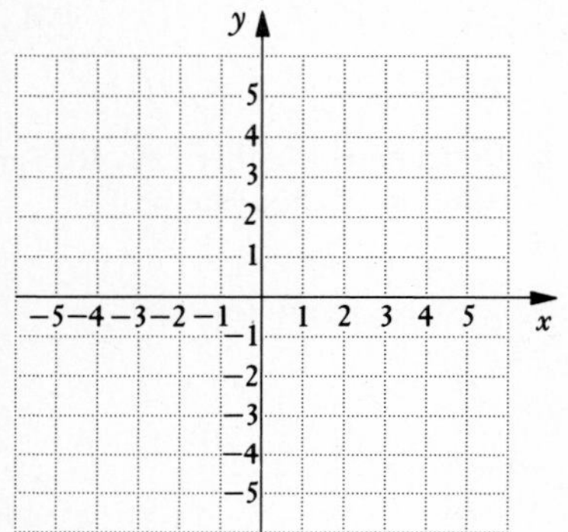

5. Solve this system by graphing:

$$y + 2x = 3,$$
$$y + 2x = -4.$$

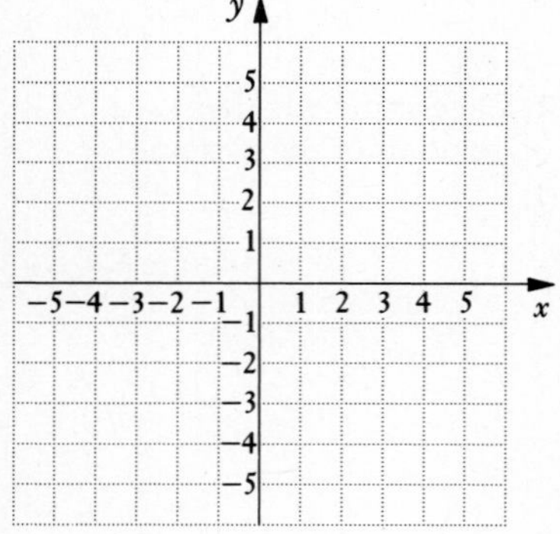

6. Classify each of the systems in Margin Exercises 3–5 as consistent or inconsistent.

ANSWERS ON PAGE A-5

All ordered pairs from line A are solutions of the first equation. All ordered pairs from line B are solutions of the second equation. The point of intersection has coordinates that make *both* equations true. The solution seems to be the point (1, 2). Graphing is not perfectly accurate, so solving by graphing may give only approximate answers.

We check the pair (1, 2) as follows.

Check:

$y - x = 1$	
$2 - 1$	1
1	TRUE

$y + x = 3$	
$2 + 1$	3
3	TRUE

The solution is (1, 2). ◄

DO EXERCISES 3 AND 4.

Sometimes the equations in a system have graphs that are parallel lines.

► **EXAMPLE 3** Solve by graphing:

$$y = -3x + 5,$$
$$y = -3x - 2.$$

We graph the equations. The lines have the same slope, -3, and different y-intercepts, so they are parallel. There is no point at which they cross, so the system has no solution.

No matter what point we try, it will *not* check in *both* equations. There is no solution. The solution set is thus the empty set, denoted $\varnothing$ or $\{\ \}$.

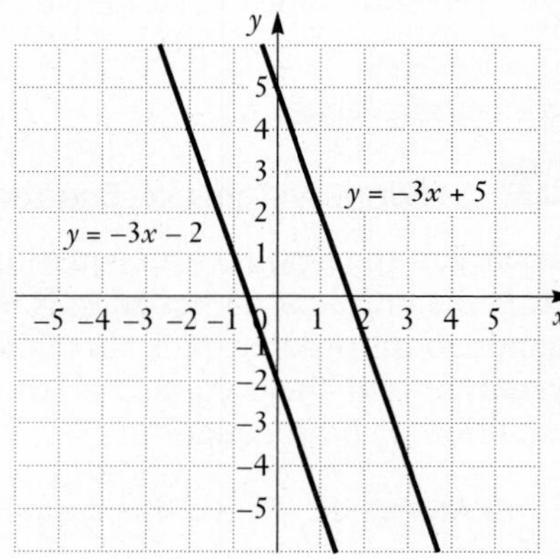

◄

A system of equations is *consistent* if it has a solution.
A system of equations is *inconsistent* if it has no solution.

The system in Example 2 is consistent. The system in Example 3 is inconsistent.

DO EXERCISES 5 AND 6.

Sometimes the equations in a system have the same graph.

► **EXAMPLE 4** Solve by graphing:

$$3y - 2x = 6,$$
$$-12y + 8x = -24.$$

We graph the equations and see that the graphs are the same. Thus any solution of one of the equations is a solution of the other. Each equation has an infinite number of solutions, two of which are shown on the graph. Each of these is also a solution of the other equation.

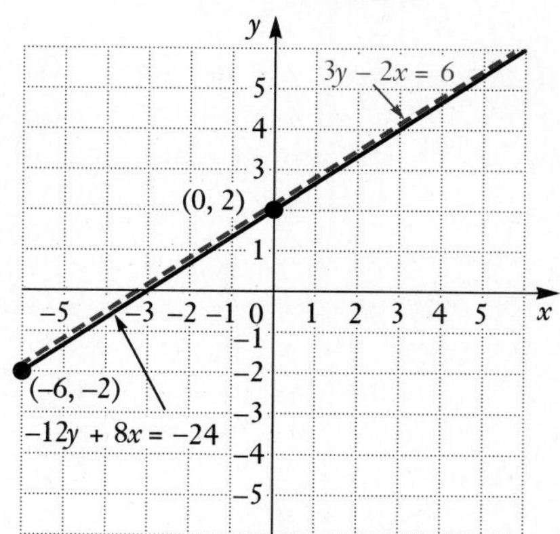

We check one such solution, (0, 2), which is the y-intercept of each equation.

$3y - 2x = 6$	
$3(2) - 2(0)$	6
$6 - 0$	
6	TRUE

$-12y + 8x = -24$	
$-12(2) + 8(0)$	-24
$-24 + 0$	
-24	TRUE

On your own, check that $(-6, -2)$ is a solution of both equations. If $(0, 2)$ is a solution and $(-6, -2)$ is a solution, then all points on the line containing them will be solutions. The system has an infinite number of solutions. ◀

A system of two equations in two variables is *dependent* if it has infinitely many solutions and *independent* if it has exactly one solution or no solutions.

The system in Example 4 is dependent. The systems in Examples 2 and 3 are independent.

When we graph a system of two equations, one of the following three things can happen.

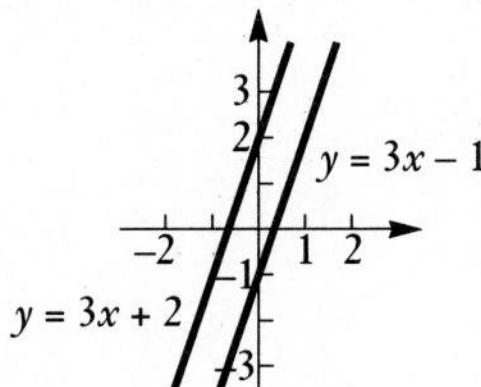

No solution.
Graphs are parallel.
Equations are *inconsistent* because there is no solution. Equations are independent.

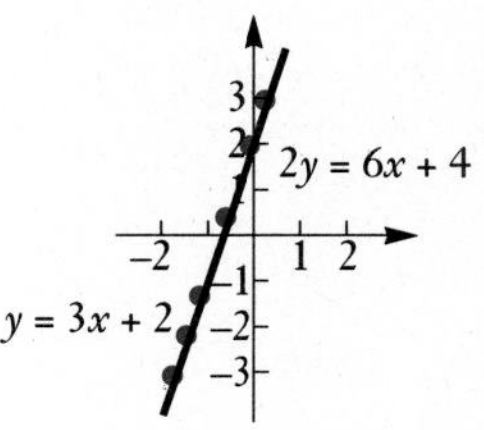

Infinitely many solutions.
Equations have the same graph. Equations are *dependent* (but they are consistent because there is at least one solution).

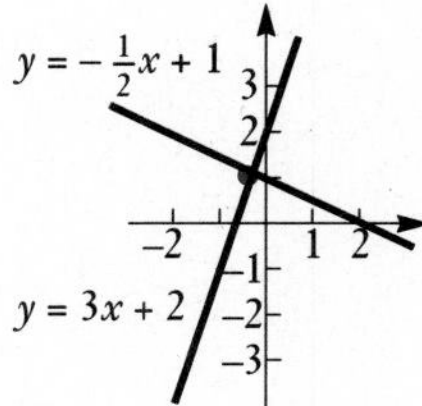

One solution.
Graphs intersect.
Equations are *consistent* (and independent).

DO EXERCISES 7 AND 8.

7. Solve this system by graphing:

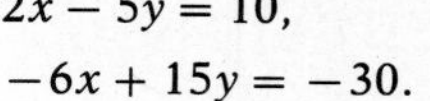

$$2x - 5y = 10,$$
$$-6x + 15y = -30.$$

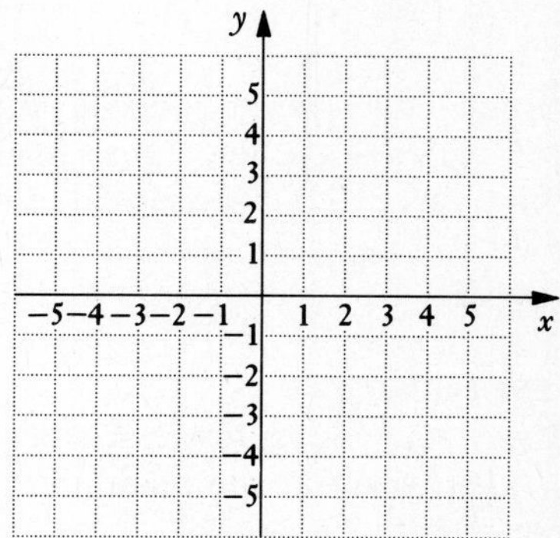

8. Classify each system in Margin Exercises 3, 4, 5, and 7 as dependent or independent.

ANSWERS ON PAGE A-5

❖ SIDELIGHTS

Study Tips: Studying for Tests and Making the Most of Tutoring Sessions

As has been stated, we will often present some tips and guidelines to enhance your learning abilities. Sometimes these tips will be focused on mathematics, but sometimes they will be more general, as is the case here where we consider test preparation and tutoring.

TEST-TAKING TIPS

Many test-taking tips have been covered on the *To the student* page at the beginning of the book. If you have not read that material, do so now. We now provide some other test-taking tips.

- *Make up your own test questions as you study.* You have probably noted by now the section and objective codes that appear throughout the book. After you have done your homework over a particular objective, write one or two questions on your own that you think might be on a test. You will be amazed at the insight this will provide. You are actually carrying out a task similar to what a teacher does in preparing an exam.
- *Ask former students for old exams.* Working such exams can be very helpful and allows you to see what various professors think is important.
- *When taking a test, read each question carefully and try to do all the questions the first time through, but pace yourself.* Answer all the questions, and mark those to recheck if you have time at the end. Very often your first hunch will be correct.
- *Try to write your test in a neat and orderly manner.* Very often your instructor tries to give you partial credit when grading an exam. If your test paper is sloppy and disorderly, it is difficult to verify the partial credit. Doing your homework in a neat and orderly manner can ease such a task on an exam. Try using an erasable pen to make your writing darker and therefore more readable.

MAKING THE MOST OF TUTORING AND HELP SESSIONS

Often you will determine that a tutoring session may be helpful. The following comments may help you to make the most of such situations.

- *Work on the topics before you go to the help or tutoring session. Do not go to such sessions with the view of yourself as an empty cup and the tutor as a magician who will instantly pour in the learning.* The primary source of your ability to learn is within you. We have seen so many students over the years go to help or tutoring sessions with no advanced preparation. You are often wasting your time and perhaps your money if you are paying for such sessions. Go to class, study the textbook, and mark trouble spots. Then use the help and tutoring sessions to deal with these difficulties most efficiently.
- *Do not be afraid to ask questions in these help and tutoring sessions*! The more you relate to your tutor, the more the tutor can help you with your difficulties.
- *Try being a tutor yourself.* Explaining a topic to someone else is often the best way to learn it.
- *What about the student who says "I could do the work at home, but on the test I made silly mistakes"*? Yes, all of us, including instructors, make silly computational mistakes in class, on homework, and on tests. But your instructor, if he or she has taught for some time, is probably aware that 90% of students who make such comments in truth do not have the depth of knowledge of the subject matter, and such silly mistakes very often are a sign that the student has not mastered the material. There is no way we can make that analysis for you. It will have to be unraveled by some careful soul searching on your part or by a conference with your instructor.

NAME SECTION DATE

EXERCISE SET 4.1

ANSWERS

a Determine whether the given ordered pair is a solution of the system of equations. Remember to use alphabetical order of variables.

1. $(1, 2)$; $4x - y = 2,$ $10x - 3y = 4$

2. $(-1, -2)$; $2x + y = -4,$ $x - y = 1$

3. $(2, 5)$; $y = 3x - 1,$ $2x + y = 4$

4. $(-1, -2)$; $x + 3y = -7,$ $3x - 2y = 12$

5. $(1, 5)$; $x + y = 6,$ $y = 2x + 3$

6. $(5, 2)$; $a + b = 7,$ $2a - 8 = b$

1. ______
2. ______
3. ______
4. ______
5. ______
6. ______

b Solve the system of equations by graphing. Then classify the system as consistent or inconsistent and as dependent or independent.

7. $x + y = 4,$ $x - y = 2$

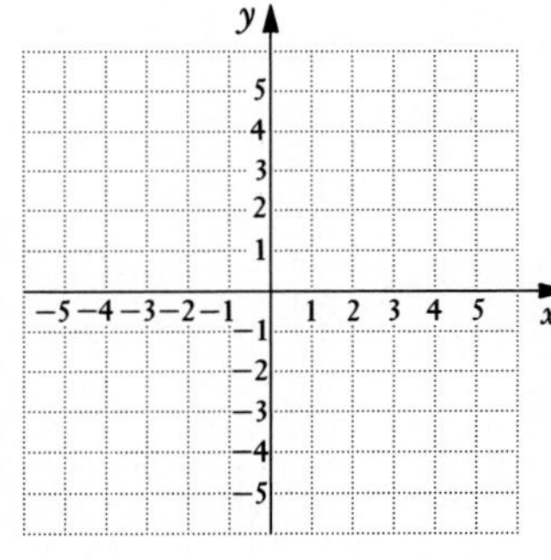

8. $x - y = 3,$ $x + y = 5$

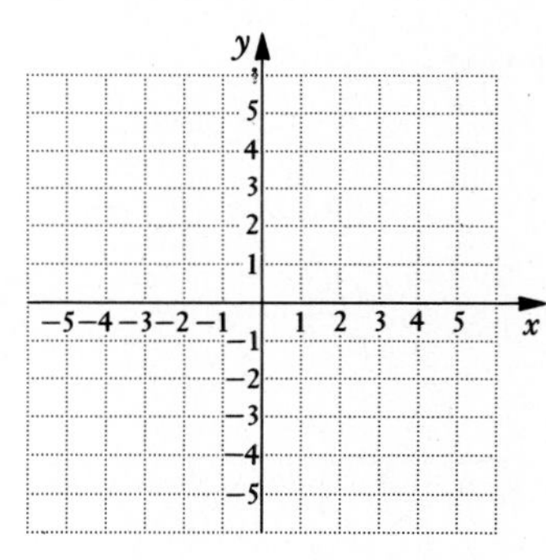

9. $2x - y = 4,$ $2x + 3y = -4$

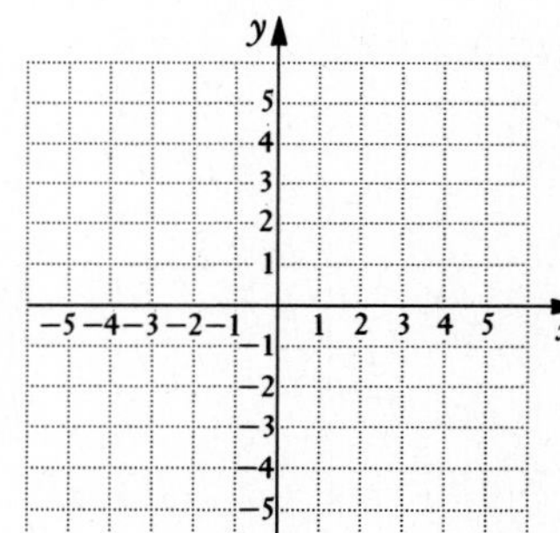

10. $3x + y = 5,$ $x - 2y = 4$

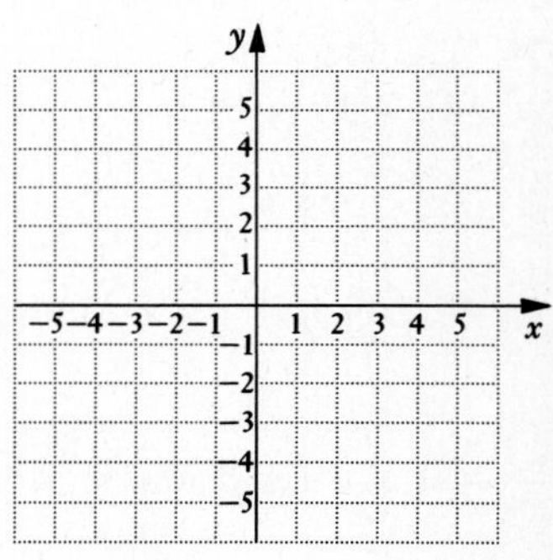

11. $2x + y = 6,$ $3x + 4y = 4$

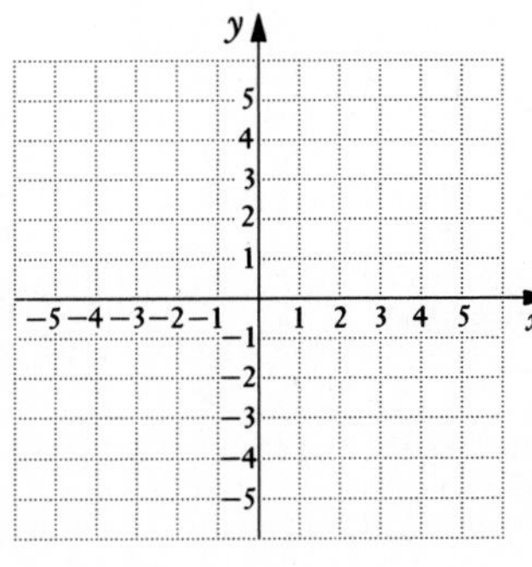

12. $2y = 6 - x,$ $3x - 2y = 6$

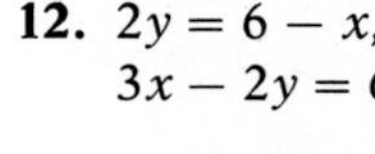

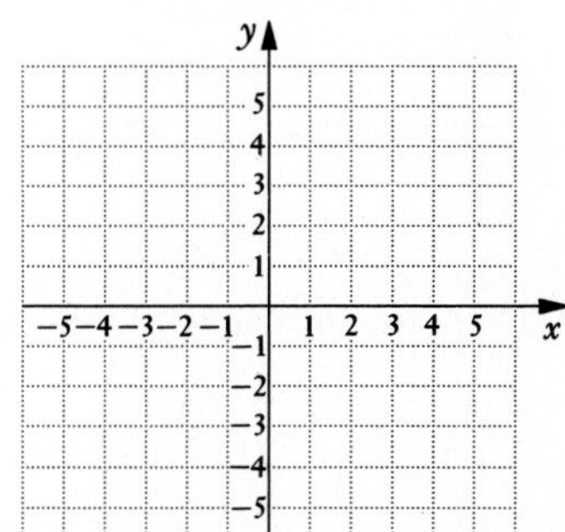

13. $a = 1 + b,$ $b = -2a + 5$

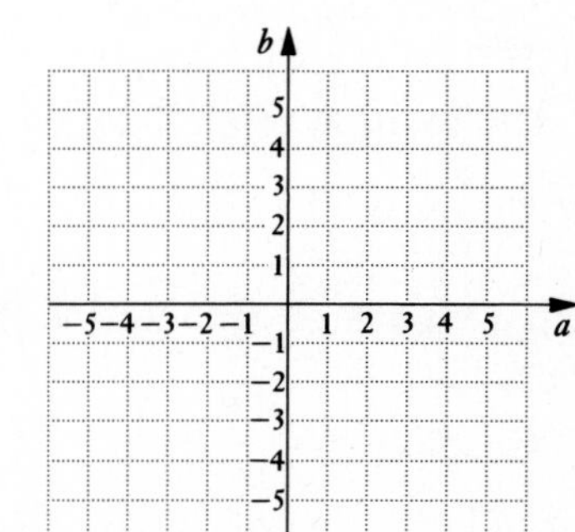

14. $x = y - 1,$ $2x = 3y$

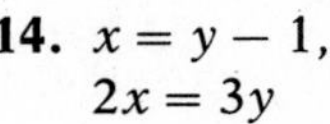

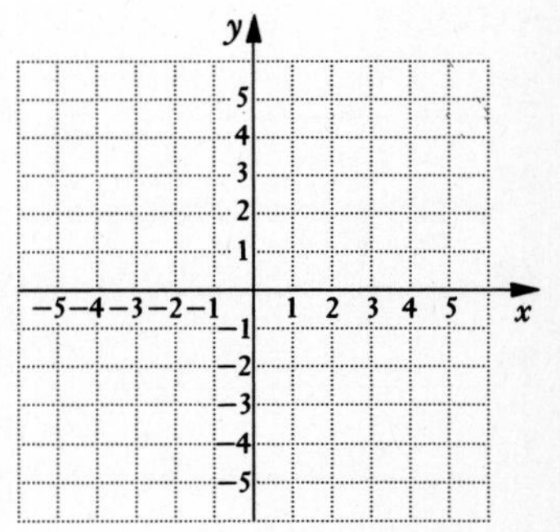

15. $2u + v = 3,$
$2u = v + 7$

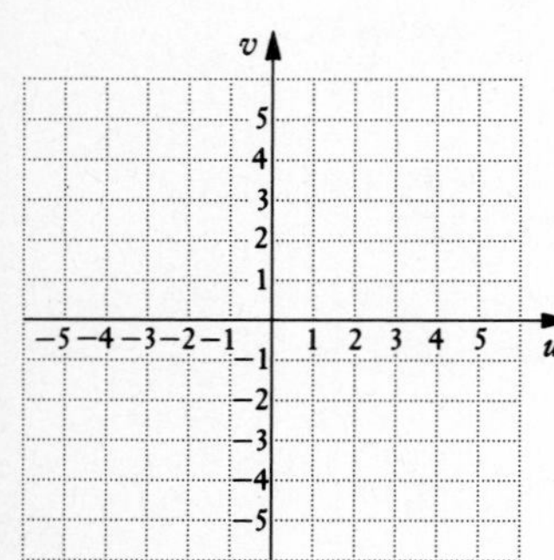

16. $2b + a = 11,$
$a - b = 5$

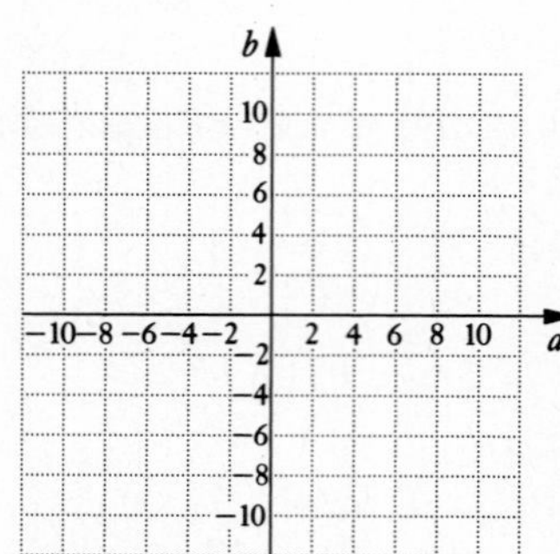

17. $y = -\frac{1}{3}x - 1,$
$4x - 3y = 18$

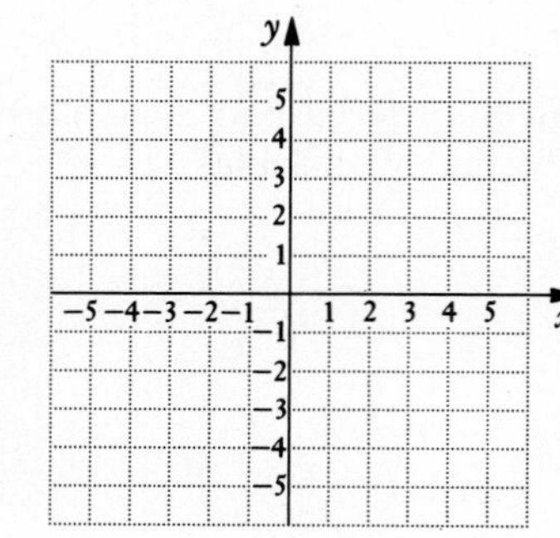

18. $y = -\frac{1}{4}x + 1,$
$2y = x - 4$

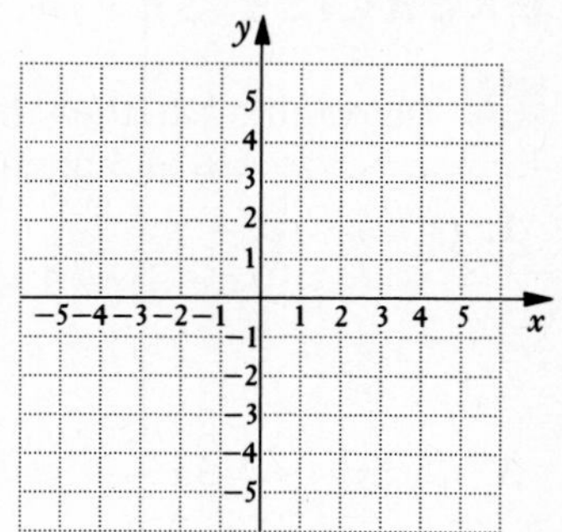

19. $6x - 2y = 2,$
$9x - 3y = 1$

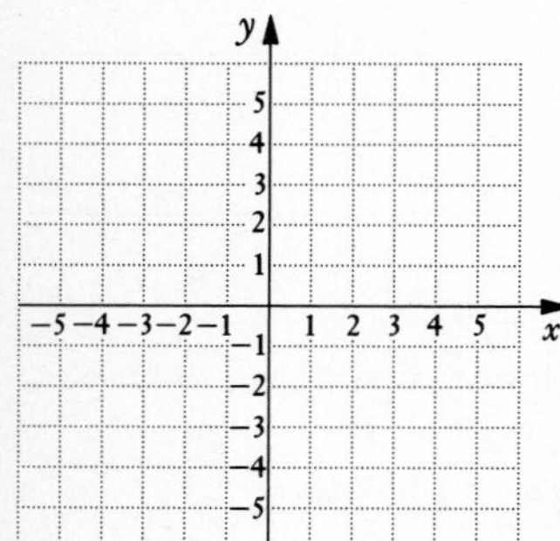

20. $y - x = 5,$
$2x - 2y = 10$

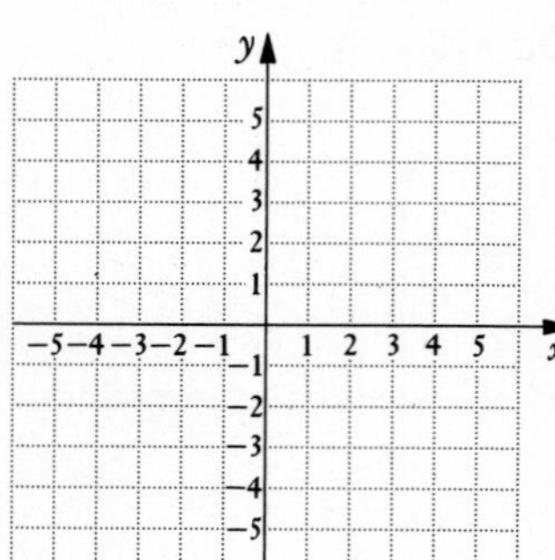

21. $x = 4,$
$y = -5$

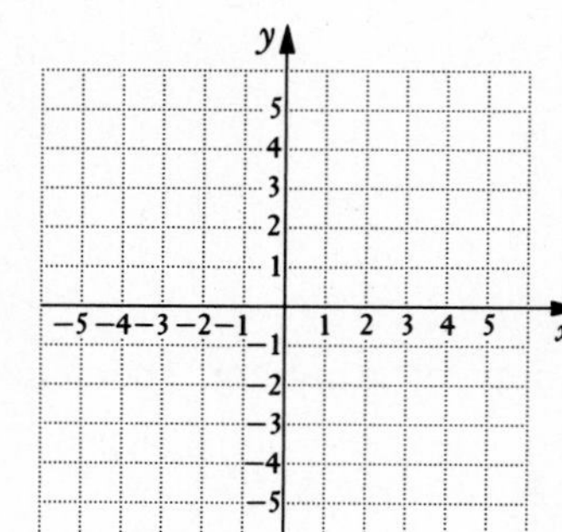

22. $x = -3,$
$y = 2$

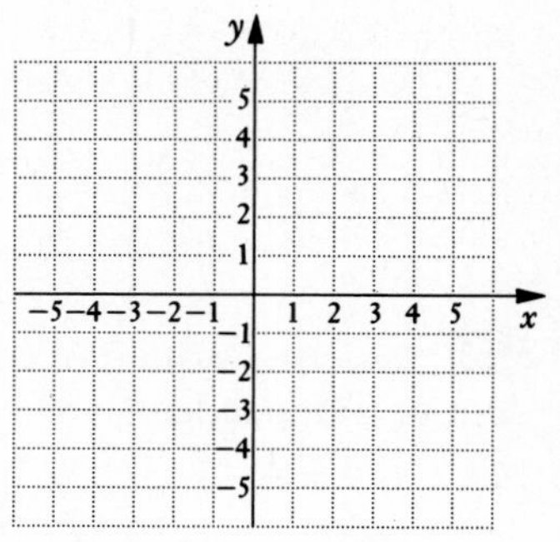

23. $y = -x - 1,$
$4x - 3y = 18$

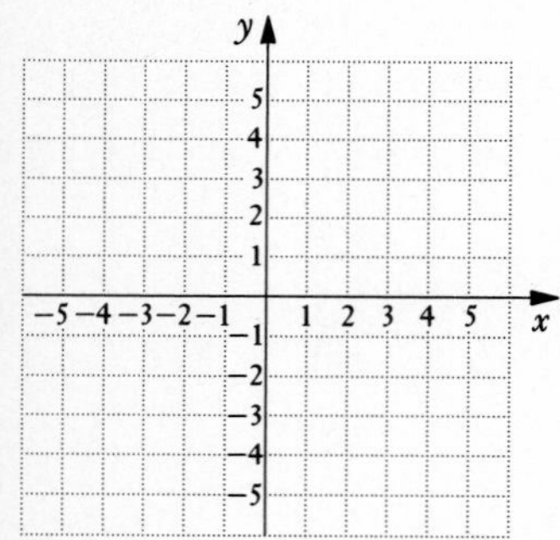

24. 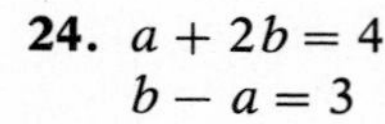
$a + 2b = 4,$
$b - a = 3$

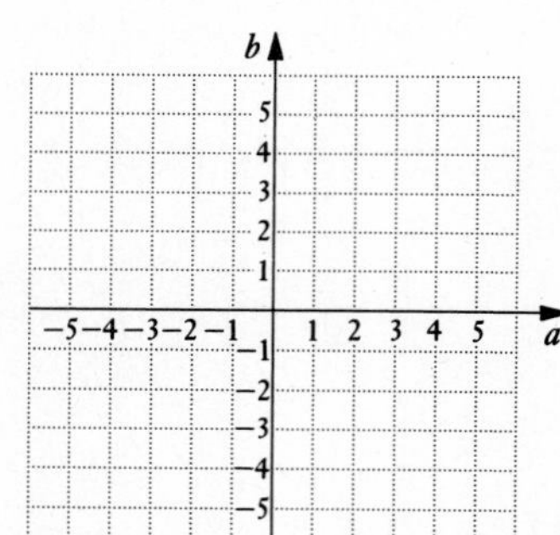

25. $2x - 3y = 6,$
$3y - 2x = -6$

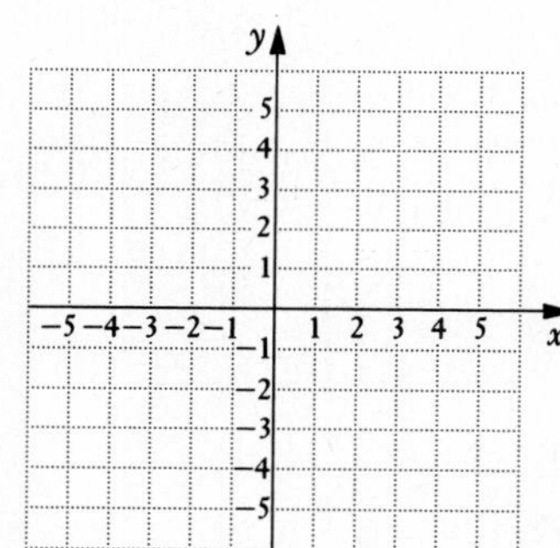

26. $y = 3 - x,$
$2x + 2y = 6$

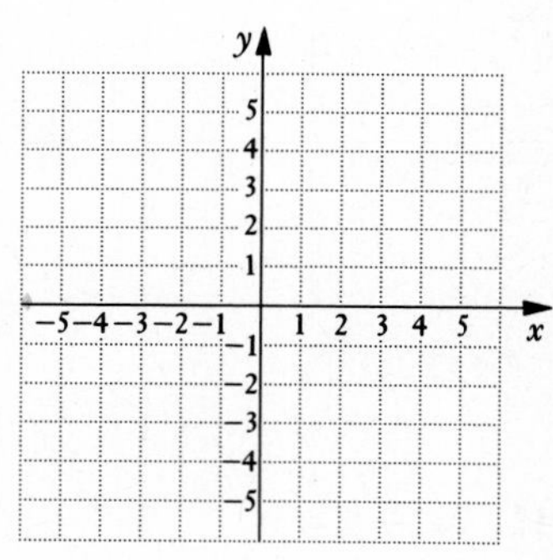

ANSWERS

27. ______

28. ______

29. ______

30. ______

31. ______

32. ______

SYNTHESIS

27. The solution of the following system is $(4, -5)$. Find A and B.

$$Ax - 6y = 13,$$
$$x - By = -8.$$

Write a system of equations with the given solution. Answers may vary.

28. $(5, 1)$

29. $(3, 6)$

30. No solution

31. Infinitely many solutions

32. Find an equation to go with $6x + 7y = -4$ so that $(-3, 2)$ is a solution of the system.

4.2 Solving by Substitution or Elimination

Consider the following system of equations:

$$5x + 9y = 2,$$
$$4x - 9y = 10.$$

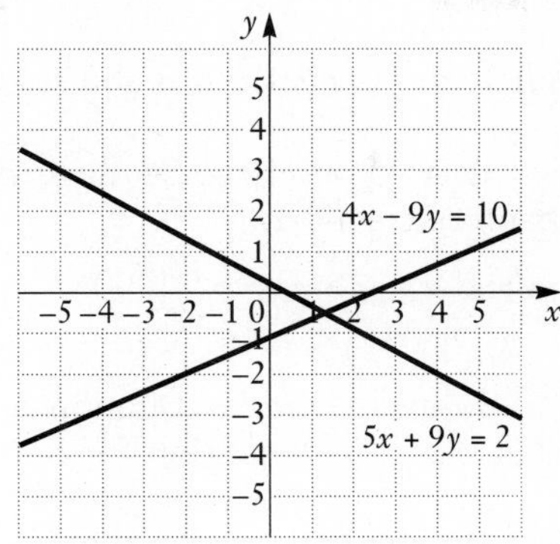

What is the solution? It is rather difficult to tell exactly. It would appear that fractions are involved. It turns out that the solution is $(\frac{4}{3}, -\frac{14}{27})$. We need techniques involving algebra to determine the solution exactly. Graphing helps us to picture the solutions of a system of equations, but solving by graphing, though useful in many applied situations, is not always fast or accurate in cases where solutions are not integers. We now learn other methods using algebra. Because they use algebra, they are called **algebraic methods.**

a The Substitution Method

One nongraphical method for solving systems is known as the **substitution method.**

▶ **EXAMPLE 1** Solve this system:

$$x + y = 4, \quad \textbf{(1)}$$
$$x = y + 1. \quad \textbf{(2)}$$

Equation (2) says that x and $y + 1$ name the same number. Thus we can substitute $y + 1$ for x in Equation (1):

$x + y = 4$ **Equation (1)**

$(y + 1) + y = 4.$ **Substituting $y + 1$ for x**

We solve this last equation, using methods learned earlier:

$(y + 1) + y = 4$

$2y + 1 = 4$ **Removing parentheses and collecting like terms**

$2y = 3$ **Subtracting 1 on both sides**

$y = \frac{3}{2}.$ **Dividing by 2**

We return to the original pair of equations and substitute $\frac{3}{2}$ for y in either of them so that we can solve for x. Calculation will be easier if we choose Equation (2):

$x = y + 1$ **Equation (2)**

$x = \frac{3}{2} + 1$ **Substituting $\frac{3}{2}$ for y**

$x = \frac{3}{2} + \frac{2}{2} = \frac{5}{2}.$

We obtain the ordered pair $(\frac{5}{2}, \frac{3}{2})$. We check to be sure that it is a solution.

OBJECTIVES

After finishing Section 4.2, you should be able to:

a Solve systems of equations in two variables by the substitution method.

b Solve systems of equations in two variables by the elimination method.

FOR EXTRA HELP

Tape 7B

Tape 5B

MAC: 4
IBM: 4

Solve by the substitution method.

1. $x + y = 6,$
$y = x + 2$

2. $y = 7 - x,$
$2x - y = 8$

(*Caution:* Use parentheses when you substitute. Then be very careful about removing them. Remember to solve for both variables.)

Solve by the substitution method.

3. $2y + x = 1,$
$y - 2x = 8$

4. $8x + 5y = 184,$
$x - y = -3$

ANSWERS ON PAGE A-5

Check:

$x + y = 4$	
$\frac{5}{2} + \frac{3}{2}$	4
$\frac{8}{2}$	
4	TRUE

$x = y + 1$	
$\frac{5}{2}$	$\frac{3}{2} + 1$
	$\frac{3}{2} + \frac{2}{2}$
	$\frac{5}{2}$ TRUE

Since $(\frac{5}{2}, \frac{3}{2})$ checks, it is the solution. ◀

This solution would have been difficult to find graphically because it involves fractions.

DO EXERCISES 1 AND 2.

Suppose neither equation of a pair has a variable alone on one side. We then solve one equation for one of the variables.

▶ **EXAMPLE 2** Solve this system:

$$2x + y = 6, \qquad (1)$$
$$3x + 4y = 4. \qquad (2)$$

First we solve one equation for one variable. Since the coefficient of y is 1 in Equation (1), it is the easier one to solve for y:

$$y = 6 - 2x. \qquad (3)$$

Then we substitute $6 - 2x$ for y in Equation (2) and solve for x:

$$3x + 4(6 - 2x) = 4 \qquad \textbf{Substituting } 6 - 2x \textbf{ for } y$$
$$3x + 24 - 8x = 4 \qquad \textbf{Multiplying to remove parentheses}$$
$$3x - 8x = 4 - 24$$
$$-5x = -20$$
$$x = 4.$$

Remember to use parentheses when you substitute. Then remove them carefully.

We return to either of the original equations, (1) or (2), or Equation (3), which we solved for y. It is generally easier to use an equation like (3), where we have solved for the specific variable. We substitute 4 for x in Equation (3):

$$y = 6 - 2x = 6 - 2(4) = 6 - 8 = -2.$$

Check:

$2x + y = 6$	
$2(4) + (-2)$	6
$8 - 2$	
6	TRUE

$3x + 4y = 4$	
$3(4) + 4(-2)$	4
$12 - 8$	
4	TRUE

Since $(4, -2)$ checks, it is the solution. ◀

DO EXERCISES 3 AND 4.

b The Elimination Method

The **elimination method** for solving systems of equations makes use of the *addition principle* for equations. Some systems are much easier to solve using the elimination method rather than the substitution method. Let us consider an example.

▶ **EXAMPLE 3** Solve this system:

$$2x - 3y = 0, \quad (1)$$
$$-4x + 3y = -1. \quad (2)$$

The key to the advantage of the elimination method for solving this system involves the $-3y$ in one equation and the $3y$ in the other. These terms are opposites. If we add the terms on the sides of the equations, these terms will add to 0, and in effect, the variable y will be "eliminated."

We will use the addition principle for equations. According to Equation (2), $-4x + 3y$ and -1 are the same number. Thus we can use a vertical form and add $-4x + 3y$ to the left side of Equation (1) and -1 to the right side:

$$\begin{array}{rl} 2x - 3y = 0 & (1) \\ -4x + 3y = -1 & (2) \\ \hline -2x + 0y = -1. & \text{Adding} \end{array}$$

We have eliminated the variable y, which is why we call this the *elimination method.* We now have an equation with just one variable, which we solve for x:

$$-2x = -1$$
$$x = \tfrac{1}{2}.$$

We now substitute $\frac{1}{2}$ for x in Equation (1) and solve for y:

$$2 \cdot \tfrac{1}{2} - 3y = 0 \quad \text{Substituting}$$
$$\left.\begin{array}{r} 1 - 3y = 0 \\ y = \tfrac{1}{3} \end{array}\right\} \quad \text{Solving for } y$$

Check:

$$\begin{array}{r|l} 2x - 3y = 0 & \\ \hline 2(\frac{1}{2}) - 3(\frac{1}{3}) & 0 \\ 1 - 1 & \\ 0 & \text{TRUE} \end{array} \qquad \begin{array}{r|l} -4x + 3y = -1 & \\ \hline -4(\frac{1}{2}) + 3(\frac{1}{3}) & -1 \\ -2 + 1 & \\ -1 & \text{TRUE} \end{array}$$

Since $(\frac{1}{2}, \frac{1}{3})$ checks, it is the solution. ◀

DO EXERCISES 5 AND 6.

Solve by the elimination method.

5. $5x + 3y = 17,$
$-5x + 2y = 3$

6. $-3a + 2b = 0,$
$3a - 4b = -1$

Note in Example 3 that a term in one equation, $-3y$, and a term in the other, $3y$, are opposites of each other. Thus their sum is 0. That fact enabled us to eliminate a variable.

In order to eliminate a variable, we sometimes use the multiplication principle to multiply one or both of the equations by a particular number before adding.

▶ **EXAMPLE 4** Solve this system:

$$3x + 3y = 15, \quad (1)$$
$$2x + 6y = 22. \quad (2)$$

If we add directly, we get $5x + 9y = 37$, but we have not eliminated a variable. However, if the $3y$ in Equation (1) were $-6y$, we could eliminate y.

ANSWERS ON PAGE A-5

7. Solve by the elimination method:

$$2y + 3x = 12,$$
$$-4y + 5x = -2.$$

8. Solve by the elimination method:

$$4x + 5y = -8,$$
$$7x + 9y = 11.$$

ANSWERS ON PAGE A-5

Thus we multiply by -2 on both sides of Equation (1) and add:

$$-6x - 6y = -30 \quad \text{Multiplying Equation (1) by } -2 \text{ on both sides}$$
$$2x + 6y = 22 \quad \text{Equation (2)}$$
$$-4x + 0 = -8 \quad \text{Adding}$$
$$x = 2. \quad \text{Solving for } x$$

Then

$$2 \cdot 2 + 6y = 22 \quad \text{Substituting 2 for } x \text{ in Equation (2)}$$
$$4 + 6y = 22$$
$$y = 3. \quad \text{Solving for } y$$

We obtain (2, 3), or $x = 2$, $y = 3$. This checks, so it is the solution. ◀

DO EXERCISE 7.

Sometimes we must multiply twice in order to make two terms opposites.

▶ **EXAMPLE 5** Solve this system:

$$2x + 3y = 17, \quad (1)$$
$$5x + 7y = 29. \quad (2)$$

We have

$$2x + 3y = 17,$$
$$5x + 7y = 29$$

From Equation (1): $10x + 15y = 85$ **Multiplying by 5**

From Equation (2): $-10x - 14y = -58$ **Multiplying by -2**

$$0 + y = 27 \quad \text{Adding}$$
$$y = 27. \quad \text{Solving for } y$$

Then

$$2x + 3 \cdot 27 = 17 \quad \text{Substituting 27 for } y \text{ in Equation (1)}$$
$$2x + 81 = 17$$
$$2x = -64 \quad \text{Solving for } x$$
$$x = -32.$$

Check:

$2x + 3y = 17$		
$2(-32) + 3(27)$	17	
$-64 + 81$		
17		TRUE

$5x + 7y = 29$		
$5(-32) + 7(27)$	29	
$-160 + 189$		
29		TRUE

We obtain $(-32, 27)$, or $x = -32$, $y = 27$, as the solution. ◀

DO EXERCISE 8.

Some systems have no solution, as we saw graphically in Section 4.1. How do we recognize such systems if we are solving using an algebraic method?

▶ **EXAMPLE 6** Solve this system:

$$y + 3x = 5, \qquad (1)$$
$$y + 3x = -2. \qquad (2)$$

If we find the slope–intercept equations for this system, we get

$$y = -3x + 5,$$
$$y = -3x - 2.$$

We graphed this system in Example 3 of Section 4.1. The lines are parallel and the system has no solution. Let us see what happens if we attempt to solve the system using the elimination method. We multiply by -1 on both sides of Equation (2) and add:

$$\begin{aligned} y + 3x &= 5 \\ -y - 3x &= 2 \quad \textbf{Multiplying by } -1 \textbf{ on both sides} \\ \hline 0 &= 7. \quad \textbf{Adding, we obtain a false equation.} \end{aligned}$$

The x-terms and the y-terms are eliminated and we end up with a *false* equation. Thus, if we obtain a false equation when solving algebraically, we know that the system has no solution. ◀

Some systems have infinitely many solutions. How can we recognize such a situation when we are solving systems using an algebraic method?

▶ **EXAMPLE 7** Solve this system:

$$3y - 2x = 6,$$
$$-12y + 8x = -24.$$

We graphed this system in Example 4 of Section 4.1. The lines are the same and the system has an infinite number of solutions. Suppose we try to solve this system using the elimination method:

$$\begin{aligned} 12y - 8x &= 24 \quad \textbf{Multiplying Equation (1) by 4} \\ -12y + 8x &= -24 \\ \hline 0 &= 0. \quad \textbf{Adding, we obtain a true equation.} \end{aligned}$$

Note that we have eliminated both variables and what remains is a true equation. It can be expressed as $0 \cdot x + 0 \cdot y = 0$, and is true for all numbers x and y. If a pair is a solution of one of the original equations, then it will be a solution of the other. The system is dependent and has an infinite number of solutions. ◀

When solving a system of two linear equations in two variables:

1. **If a false equation is obtained, such as $0 = 7$, then the system has no solution. The system is inconsistent and independent.**
2. **If a true equation is obtained, such as $0 = 0$, then the system has an infinite number of solutions. The system is consistent and dependent.**

DO EXERCISES 9 AND 10.

In carrying out the elimination method, it may help to first write the equations in the form $Ax + By = C$. When decimals or fractions occur, it also helps to *clear* before solving.

9. Solve by the substitution method:

$$y + 2x = 3,$$
$$y = -2x + 3.$$

10. Solve by the elimination method:

$$2x - 5y = 10,$$
$$-6x + 15y = -30.$$

ANSWERS ON PAGE A-5

11. Clear of decimals. Then solve.

$$0.02x + 0.03y = 0.01,$$
$$0.3x - 0.1y = 0.7$$

(*Hint:* Multiply the first equation by 100 and the second one by 10.)

12. Clear of fractions. Then solve.

$$\frac{3}{5}x + \frac{2}{3}y = \frac{1}{3},$$
$$\frac{3}{4}x - \frac{1}{3}y = \frac{1}{4}$$

ANSWERS ON PAGE A-5

▶ **EXAMPLE 8** Clear of fractions:

$$\tfrac{1}{2}x + \tfrac{2}{3}y = 1,$$
$$\tfrac{3}{4}x - \tfrac{1}{3}y = 2.$$

We multiply each equation by the LCM of the denominators.

$$6(\tfrac{1}{2}x + \tfrac{2}{3}y) = 6(1) \quad \textbf{Multiplying on both sides by 6}$$
$$12(\tfrac{3}{4}x - \tfrac{1}{3}y) = 12(2). \quad \textbf{Multiplying on both sides by 12}$$

Now we simplify.

Equation (1): $6 \cdot \tfrac{1}{2}x + 6 \cdot \tfrac{2}{3}y = 6$
$3x + 4y = 6;$

Equation (2): $12 \cdot \tfrac{3}{4}x - 12 \cdot \tfrac{1}{3}y = 24$
$9x - 4y = 24$

We can now finish solving using the elimination method. ◀

DO EXERCISES 11 AND 12.

Using the elimination method to solve systems of two equations:

1. **Write both equations in the form $Ax + By = C$.**
2. **Clear of any decimals or fractions.**
3. **Choose a variable to eliminate.**
4. **Make the chosen variable's terms opposites by multiplying one or both equations by appropriate numbers.**
5. **Eliminate a variable by adding the sides of the equations and then solve for the remaining variable.**
6. **Substitute in either of the original equations to find the value of the other variable.**

Comparing Methods

The following summary compares the graphical, substitution, and elimination methods for solving systems of equations. When deciding which method to use, consider the chart and directions from your instructor. The situation is like having a piece of wood to cut and three saws with which to cut it. The saw you use depends on the type of wood, the type of cut you are making, and how you want the wood to turn out.

Method	Strengths	Weaknesses
Graphical	Can see solution.	Inexact when solution involves numbers that are not integers.
Substitution	Works when solutions involve numbers that are not integers. Easy to use when a variable is alone on one side.	Introduces extensive computations with fractions for more complicated systems where coefficients are not 1 or -1. Cannot "see" the solution quickly.
Elimination	Works when solutions involve numbers that are not integers. Works well when coefficients are not 1 or -1, and when coefficients involve decimals or fractions. Easier to use when equations are in the form $Ax + By = C$.	Cannot "see" the solution quickly.

NAME SECTION DATE

EXERCISE SET 4.2

a Solve by the substitution method.

1. $3x + 5y = 3,$
   $x = 8 - 4y$
2. $2x - 3y = 13,$
   $y = 5 - 4x$
3. $9x - 2y = 3,$
   $3x - 6 = y$
4. $x = 3y - 3,$
   $x + 2y = 9$
5. $5m + n = 8,$
   $3m - 4n = 14$
6. $4x + y = 1,$
   $x - 2y = 16$
7. $4x + 12y = 4,$
   $-5x + y = 11$
8. $-3b + a = 7,$
   $5a + 6b = 14$

b Solve by the elimination method.

9. $x + 3y = 7,$
   $-x + 4y = 7$
10. $x + y = 9,$
    $2x - y = -3$
11. $9x + 3y = -3,$
    $2x - 3y = -8$
12. $6x - 3y = 18,$
    $6x + 3y = -12$
13. $5x + 3y = 19,$
    $2x - 5y = 11$
14. $3x + 2y = 3,$
    $9x - 8y = -2$
15. $5r - 3s = 24,$
    $3r + 5s = 28$
16. $5x - 7y = -16,$
    $2x + 8y = 26$
17. $0.3x - 0.2y = 4,$
    $0.2x + 0.3y = 1$
18. $0.7x - 0.3y = 0.5,$
    $-0.4x + 0.7y = 1.3$
19. $\frac{1}{2}x + \frac{1}{3}y = 4,$
    $\frac{1}{4}x + \frac{1}{3}y = 3$
20. $\frac{2}{3}x + \frac{1}{7}y = -11,$
    $\frac{1}{7}x - \frac{1}{3}y = -10$
21. $\frac{2}{5}x + \frac{1}{2}y = 2,$
    $\frac{1}{2}x - \frac{1}{6}y = 3$
22. $\frac{1}{3}x + \frac{1}{5}y = 7,$
    $\frac{1}{6}x - \frac{2}{5}y = -4$
23. $2x + 3y = 1,$
    $4x + 6y = 2$

ANSWERS

1. ______
2. ______
3. ______
4. ______
5. ______
6. ______
7. ______
8. ______
9. ______
10. ______
11. ______
12. ______
13. ______
14. ______
15. ______
16. ______
17. ______
18. ______
19. ______
20. ______
21. ______
22. ______
23. ______

ANSWERS

24. ______

25. ______

26. ______

27. ______

28. ______

29. ______

30. ______

31. ______

32. ______

33. ______

34. ______

35. ______

36. ______

37. ______

38. ______

39. ______

40. ______

41. ______

42. ______

24. $3x - 2y = 1,$
$-6x + 4y = -2$

25. $2x - 4y = 5,$
$2x - 4y = 6$

26. $3x - 5y = -2,$
$5y - 3x = 7$

27. $5x - 9y = 7,$
$7y - 3x = -5$

28. $a - 2b = 16,$
$b + 3 = 3a$

29. $3(a - b) = 15,$
$4a = b + 1$

30. $10x + y = 306,$
$10y + x = 90$

31. $x - \frac{1}{10}y = 100,$
$y - \frac{1}{10}x = -100$

32. $\frac{1}{8}x + \frac{3}{5}y = \frac{19}{2},$
$-\frac{3}{10}x - \frac{7}{20}y = -1$

33. $0.05x + 0.25y = 22,$
$0.15x + 0.05y = 24$

34. $1.3x - 0.2y = 12,$
$0.4x + 17y = 89$

SKILL MAINTENANCE

35. Find the slope of the line $y = 1.3x - 7$.

36. Simplify: $-9(y + 7) - 6(y - 4)$.

SYNTHESIS

Solve.

37. ▦ $3.5x - 2.1y = 106.2,$
$4.1x + 16.7y = -106.28$

38. $\frac{x+y}{2} - \frac{x-y}{5} = 1,$
$\frac{x-y}{2} + \frac{x+y}{6} = -2,$

39. Solve for x and y in terms of a and b:

$$5x + 2y = a,$$
$$x - y = b.$$

40. Determine a and b for which $(-4, -3)$ will be a solution of the system

$$ax + by = -26,$$
$$bx - ay = 7.$$

41. The points $(0, -3)$ and $(-\frac{3}{2}, 6)$ are two of the solutions of the equation $px - qy = -1$. Find p and q.

42. For $y = mx + b$, two solutions are $(1, 2)$ and $(-3, 4)$. Find m and b.

4.3 Solving Problems Using Systems of Two Equations in Two Variables

a Because you can now use systems of equations, you are in a position to translate and solve problems much more easily.

▶ **EXAMPLE 1** Two angles are supplementary. One angle is 12° less than three times the other. Find the measures of the angles.

1. *Familiarize.* Recall that two angles are supplementary if the sum of their measures is 180°. Suppose we make a guess about the answers. The measures 50° and 130° have a sum that is 180°. One angle is supposed to be 12° less than three times the other. Now $3(50°) - 12 = 150° - 12° = 138°$, which is not the other angle, 130°. Also $3(130°) - 12 = 390° - 12° = 378°$, which is not an appropriate measure for an angle. The steps we have used to see if our guess is correct help us to understand the actual steps involved in solving the problem. We let x = the measure of one angle and y = the measure of the other angle.

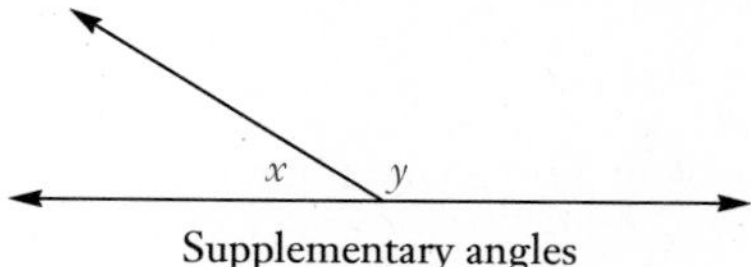

Supplementary angles

2. *Translate.* There are two statements in the problem. The fact that the two angles are supplementary can be reworded and translated as follows:

Two angles are supplementary.

Rewording: The sum of the measures is 180°.

Translating: $x + y \quad = 180$

The second statement of the problem can be translated as follows, again using x and y:

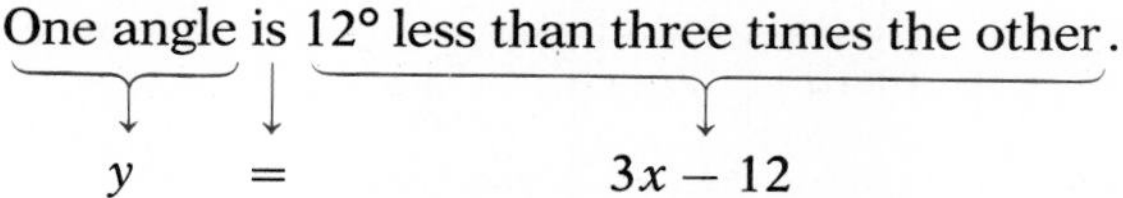

The second statement could also have been translated as $x = 3y - 12$. The problem has now been translated to a *system of equations:*

$$x + y = 180, \quad (1)$$
$$y = 3x - 12. \quad (2)$$

3. *Solve.* We solve the system of equations using algebra. Since we have a variable alone on one side, we use the substitution method:

$$x + y = 180 \quad \text{Equation (1)}$$
$$x + (3x - 12) = 180 \quad \text{Substituting } 3x - 12 \text{ for } y$$
$$4x - 12 = 180 \quad \text{Collecting like terms}$$
$$4x = 192$$
$$x = 48.$$

We return to Equation (2) and substitute 48 for x and compute y:

$$y = 3x - 12 = 3(48) - 12 = 132.$$

OBJECTIVES

After finishing Section 4.3, you should be able to:

a Solve a problem by translating to a system of two equations and solving the system.

b Solve motion problems that involve systems of equations.

FOR EXTRA HELP

Tape 8B — Tape 6A — MAC: 4 IBM: 4

1. Two angles are supplementary. One is 35° more than twice the other. Find the measures of the angles.

ANSWER ON PAGE A-5

4. *Check.* The angles seem to be 48° and 132°. The sum of the angles is 48° + 132°, or 180°, so they are supplementary. Also, three times the 48° angle minus 12° is 132°, the second angle. Thus the angles check.
5. *State.* The measure of one angle is 48° and the other is 132°. ◀

DO EXERCISE 1.

▶ **EXAMPLE 2** The manager of a store notes that one day the glove department took in $687.25 on the sale of cloth gloves and pigskin gloves. The manager knows that 20 pairs of gloves were sold and wants to reorder, but she does not know how many pairs of each kind were sold. Cloth gloves sold for $24.95 per pair, and pigskin gloves sold for $37.50 per pair. How many of each kind were sold?

1. *Familiarize.* To familiarize ourselves with the problem situation, we make a guess and do some calculations. The total number of pairs of gloves sold was 20, so we choose pairs of numbers that total 20. Let's try 12 pairs of cloth gloves and 8 pairs of pigskin gloves. How much money was taken in? The problem says that cloth gloves sold for $24.95, so the store took in

$$12(\$24.95)$$

from the cloth gloves. The store took in

$$8(\$37.50)$$

from the pigskin gloves. This makes the total taken in

$$12(\$24.95) + 8(\$37.50) = \$299.40 + \$300.00 = \$599.40.$$

Our guess is not the answer to the problem because the total taken in, according to the problem, was $687.25. Since $599.40 is smaller than $687.25, it seems reasonable to assume that more of the expensive gloves were sold than the number we guessed. We could then revise our guess, but instead let's carry out our translation.

First we list the information in a table. Let $c =$ the number of cloth gloves sold and $p =$ the number of pigskin gloves sold.

Kind of glove	Cloth	Pigskin	Totals	
Number sold	c	p	20	→ $c + p = 20$
Price	$24.95	$37.50		
Money taken in	24.95c$	37.50p$	$687.25	→ $24.95c + 37.50p = 687.25$

2. *Translate.* The total number sold was 20, so we have

$$c + p = 20.$$

The amount taken in for cloth gloves is $24.95c$, and the amount taken in for pigskin gloves is $37.50p$. These amounts are in dollars. The total was $687.25, so we have

$$24.95c + 37.50p = 687.25.$$

We can multiply on both sides by 100 in order to clear of the decimals. Thus we have the translation, a system of equations:

$$c + p = 20, \quad (1)$$
$$2495c + 3750p = 68{,}725. \quad (2)$$

3. *Solve.* What method should we use to solve the system? Since no variable appears alone and the equations are in the form $Ax + By = C$, let's use the elimination method. We eliminate c by multiplying Equation (1) by -2495 and adding to Equation (2):

$$\begin{aligned} -2495c - 2495p &= -49{,}900 && \textbf{Multiplying Equation (1) by } -2495 \\ 2495c + 3750p &= 68{,}725 \\ \hline 1255p &= 18{,}825 && \textbf{Adding} \\ p &= 15. && \textbf{Solving for } p \end{aligned}$$

To find c, we substitute 15 for p in Equation (1) and solve for c:

$$\begin{aligned} c + p &= 20 && \textbf{Equation (1)} \\ c + 15 &= 20 && \textbf{Substituting 15 for } p \\ c &= 5. && \textbf{Solving for } c \end{aligned}$$

We obtain

$$c = 5 \quad \text{and} \quad p = 15.$$

4. *Check.* We check the possible answer in the original problem. Remember that c is the number of cloth gloves and p is the number of pigskin gloves. Thus:

Number of gloves: $c + p = 5 + 15 = 20$

Money from cloth gloves: $\$24.95c = 24.95 \times 5 = \$124.75$

Money from pigskin gloves: $\$37.50p = 37.50 \times 15 = \$562.50$

Total $= \$687.25$

The numbers check.

5. *State.* The answer is that the store sold 5 pairs of cloth gloves and 15 pairs of pigskin gloves. ◀

DO EXERCISE 2.

▶ **EXAMPLE 3** *A mixture problem.* Solution A is 2% alcohol and solution B is 6% alcohol. A service station owner wants to mix the two in order to get 60 L of solution that is 3.2% alcohol. How many liters of each should the owner use?

1. *Familiarize.* We list the information in a table. We let a = the number of liters of A and b = the number of liters of B.

	Solution A	Solution B	Mixture	
Amount of solution	a liters	b liters	60 liters	$\longrightarrow a + b = 60$
Percent of alcohol	2%	6%	3.2%	
Amount of alcohol in solution	$0.02a$	$0.06b$	0.032×60, or 1.92 liters	$\longrightarrow 0.02a + 0.06b = 1.92$

To get the amount of alcohol, we multiply by the percentages.

2. *Translate.* If we add a and b in the first row, we get 60, and this gives us one equation:

$$a + b = 60.$$

2. A store sold 30 sweatshirts. They sold white ones for \$18.95 and red ones for \$19.50. They took in \$572.90. How many of each color did they sell?

ANSWER ON PAGE A-5

3. A gardener has two kinds of solutions containing weedkiller and water. One is 5% weedkiller and the other is 15% weedkiller. The gardener needs 100 L of a 12% solution and wants to make it by mixing the two. How much of each solution should be used?

Do the familiarization and translating steps by completing the following chart. Let $x =$ the number of liters of the 5% solution and $y =$ the number of liters of the 15% solution.

	5% weed-killer	15% weed-killer	Mixture
Amount of solution	x liters	y liters	
Percent of weedkiller	5%		
Amount of weedkiller in solution			

ANSWER ON PAGE A-5

If we add the amounts of alcohol in the third row, we get 1.92, and this gives us another equation:

$$0.02a + 0.06b = 1.92.$$

After clearing of decimals, we have this system:

$$a + b = 60, \quad (1)$$
$$2a + 6b = 192. \quad (2)$$

3. *Solve.* We solve the system using elimination, since the equations are in the form $Ax + By = C$ and no variable appears alone. We multiply Equation (1) by -2 and add the result to Equation (2):

$$-2a - 2b = -120 \quad \textbf{Multiplying Equation (1) by } -2$$
$$2a + 6b = 192$$
$$0 + 4b = 72 \quad \textbf{Adding}$$
$$b = 18 \quad \textbf{Solving for } b$$

$$a + 18 = 60 \quad \textbf{Substituting in Equation (1) of the system}$$
$$a = 42.$$

4. *Check.* Remember, $a =$ the number of liters of the 2% solution and $b =$ the number of liters of the 6% solution.

Total number of liters of mixture: $a + b = 42 + 18 = 60$

Amount of alcohol: $2\% \times 42 + 6\% \times 18 = 0.02 \times 42 + 0.06 \times 18 = 1.92$ liters

Percentage of alcohol in mixture: $\dfrac{1.92}{60} = 0.032$, or 3.2%

The numbers check in the original problem.

5. *State.* The answer is that the owner should use 42 L of solution A and 18 L of solution B. ◀

DO EXERCISE 3.

▶ **EXAMPLE 4** *An interest problem.* Two investments are made totaling \$4800. In the first year, they yield \$412 in simple interest. Part of the money is invested at 8% and the rest at 9%. Find the amount invested at each rate of interest.

1. *Familiarize.* Keeping information in a table will help. The columns in the table come from the formula for simple interest: $I = Prt$. We let $x =$ the number of dollars invested at 8% and $y =$ the number of dollars invested at 9%.

	First investment	Second investment	Total	
Principal, *P*	x	y	\$4800	$\longrightarrow x + y = \$4800$
Rate of interest, *r*	8%	9%		
Time, *t*	1 yr	1 yr		
Interest, *I*	$0.08x$	$0.09y$	\$412	$\longrightarrow 0.08x + 0.09y = \$412$

2. *Translate.* The total of the amounts invested is found in the first row. This gives us one equation:

$$x + y = 4800.$$

Look at the last row. The interest, or **yield,** totals \$412. This gives us a second equation:

$$8\%x + 9\%y = 412, \quad \text{or} \quad 0.08x + 0.09y = 412.$$

After we multiply on both sides to clear of decimals, we have

$$8x + 9y = 41{,}200.$$

3. *Solve.* We solve the system, again using elimination:

$$x + y = 4800,$$
$$8x + 9y = 41{,}200.$$

We find that $x = 2000$ and $y = 2800$.

4. *Check.* The sum is \$2000 + \$2800, or \$4800. The interest from \$2000 at 8% for 1 yr is 8%(\$2000), or \$160. The interest from \$2800 at 9% for 1 yr is 9%(\$2800), or \$252. The total interest is \$160 + \$252, or \$412.

5. *State.* The numbers check in the problem, so we know that \$2000 is invested at 8% and \$2800 at 9%. ◀

DO EXERCISE 4.

▶ **EXAMPLE 5** The ground floor of the John Hancock Building in Chicago is a rectangle whose perimeter is 860 ft. The length is 100 ft more than the width. Find the length and the width.

1. *Familiarize.* We make a drawing and label it. We recall, or look up, the definition of **perimeter:** $P = 2l + 2w$. Table 3 at the back of the book lists this formula. We let l = the length of the ground floor of the building and w = the width.

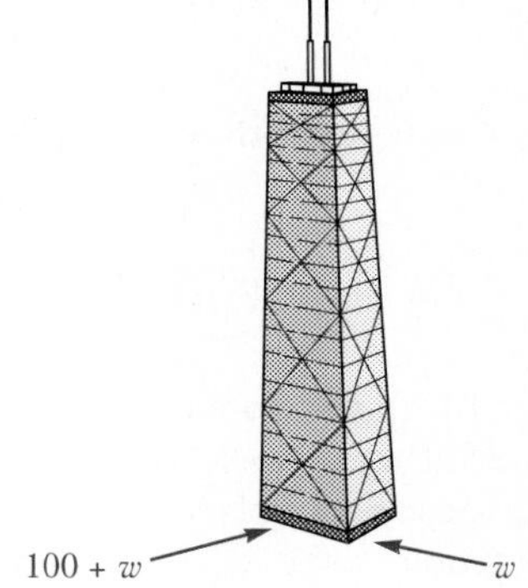

2. *Translate.* We translate as follows:

The perimeter is 860.

$$2l + 2w = 860$$

We then translate the second statement:

The length is 100 ft more than the width.

$$l = 100 + w$$

4. An investment of \$3700 is made for one year at simple interest, yielding \$297. Part of the money is invested at 7% and the rest at 9%. How much was invested at each rate?

Do the familiarization and translation steps by completing the following chart. Let x = the number of dollars invested at 7% and y = the number of dollars invested at 9%.

	First investment	Second investment	Total
Principal, *P*	x		\$3700
Rate of interest, *r*		9%	
Time, *t*	1 yr	1 yr	
Interest, *I*	$0.07x$		\$297

ANSWER ON PAGE A-5

5. The top floor of the John Hancock Building is a rectangle whose perimeter is 520 ft. The width is 60 ft less than the length. Find the width and the length of the rectangle.

ANSWER ON PAGE A-5

We now have a system of equations:

$$2l + 2w = 860, \qquad (1)$$
$$l = 100 + w. \qquad (2)$$

3. *Solve.* In this case, it is probably easier to use the substitution method, substituting $100 + w$ for l in Equation (1):

$$2(100 + w) + 2w = 860 \qquad \textbf{Substituting in Equation (1)}$$
$$200 + 2w + 2w = 860 \qquad \textbf{Multiplying to remove parentheses on the left}$$
$$200 + 4w = 860 \qquad \textbf{Collecting like terms}$$
$$4w = 660 \qquad \textbf{Subtracting 200}$$
$$w = 165. \qquad \textbf{Dividing by 4}$$

Then we substitute 165 for w in Equation (2) and solve for l:

$$l = 100 + 165 = 265.$$

4. *Check.* Consider the dimensions of 265 ft and 165 ft. The length is 100 ft more than the width. The perimeter is 2(265 ft) + 2(165 ft), or 860 ft.

5. *State.* The dimensions of 265 ft and 165 ft check in the original problem. Thus the length is 265 ft and the width is 165 ft. ◀

DO EXERCISE 5.

b Motion Problems

When a problem deals with speed, distance, and time, we can expect to use the following *motion formula.*

The Motion Formula

$$d = rt$$

Distance = Rate (or speed) · Time

From $d = rt$, we can obtain two other formulas by solving for r and t. They are

$$r = \frac{d}{t} \quad \text{and} \quad t = \frac{d}{r}.$$

In most problems involving motion, you will use one of these formulas. Thus it is important to remember at least one. You can then obtain the others as you need them by solving for a particular variable.

We have five steps for problem solving. The following tips are also helpful when solving motion problems.

Tips for Solving Motion Problems

1. **Draw a diagram using an arrow or arrows to represent distance and the direction of each object in motion.**
2. **Organize the information in a chart.**
3. **Look for as many things as you can that are the same, so you can write equations.**

► **EXAMPLE 6** A train leaves Sioux City traveling east at 30 km/h. Two hours later, another train leaves Sioux City traveling in the same direction on a parallel track at 45 km/h. At what point will the faster train overtake the slower train?

1. *Familiarize.* We first make a drawing.

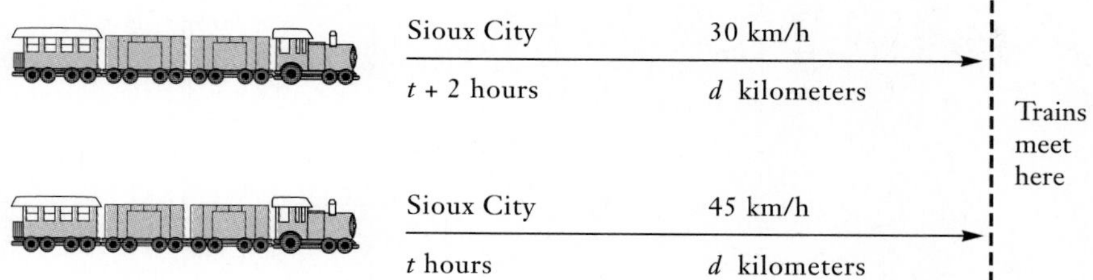

From the drawing, we see that the distances are the same. Let's call the distance d. Let t = the time traveled by the faster train. Then $t + 2$ = the time traveled by the slower train.

We organize the information in a chart, the columns of the chart being determined by the formula $d = rt$.

$d = r \cdot t$

	Distance	Rate	Time	
Slow train	d	30	$t + 2$	$\longrightarrow d = 30(t + 2)$
Fast train	d	45	t	$\longrightarrow d = 45t$

2. *Translate.* Using $d = rt$ in each row of the table, we get an equation. Thus we have a system of equations:

$$d = 30(t + 2) \qquad \textbf{(1)}$$
$$d = 45t. \qquad \textbf{(2)}$$

3. *Solve.* We solve the system:

$$45t = 30(t + 2) \qquad \textbf{Substituting } 45t \textbf{ for } d \textbf{ in Equation (1)}$$
$$45t = 30t + 60$$
$$15t = 60$$
$$t = 4.$$

The time for the faster train is 4 hr, which means that the time for the slower train is 4 + 2, or 6 hr.

4. *Check.* At 45 km/h, the faster train will travel $45 \cdot 4$, or 180 km, in 4 hr. At 30 km/h, the slower train will travel $30 \cdot 6$, or 180 km, in 6 hr. The numbers check.

5. *State.* The answer is that the trains will meet at a point 180 km east of Sioux City. ◄

DO EXERCISE 6.

6. A train leaves Barstow traveling east at 35 km/h. One hour later, a faster train leaves Barstow, also traveling east on a parallel track at 40 km/h. How far from Barstow will the faster train catch the slower one?

$d = r \cdot t$

	Distance	Rate	Time	
Slow train			t	$\to d =$
Fast train	d			$\to d =$

ANSWER ON PAGE A-5

▶ **EXAMPLE 7** A motorboat took 4 hr to make a downstream trip with a 6-mph current. The return trip against the same current took 5 hr. Find the speed of the boat in still water.

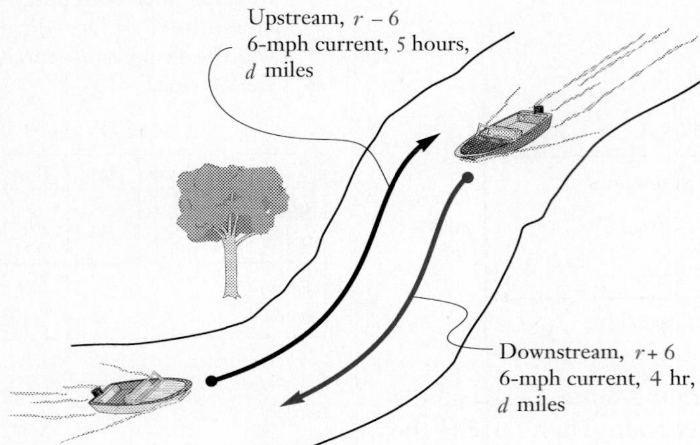

1. *Familiarize.* We first make a drawing. From the drawing, we see that the distances are the same. We let d = the distance, in miles, and r = the speed of the boat in still water, in miles per hour. Then, when the boat is traveling downstream, its speed is $r + 6$ (the current helps the boat along). When it is traveling upstream, its speed is $r - 6$ (the current holds the boat back). We can organize the information in a chart. Since the distances are the same, we use the formula $d = rt$.

$d = r \cdot t$

	Distance	Speed	Time	
Downstream	d	$r + 6$	4	$\longrightarrow d = (r + 6)4$
Upstream	d	$r - 6$	5	$\longrightarrow d = (r - 6)5$

2. *Translate.* From each row of the chart, we get an equation, $d = rt$:

$$d = 4r + 24, \quad \textbf{(1)}$$
$$d = 5r - 30. \quad \textbf{(2)}$$

3. *Solve.* Since there is a variable alone on one side of each equation, we solve the system using substitution:

$4r + 24 = 5r - 30$ **Substituting $4r + 24$ for d in Equation (2)**

$4r + 54 = 5r$ **Adding 30**

$54 = r.$ **Subtracting $4r$**

4. *Check.* If $r = 54$, then $r + 6 = 60$; and $60 \cdot 4 = 240$, the distance. If $r = 54$, then $r - 6 = 48$; and $48 \cdot 5 = 240$. In both cases, we get the same distance. Now when solving this type of problem, remember to ask yourself, "Have I found what the problem asked for?" We could solve for a certain variable but still not have solved the question of the original problem. For example, in this problem, we might have found distance when the problem wanted speed. In such a situation, we continue by substituting and solving for the other variable.

5. *State.* The speed in still water is 54 mph. ◀

DO EXERCISE 7.

7. An airplane flew for 4 hr with a 20-mph tail wind. The return flight against the same wind took 5 hr. Find the speed of the plane in still air.

$d = r \cdot t$

	Distance	Rate	Time	
With wind		$r+20$		$\to d =$
Against wind	d			$\to d =$

ANSWER ON PAGE A-5

NAME SECTION DATE

EXERCISE SET 4.3

ANSWERS

1. ______
2. ______
3. ______
4. ______
5. ______
6. ______
7. ______
8. ______
9. ______
10. ______
11. ______
12. ______

a Solve.

1. The sum of two numbers is -42. The first number minus the second is 52. What are the numbers?

2. The difference between two numbers is 11. Twice the smaller plus three times the larger is 123. What are the numbers?

3. One day a store sold 30 sweatshirts. White ones cost $9.95 and yellow ones cost $10.50. In all, $310.60 worth of sweatshirts were sold. How many of each color were sold?

4. At a club play, 117 tickets were sold. Adults' tickets cost $1.25 and children's tickets cost $0.75. In all, $129.75 was taken in. How many of each kind of ticket were sold?

5. The perimeter of a standard basketball court is 288 ft. The length is 44 ft longer than the width. Find the dimensions.

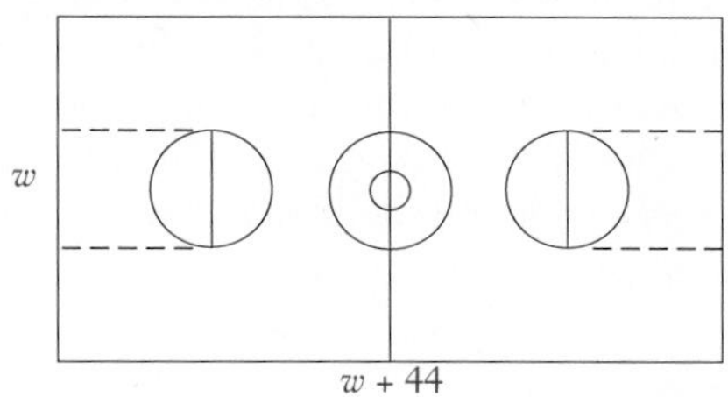

6. The perimeter of a standard tennis court, when used for doubles play, is 228 ft. The width is 42 ft less than the length. Find the dimensions.

7. Two angles are supplementary. One angle is 3° less than twice the other. Find the measures of the angles.

8. Two angles are complementary. The sum of the measure of the first angle and half the second angle is 64°. Find the measures of the angles.

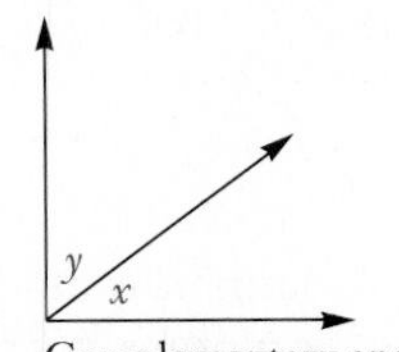

Complementary angles

9. A basketball player scored 18 times during one game. He scored a total of 30 points, two for each field goal and one for each free throw. How many field goals did he make? How many free throws?

10. At a barbecue, there were 250 dinners served. Children's plates were $1.50 each and adults' plates were $2.00 each. If the total amount of money collected was $441, how many of each type of plate was served?

11. Hockey teams receive 2 points when they win and 1 point when they tie. One season the team won a championship with 60 points. They won 9 more games than they tied. How many wins and how many ties did the team have?

12. An airplane has a total of 152 seats. The number of coach-class seats is five more than six times the number of first-class seats. How many of each type of seat are there on the plane?

ANSWERS

13. ____________

14. ____________

15. ____________

16. ____________

17. ____________

18. ____________

19. ____________

20. ____________

21. ____________

22. ____________

23. ____________

24. ____________

13. A disc jockey must play 12 commercial spots during 1 hr of a radio show. Each commercial is either 30 sec or 60 sec long. If the total commercial time during that hour is 10 min, how many 30-sec commercials were played that hour? How many 60-sec commercials?

14. A lumber company can convert logs into either lumber or plywood. In a given day, the mill turns out a total of 400 units of lumber and plywood. It makes a profit of $20 on a unit of lumber and $30 on a unit of plywood. How many of each unit must be produced and sold in order to make a profit of $11,000?

15. The difference between two numbers is 16. Three times the larger number is nine times the smaller. What are the numbers?

16. The sum of two numbers is -63. The first number minus the second is -41. What are the numbers?

17. Soybean meal is 16% protein; cornmeal is 9% protein. How many pounds of each should be mixed together in order to get a 350-lb mixture that is 12% protein?

18. A chemist has one solution that is 25% acid and a second that is 50% acid. How many liters of each should be mixed together in order to get 10 L of a solution that is 40% acid?

19. One canned juice drink is 15% orange juice; another is 5% orange juice. How many liters of each should be mixed together in order to get 10 L that is 10% orange juice?

20. Antifreeze A is 18% alcohol; antifreeze B is 10% alcohol. How many liters of each should be mixed in order to get 20 L of a mixture that is 15% alcohol?

21. Two investments are made totaling $8800. For a certain year, these investments yield $1326 in simple interest. Part of the $8800 is invested at 14% and part at 16%. Find the amount invested at each rate.

22. Two investments are made totaling $15,000. For a certain year, these investments yield $1432 in simple interest. Part of the $15,000 is invested at 9% and part at 10%. Find the amount invested at each rate.

23. $1150 is invested, part of it at 12% and part of it at 11%. The total yield is $133.75. How much was invested at each rate?

24. $27,000 is invested, part of it at 10% and part of it at 12%. The total yield is $2990. How much was invested at each rate?

25. At a concession stand, three hot dogs and five hamburgers cost \$18.50. Five hot dogs and three hamburgers cost \$16.70. Find the cost of one hot dog and the cost of one hamburger.

26. At a concession stand, five sodas and four boxes of popcorn cost \$19.90. Four sodas and five boxes of popcorn cost \$19.25. Find the cost of one soda and the cost of one box of popcorn.

27. Carlos is eight years older than his sister Maria. Four years ago, Maria was two thirds as old as Carlos was four years ago. How old are they now?

28. Paula is twelve years older than her brother Bob. Four years from now, Bob will be two thirds as old as Paula will be four years from now. How old are they now?

29. A student makes a \$9.25 purchase at the bookstore with a \$20 bill. The store has no bills and gives the change in quarters and fifty-cent pieces. There are 30 coins in all. How many of each kind are there?

30. A customer goes to a bank and gets change for a \$50 bill consisting of all \$5 bills and \$1 bills. There are 22 bills in all. How many of each kind are there?

b Solve.

31. A train leaves a station and travels north at 75 km/h. Two hours later, a second train leaves on a parallel track and travels north at 125 km/h. How far from the station will they meet? Complete the following chart to help with the familiarization.

$d \quad = \quad r \quad \cdot \quad t$

	Distance	Rate	Time	
Slow train		75	$t + 2$	$\rightarrow d = [\quad](t + 2)$
Fast train	d		t	$\rightarrow d = [\quad]t$

32. Two cars leave town at the same time traveling in opposite directions. One car travels at 80 km/h and the other at 96 km/h. In how many hours will they be 528 km apart?

33. Two motorcycles travel toward each other from Chicago and Indianapolis, which are about 350 km apart, at rates of 110 km/h and 90 km/h. They started at the same time. In how many hours will they meet?

34. Two planes travel toward each other from cities that are 780 km apart at rates of 190 km/h and 200 km/h. They started at the same time. In how many hours will they meet?

35. A motorboat took 3 hr to make a downstream trip with a 6-mph current. The return trip against the same current took 5 hr. Find the speed of the boat in still water.

36. A canoeist paddled for 4 hr with a 6-km/h current to reach a campsite. The return trip against the same current took 10 hr. Find the speed of the canoe in still water.

ANSWERS

25. ______

26. ______

27. ______

28. ______

29. ______

30. ______

31. ______

32. ______

33. ______

34. ______

35. ______

36. ______

ANSWERS

37. ______________

38. ______________

39. ______________

40. ______________

41. ______________

42. ______________

43. ______________

44. ______________

45. ______________

46. ______________

47. ______________

48. ______________

49. ______________

37. Gary computes his flight time against a head wind for a trip of 2900 mi at 5 hr. The flight would take 4 hr and 50 min if the head wind were half as great. Find the head wind and the plane's air speed.

38. A car travels from one town to another at a speed of 32 mph. If it had gone 4 mph faster, it could have made the trip in $\frac{1}{2}$ hr less time. How far apart are the towns?

39. Two airplanes start at the same time and fly toward each other from points 1000 km apart at rates of 420 km/h and 330 km/h. When will they meet?

40. A truck and a car leave a service station at the same time and travel in the same direction. The truck travels at 55 mph and the car at 40 mph. They can maintain CB radio contact within a range of 10 mi. When will they lose contact?

SYNTHESIS

41. A piece of posterboard has a perimeter of 156 in. If you cut 6 in. off the width, the length becomes four times the width. What are the dimensions of the original piece of posterboard?

42. Nancy jogs and walks to school each day. She averages 4 km/h walking and 8 km/h jogging. The distance from home to school is 6 km and she makes the trip in 1 hr. How far does she jog in a trip?

3. A limited edition of a book published by a historical society was offered for sale to its membership. The cost was one book for $12 or two books for $20. The society sold 880 books and the total amount of money taken in was $9840. How many members ordered two books?

44. The numerator of a fraction is twelve more than the denominator. The sum of the numerator and the denominator is five more than three times the denominator. What is the reciprocal of the fraction?

45. An automobile radiator contains 16 L of antifreeze and water. This mixture is 30% antifreeze. How much of this mixture should be drained and replaced with pure antifreeze so that the new mixture will be 50% antifreeze?

46. A train leaves Union Station for Central Station, 216 km away, at 9 A.M. One hour later, a train leaves Central Station for Union Station. They meet at noon. If the second train had started at 9 A.M. and the first train at 10:30 A.M., they would still have met at noon. Find the speed of each train.

47. An automobile gets 18 miles per gallon (mpg) in city driving and 24 mpg in highway driving. The car is driven 465 mi on a full tank of 23 gal of gasoline. How many miles were driven in the city and how many were driven on the highway?

48. Phil and Phyllis are siblings. Phyllis has twice as many brothers as she has sisters. Phil has the same number of brothers as sisters. How many girls and how many boys are in the family?

49. A tank at a marine exhibit contains 2000 gal of seawater. The seawater is 7.5% salt. How many gallons, to the nearest gallon, of freshwater must be added to the tank so that the mixture contains only 7% salt?

4.4 Systems of Equations in Three Variables

a Identifying Solutions

A **linear equation in three variables** is an equation equivalent to one of the type $Ax + By + Cz = D$. A solution of a system of three equations in three variables is an ordered triple (p, q, r) that makes *all three* equations true.

▶ **EXAMPLE 1** Determine whether $(\frac{3}{2}, -4, 3)$ is a solution of the following system:

$$\begin{aligned} 4x - 2y - 3z &= 5, \\ -8x - y + z &= -5, \\ 2x + y + 2z &= 5. \end{aligned}$$

We substitute $(\frac{3}{2}, -4, 3)$ into each of the three equations, using alphabetical order.

$$\begin{array}{r|l} \multicolumn{2}{c}{4x - 2y - 3z = 5} \\ \hline 4 \cdot \frac{3}{2} - 2(-4) - 3 \cdot 3 & 5 \\ 6 + 8 - 9 & \\ 5 & \end{array} \qquad \begin{array}{r|l} \multicolumn{2}{c}{-8x - y + z = -5} \\ \hline -8 \cdot \frac{3}{2} - (-4) + 3 & -5 \\ -12 + 4 + 3 & \\ -5 & \end{array}$$

$$\begin{array}{r|l} \multicolumn{2}{c}{2x + y + 2z = 5} \\ \hline 2 \cdot \frac{3}{2} + (-4) + 2 \cdot 3 & 5 \\ 3 - 4 + 6 & \\ 5 & \end{array}$$

The triple makes all three equations true, so it is a solution. ◀

DO EXERCISE 1.

b Solving Systems in Three Variables

Graphical methods for solving linear equations in three variables are unsatisfactory because a three-dimensional coordinate system is required and the graph of a linear equation in three variables is a plane. The substitution method can be used in any situation, but it is not helpful unless a variable has already been eliminated from one or more of the equations. Therefore, we will use the elimination method—essentially the same procedure for systems of three equations as for systems of two equations.

The goal is to eliminate a variable and obtain a system of two equations in two variables.

▶ **EXAMPLE 2** Solve the following system of equations:

$$\begin{aligned} x + y + z &= 4, & \textbf{(1)} \\ x - 2y - z &= 1, & \textbf{(2)} \\ 2x - y - 2z &= -1. & \textbf{(3)} \end{aligned}$$

a) We first use *any* two of the three equations to get an equation in two variables. In this case, let us use Equations (1) and (2) and add to eliminate z:

$$\begin{array}{rl} x + y + z = 4, & \textbf{(1)} \\ x - 2y - z = 1 & \textbf{(2)} \\ \hline 2x - y \phantom{{}+z} = 5. & \textbf{(4)} \quad \text{Adding} \end{array}$$

OBJECTIVES

After finishing Section 4.4, you should be able to:

- **a** Determine whether an ordered triple is a solution of a system of three equations in three variables.
- **b** Solve systems of three equations in three variables.

FOR EXTRA HELP

Tape 8A — Tape 6A — MAC: 4, IBM: 4

1. Consider the system

$$\begin{aligned} 4x + 2y + 5z &= 6, \\ 2x - y + z &= 5, \\ x + 2y - z &= 0. \end{aligned}$$

a) Determine whether $(1, 2, 3)$ is a solution.

b) Determine whether $(2, -1, 0)$ is a solution.

ANSWERS ON PAGE A-5

2. Solve. Don't forget to check.

$$\begin{aligned} 4x - y + z &= 6, \\ -3x + 2y - z &= -3, \\ 2x + y + 2z &= 3 \end{aligned}$$

ANSWER ON PAGE A-5

b) We use a different pair of equations and eliminate the *same variable* that we did in (a). Let us use Equations (1) and (3) and eliminate z. Be careful here! A common error is to eliminate a different variable the second time.

$$\begin{aligned} x + y + z &= 4 \quad &(1) \\ 2x - y - 2z &= -1 \quad &(3) \end{aligned}$$

$$\begin{aligned} 2x + 2y + 2z &= 8 \quad &\text{Multiplying Equation (1) by 2} \\ 2x - y - 2z &= -1 \\ \hline 4x + y &= 7 \quad &(5) \quad \text{Adding} \end{aligned}$$

c) Now we solve the resulting system of equations, (4) and (5). That solution will give us two of the numbers. Note that we now have two equations in two variables. Had we eliminated different variables in parts (a) and (b), this would not be the case.

$$\begin{aligned} 2x - y &= 5 \quad &(4) \\ 4x + y &= 7 \quad &(5) \\ \hline 6x &= 12 \quad &\text{Adding} \\ x &= 2 \end{aligned}$$

We can use either Equation (4) or (5) to find y. We choose Equation (5):

$$\begin{aligned} 4x + y &= 7 \quad &(5) \\ 4(2) + y &= 7 \quad &\text{Substituting 2 for } x \\ 8 + y &= 7 \\ y &= -1. \end{aligned}$$

d) We have $x = 2$ and $y = -1$. To find the value for z, we use any of the original three equations and substitute to find the third number z. Let us use Equation (1) and substitute our two numbers in it:

$$\begin{aligned} x + y + z &= 4 \quad &(1) \\ 2 + (-1) + z &= 4 \quad &\text{Substituting 2 for } x \text{ and } -1 \text{ for } y \\ 1 + z &= 4 \\ z &= 3. \end{aligned}$$

We have obtained the triple $(2, -1, 3)$. It checks in all three equations and is the solution. ◀

Using the elimination method to solve systems of three equations:

1. **Write all equations in the standard form $Ax + By + Cz = D$.**
2. **Clear of any decimals or fractions.**
3. **Choose a variable to eliminate. Then use *any* two of the three equations to get an equation in two variables.**
4. **Then use a different pair of equations and get another equation in *the same two variables.* That is, eliminate the same variable that you did in step (3).**
5. **Solve the resulting system (pair) of equations. That will give two of the numbers.**
6. **Then use any of the original three equations to find the third number.**

DO EXERCISE 2.

▶ **EXAMPLE 3** Solve the following system of equations:

$$\begin{aligned} 4x - 2y - 3z &= 5, \quad &(1)\\ -8x - y + z &= -5, \quad &(2)\\ 2x + y + 2z &= 5. \quad &(3) \end{aligned}$$

a) The equations are in standard form and have been cleared of decimals and fractions.

b) Choose a variable to eliminate. We decide to eliminate y since the y-terms are additive inverses, or opposites, of each other in Equations (2) and (3). We add:

$$\begin{aligned} -8x - y + z &= -5 \quad &(2)\\ 2x + y + 2z &= 5 \quad &(3)\\ \hline -6x + 3z &= 0. \quad &(4) \quad \textbf{Adding} \end{aligned}$$

c) We use another pair of equations to get an equation in the same two variables, x and z. That is, we eliminate the same variable y that we did in step (b). We use Equations (1) and (3) and eliminate y:

$$\begin{aligned} 4x - 2y - 3z &= 5 \quad &(1)\\ 2x + y + 2z &= 5 \quad &(3) \end{aligned}$$

$$\begin{aligned} 4x - 2y - 3z &= 5\\ 4x + 2y + 4z &= 10 \quad &\textbf{Multiplying Equation (3) by 2}\\ \hline 8x + z &= 15. \quad &(5) \quad \textbf{Adding} \end{aligned}$$

d) Now we solve the resulting system of Equations (4) and (5). That will give us two of the numbers:

$$\begin{aligned} -6x + 3z &= 0, \quad &(4)\\ 8x + z &= 15. \quad &(5) \end{aligned}$$

We multiply Equation (5) by -3. (We could multiply Equation (4) by $\frac{1}{3}$.)

$$\begin{aligned} -6x + 3z &= 0 \quad &(4)\\ -24x - 3z &= -45 \quad &\textbf{Multiplying Equation (5) by } -3\\ \hline -30x &= -45\\ x &= \tfrac{-45}{-30} = \tfrac{3}{2} \end{aligned}$$

We now use Equation (5) to find z:

$$\begin{aligned} 8x + z &= 15 \quad &(5)\\ 8(\tfrac{3}{2}) + z &= 15 \quad &\textbf{Substituting } \tfrac{3}{2} \textbf{ for } x\\ 12 + z &= 15\\ z &= 3. \end{aligned}$$

e) Then we use any of the original equations and substitute to find the third number, y. We use Equation (3) since the coefficient of y there is 1:

$$\begin{aligned} 2x + y + 2z &= 5 \quad &(3)\\ 2(\tfrac{3}{2}) + y + 2(3) &= 5 \quad &\textbf{Substituting } \tfrac{3}{2} \textbf{ for } x \textbf{ and 3 for } z\\ 3 + y + 6 &= 5\\ y + 9 &= 5\\ y &= -4. \end{aligned}$$

The solution is $(\frac{3}{2}, -4, 3)$. The check is in Example 1. ◀

DO EXERCISE 3.

3. Solve. Don't forget to check.

$$\begin{aligned} 2x + y - 4z &= 0,\\ x - y + 2z &= 5,\\ 3x + 2y + 2z &= 3 \end{aligned}$$

ANSWER ON PAGE A-5

4. Solve. Don't forget to check.

$$\begin{aligned} x + y + z &= 100, \\ x - y \quad &= -10, \\ x \quad - z &= -30 \end{aligned}$$

ANSWER ON PAGE A-5

In Example 4, which follows, certain variables have already been eliminated or are missing at the outset. In such situations, substitution can be an effective method of solution. We show *both* methods.

▶ **EXAMPLE 4** Solve:

$$\begin{aligned} x + y + z &= 180, && (1) \\ x \quad - z &= -70, && (2) \\ 2y - z &= 0. && (3) \end{aligned}$$

Method 1: Substitution. The variable z is in both Equations (2) and (3) and each equation has a missing variable. We solve Equations (2) and (3) for x and y, respectively. Then we substitute in Equation (1) to solve for x:

$x - z = -70$ **(2)**

$x = z - 70$ **Solving Equation (2) for x**

$2y - z = 0$ **(3)**

$y = \frac{1}{2}z.$ **Solving Equation (3) for y**

Now we substitute $z - 70$ for x and $\frac{1}{2}z$ for y in Equation (1) and solve for z:

$$\begin{aligned} (z - 70) + \tfrac{1}{2}z + z &= 180 && \textbf{Substituting} \\ z - 70 + \tfrac{1}{2}z + z &= 180 && \textbf{Removing parentheses} \\ 2z - 140 + z + 2z &= 360 && \textbf{Multiplying by 2 to clear of fractions} \\ 5z - 140 &= 360 \\ 5z &= 500 \\ z &= 100. && \textbf{Solving for } z \end{aligned}$$

To find x, we substitute 100 for z in the equation $x = z - 70$: $x = z - 70 = 100 - 70 = 30$. To find y, we substitute 100 for z in the equation $y = \frac{1}{2}z$: $y = \frac{1}{2}z = \frac{1}{2}(100) = 50$. The solution is (30, 50, 100).

Method 2: Elimination. We note that there is no y in Equation (2). Thus we know that at the outset, y has been eliminated from one equation. Since we need another equation with y eliminated, we use Equations (1) and (3) to eliminate y:

$x + y + z = 180$ **(1)**

$2y - z = 0$ **(3)**

$$\begin{aligned} -2x - 2y - 2z &= -360 && \textbf{Multiplying Equation (1) by } -2 \\ 2y - z &= 0 \\ \hline -2x \qquad - 3z &= -360. && \textbf{(4)} \end{aligned}$$

Now we solve the resulting system of Equations (2) and (4):

$x - z = -70$ **(2)**

$-2x - 3z = -360$ **(4)**

$$\begin{aligned} 2x - 2z &= -140 && \textbf{Multiplying Equation (2) by 2} \\ -2x - 3z &= -360 \\ \hline -5z &= -500 \\ z &= 100. \end{aligned}$$

Continuing as we did in Method 1, we get the solution (30, 50, 100). ◀

DO EXERCISE 4.

NAME SECTION DATE

EXERCISE SET 4.4

a

1. Determine whether $(1, -2, 3)$ is a solution of the system

$$x + y + z = 2,$$
$$x - 2y - z = 2,$$
$$3x + 2y + z = 2.$$

2. Determine whether $(2, -1, -2)$ is a solution of the system

$$x + y - 2z = 5,$$
$$2x - y - z = 7,$$
$$-x - 2y + 3z = 6.$$

b Solve.

3. $x + y + z = 6,$
$2x - y + 3z = 9,$
$-x + 2y + 2z = 9$

4. $2x - y + z = 10,$
$4x + 2y - 3z = 10,$
$x - 3y + 2z = 8$

5. $2x - y - 3z = -1,$
$2x - y + z = -9,$
$x + 2y - 4z = 17$

6. $x - y + z = 6,$
$2x + 3y + 2z = 2,$
$3x + 5y + 4z = 4$

7. $2x - 3y + z = 5,$
$x + 3y + 8z = 22,$
$3x - y + 2z = 12$

8. $6x - 4y + 5z = 31,$
$5x + 2y + 2z = 13,$
$x + y + z = 2$

9. $3a - 2b + 7c = 13,$
$a + 8b - 6c = -47,$
$7a - 9b - 9c = -3$

10. $x + y + z = 0,$
$2x + 3y + 2z = -3,$
$-x + 2y - 3z = -1$

11. $2x + 3y + z = 17,$
$x - 3y + 2z = -8,$
$5x - 2y + 3z = 5$

12. $2x + y - 3z = -4,$
$4x - 2y + z = 9,$
$3x + 5y - 2z = 5$

13. $2x + y + z = -2,$
$2x - y + 3z = 6,$
$3x - 5y + 4z = 7$

14. $2x + y + 2z = 11,$
$3x + 2y + 2z = 8,$
$x + 4y + 3z = 0$

15. $x - y + z = 4,$
$5x + 2y - 3z = 2,$
$3x - 7y + 4z = 8$

16. $2x + y + 2z = 3,$
$x + 6y + 3z = 4,$
$3x - 2y + z = 0$

17. $4x - y - z = 4,$
$2x + y + z = -1,$
$6x - 3y - 2z = 3$

ANSWERS

1. ______
2. ______
3. ______
4. ______
5. ______
6. ______
7. ______
8. ______
9. ______
10. ______
11. ______
12. ______
13. ______
14. ______
15. ______
16. ______
17. ______

ANSWERS

18. ______

19. ______

20. ______

21. ______

22. ______

23. ______

24. ______

25. ______

26. ______

27. ______

28. ______

29. ______

30. ______

18. $a + 2b + c = 1,$
$7a + 3b - c = -2,$
$a + 5b + 3c = 2$

19. $2r + 3s + 12t = 4,$
$4r - 6s + 6t = 1,$
$r + s + t = 1$

20. $10x + 6y + z = 7,$
$5x - 9y - 2z = 3,$
$15x - 12y + 2z = -5$

21. $4a + 9b = 8,$
$8a + 6c = -1,$
$6b + 6c = -1$

22. $3p + 2r = 11,$
$q - 7r = 4,$
$p - 6q = 1$

23. $x + y + z = 57,$
$-2x + y = 3,$
$x - z = 6$

24. $x + y + z = 105,$
$10y - z = 11,$
$2x - 3y = 7$

25. $a - 3c = 6,$
$b + 2c = 2,$
$7a - 3b - 5c = 14$

26. $L + m = 7,$
$3m + 2n = 9,$
$4L + n = 5$

SKILL MAINTENANCE

27. Solve $F = \frac{1}{2}t(c - d)$ for c.

28. Solve $F = \frac{1}{2}t(c - d)$ for d.

SYNTHESIS

Solve.

29. $w + x + y + z = 2,$
$w + 2x + 2y + 4z = 1,$
$w - x + y + z = 6,$
$w - 3x - y + z = 2$

30. $w + x - y + z = 0,$
$w - 2x - 2y - z = -5,$
$w - 3x - y + z = 4,$
$2w - x - y + 3z = 7$

4.5 Problem Solving Using Systems of Three Equations

a Many problems can be solved by first translating to a system of three equations.

► **EXAMPLE 1** The sum of three numbers is 4. The first number minus twice the second minus the third is 1. Twice the first number minus the second minus twice the third is -1. Find the numbers.

1. *Familiarize.* In this case, there are three obvious statements in the problem. The translation looks as though it can be made directly from these statements, as soon as we decide what letters to assign to the unknown numbers. Let us call the three numbers x, y, and z.
2. *Translate.* We can translate directly, from the words of the problem, as follows.

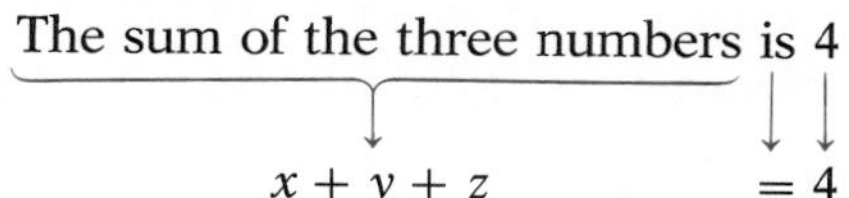

The first number minus twice the second minus the third is 1.

$x \quad - \quad 2y \quad - \quad z \quad = 1$

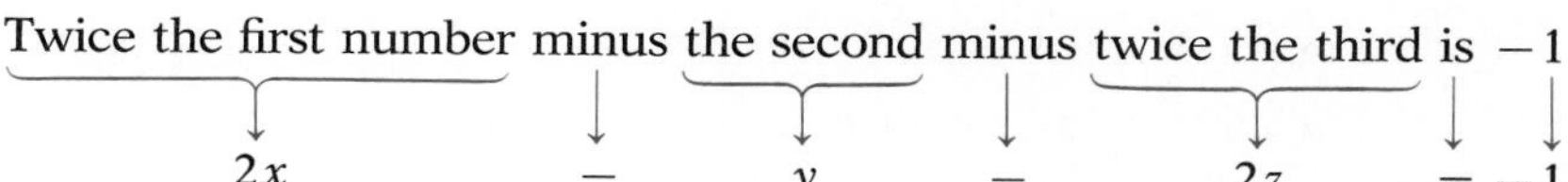

We now have a system of three equations:

$$\begin{aligned} x + y + z &= 4, \\ x - 2y - z &= 1, \\ 2x - y - 2z &= -1. \end{aligned}$$

3. *Solve.* We solved the system in Example 2 of Section 4.4. The solution is $(2, -1, 3)$.
4. *Check.* We go to the original problem. The first statement says that the sum of the three numbers is 4. That checks. The second statement says that the first number minus twice the second minus the third is 1. We calculate: $2 - 2(-1) - 3 = 1$. That checks. We leave the check of the third statement to the student.
5. *State.* The answer is that the three numbers are 2, -1, and 3. ◄

DO EXERCISE 1.

OBJECTIVE

After finishing Section 4.5, you should be able to:

a Solve problems by translating to a system of three equations in three variables.

FOR EXTRA HELP

Tape 8B

Tape 6B

MAC: 4
IBM: 4

1. The sum of three numbers is 3. The first number minus twice the second minus the third is 4. Twice the first number plus the other two is 5. Find the numbers.

ANSWER ON PAGE A-5

2. One angle of a triangle is twice as large as a second angle. The remaining angle is 20° greater than the first angle. Find the measures of the angles.

► **EXAMPLE 2** In a triangle, the largest angle is 70° greater than the smallest angle. The largest angle is twice as large as the remaining angle. Find the measure of each angle.

1. *Familiarize.* We first make a drawing. Since we don't know the size of any angle, we use x, y, and z for the measures of the angles. We let $x =$ the smallest angle, $z =$ the largest angle, and $y =$ the remaining angle.

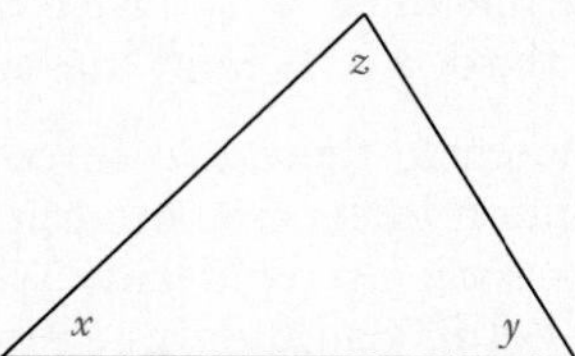

In order to solve the problem, we need to make use of a geometric fact—that is, the measures of the angles of a triangle add up to 180°.

2. *Translate.* This geometric fact about triangles gives us one equation:

$$x + y + z = 180.$$

There are two statements in the problem that we can translate directly.

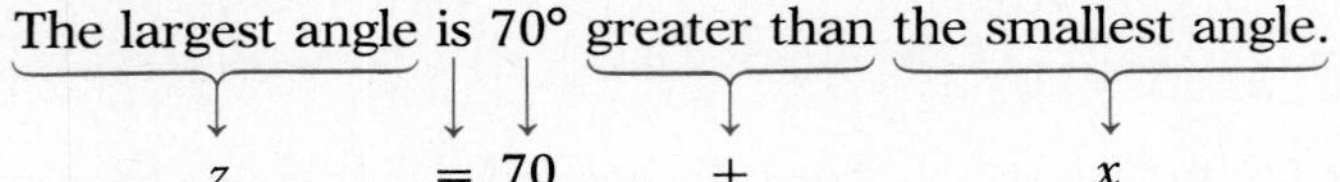

The largest angle is twice as large as the remaining angle.

$$z \quad = \quad 2y$$

We now have a system of three equations:

$$\begin{aligned} x + y + z &= 180, \\ x + 70 &= z, \\ 2y &= z; \end{aligned} \qquad \text{or} \qquad \begin{aligned} x + y + z &= 180, \\ x \qquad - z &= -70, \\ 2y - z &= 0. \end{aligned}$$

3. *Solve.* The system was solved in Example 4 of Section 4.4. The solution is (30, 50, 100).
4. *Check.* The sum of the numbers is 180, so that checks. The largest angle measures 100° and the smallest measures 30°. The largest angle is 70° greater than the smallest. The remaining angle measures 50°. The largest angle is twice as large as the remaining angle. We do have an answer to the problem.
5. *State.* The measures of the angles of the triangle are 30°, 50°, and 100°. ◄

DO EXERCISE 2.

ANSWER ON PAGE A-5

► **EXAMPLE 3** *Business.* A person receives \$391 per year in simple interest from three investments totaling \$3700. Part is invested at 8%, part at 9%, and part at 12%. There is \$1400 more invested at 12% than at 8%. Find the amount invested at each rate.

1. *Familiarize.* Let

$$x = \text{the amount invested at } 8\%,$$
$$y = \text{the amount invested at } 9\%,$$
and
$$z = \text{the amount invested at } 12\%,$$

2. *Translate.* To translate, we consider three conditions stated in the problem.

   a) The total amount of money invested is \$3700. This gives us the equation

   $$x + y + z = 3700.$$

   b) Next we consider the interest. Using the simple-interest formula $I = Prt$, we know that the interest on the 8% investment is given by $I = Prt = x(8\%)(1) = 0.08x$. Similarly, the interest on the 9% investment is $0.09y$, and on the 12% investment it is $0.12z$. The total amount of interest is given by the equation

   $$0.08x + 0.09y + 0.12z = 391.$$

   c) There is \$1400 more invested at 12% than at 8%. This gives us another equation:

   $$z = x + 1400.$$

   We now have a system of three equations:

   $$\begin{aligned} x + y + z &= 3700, \\ 0.08x + 0.09y + 0.12z &= 391, \\ z &= x + 1400. \end{aligned}$$

   Clearing the second equation of decimals and getting the third equation in the form $Ax + By + Cz = D$ gives us

   $$\begin{aligned} x + y + z &= 3700, \\ 8x + 9y + 12z &= 39{,}100, \\ -x + z &= 1400. \end{aligned}$$

3. *Solve.* We solve and get $x = 800$, $y = 700$, and $z = 2200$.
4. *Check.* The sum of \$800, \$700, and \$2200 is \$3700, so the total of the three investments checks. Also,

   $$0.08(\$800) + 0.09(\$700) + 0.12(\$2200) = \$391,$$

   so the total amount of interest checks. Finally \$2200, the amount invested at 12%, is \$1400 more than \$800, the amount invested at 8%.
5. *State.* The answer to the problem is that \$800 is invested at 8%, \$700 is invested at 9%, and \$2200 is invested at 12%. ◄

DO EXERCISE 3.

3. *Business.* A person receives \$212 per year in simple interest from three investments totaling \$2500. Part is invested at 7%, part at 8%, and part at 9%. There is \$1100 more invested at 9% than at 8%. Find the amount invested at each rate.

ANSWER ON PAGE A-5

❖ SIDELIGHTS

Careers and Their Uses of Mathematics

Students typically ask the question "Why do we have to study mathematics?" This is a question with a complex set of answers. Certainly, one answer is that you will use this mathematics in the next course. While it is a correct answer, it sometimes frustrates students, because this answer can be given in the next mathematics course, and the next one, and so on. Sometimes an answer can be given by applications like those you have seen or will see in this book. Another answer is that you are living in a society in which mathematics becomes more and more critical with each passing day. Evidence of this was provided recently by a nationwide symposium sponsored by the National Research Council's Mathematical Sciences Education Board. Results showed that "Other than demographic factors, the *strongest* predictor of earnings nine years after high school is the number of mathematics courses taken." This is a significant testimony to the need for you to take as many mathematics courses as possible.

We try to provide other answers to "Why do we have to study mathematics?" in what follows. We have listed several occupations that are attractive and popular to students. Below each occupation are listed various kinds of mathematics that are useful in that occupation.

Doctor	**Lawyer**
Equations	Equations
Percent notation	Percent notation
Graphing	Graphing
Statistics	Probability
Geometry	Statistics
Measurement	Ratio and proportion
Estimation	Area and volume
Exponents	Negative numbers
Logic	Formulas
	Calculator skills

Pilot	**Firefighter**
Equations	Percent notation
Percent notation	Graphing
Graphing	Estimation
Trigonometry	Formulas
Angles and geometry	Angles and geometry
Calculator skills	Probability
Computer skills	Statistics
Ratio and proportion	Area and geometry
Vectors	Square roots
	Exponents
	Pythagorean theorem

Accountant and businessperson	**Travel agent**
Computer skills	Whole-number skills
Calculator skills	Fraction/decimal skills
Equations	Estimation
Systems of equations	Percent notation
Formulas	Equations
Probability	Calculator skills
Statistics	Computer skills
Ratio and proportion	
Percent notation	
Estimation	

Librarian	**Machinist**
Whole-number skills	Whole-number skills
Fraction/decimal skills	Fraction/decimal skills
Estimation	Estimation
Percent notation	Percent notation
Ratio and proportion	Length, area, volume, and perimeter
Area and perimeter	Angle measures
Formulas	Geometry
Calculator skills	Pythagorean theorem
Computer skills	Square roots
	Equations
	Formulas
	Graphing
	Calculator skills
	Computer skills

Nurse	**Police officer**
Whole-number skills	Whole-number skills
Fraction/decimal skills	Fraction/decimal skills
Estimation	Estimation
Percent notation	Percent notation
Ratio and proportion	Ratio and proportion
Estimation	Geometry
Equations	Negative numbers
English/Metric measurement	Probability
Probability	Statistics
Statistics	Calculator skills
Formulas	
Exponents and scientific notation	
Calculator skills	
Computer skills	

NAME SECTION DATE

EXERCISE SET 4.5

ANSWERS

a Solve.

1. The sum of three numbers is 5. The first number minus the second plus the third is 1. The first minus the third is three more than the second. Find the numbers.

2. The sum of three numbers is 26. Twice the first minus the second is two less than the third. The third is the second minus three times the first. Find the numbers.

3. In triangle *ABC*, the measure of angle *B* is 2° more than three times the measure of angle *A*. The measure of angle *C* is 8° more than the measure of angle *A*. Find the angle measures.

4. In triangle *ABC*, the measure of angle *B* is three times the measure of angle *A*. The measure of angle *C* is 30° greater than the measure of angle *A*. Find the angle measures.

5. In a recent year, companies spent a total of $84.8 billion on newspaper, television, and radio ads. The total amount spent on television and radio ads was only $2.6 billion more than the amount spent on newspaper ads alone. The amount spent on newspaper ads was $5.1 billion more than what was spent on television ads. How much was spent on each form of advertising? (*Hint:* Let the variables represent numbers of billions of dollars.)

6. A recent basic model of a particular automobile had a cost of $12,685. The basic model with the added features of automatic transmission and power door locks was $14,070. The basic model with just air conditioning (AC) and power door locks was $13,580. A basic model with just AC and automatic transmission was $13,925. What was the individual cost of each of the three options?

1. ______

2. ______

3. ______

4. ______

5. ______

6. ______

ANSWERS

7. ______

8. ______

9. ______

10. ______

11. ______

12. ______

7. In triangle ABC, the measure of angle B is twice the measure of angle A. The measure of angle C is 80° more than that of angle A. Find the angle measures.

8. In triangle ABC, the measure of angle B is three times that of angle A. The measure of angle C is 20° more than that of angle A. Find the angle measures.

9. People are becoming more and more aware of their cholesterol levels. Recent studies indicate that a child should ingest no more than 300 mg of cholesterol per day. By eating 1 egg, 1 cupcake, and 1 slice of pizza, a child would ingest 302 mg of cholesterol. If the child ate 2 cupcakes and 3 slices of the same pizza, he or she would ingest 65 mg of cholesterol. By eating 2 eggs and 1 cupcake, the child would take in 567 mg of cholesterol. How much cholesterol is in each item?

10. A dietician works in a hospital and prepares meals under the guidance of a physician. Suppose that for a particular patient, a physician prescribes a meal to have 800 calories, 55 g of protein, and 220 mg of vitamin C. The dietician decides to prepare the meal using steak (each 3-oz serving contains 300 calories, 20 g of protein, and no vitamin C), baked potatoes (one baked potato contains 100 calories, 5 g of protein, and 20 mg of vitamin C), and brussel sprouts (one 156-g serving contains 50 calories, 5 g of protein, and 100 mg of vitamin C). How many servings of each food are required to satisfy the physician's requirements? (*Hint:* Let $s =$ the number of servings of steak, $p =$ the number of baked potatoes, and $b =$ the number of servings of brussel sprouts. Find an equation for the total number of calories, total amount of protein, and total amount of vitamin C.)

11. Repeat Exercise 10 but replace the brussel sprouts with asparagus, for which one 180-g serving contains 50 calories, 5 g of protein, and 40 mg of vitamin C. How many servings of each food are required to satisfy the physician's requirements? (Which meal would you prefer eating?)

12. Fred, Jane, and Mary made a total bowling score of 575. Fred's score was 15 more than Jane's. Mary's was 20 more than Jane's. Find the scores.

13. In the United States, the highest incidence of fraternal twin births occurs among Orientals, then blacks, and then whites. Out of every 15,400 births, the total number of fraternal twin births for all three is 739, where there are 185 more for Orientals than blacks and 231 more for Orientals than whites. How many births of fraternal twins are there for each race out of every 15,400 births?

14. The sum of the average number of times a man, a woman, and a one-year-old child cry each month is 71.7. A one-year-old cries 46.4 more times than a man. The average number of times a one-year-old cries per month is 28.3 more than the average number of times combined that a man and a woman cry. What is the average number of times per month that each cries?

15. When three pumps, A, B, and C, are running together, they can pump 3700 gal per hour. When only A and B are running, 2200 gal per hour can be pumped. When only A and C are running, 2400 gal per hour can be pumped. What is the pumping capacity of each pump?

16. Three welders, A, B, and C, can weld 37 linear feet per hour when working together. Welders A and B together can weld 22 linear feet per hour, while A and C together can weld 25 linear feet per hour. How many linear feet per hour can each weld alone?

17. One year an investment of $80,000 was made by a business club. The investment was split into three parts and lasted for one year. The first part of the investment earned 8% interest, the second 6%, and the third 9%. Total interest from the investments was $6300. The interest from the first investment was four times the interest from the second. Find the amounts of the three parts of the investment.

18. Find the year in which the first U.S. transcontinental railroad was completed. The following are some facts about the number. The sum of the digits in the year is 24. The one's digit is one more than the hundred's digit. Both the ten's and the one's digits are multiples of three.

ANSWERS

13. ______

14. ______

15. ______

16. ______

17. ______

18. ______

ANSWERS

19. ______

20. ______

21. ______

22. ______

23. ______

24. ______

19. *Golf.* On an 18-hole golf course, there are par-3 holes, par-4 holes, and par-5 holes. A golfer who shoots par on every hole has a total of 70. There are twice as many par-4 holes as there are par-5 holes. How many of each type of hole are there on the golf course?

20. *Golf.* On an 18-hole golf course, there are par-3 holes, par-4 holes, and par-5 holes. A golfer who shoots par on every hole has a total of 72. The sum of the number of par-3 holes and the number of par-5 holes is 8. How many of each type of hole are there on the golf course?

SYNTHESIS

21. Find the sum of the angle measures at the tips of the star in this figure.

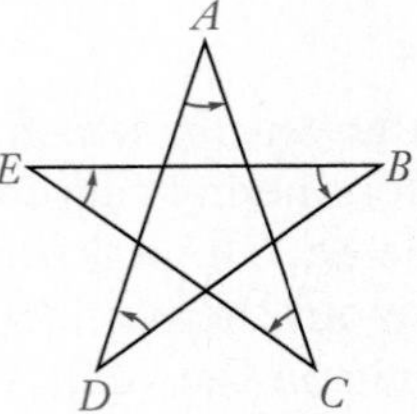

22. A theater audience of 100 people consists of men, women, and children. The ticket prices are $10 for men, $3 for women, and 50¢ for children. The total amount of money taken in is $100. How many men, women, and children are in attendance? Does there seem to be some information missing? Do some more careful reasoning.

23. Hal gives Tom as many raffle tickets as Tom has and Gary as many as Gary has. In like manner, Tom then gives Hal and Gary as many tickets as each then has. Similarly, Gary gives Hal and Tom as many tickets as each then has. If each finally has 40 tickets, with how many tickets does Tom begin?

24. At a county fair, adults' tickets sold for $5.50, senior citizens' tickets sold for $4.00, and children's tickets sold for $1.50. On the opening day, the number of children's and senior citizens' tickets sold was 30 more than half the number of adults' tickets sold. The number of senior citizens' tickets sold was 5 more than 4 times the number of children's tickets. How many of each type of ticket were sold if the total receipt from the ticket sales was $14,970?

4.6 Determinants and Cramer's Rule

OBJECTIVES

After finishing Section 4.6, you should be able to:

a Evaluate second-order determinants.

b Evaluate third-order determinants.

c Solve systems of equations using Cramer's rule.

FOR EXTRA HELP

Tape NC | Tape 6B | MAC: 4 IBM: 4

You have probably noticed by now that the elimination method primarily concerns itself with the coefficients and constants of the equations. In what follows, we learn a method for solving systems using just the coefficients and the constants.

a Evaluating Determinants

The following symbolism represents a **determinant:**

$$\begin{vmatrix} a_1 & b_1 \\ a_2 & b_2 \end{vmatrix}.$$

To evaluate a determinant, we do two multiplications and subtract.

▶ **EXAMPLE 1** Evaluate:

$$\begin{vmatrix} 2 & -5 \\ 6 & 7 \end{vmatrix}.$$

We multiply and subtract as follows:

$$\begin{vmatrix} 2 & -5 \\ 6 & 7 \end{vmatrix} = 2 \cdot 7 - 6 \cdot (-5) = 14 + 30 = 44.$$ ◀

Determinants are defined according to the pattern shown in Example 1.

The determinant $\begin{vmatrix} a_1 & b_1 \\ a_2 & b_2 \end{vmatrix}$ is defined to mean $a_1b_2 - a_2b_1$.

The value of a determinant is a *number*. In Example 1, the value is 44.

DO EXERCISES 1 AND 2.

Evaluate.

1. $\begin{vmatrix} 3 & 2 \\ 4 & 1 \end{vmatrix}$

2. $\begin{vmatrix} 5 & -2 \\ -1 & -1 \end{vmatrix}$

b Third-Order Determinants

The determinants above are called *second-order*. A *third-order determinant* is defined as follows.

$$\begin{vmatrix} a_1 & b_1 & c_1 \\ a_2 & b_2 & c_2 \\ a_3 & b_3 & c_3 \end{vmatrix} = a_1 \begin{vmatrix} b_2 & c_2 \\ b_3 & c_3 \end{vmatrix} - a_2 \begin{vmatrix} b_1 & c_1 \\ b_3 & c_3 \end{vmatrix} + a_3 \begin{vmatrix} b_1 & c_1 \\ b_2 & c_2 \end{vmatrix}$$

Note the minus sign here.

Note that the a's come from the first column.

ANSWERS ON PAGE A-5

Evaluate.

3. $\begin{vmatrix} 2 & -1 & 1 \\ 1 & 2 & -1 \\ 3 & 4 & -3 \end{vmatrix}$

4. $\begin{vmatrix} 3 & 2 & 2 \\ -2 & 1 & 4 \\ 4 & -3 & 3 \end{vmatrix}$

ANSWERS ON PAGE A-5

Note too that the second-order determinants above can be obtained by crossing out the row and the column in which the a occurs.

For a_1: $\begin{vmatrix} a_1 & b_1 & c_1 \\ a_2 & b_2 & c_2 \\ a_3 & b_3 & c_3 \end{vmatrix}$ For a_2: $\begin{vmatrix} a_1 & b_1 & c_1 \\ a_2 & b_2 & c_2 \\ a_3 & b_3 & c_3 \end{vmatrix}$

For a_3: $\begin{vmatrix} a_1 & b_1 & c_1 \\ a_2 & b_2 & c_2 \\ a_3 & b_3 & c_3 \end{vmatrix}$

▶ **EXAMPLE 2** Evaluate this third-order determinant:

$$\begin{vmatrix} -1 & 0 & 1 \\ -5 & 1 & -1 \\ 4 & 8 & 1 \end{vmatrix} = -1\begin{vmatrix} 1 & -1 \\ 8 & 1 \end{vmatrix} - (-5)\begin{vmatrix} 0 & 1 \\ 8 & 1 \end{vmatrix} + 4\begin{vmatrix} 0 & 1 \\ 1 & -1 \end{vmatrix}.$$

We calculate as follows:

$$\begin{aligned} -1\begin{vmatrix} 1 & -1 \\ 8 & 1 \end{vmatrix} - (-5)\begin{vmatrix} 0 & 1 \\ 8 & 1 \end{vmatrix} + 4\begin{vmatrix} 0 & 1 \\ 1 & -1 \end{vmatrix} &= -1[1 \cdot 1 - 8(-1)] + 5(0 \cdot 1 - 8 \cdot 1) + 4[0 \cdot (-1) - 1 \cdot 1] \\ &= -1(9) + 5(-8) + 4(-1) \\ &= -9 - 40 - 4 \\ &= -53. \end{aligned}$$

◀

DO EXERCISES 3 AND 4.

C Solving Systems by Determinants

Here is a system of two equations in two variables:

$$a_1x + b_1y = c_1,$$
$$a_2x + b_2y = c_2.$$

We form three determinants, which we call D, D_x, and D_y.

$D = \begin{vmatrix} a_1 & b_1 \\ a_2 & b_2 \end{vmatrix}$ In D, we have the coefficients of x and y.

$D_x = \begin{vmatrix} c_1 & b_1 \\ c_2 & b_2 \end{vmatrix}$ To form D_x we use D, but we replace the x-coefficients with the constants on the right side of the equations.

$D_y = \begin{vmatrix} a_1 & c_1 \\ a_2 & c_2 \end{vmatrix}$ To form D_y we use D, but we replace the y-coefficients with the constants on the right.

It is important that the replacement be done *without changing the order of the columns*. Then the solution of the system can be found as follows. This is known as **Cramer's rule:**

$$x = \frac{D_x}{D}, \qquad y = \frac{D_y}{D}.$$

▶ **EXAMPLE 3** Solve using Cramer's rule:

$$3x - 2y = 7,$$
$$3x + 2y = 9.$$

We compute D, D_x, and D_y.

$$D = \begin{vmatrix} 3 & -2 \\ 3 & 2 \end{vmatrix} = 3 \cdot 2 - 3 \cdot (-2) = 6 + 6 = 12;$$

$$D_x = \begin{vmatrix} 7 & -2 \\ 9 & 2 \end{vmatrix} = 7 \cdot 2 - 9(-2) = 14 + 18 = 32;$$

$$D_y = \begin{vmatrix} 3 & 7 \\ 3 & 9 \end{vmatrix} = 3 \cdot 9 - 3 \cdot 7 = 27 - 21 = 6.$$

Then

$$x = \frac{D_x}{D} = \frac{32}{12}, \quad \text{or} \quad \frac{8}{3} \qquad \text{and} \qquad y = \frac{D_y}{D} = \frac{6}{12} = \frac{1}{2}.$$

The solution is $(\frac{8}{3}, \frac{1}{2})$. ◀

DO EXERCISE 5.

Cramer's rule for three equations is very similar to that for two.

$$a_1x + b_1y + c_1z = d_1$$
$$a_2x + b_2y + c_2z = d_2$$
$$a_3x + b_3y + c_3z = d_3$$

$$D = \begin{vmatrix} a_1 & b_1 & c_1 \\ a_2 & b_2 & c_2 \\ a_3 & b_3 & c_2 \end{vmatrix} \qquad D_x = \begin{vmatrix} d_1 & b_1 & c_1 \\ d_2 & b_2 & c_2 \\ d_3 & b_3 & c_3 \end{vmatrix}$$

$$D_y = \begin{vmatrix} a_1 & d_1 & c_1 \\ a_2 & d_2 & c_2 \\ a_3 & d_3 & c_3 \end{vmatrix}$$

D is again the determinant of the coefficients of x, y, and z. This time we have one more determinant, D_z. We get it by using D and replacing the z-coefficients by the constants on the right:

$$D_z = \begin{vmatrix} a_1 & b_1 & d_1 \\ a_2 & b_2 & d_2 \\ a_3 & b_3 & d_3 \end{vmatrix}.$$

The solution of the system is given by

$$\mathbf{x = \frac{D_x}{D}, \qquad y = \frac{D_y}{D}, \qquad z = \frac{D_z}{D}.}$$

5. Solve using Cramer's rule.

$$20x - 15y = 75,$$
$$x + 3y = 0$$

ANSWER ON PAGE A-5

6. Solve using Cramer's rule.

$$x - 3y - 7z = 6,$$
$$2x + 3y + z = 9,$$
$$4x + y = 7$$

ANSWER ON PAGE A-5

▶ **EXAMPLE 4** Solve using Cramer's rule:

$$x - 3y + 7z = 13,$$
$$x + y + z = 1,$$
$$x - 2y + 3z = 4.$$

We compute D, D_x, D_y, and D_z:

$$D = \begin{vmatrix} 1 & -3 & 7 \\ 1 & 1 & 1 \\ 1 & -2 & 3 \end{vmatrix} = -10; \qquad D_x = \begin{vmatrix} 13 & -3 & 7 \\ 1 & 1 & 1 \\ 4 & -2 & 3 \end{vmatrix} = 20;$$

$$D_y = \begin{vmatrix} 1 & 13 & 7 \\ 1 & 1 & 1 \\ 1 & 4 & 3 \end{vmatrix} = -6; \qquad D_z = \begin{vmatrix} 1 & -3 & 13 \\ 1 & 1 & 1 \\ 1 & -2 & 4 \end{vmatrix} = -24.$$

Then

$$x = \frac{D_x}{D} = \frac{20}{-10} = -2;$$
$$y = \frac{D_y}{D} = \frac{-6}{-10} = \frac{3}{5};$$
$$z = \frac{D_z}{D} = \frac{-24}{-10} = \frac{12}{5}.$$

The solution is $(-2, \frac{3}{5}, \frac{12}{5})$. ◀

In Example 4 we would not have needed to evaluate D_z. Once we found x and y, we could have substituted them into one of the equations to find z. In practice, it is faster to use determinants to find only two of the numbers; then we find the third by substitution into an equation.

DO EXERCISE 6.

In using Cramer's rule, we divide by D. If D should be 0, we could not do so. If $D = 0$ and at least one of the other determinants is not 0, then the system does not have a solution, and we say that it is *inconsistent*. If $D = 0$ and all the other determinants are also 0, then there is an infinite set of solutions. In that case, we say that the system is *dependent*.

NAME SECTION DATE

EXERCISE SET 4.6

a Evaluate.

1. $\begin{vmatrix} 2 & 7 \\ 1 & 5 \end{vmatrix}$

2. $\begin{vmatrix} 3 & 2 \\ 2 & -3 \end{vmatrix}$

3. $\begin{vmatrix} 6 & -9 \\ 2 & 3 \end{vmatrix}$

4. $\begin{vmatrix} 3 & 2 \\ -7 & 5 \end{vmatrix}$

5. $\begin{vmatrix} -2 & -3 \\ -4 & -6 \end{vmatrix}$

6. $\begin{vmatrix} 2 & 3 \\ 1 & 1 \end{vmatrix}$

7. $\begin{vmatrix} 5 & -6 \\ 0 & 0 \end{vmatrix}$

8. $\begin{vmatrix} 0 & -3 \\ 0 & -5 \end{vmatrix}$

b Evaluate.

9. $\begin{vmatrix} 0 & 2 & 0 \\ 3 & -1 & 1 \\ 1 & -2 & 2 \end{vmatrix}$

10. $\begin{vmatrix} 3 & 0 & -2 \\ 5 & 1 & 2 \\ 2 & 0 & -1 \end{vmatrix}$

11. $\begin{vmatrix} -1 & -2 & -3 \\ 3 & 4 & 2 \\ 0 & 1 & 2 \end{vmatrix}$

12. $\begin{vmatrix} 1 & 2 & 2 \\ 2 & 1 & 0 \\ 3 & 3 & 1 \end{vmatrix}$

13. $\begin{vmatrix} 3 & 2 & -2 \\ -2 & 1 & 4 \\ -4 & -3 & 3 \end{vmatrix}$

14. $\begin{vmatrix} 2 & -1 & 1 \\ 1 & 2 & -1 \\ 3 & 4 & -3 \end{vmatrix}$

15. $\begin{vmatrix} 2 & 3 & 5 \\ 1 & 1 & 1 \\ 1 & 1 & 1 \end{vmatrix}$

16. $\begin{vmatrix} -2 & 5 & -4 \\ 3 & 6 & 6 \\ 5 & 7 & 10 \end{vmatrix}$

c Solve using Cramer's rule.

17. $3x - 4y = 6,$
$5x + 9y = 10$

18. $5x + 8y = 1,$
$3x + 7y = 5$

19. $-2x + 4y = 3,$
$3x - 7y = 1$

20. $5x - 4y = -3,$
$7x + 2y = 6$

21. $4x + 2y = 11,$
$3x - y = 2$

22. $3x - 3y = 11,$
$9x - 2y = 5$

23. $x + 4y = 8,$
$3x + 5y = 3$

24. $x + 4y = 5,$
$-3x + 2y = 13$

ANSWERS

1. ______
2. ______
3. ______
4. ______
5. ______
6. ______
7. ______
8. ______
9. ______
10. ______
11. ______
12. ______
13. ______
14. ______
15. ______
16. ______
17. ______
18. ______
19. ______
20. ______
21. ______
22. ______
23. ______
24. ______

ANSWERS

25. ______

26. ______

27. ______

28. ______

29. ______

30. ______

31. ______

32. ______

33. ______

34. ______

35. ______

36. ______

37. ______

38. ______

39. ______

40. ______

25. $\begin{aligned} 2x - 3y + 5z &= 27, \\ x + 2y - z &= -4, \\ 5x - y + 4z &= 27 \end{aligned}$

26. $\begin{aligned} x - y + 2z &= -3, \\ x + 2y + 3z &= 4, \\ 2x + y + z &= -3 \end{aligned}$

27. $\begin{aligned} r - 2s + 3t &= 6, \\ 2r - s - t &= -3, \\ r + s + t &= 6 \end{aligned}$

28. $\begin{aligned} a \quad - 3c &= 6, \\ b + 2c &= 2, \\ 7a - 3b - 5c &= 14 \end{aligned}$

29. $\begin{aligned} 4x - y - 3z &= 1, \\ 8x + y - z &= 5, \\ 2x + y + 2z &= 5 \end{aligned}$

30. $\begin{aligned} 3x + 2y + 2z &= 3, \\ x + 2y - z &= 5, \\ 2x - 4y + z &= 0 \end{aligned}$

31. $\begin{aligned} p + q + r &= 1, \\ p - 2q - 3r &= 3, \\ 4p + 5q + 6r &= 4 \end{aligned}$

32. $\begin{aligned} x + 2y - 3z &= 9, \\ 2x - y + 2z &= -8, \\ 3x - y - 4z &= 3 \end{aligned}$

SKILL MAINTENANCE

Solve.

33. $0.5x - 2.34 + 2.4x = 7.8x - 9$

34. $5x + 7x = -144$

35. Find the slope of the line $5y - 3x = 8$.

36. Find the y-intercept of the line $5y - 3x = 8$.

SYNTHESIS

Solve.

37. $\begin{vmatrix} y & -2 \\ 4 & 3 \end{vmatrix} = 44$

38. $\begin{vmatrix} 2 & x & -1 \\ -1 & 3 & 2 \\ -2 & 1 & 1 \end{vmatrix} = -12$

39. $\begin{vmatrix} m+1 & -2 \\ m-2 & 1 \end{vmatrix} = 27$

40. Show that an equation of the line through (x_1, y_1) and (x_2, y_2) can be written

$$\begin{vmatrix} x & y & 1 \\ x_1 & y_1 & 1 \\ x_2 & y_2 & 1 \end{vmatrix} = 0.$$

4.7 Systems of Inequalities

In Section 3.6, we studied the graphing of inequalities in two variables. Here we study systems of linear inequalities.

a Systems of Linear Inequalities

The following is an example of a system of two linear inequalities in two variables:

$$x + y \leq 4,$$
$$x - y < 4.$$

A **solution** of a system of linear inequalities is an ordered pair that is a solution of *both* inequalities. We now graph solutions of systems of linear inequalities. To do so, we graph each inequality and determine where the graphs overlap, or intersect.

▶ **EXAMPLE 1** Graph the solutions of the system

$$x + y \leq 4,$$
$$x - y < 4.$$

We graph the inequality $x + y \leq 4$ by first graphing the equation $x + y = 4$ using a solid line. We consider (0, 0) as a test point and find that it is a solution, so we shade all points on that side of the line using color shading. The arrows at the ends of the line also indicate the half-plane that contains the solutions.

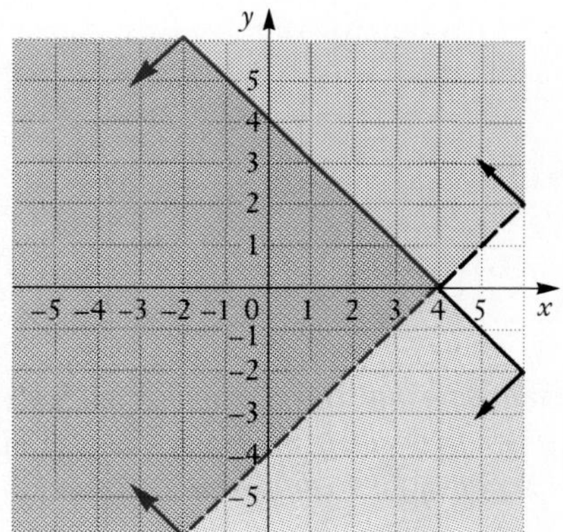

Next we graph $x - y < 4$. We begin by graphing the equation $x - y = 4$ using a dashed line and consider (0, 0) as a test point. Again, (0, 0) is a solution so we shade that side of the line using gray shading. The solution set of the system is the region that is shaded both color and gray and part of the line $x + y = 4$. ◀

DO EXERCISE 1.

OBJECTIVE

After finishing Section 4.7, you should be able to:

a Graph systems of linear inequalities.

FOR EXTRA HELP

Tape NC Tape 7A MAC: 4 IBM: 4

1. Graph:

$$x + y \geq 1,$$
$$y - x \geq 2.$$

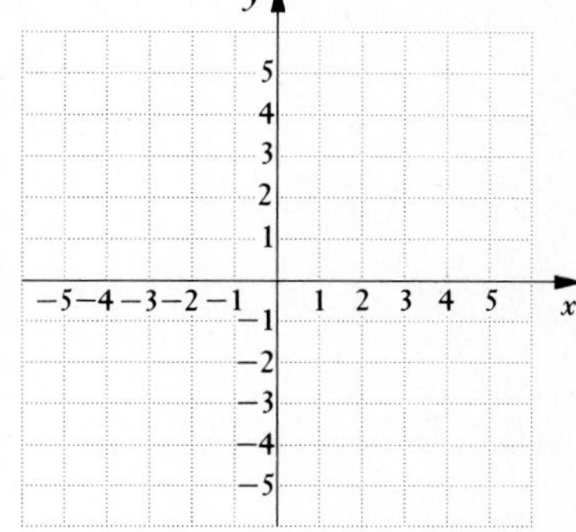

ANSWER ON PAGE A-6

2. Graph:

$$-3 \leq y < 4.$$

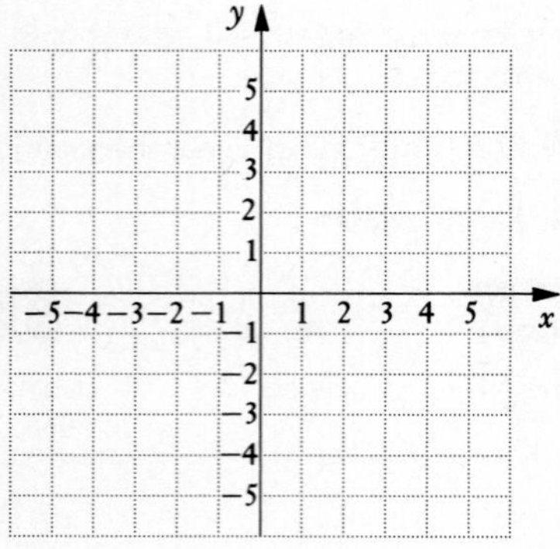

ANSWER ON PAGE A-6

▶ **EXAMPLE 2** Graph: $-2 < x \leq 5$.

This is actually a system of inequalities:

$$-2 < x,$$
$$x \leq 5.$$

We graph the equation $-2 = x$ and see that the graph of the first inequality is the half-plane to the right of the line $-2 = x$ (see the graph on the left below).

We graph the second inequality, starting with the line $x = 5$, and find that its graph is the line and also the half-plane to the left of it (see the graph on the right below).

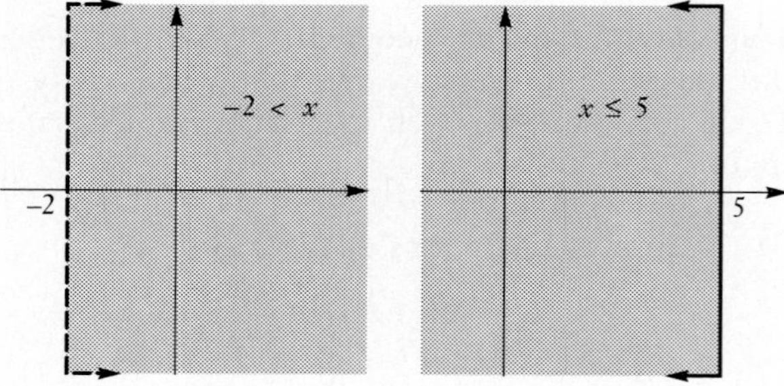

We shade the intersection of these graphs.

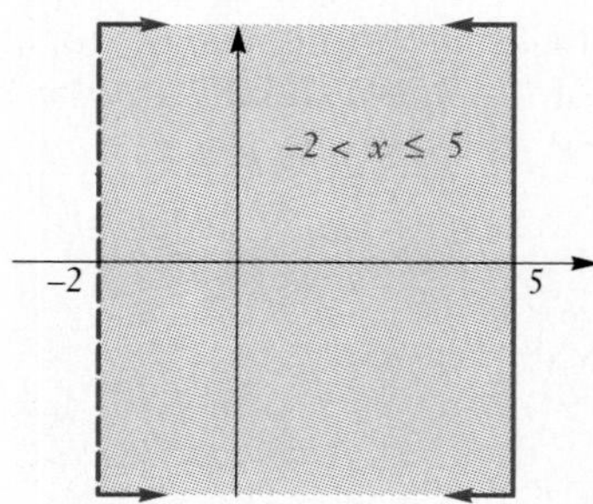

◀

DO EXERCISE 2.

A system of inequalities may have a graph that consists of a polygon and its interior. In *linear programming,* which you may study in a later course, it is important to be able to find the vertices of such a polygon.

► **EXAMPLE 3** Graph the following system of inequalities. Find the coordinates of any vertices formed.

$$6x - 2y \leq 12, \quad (1)$$
$$y - 3 \leq 0, \quad (2)$$
$$x + y \geq 0 \quad (3)$$

We graph the lines $6x - 2y = 12$, $y - 3 = 0$, and $x + y = 0$ using solid lines. The regions for each inequality are indicated by the arrows at the ends of the lines. We then note where the regions overlap and shade the region of solutions using one color.

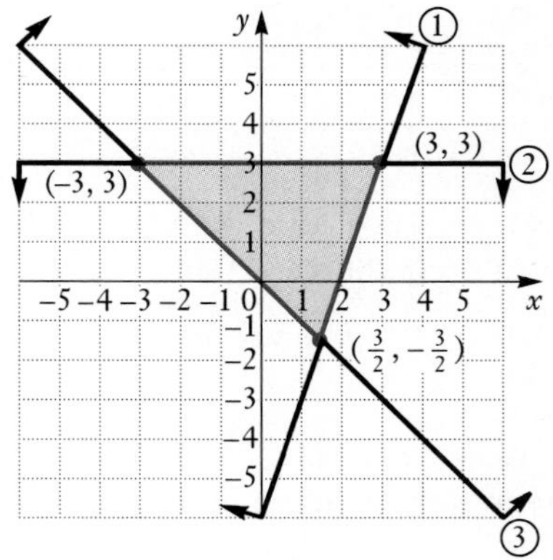

To find the vertices, we solve three different systems of equations. The system of equations from inequalities (1) and (2) is

$$6x - 2y = 12,$$
$$y - 3 = 0.$$

Solving, we obtain the vertex (3, 3).

The system of equations from inequalities (1) and (3) is

$$6x - 2y = 12,$$
$$x + y = 0.$$

Solving, we obtain the vertex $(\frac{3}{2}, -\frac{3}{2})$.

The system of equations from inequalities (2) and (3) is

$$y - 3 = 0,$$
$$x + y = 0.$$

Solving, we obtain the vertex $(-3, 3)$. ◄

DO EXERCISE 3.

► **EXAMPLE 4** Graph the following system of inequalities. Find the coordinates of any vertices formed.

$$x + y \leq 16, \quad (1)$$
$$3x + 6y \leq 60, \quad (2)$$
$$x \geq 0, \quad (3)$$
$$y \geq 0 \quad (4)$$

We graph each inequality using solid lines. The regions for each inequality are indicated by the arrows at the ends of the lines. We then note where the regions overlap and shade the region of solutions using one color.

3. Graph the system of inequalities. Find the coordinates of any vertices formed.

$$5x + 6y \leq 30,$$
$$0 \leq y \leq 3,$$
$$0 \leq x \leq 4$$

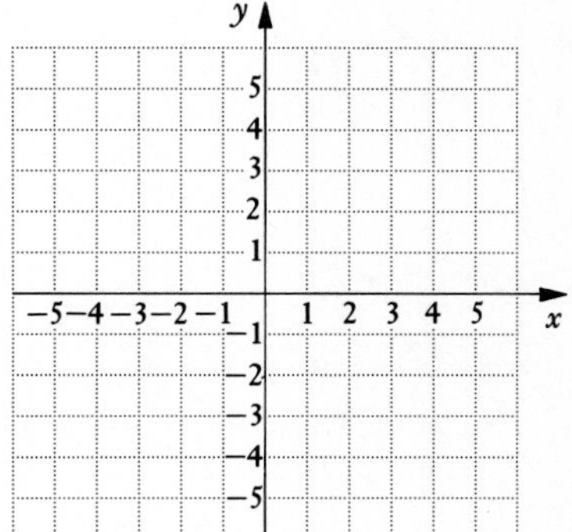

ANSWER ON PAGE A-6

4. Graph the system of inequalities. Find the coordinates of any vertices formed.

$$2x + 4y \leq 8,$$
$$x + y \leq 3,$$
$$x \geq 0,$$
$$y \geq 0$$

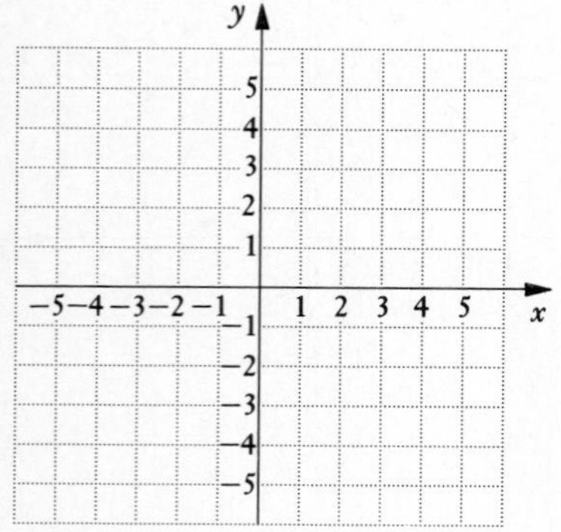

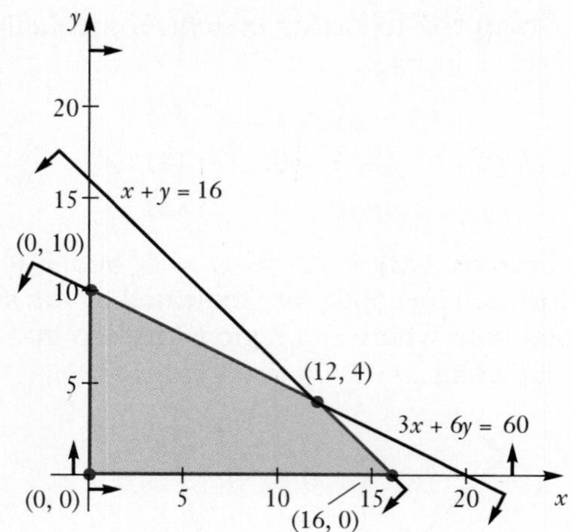

To find the vertices, we solve four different systems of equations. The system of equations from inequalities (1) and (2) is

$$x + y = 16,$$
$$3x + 6y = 60.$$

Solving, we obtain the vertex (12, 4).

The system of equations from inequalities (1) and (4) is

$$x + y = 16,$$
$$y = 0.$$

Solving, we obtain the vertex (16, 0).

The system of equations from inequalities (3) and (4) is

$$x = 0,$$
$$y = 0.$$

The vertex is obviously (0, 0).

The system of equations from inequalities (2) and (3) is

$$3x + 6y = 60,$$
$$x = 0.$$

Solving, we obtain the vertex (0, 10). ◀

DO EXERCISE 4.

ANSWER ON PAGE A-6

NAME SECTION DATE

EXERCISE SET 4.7

a Graph the system of inequalities. Find the coordinates of any vertices formed.

1. $y < x,$
$y > -x + 3$

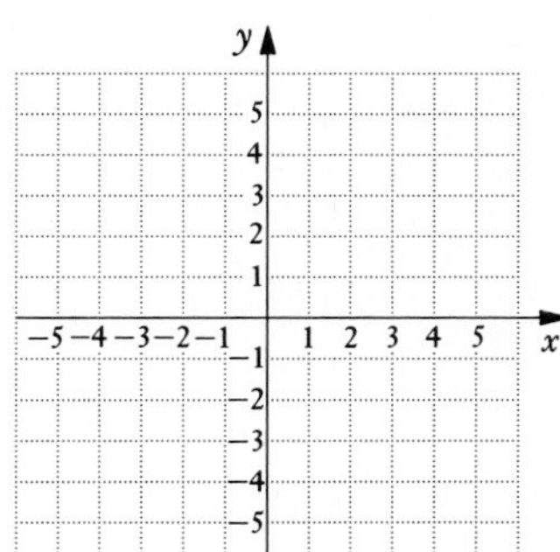

2. $y > x,$
$y < -x + 1$

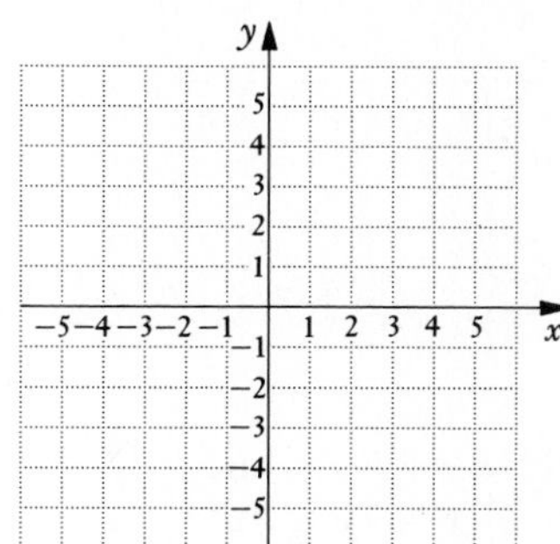

3. $y \geq x,$
$y \leq -x + 4$

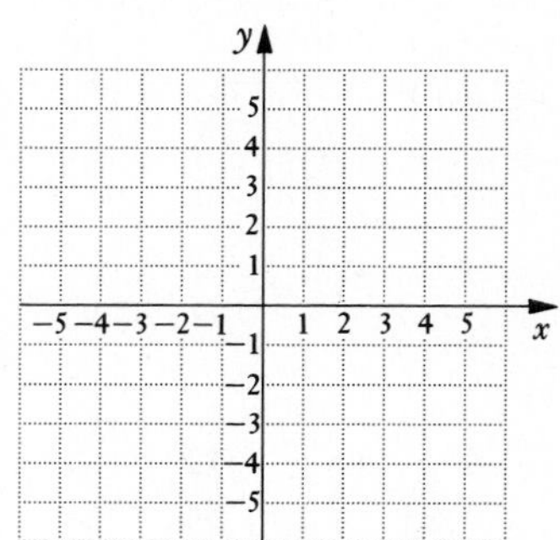

4. $y \geq x,$
$y \leq -x + 2$

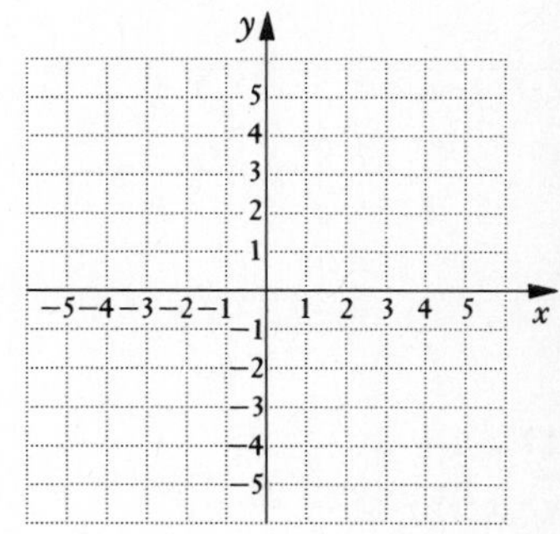

5. $y \geq -2,$
$x \geq 1$

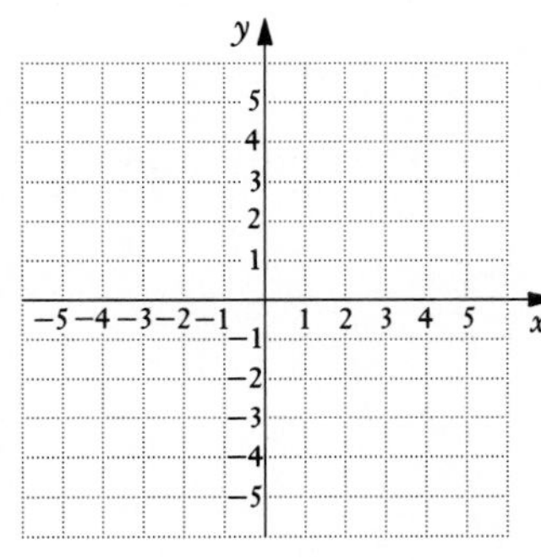

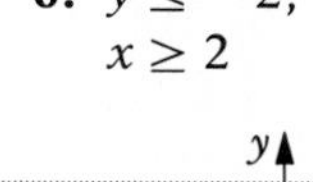

6. $y \leq -2,$
$x \geq 2$

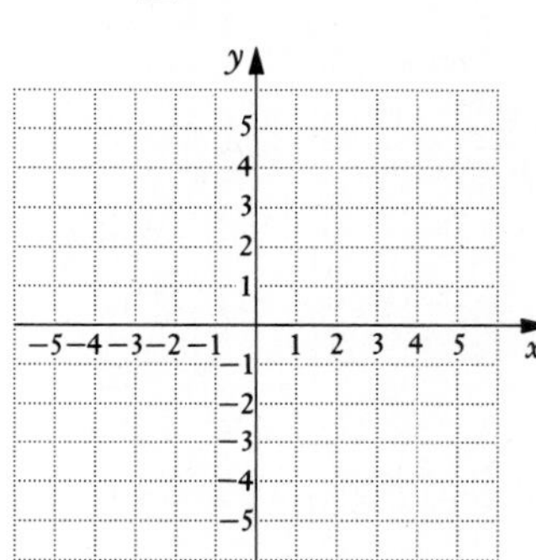

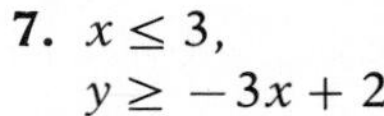

7. $x \leq 3,$
$y \geq -3x + 2$

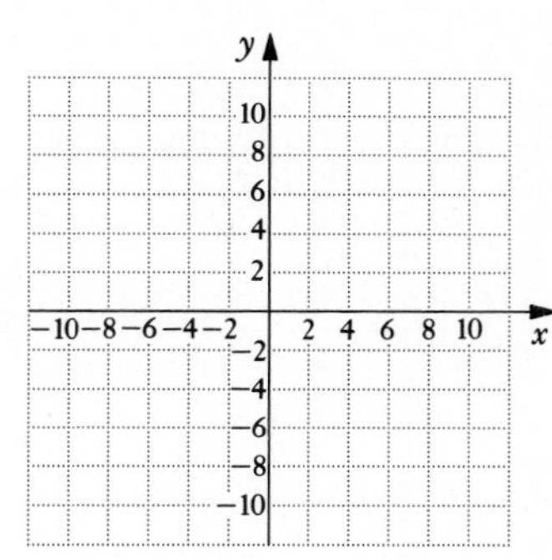

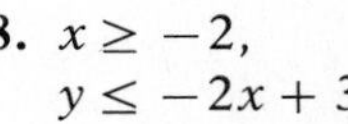

8. $x \geq -2,$
$y \leq -2x + 3$

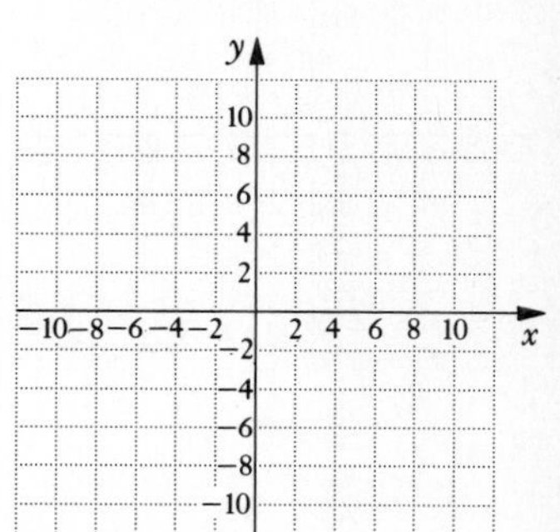

9. $y \geq -2,$
$y \geq x + 3$

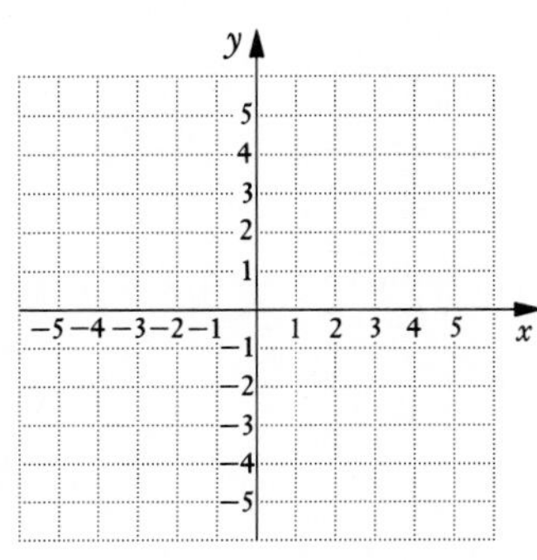

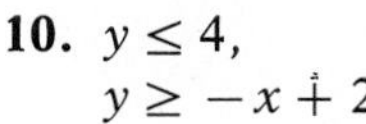

10. $y \leq 4,$
$y \geq -x + 2$

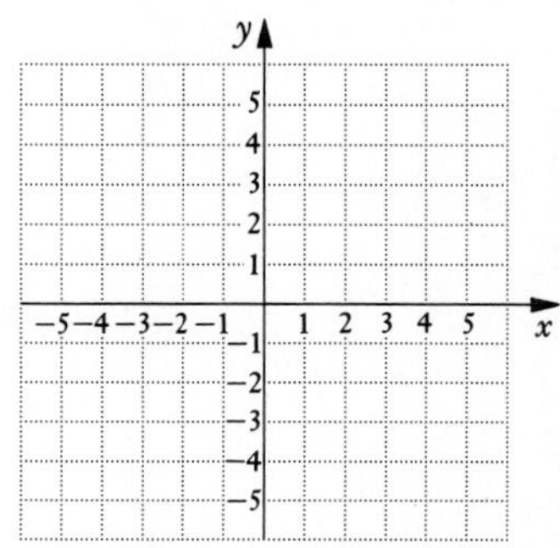

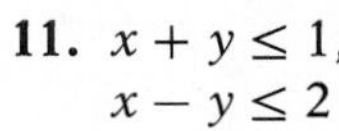

11. $x + y \leq 1,$
$x - y \leq 2$

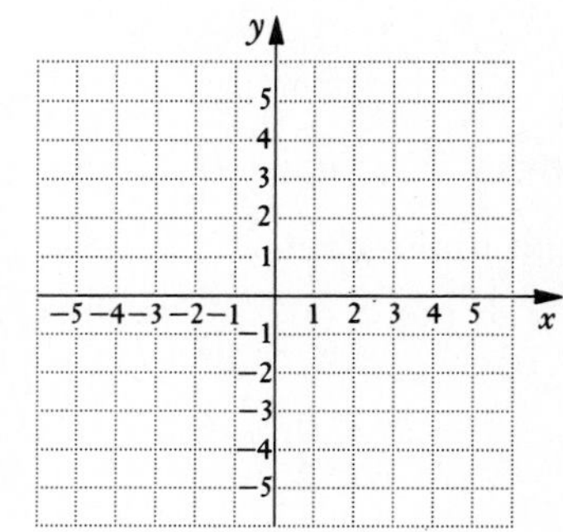

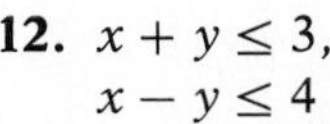

12. $x + y \leq 3,$
$x - y \leq 4$

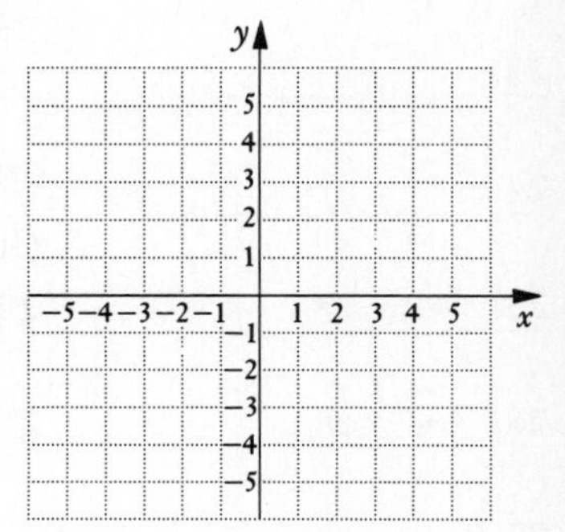

13. $y - 2x \geq 1,$
$y - 2x \leq 3$

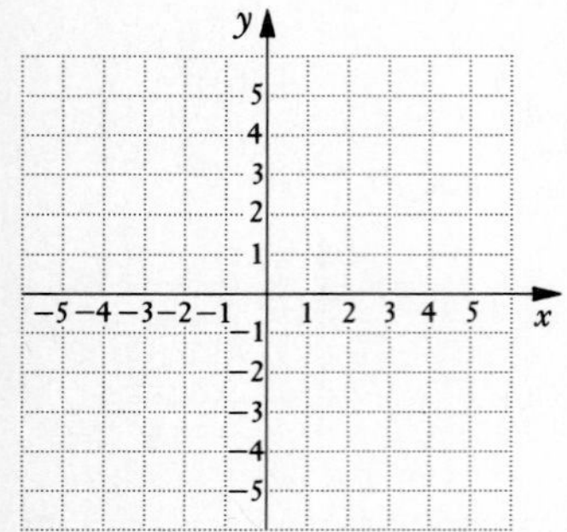

14. $y + 3x \geq 0,$
$y + 3x \leq 2$

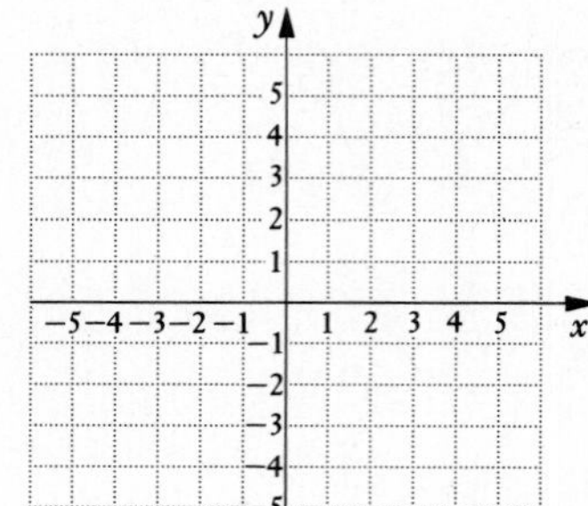

15. $y \leq 2x + 1,$
$y \geq -2x + 1,$
$x \leq 2$

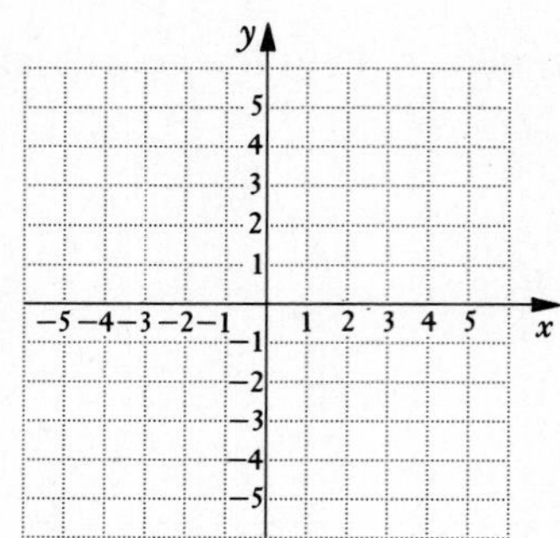

16. $x - y \leq 2,$
$x + 2y \geq 8,$
$y \leq 4$

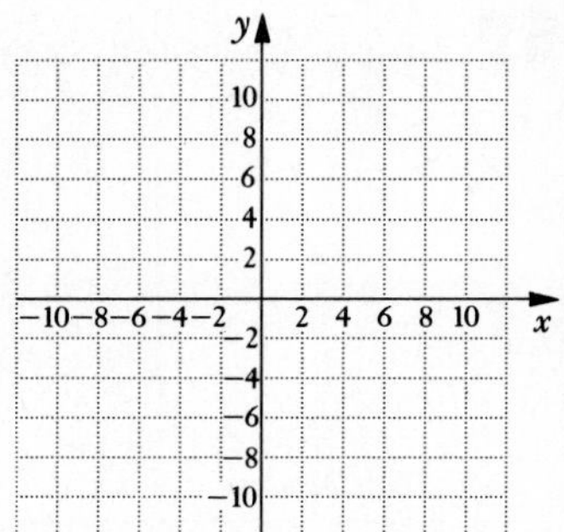

17. $x + 2y \leq 12,$
$2x + y \leq 12,$
$x \geq 0,$
$y \geq 0$

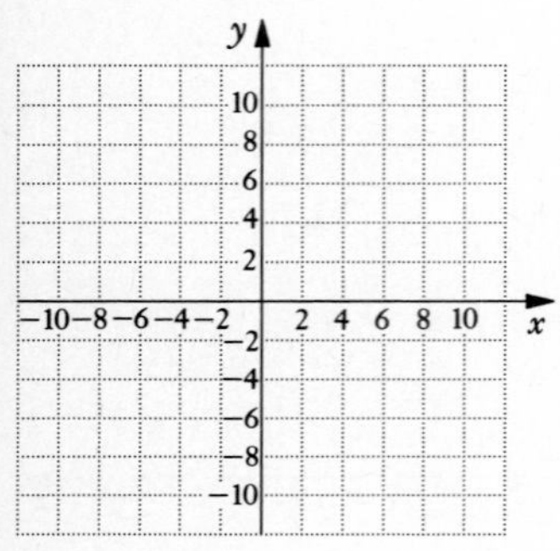

$4y - 3x \geq -12,$
$4y + 3x \geq -36,$
$y \leq 0,$
$x \leq 0$

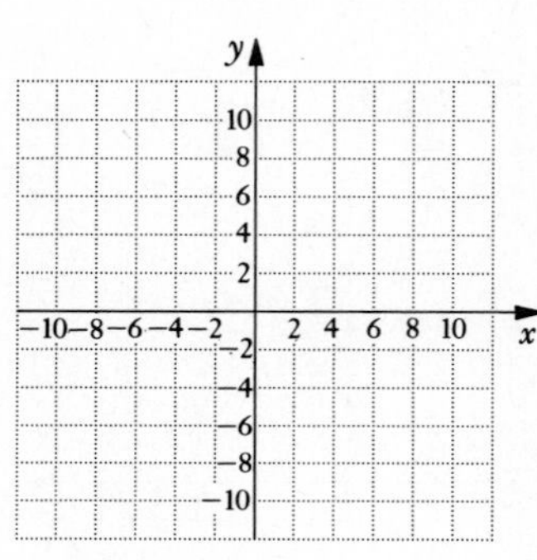

19. $8x + 5y \leq 40,$
$x + 2y \leq 8,$
$x \geq 0,$
$y \geq 0$

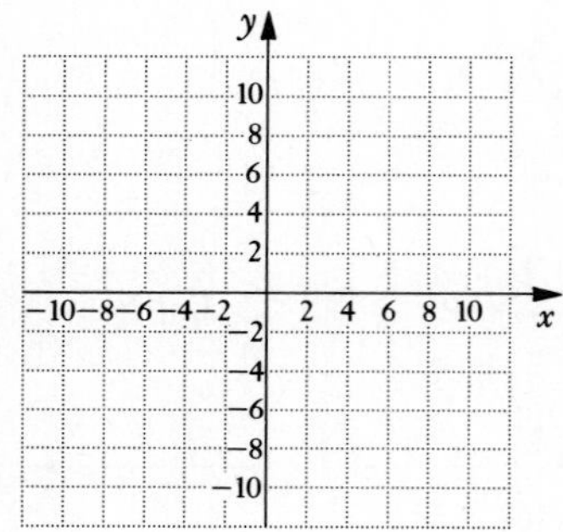

20. $y - x \geq 1,$
$y - x \leq 3,$
$2 \leq x \leq 5$

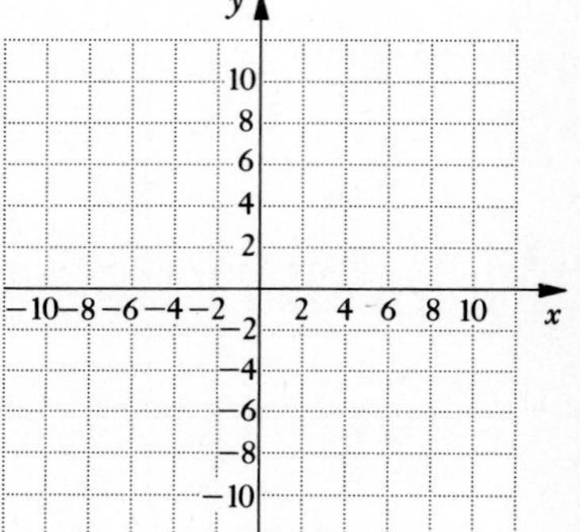

ANSWERS

21. ______

22. ______

23. ______

24. ______

25. See graph.

26. See graph.

27. See graph.

28. See graph.

SKILL MAINTENANCE

21. Simplify: $\dfrac{4^3 + 5 \cdot 6 - 7 \cdot 8}{|3 - 5|^5 + 4(9 - 2)}$.

22. Simplify: $-7x - 8(9 - 4x) + 21x$.

Solve.

23. $5(3x - 4) = -2(x + 5)$

24. $4(3x + 4) = 2 - x$

SYNTHESIS

Graph the system.

25. $x + y \geq 5,$
$x + y \leq -3$

26. $x + y \leq 8,$
$x + y \leq -2$

27. $x - 2y \leq 0,$
$-2x + y \leq 2,$
$x \leq 2,$
$y \leq 2,$
$x + y \leq 4$

28. $x + y \geq 1,$
$-x + y \leq 2,$
$x \leq 4,$
$y \geq 0,$
$y \leq 4,$
$x \leq 2$

SUMMARY AND REVIEW: CHAPTER 4

IMPORTANT PROPERTIES AND FORMULAS

Second-order determinant:

$$\begin{vmatrix} a_1 & b_1 \\ a_2 & b_2 \end{vmatrix} = a_1b_2 - a_2b_1$$

Third-order determinant:

$$\begin{vmatrix} a_1 & b_1 & c_1 \\ a_2 & b_2 & c_2 \\ a_3 & b_3 & c_3 \end{vmatrix} = a_1\begin{vmatrix} b_2 & c_2 \\ b_3 & c_3 \end{vmatrix} - a_2\begin{vmatrix} b_1 & c_1 \\ b_3 & c_3 \end{vmatrix} + a_3\begin{vmatrix} b_1 & c_1 \\ b_2 & c_2 \end{vmatrix}$$

REVIEW EXERCISES

The review sections and objectives to be tested in addition to the material in this chapter are [1.5c], [2.1d], [2.3a], and [3.3a, b].

Solve the system by graphing. Classify the system as consistent or inconsistent and dependent or independent.

1. $4x - y = -9,$
$2x - 2y = -6$

2. $15x + 10y = -20,$
$3x + 2y = -4$

3. $y - 2x = 4,$
$y - 2x = 5$

Solve by the substitution method.

4. $7x - 4y = 6,$
$y - 3x = -2$

5. $y = x + 2,$
$y - x = 8$

6. $9x - 6y = 2,$
$x = 4y + 5$

Solve by the elimination method.

7. $8x - 2y = 10,$
$-4y - 3x = -17$

8. $4x - 7y = 18,$
$9x + 14y = 40$

9. $3x - 5y = -4,$
$5x - 3y = 4$

Solve.

10. Jimmy has $20.00 to spend at the store. He can spend all the money for two record albums and a poster, or he can buy one record album and two posters and have $1.00 left over. What is the price of a record album? What is the price of a poster?

11. A train leaves town at noon traveling north at 44 mph. One hour later, another train travels north on a parallel track at 52 mph. How many hours will the second train travel before it overtakes the first train?

12. A beaker of alcohol contains a solution that is 30% alcohol. In another beaker is a solution that is 50% alcohol. How much of each can be added to obtain 40 L of a solution that is 45% alcohol?

Solve.

13. $x + 2y + z = 10,$
$2x - y + z = 8,$
$3x + y + 4z = 2$

14. $3x + 2y + z = 3,$
$6x - 4y - 2z = -34,$
$-x + 3y - 3z = 14$

15. $2x - 5y - 2z = -4,$
$7x + 2y - 5z = -6,$
$-2x + 3y + 2z = 4$

16. $-5x + 5y = -6,$
$2x - 2y = 4$

17. $3x + y = 2,$
$x + 3y + z = 0,$
$x + z = 2$

18. $3x + 4y = 6,$
$1.5x - 3 = -2y$

19. $x + y + 2z = 1,$
$x - y + z = 1,$
$x + 2y + z = 2$

Solve.

20. In triangle ABC, the measure of angle A is four times the measure of angle C and the measure of angle B is 45° more than the measure of angle C. What are the measures of the angles of the triangle?

21. Find the three-digit number in which the sum of the digits is 11, the ten's digit is three less than the sum of the hundred's and unit's digits, and the unit's digit is five less than the hundred's digit.

22. Lynn has \$194 in her purse, consisting of \$20, \$5, and \$1 bills. The number of \$1 bills is one less than the total number of \$20 and \$5 bills. If she has 39 bills in her purse, how many of each denomination does she have?

Evaluate.

23. $\begin{vmatrix} -2 & 4 \\ -3 & 5 \end{vmatrix}$

24. $\begin{vmatrix} 2 & 3 & 0 \\ 1 & 4 & -2 \\ 2 & -1 & 5 \end{vmatrix}$

Solve using Cramer's rule. Show your work.

25. $2x + 3y = 6,$
$x - 4y = 14$

26. $2x + y + z = -2,$
$2x - y + 3z = 6,$
$3x - 5y + 4z = 7$

27. $3x + 4y = -13,$
$5x + 6y = 8$

28. $x + 4y - z = -5,$
$3x + 2y + 2z = 15,$
$x - 2y + z = 8$

Graph. Find the coordinates of any vertices formed.

29. $y \geq -3,$
$x \geq 2$

30. $x + 3y \geq -1,$
$x + 3y \leq 4$

31. $x - 3y \leq 3,$
$x + 3y \geq 9,$
$y \leq 6$

SKILL MAINTENANCE

32. Solve: $4x - 5x + 8 = -9x + 2x.$

33. Solve $Q = at - 4t$ for t.

34. Simplify: $-8x + 10(x - 4) - 12(2x - 8) + 16.$

35. Find the slope of the line $5x - 8y = 40.$

SYNTHESIS

36. Solve:

$$\begin{vmatrix} -3 & x & -4 \\ -2 & 5 & 3 \\ -4 & 3 & 2 \end{vmatrix} = 25.$$

❖ THINKING IT THROUGH

1. Briefly compare the strengths and weaknesses of the graphical, substitution, and elimination methods as applied to the solution of systems of two linear equations in two variables.
2. List a system of equations with no solution. (Answers may vary.)
3. List a system of equations with infinitely many solutions. (Answers may vary.)
4. Explain the advantages of using a system of equations to solve certain kinds of problems.

NAME SECTION DATE

TEST: CHAPTER 4

Solve the system by graphing. Classify the system as consistent or inconsistent and dependent or independent.

1. $y = 3x + 7,$
$3x + 2y = -4$

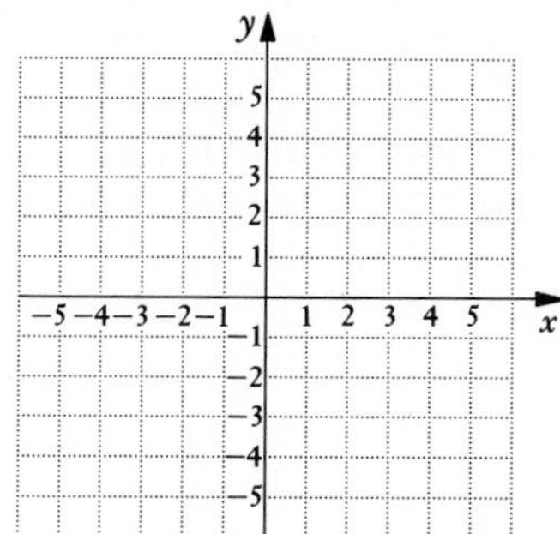

2. $y = 3x + 4,$
$y = 3x - 2$

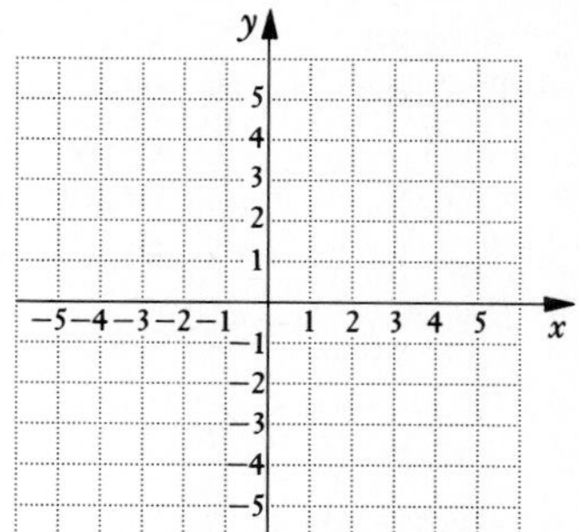

3. $y - 3x = 6,$
$6x - 2y = -12$

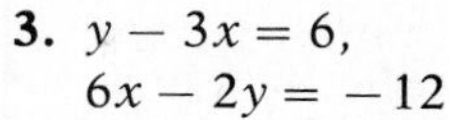

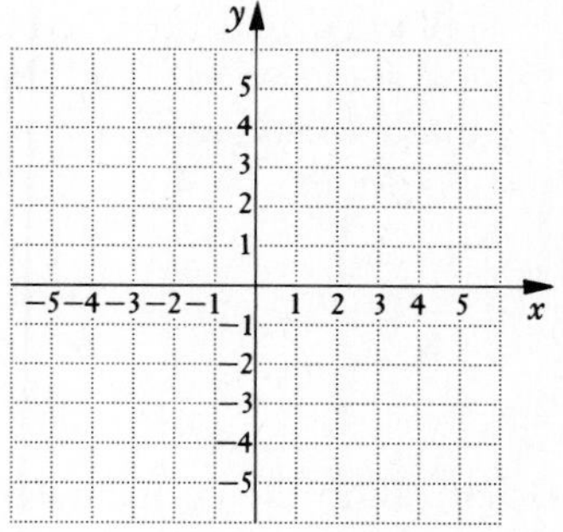

Solve by the substitution method.

4. $x + 3y = -8,$
$4x - 3y = 23$

5. $2x + 4y = -6,$
$y = 3x - 9$

Solve by the elimination method.

6. $4x - 6y = 3,$
$6x - 4y = -3$

7. $4y + 2x = 18,$
$3x + 6y = 26$

8. The perimeter of a rectangle is 96. The length of the rectangle is six less than twice the width. Find the dimensions of the rectangle.

9. A fast-food restaurant sold 132 chicken dinners during one day. Two-piece dinners sold for \$4.50, and three-piece dinners sold for \$5.50. The restaurant's total receipts that day from chicken dinners were \$656. How many of each size dinner did they sell?

Solve.

10. $-3x + y - 2z = 8,$
$-x + 2y - z = 5,$
$2x + y + z = -3$

11. $6x + 2y - 4z = 15,$
$-3x - 4y + 2z = -6,$
$4x - 6y + 3z = 8$

12. $2x + 2y = 0,$
$4x + 4z = 4,$
$2x + y + z = 2$

13. $3x + 3z = 0,$
$2x + 2y = 2,$
$3y + 3z = 3$

ANSWERS

1. ______
2. ______
3. ______
4. ______
5. ______
6. ______
7. ______
8. ______
9. ______
10. ______
11. ______
12. ______
13. ______

ANSWERS

14. ______

15. ______

16. ______

17. ______

18. ______

19. ______

20. ______

21. ______

22. ______

23. ______

24. ______

Solve using Cramer's rule. Show your work.

14. $7x - 8y = 10,$
$9x + 5y = -2$

15. $x + 3y - 3z = 12,$
$3x - y + 4z = 0,$
$-x + 2y - z = 1$

16. A plane flew for 5 hr with a 20-km/h tail wind and returned in 7 hr against the same wind. Find the speed of the plane in still air.

17. Mixture A is 34% salt and the rest water. Mixture B is 61% salt and the rest water. How many pounds of each mixture would be needed in order to obtain 120 lb of a mixture that is 50% salt?

Graph. Find the coordinates of any vertices formed.

18. $x + y \geq 3,$
$x - y \geq 5$

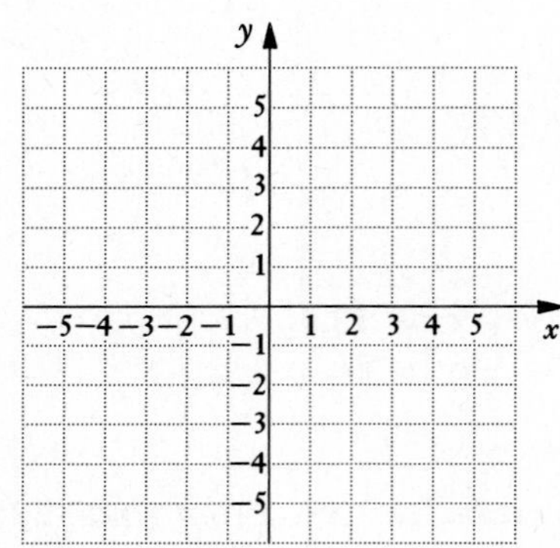

19. $2y - x \geq -7,$
$2y + 3x \leq 15,$
$y \leq 0,$
$x \leq 0$

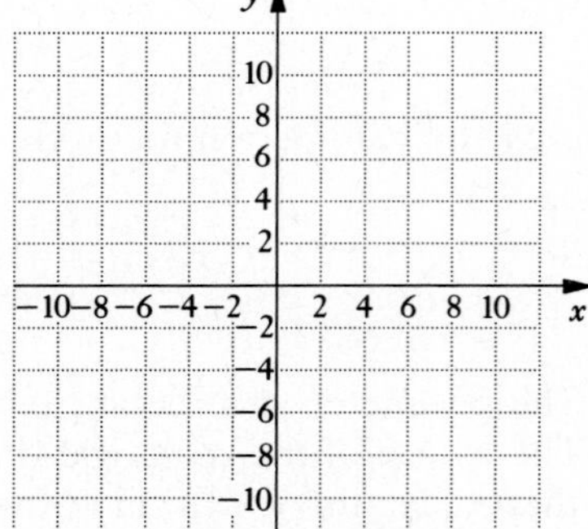

SKILL MAINTENANCE

20. Find the slope of the line containing the points $(-3, 5)$ and $(2, -6)$.

21. Solve $P = 4a - 3b$ for a.

22. Solve: $-3x - 5 + 6x = 8x - 14$.

23. Simplify: $-8(t + 8) - 12(t - 4)$.

SYNTHESIS

24. The graph of the equation $y = mx + b$ contains the points $(-1, 3)$ and $(-2, -4)$. Find m and b.

CUMULATIVE REVIEW: CHAPTERS 1–4

Evaluate for $a = 2$ and $b = -5$.

1. $\frac{a+b}{3}$

2. $ab - 2a$

Simplify.

3. $|-13|$

4. $\left|\frac{0}{2}\right|$

5. $(-2.1)(3.8)(-11.0)$

6. $\left(\frac{5}{9}\right) \div \left(-\frac{7}{3}\right)$

7. $9x - 3(2x - 11)$

8. $[5(6-3)+2] - [4(5-6)+11]$

9. $\frac{-10a^7b^{-11}}{25a^{-4}b^{22}}$

10. $4b + 2 - [7 - 6(2b + 1)]$

11. $\frac{y^4}{y^{-6}}$

Solve.

12. $6y - 5(3y - 4) = 10$

13. $-3 + 5x = 2x + 15$

14. $A = \pi r^2 h$, for h

15. $L = \frac{1}{3}m(k + p)$, for p

16. $5x + 8 > 2x + 5$

17. $2x - 10 \leq -4$ *or* $x - 4 \geq 3$

18. $-12 \leq -3x + 1 < 0$

19. $|8y - 3| \geq 15$

20. $|x + 1| = 4$

Graph on a plane.

21. $3y = 9$

22. $y = -\frac{1}{2}x - 3$

23. $3x - 1 = y$

24. $y > 3x - 4$

25. $3x + 5y = 15$

26. $2x - y \leq 6$

27. Solve this system of equations by graphing: $2x - y = 7$,
$x + 3y = 0$.

Solve.

28. $3x + 4y = 4$,
$x - 2y = 2$

29. $3x + y = 4$,
$6x - y = 5$

30. $4x + 3y = -2$,
$2x - 5y = -12$

31. $$\begin{aligned} 2x + 5y - 3z &= -11, \\ -5x + 3y - 2z &= -7, \\ 3x - 2y + 5z &= 12 \end{aligned}$$

32. $$\begin{aligned} x - y + z &= 1, \\ 2x + y + z &= 3, \\ x + y - 2z &= 4 \end{aligned}$$

Solve using Cramer's rule. Show your work.

33. $$\begin{aligned} 9x - 6y &= -7, \\ -3x + 12y &= 4 \end{aligned}$$

34. $$\begin{aligned} 2x + y &= 0, \\ x + 2y + z &= 0, \\ x + z &= 1 \end{aligned}$$

Evaluate.

35. $\begin{vmatrix} -7 & 2 \\ -5 & 3 \end{vmatrix}$

36. $\begin{vmatrix} -2 & 1 & 0 \\ -6 & 2 & -1 \\ 3 & -1 & 4 \end{vmatrix}$

Graph.

37. $x + y < -3,$
$x - y < 1$

38. $x \geq 0,$
$y \leq x - 2$

39. Find the slope and the y-intercept of $5y - 4x = 20$.

40. Find the slope and the y-intercept of $2x + 4y = 7$.

41. Find an equation of the line with slope -3 and containing the point $(5, 2)$.

42. Find an equation of the line parallel to $3x - 9y = 2$ that contains the point $(-6, 2)$.

43. A piece of wire 10 m long is to be cut into two pieces, one of them two thirds as long as the other. How should the wire be cut?

44. Japan has the most expensive movie ticket prices in the world. On the average, a ticket in Japan costs \$6.09 more than a ticket in the United States. Two tickets in the United States cost \$0.84 less than one ticket in Japan. What is the average price of a movie ticket in each country?

45. There are four more nickels than dimes in a piggy bank. The total amount of money in the bank is \$2.45. How many of each type of coin are in the bank?

46. One month a family spent \$340 for electricity, rent, and telephone. The electric bill was one fourth of the rent, and the rent was \$200 more than the phone bill. How much was the electric bill?

47. In the United States, the number N, in millions, of bicyclists t years since 1983 is approximated by $N = 3t + 72$. When will there be more than 100,000,000 bicyclists?

48. Linda has a total of 225 on three tests. The sum of the scores on the first and second tests exceeds her third score by 61. Her first score exceeds her second by 6. Find the three scores.

SYNTHESIS

49. Tammy's age is the sum of the ages of Carmen and Dennis. Carmen's age is two more than the sum of the ages of Dennis and Mark. Dennis's age is four times Mark's age. The sum of all four ages is 42. How old is Tammy?

50. Two solutions to the equation $y = mx + b$ are $(5, -3)$ and $(-4, 2)$. Find m and b.

INTRODUCTION A polynomial is a type of algebraic expression that contains one or more terms. In this chapter, you will learn to manipulate polynomials and to use them to solve problems.

The review sections to be tested in addition to the material in this chapter are 2.6, 3.2, 4.4, and 4.5. ❖

Polynomials

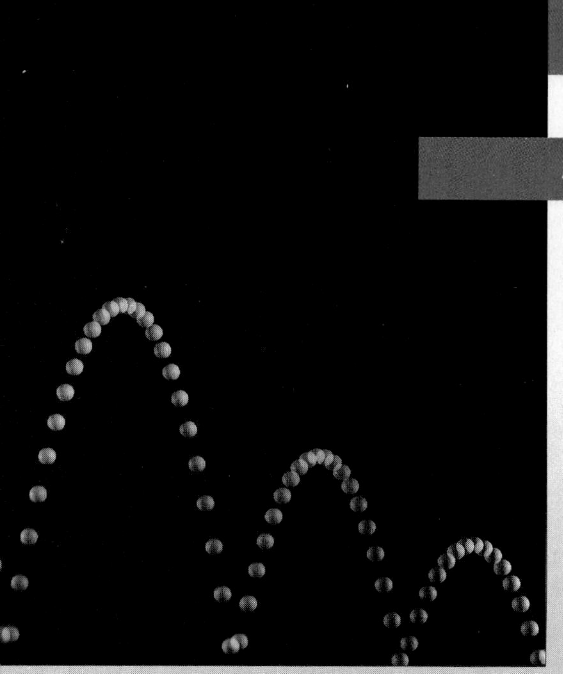

AN APPLICATION

Suppose that an object is thrown upward with an initial velocity of 80 ft/sec from a height of 224 ft. Find its height after t seconds.

THE MATHEMATICS

The height h after t seconds is given by

$$h = \underbrace{-16t^2 + 80t + 224.}_{\text{This is a polynomial.}}$$

We find the height after 2 sec by evaluating the polynomial when $t = 2$:

$$h = -16(2)^2 + 80(2) + 224 = 320 \text{ ft.}$$

❖ POINTS TO REMEMBER: CHAPTER 5

Product Law of Exponents:	$a^n a^m = a^{n+m}$
Distributive Laws:	$a(b + c) = ab + ac, \quad a(b - c) = ab - ac$
Meaning of a Factor:	If $N = ab$, then a and b are factors of N.
Equation-Solving Skills:	Section 2.1
Pythagorean Theorem:	The sum of the squares of the legs of a right triangle is equal to the square of the hypotenuse: $a^2 + b^2 = c^2$.

PRETEST: CHAPTER 5

1. Evaluate $P = 2x^2 - 3x + 1$ when $x = -1$.

2. Add: $(13x^2y - 4xy^2 + 3xy) + (4xy^2 - 7x^2y - 2xy)$.

3. Subtract:
$(5m^3 - 3m^2 + 6m + 3) - (6m - 9 - m^2 + 4m^2)$.

4. Arrange the polynomial $4xy^5 - 3x^6y^2 + x^2y^3 - 2y$ in descending powers of y.

Multiply.

5. $(x^2 - 1)(x^2 - 2x + 1)$

6. $(2y + 5z)(4y - z)$

7. $(a + 3b)(a - 3b)$

8. $(5t - 3m^2)^2$

Factor.

9. $4x^2 + 4x - 3$

10. $50m^2 + 40m + 8$

11. $4t^6 + 4t^3$

12. $a^2 + 6a + 8$

13. $x^2 - 49y^2$

14. $y^3 + 3y^2 + 4y + 12$

15. Solve: $6x + 8 = 9x^2$.

16. The square of a number is 35 more than twice the number. Find the number.

17. Divide: $(x^2 + 14x - 28) \div (x - 3)$.

18. Divide using synthetic division:
$(x^4 - x^3 + x^2 - 16) \div (x + 2)$.

5.1 Polynomials: Addition and Subtraction

A **polynomial** is a type of algebraic expression.

a Polynomial Expressions

The following are examples of **monomials:**

$$0, \quad -3, \quad z, \quad 8x, \quad -7y^2, \quad 4a^2b^3, \quad 1.3p^4q^5r^7.$$

Each expression is a constant or a constant times some variable or variables to powers that are nonnegative integers.

> **A *polynomial* is a monomial or a combination of sums and/or differences of monomials.**

Expressions like these are called **polynomials in one variable:**

$$5x^2, \quad 8a, \quad 2, \quad 2x + 3, \quad -7x + 5, \quad 2y^2 + 5y - 3,$$
$$5a^4 - 3a^2 + \tfrac{1}{4}a - 8, \quad b^6 + 3b^5 - 8b + 7b^4 + \tfrac{1}{2}.$$

Expressions like these are called **polynomials in several variables:**

$$5a - ab^2 + 7b + 2, \quad 9xy^2z - 4x^3z + (-14x^4y^2) + 9, \quad 15x^3y^2.$$

The following are algebraic expressions that are not polynomials:

$$(1)\ \frac{y^2 - 3}{y^2 + 4}, \quad (2)\ 4a^{-3}b^4, \quad (3)\ \frac{2xy}{x^3 - y^3}.$$

Expressions (1) and (3) are not polynomials because they represent quotients. Expression (2) is not a polynomial because x is raised to a power that is a negative integer.

The polynomial $5x^3y - 7xy^2 + 2$ has three **terms:**

$$5x^3y, \quad -7xy^2, \quad \text{and} \quad 2.$$

The **coefficients** of the terms are 5, -7, and 2.

The **degree of a term** is the sum of the exponents of the variables, if there are variables. The degree of a constant term is 0, except when the constant term is 0. Mathematicians agree that the polynomial 0 has no degree. This is because we can express 0 as $0 = 0x^5 = 0x^8$, and so on, using any exponent we wish. The **degree of a polynomial** is the same as its term of highest degree.

The **leading term** of a polynomial is the term of highest degree. Its coefficient is called the **leading coefficient.**

OBJECTIVES

After finishing Section 5.1, you should be able to:

- **a** Determine the degree of each term and the degree of a polynomial; identify terms, coefficients, monomials, binomials, and trinomials; arrange polynomials in ascending or descending order; and determine the leading coefficient.
- **b** Evaluate polynomials for specified values of the variables.
- **c** Collect like terms.
- **d** Add polynomials.
- **e** Write an equivalent expression for the additive inverse, or opposite, of a polynomial.
- **f** Subtract polynomials.

FOR EXTRA HELP

Tape 8C

Tape 7A

MAC: 5
IBM: 5

1. Identify the terms and the leading term.

$-92x^5 - 8x^4 + x^2 + 5$

2. Identify the coefficient of each term and the leading coefficient.

$5x^3y - 4xy^2 - 2x^3 + xy - y - 5$

3. Determine the degree of each term and the degree of the polynomial. Then find the leading term and the leading coefficient.

a) $6x^2 - 5x^3 + 2x - 7$

b) $2y - 4 - 5x + 7x^2y^3z^2 + 5xy^2$

4. Consider the following polynomials.

a) $3x^2 - 2$
b) $5x^3 + 9x - 3$
c) $4x^2$
d) $-7y$
e) -3
f) $8x^3 - 2x^2$
g) $-4y^2 - 5 - 5y$
h) $5 - 3x$

Identify the monomials, the binomials, and the trinomials.

5. a) Arrange in ascending order:

$5 - 6x^2 + 7x^3 - x^4 + 10x.$

b) Arrange in descending order:

$5 - 6x^2 + 7x^3 - x^4 + 10x.$

ANSWERS ON PAGE A-6

▶ **EXAMPLE 1** Determine the degree of each term and the degree of the polynomial. Then find the leading term and the leading coefficient.

$$2x^3 + 8x^2 - 17x - 3$$

Term	$2x^3$	$8x^2$	$-17x$	-3
Degree	3	2	1	0
Degree of polynomial	3			
Leading term	$2x^3$			
Leading coefficient	2			

◀

▶ **EXAMPLE 2** Determine the degree of each term and the degree of the polynomial. Then find the leading term and the leading coefficient.

$$6x^2 + 8x^2y^3 - 17xy - 24xy^2z^4 + 2y + 3$$

Term	$6x^2$	$8x^2y^3$	$-17xy$	$-24xy^2z^4$	$2y$	3
Degree	2	5	2	7	1	0
Degree of polynomial	7					
Leading term	$-24xy^2z^4$					
Leading coefficient	-24					

◀

DO EXERCISES 1–3.

The following are some names for certain kinds of polynomials.

Type	Definition	Examples
Monomial	A polynomial of one term	$4,\ -3p,\ 5x^2,\ -7a^2b^3,\ 0,\ xyz$
Binomial	A polynomial of two terms	$2x + 7,\ a - 3b,\ 5x^2 + 7y^3$
Trinomial	A polynomial of three terms	$x^2 - 7x + 12,\ 4a^2 + 2ab + b^2$

We generally arrange polynomials in one variable so that the exponents *decrease* from left to right, which is **descending order.** Sometimes they may be written so that the exponents *increase* from left to right, which is **ascending order.** Generally, if an exercise is written in one kind of order, we write the answer in that same order.

▶ **EXAMPLE 3** Arrange in descending order: $12 + x^2 - 7x.$

$$12 + x^2 - 7x = x^2 - 7x + 12$$ ◀

▶ **EXAMPLE 4** Arrange in ascending order: $12 + x^2 - 7x.$

$$12 + x^2 - 7x = 12 - 7x + x^2$$ ◀

DO EXERCISES 4 AND 5.

We can also arrange the terms of polynomials in several variables in ascending or descending order, with respect to the powers of one of the variables.

► **EXAMPLE 5** Arrange in ascending powers of x:

$$x^4 + 2 - 5x^2 + 3x^3y + 7xy^2.$$

We have

$$x^4 + 2 - 5x^2 + 3x^3y + 7xy^2 = 2 + 7xy^2 - 5x^2 + 3x^3y + x^4.$$ ◄

► **EXAMPLE 6** Arrange in descending powers of x:

$$y^4 + 2 - 5x^2 + 3x^3y + 9xy.$$

We have

$$y^4 + 2 - 5x^2 + 3x^3y + 9xy = 3x^3y - 5x^2 + 9xy + y^4 + 2.$$ ◄

DO EXERCISE 6.

b Evaluating Polynomials

We may often represent a polynomial with a single letter, such as

$$P = 5x^7 + 3x^5 - 4x^2 - 5.$$

To find the values of a polynomial, we substitute a number for each occurrence of the variable. Then we calculate, using our rules for order of operations (Section 1.5). In such a way, we **evaluate** the polynomial.

► **EXAMPLE 7** Evaluate the polynomial $P = -x^2 + 4x - 1$ when $x = 2$, $x = 10$, and $x = -10$.

When $x = 2$, $P = -x^2 + 4x - 1 = -(2)^2 + 4(2) - 1 = -4 + 8 - 1 = 3$.

When $x = 10$, $P = -(10)^2 + 4(10) - 1 = -100 + 40 - 1 = -61$.

When $x = -10$, $P = -(-10)^2 + 4(-10) - 1 = -100 - 40 - 1 = -141$. ◄

DO EXERCISE 7.

Polynomials are useful in many real-world applications.

► **EXAMPLE 8** The local supermarket is having a guessing contest with a free trip to Florida as the prize. The object of the contest is to guess the number of oranges in a pile like the one shown here, where each layer in the pile is an equilateral triangle. You can use mathematics to win!

6. **a)** Arrange in ascending powers of y:

$$5x^4y - 3y^2 + 3x^2y^3 + x^3 - 5.$$

b) Arrange in descending powers of y:

$$5x^4y - 3y^2 + 3x^2y^3 + x^3 - 5.$$

7. Evaluate the polynomial $P = x^2 - 2x + 5$, when the variable has the given values.

a) $x = 0$

b) $x = 4$

c) $x = -2$

8. The cost C, in cents per kilometer, of operating an automobile at speed s, in kilometers per hour, is approximated by the polynomial $C = 0.002s^2 - 0.21s + 15$. How much does it cost to operate at 80 km/h?

ANSWERS ON PAGE A-6

Collect like terms.

9. $3y - 4x + 6xy^2 - 2xy^2$

10. $3xy^3 + 2x^3y + 5xy^3 - 3x^2y - 6x^2y$

Add.

11. $(3x^3 + 4x^2 - 7x - 2) + (-7x^3 - 2x^2 + 3x + 4)$

12. $(7y^5 - 5) + (3y^5 - 4y^2 + 10)$

13. $(5p^2q^4 - 2p^2q^2 - 3q) + (-6p^2q^2 + 3q + 5)$

ANSWERS ON PAGE A-6

If the layers are equilateral triangles and all the spheres are the same size, we can use a particular polynomial to tell the count. The number N is given by the polynomial

$$N = \tfrac{1}{6}x^3 + \tfrac{1}{2}x^2 + \tfrac{1}{3}x,$$

where $x =$ the number of layers and $N =$ the number of spheres. Find the number of spheres in a pile with 9 layers.

We substitute 9 for x and calculate:

$$\begin{aligned} N &= \tfrac{1}{6}x^3 + \tfrac{1}{2}x^2 + \tfrac{1}{3}x \\ &= \tfrac{1}{6}(9)^3 + \tfrac{1}{2}(9)^2 + \tfrac{1}{3}(9) \\ &= \tfrac{1}{6} \cdot 729 + \tfrac{1}{2} \cdot 81 + 3 \\ &= 121.5 + 40.5 + 3 \\ &= 165. \end{aligned}$$

The answer is that when there are 9 layers, there will be 165 spheres. (Have a nice trip!) ◀

DO EXERCISE 8 ON THE PRECEDING PAGE.

c Collecting Like Terms

If two terms of a polynomial have the same variables raised to the same powers, the terms are called **similar,** or **like terms.** Similar terms can be "combined" or "collected" using the distributive laws. We show the use of the distributive law here for understanding, but in practice you should try to collect like terms mentally, writing only the answer.

▶ **EXAMPLES** Collect like terms.

9. $3x^2 - 4y + 2x^2 = 3x^2 + 2x^2 - 4y$ **Rearranging using the commutative and associative laws**

$= (3 + 2)x^2 - 4y$ **Using a distributive law**

$= 5x^2 - 4y$

10. $4x^3 + 5x - 4x^2 - 2x^3 + 5x^2 = 2x^3 + x^2 + 5x$

11. $3x^2y + 5xy^2 - 3x^2y - xy^2 = 4xy^2$ ◀

DO EXERCISES 9 AND 10.

d Addition

We can find the sum of two polynomials by writing a plus sign between them and then collecting like terms. Ordinarily this can be done mentally.

▶ **EXAMPLE 12** Add: $(-3x^3 + 2x - 4) + (4x^3 + 3x^2 + 2)$.

$$(-3x^3 + 2x - 4) + (4x^3 + 3x^2 + 2) = x^3 + 3x^2 + 2x - 2$$ ◀

Using columns is often helpful. To do so, we write the polynomials one under the other, with like terms under one another, and leave spaces for missing terms. Let us do the addition in Example 12 using columns.

$$\begin{array}{rrrr} -3x^3 & & +\,2x & -\,4 \\ 4x^3 & +\,3x^2 & & +\,2 \\ \hline x^3 & +\,3x^2 & +\,2x & -\,2 \end{array}$$

▶ **EXAMPLE 13** Add: $4ax^2 + 4bx - 5$ and $-6ax^2 + 5bx + 8$.

$$\begin{array}{r} 4ax^2 + 4bx - 5 \\ -6ax^2 + 5bx + 8 \\ \hline -2ax^2 + 9bx + 3 \end{array}$$ ◀

▶ **EXAMPLE 14** Add: $13x^3y + 3x^2y - 5y$ and $x^3y + 4x^2y - 3xy + 3y$.

$$(13x^3y + 3x^2y - 5y) + (x^3y + 4x^2y - 3xy + 3y)$$
$$= 14x^3y + 7x^2y - 3xy - 2y$$ ◀

DO EXERCISES 11–13 ON THE PRECEDING PAGE.

e Opposites

If the sum of two polynomials is 0, they are called **additive inverses,** or **opposites** of each other. For example,

$$(3x^2 - 5x + 2) + (-3x^2 + 5x - 2) = 0,$$

so the opposite of $(3x^2 - 5x + 2)$ is $(-3x^2 + 5x - 2)$. We can say the same thing using algebraic symbolism, as follows:

The opposite of $(3x^2 - 5x + 2)$ is $(-3x^2 + 5x - 2)$.

$$- \quad (3x^2 - 5x + 2) = -3x^2 + 5x - 2$$

> **The *opposite* of a polynomial *P* can be symbolized by $-P$ or by replacing each term with its opposite. The two expressions for the opposite are equivalent.**

▶ **EXAMPLE 15** Write two equivalent expressions for the opposite of

$$7xy^2 - 6xy - 4y + 3.$$

a) $-(7xy^2 - 6xy - 4y + 3)$ **Writing an inverse sign in front**

b) $-7xy^2 + 6xy + 4y - 3$ **Writing the opposite of each term** ◀

DO EXERCISES 14–16.

f Subtraction

To subtract one polynomial from another, we add the opposite of the polynomial being subtracted.

▶ **EXAMPLE 16** Subtract: $(-9x^5 + 2x^2 + 4) - (2x^5 + 4x^3 - 3x^2)$.

$$(-9x^5 + 2x^2 + 4) - (2x^5 + 4x^3 - 3x^2)$$
$$= (-9x^5 + 2x^2 + 4) + (-2x^5 - 4x^3 + 3x^2)$$ **Adding the opposite of the polynomial being subtracted**
$$= -11x^5 - 4x^3 + 5x^2 + 4$$ ◀

After some practice, you will find that you can skip some steps, by mentally taking the opposite of each term and then collecting like terms. Eventually, all you will write is the answer.

DO EXERCISES 17–19.

Write two equivalent expressions for the additive inverse, or opposite.

14. $4x^3 - 5x^2 + \frac{1}{4}x - 10$

15. $8xy^2 - 4x^3y^2 - 9x - \frac{1}{5}$

16. $-9y^5 - 8y^4 + \frac{1}{2}y^3 - y^2 + y - 1$

Subtract.

17. $(6x^2 + 4) - (3x^2 - 1)$

18. $(9y^3 - 2y - 4) - (-5y^3 - 8)$

19. $(-3p^2 + 5p - 4) - (-4p^2 + 11p - 2)$

ANSWERS ON PAGE A-6

Subtract.

20. $(2y^5 - y^4 + 3y^3 - y^2 - y - 7) - (-y^5 + 2y^4 - 2y^3 + y^2 - y - 4)$

21. $(4p^4q - 5p^3q^2 + p^2q^3 + 2q^4) - (-5p^4q + 5p^3q^2 - 3p^2q^3 - 7q^4)$

22. $\left(\frac{3}{2}y^3 - \frac{1}{2}y^2 + 0.3\right) - \left(\frac{1}{2}y^3 + \frac{1}{2}y^2 - \frac{4}{3}y + 0.2\right)$

ANSWERS ON PAGE A-6

We can also use columns for subtraction. We mentally change the signs of the polynomial being subtracted.

▶ **EXAMPLE 17** Subtract:

$$(4x^2y - 6x^3y^2 + x^2y^2) - (4x^2y + x^3y^2 + 3x^2y^3 - 8x^2y^2).$$

Write: (Subtract)

$$\begin{array}{l} 4x^2y - 6x^3y^2 \qquad\qquad\quad + x^2y^2 \\ \underline{-(4x^2y + x^3y^2 + 3x^2y^3 - 8x^2y^2)} \end{array}$$

Think: (Add)

$$\begin{array}{r} 4x^2y - 6x^3y^2 \qquad\qquad\quad + x^2y^2 \\ \underline{-4x^2y - x^3y^2 - 3x^2y^3 + 8x^2y^2} \\ -7x^3y^2 - 3x^2y^3 + 9x^2y^2 \end{array}$$

Mentally, take the opposite of each term and add. ◀

As with addition, you should avoid the use of columns as much as possible. After sufficient experience, you may be able to write only the answer.

DO EXERCISES 20–22.

❖ SIDELIGHTS

Factors and Sums

To *factor* a number is to express it as a product. Since $15 = 5 \cdot 3$, we say that 15 is *factored* and that 5 and 3 are *factors* of 15. In the table below, the top number has been factored in such a way that the sum of the factors is the bottom number. For example, in the first column, 56 has been factored as $7 \cdot 8$, and $7 + 8 = 15$, the bottom number. Such thinking is important in algebra when we factor trinomials of the type $x^2 + bx + c$.

Product	56	63	36	72	−140	−96	48	168	110			
Factor	7									−9	−24	−3
Factor	8									−10	18	
Sum	15	16	−20	−38	−4	4	−14	−29	−21			18

EXERCISE

Find the missing numbers in the table.

NAME SECTION DATE

EXERCISE SET 5.1

a Determine the degree of each term and the degree of the polynomial. Then find the leading term and the leading coefficient.

1. $-11x^4 - x^3 + x^2 + 3x - 9$

2. $t^3 - 3t^2 + t + 1$

3. $y^3 + 2y^7 + x^2y^4 - 8$

4. $u^2 + 3v^5 - u^3v^4 - 7$

5. $a^5 + 4a^2b^4 + 6ab + 4a - 3$

6. $8p^6 + 2p^4t^4 - 7p^3t + 5p^2 - 14$

Arrange in descending powers of y.

7. $23 - 4y^3 + 7y - 6y^2$

8. $5 - 8y + 6y^2 + 11y^3 - 18y^4$

9. $x^2y^2 + x^3y - xy^3 + 1$

10. $x^3y - x^2y^2 + xy^3 + 6$

11. $3ay - 5a^5y^5 - 4a^2y^3$

12. $by^6 - 4b^7y^2 + 7ay^5 - 4y - 5b$

Arrange in ascending powers of x.

13. $4x + 12 + 3x^4 - 5x^2$

14. $-5x^2 + 10x + 5$

15. $-9x^3y + 3xy^3 + x^2y^2 + 2x^4$

16. $5x^2y^2 - 9xy + 8x^3y^2 - 5x^4$

17. $4ax - 7ab + 4x^6 - 7ax^2$

18. $5xy^8 - 3ax^5 + 4ax^3 - 12a + 5x^5$

b Evaluate the polynomial for the specified values of the variable.

19. $P = 4x^2 - 3x + 2$, when $x = 4$; when $x = 0$

20. $Q = -5x^3 + 7x^2 - 12$, when $x = 3$; when $x = -1$

21. $P = 8y^3 - 12y - 5$, when $y = -2$; when $y = \frac{1}{3}$

22. $Q = 9y^3 + 8y^2 - 4y - 9$, when $y = -3$; when $y = 0$

ANSWERS

1. ________
2. ________
3. ________
4. ________
5. ________
6. ________
7. ________
8. ________
9. ________
10. ________
11. ________
12. ________
13. ________
14. ________
15. ________
16. ________
17. ________
18. ________
19. ________
20. ________
21. ________
22. ________

ANSWERS

23. ______

24. ______

25. ______

26. ______

27. ______

28. ______

29. ______

30. ______

31. ______

32. ______

33. ______

34. ______

35. ______

36. ______

37. ______

38. ______

23. *Number of games in a league.* If there are n teams in a league and each team plays each other once in a season, the total number of games played is given by the polynomial $N = \frac{1}{2}n^2 - \frac{1}{2}n$. Find the number of games played when there are 8 teams; when there are 20 teams.

24. *Number of games in a league.* If there are n teams in a league and each team plays each other twice in a season, the total number of games played is given by the polynomial function $N = n^2 - n$. Find the number of games played when there are 12 teams in a league and each team plays each other team twice.

Surface area of a right circular cylinder. The surface area of a right circular cylinder is given by the polynomial

$$2\pi rh + 2\pi r^2,$$

where h = the height and r = the radius of the base.

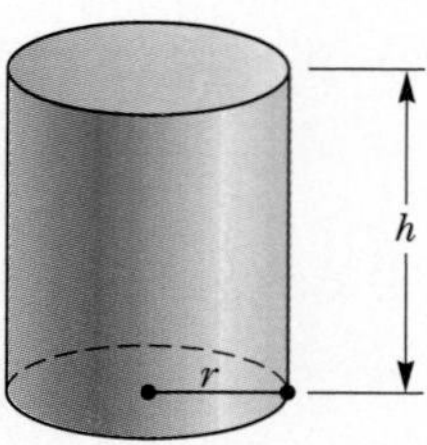

25. A 12-oz beverage can has height 4.7 in. and radius 1.2 in. Evaluate the polynomial for $h = 4.7$ and $r = 1.2$ to find the surface area of the can. Use 3.14 as an approximation for π.

26. A 16-oz beverage can has height 6.3 in. and radius 1.2 in. Evaluate the polynomial for $h = 6.3$ and $r = 1.2$ to find the surface area of the can. Use 3.14 as an approximation for π.

Total revenue. An electronics firm is marketing a new kind of stereo. *Total revenue* is the total amount of money taken in. The firm determines that when it sells x stereos, the total revenue R, in dollars, is given by the polynomial

$$R = 280x - 0.4x^2.$$

27. What is the total revenue from the sale of 200 stereos?

28. What is the total revenue from the sale of 450 stereos?

Total cost. Total cost is the total cost of producing x stereos. The electronics firm determines that the total cost C, in dollars, of producing x stereos is given by the polynomial

$$C = 7000 + 0.6x^2.$$

29. What is the total cost of producing 200 stereos?

30. What is the total cost of producing 450 stereos?

Total profit. Total profit P is defined as *total revenue R minus total cost C*, and is given by

$$P = R - C.$$

For each of the following, find the total profit P.

31. $R = 280x - 0.4x^2,\ C = 7000 + 0.6x^2$

32. $R = 280x - 0.7x^2,\ C = 8000 + 0.5x^2$

c Collect like terms.

33. $6x^2 - 7x^2 + 3x^2$

34. $-2y^2 - 7y^2 + 5y^2$

35. $5x - 4y - 2x + 5y$

36. $4a - 9b - 6a + 3b$

37. $5a + 7 - 4 + 2a - 6a + 3$

38. $9x + 12 - 8 - 7x + 5x + 10$

39. $3a^2b + 4b^2 - 9a^2b - 6b^2$

40. $5x^2y^2 + 4x^3 - 8x^2y^2 - 12x^3$

41. $8x^2 - 3xy + 12y^2 + x^2 - y^2 + 5xy + 4y^2$

42. $a^2 - 2ab + b^2 + 9a^2 + 5ab - 4b^2 + a^2$

43. $4x^2y - 3y + 2xy^2 - 5x^2y + 7y + 7xy^2$

44. $3xy^2 + 4xy - 7xy^2 + 7xy + x^2y$

d Add.

45. $(3x^2 + 5y^2 + 6) + (2x^2 - 3y^2 - 1)$

46. $(9y^2 + 8y - 4) + (12y^2 - 5y + 8)$

47. $(2a - c + 3b) + (4a - 2b + 2c)$

48. $(5x + 2z - 4y) + (9x + 12y - 8z)$

49. $(a^2 - 3b^2 + 4c^2) + (-5a^2 + 2b^2 - c^2)$

50. $(x^2 - 5y^2 - 9z^2) + (-6x^2 + 9y^2 - 2z^2)$

51. $(x^2 + 2x - 3xy - 7) + (-3x^2 - x + 2xy + 6)$

52. $(3a^2 - 2b + ab + 6) + (-a^2 + 5b - 5ab - 2)$

53. $(7x^2y - 3xy^2 + 4xy) + (-2x^2y - xy^2 + xy)$

54. $(7ab - 3ac + 5bc) + (13ab - 15ac - 8bc)$

55. $(2r^2 + 12r - 11) + (6r^2 - 2r + 4) + (r^2 - r - 2)$

56. $(5x^2 + 19x - 23) + (-7x^2 - 11x + 12) + (-x^2 - 9x + 8)$

57. Add: $(\frac{2}{3}xy + \frac{5}{6}xy^2 + 5.1x^2y) + (-\frac{4}{5}xy + \frac{3}{4}xy^2 - 3.4x^2y)$.

58. Add: $(\frac{1}{8}xy - \frac{3}{5}x^3y^2 + 4.3y^3) + (-\frac{1}{3}xy - \frac{3}{4}x^3y^2 - 2.9y^3)$.

e Write two equivalent expressions for the opposite of the polynomial.

59. $5x^3 - 7x^2 + 3x - 6$

60. $-8y^4 - 18y^3 + 4y - 9$

61. $-12y^5 + 4ay^4 - 7by^2$

62. $7ax^3y^2 - 8by^4 - 7abx - 12ay$

ANSWERS

39. ______

40. ______

41. ______

42. ______

43. ______

44. ______

45. ______

46. ______

47. ______

48. ______

49. ______

50. ______

51. ______

52. ______

53. ______

54. ______

55. ______

56. ______

57. ______

58. ______

59. ______

60. ______

61. ______

62. ______

ANSWERS

63.

64.

65.

66.

67.

68.

69.

70.

71.

72.

73.

74.

75.

76.

77.

78.

79.

80.

81.

82. See graph.

83.

84.

85.

86.

87.

f Subtract.

63. $(8x - 4) - (-5x + 2)$

64. $(9y + 3) - (-4y - 2)$

65. $(-3x^2 + 2x + 9) - (x^2 + 5x - 4)$

66. $(-9y^2 + 4y + 8) - (4y^2 + 2y - 3)$

67. $(5a + c - 2b) - (3a + 2b - 2c)$

68. $(z + 8x - 4y) - (4x + 6y - 3z)$

69. $(3x^2 - 2x - x^3) - (5x^2 - x^3 - 8x)$

70. $(8y^2 - 4y^3 - 3y) - (3y^2 - 9y - 7y^3)$

71. $(5a^2 + 4ab - 3b^2) - (9a^2 - 4ab + 2b^2)$

72. $(9y^2 - 14yz - 8z^2) - (12y^2 - 8yz + 4z^2)$

73. Find $P - Q$ given that $P = 2 - 3y$ and $Q = 6y - 7$.

74. Find $P - Q$ given that $P = 4 + 8x^2 - 5x$ and $Q = -2x^2 + 3x - 2$.

Subtract.

75. $(6ab - 4a^2b + 6ab^2) - (3ab^2 - 10ab - 12a^2b)$

76. $(10xy - 4x^2y^2 - 3y^3) - (-9x^2y^2 + 4y^3 - 7xy)$

77. $(0.09y^4 - 0.052y^3 + 0.93) - (0.03y^4 - 0.084y^3 + 0.94y^2)$

78. $(1.23x^4 - 3.122x^3 + 1.11x) - (0.79x^4 - 8.734x^3 + 0.04x^2 + 6.71x)$

79. $(\frac{5}{8}x^4 - \frac{1}{4}x^2 - \frac{1}{2}) - (-\frac{3}{8}x^4 + \frac{3}{4}x^2 + \frac{1}{2})$

80. $(\frac{5}{6}y^4 - \frac{1}{2}y^2 - 7.8y + \frac{1}{3}) - (-\frac{3}{8}y^4 + \frac{3}{4}y^2 + 3.4y - \frac{1}{5})$

SKILL MAINTENANCE

81. Solve: $|4 - 2x| < 18$.

82. Graph: $3y - 4x = 12$.

SYNTHESIS

83. Find a polynomial that gives the surface area of a box like this one, with an open top and dimensions as shown.

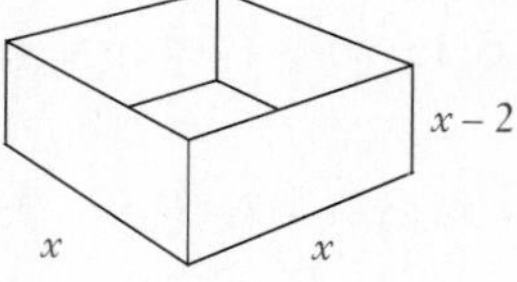

Add. Assume that the exponents are natural numbers.

84. $(2x^{2a} + 4x^a + 3) + (6x^{2a} + 3x^a + 4)$

85. $(47x^{4a} + 3x^{3a} + 22x^{2a} + x^a + 1) + (37x^{3a} + 8x^{2a} + 3)$

Subtract. Assume that the exponents are natural numbers.

86. $(3x^{6a} - 5x^{5a} + 4x^{3a} + 8) - (2x^{6a} + 4x^{4a} + 3x^{3a} + 2x^{2a})$

87. $(2x^{5b} + 4x^{4b} + 3x^{3b} + 8) - (x^{5b} + 2x^{3b} + 6x^{2b} + 9x^b + 8)$

5.2 Multiplication of Polynomials; Special Products

a Multiplication of Any Two Polynomials

Multiplying Monomials

To multiply monomials, we first multiply their coefficients. Then we multiply the variables using the rules for exponents and the commutative and associative laws as studied in Chapter 1.

► **EXAMPLES** Multiply and simplify.

1. $(-8x^4y^7)(5x^3y^2) = -8 \cdot 5 \cdot x^4 \cdot x^3 \cdot y^7 \cdot y^2$
$= -40x^{4+3}y^{7+2}$ **Adding exponents**
$= -40x^7y^9$

2. $(3x^2yz^5)(-6x^5y^{10}z^2) = 3 \cdot (-6) \cdot x^2 \cdot x^5 \cdot y \cdot y^{10} \cdot z^5 \cdot z^2$
$= -18x^7y^{11}z^7$ ◄

DO EXERCISES 1–3.

Multiplying Monomials and Binomials

The distributive law is the basis for multiplying polynomials other than monomials. We first multiply a monomial and a binomial.

► **EXAMPLE 3** Multiply: $2x(3x - 5)$.

$2x \cdot (3x - 5) = 2x \cdot 3x - 2x \cdot 5$ **Using a distributive law**
$= 6x^2 - 10x$ **Multiplying monomials** ◄

► **EXAMPLE 4** Multiply: $3a^2b(a^2 - b^2)$.

$3a^2b \cdot (a^2 - b^2) = 3a^2b \cdot a^2 - 3a^2b \cdot b^2$ **Using a distributive law**
$= 3a^4b - 3a^2b^3$ ◄

DO EXERCISES 4 AND 5.

Next we multiply two binomials. To do so, we use the distributive law twice, first considering one of the binomials as a single expression and multiplying it by each term of the other binomial.

► **EXAMPLE 5** Multiply: $(3y^2 + 4)(y^2 - 2)$.

$(3y^2 + 4)(y^2 - 2) = \underbrace{3y^2(y^2 - 2)}_{(a)} + \underbrace{4(y^2 - 2)}_{(b)}$. **Using a distributive law**

We consider parts (a) and (b) separately.

a) $3y^2(y^2 - 2) = 3y^2 \cdot y^2 - 3y^2 \cdot 2$ **Using a distributive law**
$= 3y^4 - 6y^2$ **Multiplying the monomials**

b) $4(y^2 - 2) = 4 \cdot y^2 - 4 \cdot 2$ **Using a distributive law**
$= 4y^2 - 8$ **Multiplying the monomials**

We collect like terms if we can:

$$(3y^2 + 4)(y^2 - 2) = (3y^4 - 6y^2) + (4y^2 - 8) = 3y^4 - 2y^2 - 8.$$ ◄

DO EXERCISES 6 AND 7.

OBJECTIVES

After finishing Section 5.2, you should be able to:

a. Multiply any two polynomials.
b. Use the FOIL method to multiply two binomials.
c. Use a rule to square a binomial.
d. Use a rule to multiply a sum and a difference of the same two terms.

FOR EXTRA HELP

Tape 9A

Tape 7B

MAC: 5
IBM: 5

Multiply.

1. $(9y^2)(-2y)$
2. $(4x^3y)(6x^5y^2)$
3. $(-5xy^7z^4)(18x^3y^2z^8)$

Multiply.

4. $(-3y)(2y + 6)$
5. $(2xy)(4y^2 - 5)$

Multiply.

6. $(5x^2 - 4)(x + 3)$
7. $(2y + 3)(3y - 4)$

ANSWERS ON PAGE A-6

Multiplying Any Two Polynomials

To find a quick way to multiply any two polynomials, let us consider another example.

▶ **EXAMPLE 6** Multiply: $(p + 2)(p^4 - 2p^3 + 3)$.

By the distributive law, we have

$$\begin{aligned}(p+2)(p^4 - 2p^3 + 3) &= (p+2)(p^4) - (p+2)(2p^3) + (p+2)(3)\\ &= p(p^4) + 2(p^4) - p(2p^3) - 2(2p^3) + p(3) + 2(3)\\ &= p^5 + 2p^4 - 2p^4 - 4p^3 + 3p + 6\\ &= p^5 - 4p^3 + 3p + 6. \quad \text{Collecting like terms}\end{aligned}$$

◀

DO EXERCISES 8 AND 9.

From the preceding examples, we can see how to multiply any two polynomials.

To multiply any two polynomials P and Q, select one of the polynomials, say P. Then multiply each term of P by every term of Q and collect like terms.

We can use columns for long multiplications. We multiply each term at the top by every term at the bottom, keeping like terms in columns and adding spaces for missing terms. Then we add.

▶ **EXAMPLE 7** Multiply: $(5x^3 + x - 4)(-2x^2 + 3x + 6)$.

$$\begin{array}{rrrrrrl} & & 5x^3 & & + x & - 4 & \\ & & & -2x^2 & + 3x & + 6 & \\ \hline & & 30x^3 & & + 6x & - 24 & \text{Multiplying by 6}\\ & 15x^4 & & + 3x^2 & - 12x & & \text{Multiplying by } 3x\\ -10x^5 & & - 2x^3 & + 8x^2 & & & \text{Multiplying by } -2x^2\\ \hline -10x^5 & + 15x^4 & + 28x^3 & + 11x^2 & - 6x & - 24 & \end{array}$$

◀

DO EXERCISES 10 AND 11.

b Product of Two Binomials: FOIL

We now consider what are called **special product rules.** These are faster ways to multiply in certain situations.

Let us find a faster *special product rule* for finding the product of two binomials. Consider $(x + 7)(x + 4)$. We multiply each term of one by every term of the other. Then we collect like terms.

$$(x + 7)(x + 4) = x \cdot x + 4x + 7x + 7 \cdot 4 = x^2 + 11x + 28$$

This also shows a special product rule for multiplying two binomials:

$$(x + 7)(x + 4) = \underbrace{x \cdot x}_{\text{First terms}} + \underbrace{4x}_{\text{Outside terms}} + \underbrace{7x}_{\text{Inside terms}} + \underbrace{7(4)}_{\text{Last terms}}$$

This method of multiplying is called the **FOIL method.**

Multiply.

8. $(p - 3)(p^3 + 4p^2 - 5)$

9. $(2x^3 + 4x - 5)(x - 4)$

Multiply.

10. $(-4x^3 - 2x + 1)(-2x^2 - 3x + 6)$

11. $(a^2 - 2ab + b^2)(a^3 + 3ab - b^2)$

ANSWERS ON PAGE A-6

The FOIL Method

To multiply two binomials $A + B$ and $C + D$, multiply the First terms AC, the Outside terms AD, the Inside terms BC, and then the Last terms BD. Then collect like terms, if possible.

$$(A + B)(C + D) = AC + AD + BC + BD$$

1. **Multiply First terms: AC.**
2. **Multiply Outside terms: AD.**
3. **Multiply Inside terms: BC.**
4. **Multiply Last terms: BD.**

↓
FOIL

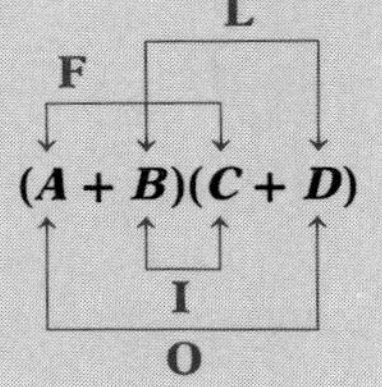

▶ **EXAMPLES** Multiply.

8. $(x + 5)(x - 8) = \overset{F}{x^2} \overset{O}{- 8x} \overset{I}{+ 5x} \overset{L}{- 40}$
$= x^2 - 3x - 40$ **Collecting like terms**

We write the result in descending order since the original binomials are in descending order.

9. $(3xy + 2x)(x^2 + 2xy^2) = \overset{F}{3x^3y} + \overset{O}{6x^2y^3} + \overset{I}{2x^3} + \overset{L}{4x^2y^2}$
10. $(2x - 3)(y + 2) = 2xy + 4x - 3y - 6$
11. $(2x + 3y)(x - 4y) = 2x^2 - 8xy + 3xy - 12y^2 = 2x^2 - 5xy - 12y^2$ ◀

DO EXERCISES 12–14.

C Squares of Binomials

We can use the FOIL method to develop special product rules for the square of a binomial:

$$(A + B)^2 = (A + B)(A + B) = A^2 + AB + AB + B^2 = A^2 + 2AB + B^2$$

$$(A - B)^2 = (A - B)(A - B) = A^2 - AB - AB + B^2 = A^2 - 2AB + B^2$$

$$\mathbf{(A + B)^2 = A^2 + 2AB + B^2;}$$
$$\mathbf{(A - B)^2 = A^2 - 2AB + B^2}$$

The square of a binomial is the square of the first term, plus twice the product of the two terms, plus the square of the last term.

▶ **EXAMPLES** Multiply.

It can be helpful to memorize the words of the rules and say them while you are calculating.

$(A - B)^2 = A^2 - 2\,A\,B + B^2$

12. $(y - 5)^2 = y^2 - 2(y)(5) + 5^2 = y^2 - 10y + 25$
13. $(2x + 3y)^2 = (2x)^2 + 2(2x)(3y) + (3y)^2 = 4x^2 + 12xy + 9y^2$
14. $(3x^2 - 5xy^2)^2 = (3x^2)^2 - 2(3x^2)(5xy^2) + (5xy^2)^2$
$= 9x^4 - 30x^3y^2 + 25x^2y^4$
15. $(\frac{1}{2}a^2 - b^3)^2 = (\frac{1}{2}a^2)^2 - 2(\frac{1}{2}a^2)(b^3) + (b^3)^2 = \frac{1}{4}a^4 - a^2b^3 + b^6$ ◀

DO EXERCISES 15–18.

Multiply.

12. $(y - 4)(y + 10)$

13. $(p + 5q)(2p - 3q)$

14. $(x^2y + 2x)(xy^2 + y^2)$

Multiply.

15. $(a - b)^2$

16. $(x + 8)^2$

17. $(3x - 7)^2$

18. $(m^3 + \frac{1}{4}n)^2$

ANSWERS ON PAGE A-6

Multiply.

19. $(x + 8)(x - 8)$

20. $(4y - 7)(4y + 7)$

Multiply.

21. $(2.8a + 4.1b)(2.8a - 4.1b)$

22. $\left(3w - \frac{3}{5}q^2\right)\left(3w + \frac{3}{5}q^2\right)$

Multiply.

23. $(7x^2y + 2y)(-2y + 7x^2y)$

24. $(2x + 3 - 5y)(2x + 3 + 5y)$

25. Multiply:
$(3x + 2y)(3x - 2y)(9x^2 + 4y^2)$.

ANSWERS ON PAGE A-6

d Products of Sums and Differences

Another special case of a product of two binomials is the product of a sum and a difference. Note the following:

$$(A + B)(A - B) = \underset{\text{F}}{A^2} - \underset{\text{O}}{AB} + \underset{\text{I}}{AB} - \underset{\text{L}}{B^2} = A^2 - B^2$$

> $$(A + B)(A - B) = A^2 - B^2$$
> **The product of the sum and difference of the same two terms is the square of the first term minus the square of the second term.**

► **EXAMPLES** Multiply.

$(A + B)(A - B) = A^2 - B^2$

16. $(y + 5)(y - 5) = y^2 - 5^2 = y^2 - 25$ ← Say the rule as you work.

17. $(2xy^2 + 3x)(2xy^2 - 3x) = (2xy^2)^2 - (3x)^2 = 4x^2y^4 - 9x^2$

18. $(0.2t - 1.4m)(0.2t + 1.4m) = (0.2t)^2 - (1.4m)^2 = 0.04t^2 - 1.96m^2$

19. $(\frac{2}{3}n - m^2)(\frac{2}{3}n + m^2) = (\frac{2}{3}n)^2 - (m^2)^2 = \frac{4}{9}n^2 - m^4$ ◄

DO EXERCISES 19–22.

► **EXAMPLES** Multiply.

20. $(5y + 4 + 3x)(5y + 4 - 3x) = (5y + 4)^2 - (3x)^2$
$= 25y^2 + 40y + 16 - 9x^2$

Here we treat $5y + 4$ as the first expression A and $3x$ as the second B.

21. $(3xy^2 + 4y)(-3xy^2 + 4y) = -(3xy^2)^2 + (4y)^2 = 16y^2 - 9x^2y^4$ ◄

DO EXERCISES 23 AND 24.

Try to multiply polynomials mentally, even when several types are mixed. First check to see what types of polynomials are to be multiplied. Then use the quickest method. Sometimes we might use more than one method.

► **EXAMPLE 22** Multiply: $(a - 5b)(a + 5b)(a^2 - 25b^2)$.

We first note that $a - 5b$ and $a + 5b$ can be multiplied using the rule $(A - B)(A + B) = A^2 - B^2$. Then we square, using $(A - B)^2 = A^2 - 2AB + B^2$.

$$\begin{aligned}(a - 5b)(a + 5b)(a^2 - 25b^2) &= (a^2 - 25b^2)(a^2 - 25b^2) = (a^2 - 25b^2)^2 \\ &= (a^2)^2 - 2(a^2)(25b^2) + (25b^2)^2 \\ &= a^4 - 50a^2b^2 + 625b^4\end{aligned}$$ ◄

DO EXERCISE 25.

NAME SECTION DATE

EXERCISE SET 5.2

a Multiply.

1. $2y^2 \cdot 5y$
2. $-3x^2 \cdot 2xy$
3. $5x(-4x^2y)$
4. $-3ab^2(2a^2b^2)$
5. $(2x^3y^2)(-5x^2y^4)$
6. $(7a^2bc^4)(-8ab^3c^2)$
7. $2x(3 - x)$
8. $4a(a^2 - 5a)$
9. $3ab(a + b)$
10. $2xy(2x - 3y)$
11. $5cd(3c^2d - 5cd^2)$
12. $a^2(2a^2 - 5a^3)$
13. $(5x + 2)(3x - 1)$
14. $(2a - 3b)(4a - b)$
15. $(s + 3t)(s - 3t)$
16. $(y + 4)(y - 4)$
17. $(x - y)(x - y)$
18. $(a + 2b)(a + 2b)$
19. $(y + 8x)(2y - 7x)$
20. $(x + y)(x - 2y)$
21. $(a^2 - 2b^2)(a^2 - 3b^2)$
22. $(2m^2 - n^2)(3m^2 - 5n^2)$
23. $(x - 4)(x^2 + 4x + 16)$
24. $(y + 3)(y^2 - 3y + 9)$
25. $(x + y)(x^2 - xy + y^2)$
26. $(a - b)(a^2 + ab + b^2)$

ANSWERS

1. ____
2. ____
3. ____
4. ____
5. ____
6. ____
7. ____
8. ____
9. ____
10. ____
11. ____
12. ____
13. ____
14. ____
15. ____
16. ____
17. ____
18. ____
19. ____
20. ____
21. ____
22. ____
23. ____
24. ____
25. ____
26. ____

ANSWERS

27. ______

28. ______

29. ______

30. ______

31. ______

32. ______

33. ______

34. ______

35. ______

36. ______

37. ______

38. ______

39. ______

40. ______

41. ______

42. ______

43. ______

44. ______

45. ______

46. ______

47. ______

48. ______

49. ______

27. $(a^2 + a - 1)(a^2 + 4a - 5)$

28. $(x^2 - 2x + 1)(x^2 + x + 2)$

29. $(4a^2b - 2ab + 3b^2)(ab - 2b + a)$

30. $(2x^2 + y^2 - 2xy)(x^2 - 2y^2 - xy)$

31. $(x + \frac{1}{4})(x + \frac{1}{4})$

32. $(b - \frac{1}{3})(b - \frac{1}{3})$

33. $(1.3x - 4y)(2.5x + 7y)$

34. $(40a - 0.24b)(0.3a + 10b)$

b **c** Multiply.

35. $(a + 2)(a + 3)$

36. $(x + 5)(x + 8)$

37. $(y + 3)(y - 2)$

38. $(y - 4)(y + 7)$

39. $(2a + \frac{1}{3})^2$

40. $(3c - \frac{1}{2})^2$

41. $(x - 2y)^2$

42. $(2s + 3t)^2$

43. $(b - \frac{1}{3})(b - \frac{1}{2})$

44. $(x - \frac{1}{2})(x - \frac{1}{4})$

45. $(2x + 9)(x + 2)$

46. $(3b + 2)(2b - 5)$

47. $(20a - 0.16b)^2$

48. $(10p^2 + 2.3q)^2$

49. $(2x - 3y)(2x + y)$

50. $(2a - 3b)(2a - b)$

51. $(x + 3)^2$

52. $(y - 7)^2$

53. $(2x^2 - 3y^2)^2$

54. $(3s^2 + 4t^2)^2$

55. $(a^2b^2 + 1)^2$

56. $(x^2y - xy^2)^2$

57. $(0.1a^2 - 5b)^2$

58. $(6p + 0.45q^2)^2$

d Multiply.

59. $(c + 2)(c - 2)$

60. $(x - 3)(x + 3)$

61. $(2a + 1)(2a - 1)$

62. $(3 - 2x)(3 + 2x)$

63. $(3m - 2n)(3m + 2n)$

64. $(3x + 5y)(3x - 5y)$

65. $(x^2 + yz)(x^2 - yz)$

66. $(2a^2 + 5ab)(2a^2 - 5ab)$

67. $(-mn + m^2)(mn + m^2)$

68. $(1.6 + pq)(-1.6 + pq)$

69. $(\frac{1}{2}p - \frac{2}{3}q)(\frac{1}{2}p + \frac{2}{3}q)$

70. $(\frac{3}{5}ab + 4c)(\frac{3}{5}ab - 4c)$

71. $(x + 1)(x - 1)(x^2 + 1)$

72. $(y - 2)(y + 2)(y^2 + 4)$

73. $(a - b)(a + b)(a^2 - b^2)$

74. $(2x - y)(2x + y)(4x^2 - y^2)$

ANSWERS

50. ______

51. ______

52. ______

53. ______

54. ______

55. ______

56. ______

57. ______

58. ______

59. ______

60. ______

61. ______

62. ______

63. ______

64. ______

65. ______

66. ______

67. ______

68. ______

69. ______

70. ______

71. ______

72. ______

73. ______

74. ______

ANSWERS

75. ____
76. ____
77. ____
78. ____
79. ____
80. ____
81. ____
82. ____
83. ____
84. ____
85. ____
86. ____
87. ____
88. ____
89. ____
90. ____
91. ____
92. ____
93. ____
94. ____
95. ____
96. ____
97. ____
98. ____
99. ____
100. ____
101. ____
102. ____

75. $(a + b + 1)(a + b - 1)$

76. $(m + n + 2)(m + n - 2)$

77. $(2x + 3y + 4)(2x + 3y - 4)$

78. $(3a - 2b + c)(3a - 2b - c)$

79. Suppose P dollars is invested in a savings account at interest rate i, compounded annually, for 2 years. The amount A in the account after 2 years is given by

$$A = P(1 + i)^2.$$

Find an equivalent expression for A.

80. Suppose P dollars is invested in a savings account at interest rate i, compounded semiannually, for 1 year. The amount A in the account after 1 year is given by

$$A = P\left(1 + \frac{i}{2}\right)^2.$$

Find an equivalent expression for A.

SKILL MAINTENANCE

81. In a factory there are three machines A, B, and C. When all three are running, they produce 222 suitcases per day. If A and B work but C does not, they produce 159 suitcases per day. If B and C work but A does not, they produce 147 suitcases. What is the daily production of each machine?

82. There are 50 dimes in a roll of dimes, 40 nickels in a roll of nickels, and 40 quarters in a roll of quarters. A student has 13 rolls of coins that have a total value of \$89. There are three more rolls of dimes than nickels. How many of each type of roll of coins does the student have?

SYNTHESIS

Multiply. Assume that variables in exponents represent natural numbers.

83. $(6y)^2(-\frac{1}{3}x^2y^3)^3$

84. $[(a^{2n})^{2n}]^4$

85. $(-r^6s^2)^3\left(-\frac{r^2}{6}\right)^2(9s^4)$

86. $(z^{n^2})^{n^3}(z^{4n^3})^{n^2}$

87. $(a^xb^{2y})(\frac{1}{2}a^{3x}b)^2$

88. $(-8s^3t)(2s^5 - 3s^3t^4 + st^7 - t^{10})$

89. $y^3z^n(y^{3n}z^3 - 4yz^{2n})$

90. $[(2x - 1)^2 - 1]^2$

91. $(y - 1)^6(y + 1)^6$

92. $(r^2 + s^2)^2(r^2 + 2rs + s^2)(r^2 - 2rs + s^2)$

93. $(3x^5 - \frac{5}{11})^2$

94. $(\frac{2}{3}x + \frac{1}{3}y + 1)(\frac{2}{3}x - \frac{1}{3}y - 1)$

95. $(x - \frac{1}{7})(x^2 + \frac{1}{7}x + \frac{1}{49})$

96. $(4x^2 + 2xy + y^2)(4x^2 - 2xy + y^2)$

97. $(x^2 - 7x + 12)(x^2 + 7x + 12)$

98. $(x^a + y^b)(x^a - y^b)(x^{2a} + y^{2b})$

99. $[a - (b - 1)][(b - 1)^2 + a(b - 1) + a^2]$

100. $(x - 1)(x^2 + x + 1)(x^3 + 1)$

101. $(x^{a-b})^{a+b}$

102. $(M^{x+y})^{x+y}$

5.3 Common Factors and Factoring by Grouping

Factoring is the reverse of multiplying. To **factor** an expression means to write an equivalent expression that is a product. To factor, we generally use the distributive laws.

a Terms with Common Factors

When factoring, you should first look for factors common to all the terms of an expression.

► **EXAMPLE 1** Factor out a common factor: $4y^2 - 8$.

$$4y^2 - 8 = 4 \cdot y^2 - 4 \cdot 2 \quad \text{Noting that 4 is a common factor}$$
$$= 4(y^2 - 2)$$ ◄

In some cases, there is more than one common factor. In Example 2 below, for instance, 5 is a common factor, x^3 is a common factor, and $5x^3$ is a common factor. If there is more than one common factor, we usually choose the one with the largest coefficient and the greatest exponent. In Example 2, the greatest common factor is $5x^3$.

► **EXAMPLES** Factor out a common factor.

2. $5x^4 - 20x^3 = 5x^3(x - 4)$ **Try to write your answer directly. Multiply mentally to check your answer.**
3. $12x^2y - 20x^3y = 4x^2y(3 - 5x)$
4. $10p^6q^2 - 4p^5q^3 + 2p^4q^4 = 2p^4q^2(5p^2 - 2pq + q^2)$ ◄

The polynomials in Examples 1–4 have been **factored completely.** They cannot be factored further. The factors in the resulting factorization are said to be **prime polynomials.**

DO EXERCISES 1–4.

Consider the polynomial

$$-3x^2 + 6.$$

We can factor this as

$$-3x^2 + 6 = 3(-x^2 + 2),$$

or as

$$-3x^2 + 6 = -3(x^2 - 2).$$

In certain situations, the latter factorization will be helpful. It allows the leading coefficient of the factor $x^2 - 2$ to be 1.

► **EXAMPLES** Factor out a common factor with a negative coefficient.

5. $-4x - 24 = -4(x + 6)$
6. $-2x^2 + 6x - 10 = -2(x^2 - 3x + 5)$ ◄

DO EXERCISES 5 AND 6.

OBJECTIVES

After finishing Section 5.3, you should be able to:

a Factor polynomials whose terms have a common factor.

b Factor certain polynomials with four terms by grouping.

FOR EXTRA HELP

Tape 9B

Tape 7B

MAC: 5
IBM: 5

Factor out a common factor.

1. $3x^2 - 6$

2. $4x^5 - 8x^3$

3. $9y^4 - 15y^3 + 3y^2$

4. $6x^2y - 21x^3y^2 + 3x^2y^3$

Factor out a common factor with a negative coefficient.

5. $-8x + 32$

6. $-3x^2 - 15x + 9$

ANSWERS ON PAGE A-6

Factor by grouping.

7. $(p + q)(x + 2) + (p + q)(x + y)$

8. $(y + 3)(y - 21) + (y + 3)(y + 10)$

9. $x^3 + 5x^2 + 4x - 20$

10. $5y^3 + 2y^2 - 10y - 4$

ANSWERS ON PAGE A-6

b Factoring by Grouping

In expressions of four or more terms, there may be a common binomial factor. We proceed as in the following examples.

▶ **EXAMPLE 7** Factor: $(a - b)(x + 5) + (a - b)(x - y^2)$.

$$\begin{aligned}(a - b)(x + 5) + (a - b)(x - y^2) &= (a - b)[(x + 5) + (x - y^2)] \\ &= (a - b)(2x + 5 - y^2)\end{aligned}$$ ◀

▶ **EXAMPLE 8** Factor: $y^3 + 3y^2 + 4y + 12$.

$$\begin{aligned}y^3 + 3y^2 + 4y + 12 &= (y^3 + 3y^2) + (4y + 12) && \textbf{Grouping} \\ &= y^2(y + 3) + 4(y + 3) && \textbf{Factoring out common factors} \\ &= (y^2 + 4)(y + 3) && \textbf{Factoring out } y + 3\end{aligned}$$ ◀

In this example, we factored two parts of the expression. Then we factored as in Example 7.

▶ **EXAMPLE 9** Factor: $4x^3 - 15 + 20x^2 - 3x$.

$$\begin{aligned}4x^3 - 15 + 20x^2 - 3x &= 4x^3 + 20x^2 - 3x - 15 && \textbf{Grouping} \\ &= 4x^2(x + 5) - 3(x + 5) && \textbf{Factoring out common factors} \\ &= (4x^2 - 3)(x + 5) && \textbf{Factoring out } x + 5\end{aligned}$$ ◀

▶ **EXAMPLE 10** Factor: $ax^2 + ay + bx^2 + by$.

$$\begin{aligned}ax^2 + ay + bx^2 + by &= a(x^2 + y) + b(x^2 + y) \\ &= (a + b)(x^2 + y)\end{aligned}$$ ◀

Not all polynomials with four terms can be factored by grouping. An example is

$$x^3 + x^2 + 3x - 3.$$

Note that neither $x^2(x + 1) + 3(x - 1)$ nor $x(x^2 + 3) + (x^2 - 3)$ nor any other groupings allow us to factor out a common binomial.

DO EXERCISES 7–10.

NAME SECTION DATE

EXERCISE SET 5.3

a Factor.

1. $4a^2 + 2a$
2. $6y^2 + 3y$
3. $y^3 + 9y^2$
4. $x^3 + 8x^2$
5. $6x^2 - 3x^4$
6. $8y^2 + 4y^4$
7. $4x^2y - 12xy^2$
8. $5x^2y^3 + 15x^3y^2$
9. $3y^2 - 3y - 9$
10. $5x^2 - 5x + 15$
11. $4ab - 6ac + 12ad$
12. $8xy + 10xz - 14xw$
13. $10a^4 + 15a^2 - 25a - 30$
14. $12t^5 - 20t^4 + 8t^2 - 16$

Factor out a factor with a negative coefficient.

15. $-3x + 12$
16. $-5x - 40$
17. $-6y - 72$
18. $-8t + 72$
19. $-2x^2 + 4x - 12$
20. $-2x^2 + 12x + 40$
21. $-3y^2 + 24y$
22. $-7x^2 - 56y$
23. $-3y^3 + 12y^2 - 15y$
24. $-4m^4 - 32m^3 + 64m$
25. $-x^2 + 3x - 7$
26. $-p^3 - 4p^2 + 11$
27. $-a^4 + 2a^3 - 13a$
28. $-m^3 - m^2 + m - 2$

b Factor.

29. $a(b - 2) + c(b - 2)$
30. $a(x^2 - 3) - 2(x^2 - 3)$
31. $(x - 2)(x + 5) + (x - 2)(x + 8)$
32. $(m - 4)(m + 3) + (m - 4)(m - 3)$
33. $a^2(x - y) + a^2(x - y)$
34. $3x^2(x - 6) + 3x^2(x - 6)$

1. ______
2. ______
3. ______
4. ______
5. ______
6. ______
7. ______
8. ______
9. ______
10. ______
11. ______
12. ______
13. ______
14. ______
15. ______
16. ______
17. ______
18. ______
19. ______
20. ______
21. ______
22. ______
23. ______
24. ______
25. ______
26. ______
27. ______
28. ______
29. ______
30. ______
31. ______
32. ______
33. ______
34. ______

ANSWERS

35. ______

36. ______

37. ______

38. ______

39. ______

40. ______

41. ______

42. ______

43. ______

44. ______

45. ______

46. ______

47. ______

48. ______

49. ______

50. ______

51. ______

52. ______

53. ______

54. ______

35. $ac + ad + bc + bd$

36. $xy + xz + wy + wz$

37. $b^3 - b^2 + 2b - 2$

38. $y^3 - y^2 + 3y - 3$

39. $y^3 - 8y^2 + y - 8$

40. $t^3 + 6t^2 - 2t - 12$

41. $24x^3 - 36x^2 + 72x - 108$

42. $10a^4 + 15a^2 - 25a - 30$

43. $x^6 + x^5 - x^3 + x^2$

44. $y^4 - y^3 + y^2 + y$

45. $2y^4 + 6y^2 + 5y^2 + 15$

46. $2xy + x^2y - 6 - 3x$

SKILL MAINTENANCE

47. A person receives \$244 per year in simple interest from three investments totaling \$3500. Part is invested at $5\frac{1}{2}\%$, part at 7%, and part at 8%. There is \$1100 more invested at 8% than at 7%. Find the amount invested at each rate.

48. One angle of a triangle is four times another angle. The third angle is 75°. What are the measures of the first two angles?

SYNTHESIS

Factor. Assume that all exponents are natural numbers.

49. $4y^{4a} + 12y^{2a} + 10y^{2a} + 30$

50. $7y^{2a+b} - 5y^{a+b} + 3y^{a+2b}$

51. *Total revenue.* An electronics firm is marketing a new kind of stereo. Total revenue is the total amount of money taken in. The firm determines that when it sells x stereos, the total revenue R is given by

$$R = 280x - 0.4x^2 \text{ dollars.}$$

Find an equivalent expression for R by factoring out $0.4x$.

52. *Total cost.* The electronics firm determines that the total cost C of producing x stereos is given by

$$C = 7000x + 0.6x^2.$$

Find an equivalent expression for C by factoring out $0.6x$.

Number of diagonals. The number of diagonals D of a polygon having n sides is given by

$$D = \tfrac{1}{2}n^2 - \tfrac{3}{2}n.$$

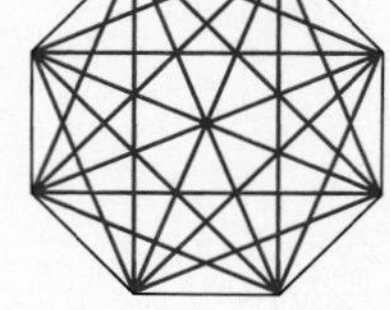

53. Find the number of diagonals in a polygon with 8 sides.

54. Find an equivalent expression for D by factoring out a common factor.

5.4 Factoring Trinomials $x^2 + bx + c$ and $ax^2 + bx + c, a \neq 1$

We begin our study of the factoring of trinomials with a trial-and-error method for factoring a general trinomial of the type $x^2 + bx + c$. Then we consider trinomials of the type $ax^2 + bx + c$, where the leading coefficient a is not 1.

a Factoring Trinomials of the Type $x^2 + bx + c$

To try to factor trinomials of the type $x^2 + bx + c$, we can use a trial-and-error procedure.

Constant Term Positive

Recall the FOIL method of multiplying two binomials:

$$\begin{aligned}(x+3)(x+5) &= \overset{F}{x^2} + \overset{O}{5x} + \overset{I}{3x} + \overset{L}{15} \\ &= x^2 + 8x + 15.\end{aligned}$$

The product is a trinomial. In this example, the leading term has a coefficient of 1. The constant term is positive. To factor $x^2 + 8x + 15$, we think of FOIL in reverse. We multiplied x times x to get the first term of the trinomial. Thus the first term of each binomial factor is x. We want to find numbers p and q such that

$$x^2 + 8x + 15 = (x + p)(x + q).$$

To get the middle term and the last term of the trinomial, we look for two numbers whose product is 15 and whose sum is 8. In this case, we know beforehand that the numbers are 3 and 5. Thus the factorization is

$$(x + 3)(x + 5), \quad \text{or} \quad (x + 5)(x + 3)$$

by the commutative law of multiplication. In general,

$$(x + p)(x + q) = x^2 + (p + q)x + pq.$$

To factor, we can use this equation in reverse.

► **EXAMPLE 1** Factor: $x^2 + 9x + 8$.

Think of FOIL in reverse. The first term of each factor is x. We are looking for numbers p and q such that

$$x^2 + 9x + 8 = (x + p)(x + q).$$

We look for two numbers whose product is 8 and whose sum is 9. Since both 8 and 9 are positive, we need consider only positive factors.

Pairs of factors	Sums of factors
2, 4	6
1, 8	9 ← The numbers we need are 1 and 8.

The factorization is $(x + 1)(x + 8)$. We can check by multiplying to see whether we get the original trinomial. ◄

DO EXERCISES 1 AND 2.

OBJECTIVES

After finishing Section 5.4, you should be able to:

a Factor trinomials of the type $x^2 + bx + c$.

b Factor trinomials of the type $ax^2 + bx + c, a \neq 1$.

FOR EXTRA HELP

Tape 9C | Tape 8A | MAC: 5 IBM: 5

Factor. Check by multiplying.

1. $x^2 + 5x + 6$

2. $y^2 + 7y + 10$

ANSWERS ON PAGE A-6

Factor.

3. $m^2 - 8m + 12$

4. $24 - 11t + t^2$

ANSWERS ON PAGE A-6

When the constant term of a trinomial is positive, we look for two factors with the same sign. The sign is that of the middle term.

▶ **EXAMPLE 2** Factor: $y^2 - 9y + 20$.

Since the constant term is positive and the coefficient of the middle term is negative, we look for a factorization of 20 in which both factors are negative. Their sum must be -9.

Pairs of factors	Sums of factors
$-1, -20$	-21
$-2, -10$	-12
$-4, -5$	-9 ← The numbers we need are -4 and -5.

The factorization is $(y - 4)(y - 5)$. ◀

DO EXERCISES 3 AND 4.

Constant Term Negative

When the constant term is negative, we look for two factors whose product is negative. One of them must be positive and the other negative. Their sum must be the coefficient of the *middle* term.

▶ **EXAMPLE 3** Factor: $x^3 - x^2 - 30x$.

Always look first for a common factor. This time there is one, x. We first factor it out:

$$x^3 - x^2 - 30x = x(x^2 - x - 30).$$

Now we consider $x^2 - x - 30$. Since the constant term is negative, we look for a factorization of -30 in which one factor is positive and one factor is negative. The sum of the factors must be -1, so the negative factor must have the larger absolute value. Thus we consider only pairs of factors in which the negative factor has the larger absolute value.

Pairs of factors	Sums of factors
$1, -30$	-29
$2, -15$	-13
$3, -10$	-7
$5, -6$	-1 ← The numbers we want are 5 and -6.

The factorization of $x^2 - x - 30$ is

$$(x + 5)(x - 6).$$

The factorization of the original trinomial is

$$x(x + 5)(x - 6).$$ ◀

▶ **EXAMPLE 4** Factor: $x^2 + 17x - 110$.

Since the constant term is negative, we look for a factorization of -110 in which one factor is positive and one factor is negative. Their sum must be 17, so the positive factor must have the larger absolute value. Thus we consider only pairs of factors in which the positive term has the larger absolute value.

Pairs of factors	Sums of factors
-1, 110	109
-2, 55	53
-5, 22	17 ← The numbers we need are -5 and 22.
-10, 11	1

The factorization is

$$(x - 5)(x + 22).$$ ◀

Some trinomials are not factorable.

▶ **EXAMPLE 5** Factor: $x^2 - x - 7$.

There are no factors of -7 whose sum is -1. This trinomial is *not* factorable into binomials. ◀

> **To factor $x^2 + bx + c$:**
> 1. **First arrange in descending order. Use a trial-and-error procedure that looks for factors of c whose sum is b.**
> 2. **If c is positive, the signs of the factors are the same as the sign of b.**
> 3. **If c is negative, one factor is positive and the other is negative. If the sum of the two factors is the opposite of b, changing the signs of each factor will give the desired factors whose sum is b.**
> 4. **Check your result by multiplying.**

DO EXERCISES 5–8.

The procedure considered here can also be applied to a trinomial with more than one variable.

▶ **EXAMPLE 6** Factor: $x^2 - 2xy - 48y^2$.

We look for numbers p and q such that

$$x^2 - 2xy - 48y^2 = (x + py)(x + qy).$$

Our thinking is much the same as if we were factoring $x^2 - 2x - 48$. We look for factors of -48 whose sum is -2. Those factors are 6 and -8. Then

$$x^2 - 2xy - 48y^2 = (x + 6y)(x - 8y).$$

We check by multiplying. ◀

DO EXERCISES 9 AND 10.

Sometimes a trinomial like $x^4 + 2x^2 - 15$ can be factored using the following method. We can first think of the trinomial as $(x^2)^2 + 2x^2 - 15$, or we can make a substitution (perhaps, just mentally), letting $u = x^2$. Then the trinomial becomes

$$u^2 + 2u - 15.$$

We factor this trinomial and if a factorization is found, we replace all occurrences of u by x^2.

Factor.

5. $x^3 + 4x^2 - 12x$

6. $y^2 - 4y - 12$

7. $x^2 - 110 - x$

8. $x^2 + x - 5$

Factor.

9. $x^2 - 5xy + 6y^2$

10. $p^2 - 6pq - 16q^2$

ANSWERS ON PAGE A-6

Factor.

11. $x^4 - 11x^2 + 18$

12. $p^6 + p^3 - 6$

ANSWERS ON PAGE A-6

► **EXAMPLE 7** Factor: $x^4 + 2x^2 - 15$.

We let $u = x^2$. Then consider $u^2 + 2u - 15$. The constant term is negative and the middle term is positive. Thus we look for pairs of factors of -15, one positive and one negative, such that the positive factor has the larger absolute value and the sum of the factors is 2.

Pairs of factors	Sums of factors
$-1, 15$	14
$-3, 5$	2

The numbers we want are -3 and 5, and the desired factorization of $u^2 + 2u - 15$ is

$$(u - 3)(u + 5).$$

Replacing u by x^2, we obtain the following factorization of the original trinomial:

$$(x^2 - 3)(x^2 + 5).$$

With practice, you will make such a substitution mentally. ◄

DO EXERCISES 11 AND 12.

b Factoring Trinomials of the Type $ax^2 + bx + c$, $a \neq 1$

Now we consider trinomials in which the coefficient a of the leading term, ax^2, is not 1. We consider two methods. Use the one that works best for you or the one that your instructor chooses for you. Both methods involve trial and error, but the first requires trial and error in only one step.

Method 1: The FOIL Method

We first consider the **FOIL method** for factoring trinomials of the type

$$ax^2 + bx + c, \quad \text{where } a \neq 1.$$

Consider the following multiplication.

$$\begin{aligned} & \quad\;\; \text{F} \quad\;\; \text{O} \quad\;\; \text{I} \quad\;\; \text{L} \\ (2x + 5)(3x + 4) &= 6x^2 + \underbrace{8x + 15x} + 20 \\ &= 6x^2 + \quad 23x \quad + 20 \end{aligned}$$

To factor $6x^2 + 23x + 20$, we do the reverse of what we just did. We look for two binomials $rx + p$ and $sx + q$ whose product is this trinomial. The product of the First terms must be $6x^2$. The product of the Outside terms plus the product of the Inside terms must be $23x$. The product of the Last terms must be 20. We know from the preceding discussion that the answer is

$$(2x + 5)(3x + 4).$$

Generally, however, finding such an answer is a trial-and-error process. It turns out that

$$(-2x - 5)(-3x - 4)$$

is also a correct answer, but we usually choose an answer in which the first coefficients are positive.

We use the following method.

To factor $ax^2 + bx + c$, $a \neq 1$, using the FOIL method:

1. **Factor out a common factor, if any.**
2. **Factor the term ax^2. This gives possibilities for r and s:**

$$(rx + p)(sx + q)$$
$$rx \cdot sx = ax^2.$$

3. **Factor the last term, c. This gives possibilities for p and q:**

$$(rx + p)(sx + q)$$
$$p \cdot q = c.$$

4. **Look for combinations of factors from steps (2) and (3) for which the sum of their products is the middle term, bx:**

$$rx \cdot q$$
$$(rx + p)(sx + q)$$
$$p \cdot sx. \qquad rx \cdot q + p \cdot sx \stackrel{?}{=} bx$$

▶ **EXAMPLE 8** Factor: $3x^2 + 10x - 8$.

1. First we factor out a common factor, if any. There is none (other than 1 or -1).
2. Next we factor the first term, $3x^2$. The only possibility for factors is $3x \cdot x$. The desired factorization is then of the form

$$(3x + \square)(x + \square),$$

where we must determine other numbers for the blanks.

3. We then factor the last term, -8, which is negative. The possibilities are $(-8)(1)$, $8(-1)$, $2(-4)$, and $(-2)(4)$.
4. From steps (2) and (3), we see that there are 8 possibilities for factorization. We look for combinations such that the sum of their products is the middle term, $10x$:

$3x$, $-8x$: $(3x - 8)(x + 1) = 3x^2 - 5x - 8$; Wrong middle term

$-3x$, $8x$: $(3x + 8)(x - 1) = 3x^2 + 5x - 8$; Wrong middle term

$-12x$, $2x$: $(3x + 2)(x - 4) = 3x^2 - 10x - 8$; Wrong middle term

$12x$, $-2x$: $(3x - 2)(x + 4) = 3x^2 + 10x - 8$ Correct middle term!

We need not consider any other possibilities. The factorization is $(3x - 2)(x + 4)$. ◀

DO EXERCISES 13 AND 14.

▶ **EXAMPLE 9** Factor: $6x^6 - 19x^5 + 10x^4$.

1. We first factor out a common factor, if any. The expression x^4 is common to all terms, so we factor it out: $x^4(6x^2 - 19x + 10)$.
2. Now we factor the trinomial $6x^2 - 19x + 10$. We factor the first term, $6x^2$, and get $6x$, x and $3x$, $2x$. We then have these as possibilities for factorizations: $(3x + \square)(2x + \square)$ or $(6x + \square)(x + \square)$.

Factor.

13. $3x^2 - 13x - 56$

14. $3x^2 + 5x + 2$

ANSWERS ON PAGE A-6

Factor.

15. $24y^2 - 46y + 10$

16. $20x^5 - 46x^4 + 24x^3$

3. We next factor the last term, 10, which is positive. The possibilities are

$$10, 1 \quad \text{and} \quad -10, -1 \qquad \text{and} \qquad 5, 2 \quad \text{and} \quad -5, -2.$$

4. From steps (2) and (3), we see that there are 16 possibilities for factorization. We look for combinations of factors from steps (2) and (3) such that the sum of their products is the middle term, $-19x$. The sign of the middle term is negative, but the sign of the last term, 10, is positive. Thus the signs of both factors of the last term, 10, must be negative. We see from our list of factors in step (3) that we can use only $-10, -1$ and $-5, -2$ as possibilities. This reduces the possibilities for factorization to 8. We start by using these factors with $(3x + \square)(2x + \square)$. Should we not find the correct factorization, we will consider $(6x + \square)(x + \square)$.

$$(3x - 10)(2x - 1) = 6x^2 - 23x + 10;$$
Outer product: $-3x$; inner product: $-20x$. Wrong middle term

$$(3x - 1)(2x - 10) = 6x^2 - 32x + 10$$
Outer product: $-30x$; inner product: $-2x$. Wrong middle term

$$(3x - 5)(2x - 2) = 6x^2 - 16x + 10;$$
Outer product: $-6x$; inner product: $-10x$. Wrong middle term

$$(3x - 2)(2x - 5) = 6x^2 - 19x + 10$$
Outer product: $-15x$; inner product: $-4x$. Correct middle term!

We have a correct answer. We need not consider $(6x + \square)(x + \square)$.

Look again at the possibility: $(3x - 5)(2x - 2)$. Without multiplying, we can reject such a possibility, noting that

$$(3x - 5)(2x - 2) = 2(3x - 5)(x - 1).$$

The expression $2x - 2$ has a common factor, 2. But we removed the largest common factor before we began. If this expression were a factorization, then 2 would have to be a common factor along with x^4. Thus, as we saw when we multiplied, $(3x - 5)(2x - 2)$ cannot be part of the factorization of the original trinomial. We can now eliminate factorizations that have a common factor, given that we factored out the largest common factor at the outset.

The factorization of $6x^2 - 19x + 10$ is $(3x - 2)(2x - 5)$. But do not forget the common factor! We must include it in order to get a complete factorization of the original trinomial:

$$6x^4 - 19x^5 + 10x^4 = x^4(3x - 2)(2x - 5).$$ ◀

Here is another tip that might speed up your factoring. Suppose in Example 9 that we considered the possibility

$$(3x + 2)(2x + 5) = 6x^2 + 19x + 10.$$

We might have tried this before noticing that using all plus signs would give us a plus sign for the middle term. If we change *both* signs, however, we get the correct answer before including the common factor:

$$(3x - 2)(2x - 5) = 6x^2 - 19x + 10.$$

DO EXERCISES 15 AND 16.

ANSWERS ON PAGE A-6

Tips for Factoring $ax^2 + bx + c$, $a \neq 1$, Using the FOIL Method

1. **If the largest common factor has been factored out of the original trinomial, then no binomial factor can have a common factor (other than 1 or -1).**
2. **a) If all the signs of all the terms are positive, then the signs of all the terms of the binomial factors are positive.**
   b) If a and c are positive and b is negative, then the signs of the factors of c are negative.
   c) If a is positive and c is negative, then the signs of the factors of c will be opposites.
3. **Be systematic about your trials. Keep track of those possibilities that you have tried and those that you have not.**

Keep in mind that this method of factoring trinomials of the type $ax^2 + bx + c$ involves trial and error. As you practice, you will find that you can make better and better guesses.

DO EXERCISES 17 AND 18.

Factor.

17. $3x^2 + 19x + 20$

18. $16x^2 - 12 + 16x$

Method 2: The Grouping Method

The second method for factoring trinomials of the type $ax^2 + bx + c$, $a \neq 1$, is known as the **grouping method.** It involves not only trial and error and FOIL but also factoring by grouping. We know how to factor the trinomial $x^2 + 7x + 10$. We look for factors of the constant term, 10, whose sum is the coefficient, 7, of the middle term:

$$x^2 + 7x + 10$$

(1) Factor: $10 = 2 \cdot 5$.
(2) Sum of factors: $2 + 5 = 7$.

But what happens when the leading coefficient is not 1? Consider the trinomial $6x^2 + 23x + 20$. The method we use is similar to what we used for the preceding trinomial, but we need two more steps. The method is outlined as follows.

To factor $ax^2 + bx + c$, $a \neq 1$, using the grouping method:

1. **Factor out a common factor, if any.**
2. **Multiply the leading coefficient a and the constant c.**
3. **Try to factor the product ac so that the sum of the factors is b. That is, find integers p and q such that $pq = ac$ and $p + q = b$.**
4. **Split the middle term. That is, write it as a sum using the factors found in step (3).**
5. **Factor by grouping.**

▶ **EXAMPLE 10** Factor: $6x^2 + 23x + 20$.

1. First we factor out a common factor, if any. There is none (other than 1 or -1).
2. We multiply the leading coefficient 6 and the constant 20: $6 \cdot 20 = 120$.
3. Then we look for a factorization of 120 in which the sum of the factors is the coefficient, 23, of the middle term.

ANSWERS ON PAGE A-6

Factor.

19. $4x^2 + 4x - 3$

20. $4x^2 + 37x + 9$

Factor.

21. $10y^4 - 7y^3 - 12y^2$

22. $36a^3 + 21a^2 + a$

ANSWERS ON PAGE A-6

Pairs of factors	Sums of factors
1, 120	121
2, 60	62
3, 40	43
4, 30	34
5, 24	29
6, 20	26
8, 15	23 ← **8 + 15 = 23**
10, 12	22

4. Next we split the middle term as a sum or difference using the factors found in step (3): $6x^2 + 23x + 20 = 6x^2 + 8x + 15x + 20$.
5. We factor by grouping as follows:

$$\begin{aligned} 6x^2 + 23x + 20 &= 6x^2 + 8x + 15x + 20 \\ &= 2x(3x + 4) + 5(3x + 4) \\ &= (2x + 5)(3x + 4). \end{aligned}$$

Factoring by grouping; see Section 5.3

It does not matter which way we split the middle term so long as we split it using the factors found in step (3). We still get the same factorization, although the factors may be in a different order. Note the following:

$$\begin{aligned} 6x^2 + 23x + 20 &= 6x^2 + 15x + 8x + 20 \\ &= 3x(2x + 5) + 4(2x + 5) \\ &= (3x + 4)(2x + 5) \end{aligned}$$

Check by multiplying: $(3x + 4)(2x + 5) = 6x^2 + 23x + 20$. ◀

DO EXERCISES 19 AND 20.

▶ **EXAMPLE 11** Factor: $6x^4 - 116x^3 - 80x^2$.

1. First we factor out a common factor, if any. The expression $2x^2$ is common to all three terms: $2x^2(3x^2 - 58x - 40)$.
2. Now we factor the trinomial $3x^2 - 58x - 40$. We multiply the leading coefficient and the constant, 3 and -40: $3(-40) = -120$.
3. Next we try to factor -120 so that the sum of the factors is -58.

Pairs of factors	Sums of factors
−1, 120	119
1, −120	−119
−2, 60	58
2, −60	−58

Pairs of factors	Sums of factors
−8, 15	7
8, −15	−7
−10, 12	2
10, −12	−2

4. We split the middle term, $-58x$, as follows: $-58x = 2x - 60x$.
5. We then factor by grouping:

$$\begin{aligned} 3x^2 - 58x - 40 &= 3x^2 + 2x - 60x - 40 \\ &= x(3x + 2) - 20(3x + 2) \\ &= (x - 20)(3x + 2). \end{aligned}$$

Substituting $2x - 60x$ for $-58x$

The factorization of $3x^2 - 58x - 40$ is $(x - 20)(3x + 2)$. But don't forget the common factor! We must include it to get a factorization of the original trinomial:

$$6x^4 - 116x^3 - 80x^2 = 2x^2(x - 20)(3x + 2).$$ ◀

DO EXERCISES 21 AND 22.

NAME SECTION DATE

EXERCISE SET 5.4

a Factor.

1. $x^2 + 9x + 20$
2. $x^2 + 8x + 15$
3. $t^2 - 8t + 15$
4. $y^2 - 12y + 27$
5. $x^2 - 27 - 6x$
6. $t^2 - 15 - 2t$
7. $2y^2 - 16y + 32$
8. $2a^2 - 20a + 50$
9. $p^2 + 3p - 54$
10. $m^2 + m - 72$
11. $14x + x^2 + 45$
12. $12y + y^2 + 32$
13. $y^2 + 2y - 63$
14. $p^2 + 3p - 40$
15. $t^2 - 11t + 28$
16. $y^2 - 14y + 45$
17. $3x + x^2 - 10$
18. $x + x^2 - 6$
19. $x^2 + 5x + 6$
20. $y^2 + 8y + 7$
21. $56 + x - x^2$
22. $32 + 4y - y^2$
23. $32y + 4y^2 - y^3$
24. $56x + x^2 - x^3$
25. $x^4 + 11x^2 - 80$
26. $y^4 + 5y^2 - 84$
27. $x^2 - 3x + 7$

ANSWERS

1. ______
2. ______
3. ______
4. ______
5. ______
6. ______
7. ______
8. ______
9. ______
10. ______
11. ______
12. ______
13. ______
14. ______
15. ______
16. ______
17. ______
18. ______
19. ______
20. ______
21. ______
22. ______
23. ______
24. ______
25. ______
26. ______
27. ______

ANSWERS

28. ______
29. ______
30. ______
31. ______
32. ______
33. ______
34. ______
35. ______
36. ______
37. ______
38. ______
39. ______
40. ______
41. ______
42. ______
43. ______
44. ______
45. ______
46. ______
47. ______
48. ______
49. ______
50. ______
51. ______
52. ______
53. ______
54. ______

28. $x^2 + 12x + 13$

29. $x^2 + 12xy + 27y^2$

30. $p^2 - 5pq - 24q^2$

31. $x^2 - 14x + 49$

32. $y^2 + 8y + 16$

33. $x^4 + 50x^2 + 49$

34. $p^4 + 80p^2 + 79$

35. $x^6 - x^3 - 42$

36. $x^8 - 7x^4 + 10$

b Factor.

37. $3x^2 - 16x - 12$

38. $6x^2 - 5x - 25$

39. $6x^3 - 15x - x^2$

40. $10y^3 - 12y - 7y^2$

41. $3a^2 - 10a + 8$

42. $12a^2 - 7a + 1$

43. $35y^2 + 34y + 8$

44. $9a^2 + 18a + 8$

45. $4t + 10t^2 - 6$

46. $8x + 30x^2 - 6$

47. $8x^2 - 16 - 28x$

48. $18x^2 - 24 - 6x$

49. $12x^3 - 31x^2 + 20x$

50. $15x^3 - 19x^2 - 10x$

51. $14x^4 - 19x^3 - 3x^2$

52. $70x^4 - 68x^3 + 16x^2$

53. $3a^2 - a - 4$

54. $6a^2 - 7a - 10$

55. $9x^2 + 15x + 4$

56. $6y^2 - y - 2$

57. $3 + 35z - 12z^2$

58. $8 - 6a - 9a^2$

59. $-4t^2 - 4t + 15$

60. $-12a^2 + 7a - 1$

61. $3x^3 - 5x^2 - 2x$

62. $18y^3 - 3y^2 - 10y$

63. $24x^2 - 2 - 47x$

64. $15y^2 - 10 - 15y$

65. $21x^2 + 37x + 12$

66. $10y^2 + 23y + 12$

67. $40x^4 + 16x^2 - 12$

68. $24y^4 + 2y^2 - 15$

69. $12a^2 - 17ab + 6b^2$

70. $20p^2 - 23pq + 6q^2$

71. $2x^2 + xy - 6y^2$

72. $8m^2 - 6mn - 9n^2$

73. $6x^2 - 29xy + 28y^2$

74. $10p^2 + 7pq - 12q^2$

75. $9x^2 - 30xy + 25y^2$

76. $4p^2 + 12pq + 9q^2$

77. $3x^6 + x^3 - 2$

78. $2p^8 + 11p^4 + 15$

ANSWERS

55. ______

56. ______

57. ______

58. ______

59. ______

60. ______

61. ______

62. ______

63. ______

64. ______

65. ______

66. ______

67. ______

68. ______

69. ______

70. ______

71. ______

72. ______

73. ______

74. ______

75. ______

76. ______

77. ______

78. ______

ANSWERS

79. a) ______

b) ______

80. a) ______

b) ______

81. ______

82. ______

83. ______

84. ______

85. ______

86. ______

87. ______

88. ______

89. ______

90. ______

91. ______

92. ______

93. ______

94. ______

95. ______

96. ______

97. ______

98. ______

99. ______

100. ______

79. *Height of a thrown object.* Suppose that an object is thrown upward with an initial velocity of 80 ft/sec from a height of 224 ft. Its height h after t seconds is given by

$$h = -16t^2 + 80t + 224.$$

a) What is the height of the object after 0 sec? 1 sec? 3 sec? 4 sec? 6 sec?

b) Find an equivalent expression for h by factoring.

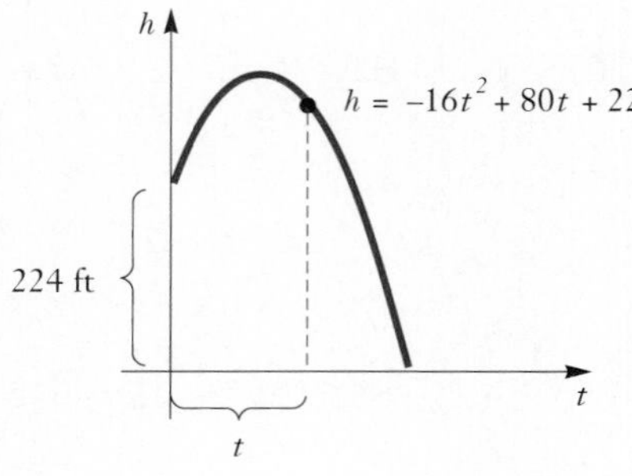

80. *Height of a thrown object.* Suppose that an object is thrown upward with an initial velocity of 96 ft/sec from a height of 880 ft. Its height h after t seconds is given by

$$h = -16t^2 + 96t + 880.$$

a) What is the height of the object after 0 sec? 1 sec? 3 sec? 8 sec? 10 sec?

b) Find an equivalent expression for h by factoring.

SYNTHESIS

Factor. Assume that variables in exponents represent positive integers.

81. $p^2q^2 + 7pq + 12$

82. $2x^4y^6 - 3x^2y^3 - 20$

83. $x^2 - \frac{4}{25} + \frac{3}{5}x$

84. $y^2 - \frac{8}{49} + \frac{2}{7}y$

85. $y^2 + 0.4y - 0.05$

86. $t^2 + 0.6t - 0.27$

87. $7a^2b^2 + 6 + 13ab$

88. $9x^2y^2 - 4 + 5xy$

89. $3x^2 + 12x - 495$

90. $15t^3 - 60t^2 - 315t$

91. $216x + 78x^2 + 6x^3$

92. $x^{2a} + 5x^a - 24$

93. $x^2 + ax + bx + ab$

94. $4x^{2a} - 4x^a - 3$

95. $bdx^2 + adx + bcx + ac$

96. $\frac{1}{4}p^2 - \frac{2}{5}p + \frac{4}{25}$

97. $(x + 3)^2 - 2(x + 3) - 35$

98. $6(x - 7)^2 + 13(x - 7) - 5$

99. Find all integers m for which $x^2 + mx + 75$ can be factored.

100. One of the factors of $x^2 - 345x - 7300$ is $x + 20$. Find the other factor.

5.5 Factoring Trinomial Squares and Differences of Squares

We now introduce a faster procedure to factor trinomials that are squares of binomials and binomials that are differences of squares.

a Trinomial Squares

Consider the trinomial

$$x^2 + 6x + 9.$$

To factor it, we can use the method considered in the preceding section. We look for factors of 9 whose sum is 6. We see that these factors are 3 and 3 and the factorization is

$$x^2 + 6x + 9 = (x + 3)(x + 3) = (x + 3)^2.$$

Note that the result is the square of a binomial. We also say that $x^2 + 6x + 9$ is a **trinomial square**. We can certainly use the trial-and-error procedure to factor trinomial squares, but we want to develop an even faster procedure.

In order to do so, we must first be able to recognize when a trinomial is a square.

> **How to recognize a trinomial square:**
>
> **a) Two of the terms must be squares, such as A^2 and B^2.**
>
> **b) There must be no minus sign before A^2 or B^2.**
>
> **c) Multiplying A and B (which are the square roots of these expressions) and doubling the result should give the remaining term, $2AB$, or its opposite, $-2AB$.**

▶ **EXAMPLES** Determine whether the polynomial is a trinomial square.

1. $x^2 + 10x + 25$

a) Two terms are squares: x^2 and 25.

b) There is no minus sign before either x^2 or 25.

c) If we multiply the square roots, x and 5, and double the product, we get $10x$, the remaining term.

Thus the trinomial is a square.

2. $4x + 16 + 3x^2$

a) Only one term, 16, is a square ($3x^2$ is not a square because 3 is not a perfect square integer and $4x$ is not a square because x is not a square).

Therefore, the trinomial is not a square.

3. $100y^2 + 81 - 180y$

(It can help to first write this in descending order: $100y^2 - 180y + 81$.)

a) Two of the terms, $100y^2$ and 81, are squares.

b) There is no minus sign before either $100y^2$ or 81.

c) If we multiply the square roots, $10y$ and 9, and double the product, we get the opposite of the remaining term: $2(10y)(9) = 180y$, which is the opposite of $-180y$.

Thus, $100y^2 + 81 - 180y$ is a trinomial square. ◀

OBJECTIVES

After finishing Section 5.5, you should be able to:

a Factor trinomial squares.

b Factor differences of squares.

c Factor certain polynomials with four terms by grouping and possibly using the factoring of a trinomial square or the difference of squares.

FOR EXTRA HELP

Tape NC

Tape 8A

MAC: 5
IBM: 5

1. Which of the following are trinomial squares?

a) $x^2 + 6x + 9$

b) $x^2 - 8x + 16$

c) $x^2 + 6x + 11$

d) $4x^2 + 25 - 20x$

e) $16x^2 - 20x + 25$

f) $16 + 14x + 5x^2$

g) $x^2 + 8x - 16$

h) $x^2 - 8x - 16$

ANSWERS ON PAGE A-6

Factor.

2. $x^2 + 14x + 49$

3. $9y^2 - 30y + 25$

4. $16x^2 + 72xy + 81y^2$

5. $16x^4 - 40x^2y^3 + 25y^6$

Factor.

6. $-8a^2 + 24ab - 18b^2$

7. $3a^2 - 30ab + 75b^2$

ANSWERS ON PAGE A-6

DO EXERCISE 1 ON THE PRECEDING PAGE.

The factors of a trinomial square are two identical binomials. We use the following equations.

$$A^2 + 2AB + B^2 = (A + B)^2;$$
$$A^2 - 2AB + B^2 = (A - B)^2$$

▶ **EXAMPLE 4** Factor: $x^2 - 10x + 25$.

$x^2 - 10x + 25 = (x - 5)^2$ **We find the square terms and write their square roots with a minus sign between them.**

Note the sign! ◀

▶ **EXAMPLE 5** Factor: $16y^2 + 49 + 56y$.

$16y^2 + 49 + 56y = 16y^2 + 56y + 49$ **Using the commutative law of addition**

$= (4y + 7)^2$ **We find the square terms and write their square roots with a plus sign between them.** ◀

▶ **EXAMPLE 6** Factor: $-20xy + 4y^2 + 25x^2$.

We have

$-20xy + 4y^2 + 25x^2 = 4y^2 - 20xy + 25x^2$ **Writing descending order in *y***

$= (2y - 5x)^2$.

This square can also be expressed as

$$25x^2 - 20xy + 4y^2 = (5x - 2y)^2.$$ ◀

DO EXERCISES 2–5.

In factoring, we must always remember to look *first* for a factor common to all the terms.

▶ **EXAMPLE 7** Factor: $2x^2 - 12xy + 18y^2$.

Always remember to look first for a common factor. This time there is a common factor, 2.

$2x^2 - 12xy + 18y^2 = 2(x^2 - 6xy + 9y^2)$ **Removing the common factor**

$= 2(x - 3y)^2$ **Factoring the trinomial square** ◀

▶ **EXAMPLE 8** Factor: $-4y^2 - 144y^8 + 48y^5$.

$-4y^2 - 144y^8 + 48y^5$

$= -4y^2(1 + 36y^6 - 12y^3)$ **Removing the common factor**

$= -4y^2(1 - 12y^3 + 36y^6)$ **Changing order**

$= -4y^2(1 - 6y^3)^2$ **Factoring the trinomial square** ◀

DO EXERCISES 6 AND 7.

b Differences of Squares

The following is a difference of squares:

$$x^2 - 9.$$

To factor a difference of two expressions that are squares, we can use a pattern for multiplying a sum and a difference that we used earlier.

> $$A^2 - B^2 = (A + B)(A - B)$$
> **To factor a difference of two squares, write the square root of the first expression *plus* the square root of the second *times* the square root of the first *minus* the square root of the second.**

You should memorize this rule, and then say it as you work.

▶ **EXAMPLE 9** Factor: $x^2 - 9$.

$$x^2 - 9 = x^2 - 3^2 = (x + 3)(x - 3)$$ ◀

▶ **EXAMPLE 10** Factor: $25y^6 - 49x^2$.

$$\begin{array}{ccccc} A^2 & - B^2 & = (A + B)(A - B) \\ \downarrow & \downarrow & \downarrow\ \downarrow\ \downarrow\ \downarrow \\ 25y^6 - 49x^2 = (5y^3)^2 & - (7x)^2 & = (5y^3 + 7x)(5y^3 - 7x) \end{array}$$ ◀

DO EXERCISES 8 AND 9.

Common factors should always be removed. Removing common factors actually eases the factoring process because the type of factoring to be done becomes clearer.

▶ **EXAMPLE 11** Factor: $5 - 5x^2y^6$.

There is a common factor, 5.

$$\begin{aligned} 5 - 5x^2y^6 &= 5(1 - x^2y^6) && \text{Removing the common factor} \\ &= 5[1^2 - (xy^3)^2] \\ &= 5(1 + xy^3)(1 - xy^3) && \text{Factoring the difference of squares} \end{aligned}$$ ◀

▶ **EXAMPLE 12** Factor: $2x^4 - 8y^4$.

There is a common factor, 2.

$$\begin{aligned} 2x^4 - 8y^4 &= 2(x^4 - 4y^4) && \text{Removing the common factor} \\ &= 2[(x^2)^2 - (2y^2)^2] \\ &= 2(x^2 + 2y^2)(x^2 - 2y^2) && \text{Factoring the difference of squares} \end{aligned}$$ ◀

Factor.

8. $y^2 - 4$

9. $49x^4 - 25y^{10}$

ANSWERS ON PAGE A-6

Factor.

10. $25x^2y^2 - 4a^2$

11. $9x^2 - 16y^2$

12. $20x^2 - 5y^2$

13. $81x^4y^2 - 16y^2$

14. Factor: $a^3 + a^2 - 16a - 16$.

Factor.

15. $x^2 + 2x + 1 - p^2$

16. $y^2 - 8y + 16 - 9m^2$

17. $x^2 + 8x + 16 - 100t^2$

18. $64p^2 - (x^2 + 8x + 16)$

ANSWERS ON PAGE A-6

▶ **EXAMPLE 13** Factor: $16x^4y - 81y$.

There is a common factor, y.

$$\begin{aligned} 16x^4y - 81y &= y(16x^4 - 81) && \textbf{Removing the common factor} \\ &= y[(4x^2)^2 - 9^2] \\ &= y(4x^2 + 9)(4x^2 - 9) && \textbf{Factoring the difference of squares} \\ &= y(4x^2 + 9)(2x + 3)(2x - 3) && \textbf{Factoring } 4x^2 - 9\textbf{, which is also a difference of squares} \end{aligned}$$ ◀

In Example 13, it may be tempting to try to factor $(4x^2 + 9)$. Note that it is a sum of two expressions that are squares. If the greatest common factor of a sum of squares has been removed, then a sum of squares cannot be factored using real numbers. Note also that one of the factors, $4x^2 - 9$, could be factored further. Whenever that is possible, you should do so. That way you will be factoring *completely*.

DO EXERCISES 10–13.

C More Factoring by Grouping

Sometimes when factoring a polynomial with four terms, we might get a factor that can be factored further using other methods we have learned.

▶ **EXAMPLE 14** Factor: $x^3 + 3x^2 - 4x - 12$.

$$\begin{aligned} x^3 + 3x^2 - 4x - 12 &= x^2(x + 3) - 4(x + 3) \\ &= (x^2 - 4)(x + 3) \\ &= (x + 2)(x - 2)(x + 3) \end{aligned}$$ ◀

DO EXERCISE 14.

A difference of squares can have more than two terms. For example, one of the squares may be a trinomial. We can factor by a type of grouping.

▶ **EXAMPLE 15** Factor: $x^2 + 6x + 9 - y^2$.

$$\begin{aligned} x^2 + 6x + 9 - y^2 &= (x^2 + 6x + 9) - y^2 && \textbf{Grouping as a trinomial minus } y^2 \textbf{ to show a difference of squares} \\ &= (x + 3)^2 - y^2 \\ &= (x + 3 + y)(x + 3 - y) \end{aligned}$$ ◀

DO EXERCISES 15–18.

NAME SECTION DATE

EXERCISE SET 5.5

a Factor.

1. $y^2 - 6y + 9$
2. $x^2 - 8x + 16$
3. $x^2 + 14x + 49$
4. $x^2 + 16x + 64$
5. $x^2 + 1 + 2x$
6. $x^2 + 1 - 2x$
7. $y^2 + 36 - 12y$
8. $y^2 + 36 + 12y$
9. $-18y^2 + y^3 + 81y$
10. $24a^2 + a^3 + 144a$
11. $12a^2 + 36a + 27$
12. $20y^2 + 100y + 125$
13. $2x^2 - 40x + 200$
14. $32x^2 + 48x + 18$
15. $1 - 8d + 16d^2$
16. $64 + 25y^2 - 80y$
17. $y^4 - 8y^2 + 16$
18. $a^4 - 10a^2 + 25$
19. $0.25x^2 + 0.30x + 0.09$
20. $0.04x^2 - 0.28x + 0.49$
21. $p^2 - 2pq + q^2$
22. $m^2 + 2mn + n^2$
23. $a^2 + 4ab + 4b^2$
24. $49p^2 - 14pq + q^2$
25. $25a^2 - 30ab + 9b^2$
26. $49p^2 - 84pq + 36q^2$
27. $y^6 + 26y^3 + 169$
28. $p^6 - 10p^3 + 25$
29. $16x^{10} - 8x^5 + 1$
30. $9x^{10} + 12x^5 + 4$
31. $x^4 + 2x^2y^2 + y^4$
32. $p^6 - 2p^3q^4 + q^8$

b Factor.

33. $x^2 - 16$
34. $y^2 - 9$
35. $p^2 - 49$

ANSWERS

1. ______
2. ______
3. ______
4. ______
5. ______
6. ______
7. ______
8. ______
9. ______
10. ______
11. ______
12. ______
13. ______
14. ______
15. ______
16. ______
17. ______
18. ______
19. ______
20. ______
21. ______
22. ______
23. ______
24. ______
25. ______
26. ______
27. ______
28. ______
29. ______
30. ______
31. ______
32. ______
33. ______
34. ______
35. ______

ANSWERS

36. ______
37. ______
38. ______
39. ______
40. ______
41. ______
42. ______
43. ______
44. ______
45. ______
46. ______
47. ______
48. ______
49. ______
50. ______
51. ______
52. ______
53. ______
54. ______
55. ______
56. ______
57. ______
58. ______
59. ______
60. ______
61. ______
62. ______
63. ______
64. ______
65. ______
66. ______
67. ______
68. ______
69. ______
70. ______
71. ______
72. ______
73. ______
74. ______

36. $m^2 - 64$ **37.** $p^2q^2 - 25$ **38.** $a^2b^2 - 81$

39. $6x^2 - 6y^2$ **40.** $8x^2 - 8y^2$ **41.** $4xy^4 - 4xz^4$

42. $25ab^4 - 25az^4$ **43.** $4a^3 - 49a$ **44.** $9x^3 - 25x$

45. $3x^8 - 3y^8$ **46.** $9a^4 - a^2b^2$ **47.** $9a^4 - 25a^2b^4$

48. $16x^6 - 121x^2y^4$ **49.** $\frac{1}{25} - x^2$ **50.** $\frac{1}{16} - y^2$

51. $0.04x^2 - 0.09y^2$ **52.** $0.01x^2 - 0.04y^2$

C Factor.

53. $m^3 - 7m^2 - 4m + 28$ **54.** $x^3 + 8x^2 - x - 8$ **55.** $a^3 - ab^2 - 2a^2 + 2b^2$

56. $p^2q - 25q + 3p^2 - 75$ **57.** $(a + b)^2 - 100$ **58.** $(p - 7)^2 - 144$

59. $a^2 + 2ab + b^2 - 9$ **60.** $x^2 - 2xy + y^2 - 25$ **61.** $r^2 - 2r + 1 - 4s^2$

62. $c^2 + 4cd + 4d^2 - 9p^2$ **63.** $2m^2 + 4mn + 2n^2 - 50b^2$ **64.** $12x^2 + 12x + 3 - 3y^2$

65. $9 - (a^2 + 2ab + b^2)$ **66.** $16 - (x^2 - 2xy + y^2)$

SYNTHESIS

Factor. Assume that variables in exponents represent positive integers.

67. $\frac{1}{4}p^2 - \frac{2}{5}p + \frac{4}{25}$ **68.** $x^4y^4 - 8x^2y^2 + 16$ **69.** $x^{4a} - y^{2b}$

70. $5c^{100} - 80d^{100}$ **71.** $9x^{2n} - 6x^n + 1$ **72.** $y^8 - 256$

73. $(3a + 4)^2 - 49b^2$ **74.** $x^{16} - 1$

5.6 Factoring Sums or Differences of Cubes

a Although a sum of two squares cannot be factored using real-number coefficients, a sum of two cubes can. In this section, we develop the patterns for factoring a sum or a difference of two expressions that are cubes.

Consider the following products:

$$\begin{aligned}(A+B)(A^2-AB+B^2) &= A(A^2-AB+B^2)+B(A^2-AB+B^2)\\ &= A^3-A^2B+AB^2+A^2B-AB^2+B^3\\ &= A^3+B^3\end{aligned}$$

and

$$\begin{aligned}(A-B)(A^2+AB+B^2) &= A(A^2+AB+B^2)-B(A^2+AB+B^2)\\ &= A^3+A^2B+AB^2-A^2B-AB^2-B^3\\ &= A^3-B^3.\end{aligned}$$

The above equations (reversed) show how we can factor a sum or a difference of two cubes.

$$\mathbf{A^3+B^3=(A+B)(A^2-AB+B^2),}$$
$$\mathbf{A^3-B^3=(A-B)(A^2+AB+B^2)}$$

Note that what we are considering here is a sum or a difference of cubes. We are not cubing a binomial. For example, $(A+B)^3$ is *not* the same as A^3+B^3.

The table of cubes in the margin will help in the following examples.

▶ **EXAMPLE 1** Factor: x^3-27.

We have

$$x^3-27=x^3-3^3.$$

In one set of parentheses, we write the cube root of the first term, x. Then we write the cube root of the second term, -3. This gives us the expression $x-3$:

$$(x-3)(\qquad\qquad)$$

To get the next factor, we think of $x-3$ and do the following:

Square the first term: x^2.
Multiply the terms and then change the sign: $3x$.
Square the second term: 9.

$$(x-3)(x^2+3x+9).$$

Note that we cannot factor x^2+3x+9. It is not a trinomial square nor can it be factored by trial and error. ◀

DO EXERCISES 1 AND 2.

▶ **EXAMPLE 2** Factor: $125x^3+y^3$.

We have

$$125x^3+y^3=(5x)^3+y^3.$$

In one set of parentheses, we write the cube root of the first term, $5x$. Then we write a plus sign, and then the cube root of the second term, y:

$$(5x+y)(\qquad\qquad)$$

OBJECTIVE

After finishing Section 5.6, you should be able to:

a Factor sums and differences of two cubes.

FOR EXTRA HELP

Tape 10A | Tape 8B | MAC: 5 IBM: 5

N	N^3
0.2	0.008
0.1	0.001
0	0
1	1
2	8
3	27
4	64
5	125
6	216
7	343
8	512
9	729
10	1000

Factor.

1. x^3-8

2. $64-y^3$

ANSWERS ON PAGE A-6

Factor.

3. $27x^3 + y^3$

4. $8y^3 + z^3$

Factor.

5. $m^6 - n^6$

6. $16x^7y + 54xy^7$

7. $729x^6 - 64y^6$

8. $x^3 - 0.027$

ANSWERS ON PAGE A-6

To get the next factor, we think of $5x + y$ and do the following:

Square the first term: $25x^2$.
Multiply the terms and then change the sign: $-5xy$.
Square the second term: y^2.

$(5x + y)(25x^2 - 5xy + y^2)$. ◀

DO EXERCISES 3 AND 4.

▶ **EXAMPLE 3** Factor: $128y^7 - 250x^6y$.

We first look for a common factor:

$$\begin{aligned}128y^7 - 250x^6y &= 2y(64y^6 - 125x^6) = 2y[(4y^2)^3 - (5x^2)^3]\\ &= 2y(4y^2 - 5x^2)(16y^4 + 20x^2y^2 + 25x^4).\end{aligned}$$ ◀

▶ **EXAMPLE 4** Factor: $a^6 - b^6$.

We can express this polynomial as a difference of squares:

$$(a^3)^2 - (b^3)^2.$$

We factor as follows:

$$(a^3 + b^3)(a^3 - b^3).$$

One factor is a sum of two cubes, and the other factor is a difference of two cubes. We factor them:

$$(a + b)(a^2 - ab + b^2)(a - b)(a^2 + ab + b^2).$$

We have now factored completely. ◀

In Example 4, had we thought of factoring first as a difference of two cubes, we would have had

$$\begin{aligned}(a^2)^3 - (b^2)^3 &= (a^2 - b^2)(a^4 + a^2b^2 + b^4)\\ &= (a + b)(a - b)(a^4 + a^2b^2 + b^4).\end{aligned}$$

In this case, we might have missed some factors; $a^4 + a^2b^2 + b^4$ can be factored as $(a^2 - ab + b^2)(a^2 + ab + b^2)$, but we probably would not have known to do such factoring.

▶ **EXAMPLE 5** Factor: $64a^6 - 729b^6$.

We have

$$\begin{aligned}64a^6 - 729b^6 &= (8a^3 - 27b^3)(8a^3 + 27b^3) \quad \textbf{Factoring a difference of squares}\\ &= [(2a)^3 - (3b)^3][(2a)^3 + (3b)^3].\end{aligned}$$

Each factor is a sum or a difference of cubes. We factor each:

$$= (2a - 3b)(4a^2 + 6ab + 9b^2)(2a + 3b)(4a^2 - 6ab + 9b^2).$$ ◀

Remember the following:

Sum of cubes:	$A^3 + B^3 = (A + B)(A^2 - AB + B^2)$;
Difference of cubes:	$A^3 - B^3 = (A - B)(A^2 + AB + B^2)$;
Difference of squares:	$A^2 - B^2 = (A + B)(A - B)$;
Sum of squares:	$A^2 + B^2$ **cannot be factored using real numbers if the largest common factor has been removed.**

DO EXERCISES 5–8.

NAME SECTION DATE

EXERCISE SET 5.6

a Factor.

1. $x^3 + 8$
2. $c^3 + 27$
3. $y^3 - 64$
4. $z^3 - 1$
5. $w^3 + 1$
6. $x^3 + 125$
7. $8a^3 + 1$
8. $27x^3 + 1$
9. $y^3 - 8$
10. $p^3 - 27$
11. $8 - 27b^3$
12. $64 - 125x^3$
13. $64y^3 + 1$
14. $125x^3 + 1$
15. $8x^3 + 27$
16. $27y^3 + 64$
17. $a^3 - b^3$
18. $x^3 - y^3$
19. $a^3 + \frac{1}{8}$
20. $b^3 + \frac{1}{27}$
21. $2y^3 - 128$
22. $3z^3 - 3$
23. $24a^3 + 3$
24. $54x^3 + 2$
25. $rs^3 + 64r$
26. $ab^3 + 125a$
27. $5x^3 - 40z^3$

ANSWERS

1. ____
2. ____
3. ____
4. ____
5. ____
6. ____
7. ____
8. ____
9. ____
10. ____
11. ____
12. ____
13. ____
14. ____
15. ____
16. ____
17. ____
18. ____
19. ____
20. ____
21. ____
22. ____
23. ____
24. ____
25. ____
26. ____
27. ____

ANSWERS

28.

29.

30.

31.

32.

33.

34.

35.

36.

37.

38.

39. See graph.

40.

41.

42.

43.

44.

45.

46.

47.

48.

49.

50.

51.

52.

53.

54.

55.

56.

28. $2y^3 - 54z^3$ **29.** $x^3 + 0.001$ **30.** $y^3 + 0.125$

31. $64x^6 - 8t^6$ **32.** $125c^6 - 8d^6$ **33.** $2y^4 - 128y$

34. $3z^5 - 3z^2$ **35.** $z^6 - 1$ **36.** $t^6 + 1$

37. $t^6 + 64y^6$ **38.** $p^6 - q^6$

SKILL MAINTENANCE

39. Graph: $5x = 10 - 2y$.

Solve.

40. $|x| = 27$ **41.** $|5x - 6| \leq 39$ **42.** $|5x - 6| > 39$

SYNTHESIS

Consider these polynomials:

$$(a + b)^3; \quad a^3 + b^3; \quad (a + b)(a^2 - ab + b^2);$$
$$(a + b)(a^2 + ab + b^2); \quad (a + b)(a - b)(a - b).$$

43. Evaluate each polynomial when $a = -2$ and $b = 3$.

44. Evaluate each polynomial when $a = 4$ and $b = -1$.

Factor. Assume that variables in exponents represent natural numbers.

45. $x^{6a} + y^{3b}$ **46.** $a^3x^3 - b^3y^3$ **47.** $3x^{3a} + 24y^{3b}$

48. $\frac{8}{27}x^3 + \frac{1}{64}y^3$ **49.** $\frac{1}{24}x^3y^3 + \frac{1}{3}z^3$ **50.** $\frac{1}{16}x^{3a} + \frac{1}{2}y^{6a}z^{9b}$

51. $7x^3 - \frac{7}{8}$ **52.** $[(c - d)^3 - d^3]^2$ **53.** $(x + y)^3 - x^3$

54. $(1 - x)^3 + (x - 1)^6$ **55.** $(a + 2)^3 - (a - 2)^3$ **56.** $y^4 - 8y^3 - y + 8$

5.7 Factoring: A General Strategy

a Once you know the kind of expression you have to factor, you can do so without too much difficulty. Below is a general strategy for factoring.

> **a) Always look for a common factor (other than 1 or -1). Factor out the greatest common factor.**
>
> **b) Then look at the number of terms.**
>
> ***Two terms:*** **Try factoring as a difference of squares first. Next, try factoring as a sum or difference of cubes. Do *not* try to factor a *sum* of squares.**
>
> ***Three terms:*** **Determine whether the trinomial is a square. If it is, you know how to factor. If not, try trial and error.**
>
> ***Four or more terms:*** **Try factoring by grouping and removing a common binomial factor. Next, try grouping into a difference of squares, one of which is a trinomial.**
>
> **c) Always *factor completely.* If a factor with more than one term can be factored, you should factor it.**

▶ **EXAMPLE 1** Factor: $10a^2x - 40b^2x$.

a) We look first for a common factor:

$$10x(a^2 - 4b^2). \quad \text{Factoring out the largest common factor}$$

b) The factor $a^2 - 4b^2$ has only two terms. It is a difference of squares. We factor it, keeping the common factor: $10x(a + 2b)(a - 2b)$.

c) Have we factored completely? Yes, because none of the factors with more than one term can be factored further into a polynomial of smaller degree. ◀

▶ **EXAMPLE 2** Factor: $x^6 - y^6$.

a) We look for a common factor. There isn't one (other than 1 or -1).

b) There are only two terms. It is a difference of squares: $(x^3)^2 - (y^3)^2$. We factor it: $(x^3 + y^3)(x^3 - y^3)$. One factor is a sum of two cubes, and the other factor is a difference of two cubes. We factor them:

$$(x + y)(x^2 - xy + y^2)(x - y)(x^2 + xy + y^2).$$

c) We have factored completely because none of the factors can be factored further into a polynomial of smaller degree. ◀

DO EXERCISES 1–3.

▶ **EXAMPLE 3** Factor: $10x^6 + 40y^2$.

a) We remove the largest common factor: $10(x^6 + 4y^2)$.

b) In the parentheses, there are two terms, a sum of squares, which cannot be factored. ◀

OBJECTIVE

After finishing Section 5.7, you should be able to:

a Factor polynomials using any of the methods considered in this chapter.

FOR EXTRA HELP

Tape 10B

Tape 8B

MAC: 5
IBM: 5

Factor completely.

1. $3y^3 - 12x^2y$

2. $7a^3 - 7$

3. $64x^6 - 729y^6$

ANSWERS ON PAGE A-6

Factor.

4. $3x - 6 - bx^2 + 2bx$

5. $5y^4 + 20x^6$

6. $6x^2 - 3x - 18$

7. $a^3 - ab^2 - a^2b + b^3$

8. $3x^2 + 18ax + 27a^2$

9. $2x^2 - 20x + 50 - 18b^2$

ANSWERS ON PAGE A-7

▶ **EXAMPLE 4** Factor: $2x^2 + 50a^2 - 20ax$.

a) We remove the largest common factor: $2(x^2 + 25a^2 - 10ax)$.

b) In the parentheses, there are three terms. The trinomial is a square. We factor it: $2(x - 5a)^2$.

c) None of the factors with more than one term can be factored further. ◀

▶ **EXAMPLE 5** Factor: $6x^2 - 20x - 16$.

a) We remove the largest common factor: $2(3x^2 - 10x - 8)$.

b) In the parentheses, there are three terms. The trinomial is not a square. We factor by trial: $2(x - 4)(3x + 2)$.

c) We cannot factor further. ◀

▶ **EXAMPLE 6** Factor: $3x + 12 + ax^2 + 4ax$.

a) There is no common factor (other than 1 or -1).

b) There are four terms. We try grouping to remove a common binomial factor:

$$3(x + 4) + ax(x + 4) \quad \textbf{Factoring two grouped binomials}$$
$$= (3 + ax)(x + 4). \quad \textbf{Removing the common binomial factor}$$

c) None of the factors with more than one term can be factored further. ◀

▶ **EXAMPLE 7** Factor: $y^2 - 9a^2 + 12y + 36$.

a) There is no common factor (other than 1 or -1).

b) There are four terms. We try grouping to remove a common binomial factor, but that is not possible. We try grouping as a difference of squares:

$$(y^2 + 12y + 36) - 9a^2 = (y + 6)^2 - (3a)^2$$
$$= (y + 6 + 3a)(y + 6 - 3a). \quad \textbf{Factoring the difference of squares}$$

c) No factor with more than one term can be factored further. ◀

▶ **EXAMPLE 8** Factor: $x^3 - xy^2 + x^2y - y^3$.

a) There is no common factor (other than 1 or -1).

b) There are four terms. We try grouping to remove a common binomial factor:

$$x(x^2 - y^2) + y(x^2 - y^2) \quad \textbf{Factoring two grouped binomials}$$
$$= (x + y)(x^2 - y^2). \quad \textbf{Removing the common binomial factor}$$

c) The factor $x^2 - y^2$ can be factored further, giving

$$(x + y)(x + y)(x - y). \quad \textbf{Factoring a difference of squares}$$

None of the factors with more than one term can be factored further, so we have factored completely. ◀

DO EXERCISES 4–9.

NAME SECTION DATE

EXERCISE SET 5.7

a Factor completely.

1. $x^2 - 144$
2. $y^2 - 81$
3. $2x^2 + 11x + 12$
4. $8a^2 + 18a - 5$
5. $3x^4 - 12$
6. $2xy^2 - 50x$
7. $a^2 + 25 + 10a$
8. $p^2 + 64 + 16p$
9. $2x^2 - 10x - 132$
10. $3y^2 - 15y - 252$
11. $9x^2 - 25y^2$
12. $16a^2 - 81b^2$
13. $m^6 - 1$
14. $64t^6 - 1$
15. $x^2 + 6x - y^2 + 9$
16. $t^2 + 10t - p^2 + 25$
17. $250x^3 - 128y^3$
18. $27a^3 - 343b^3$
19. $8m^3 + m^6 - 20$
20. $-37x^2 + x^4 + 36$
21. $ac + cd - ab - bd$
22. $xw - yw + xz - yz$
23. $4c^2 - 4cd + d^2$
24. $70b^2 - 3ab - a^2$
25. $-7x^2 + 2x^3 + 4x - 14$
26. $9m^2 + 3m^3 + 8m + 24$
27. $2x^3 + 6x^2 - 8x - 24$
28. $3x^3 + 6x^2 - 27x - 54$
29. $16x^3 + 54y^3$
30. $250a^3 + 54b^3$

ANSWERS

1. ______
2. ______
3. ______
4. ______
5. ______
6. ______
7. ______
8. ______
9. ______
10. ______
11. ______
12. ______
13. ______
14. ______
15. ______
16. ______
17. ______
18. ______
19. ______
20. ______
21. ______
22. ______
23. ______
24. ______
25. ______
26. ______
27. ______
28. ______
29. ______
30. ______

ANSWERS

31. ______
32. ______
33. ______
34. ______
35. ______
36. ______
37. ______
38. ______
39. ______
40. ______
41. ______
42. ______
43. ______
44. ______
45. ______
46. ______
47. ______
48. ______
49. ______
50. ______
51. ______
52. ______
53. ______
54. ______
55. ______
56. ______
57. ______
58. ______
59. ______
60. ______
61. ______
62. ______
63. ______
64. ______

31. $36y^2 - 35 + 12y$

32. $2b - 28a^2b + 10ab$

33. $a^8 - b^8$

34. $2x^4 - 32$

35. $a^3b - 16ab^3$

36. $x^3y - 25xy^3$

37. $a(b - 2) + c(b - 2)$

38. $(x - 2)(x + 5) + (x - 2)(x + 8)$

39. $5x^3 - 5x^2y - 5xy^2 + 5y^3$

40. $a^3 - ab^2 + a^2b - b^3$

41. $42ab + 27a^2b^2 + 8$

42. $-23xy + 20x^2y^2 + 6$

43. $8y^4 - 125y$

44. $64p^4 - p$

SKILL MAINTENANCE

Solve.

45. $2x - 3y + 4z = 10,$
$4x + 6y - 4z = -5,$
$-8x - 9y + 8z = -2$

46. $3x - 2y + 6z = 12,$
$6x + 2y - 3z = -9,$
$-9x + 4y - 9z = 10$

SYNTHESIS

Factor. Assume that variables in exponents represent natural numbers.

47. $30y^4 - 97xy^2 + 60x^2$

48. $3x^2y^2z + 25xyz^2 + 28z^3$

49. $5x^3 - \frac{5}{27}$

50. $-16 + 17(5 - y^2) - (5 - y^2)^2$

51. $(x - p)^2 - p^2$

52. $x^4 - 50x^2 + 49$

53. $(y - 1)^4 - (y - 1)^2$

54. $s^6 - 729t^6$

55. $x^6 - 2x^5 + x^4 - x^2 + 2x - 1$

56. $27x^{6s} + 64y^{3t}$

57. $4x^2 + 4xy + y^2 - r^2 + 6rs - 9s^2$

58. $c^4d^4 - a^{16}$

59. $c^{2w+1} + 2c^{w+1} + c$

60. $24x^{2a} - 6$

61. $y^9 - y$

62. $1 - \dfrac{x^{27}}{1000}$

63. $3(x + 1)^2 + 9(x + 1) - 12$

64. $8(a - 3)^2 - 64(a - 3) + 128$

5.8 Solving Equations and Problems Using Polynomials

OBJECTIVES

After finishing Section 5.8, you should be able to:

a Solve equations by factoring and using the principle of zero products.

b Solve problems by translating to an equation and then solving the equation using factoring and the principle of zero products.

FOR EXTRA HELP

Tape 10C

Tape 9A

MAC: 5
IBM: 5

We now introduce a new equation-solving procedure and use it along with factoring to solve *quadratic equations.*

A *quadratic equation* is an equation equivalent to one of the standard form

$$ax^2 + bx + c = 0,$$

where a, b, and c are all real-number coefficients and $a > 0$. The trinomial on the left is of second degree.

a The Principle of Zero Products

When we multiply two or more numbers, the product will be 0 if one of the factors is 0. Conversely, if a product is 0, then at least one of the factors must be 0. This property of real numbers gives us another principle for solving equations.

The Principle of Zero Products

For any real numbers a and b:

If $ab = 0$, then $a = 0$ or $b = 0$ (or both). If $a = 0$ or $b = 0$, then $ab = 0$.

To use this principle in solving equations, we make sure that there is a 0 on one side of the equation and a factorization on the other side.

► **EXAMPLE 1** Solve: $x^2 - x = 6$.

In order to use the principle of zero products, we must have 0 on one side of the equation, so we subtract 6 on both sides:

$$x^2 - x - 6 = 0. \quad \text{Getting 0 on one side}$$

We need a factorization on the other side, so we factor the polynomial:

$$(x - 3)(x + 2) = 0. \quad \text{Factoring}$$

We now set each factor equal to 0 (this is a use of the principle of zero products):

$$x - 3 = 0 \quad \text{or} \quad x + 2 = 0. \quad \text{Using the principle of zero products}$$

This gives us two simple linear equations. We solve them separately,

$$x = 3 \quad \text{or} \quad x = -2,$$

and check as follows.

Check:

$x^2 - x = 6$	
$3^2 - 3$	6
$9 - 3$	
6	

$x^2 - x = 6$	
$(-2)^2 - (-2)$	6
$4 + 2$	
6	

The numbers 3 and -2 are both solutions. ◄

1. Solve: $x^2 + 8 = 6x$.

2. Solve: $5y + 2y^2 = 3$.

3. Solve: $8b^2 = 16b$.

ANSWERS ON PAGE A-7

To solve an equation using the principle of zero products:

1. **Obtain a 0 on one side of the equation.**
2. **Factor the other side.**
3. **Set each factor equal to 0.**
4. **Solve the resulting equations.**

DO EXERCISE 1.

When you solve an equation using the principle of zero products, you may wish to check by substitution as we did in Example 1. Such a check will detect errors in solving.

CAUTION! When you are using the principle of zero products, it is important that you be sure there is a 0 on one side of the equation. If neither side of the equation is 0, the procedure will not work.

For example, consider $x^2 - x = 6$ in Example 1 as

$$x(x - 1) = 6.$$

Suppose we reasoned as follows, setting factors equal to 6:

$$x = 6 \quad \text{or} \quad x - 1 = 6$$
$$x = 7.$$

Neither 6 nor 7 check, as shown below:

$x(x-1) = 6$	
$6(6-1)$	6
$6(5)$	
30	FALSE

$x(x-1) = 6$	
$7(7-1)$	6
$7(6)$	
42	FALSE

▶ **EXAMPLE 2** Solve: $7y + 3y^2 = -2$.

Since there must be a 0 on one side of the equation, we add 2 to get 0 on one side and arrange in descending order. Then we factor and use the principle of zero products.

$$7y + 3y^2 = -2$$
$$3y^2 + 7y + 2 = 0$$
$$(3y + 1)(y + 2) = 0 \quad \textbf{Factoring}$$
$$3y + 1 = 0 \quad \text{or} \quad y + 2 = 0 \quad \textbf{Using the principle of zero products}$$
$$y = -\tfrac{1}{3} \quad \text{or} \quad y = -2$$

The solutions are $-\frac{1}{3}$ and -2. ◀

DO EXERCISE 2.

▶ **EXAMPLE 3** Solve: $5b^2 = 10b$.

$$5b^2 = 10b$$
$$5b^2 - 10b = 0 \quad \textbf{Getting 0 on one side}$$
$$5b(b - 2) = 0 \quad \textbf{Factoring}$$
$$5b = 0 \quad \text{or} \quad b - 2 = 0 \quad \textbf{Using the principle of zero products}$$
$$b = 0 \quad \text{or} \quad b = 2$$

The solutions are 0 and 2. ◀

DO EXERCISE 3.

► **EXAMPLE 4** Solve: $x^2 - 6x + 9 = 0$.

$$x^2 - 6x + 9 = 0$$
$$(x - 3)(x - 3) = 0 \quad \text{Factoring}$$
$$x - 3 = 0 \quad \text{or} \quad x - 3 = 0 \quad \text{Using the principle of zero products}$$
$$x = 3 \quad \text{or} \quad x = 3$$

There is only one solution, 3. ◄

DO EXERCISE 4.

► **EXAMPLE 5** Solve: $3x^3 - 9x^2 = 30x$.

$$3x^3 - 9x^2 = 30x$$
$$3x^3 - 9x^2 - 30x = 0 \quad \text{Getting 0 on one side}$$
$$3x(x^2 - 3x - 10) = 0 \quad \text{Factoring out a common factor}$$
$$3x(x + 2)(x - 5) = 0 \quad \text{Factoring the trinomial}$$
$$3x = 0 \quad \text{or} \quad x + 2 = 0 \quad \text{or} \quad x - 5 = 0 \quad \text{Using the principle of zero products}$$
$$x = 0 \quad \text{or} \quad x = -2 \quad \text{or} \quad x = 5$$

The solutions are 0, -2, and 5. ◄

DO EXERCISE 5.

b Problem Solving

Some problems can be translated to quadratic equations, some of which we can now solve. The problem-solving process is the same as for other kinds of equations, except that we now can solve equations by factoring and using the principle of zero products.

► **EXAMPLE 6** There are two square flower gardens on the campus mall. The length of a side of one garden is 4 ft less than the length of a side of the other. The area of the larger garden is 56 ft^2 greater than that of the smaller. Find the dimensions of the gardens.

1. *Familiarize.* We draw a picture and label it.

We let $x =$ the length of a side of the larger garden. Then $x - 4 =$ the length of the side of the smaller garden. We note that the larger garden has area 56 greater than the smaller garden.

2. *Translate.* We use the following statement:

Area of large garden is 56 greater than area of small garden.

$$x^2 \quad = 56 \quad + \quad (x - 4)^2$$

4. Solve: $25 + x^2 = -10x$.

5. Solve: $x^3 + x^2 = 6x$.

ANSWERS ON PAGE A-7

6. The square of a number minus twice the number is 48. Find the number.

7. The width of a rectangle is 5 cm less than the length. The area is 24 cm^2. Find the dimensions.

ANSWERS ON PAGE A-7

3. *Solve.* We solve the equation:

$$x^2 = 56 + (x - 4)^2$$

$$x^2 = 56 + x^2 - 8x + 16 \quad \textbf{Squaring a binomial}$$

$$8x = 72 \quad \textbf{Collecting like terms and simplifying}$$

$$x = 9.$$

4. *Check.* Since we used x to represent the length of a side of the larger garden, it follows that the smaller garden has sides of length $x - 4$, or 5. The areas of the gardens are then 5^2, or 25 ft^2, and 9^2, or 81 ft^2. The difference is 56 ft^2, so we do have a solution of the problem.

5. *State.* The smaller garden has sides of length 5 ft, and the larger garden has sides of length 9 ft. ◀

DO EXERCISE 6.

Example 6 is interesting in that the equation turned out to be linear, although initially it didn't seem as though it would. In the following example, that does not happen and we will need to use the principle of zero products.

▶ **EXAMPLE 7** The width of a rectangular swimming pool is 2 m less than the length. The area is 15 m^2. Find the dimensions of the pool.

1. *Familiarize.* We make a drawing and label the length and the width.

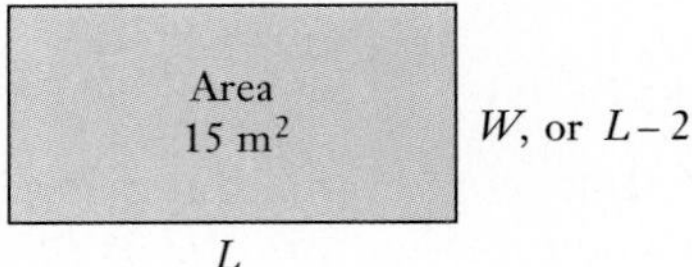

We let L = the length of the square and W = the width. We recall the formula for the area of a rectangle:

$$A = LW.$$

2. *Translate.* We can now translate to a system of two equations:

$$15 = LW, \quad \textbf{Substituting 15 for } \boldsymbol{A}$$

$$W = L - 2.$$

3. *Solve.* We solve the system of equations. Substituting from the second equation into the first, we obtain

$$15 = L(L - 2).$$

We now have an equation in one variable, which we can solve:

$$15 = L^2 - 2L \quad \textbf{Multiplying on the right}$$

$$0 = L^2 - 2L - 15 \quad \textbf{Getting 0 on one side}$$

$$0 = (L - 5)(L + 3) \quad \textbf{Factoring}$$

$$L - 5 = 0 \quad \text{or} \quad L + 3 = 0 \quad \textbf{Using the principle of zero products}$$

$$L = 5 \quad \text{or} \quad L = -3.$$

4. *Check.* We check the possible solutions in the original problem. The number -3 is not a solution because length cannot be negative in this problem. If the length is 5 and the width is 2 m less, or 3, then the area will be $3 \cdot 5$, or 15 m^2. We have a solution.

5. *State.* The answer is that the pool has length 5 m and width 3 m. ◀

DO EXERCISE 7.

The following problem involves the **Pythagorean theorem,** which relates the lengths of the sides of a right triangle. A **right triangle** has a 90° angle. The side opposite the 90° angle is called the **hypotenuse.** The other sides are called **legs.**

The Pythagorean Theorem

The sum of the squares of the legs of a right triangle is equal to the square of the hypotenuse:

$$a^2 + b^2 = c^2.$$

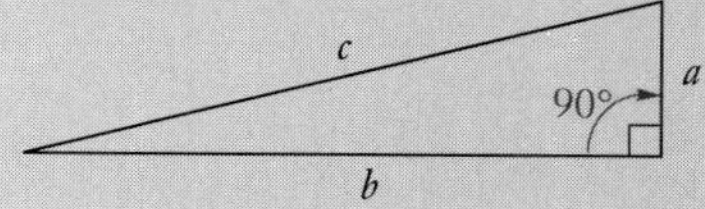

► **EXAMPLE 8** The lengths of the sides of a right triangle are three consecutive integers. Find their lengths.

1. *Familiarize.* We first make a drawing. We let

$$x = \text{the length of the first side.}$$

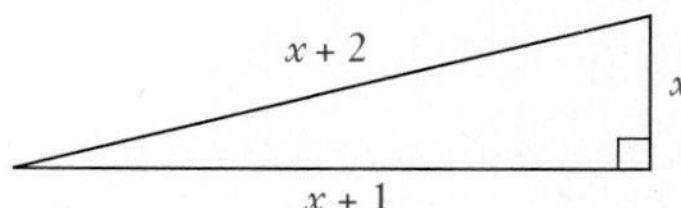

Since the lengths are consecutive integers, we know that

$$x + 1 = \text{the length of the second side}$$

and

$$x + 2 = \text{the length of the third side.}$$

The hypotenuse is always the longest side of a right triangle, so we know it is the side with length $x + 2$.

2. *Translate.* Applying the Pythagorean theorem, we obtain the following translation:

$$a^2 + b^2 = c^2$$
$$x^2 + (x + 1)^2 = (x + 2)^2.$$

3. *Solve.* We solve the equation as follows:

$$x^2 + (x^2 + 2x + 1) = x^2 + 4x + 4 \quad \textbf{Squaring the binomials}$$
$$2x^2 + 2x + 1 = x^2 + 4x + 4 \quad \textbf{Collecting like terms}$$
$$x^2 - 2x - 3 = 0 \quad \textbf{Subtracting } 4x \textbf{ and 4 to get 0 on one side}$$
$$(x - 3)(x + 1) = 0 \quad \textbf{Factoring}$$
$$x - 3 = 0 \quad \text{or} \quad x + 1 = 0$$
$$x = 3 \quad \text{or} \quad x = -1.$$

4. *Check.* The integer -1 cannot be a length of a side because it is negative. When $x = 3$, $x + 1 = 4$, and $x + 2 = 5$; and $3^2 + 4^2 = 5^2$. So 3, 4, and 5 check.

5. *State.* The consecutive integers 3, 4, and 5 are the lengths of the sides. ◄

DO EXERCISE 8.

8. One leg of a right triangle has length 5 m. The lengths of the other sides are consecutive integers. Find the lengths of the other sides of the triangle.

ANSWER ON PAGE A-7

❖ SIDELIGHTS

Computer–Calculator Exercises: Graphing and Solving Polynomial Equations

There are many computer software packages and calculators that can graph equations. Below is such an example. It is the graph of $y = x^3 - 3x - 1$.

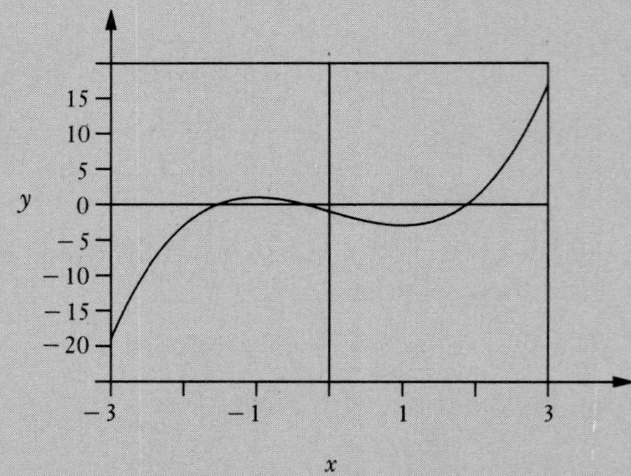

Generally such computer software packages and calculators can also be used to solve an equation like $x^3 - 3x - 1 = 0$. Solutions occur where the graph crosses the x-axis. For this equation, the solutions are approximately -1.53, -0.35, and 1.88.

EXERCISES

Use a computer software package or a graphing calculator to graph each equation. Then approximate the values of x for which $y = 0$.

1. $y = x^3 - 3x + 2$

2. $y = x^4 - 2x^2$

3. $y = \frac{1}{3}x^3 - \frac{1}{2}x^2 - 2x + 1$

4. $y = 280x - 0.4x^2$

5. $y = 0.1x^4 - x^2 + 0.4$

6. $y = 2x^4 - 4x^2 + 2$

7. $x^4 + x^3 - 4x^2 - 2x + 4$

8. $y = x^4 - x^3 - 11x^2 + 9x + 18$

NAME SECTION DATE

EXERCISE SET 5.8

a Solve.

1. $x^2 + 3x = 28$
2. $y^2 - 4y = 45$
3. $y^2 + 16 = 8y$
4. $r^2 + 1 = 2r$
5. $x^2 - 12x + 36 = 0$
6. $y^2 + 16y + 64 = 0$
7. $9x + x^2 + 20 = 0$
8. $8y + y^2 + 15 = 0$
9. $x^2 + 8x = 0$
10. $t^2 + 9t = 0$
11. $x^2 - 9 = 0$
12. $p^2 - 16 = 0$
13. $z^2 = 36$
14. $y^2 = 81$
15. $y^2 + 2y = 63$
16. $a^2 + 3a = 40$
17. $32 + 4x - x^2 = 0$
18. $27 + 12t + t^2 = 0$
19. $3b^2 + 8b + 4 = 0$
20. $9y^2 + 15y + 4 = 0$
21. $8y^2 - 10y + 3 = 0$
22. $4x^2 + 11x + 6 = 0$
23. $6z - z^2 = 0$
24. $8y - y^2 = 0$

ANSWERS

1. ______
2. ______
3. ______
4. ______
5. ______
6. ______
7. ______
8. ______
9. ______
10. ______
11. ______
12. ______
13. ______
14. ______
15. ______
16. ______
17. ______
18. ______
19. ______
20. ______
21. ______
22. ______
23. ______
24. ______

ANSWERS

25. ______

26. ______

27. ______

28. ______

29. ______

30. ______

31. ______

32. ______

33. ______

34. ______

35. ______

36. ______

37. ______

38. ______

39. ______

40. ______

25. $12z^2 + z = 6$

26. $6x^2 - 7x = 10$

27. $5x^2 - 20 = 0$

28. $6y^2 - 54 = 0$

29. $21r^2 + r - 10 = 0$

30. $12a^2 - 5a - 28 = 0$

31. $15y^2 = 3y$

32. $18x^2 = 9x$

33. $14 = x(x - 5)$

34. $x(x - 5) = 24$

35. $2x^3 - 2x^2 = 12x$

36. $50y + 5y^3 = 35y^2$

37. $2x^3 = 128x$

38. $147y = 3y^3$

b Solve.

39. If 4 times the square of a number is 45 more than 8 times the number, what is the number?

40. The square of a number plus the number is 132. What is the number?

41. The length of the top of a workbench is 4 cm greater than the width. The area is 96 cm^2. Find the length and the width.

42. A flower bed is to be 3 m longer than it is wide. The flower bed will have an area of 108 m^2. What will its dimensions be?

43. The sum of the squares of two consecutive odd positive integers is 202. Find the integers.

44. The sum of the squares of two consecutive odd positive integers is 394. Find the integers.

45. If the sides of a square are lengthened by 4 cm, the area becomes 49 cm^2. Find the length of a side of the original square.

46. If the sides of a square are lengthened by 6 m, the area becomes 144 m^2. Find the length of a side of the original square.

47. The base of a triangle is 9 cm greater than the height. The area is 56 cm^2. Find the height and the base.

48. The base of a triangle is 5 cm less than the height. The area is 18 cm^2. Find the height and the base.

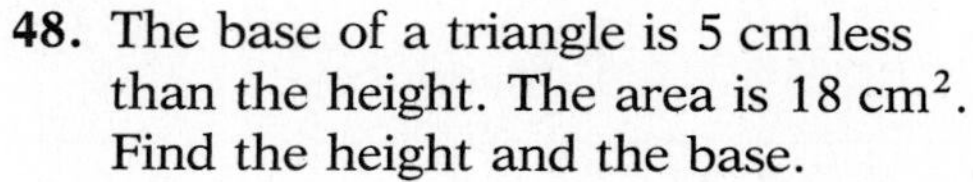

49. The perimeter of a square is 4 more than its area. Find the length of a side.

50. The area of a square is 12 more than its perimeter. Find the length of a side.

51. Find three consecutive integers such that the product of the first and third minus the second is 1 more than 10 times the third.

52. Find three consecutive integers such that four times the square of the third minus three times the square of the first minus 41 is twice the square of the second.

ANSWERS

41. ____________

42. ____________

43. ____________

44. ____________

45. ____________

46. ____________

47. ____________

48. ____________

49. ____________

50. ____________

51. ____________

52. ____________

ANSWERS

53. ______
54. ______
55. ______
56. ______
57. ______
58. ______
59. ______
60. ______
61. ______
62. ______
63. ______
64. ______
65. ______
66. ______
67. ______
68. ______
69. ______

53. One leg of a right triangle has length 9 m. The other sides have lengths that are consecutive integers. Find these lengths.

54. One leg of a right triangle is 10 cm. The other sides have lengths that are consecutive even integers. Find these lengths.

55. Suppose that an object is thrown upward with an initial velocity of 80 ft/sec from a height of 224 ft. Its height h after t seconds is given by

$$h = -16t^2 + 80t + 224.$$

After what amount of time will the object reach the ground? That is, for what positive value of t does $h = 0$?

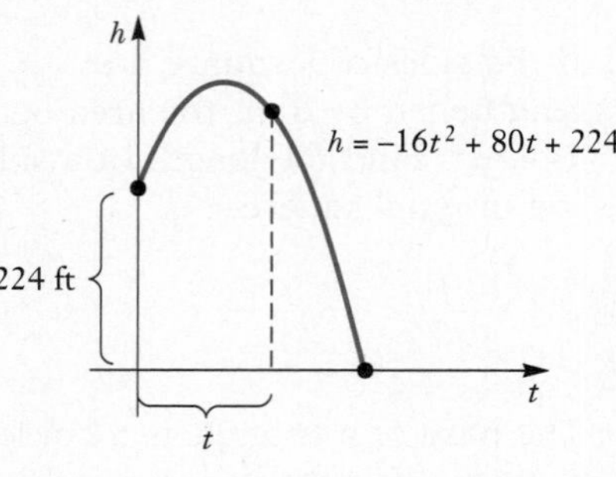

56. Suppose that an object is thrown upward with an initial velocity of 96 ft/sec from a height of 880 ft. Its height h after t seconds is given by

$$h = -16t^2 + 96t + 880.$$

After what amount of time will the object reach the ground? That is, for what positive value of t does $h = 0$?

SKILL MAINTENANCE

Solve.

57. $2x - 14y + 10z = 100,$
$5y - 8z = 80,$
$4z = 64$

58. $x + y = 0,$
$z - y = -2,$
$x - z = 6$

SYNTHESIS

Solve.

59. $x(x + 8) = 16(x - 1)$

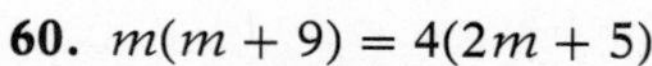

60. $m(m + 9) = 4(2m + 5)$

61. $(a - 5)^2 = 36$

62. $(x - 6)^2 = 81$

63. $(3x^2 - 7x - 20)(x - 5) = 0$

64. $3x^3 + 6x^2 - 27x - 54 = 0$

65. $2x^3 + 6x^2 = 8x + 24$

66. A square and an equilateral triangle have the same perimeter. The area of the square is 9 cm^2. What is the length of a side of the triangle?

67. The sum of two numbers is 17 and the sum of their squares is 205. Find the numbers.

68. A rectangular piece of tin is twice as long as it is wide. Squares 2 cm on a side are cut out of each corner and the ends are turned up to make a box whose volume is 480 cm^3. What are the dimensions of the piece of tin?

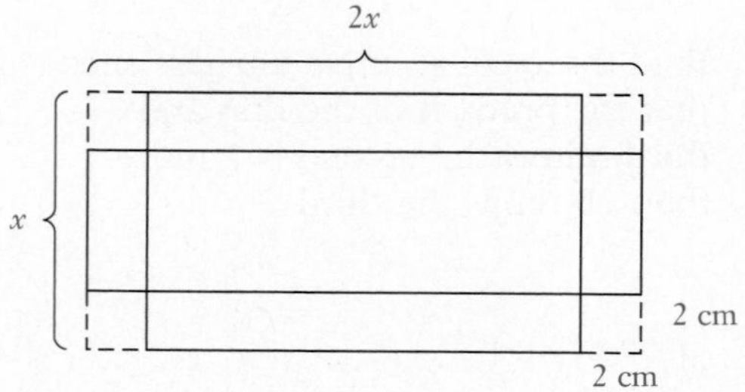

69. The hypotenuse of a right triangle is 3 cm longer than one of its legs and 6 cm more than the other leg. What is the area of the triangle?

5.9 Division of Polynomials

As we consider division of polynomials, you will see that such division is similar to what is done in arithmetic.

a Divisor a Monomial

We first consider division by a monomial. When we are dividing a monomial by a monomial, we can use our rules of exponents and subtract exponents when variables in the bases are the same. (We studied this in Section 1.6.) For example,

$$\frac{45x^{10}}{3x^4} = 15x^{10-4} = 15x^6, \qquad \frac{48a^2b^5}{-3ab^2} = \frac{48}{-3}a^{2-1}b^{5-2} = -16ab^3.$$

When we are dividing a monomial into a polynomial, we break up the division into an addition of quotients of monomials. To do so, we use the rule for addition using fractional notation in reverse. That is, since

$$\frac{A}{C} + \frac{B}{C} = \frac{A+B}{C}, \quad \text{we know that} \quad \frac{A+B}{C} = \frac{A}{C} + \frac{B}{C}.$$

▶ **EXAMPLE 1** Divide $12x^3 + 8x^2 + x + 4$ by $4x$.

$$\frac{12x^3 + 8x^2 + x + 4}{4x} \qquad \text{Writing a fractional expression}$$

$$= \frac{12x^3}{4x} + \frac{8x^2}{4x} + \frac{x}{4x} + \frac{4}{4x} \qquad \text{Doing the reverse of adding*}$$

$$= 3x^2 + 2x + \frac{1}{4} + \frac{1}{x} \qquad \text{Doing the four indicated divisions}$$ ◀

DO EXERCISE 1.

▶ **EXAMPLE 2** Divide: $(8x^4y^5 - 3x^3y^4 + 5x^2y^3) \div x^2y^3$.

$$\frac{8x^4y^5 - 3x^3y^4 + 5x^2y^3}{x^2y^3} = \frac{8x^4y^5}{x^2y^3} - \frac{3x^3y^4}{x^2y^3} + \frac{5x^2y^3}{x^2y^3}$$

$$= 8x^2y^2 - 3xy + 5$$ ◀

You should try to write only the answer.

To divide a polynomial by a monomial, we can divide each term by the monomial.

DO EXERCISES 2 AND 3.

---

* To see this, do the addition

$$\frac{12x^3}{4x} + \frac{8x^2}{4x} + \frac{x}{4x} + \frac{4}{4x}$$

and you will get the expression

$$\frac{12x^3 + 8x^2 + x + 4}{4x}.$$

OBJECTIVES

After finishing Section 5.9 you should be able to:

a Divide a polynomial by a monomial.

b Divide a polynomial by a divisor that is not a monomial, and if there is a remainder, express the result in two ways.

c Use synthetic division to divide a polynomial by a binomial of the type $x - a$.

FOR EXTRA HELP

Tape 10D | Tape 9A | MAC: 5 IBM: 5

1. Divide:

$$\frac{x^3 + 16x^2 + 6x}{2x}.$$

Divide.

2. $(15y^5 - 6y^4 + 18y^3) \div 3y^2$

3. $(x^4y^3 + 10x^3y^2 + 16x^2y) \div 2x^2y$

ANSWERS ON PAGE A-7

4. Divide and check:

$x - 2\,\overline{)\,x^2 + 3x - 10}$.

b Divisor Not a Monomial

When the divisor is not a monomial, we use a procedure very much like long division in arithmetic.

▶ **EXAMPLE 3** Divide $x^2 + 5x + 8$ by $x + 3$.

We have

$$\begin{array}{r l} x & \leftarrow \text{Divide the first term of the dividend by the first term of the divisor: } x^2/x = x. \\ x + 3\,\overline{)\,x^2 + 5x + 8} & \\ x^2 + 3x & \leftarrow \text{Multiply } x \text{ above by the divisor.} \\ \hline 2x & \leftarrow \text{Subtract: } (x^2 + 5x) - (x^2 + 3x) = x^2 + 5x - x^2 - 3x = 2x. \end{array}$$

We now "bring down" the other terms of the dividend—in this case, 8.

$$\begin{array}{r l} x + 2 & \leftarrow \text{Divide the first term by the first term: } 2x/x = 2. \\ x + 3\,\overline{)\,x^2 + 5x + 8} & \\ x^2 + 3x & \\ \hline 2x + 8 & \leftarrow \text{The 8 has been "brought down."} \\ 2x + 6 & \leftarrow \text{Multiply 2 by the divisor.} \\ \hline 2 & \leftarrow \text{Subtract: } (2x + 8) - (2x + 6) = 2x + 8 - 2x - 6 = 2. \end{array}$$

The answer is $x + 2$ with $\text{R} = 2$, or

$$x + 2 + \frac{2}{x + 3}.$$

This expression is the remainder over the divisor.

To check, we multiply the quotient by the divisor and add the remainder to see if we get the dividend.

$$\underbrace{(x + 3)}_{\text{Divisor}} \quad \underbrace{(x + 2)}_{\text{Quotient}} \quad + \quad \underbrace{2}_{\text{Remainder}} \quad = (x^2 + 5x + 6) + 2 = \underbrace{x^2 + 5x + 8}_{\text{Dividend}}$$

The answer checks. ◀

DO EXERCISE 4.

> Always remember the following when dividing polynomials:
> **1.** Arrange polynomials in descending order.
> **2.** If there are missing terms in the dividend, either write them with 0 coefficients or leave space for them.

▶ **EXAMPLE 4** Divide: $(125y^3 - 8) \div (5y - 2)$.

a)

$$\begin{array}{r} 25y^2 + 10y + 4 \\ 5y - 2\,\overline{)\,125y^3 + 0y^2 + 0y - 8} \\ 125y^3 - 50y^2 \\ \hline 50y^2 + 0y \\ 50y^2 - 20y \\ \hline 20y - 8 \\ 20y - 8 \\ \hline 0 \end{array}$$

When there are missing terms, we can write them in, as in (a), or leave space for them, as in (b).

This subtraction is $125y^3 - (125y^3 - 50y^2)$. We get $50y^2$.

ANSWER ON PAGE A-7

b)

$$
\begin{array}{r}
25y^2 + 10y + 4 \\
5y - 2\,\overline{)\,125y^3 \qquad\qquad\quad - 8}\\
\underline{125y^3 - 50y^2} \qquad\quad\\
50y^2 \qquad\quad\\
\underline{50y^2 - 20y} \qquad\\
20y - 8\\
\underline{20y - 8}\\
0
\end{array}
$$

This subtraction is $50y^2 - (50y^2 - 20y)$. We get $20y$.

The answer is $25y^2 + 10y + 4$. ◀

DO EXERCISE 5.

5. Divide and check:
$(9y^4 + 14y^2 - 8) \div (3y + 2)$.

▶ **EXAMPLE 5** Divide: $(x^4 - 9x^2 - 5) \div (x - 2)$.

$$
\begin{array}{r}
x^3 + 2x^2 - 5x - 10 \\
x - 2\,\overline{)\,x^4 \qquad - 9x^2 \qquad\qquad - 5}\\
\underline{x^4 - 2x^3} \qquad\qquad\qquad\qquad\\
2x^3 - 9x^2 \qquad\qquad\quad\\
\underline{2x^3 - 4x^2} \qquad\qquad\quad\\
- 5x^2 \qquad\qquad\quad\\
\underline{- 5x^2 + 10x} \qquad\quad\\
- 10x - 5\\
\underline{- 10x + 20}\\
- 25
\end{array}
$$

The first subtraction is $x^4 - (x^4 - 2x^3)$.

The second subtraction is $(2x^3 - 9x^2) - (2x^3 - 4x^2)$.

The answer is $x^3 + 2x^2 - 5x - 10$ with R $= -25$, or

$$x^3 + 2x^2 - 5x - 10 + \frac{-25}{x - 2}.$$ ◀

DO EXERCISES 6 AND 7.

Divide and check.

6. $(y^3 - 11y^2 + 6) \div (y - 3)$

7. $(x^3 + 9x^2 - 5) \div (x - 1)$

When dividing, we may "come out even" (have a remainder of 0) or we may not. If not, how long should we keep working? We continue until the degree of the remainder is less than the degree of the divisor, as in the next example. The answer can be given by writing the quotient and remainder or by using a fractional expression, as you saw in Example 5 and as you will see in the next example.

▶ **EXAMPLE 6** Divide: $(6x^3 + 9x^2 - 5) \div (x^2 - 2x)$.

$$
\begin{array}{r}
6x + 21 \qquad\qquad\quad\\
x^2 - 2x\,\overline{)\,6x^3 + 9x^2 + 0x - 5}\\
\underline{6x^3 - 12x^2} \qquad\qquad\quad\\
21x^2 + 0x \qquad\\
\underline{21x^2 - 42x} \qquad\\
42x - 5
\end{array}
$$

Again we have a missing term, so we can write it in.

The degree of the remainder is less than the degree of the divisor, so we are finished.

The answer is $6x + 21$ with R $= 42x - 5$, or

$$6x + 21 + \frac{42x - 5}{x^2 - 2x}.$$ ◀

DO EXERCISE 8.

8. Divide and check:
$(y^3 - 11y^2 + 6) \div (y^2 - 3)$.

ANSWERS ON PAGE A-7

C Synthetic Division

To divide a polynomial by a binomial of the type $x - a$, we can streamline the usual procedure by a process called **synthetic division.**

Compare the following. In **A** we perform a division. In **B** we do not write the variables.

A.

$$\begin{array}{r}
4x^2 + 5x + 11 \\
x - 2\,\overline{)\,4x^3 - 3x^2 + \quad x + \ 7} \\
\underline{4x^3 - 8x^2} \qquad\qquad \\
5x^2 + \quad x \qquad \\
\underline{5x^2 - 10x} \qquad \\
11x + \ 7 \\
\underline{11x - 22} \\
29
\end{array}$$

B.

$$\begin{array}{r}
4 + 5 + 11 \\
1 - 2\,\overline{)\,4 - 3 + \ 1 + \ 7} \\
\underline{4 - 8} \qquad\quad \\
5 + \ 1 \quad \\
\underline{5 - 10} \quad \\
11 + \ 7 \\
\underline{11 - 22} \\
29
\end{array}$$

In **B** there is still some duplication of writing. Also, since we can subtract by adding the opposite, we can use 2 instead of -2 and then add instead of subtracting.

C. *Synthetic Division*

a) $\begin{array}{r|rrrr} 2 & 4 & -3 & +1 & +7 \\ & & & & \\ \hline & 4 & & & \end{array}$

Write the 2 instead of -2 of the divisor $x - 2$ and the coefficients of the dividend.

Bring down the first coefficient.

b) $\begin{array}{r|rrrr} 2 & 4 & -3 & +1 & +7 \\ & & 8 & & \\ \hline & 4 & 5 & & \end{array}$

Multiply 4 by 2 to get 8. Add 8 and -3.

c) $\begin{array}{r|rrrr} 2 & 4 & -3 & +1 & +7 \\ & & 8 & 10 & \\ \hline & 4 & 5 & 11 & \end{array}$

Multiply 5 by 2 to get 10. Add 10 and 1.

d) $\begin{array}{r|rrr|r} 2 & 4 & -3 & +1 & +7 \\ & & 8 & 10 & 22 \\ \hline & 4 & 5 & 11 & 29 \end{array}$

Multiply 11 by 2 to get 22. Add 22 and 7.

Quotient Remainder

The last number, 29, is the remainder. The other numbers are the coefficients of the quotient, with that of the term of highest degree first, as follows.

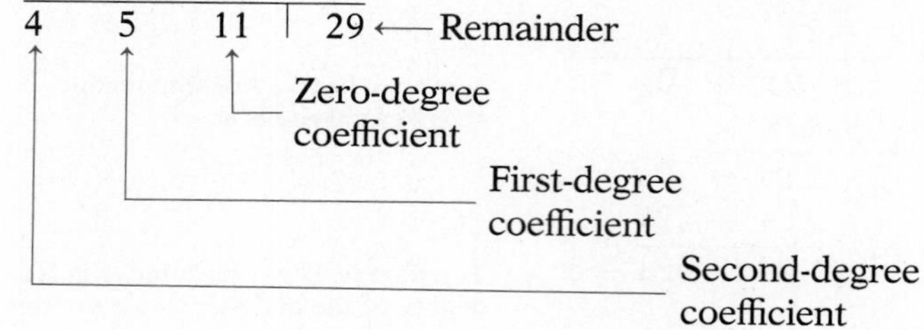

The answer is $4x^2 + 5x + 11$ with R = 29, or $4x^2 + 5x + 11 + \dfrac{29}{x - 2}$.

> It is important to remember that in order for this method to work, the divisor must be of the form $x - a$, that is, a variable minus a constant. The coefficient of the variable must be 1.

▶ **EXAMPLE 7** Use synthetic division to divide:

$$(x^3 + 6x^2 - x - 30) \div (x - 2).$$

We have

$$\begin{array}{r|rrrr} 2 & 1 & 6 & -1 & -30 \\ & & 2 & 16 & 30 \\ \hline & 1 & 8 & 15 & 0 \end{array}$$

The answer is $x^2 + 8x + 15$ with $R = 0$, or just $x^2 + 8x + 15$. ◀

DO EXERCISE 9.

When there are missing terms, be sure to write 0's for the coefficients of the missing terms.

▶ **EXAMPLES** Use synthetic division to divide.

8. $(2x^3 + 7x^2 - 5) \div (x + 3)$

There is no x-term, so we must write a 0 for the coefficient. Note that $x + 3 = x - (-3)$, so we write -3 in the left corner.

$$\begin{array}{r|rrrr} -3 & 2 & 7 & 0 & -5 \\ & & -6 & -3 & 9 \\ \hline & 2 & 1 & -3 & 4 \end{array}$$

The answer is $2x^2 + x - 3$ with $R = 4$, or $2x^2 + x - 3 + \dfrac{4}{x+3}$.

9. $(x^3 + 4x^2 - x - 4) \div (x + 4)$

$$\begin{array}{r|rrrr} -4 & 1 & 4 & -1 & -4 \\ & & -4 & 0 & 4 \\ \hline & 1 & 0 & -1 & 0 \end{array}$$

The answer is $x^2 - 1$.

10. $(x^4 - 1) \div (x - 1)$

$$\begin{array}{r|rrrrr} 1 & 1 & 0 & 0 & 0 & -1 \\ & & 1 & 1 & 1 & 1 \\ \hline & 1 & 1 & 1 & 1 & 0 \end{array}$$

The answer is $x^3 + x^2 + x + 1$.

11. $(8x^5 - 6x^3 + x - 8) \div (x + 2)$

$$\begin{array}{r|rrrrrr} -2 & 8 & 0 & -6 & 0 & 1 & -8 \\ & & -16 & 32 & -52 & 104 & -210 \\ \hline & 8 & -16 & 26 & -52 & 105 & -218 \end{array}$$

The answer is $8x^4 - 16x^3 + 26x^2 - 52x + 105$ with $R = -218$, or

$$8x^4 - 16x^3 + 26x^2 - 52x + 105 + \frac{-218}{x+2}. \quad ◀$$

DO EXERCISES 10 AND 11.

9. Use synthetic division to divide:

$(2x^3 - 4x^2 + 8x - 8) \div (x - 3)$.

Use synthetic division to divide.

10. $(x^3 - 2x^2 + 5x - 4) \div (x + 2)$

11. $(y^3 + 1) \div (y + 1)$

ANSWERS ON PAGE A-7

❖ SIDELIGHTS

Handling Dimension Symbols (Part I)*

In many applications, we add, subtract, multiply, and divide quantities having units, or dimensions, such as ft, km, sec, hr, etc. For example, to find average speed, we divide total distance by total time. What results is notation very much like a rational expression.

EXAMPLE 1 A car travels 150 km in 2 hr. What is its average speed?

$$\text{Speed} = \frac{150 \text{ km}}{2 \text{ hr}} \quad \text{or} \quad 75 \frac{\text{km}}{\text{hr}}$$

(The standard abbreviation for km/hr is km/h, but it does not suit our present discussion well.)

The symbol km/hr makes it look as if we are dividing kilometers by hours. It may be argued that we can divide only numbers. Nevertheless, we treat dimension symbols, such as km, ft, and hr, as if they were numerals or variables, obtaining correct results mechanically.

EXAMPLE 2 Compare

$$\frac{150x}{2y} = \frac{150}{2} \cdot \frac{x}{y} = 75\,\frac{x}{y}$$

with

$$\frac{150 \text{ km}}{2 \text{ hr}} = \frac{150}{2}\,\frac{\text{km}}{\text{hr}} = 75\,\frac{\text{km}}{\text{hr}}.$$

EXAMPLE 3 Compare

$$3x + 2x = (3 + 2)x = 5x$$

with

$$3 \text{ ft} + 2 \text{ ft} = (3 + 2) \text{ ft} = 5 \text{ ft}.$$

EXAMPLE 4 Compare

$$5x \cdot 3x = 15x^2$$

with

$$5 \text{ ft} \cdot 3 \text{ ft} = 15 \text{ ft}^2 \text{ (square feet)}.$$

EXAMPLE 5 Compare

$$5x \cdot 8y = 40xy$$

with

$$5 \text{ men} \cdot 8 \text{ hours} = 40 \text{ man-hours}.$$

If 5 men work 8 hours, the total amount of labor is 40 man-hours, which is the same as 4 men working 10 hours.

EXAMPLE 6 Compare

$$\frac{300x \cdot 240y}{15t} = 4800\,\frac{xy}{t}$$

with

$$\frac{300 \text{ kW} \cdot 240 \text{ hr}}{15 \text{ da}} = 4800\,\frac{\text{kW-hr}}{\text{da}}.$$

If an electrical device uses 300 kilowatts for 240 hours over a period of 15 days, its rate of usage of energy is 4800 kilowatt-hours per day. The standard abbreviation for kilowatt-hours is kWh.

These "multiplications" and "divisions" can have humorous interpretations. For example,

$$2 \text{ barns} \cdot 4 \text{ dances} = 8 \text{ barn-dances},$$

$$2 \text{ dances} \cdot 4 \text{ dances} = 8 \text{ dances}^2 \text{ (8 square dances)},$$

and

$$\text{Ice} \cdot \text{Ice} \cdot \text{Ice} = \text{Ice}^3 \text{ (Ice cubed)}.$$

However, the fact that such amusing examples exist causes us no trouble, since they do not come up in practice.

EXERCISES

Add these measures.

1. 45 ft + 23 ft
2. 55 km/hr + 27 km/hr
3. 17 g + 28 g
4. 3.4 lb + 5.2 lb

Find average speeds, given total distance and total time.

5. 90 mi, 6 hr
6. 640 km, 20 hr
7. 9.9 m, 3 sec
8. 76 ft, 4 min

Perform these calculations.

9. $\dfrac{3 \text{ in.} \cdot 8 \text{ lb}}{6 \text{ sec}}$
10. $\dfrac{60 \text{ men} \cdot 8 \text{ hr}}{20 \text{ da}}$
11. $36 \text{ ft} \cdot \dfrac{1 \text{ yd}}{3 \text{ ft}}$
12. $55\,\dfrac{\text{mi}}{\text{hr}} \cdot 4 \text{ hr}$
13. $5 \text{ ft}^3 + 11 \text{ ft}^3$
14. $\dfrac{3 \text{ lb}}{14 \text{ ft}} \cdot \dfrac{7 \text{ lb}}{6 \text{ ft}}$
15. Divide \$4850 by 5 days.
16. Divide \$25.60 by 8 hr.

---

* For more on handling dimension symbols, see p. 594.

NAME SECTION DATE

EXERCISE SET 5.9

ANSWERS

1.
2.
3.
4.
5.
6.
7.
8.
9.
10.
11.
12.
13.
14.
15.
16.
17.
18.
19.
20.

a Divide.

1. $\dfrac{30x^8 - 15x^6 + 40x^4}{5x^4}$

2. $\dfrac{24y^6 + 18y^5 - 36y^2}{6y^2}$

3. $(9x^3y^4 - 18x^2y^3 + 27xy^2) \div 9xy$

4. $(24a^4b^3 + 28a^3b^2 - 20a^2b) \div 2a^2b$

b Divide.

5. $(x^2 + 10x + 21) \div (x + 3)$

6. $(y^2 - 8y + 16) \div (y - 4)$

7. $(a^2 - 8a - 16) \div (a + 4)$

8. $(y^2 - 10y - 25) \div (y - 5)$

9. $(x^2 + 7x + 14) \div (x + 5)$

10. $(t^2 - 7t - 9) \div (t - 3)$

11. $(4y^3 + 6y^2 + 14) \div (2y + 4)$

12. $(6x^3 - x^2 - 10) \div (3x + 4)$

13. $(10y^3 + 6y^2 - 9y + 10) \div (5y - 2)$

14. $(6x^3 - 11x^2 + 11x - 2) \div (2x - 3)$

15. $(2x^4 - x^3 - 5x^2 + x - 6) \div (x^2 + 2)$

16. $(3x^4 + 2x^3 - 11x^2 - 2x + 5) \div (x^2 - 2)$

17. $(2x^5 - x^4 + 2x^3 - x) \div (x^2 - 3x)$

18. $(2x^5 + 3x^3 + x^2 - 4) \div (x^2 + x)$

c Use synthetic division to divide.

19. $(x^3 - 2x^2 + 2x - 5) \div (x - 1)$

20. $(x^3 - 2x^2 + 2x - 5) \div (x + 1)$

ANSWERS

21. ______

22. ______

23. ______

24. ______

25. ______

26. ______

27. ______

28. ______

29. ______

30. ______

31. ______

32. ______

33. ______

34. ______

35. See graph.

36. See graph.

37. ______

38. ______

39. ______

40. ______

41. ______

42. ______

43. ______

44. ______

21. $(a^2 + 11a - 19) \div (a + 4)$

22. $(a^2 + 11a - 19) \div (a - 4)$

23. $(x^3 - 7x^2 - 13x + 3) \div (x - 2)$

24. $(x^3 - 7x^2 - 13x + 3) \div (x + 2)$

25. $(3x^3 + 7x^2 - 4x + 3) \div (x + 3)$

26. $(3x^3 + 7x^2 - 4x + 3) \div (x - 3)$

27. $(y^3 - 3y + 10) \div (y - 2)$

28. $(x^3 - 2x^2 + 8) \div (x + 2)$

29. $(3x^4 - 25x^2 - 18) \div (x - 3)$

30. $(6y^4 + 15y^3 + 28y + 6) \div (y + 3)$

31. $(x^3 - 27) \div (x - 3)$

32. $(y^3 + 27) \div (y + 3)$

33. $(y^4 - 16) \div (y - 2)$

34. $(x^5 - 32) \div (x - 2)$

SKILL MAINTENANCE

Graph.

35. $2x - 3y = 6$

36. $x = -2$

Solve.

37. $35t^2 + 18t = 8$

38. $16x^2 + 56x + 49 = 0$

SYNTHESIS

Divide.

39. $(2x^6 + 5x^4 - x^3 + 1) \div (x^2 + x + 1)$

40. $(2y^5 + y^3 - 2y - 3) \div (y^2 - 3y + 1)$

41. $(x^4 - x^3y + x^2y^2 + 2x^2y - 2xy^2 + 2y^3) \div (x^2 - xy + y^2)$

42. $(x^4 - y^4) \div (x - y)$

43. $(a^7 + b^7) \div (a + b)$

44. Find k so that when $x^3 - kx^2 + 3x + 7k$ is divided by $x + 2$, the remainder will be 0.

SUMMARY AND REVIEW: CHAPTER 5

IMPORTANT PROPERTIES AND FORMULAS

Factoring Formulas: $A^2 - B^2 = (A + B)(A - B)$, $A^2 + 2AB + B^2 = (A + B)^2$, $A^2 - 2AB + B^2 = (A - B)^2$, $A^3 + B^3 = (A + B)(A^2 - AB + B^2)$, $A^3 - B^3 = (A - B)(A^2 + AB + B^2)$

The Principle of Zero Products: For any real numbers a and b: If $ab = 0$, then $a = 0$ or $b = 0$. If $a = 0$ or $b = 0$, then $ab = 0$.

REVIEW EXERCISES

The review sections and objectives to be tested in addition to the material in this chapter are [4.4a], [4.5a], [3.2a], and [2.6e].

1. Evaluate $P = x^3 - x^2 + 4x$ when $x = 0$ and $x = -1$.

2. Evaluate $P = 4 - 2x - x^2$ when $x = -2$ and $x = 5$.

3. Given the polynomial $3x^6y - 7x^8y^3 + 2x^3 - 3x^2$:

a) Determine the degree of each term and the degree of the polynomial.

b) Determine the leading term and the leading coefficient.

c) Arrange in ascending powers of x.

d) Arrange in descending powers of y.

Collect like terms.

4. $4x^2y - 3xy^2 - 5x^2y + xy^2$

5. $3ab - 10 + 5ab^2 - 2ab + 7ab^2 + 14$

Add, subtract, or multiply.

6. $(-6x^3 - 4x^2 + 3x + 1) + (5x^3 + 2x + 6x^2 + 1)$

7. $(4x^3 - 2x^2 - 7x + 5) + (8x^2 - 3x^3 - 9 + 6x)$

8. $(-9xy^2 - xy + 6x^2y) + (-5x^2y - xy + 4xy^2) + (12x^2y - 3xy^2 + 6xy)$

9. $(3x - 5) - (-6x + 2)$

10. $(4a - b + 3c) - (6a - 7b - 4c)$

11. $(9p^2 - 4p + 4) - (-7p^2 + 4p + 4)$

12. $(6x^2 - 4xy + y^2) - (2y^2 + 3xy - 2y^2)$

13. $(3x^2y)(-6xy^3)$

14. $(x^4 - 2x^2 + 3)(x^4 + x^2 - 1)$

15. $(4ab + 3c)(2ab - c)$

16. $(2x + 5y)(2x - 5y)$

17. $(2x - 5y)^2$

18. $(5x^2 - 7x + 3)(4x^2 + 2x - 9)$

19. $(x^2 + 4y^3)^2$

20. $(x - 5)(x^2 + 5x + 25)$

21. $(x - \frac{1}{3})(x - \frac{1}{6})$

Factor.

22. $9y^4 - 3y^2$

23. $15x^4 - 18x^3 + 21x^2 - 9x$

24. $a^2 - 12a + 27$

25. $3m^2 + 14m + 8$

26. $25x^2 + 20x + 4$

27. $4y^2 - 16$

28. $ax + 2bx - ay - 2by$

29. $4x^4 + 4x^2 + 20$

30. $27x^3 - 8$

31. $0.064b^3 - 0.125c^3$

32. $y^5 - y$

33. $2z^8 - 16z^6$

34. $54x^6y - 2y$

35. $1 + a^3$

36. $36x^2 - 120x + 100$

37. $6t^2 + 17pt + 5p^2$

38. $x^3 + 2x^2 - 9x - 18$

39. $a^2 - 2ab + b^2 - 4t^2$

Solve.

40. $x^2 - 20x = -100$

41. $6b^2 - 13b + 6 = 0$

42. $8y^2 + 5 = 14y$

43. The area of a square is five more than four times the length of a side. What is the length of a side of the square?

44. The sum of the squares of three consecutive odd numbers is 83. Find the numbers.

45. A photograph is 3 in. longer than it is wide. When a 2-in. border is placed around the photograph, the total area of the photograph and the border is 108 in^2. Find the dimensions of the photograph.

Divide.

46. $(20r^2s^3 - 15r^2s^2 - 10r^3s^3) \div 5r^2s$

47. $(y^3 - 64) \div (y - 4)$

48. $(4x^3 + 3x^2 - 5x - 2) \div (x^2 + 1)$

Divide using synthetic division. Show your work.

49. $(x^3 + 3x^2 + 2x - 6) \div (x - 3)$

50. $(4x^3 + 6x^2 - 5) \div (x + 3)$

SKILL MAINTENANCE

51. Solve: $\begin{aligned} 3x + 2y + z &= 3, \\ 2x - y + 2z &= 16, \\ x + y - z &= -9. \end{aligned}$

52. Graph: $5y - 4x = -12$.

Solve.

53. $|10 - 3x| \leq 14$

54. $|10 - 3x| \geq 14$

55. There are three machines A, B, and C in a factory. When all three work, they produce 287 screws per hour. When only A and C work, they produce 197 screws per hour. When only A and B work, they produce 202 screws per hour. How many screws per hour can each produce alone?

SYNTHESIS

56. Factor: $128x^6 - 2y^6$.

57. Solve: $64x^3 = x$.

❖ THINKING IT THROUGH

Explain the error in each of the following.

1. $(a + 3)^2 = a^2 + 9$
2. $a^3 + b^3 = (a + b)(a^2 - 2ab + b^2)$
3. $(a - b)(a - b) = a^2 - b^2$
4. $(x + 3)(x - 4) = x^2 - 12$
5. $(p + 7)(p - 7) = p^2 + 49$
6. $(t - 3)^2 = t^2 - 9$
7. What law of real numbers is most important to our learning quick ways to multiply and factor polynomials?
8. In this chapter, we learned to solve equations that we could not have solved in Chapter 2. What new kinds of equations are we able to solve, how do the solutions differ from before, and how is the procedure different?

NAME SECTION DATE

TEST: CHAPTER 5

1. Evaluate $P = 2x^3 + 3x^2 - x + 4$ when $x = 0$ and $x = -2$.

Given the polynomial $3xy^3 - 4x^2y + 5x^5y^4 - 2x^4y$:

2. Determine the degree of the polynomial.

3. Determine the leading term.

4. Arrange in descending powers of x.

5. Collect like terms: $5xy - 2xy^2 - 2xy + 5xy^2$

Add.

6. $-6x^3 + 3x^2 - 4y$ and $3x^3 - 2y - 7y^2$

7. $4a^3 - 2a^2 + 6a - 5$ and $3a^3 - 3a + 2 - 4a^2$

8. $5m^3 - 4m^2n - 6mn^2 - 3n^3$ and $9mn^2 - 4n^3 + 2m^3 + 6m^2n$

Subtract.

9. $(9a - 4b) - (3a + 4b)$

10. $(4x^2 - 3x + 7) - (-3x^2 + 4x - 6)$

11. $(6y^2 - 2y - 5y^3) - (4y^2 - 7y - 6y^3)$

Multiply.

12. $(-4x^2y)(-16xy^2)$

13. $(6a - 5b)(2a + b)$

14. $(x - y)(x^2 - xy - y^2)$

15. $(3m^2 + 4m - 2)(-m^2 - 3m + 5)$

16. $(4y - 9)^2$

17. $(x - 2y)(x + 2y)$

ANSWERS

1. ____
2. ____
3. ____
4. ____
5. ____
6. ____
7. ____
8. ____
9. ____
10. ____
11. ____
12. ____
13. ____
14. ____
15. ____
16. ____
17. ____

ANSWERS

18. ______
19. ______
20. ______
21. ______
22. ______
23. ______
24. ______
25. ______
26. ______
27. ______
28. ______
29. ______
30. ______
31. ______
32. ______
33. ______
34. ______
35. ______
36. ______
37. ______
38. ______
39. ______
40. ______
41. ______
42. ______
43. a) ______
b) ______
44. ______

Factor.

18. $9x^2 + 7x$

19. $24y^3 + 16y^2$

20. $y^3 + 5y^2 - 4y - 20$

21. $p^2 - 12p - 28$

22. $12m^2 + 20m + 3$

23. $9y^2 - 25$

24. $3r^3 - 3$

25. $9x^2 + 25 - 30x$

26. $(z + 1)^2 - b^2$

27. $x^8 - y^8$

28. $y^2 + 8y + 16 - 100t^2$

29. $20a^2 - 5b^2$

30. $24x^2 - 46x + 10$

31. $16a^7b + 54ab^7$

Solve.

32. $x^2 - 18 = 3x$

33. $5y^2 - 125 = 0$

34. $2x^2 + 21 = -17x$

35. A photograph is 3 cm longer than it is wide. Its area is 40 cm^2. Find the length and the width.

36. Divide: $(x^2 - 10x - 25) \div (x - 3)$.

37. Divide using synthetic division:
$(6y^4 + 15y^3 - 28y + 6) \div (y + 2)$.

SKILL MAINTENANCE

Solve.

38. $|3x + 8| < 10$

39. $|3x + 8| > 10$

40. Graph: $x - 3y = -6$.

41. Solve:

$$\begin{aligned} 2x - y + z &= 9, \\ x - y + z &= 4, \\ x + 2y - z &= 5. \end{aligned}$$

42. There are 70 questions on a test. The questions are either multiple-choice, true–false, or fill-in. There are twice as many true–false as fill-in and five more multiple-choice than true–false. How many of each type of question are there on the test?

SYNTHESIS

43. **a)** Multiply: $(x^2 + x + 1)(x^3 - x^2 + 1)$.
b) Factor: $x^5 + x + 1$.

44. Factor: $6x^{2n} - 7x^n - 20$.

CUMULATIVE REVIEW: CHAPTERS 1–5

1. Evaluate $\frac{2m - n}{4}$ when $m = 3$ and $n = 2$.

Simplify.

2. $|0|$

3. $-8 - (-4)$

4. $-3a + (6a - 1) - 4(a + 1)$

5. $2[5(3x + 4) - 2x] - [7(3x + 4) - 8]$

6. $[(-3a^{-6}b^2)^5]^{-2}$

7. $(x^2 + 4x - xy - 9) + (-3x^2 - 3x + 8)$

8. $(6x^2 - 3x + 2x^3) - (8x^2 - 9x + 2x^3)$

9. $(a^2 - a - 3) \cdot (a^2 + 2a - 3)$

10. $(x + 4)(x + 9)$

Solve.

11. $8 - 3x = 6x - 10$

12. $\frac{1}{2}x - 3 = \frac{7}{2}$

13. $A = \frac{1}{2}h(a + b)$ for b

14. $6x - 1 \leq 3(5x + 2)$

15. $4x - 3 < 2$ or $x - 3 > 1$

16. $|2x - 3| < 7$

17. $x + y + z = -5,$ $x - z = 10,$ $y - z = 12$

18. $2x + 5y = -2,$ $5x + 3y = 14$

19. $3x - y = 7,$ $2x + 2y = 5$

20. $x + 2y - z = 0,$ $3x + y - 2z = -1,$ $x - 4y + z = -2$

Solve using Cramer's rule. Show your work.

21. $2x + 3y = 2,$ $-4x + 3y = -7$

22. $-3a + 2b = 0,$ $3a - 4b = -1$

23. $2x + 2y - 4z = 1,$ $-2x - 4y + 8z = -1,$ $4x + 4y + 4z = 5$

Solve.

24. $11x + x^2 + 24 = 0$

25. $2x^2 - 15x = -7$

26. Evaluate: $\begin{vmatrix} 3 & 2 & -4 \\ -1 & 3 & 0 \\ 0 & 1 & 5 \end{vmatrix}$.

Factor.

27. $3x^3 - 12x^2$

28. $2x^4 + x^3 + 2x + 1$

29. $x^2 + 5x - 14$

30. $20a^2 - 23a + 6$

31. $4x^2 - 25$

32. $2x^2 - 28x + 98$

33. $a^3 + 64$

34. $8x^3 - 1$

35. $4a^3 + a^6 - 12$

36. $4x^4y^2 - x^2y^4$

Graph.

37. $y = -2x$

38. $y = \frac{1}{2}x$

39. $4y + 3x = 12 + 3x$

40. $6y + 24 = 0$

41. $y > x + 6$

42. $2x + y \leq 2$

Graph. List the vertices.

43. $y \geq -x,$
$y \leq 2x + 1$

44. $2x + 3y \leq 6,$
$5x - 5y \leq 15,$
$x \geq 0$

45. Write an equation of the line containing the point (3, 7) and parallel to the line $x + 2y = 6$.

46. Write an equation of the line containing the point (3, −2) and perpendicular to the line $3x + 4y = 5$.

47. Find an equation of the line containing the points (−1, 4) and (−1, 0).

48. Find an equation of the line with slope −3 and through the point (2, 1).

49. A family with twins did an experiment on the cost of diapering their two children. Disposables were used on one child and cloth diapers from a diaper service were used on the other child. After the twins were out of diapers, calculations showed that for both children $2886 was spent on diapering. Using disposables cost $546 more than using the diaper service. How much did each cost?

50. When filing your income tax, you may claim a casualty loss only if it exceeds 10% of your adjusted gross income plus $100. If the loss was $1500, can you claim it if your adjusted gross income is $13,500?

51. Caracas, Venezuela, has the most inexpensive gas in the world. The total cost of 10 gal of gas in Caracas and 5 gal of gas in Lagos, Nigeria, is $2.90. The difference in the price per gallon in each location is $0.16. What is the price per gallon in each city?

52. In a factory there are three polishing machines, A, B, and C. When all three of them are working, 5700 lenses can be polished in one week. When only A and B are working, 3400 lenses can be polished in one week. When only B and C are working, 4200 lenses can be polished in one week. How many lenses can be polished in a week by each machine?

SYNTHESIS

53. Solve: $|x + 1| \leq |x - 3|$.

54. A piece of copper pipe has a length of 100 cm at 18°C. At 20°C the length of pipe changes to 100.00356 cm. Find the length of pipe at 40°C and at 0°C.

55. Bert and Sally Ng have been mathematics professors at a state university. Together they have 46 years of service. Two years ago, Bert had taught 2.5 times as many years as Sally. How long has each taught at the university?

56. Debbie bought three record albums. She paid full price for the first, and then received a 10% discount on the second and a 20% discount on the third. The second album's discount price was twice the price of the first album. The third album's original price was twice the discount price of the second. Debbie paid a total of $27.90 for the three albums. What did each cost originally?

INTRODUCTION A rational expression is the ratio of two polynomials. In this chapter, we learn to manipulate rational expressions and solve rational equations. The result will be an ability to solve problems like the one below and others that we could not have solved before.

The review sections to be tested in addition to the material in this chapter are 4.3, 4.6, 5.4, and 5.6. ❖

Rational Expressions and Equations

AN APPLICATION

A boat travels 246 mi downstream in the same time that it takes to travel 90 mi upstream. The speed of the current in the stream is 5.5 mph. Find the speed of the boat in still water.

THE MATHEMATICS

Let $b =$ the speed of the boat in still water. Then the problem translates to the following equation.

These are rational expressions.

$$\underbrace{\frac{246}{b+5.5} = \frac{90}{b-5.5}}$$

This is a rational equation.

❖ POINTS TO REMEMBER: CHAPTER 6

Motion Formulas: $d = rt, \quad t = \frac{d}{r}, \quad r = \frac{d}{t}$

Equation-Solving Skills: Section 2.1
Formula-Solving Skills: Section 2.3

PRETEST: CHAPTER 6

1. Find the LCM for $x^3 - 2x^2$ and $2x^2 - 6x + 4$.

Perform the indicated operations and simplify.

2. $\frac{5}{y^2 + 8y + 16} - \frac{4}{y^2 + 9y + 20}$

3. $\frac{a - 1}{1 - a} + \frac{3a + 3}{a^2 - 1}$

4. $\frac{2}{y + 8} + \frac{24}{y^2 - 64} - \frac{3}{2y - 16}$

5. $\frac{y^2 - 8y - 33}{y^2 - 6y + 8} \div \frac{y^2 + y - 6}{y^2 - 15y + 44}$

6. $\frac{y^2 + 6y + 9}{3y^2 - 27} \cdot \frac{2y - 6}{y^2 + 4y + 3}$

7. Simplify: $\dfrac{4 - \frac{4}{y}}{4 + \frac{4}{y}}$.

Solve.

8. $\frac{x}{11} + \frac{x}{6} = 2$

9. $\frac{3}{y + 7} - \frac{6}{y^2 + 10y + 21} = \frac{4}{y + 3}$

10. Solve for b: $M = \dfrac{ab + c}{ab}$.

11. Find an equation of variation in which y varies jointly as x and the square of z and inversely as w, and $y = 1$ when $x = 5$, $z = 4$, and $w = 20$.

12. The amount of concrete necessary to construct a patio varies directly as the area of the patio. It takes 4.5 yd^3 of concrete to construct a 200-ft^2 patio. How many cubic yards of concrete will be needed to construct a 250-ft^2 patio?

13. Airplane A travels 50 km/h faster than airplane B. Airplane A travels 875 km in the same time that it takes airplane B to travel 700 km. Find the speed of each plane.

14. One typist can type a report in 4 hr. Another typist can type the same report in 3 hr. How long would it take them to type the same report working together?

6.1 Multiplying, Dividing, and Simplifying Rational Expressions

a Rational Expressions and Replacements

Rational numbers are quotients of integers. The following are rational numbers:

$$\frac{4}{5}, \quad \frac{-7}{8}, \quad \frac{12}{-5}, \quad \frac{634}{1}.$$

The following are called **rational expressions,** or **fractional expressions.** They are quotients of polynomials.

$$\frac{7}{8}, \quad \frac{8}{y+5}, \quad \frac{t^4-5t}{t^2-3t-28}, \quad \frac{x^2+7xy-4}{x^3-y^3}$$

Rational expressions indicate division. For example,

$$\frac{7}{8} \text{ means } 7 \div 8 \quad \text{and} \quad \frac{2x+y}{x-y} \text{ means } (2x+y) \div (x-y).$$

Because division is indicated by a rational expression, replacement of a variable by a number that makes the denominator 0 cannot be allowed. For example, in

$$\frac{8}{x+5},$$

-5 is not a meaningful replacement because it allows the denominator to be 0. The meaningful replacements are all those real numbers except -5; that is, the replacements are restricted to all x such that $x \neq -5$.

▶ **EXAMPLE 1** Find the meaningful replacements in

$$\frac{t^4-5t}{t^2-3t-28}.$$

The meaningful replacements are all those real numbers for which the denominator is not 0. To find them, we first find those that do make the denominator 0. We set the denominator equal to 0 and solve:

$$t^2-3t-28=0$$
$$(t-7)(t+4)=0 \quad \text{Factoring}$$
$$t-7=0 \quad \text{or} \quad t+4=0 \quad \text{Principle of zero products}$$
$$t=7 \quad \text{or} \quad t=-4.$$

The meaningful replacements are all real numbers except 7 and -4; that is, the replacements are restricted to all t such that $t \neq 7$ and $t \neq -4$. ◀

DO EXERCISES 1 AND 2.

OBJECTIVES

After finishing Section 6.1, you should be able to:

a Find the meaningful replacements in a rational expression.

b Multiply a rational expression by 1, using an expression like A/A.

c Simplify rational expressions.

d Multiply rational expressions and simplify.

e Divide rational expressions and simplify.

FOR EXTRA HELP

Tape 11A | Tape 9B | MAC: 6 IBM: 6

Find the meaningful replacements in the expression.

1. $\dfrac{y^2-4y+9}{2y+5}$

2. $\dfrac{x^2-9}{x^2-7x+10}$

ANSWERS ON PAGE A-7

Multiply.

3. $\frac{3x+2y}{5x+4y} \cdot \frac{x}{x}$

4. $\frac{2x^2-y}{3x+4} \cdot \frac{3x+2}{3x+2}$

5. $\frac{-1}{-1} \cdot \frac{2a-5}{a-b}$

ANSWERS ON PAGE A-7

b Multiplying by 1

To multiply two rational expressions, multiply numerators and multiply denominators.

For example,

$$\frac{x+3}{y-4} \cdot \frac{x^3}{y+5} = \frac{(x+3)x^3}{(y-4)(y+5)}.$$ **Multiplying numerators and multiplying denominators**

Note that we do not carry out the multiplications, because it is easier to simplify if we do not. In order to learn to simplify, we must first consider multiplying by 1. Recall that 1 is the identity for multiplication—multiplying any number a by 1 gives the same number a.

Any rational expression with the same numerator and denominator names the number 1:

$$\frac{y+5}{y+5}, \quad \frac{4x^2-5}{4x^2-5}, \quad \frac{-1}{-1}.$$ **All name the number 1 for all meaningful replacements.**

We can multiply by 1 to get equivalent expressions. For example, let us multiply $(x+y)/5$ by 1:

$$\frac{x+y}{5} \cdot \frac{x-y}{x-y} = \frac{(x+y)(x-y)}{5(x-y)}.$$ **Multiplying by $\frac{x-y}{x-y}$, which is 1**

We know that

$$\frac{x+y}{5} \quad \text{and} \quad \frac{(x+y)(x-y)}{5(x-y)}$$

are equivalent. This means that they will name the same number for all replacements that do not make a denominator 0.

► **EXAMPLES** Multiply to obtain equivalent expressions.

2. $\frac{x^2+3}{x-1} \cdot \frac{x+1}{x+1} = \frac{(x^2+3)(x+1)}{(x-1)(x+1)}$

3. $\frac{-1}{-1} \cdot \frac{x-4}{x-y} = \frac{-1 \cdot (x-4)}{-1 \cdot (x-y)}$ ◄

DO EXERCISES 3–5.

c Simplifying Rational Expressions

We can simplify rational expressions by "removing" factors that are equal to 1. We first factor the numerator and the denominator and then factor the rational expression, so that a factor is equal to 1.

▶ **EXAMPLE 4** Simplify by removing a factor equal to 1: $\frac{120}{320}$.

$\frac{120}{320} = \frac{40 \cdot 3}{40 \cdot 8}$ **Factoring the numerator and the denominator**

$= \frac{40}{40} \cdot \frac{3}{8}$ **Factoring the rational expression**

$= 1 \cdot \frac{3}{8}$ **$\frac{40}{40} = 1$**

$= \frac{3}{8}$ **Removing a factor of 1** ◀

DO EXERCISE 6.

▶ **EXAMPLES** Simplify by removing a factor equal to 1.

5. $\frac{5x^2}{x} = \frac{5x \cdot x}{1 \cdot x}$ **Factoring the numerator and the denominator**

$= \frac{5x}{1} \cdot \frac{x}{x}$ **Factoring the rational expression**

$= 5x \cdot 1$ **$\frac{x}{x} = 1$ for all meaningful replacements**

$= 5x$ **Removing a factor of 1**

In this example we supplied a 1 in the denominator. This can always be done, but it is not necessary.

6. $\frac{4a + 8}{2} = \frac{2(2a + 4)}{2 \cdot 1}$ **Factoring the numerator and the denominator**

$= \frac{2}{2} \cdot \frac{2a + 4}{1}$ **Factoring the rational expression**

$= \frac{2a + 4}{1}$ **Removing a factor of 1**

$= 2a + 4$

A common error here is to get $2a + 8$. This happens if you take an incorrect shortcut. You should go through the *entire* process of removing a factor of 1. ◀

DO EXERCISES 7 AND 8.

▶ **EXAMPLES** Simplify by removing a factor equal to 1.

7. $\frac{2x^2 + 4x}{6x^2 + 2x} = \frac{2x(x + 2)}{2x(3x + 1)}$ **Factoring the numerator and the denominator**

$= \frac{2x}{2x} \cdot \frac{x + 2}{3x + 1}$ **Factoring the rational expression**

$= \frac{x + 2}{3x + 1}$ **Removing a factor of 1**

8. $\frac{x^2 - 1}{2x^2 - x - 1} = \frac{(x - 1)(x + 1)}{(2x + 1)(x - 1)}$ **Factoring the numerator and the denominator**

$= \frac{x - 1}{x - 1} \cdot \frac{x + 1}{2x + 1}$ **Factoring the rational expression**

$= \frac{x + 1}{2x + 1}$ **Removing a factor of 1**

---

6. Simplify by removing a factor of 1:

$\frac{128}{160}$.

Simplify by removing a factor of 1.

7. $\frac{7x^2}{x}$

8. $\frac{6a + 9}{3}$

ANSWERS ON PAGE A-7

Simplify by removing a factor of 1.

9. $\dfrac{6x^2 + 4x}{4x^2 + 8x}$

10. $\dfrac{2y^2 + 6y + 4}{y^2 - 1}$

ANSWERS ON PAGE A-7

9. $\dfrac{9x^2 + 6xy - 3y^2}{12x^2 - 12y^2} = \dfrac{3(x + y)(3x - y)}{3(4)(x + y)(x - y)}$ Factoring the numerator and the denominator

$= \dfrac{3(x + y)}{3(x + y)} \cdot \dfrac{3x - y}{4(x - y)}$ Factoring the rational expression

$= \dfrac{3x - y}{4(x - y)}$ Removing a factor of 1

For purposes of later work, we usually do not multiply out the numerator and the denominator. ◀

CANCELING. Canceling is a shortcut that you may have used for removing a factor of 1 when working with fractional notation or rational expressions. With great concern, we mention it as a possibility of speeding up your work here. Canceling may be done for removing factors of 1 only in products. It may *not* be done in sums or when adding expressions together. Our concern is that canceling be done with care and understanding. Example 9 might have been done faster as follows:

$\dfrac{9x^2 + 6xy - 3y^2}{12x^2 - 12y^2} = \dfrac{\cancel{3}\cancel{(x + y)}(3x - y)}{\cancel{3}(4)\cancel{(x + y)}(x - y)}$ When a factor of 1 is noted, it is "canceled" as shown.

$= \dfrac{3x - y}{4(x - y)}.$ Removing a factor of 1: $\dfrac{3(x + y)}{3(x + y)} = 1$

CAUTION! The difficulty with canceling is that it can be applied incorrectly in situations such as the following:

$\dfrac{2 + 3}{\cancel{2}} = 3,$ $\dfrac{\cancel{4} + 1}{\cancel{4} + 2} = \dfrac{1}{2},$ $\dfrac{1\cancel{5}}{\cancel{5}4} = \dfrac{1}{4}.$

Wrong! Wrong! Wrong!

In each of these situations, the expressions canceled were *not* factors of 1. Factors are parts of products. For example, in $2 \cdot 3$, 2 and 3 are factors, but in $2 + 3$, 2 and 3 are *not* factors. **If you can't factor, you can't cancel! If in doubt, don't cancel!**

DO EXERCISES 9 AND 10.

d Multiplying and Simplifying

After multiplying, you should usually simplify if possible.

► **EXAMPLES** Multiply. Then simplify by removing a factor of 1.

10. $\dfrac{x + 2}{x - 3} \cdot \dfrac{x^2 - 4}{x^2 + x - 2} = \dfrac{(x + 2)(x^2 - 4)}{(x - 3)(x^2 + x - 2)}$ Multiplying the numerators and the denominators

$= \dfrac{(x + 2)(x + 2)(x - 2)}{(x - 3)(x + 2)(x - 1)}$ Factoring the numerators and the denominators

$= \dfrac{(x + 2)\cancel{(x + 2)}(x - 2)}{(x - 3)\cancel{(x + 2)}(x - 1)}$ Removing a factor of 1: $\dfrac{x + 2}{x + 2} = 1$

$= \dfrac{(x + 2)(x - 2)}{(x - 3)(x - 1)}$ Simplifying

We may not always multiply out the numerator and the denominator.

11. $\frac{a^3 - b^3}{a^2 - b^2} \cdot \frac{a^2 + 2ab + b^2}{a^2 + ab + b^2}$

$= \frac{(a^3 - b^3)(a^2 + 2ab + b^2)}{(a^2 - b^2)(a^2 + ab + b^2)}$

$= \frac{(a - b)(a^2 + ab + b^2)(a + b)(a + b)}{(a - b)(a + b)(a^2 + ab + b^2) \cdot 1}$ **Factoring the numerators and the denominators**

$= \frac{\cancel{(a - b)}\cancel{(a^2 + ab + b^2)}\cancel{(a + b)}(a + b)}{\cancel{(a - b)}\cancel{(a + b)}\cancel{(a^2 + ab + b^2)} \cdot 1}$ **Removing a factor of 1:** $\frac{(a - b)(a^2 + ab + b^2)(a + b)}{(a - b)(a^2 + ab + b^2)(a + b)} = 1$

$= \frac{a + b}{1}$ **Simplifying**

$= a + b$ ◀

DO EXERCISES 11 AND 12.

e Dividing and Simplifying

Two expressions are reciprocals (or multiplicative inverses) of each other if their product is 1. To find the reciprocal of a rational expression, we interchange the numerator and the denominator.

The reciprocal of $\frac{x + 2y}{x + y - 1}$ is $\frac{x + y - 1}{x + 2y}$.

The reciprocal of $y - 8$ is $\frac{1}{y - 8}$.

DO EXERCISES 13–15.

We divide with rational expressions the way we did in arithmetic. We multiply by the reciprocal of the divisor.

We sometimes say, "invert the divisor and multiply."

▶ **EXAMPLES** Divide. Simplify by removing a factor of 1 if possible.

12. $\frac{x - 2}{x + 1} \div \frac{x + 5}{x - 3} = \frac{x - 2}{x + 1} \cdot \frac{x - 3}{x + 5}$ **Multiplying by the reciprocal of the divisor**

$= \frac{(x - 2)(x - 3)}{(x + 1)(x + 5)}$

Multiply. Simplify by removing a factor of 1 if possible.

11. $\frac{(x - y)^3}{x + y} \cdot \frac{3x + 3y}{x^2 - y^2}$

12. $\frac{a^3 + b^3}{a^2 - b^2} \cdot \frac{a^2 - 2ab + b^2}{a^2 - ab + b^2}$

Find the reciprocal.

13. $\frac{x + 3}{x - 5}$

14. $x + 7$

15. $\frac{1}{y^3 - 9}$

ANSWERS ON PAGE A-7

Divide. Simplify by removing a factor of 1 if possible.

16. $\dfrac{x^2+7x+10}{2x-4} \div \dfrac{x^2-3x-10}{x-2}$

17. $\dfrac{a^2-b^2}{ab} \div \dfrac{a^2-2ab+b^2}{2a^2b^2}$

18. Perform the indicated operations and simplify:

$$\left[\frac{a^3+8}{a-2} \div (a^2-2a+4)\right] \cdot (a-2)^2$$

ANSWERS ON PAGE A-7

13. $\dfrac{a^2-1}{a-1} \div \dfrac{a^2-2a+1}{a+1}$

$= \dfrac{a^2-1}{a-1} \cdot \dfrac{a+1}{a^2-2a+1}$ **Multiplying by the reciprocal of the divisor**

$= \dfrac{(a^2-1)(a+1)}{(a-1)(a^2-2a+1)}$ **Multiplying the numerators and the denominators**

$= \dfrac{(a+1)(a-1)(a+1)}{(a-1)(a-1)(a-1)}$ **Factoring the numerator and the denominator**

$= \dfrac{(a+1)\cancel{(a-1)}(a+1)}{(a-1)\cancel{(a-1)}(a-1)}$ **Removing a factor of 1: $\dfrac{a-1}{a-1}=1$**

$= \dfrac{(a+1)(a-1)}{(a-1)(a-1)}$ **Simplifying** ◀

DO EXERCISES 16 AND 17.

▶ **EXAMPLE 14** Perform the indicated operations and simplify:

$$\left[\frac{c^3-d^3}{(c+d)^2} \div (c-d)\right] \cdot (c+d).$$

We have

$$\left[\frac{c^3-d^3}{(c+d)^2} \div (c-d)\right] \cdot (c+d)$$

$$= \frac{c^3-d^3}{(c+d)^2} \cdot \frac{1}{c-d} \cdot (c+d)$$

$$= \frac{(c-d)(c^2+cd+d^2)(c+d)}{(c+d)(c+d)(c-d)}$$

$$= \frac{\cancel{(c-d)}(c^2+cd+d^2)\cancel{(c+d)}}{(c+d)\cancel{(c+d)}\cancel{(c-d)}} \qquad \frac{(c-d)(c+d)}{(c-d)(c+d)} = 1$$

$$= \frac{c^2+cd+d^2}{c+d}.$$ ◀

Keep in mind that the procedures we learn in this chapter are by their nature rather long. It may help you to write out lots of steps as you do the problems. If you have difficulty, consider taking a clean sheet of paper and starting over.

DO EXERCISE 18.

NAME SECTION DATE

EXERCISE SET 6.1

a Find the meaningful replacements in the expression.

1. $\dfrac{3a^2-16}{5a-20}$

2. $\dfrac{t^2+t+102}{3t-39}$

3. $\dfrac{y^2-y-2}{y^2-15y+54}$

4. $\dfrac{x^3-x^2+x+2}{x^2+11x+28}$

b Multiply to obtain equivalent expressions. Do not simplify.

5. $\dfrac{3x}{3x}\cdot\dfrac{x+1}{x+3}$

6. $\dfrac{4-y^2}{6-y}\cdot\dfrac{-1}{-1}$

7. $\dfrac{t-3}{t+2}\cdot\dfrac{t+3}{t+3}$

8. $\dfrac{p-4}{p-5}\cdot\dfrac{p+5}{p+5}$

c Simplify by removing factors of 1.

9. $\dfrac{9y^2}{15y}$

10. $\dfrac{6x^3}{18x^2}$

11. $\dfrac{16p^3}{24p^7}$

12. $\dfrac{48t^5}{56t^{11}}$

13. $\dfrac{2a-6}{2}$

14. $\dfrac{3a-6}{3}$

15. $\dfrac{6x+9}{24}$

16. $\dfrac{30a-25}{45}$

17. $\dfrac{4y-12}{4y+12}$

18. $\dfrac{8x+16}{8x-16}$

ANSWERS

1. ______

2. ______

3. ______

4. ______

5. ______

6. ______

7. ______

8. ______

9. ______

10. ______

11. ______

12. ______

13. ______

14. ______

15. ______

16. ______

17. ______

18. ______

ANSWERS

19. ______

20. ______

21. ______

22. ______

23. ______

24. ______

25. ______

26. ______

27. ______

28. ______

29. ______

30. ______

31. ______

32. ______

19. $\dfrac{t^2 - 16}{t^2 - 8t + 16}$

20. $\dfrac{p^2 - 25}{p^2 + 10p + 25}$

21. $\dfrac{x^2 - 9x + 8}{x^2 + 3x - 4}$

22. $\dfrac{y^2 + 8y - 9}{y^2 - 5y + 4}$

23. $\dfrac{a^3 - b^3}{a^2 - b^2}$

24. $\dfrac{x^2 - y^2}{x^3 + y^3}$

d Multiply and simplify.

25. $\dfrac{x^4}{3x + 6} \cdot \dfrac{5x + 10}{5x^7}$

26. $\dfrac{10t}{6t - 12} \cdot \dfrac{20t - 40}{30t^3}$

27. $\dfrac{x^2 - 16}{x^2} \cdot \dfrac{x^2 - 4x}{x^2 - x - 12}$

28. $\dfrac{y^2 + 10y + 25}{y^2 - 9} \cdot \dfrac{y^2 - 3y}{y + 5}$

29. $\dfrac{y^2 - 16}{2y + 6} \cdot \dfrac{y + 3}{y - 4}$

30. $\dfrac{m^2 - n^2}{4m + 4n} \cdot \dfrac{m + n}{m - n}$

31. $\dfrac{x^2 - 2x - 35}{2x^3 - 3x^2} \cdot \dfrac{4x^3 - 9x}{7x - 49}$

32. $\dfrac{y^2 - 10y + 9}{y^2 - 1} \cdot \dfrac{y + 4}{y^2 - 5y - 36}$

33. $\dfrac{c^3+8}{c^2-4}\cdot\dfrac{c^2-4c+4}{c^2-2c+4}$

34. $\dfrac{x^3-27}{x^2-9}\cdot\dfrac{x^2-6x+9}{x^2+3x+9}$

35. $\dfrac{x^2-y^2}{x^3-y^3}\cdot\dfrac{x^2+xy+y^2}{x^2+2xy+y^2}$

36. $\dfrac{4x^2-9y^2}{8x^3-27y^3}\cdot\dfrac{4x^2+6xy+9y^2}{4x^2+12xy+9y^2}$

e Divide and simplify.

37. $\dfrac{12x^8}{3y^4}\div\dfrac{16x^3}{6y}$

38. $\dfrac{9a^7}{8b^2}\div\dfrac{12a^2}{24b^7}$

39. $\dfrac{3y+15}{y}\div\dfrac{y+5}{y}$

40. $\dfrac{6x+12}{x}\div\dfrac{x+2}{x^3}$

41. $\dfrac{y^2-9}{y}\div\dfrac{y+3}{y+2}$

42. $\dfrac{x^2-4}{x}\div\dfrac{x-2}{x+4}$

43. $\dfrac{4a^2-1}{a^2-4}\div\dfrac{2a-1}{a-2}$

44. $\dfrac{25x^2-4}{x^2-9}\div\dfrac{5x-2}{x+3}$

45. $\dfrac{x^2-16}{x^2-10x+25}\div\dfrac{3x-12}{x^2-3x-10}$

46. $\dfrac{y^2-36}{y^2-8y+16}\div\dfrac{3y-18}{y^2-y-12}$

ANSWERS

33. ______

34. ______

35. ______

36. ______

37. ______

38. ______

39. ______

40. ______

41. ______

42. ______

43. ______

44. ______

45. ______

46. ______

ANSWERS

47. ____________

48. ____________

49. ____________

50. ____________

51. ____________

52. ____________

53. ____________

54. ____________

55. ____________

56. ____________

57. ____________

58. ____________

59. ____________

60. ____________

61. ____________

62. ____________

47. $\dfrac{y^3 + 3y}{y^2 - 9} \div \dfrac{y^2 + 5y - 14}{y^2 + 4y - 21}$

48. $\dfrac{a^3 + 4a}{a^2 - 16} \div \dfrac{a^2 + 8a + 15}{a^2 + a - 20}$

49. $\dfrac{x^3 - 64}{x^3 + 64} \div \dfrac{x^2 - 16}{x^2 - 4x + 16}$

50. $\dfrac{8y^3 + 27}{64y^3 - 1} \div \dfrac{4y^2 - 9}{16y^2 + 4y + 1}$

Perform the indicated operations and simplify.

51. $\left[\dfrac{r^2 - 4s^2}{r + 2s} \div (r + 2s)\right] \cdot \dfrac{2s}{r - 2s}$

52. $\left[\dfrac{d^2 - d}{d^2 - 6d + 8} \cdot \dfrac{d - 2}{d^2 + 5d}\right] \div \dfrac{5d}{d^2 - 9d + 20}$

SKILL MAINTENANCE

53. A basketball player scored 15 times during one game. She scored a total of 27 points, two for each field goal and one for each free throw. How many field goals did the player make? How many free throws did she make?

54. An airplane has a total of 114 seats. The number of coach seats is six more than five times the number of first-class seats. How many of each type of seat are there on the plane?

55. The perimeter of a rectangular field is 628 m. The length of the field is 6 m greater than the width. Find the area of the field.

56. At a barbecue there were 250 dinners served. The cost of a dinner was \$1.50 each for children and \$4.00 each for adults. The total amount of money collected for dinners at the barbecue was \$705. How many of each type of plate was served?

SYNTHESIS

Simplify.

57. $\dfrac{x(x + 1) - 2(x + 3)}{(x + 1)(x + 2)(x + 3)}$

58. $\dfrac{2x - 5(x + 2) - (x - 2)}{x^2 - 4}$

59. $\dfrac{m^2 - t^2}{m^2 + t^2 + m + t + 2mt}$

60. $\dfrac{a^3 - 2a^2 + 2a - 4}{a^3 - 2a^2 - 3a + 6}$

61. $\dfrac{x^3 + x^2 - y^3 - y^2}{x^2 - 2xy + y^2}$

62. $\dfrac{x^5 - x^3 + x^2 - 1 - (x^3 - 1)(x + 1)^2}{(x^2 - 1)^2}$

6.2 LCM's, LCD's, Addition, and Subtraction

a Finding LCM's by Factoring

To add rational expressions when denominators are different, we first find a common denominator. Let us review the procedure in arithmetic first. To do the addition

$$\frac{5}{42} + \frac{7}{12},$$

we find a common denominator. We look for the least common multiple (LCM) of both 42 and 12. That number becomes the least common denominator (LCD).

To find the LCM, we factor both numbers completely (into primes).

$42 = 2 \cdot 3 \cdot 7$ ← Any multiple of 42 has these factors.

$12 = 2 \cdot 2 \cdot 3$ ← Any multiple of 12 has these factors.

The LCM is the number that has 2 as a factor twice, 3 as a factor once, and 7 as a factor once. The LCM is $2 \cdot 2 \cdot 3 \cdot 7$, or 84.

To obtain the LCM, use each factor the greatest number of times that it occurs in any one prime factorization.

▶ **EXAMPLE 1** Find the LCM of 18 and 24.

$18 = 3 \cdot 3 \cdot 2$
$24 = 2 \cdot 2 \cdot 2 \cdot 3$

The LCM is $3 \cdot 3 \cdot 2 \cdot 2 \cdot 2$, or 72. ◀

DO EXERCISES 1 AND 2.

Now let us return to adding $\frac{5}{42}$ and $\frac{7}{12}$:

$$\frac{5}{42} + \frac{7}{12} = \frac{5}{2 \cdot 3 \cdot 7} + \frac{7}{2 \cdot 2 \cdot 3}. \quad \text{Factoring the denominators}$$

The LCD is $2 \cdot 2 \cdot 3 \cdot 7$. To get this LCD in the first denominator, we need a 2. In the second denominator, we need a 7. We multiply by 1, as follows:

$$\frac{5}{2 \cdot 3 \cdot 7} \cdot \frac{2}{2} + \frac{7}{2 \cdot 2 \cdot 3} \cdot \frac{7}{7} = \frac{10}{2 \cdot 2 \cdot 3 \cdot 7} + \frac{49}{2 \cdot 2 \cdot 3 \cdot 7}$$
$$= \frac{59}{2 \cdot 2 \cdot 3 \cdot 7} = \frac{59}{84}.$$

Multiplying the first fraction by $\frac{2}{2}$ gave us an equivalent fraction with a denominator that is the LCD. Multiplying the second fraction by $\frac{7}{7}$ also gave us an equivalent fraction with a denominator that is the LCD. Now that we have a common denominator, we can add the numerators.

DO EXERCISES 3 AND 4.

OBJECTIVES

After finishing Section 6.2, you should be able to:

a Find the LCM of several algebraic expressions by factoring.
b Add and subtract rational expressions.
c Simplify combined additions and subtractions of rational expressions.

FOR EXTRA HELP

Tape 11B | Tape 9B | MAC: 6 IBM: 6

Find the LCM by factoring.

1. 18, 30

2. 12, 18, 24

Add, first finding the LCD of the denominators.

3. $\frac{5}{12} + \frac{11}{30}$

4. $\frac{7}{12} + \frac{13}{18} + \frac{1}{24}$

ANSWERS ON PAGE A-7

Find the LCM.

5. $a^2b^2, \quad 5a^3b$

6. $y^2 + 7y + 12, \quad y^2 + 8y + 16,$
$y + 4$

7. $x^2 - 9, \quad x^3 - x^2 - 6x, \quad 2x^2$

8. $a^2 - b^2, \quad 2b - 2a$

ANSWERS ON PAGE A-7

► **EXAMPLES**

2. Find the LCM of $12xy^2$ and $15x^3y$.

We factor each expression completely.

$$12xy^2 = 2 \cdot 2 \cdot 3 \cdot x \cdot y \cdot y \qquad 15x^3y = 3 \cdot 5 \cdot x \cdot x \cdot x \cdot y$$

$$\text{LCM} = 2 \cdot 2 \cdot 3 \cdot y \cdot y \cdot 5 \cdot x \cdot x \cdot x = 60x^3y^2$$

Remember: To find the LCM, use each factor the greatest number of times that it occurs in any one prime factorization.

3. Find the LCM of $x^2 + 2x + 1$, $5x^2 - 5x$, and $x^2 - 1$.

$$\left.\begin{aligned} x^2 + 2x + 1 &= (x + 1)(x + 1); \\ 5x^2 - 5x &= 5x(x - 1); \\ x^2 - 1 &= (x + 1)(x - 1) \end{aligned}\right\} \text{Factoring}$$

$$\text{LCM} = 5x(x + 1)(x + 1)(x - 1)$$

4. Find the LCM of $x^2 - y^2$, $x^3 + y^3$, and $x^2 + 2xy + y^2$.

$$\begin{aligned} x^2 - y^2 &= (x - y)(x + y); \\ x^3 + y^3 &= (x + y)(x^2 - xy + y^2); \\ x^2 + 2xy + y^2 &= (x + y)(x + y) \end{aligned}$$

$$\text{LCM} = (x - y)(x + y)(x + y)(x^2 - xy + y^2)$$ ◄

The opposite, or additive inverse, of an LCM is also an LCM. For example, if $(x + 2)(x - 3)$ is an LCM, then $-(x + 2)(x - 3)$ is also an LCM. We can name the latter $(x + 2)(-1)(x - 3)$, or $(x + 2)(3 - x)$. If, when we are finding LCM's, factors that are opposites occur, we do not use them both. For example if $a - b$ occurs in one factorization and $b - a$ occurs in another, we do not use them both, since they are opposites.

► **EXAMPLE 5** Find the LCM of $x^2 - y^2$ and $3y - 3x$.

$$x^2 - y^2 = (x + y)(x - y)$$

$$3y - 3x = 3(y - x), \text{ or } -3(x - y)$$

We can use $(x - y)$ or $(y - x)$, but we do not use both.

$$\text{LCM} = 3(x + y)(x - y), \quad \text{or } 3(x + y)(y - x), \quad \text{or } -3(x + y)(x - y)$$ ◄

DO EXERCISES 5–8.

b Adding and Subtracting Rational Expressions

When denominators are the same, add or subtract the numerators and keep the same denominator.

▶ **EXAMPLE 6** Add: $\dfrac{3+x}{x}+\dfrac{4}{x}$.

$$\frac{3+x}{x}+\frac{4}{x}=\frac{7+x}{x}$$ ← This expression does *not* simplify to 7. ◀

Example 6 shows that

$$\frac{3+x}{x}+\frac{4}{x} \quad \text{and} \quad \frac{7+x}{x}$$

are equivalent expressions. They name the same number for all replacements except 0.

▶ **EXAMPLE 7** Add: $\dfrac{4x^2-5xy}{x^2-y^2}+\dfrac{2xy-y^2}{x^2-y^2}$.

$$\frac{4x^2-5xy}{x^2-y^2}+\frac{2xy-y^2}{x^2-y^2}=\frac{4x^2-3xy-y^2}{x^2-y^2}$$ **Adding the numerators**

$$=\frac{(4x+y)(x-y)}{(x+y)(x-y)}$$ **Factoring the numerator and the denominator**

$$=\frac{(4x+y)\cancel{(x-y)}}{(x+y)\cancel{(x-y)}}$$ **Removing a factor of 1: $\dfrac{x-y}{x-y}=1$**

$$=\frac{4x+y}{x+y}$$ ◀

DO EXERCISES 9 AND 10.

▶ **EXAMPLE 8** Subtract: $\dfrac{4x+5}{x+3}-\dfrac{x-2}{x+3}$.

$$\frac{4x+5}{x+3}-\frac{x-2}{x+3}=\frac{4x+5-(x-2)}{x+3}$$ **Subtracting numerators**

$$=\frac{4x+5-x+2}{x+3}$$

$$=\frac{3x+7}{x+3}$$ ◀

A common error: Forgetting these parentheses. If you forget them, you will be subtracting only *part* of the numerator, $x-2$.

DO EXERCISES 11 AND 12.

When one denominator is the opposite, or additive inverse, of the other, multiply one expression by $-1/-1$. This gives a common denominator.

Add.

9. $\dfrac{5+y}{y}+\dfrac{7}{y}$

10. $\dfrac{2x^2+5x-9}{x-5}+\dfrac{x^2-19x+4}{x-5}$

Subtract.

11. $\dfrac{a}{b+2}-\dfrac{b}{b+2}$

12. $\dfrac{4y+7}{x^2+y^2}-\dfrac{3y-5}{x^2+y^2}$

ANSWERS ON PAGE A-7

Add.

13. $\dfrac{b}{3b} + \dfrac{b^3}{-3b}$

14. $\dfrac{3x^2 + 4}{x - 5} + \dfrac{x^2 - 7}{5 - x}$

Subtract.

15. $\dfrac{3}{4y} - \dfrac{7x}{-4y}$

16. $\dfrac{4x^2}{2x - y} - \dfrac{7x^2}{y - 2x}$

ANSWERS ON PAGE A-7

▶ **EXAMPLE 9** Add: $\dfrac{a}{2a} + \dfrac{a^3}{-2a}$.

$$\frac{a}{2a} + \frac{a^3}{-2a} = \frac{a}{2a} + \frac{-1}{-1} \cdot \frac{a^3}{-2a}$$ **Multiplying by $\frac{-1}{-1}$** ← This is equal to 1 (not -1).

$$= \frac{a}{2a} + \frac{-a^3}{2a}$$

$$= \frac{a - a^3}{2a}$$ **Adding numerators**

$$= \frac{a(1 - a^2)}{2a}$$ **Factoring**

$$= \frac{\not{a}(1 - a^2)}{2\not{a}}$$ **Removing a factor of 1: $\frac{a}{a} = 1$**

$$= \frac{1 - a^2}{2}$$ ◀

▶ **EXAMPLE 10** Subtract: $\dfrac{x^2}{5y} - \dfrac{x^3}{-5y}$.

$$\frac{x^2}{5y} - \frac{x^3}{-5y} = \frac{x^2}{5y} - \frac{-1}{-1} \cdot \frac{x^3}{-5y}$$ **Multiplying by $\frac{-1}{-1}$**

$$= \frac{x^2}{5y} - \frac{-x^3}{5y} = \frac{x^2 - (-x^3)}{5y}$$ ← Don't forget these parentheses!

$$= \frac{x^2 + x^3}{5y}$$ ◀

▶ **EXAMPLE 11** Subtract: $\dfrac{5x}{x - 2y} - \dfrac{3y - 7}{2y - x}$.

$$\frac{5x}{x - 2y} - \frac{3y - 7}{2y - x} = \frac{5x}{x - 2y} - \frac{-1}{-1} \cdot \frac{3y - 7}{2y - x}$$

$$= \frac{5x}{x - 2y} - \frac{-3y + 7}{x - 2y}$$ Remember: $-1(2y - x) = -(2y - x) = (x - 2y)$.

$$= \frac{5x - (-3y + 7)}{x - 2y}$$ **Subtracting numerators**

$$= \frac{5x + 3y - 7}{x - 2y}$$ ◀

DO EXERCISES 13–16.

When denominators are different, but not additive inverses of each other, we first find the least common denominator, the LCD, and then add or subtract the numerators.

▶ **EXAMPLE 12** Add: $\frac{2a}{5} + \frac{3b}{2a}$.

We first find the LCD:

$$\begin{matrix} 5 \\ 2a \end{matrix} \quad \text{The LCD is } 5 \cdot 2a, \text{ or } 10a.$$

Now we multiply each expression by 1. We choose whatever symbol for 1 will give us the LCD in each denominator. In this case, we use $2a/2a$ and 5/5:

$$\frac{2a}{5} \cdot \frac{2a}{2a} + \frac{3b}{2a} \cdot \frac{5}{5} = \frac{4a^2}{10a} + \frac{15b}{10a} = \frac{4a^2 + 15b}{10a}.$$

Multiplying by $2a/2a$ in the first term gave us a denominator of $10a$. Multiplying by $\frac{5}{5}$ in the second term also gave us a denominator of $10a$. ◀

▶ **EXAMPLE 13** Add: $\frac{3x^2 + 3xy}{x^2 - y^2} + \frac{2 - 3x}{x - y}$.

We first find the LCD of the denominators:

$$\left.\begin{aligned} x^2 - y^2 &= (x + y)(x - y) \\ x - y &= x - y \end{aligned}\right\} \quad \text{The LCD is } (x + y)(x - y).$$

We now multiply by 1 to get the LCD in the second expression. Then we add and simplify if possible.

$$\frac{3x^2 + 3xy}{(x + y)(x - y)} + \frac{2 - 3x}{x - y} \cdot \frac{x + y}{x + y} \quad \textbf{Multiplying by 1 to get the LCD}$$

$$= \frac{3x^2 + 3xy}{(x + y)(x - y)} + \frac{(2 - 3x)(x + y)}{(x - y)(x + y)}$$

$$= \frac{3x^2 + 3xy}{(x + y)(x - y)} + \frac{2x + 2y - 3x^2 - 3xy}{(x - y)(x + y)} \quad \textbf{Multiplying in the numerator}$$

$$= \frac{3x^2 + 3xy + 2x + 2y - 3x^2 - 3xy}{(x + y)(x - y)} \quad \textbf{Adding the numerators}$$

$$= \frac{2x + 2y}{(x + y)(x - y)} = \frac{2(x + y)}{(x + y)(x - y)} \quad \textbf{Combining like terms and then factoring the numerator}$$

$$= \frac{2\cancel{(x + y)}}{\cancel{(x + y)}(x - y)} \quad \textbf{Removing a factor of 1: } \frac{x + y}{x + y} = 1$$

$$= \frac{2}{x - y}$$ ◀

DO EXERCISES 17 AND 18.

Add.

17. $\frac{3x}{7} + \frac{4y}{3x}$

18. $\frac{2xy - 2x^2}{x^2 - y^2} + \frac{2x + 3}{x + y}$

ANSWERS ON PAGE A-7

Subtract.

19. $\dfrac{a}{a+3} - \dfrac{a-4}{a}$

20. $\dfrac{4y-5}{y^2-7y+12} - \dfrac{y+7}{y^2+2y-15}$

21. Do this calculation:

$$\frac{8x}{x^2-1} + \frac{2}{1-x} - \frac{4}{x+1}.$$

▶ **EXAMPLE 14** Subtract: $\dfrac{2y+1}{y^2-7y+6} - \dfrac{y+3}{y^2-5y-6}$.

$$\frac{2y+1}{y^2-7y+6} - \frac{y+3}{y^2-5y-6}$$

$$= \frac{2y+1}{(y-6)(y-1)} - \frac{y+3}{(y-6)(y+1)}$$ **The LCD is $(y-6)(y-1)(y+1)$.**

$$= \frac{2y+1}{(y-6)(y-1)} \cdot \frac{y+1}{y+1} - \frac{y+3}{(y-6)(y+1)} \cdot \frac{y-1}{y-1}$$ **Multiplying by 1 to get the LCD**

$$= \frac{(2y+1)(y+1)-(y+3)(y-1)}{(y-6)(y-1)(y+1)}$$ **Subtracting the numerators**

$$= \frac{2y^2+3y+1-(y^2+2y-3)}{(y-6)(y-1)(y+1)}$$ **Multiplying and removing parentheses**

$$= \frac{2y^2+3y+1-y^2-2y+3}{(y-6)(y-1)(y+1)} = \frac{y^2+y+4}{(y-6)(y-1)(y+1)}$$ ◀

We usually do not multiply out a numerator or a denominator. Not doing so will be helpful when we solve equations.

DO EXERCISES 19 AND 20.

C Combined Additions and Subtractions

▶ **EXAMPLE 15** Perform the indicated operations and simplify.

$$\frac{2x}{x^2-4} + \frac{5}{2-x} - \frac{1}{2+x}$$

$$= \frac{2x}{(x-2)(x+2)} + \frac{5}{2-x} - \frac{1}{2+x}$$

$$= \frac{2x}{(x-2)(x+2)} + \frac{-1}{-1} \cdot \frac{5}{(2-x)} - \frac{1}{x+2}$$ **Multiplying by $\frac{-1}{-1}$**

$$= \frac{2x}{(x-2)(x+2)} + \frac{-5}{x-2} - \frac{1}{x+2}$$ **The LCD is $(x-2)(x+2)$.**

$$= \frac{2x}{(x-2)(x+2)} + \frac{-5}{x-2} \cdot \frac{x+2}{x+2} - \frac{1}{x+2} \cdot \frac{x-2}{x-2}$$ **Multiplying by 1 to get the LCD**

$$= \frac{2x-5(x+2)-(x-2)}{(x-2)(x+2)}$$ **Adding and subtracting the numerators**

$$= \frac{2x-5x-10-x+2}{(x-2)(x+2)}$$ **Removing parentheses and simplifying**

$$= \frac{-4x-8}{(x-2)(x+2)}$$

$$= \frac{-4\cancel{(x+2)}}{(x-2)\cancel{(x+2)}}$$ **Removing a factor of 1: $\frac{x+2}{x+2} = 1$**

$$= \frac{-4}{x-2}, \quad \text{or} \quad -\frac{4}{x-2}$$

Another correct answer is $4/(2-x)$. It is found by multiplying by $-1/-1$. ◀

ANSWERS ON PAGE A-7

DO EXERCISE 21.

NAME SECTION DATE

EXERCISE SET 6.2

a Find the LCM by factoring.

1. 12, 18 **2.** 15, 20 **3.** 18, 48 **4.** 45, 54

5. 24, 36 **6.** 30, 75 **7.** 9, 15, 5 **8.** 27, 35, 63

Add, first finding the LCD.

9. $\frac{5}{6} + \frac{4}{15}$ **10.** $\frac{5}{36} + \frac{5}{24}$ **11.** $\frac{7}{12} + \frac{11}{18}$

12. $\frac{11}{30} + \frac{19}{75}$ **13.** $\frac{3}{4} + \frac{7}{30} + \frac{1}{16}$ **14.** $\frac{5}{8} + \frac{7}{12} + \frac{11}{40}$

Find the LCM.

15. $12x^2y$, $4xy$ **16.** $18r^2s$, $12rs^3$ **17.** $y^2 - 9$, $3y + 9$

18. $a^2 - b^2$, $ab + b^2$ **19.** $15ab^2$, $3ab$, $10a^3b$ **20.** $6x^2y^2$, $9x^3y$, $15y^3$

21. $5y - 15$, $y^2 - 6y + 9$ **22.** $x^2 + 10x + 25$, $x^2 + 2x - 15$

23. $x^2 - 4$, $2 - x$ **24.** $y^2 - 9$, $3 - y$

25. $2r^2 - 5r - 12$, $3r^2 - 13r + 4$, $r^2 - 16$

26. $2x^2 - 5x - 3$, $2x^2 - x - 1$, $x^2 - 6x + 9$

27. $x^5 + 4x^3$, $x^3 - 4x^2 + 4x$ **28.** $9x^3 + 9x^2 - 18x$, $6x^5 + 24x^4 + 24x^3$

29. $x^5 - 2x^4 + x^3$, $2x^3 + 2x$, $5x + 5$ **30.** $x^5 - 4x^4 + 4x^3$, $3x^2 - 12$, $2x + 4$

ANSWERS

1. ______
2. ______
3. ______
4. ______
5. ______
6. ______
7. ______
8. ______
9. ______
10. ______
11. ______
12. ______
13. ______
14. ______
15. ______
16. ______
17. ______
18. ______
19. ______
20. ______
21. ______
22. ______
23. ______
24. ______
25. ______
26. ______
27. ______
28. ______
29. ______
30. ______

ANSWERS

31. ______
32. ______
33. ______
34. ______
35. ______
36. ______
37. ______
38. ______
39. ______
40. ______
41. ______
42. ______
43. ______
44. ______
45. ______
46. ______
47. ______
48. ______

b Add or subtract. Simplify by removing a factor of 1 when possible.

31. $\frac{a-3b}{a+b}+\frac{a+5b}{a+b}$

32. $\frac{x-5y}{x+y}+\frac{x+7y}{x+y}$

33. $\frac{4y+3}{y-2}-\frac{y-2}{y-2}$

34. $\frac{3t+2}{t-4}-\frac{t-4}{t-4}$

35. $\frac{a^2}{a-b}+\frac{b^2}{b-a}$

36. $\frac{r^2}{r-s}+\frac{s^2}{s-r}$

37. $\frac{3}{x}-\frac{8}{-x}$

38. $\frac{2}{a}-\frac{5}{-a}$

39. $\frac{2x-10}{x^2-25}-\frac{5-x}{25-x^2}$

40. $\frac{y-9}{y^2-16}-\frac{7-y}{16-y^2}$

Add or subtract. Simplify by removing a factor of 1 when possible. If a denominator has three or more factors (other than monomials), leave it factored.

41. $\frac{y-2}{y+4}+\frac{y+3}{y-5}$

42. $\frac{x-2}{x+3}+\frac{x+2}{x-4}$

43. $\frac{4xy}{x^2-y^2}+\frac{x-y}{x+y}$

44. $\frac{5ab}{a^2-b^2}+\frac{a+b}{a-b}$

45. $\frac{9x+2}{3x^2-2x-8}+\frac{7}{3x^2+x-4}$

46. $\frac{3y+2}{2y^2-y-10}+\frac{8}{2y^2-7y+5}$

47. $\frac{4}{x+1}+\frac{x+2}{x^2-1}+\frac{3}{x-1}$

48. $\frac{-2}{y+2}+\frac{5}{y-2}+\frac{y+3}{y^2-4}$

49. $\dfrac{x-1}{3x+15} - \dfrac{x+3}{5x+25}$

50. $\dfrac{y-2}{4y+8} - \dfrac{y+6}{5y+10}$

51. $\dfrac{5ab}{a^2-b^2} - \dfrac{a-b}{a+b}$

52. $\dfrac{6xy}{x^2-y^2} - \dfrac{x+y}{x-y}$

53. $\dfrac{3y}{y^2-7y+10} - \dfrac{2y}{y^2-8y+15}$

54. $\dfrac{5x}{x^2-6x+8} - \dfrac{3x}{x^2-x-12}$

55. $\dfrac{y}{y^2-y-20} + \dfrac{2}{y+4}$

56. $\dfrac{6}{y^2+6y+9} + \dfrac{5}{y^2-9}$

57. $\dfrac{3y+2}{y^2+5y-24} + \dfrac{7}{y^2+4y-32}$

58. $\dfrac{3y+2}{y^2-7y+10} + \dfrac{2y}{y^2-8y+15}$

59. $\dfrac{3x-1}{x^2+2x-3} - \dfrac{x+4}{x^2-9}$

60. $\dfrac{3p-2}{p^2+2p-24} - \dfrac{p-3}{p^2-16}$

ANSWERS

49. ______

50. ______

51. ______

52. ______

53. ______

54. ______

55. ______

56. ______

57. ______

58. ______

59. ______

60. ______

ANSWERS

61.

62.

63.

64.

65.

66.

67.

68.

69.

70.

71.

72.

73. See graph.

74. See graph.

75.

76.

77.

78.

79.

80.

81.

82.

83.

84.

C Perform the indicated operations and simplify.

61. $\frac{1}{x+1} - \frac{x}{x-2} + \frac{x^2+2}{x^2-x-2}$

62. $\frac{2}{y+3} - \frac{y}{y-1} + \frac{y^2+2}{y^2+2y-3}$

63. $\frac{x-1}{x-2} - \frac{x+1}{x+2} + \frac{x-6}{x^2-4}$

64. $\frac{y-3}{y-4} - \frac{y+2}{y+4} + \frac{y-7}{y^2-16}$

65. $\frac{4x}{x^2-1} + \frac{3x}{1-x} - \frac{4}{x-1}$

66. $\frac{5y}{1-2y} - \frac{2y}{2y+1} + \frac{3}{4y^2-1}$

67. $\frac{5}{3-2x} - \frac{3}{2x-3} + \frac{x-3}{2x^2-x-3}$

68. $\frac{2r}{r^2-s^2} - \frac{1}{r+s} + \frac{1}{s-r}$

69. $\frac{3}{2c-1} - \frac{1}{c+2} + \frac{5}{2c^2+3c-2}$

70. $\frac{3y-1}{2y^2+y-3} - \frac{2-y}{y-1} - \frac{y}{1-y}$

71. $\frac{1}{x+y} + \frac{1}{y-x} - \frac{2x}{x^2-y^2}$

72. $\frac{1}{b-a} + \frac{1}{a+b} - \frac{2b}{a^2-b^2}$

SKILL MAINTENANCE

73. Graph: $2x - 3y > 6$.

74. Graph: $5x + 3y \leq 15$.

Factor.

75. $12y^4 - 15y^3 + 3y^2$

76. $x^2 - 25$

77. $x^2 - 14x + 49$

SYNTHESIS

Find the LCM.

78. 18, 42, 82, 120, 300, 700

79. $x^8 - x^4,\ x^5 - x^2,\ x^5 - x^3,\ x^5 + x^2$

80. The LCM of two expressions is $8a^4b^7$. One of the expressions is $2a^3b^7$. List all possibilities for the other expression.

Perform the indicated operations and simplify.

81. $\frac{b-c}{a-(b-c)} - \frac{b-a}{(b-a)-c}$

82. $\frac{x+y+1}{y-(x+1)} + \frac{x+y-1}{x-(y-1)} - \frac{x-y-1}{1-(y-x)}$

83. $\frac{x^2}{3x^2-5x-2} - \frac{2x}{3x+1} \cdot \frac{1}{x-2}$

84. $\frac{x}{x^4-y^4} - \frac{1}{x^2+2xy+y^2}$

6.3 Solving Rational Equations

a In Sections 6.1 and 6.2, we studied operations with rational **expressions.** These are expressions that do not have an equals sign. We cannot clear expressions of fractions other than occasionally when simplifying by removing a factor of 1. In this section, we are studying rational **equations.** Equations do have an equals sign, and we can clear of fractions as we did in Chapter 2. The difference is that now we may have variables in denominators.

A **rational,** or **fractional, equation** is an equation containing one or more rational expressions. Here are some examples:

$$\frac{2}{3}-\frac{5}{6}=\frac{1}{x}, \qquad x+\frac{6}{x}=5, \qquad \frac{x^2}{x-2}=\frac{4}{x-2}.$$

> **To solve a rational equation, our first step is to clear the equation of fractions. We use the multiplication principle as in Chapter 2. We multiply both sides of the equation by the LCM of all the denominators. Then we carry out the equation-solving process as we learned it in Chapter 2.**

▶ **EXAMPLE 1** Solve: $\frac{2}{3}-\frac{5}{6}=\frac{1}{x}$.

The LCM of all the denominators is $6x$, or $2\cdot 3\cdot x$. Using the multiplication principle of Chapter 2, we multiply on both sides of the equation by the LCM.

$$(2\cdot 3\cdot x)\cdot\left(\frac{2}{3}-\frac{5}{6}\right)=(2\cdot 3\cdot x)\cdot\frac{1}{x}$$ **Multiplying by the LCM. We are *not* multiplying by 1.**

$$2\cdot 3\cdot x\cdot\frac{2}{3}-2\cdot 3\cdot x\cdot\frac{5}{6}=2\cdot 3\cdot x\cdot\frac{1}{x}$$ **Multiplying to remove parentheses**

When clearing of fractions, be sure to multiply *every* term in the equation by the LCM.

$$2\cdot x\cdot 2-x\cdot 5=2\cdot 3$$ **Simplifying. Note that we have now cleared of the fractions.**

$$4x-5x=6$$
$$-x=6$$
$$-1\cdot x=6$$
$$x=-6$$

Check:

$\frac{2}{3}-\frac{5}{6}$	$=\frac{1}{x}$	
$\frac{2}{3}-\frac{5}{6}$	$\frac{1}{-6}$	
$\frac{4}{6}-\frac{5}{6}$	$-\frac{1}{6}$	
$-\frac{1}{6}$		TRUE

The solution is -6.

DO EXERCISE 1. ◀

OBJECTIVE

After finishing Section 6.3, you should be able to:

a Solve rational equations.

FOR EXTRA HELP

Tape 11C

Tape 10A

MAC: 6
IBM: 6

1. Solve: $\frac{2}{3}+\frac{5}{6}=\frac{1}{x}$.

ANSWER ON PAGE A-7

2. Solve: $\dfrac{y-4}{5} - \dfrac{y+7}{2} = 5$.

► **EXAMPLE 2** Solve: $\dfrac{x+1}{2} - \dfrac{x-3}{3} = 3$.

The LCM of all the denominators is $2 \cdot 3$, or 6. We multiply on both sides of the equation by the LCM.

$$2 \cdot 3\left(\frac{x+1}{2} - \frac{x-3}{3}\right) = 2 \cdot 3 \cdot 3 \quad \text{Multiplying on both sides by the LCM}$$

$$2 \cdot 3 \cdot \frac{x+1}{2} - 2 \cdot 3 \cdot \frac{x-3}{3} = 2 \cdot 3 \cdot 3 \quad \text{Multiplying to remove parentheses}$$

$$3(x+1) - 2(x-3) = 18 \quad \text{Simplifying}$$

$$\left.\begin{aligned} 3x + 3 - 2x + 6 &= 18 \\ x + 9 &= 18 \end{aligned}\right\} \quad \text{Multiplying and collecting like terms}$$

$$x = 9$$

Check:

$$\begin{array}{c|c} \dfrac{x+1}{2} - \dfrac{x-3}{3} & = 3 \\ \hline \dfrac{9+1}{2} - \dfrac{9-3}{3} & 3 \\ 5 - 2 & \\ 3 & \text{TRUE} \end{array}$$

Clearing of fractions is a valid procedure only when solving equations, *not* when adding, subtracting, multiplying, or dividing rational expressions.

The solution is 9. ◄

DO EXERCISE 2.

When we multiply by an expression with a variable, we may not get equivalent equations. Thus we must *always* check possible solutions in the original equation.

1. **If you have carried out all algebraic procedures correctly, you need only check to see if a number is a meaningful replacement in all parts of the original equation.**
2. **To be sure that no computational errors have been made and that you indeed have a solution, a complete check is necessary, as we did in Chapter 2.**

► **EXAMPLE 3** Solve: $\dfrac{2x}{x-3} - \dfrac{6}{x} = \dfrac{18}{x^2 - 3x}$.

The LCM of the denominators is $x(x-3)$. We multiply by $x(x-3)$.

$$x(x-3)\left(\frac{2x}{x-3} - \frac{6}{x}\right) = x(x-3)\left(\frac{18}{x^2-3x}\right) \quad \text{Multiplying on both sides by the LCM}$$

$$x(x-3) \cdot \frac{2x}{x-3} - x(x-3) \cdot \frac{6}{x} = x(x-3) \cdot \frac{18}{x^2-3x} \quad \text{Multiplying to remove parentheses}$$

$$2x^2 - 6(x-3) = 18 \quad \text{Simplifying}$$

$$2x^2 - 6x + 18 = 18$$

$$2x^2 - 6x = 0$$

$$2x(x-3) = 0$$

$$2x = 0 \quad \text{or} \quad x - 3 = 0 \quad \text{Using the principle of zero products}$$

$$x = 0 \quad \text{or} \quad x = 3$$

ANSWER ON PAGE A-7

The numbers 0 and 3 are possible solutions. We look at the original equation and see that neither is a meaningful replacement. We can also carry out a check, as follows.

Check:

For 0:

$$\frac{2x}{x-3} - \frac{6}{x} = \frac{18}{x^2-3x}$$

$$\frac{2(0)}{0-3} - \frac{6}{0} \;\Big|\; \frac{18}{0^2-3(0)} \quad \text{FALSE}$$

For 3:

$$\frac{2x}{x-3} - \frac{6}{x} = \frac{18}{x^2-3x}$$

$$\frac{2(3)}{3-3} - \frac{6}{3} \;\Big|\; \frac{18}{3^2-3(3)}$$

$$\frac{6}{0} - 2 \;\Big|\; \frac{18}{0} \quad \text{FALSE}$$

The equation has *no solution.* ◀

DO EXERCISE 3.

► **EXAMPLE 4** Solve: $\dfrac{x^2}{x-2} = \dfrac{4}{x-2}$.

The LCM of the denominators is $x - 2$. We multiply by $x - 2$.

$$(x-2) \cdot \frac{x^2}{x-2} = (x-2) \cdot \frac{4}{x-2}$$

$$x^2 = 4 \qquad \textbf{Simplifying}$$

$$x^2 - 4 = 0$$

$$(x+2)(x-2) = 0$$

$$x = -2 \quad \text{or} \quad x = 2 \qquad \textbf{Using the principle of zero products}$$

Check:

For 2:

$$\frac{x^2}{x-2} = \frac{4}{x-2}$$

$$\frac{2^2}{2-2} \;\Big|\; \frac{4}{2-2}$$

$$\frac{4}{0} \;\Big|\; \frac{4}{0}$$

For -2:

$$\frac{x^2}{x-2} = \frac{4}{x-2}$$

$$\frac{(-2)^2}{-2-2} \;\Big|\; \frac{4}{-2-2}$$

$$\frac{4}{-4} \;\Big|\; \frac{4}{-4}$$

The number -2 is a solution, but 2 is not (it results in division by 0). ◀

► **EXAMPLE 5** Solve: $x + \dfrac{6}{x} = 5$.

The LCM of the denominators is x. We multiply on both sides by x.

$$x\left(x + \frac{6}{x}\right) = x \cdot 5 \qquad \textbf{Multiplying on both sides by } x$$

$$x \cdot x + x \cdot \frac{6}{x} = 5x$$

$$x^2 + 6 = 5x \qquad \textbf{Simplifying}$$

$$x^2 - 5x + 6 = 0 \qquad \textbf{Getting 0 on one side}$$

$$(x-3)(x-2) = 0 \qquad \textbf{Factoring}$$

$$x = 3 \quad \text{or} \quad x = 2 \qquad \textbf{Using the principle of zero products}$$

3. Solve:

$$\frac{4x}{x+5} + \frac{20}{x} = \frac{100}{x^2+5x}.$$

ANSWER ON PAGE A-7

Solve.

4. $\frac{x^2}{x-3} = \frac{9}{x-3}$

5. $x - \frac{12}{x} = 1$

Check: For 3: For 2:

$$x + \frac{6}{x} = 5 \qquad x + \frac{6}{x} = 5$$

$3 + \frac{6}{3}$	5
5	

$2 + \frac{6}{2}$	5
5	

The solutions are 2 and 3. ◀

DO EXERCISES 4 AND 5.

▶ **EXAMPLE 6** Solve: $\frac{2}{x-1} = \frac{3}{x+1}$.

The LCM of the denominators is $(x-1)(x+1)$. We multiply by $(x-1)(x+1)$.

$$(x-1)(x+1) \cdot \frac{2}{(x-1)} = (x-1)(x+1) \cdot \frac{3}{x+1} \qquad \textbf{Multiplying}$$

$$2(x+1) = 3(x-1) \qquad \textbf{Simplifying}$$

$$2x + 2 = 3x - 3$$

$$5 = x$$

We leave the check to the student. The number 5 checks and is the solution. ◀

▶ **EXAMPLE 7** Solve: $\frac{2}{x+5} + \frac{1}{x-5} = \frac{16}{x^2-25}$.

The LCM is of the denominators $(x+5)(x-5)$. We multiply by $(x+5)(x-5)$.

$$(x+5)(x-5) \cdot \left[\frac{2}{x+5} + \frac{1}{x-5}\right] = (x+5)(x-5) \cdot \frac{16}{x^2-25}$$

$$(x+5)(x-5) \cdot \frac{2}{x+5} + (x+5)(x-5) \cdot \frac{1}{x-5} = (x+5)(x-5) \cdot \frac{16}{x^2-25}$$

$$2(x-5) + (x+5) = 16$$

$$2x - 10 + x + 5 = 16$$

$$3x = 21$$

$$x = 7$$

Check: $\frac{2}{x+5} + \frac{1}{x-5} = \frac{16}{x^2-25}$

$\frac{2}{7+5} + \frac{1}{7-5}$	$\frac{16}{7^2-25}$
$\frac{2}{12} + \frac{1}{2}$	$\frac{16}{49-25}$
$\frac{8}{12}$	$\frac{16}{24}$
$\frac{2}{3}$	$\frac{2}{3}$

Note: In this section, we have introduced a new use of the LCM. Before, you used the LCM in adding or subtracting rational expressions. *Now* we have equations. There are equals signs. We clear of fractions by multiplying on both sides of the equation by the LCM. This eliminates the denominators. *Do not* make the mistake of trying to "clear of fractions" when you do not have an equation!

The solution is 7. ◀

DO EXERCISES 6 AND 7.

Solve.

6. $\frac{2}{x-1} = \frac{3}{x+2}$

7. $\frac{2}{x^2-9} + \frac{5}{x-3} = \frac{3}{x+3}$

ANSWERS ON PAGE A-7

NAME SECTION DATE

EXERCISE SET 6.3

a Solve.

1. $\frac{2}{5}+\frac{7}{8}=\frac{y}{20}$

2. $\frac{4}{5}+\frac{1}{3}=\frac{t}{9}$

3. $\frac{1}{3}-\frac{5}{6}=\frac{1}{x}$

4. $\frac{5}{8}-\frac{2}{5}=\frac{1}{y}$

5. $\frac{x}{3}-\frac{x}{4}=12$

6. $\frac{y}{5}-\frac{y}{3}=15$

7. $y+\frac{5}{y}=-6$

8. $x+\frac{4}{x}=-5$

9. $\frac{4}{z}+\frac{2}{z}=3$

10. $\frac{4}{3y}-\frac{3}{y}=\frac{10}{3}$

11. $\frac{x-3}{x+2}=\frac{1}{5}$

12. $\frac{y-5}{y+1}=\frac{3}{5}$

13. $\frac{3}{y+1}=\frac{2}{y-3}$

14. $\frac{4}{x-1}=\frac{3}{x+2}$

15. $\frac{y-1}{y-3}=\frac{2}{y-3}$

16. $\frac{x-2}{x-4}=\frac{2}{x-4}$

17. $\frac{x+1}{x}=\frac{3}{2}$

18. $\frac{y+2}{y}=\frac{5}{3}$

19. $\frac{2}{x}-\frac{3}{x}+\frac{4}{x}=5$

20. $\frac{4}{y}-\frac{6}{y}+\frac{8}{y}=8$

21. $\frac{1}{2}-\frac{4}{9x}=\frac{4}{9}-\frac{1}{6x}$

22. $-\frac{1}{3}-\frac{5}{4y}=\frac{3}{4}-\frac{1}{6y}$

23. $\frac{60}{x}-\frac{60}{x-5}=\frac{2}{x}$

24. $\frac{50}{y}-\frac{50}{y-2}=\frac{4}{y}$

25. $\frac{7}{5x-2}=\frac{5}{4x}$

26. $\frac{5}{y+4}=\frac{3}{y-2}$

ANSWERS

1. ______
2. ______
3. ______
4. ______
5. ______
6. ______
7. ______
8. ______
9. ______
10. ______
11. ______
12. ______
13. ______
14. ______
15. ______
16. ______
17. ______
18. ______
19. ______
20. ______
21. ______
22. ______
23. ______
24. ______
25. ______
26. ______

ANSWERS

27. ______

28. ______

29. ______

30. ______

31. ______

32. ______

33. ______

34. ______

35. ______

36. ______

37. ______

38. ______

39. ______

40. ______

41. ______

42. ______

43. ______

44. ______

45. ______

46. ______

47. ______

48. ______

49. ______

27. $\frac{x}{x-2}+\frac{x}{x^2-4}=\frac{x+3}{x+2}$

28. $\frac{3}{y-2}+\frac{2y}{4-y^2}=\frac{5}{y+2}$

29. $\frac{6}{x^2-4x+3}-\frac{1}{x-3}=\frac{1}{4x-4}$

30. $\frac{8}{x^2-25}-\frac{2}{x-5}=\frac{1}{2x+10}$

31. $\frac{1}{4y^2-36}+\frac{2}{y-3}=\frac{5}{y+3}$

32. $\frac{7}{x-2}-\frac{8}{x+5}=\frac{1}{2x^2+6x-20}$

33. $\frac{a}{2a-6}-\frac{3}{a^2-6a+9}=\frac{a-2}{3a-9}$

34. $\frac{2}{x+4}+\frac{2x-1}{x^2+2x-8}=\frac{1}{x-2}$

35. $\frac{2x+3}{x-1}=\frac{10}{x^2-1}+\frac{2x-3}{x+1}$

36. $\frac{y}{y+1}+\frac{3y+5}{y^2+4y+3}=\frac{2}{y+3}$

37. $\frac{4}{x+3}+\frac{7}{x^2-3x+9}=\frac{108}{x^3+27}$

38. $\frac{3x}{x+2}+\frac{72}{x^3+8}=\frac{24}{x^2-2x+4}$

39. $\frac{3x}{x+2}+\frac{6}{x}+4=\frac{12}{x^2+2x}$

40. $\frac{5x}{x-7}-\frac{35}{x+7}=\frac{490}{x^2-49}$

41. $\frac{2x-14}{x^2+3x-28}+\frac{2-x}{4-x}-\frac{x+3}{x+7}=0$

42. $\frac{x-1}{3}+\frac{6x+1}{15}+\frac{2(x-2)}{13-7x}=\frac{2(x+2)}{5}$

SKILL MAINTENANCE

Factor.

43. $4x^2-5x-51$

44. $4t^3+500$

45. $1-t^6$

SYNTHESIS

Solve.

46. $\left(\frac{1}{1+x}+\frac{x}{1-x}\right)\div\left(\frac{x}{1+x}-\frac{1}{1-x}\right)=-1$

47. $\frac{x+3}{x+2}-\frac{x+4}{x+3}=\frac{x+5}{x+4}-\frac{x+6}{x+5}$

48. $\frac{7}{x-9}-\frac{7}{x}=\frac{63}{x^2-9x}$

49. $\frac{36}{x+4}-\frac{27}{x+3}=\frac{9x}{x^2+7x+12}$

6.4 Solving Problems

a Work Problems

▶ **EXAMPLE 1** Lon can mow a lawn in 4 hr. Penny can mow the same lawn in 5 hr. How long would it take both of them, working together, to mow the lawn?

1. *Familiarize.* We familiarize ourselves with the problem by considering two *incorrect* ways of translating the problem to mathematical language.
   a) A common *incorrect* way to translate the problem is to add the two times:

   $$4 \text{ hr} + 5 \text{ hr} = 9 \text{ hr}.$$

   Now think about this. Lon can do the job alone in 4 hr. If Lon and Penny work together, whatever time it takes them should be *less* than 4 hr. Thus we reject 9 hr as a solution, but we do have a partial check on any answer we get. The answer should be less than 4 hr.

   b) Another *incorrect* way to translate the problem is as follows. Suppose the two people split up the mowing job in such a way that Lon does half the mowing and Penny does the other half. Then

   $$\text{Lon mows } \tfrac{1}{2} \text{ the lawn in } \tfrac{1}{2}(4 \text{ hr}), \text{ or } 2 \text{ hr},$$

   and

   $$\text{Penny mows } \tfrac{1}{2} \text{ the lawn in } \tfrac{1}{2}(5 \text{ hr}), \text{ or } 2\tfrac{1}{2} \text{ hr}.$$

   But time is wasted since Lon would get done $\frac{1}{2}$ hr earlier than Penny. In effect, they have not worked together to get the job done as fast as possible. If Lon helps Penny after completing his half, the entire job could be done in a time somewhere between 2 hr and $2\frac{1}{2}$ hr.

   We proceed to a translation by considering how much of the job is finished in 1 hr, 2 hr, 3 hr, and so on. It takes Lon 4 hr to do the mowing job alone. Then in 1 hr, he can do $\frac{1}{4}$ of the job. It takes Penny 5 hr to do the job alone. Then in 1 hr, she can do $\frac{1}{5}$ of the job. Working together, they can do

   $$\tfrac{1}{4} + \tfrac{1}{5}, \text{ or } \tfrac{9}{20} \text{ of the job in 1 hr.}$$

   In 2 hr, Lon can do $2(\frac{1}{4})$ of the job and Penny can do $2(\frac{1}{5})$ of the job. Working together, they can do

   $$2(\tfrac{1}{4}) + 2(\tfrac{1}{5}), \text{ or } \tfrac{9}{10} \text{ of the job in 2 hr.}$$

   Continuing this reasoning, we can form a table like the following one.

	Fraction of the job completed		
Time	**Lon**	**Penny**	**Together**
1 hr	$\frac{1}{4}$	$\frac{1}{5}$	$\frac{1}{4} + \frac{1}{5}$, or $\frac{9}{20}$
2 hr	$2(\frac{1}{4})$	$2(\frac{1}{5})$	$2(\frac{1}{4}) + 2(\frac{1}{5})$, or $\frac{9}{10}$
3 hr	$3(\frac{1}{4})$	$3(\frac{1}{5})$	$3(\frac{1}{4}) + 3(\frac{1}{5})$, or $1\frac{7}{20}$
t hr	$t(\frac{1}{4})$	$t(\frac{1}{5})$	$t(\frac{1}{4}) + t(\frac{1}{5})$

OBJECTIVES

After finishing Section 6.4, you should be able to:

- **a** Solve work problems and certain basic problems involving rational equations.
- **b** Solve problems involving proportions.
- **c** Solve motion problems.

FOR EXTRA HELP

Tape 12A

Tape 10A

MAC: 6
IBM: 6

1. Fred does a certain typing job in 6 hr. Fran can do the same job in 4 hr. How long would it take them to do the same amount of typing working together?

ANSWER ON PAGE A-7

From the table, we see that if they work 3 hr, the fraction of the job that they will finish is $1\frac{7}{20}$, which is more of the job than needs to be done. We also see that the answer is somewhere between 2 hr and 3 hr. What we want is a number t such that the fraction of the job that gets completed is 1; that is, the job is just completed—not more ($1\frac{7}{20}$) and not less ($\frac{9}{10}$).

2. *Translate.* From the table, we see that the time we want is some number t for which

$$t\left(\frac{1}{4}\right) + t\left(\frac{1}{5}\right) = 1, \quad \text{or} \quad \frac{t}{4} + \frac{t}{5} = 1,$$

where 1 represents the idea that the entire job is completed in time t.

3. *Solve.* We solve the equation:

$$\frac{t}{4} + \frac{t}{5} = 1$$

$$20\left(\frac{t}{4} + \frac{t}{5}\right) = 20 \cdot 1 \qquad \textbf{The LCM is } 2 \cdot 2 \cdot 5\textbf{, or } 20.$$

$$20 \cdot \frac{t}{4} + 20 \cdot \frac{t}{5} = 20$$

$$5t + 4t = 20$$

$$9t = 20$$

$$t = \frac{20}{9}, \quad \text{or } 2\frac{2}{9} \text{ hr.}$$

4. *Check.* The check can be done by using $\frac{20}{9}$ for t and substituting into the original equation:

$$\frac{20}{9}\left(\frac{1}{4}\right) + \frac{20}{9}\left(\frac{1}{5}\right) = \frac{5}{9} + \frac{4}{9} = \frac{9}{9} = 1.$$

We also have a partial check in what we learned from our familiarization. The answer, $2\frac{2}{9}$ hr, is between 2 hr and 3 hr (see the table), and it is less than 4 hr, the time it takes Lon working alone.

5. *State.* It takes $2\frac{2}{9}$ hr for them to do the job working together. ◀

The following principle is helpful when solving work problems such as the one in Example 1.

The Work Principle

Suppose that a = the time it takes person A to do a job, b = the time it takes person B to do the same job, and t = the time it takes them to do the same job working together. Then

$$\frac{t}{a} + \frac{t}{b} = 1.$$

DO EXERCISE 1.

▶ **EXAMPLE 2** At a factory, smokestack A pollutes the air twice as fast as smokestack B. When the stacks operate together, they yield a certain amount of pollution in 15 hr. Find the time it would take each to yield that same amount of pollution operating alone.

1. *Familiarize.* Let a = the number of hours it takes A to yield the pollution. Then $2a$ = the number of hours it takes B to yield the same amount of pollution. Now

$\frac{15}{a}$ is the fraction of the pollution produced by A in 15 hr working alone

and

$\frac{15}{2a}$ is the fraction of the pollution produced by B in 15 hr working alone.

> Common sense will help us with this reasoning. The slower smokestack will take *longer* to generate pollution than the faster stack. Since B is slower, the time required is longer. Thus, $2a$ is the time required for B.

2. *Translate.* Using the work principle, we get the following equation:

$$\frac{15}{a} + \frac{15}{2a} = 1.$$

3. *Solve.* We solve the equation:

$$2a\left(\frac{15}{a} + \frac{15}{2a}\right) = 2a \cdot 1$$ **Multiplying on both sides by the LCM of the denominators to clear of fractions**

$$\frac{30a}{a} + \frac{30a}{2a} = 2a$$

$$30 + 15 = 2a$$

$$45 = 2a$$

$$a = \frac{45}{2}, \quad \text{or } 22\frac{1}{2}.$$

4. *Check.* Thus if the number checks, the answer will be that A takes a, or $22\frac{1}{2}$ hr, and B takes $2a$, or 45 hr. Part of the check is obvious: B will take twice as long, so A pollutes twice as fast. We can complete the check by verifying the work principle:

$$\frac{15}{22\frac{1}{2}} + \frac{15}{45} = \frac{15}{1} \cdot \frac{2}{45} + \frac{15}{45} = \frac{30}{45} + \frac{15}{45} = \frac{45}{45} = 1.$$

5. *State.* Thus A takes $22\frac{1}{2}$ hr, working alone, and B takes 45 hr, working alone, to yield the same amount of pollution. ◀

DO EXERCISE 2.

2. Two pipes carry water to the same tank. Pipe A, working alone, can fill the tank three times as fast as pipe B. Together the pipes can fill the tank in 24 hr. Find the time each would take to fill the tank alone.

b Proportion Problems

Any rational expression a/b represents a **ratio.** Percent can be considered a ratio. For example, 67% is the ratio of 67 to 100, or 67/100. The ratio of two different kinds of measure is called a **rate.** Speed is an example of a rate. Florence Griffith Joyner set a world record in a recent Olympics with a time of 10.49 sec in the 100-m dash. Her speed, or rate, was

$$\frac{100 \text{ m}}{10.49 \text{ sec}}, \quad \text{or } 9.5\,\frac{\text{m}}{\text{sec}}.$$ **Rounded to the nearest tenth**

> **An equality of ratios, $A/B = C/D$, is called a *proportion.* The numbers named in a true proportion are said to be *proportional.***

ANSWER ON PAGE A-7

3. A pitcher gave up 71 earned runs in 285 innings in a recent year. At this rate, how many runs did the pitcher give up every 9 innings? (There are 9 innings in a baseball game.)

ANSWER ON PAGE A-7

We can use proportions to solve applied problems by expressing a ratio in two ways. For example, suppose it takes 8 gal of gas to drive 120 mi, and we wish to find how much will be required to go 550 mi. If we assume that the car uses gas at the same rate throughout the trip, the ratios are the same, and we can write a proportion.

$$\begin{matrix}\text{Gas} \longrightarrow \\ \text{Miles} \longrightarrow\end{matrix} \frac{8}{120} = \frac{x}{550} \begin{matrix}\longleftarrow \text{Gas} \\ \longleftarrow \text{Miles}\end{matrix}$$

To solve this proportion, we multiply by 550 to get x alone on one side:

$$550 \cdot \frac{8}{120} = 550 \cdot \frac{x}{550}$$
$$\frac{550 \cdot 8}{120} = x$$
$$36.67 \approx x.$$

Thus, 36.67 gal will be required to go 550 mi. (Note that we could have multiplied by the LCM of 120 and 550, which is 66,000, but that would have been more complicated.) We can also use **cross products** to solve proportions:

$$\frac{8}{120} = \frac{x}{550}$$ **If $\frac{A}{B} = \frac{C}{D}$, then $AD = BC$.**

$$8 \cdot 550 = 120 \cdot x$$ **$8 \cdot 550$ and $120 \cdot x$ are called *cross products.***

$$\frac{8 \cdot 550}{120} = x$$
$$36.67 = x.$$

We can verify this method using the multiplication principle, multiplying on both sides of the proportion by 550 and then by 120, but we will not do so here.

► **EXAMPLE 3** *Earned-run average.* A pitcher gave up 76 earned runs in 198 innings in a recent year. At this rate, how many runs did the pitcher give up every 9 innings? The answer to this question is the pitcher's **earned-run average, ERA.** (An "earned" run is a run scored without the benefit of errors.)

Let E = the number of earned runs that the pitcher gives up every 9 innings. Then we translate to a proportion:

$$\text{Runs each 9 innings} \longrightarrow \frac{E}{9} = \frac{76}{198}. \begin{matrix}\longleftarrow \text{Runs (total)} \\ \longleftarrow \text{Innings pitched (total)}\end{matrix}$$

We solve the proportion:

$$9 \cdot \frac{E}{9} = 9 \cdot \frac{76}{198}$$ **Multiplying by 9 to get E alone**

$$E = \frac{9 \cdot 76}{198}$$
$$E \approx 3.45.$$ **Rounded to the nearest hundredth**

The pitcher gives up 3.45 earned runs every 9 innings pitched. ◄

DO EXERCISE 3.

▶ **EXAMPLE 4** *Estimating wildlife populations.* To determine the number of deer in a forest, a conservationist catches 612 deer, tags them, and lets them loose. Later, 244 deer are caught, and 72 of them are found to be tagged. Estimate how many deer are in the forest.

Let $D =$ the number of deer in the forest. Then we translate to a proportion:

$$\text{Deer tagged originally} \longrightarrow \frac{612}{D} = \frac{72}{244} \begin{array}{l} \longleftarrow \text{Tagged deer caught later} \\ \longleftarrow \text{Deer caught later} \end{array}$$
Deer in forest $\longrightarrow D$

We solve the proportion. This time we multiply by the LCM, which is $244D$:

$$244D \cdot \frac{612}{D} = 244D \cdot \frac{72}{244} \qquad \textbf{Multiplying by } \mathbf{244D}$$

$$244 \cdot 612 = D \cdot 72$$

$$\frac{244 \cdot 612}{72} = D \qquad \textbf{Dividing by 72}$$

$$2074 = D.$$

We estimate that there are about 2074 deer in the forest. ◀

DO EXERCISE 4.

4. *Estimating wildlife populations.* To determine the number of fish in a lake, a conservationist catches 225 fish, tags them, and throws them back into the lake. Later, 108 fish are caught, and 15 of them are found to be tagged. Estimate how many fish are in the lake.

C Motion Problems

Problems dealing with distance, speed, and time are called **motion problems.** To translate them, we use either the basic motion formula, $d = rt$, or either of two formulas $r = d/t$ or $t = d/r$, which can be derived from $d = rt$.

▶ **EXAMPLE 5** A car leaves Sioux City traveling east with the cruise control set at 40 mph. Two hours later, another car leaves Sioux City in the same direction with the cruise control set at 55 mph. How far from Sioux City will the faster car catch the slower one?

1. *Familiarize.* We first make a drawing.

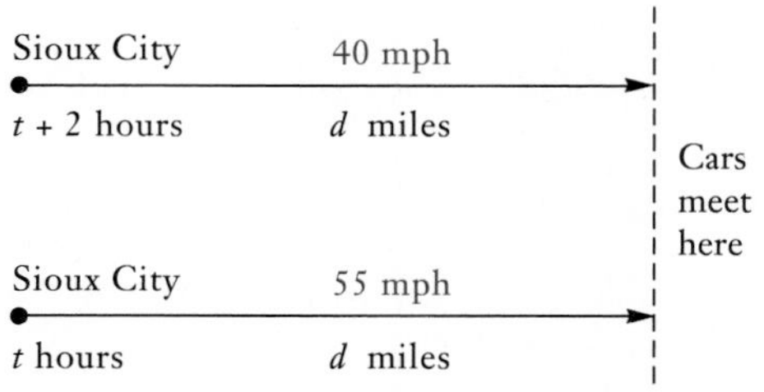

From the drawing, we see that the distances are the same. We let $d =$ the distance. We don't know the times, so we let $t =$ the number of hours traveled by the faster car. Then the time for the slower one will be $t + 2$. We can organize the information in a table.

	Distance	Speed	Time	
Slow car	d	40	$t + 2$	$\longrightarrow d = 40(t + 2)$
Fast car	d	55	t	$\longrightarrow d = 55t$

2. *Translate.* Using $d = rt$ in each row of the table, we get an equation. Thus we get a system of two equations:

$$d = 40(t + 2) \quad \text{and} \quad d = 55t. \qquad \textbf{This is the translation.}$$

ANSWER ON PAGE A-7

5. A train leaves Georgetown traveling east at 35 mph. One hour later, a faster train leaves Georgetown, also traveling east on a parallel track at 40 mph. How far from Georgetown will the faster train catch the slower one?

3. *Solve.* We solve:

$$\begin{aligned} 40(t+2) &= 55t \quad \textbf{Using substitution} \\ 40t + 80 &= 55t \\ 80 &= 15t \\ t &= 5\tfrac{1}{3}. \end{aligned}$$

4. *Check.* At 55 mph, the faster car will go $55(5\frac{1}{3})$, or $293\frac{1}{3}$ mi in $5\frac{1}{3}$ hr. At 40 mph, the slower car will go $40(5\frac{1}{3} + 2)$, or $293\frac{1}{3}$ mi in $7\frac{1}{3}$ hr. We have a check.
5. *State.* The faster car will catch the slower $293\frac{1}{3}$ mi from Sioux City. ◀

DO EXERCISE 5.

▶ **EXAMPLE 6** An airplane flies 1062 mi with the wind. In the same amount of time, it can fly 738 mi against the wind. The speed of the plane in still air is 200 mph. Find the speed of the wind.

1. *Familiarize.* We first make a drawing. We let w = the speed of the wind and t = the time, and then organize the facts in a chart.

1062 mi

$200 + w$ (The wind pushes the plane and increases the speed over the ground.)

738 mi

$200 - w$ (The wind slows down the plane and decreases the speed over the ground.)

6. A boat travels 246 mi downstream in the same time that it takes to travel 180 mi upstream. The speed of the current in the stream is 5.5 mph. Find the speed of the boat in still water.

	Distance	Speed	Time
With wind	1062	$200 + w$	t
Against wind	738	$200 - w$	t

2. *Translate.* Using $t = d/r$, we get a system of equations:

$$t = \frac{1062}{200 + w} \quad \text{and} \quad t = \frac{738}{200 - w}. \qquad \textbf{This is the translation.}$$

3. *Solve.* We solve:

$$\begin{aligned} \frac{1062}{200 + w} &= \frac{738}{200 - w} \qquad \textbf{Using substitution} \\ (200 + w)(200 - w)\left(\frac{1062}{200 + w}\right) &= (200 + w)(200 - w)\left(\frac{738}{200 - w}\right) \\ (200 - w)1062 &= (200 + w)738 \\ 212{,}400 - 1062w &= 147{,}600 + 738w \\ 64{,}800 &= 1800w \\ w &= 36. \end{aligned}$$

4. *Check.* With the wind, the speed of the plane is 236 mph. Dividing the distance, 1062 mi, by the speed, 236 mph, we get 4.5 hr. Against the wind, the speed of the plane is 164 mph. Dividing the distance, 738 mi, by the speed, 164 mph, we get 4.5 hr. The answer checks.
5. *State.* The speed of the wind is 36 mph. ◀

ANSWERS ON PAGE A-7

DO EXERCISE 6.

NAME SECTION DATE

EXERCISE SET 6.4

ANSWERS

a Solve.

1. The sum of a number and 21 times its reciprocal is -10. Find the number.

2. The sum of a number and 6 times its reciprocal is -5. Find the number.

3. The reciprocal of 5 plus the reciprocal of 6 is the reciprocal of what number?

4. The reciprocal of 8 plus the reciprocal of 7 is the reciprocal of what number?

5. Wilma, an experienced shipping clerk, can fill a certain order in 5 hr. Willy, a new clerk, needs 9 hr to do the same job. Working together, how long will it take them to fill the order?

6. Paul can paint a room in 4 hr. Paula can paint the same room in 3 hr. Working together, how long will it take them to paint the room?

7. A swimming pool can be filled in 12 hr if water enters through a pipe alone, or in 30 hr if water enters through a hose alone. If water is entering through both the pipe and the hose, how long will it take to fill the pool?

8. A tank can be filled in 18 hr by pipe A alone and in 22 hr by pipe B alone. How long will it take to fill the tank if both pipes are working?

9. Bill can clear a lot in 5.5 hr. His partner can do the same job in 7.5 hr. How long will it take them to clear the lot working together?

Hint: You may find that multiplying on both sides of your equation by $\frac{1}{10}$ will clear of decimals.

10. One printing press can print an order of booklets in 4.5 hr. Another press can do the same job in 5.5 hr. How long will it take if both presses are used?

1. ______

2. ______

3. ______

4. ______

5. ______

6. ______

7. ______

8. ______

9. ______

10. ______

ANSWERS

11. ________ ________

12. ________

13. ________

14. ________

15. ________

16. ________

17. ________

18. ________

19. ________

20. ________

11. A can paint the neighbor's house 4 times as fast as B. The year they worked together it took them 8 days. How long would it take each to paint the house alone?

12. A can deliver papers 3 times as fast as B. If they work together, it takes them 1 hr. How long would it take each to deliver the papers alone?

b Solve.

13. Light travels 930,000 mi in 5 sec. How far can light travel in 60 sec?

14. It is known that a black racer snake can travel 46 km in 20 hr. How far can a black racer travel in 3 hr?

15. Last season, a major league baseball player got 120 hits in 300 times at bat. If the player expects to bat 500 times in the entire season with the same ratio of hits to at-bats, how many hits will the player have?

16. The coffee beans from 14 trees are required to produce 7.7 kg of coffee (this is the average amount that each person in the United States drinks each year). How many trees are required to produce 638 kg of coffee?

17. A student traveled 234 km in 14 days. At this same ratio, how far would the student travel in 56 days?

18. 10 cm^3 of a normal specimen of human blood contains 1.2 g of hemoglobin. How many grams does 32 cm^3 of the same blood contain?

19. To determine the number of deer in a game preserve, a conservationist catches 318 deer, tags them, and lets them loose. Later, 168 deer are caught; 56 of them are tagged. How many deer are in the preserve?

20. To determine the number of trout in a lake, a conservationist catches 112 trout, tags them, and throws them back into the lake. Later, 82 trout are caught; 32 of them are tagged. How many trout are in the lake?

21. The ratio of the weight of an object on Mars to the weight of an object on earth is 0.4 to 1.

a) How much will a 12-T rocket weigh on Mars?
b) How much will a 120-lb astronaut weigh on Mars?

22. The ratio of the weight of an object on the moon to the weight of an object on earth is 0.16 to 1.

a) How much will a 12-T rocket weigh on the moon?
b) How much will a 180-lb astronaut weigh on the moon?

23. Consider the numbers 1, 2, 3, and 5. If the same number is added to each of the numbers, it is found that the ratio of the first new number to the second is the same as the ratio of the third new number to the fourth. Find the number.

24. A rope is 28 ft long. How can the rope be cut in such a way that the ratio of the resulting two segments is 3 to 5?

C Solve.

25. The speed of a stream is 3 mph. A boat travels 4 mi upstream in the same time it takes to travel 10 mi downstream. What is the speed of the boat in still water?

26. The speed of a stream is 4 mph. A boat travels 6 mi upstream in the same time it takes to travel 12 mi downstream. What is the speed of the boat in still water?

27. A train leaves a station and travels north at 75 km/h. Two hours later, a second train leaves on a parallel track traveling north at 125 km/h. How far from the station will the second train overtake the first train?

28. A private airplane leaves an airport and flies due east at 180 km/h. Two hours later, a jet leaves the same airport and flies due east at 900 km/h. How far from the airport will the jet overtake the private plane?

29. The speed of train A is 12 mph slower than the speed of train B. Train A travels 230 mi in the same time it takes train B to travel 290 mi. Find the speed of each train.

30. The speed of a passenger train is 14 mph faster than the speed of a freight train. The passenger train travels 400 mi in the same time it takes the freight train to travel 330 mi. Find the speed of each train.

ANSWERS

21. a) ____________

b) ____________

22. a) ____________

b) ____________

23. ____________

24. ____________

25. ____________

26. ____________

27. ____________

28. ____________

29. ____________

30. ____________

ANSWERS

31. ______

32. ______

33. ______

34. ______

35. ______

36. ______

37. ______

38. ______

39. ______

40. ______

41. ______

42. ______

43. ______

44. ______

45. ______

31. A boat can move at a speed of 15 km/h in still water. The boat travels 140 km downstream in a river in the same time it takes to travel 35 km upstream. What is the speed of the river?

32. A paddleboat can move at a speed of 2 km/h in still water. The boat is paddled 4 km downstream in a river in the same time it takes to go 1 km upstream. What is the speed of the river?

SYNTHESIS

33. An automobile gets 22.5 miles per gallon (mpg) in city driving and 30 mpg in highway driving. The car is driven 465 mi on a full tank of 18.4 gal of gasoline. How many miles were driven in the city and how many were driven on the highway?

34. A woman drives to work at 45 mph and arrives 1 min early. She drives to work at 40 mph and arrives 1 min late. How far does she live from work?

35. At what time after 4:00 will the minute hand and the hour hand of a clock first be in the same position?

36. At what time after 10:30 will the hands of a clock first be perpendicular?

37. A boat travels 96 km downstream in 4 hr. It travels 28 km upstream in 7 hr. Find the speed of the boat and the speed of the stream.

38. An airplane carries enough fuel for 6 hr of flight time, and its speed in still air is 240 mph. It leaves an airport against a wind of 40 mph and returns to the same airport with a wind of 40 mph. How far can it fly under those conditions without refueling?

39. A motor boat travels 3 times as fast as the current. A trip up the river and back takes 10 hr, and the total distance of the trip is 100 km. Find the speed of the current.

40. An employee drives to work at 50 mph and arrives 1 min late. The employee drives to work at 60 mph and arrives 5 min early. How far does the employee live from work?

Average speed is defined as *total distance divided by total time.*

41. A driver went 200 km. For the first half of the distance of a trip, the driver traveled at a speed of 40 km/h. For the second half of the distance of a trip, the driver traveled at a speed of 60 km/h. What was the average speed for the entire trip? (It is *not* 50 km/h.)

42. For half of the distance of a trip, a driver travels at 40 mph. What speed would the driver have to travel for the last half of the distance of a trip so that the average speed for the entire trip would be 45 mph?

43. Three trucks A, B, and C, working together, can move a load of sand in t hours. When working alone, it takes A 1 extra hour to move the sand, B 6 extra hours, and C t extra hours. Find t.

44. Explain how you might go about estimating the number of words in a book using a proportion.

45. A computer takes 1.23×10^{-6} sec to do one operation. How many of the same operations can it do in one minute?

6.5 Formulas

a Solving formulas is important in business, technology, science, and engineering.

▶ **EXAMPLE 1** *Earned-run average.* The formula $E = 9R/I$ tells how to calculate a pitcher's earned-run average (ERA). In this formula, E is the ERA, R is the number of earned runs, and I is the number of innings pitched. Solve this formula for I.

We multiply by the LCM, which is I:

$$E = \frac{9R}{I}$$

$$I \cdot E = I \cdot \frac{9R}{I}$$

$$I \cdot E = 9R.$$

> In solving formulas, we proceed as though we were solving equations with numbers. In this case, we have a rational equation, so the first thing we do is clear of fractions.

Next we divide by E: $\frac{I \cdot E}{E} = \frac{9R}{E}$

$$I = \frac{9R}{E}.$$

From the formula solved for I, we can calculate the number of innings a pitcher pitched if we know the earned-run average and the number of earned runs given up. ◀

DO EXERCISE 1.

▶ **EXAMPLE 2** *Resistance.* The formula $\frac{1}{R} = \frac{1}{r_1} + \frac{1}{r_2}$ gives the resistance R of two resistors r_1 and r_2 connected in parallel.* Solve it for r_1.

We multiply by the LCM, which is Rr_1r_2:

$$Rr_1r_2 \cdot \frac{1}{R} = Rr_1r_2 \cdot \left[\frac{1}{r_1} + \frac{1}{r_2}\right] \quad \text{Multiplying by the LCM}$$

$$Rr_1r_2 \cdot \frac{1}{R} = Rr_1r_2 \cdot \frac{1}{r_1} + Rr_1r_2 \cdot \frac{1}{r_2} \quad \text{Multiplying to remove parentheses}$$

$$r_1r_2 = Rr_2 + Rr_1. \quad \text{Simplifying by removing factors of 1}$$

One might be tempted at this point to multiply by $1/r_2$ to get r_1 alone on the left, *but* note that there is an r_1 on the right. We must get all the terms involving r_1 on the *same side* of the equation.

$$r_1r_2 - Rr_1 = Rr_2 \quad \text{Subtracting } Rr_1$$

$$r_1(r_2 - R) = Rr_2 \quad \text{Factoring out } r_1$$

$$r_1 = \frac{Rr_2}{r_2 - R} \quad \text{Dividing by } r_2 - R \text{ to get } r_1 \text{ alone}$$ ◀

DO EXERCISE 2.

---

* Note that R, r_1, and r_2 are all different variables. It is common to use subscripts, as in r_1 and r_2, to distinguish variables. In many applications, this helps to keep ideas straight.

OBJECTIVE

After finishing Section 6.5, you should be able to:

a Solve a formula for a given letter.

FOR EXTRA HELP

Tape 12B | Tape 10B | MAC: 6 IBM: 6

1. *Boyle's law.* The formula
$$\frac{PV}{T} = k$$
relates the pressure, volume, and temperature of a gas. Solve it for T. (*Hint:* Begin by clearing of fractions.)

2. *Work formula.* The formula
$$\frac{1}{t} = \frac{1}{u} + \frac{1}{v}$$
gives the total time t for some work to be done by two workers whose individual times are u and v. Solve it for t. (*Hint:* Begin by clearing of fractions. Then multiply to remove parentheses. Get all terms with t alone on one side. Then factor out t.)

ANSWERS ON PAGE A-7

3. Solve $V = \dfrac{A}{n(T-t)}$ for T.

4. Solve $F = \dfrac{kmM}{d^2}$ for M.

5. Solve $L = \dfrac{t}{5}(r-w)$ for t.

ANSWERS ON PAGE A-7

To solve a formula, we do the same things we would do to solve an equation with numbers. For rational formulas, the procedure is as follows.

To solve a rational formula:

1. **Clear of fractions.**
2. **Multiply to remove parentheses, if necessary.**
3. **Get all terms with the unknown alone on one side of the equation.**
4. **Factor out the unknown.**
5. **Use the multiplication principle to get the unknown alone on one side.**

▶ **EXAMPLE 3** Solve the formula $I = \dfrac{pT}{M + pn}$ for p.

We first clear of fractions by multiplying by the LCM, $M + pn$:

$$I = \frac{pT}{M+pn}$$

$$I(M+pn) = \frac{pT}{M+pn}(M+pn)$$

$$IM + Ipn = pT.$$

We must now get all terms containing p alone on one side:

$$IM = pT - Ipn \qquad \textbf{Subtracting } Ipn$$

$$IM = p(T - In) \qquad \textbf{Factoring out the letter } p$$

$$\frac{IM}{T - In} = p. \qquad \textbf{Dividing by } T - In$$

We now have p alone on one side, and p does *not* appear on the other side, so we have solved the formula for p. ◀

DO EXERCISE 3.

▶ **EXAMPLE 4** *Telephone calls.* The formula $N = \dfrac{kP_1P_2}{d^2}$ is used to find the number N of telephone calls between two cities of populations P_1 and P_2 at a distance d from each other. Solve it for P_1.

Note that P_1 and P_2 are different variables (or constants). So, don't forget to write the subscripts, 1 and 2. We first clear of fractions:

$$d^2N = kP_1P_2. \qquad \textbf{Multiplying by } d^2$$

Now there is no rational expression in the equation. Next,

$$\frac{d^2N}{kP_2} = P_1. \qquad \textbf{Dividing by } kP_2 \textbf{ to get } P_1 \textbf{ alone}$$ ◀

▶ **EXAMPLE 5** Solve $P = \dfrac{k}{4}(a + b)$ for k.

$$4P = k(a+b) \qquad \textbf{Multiplying by 4}$$

$$\frac{4P}{a+b} = k \qquad \textbf{Dividing by } a + b$$ ◀

DO EXERCISES 4 AND 5.

NAME SECTION DATE

EXERCISE SET 6.5

ANSWERS

a Solve.

1. $\frac{W_1}{W_2} = \frac{d_1}{d_2}$ for d_1

2. $\frac{W_1}{W_2} = \frac{d_1}{d_2}$ for W_2

3. $s = \frac{(v_1 + v_2)t}{2}$ for t

4. $s = \frac{(v_1 + v_2)t}{2}$ for v_1

5. $\frac{1}{R} = \frac{1}{r_1} + \frac{1}{r_2}$ for r_2

6. $\frac{1}{R} = \frac{1}{r_1} + \frac{1}{r_2}$ for R

7. $R = \frac{gs}{g + s}$ for s

8. $I = \frac{2V}{V + 2r}$ for V

9. $\frac{1}{p} + \frac{1}{q} = \frac{1}{f}$ for p

10. $\frac{1}{p} + \frac{1}{q} = \frac{1}{f}$ for f

1. ________

2. ________

3. ________

4. ________

5. ________

6. ________

7. ________

8. ________

9. ________

10. ________

ANSWERS

11. ______

12. ______

13. ______

14. ______

15. ______

16. ______

17. ______

18. ______

19. See graph.

20. ______

21. ______

22. ______

23. ______

24. ______

25. ______

26. ______

11. $I = \dfrac{nE}{E + nr}$ for E

12. $I = \dfrac{nE}{E + nr}$ for n

13. $S = \dfrac{H}{m(t_1 - t_2)}$ for H

14. $S = \dfrac{H}{m(t_1 - t_2)}$ for t_1

15. $\dfrac{E}{e} = \dfrac{R + r}{r}$ for e

16. $\dfrac{E}{e} = \dfrac{R + r}{r}$ for r

17. $A = P(1 + rt)$ for r

18. $V = \dfrac{1}{3}\pi h^2(3R - h)$ for R

SKILL MAINTENANCE

19. Graph on a plane: $6x - y < 6$.

20. Factor: $p^3 - 27q^3$.

21. Factor: $t^3 + 8b^3$.

22. Factor: $a^3 - 8b^3$.

SYNTHESIS

Person A can do a job in a hours and person B can do the same job in b hours. Working together, they can do the same job in t hours.

23. Find a formula relating a, b, and t.

24. Solve the formula for t.

25. Solve the formula for a.

26. Solve the formula for b.

6.6 Complex Rational Expressions

a A **complex rational expression,** or **complex fractional expression,** is a rational expression that has one or more rational expressions within its numerator or denominator. Here are some examples:

$$\frac{1 + \frac{y}{x}}{x}, \quad \frac{\frac{x+y}{3}}{\frac{5x}{x+2}}, \quad \frac{\frac{1}{a} + \frac{1}{b}}{\frac{1}{a^3} + \frac{1}{b^3}}.$$

These are rational expressions within the complex rational expression.

There are two methods that can be used to simplify complex rational expressions. We will consider both of them. Use the one that works best for you or the one that your instructor directs you to use.

Method 1: Multiplying by the LCM of All the Denominators

> ***Method 1.* To simplify a complex rational expression:**
>
> 1. **First, find the LCM of all the denominators of all the expressions occurring within both the numerator and the denominator of the complex rational expression. Let a = the LCM.**
> 2. **Multiply by 1 using a/a.**
> 3. **Simplify.**

► **EXAMPLE 1** Simplify: $\dfrac{x + \frac{1}{5}}{x - \frac{1}{3}}$.

We first find the LCM of all the denominators of all the rational expressions occurring in both the numerator and the denominator of the complex rational expression. The denominators are 3 and 5. The LCM of these denominators is $3 \cdot 5$, or 15. We multiply by 15/15.

$$\frac{x + \frac{1}{5}}{x - \frac{1}{3}} = \left(\frac{x + \frac{1}{5}}{x - \frac{1}{3}}\right) \cdot \frac{15}{15}$$ **Multiplying by 1**

$$= \frac{\left(x + \frac{1}{5}\right) \cdot 15}{\left(x - \frac{1}{3}\right) \cdot 15}$$ **Multiplying the numerators and the denominators**

$$= \frac{15x + \frac{1}{5} \cdot 15}{15x - \frac{1}{3} \cdot 15}$$ **Carrying out the multiplications using the distributive laws**

$$= \frac{15x + 3}{15x - 5}$$ **No further simplifying is possible.** ◄

OBJECTIVE

After finishing Section 6.6, you should be able to:

a Simplify complex rational expressions.

FOR EXTRA HELP

Tape 12C

Tape 10B

MAC: 6
IBM: 6

Simplify. Use method 1.

1. $\dfrac{y+\dfrac{1}{2}}{y-\dfrac{1}{7}}$

2. $\dfrac{1-\dfrac{1}{x}}{1-\dfrac{1}{x^2}}$

ANSWERS ON PAGE A-7

▶ **EXAMPLE 2** Simplify: $\dfrac{1+\dfrac{1}{x}}{1-\dfrac{1}{x^2}}$.

We first find the LCM of all the denominators of all the rational expressions occurring in both the numerator and the denominator of the complex rational expression. The denominators are x and x^2. The LCM of these denominators is x^2. We multiply by x^2/x^2.

$$\frac{1+\frac{1}{x}}{1-\frac{1}{x^2}} = \left(\frac{1+\frac{1}{x}}{1-\frac{1}{x^2}}\right)\cdot\frac{x^2}{x^2} \qquad \textbf{Multiplying by 1}$$

$$= \frac{\left(1+\frac{1}{x}\right)\cdot x^2}{\left(1-\frac{1}{x^2}\right)\cdot x^2} \qquad \textbf{Multiplying the numerators and the denominators}$$

$$= \frac{x^2+\frac{1}{x}\cdot x^2}{x^2-\frac{1}{x^2}\cdot x^2} \qquad \textbf{Carrying out the multiplications using the distributive laws}$$

$$= \frac{x^2+x}{x^2-1}$$

$$= \frac{x(x+1)}{(x+1)(x-1)} \qquad \textbf{Factoring}$$

$$= \frac{x\cancel{(x+1)}}{\cancel{(x+1)}(x-1)} \qquad \textbf{Removing a factor of 1: } \frac{x+1}{x+1}=1$$

$$= \frac{x}{x-1}$$ ◀

DO EXERCISES 1 AND 2.

▶ **EXAMPLE 3** Simplify: $\dfrac{\dfrac{1}{a}+\dfrac{1}{b}}{\dfrac{1}{a^3}+\dfrac{1}{b^3}}$.

We first find the LCM of all the denominators of all the rational expressions occurring in both the numerator and the denominator of the complex rational expression. The denominators are a, b, a^3, and b^3. The LCM of these denominators is a^3b^3. We multiply by a^3b^3/a^3b^3.

$$\frac{\frac{1}{a}+\frac{1}{b}}{\frac{1}{a^3}+\frac{1}{b^3}} = \left(\frac{\frac{1}{a}+\frac{1}{b}}{\frac{1}{a^3}+\frac{1}{b^3}}\right)\cdot\frac{a^3b^3}{a^3b^3} \qquad \textbf{Multiplying by 1}$$

$$= \frac{\left(\frac{1}{a}+\frac{1}{b}\right)\cdot a^3b^3}{\left(\frac{1}{a^3}+\frac{1}{b^3}\right)\cdot a^3b^3} \qquad \textbf{Multiplying the numerators and the denominators}$$

Then

$$= \frac{\frac{1}{a} \cdot a^3b^3 + \frac{1}{b} \cdot a^3b^3}{\frac{1}{a^3} \cdot a^3b^3 + \frac{1}{b^3} \cdot a^3b^3}$$ **Carrying out the multiplications using a distributive law**

$$= \frac{a^2b^3 + a^3b^2}{b^3 + a^3} = \frac{a^2b^2(b + a)}{(b + a)(a^2 - ab + b^2)}$$ **Factoring**

$$= \frac{a^2b^2\cancel{(b + a)}}{\cancel{(b + a)}(a^2 - ab + b^2)}$$ **Removing a factor of 1:** $\frac{b + a}{b + a} = 1$

$$= \frac{a^2b^2}{a^2 - ab + b^2}$$ ◀

DO EXERCISES 3 AND 4.

Simplify. Use method 1.

3. $\dfrac{\frac{1}{a} + \frac{1}{b}}{\frac{1}{a} - \frac{1}{b}}$

4. $\dfrac{\frac{1}{a} - \frac{1}{b}}{\frac{1}{a^3} - \frac{1}{b^3}}$

Method 2: Adding in the Numerator and Denominator

Method 2. **To simplify a complex rational expression:**

1. **Add or subtract, as necessary, to get a single rational expression in the numerator.**
2. **Add or subtract, as necessary, to get a single rational expression in the denominator.**
3. **Divide the numerator by the denominator.**
4. **If possible, simplify by removing a factor of 1.**

We will redo Examples 1–3 using this method.

► **EXAMPLE 4** Simplify: $\dfrac{x + \frac{1}{5}}{x - \frac{1}{3}}$.

$$\frac{x + \frac{1}{5}}{x - \frac{1}{3}} = \frac{x \cdot \frac{5}{5} + \frac{1}{5}}{x - \frac{1}{3}} = \frac{\frac{5x + 1}{5}}{x - \frac{1}{3}}$$ **Finding the LCM in the numerator, multiplying by 1, and adding**

$$= \frac{\frac{5x + 1}{5}}{x \cdot \frac{3}{3} - \frac{1}{3}} = \frac{\frac{5x + 1}{5}}{\frac{3x - 1}{3}}$$ **Finding the LCM in the denominator, multiplying by 1, and subtracting**

$$= \frac{5x + 1}{5} \cdot \frac{3}{3x - 1}$$ **Multiplying by the reciprocal of the denominator**

$$= \frac{15x + 3}{15x - 5}$$ **Multiplying the numerators and the denominators. Simplifying by removing a factor of 1 is not possible in this case.**

If you feel more comfortable doing so, you can always write denominators of 1 where there are no denominators. In this case, you could start out by writing

$$\frac{\frac{x}{1} + \frac{1}{5}}{\frac{x}{1} - \frac{1}{3}}$$ ◀

ANSWERS ON PAGE A-7

Simplify. Use method 2.

5. $\dfrac{y+\dfrac{1}{2}}{y-\dfrac{1}{7}}$

6. $\dfrac{1-\dfrac{1}{x}}{1-\dfrac{1}{x^2}}$

▶ **EXAMPLE 5** Simplify: $\dfrac{1+\dfrac{1}{x}}{1-\dfrac{1}{x^2}}$.

$$\frac{1+\frac{1}{x}}{1-\frac{1}{x^2}} = \frac{\frac{x}{x}+\frac{1}{x}}{\frac{x^2}{x^2}-\frac{1}{x^2}}$$ **Finding the LCM and multiplying by 1**

$$= \frac{\frac{x+1}{x}}{\frac{x^2-1}{x^2}}$$ **Adding in the numerator and subtracting in the denominator**

$$= \frac{x+1}{x}\cdot\frac{x^2}{x^2-1}$$ **Multiplying by the reciprocal of the denominator**

$$= \frac{(x+1)\cdot x\cdot x}{x(x-1)(x+1)}$$ **Removing a factor of 1: $\dfrac{x(x+1)}{x(x+1)} = 1$**

$$= \frac{x}{x-1}$$ ◀

DO EXERCISES 5 AND 6.

Simplify. Use method 2.

7. $\dfrac{\dfrac{1}{a}+\dfrac{1}{b}}{\dfrac{1}{a}-\dfrac{1}{b}}$

8. $\dfrac{\dfrac{1}{a}-\dfrac{1}{b}}{\dfrac{1}{a^3}-\dfrac{1}{b^3}}$

▶ **EXAMPLE 6** Simplify: $\dfrac{\dfrac{1}{a}+\dfrac{1}{b}}{\dfrac{1}{a^3}+\dfrac{1}{b^3}}$.

$$\frac{\frac{1}{a}+\frac{1}{b}}{\frac{1}{a^3}+\frac{1}{b^3}} = \frac{\frac{1}{a}\cdot\frac{b}{b}+\frac{1}{b}\cdot\frac{a}{a}}{\frac{1}{a^3}\cdot\frac{b^3}{b^3}+\frac{1}{b^3}\cdot\frac{a^3}{a^3}} = \frac{\frac{b}{ab}+\frac{a}{ab}}{\frac{b^3}{a^3b^3}+\frac{a^3}{a^3b^3}}$$

$$= \frac{\frac{b+a}{ab}}{\frac{b^3+a^3}{a^3b^3}}$$ **Adding in the numerator and the denominator**

$$= \frac{b+a}{ab}\cdot\frac{a^3b^3}{b^3+a^3}$$ **Multiplying by the reciprocal of the denominator**

$$= \frac{(b+a)a^3b^3}{ab(b^3+a^3)}$$

$$= \frac{(b+a)\cdot ab\cdot a^2b^2}{ab(b+a)(b^2-ab+a^2)}$$ **Removing a factor of 1: $\dfrac{ab(b+a)}{ab(b+a)} = 1$**

$$= \frac{a^2b^2}{b^2-ab+a^2}$$ ◀

DO EXERCISES 7 AND 8.

ANSWERS ON PAGE A-7

NAME SECTION DATE

EXERCISE SET 6.6

ANSWERS

a Simplify. Use either method 1 or method 2 as you wish or as directed by your instructor.

1. $\dfrac{\dfrac{1}{x}+4}{\dfrac{1}{x}-3}$

2. $\dfrac{\dfrac{1}{y}+7}{\dfrac{1}{y}-5}$

3. $\dfrac{x-\dfrac{1}{x}}{x+\dfrac{1}{x}}$

4. $\dfrac{y+\dfrac{1}{y}}{y-\dfrac{1}{y}}$

5. $\dfrac{\dfrac{3}{x}+\dfrac{4}{y}}{\dfrac{4}{x}-\dfrac{3}{y}}$

6. $\dfrac{\dfrac{2}{y}+\dfrac{5}{z}}{\dfrac{1}{y}-\dfrac{4}{z}}$

7. $\dfrac{\dfrac{x^2-y^2}{xy}}{\dfrac{x-y}{y}}$

8. $\dfrac{\dfrac{a^2-b^2}{ab}}{\dfrac{a-b}{b}}$

9. $\dfrac{a-\dfrac{3a}{b}}{b-\dfrac{b}{a}}$

10. $\dfrac{1-\dfrac{2}{3x}}{x-\dfrac{4}{9x}}$

11. $\dfrac{\dfrac{1}{a}+\dfrac{1}{b}}{\dfrac{a^2-b^2}{ab}}$

12. $\dfrac{\dfrac{1}{x}+\dfrac{1}{y}}{\dfrac{x^2-y^2}{xy}}$

13. $\dfrac{\dfrac{1}{x+h}-\dfrac{1}{x}}{h}$

It may help you to write h as $\dfrac{h}{1}$.

14. $\dfrac{\dfrac{1}{a-h}-\dfrac{1}{a}}{h}$

1. ______
2. ______
3. ______
4. ______
5. ______
6. ______
7. ______
8. ______
9. ______
10. ______
11. ______
12. ______
13. ______
14. ______

ANSWERS

15. ______

16. ______

17. ______

18. ______

19. ______

20. ______

21. ______

22. ______

23. ______

24. ______

25. ______

26. ______

27. ______

28. ______

29. ______

30. ______

15. $\dfrac{\dfrac{y^2 - y - 6}{y^2 - 5y - 14}}{\dfrac{y^2 + 6y + 5}{y^2 - 6y - 7}}$

16. $\dfrac{\dfrac{x^2 - x - 12}{x^2 - 2x - 15}}{\dfrac{x^2 + 8x + 12}{x^2 - 5x - 14}}$

17. $\dfrac{\dfrac{1}{x+2} + \dfrac{4}{x-3}}{\dfrac{2}{x-3} - \dfrac{7}{x+2}}$

18. $\dfrac{\dfrac{1}{y-4} + \dfrac{1}{y+5}}{\dfrac{6}{y+5} + \dfrac{2}{y-4}}$

19. $\dfrac{\dfrac{6}{x^2-4} - \dfrac{5}{x+2}}{\dfrac{7}{x^2-4} - \dfrac{4}{x-2}}$

20. $\dfrac{\dfrac{1}{x^2-1} + \dfrac{5}{x^2-5x+4}}{\dfrac{1}{x^2-1} + \dfrac{2}{x^2+3x+2}}$

21. $\dfrac{\dfrac{1}{a^2} - \dfrac{1}{b^2}}{\dfrac{1}{a^3} + \dfrac{1}{b^3}}$

22. $\dfrac{\dfrac{1}{x^2} - \dfrac{1}{y^2}}{\dfrac{1}{x^3} - \dfrac{1}{y^3}}$

SYNTHESIS

Simplify.

23. $\dfrac{5x^{-1} - 5y^{-1} + 10x^{-1}y^{-1}}{6x^{-1} - 6y^{-1} + 12x^{-1}y^{-1}}$

24. $\dfrac{\dfrac{4a}{2a^2 - a - 1} - \dfrac{4}{a-1}}{\dfrac{1}{a-1} + \dfrac{6}{2a+1}}$

25. $2 + \cfrac{2}{2 + \cfrac{2}{2 + \cfrac{2}{2 + \cfrac{2}{x}}}}$

26. $\left[\dfrac{\dfrac{x+3}{x-3} + 1}{\dfrac{x+3}{x-3} - 1}\right]^8$

27. $(a^2 - ab + b^2)^{-1}(a^2b^{-1} + b^2a^{-1})(a^{-2} - b^{-2})(a^{-2} + 2a^{-1}b^{-1} + b^{-2})^{-1}$

Find the reciprocal of each of the following and simplify.

28. $x^2 - \dfrac{1}{x}$

29. $\dfrac{1 - \dfrac{1}{a}}{a - 1}$

30. $\dfrac{a^3 + b^3}{a + b}$

6.7 Variation: Direct, Inverse, and Joint

a Direct Variation

A plumber earns \$9 per hour. In 1 hour, \$9 is earned. In 2 hours, \$18 is earned. In 3 hours, \$27 is earned, and so on. This creates a set of ordered pairs of numbers, all having the same ratio:

$$(1, 9), \quad (2, 18), \quad (3, 27), \quad (4, 36), \quad \text{and so on.}$$

The ratio of earnings to time is $\frac{9}{1}$ in every case.

Whenever a situation gives rise to pairs of numbers in which the ratio is constant, we say that there is **direct variation.** Here the earnings *vary directly* as the time:

$$\frac{E}{t} = 9 \text{ (a constant)}, \quad \text{or} \quad E = 9t.$$

Note that the constant, 9, is positive. Note also that when one variable increases, so does the other. When one decreases, so does the other.

> **Whenever a situation gives rise to a relation among variables $y = kx$, where k is a positive constant, we say that there is *direct variation,* or that *y varies directly as x,* or that *y is directly proportional to x.* The number k is called the *variation constant,* or the *constant of proportionality.***

The graph of $y = kx$, $k > 0$, always goes through the origin and rises from left to right. Note that as x increases, y increases. The constant k is also the slope of the line.

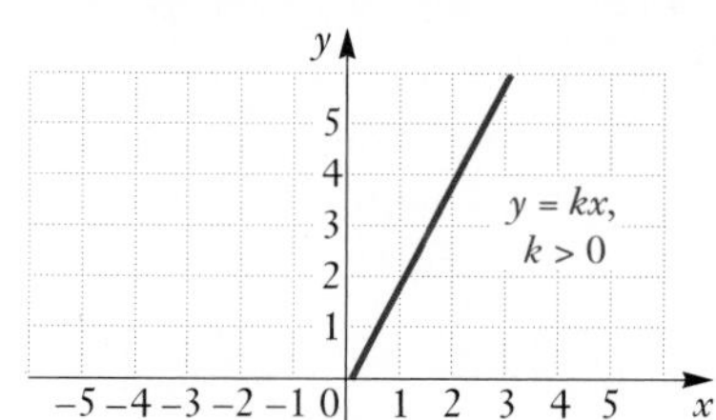

► **EXAMPLE 1** Find the variation constant and an equation of variation where y varies directly as x and $y = 32$ when $x = 2$.

We know that (2, 32) is a solution of $y = kx$. Thus,

$$32 = k \cdot 2 \qquad \textbf{Substituting}$$

$$\frac{32}{2} = k, \quad \text{or} \quad k = 16. \qquad \textbf{Solving for } k$$

The variation constant is 16. The equation of variation is $y = 16x$. ◄

DO EXERCISES 1 AND 2.

OBJECTIVES

After finishing Section 6.7, you should be able to:

- **a** Given a description of direct variation, write an equation of variation.
- **b** Solve problems involving direct variation.
- **c** Given a description of inverse variation, write an equation of variation.
- **d** Solve problems involving inverse variation.
- **e** Given a situation of variation, write an equation of variation or a proportion.
- **f** Solve problems involving variation.

FOR EXTRA HELP

Tape 12D

Tape 11A

MAC: 6
IBM: 6

1. Find the variation constant and an equation of variation where y varies directly as x and $y = 8$ when $x = 20$.

2. Find the variation constant and an equation of variation where y varies directly as x and $y = 5.6$ when $x = 8$.

ANSWERS ON PAGE A-7

3. *Ohm's law* states that the voltage V in an electric circuit varies directly as the number of amperes I of electric current in the circuit. If the voltage is 10 volts when the current is 3 amperes, what is the voltage when the current is 15 amperes?

4. *An ecology problem.* The amount of garbage G produced in the United States varies directly as the number of people N who produce the garbage. It is known that 50 tons of garbage is produced by 200 people in 1 year. The population of San Francisco is 705,000. How much garbage is produced by San Francisco in 1 year?

ANSWERS ON PAGE A-7

b Direct-Variation Problems

▶ **EXAMPLE 2** The number of centimeters W of water produced from melting snow varies directly as S, the number of centimeters of snow. Meteorologists have found that 150 cm of snow will melt to 16.8 cm of water. How many centimeters of water will 200 cm of snow melt to?

We first find the variation constant using the data and then find an equation of variation:

$$W = kS \quad \textbf{W varies directly as S.}$$
$$16.8 = k \cdot 150 \quad \textbf{Substituting}$$
$$\frac{16.8}{150} = k \quad \textbf{Solving for k}$$
$$0.112 = k. \quad \textbf{This is the variation constant.}$$

The equation of variation is $W = 0.112S$.

Next, we use the equation to find how many centimeters of water will result from melting 200 cm of snow:

$$W = 0.112S$$
$$W = 0.112(200) \quad \textbf{Substituting}$$
$$W = 22.4.$$

Thus, 200 cm of snow will melt to 22.4 cm of water. ◀

DO EXERCISES 3 AND 4.

c Inverse Variation

A bus is traveling a distance of 20 mi. At a speed of 20 mph, it will take 1 hr. At 40 mph, it will take $\frac{1}{2}$ hr. At 60 mph, it will take $\frac{1}{3}$ hr, and so on. This gives rise to a set of pairs of numbers, all having the same product:

$$(20, 1), \quad (40, \tfrac{1}{2}), \quad (60, \tfrac{1}{3}), \quad (80, \tfrac{1}{4}), \quad \text{and so on.}$$

Whenever a situation gives rise to pairs of numbers whose product is constant, we say that there is **inverse variation.** Here the time *varies inversely* as the speed:

$$rt = 20 \text{ (a constant)}, \quad \text{or} \quad t = \frac{20}{r}.$$

Note that the constant, 20, is positive. Note also that when one variable increases, the other decreases.

Whenever a situation gives rise to a relation among variables $y = k/x$, where k is a positive constant, we say that there is *inverse variation*, or that *y varies inversely as x*, or that *y is inversely proportional to x*. The number k is called the *variation constant*.

Although we will not study such graphs until Chapter 9, it is helpful to look at the graph of $y = k/x$, $k > 0$. The graph is like the one shown at the right for positive values of x. Note that as x increases, y decreases.

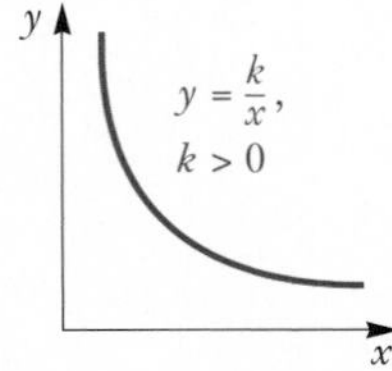

▶ **EXAMPLE 3** Find the variation constant and an equation of variation where y varies inversely as x and $y = 32$ when $x = 0.2$.

We know that (0.2, 32) is a solution of $y = k/x$. We substitute:

$$y = \frac{k}{x}$$

$$32 = \frac{k}{0.2} \quad \textbf{Substituting}$$

$$(0.2)32 = k$$

$$6.4 = k. \quad \textbf{Solving}$$

The variation constant is 6.4. The equation of variation is $y = 6.4/x$. ◀

DO EXERCISE 5.

d Inverse-Variation Problems

▶ **EXAMPLE 4** The time t required to do a certain job varies inversely as the number of people P who work on the job (assuming that all do the same amount of work). It takes 4 hr for 12 people to erect some football bleachers. How long would it take 3 people to do the same job?

We first find the variation constant using the data and then find an equation of variation:

$$t = \frac{k}{P} \quad \textbf{\textit{t} varies inversely as \textit{P}.}$$

$$4 = \frac{k}{12} \quad \textbf{Substituting}$$

$$48 = k. \quad \textbf{Solving for \textit{k}, the variation constant}$$

The equation of variation is $t = 48/P$.

Next, we use the equation to find the time it would take 3 people to do the job:

$$t = \frac{48}{P}$$

$$t = \frac{48}{3} \quad \textbf{Substituting}$$

$$t = 16.$$

It would take 16 hr for 3 people to do the job. ◀

DO EXERCISE 6.

5. Find the variation constant and an equation of variation where y varies inversely as x and $y = 0.012$ when $x = 50$.

6. *Time and speed.* The time t required to drive a fixed distance varies inversely as the speed r. It takes 5 hr at 60 km/h to drive a fixed distance. How long would it take to drive that same distance at 40 km/h?

ANSWERS ON PAGE A-7

7. Find an equation of variation where y varies directly as the square of x and $y = 175$ when $x = 5$.

8. Find an equation of variation where y varies inversely as the square of x and $y = \frac{1}{4}$ when $x = 6$.

e Other Kinds of Variation

We now look at other kinds of variation. Consider the equation for the area of a circle, in which A and r are variables and π is a constant:

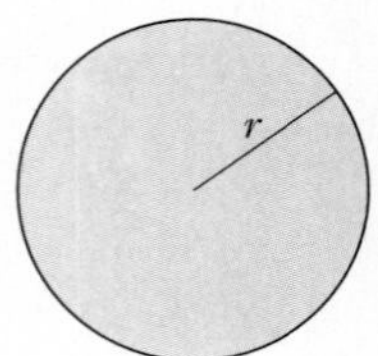

$$A = \pi r^2.$$

We say that the area *varies directly* as the square of the radius.

> **y varies directly as the nth power of x if there is some positive constant k such that $y = kx^n$.**

▶ **EXAMPLE 5** Find an equation of variation where y varies directly as the square of x and $y = 12$ when $x = 2$.

We write an equation of variation and find k:

$$y = kx^2$$
$$12 = k \cdot 2^2 \qquad \textbf{Substituting}$$
$$12 = k \cdot 4$$
$$3 = k.$$

Thus, $y = 3x^2$. ◀

DO EXERCISE 7.

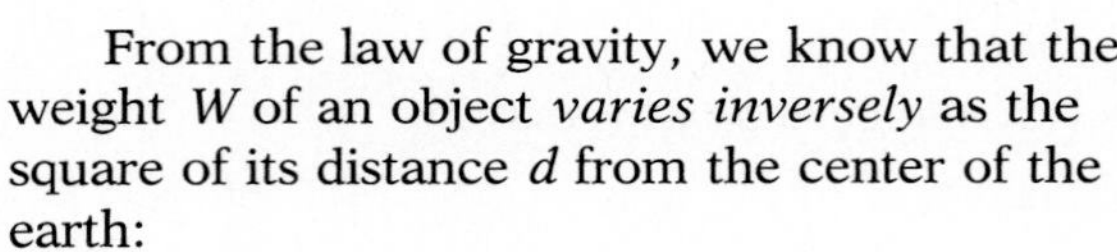

From the law of gravity, we know that the weight W of an object *varies inversely* as the square of its distance d from the center of the earth:

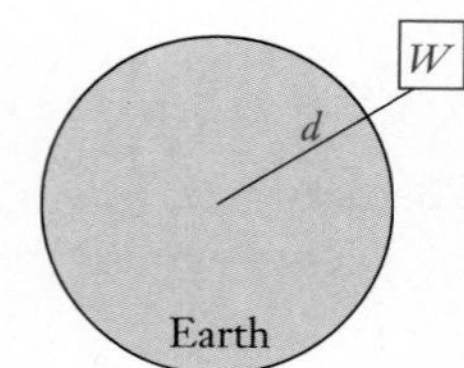

$$W = \frac{k}{d^2}.$$

> **y varies inversely as the nth power of x if there is some positive constant k such that $y = k/x^n$.**

▶ **EXAMPLE 6** Find an equation of variation where W varies inversely as the square of d and $W = 3$ when $d = 5$.

$$W = \frac{k}{d^2}$$
$$3 = \frac{k}{5^2} \qquad \textbf{Substituting}$$
$$3 = \frac{k}{25}$$
$$75 = k$$

Thus, $W = \dfrac{75}{d^2}$. ◀

DO EXERCISE 8.

ANSWERS ON PAGE A-7

Consider the equation for the area A of a triangle with height h and base b:

$$A = \frac{1}{2}bh.$$

We say that the area *varies jointly* as the height and the base.

> **y varies jointly as x and z if there is some positive constant k such that $y = kxz$.**

► **EXAMPLE 7** Find an equation of variation where y varies jointly as x and z and $y = 42$ when $x = 2$ and $z = 3$.

$$\begin{aligned} y &= kxz \\ 42 &= k \cdot 2 \cdot 3 \qquad \textbf{Substituting} \\ 42 &= k \cdot 6 \\ 7 &= k \end{aligned}$$

Thus, $y = 7xz$. ◄

DO EXERCISE 9.

The equation

$$y = k \cdot \frac{xz^2}{w}$$

asserts that y varies jointly as x and the square of z, and inversely as w.

► **EXAMPLE 8** Find an equation of variation where y varies jointly as x and z and inversely as the square of w, and $y = 105$ when $x = 3$, $z = 20$, and $w = 2$.

$$\begin{aligned} y &= k \cdot \frac{xz}{w^2} \\ 105 &= k \cdot \frac{3 \cdot 20}{2^2} \qquad \textbf{Substituting} \\ 105 &= k \cdot 15 \\ 7 &= k \end{aligned}$$

Thus, $y = 7 \cdot \dfrac{xz}{w^2}$. ◄

DO EXERCISE 10.

9. Find an equation of variation where y varies jointly as x and z and $y = 65$ when $x = 10$ and $z = 13$.

10. Find an equation of variation where y varies jointly as x and the square of z and inversely as w, and $y = 80$ when $x = 4$, $z = 10$, and $w = 25$.

ANSWERS ON PAGE A-7

f Solving Variation Problems

Many problem situations can be described with equations of variation.

▶ **EXAMPLE 9** The volume of wood V in a tree varies jointly as the height h and the square of the girth g (girth is distance around). If the volume of a redwood tree is 216 m³ when the height is 30 m and the girth is 1.5 m, what is the height of a tree whose volume is 960 m³ and girth is 2 m?

We first find k using the first set of data. Then we solve for h using the second set of data.

$$V = khg^2$$
$$216 = k \cdot 30 \cdot 1.5^2$$
$$3.2 = k$$

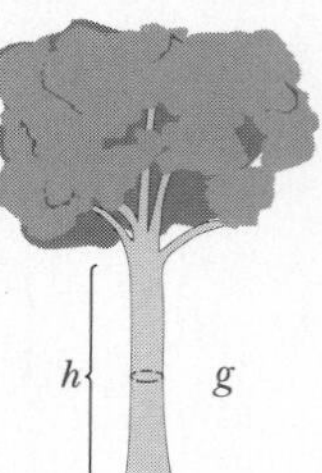

Then the equation of variation is $V = 3.2hg^2$. We substitute the second set of data into the equation:

$$960 = 3.2 \cdot h \cdot 2^2$$
$$75 = h.$$

Therefore, the height is 75 m. ◀

▶ **EXAMPLE 10** The intensity I of a TV signal varies inversely as the square of the distance d from the transmitter. If the intensity is 23 watts per square meter (W/m²) at a distance of 2 km, what is the intensity at a distance of 6 km?

We first find k using the first set of data. Then we solve for I using the second set of data.

$$I = \frac{k}{d^2}$$
$$23 = \frac{k}{2^2}$$
$$92 = k$$

Then the equation of variation is $I = 92/d^2$. We substitute the second distance into the equation:

$$I = \frac{92}{d^2} = \frac{92}{6^2} \approx 2.56.$$ **Rounded to the nearest hundredth**

Therefore, at 6 km, the intensity is about 2.56 W/m². ◀

DO EXERCISES 11 AND 12.

11. *Distance of a dropped object.* The distance s that an object falls when dropped from some point above the ground varies directly as the square of the time t that it falls. If the object falls 19.6 m in 2 sec, how far will the object fall in 10 sec?

12. *Resistance and length.* At a fixed temperature, the resistance R of a wire varies directly as the length l and inversely as the square of its diameter d. If the resistance is 0.1 ohm when the diameter is 1 mm and the length is 50 cm, what is the resistance when the length is 2000 cm and the diameter is 2 mm?

ANSWERS ON PAGE A-7

NAME SECTION DATE

EXERCISE SET 6.7

ANSWERS

a Find the variation constant and an equation of variation where y varies directly as x and the following are true.

1. $y = 24$ when $x = 3$

2. $y = 5$ when $x = 12$

3. $y = 3.6$ when $x = 1$

4. $y = 2$ when $x = 5$

5. $y = 0.8$ when $x = 0.5$

6. $y = 0.6$ when $x = 0.4$

b Solve.

7. *Ohm's law.* The electric current I, in amperes, in a circuit varies directly as the voltage V. When 12 volts are applied, the current is 4 amperes. What is the current when 18 volts are applied?

8. *Hooke's law.* Hooke's law states that the distance d that a spring is stretched by a hanging object varies directly as the weight w of the object. If the distance is 40 cm when the weight is 3 kg, what is the distance when the weight is 5 kg?

9. *Weekly allowance.* According to Fidelity Investments *Investment Vision Magazine*, the average weekly allowance A of children varies directly as their grade level, G. It is known that the average allowance of a 9th grade student is \$9.66 per week. What then is the average weekly allowance of a 4th grade student?

10. The number N of aluminum cans used each year varies directly as the number of people using the cans. If 250 people use 60,000 cans in one year, how many cans are used each year in Dallas, which has a population of 850,000?

11. The number of kilograms of water W in a human body varies directly as the total weight. A person weighing 96 kg contains 64 kg of water. How many kilograms of water are in a person weighing 75 kg?

12. The weight M of an object on Mars varies directly as its weight E on earth. A person who weighs 209 lb on earth weighs 83.6 lb on Mars. How much would a 220-lb person weigh on Mars?

1. ______
2. ______
3. ______
4. ______
5. ______
6. ______
7. ______
8. ______
9. ______
10. ______
11. ______
12. ______

ANSWERS

13. ______

14. ______

15. ______

16. ______

17. ______

18. ______

19. ______

20. ______

21. ______

22. ______

23. ______

24. ______

25. ______

26. ______

27. ______

28. ______

c Find the variation constant and an equation of variation where y varies inversely as x and the following are true.

13. $y = 6$ when $x = 10$

14. $y = 16$ when $x = 4$

15. $y = 12$ when $x = 3$

16. $y = 9$ when $x = 5$

17. $y = 0.4$ when $x = 0.8$

18. $y = 1.5$ when $x = 0.3$

d Solve.

19. *Electrical current.* The current I in an electrical conductor varies inversely as the resistance R of the conductor. If the current is $\frac{1}{2}$ ampere when the resistance is 240 ohms, what is the current when the resistance is 540 ohms?

20. *Pump time.* The time t required to empty a tank varies inversely as the rate r of pumping. If a pump can empty a tank in 45 min at the rate of 600 kL per minute, how long will it take the pump to empty the same tank at the rate of 1000 kL per minute?

21. *Beam weight.* The weight W that a horizontal beam can support varies inversely as the length L of the beam. Suppose that an 8-m beam can support 1200 kg. How many kilograms can a 14-m beam support?

22. *Radio waves.* The wavelength W of a radio wave varies inversely as its frequency F. A wave with a frequency of 1200 kHz (kilohertz) has a length of 300 m. What is the length of a wave with a frequency of 800 kHz?

e Find an equation of variation in which the following are true.

23. y varies directly as the square of x and $y = 0.15$ when $x = 0.1$.

24. y varies directly as the square of x and $y = 6$ when $x = 3$.

25. y varies inversely as the square of x and $y = 0.15$ when $x = 0.1$.

26. y varies inversely as the square of x and $y = 6$ when $x = 3$.

27. y varies jointly as x and z and $y = 56$ when $x = 7$ and $z = 8$.

28. y varies directly as x and inversely as z and $y = 4$ when $x = 12$ and $z = 15$.

29. y varies jointly as x and the square of z and $y = 105$ when $x = 14$ and $z = 5$.

30. y varies jointly as x and z and inversely as w and $y = \frac{3}{2}$ when $x = 2$, $z = 3$, and $w = 4$.

31. y varies jointly as x and z and inversely as the product of w and p, and $y = \frac{3}{28}$ when $x = 3$, $z = 10$, $w = 7$, and $p = 8$.

32. y varies jointly as x and z and inversely as the square of w, and $y = \frac{12}{5}$ when $x = 16$, $z = 3$, and $w = 5$.

f Solve.

33. *Stopping distance of a car.* The stopping distance d of a car after the brakes are applied varies directly as the square of the speed r. If a car traveling 60 mph can stop in 200 ft, how many feet will it take the same car to stop when it is traveling 80 mph?

34. *Area of a cube.* The area of a cube varies directly as the square of the length of a side. If a cube has an area of 168.54 in^2 when the length of a side is 5.3 in., what will the area be when the length of a side is 10.2 in.?

35. *Weight of an astronaut.* The weight W of an object varies inversely as the square of the distance d from the center of the earth. At sea level (3978 mi from the center of the earth), an astronaut weighs 220 lb. Find her weight when she is 200 mi above the surface of the earth and the spacecraft is not in motion.

36. *Intensity of light.* The intensity I of light from a lightbulb varies inversely as the square of the distance d from the bulb. Suppose that I is 90 W/m^2 when the distance is 5 m. Find the intensity at a distance of 10 m.

37. *Earned-run average.* A pitcher's earned-run average A varies directly as the number R of earned runs allowed and inversely as the number I of innings pitched. In a recent year, Tom Seaver had an earned-run average of 2.92. He gave up 85 earned runs in 262 innings. How many earned runs would he have given up had he pitched 300 innings with the same average? Round to the nearest whole number.

38. *Boyle's law: Volume of a gas.* The volume V of a given mass of a gas varies directly as the temperature T and inversely as the pressure P. If $V = 231\ \text{cm}^3$ when $T = 42°$ and $P = 20\ \text{kg/cm}^2$, what is the volume when $T = 30°$ and $P = 15\ \text{kg/cm}^2$?

ANSWERS

29. ______

30. ______

31. ______

32. ______

33. ______

34. ______

35. ______

36. ______

37. ______

38. ______

ANSWERS

39. ______

40. a) ______

b) ______

41. ______

42. ______

43. ______

44. a) ______

b) ______

c) ______

d) ______

e) ______

45. ______

46. ______

47. ______

48. ______

39. *Water flow.* The amount Q of water emptied by a pipe varies directly as the square of the diameter d. A pipe 5 in. in diameter will empty 225 gal of water over a fixed time period. Assuming the same kind of flow, how many gallons of water are emptied in the same amount of time by a pipe that is 9 in. in diameter?

40. *Weight, volume, and diameter of a sphere.* The weight W of a sphere of a given kind of material varies directly as its volume V, and its volume V varies directly as the cube of its diameter.

a) Find an equation of variation relating the weight W to the diameter d.

b) A sphere that is 5 in. in diameter is known to weigh 25 lb. Find the weight of a sphere that is 8 in. in diameter.

SKILL MAINTENANCE

41. There are 75 questions on a college entrance examination. Two points are awarded for each correct answer and one-half point is deducted for each wrong answer. A score of 100 indicates how many correct and how many wrong answers?

42. A pentagon with all five sides congruent has the same perimeter as an octagon with all eight sides congruent. One side of the pentagon is 2 less than 3 times one side of the octagon. Find the perimeters.

SYNTHESIS

43. A loan of \$9000 earns simple interest of \$1665 in one year. Find an equation of variation that describes the simple interest I in terms of the loan principal P.

44. In each of the following equations, state whether y varies directly as x, inversely as x, or neither directly nor inversely as x.

a) $7xy = 14$

b) $x - 2y = 12$

c) $-2x + 3y = 0$

d) $x = \frac{3}{4}y$

e) $\frac{x}{y} = 2$

45. The area of a circle varies directly as the square of the length of a diameter. What is the variation constant?

46. A jar of peanut butter in the shape of a right circular cylinder is 4 in. high and 3 in. in diameter and sells for \$1.20. Assuming the same ratio of peanut butter to cost, what is the cost of a jar that is 6 in. high and 6 in. in diameter?

Describe, in words, the variation given by the equation.

47. $Q = \frac{kp^2}{q^3}$

48. $W = \frac{km_1M_1}{d^2}$

SUMMARY AND REVIEW: CHAPTER 6

IMPORTANT PROPERTIES AND FORMULAS

Direct Variation: $y = kx$ *Inverse Variation:* $y = \frac{k}{x}$ *Joint Variation:* $y = kxy$

REVIEW EXERCISES

The review sections and objectives to be tested in addition to the material in this chapter are [4.3a], [4.6b], [5.4a, b], and [5.6a].

Find the LCM.

1. $6x^3, \quad 16x^2$
2. $x^2 - 49, \quad 3x + 1$
3. $x^2 + x - 20, \quad x^2 + 3x - 10$

Perform the indicated operations and simplify.

Caution! Do not make the mistake of trying to clear of fractions when you *do not* have an equation.

4. $\frac{y^2 - 64}{2y + 10} \cdot \frac{y + 5}{y + 8}$
5. $\frac{x^3 - 8}{x^2 - 25} \cdot \frac{x^2 + 10x + 25}{x^2 + 2x + 4}$
6. $\frac{9a^2 - 1}{a^2 - 9} \div \frac{3a + 1}{a + 3}$
7. $\frac{x^3 - 64}{x^2 - 16} \div \frac{x^2 + 5x + 6}{x^2 - 3x - 18}$
8. $\frac{x}{x^2 + 5x + 6} - \frac{2}{x^2 + 3x + 2}$
9. $\frac{9xy}{x^2 - y^2} + \frac{x + y}{x - y}$
10. $\frac{2x^2}{x - y} + \frac{2y^2}{x + y}$
11. $\frac{3}{y + 4} - \frac{y}{y - 1} + \frac{y^2 + 3}{y^2 + 3y - 4}$

Solve.

12. $\frac{6}{x} + \frac{4}{x} = 5$
13. $\frac{x}{7} + \frac{x}{4} = 1$
14. $\frac{5}{3x + 2} = \frac{3}{2x}$
15. $\frac{4x}{x + 1} + \frac{4}{x} + 9 = \frac{4}{x^2 + x}$
16. $\frac{90}{x^2 - 3x + 9} - \frac{5x}{x + 3} = \frac{405}{x^3 + 27}$
17. $\frac{2}{x - 3} + \frac{1}{4x + 20} = \frac{1}{x^2 + 2x - 15}$

Solve.

18. Danielle can paint the outside of the house in 12 hr. Bill can paint the same house in 9 hr. How long would it take them working together to paint the house?

19. A camper gets on an interstate highway and sets its cruise control to a speed of 45 mph. Three hours later, a car gets on the same interstate at the same place and sets its cruise control to a speed of 55 mph. After what distance will the car overtake the camper?

20. The speed of a stream is 6 mph. A boat travels 50 mi downstream in the same time that it takes to travel 30 mi upstream. What is the speed of the boat in still water?

21. Solve $R = \dfrac{gs}{g + s}$ for s; for g.

22. Solve $S = \dfrac{H}{m(t_1 - t_2)}$ for m; for t_1.

23. Find an equation of variation where y varies directly as x and $y = 100$ when $x = 25$.

24. Find an equation of variation where y varies inversely as x and $y = 100$ when $x = 25$.

Solve.

25. The pitch P of a musical tone varies inversely as its wavelength W. One tone has a pitch of 330 vibrations per second and a wavelength of 3.2 ft. Find the wavelength of another tone that has a pitch of 550 vibrations per second.

26. The score N on a test varies directly as the number of correct responses a. A student who gets 28 questions correct earns a score of 87. What is the score of a student who gets 25 questions correct?

27. The power P expended by heat in an electric circuit of fixed resistance varies directly as the square of the current C in the circuit. A circuit expends 180 watts when a current of 6 amperes is flowing. What is the heat expended when the current is 10 amperes?

28. A warning dye is used by people in lifeboats to aid searching airplanes. The radius r of the circle formed by the dye varies directly as the square root of the volume V. It is found that 4 L of dye will spread to a circle of radius 5 m. To what radius will 9 L spread?

Simplify.

29. $\dfrac{3 + \dfrac{3}{y}}{4 + \dfrac{4}{y}}$

30. $\dfrac{\dfrac{2}{a} + \dfrac{2}{b}}{\dfrac{4}{a^3} + \dfrac{4}{b^3}}$

31. $\dfrac{\dfrac{y^2 + 4y - 77}{y^2 - 10y + 25}}{\dfrac{y^2 - 5y - 14}{y^2 - 25}}$

SKILL MAINTENANCE

Graph on a plane.

32. $y - 2x \geq 4$

33. $x > -3$

34. Factor: $125x^3 - 8y^3$.

35. Factor: $6x^2 + 29x - 42$.

36. A side of a square is five less than a side of an equilateral triangle. The perimeter of the square is the same as the perimeter of the triangle. Find the length of a side of the square and the length of a side of the triangle.

SYNTHESIS

37. Solve: $\dfrac{5}{x - 13} - \dfrac{5}{x} = \dfrac{65}{x^2 - 13x}$.

38. Find the reciprocal and simplify: $\dfrac{a - b}{a^3 - b^3}$.

❖ THINKING IT THROUGH

1. Explain at least three different uses of the LCM studied in this chapter.
2. You have learned to solve new kinds of equations in this chapter. Explain how the equations differ from those you have studied previously and how the equation-solving process differs.
3. Explain the difference between a rational expression and a rational equation.

NAME SECTION DATE

TEST: CHAPTER 6

ANSWERS

1. Find the LCM for $x^2 + x - 6$ and $x^2 + 8x + 15$.

Perform the indicated operations and simplify.

2. $\dfrac{2x^2 + 20x + 50}{x^2 - 4} \cdot \dfrac{x + 2}{x + 5}$

3. $\dfrac{x}{x^2 + 11x + 30} - \dfrac{5}{x^2 + 9x + 20}$

4. $\dfrac{y^2 - 16}{2y + 6} \div \dfrac{y - 4}{y + 3}$

5. $\dfrac{x^2}{x - y} + \dfrac{y^2}{y - x}$

6. $\dfrac{1}{x + 1} - \dfrac{x + 2}{x^2 - 1} + \dfrac{3}{x - 1}$

7. $\dfrac{a}{a - b} + \dfrac{b}{a^2 + ab + b^2} - \dfrac{2}{a^3 - b^3}$

Solve.

8. $\dfrac{2}{x - 1} = \dfrac{3}{x + 3}$

9. $\dfrac{7x}{x + 3} + \dfrac{21}{x - 3} = \dfrac{126}{x^2 - 9}$

10. $\dfrac{2x}{x - 7} - \dfrac{14}{x} = \dfrac{98}{x^2 - 7x}$

11. $\dfrac{1}{3x - 6} - \dfrac{1}{x^2 - 4} = \dfrac{3}{x + 2}$

Solve.

12. Two pipes are filling a tank. One of them can fill the tank alone in 6 hr. The other can fill the tank alone in 4 hr. How long will it take them together to fill the tank?

13. A boat can go upstream a certain distance in 6 hr. It can go that same distance downstream in 2 hr and 15 min. The speed of the stream is 5 km/hr. What is the speed of the boat in still water?

1. ______

2. ______

3. ______

4. ______

5. ______

6. ______

7. ______

8. ______

9. ______

10. ______

11. ______

12. ______

13. ______

ANSWERS

14. ______

15. ______

16. ______

17. ______

18. ______

19. ______

20. ______

21. ______

22. ______

23. ______

24. ______

25. ______

26. ______

27. ______

28. ______

29. ______

14. A boat can travel at a speed of 12 km/h in still water. The boat travels 50 km downstream in a river in the same time it takes to travel 10 km upstream. What is the speed of the river current?

15. Bob can unload a truckload of melons in 5 hr. Working together, it takes Bob and Jack 2 hr to do the same job. How long would it take Jack to unload the truck by himself?

16. Solve $m = \dfrac{2t}{r+t}$ for t.

17. Find an equation of variation where Q varies jointly as x and y and $Q = 25$ when $x = 2$ and $y = 5$.

Solve.

18. Income I varies directly as time t worked. A job pays $275 for 40 hr of work. What does it pay for 72 hr?

19. The time t required to drive a certain distance varies inversely as the speed s. You can make a trip in $3\frac{1}{2}$ hr at 45 mph. How long will it take you to make the same trip at 60 mph?

20. The area of a balloon varies directly as the square of its radius. The area is 3.4 cm^2 when the radius is 5 cm. What is the area when the radius is 7 cm?

21. Simplify: $\dfrac{1 - \dfrac{1}{x^2}}{1 - \dfrac{1}{x}}$.

22. Simplify: $\dfrac{\dfrac{1}{a^3} + \dfrac{1}{b^3}}{\dfrac{1}{a} + \dfrac{1}{b}}$.

SKILL MAINTENANCE

Factor.

23. $64a^3 + 1$

24. $16t^2 - 24t - 72$

25. Factor: $p^2 + 2pq - 15q^2$.

26. Graph on a plane: $2x + 5y > 10$.

27. A disc jockey must play 12 commercial spots during 1 hr of a radio show. Each commercial is either 30 sec or 60 sec long. The total commercial time during that hour is 9 min. How many 30-sec commercials were played? How many 60-sec commercials were played?

SYNTHESIS

28. Solve: $\dfrac{6}{x-15} - \dfrac{6}{x} = \dfrac{90}{x^2 - 15x}$.

29. Find the LCM for $1 - t^6$ and $1 + t^6$.

CUMULATIVE REVIEW: CHAPTERS 1–6

1. Evaluate $\frac{4a}{7b}$ when $a = 14$ and $b = 6$.

Simplify.

2. $|-12|$

3. $-\frac{1}{3} \div \frac{3}{8}$

4. $12b - [9 - 7(5b - 6)]$

5. $5^3 \div \frac{1}{2} - 2(3^2 + 2^2)$

6. $\left(\frac{2x^3y^{-6}}{-4y^{-2}}\right)^2$

7. $(6p^2 - 2p + 5) - (-10p^2 + 6p + 5)$

8. $(6m - n)^2$

9. $(3a - 4b)(5a + 2b)$

10. $\frac{y^2 - 4}{3y + 33} \cdot \frac{y + 11}{y + 2}$

11. $\frac{9x^2 - 25}{x^2 - 16} \div \frac{3x + 5}{x - 4}$

12. $\frac{2x + 1}{4x - 12} - \frac{x - 2}{5x - 15}$

13. $\frac{2}{x + 2} + \frac{3}{x - 2} - \frac{x + 1}{x^2 - 4}$

14. $\dfrac{\frac{1}{x} + \frac{2}{y}}{\frac{3}{x} - \frac{1}{y}}$

15. $\dfrac{1 - \frac{2}{y^2}}{1 - \frac{1}{y^3}}$

16. $(2x^3 - 7x^2 + x - 3) \div (x + 2)$

Solve.

17. $9y - (5y - 3) = 33$

18. $F = \frac{9}{5}C + 32$ for C

19. $-3 < -2x - 6 < 0$

20. $|x| \geq 2.1$

21. $4x - 2y = 6,$
$6x - 3y = 9$

22. $x + 2y - 2z = 9,$
$2x - 3y + 4z = -4,$
$5x - 4y + 2z = 5$

23. $8x = 1 + 16x^2$

24. $14 + 3x = 2x^2$

25. $\frac{3x}{x - 2} - \frac{6}{x + 2} = \frac{24}{x^2 - 4}$

26. $\frac{6}{x - 5} = \frac{2}{2x}$

27. $P = \frac{3a}{a + b}$ for a

Solve using Cramer's rule.

28. $4x + 5y = -3,$
$x = 1 - 3y$

29. $x + 6y + 4z = -2,$
$4x + 4y + z = 2,$
$3x + 2y - 4z = 5$

Factor.

30. $4x^3 + 18x^2$

31. $8a^3 - 4a^2 - 6a + 3$

32. $x^2 + 8x - 84$

33. $6x^2 + 11x - 10$

34. $16y^2 - 81$

35. $t^2 - 16t + 64$

36. $64x^3 + 8$

37. $0.027b^3 - 0.008c^3$

38. $x^6 - x^2$

39. $20x^2 + 7x - 3$

Graph.

40. $y = -5x + 4$

41. $3x - 18 = 0$

42. $x + 3y < 4$

43. $x + y \geq 4,$
$x - y > 1$

44. Find an equation of the line with slope $-\frac{1}{2}$ passing through the point $(2, -2)$.

45. Find an equation of the line that is perpendicular to the line $2x + y = 5$ and passes through the point $(3, -1)$.

46. Medical expenses can be deducted on your tax return only if they exceed $7\frac{1}{2}\%$ of your adjusted gross income. If your expenses were \$3400, what restriction is there on your adjusted gross income (assuming no expenses can be deducted)?

47. According to the Census Bureau, 19.7% of housing is in the West, 24.6% in the Midwest, and 34.8% in the South. What percent is in the Northeast?

48. The residential taxes per \$100 of property value are \$2.43 more in Detroit than in Chicago. The taxes for \$10,000 of property in Detroit and \$20,000 of property in Chicago total \$741. What are the residential property taxes per \$100 of property value in each city?

49. A hockey team played 81 games in a season. They won 1 fewer game than 3 times the number of ties and lost 8 fewer games than they won. How many games did they win? lose? tie?

50. Dave can paint the outside of his house in 12 hr. Bill can paint the same house in 9 hr. How long would it take them to paint the house together?

51. The cost c of an insurance policy varies directly as the age a of the insured. A 32-year-old person pays an annual premium of \$152. What is the age of a person who pays \$285?

52. Find the LCM for 15, 38, and 65.

SYNTHESIS

53. The graph of $y = ax^2 + bx + c$ contains the three points $(4, 2)$, $(2, 0)$, and $(1, 2)$. Find a, b, and c.

54. Solve:

$$\frac{18}{x-9} + \frac{10}{x+5} = \frac{28x}{x^2 - 4x - 45}.$$

55. Solve: $16x^3 = x$.

INTRODUCTION You have probably already studied square roots. In this chapter, we introduce cube roots, fourth roots, and so on. We study them in connection with radical expressions and solve problems involving these roots, such as the one below. We also define expressions with rational exponents.

The review sections to be tested in addition to the material in this chapter are 5.8, 6.1, and 6.3. ❖

Radical Expressions and Equations

AN APPLICATION

Often we see road-pavement messages of various sizes. If the letters are too small, they cannot be read in time to heed their warnings. If the letters are too large, they are expensive to maintain and can be a hazard when the road is wet. What length L will make a road-pavement message most readable from a distance d when the eyes are at a height h from the surface of the road?

THE MATHEMATICS

In a psychological study, it was determined that the proper length L of letters painted on pavement is given by

$$L = \frac{0.000169d^{2.27}}{h},$$

This is a rational exponent.

where d is the distance, in feet, of a car from the lettering and h is the height, in feet, of the eye from the road.

❖ POINTS TO REMEMBER: CHAPTER 7

Pythagorean Theorem: In a right triangle, the sums of the squares of the legs is equal to the square of the hypotenuse: $a^2 + b^2 = c^2$.

Product Rule of Exponents: $a^m a^n = a^{m+n}$

Quotient Rule of Exponents: $\dfrac{a^m}{a^n} = a^{m-n}$

Power Rule of Exponents: $(a^m)^n = a^{mn}$

Raising a Product to a Power: $(ab)^n = a^n b^n$

Raising a Quotient to a Power: $\left(\dfrac{a}{b}\right)^n = \dfrac{a^n}{b^n}$

PRETEST: CHAPTER 7

Simplify. Assume that letters can represent *any* real number.

1. $\sqrt{t^2}$

2. $\sqrt[3]{27x^3}$

3. $\sqrt[12]{y^{12}}$

In the remaining questions on this Pretest, assume that all expressions under the radical represent positive numbers. Simplify.

4. $(\sqrt[3]{9a^2b})^2$

5. $\sqrt{45} - 3\sqrt{125} + 4\sqrt{80}$

Multiply and simplify.

6. $\sqrt{18x^2}\sqrt{8x^3}$

7. $(2\sqrt{6} - 1)^2$

8. Divide. Then simplify by taking roots if possible.

$$\frac{\sqrt{52a^4}}{\sqrt{13a^3}}$$

9. Rationalize the denominator:

$$\frac{x - \sqrt{z}}{3x + \sqrt{z}}.$$

Solve.

10. $\sqrt{2x - 1} = 5$

11. $\sqrt[3]{1 - 6x} = 2$

12. $\sqrt{3x + 1} - \sqrt{2x} = 1$

In Questions 13 and 14, give an exact answer and an approximation to three decimal places.

13. In a right triangle with leg $b = 5$ and hypotenuse $c = 8$, find the length of leg a.

14. The diagonal of a square has length 8 ft. Find the length of a side.

15. Use rational exponents to write a single radical expression:

$$\sqrt{x}\sqrt[3]{x - 4}.$$

16. Determine whether $-1 + i$ is a solution of $x^2 + 2x + 4 = 0$.

17. Subtract: $(-2 + 7i) - (3 - 6i)$.

18. Multiply: $(4 + 3i)(3 - 4i)$.

19. Divide: $\dfrac{-4 + i}{3 - 2i}$.

20. Simplify: i^{87}.

7.1 Introduction to Roots and Radical Expressions

a Square Roots

When we raise a number to the second power, we have **squared** the number. Sometimes we may need to find the number that was squared. We call this process finding a square root of a number.

The number c is a *square root* of a if $c^2 = a$.

For example:

5 is a square root of 25 because $5^2 = 5 \cdot 5 = 25$;

-5 is a square root of 25 because $(-5)^2 = (-5)(-5) = 25$;

-4 does not have a real square root because there is no real number b such that $b^2 = -4$.

Every positive real number has two real-number square roots. The number 0 has just one square root, 0 itself. Negative numbers do not have real-number square roots.*

▶ **EXAMPLE 1** Find the two square roots of 64.

The square roots of 64 are 8 and -8 because $8^2 = 64$ and $(-8)^2 = 64$. ◀

DO EXERCISES 1–3.

The *principal square root* of a nonnegative number is its nonnegative square root. The symbol $\sqrt{a}$ represents the principal square root of a. To name the negative square root of a, we can write $-\sqrt{a}$.

▶ **EXAMPLES** Simplify.

Remember: $\sqrt{\ }$ indicates the principal (nonnegative) square root.

2. $\sqrt{25} = 5$
3. $\sqrt{\frac{25}{64}} = \frac{5}{8}$
4. $-\sqrt{64} = -8$ Since $\sqrt{64} = 8$, $-\sqrt{64} = -8$.
5. $\sqrt{0} = 0$
6. $\sqrt{0.0049} = 0.07$
7. $\sqrt{-49}$ Does not exist as a real number. Negative numbers do not have real-number square roots. ◀

DO EXERCISES 4–10.

---

* Later, we will see that there is a number system other than the real-number system, in which negative numbers do have square roots.

OBJECTIVES

After finishing Section 7.1, you should be able to:

a. Find principal square roots and their opposites.
b. Simplify radical expressions with perfect-square radicands.
c. Find cube roots, simplifying certain expressions.
d. Simplify expressions involving odd and even roots.

FOR EXTRA HELP

Tape 13A

Tape 11A

MAC: 7
IBM: 7

Find the square roots.

1. 9
2. 36
3. 121

Simplify.

4. $\sqrt{1}$
5. $\sqrt{36}$
6. $\sqrt{\frac{81}{100}}$
7. $\sqrt{0.0064}$

Find the following.

8. a) $\sqrt{16}$
   b) $-\sqrt{16}$
   c) $\sqrt{-16}$
9. a) $\sqrt{49}$
   b) $-\sqrt{49}$
   c) $\sqrt{-49}$
10. a) $\sqrt{144}$
    b) $-\sqrt{144}$
    c) $\sqrt{-144}$

ANSWERS ON PAGE A-7

Identify the radicand.

11. $\sqrt{28 + x}$

12. $\sqrt{\dfrac{y}{y + 3}}$

Find the following. Assume that letters can represent any real number.

13. $\sqrt{y^2}$

14. $\sqrt{(-24)^2}$

15. $\sqrt{(5y)^2}$

16. $\sqrt{16y^2}$

17. $\sqrt{(x + 7)^2}$

18. $\sqrt{4(x - 2)^2}$

19. $\sqrt{49(y + 5)^2}$

20. $\sqrt{x^2 - 6x + 9}$

ANSWERS ON PAGE A-7

The symbol $\sqrt{\ }$ is called a *radical*. An expression written with a radical is called a *radical expression*. The expression written under the radical is called the *radicand*.

These are radical expressions:

$$\sqrt{5}, \quad \sqrt{a}, \quad -\sqrt{5x}, \quad \sqrt{y^2 + 7}.$$

► **EXAMPLE 8** Identify the radicand in $\sqrt{x^2 - 9}$.

The radicand in $\sqrt{x^2 - 9}$ is $x^2 - 9$. ◄

DO EXERCISES 11 AND 12.

b Finding $\sqrt{a^2}$

In the expression $\sqrt{a^2}$, the radicand is a perfect square.

Suppose $a = 5$. Then we have $\sqrt{5^2}$, which is $\sqrt{25}$, or 5.

Suppose $a = -5$. Then we have $\sqrt{(-5)^2}$, which is $\sqrt{25}$, or 5.

Suppose $a = 0$. Then we have $\sqrt{0^2}$, which is $\sqrt{0}$, or 0.

The symbol $\sqrt{a^2}$ does not represent a negative number. It represents the principal square root of a^2. Note that if a represents a positive number or 0, then $\sqrt{a^2}$ represents a. If a is negative, then $\sqrt{a^2}$ represents the opposite of a. In all cases, the radical expression represents the absolute value of a.

For any real number a, $\sqrt{a^2} = |a|$. The principal (nonnegative) square root of a^2 is the absolute value of a.

► **EXAMPLES** Find the following. Assume that letters can represent any real number.

9. $\sqrt{(-16)^2} = |-16|$, or 16

10. $\sqrt{(3b)^2} = |3b|$, or $3|b|$ ←

$|3b|$ can be simplified to $3|b|$ because the absolute value of any product is the product of the absolute values. That is, $|a \cdot b| = |a| \cdot |b|$. In this case, $|3b| = |3| \cdot |b|$, or $3|b|$.

11. $\sqrt{(x - 1)^2} = |x - 1|$

12. $\sqrt{x^2 + 8x + 16} = \sqrt{(x + 4)^2}$

$= |x + 4|$ — $|x + 4|$ is *not* the same as $|x| + 4$. ◄

DO EXERCISES 13–20.

c Cube Roots

The number c is the cube root of a if its third power is a, that is, if $c^3 = a$. For example:

2 is the cube root of 8 because $2^3 = 2 \cdot 2 \cdot 2 = 8$;

-4 is the cube root of -64 because $(-4)^3 = (-4)(-4)(-4) = -64$.

We used the word "the" with cube roots because of the following.

Every real number has exactly one cube root in the system of real numbers.

The symbol $\sqrt[3]{a}$ represents the cube root of a.

▶ **EXAMPLES** Find the following.

13. $\sqrt[3]{8} = 2$

14. $\sqrt[3]{-27} = -3$

15. $\sqrt[3]{-\frac{216}{125}} = -\frac{6}{5}$

16. $\sqrt[3]{x^3} = x$

17. $\sqrt[3]{-8y^3} = -2y$ ◀

No absolute-value signs are needed when finding cube roots because a real number has just one cube root. The real-number cube root of a positive number is positive. The real-number cube root of a negative number is negative. The cube root of 0 is 0.

DO EXERCISES 21–24.

d Odd and Even *k*th Roots

The 5th root of a number a is the number c for which $c^5 = a$. There are also 7th roots, 9th roots, and so on. Whenever the number k in $\sqrt[k]{\ }$ is an odd number, we say that we are taking an **odd root.** The number k is called the **index.**

Every number has just one real-number odd root. If the number is positive, then the root is positive. If the number is negative, then the root is negative. If the number is 0, then the root is 0.

▶ **EXAMPLES** Find the following.

18. $\sqrt[5]{32} = 2$

19. $\sqrt[5]{-32} = -2$

20. $-\sqrt[5]{32} = -2$

21. $-\sqrt[5]{-32} = -(-2) = 2$

22. $\sqrt[7]{x^7} = x$

23. $\sqrt[9]{(x-1)^9} = x - 1$ ◀

Absolute-value signs are never needed when finding odd roots.

DO EXERCISES 25 AND 26.

Find the following.

21. $\sqrt[3]{-64}$

22. $\sqrt[3]{27y^3}$

23. $\sqrt[3]{8(x+2)^3}$

24. $\sqrt[3]{-\frac{343}{64}}$

Find the following.

25. $\sqrt[5]{-32x^5}$

26. $\sqrt[7]{(3x+2)^7}$

ANSWERS ON PAGE A-7

Find the following. Assume that letters can represent any real number.

27. a) $\sqrt[4]{81}$

b) $-\sqrt[4]{81}$

c) $\sqrt[4]{-81}$

28. $\sqrt[4]{16(x-2)^4}$

29. $\sqrt[6]{x^6}$

30. $\sqrt[8]{(x+3)^8}$

ANSWERS ON PAGE A-7

When the index k in $\sqrt[k]{}$ is an even number, we say that we are taking an **even root.** Every positive real number has two real-number kth roots when k is even. One of those roots is positive and one is negative. Negative real numbers do not have real-number kth roots when k is even. When we are finding even kth roots, absolute-value signs are sometimes necessary, as they are with square roots. When the index is 2, we do not write it.

► **EXAMPLES** Find the following. Assume that letters can represent any real number.

24. $\sqrt[4]{16} = 2$

25. $-\sqrt[4]{16} = -2$

26. $\sqrt[4]{-16}$ Does not exist as a real number.

27. $\sqrt[4]{81x^4} = 3|x|$

28. $\sqrt[6]{(y+7)^6} = |y+7|$

29. $\sqrt{81y^2} = 9|y|$ ◄

For any real number a:

a) $\sqrt[k]{a^k} = |a|$ **when k is even. We use absolute value when k is even unless a is nonnegative.**

b) $\sqrt[k]{a^k} = a$ **when k is odd. We do not use absolute value when k is odd.**

DO EXERCISES 27–30.

❖ SIDELIGHTS

Wind Chill Temperature

Calculators are often used to approximate square roots. For example, using a calculator, we can approximate $\sqrt{73}$ as follows:

$$\sqrt{73} \approx 8.544003745.$$

Different calculators may give different numbers of digits in their readouts.

We can use approximations of square roots to consider an application involving the effect of wind on the feeling of cold in the winter. In cold weather we feel colder when there is wind than when there is not. The *wind chill temperature* is what the temperature would have to be with no wind in order to give the same chilling effect. A formula for finding the wind chill temperature is

$$T_w = 91.4 - \frac{(10.45 + 6.68\sqrt{v} - 0.447v)(457 - 5T)}{110},$$

where T is the actual temperature given by a thermometer, in degrees Fahrenheit, and v is the wind speed, in miles per hour.

EXERCISES

Use a calculator to find the wind chill temperature in each case. Round to the nearest degree.

1. $T = 30°\text{F}$, $v = 25$ mph

2. $T = 10°\text{F}$, $v = 25$ mph

3. $T = 20°\text{F}$, $v = 20$ mph

4. $T = 20°\text{F}$, $v = 40$ mph

5. $T = -10°\text{F}$, $v = 30$ mph

6. $T = -30°\text{F}$, $v = 30$ mph

NAME SECTION DATE

EXERCISE SET 7.1

a Find the square roots.

1. 16 **2.** 225 **3.** 144 **4.** 9 **5.** 400 **6.** 81

Find the following.

7. $-\sqrt{\frac{49}{36}}$ **8.** $-\sqrt{\frac{361}{9}}$ **9.** $\sqrt{196}$ **10.** $\sqrt{441}$ **11.** $-\sqrt{\frac{16}{81}}$

12. $-\sqrt{\frac{81}{144}}$ **13.** $\sqrt{0.09}$ **14.** $\sqrt{0.36}$ **15.** $-\sqrt{0.0049}$ **16.** $\sqrt{0.0144}$

Identify the radicand.

17. $5\sqrt{p^2+4}$ **18.** $-7\sqrt{y^2-8}$ **19.** $x^2y^2\sqrt{\frac{x}{y+4}}$ **20.** $a^2b^3\sqrt{\frac{a}{a^2-b}}$

b Find the following. Assume that letters can represent any real number.

21. $\sqrt{16x^2}$ **22.** $\sqrt{25t^2}$

23. $\sqrt{(-7c)^2}$ **24.** $\sqrt{(-6b)^2}$

25. $\sqrt{(a+1)^2}$ **26.** $\sqrt{(5-b)^2}$

27. $\sqrt{x^2-4x+4}$ **28.** $\sqrt{y^2+16y+64}$

29. $\sqrt{4x^2+28x+49}$ **30.** $\sqrt{9x^2-30x+25}$

c Simplify.

31. $\sqrt[3]{27}$ **32.** $-\sqrt[3]{64}$ **33.** $\sqrt[3]{-64x^3}$

ANSWERS

1. ____
2. ____
3. ____
4. ____
5. ____
6. ____
7. ____
8. ____
9. ____
10. ____
11. ____
12. ____
13. ____
14. ____
15. ____
16. ____
17. ____
18. ____
19. ____
20. ____
21. ____
22. ____
23. ____
24. ____
25. ____
26. ____
27. ____
28. ____
29. ____
30. ____
31. ____
32. ____
33. ____

ANSWERS

34. ______
35. ______
36. ______
37. ______
38. ______
39. ______
40. ______
41. ______
42. ______
43. ______
44. ______
45. ______
46. ______
47. ______
48. ______
49. ______
50. ______
51. ______
52. ______
53. ______
54. ______
55. ______
56. ______
57. ______
58. ______
59. ______
60. ______
61. a) ______
b) ______
c) ______
d) ______

34. $\sqrt[3]{-125y^3}$ **35.** $\sqrt[3]{-216}$ **36.** $-\sqrt[3]{-1000}$

37. $\sqrt[3]{0.343(x+1)^3}$ **38.** $\sqrt[3]{0.000008(y-2)^3}$

d Find the following. Assume that letters can represent any real number.

39. $\sqrt[4]{625}$ **40.** $-\sqrt[4]{256}$ **41.** $\sqrt[5]{-1}$

42. $\sqrt[5]{-32}$ **43.** $\sqrt[5]{-\dfrac{32}{243}}$ **44.** $\sqrt[5]{-\dfrac{1}{32}}$

45. $\sqrt[6]{x^6}$ **46.** $\sqrt[8]{y^8}$ **47.** $\sqrt[4]{(5a)^4}$

48. $\sqrt[4]{(7b)^4}$ **49.** $\sqrt[10]{(-6)^{10}}$ **50.** $\sqrt[12]{(-10)^{12}}$

51. $\sqrt[414]{(a+b)^{414}}$ **52.** $\sqrt[1976]{(2a+b)^{1976}}$ **53.** $\sqrt[7]{y^7}$

54. $\sqrt[3]{(-6)^3}$ **55.** $\sqrt[5]{(x-2)^5}$ **56.** $\sqrt[9]{(2xy)^9}$

SKILL MAINTENANCE

Solve.

57. $x^2 + x - 2 = 0$ **58.** $x^2 + x = 0$

59. $4x^2 - 49 = 0$ **60.** $2x^2 - 26x + 72 = 0$

SYNTHESIS

61. *Parking.* A parking lot has attendants to park the cars. The number N of temporary stalls needed for waiting cars before attendants can get to them is given by the formula $N = 2.5\sqrt{A}$, where A is the number of arrivals in peak hours. Find the number of spaces needed when the average number of arrivals in peak hours is **(a)** 25; **(b)** 36; **(c)** 49; **(d)** 64.

7.2 Multiplying and Simplifying with Radical Expressions

a Simplifying by Factoring

Note that $\sqrt{4}\sqrt{25} = 2 \cdot 5 = 10$. Also $\sqrt{4 \cdot 25} = \sqrt{100} = 10$. Likewise,

$$\sqrt[3]{27}\,\sqrt[3]{8} = 3 \cdot 2 = 6 \quad \text{and} \quad \sqrt[3]{27 \cdot 8} = \sqrt[3]{216} = 6.$$

These examples suggest the following.

> The Product Rule for Radicals
>
> **For any nonnegative real numbers a and b and any index k,**
>
> $$\sqrt[k]{a} \cdot \sqrt[k]{b} = \sqrt[k]{a \cdot b}.$$
>
> **(To multiply, we multiply the radicands.)**

Note that the index k must be the same throughout.

► **EXAMPLES** Multiply.

1. $\sqrt{3} \cdot \sqrt{5} = \sqrt{3 \cdot 5} = \sqrt{15}$
2. $\sqrt{x+2}\,\sqrt{x-2} = \sqrt{(x+2)(x-2)} = \sqrt{x^2 - 4}$
3. $\sqrt[3]{4}\,\sqrt[3]{5} = \sqrt[3]{4 \cdot 5} = \sqrt[3]{20}$
4. $\sqrt[4]{\dfrac{y}{5}}\,\sqrt[4]{\dfrac{7}{x}} = \sqrt[4]{\dfrac{y}{5} \cdot \dfrac{7}{x}} = \sqrt[4]{\dfrac{7y}{5x}}$

A common error is to omit the index in the answer. ◄

DO EXERCISES 1–4.

Reading the product rule from right to left, we have

$$\sqrt[k]{ab} = \sqrt[k]{a} \cdot \sqrt[k]{b}.$$

This shows a way to factor and thus simplify radical expressions. Consider $\sqrt{20}$. The number 20 has the factor 4, which is a perfect square. Therefore,

$$\begin{aligned} \sqrt{20} &= \sqrt{4 \cdot 5} && \text{Factoring the radicand (4 is a perfect square)} \\ &= \sqrt{4} \cdot \sqrt{5} && \text{Factoring into two radicals} \\ &= 2\sqrt{5}. && \text{Taking the square root of 4} \end{aligned}$$

> **To simplify a radical expression by factoring, look for the largest factors of the radicand that are perfect kth powers (where k is the index). Then take the kth root of the resulting factors. A radical expression, with index k, is *simplified* when its radicand has no factors that are perfect k powers.**

OBJECTIVES

After finishing Section 7.2, you should be able to:

- a Simplify radical expressions by factoring, assuming that all expressions under the radical represent nonnegative numbers.
- b Multiply and simplify radical expressions.
- c Approximate expressions using a calculator or a square root table.

FOR EXTRA HELP

Tape 13A | Tape 11B | MAC: 7 IBM: 7

Multiply.

1. $\sqrt{19}\,\sqrt{7}$

2. $\sqrt{x+2y}\,\sqrt{x-2y}$

3. $\sqrt[4]{403}\,\sqrt[4]{7}$

4. $\sqrt[3]{8x}\,\sqrt[3]{x^4+5}$

ANSWERS ON PAGE A-7

Simplify by factoring.

5. $\sqrt{32}$

6. $\sqrt[3]{80}$

Simplify by factoring. Assume that all expressions under radicals represent nonnegative numbers. Thus no absolute-value signs will be needed.

7. $\sqrt{300}$

8. $\sqrt{36y^2}$

9. $\sqrt{3x^2 + 12x + 12}$

10. $\sqrt{12ab^3c^2}$

11. $\sqrt[3]{16}$

12. $\sqrt[3]{81x^4y^8}$

13. $\sqrt[3]{(a + b)^4}$

ANSWERS ON PAGE A-7

▶ **EXAMPLES** Simplify by factoring.

5. $\sqrt{50} = \sqrt{25 \cdot 2} = \sqrt{25} \cdot \sqrt{2} = 5\sqrt{2}$

This factor is a perfect square. (points to 25)

6. $\sqrt[3]{32} = \sqrt[3]{8 \cdot 4} = \sqrt[3]{8} \cdot \sqrt[3]{4} = 2\sqrt[3]{4}$

This factor is a perfect cube (third power). (points to 8) ◀

DO EXERCISES 5 AND 6.

> **In many situations, expressions under radicals never represent negative numbers. In such cases, absolute-value notation is not necessary. For this reason, we will henceforth assume that *all expressions under radicals are nonnegative.***

▶ **EXAMPLES** Simplify by factoring. Assume that all expressions under radicals represent nonnegative numbers.

7. $\sqrt{5x^2} = \sqrt{x^2 \cdot 5}$ **Factoring the radicand**

$= \sqrt{x^2} \cdot \sqrt{5}$ **Factoring into two radicals**

$= x \cdot \sqrt{5}$ **Taking the square root of x^2**

8. $\sqrt{2x^2 - 4x + 2} = \sqrt{2(x-1)^2}$

$= \sqrt{(x-1)^2} \cdot \sqrt{2}$

$= (x-1) \cdot \sqrt{2}$

Absolute value is not needed because expressions under radicals are not negative.

9. $\sqrt{216x^5y^3} = \sqrt{36 \cdot 6 \cdot x^4 \cdot x \cdot y^2 \cdot y}$

$= \sqrt{36 \cdot x^4 \cdot y^2 \cdot 6 \cdot x \cdot y}$ **Factoring the radicand**

$= \sqrt{36}\sqrt{x^4}\sqrt{y^2}\sqrt{6xy}$ **Factoring into several radicals**

$= 6x^2y\sqrt{6xy}$ **Taking square roots**

Note: Had we not seen that $216 = 36 \cdot 6$, where 36 is the largest square factor of 216, we could have found the prime factorization

$$2 \cdot 2 \cdot 2 \cdot 3 \cdot 3 \cdot 3.$$

Each pair of factors makes a square, so

$$\sqrt{2 \cdot 2 \cdot 2 \cdot 3 \cdot 3 \cdot 3} = \sqrt{2^2 \cdot 3^2 \cdot 2 \cdot 3} = 2 \cdot 3\sqrt{2 \cdot 3}.$$

10. $\sqrt[3]{16a^7b^{11}} = \sqrt[3]{8 \cdot 2 \cdot a^6 \cdot a \cdot b^9 \cdot b^2}$ **Factoring the radicand. We look for the largest powers that are multiples of 3.**

$= \sqrt[3]{8} \cdot \sqrt[3]{a^6} \cdot \sqrt[3]{b^9} \cdot \sqrt[3]{2ab^2}$ **Factoring into radicals**

$= 2a^2b^3\sqrt[3]{2ab^2}$ **Taking cube roots** ◀

DO EXERCISES 7–13.

b Multiplying and Simplifying

Sometimes after we multiply we can then simplify by factoring.

▶ **EXAMPLES** Multiply and then simplify by factoring. Assume that all expressions under radicals represent nonnegative numbers.

11. $\sqrt{15}\sqrt{6} = \sqrt{15 \cdot 6} = \sqrt{90} = \sqrt{9 \cdot 10} = 3\sqrt{10}$

12. $3\sqrt[3]{25} \cdot 2\sqrt[3]{5} = 6 \cdot \sqrt[3]{25 \cdot 5}$ $= 6 \cdot \sqrt[3]{125}$ **Multiplying radicands**

$= 6 \cdot 5$, or 30   **Taking the cube root of 125**

13. $\sqrt[3]{18y^3}\sqrt[3]{4x^2} = \sqrt[3]{18y^3 \cdot 4x^2} = \sqrt[3]{72y^3x^2}$ **Multiplying radicands**

$= \sqrt[3]{8y^3 \cdot 9x^2}$ **Factoring the radicand**

$= \sqrt[3]{8y^3}\sqrt[3]{9x^2}$ **Factoring into two radicals**

$= 2y\sqrt[3]{9x^2}$ **Taking the cube root** ◀

DO EXERCISES 14–18.

c Approximating Square Roots

We often need to use rational numbers to approximate square roots that are irrational. Such approximations can be found using a table such as Table 1 on p. 641. They can also be found on a calculator with a square root key. For example, if we were to approximate $\sqrt{37}$ using a calculator, we might get

$\sqrt{37} \approx 6.082762530.$ **Using a calculator with a 10-digit readout**

Different calculators give different numbers of digits in their readouts. This may cause some variance in their answers. We might round to the third decimal place. Then

$\sqrt{37} \approx 6.083.$ **This can also be found in Table 1.**

DO EXERCISES 19 AND 20.

Now consider $\sqrt{275}$. To approximate such a root, we can use a calculator, or we can also factor and use Table 1. Different procedures can lead to variance in approximations.

▶ **EXAMPLE 14** Use your calculator or Table 1 to approximate $\sqrt{275}$. Round to three decimal places.

Using a calculator gives us

$$\sqrt{275} \approx 16.58312395 \approx 16.583.$$

Using factoring and Table 1, we get

$\sqrt{275} = \sqrt{25 \cdot 11}$ **Factoring the radicand**

$= \sqrt{25} \cdot \sqrt{11}$ **Factoring into two radicals**

$= 5\sqrt{11}$ **Simplifying**

$\approx 5 \times 3.317$ **Using Table 1**

$= 16.585.$

Multiply and then simplify by factoring. Assume that all expressions under radicals represent nonnegative numbers.

14. $\sqrt{3}\sqrt{6}$

15. $\sqrt{18y}\sqrt{14y}$

16. $\sqrt[3]{3x^2y}\sqrt[3]{36x}$

17. $\sqrt{7a}\sqrt{21b}$

18. $\sqrt[3]{2(y+5)}\sqrt[3]{4(y+5)^4}$

Use your calculator or Table 1 to approximate the square roots. Round to three decimal places.

19. $\sqrt{23}$

20. $\sqrt{97}$

ANSWERS ON PAGE A-7

Use a calculator or Table 1 to approximate to the nearest thousandth.

21. $\sqrt{160}$

22. $\sqrt{341}$

Use a calculator or Table 1 to approximate to the nearest thousandth.

23. $\dfrac{18-\sqrt{810}}{3}$

24. $\dfrac{6+\sqrt{450}}{3}$

ANSWERS ON PAGE A-7

Note the variance in the answers. Because calculators are so much a part of our everyday life, the answers at the back of the book are found using a calculator. If you use another method, such as the one above, and you get an answer slightly different from the one given at the back, keep in mind that your work may not be wrong. You may have used a method that gives a different answer. ◀

DO EXERCISES 21 AND 22.

▶ **EXAMPLE 15** Approximate to the nearest thousandth: $\dfrac{16-\sqrt{640}}{4}$.

Method 1. Using a calculator, we get

$$\begin{aligned}\frac{16-\sqrt{640}}{4} &\approx \frac{16-25.298221128}{4}\\ &= \frac{-9.298221280}{4}\\ &= -2.324555320\\ &\approx -2.325.\end{aligned}$$

Method 2. Using factoring and Table 1 gives us

$$\begin{aligned}\frac{16-\sqrt{640}}{4} &= \frac{16-\sqrt{64\cdot 10}}{4} && \textbf{Factoring the radicand}\\ &= \frac{16-\sqrt{64}\cdot\sqrt{10}}{4} && \textbf{Factoring into two radicals}\\ &= \frac{16-8\sqrt{10}}{4}\\ &= \frac{4(4-2\sqrt{10})}{4} = 4-2\sqrt{10} && \textbf{Removing a factor of 1}\\ &\approx 4-2(3.162) && \textbf{Using Table 1}\\ &= 4-6.324\\ &= -2.324.\end{aligned}$$

A common error here is to divide 4 into 16 but *not* into 8, and get $4-8\sqrt{10}$. Always remember to *remove a factor of* 1.

Note again the variance in answers. ◀

DO EXERCISES 23 AND 24.

There are situations in which we may not be able to approximate a root using a calculator because the number does not fit. In such a case, we can again use factoring. For example,

$$\begin{aligned}\sqrt{0.000000005768} &= \sqrt{57.68\times 10^{-10}}\\ &= \sqrt{57.68}\times\sqrt{10^{-10}}\\ &= \sqrt{57.68}\times\sqrt{(10^{-5})^2}\\ &\approx 7.595\times 10^{-5} && \textbf{Using a calculator}\\ &= 0.00007595.\end{aligned}$$

NAME SECTION DATE

EXERCISE SET 7.2

a Simplify by factoring. Assume that all expressions under radicals represent nonnegative numbers.

1. $\sqrt{8}$ **2.** $\sqrt{18}$ **3.** $\sqrt{24}$ **4.** $\sqrt{20}$ **5.** $\sqrt{40}$

6. $\sqrt{90}$ **7.** $\sqrt{180x^4}$ **8.** $\sqrt{175y^6}$ **9.** $\sqrt[3]{54x^8}$ **10.** $\sqrt[3]{40y^3}$

11. $\sqrt[3]{80t^8}$ **12.** $\sqrt[3]{108x^5}$ **13.** $\sqrt[4]{80}$ **14.** $\sqrt[4]{32}$

15. $\sqrt[4]{243x^8y^{10}}$ **16.** $\sqrt[4]{162c^4d^6}$ **17.** $\sqrt[3]{(x+y)^4}$ **18.** $\sqrt[3]{(p-q)^5}$

19. $\sqrt[3]{-24x^4y^5}$ **20.** $\sqrt[3]{-250a^2b^6}$ **21.** $\sqrt[5]{96x^7y^{15}}$ **22.** $\sqrt[5]{p^{14}q^9r^{23}}$

b Multiply and simplify by factoring. Assume that all expressions under radicals represent nonnegative numbers.

23. $\sqrt{15}\sqrt{6}$ **24.** $\sqrt{2}\sqrt{32}$ **25.** $\sqrt[3]{3}\sqrt[3]{18}$ **26.** $\sqrt[3]{2}\sqrt[3]{20}$

27. $\sqrt{45}\sqrt{60}$ **28.** $\sqrt{24}\sqrt{75}$ **29.** $\sqrt{5b^3}\sqrt{10c^4}$ **30.** $\sqrt{2x^3y}\sqrt{12xy}$

31. $\sqrt[3]{y^4}\sqrt[3]{16y^5}$ **32.** $\sqrt[3]{5^2t^4}\sqrt[3]{5^4t^6}$ **33.** $\sqrt[3]{(b+3)^4}\sqrt[3]{(b+3)^2}$

ANSWERS

1. ___
2. ___
3. ___
4. ___
5. ___
6. ___
7. ___
8. ___
9. ___
10. ___
11. ___
12. ___
13. ___
14. ___
15. ___
16. ___
17. ___
18. ___
19. ___
20. ___
21. ___
22. ___
23. ___
24. ___
25. ___
26. ___
27. ___
28. ___
29. ___
30. ___
31. ___
32. ___
33. ___

ANSWERS

34. ______
35. ______
36. ______
37. ______
38. ______
39. ______
40. ______
41. ______
42. ______
43. ______
44. ______
45. ______
46. ______
47. ______
48. ______
49. ______
50. ______
51. ______
52. ______
53. a) ______
b) ______
c) ______
54. ______
55. ______
56. ______
57. ______
58. ______
59. ______
60. ______

34. $\sqrt[3]{(x+y)^3}\sqrt[3]{(x+y)^5}$ **35.** $\sqrt{12a^3b}\sqrt{8a^4b^2}$ **36.** $\sqrt{18x^2y^3}\sqrt{6xy^4}$

37. $\sqrt[4]{16}\cdot\sqrt[4]{64}$ **38.** $\sqrt[5]{64}\cdot\sqrt[5]{16}$

39. $\sqrt[4]{10a^3}\cdot\sqrt[4]{8a^2}$ **40.** $\sqrt[4]{18x^3}\cdot\sqrt[4]{9x^2}$

41. $\sqrt{30x^3y^4}\cdot\sqrt{18x^2y^5}$ **42.** $\sqrt[5]{64x^8y^{11}}\cdot\sqrt[5]{128x^9y^{14}}$

43. $\sqrt[5]{a^3(b-c)^7}\cdot\sqrt[5]{a^8(b-c)^{11}}$ **44.** $\sqrt[5]{x^4(b+c)^7}\cdot\sqrt[5]{x^4(b+c)^9}$

C Approximate to the nearest thousandth using a calculator or Table 1.

45. $\sqrt{180}$ **46.** $\sqrt{124}$ **47.** $\dfrac{8+\sqrt{480}}{4}$ **48.** $\dfrac{12-\sqrt{450}}{3}$

49. $\dfrac{16-\sqrt{48}}{20}$ **50.** $\dfrac{25-\sqrt{250}}{10}$ **51.** $\dfrac{24+\sqrt{128}}{8}$ **52.** $\dfrac{96-\sqrt{90}}{12}$

53. *Speed of a skidding car.* After an accident, police can estimate the speed that a car was traveling by measuring its skid marks. The formula

$$r = 2\sqrt{5L}$$

can be used, where r is the speed, in miles per hour, and L is the length of the skid marks, in feet. Estimate the speed of a car that left skid marks **(a)** 20 ft long; **(b)** 70 ft long; **(c)** 90 ft long.

SKILL MAINTENANCE

Solve.

54. $9x^2 - 15x = 0$ **55.** $x^2 - \dfrac{2}{3}x = 0$

56. Multiply and simplify: $\dfrac{x^3-y^3}{x+y}\cdot\dfrac{x^2-y^2}{x^2+xy+y^2}$.

SYNTHESIS

Multiply and simplify.

57. $\sqrt{1.6\times 10^3}\sqrt{36\times 10^{-8}}$ **58.** $\sqrt{968}\sqrt{1014}$ **59.** $\sqrt[3]{48}\sqrt[3]{63}\sqrt[3]{196}$

60. Solve for t: $\sqrt[3]{2x^{t+3}}\cdot\sqrt[3]{32x^t} = 4x^9$.

7.3 Dividing and Simplifying with Radical Expressions

a Dividing Radical Expressions

Note that $\sqrt[3]{\frac{27}{8}} = \frac{3}{2}$ and that $\frac{\sqrt[3]{27}}{\sqrt[3]{8}} = \frac{3}{2}$. This example suggests the following.

> The Quotient Rule for Radicals
>
> **For any nonnegative number a and any positive number b, and any index k,**
>
> $$\frac{\sqrt[k]{a}}{\sqrt[k]{b}} = \sqrt[k]{\frac{a}{b}}.$$
>
> **(To divide, we divide the radicands. After doing this, we can sometimes simplify by taking roots.)**

▶ **EXAMPLES** Divide. Then simplify by taking roots, if possible. Assume that all expressions under radicals represent positive numbers.

1. $\frac{\sqrt{80}}{\sqrt{5}} = \sqrt{\frac{80}{5}} = \sqrt{16} = 4$ — We divide the radicands.

2. $\frac{3\sqrt{2}}{5\sqrt{3}} = \frac{3}{5} \cdot \frac{\sqrt{2}}{\sqrt{3}} = \frac{3}{5} \cdot \sqrt{\frac{2}{3}}$

3. $\frac{5\sqrt[3]{32}}{\sqrt[3]{2}} = 5\sqrt[3]{\frac{32}{2}} = 5\sqrt[3]{16} = 5\sqrt[3]{8 \cdot 2} = 5\sqrt[3]{8}\sqrt[3]{2} = 5 \cdot 2\sqrt[3]{2} = 10\sqrt[3]{2}$

4. $\frac{\sqrt{72xy}}{2\sqrt{2}} = \frac{1}{2}\frac{\sqrt{72xy}}{\sqrt{2}} = \frac{1}{2}\sqrt{\frac{72xy}{2}} = \frac{1}{2}\sqrt{36xy} = \frac{1}{2}\sqrt{36}\sqrt{xy}$

   $= \frac{1}{2} \cdot 6\sqrt{xy} = 3\sqrt{xy}$

5. $\frac{\sqrt[4]{33a^5b^3}}{\sqrt[4]{2b^{-1}}} = \sqrt[4]{\frac{33a^5b^3}{2b^{-1}}} = \sqrt[4]{\frac{33}{2}a^5b^4} = \sqrt[4]{a^4b^4}\sqrt[4]{\frac{33}{2}a} = ab\sqrt[4]{\frac{33}{2}a}$ ◀

DO EXERCISES 1–5.

b Roots of Quotients

We can reverse the quotient rule to simplify a quotient. We simplify the root of a quotient by taking the root of the numerator and of the denominator separately.

> **For any nonnegative number a and any positive number b, and any index k,**
>
> $$\sqrt[k]{\frac{a}{b}} = \frac{\sqrt[k]{a}}{\sqrt[k]{b}}.$$
>
> **(We can take the kth root of the numerator and of the denominator separately.)**

OBJECTIVES

After finishing Section 7.3, you should be able to:

- a Divide and simplify radical expressions.
- b Simplify radical expressions having a quotient for a radicand.
- c Calculate combinations of roots and powers.

FOR EXTRA HELP

Tape 13B | Tape 11B | MAC: 7 IBM: 7

Divide. Then simplify by taking roots, if possible. Assume that all expressions under radicals represent positive numbers.

1. $\frac{\sqrt{75}}{\sqrt{3}}$

2. $\frac{14\sqrt{128xy}}{2\sqrt{2}}$

3. $\frac{\sqrt{50a^3}}{\sqrt{2a}}$

4. $\frac{4\sqrt[3]{250}}{7\sqrt[3]{2}}$

5. $\frac{\sqrt[3]{8a^3b}}{\sqrt[3]{27b^{-2}}}$

ANSWERS ON PAGE A-7

Simplify by taking roots of the numerator and the denominator. Assume that all expressions under radicals represent positive numbers.

6. $\sqrt{\dfrac{25}{36}}$

7. $\sqrt{\dfrac{x^2}{100}}$

8. $\sqrt[3]{\dfrac{54x^5}{125}}$

Calculate as shown. Then use the power–root rule to calculate another way.

9. $(\sqrt[3]{125})^2$

10. $(\sqrt{6y})^3$

11. $\sqrt{(2a^3b)^3}$

ANSWERS ON PAGE A-7

► **EXAMPLES** Simplify by taking roots of the numerator and the denominator. Assume that all expressions under radicals represent positive numbers.

6. $\sqrt[3]{\dfrac{27}{125}} = \dfrac{\sqrt[3]{27}}{\sqrt[3]{125}} = \dfrac{3}{5}$ — We take the cube root of the numerator and of the denominator.

7. $\sqrt{\dfrac{25}{y^2}} = \dfrac{\sqrt{25}}{\sqrt{y^2}} = \dfrac{5}{y}$ — We take the square root of the numerator and of the denominator.

8. $\sqrt{\dfrac{16x^3}{y^4}} = \dfrac{\sqrt{16x^3}}{\sqrt{y^4}} = \dfrac{\sqrt{16x^2 \cdot x}}{\sqrt{y^4}} = \dfrac{4x\sqrt{x}}{y^2}$

9. $\sqrt[3]{\dfrac{27y^5}{343x^3}} = \dfrac{\sqrt[3]{27y^5}}{\sqrt[3]{343x^3}} = \dfrac{\sqrt[3]{27y^3 \cdot y^2}}{\sqrt[3]{343x^3}} = \dfrac{\sqrt[3]{27y^3} \cdot \sqrt[3]{y^2}}{\sqrt[3]{343x^3}} = \dfrac{3y\sqrt[3]{y^2}}{7x}$ ◄

We are assuming that no expression represents 0 or a negative number. Thus we need not be concerned about zero denominators.

DO EXERCISES 6–8.

C Powers and Roots Combined

Consider the following:

$$\sqrt[3]{8^2} = \sqrt[3]{64} = 4; \qquad (\sqrt[3]{8})^2 = (2)^2 = 4.$$

This suggests another important property of radical expressions.

> **The Power–Root Rule**
>
> **For any real number a and any index k for which $\sqrt[k]{a}$ exists, and any natural number m,**
>
> $$\sqrt[k]{a^m} = (\sqrt[k]{a})^m.$$
>
> **(We can raise to a power and then take a root, or we can take a root and then raise to a power.)**

In some cases, one way of calculating is easier than the other.

► **EXAMPLES** Calculate as shown. Then use the power–root rule to calculate another way. Assume that all expressions under radicals are positive.

10. a) $\sqrt[3]{27^2} = \sqrt[3]{729} = 9$ **Using a calculator or Table 1**

b) $(\sqrt[3]{27})^2 = (3)^2 = 9$

11. a) $\sqrt[3]{2^6} = \sqrt[3]{64} = 4$

b) $(\sqrt[3]{2})^6 = \sqrt[3]{2}\sqrt[3]{2}\sqrt[3]{2}\sqrt[3]{2}\sqrt[3]{2}\sqrt[3]{2} = 2 \cdot 2 = 4$

12. a) $\sqrt{(5x)^3} = \sqrt{5^3x^3} = \sqrt{5^2 \cdot x^2 \cdot 5 \cdot x} = \sqrt{5^2x^2}\sqrt{5x} = 5x\sqrt{5x}$

b) $(\sqrt{5x})^3 = \sqrt{5x}\sqrt{5x}\sqrt{5x} = 5x\sqrt{5x}$

13. a) $\sqrt{(5a^2b^3)^3} = \sqrt{125a^6b^9} = \sqrt{5 \cdot 5 \cdot 5 \cdot a^3 \cdot a^3 \cdot b^4 \cdot b^4 \cdot b}$
$= 5a^3b^4\sqrt{5b}$

b) $(\sqrt{5a^2b^3})^3 = \sqrt{5a^2b^3} \cdot \sqrt{5a^2b^3} \cdot \sqrt{5a^2b^3} = 5a^2b^3\sqrt{5a^2b^3}$
$= 5a^2b^3\sqrt{a^2b^2 \cdot 5b} = 5a^2b^3\sqrt{a^2b^2}\sqrt{5b}$
$= 5a^2b^3 \cdot ab\sqrt{5b} = 5a^3b^4\sqrt{5b}$ ◄

DO EXERCISES 9–11.

NAME SECTION DATE

EXERCISE SET 7.3

a Divide. Then simplify by taking roots, if possible. Assume that all expressions under radicals represent positive numbers.

1. $\dfrac{\sqrt{21a}}{\sqrt{3a}}$ **2.** $\dfrac{\sqrt{28y}}{\sqrt{4y}}$ **3.** $\dfrac{\sqrt[3]{54}}{\sqrt[3]{2}}$ **4.** $\dfrac{\sqrt[3]{40}}{\sqrt[3]{5}}$

5. $\dfrac{\sqrt{40xy^3}}{\sqrt{8x}}$ **6.** $\dfrac{\sqrt{56ab^3}}{\sqrt{7a}}$ **7.** $\dfrac{\sqrt[3]{96a^4b^2}}{\sqrt[3]{12a^2b}}$ **8.** $\dfrac{\sqrt[3]{189x^5y^7}}{\sqrt[3]{7x^2y^2}}$

9. $\dfrac{\sqrt{144xy}}{2\sqrt{2}}$ **10.** $\dfrac{\sqrt{75ab}}{3\sqrt{3}}$ **11.** $\dfrac{\sqrt[4]{48x^9y^{13}}}{\sqrt[4]{3xy^5}}$ **12.** $\dfrac{\sqrt[5]{64a^{11}b^{28}}}{\sqrt[5]{2ab^2}}$

13. $\dfrac{\sqrt{x^3 - y^3}}{\sqrt{x - y}}$ **14.** $\dfrac{\sqrt{r^3 + s^3}}{\sqrt{r + s}}$

Hint: To divide $x^3 - y^3$ by $x - y$, factor $x^3 - y^3$. Then simplify the rational expression by removing a factor of 1.

b Simplify by taking roots of the numerator and the denominator. Assume that all expressions under radicals represent positive numbers.

15. $\sqrt{\dfrac{16}{25}}$ **16.** $\sqrt{\dfrac{100}{81}}$ **17.** $\sqrt[3]{\dfrac{64}{27}}$ **18.** $\sqrt[3]{\dfrac{343}{512}}$ **19.** $\sqrt{\dfrac{49}{y^2}}$

20. $\sqrt{\dfrac{121}{x^2}}$ **21.** $\sqrt{\dfrac{25y^3}{x^4}}$ **22.** $\sqrt{\dfrac{36a^5}{b^6}}$ **23.** $\sqrt[3]{\dfrac{8x^5}{27y^3}}$ **24.** $\sqrt[3]{\dfrac{64x^7}{216y^6}}$

25. $\sqrt[4]{\dfrac{81x^4}{16}}$ **26.** $\sqrt[4]{\dfrac{625a^8}{b^4}}$ **27.** $\sqrt[4]{\dfrac{p^5q^8}{r^{12}}}$ **28.** $\sqrt[4]{\dfrac{a^{11}b^{13}}{c^8}}$

ANSWERS

1. ______
2. ______
3. ______
4. ______
5. ______
6. ______
7. ______
8. ______
9. ______
10. ______
11. ______
12. ______
13. ______
14. ______
15. ______
16. ______
17. ______
18. ______
19. ______
20. ______
21. ______
22. ______
23. ______
24. ______
25. ______
26. ______
27. ______
28. ______

ANSWERS

29. ______

30. ______

31. ______

32. ______

33. ______

34. ______

35. ______

36. ______

37. ______

38. ______

39. ______

40. ______

41. ______

42. ______

43. ______

44. ______

45. ______

46. ______

47. ______

48. a) ______

b) ______

c) ______

49. ______

50. ______

51. ______

29. $\sqrt[5]{\frac{32x^8}{y^{10}}}$ **30.** $\sqrt[5]{\frac{32b^{10}}{243a^{20}}}$ **31.** $\sqrt[6]{\frac{x^{13}}{y^6z^{12}}}$ **32.** $\sqrt[6]{\frac{p^9q^{24}}{r^{18}}}$

C Calculate as shown. Then use $\sqrt[k]{a^m} = (\sqrt[k]{a})^m$ to calculate another way. Assume that all expressions under radicals represent positive numbers.

33. $\sqrt{(6a)^3}$ **34.** $\sqrt{(7y)^3}$ **35.** $(\sqrt[3]{16b^2})^2$ **36.** $(\sqrt[3]{25r^2})^2$

37. $\sqrt{(18a^2b)^3}$ **38.** $\sqrt{(12x^2y)^3}$ **39.** $(\sqrt[3]{12c^2d})^2$ **40.** $(\sqrt[3]{9x^2y})^2$

41. $\sqrt[3]{(7x^2y)^2}$ **42.** $\sqrt[3]{(4ab^2)^2}$ **43.** $(\sqrt[4]{81xy^5})^2$ **44.** $(\sqrt[4]{16a^4b^8})^2$

SKILL MAINTENANCE

Solve.

45. $\frac{12x}{x-4} - \frac{3x^2}{x+4} = \frac{384}{x^2-16}$ **46.** $\frac{2}{3} + \frac{1}{t} = \frac{4}{5}$

47. The width of a rectangle is one fourth the length. The area is twice the perimeter. Find the dimensions of the rectangle.

SYNTHESIS

48. *Pendulums.* The *period* of a pendulum is the time it takes to complete one cycle, swinging to and fro. If a pendulum consists of a ball on a string, the period T is given by the formula

$$T = 2\pi\sqrt{\frac{L}{980}},$$

where T is in seconds and L is the length of the pendulum in centimeters. Find the period of a pendulum of length **(a)** 65 cm; **(b)** 98 cm; **(c)** 120 cm. Use 3.14 for π.

Divide and simplify. Assume that all expressions under radicals represent positive numbers.

49. $\frac{7\sqrt{a^2b}\sqrt{25xy}}{5\sqrt{a^{-4}b^{-1}}\sqrt{49x^{-1}y^{-3}}}$ **50.** $\frac{(\sqrt[3]{81mn^2})^2}{(\sqrt[3]{mn})^2}$ **51.** $\frac{\sqrt{44x^2y^9z}\sqrt{22y^9z^6}}{(\sqrt{11xy^8z^2})^2}$

7.4 Addition, Subtraction, and More Multiplication

a Addition and Subtraction Involving Radicals

Any two real numbers can be added. For instance, the sum of 7 and $\sqrt{3}$ can be expressed as

$$7 + \sqrt{3}.$$

We cannot simplify this name for the sum. However, when we have **like radicals** (radicals having the same index and radicand), we can use the distributive laws to simplify by collecting like radical terms.

► **EXAMPLES** Add or subtract. Simplify by collecting like radical terms, if possible.

1. $6\sqrt{7} + 4\sqrt{7} = (6 + 4)\sqrt{7}$ **Using a distributive law (factoring out $\sqrt{7}$)**
   $= 10\sqrt{7}$

2. $8\sqrt[3]{2} - 7x\sqrt[3]{2} + 5\sqrt[3]{2} = (8 - 7x + 5)\sqrt[3]{2}$ **Factoring out $\sqrt[3]{2}$**
   $= (13 - 7x)\sqrt[3]{2}$ These parentheses *are* necessary!

3. $6\sqrt[5]{4x} + 4\sqrt[5]{4x} - \sqrt[3]{4x} = (6 + 4)\sqrt[5]{4x} - \sqrt[3]{4x}$
   $= 10\sqrt[5]{4x} - \sqrt[3]{4x}$

   Note that these expressions have the *same* radicand, but they are *not* like radicals since they do not have the same indices (sing., index). ◄

DO EXERCISES 1 AND 2.

Sometimes we need to simplify radicals by factoring in order to obtain terms with like radicals.

► **EXAMPLES** Add or subtract. Simplify by collecting like radical terms, if possible.

4. $3\sqrt{8} - 5\sqrt{2} = 3(\sqrt{4 \cdot 2}) - 5\sqrt{2}$ **Factoring 8**
   $= 3(\sqrt{4} \cdot \sqrt{2}) - 5\sqrt{2}$ **Factoring $\sqrt{4 \cdot 2}$ into two radicals**
   $= 3(2\sqrt{2}) - 5\sqrt{2}$ **Taking the square root of 4**
   $= 6\sqrt{2} - 5\sqrt{2}$
   $= (6 - 5)\sqrt{2}$ **Collecting like radical terms**
   $= \sqrt{2}$

5. $5\sqrt{2} - 4\sqrt{3}$ No simplification possible

6. $5\sqrt[3]{16y^4} + 7\sqrt[3]{2y} = 5\sqrt[3]{8y^3 \cdot 2y} + 7\sqrt[3]{2y}$
   $= 5\sqrt[3]{8y^3} \cdot \sqrt[3]{2y} + 7\sqrt[3]{2y}$ **Factoring the first radical**
   $= 5 \cdot 2y \cdot \sqrt[3]{2y} + 7\sqrt[3]{2y}$ **Taking the cube root**
   $= 10y\sqrt[3]{2y} + 7\sqrt[3]{2y}$
   $= (10y + 7)\sqrt[3]{2y}$ **Collecting like radical terms**

   Note that parentheses are necessary here. ◄

DO EXERCISES 3–5.

OBJECTIVES

After finishing Section 7.4, you should be able to:

a Add or subtract with radical notation and simplify.

b Multiply expressions involving radicals in which some factors contain more than one term.

FOR EXTRA HELP

Tape 13C Tape 12A MAC: 7 IBM: 7

Add or subtract. Simplify by collecting like radical terms, if possible.

1. $5\sqrt{2} + 8\sqrt{2}$

2. $7\sqrt[4]{5x} + 3\sqrt[4]{5x} - \sqrt{7}$

Add or subtract. Simplify by collecting like radical terms, if possible.

3. $7\sqrt{45} - 2\sqrt{5}$

4. $3\sqrt[3]{y^5} + 4\sqrt[3]{y^2} + \sqrt[3]{8y^6}$

5. $\sqrt{25x - 25} - \sqrt{9x - 9}$

ANSWERS ON PAGE A-8

b More Multiplication with Radicals

To multiply expressions in which some factors contain more than one term, we use the procedures for multiplying polynomials.

▶ **EXAMPLES** Multiply.

7. $\sqrt{3}(x - \sqrt{5}) = \sqrt{3} \cdot x - \sqrt{3} \cdot \sqrt{5}$ Using a distributive law
$= x\sqrt{3} - \sqrt{15}$ Multiplying radicals

8. $\sqrt[3]{y}(\sqrt[3]{y^2} + \sqrt[3]{2}) = \sqrt[3]{y} \cdot \sqrt[3]{y^2} + \sqrt[3]{y} \cdot \sqrt[3]{2}$ Using a distributive law
$= \sqrt[3]{y^3} + \sqrt[3]{2y}$ Multiplying radicals
$= y + \sqrt[3]{2y}$ Simplifying $\sqrt[3]{y^3}$ ◀

DO EXERCISES 6 AND 7.

▶ **EXAMPLE 9** Multiply: $(4\sqrt{3} + \sqrt{2})(\sqrt{3} - 5\sqrt{2})$.

$$\begin{aligned}(4\sqrt{3} + \sqrt{2})(\sqrt{3} - 5\sqrt{2}) &= \overset{F}{4(\sqrt{3})^2} - \overset{O}{20\sqrt{3} \cdot \sqrt{2}} + \overset{I}{\sqrt{2} \cdot \sqrt{3}} - \overset{L}{5(\sqrt{2})^2} \\ &= 4 \cdot 3 - 20\sqrt{6} + \sqrt{6} - 5 \cdot 2 \\ &= 12 - 20\sqrt{6} + \sqrt{6} - 10 \\ &= 2 - 19\sqrt{6}\end{aligned}$$ ◀

▶ **EXAMPLE 10** Multiply: $(\sqrt{a} + \sqrt{3})(\sqrt{b} + \sqrt{3})$. Assume that all expressions under radicals represent nonnegative numbers.

$$\begin{aligned}(\sqrt{a} + \sqrt{3})(\sqrt{b} + \sqrt{3}) &= \sqrt{a}\sqrt{b} + \sqrt{a}\sqrt{3} + \sqrt{3}\sqrt{b} + \sqrt{3}\sqrt{3} \\ &= \sqrt{ab} + \sqrt{3a} + \sqrt{3b} + 3\end{aligned}$$ ◀

DO EXERCISES 8 AND 9.

▶ **EXAMPLE 11** Multiply: $(\sqrt{5} + \sqrt{7})(\sqrt{5} - \sqrt{7})$.

$(\sqrt{5} + \sqrt{7})(\sqrt{5} - \sqrt{7}) = (\sqrt{5})^2 - (\sqrt{7})^2$ This is now a difference of two squares.
$= 5 - 7$
$= -2$ ◀

▶ **EXAMPLE 12** Multiply: $(\sqrt{a} + \sqrt{b})(\sqrt{a} - \sqrt{b})$. Assume that $a \geq 0$ and $b \geq 0$.

$$\begin{aligned}(\sqrt{a} + \sqrt{b})(\sqrt{a} - \sqrt{b}) &= (\sqrt{a})^2 - (\sqrt{b})^2 \\ &= a - b\end{aligned}$$ ◀

Expressions of the form $\sqrt{a} + \sqrt{b}$ and $\sqrt{a} - \sqrt{b}$ are called **conjugates.** Their product is always an expression that has no radicals.

DO EXERCISES 10 AND 11.

▶ **EXAMPLE 13** Multiply: $(\sqrt{3} + x)^2$.

$(\sqrt{3} + x)^2 = (\sqrt{3})^2 + 2x\sqrt{3} + x^2$ Squaring a binomial
$= 3 + 2x\sqrt{3} + x^2$ ◀

DO EXERCISES 12 AND 13.

Multiply. Assume that all expressions under radicals represent nonnegative numbers.

6. $\sqrt{2}(5\sqrt{3} + 3\sqrt{7})$

7. $\sqrt[3]{a^2}(\sqrt[3]{3a} - \sqrt[3]{2})$

Multiply. Assume that all expressions under radicals represent nonnegative numbers.

8. $(\sqrt{3} - 5\sqrt{2})(2\sqrt{3} + \sqrt{2})$

9. $(\sqrt{a} + 2\sqrt{3})(3\sqrt{b} - 4\sqrt{3})$

Multiply. Assume that all expressions under radicals represent nonnegative numbers.

10. $(\sqrt{2} + \sqrt{5})(\sqrt{2} - \sqrt{5})$

11. $(\sqrt{p} - \sqrt{q})(\sqrt{p} + \sqrt{q})$

Multiply.

12. $(2\sqrt{5} - y)^2$

13. $(3\sqrt{6} + 2)^2$

ANSWERS ON PAGE A-8

NAME SECTION DATE

EXERCISE SET 7.4

ANSWERS

a Add or subtract. Simplify by collecting like radical terms if possible, assuming that all expressions under radicals represent nonnegative numbers.

1. $6\sqrt{3} + 2\sqrt{3}$

2. $8\sqrt{5} + 9\sqrt{5}$

3. $9\sqrt[3]{5} - 6\sqrt[3]{5}$

4. $14\sqrt[5]{2} - 6\sqrt[5]{2}$

5. $4\sqrt[3]{y} + 9\sqrt[3]{y}$

6. $6\sqrt[4]{t} - 3\sqrt[4]{t}$

7. $8\sqrt{2} - 6\sqrt{2} + 5\sqrt{2}$

8. $2\sqrt{6} + 8\sqrt{6} - 3\sqrt{6}$

9. $4\sqrt[3]{3} - \sqrt{5} + 2\sqrt[3]{3} + \sqrt{5}$

10. $5\sqrt{7} - 8\sqrt[4]{11} + \sqrt{7} + 9\sqrt[4]{11}$

11. $8\sqrt{27} - 3\sqrt{3}$

12. $9\sqrt{50} - 4\sqrt{2}$

13. $8\sqrt{45} + 7\sqrt{20}$

14. $9\sqrt{12} + 16\sqrt{27}$

15. $18\sqrt{72} + 2\sqrt{98}$

16. $12\sqrt{45} - 8\sqrt{80}$

17. $3\sqrt[3]{16} + \sqrt[3]{54}$

18. $\sqrt[3]{27} - 5\sqrt[3]{8}$

1. ______
2. ______
3. ______
4. ______
5. ______
6. ______
7. ______
8. ______
9. ______
10. ______
11. ______
12. ______
13. ______
14. ______
15. ______
16. ______
17. ______
18. ______

ANSWERS

19. ______

20. ______

21. ______

22. ______

23. ______

24. ______

25. ______

26. ______

27. ______

28. ______

29. ______

30. ______

31. ______

32. ______

33. ______

34. ______

35. ______

36. ______

19. $2\sqrt{128} - \sqrt{18} + 4\sqrt{32}$

20. $5\sqrt{50} - 2\sqrt{18} + 9\sqrt{32}$

21. $\sqrt{5a} + 2\sqrt{45a^3}$

22. $4\sqrt{3x^3} - \sqrt{12x}$

23. $\sqrt[3]{24x} - \sqrt[3]{3x^4}$

24. $\sqrt[3]{54x} - \sqrt[3]{2x^4}$

25. $\sqrt{8y - 8} + \sqrt{2y - 2}$

26. $\sqrt{12t + 12} + \sqrt{3t + 3}$

27. $\sqrt{x^3 - x^2} + \sqrt{9x - 9}$

28. $\sqrt{4x - 4} - \sqrt{x^3 - x^2}$

29. $5\sqrt[3]{32} - \sqrt[3]{108} + 2\sqrt[3]{256}$

30. $3\sqrt[3]{8x} - 4\sqrt[3]{27x} + 2\sqrt[3]{64x}$

b Multiply.

31. $\sqrt{6}(2 - 3\sqrt{6})$

32. $\sqrt{3}(4 + \sqrt{3})$

33. $\sqrt{2}(\sqrt{3} - \sqrt{5})$

34. $\sqrt{5}(\sqrt{5} - \sqrt{2})$

35. $\sqrt{3}(2\sqrt{5} - 3\sqrt{4})$

36. $\sqrt{2}(3\sqrt{10} - 2\sqrt{2})$

37. $\sqrt[3]{2}(\sqrt[3]{4} - 2\sqrt[3]{32})$

38. $\sqrt[3]{3}(\sqrt[3]{9} - 4\sqrt[3]{21})$

39. $\sqrt[3]{a}(\sqrt[3]{2a^2} + \sqrt[3]{16a^2})$

40. $\sqrt[3]{x}(\sqrt[3]{3x^2} - \sqrt[3]{81x^2})$

41. $(\sqrt{3} - \sqrt{2})(\sqrt{3} + \sqrt{2})$

42. $(\sqrt{5} + \sqrt{6})(\sqrt{5} - \sqrt{6})$

43. $(\sqrt{8} + 2\sqrt{5})(\sqrt{8} - 2\sqrt{5})$

44. $(\sqrt{18} + 3\sqrt{7})(\sqrt{18} - 3\sqrt{7})$

45. $(7 + \sqrt{5})(7 - \sqrt{5})$

46. $(4 - \sqrt{3})(4 + \sqrt{3})$

47. $(2 - \sqrt{3})(2 + \sqrt{3})$

48. $(11 - \sqrt{2})(11 + \sqrt{2})$

49. $(\sqrt{8} + \sqrt{5})(\sqrt{8} - \sqrt{5})$

50. $(\sqrt{6} - \sqrt{7})(\sqrt{6} + \sqrt{7})$

51. $(3 + 2\sqrt{7})(3 - 2\sqrt{7})$

52. $(6 - 3\sqrt{2})(6 + 3\sqrt{2})$

ANSWERS

37. ____________

38. ____________

39. ____________

40. ____________

41. ____________

42. ____________

43. ____________

44. ____________

45. ____________

46. ____________

47. ____________

48. ____________

49. ____________

50. ____________

51. ____________

52. ____________

ANSWERS

53. ______

54. ______

55. ______

56. ______

57. ______

58. ______

59. ______

60. ______

61. ______

62. ______

63. ______

64. ______

65. ______

66. ______

67. ______

68. ______

69. ______

70. ______

71. ______

72. ______

73. ______

74. ______

For the following exercises, assume that all expressions under radicals represent nonnegative numbers.

53. $(\sqrt{a} + \sqrt{b})(\sqrt{a} - \sqrt{b})$

54. $(\sqrt{x} - \sqrt{y})(\sqrt{x} + \sqrt{y})$

55. $(3 - \sqrt{5})(2 + \sqrt{5})$

56. $(2 + \sqrt{6})(4 - \sqrt{6})$

57. $(\sqrt{3} + 1)(2\sqrt{3} + 1)$

58. $(4\sqrt{3} + 5)(\sqrt{3} - 2)$

59. $(2\sqrt{7} - 4\sqrt{2})(3\sqrt{7} + 6\sqrt{2})$

60. $(4\sqrt{5} + 3\sqrt{3})(3\sqrt{5} - 4\sqrt{3})$

61. $(\sqrt{a} + \sqrt{2})(\sqrt{a} + \sqrt{3})$

62. $(2 - \sqrt{x})(1 - \sqrt{x})$

63. $(2\sqrt[3]{3} + \sqrt[3]{2})(\sqrt[3]{3} - 2\sqrt[3]{2})$

64. $(3\sqrt[4]{7} + \sqrt[4]{6})(2\sqrt[4]{9} - 3\sqrt[4]{6})$

65. $(2 + \sqrt{3})^2$

66. $(\sqrt{5} + 1)^2$

67. $(\sqrt[5]{9} - \sqrt[5]{3})(\sqrt[5]{8} + \sqrt[5]{27})$

68. $(\sqrt[3]{8x} - \sqrt[3]{5y})^2$

SYNTHESIS

Multiply and simplify.

69. $\sqrt{9 + 3\sqrt{5}}\sqrt{9 - 3\sqrt{5}}$

70. $(\sqrt{x + 2} - \sqrt{x - 2})^2$

71. $(\sqrt{3} + \sqrt{5} - \sqrt{6})^2$

72. $\sqrt[3]{y}(1 - \sqrt[3]{y})(1 + \sqrt[3]{y})$

73. $(\sqrt[3]{9} - 2)(\sqrt[3]{9} + 4)$

74. $[\sqrt{3 + \sqrt{2 + \sqrt{1}}}]^4$

7.5 Rationalizing Numerators and Denominators

a Rationalizing Denominators

Sometimes in mathematics it is useful to find an equivalent expression without a radical in the denominator. This provides a standard notation for expressing results. The procedure for finding such an expression is called **rationalizing a denominator.** We carry this out by multiplying by 1 in either of two ways.

One way is to multiply by 1 under the radical to make the denominator of the radicand a perfect power.

▶ **EXAMPLE 1** Rationalize the denominator: $\sqrt{\frac{7}{3}}$.

We multiply by 1 under the radical, using $\frac{3}{3}$. We do this so that the denominator of the radicand will be a perfect power.

$$\sqrt{\frac{7}{3}} = \sqrt{\frac{7}{3} \cdot \frac{3}{3}} = \sqrt{\frac{21}{9}} = \frac{\sqrt{21}}{\sqrt{9}} = \frac{\sqrt{21}}{3}$$ ◀

DO EXERCISE 1.

▶ **EXAMPLE 2** Rationalize the denominator: $\sqrt[3]{\frac{7}{9}}$.

$$\sqrt[3]{\frac{7}{9}} = \sqrt[3]{\frac{7}{3 \cdot 3} \cdot \frac{3}{3}}$$ **Multiplying by $\frac{3}{3}$ to make the denominator of the radicand a perfect cube**

$$= \sqrt[3]{\frac{21}{3 \cdot 3 \cdot 3}} = \frac{\sqrt[3]{21}}{\sqrt[3]{3^3}} = \frac{\sqrt[3]{21}}{3}$$ ◀

DO EXERCISE 2.

Another way to rationalize a denominator is to multiply by 1 outside the radical in order to make the denominator a perfect power.

▶ **EXAMPLE 3** Rationalize the denominator: $\sqrt{\frac{7}{3}}$.

$$\sqrt{\frac{7}{3}} = \frac{\sqrt{7}}{\sqrt{3}} \cdot \frac{\sqrt{3}}{\sqrt{3}} = \frac{\sqrt{7} \cdot \sqrt{3}}{\sqrt{3} \cdot \sqrt{3}} = \frac{\sqrt{21}}{(\sqrt{3})^2} = \frac{\sqrt{21}}{3}$$ ◀

DO EXERCISE 3.

OBJECTIVES

After finishing Section 7.5, you should be able to:

a Rationalize the denominator of a radical expression.

b Rationalize the numerator of a radical expression.

c Rationalize denominators or numerators having two terms.

FOR EXTRA HELP

Tape 13D | Tape 12A | MAC: 7 IBM: 7

1. Rationalize the denominator. Multiply by 1 under the radical.

$$\sqrt{\frac{2}{5}}$$

2. Rationalize the denominator. Multiply by 1 under the radical.

$$\sqrt[3]{\frac{5}{4}}$$

3. Rationalize the denominator. Multiply by 1 outside the radical.

$$\sqrt{\frac{2}{5}}$$

ANSWERS ON PAGE A-8

4. Rationalize the denominator. Multiply by 1 outside the radical.

$$\sqrt{\frac{4a}{3b}}$$

▶ **EXAMPLE 4** Rationalize the denominator: $\sqrt{\frac{2a}{5b}}$. Assume that all expressions under radicals represent positive numbers.

$$\sqrt{\frac{2a}{5b}} = \frac{\sqrt{2a}}{\sqrt{5b}} \quad \textbf{Converting to a quotient of radicals}$$

$$= \frac{\sqrt{2a}}{\sqrt{5b}} \cdot \frac{\sqrt{5b}}{\sqrt{5b}} \quad \textbf{Multiplying by 1}$$

$$= \frac{\sqrt{10ab}}{(\sqrt{5b})^2} \quad \textbf{The denominator is a perfect power.}$$

$$= \frac{\sqrt{10ab}}{5b}$$ ◀

DO EXERCISE 4.

Rationalize the denominator.

5. $\frac{\sqrt[4]{7}}{\sqrt[4]{2}}$

6. $\sqrt[3]{\frac{3x^5}{2y}}$

▶ **EXAMPLE 5** Rationalize the denominator: $\frac{\sqrt[3]{a}}{\sqrt[3]{9x}}$.

We have

$$\frac{\sqrt[3]{a}}{\sqrt[3]{9x}} = \frac{\sqrt[3]{a}}{\sqrt[3]{9x}} \cdot \frac{\sqrt[3]{3x^2}}{\sqrt[3]{3x^2}} \quad \textbf{Multiplying by 1}$$

To choose the symbol for 1, we look at the radicand $9x$. This is $3 \cdot 3 \cdot x$. To make it a cube, we need another 3 and two more x's. Thus we multiply by $\sqrt[3]{3x^2}/\sqrt[3]{3x^2}$. We have

$$\frac{\sqrt[3]{3ax^2}}{\sqrt[3]{27x^3}}, \quad \text{which simplifies to} \quad \frac{\sqrt[3]{3ax^2}}{3x}.$$ ◀

DO EXERCISES 5 AND 6.

▶ **EXAMPLE 6** Rationalize the denominator: $\frac{3x}{\sqrt[5]{2x^2y^3}}$.

$$\frac{3x}{\sqrt[5]{2x^2y^3}} \cdot \frac{\sqrt[5]{16x^3y^2}}{\sqrt[5]{16x^3y^2}} = \frac{3x\sqrt[5]{16x^3y^2}}{\sqrt[5]{32x^5y^5}} = \frac{3x\sqrt[5]{16x^3y^2}}{2xy} = \frac{3\sqrt[5]{16x^3y^2}}{2y}$$ ◀

DO EXERCISE 7.

7. Rationalize the denominator.

$$\frac{7x}{\sqrt[3]{4xy^5}}$$

b Rationalizing Numerators

Sometimes in calculus it is necessary to rationalize a numerator. We use a similar procedure.

▶ **EXAMPLE 7** Rationalize the numerator: $\sqrt{\frac{7}{5}}$.

$$\sqrt{\frac{7}{5}} = \frac{\sqrt{7}}{\sqrt{5}} = \frac{\sqrt{7}}{\sqrt{5}} \cdot \frac{\sqrt{7}}{\sqrt{7}} \quad \textbf{Multiplying by 1}$$

$$= \frac{\sqrt{49}}{\sqrt{35}} \quad \textbf{The radicand in the numerator is a perfect square.}$$

$$= \frac{7}{\sqrt{35}}$$ ◀

ANSWERS ON PAGE A-8

▶ **EXAMPLE 8** Rationalize the numerator: $\dfrac{\sqrt[3]{4a^2}}{\sqrt[3]{5b^5}}$.

$$\frac{\sqrt[3]{4a^2}}{\sqrt[3]{5b^5}} = \frac{\sqrt[3]{4a^2}}{\sqrt[3]{5b^5}} \cdot \frac{\sqrt[3]{2a}}{\sqrt[3]{2a}} \quad \textbf{Multiplying by 1}$$

$$= \frac{2a}{\sqrt[3]{5b^5 \cdot 2a}}$$

$$= \frac{2a}{\sqrt[3]{b^3 \cdot 10ab^2}}$$

$$= \frac{2a}{b\sqrt[3]{10ab^2}}$$ ◀

DO EXERCISES 8 AND 9.

Rationalize the numerator.

8. $\sqrt{\dfrac{11}{6}}$

9. $\dfrac{\sqrt[3]{3a^4}}{\sqrt[3]{2c^2}}$

C Rationalizing When There Are Two Terms

DO EXERCISES 10 AND 11.

Multiply.

10. $(c - \sqrt{b})(c + \sqrt{b})$

11. $(\sqrt{a} + \sqrt{b})(\sqrt{a} - \sqrt{b})$

Pairs of expressions like $c - \sqrt{b}$, $c + \sqrt{b}$ and $\sqrt{a} - \sqrt{b}$, $\sqrt{a} + \sqrt{b}$ are called **conjugates.** The product of a pair of conjugates has no radicals in it. Thus when we wish to rationalize a denominator that has two terms and one or more of them involves a square-root radical, we multiply by 1 using the conjugate of the denominator to write a symbol for 1.

▶ **EXAMPLES** What symbol for 1 would you use to rationalize the denominator?

	Expression	*Symbol for 1*
9.	$\dfrac{3}{x + \sqrt{7}}$	$\dfrac{x - \sqrt{7}}{x - \sqrt{7}}$
10.	$\dfrac{\sqrt{7} + 4}{3 - 2\sqrt{5}}$	$\dfrac{3 + 2\sqrt{5}}{3 + 2\sqrt{5}}$

Change the operation sign to obtain the conjugate. Use the conjugate for the numerator and denominator of the symbol for 1. ◀

DO EXERCISES 12 AND 13.

What symbol for 1 would you use to rationalize the denominator?

12. $\dfrac{\sqrt{5} + 1}{\sqrt{3} - y}$

13. $\dfrac{1}{\sqrt{2} + \sqrt{3}}$

▶ **EXAMPLE 11** Rationalize the denominator: $\dfrac{4 + \sqrt{2}}{\sqrt{5} - \sqrt{2}}$.

$$\frac{4 + \sqrt{2}}{\sqrt{5} - \sqrt{2}} = \frac{4 + \sqrt{2}}{\sqrt{5} - \sqrt{2}} \cdot \frac{\sqrt{5} + \sqrt{2}}{\sqrt{5} + \sqrt{2}}$$ **Multiplying by 1, using the conjugate of $\sqrt{5} - \sqrt{2}$, which is $\sqrt{5} + \sqrt{2}$**

$$= \frac{(4 + \sqrt{2})(\sqrt{5} + \sqrt{2})}{(\sqrt{5} - \sqrt{2})(\sqrt{5} + \sqrt{2})}$$ **Multiplying numerators and denominators**

$$= \frac{4\sqrt{5} + 4\sqrt{2} + \sqrt{2}\sqrt{5} + (\sqrt{2})^2}{(\sqrt{5})^2 - (\sqrt{2})^2}$$ **Using FOIL**

$$= \frac{4\sqrt{5} + 4\sqrt{2} + \sqrt{10} + 2}{5 - 2}$$ **Squaring in the denominator**

$$= \frac{4\sqrt{5} + 4\sqrt{2} + \sqrt{10} + 2}{3}$$ ◀

ANSWERS ON PAGE A-8

Rationalize the denominator.

14. $\dfrac{5+\sqrt{2}}{1-\sqrt{2}}$

15. $\dfrac{14}{3+\sqrt{2}}$

What symbol for 1 would you use to rationalize the numerator?

16. $\dfrac{5-\sqrt{7}}{\sqrt{6}-3}$

17. $\dfrac{\sqrt{a}+\sqrt{b}}{\sqrt{c}-\sqrt{q}}$

18. Rationalize the numerator:
$$\frac{3+\sqrt{5}}{\sqrt{2}-\sqrt{6}}.$$

ANSWERS ON PAGE A-8

▶ **EXAMPLE 12** Rationalize the denominator: $\dfrac{4}{\sqrt{3}+x}$.

$$\begin{aligned}\frac{4}{\sqrt{3}+x} &= \frac{4}{\sqrt{3}+x}\cdot\frac{\sqrt{3}-x}{\sqrt{3}-x}\\ &= \frac{4(\sqrt{3}-x)}{(\sqrt{3}+x)(\sqrt{3}-x)}\\ &= \frac{4(\sqrt{3}-x)}{(\sqrt{3})^2-x^2}\\ &= \frac{4(\sqrt{3}-x)}{3-x^2}\\ &= \frac{4\sqrt{3}-4x}{3-x^2}\end{aligned}$$ ◀

DO EXERCISES 14 AND 15.

We can also rationalize a numerator with more than one term using the conjugate of the numerator.

▶ **EXAMPLES** What symbol for 1 would you use to rationalize the numerator?

	Expression	*Symbol for 1*
13.	$\dfrac{\sqrt{5}+\sqrt{2}}{1+\sqrt{3}}$	$\dfrac{\sqrt{5}-\sqrt{2}}{\sqrt{5}-\sqrt{2}}$
14.	$\dfrac{x-\sqrt{7}}{3}$	$\dfrac{x+\sqrt{7}}{x+\sqrt{7}}$
15.	$\dfrac{3-2\sqrt{5}}{\sqrt{7}+4}$	$\dfrac{3+2\sqrt{5}}{3+2\sqrt{5}}$

Change the operation sign to obtain the conjugate. Use the conjugate for the numerator and denominator of the symbol for 1. ◀

DO EXERCISES 16 AND 17.

▶ **EXAMPLE 16** Rationalize the numerator: $\dfrac{4+\sqrt{2}}{\sqrt{5}-\sqrt{2}}$.

$$\begin{aligned}\frac{4+\sqrt{2}}{\sqrt{5}-\sqrt{2}} &= \frac{4+\sqrt{2}}{\sqrt{5}-\sqrt{2}}\cdot\frac{4-\sqrt{2}}{4-\sqrt{2}}\\ &= \frac{16-(\sqrt{2})^2}{4\sqrt{5}-\sqrt{5}\sqrt{2}-4\sqrt{2}+(\sqrt{2})^2}\\ &= \frac{14}{4\sqrt{5}-\sqrt{10}-4\sqrt{2}+2}\end{aligned}$$ ◀

DO EXERCISE 18.

NAME SECTION DATE

EXERCISE SET 7.5

a Rationalize the denominator. Assume that all expressions under radicals represent positive numbers.

1. $\sqrt{\frac{6}{5}}$

2. $\sqrt{\frac{11}{6}}$

3. $\sqrt{\frac{10}{3}}$

4. $\sqrt{\frac{22}{7}}$

5. $\frac{6\sqrt{5}}{5\sqrt{3}}$

6. $\frac{2\sqrt{3}}{5\sqrt{2}}$

7. $\sqrt[3]{\frac{16}{9}}$

8. $\sqrt[3]{\frac{3}{9}}$

9. $\frac{\sqrt[3]{3a}}{\sqrt[3]{5c}}$

10. $\frac{\sqrt[3]{7x}}{\sqrt[3]{3y}}$

11. $\frac{\sqrt[3]{2y^4}}{\sqrt[3]{6x^4}}$

12. $\frac{\sqrt[3]{3a^4}}{\sqrt[3]{7b^2}}$

13. $\frac{1}{\sqrt[3]{xy}}$

14. $\frac{1}{\sqrt[4]{ab}}$

15. $\sqrt{\frac{5x}{18}}$

16. $\sqrt{\frac{3a}{50}}$

17. $\sqrt[3]{\frac{4}{5x^5y^2}}$

18. $\sqrt[3]{\frac{7c}{100ab^5}}$

ANSWERS

1. ______

2. ______

3. ______

4. ______

5. ______

6. ______

7. ______

8. ______

9. ______

10. ______

11. ______

12. ______

13. ______

14. ______

15. ______

16. ______

17. ______

18. ______

ANSWERS

19. ______

20. ______

21. ______

22. ______

23. ______

24. ______

25. ______

26. ______

27. ______

28. ______

29. ______

30. ______

31. ______

32. ______

33. ______

34. ______

35. ______

36. ______

19. $\sqrt[4]{\dfrac{1}{8x^7y^3}}$ **20.** $\sqrt[5]{\dfrac{1}{8x^7y^3}}$

21. $\dfrac{2x}{\sqrt[5]{18x^8y^6}}$ **22.** $\dfrac{17x}{\sqrt[4]{9x^7y^6}}$

b Rationalize the numerator. Assume that all expressions under radicals represent positive numbers.

23. $\dfrac{\sqrt{7}}{\sqrt{3x}}$ **24.** $\dfrac{\sqrt{6}}{\sqrt{5x}}$ **25.** $\sqrt{\dfrac{14}{21}}$

26. $\sqrt{\dfrac{12}{15}}$ **27.** $\dfrac{4\sqrt{13}}{3\sqrt{7}}$ **28.** $\dfrac{5\sqrt{21}}{2\sqrt{6}}$

29. $\dfrac{\sqrt[3]{7}}{\sqrt[3]{2}}$ **30.** $\dfrac{\sqrt[3]{5}}{\sqrt[3]{4}}$ **31.** $\sqrt{\dfrac{7x}{3y}}$

32. $\sqrt{\dfrac{6a}{2b}}$ **33.** $\dfrac{\sqrt[3]{5y^4}}{\sqrt[3]{6x^5}}$ **34.** $\dfrac{\sqrt[3]{3a^5}}{\sqrt[3]{7b^2}}$

35. $\dfrac{\sqrt{ab}}{3}$ **36.** $\dfrac{\sqrt{xy}}{5}$

c Rationalize the denominator. Assume that all expressions under radicals represent positive numbers.

37. $\dfrac{5}{8 - \sqrt{6}}$

38. $\dfrac{7}{9 + \sqrt{10}}$

39. $\dfrac{-4\sqrt{7}}{\sqrt{5} - \sqrt{3}}$

40. $\dfrac{34\sqrt{5}}{2\sqrt{5} - \sqrt{3}}$

41. $\dfrac{\sqrt{5} - 2\sqrt{6}}{\sqrt{3} - 4\sqrt{5}}$

42. $\dfrac{\sqrt{6} - 3\sqrt{5}}{\sqrt{3} - 2\sqrt{7}}$

43. $\dfrac{\sqrt{x} - \sqrt{y}}{\sqrt{x} + \sqrt{y}}$

44. $\dfrac{\sqrt{a} + \sqrt{b}}{\sqrt{a} - \sqrt{b}}$

45. $\dfrac{5\sqrt{3} - 3\sqrt{2}}{3\sqrt{2} - 2\sqrt{3}}$

46. $\dfrac{7\sqrt{2} + 4\sqrt{3}}{4\sqrt{3} - 3\sqrt{2}}$

47. $\dfrac{\sqrt{x} - 2\sqrt{y}}{2\sqrt{x} + \sqrt{y}}$

48. $\dfrac{5t - \sqrt{c}}{4t - \sqrt{c}}$

Rationalize the numerator. Assume that all expressions under radicals represent positive numbers.

49. $\dfrac{\sqrt{3} + 5}{8}$

50. $\dfrac{3 - \sqrt{2}}{5}$

51. $\dfrac{\sqrt{3} - 5}{\sqrt{2} + 5}$

52. $\dfrac{\sqrt{6} - 3}{\sqrt{3} + 7}$

ANSWERS

37. ______

38. ______

39. ______

40. ______

41. ______

42. ______

43. ______

44. ______

45. ______

46. ______

47. ______

48. ______

49. ______

50. ______

51. ______

52. ______

ANSWERS

53. ____

54. ____

55. ____

56. ____

57. ____

58. ____

59. ____

60. ____

61. ____

62. ____

63. ____

64. ____

65. ____

66. ____

67. ____

68. ____

69. ____

70. ____

71. ____

72. ____

73. ____

74. ____

75. ____

76. ____

53. $\dfrac{\sqrt{x}-\sqrt{y}}{\sqrt{x}+\sqrt{y}}$

54. $\dfrac{\sqrt{x}+\sqrt{y}}{\sqrt{x}-\sqrt{y}}$

55. $\dfrac{4\sqrt{6}-5\sqrt{3}}{2\sqrt{3}+7\sqrt{6}}$

56. $\dfrac{8\sqrt{2}+5\sqrt{3}}{5\sqrt{3}-7\sqrt{2}}$

57. $\dfrac{\sqrt{2}+3\sqrt{x}}{\sqrt{2}-\sqrt{x}}$

58. $\dfrac{\sqrt{7}-2\sqrt{x}}{\sqrt{7}+\sqrt{x}}$

59. $\dfrac{a\sqrt{b}+c}{\sqrt{b}+c}$

60. $\dfrac{b+a\sqrt{c}}{a-\sqrt{c}}$

SKILL MAINTENANCE

Solve.

61. $\dfrac{1}{2}-\dfrac{1}{3}=\dfrac{1}{t}$

62. $\dfrac{5}{x-1}+\dfrac{9}{x^2+x+1}=\dfrac{15}{x^3-1}$

Divide and simplify.

63. $\dfrac{1}{x^3-y^3}\div\dfrac{1}{(x-y)(x^2+xy+y^2)}$

64. $\dfrac{2x^2-x-6}{x^2+4x+3}\div\dfrac{2x^2+x-3}{x^2-1}$

SYNTHESIS

Assume that all expressions under radicals in Exercises 65–76 represent positive numbers.

Rationalize the denominator.

65. $\dfrac{\sqrt{5}+\sqrt{10}-\sqrt{6}}{\sqrt{50}}$

66. $\dfrac{3\sqrt{y}+4\sqrt{yz}}{5\sqrt{y}-2\sqrt{z}}$

67. $\dfrac{b+\sqrt{b}}{1+b+\sqrt{b}}$

68. $\dfrac{12}{3-\sqrt{3-y}}$

69. $\dfrac{36a^2b}{\sqrt[3]{6a^2b}}$

Rationalize the numerator.

70. $\dfrac{\sqrt{y+18}-\sqrt{y}}{18}$

71. $\dfrac{\sqrt{x+6}-5}{\sqrt{x+6}+5}$

72. $\dfrac{\sqrt{x+h}-\sqrt{x}}{h}$

Simplify. (*Hint:* Rationalize the denominator.)

73. $\sqrt{a^2-3}-\dfrac{a^2}{\sqrt{a^2-3}}$

74. $5\sqrt{\dfrac{x}{y}}+4\sqrt{\dfrac{y}{x}}-\dfrac{3}{\sqrt{xy}}$

75. $\dfrac{\dfrac{1}{\sqrt{w}}-\sqrt{w}}{\dfrac{\sqrt{w}+1}{\sqrt{w}}}$

76. $\dfrac{1}{4+\sqrt{3}}+\dfrac{1}{\sqrt{3}}+\dfrac{1}{\sqrt{3}-4}$

7.6 Rational Numbers as Exponents

a Rational Exponents

Expressions like $a^{1/2}$, $5^{-1/4}$, and $(2y)^{4/5}$ have not yet been defined. We will define such expressions so that the usual properties of exponents hold.

Consider $a^{1/2} \cdot a^{1/2}$. If we still want to multiply by adding exponents, it must follow that $a^{1/2} \cdot a^{1/2} = a^{1/2+1/2}$, or a^1. Thus we should define $a^{1/2}$ to be a square root of a. Similarly, $a^{1/3} \cdot a^{1/3} \cdot a^{1/3} = a^{1/3+1/3+1/3}$, or a^1, so $a^{1/3}$ should be defined to mean $\sqrt[3]{a}$.

For any nonnegative real number a and any index n, $a^{1/n}$ means $\sqrt[n]{a}$ (the nonnegative nth root of a).

Whenever we use rational exponents, we assume that the bases are nonnegative.

▶ **EXAMPLES** Rewrite without rational exponents.

1. $x^{1/2} = \sqrt{x}$
2. $27^{1/3} = \sqrt[3]{27}$, or 3
3. $(abc)^{1/5} = \sqrt[5]{abc}$ ◀

DO EXERCISES 1–5.

▶ **EXAMPLES** Rewrite with rational exponents.

4. $\sqrt[5]{7xy} = (7xy)^{1/5}$ — We need parentheses around the radicand here.
5. $\sqrt[7]{\frac{x^3y}{9}} = \left(\frac{x^3y}{9}\right)^{1/7}$ ◀

DO EXERCISES 6–8.

How should we define $a^{2/3}$? If the usual properties of exponents are to hold, we have $a^{2/3} = (a^{1/3})^2$, or $(\sqrt[3]{a})^2$, or $\sqrt[3]{a^2}$. We define this accordingly.

For any natural numbers m and n ($n \neq 1$) and any nonnegative real number a,

$$a^{m/n} \text{ means } \sqrt[n]{a^m}, \text{ or } (\sqrt[n]{a})^m.$$

▶ **EXAMPLES** Rewrite without rational exponents.

6. $(27)^{2/3} = \sqrt[3]{27^2}$, or $(\sqrt[3]{27})^2$ — This is easier to compute using $(\sqrt[3]{27})^2$.
   $= 3^2$, or 9
7. $4^{3/2} = \sqrt[2]{4^3}$, or $(\sqrt[2]{4})^3$
   $= 2^3$, or 8 ◀

DO EXERCISES 9–11.

OBJECTIVES

After finishing Section 7.6, you should be able to:

- **a** Write expressions with or without rational exponents.
- **b** Write expressions without negative exponents.
- **c** Use the laws of exponents with rational exponents.
- **d** Use rational exponents to simplify radical expressions.

FOR EXTRA HELP

Tape 14A | Tape 12B | MAC: 7 IBM: 7

Rewrite without rational exponents.

1. $y^{1/4}$
2. $(3a)^{1/2}$
3. $16^{1/4}$
4. $(125)^{1/3}$
5. $(a^3b^2c)^{1/5}$

Rewrite with rational exponents.

6. $\sqrt[3]{19}$
7. $\sqrt{abc}$
8. $\sqrt[5]{\frac{x^2y}{16}}$

Rewrite without rational exponents.

9. $x^{3/2}$
10. $8^{2/3}$
11. $4^{5/2}$

ANSWERS ON PAGE A-8

Rewrite with rational exponents.

12. $(\sqrt[3]{7abc})^4$

13. $\sqrt[5]{6^7}$

Rewrite with positive exponents.

14. $16^{-1/4}$

15. $(3xy)^{-7/8}$

Use the laws of exponents to simplify.

16. $7^{1/3} \cdot 7^{3/5}$

17. $\dfrac{5^{7/6}}{5^{5/6}}$

18. $(9^{3/5})^{2/3}$

ANSWERS ON PAGE A-8

▶ **EXAMPLES** Rewrite with rational exponents.

8. $\sqrt[3]{9^4} = 9^{4/3}$

9. $(\sqrt[4]{7xy})^5 = (7xy)^{5/4}$ ◀

DO EXERCISES 12 AND 13.

b Negative Rational Exponents

Negative rational exponents have a meaning similar to that of negative integer exponents.

> **For any rational number m/n and any positive real number a,**
>
> $$a^{-m/n} \quad \text{means} \quad \frac{1}{a^{m/n}},$$
>
> **that is, $a^{m/n}$ and $a^{-m/n}$ are reciprocals.**

▶ **EXAMPLES** Rewrite with positive exponents

Don't make the mistake of thinking that a negative exponent means that the entire expression is negative.

10. $4^{-1/2} = \dfrac{1}{4^{1/2}}$ **$4^{-1/2}$ is the reciprocal of $4^{1/2}$.**

Since $4^{1/2} = \sqrt{4} = 2$, the answer simplifies to $\frac{1}{2}$.

11. $(5xy)^{-4/5} = \dfrac{1}{(5xy)^{4/5}}$ **$(5xy)^{-4/5}$ is the reciprocal of $(5xy)^{4/5}$.** ◀

DO EXERCISES 14 AND 15.

c Laws of Exponents

The same laws hold for rational-number exponents as for integer exponents. We list them for review.

> **For any real number a and any rational exponents m and n:**
>
> 1. $a^m \cdot a^n = a^{m+n}$ **In multiplying, we can add exponents if the bases are the same.**
> 2. $\dfrac{a^m}{a^n} = a^{m-n}$ **In dividing, we can subtract exponents if the bases are the same.**
> 3. $(a^m)^n = a^{m \cdot n}$ **To raise a power to a power, we can multiply the exponents.**

▶ **EXAMPLES** Use the laws of exponents to simplify.

12. $3^{1/5} \cdot 3^{3/5} = 3^{1/5+3/5} = 3^{4/5}$ **Adding exponents**

13. $\dfrac{7^{1/4}}{7^{1/2}} = 7^{1/4-1/2} = 7^{1/4-2/4} = 7^{-1/4}$ **Subtracting exponents**

14. $(7.2^{2/3})^{3/4} = 7.2^{2/3 \cdot 3/4} = 7.2^{6/12}$ $= 7.2^{1/2}$ **Multiplying exponents** ◀

DO EXERCISES 16–18.

d Simplifying Radical Expressions

Rational exponents can be used to simplify some radical expressions. The procedure is as follows.

> **1. Convert radical expressions to exponential expressions.**
> **2. Use arithmetic and the laws of exponents to simplify.**
> **3. Convert back to radical notation when appropriate.**
> ***Important:* This works only when all expressions under radicals are nonnegative since rational exponents are not defined otherwise. No absolute-value signs will be needed.**

▶ **EXAMPLES** Use rational exponents to simplify.

15. $\sqrt[6]{x^3} = x^{3/6}$ **Converting to an exponential expression**
$= x^{1/2}$ **Using arithmetic to simplify the exponent**
$= \sqrt{x}$ **Converting back to radical notation**

16. $\sqrt[6]{4} = 4^{1/6}$ **Converting to exponential notation**
$= (2^2)^{1/6}$ **Renaming 4 as 2^2**
$= 2^{2/6}$ **Using the third law of exponents**
$= 2^{1/3}$ **Using arithmetic to simplify the exponent**
$= \sqrt[3]{2}$ **Converting back to radical notation** ◀

DO EXERCISES 19–21.

▶ **EXAMPLE 17** Use rational exponents to simplify: $\sqrt[8]{a^2b^4}$.

$\sqrt[8]{a^2b^4} = (a^2b^4)^{1/8}$ **Converting to exponential notation**
$= a^{2/8} \cdot b^{4/8}$ **Using the third law of exponents**
$= a^{1/4} \cdot b^{1/2}$ **Using arithmetic to simplify the exponents**
$= a^{1/4} \cdot b^{2/4}$ **Converting the exponents to fractions that have the least common denominator of the fractions**
$= (ab^2)^{1/4}$ **Using the third law of exponents (in reverse)**
$= \sqrt[4]{ab^2}$ **Converting back to radical notation** ◀

DO EXERCISES 22–24.

We can use properties of rational exponents to write a single radical expression for a product or quotient.

▶ **EXAMPLES** Use rational exponents to write a single radical expression.

18. $\sqrt[3]{5} \cdot \sqrt{2} = 5^{1/3} \cdot 2^{1/2}$ **Converting to exponential notation**
$= 5^{2/6} \cdot 2^{3/6}$ **Rewriting so that exponents have a common denominator**
$= (5^2 \cdot 2^3)^{1/6}$ **Using the third law of exponents**
$= \sqrt[6]{5^2 \cdot 2^3}$ **Converting back to radical notation**
$= \sqrt[6]{200}$ **Multiplying under the radical**

Use rational exponents to simplify.

19. $\sqrt[4]{a^2}$

20. $\sqrt[4]{x^4}$

21. $\sqrt[6]{8}$

Use rational exponents to simplify.

22. $\sqrt[5]{a^5b^{10}}$

23. $\sqrt[4]{x^4y^{12}}$

24. $\sqrt[12]{x^3y^6}$

ANSWERS ON PAGE A-8

Use rational exponents to write a single radical expression.

25. $\sqrt[4]{7} \cdot \sqrt{3}$

26. $\dfrac{\sqrt[4]{(a-b)^5}}{(a-b)}$

Write a single radical expression.

27. $x^{2/3}y^{1/2}z^{5/6}$

28. $\dfrac{a^{1/2}b^{3/8}}{a^{1/4}b^{1/8}}$

ANSWERS ON PAGE A-8

19. $\sqrt{x-2} \cdot \sqrt[4]{3y} = (x-2)^{1/2}(3y)^{1/4}$ **Converting to exponential notation**

$= (x-2)^{2/4}(3y)^{1/4}$ **Writing exponents with a common denominator**

$= [(x-2)^2(3y)]^{1/4}$ **Using the third law of exponents (in reverse)**

$= \sqrt[4]{(x^2-4x+4) \cdot 3y}$ **Converting back to radical notation**

$= \sqrt[4]{3x^2y - 12xy + 12y}$ **Multiplying under the radical**

20. $\dfrac{\sqrt[4]{(x+y)^3}}{\sqrt{x+y}} = \dfrac{(x+y)^{3/4}}{(x+y)^{1/2}}$ **Converting to exponential notation**

$= (x+y)^{3/4-1/2}$ **Using the second law of exponents**

$= (x+y)^{1/4}$ **Subtracting exponents**

$= \sqrt[4]{x+y}$ **Converting back to radical notation** ◀

DO EXERCISES 25 AND 26.

► **EXAMPLE 21** Write a single radical expression for $a^{1/2}b^{-1/2}c^{5/6}$.

$a^{1/2}b^{-1/2}c^{5/6} = a^{3/6}b^{-3/6}c^{5/6}$ **Rewriting exponents with a common denominator**

$= (a^3b^{-3}c^5)^{1/6}$ **Using the third law of exponents**

$= \sqrt[6]{a^3b^{-3}c^5}$ **Converting to radical notation** ◀

DO EXERCISES 27 AND 28.

We have now seen several different methods of simplifying radical expressions. We list them here for reference.

Methods of Simplifying Radical Expressions

1. ***Simplifying by factoring.*** **We factor the radicand, looking for factors that are perfect powers.**
   Example: $\sqrt[3]{16} = \sqrt[3]{8}\,\sqrt[3]{2} = 2\sqrt[3]{2}$
2. ***Rationalizing denominators.*** **Radical expressions are usually considered simpler if there are no radicals in the denominator.**
   Example: $\dfrac{1}{\sqrt{2}} = \dfrac{1}{\sqrt{2}} \cdot \dfrac{\sqrt{2}}{\sqrt{2}} = \dfrac{\sqrt{2}}{2}$
3. ***Collecting like radical terms.***
   Example: $\sqrt{8} + 3\sqrt{2} = \sqrt{4} \cdot \sqrt{2} + 3\sqrt{2} = 2\sqrt{2} + 3\sqrt{2} = 5\sqrt{2}$
4. ***Using rational exponents to simplify.*** **We convert to exponential notation and then use arithmetic and the laws of exponents to simplify the exponents. Then we convert back to radical notation.**
   ***Caution*! This works only when there are only nonnegative expressions under radicals.**

NAME SECTION DATE

EXERCISE SET 7.6

a Rewrite without exponents.

1. $x^{1/4}$
2. $y^{1/5}$
3. $(8)^{1/3}$
4. $(16)^{1/2}$
5. $(a^2b^2)^{1/5}$
6. $(x^3y^3)^{1/4}$
7. $16^{3/4}$
8. $4^{7/2}$

Rewrite with rational exponents.

9. $\sqrt[3]{20}$
10. $\sqrt[3]{19}$
11. $\sqrt[5]{xy^2z}$
12. $\sqrt[7]{x^3y^2z^2}$
13. $(\sqrt{3mn})^3$
14. $(\sqrt[3]{7xy})^4$
15. $(\sqrt[7]{8x^2y})^5$
16. $(\sqrt[6]{2a^5b})^7$

b Rewrite with positive exponents.

17. $x^{-1/3}$
18. $y^{-1/4}$
19. $\dfrac{1}{x^{-2/3}}$
20. $\dfrac{1}{x^{-5/6}}$

c Use the laws of exponents to simplify.

21. $5^{3/4} \cdot 5^{1/8}$
22. $11^{2/3} \cdot 11^{1/2}$
23. $\dfrac{7^{5/8}}{7^{3/8}}$
24. $\dfrac{9^{9/11}}{9^{7/11}}$
25. $\dfrac{8.3^{3/4}}{8.3^{2/5}}$
26. $\dfrac{3.9^{3/5}}{3.9^{1/4}}$
27. $(10^{3/8})^{2/5}$
28. $(5^{5/9})^{3/7}$
29. $a^{2/3} \cdot a^{5/6}$
30. $x^{3/4} \cdot x^{5/12}$
31. $(a^{2/3} \cdot b^{5/8})^4$
32. $(x^{-1/3} \cdot y^{-2/5})^{-15}$

d Use rational exponents to simplify.

33. $\sqrt[6]{a^4}$
34. $\sqrt[6]{y^2}$
35. $\sqrt[3]{8y^6}$
36. $\sqrt{x^4y^6}$
37. $\sqrt[4]{32}$
38. $\sqrt[8]{81}$
39. $\sqrt[6]{4x^2}$
40. $\sqrt[4]{16x^4y^2}$

ANSWERS

1. ______
2. ______
3. ______
4. ______
5. ______
6. ______
7. ______
8. ______
9. ______
10. ______
11. ______
12. ______
13. ______
14. ______
15. ______
16. ______
17. ______
18. ______
19. ______
20. ______
21. ______
22. ______
23. ______
24. ______
25. ______
26. ______
27. ______
28. ______
29. ______
30. ______
31. ______
32. ______
33. ______
34. ______
35. ______
36. ______
37. ______
38. ______
39. ______
40. ______

ANSWERS
41. ______
42. ______
43. ______
44. ______
45. ______
46. ______
47. ______
48. ______
49. ______
50. ______
51. ______
52. ______
53. ______
54. ______
55. ______
56. ______
57. ______
58. ______
59. ______
60. ______
61. ______
62. ______
63. ______
64. ______
65. ______
66. ______
67. ______
68. ______
69. ______
70. ______
71. ______
72. ______
73. ______
74. ______
75. ______
76. ______
77. ______

41. $\sqrt[5]{32c^{10}d^{15}}$ **42.** $\sqrt[4]{16x^{12}y^{16}}$ **43.** $\sqrt[6]{\frac{m^{12}n^{24}}{64}}$ **44.** $\sqrt[5]{\frac{x^{15}y^{20}}{32}}$

45. $\sqrt[8]{r^4s^2}$ **46.** $\sqrt[12]{64t^6s^6}$ **47.** $\sqrt[3]{27a^3b^9}$ **48.** $\sqrt[4]{81x^8y^8}$

49. $\sqrt[5]{32x^{15}y^{40}}$ **50.** $\sqrt[3]{1000a^9b^{18}}$ **51.** $\sqrt[4]{64p^{12}q^{32}}$ **52.** $\sqrt[4]{81x^{48}y^{84}}$

Use rational exponents to write a single radical expression.

53. $\sqrt[3]{3}\sqrt{3}$ **54.** $\sqrt[3]{7}\cdot\sqrt[4]{5}$ **55.** $\sqrt{x}\sqrt[3]{2x}$ **56.** $\sqrt[3]{y}\sqrt[5]{3y}$

57. $\sqrt{x}\sqrt[3]{x-2}$ **58.** $\sqrt[4]{3x}\sqrt{y+4}$ **59.** $\frac{\sqrt[3]{(a+b)^2}}{\sqrt{(a+b)}}$ **60.** $\frac{\sqrt[3]{(x+y)^2}}{\sqrt[4]{(x+y)^3}}$

61. $a^{2/3}\cdot b^{3/4}$ **62.** $x^{1/3}\cdot y^{1/4}\cdot z^{1/6}$ **63.** $\frac{s^{7/12}\cdot t^{7/6}}{s^{1/3}\cdot t^{-1/6}}$ **64.** $\frac{x^{8/15}\cdot y^{7/5}}{x^{1/3}\cdot y^{-1/5}}$

65. $\sqrt[4]{x^3y^5}\cdot\sqrt{xy}$ **66.** $\sqrt[5]{a^3b}\sqrt{ab}$ **67.** $\sqrt{a^4b^3c^4}\sqrt[3]{ab^2c}$ **68.** $\sqrt[3]{xy^2z}\sqrt{x^3yz^2}$

69. $\left(\frac{p^{3/4}q^{-2/3}}{p^{7/8}q^{-3/4}}\right)^2$ **70.** $\left(\frac{c^{-4/5}d^{5/9}}{c^{3/10}d^{1/6}}\right)^3$

SYNTHESIS

Use rational exponents to write a single radical expression and simplify.

71. $\sqrt{x^5\sqrt[3]{x^4}}$ **72.** $\frac{\sqrt{(a+b)^3}\sqrt[3]{(a+b)^2}}{\sqrt[4]{a+b}}$ **73.** $\sqrt[4]{\sqrt[3]{8x^3y^6}}$

Simplify.

74. $(-\sqrt[4]{7}\sqrt[3]{w})^{12}$ **75.** $\sqrt[12]{p^2+2pq+q^2}$

76. $\frac{1}{\sqrt[3]{3}-\sqrt[3]{2}}$ **77.** $[\sqrt[10]{\sqrt[5]{x^{15}}}]^5[\sqrt[5]{\sqrt[10]{x^{15}}}]^5$

7.7 Solving Radical Equations

a The Principle of Powers

A **radical equation** has variables in one or more radicands. These are radical equations:

$$\sqrt[3]{2x} + 1 = 5, \qquad \sqrt{x} + \sqrt{4x - 2} = 7.$$

To solve such equations, we need a new principle. Suppose an equation $a = b$ is true. If we square both sides, we get another true equation: $a^2 = b^2$. This can be generalized.

> The Principle of Powers
>
> **If an equation $a = b$ is true, then $a^n = b^n$ is true for any natural number n.**

If an equation $a^n = b^n$ is true, it *may not* be true that $a = b$. For example, $3^2 = (-3)^2$ is true, but $3 = -3$ is not true.

► **EXAMPLE 1** Solve: $\sqrt{x} - 3 = 4$.

$$\begin{aligned} \sqrt{x} - 3 &= 4 \\ \sqrt{x} &= 7 && \text{Adding to isolate the radical} \\ (\sqrt{x})^2 &= 7^2 && \text{Using the principle of powers} \\ x &= 49 \end{aligned}$$

Check:

$\sqrt{x} - 3 = 4$	
$\sqrt{49} - 3$	4
$7 - 3$	
4	TRUE

The solution is 49. ◄

> **The principle of powers does not always give equivalent equations. For this reason, a check is a must!**

► **EXAMPLE 2** Solve: $\sqrt{x} = -3$.

We might observe at the outset that this equation has no solution because the principal square root of a number is never negative. Let us continue as above for comparison:

$$\begin{aligned} (\sqrt{x})^2 &= (-3)^2 && \text{Using the principle of powers (squaring)} \\ x &= 9 \end{aligned}$$

Check:

$\sqrt{x} = -3$	
$\sqrt{9}$	-3
3	FALSE

The number 9 does *not* check. Thus the equation $\sqrt{x} = -3$ has no real-number solution. ◄

OBJECTIVES

After finishing Section 7.7, you should be able to:

a Solve radical equations with one radical term.

b Solve radical equations with two radical terms.

FOR EXTRA HELP

Tape 14B

Tape 12B

MAC: 7
IBM: 7

Solve.

1. $\sqrt{x} - 7 = 3$

2. $\sqrt{x} = -2$

Solve.

3. $x + 2 = \sqrt{2x + 7}$

4. $x + 1 = 3\sqrt{x - 1}$

ANSWERS ON PAGE A-8

Note in Example 2 that the equation $x = 9$ has solution 9, but that $\sqrt{x} = -3$ has *no* solution. Thus the equations $x = 9$ and $\sqrt{x} = -3$ are *not* equivalent.

DO EXERCISES 1 AND 2.

To solve an equation with a radical term, we first isolate the radical term on one side of the equation. Then we use the principle of powers.

▶ **EXAMPLE 3** Solve: $x - 7 = 2\sqrt{x + 1}$.

The radical term is already isolated. We proceed with the principle of powers:

$$x - 7 = 2\sqrt{x + 1}$$

$$(x - 7)^2 = (2\sqrt{x + 1})^2 \quad \text{Using the principle of powers (squaring)}$$

$$x^2 - 14x + 49 = 2^2(\sqrt{x + 1})^2 \quad \text{Squaring the binomial on the left; raising a product to a power on the right}$$

$$x^2 - 14x + 49 = 4(x + 1)$$

$$x^2 - 14x + 49 = 4x + 4$$

$$x^2 - 18x + 45 = 0$$

$$(x - 3)(x - 15) = 0 \quad \text{Factoring}$$

$$x - 3 = 0 \quad \text{or} \quad x - 15 = 0 \quad \text{Using the principle of zero products}$$

$$x = 3 \quad \text{or} \quad x = 15.$$

The possible solutions are 3 and 15. Let us check.

For 3:

$x - 7$	$= 2\sqrt{x + 1}$	
$3 - 7$	$2\sqrt{3 + 1}$	
-4	$2\sqrt{4}$	
	$2(2)$	
	4	FALSE

For 15:

$x - 7$	$= 2\sqrt{x + 1}$	
$15 - 7$	$2\sqrt{15 + 1}$	
8	$2\sqrt{16}$	
	$2(4)$	
	8	TRUE

The number 3 does *not* check, but the number 15 does check. The solution is 15. ◀

DO EXERCISES 3 AND 4.

▶ **EXAMPLE 4** Solve: $x = \sqrt{x + 7} + 5$.

$$x = \sqrt{x + 7} + 5$$

$$x - 5 = \sqrt{x + 7} \quad \text{Subtracting 5 to isolate the radical term}$$

$$(x - 5)^2 = (\sqrt{x + 7})^2 \quad \text{Using the principle of powers (squaring both sides)}$$

$$x^2 - 10x + 25 = x + 7$$

$$x^2 - 11x + 18 = 0$$

$$(x - 9)(x - 2) = 0 \quad \text{Factoring}$$

$$x = 9 \quad \text{or} \quad x = 2 \quad \text{Using the principle of zero products}$$

The possible solutions are 9 and 2. Let us check.

For 9:

x	$= \sqrt{x + 7} + 5$	
9	$\sqrt{9 + 7} + 5$	
	9	TRUE

For 2:

x	$= \sqrt{x + 7} + 5$	
2	$\sqrt{2 + 7} + 5$	
	8	FALSE

Since 9 checks but 2 does not, the solution is 9. ◀

► **EXAMPLE 5** Solve: $\sqrt[3]{2x+1}+5=0$.

$$\sqrt[3]{2x+1}+5=0$$
$$\sqrt[3]{2x+1}=-5 \quad \text{Subtracting 5; this isolates the radical term}$$
$$(\sqrt[3]{2x+1})^3=(-5)^3 \quad \text{Using the principle of powers (raising to the third power)}$$
$$2x+1=-125$$
$$2x=-126 \quad \text{Subtracting 1}$$
$$x=-63$$

Check:

$\sqrt[3]{2x+1}+5=0$	
$\sqrt[3]{2\cdot(-63)+1}+5$	0
$\sqrt[3]{-125}+5$	
$-5+5$	
0	TRUE

The solution is -63. ◄

DO EXERCISES 5 AND 6.

Solve.

5. $x=\sqrt{x+5}+1$

6. $\sqrt[4]{x-1}-2=0$

b Equations with Two Radical Terms

A general strategy for solving equations with two radical terms is as follows.

1. **Isolate one of the radical terms.**
2. **Use the principle of powers.**
3. **If a radical remains, perform steps (1) and (2) again.**
4. **Check possible solutions.**

► **EXAMPLE 6** Solve: $\sqrt{x-3}+\sqrt{x+5}=4$.

$$\sqrt{x-3}+\sqrt{x+5}=4$$
$$\sqrt{x-3}=4-\sqrt{x+5} \quad \text{Subtracting } \sqrt{x+5}\text{; this isolates one of the radical terms}$$
$$(\sqrt{x-3})^2=(4-\sqrt{x+5})^2 \quad \text{Using the principle of powers (squaring both sides)}$$

Here we are squaring a binomial. We square 4, then find twice the product of 4 and $\sqrt{x+5}$, and then the square of $\sqrt{x+5}$. (See the rule in Section 5.2.)

$$x-3=16-8\sqrt{x+5}+(x+5)$$
$$-3=21-8\sqrt{x+5} \quad \text{Subtracting } x \text{ and collecting like terms}$$
$$-24=-8\sqrt{x+5} \quad \text{Isolating the remaining radical term}$$
$$3=\sqrt{x+5} \quad \text{Dividing by } -8$$
$$3^2=(\sqrt{x+5})^2 \quad \text{Squaring}$$
$$9=x+5$$
$$4=x$$

The number 4 checks and is the solution. ◄

ANSWERS ON PAGE A-8

Solve.

7. $\sqrt{x} - \sqrt{x-5} = 1$

8. $\sqrt{2x-5} - 2 = \sqrt{x-2}$

9. Solve:
$\sqrt{3x+1} - 1 - \sqrt{x+4} = 0.$

ANSWERS ON PAGE A-8

A common error in solving equations like $\sqrt{x-3} + \sqrt{x+5} = 4$ is to square the left side, obtaining $(x-3) + (x+5)$. That is wrong because the square of a sum is *not* the sum of the squares.
Example: $\sqrt{9} + \sqrt{16} = 7$, but $9 + 16 = 25$, not 7^2.

► **EXAMPLE 7** Solve: $\sqrt{2x-5} = 1 + \sqrt{x-3}$.

$$\sqrt{2x-5} = 1 + \sqrt{x+3}$$
$$(\sqrt{2x-5})^2 = (1 + \sqrt{x-3})^2 \quad \text{One radical is already isolated; we square both sides.}$$
$$2x-5 = 1 + 2\sqrt{x-3} + (\sqrt{x-3})^2$$
$$2x-5 = 1 + 2\sqrt{x-3} + (x-3)$$
$$x-3 = 2\sqrt{x-3} \quad \text{Isolating the remaining radical term}$$
$$(x-3)^2 = (2\sqrt{x-3})^2 \quad \text{Squaring both sides}$$
$$x^2 - 6x + 9 = 4(x-3)$$
$$x^2 - 6x + 9 = 4x - 12$$
$$x^2 - 10x + 21 = 0$$
$$(x-7)(x-3) = 0 \quad \text{Factoring}$$
$$x = 7 \quad \text{or} \quad x = 3 \quad \text{Using the principle of zero products}$$

The numbers 7 and 3 check and are the solutions. ◄

DO EXERCISES 7 AND 8.

► **EXAMPLE 8** Solve: $\sqrt{x+2} - \sqrt{2x+2} + 1 = 0$.

We first isolate one radical.

$$\sqrt{x+2} - \sqrt{2x+2} + 1 = 0$$
$$\sqrt{x+2} = \sqrt{2x+2} - 1 \quad \text{Adding } \sqrt{2x+2} \text{ and subtracting 1 to isolate a radical}$$
$$(\sqrt{x+2})^2 = (\sqrt{2x+2} - 1)^2 \quad \text{Squaring both sides}$$
$$x+2 = (\sqrt{2x+2})^2 - 2\sqrt{2x+2} + 1$$
$$x+2 = 2x+2 - 2\sqrt{2x+2} + 1$$
$$-x-1 = -2\sqrt{2x+2} \quad \text{Isolating the remaining radical}$$
$$x+1 = 2\sqrt{2x+2} \quad \text{Multiplying by } -1$$
$$(x+1)^2 = (2\sqrt{2x+2})^2 \quad \text{Squaring both sides}$$
$$x^2 + 2x + 1 = 4(2x+2)$$
$$x^2 + 2x + 1 = 8x + 8$$
$$x^2 - 6x - 7 = 0$$
$$(x-7)(x+1) = 0 \quad \text{Factoring}$$
$$x-7 = 0 \quad \text{or} \quad x+1 = 0 \quad \text{Using the principle of zero products}$$
$$x = 7 \quad \text{or} \quad x = -1$$

The check is left to the student. The number 7 checks, but -1 does not. The solution is 7. ◄

DO EXERCISE 9.

NAME SECTION DATE

EXERCISE SET 7.7

a Solve.

1. $\sqrt{4x - 5} = 1$
2. $\sqrt{3x + 8} = 10$
3. $\sqrt{5x} + 2 = 9$
4. $\sqrt{2x} - 3 = 9$
5. $\sqrt{y + 1} - 5 = 8$
6. $\sqrt{x - 2} - 7 = 4$
7. $\sqrt{3y + 1} = 9$
8. $\sqrt{2y + 1} = 13$
9. $\sqrt[3]{x} = -3$
10. $\sqrt[3]{y} = -4$
11. $\sqrt{x + 2} = -4$
12. $\sqrt{y - 3} = -2$
13. $\sqrt[3]{x + 5} = 2$
14. $\sqrt[3]{x - 2} = 3$
15. $\sqrt[4]{y - 3} = 2$
16. $\sqrt[4]{x + 3} = 3$
17. $\sqrt[3]{6x + 9} + 8 = 5$
18. $\sqrt[3]{3y + 6} + 2 = 3$
19. $8 = \frac{1}{\sqrt{x}}$
20. $\frac{1}{\sqrt{y}} = 3$

b Solve.

21. $\sqrt{3y + 1} = \sqrt{2y + 6}$
22. $\sqrt{5x - 3} = \sqrt{2x + 3}$
23. $\sqrt{y - 5} + \sqrt{y} = 5$
24. $\sqrt{x - 9} + \sqrt{x} = 1$

ANSWERS

1. ______
2. ______
3. ______
4. ______
5. ______
6. ______
7. ______
8. ______
9. ______
10. ______
11. ______
12. ______
13. ______
14. ______
15. ______
16. ______
17. ______
18. ______
19. ______
20. ______
21. ______
22. ______
23. ______
24. ______

ANSWERS

25.

26.

27.

28.

29.

30.

31.

32.

33.

34.

35.

36.

37.

38.

39.

40.

41.

42.

43.

44.

45.

46.

47.

48.

49.

50.

51.

52.

53.

54.

25. $3 + \sqrt{z-6} = \sqrt{z+9}$

26. $\sqrt{4x-3} = 2 + \sqrt{2x-5}$

27. $\sqrt{20-x} + 8 = \sqrt{9-x} + 11$

28. $4 + \sqrt{10-x} = 6 + \sqrt{4-x}$

29. $\sqrt{4y+1} - \sqrt{y-2} = 3$

30. $\sqrt{y+15} - \sqrt{2y+7} = 1$

31. $\sqrt{x+2} + \sqrt{3x+4} = 2$

32. $\sqrt{6x+7} - \sqrt{3x+3} = 1$

33. $\sqrt{3x-5} + \sqrt{2x+3} + 1 = 0$

34. $\sqrt{2m-3} + 2 - \sqrt{m+7} = 0$

35. $2\sqrt{t-1} - \sqrt{3t-1} = 0$

36. $3\sqrt{2y+3} - \sqrt{y+10} = 0$

SKILL MAINTENANCE

37. Solve: $x^2 + 2.8x = 0$.

38. The base of a triangle is 2 in. longer than the height. The area is $31\frac{1}{2}$ in^2. Find the height and the base.

SYNTHESIS

Solve.

39. $\sqrt[3]{\frac{z}{4}} - 10 = 2$

40. $\sqrt[4]{z^2 + 17} = 3$

41. $\sqrt{\sqrt{y+49} - \sqrt{y}} = \sqrt{7}$

42. $\sqrt[3]{x^2 + x + 15} - 3 = 0$

43. $\sqrt{\sqrt{x^2 + 9x + 34}} = 2$

44. $\sqrt{8-b} = b\sqrt{8-b}$

45. $\sqrt{x-2} - \sqrt{x+2} + 2 = 0$

46. $6\sqrt{y} + 6y^{-1/2} = 37$

47. $\sqrt{a^2 + 30a} = a + \sqrt{5a}$

48. $\sqrt{\sqrt{x} + 4} = \sqrt{x} - 2$

49. $\dfrac{x-1}{\sqrt{x^2 + 3x + 6}} = \dfrac{1}{4}$

50. $\sqrt{x+1} - \dfrac{2}{\sqrt{x+1}} = 1$

51. $\sqrt{y^2 + 6} + y - 3 = 0$

52. $2\sqrt{x-1} - \sqrt{3x-5} = \sqrt{x-9}$

53. $\sqrt{y+1} - \sqrt{2y-5} = \sqrt{y-2}$

54. Evaluate: $\sqrt{7 + 4\sqrt{3}} - \sqrt{7 - 4\sqrt{3}}$.

7.8 Applications Involving Powers and Roots

a There are many kinds of problems that involve powers and roots. Many also involve right triangles and the Pythagorean theorem, which we studied in Section 5.8.

▶ **EXAMPLE 1** You and a friend have decided to plant a vegetable garden in the backyard. You decide that it will be a 30-ft by 40-ft rectangle and begin to lay it out using string. You soon realize that it is difficult to form the right angles and that it would be helpful to know the length of a diagonal. Find the length of a diagonal.

Using the Pythagorean theorem, $a^2 + b^2 = c^2$, we substitute 30 for a, 40 for b, and solve for c:

$$a^2 + b^2 = c^2$$
$$30^2 + 40^2 = c^2 \quad \textbf{Substituting}$$
$$900 + 1600 = c^2$$
$$2500 = c^2$$
$$\sqrt{2500} = c$$
$$50 = c.$$

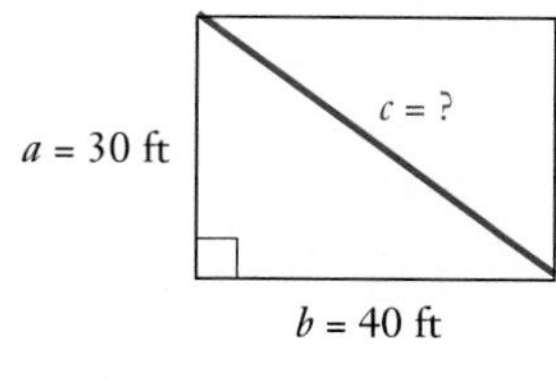

The length of the hypotenuse, or the diagonal, is 50 ft. Knowing this measurement would help in laying out the garden. Construction workers often use a procedure like this to lay out a right angle. ◀

▶ **EXAMPLE 2** Find the length of the hypotenuse in this right triangle. Give an exact answer and an approximation to three decimal places.

$$7^2 + 4^2 = c^2 \quad \textbf{Substituting in the Pythagorean equation}$$
$$49 + 16 = c^2$$
$$65 = c^2$$

Exact answer: $c = \sqrt{65}$

Approximation: $c \approx 8.062$ **Using a calculator** ◀

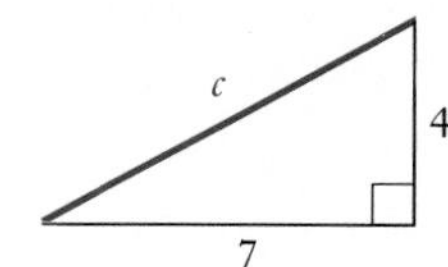

▶ **EXAMPLE 3** Find the length of the leg of this right triangle. Give an exact answer and an approximation to three decimal places.

$$1^2 + b^2 = (\sqrt{11})^2 \quad \textbf{Substituting in the Pythagorean equation}$$
$$1 + b^2 = 11$$
$$b^2 = 10$$

Exact answer: $b = \sqrt{10}$

Approximation: $b \approx 3.162$ ◀

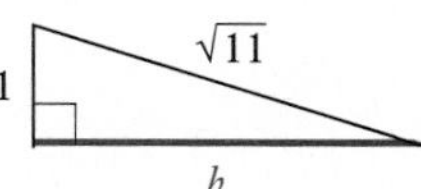

DO EXERCISES 1–3.

OBJECTIVE

After finishing Section 7.8, you should be able to:

a Given two of the lengths of the sides of a right triangle, find the length of the third side and solve applications involving powers and roots.

FOR EXTRA HELP

Tape 14C | Tape 13A | MAC: 7 IBM: 7

1. Find the length of the hypotenuse of this right triangle. Give an exact answer and an approximation to three decimal places.

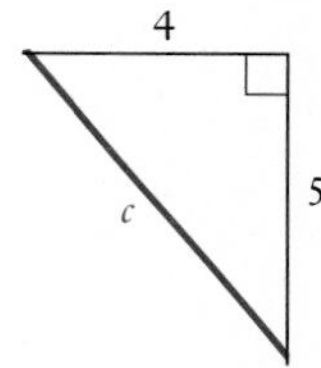

2. Find the length of the leg of this right triangle. Give an exact answer and an approximation to three decimal places.

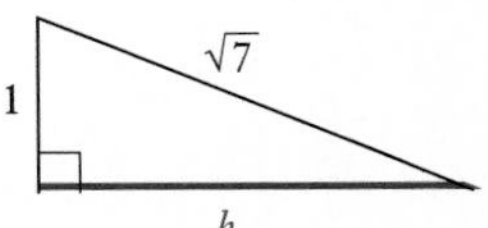

3. Find the length of the hypotenuse of this right triangle. Give an exact answer and an approximation to three decimal places.

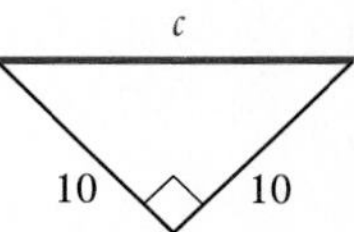

ANSWERS ON PAGE A-8

4. Suppose the catcher in Example 4 makes a throw to second base. How far is that throw? Give an exact answer and an approximation to three decimal places.

▶ **EXAMPLE 4** *Baseball diamond.* A baseball diamond is actually a square 90 ft on a side. Suppose a catcher fields a bunt along the third-base line 10 ft from home plate. How far would the catcher have to throw the ball to first base? Give an exact answer and an approximation to three decimal places.

We first make a drawing and let d represent the distance to first base. We see that we have a right triangle, where the length of the leg from home to first base is 90 ft. The length of the leg from home to where the catcher fields the ball is 10 ft. We substitute these values into the Pythagorean equation to find d:

$$d^2 = 90^2 + 10^2$$
$$d^2 = 8100 + 100$$
$$d^2 = 8200$$
$$d = \sqrt{8200}.$$

Exact answer: $d = \sqrt{8200}$

Approximation: $c \approx 90.554$ ft

Remember: Answers at the back of the book have been found using a calculator.

If you use Table 1 to find an approximation, you will need to simplify before finding an approximation in the table:

$$d = \sqrt{8200} = \sqrt{100 \cdot 82} = 10\sqrt{82}$$
$$\approx 10(9.055) = 90.550 \text{ ft.}$$

Note that we get a variance in the third decimal place. ◀

DO EXERCISE 4.

5. Referring to Example 5, find L given that $h = 3$ ft and $d = 180$ ft. You will need a calculator with an exponential key $\boxed{y^x}$.

▶ **EXAMPLE 5** ▦ *Road-pavement messages.* In a psychological study, it was determined that the proper length L of the letters of a word painted on pavement is given by

$$L = \frac{0.000169d^{2.27}}{h},$$

where d is the distance of a car from the lettering and h is the height of the eye above the road. All units are in feet. According to the study and this formula, if a person h feet above the road is to read a message d feet away, that message will be the most readable if the length of the letters is L.

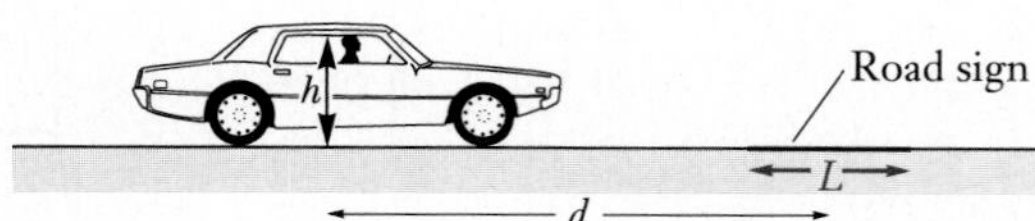

Find L, given that $h = 4$ ft and $d = 180$ ft. You will need a calculator with an exponential key $\boxed{y^x}$.

We substitute 4 for h and 180 for d and calculate L using a calculator with an exponential key $\boxed{y^x}$:

$$L = \frac{0.000169(180)^{2.27}}{4} \approx \frac{0.000169(131{,}664.4674)}{4}$$
$$\approx 5.6 \text{ ft.}$$ ◀

DO EXERCISE 5.

NAME SECTION DATE

EXERCISE SET 7.8

a In a right triangle, find the length of the side not given. Give an exact answer and an approximation to three decimal places.

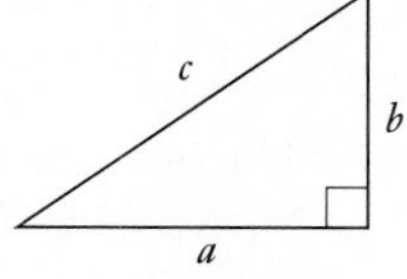

1. $a = 3,\ b = 5$

2. $a = 8,\ b = 10$

3. $a = 12,\ b = 12$

4. $a = 10,\ b = 10$

5. $b = 12,\ c = 13$

6. $a = 5,\ c = 12$

7. $c = 6,\ a = \sqrt{5}$

8. $c = 8,\ a = 4\sqrt{3}$

9. $b = 1,\ c = \sqrt{13}$

10. $a = 1,\ c = \sqrt{20}$

11. $a = 1,\ c = \sqrt{n}$

12. $c = 2,\ a = \sqrt{n}$

In the following problems, give an exact answer and, where appropriate, an approximation to three decimal places.

13. *Guy wire.* How long is a guy wire reaching from the top of a 15-ft pole to a point on the ground 10 ft from the pole?

14. *Softball diamond.* A slow-pitch softball diamond is actually a square 65 ft on a side. How far is it from home to second base?

15. Triangle ABC has sides of lengths 25 ft, 25 ft, and 30 ft. Triangle PQR has sides of lengths 25 ft, 25 ft, and 40 ft. Which triangle has the greater area and by how much?

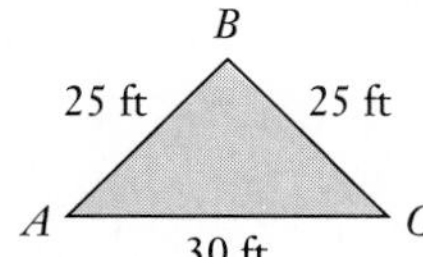

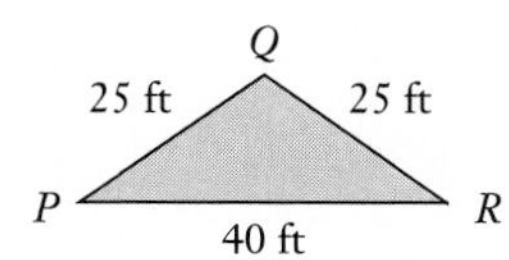

16. *Bridge expansion.* During the summer heat, a 2-mi bridge expands 2 ft in length. Assuming the bulge occurs straight up the middle, how high is the bulge? (The answer may surprise you. In reality, bridges are built with expansion spaces to avoid such buckling.)

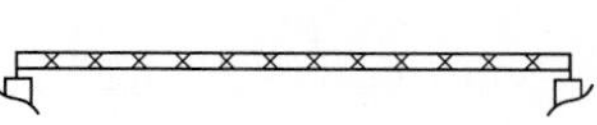

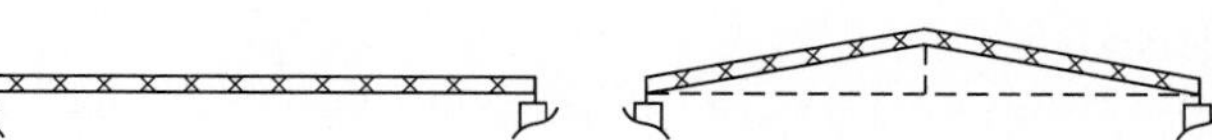

17. The two equal sides of an isosceles right triangle are of length s. Find a formula for the length of the hypotenuse.

18. Each side of a regular octagon has length s. Find a formula for the distance d between the parallel sides of the octagon.

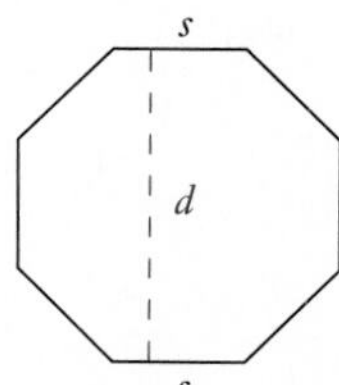

ANSWERS

1. ______
2. ______
3. ______
4. ______
5. ______
6. ______
7. ______
8. ______
9. ______
10. ______
11. ______
12. ______
13. ______
14. ______
15. ______
16. ______
17. ______
18. ______

ANSWERS

19. ______

20. ______

21. ______

22. ______

23. ______

24. ______

25. ______

26. ______

27. ______

28. ______

29. ______

30. a) ______

b) ______

c) ______

d) ______

19. ▦ *Road-pavement messages.* Using the formula of Example 5, find the length L of a road-pavement message when $h = 4$ ft and $d = 200$ ft.

20. ▦ *Road-pavement messages.* Using the formula of Example 5, find the length L of a road-pavement message when $h = 8$ ft and $d = 300$ ft.

21. The diagonal of a square has length $8\sqrt{2}$ ft. Find the length of a side of the square.

22. The length and the width of a rectangle are given by consecutive integers. The area of the rectangle is 90 cm². Find the length of a diagonal of the rectangle.

23. *Television sets.* What does it mean to refer to a 20-in. TV set or a 25-in. TV set? Such units refer to the diagonal of the screen. A 20-in. TV set also has a width of 16 in. What is its height?

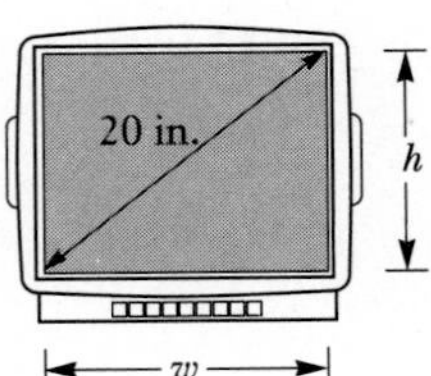

24. *Television sets.* A 25-in. TV set has a screen with a height of 15 in. What is its width?

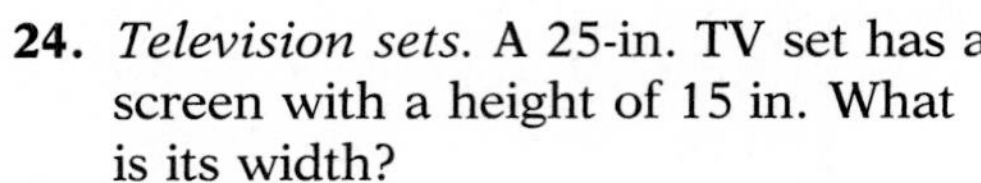

25. Find all points on the y-axis of a Cartesian coordinate system that are 5 units from the point (3, 0).

26. Find all points on the x-axis of a Cartesian coordinate system that are 5 units from the point (0, 4).

SKILL MAINTENANCE

Solve.

27. $x^2 - 11x + 24 = 0$

28. $2x^2 + 11x - 21 = 0$

SYNTHESIS

29. The area of square $PQRS$ is 100 ft². A, B, C, and D are midpoints of the sides on which they lie. Find the area of square $ABCD$.

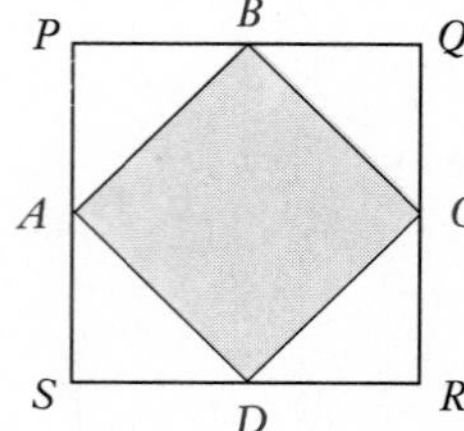

30. A *Pythagorean triple* is a set of three numbers a, b, c for which $a^2 + b^2 = c^2$. Determine whether each of the following is a Pythagorean triple.

a) 3, 4, 5 **b)** 10, 24, 25
c) 36, 48, 49 **d)** 36, 48, 60

7.9 The Complex Numbers

a Imaginary and Complex Numbers

Negative numbers do not have square roots in the real-number system. Mathematicians have invented a larger number system that contains the real-number system, but is such that negative numbers have square roots. That system is called the **complex-number system.** We begin by making up a number that is a square root of -1. We call this new number i.

We define the number $i = \sqrt{-1}$. That is, $i = \sqrt{-1}$ and $i^2 = -1$.

To express roots of negative numbers in terms of i, we can use the fact that in the complex numbers, $\sqrt{-p} = \sqrt{-1}\sqrt{p}$ when p is a positive real number.

► **EXAMPLES** Express in terms of i.

1. $\sqrt{-7} = \sqrt{-1 \cdot 7} = \sqrt{-1} \cdot \sqrt{7} = i\sqrt{7}$, or $\sqrt{7}\,i$ ← i is *not* under the radical.
2. $\sqrt{-16} = \sqrt{-1 \cdot 16} = \sqrt{-1} \cdot \sqrt{16} = i \cdot 4 = 4i$
3. $-\sqrt{-13} = -\sqrt{-1 \cdot 13} = -\sqrt{-1} \cdot \sqrt{13} = -i\sqrt{13}$, or $-\sqrt{13}\,i$
4. $-\sqrt{-64} = -\sqrt{-1 \cdot 64} = -\sqrt{-1} \cdot \sqrt{64} = -i \cdot 8 = -8i$
5. $\sqrt{-48} = \sqrt{-1 \cdot 48} = \sqrt{-1} \cdot \sqrt{48} = i\sqrt{48} = i \cdot 4\sqrt{3} = 4\sqrt{3}\,i$, or $4i\sqrt{3}$ ◄

DO EXERCISES 1–5.

An *imaginary* number* is a number that can be named *bi*, where *b* is some real number and $b \neq 0$.

To form the system of **complex numbers,** we take the imaginary numbers and the real numbers and all possible sums of real and imaginary numbers. These are complex numbers:

$$7 - 4i, \qquad -\pi + 19i, \qquad 37, \qquad i\sqrt{8}.$$

A *complex number* is any number that can be named $a + bi$, where a and b are any real numbers. (Note that a and b both can be 0.)

* Don't let the name "imaginary" fool you. The imaginary numbers are very important in such fields as engineering and the physical sciences.

OBJECTIVES

After finishing Section 7.9, you should be able to:

- **a** Express imaginary numbers as *bi*, where *b* is a nonzero real number, and complex numbers as $a + bi$, where a and b are real numbers.
- **b** Add and subtract complex numbers.
- **c** Multiply complex numbers.
- **d** Write expressions involving powers of i in the form $a + bi$.
- **e** Find conjugates of complex numbers and divide complex numbers.
- **f** Determine whether a given complex number is a solution of an equation.

FOR EXTRA HELP

Tape 14D

Tape 13A

MAC: 7
IBM: 7

Express in terms of i.

1. $\sqrt{-5}$
2. $\sqrt{-25}$
3. $-\sqrt{-11}$
4. $-\sqrt{-36}$
5. $\sqrt{-54}$

ANSWERS ON PAGE A-8

Add or subtract.

6. $(7 + 4i) + (8 - 7i)$

7. $(-5 - 6i) + (-7 + 12i)$

8. $(8 + 3i) - (5 + 8i)$

9. $(5 - 4i) - (-7 + 3i)$

ANSWERS ON PAGE A-8

Since $0 + bi = bi$, every imaginary number is a complex number. Similarly, $a + 0i = a$, so every real number is a complex number. The relationships among various real and complex numbers is shown in the following diagram.

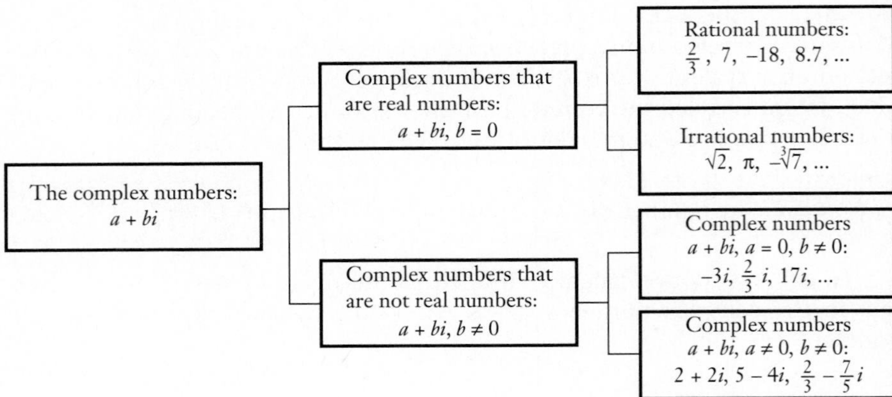

It is important to keep in mind some comparisons between numbers that have real-number roots and those that have complex-number roots that are not real. For example, $\sqrt{-48}$ is a complex number that is not a real number because we are taking the square root of a negative number. **But,** $\sqrt[3]{-125}$ is a real number because we are taking the cube root of a negative number and **any** real number has a cube root that is a real number.

b Addition and Subtraction

The complex numbers follow the commutative and associative laws of addition. Thus we can add and subtract them as we do binomials in real numbers.

► **EXAMPLES** Add or subtract.

6. $(8 + 6i) + (3 + 2i) = (8 + 3) + (6 + 2)i$ **Collecting the real parts and the imaginary parts**

$= 11 + 8i$

7. $(3 + 2i) - (5 - 2i) = (3 - 5) + [2 - (-2)]i$

$= -2 + 4i$ ◄

DO EXERCISES 6–9.

c Multiplication

The complex numbers obey the commutative, associative, and distributive laws. But although the property $\sqrt{a}\sqrt{b} = \sqrt{ab}$ does *not* hold in general, it does hold when $a = -1$ and b is a positive real number. To multiply square roots of negative real numbers, we first express them in terms of i. For example,

$$\sqrt{-2} \cdot \sqrt{-5} = \sqrt{-1} \cdot \sqrt{2} \cdot \sqrt{-1}\sqrt{5} = i\sqrt{2} \cdot i\sqrt{5}$$
$$= i^2\sqrt{10} = -\sqrt{10} \quad \text{is correct!}$$

But

$$\sqrt{-2} \cdot \sqrt{-5} = \sqrt{(-2)(-5)} = \sqrt{10} \quad \text{is wrong!}$$

Keeping this and the fact that $i^2 = -1$ in mind, we multiply in much the same way that we do with real numbers.

▶ **EXAMPLES** Multiply.

8. $\sqrt{-49} \cdot \sqrt{-16} = \sqrt{-1} \cdot \sqrt{49} \cdot \sqrt{-1} \cdot \sqrt{16}$
$= i \cdot 7 \cdot i \cdot 4$
$= i^2(28)$
$= (-1)(28)$ $\quad i^2 = -1$
$= -28$

9. $\sqrt{-3} \cdot \sqrt{-7} = \sqrt{-1} \cdot \sqrt{3} \cdot \sqrt{-1} \cdot \sqrt{7}$
$= i \cdot \sqrt{3} \cdot i \cdot \sqrt{7}$
$= i^2(\sqrt{21})$
$= (-1)\sqrt{21}$ $\quad i^2 = -1$
$= -\sqrt{21}$

10. $-2i \cdot 5i = -10 \cdot i^2$
$= (-10)(-1)$ $\quad i^2 = -1$
$= 10$

11. $(-4i)(3 - 5i) = (-4i) \cdot 3 - (-4i)(5i)$ **Using a distributive law**
$= -12i + 20i^2$
$= -12i + 20(-1)$ $\quad i^2 = -1$
$= -12i - 20$
$= -20 - 12i$

12. $(1 + 2i)(1 + 3i) = 1 + 3i + 2i + 6i^2$ **Multiplying each term of one number by every term of the other (FOIL)**
$= 1 + 3i + 2i - 6$ $\quad i^2 = -1$
$= -5 + 5i$ **Collecting like terms**

13. $(3 - 2i)^2 = 3^2 - 2(3)(2i) + (2i)^2$
$= 9 - 12i + 4i^2$
$= 9 - 12i - 4$
$= 5 - 12i$ ◀

DO EXERCISES 10–17.

Multiply.

10. $\sqrt{-25} \cdot \sqrt{-4}$

11. $\sqrt{-2} \cdot \sqrt{-17}$

12. $-6i \cdot 7i$

13. $-3i(4 - 3i)$

14. $5i(-5 + 7i)$

15. $(1 + 3i)(1 + 5i)$

16. $(3 - 2i)(1 + 4i)$

17. $(3 + 2i)^2$

ANSWERS ON PAGE A-8

d Powers of *i*

We now want to simplify certain expressions involving powers of i. To do so, we first see how to simplify powers of i. Simplifying powers of i can be done by using the fact that $i^2 = -1$ and expressing the given power of i in terms of i^2. Consider the following:

$$i,$$
$$i^2 = -1,$$
$$i^3 = i^2 \cdot i = (-1)i = -i,$$
$$i^4 = (i^2)^2 = (-1)^2 = 1,$$
$$i^5 = i^4 \cdot i = (i^2)^2 \cdot i = (-1)^2 \cdot i = i,$$
$$i^6 = (i^2)^3 = (-1)^3 = -1.$$

Note that the powers of i cycle themselves through the values i, -1, $-i$, and 1.

Simplify.

18. i^{47}

19. i^{68}

20. i^{85}

21. i^{90}

Simplify.

22. $8 - i^5$

23. $7 + 4i^2$

24. $6i^{11} + 7i^{14}$

25. $i^{34} - i^{55}$

Find the conjugate.

26. $6 + 3i$

27. $-9 - 5i$

28. $\pi - \frac{1}{4}i$

ANSWERS ON PAGE A-8

► **EXAMPLES** Simplify.

14. $i^{37} = i^{36} \cdot i = (i^2)^{18} \cdot i = (-1)^{18} \cdot i = 1 \cdot i = i$

15. $i^{58} = (i^2)^{29} = (-1)^{29} = -1$

16. $i^{75} = i^{74} \cdot i = (i^2)^{37} \cdot i = (-1)^{37} \cdot i = -1 \cdot i = -i$

17. $i^{80} = (i^2)^{40} = (-1)^{40} = 1$ ◄

DO EXERCISES 18–21.

Now let us simplify other expressions.

► **EXAMPLES** Simplify to the form $a + bi$.

18. $8 - i^2 = 8 - (-1) = 8 + 1 = 9$

19. $17 + 6i^3 = 17 + 6 \cdot i^2 \cdot i = 17 + 6(-1)i = 17 - 6i$

20. $i^{22} - 67i^2 = (i^2)^{11} - 67(-1) = (-1)^{11} + 67 = -1 + 67 = 66$

21. $i^{23} + i^{48} = (i^{22}) \cdot i + (i^2)^{24} = (i^2)^{11} \cdot i + (-1)^{24} = (-1)^{11} \cdot i + (-1)^{24}$
$= -i + 1 = 1 - i$ ◄

DO EXERCISES 22–25.

e Conjugates and Division

Conjugates of complex numbers are defined as follows.

The *conjugate* of a complex number $a + bi$ is $a - bi$, and the *conjugate* of $a - bi$ is $a + bi$.

► **EXAMPLES** Find the conjugate.

22. $5 + 7i$ The conjugate is $5 - 7i$.

23. $14 - 3i$ The conjugate is $14 + 3i$.

24. $-3 - 9i$ The conjugate is $-3 + 9i$.

25. $4i$ The conjugate is $-4i$. ◄

DO EXERCISES 26–28.

When we multiply a complex number by its conjugate, we get a real number.

► **EXAMPLES** Multiply.

26.
$$\begin{aligned}(5 + 7i)(5 - 7i) &= 5^2 - (7i)^2 && \text{Using } (A + B)(A - B) = A^2 - B^2\\ &= 25 - 49i^2\\ &= 25 - 49(-1) && i^2 = -1\\ &= 25 + 49\\ &= 74\end{aligned}$$

27.
$$\begin{aligned}(2 - 3i)(2 + 3i) &= 2^2 - (3i)^2\\ &= 4 - 9i^2\\ &= 4 - 9(-1) && i^2 = -1\\ &= 4 + 9\\ &= 13\end{aligned}$$ ◄

DO EXERCISES 29 AND 30.

We use conjugates in dividing complex numbers.

▶ **EXAMPLE 28** Divide and simplify to the form $a + bi$: $\dfrac{-5 + 9i}{1 - 2i}$.

$$\frac{-5 + 9i}{1 - 2i} \cdot \frac{1 + 2i}{1 + 2i} = \frac{(-5 + 9i)(1 + 2i)}{(1 - 2i)(1 + 2i)}$$

Multiplying by 1 using the conjugate of the denominator in the symbol for 1

$$= \frac{-5 - 10i + 9i + 18i^2}{1^2 - 4i^2}$$

$$= \frac{-5 - i - 18}{1 - 4(-1)}$$

$$= \frac{-23 - i}{5}$$

$$= -\frac{23}{5} - \frac{1}{5}i$$ ◀

Note the similarity between the preceding example and rationalizing denominators. In both cases, we used the conjugate of the denominator to write another name for 1. In Example 28, the symbol for the number 1 was chosen using the conjugate of the divisor, $1 - 2i$.

▶ **EXAMPLE 29** What symbol for 1 would you use to divide?

Division to be done	*Symbol for 1*
$\dfrac{3 + 5i}{4 + 3i}$	$\dfrac{4 - 3i}{4 - 3i}$

◀

▶ **EXAMPLE 30** Divide and simplify to the form $a + bi$: $\dfrac{3 + 5i}{4 + 3i}$.

$$\frac{3 + 5i}{4 + 3i} \cdot \frac{4 - 3i}{4 - 3i} = \frac{(3 + 5i)(4 - 3i)}{(4 + 3i)(4 - 3i)}$$

Multiplying by 1

$$= \frac{12 - 9i + 20i - 15i^2}{4^2 - 9i^2}$$

$$= \frac{12 + 11i - 15(-1)}{16 - 9(-1)}$$

$$= \frac{27 + 11i}{25}$$

$$= \frac{27}{25} + \frac{11}{25}i$$ ◀

DO EXERCISES 31 AND 32.

Multiply.

29. $(7 - 2i)(7 + 2i)$

30. $(-3 - i)(-3 + i)$

Divide and simplify to the form $a + bi$.

31. $\dfrac{6 + 2i}{1 - 3i}$

32. $\dfrac{2 + 3i}{-1 + 4i}$

ANSWERS ON PAGE A-8

33. Determine whether $-i$ is a solution of $x^2 + 1 = 0$.

34. Determine whether $1 - i$ is a solution of $x^2 - 2x + 2 = 0$.

ANSWERS ON PAGE A-8

f Solutions of Equations

The equation $x^2 + 1 = 0$ has no real-number solution, but it has two nonreal complex solutions.

▶ **EXAMPLE 31** Determine whether i is a solution of the equation $x^2 + 1 = 0$.

We substitute i for x in the equation.

$$\begin{array}{r|l} x^2 + 1 & = 0 \\ \hline i^2 + 1 & 0 \\ -1 + 1 & \\ 0 & \end{array}$$

The number i is a solution. ◀

DO EXERCISE 33.

Any equation consisting of a polynomial in one variable on one side and 0 on the other has complex-number solutions (some may be real). It is not always easy to find the solutions, but they always exist.

▶ **EXAMPLE 32** Determine whether $1 + i$ is a solution of the equation $x^2 - 2x + 2 = 0$.

We substitute $1 + i$ for x in the equation.

$$\begin{array}{r|l} x^2 - 2x + 2 & = 0 \\ \hline (1 + i)^2 - 2(1 + i) + 2 & 0 \\ 1 + 2i + i^2 - 2 - 2i + 2 & \\ 1 + 2i - 1 - 2 - 2i + 2 & \\ (1 - 1 - 2 + 2) + (2 - 2)i & \\ 0 + 0i & \\ 0 & \end{array}$$

$1 + i$ is a solution. ◀

▶ **EXAMPLE 33** Determine whether $2i$ is a solution of $x^2 + 3x - 4 = 0$.

$$\begin{array}{r|l} x^2 + 3x - 4 & = 0 \\ \hline (2i)^2 + 3(2i) - 4 & 0 \\ 4i^2 + 6i - 4 & \\ -4 + 6i - 4 & \\ -8 + 6i & \end{array}$$

$2i$ is not a solution. ◀

DO EXERCISE 34.

NAME SECTION DATE

EXERCISE SET 7.9

a Express in terms of i.

1. $\sqrt{-15}$
2. $\sqrt{-17}$
3. $\sqrt{-16}$
4. $\sqrt{-25}$
5. $-\sqrt{-12}$
6. $-\sqrt{-20}$
7. $\sqrt{-3}$
8. $\sqrt{-4}$
9. $\sqrt{-81}$
10. $\sqrt{-27}$
11. $\sqrt{-98}$
12. $-\sqrt{-18}$
13. $-\sqrt{-49}$
14. $-\sqrt{-125}$
15. $4-\sqrt{-60}$
16. $6-\sqrt{-84}$
17. $\sqrt{-4}+\sqrt{-12}$
18. $-\sqrt{-76}+\sqrt{-125}$

b Add or subtract and simplify.

19. $(3+2i)+(5-i)$
20. $(-2+3i)+(7+8i)$
21. $(4-3i)+(5-2i)$
22. $(-2-5i)+(1-3i)$
23. $(9-i)+(-2+5i)$
24. $(6+4i)+(2-3i)$
25. $(3-i)-(5+2i)$
26. $(-2+8i)-(7+3i)$
27. $(4-2i)-(5-3i)$

ANSWERS

1. ______
2. ______
3. ______
4. ______
5. ______
6. ______
7. ______
8. ______
9. ______
10. ______
11. ______
12. ______
13. ______
14. ______
15. ______
16. ______
17. ______
18. ______
19. ______
20. ______
21. ______
22. ______
23. ______
24. ______
25. ______
26. ______
27. ______

ANSWERS

28. ______

29. ______

30. ______

31. ______

32. ______

33. ______

34. ______

35. ______

36. ______

37. ______

38. ______

39. ______

40. ______

41. ______

42. ______

43. ______

44. ______

45. ______

46. ______

47. ______

48. ______

49. ______

50. ______

51. ______

52. ______

28. $(-2 - 3i) - (1 - 5i)$

29. $(9 + 5i) - (-2 - i)$

30. $(6 - 3i) - (2 + 4i)$

C Multiply.

31. $\sqrt{-36} \cdot \sqrt{-9}$

32. $\sqrt{-16} \cdot \sqrt{-64}$

33. $\sqrt{-5} \cdot \sqrt{-2}$

34. $\sqrt{-11} \cdot \sqrt{-3}$

35. $-3i \cdot 7i$

36. $8i \cdot 5i$

37. $5i(4 - 7i)$

38. $-2i(-3 + 5i)$

39. $(3 + 2i)(1 + i)$

40. $(4 + 3i)(2 + 5i)$

41. $(2 + 3i)(6 - 2i)$

42. $(5 + 6i)(2 - i)$

43. $(6 - 5i)(3 + 4i)$

44. $(5 - 6i)(2 + 5i)$

45. $(7 - 2i)(2 - 6i)$

46. $(-4 + 5i)(3 - 4i)$

47. $(3 - 2i)^2$

48. $(5 - 2i)^2$

49. $(2 + 3i)^2$

50. $(4 + 2i)^2$

51. $(-2 + 3i)^2$

52. $(-5 - 2i)^2$

d Simplify.

53. i^7 **54.** i^{11} **55.** i^{24} **56.** i^{35}

57. i^{42} **58.** i^{64} **59.** i^9 **60.** $(-i)^{71}$

61. i^6 **62.** $(-i)^4$ **63.** $(5i)^3$ **64.** $(-3i)^5$

Simplify to the form $a + bi$.

65. $7 + i^4$ **66.** $-18 + i^3$ **67.** $i^4 - 26i$

68. $i^5 + 37i$ **69.** $i^2 + i^4$ **70.** $5i^5 + 4i^3$

71. $i^5 + i^7$ **72.** $i^{84} - i^{100}$ **73.** $1 + i + i^2 + i^3 + i^4$

74. $i - i^2 + i^3 - i^4 + i^5$ **75.** $5 - \sqrt{-64}$ **76.** $\sqrt{-12} + 36i$

77. $\dfrac{8 - \sqrt{-24}}{4}$ **78.** $\dfrac{9 + \sqrt{-9}}{3}$

e Divide and simplify to the form $a + bi$.

79. $\dfrac{3 + 2i}{2 + i}$ **80.** $\dfrac{4 + 5i}{5 - i}$ **81.** $\dfrac{5 - 2i}{2 + 5i}$ **82.** $\dfrac{3 - 2i}{4 + 3i}$

ANSWERS

53. ______

54. ______

55. ______

56. ______

57. ______

58. ______

59. ______

60. ______

61. ______

62. ______

63. ______

64. ______

65. ______

66. ______

67. ______

68. ______

69. ______

70. ______

71. ______

72. ______

73. ______

74. ______

75. ______

76. ______

77. ______

78. ______

79. ______

80. ______

81. ______

82. ______

ANSWERS

83. ______
84. ______
85. ______
86. ______
87. ______
88. ______
89. ______
90. ______
91. ______
92. ______
93. ______
94. ______
95. ______
96. ______
97. ______
98. ______
99. ______
100. ______
101. ______
102. ______
103. ______
104. ______
105. ______
106. ______
107. ______
108. ______
109. ______
110. ______
111. ______

83. $\frac{8-3i}{7i}$ **84.** $\frac{3+8i}{5i}$ **85.** $\frac{4}{3+i}$ **86.** $\frac{6}{2-i}$

87. $\frac{2i}{5-4i}$ **88.** $\frac{8i}{6+3i}$ **89.** $\frac{4}{3i}$ **90.** $\frac{5}{6i}$

91. $\frac{2-4i}{8i}$ **92.** $\frac{5+3i}{i}$ **93.** $\frac{6+3i}{6-3i}$ **94.** $\frac{4-5i}{4+5i}$

f Determine whether the complex number is a solution of the equation.

95. $1+2i$;
$x^2-2x+5=0$

96. $1-2i$;
$x^2-2x+5=0$

97. $1-i$;
$x^2+2x+2=0$

98. $2+i$;
$x^2-4x-5=0$

SKILL MAINTENANCE

Solve.

99. $\frac{196}{x^2-7x+49}-\frac{2x}{x+7}=\frac{2058}{x^3+343}$ **100.** $\frac{5}{t}-\frac{3}{2}=\frac{4}{7}$

SYNTHESIS

101. Evaluate $\frac{z^4-z^2}{z-1}$ when $z=2i$. **102.** Evaluate $\frac{1}{w-w^2}$ when $w=\frac{1-i}{10}$.

Express in terms of i.

103. $\frac{1}{8}(-24-\sqrt{-1024})$ **104.** $12\sqrt{-\frac{1}{32}}$ **105.** $7\sqrt{-64}-9\sqrt{-256}$

Simplify.

106. $\frac{i^5+i^6+i^7+i^8}{(1-i)^4}$ **107.** $(1-i)^3(1+i)^3$ **108.** $\frac{5-\sqrt{5}i}{\sqrt{5}i}$

109. $\frac{6}{1+\frac{3}{i}}$ **110.** $\left(\frac{1}{2}-\frac{1}{3}i\right)^2-\left(\frac{1}{2}+\frac{1}{3}i\right)^2$ **111.** $\frac{i-i^{38}}{1+i}$

SUMMARY AND REVIEW: CHAPTER 7

IMPORTANT PROPERTIES AND FORMULAS

$\sqrt{a^2} = |a|$; $\sqrt[k]{a^k} = |a|$, when k is even; $\sqrt[k]{a^k} = a$, when k is odd;

$\sqrt[k]{ab} = \sqrt[k]{a} \cdot \sqrt[k]{b}$; $\sqrt[k]{\dfrac{a}{b}} = \dfrac{\sqrt[k]{a}}{\sqrt[k]{b}}$; $\sqrt[k]{a^m} = (\sqrt[k]{a})^m$;

$a^{1/n} = \sqrt[n]{a}$; $a^{m/n} = \sqrt[n]{a^m} = (\sqrt[n]{a})^m$; $a^{-m/n} = \dfrac{1}{a^{m/n}}$

Principle of Powers: If $a = b$ is true, then $a^n = b^n$ is true.
Pythagorean Theorem: $a^2 + b^2 = c^2$, in a right triangle.
$i = \sqrt{-1}$, $i^2 = -1$
Imaginary Numbers: bi, $i^2 = -1$
Complex Numbers: $a + bi$, $i^2 = -1$
Conjugates: $a + bi$, $a - bi$

REVIEW EXERCISES

The review sections and objectives to be tested in addition to the material in this chapter are [5.8a, b]; [6.1d, e], and [6.3a].

Simplify. Assume that letters can represent *any* real number.

1. $\sqrt{81a^2}$
2. $\sqrt{(c-3)^2}$
3. $\sqrt{x^2 - 6x + 9}$
4. $\sqrt{4x^2 + 4x + 1}$
5. $\sqrt[5]{-32}$
6. $\sqrt[3]{-\dfrac{1}{27}}$
7. $\sqrt[10]{x^{10}}$
8. $-\sqrt[13]{(-3)^{13}}$

Use a calculator or Table 1. Approximate to three decimal places.

9. $\sqrt{245}$
10. $\sqrt{112}$
11. $\dfrac{14 - \sqrt{245}}{7}$

Multiply and simplify by factoring. Simplify by collecting like radical terms if possible, assuming that all expressions under radicals represent nonnegative numbers.

12. $\sqrt{3x^2}\sqrt{6y^3}$
13. $\sqrt[3]{a^5 b}\sqrt[3]{27b}$

Divide. Then simplify by taking roots, if possible.

14. $\dfrac{\sqrt[3]{60xy^3}}{\sqrt[3]{10x}}$
15. $\dfrac{\sqrt{75x}}{2\sqrt{3}}$

Simplify.

16. $(\sqrt{8xy^2})^2$
17. $(\sqrt[3]{4a^2 b})^2$

Add, subtract, or multiply. Simplify by collecting like radical terms if possible, assuming that all expressions under radicals represent nonnegative numbers.

18. $12\sqrt[3]{135} - 3\sqrt[3]{40}$
19. $\sqrt{50} + 2\sqrt{18} + \sqrt{32}$
20. $(\sqrt[3]{27} - \sqrt[3]{2})(\sqrt[3]{27} + \sqrt[3]{2})$
21. $(\sqrt{5} - 3\sqrt{8})(\sqrt{5} + 2\sqrt{8})$
22. $(1 - \sqrt{7})^2$

23. Rationalize the denominator. Assume that a and b represent positive numbers.

$$\frac{2\sqrt{a}+\sqrt{b}}{\sqrt{a}+\sqrt{b}}$$

24. Rationalize the numerator of the expression in Exercise 23.

25. Rewrite with rational exponents: $(\sqrt[5]{8x^6y^2})^4$.

26. Rewrite without rational exponents: $(5a)^{3/4}$.

Use rational exponents to write a single radical expression.

27. $x^{1/3} \cdot y^{1/4}$

28. $\sqrt[4]{x}\sqrt[3]{(x-3)}$

Solve.

29. $\sqrt{3x-3} = 1 + \sqrt{x}$

30. $\sqrt[4]{x+3} = 2$

In a right triangle, find the length of the side not given. Give an exact answer and an answer to three decimal places.

31. $a = 7,\ b = 24$

32. $a = 2,\ c = 5\sqrt{2}$

33. A guy wire is attached to a pole at a point that is 9 ft from the base. It reaches out to a point on the ground 20 ft from the base of the pole. How long is the wire?

34. Express in terms of i and simplify: $-\sqrt{-8}$.

35. Add: $(-4 + 3i) + (2 - 12i)$.

36. Subtract: $(4 - 7i) - (3 - 8i)$.

Multiply.

37. $(2 + 5i)(2 - 5i)$

38. i^{13}

39. $(6 - 3i)(2 - i)$

Divide.

40. $\dfrac{-3 + 2i}{5i}$

41. $\dfrac{6 - 3i}{2 - i}$

42. Determine whether $1 + i$ is a solution of $x^2 + x + 2 = 0$.

SKILL MAINTENANCE

43. Find three consecutive positive integers such that the product of the first and second integers is 26 less than the product of the second and third integers.

44. Solve: $\dfrac{7}{x+2} + \dfrac{5}{x^2 - 2x + 4} = \dfrac{84}{x^3 + 8}$.

45. A kitchen floor is 3 ft longer than it is wide. The area of the floor is 180 ft^2. Find the dimensions of the kitchen.

46. Multiply and simplify:

$$\frac{x^2 + 3x}{x^2 - y^2} \cdot \frac{x^2 - xy + 2x - 2y}{x^2 - 9}.$$

SYNTHESIS

47. Solve: $\sqrt{11x + \sqrt{6 + x}} = 6$.

48. Simplify: $\sqrt{432} - \sqrt{6125} + \sqrt{845} - \sqrt{4800}$.

❖ THINKING IT THROUGH

1. We learned to solve a new kind of equation in this chapter. Explain how the procedure for solving this kind of equation differs from others we have solved.
2. Explain why $\sqrt{x^2} = |x|$, when x is considered to be an arbitrary real number.
3. Explain the difference between a complex number and a real number. Give two examples of complex numbers that are not real numbers.

NAME SECTION DATE

TEST: CHAPTER 7

ANSWERS

In Questions 1–4, assume that letters can represent *any* real number. Simplify.

1. $\sqrt{36y^2}$

2. $\sqrt{x^2 + 10x + 25}$

3. $\sqrt[3]{-8}$

4. $\sqrt[10]{(-4)^{10}}$

5. Use a calculator or Table 1 to approximate $\sqrt{148}$ to three decimal places.

Assume henceforth on this test that *all* expressions under radicals represent nonnegative numbers.

6. Multiply and simplify by factoring:
$$\sqrt[3]{x^4}\,\sqrt[3]{8x^5}.$$

7. Divide. Then simplify by taking roots, if possible.
$$\sqrt{\frac{20x^3y}{4y}}$$

8. Simplify: $(\sqrt[3]{16a^2b})^2$.

9. Add. Then simplify by collecting like terms.
$$3\sqrt{128} + 2\sqrt{18} + 2\sqrt{32}$$

10. Multiply and simplify:
$$(\sqrt{20} + 2\sqrt{5})(\sqrt{20} - 3\sqrt{5}).$$

11. Rationalize the denominator:
$$\frac{\sqrt{a} + \sqrt{b}}{\sqrt{a} - \sqrt{b}}.$$

12. Rewrite with rational exponents:
$$(\sqrt{5xy^2})^5.$$

13. Use rational exponents to write a single radical expression:
$$\sqrt[4]{2y}\,\sqrt{x - 3}.$$

1. ______
2. ______
3. ______
4. ______
5. ______
6. ______
7. ______
8. ______
9. ______
10. ______
11. ______
12. ______
13. ______

ANSWERS

14. ________

15. ________

16. ________

17. ________

18. ________

19. ________

20. ________

21. ________

22. ________

23. ________

24. ________

25. ________

26. ________

27. ________

28. ________

29. ________

30. ________

31. ________

14. Solve:
$$\sqrt{y-6} = \sqrt{y+9} - 3.$$

15. Express in terms of i and simplify:
$$\sqrt{-18}.$$

16. Subtract: $(5 + 8i) - (-2 + 3i)$.

17. Multiply: $(1 - i)^2$.

18. Simplify: i^{95}.

19. Express in the form $a + bi$:
$$(8i)^3 - i^{16}.$$

20. Divide: $\dfrac{-7 + 14i}{6 - 8i}$.

21. Determine whether $1 + 2i$ is a solution of $x^2 + 2x + 5 = 0$.

In a right triangle, find the length of the side not given. Give an exact answer and an approximation to three decimal places.

22. $a = 7,\ b = 7$

23. $a = 1,\ c = \sqrt{5}$

24. A baseball diamond is actually a square 90 ft on a side. How far would a catcher have to throw a ball from home to second base?

SKILL MAINTENANCE

25. Solve: $x^2 - 10x + 25 = 0$.

26. Divide and simplify:
$$\frac{x^3 - 27}{x^2 - 16} \div \frac{x^2 + 3x + 9}{x + 4}.$$

27. Solve: $\dfrac{11x}{x+3} + \dfrac{33}{x} + 12 = \dfrac{99}{x^2 + 3x}$.

28. Find two consecutive even integers whose product is 288.

SYNTHESIS

29. Solve:
$$\sqrt{2x-2} + \sqrt{7x+4} = \sqrt{13x+10}.$$

30. Simplify: $\dfrac{1 - 4i}{4i(1 + 4i)^{-1}}$.

31. Simplify: $\sqrt{1250x^3y} - \sqrt{1800xy^3} - \sqrt{162x^3y^3}$.

CUMULATIVE REVIEW: CHAPTERS 1–7

Simplify.

1. $-\frac{3}{5}\left(-\frac{7}{10}\right)$

2. $36.2 - 73.4$

3. $7c - [4 - 3(6c - 9)]$

4. $6^2 - 3^3 \cdot 2 + 3 \cdot 8^2$

5. $\left(\frac{4x^4y^{-5}}{3x^{-3}}\right)^3$

6. $(2x^2 - 3x + 1) + (6x - 3x^3 + 7x^2 - 4)$

7. $(2x^2 - y)^2$

8. $(5x^2 - 2x + 1)(3x^2 + x - 2)$

9. $\frac{x^3 + 64}{x^2 - 49} \cdot \frac{x^2 - 14x + 49}{x^2 - 4x + 16}$

10. $\frac{x}{x+2} + \frac{1}{x-3} - \frac{x^2 - 2}{x^2 - x - 6}$

11. $\dfrac{\frac{y^2 - 5y - 6}{y^2 - 7y - 18}}{\frac{y^2 + 3y + 2}{y^2 + 4y + 4}}$

12. $(y^3 + 3y^2 - 5) \div (y + 2)$

13. $\sqrt[3]{-8x^3}$

14. $\sqrt{16x^2 - 32x + 16}$

15. $9\sqrt{75} + 6\sqrt{12}$

16. $\sqrt{2xy^2} \cdot \sqrt{8xy^3}$

17. $\frac{3\sqrt{5}}{\sqrt{6} - \sqrt{3}}$

18. $\sqrt[6]{\frac{m^{12}n^{24}}{64}}$

19. $6^{2/9} \cdot 6^{2/3}$

20. $(6 + i) - (3 - 4i)$

21. $\frac{2 - i}{6 + 5i}$

Solve.

22. $\frac{1}{5} + \frac{3}{10}x = \frac{4}{5}$

23. $M = \frac{1}{8}(c - 3)$ for c

24. $3a - 4 < 10 + 5a$

25. $-8 < x + 2 < 15$

26. $|3x - 6| = 2$

27. $3x + 5y = 30,$
$5x + 3y = 34$

28. $3x + 2y - z = 7,$
$-x + y + 2z = 9,$
$5x + 5y + z = -1$

29. $625 = 49y^2$

30. $\frac{6x}{x-5} - \frac{300}{x^2 + 5x + 25} = \frac{2250}{x^3 - 125}$

31. $\frac{3x^2}{x+2} + \frac{5x - 22}{x - 2} = \frac{-48}{x^2 - 4}$

32. $I = \frac{nE}{R + nr}$ for r

33. $\sqrt{4x + 1} - 2 = 3$

34. $2\sqrt{1 - x} = \sqrt{5}$

Solve using Cramer's rule.

35. $6x - 2y = -3,$
$4x + y = 5$

36. $x + 6y - 4z = 0,$
$2x - 3y + 8z = -9,$
$-x + 9y - 4z = 7$

Graph.

37. $y = -\frac{2}{3}x + 2$

38. $4x - 2y = 8$

39. $4x \geq 5y + 20$

40. $y \geq -3,$
$y \leq 2x + 3$

Factor.

41. $12x^2y^2 - 30xy^3$

42. $3x^2 - 17x - 28$

43. $y^2 - y - 132$

44. $27y^3 + 8$

45. $4x^2 - 625$

46. The average cost of a movie ticket in Finland is \$1.54 more than the average cost of a movie ticket in Sweden. The sum of the average costs is \$14.44. How much does a movie ticket cost in each country?

47. Find the slope and the y-intercept of the line $3x - 2y = 8$.

48. Find an equation for the line perpendicular to the line $3x - y = 5$ and passing through (1, 4).

49. The University of Pittsburgh Medical Center and the Medical College of Virginia have together performed 669 heart transplants. The University of Pittsburgh has performed 33 more than twice the number of transplants at the Medical College of Virginia. How many transplants has each hospital performed?

50. One tractor can plow a field in 3 hr. Another tractor can plow the field in 1.5 hr. How long should it take them to plow the field together?

51. The height h of triangles of fixed area varies inversely as the base b. Suppose the height is 100 ft when the base is 20 ft. Find the height when the base is 16 ft. What is the fixed area?

SYNTHESIS

52. ▦ *The Pythagorean formula in three dimensions.* The length d of a diagonal of a rectangular box is given by $d = \sqrt{a^2 + b^2 + c^2}$, where a, b, and c are the lengths of the sides. Find the length of a diagonal of a box whose sides have lengths 2 ft, 4 ft, and 5 ft.

53. Graph: $|x + y| \leq 1$.

54. Solve: $\dfrac{x + \sqrt{x+1}}{x - \sqrt{x+1}} = \dfrac{5}{11}$.

55. Factor: $x^2 - 3x + \dfrac{5}{4}$.

INTRODUCTION We began our study of quadratic equations in Chapter 5. Here we extend our equation-solving skills to those quadratic equations that do not lend themselves to solving easily by factoring. We consider real-number and complex-number solutions and learn to solve problems and to graph quadratic equations.

The review sections to be tested in addition to the material in this chapter are 4.3, 6.2, 7.2, and 7.7. ❖

Quadratic Equations

AN APPLICATION

The height of the World Trade Center in New York is 1377 ft (excluding television towers and antennas). How long would it take an object to fall from the top?

THE MATHEMATICS

Let $t =$ the time, in seconds, of the fall. To find the answer to the problem, we solve the equation

$$\underbrace{1377 = 16t^2}_{\uparrow}.$$

This is a *quadratic equation.*

❖ POINTS TO REMEMBER: CHAPTER 8

Every positive real number has two real-number square roots.
Factoring skills (Chapter 5)
Skills at manipulating square-root symbolism (Chapter 7)
Motion Formula: $d = rt$

Work Formula: $\frac{t}{a} + \frac{t}{b} = 1$

Pythagorean Theorem: $a^2 + b^2 = c^2$

PRETEST: CHAPTER 8

Solve.

1. $5x^2 + 15x = 0$

2. $y^2 + 4y + 8 = 0$

3. $x^2 - 10x + 25 = 0$

4. $3x^4 - 7x^2 + 2 = 0$

5. $\frac{2x}{2x+1} + \frac{x+1}{2x-1} = \frac{6}{4x^2-1}$

6. Solve. Give the exact solution and approximate the solutions to the nearest tenth.
$$2x^2 + 4x - 1 = 0$$

7. Solve for T:
$$W = \sqrt{\frac{1}{RT}}.$$

8. Write a quadratic equation having solutions $\frac{2}{3}$ and -2.

9. For $y = 2(x-3)^2 - 2$:
   a) Find the vertex.
   b) Find the line of symmetry.
   c) Graph the equation.

10. Find the x-intercepts: $y = x^2 - 6x + 4$.

11. Find the quadratic equation that fits these data points: (0, 1), (1, 0), and (2, 7).

12. Find three consecutive even integers such that twice the product of the first two is equal to the square of the third plus five times the third.

13. During the first part of a trip, a car travels 100 mi at a certain speed. It travels 120 mi on the second part of the trip at a speed 8 mph faster. The total time for the trip is 5 hr. Find the speed of the car during each part of the trip.

14. Find the area of the largest rectangular region that can be enclosed using 300 ft of fencing.

Solve.

15. $x(x-3)(x+5) < 0$

16. $\frac{x-2}{x+3} \geq 0$

8.1 The Basics of Solving Quadratic Equations

The following are **quadratic equations.** They contain polynomials of second degree:

$$5x^2 + 8x - 2 = 0, \qquad 3t^2 - \tfrac{1}{4}t = 9, \qquad 7y^2 = -8y, \qquad 2m^2 = 0.$$

The quadratic equation

$$5x^2 + 8x - 2 = 0$$

is said to be in **standard form.** The quadratic equation

$$5x^2 = 2 - 8x$$

is equivalent to the preceding, but it is *not* in standard form.

> **An equation of the type $ax^2 + bx + c = 0$, where a, b, and c are real-number constants and $a > 0$, is called the *standard form of a quadratic equation.***

Suppose we are considering an equation like $-5x^2 + 4x - 7 = 0$. It is not in standard form. We can find an equivalent equation that is in standard form by multiplying on both sides by -1:

$$-1(-5x^2 + 4x - 7) = -1(0)$$
$$5x^2 - 4x + 7 = 0.$$

We require $a > 0$ in the standard form to allow a smoother proof of the quadratic formula, which we consider later.

a The Principle of Square Roots

In Section 5.8, we studied the use of factoring and the principle of zero products to solve certain quadratic equations. Let us review that procedure and introduce a new one.

► **EXAMPLE 1** Solve: $3x^2 = 2 - x$.

We first find standard form. Then we factor and use the principle of zero products.

$$3x^2 = 2 - x$$
$$3x^2 + x - 2 = 0 \qquad \text{Adding } x \text{ and subtracting 2}$$
$$(3x - 2)(x + 1) = 0 \qquad \text{Factoring}$$
$$3x - 2 = 0 \quad or \quad x + 1 = 0 \qquad \text{Using the principle of zero products}$$
$$3x = 2 \quad or \quad x = -1$$
$$x = \tfrac{2}{3} \quad or \quad x = -1$$

Check: For -1:

$3x^2 + x - 2 = 0$	
$3(-1)^2 + (-1) - 2$	0
$3 - 1 - 2$	
0	TRUE

For $\frac{2}{3}$:

$3x^2 + x - 2 = 0$	
$3(\frac{2}{3})^2 + (\frac{2}{3}) - 2$	0
$3 \cdot \frac{4}{9} + \frac{2}{3} - 2$	
$\frac{12}{9} + \frac{6}{9} - 2$	
$\frac{18}{9} - 2$	
0	TRUE

OBJECTIVES

After finishing Section 8.1, you should be able to:

- **a** Solve quadratic equations using the principle of square roots.
- **b** Solve quadratic equations by completing the square.
- **c** Solve problems using quadratic equations.

FOR EXTRA HELP

Tape 15A

Tape 13B

MAC: 8
IBM: 8

1. Solve: $5x^2 = 8x - 3$.

2. Solve: $x^2 = 16$.

3. Solve: $5x^2 = 15$.

ANSWERS ON PAGE A-8

The solutions are -1 and $\frac{2}{3}$. ◀

DO EXERCISE 1.

▶ **EXAMPLE 2** Solve: $x^2 = 25$.

We first find standard form. Then we factor and use the principle of zero products.

$$x^2 = 25$$
$$x^2 - 25 = 0 \quad \text{Subtracting 25}$$
$$(x-5)(x+5) = 0 \quad \text{Factoring}$$
$$x - 5 = 0 \quad or \quad x + 5 = 0 \quad \text{Using the principle of zero products}$$
$$x = 5 \quad or \quad x = -5$$

The solutions are 5 and -5. We leave the check to the student. ◀

DO EXERCISE 2.

Consider the equation $x^2 = 25$ again. We know from Chapter 7 that the number 25 has two real-number square roots, namely, 5 and -5. Note that these are the solutions of the equation in Example 2. This exemplifies the principle of square roots, which provides a quick method for solving equations of the type $x^2 = k$.

> The Principle of Square Roots
>
> **The equation $x^2 = k$ has two real-number solutions when $k > 0$. The solutions are $\sqrt{k}$ and $-\sqrt{k}$.**
>
> **The equation $x^2 = 0$ has 0 as its only solution.**
>
> **The equation $x^2 = k$ has two nonreal complex solutions when $k < 0$.**

▶ **EXAMPLE 3** Solve: $3x^2 = 6$.

We have

$$3x^2 = 6$$
$$x^2 = 2 \quad \text{Dividing by 3}$$
$$x = \sqrt{2} \quad or \quad x = -\sqrt{2}. \quad \text{Using the principle of square roots}$$

We often use the symbol $\pm\sqrt{2}$ to represent both of the numbers $\sqrt{2}$ and $-\sqrt{2}$. We check as follows.

For $\sqrt{2}$:

$3x^2 = 6$	
$3(\sqrt{2})^2$	6
$3 \cdot 2$	
6	TRUE

For $-\sqrt{2}$:

$3x^2 = 6$	
$3(-\sqrt{2})^2$	6
$3 \cdot 2$	
6	TRUE

The solutions are $\sqrt{2}$ and $-\sqrt{2}$, or $\pm\sqrt{2}$. ◀

DO EXERCISE 3.

Sometimes we rationalize denominators to simplify answers.

▶ **EXAMPLE 4** Solve: $-5x^2 + 2 = 0$.

$$-5x^2 + 2 = 0$$

$$x^2 = \frac{2}{5} \quad \text{Subtracting 2 and dividing by } -5$$

$$x = \sqrt{\frac{2}{5}} \quad or \quad x = -\sqrt{\frac{2}{5}} \quad \text{Using the principle of square roots}$$

$$x = \sqrt{\frac{2}{5} \cdot \frac{5}{5}} \quad or \quad x = -\sqrt{\frac{2}{5} \cdot \frac{5}{5}} \quad \text{Rationalizing the denominators}$$

$$x = \frac{\sqrt{10}}{5} \quad or \quad x = -\frac{\sqrt{10}}{5}$$

Check: We check both numbers at once, since there is only an x^2-term in the equation.

$$\begin{array}{r|l} -5x^2 + 2 & = 0 \\ \hline -5\left(\pm\frac{\sqrt{10}}{5}\right)^2 + 2 & 0 \\ -5\left(\frac{10}{25}\right) + 2 & \\ -2 + 2 & \\ 0 & \text{TRUE} \end{array}$$

The solutions are $\frac{\sqrt{10}}{5}$ and $-\frac{\sqrt{10}}{5}$, or $\pm\frac{\sqrt{10}}{5}$. ◀

DO EXERCISE 4.

Sometimes we get solutions that are complex numbers.

▶ **EXAMPLE 5** Solve: $4x^2 + 9 = 0$.

$$4x^2 + 9 = 0$$

$$x^2 = -\frac{9}{4} \quad \text{Subtracting 9 and dividing by 4}$$

$$x = \sqrt{-\frac{9}{4}} \quad or \quad x = -\sqrt{-\frac{9}{4}} \quad \text{Using the principle of square roots}$$

$$x = \frac{3}{2}i \quad or \quad x = -\frac{3}{2}i \quad \text{Simplifying}$$

Check:

$$\begin{array}{r|l} 4x^2 + 9 & = 0 \\ \hline 4\left(\pm\frac{3}{2}i\right)^2 + 9 & 0 \\ 4\left(-\frac{9}{4}\right) + 9 & \\ -9 + 9 & \\ 0 & \text{TRUE} \end{array}$$

The solutions are $\frac{3}{2}i$ and $-\frac{3}{2}i$, or $\pm\frac{3}{2}i$. ◀

DO EXERCISE 5.

4. Solve: $-3x^2 + 8 = 0$.

5. Solve: $2x^2 + 1 = 0$.

ANSWERS ON PAGE A-8

6. Solve: $(x-1)^2 = 5$.

7. Solve: $x^2 + 16x + 64 = 11$.

ANSWERS ON PAGE A-8

The equation $(x-2)^2 = 7$ can also be solved using the principle of square roots.

▶ **EXAMPLE 6** Solve: $(x-2)^2 = 7$.

We have

$$(x-2)^2 = 7$$
$$x - 2 = \sqrt{7} \quad or \quad x - 2 = -\sqrt{7} \qquad \text{Using the principle of square roots}$$
$$x = 2 + \sqrt{7} \quad or \quad x = 2 - \sqrt{7}.$$

We leave the checks to the student. The solutions are $2 + \sqrt{7}$ and $2 - \sqrt{7}$, or $2 \pm \sqrt{7}$. ◀

DO EXERCISE 6.

In Example 6, the left side of the equation is the square of a binomial. If we can express an equation in such a form, we can proceed as we did in Example 6.

▶ **EXAMPLE 7** Solve: $x^2 + 6x + 9 = 2$.

We have

$$x^2 + 6x + 9 = 2 \qquad \text{The left side is the square of a binomial.}$$
$$(x+3)^2 = 2$$
$$x + 3 = \sqrt{2} \quad or \quad x + 3 = -\sqrt{2} \qquad \text{Using the principle of square roots}$$
$$x = -3 + \sqrt{2} \quad or \quad x = -3 - \sqrt{2}.$$

The solutions are $-3 + \sqrt{2}$ and $-3 - \sqrt{2}$, or $-3 \pm \sqrt{2}$. ◀

DO EXERCISE 7.

b Completing the Square

We have seen that we can solve a quadratic equation like $(x-2)^2 = 7$ by using the principle of square roots. We also noted that we can solve an equation like $x^2 + 6x + 9 = 2$ in like manner because the expression on the left side is the square of a binomial, $(x+3)^2$. This second procedure is the basis for a method of solving quadratic equations called **completing the square.** It can be used to solve any quadratic equation.

Suppose we have the following quadratic equation:

$$x^2 + 14x = 4.$$

If we could add a constant to both sides of the equation that would make the expression on the left the square of a binomial, we could then solve the equation using the principle of square roots.

How can we determine what to add to $x^2 + 14x$ to construct the square of a binomial? We want to find a number a such that the following equation is satisfied:

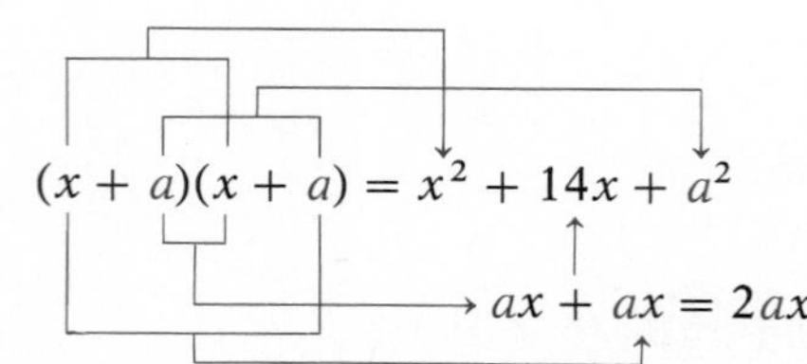

Thus a is such that $2ax = 14x$. Solving for a, we get

$$a = \frac{14x}{2x} = \frac{14}{2} = 7.$$

That is, a is half of the coefficient of x in $x^2 + 14x$. Since $a^2 = (\frac{14}{2})^2 = 7^2 = 49$, we add 49 to our original expression:

$x^2 + 14x + 49$ is the square of $x + 7$;

that is, $x^2 + 14x + 49 = (x + 7)^2$.

To *complete the square* of an expression like $x^2 + bx$, we take half the coefficient of x, $b/2$, and square. Then we add that number, which is $(b/2)^2$.

Returning to solving our original equation, we first add 49 on both sides to *complete the square.* Then we solve as follows.

$$x^2 + 14x = 4$$

$$x^2 + 14x + 49 = 4 + 49 \quad \textbf{Adding 49: } (\tfrac{14}{2})^2 = 7^2 = 49$$

$$(x + 7)^2 = 53$$

$$x + 7 = \sqrt{53} \quad or \quad x + 7 = -\sqrt{53} \quad \textbf{Using the principle of square roots}$$

$$x = -7 + \sqrt{53} \quad or \quad x = -7 - \sqrt{53}$$

The solutions are $-7 \pm \sqrt{53}$.

We have seen that a quadratic equation $(x + k)^2 = d$ can be solved using the principle of square roots. Any equation, such as $x^2 - 6x + 8 = 0$, can be put in this form by completing the square. Then we can solve as before.

▶ **EXAMPLE 8** Solve: $x^2 - 6x + 8 = 0$.

We have

$$x^2 - 6x + 8 = 0$$

$$x^2 - 6x = -8. \quad \textbf{Subtracting 8}$$

We take half of -6 and square it, to get 9. Then we add 9 on *both* sides of the equation. This makes the left side the square of a binomial, $x - 3$. We have now *completed the square* and can continue to solve:

$$x^2 - 6x + 9 = -8 + 9 \quad \textbf{Adding 9}$$

$$(x - 3)^2 = 1$$

$$x - 3 = 1 \quad or \quad x - 3 = -1 \quad \textbf{Using the principle of square roots}$$

$$x = 4 \quad or \quad x = 2.$$

The solutions are 2 and 4. ◀

This method of solving is called *completing the square.* Example 8 can be solved more easily by factoring. We solved it by completing the square to illustrate that we can use this method to solve any quadratic equation.

DO EXERCISES 8 AND 9.

Solve.

8. $x^2 + 6x + 8 = 0$

9. $x^2 - 8x - 20 = 0$

ANSWERS ON PAGE A-8

10. Solve by completing the square: $x^2 + 6x - 1 = 0$.

▶ **EXAMPLE 9** Solve $x^2 + 4x - 7 = 0$ by completing the square.

$$x^2 + 4x - 7 = 0$$

$$x^2 + 4x = 7 \qquad \textbf{Adding 7}$$

$$x^2 + 4x + 4 = 7 + 4 \qquad \textbf{Adding 4: } (\tfrac{4}{2})^2 = (2)^2 = 4$$

$$(x + 2)^2 = 11$$

$$x + 2 = \sqrt{11} \quad or \quad x + 2 = -\sqrt{11} \qquad \textbf{Using the principle of square roots}$$

$$x = -2 + \sqrt{11} \quad or \quad x = -2 - \sqrt{11}.$$

The solutions are $-2 \pm \sqrt{11}$. ◀

DO EXERCISE 10.

When the coefficient of x^2 is not 1, we can make it 1, as shown in the following example.

▶ **EXAMPLE 10** Solve $2x^2 = 3x - 7$ by completing the square.

We first obtain standard form. We then multiply on both sides by $\frac{1}{2}$ to make the coefficient of x^2 equal to 1.

$$2x^2 = 3x - 7$$

$$2x^2 - 3x + 7 = 0 \qquad \textbf{Finding standard form}$$

$$\frac{1}{2}(2x^2 - 3x + 7) = \frac{1}{2} \cdot 0 \qquad \textbf{Multiplying by } \frac{1}{2} \textbf{ to make the } x^2\textbf{-coefficient 1}$$

$$x^2 - \frac{3}{2}x + \frac{7}{2} = 0$$

$$x^2 - \frac{3}{2}x = -\frac{7}{2} \qquad \textbf{Subtracting } \frac{7}{2}$$

$$x^2 - \frac{3}{2}x + \frac{9}{16} = -\frac{7}{2} + \frac{9}{16} \qquad \textbf{Adding } \frac{9}{16}\textbf{: } \left[\frac{1}{2}\left(-\frac{3}{2}\right)\right]^2 = \left[-\frac{3}{4}\right]^2 = \frac{9}{16}$$

$$\left(x - \frac{3}{4}\right)^2 = -\frac{56}{16} + \frac{9}{16} \qquad \textbf{Finding a common denominator}$$

$$\left(x - \frac{3}{4}\right)^2 = -\frac{47}{16}$$

$$x - \frac{3}{4} = \sqrt{-\frac{47}{16}} \quad or \quad x - \frac{3}{4} = -\sqrt{-\frac{47}{16}} \qquad \textbf{Using the principle of square roots}$$

$$x - \frac{3}{4} = i\sqrt{\frac{47}{16}} \quad or \quad x - \frac{3}{4} = -i\sqrt{\frac{47}{16}} \qquad \sqrt{-1} = i$$

$$x = \frac{3}{4} + \frac{i\sqrt{47}}{4} \quad or \quad x = \frac{3}{4} - \frac{i\sqrt{47}}{4}$$

The solutions are $\frac{3 \pm i\sqrt{47}}{4}$.

ANSWER ON PAGE A-8

Solving by Completing the Square

To solve an equation $ax^2 + bx + c = 0$ by completing the square:

1. **If $a \neq 1$, multiply by $1/a$ so that the coefficient of x^2 is 1.**
2. **If the coefficient of x^2 is 1, add or subtract so that the equation is in the form**

$$x^2 + bx = -c, \quad \text{or} \quad x^2 + \frac{b}{a}x = -\frac{c}{a} \text{ if step (1) has been applied.}$$

3. **Take half of the coefficient of x and square it. Add the result on both sides of the equation.**
4. **Express the side with the variables as the square of a binomial.**
5. **Use the principle of square roots and complete the solution.**

DO EXERCISE 11.

11. Solve by completing the square:

$$3x^2 - 2x = 7.$$

C Problem Solving

If you put money in a savings account, the bank will pay you interest. At the end of a year, the bank will start paying you interest on both the original amount and the interest that you have already earned. This is called **compounding interest annually.**

The Compound-Interest Formula

If an amount of money P is invested at interest rate r, compounded annually, then in t years, it will grow to the amount A given by

$$A = P(1 + r)^t.$$

We can use quadratic equations to solve certain interest problems.

▶ **EXAMPLE 11** $1000 invested at 8.4%, compounded annually, for 2 years will grow to what amount?

We have

$$A = P(1 + r)^t$$

$$A = 1000(1 + 0.084)^2 \qquad \textbf{Substituting into the formula}$$

$$A = 1000(1.084)^2$$

$$A = 1000(1.175056)$$

$$A \approx 1175.06. \qquad \textbf{Computing and rounding}$$

The amount is $1175.06. ◀

▶ **EXAMPLE 12** $4000 is invested at interest rate r, compounded annually. In 2 years, it grows to $4410. What is the interest rate?

We know that $4000 is originally invested. Thus P is $4000. That amount grows to $4410 in 2 years. Thus when t is 2, A is $4410. We substitute 4000 for P, 4410 for A, and 2 for t in the formula, and solve the resulting equation for r:

$$A = P(1 + r)^t$$

$$4410 = 4000(1 + r)^2$$

$$\tfrac{4410}{4000} = (1 + r)^2$$

$$\tfrac{441}{400} = (1 + r)^2.$$

ANSWER ON PAGE A-8

12. Suppose that \$3000 is invested at 9.6%, compounded annually, for 2 years. To what amount will the investment grow?

13. Suppose that \$2500 is invested at interest rate r, compounded annually. In 2 years, the investment grows to \$3600. What is the interest rate?

14. The Sears Tower in Chicago is 1451 ft tall. How long would it take an object to fall freely from the top?

ANSWERS ON PAGE A-8

We then have

$$\sqrt{\tfrac{441}{400}} = 1 + r \quad or \quad -\sqrt{\tfrac{441}{400}} = 1 + r \qquad \text{Using the principle of square roots}$$
$$\tfrac{21}{20} = 1 + r \quad or \quad -\tfrac{21}{20} = 1 + r \qquad \text{Simplifying}$$
$$-\tfrac{20}{20} + \tfrac{21}{20} = r \quad or \quad -\tfrac{20}{20} - \tfrac{21}{20} = r$$
$$\tfrac{1}{20} = r \quad or \quad -\tfrac{41}{20} = r.$$

Since the interest rate cannot be negative, we have

$$\tfrac{1}{20} = r$$
$$r = 0.05, \quad \text{or } 5\%.$$

The interest rate must be 5%. ◀

DO EXERCISES 12 AND 13.

▶ **EXAMPLE 13** The formula $s = 16t^2$ is used to approximate the distance s, in feet, that an object falls freely from rest in t seconds. The RCA Building in New York City is 850 ft tall. How long will it take an object to fall from the top?

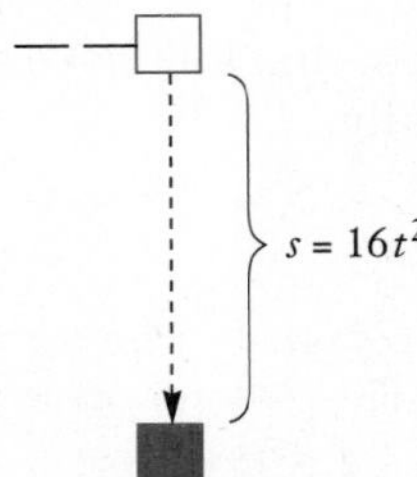

Using the formula $s = 16t^2$, we see that we know s and want to find t. We substitute 850 for s and solve for t:

$$s = 16t^2$$
$$850 = 16t^2 \qquad \text{Substituting 850 for } s$$
$$\frac{850}{16} = t^2 \qquad \text{Solving for } t^2$$
$$53.125 = t^2$$
$$\sqrt{53.125} = t \qquad \text{Using the principle of square roots; rejecting the negative square root since } t \text{ is not negative in this problem}$$
$$7.3 \approx t. \qquad \text{Using a calculator to find the square root and rounding to the nearest tenth}$$

It takes about 7.3 sec for an object to fall freely from the top of the RCA Building. ◀

DO EXERCISE 14.

NAME SECTION DATE

EXERCISE SET 8.1

a Solve.

1. $4x^2 = 20$

2. $3x^2 = 21$

3. $25x^2 + 4 = 0$

4. $9x^2 + 16 = 0$

5. $2x^2 - 3 = 0$

6. $3x^2 - 7 = 0$

7. $(x + 2)^2 = 49$

8. $(x - 1)^2 = 6$

9. $(x - 3)^2 = 21$

10. $(x + 3)^2 = 6$

11. $(x - 13)^2 = 8$

12. $(x - 13)^2 = 64$

13. $(x - 7)^2 = -4$

14. $(x + 1)^2 = -9$

15. $(x - 9)^2 = 34$

16. $(t - 2)^2 = 25$

17. $(x - \frac{3}{2})^2 = \frac{7}{2}$

18. $(y + \frac{3}{4})^2 = \frac{17}{16}$

19. $x^2 + 6x + 9 = 64$

20. $x^2 + 10x + 25 = 100$

21. $y^2 - 14y + 49 = 4$

22. $p^2 - 8p + 16 = 1$

b Solve by completing the square. Show your work.

23. $x^2 + 2x = 5$

24. $x^2 + 4x = 2$

25. $x^2 - 18x = 10$

26. $x^2 - 22x = 11$

27. $x^2 + x = 1$

28. $x^2 - x = 3$

29. $t^2 - 5t = 7$

30. $y^2 + 9y = 8$

31. $x^2 + \frac{3}{2}x = 3$

32. $x^2 - \frac{4}{3}x = \frac{2}{3}$

33. $m^2 - \frac{9}{2}m = \frac{3}{2}$

34. $r^2 + \frac{2}{5}r = \frac{4}{5}$

35. $x^2 + 6x - 16 = 0$

36. $x^2 - 8x + 15 = 0$

37. $x^2 + 22x + 102 = 0$

38. $x^2 + 18x + 74 = 0$

39. $x^2 - 10x - 4 = 0$

40. $x^2 + 10x - 4 = 0$

ANSWERS

1.
2.
3.
4.
5.
6.
7.
8.
9.
10.
11.
12.
13.
14.
15.
16.
17.
18.
19.
20.
21.
22.
23.
24.
25.
26.
27.
28.
29.
30.
31.
32.
33.
34.
35.
36.
37.
38.
39.
40.

ANSWERS

41.

42.

43.

44.

45.

46.

47.

48.

49.

50.

51.

52.

53.

54.

55.

56.

57.

58.

59.

60.

61.

62.

63. See graph.

64.

65.

66.

67.

68.

69.

70.

71.

72.

41. $x^2 + 7x - 2 = 0$ **42.** $x^2 - 7x - 2 = 0$ **43.** $x^2 - 3x - 28 = 0$

44. $x^2 + 3x - 28 = 0$ **45.** $x^2 - \frac{3}{2}x - \frac{1}{2} = 0$ **46.** $x^2 + \frac{3}{2}x - 2 = 0$

47. $2x^2 - 3x - 17 = 0$ **48.** $2x^2 + 3x - 1 = 0$ **49.** $3x^2 - 4x - 1 = 0$

50. $3x^2 + 4x - 3 = 0$ **51.** $x^2 + x + 2 = 0$ **52.** $x^2 - x + 1 = 0$

53. $x^2 - 4x + 13 = 0$ **54.** $x^2 - 6x + 13 = 0$

C Use $A = P(1 + r)^t$ for Exercises 55–60. What is the interest rate?

55. \$2560 grows to \$2890 in 2 years **56.** \$1000 grows to \$1210 in 2 years

57. \$1000 grows to \$1440 in 2 years **58.** \$2560 grows to \$3610 in 2 years

59. \$6250 grows to \$6760 in 2 years **60.** \$6250 grows to \$7290 in 2 years

The formula $s = 16t^2$ is used to approximate the distance s, in feet, that an object falls freely from rest in t seconds. Use the formula for Exercises 61 and 62.

61. The height of the World Trade Center in New York is 1377 ft (excluding television towers and antennas). How long would it take an object to fall from the top?

62. The John Hancock Building in Chicago is 1107 ft tall. How long would it take an object to fall from the top?

SKILL MAINTENANCE

63. Graph: $y = 2x + 1$.

64. Simplify: $\sqrt{88}$.

65. Approximate to the nearest tenth: $14 - \sqrt{88}$.

66. Rationalize the denominator: $\sqrt{\frac{2}{5}}$.

SYNTHESIS

Find b such that the trinomial is a square.

67. $x^2 + bx + 64$ **68.** $x^2 + bx + 75$

Solve.

69. $x(2x^2 + 9x - 56)(3x + 10) = 0$ **70.** $(x - \frac{1}{3})(x - \frac{1}{3}) + (x - \frac{1}{3})(x + \frac{2}{9}) = 0$

71. Boats A and B leave the same point at the same time at right angles. B travels 7 km/h slower than A. After 4 hr they are 68 km apart. Find the speed of each boat.

72. Find three consecutive integers such that the square of the first plus the product of the other two is 67.

8.2 The Quadratic Formula

There are at least two reasons for learning to complete the square. One is to enhance your ability to graph certain second-degree equations, which you will encounter in Chapter 9. The other is to prove a general formula that can be used to solve quadratic equations.

a Solving Using the Quadratic Formula

Each time you solve by completing the square, the procedure is the same. In situations like this in mathematics, when we do about the same kind of procedure many times, we look for a formula to speed up our work. Consider any quadratic equation in standard form:

$$ax^2 + bx + c = 0, \quad a > 0.$$

Let's solve by completing the square. As we carry out the steps, compare them with Example 10 in the preceding section.

$$x^2 + \frac{b}{a}x + \frac{c}{a} = 0 \qquad \text{Multiplying by } \frac{1}{a}$$

$$x^2 + \frac{b}{a}x = -\frac{c}{a} \qquad \text{Subtracting } \frac{c}{a}$$

Half of b/a is $b/2a$. The square is $b^2/4a^2$. We add $b^2/4a^2$ on both sides:

$$x^2 + \frac{b}{a}x + \frac{b^2}{4a^2} = -\frac{c}{a} + \frac{b^2}{4a^2} \qquad \text{Adding } \frac{b^2}{4a^2}$$

$$\left(x + \frac{b}{2a}\right)^2 = -\frac{4ac}{4a^2} + \frac{b^2}{4a^2} \qquad \text{Factoring the left side and finding a common denominator on the right}$$

$$\left(x + \frac{b}{2a}\right)^2 = \frac{b^2 - 4ac}{4a^2}$$

$$x + \frac{b}{2a} = \sqrt{\frac{b^2 - 4ac}{4a^2}} \quad or \quad x + \frac{b}{2a} = -\sqrt{\frac{b^2 - 4ac}{4a^2}}. \qquad \text{Using the principle of square roots}$$

Since $a > 0$, $\sqrt{4a^2} = 2a$, so we can simplify as follows:

$$x + \frac{b}{2a} = \frac{\sqrt{b^2 - 4ac}}{2a} \quad \text{or} \quad x + \frac{b}{2a} = -\frac{\sqrt{b^2 - 4ac}}{2a}.$$

Thus,

$$x = -\frac{b}{2a} + \frac{\sqrt{b^2 - 4ac}}{2a} \quad \text{or} \quad x = -\frac{b}{2a} - \frac{\sqrt{b^2 - 4ac}}{2a},$$

so

$$x = -\frac{b}{2a} \pm \frac{\sqrt{b^2 - 4ac}}{2a},$$

or

$$x = \frac{-b \pm \sqrt{b^2 - 4ac}}{2a}.$$

> The Quadratic Formula
>
> **The solutions of $ax^2 + bx + c = 0$ are given by**
>
> $$x = \frac{-b \pm \sqrt{b^2 - 4ac}}{2a}.$$

OBJECTIVES

After finishing Section 8.2, you should be able to:

- **a** Solve quadratic equations using the quadratic formula.
- **b** Find approximate solutions using a calculator or a square-root table.

FOR EXTRA HELP

Tape 15B

Tape 13B

MAC: 8
IBM: 8

1. Consider the equation

$$2x^2 = 4 + 7x.$$

a) Solve using the quadratic formula.

b) Solve by factoring.

ANSWERS ON PAGE A-8

Note that the formula also holds when $a < 0$. A similar proof would show this, but we will not consider it here.

▶ **EXAMPLE 1** Solve $5x^2 + 8x = -3$ using the quadratic formula.

We first find standard form and determine a, b, and c:

$$5x^2 + 8x + 3 = 0;$$
$$a = 5, \quad b = 8, \quad c = 3.$$

We then use the quadratic formula:

$$x = \frac{-b \pm \sqrt{b^2 - 4ac}}{2a}$$

$$x = \frac{-8 \pm \sqrt{8^2 - 4 \cdot 5 \cdot 3}}{2 \cdot 5} \quad \text{Substituting}$$

Be sure to write the fraction bar all the way across.

$$x = \frac{-8 \pm \sqrt{64 - 60}}{10}$$

$$x = \frac{-8 \pm \sqrt{4}}{10}$$

$$x = \frac{-8 \pm 2}{10}$$

$$x = \frac{-8 + 2}{10} \quad or \quad x = \frac{-8 - 2}{10}$$

$$x = \frac{-6}{10} \quad or \quad x = \frac{-10}{10}$$

$$x = -\frac{3}{5} \quad or \quad x = -1.$$

The solutions are $-\frac{3}{5}$ and -1. ◀

It turns out that we could have solved the equation in Example 1 by factoring as follows. This would have been easier:

$$5x^2 + 8x + 3 = 0$$
$$(5x + 3)(x + 1) = 0$$
$$5x + 3 = 0 \quad or \quad x + 1 = 0$$
$$5x = -3 \quad or \quad x = -1$$
$$x = -\tfrac{3}{5} \quad or \quad x = -1.$$

To solve a quadratic equation:

1. **Check for the form $ax^2 = p$ or $(x + k)^2 = d$. If it is in either of these forms, use the principle of square roots as in Section 8.1.**
2. **If it is not in the form of step (1), write it in standard form $ax^2 + bx + c = 0$ with a and b nonzero.**
3. **Then try factoring.**
4. **If it is not possible to factor or factoring seems difficult, use the quadratic formula.**

The solutions of a quadratic equation can always be found using the quadratic formula. They cannot always be found by factoring.

DO EXERCISE 1.

▶ **EXAMPLE 2** Solve: $5x^2 - 8x = 3$.

We first find standard form and determine a, b, and c:

$$5x^2 - 8x - 3 = 0;$$
$$a = 5, \quad b = -8, \quad c = -3$$

We then substitute into the quadratic formula:

$$x = \frac{-(-8) \pm \sqrt{(-8)^2 - 4 \cdot 5 \cdot (-3)}}{2 \cdot 5}$$
$$= \frac{8 \pm \sqrt{64 + 60}}{10} = \frac{8 \pm \sqrt{124}}{10} = \frac{8 \pm \sqrt{4 \cdot 31}}{10}$$
$$= \frac{8 \pm 2\sqrt{31}}{10} = \frac{2(4 \pm \sqrt{31})}{2 \cdot 5} = \frac{4 \pm \sqrt{31}}{5}.$$

↑

> CAUTION! To avoid a common error in simplifying, remember to *factor the numerator and the denominator* and then remove a factor of 1.

The checking of possible solutions such as these is cumbersome. When the quadratic formula is being used to solve a quadratic equation in standard form, no numbers are ever obtained that are not solutions of the original equation unless a mistake has been made. Therefore, it is usually better to check your work, rather than to check by substituting into the original equation.

The solutions are

$$\frac{4 + \sqrt{31}}{5} \quad \text{and} \quad \frac{4 - \sqrt{31}}{5}.$$ ◀

DO EXERCISE 2.

Some quadratic equations have solutions that are nonreal complex numbers.

▶ **EXAMPLE 3** Solve: $x^2 + x + 1 = 0$.

We have

$$a = 1, \quad b = 1, \quad c = 1$$
$$x = \frac{-b \pm \sqrt{b^2 - 4ac}}{2a}$$
$$= \frac{-1 \pm \sqrt{1^2 - 4 \cdot 1 \cdot 1}}{2 \cdot 1} = \frac{-1 \pm \sqrt{1 - 4}}{2}$$
$$= \frac{-1 \pm \sqrt{-3}}{2} = \frac{-1 \pm i\sqrt{3}}{2}.$$

The solutions are

$$\frac{-1 + i\sqrt{3}}{2} \quad \text{and} \quad \frac{-1 - i\sqrt{3}}{2}.$$

The solutions can also be expressed in the form

$$-\frac{1}{2} + i\frac{\sqrt{3}}{2} \quad \text{and} \quad -\frac{1}{2} - i\frac{\sqrt{3}}{2}.$$ ◀

DO EXERCISE 3.

2. Solve using the quadratic formula:

$$3x^2 + 2x = 7.$$

3. Solve: $x^2 - x + 2 = 0$.

ANSWERS ON PAGE A-8

4. Solve: $3 = \frac{5}{x} + \frac{2}{x^2}$.

5. Approximate to the nearest tenth:

$$\frac{8 \pm \sqrt{47}}{7}.$$

6. Approximate the solutions to Margin Exercise 2. Round to the nearest tenth.

ANSWERS ON PAGE A-8

▶ **EXAMPLE 4** Solve: $2 + \frac{7}{x} = \frac{4}{x^2}$.

We first find standard form:

$$2x^2 + 7x = 4 \qquad \text{Multiplying by } x^2\text{, the LCM of the denominators}$$

$$2x^2 + 7x - 4 = 0 \qquad \text{Subtracting 4}$$

$$a = 2, \quad b = 7, \quad c = -4$$

$$x = \frac{-7 \pm \sqrt{7^2 - 4 \cdot 2 \cdot (-4)}}{2 \cdot 2}$$

$$x = \frac{-7 \pm \sqrt{49 + 32}}{4}$$

$$x = \frac{-7 \pm \sqrt{81}}{4} = \frac{-7 \pm 9}{4}$$

$$x = \frac{-7 + 9}{4} = \frac{1}{2} \quad \text{or} \quad x = \frac{-7 - 9}{4} = -4$$

The quadratic formula always gives correct results when we start with the standard form. In such cases, we need check only to detect errors. In this case, since we started with a rational equation, we *do* need to check. We cleared of fractions before obtaining standard form, and this step could possibly introduce numbers that do not check in the original equation. At least we need to show that neither of the numbers makes a denominator 0. Since neither of them does, the solutions are $\frac{1}{2}$ and -4. ◀

DO EXERCISE 4.

b Approximating Solutions

A calculator or a square-root table can be used to approximate solutions.

▶ **EXAMPLE 5** Approximate to the nearest tenth the solutions of the equation in Example 2.

Using a calculator or a square root table, we find that

$$\sqrt{31} \approx 5.567764363.$$

Then

$$\frac{4 + \sqrt{31}}{5} \approx \frac{4 + 5.567764363}{5}$$

$$\approx \frac{9.567764363}{5} \qquad \text{Adding}$$

$$\approx 1.913552873 \qquad \text{Dividing}$$

$$\approx 1.9; \qquad \text{Rounding to the nearest tenth}$$

$$\frac{4 - \sqrt{31}}{5} \approx \frac{4 - 5.567764363}{5}$$

$$\approx \frac{-1.567764363}{5} \qquad \text{Subtracting}$$

$$\approx -0.313552873 \qquad \text{Dividing}$$

$$\approx -0.3. \qquad \text{Rounding to the nearest tenth}$$ ◀

We could also use Table 1, but there might be a variance in the answers.

DO EXERCISES 5 AND 6.

NAME SECTION DATE

EXERCISE SET 8.2

a Solve.

1. $x^2 + 6x + 4 = 0$
2. $x^2 - 6x - 4 = 0$
3. $3p^2 = -8p - 5$
4. $3u^2 = 18u - 6$
5. $x^2 - x + 1 = 0$
6. $x^2 + x + 2 = 0$
7. $x^2 + 13 = 4x$
8. $x^2 + 13 = 6x$
9. $r^2 + 3r = 8$
10. $h^2 + 4 = 6h$
11. $1 + \frac{2}{x} + \frac{5}{x^2} = 0$
12. $1 + \frac{5}{x^2} = \frac{2}{x}$
13. $3x + x(x - 2) = 0$
14. $4x + x(x - 3) = 0$
15. $14x^2 + 9x = 0$
16. $19x^2 + 8x = 0$
17. $25x^2 - 20x + 4 = 0$
18. $36x^2 + 84x + 49 = 0$
19. $4x(x - 2) - 5x(x - 1) = 2$
20. $3x(x + 1) - 7x(x + 2) = 6$
21. $14(x - 4) - (x + 2) = (x + 2)(x - 4)$
22. $11(x - 2) + (x - 5) = (x + 2)(x - 6)$
23. $5x^2 = 13x + 17$
24. $25x = 3x^2 + 28$

ANSWERS

1. ______
2. ______
3. ______
4. ______
5. ______
6. ______
7. ______
8. ______
9. ______
10. ______
11. ______
12. ______
13. ______
14. ______
15. ______
16. ______
17. ______
18. ______
19. ______
20. ______
21. ______
22. ______
23. ______
24. ______

ANSWERS

25. ______
26. ______
27. ______
28. ______
29. ______
30. ______
31. ______
32. ______
33. ______
34. ______
35. ______
36. ______
37. ______
38. ______
39. ______
40. ______
41. ______
42. ______
43. ______
44. ______
45. ______
46. ______
47. ______
48. ______
49. ______
50. ______
51. ______
52. ______
53. ______

25. $x^2 + 5 = 2x$

26. $x^2 + 5 = 4x$

27. $x + \frac{1}{x} = \frac{13}{6}$

28. $\frac{3}{x} + \frac{x}{3} = \frac{5}{2}$

29. $\frac{1}{x} + \frac{1}{x+3} = \frac{1}{2}$

30. $\frac{1}{x} + \frac{1}{x+4} = \frac{1}{7}$

31. $(2t - 3)^2 + 17t = 15$

32. $2y^2 - (y + 2)(y - 3) = 12$

33. $(x - 2)^2 + (x + 1)^2 = 0$

34. $(x + 3)^2 + (x - 1)^2 = 0$

35. $x^3 - 8 = 0$
(*Hint:* Factor the difference of cubes. Then use the quadratic formula.)

36. $x^3 + 1 = 0$

b Use a calculator or Table 1 to approximate solutions to the nearest tenth.

37. $x^2 + 4x - 7 = 0$

38. $x^2 + 6x + 4 = 0$

39. $x^2 - 6x + 4 = 0$

40. $x^2 - 4x + 1 = 0$

41. $2x^2 - 3x - 7 = 0$

42. $3x^2 - 3x - 2 = 0$

43. $5x^2 = 3 + 8x$

44. $2y^2 + 2y - 3 = 0$

SKILL MAINTENANCE

45. A store has coffee A worth \$1.50 per pound and coffee B worth \$2.50 per pound. It wants to mix the coffees to obtain 50 lb of a coffee worth \$1.90 per pound. How much of each kind of coffee should be used?

SYNTHESIS

Solve.

46. ▦ $2.2x^2 + 0.5x - 1 = 0$

47. ▦ $5.33x^2 - 8.23x - 3.24 = 0$

48. $\frac{5}{x} + \frac{x}{4} = \frac{11}{7}$

49. $2x^2 - x - \sqrt{5} = 0$

50. $\sqrt{3}x^2 + 6x + \sqrt{3} = 0$

51. $ix^2 - x - 1 = 0$

52. $(1 + \sqrt{3})x^2 - (3 + 2\sqrt{3})x + 3 = 0$

53. $\frac{x}{x+1} = 4 + \frac{1}{3x^2 - 3}$

8.3 The Discriminant and Solutions to Quadratic Equations

a The Discriminant

From the quadratic formula, we know that the solutions x_1 and x_2 of a quadratic equation are given by

$$x_1 = \frac{-b + \sqrt{b^2 - 4ac}}{2a} \quad \text{and} \quad x_2 = \frac{-b - \sqrt{b^2 - 4ac}}{2a}.$$

The expression $b^2 - 4ac$ shows the nature of the solutions. This expression is called the **discriminant.** If it is 0, then it doesn't matter whether we choose the plus or minus sign in the formula; hence there is just one real solution. If the discriminant is positive, there will be two real solutions. If it is negative, we will be taking the square root of a negative number; hence there will be two nonreal complex-number solutions, and they will be complex conjugates. We summarize.

Discriminant $b^2 - 4ac$	Nature of solutions
0	Only one solution; it is a real number
Positive	Two different real-number solutions
Negative	Two different nonreal complex-number solutions (complex conjugates)

The discriminant also gives information about solving. When the discriminant is a perfect square, you can solve the equation by factoring, not needing the quadratic formula.

► **EXAMPLE 1** Determine the nature of the solutions of $9x^2 - 12x + 4 = 0$.

We have

$$a = 9, \quad b = -12, \quad c = 4.$$

We compute the discriminant:

$$\begin{aligned} b^2 - 4ac &= (-12)^2 - 4 \cdot 9 \cdot 4 \\ &= 144 - 144 \\ &= 0. \end{aligned}$$

There is just one solution, and it is a real number. Since 0 is a perfect square, the equation can be solved by factoring. ◄

► **EXAMPLE 2** Determine the nature of the solutions of $x^2 + 5x + 8 = 0$.

We have

$$a = 1, \quad b = 5, \quad c = 8.$$

We compute the discriminant:

$$\begin{aligned} b^2 - 4ac &= 5^2 - 4 \cdot 1 \cdot 8 \\ &= 25 - 32 \\ &= -7. \end{aligned}$$

Since the discriminant is negative, there are two nonreal complex-number solutions. The equation cannot be solved by factoring because -7 is not a perfect square. ◄

OBJECTIVES

After finishing Section 8.3, you should be able to:

a Determine the nature of the solutions of a quadratic equation with real-number coefficients without solving it.

b Write a quadratic equation having two numbers specified as solutions.

FOR EXTRA HELP

Tape NC

Tape 14A

MAC: 8
IBM: 8

Determine the nature of the solutions without solving.

1. $x^2 + 5x - 3 = 0$

2. $9x^2 - 6x + 1 = 0$

3. $3x^2 - 2x + 1 = 0$

Find a quadratic equation having the following solutions.

4. -4 and $\frac{5}{3}$

5. $5i$ and $-5i$

6. $-2\sqrt{2}$ and $\sqrt{2}$

ANSWERS ON PAGE A-8

▶ **EXAMPLE 3** Determine the nature of the solutions of $x^2 + 5x + 6 = 0$.

We have

$$a = 1, \quad b = 5, \quad c = 6;$$
$$b^2 - 4ac = 5^2 - 4 \cdot 1 \cdot 6 = 1.$$

Since the discriminant is positive and a perfect square, there are two solutions and they are real numbers. The equation can be solved by factoring since the discriminant is a perfect square. ◀

DO EXERCISES 1–3.

b Writing Equations from Solutions

We know by the principle of zero products that $(x - 2)(x + 3) = 0$ has solutions 2 and -3. If we know the solutions of an equation, we can write the equation, using this principle in reverse.

▶ **EXAMPLE 4** Find a quadratic equation whose solutions are 3 and $-\frac{2}{5}$.

We have

$$x = 3 \quad or \quad x = -\tfrac{2}{5}$$
$$x - 3 = 0 \quad or \quad x + \tfrac{2}{5} = 0 \qquad \text{Getting the 0's on one side}$$
$$(x - 3)(x + \tfrac{2}{5}) = 0 \qquad \text{Using the principle of zero products (multiplying)}$$
$$x^2 + \tfrac{2}{5}x - 3x - 3 \cdot \tfrac{2}{5} = 0 \qquad \text{Using FOIL}$$
$$x^2 - \tfrac{13}{5}x - \tfrac{6}{5} = 0, \quad \text{or}$$
$$5x^2 - 13x - 6 = 0. \qquad \text{Collecting like terms and multiplying by 5}$$ ◀

▶ **EXAMPLE 5** Write a quadratic equation whose solutions are $2i$ and $-2i$.

We have

$$x = 2i \quad or \quad x = -2i$$
$$x - 2i = 0 \quad or \quad x + 2i = 0 \qquad \text{Getting the 0's on one side}$$
$$(x - 2i)(x + 2i) = 0 \qquad \text{Using the principle of zero products (multiplying)}$$
$$x^2 + 2ix - 2ix - (2i)^2 = 0$$
$$x^2 - (2i)^2 = 0 \qquad \text{Using FOIL}$$
$$x^2 - 4i^2 = 0$$
$$x^2 + 4 = 0.$$ ◀

▶ **EXAMPLE 6** Write a quadratic equation whose solutions are $\sqrt{3}$ and $-2\sqrt{3}$.

We have

$$x = \sqrt{3} \quad or \quad x = -2\sqrt{3}$$
$$x - \sqrt{3} = 0 \quad or \quad x + 2\sqrt{3} = 0 \qquad \text{Getting the 0's on one side}$$
$$(x - \sqrt{3})(x + 2\sqrt{3}) = 0 \qquad \text{Using the principle of zero products}$$
$$x^2 + 2\sqrt{3}x - \sqrt{3}x - 2(\sqrt{3})^2 = 0 \qquad \text{Using FOIL}$$
$$x^2 + \sqrt{3}x - 6 = 0. \qquad \text{Collecting like terms}$$ ◀

DO EXERCISES 4–6.

NAME SECTION DATE

EXERCISE SET 8.3

ANSWERS

a Determine the nature of the solutions of the equation.

1. $x^2 - 6x + 9 = 0$

2. $x^2 + 10x + 25 = 0$

3. $x^2 + 7 = 0$

4. $x^2 + 2 = 0$

5. $x^2 - 2 = 0$

6. $x^2 - 5 = 0$

7. $4x^2 - 12x + 9 = 0$

8. $4x^2 + 8x - 5 = 0$

9. $x^2 - 2x + 4 = 0$

10. $x^2 + 3x + 4 = 0$

11. $9t^2 - 3t = 0$

12. $4m^2 + 7m = 0$

13. $y^2 = \frac{1}{2}y + \frac{3}{5}$

14. $y^2 + \frac{9}{4} = 4y$

15. $4x^2 - 4\sqrt{3}x + 3 = 0$

16. $6y^2 - 2\sqrt{3}y - 1 = 0$

1. ______
2. ______
3. ______
4. ______
5. ______
6. ______
7. ______
8. ______
9. ______
10. ______
11. ______
12. ______
13. ______
14. ______
15. ______
16. ______

ANSWERS

17. ____
18. ____
19. ____
20. ____
21. ____
22. ____
23. ____
24. ____
25. ____
26. ____
27. ____
28. ____
29. ____
30. a) ____
b) ____
31. a) ____
b) ____
32. ____
33. ____

b Write a quadratic equation having the given numbers as solutions.

17. $-11,\ 9$

18. $-4,\ 4$

19. 7, only solution
[*Hint:* It must be a double solution, that is, $(x-7)(x-7)=0$.]

20. -5, only solution

21. $-\frac{2}{5},\ \frac{6}{5}$

22. $-\frac{1}{4},\ -\frac{1}{2}$

23. $\frac{c}{2},\ \frac{d}{2}$

24. $\frac{k}{3},\ \frac{m}{4}$

25. $\sqrt{2},\ 3\sqrt{2}$

26. $-\sqrt{3},\ 2\sqrt{3}$

SKILL MAINTENANCE

27. During a one-hour television show, there were 12 commercials. Some of the commercials were 30 sec long and the others were 60 sec long. The amount of time for 30-sec commercials was 6 min less than the total number of minutes of commercial time during the show. How many 30-sec commercials were used? How many 60-sec commercials were used?

SYNTHESIS

Prove.

28. The sum of the solutions of $ax^2 + bx + c = 0$ is $-b/a$.

29. The product of the solutions of $ax^2 + bx + c = 0$ is c/a.

For each equation under the given condition, **(a)** find k and **(b)** find the other solution.

30. $kx^2 - 17x + 33 = 0$; one solution is 3.

31. $kx^2 - 2x + k = 0$; one solution is -3.

32. Find k if $kx^2 - 4x + (2k-1) = 0$ and the product of the solutions is 3.

33. Find a quadratic equation for which the sum of the solutions is $\sqrt{3}$ and the product is 8.

8.4 Equations Reducible to Quadratic

a Certain equations that are not really quadratic can be thought of in such a way that they can be solved as quadratic. For example, consider this fourth-degree equation.

$$\begin{array}{llll} x^4 & -\ 9x^2 & +\ 8 = 0 & \\ \downarrow & \quad\downarrow & \ \downarrow \ \ \downarrow & \\ (x^2)^2 & -\ 9(x^2) & +\ 8 = 0 & \text{Thinking of } x^4 \text{ as } (x^2)^2 \\ \downarrow & \quad\downarrow & \ \downarrow \ \ \downarrow & \\ u^2 & -\ 9u & +\ 8 = 0 & \text{To make this clearer, write } u \text{ instead of } x^2. \end{array}$$

The equation $u^2 - 9u + 8 = 0$ can be solved by factoring or by the quadratic formula. After that, we can find x by remembering that $x^2 = u$. Equations that can be solved like this are said to be **reducible to quadratic.**

▶ **EXAMPLE 1** Solve: $x^4 - 9x^2 + 8 = 0$.

Let $u = x^2$. Then we solve the equation found by substituting u for x^2:

$$u^2 - 9u + 8 = 0$$
$$(u - 8)(u - 1) = 0 \quad \text{Factoring}$$
$$u - 8 = 0 \quad or \quad u - 1 = 0 \quad \text{Using the principle of zero products}$$
$$u = 8 \quad or \quad u = 1.$$

Now we substitute x^2 for u and solve these equations:

$$x^2 = 8 \quad or \quad x^2 = 1$$
$$x = \pm\sqrt{8} \quad or \quad x = \pm 1$$
$$x = \pm 2\sqrt{2} \quad or \quad x = \pm 1.$$

To check, first note that when $x = 2\sqrt{2}$, $x^2 = 8$ and $x^4 = 64$. Also, when $x = -2\sqrt{2}$, $x^2 = 8$ and $x^4 = 64$. Similarly, when $x = 1$, $x^2 = 1$ and $x^4 = 1$, and when $x = -1$, $x^2 = 1$ and $x^4 = 1$. Thus instead of making four checks we need make only two.

Check:

For $\pm 2\sqrt{2}$:

$x^4 - 9x^2 + 8 = 0$	
$(\pm 2\sqrt{2})^4 - 9(\pm 2\sqrt{2})^2 + 8$	0
$64 - 9 \cdot 8 + 8$	
0	

For ± 1:

$x^4 - 9x^2 + 8 = 0$	
$(\pm 1)^4 - 9(\pm 1)^2 + 8$	0
$1 - 9 + 8$	
0	

A common error is to solve for u and then forget to solve for x. Remember that you must find values for the *original* variable!

The solutions are 1, -1, $2\sqrt{2}$, and $-2\sqrt{2}$. ◀

A check by substituting is necessary when solving equations reducible to quadratic. Reducing the equation to quadratic can sometimes introduce numbers that are not solutions of the original equation.

DO EXERCISE 1.

OBJECTIVE

After finishing Section 8.4, you should be able to:

a Solve equations that are reducible to quadratic.

FOR EXTRA HELP

Tape 15C Tape 14A MAC: 8 IBM: 8

1. Solve: $x^4 - 10x^2 + 9 = 0$.

ANSWER ON PAGE A-8

2. Solve $x + 3\sqrt{x} - 10 = 0$. Be sure to check.

▶ **EXAMPLE 2** Solve: $x - 3\sqrt{x} - 4 = 0$.

Let $u = \sqrt{x}$. Then we solve the equation found by substituting u for $\sqrt{x}$ (and, of course, u^2 for x):

$$u^2 - 3u - 4 = 0$$
$$(u - 4)(u + 1) = 0$$
$$u = 4 \quad \textit{or} \quad u = -1.$$

Now we substitute $\sqrt{x}$ for u and solve these equations:

$$\sqrt{x} = 4 \quad \text{or} \quad \sqrt{x} = -1.$$

Squaring the first equation, we get $x = 16$. The second equation has no real solution since principal square roots are never negative.

The number 16 checks and is the solution. We will not be considering nonreal solutions in this section. ◀

DO EXERCISE 2.

3. Solve: $(x^2 - x)^2 - 14(x^2 - x) + 24 = 0$.

▶ **EXAMPLE 3** Solve: $(x^2 - 1)^2 - (x^2 - 1) - 2 = 0$.

Let $u = x^2 - 1$. Then we solve the equation found by substituting u for $x^2 - 1$:

$$u^2 - u - 2 = 0$$
$$(u - 2)(u + 1) = 0$$
$$u = 2 \quad \textit{or} \quad u = -1.$$

Now we substitute $x^2 - 1$ for u and solve these equations:

$$x^2 - 1 = 2 \quad \textit{or} \quad x^2 - 1 = -1$$
$$x^2 = 3 \quad \textit{or} \quad x^2 = 0$$
$$x = \pm\sqrt{3} \quad \textit{or} \quad x = 0.$$

The numbers $\sqrt{3}$, $-\sqrt{3}$, and 0 check. They are the solutions. ◀

DO EXERCISE 3.

4. Solve: $x^{-2} + x^{-1} - 6 = 0$.

▶ **EXAMPLE 4** Solve: $y^{-2} - y^{-1} - 2 = 0$.

Let $u = y^{-1}$. Then we solve the equation found by substituting u for y^{-1} and u^2 for y^{-2}:

$$u^2 - u - 2 = 0$$
$$(u - 2)(u + 1) = 0$$
$$u = 2 \quad \textit{or} \quad u = -1.$$

Now we substitute y^{-1} or $1/y$ for u and solve these equations:

$$\frac{1}{y} = 2 \quad \text{or} \quad \frac{1}{y} = -1.$$

Solving, we get

$$y = \frac{1}{2} \quad \text{or} \quad y = \frac{1}{(-1)} = -1.$$

The numbers $\frac{1}{2}$ and -1 both check. They are the solutions. ◀

ANSWERS ON PAGE A-8

DO EXERCISE 4.

NAME SECTION DATE

EXERCISE SET 8.4

a Solve.

1. $x^4 - 10x^2 + 25 = 0$

2. $x^4 - 3x^2 + 2 = 0$

3. $x - 10\sqrt{x} + 9 = 0$

4. $2x - 9\sqrt{x} + 4 = 0$

5. $(x^2 - 6x)^2 - 2(x^2 - 6x) - 35 = 0$

6. $(x^2 + 5x)^2 + 2(x^2 + 5x) - 24 = 0$

7. $x^{-2} - x^{-1} - 6 = 0$

8. $4x^{-2} - x^{-1} - 5 = 0$

9. $(1 + \sqrt{x})^2 + (1 + \sqrt{x}) - 6 = 0$

10. $(2 + \sqrt{x})^2 - 3(2 + \sqrt{x}) - 10 = 0$

11. $(y^2 - 5y)^2 - 2(y^2 - 5y) - 24 = 0$

12. $(2t^2 + t)^2 - 4(2t^2 + t) + 3 = 0$

13. $w^4 - 4w^2 - 2 = 0$

14. $t^4 - 5t^2 + 5 = 0$

15. $2x^{-2} + x^{-1} - 1 = 0$

16. $m^{-2} + 9m^{-1} - 10 = 0$

17. $6x^4 - 19x^2 + 15 = 0$

18. $6x^4 - 17x^2 + 5 = 0$

ANSWERS

1. ________
2. ________
3. ________
4. ________
5. ________
6. ________
7. ________
8. ________
9. ________
10. ________
11. ________
12. ________
13. ________
14. ________
15. ________
16. ________
17. ________
18. ________

ANSWERS

19. ______
20. ______
21. ______
22. ______
23. ______
24. ______
25. ______
26. ______
27. ______
28. ______
29. ______
30. ______
31. ______
32. ______
33. ______
34. ______
35. ______
36. ______
37. ______
38. ______

19. $x^{2/3} - 4x^{1/3} - 5 = 0$

20. $x^{2/3} + 2x^{1/3} - 8 = 0$

21. $\left(\frac{x+3}{x-3}\right)^2 - \left(\frac{x+3}{x-3}\right) - 6 = 0$

22. $\left(\frac{x-4}{x+1}\right)^2 - 2\left(\frac{x-4}{x+1}\right) - 35 = 0$

23. $9\left(\frac{x+2}{x+3}\right)^2 - 6\left(\frac{x+2}{x+3}\right) + 1 = 0$

24. $16\left(\frac{x-1}{x-8}\right)^2 + 8\left(\frac{x-1}{x-8}\right) + 1 = 0$

25. $\left(\frac{y^2-1}{y}\right)^2 - 4\left(\frac{y^2-1}{y}\right) - 12 = 0$

26. $\left(\frac{x^2-2}{x}\right)^2 - 7\left(\frac{x^2-2}{x}\right) - 18 = 0$

SKILL MAINTENANCE

27. Multiply and simplify: $\sqrt{3x^2}\sqrt{3x^3}$.

28. Solution A is 18% alcohol and solution B is 45% alcohol. How much of each should be mixed together to get 12 L of a solution that is 36% alcohol?

29. Subtract: $\frac{x+1}{x-1} - \frac{x+1}{x^2+x+1}$.

30. Add: $\frac{x+1}{x-1} + \frac{x+1}{x^2+x+1}$.

SYNTHESIS

Solve. Check possible solutions by substituting into the original equation.

31. ▦ $6.75x - 35\sqrt{x} - 5.36 = 0$

32. ▦ $\pi x^4 - \pi^2 x^2 - \sqrt{99.3} = 0$

33. $\frac{x}{x-1} - 6\sqrt{\frac{x}{x-1}} - 40 = 0$

34. $\left(\sqrt{\frac{x}{x-3}}\right)^2 - 24 = 10\sqrt{\frac{x}{x-3}}$

35. $\sqrt{x-3} - \sqrt[4]{x-3} = 12$

36. $a^3 - 26a^{3/2} - 27 = 0$

37. $x^6 - 28x^3 + 27 = 0$

38. $x^6 + 7x^3 - 8 = 0$

8.5 Formulas and Applied Problems

a To solve a formula for a certain letter, we use the principles for solving equations to get that letter alone on one side. When square roots appear, we can usually eliminate the radical signs by squaring both sides.

► **EXAMPLE 1** *A pendulum formula.* Solve $T = 2\pi\sqrt{\frac{L}{g}}$ for L.

$$T = 2\pi\sqrt{\frac{L}{g}}$$

$$T^2 = \left(2\pi\sqrt{\frac{L}{g}}\right)^2 \quad \text{Using the principle of powers (squaring)}$$

$$T^2 = 2^2\pi^2\frac{L}{g}$$

$$gT^2 = 4\pi^2 L \quad \text{Clearing of fractions}$$

$$\frac{gT^2}{4\pi^2} = L \quad \text{Multiplying by } \frac{1}{4\pi^2}\text{, or dividing by } 4\pi^2$$

We now have L alone on one side, so the formula is solved for L. ◄

DO EXERCISE 1.

In most formulas, the letters represent nonnegative numbers, so we need not use absolute-value signs when taking square roots.

► **EXAMPLE 2** *A right triangle formula.* Solve $c^2 = a^2 + b^2$ for a.

$$c^2 - b^2 = a^2 \quad \text{Subtracting } b^2 \text{ to get } a^2 \text{ alone}$$

$$\sqrt{c^2 - b^2} = a \quad \text{Taking the square root}$$ ◄

► **EXAMPLE 3** *A motion formula.* Solve $s = gt + 16t^2$ for t.

This time we use the quadratic formula to get t alone on one side of the equation:

$$16t^2 + gt - s = 0 \quad \text{Writing the equation in standard form}$$

$$a = 16, \quad b = g, \quad c = -s$$

$$t = \frac{-g \pm \sqrt{g^2 - 4 \cdot 16 \cdot (-s)}}{2 \cdot 16}. \quad \text{Using the quadratic formula}$$

Since taking the negative square root would result in a negative answer, we take the positive one:

$$t = \frac{-g + \sqrt{g^2 + 64s}}{32}.$$ ◄

DO EXERCISES 2 AND 3.

The following list of steps should help you when solving formulas for a given letter. Try to remember that when solving a formula, you do the same things you would do to solve any equation. The same principles hold.

OBJECTIVES

After finishing Section 8.5, you should be able to:

a Solve a formula for a given letter.

b Solve applied problems involving quadratic equations.

FOR EXTRA HELP

Tape NC

Tape 14B

MAC: 8
IBM: 8

1. Solve $A = \sqrt{\frac{w_1}{w_2}}$ for w_2.

2. Solve $V = \pi r^2 h$ for r.

3. Solve $Ls^2 + Rs = -\frac{1}{C}$ for s.

ANSWERS ON PAGE A-8

4. Solve for b: $\dfrac{b}{\sqrt{a^2 - b^2}} = p.$

To solve a formula for a letter, say b:

1. **Clear of fractions and use the principle of powers, as needed, to eliminate radicals and until b does not appear in any denominator. (In some cases you may clear of fractions first, and in some cases you may use the principle of powers first. But do these until radicals are gone and b is not in any denominator.)**
2. **Collect all terms with b^2 in them. Also collect all terms with b in them.**
3. **If b^2 does not appear, you can finish by using just the addition and multiplication principles. Get all terms containing b on one side of the equation; then factor out b. Dividing on both sides will then get b alone.**
4. **If b^2 appears but b does not appear to the first power, solve the equation for b^2. Then take the square root on both sides.**
5. **If there are terms containing both b and b^2, put the equation in standard form and use the quadratic formula.**

▶ **EXAMPLE 4** Solve the following formula for a: $q = \dfrac{a}{\sqrt{a^2 + b^2}}$.

In this case, we could either clear of fractions first or use the principle of powers first. Let us clear of fractions. We then have

$$q\sqrt{a^2 + b^2} = a.$$

Now we square both sides and then continue:

$$(q\sqrt{a^2 + b^2})^2 = a^2 \quad \text{Squaring}$$

Don't forget to square both q and $\sqrt{a^2 + b^2}$.

$$q^2(a^2 + b^2) = a^2$$

$$q^2a^2 + q^2b^2 = a^2$$

$$q^2b^2 = a^2 - q^2a^2 \quad \text{Getting all } a^2\text{-terms together}$$

$$q^2b^2 = a^2(1 - q^2) \quad \text{Factoring out } a^2$$

$$\frac{q^2b^2}{1 - q^2} = a^2 \quad \text{Multiplying by } \frac{1}{1 - q^2}\text{, or dividing by } 1 - q^2$$

$$\sqrt{\frac{q^2b^2}{1 - q^2}} = a \quad \text{Taking the square root}$$

$$\frac{qb}{\sqrt{1 - q^2}} = a. \quad \text{Simplifying}$$

◀

DO EXERCISE 4.

b Applied Problems

Sometimes when we translate a problem to mathematical language, we get a quadratic equation. If so, we proceed as usual to solve the problem. That is, we solve the equation and then check possible answers in the original problem.

▶ **EXAMPLE 5** A rectangular garden is 60 ft by 80 ft. Part of the garden is torn up to install a sidewalk of uniform width around it. The area of the new garden is $\frac{1}{6}$ of the old area. How wide is the sidewalk?

ANSWER ON PAGE A-8

1. *Familiarize.* We first make a drawing and label it with the known information. We don't know how wide the sidewalk is, so we have called its width x.

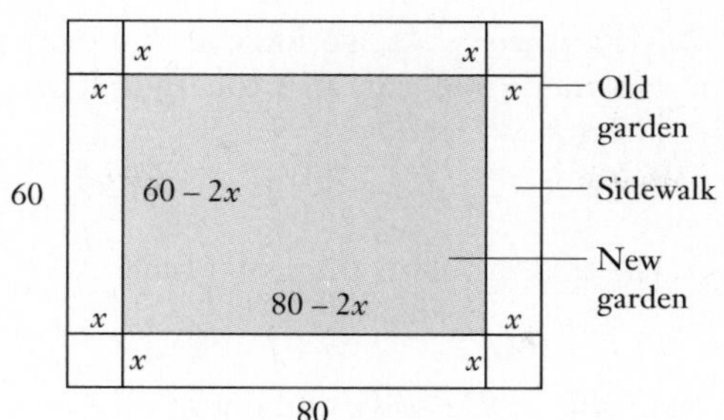

2. *Translate.* Remember, the area of a rectangle is $l \times w$ (length times width). Then:

   Area of old garden $= 60 \cdot 80$;

   Area of new garden $= (60 - 2x)(80 - 2x)$.

   Since the new garden is $\frac{1}{6}$ of the old, we have

$$(60 - 2x)(80 - 2x) = \tfrac{1}{6} \cdot 60 \cdot 80.$$

3. *Solve.* We solve the equation:

$$4800 - 120x - 160x + 4x^2 = 800 \quad \textbf{Multiplying binomials on the left}$$

$$4x^2 - 280x + 4000 = 0 \quad \textbf{Collecting like terms}$$

$$x^2 - 70x + 1000 = 0 \quad \textbf{Multiplying on both sides by } \tfrac{1}{4}\textbf{, or dividing by 4}$$

$$(x - 20)(x - 50) = 0 \quad \textbf{Factoring}$$

$$x = 20 \quad or \quad x = 50. \quad \textbf{Using the principle of zero products}$$

4. *Check.* We check in the original problem. We see that 50 is not a solution because when $x = 50$, $60 - 2x = -40$, and the width of the garden cannot be negative.

   If the sidewalk is 20 ft wide, then the garden itself will have length $80 - 2 \cdot 20$, or 40 ft. The width will be $60 - 2 \cdot 20$, or 20 ft. The new area is thus $20 \cdot 40$, or 800 ft^2. The old area was $60 \cdot 80$, or 4800 ft^2. The new area of 800 ft^2 is $\frac{1}{6}$ of this, so the number 20 checks.

5. *State.* The sidewalk is 20 ft wide. ◄

DO EXERCISE 5.

► **EXAMPLE 6** A wire is stretched from the ground to the top of a building, as shown below. The wire is 20 ft long. The height of the building is 4 ft greater than the distance d from the building. Find the distance d and the height of the building.

1. *Familiarize.* We first make a drawing and label it. We do not know d or the height of the building, but the height must be $d + 4$.

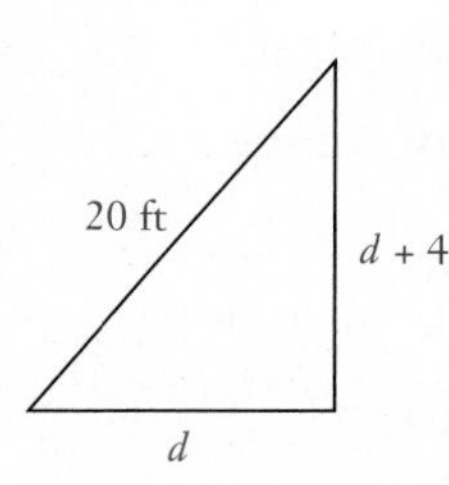

5. An open box is to be made from a 10-ft by 20-ft rectangular piece of cardboard by cutting a square from each corner. The area of the bottom of the box is to be 96 ft^2. What is the length of the sides of the squares that are cut from the corners?

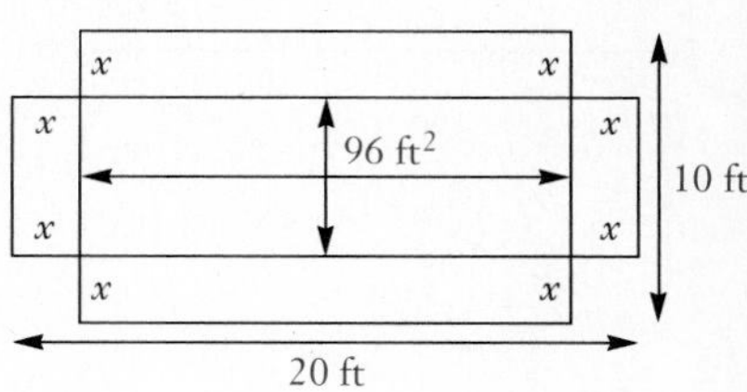

ANSWER ON PAGE A-8

6. Three towns A, B, and C are situated as shown. The roads at A form a right angle. The distance from A to B is 2 mi less than the distance from A to C. The distance from B to C is 10 mi. Find the distance from A to B and the distance from A to C.

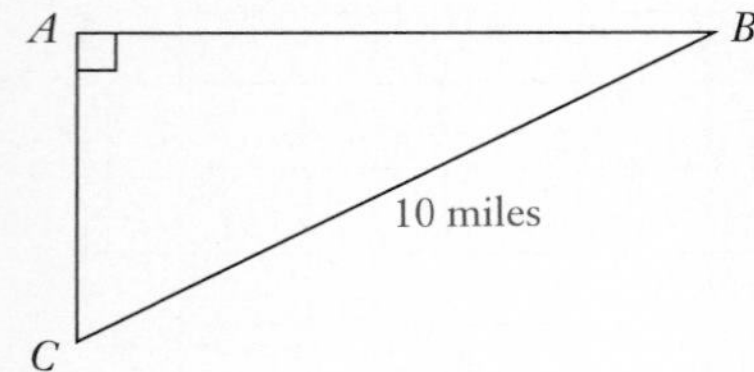

7. Three towns A, B, and C are situated as shown. The roads at A form a right angle. The distance from A to B is 2 mi less than the distance from A to C. The distance from B to C is 8 mi. Find the distance from A to B and the distance from A to C. Find exact and approximate answers to the nearest hundredth of a mile.

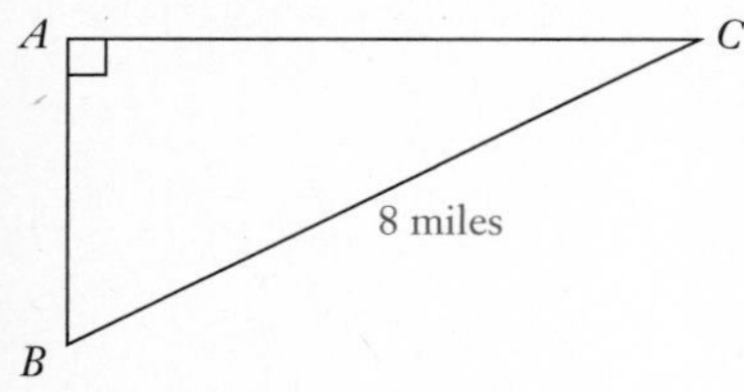

ANSWERS ON PAGE A-8

2. *Translate.* As we look at the drawing, we see that a right triangle is formed. We can use the Pythagorean equation, which we studied in Chapter 7:

$$c^2 = a^2 + b^2.$$

In this problem then, we have

$$20^2 = d^2 + (d + 4)^2.$$

3. *Solve.* We solve the equation:

$$\begin{aligned}
400 &= d^2 + d^2 + 8d + 16 && \textbf{Squaring}\\
2d^2 + 8d - 384 &= 0 && \textbf{Finding standard form}\\
d^2 + 4d - 192 &= 0 && \textbf{Multiplying by } \tfrac{1}{2}\textbf{, or dividing by 2}\\
(d + 16)(d - 12) &= 0 && \textbf{Factoring}\\
d + 16 = 0 \quad &or \quad d - 12 = 0 && \textbf{Using the principle of zero products}\\
d = -16 \quad &or \quad d = 12.
\end{aligned}$$

4. *Check.* We know that -16 is not an answer because distances are not negative. The number 12 checks (we leave the check to you).

5. *State.* The answer is that $d = 12$ ft, and the height of the building is 16 ft. ◀

DO EXERCISE 6.

► **EXAMPLE 7** Suppose that the wire in Example 6 has length 10 ft. Find the distance d and the height of the building.

Using the same reasoning that we did in Example 6, we translate the problem to the equation

$$10^2 = d^2 + (d + 4)^2.$$

We solve as follows. Note that the quadratic equation we get is not easily factored, so we use the quadratic formula:

$$\begin{aligned}
100 &= d^2 + d^2 + 8d + 16 && \textbf{Squaring}\\
2d^2 + 8d - 84 &= 0 && \textbf{Finding standard form}\\
d^2 + 4d - 42 &= 0 && \textbf{Multiplying by } \tfrac{1}{2}\textbf{, or dividing by 2}
\end{aligned}$$

$$d = \frac{-b \pm \sqrt{b^2 - 4ac}}{2a} = \frac{-4 \pm \sqrt{4^2 - 4(1)(-42)}}{2(1)}$$

$$= \frac{-4 \pm \sqrt{16 + 168}}{2} = \frac{-4 \pm \sqrt{184}}{2} = \frac{-4 \pm \sqrt{4(46)}}{2}$$

$$= \frac{-4 \pm 2\sqrt{46}}{2} = -2 \pm \sqrt{46}.$$

Since $\sqrt{46} > 2$, the use of $-\sqrt{46}$ would give a negative distance d. It follows that d is given by $d = -2 + \sqrt{46}$ and the height of the building is $d + 4$, or $2 + \sqrt{46}$. Using a calculator, we find that $d = -2 + 6.782 \approx 4.782$ ft, and the height $= 2 + \sqrt{46} \approx 2 + 6.782 = 8.782$ ft. ◀

DO EXERCISE 7.

NAME SECTION DATE

EXERCISE SET 8.5

ANSWERS

a Solve the formula for the given letter. Assume that all variables represent non-negative numbers.

1. $A = 6s^2$, for s
(Area of a cube)

2. $A = 4\pi r^2$, for r
(Area of a sphere)

3. $F = \dfrac{Gm_1 m_2}{r^2}$, for r
(Law of gravity)

4. $N = \dfrac{kQ_1 Q_2}{s^2}$, for s
(Number of phone calls between two cities)

5. $E = mc^2$, for c
(Energy–mass relationship)

6. $A = \pi r^2$, for r
(Area of a circle)

7. $a^2 + b^2 = c^2$, for b
(Pythagorean formula in two dimensions)

8. $a^2 + b^2 + c^2 = d^2$, for c
(Pythagorean formula in three dimensions)

9. $N = \dfrac{k^2 - 3k}{2}$, for k
(Number of diagonals of a polygon)

10. $s = v_0 t + \dfrac{gt^2}{2}$, for t
(A motion formula)

1. ________

2. ________

3. ________

4. ________

5. ________

6. ________

7. ________

8. ________

9. ________

10. ________

ANSWERS

11. ______

12. ______

13. ______

14. ______

15. ______

16. ______

17. ______

18. ______

19. ______

20. ______

11. $A = 2\pi r^2 + 2\pi rh$, for r
(Area of a cylinder)

12. $A = \pi r^2 + \pi rs$, for r
(Area of a cone)

13. $T = 2\pi\sqrt{\dfrac{L}{g}}$, for g
(A pendulum formula)

14. $W = \sqrt{\dfrac{1}{LC}}$, for L
(An electricity formula)

15. $P_1 - P_2 = \dfrac{32LV}{gD^2}$, for D

16. $N + p = \dfrac{6.2A^2}{pR^2}$, for R

17. $m = \dfrac{m_0}{\sqrt{1 - \dfrac{v^2}{c^2}}}$, for v
(A relativity formula)

18. Solve the formula given in Exercise 17 for c.

b Solve.

19. The width of a rectangle is 4 ft less than the length. The area is 12 ft^2. Find the length and the width.

20. The width of a rectangle is 5 m less than the length. The area is 24 m^2. Find the length and the width.

21. The length of a rectangle is twice the width. The area is 288 yd^2. Find the length and the width.

22. The length of a rectangle is twice the width. The area is 338 cm^2. Find the length and the width.

23. The outside of a picture frame measures 14 in. by 20 in.; 160 in^2 of picture shows. Find the width of the frame.

24. The outside of a picture frame measures 12 cm by 20 cm; 84 cm^2 of picture shows. Find the width of the frame.

25. The hypotenuse of a right triangle is 26 ft long. The length of one leg is 14 ft more than the other. Find the lengths of the legs.

26. The hypotenuse of a right triangle is 25 m long. The length of one leg is 17 m less than the other. Find the lengths of the legs.

27. The hypotenuse of a right triangle is 5 ft long. One leg is 1 ft shorter than the other. Find the lengths of the legs.

28. The hypotenuse of a right triangle is 13 m long. One leg is 7 m longer than the other. Find the lengths of the legs.

29. A student opens a mathematics book to two facing pages. The product of the page numbers is 2756. Find the page numbers.

30. A student opens a mathematics book to two facing pages. The product of the page numbers is 1806. Find the page numbers.

ANSWERS

21. ______

22. ______

23. ______

24. ______

25. ______

26. ______

27. ______

28. ______

29. ______

30. ______

ANSWERS

31. ______

32. ______

33. ______

34. ______

35. ______

36. ______

37. ______

38. ______

39. ______

40. ______

41. ______

Solve. Find exact and approximate answers rounded to the nearest tenth.

31. The width of a rectangle is 4 ft less than the length. The area is 10 ft^2. Find the length and the width.

32. The width of a rectangle is 5 m less than the length. The area is 20 m^2. Find the length and the width.

33. The length of a rectangle is twice the width. The area is 256 yd^2. Find the length and the width.

34. The length of a rectangle is twice the width. The area is 328 cm^2. Find the length and the width.

35. The outside of a picture frame measures 14 in. by 20 in.; 100 in^2 of picture shows. Find the width of the frame.

36. The outside of a picture frame measures 12 cm by 20 cm; 80 cm^2 of picture shows. Find the width of the frame.

37. The hypotenuse of a right triangle is 24 ft long. The length of one leg is 14 ft more than the other. Find the lengths of the legs.

38. The hypotenuse of a right triangle is 22 m long. The length of one leg is 10 m less than the other. Find the lengths of the legs.

SKILL MAINTENANCE

39. Express in terms of i: $\sqrt{-20}$.

40. Subtract:

$$\frac{x}{x^2 + 17x + 72} - \frac{8}{x^2 + 15x + 56}.$$

41. Solve: $\sqrt{x^2} = -20$.

8.6 Rational Equations and Applied Problems

a Rational Equations That Become Quadratic

In a rational equation, when we clear of fractions, we sometimes get a quadratic equation.

► **EXAMPLE 1** Solve: $\dfrac{14}{x+2} - \dfrac{1}{x-4} = 1.$

We multiply by the LCM of the denominators: $(x + 2)(x - 4)$.

$$(x+2)(x-4) \cdot \left[\frac{14}{x+2} - \frac{1}{x-4}\right] = (x+2)(x-4) \cdot 1$$

$$(x+2)(x-4) \cdot \frac{14}{x+2} - (x+2)(x-4) \cdot \frac{1}{x-4} = (x+2)(x-4) \cdot 1$$ Using a distributive law

$$14(x-4) - (x+2) = (x+2)(x-4)$$ Simplifying by removing factors of 1

$$14x - 56 - x - 2 = x^2 - 2x - 8$$ Multiplying

$$13x - 58 = x^2 - 2x - 8$$ Collecting like terms

$$0 = x^2 - 15x + 50$$ Writing standard form

$$0 = (x-5)(x-10)$$ Factoring

$x = 5$ *or* $x = 10$ Using the principle of zero products

Since we started with a fractional equation, we might have introduced some numbers that are not solutions of the original equation when we cleared of fractions. This time, we must check.

Check: For 5:

$$\frac{14}{x+2} - \frac{1}{x-4} = 1$$

$\dfrac{14}{5+2} - \dfrac{1}{5-4}$	1
$\dfrac{14}{7} - 1$	
$2 - 1$	
1	

For 10:

$$\frac{14}{x+2} - \frac{1}{x-4} = 1$$

$\dfrac{14}{10+2} - \dfrac{1}{10-4}$	1
$\dfrac{14}{12} - \dfrac{1}{6}$	
$\dfrac{14}{12} - \dfrac{2}{12}$	
1	

The solutions are 5 and 10. ◄

OBJECTIVES

After finishing Section 8.6, you should be able to:

a Solve rational equations that become quadratic after clearing of fractions.

b Solve problems that translate to this type of equation.

FOR EXTRA HELP

Tape 16A

Tape 14B

MAC: 8
IBM: 8

1. Solve: $\dfrac{24}{r-2} + \dfrac{24}{r+2} = 5.$

2. It takes Red 9 hr longer to build a wall than it takes Mort. If they work together, they can build the wall in 20 hr. Working alone, how long would it take Mort to build the wall?

ANSWERS ON PAGE A-8

If for a rational equation, clearing of fractions produces a quadratic equation, we solve that resulting equation as we would solve any quadratic equation. However, it is necessary to check possible solutions by substituting into the original equation. Assuming you performed the operations correctly, it is sufficient to see if possible solutions produce zero denominators.

DO EXERCISE 1.

b Applied Problems

Some problems translate to equations like that in Example 1.

▶ **EXAMPLE 2** Fran and Sally work together and mow a lawn in 4 hr. It would take Sally 6 hr more than Fran to do the job alone. How long would each need to do the job working alone?

1. *Familiarize.* Let $x =$ the time it takes Fran to work alone. Then $x + 6 =$ the amount of time it takes Sally to do the work alone.
2. *Translate.* Recalling the work formula from Section 6.4, we see that they can do $(4/x) + [4/(x + 6)]$ of the job working together. Thus we get the equation

$$\frac{4}{x} + \frac{4}{x+6} = 1.$$

3. *Solve.*

$$x(x+6)\cdot\left[\frac{4}{x} + \frac{4}{x+6}\right] = x(x+6)\cdot 1 \qquad \textbf{Multiplying by the LCM}$$

$$x(x+6)\cdot\frac{4}{x} + x(x+6)\cdot\frac{4}{x+6} = x(x+6)\cdot 1$$

$$4(x+6) + 4x = x(x+6) \qquad \textbf{Simplifying}$$

$$4x + 24 + 4x = x^2 + 6x$$

$$8x + 24 = x^2 + 6x$$

$$0 = x^2 - 2x - 24 \qquad \textbf{Standard form}$$

$$0 = (x-6)(x+4) \qquad \textbf{Factoring}$$

$$x = 6 \quad or \quad x = -4 \qquad \textbf{Using the principle of zero products}$$

4. *Check.* Since negative time has no meaning in this problem, -4 is not a solution. Let's see if 6 checks. This is the time for Fran to do the job alone. Then Sally would take 12 hr alone. Thus Fran would do $\frac{4}{6}$ of the job in 4 hr and Sally would do $\frac{4}{12}$ of it. Thus in 4 hr they would do $\frac{4}{6} + \frac{4}{12}$, or $\frac{8}{12} + \frac{4}{12}$ of the job. This is all of it, so the numbers check.
5. *State.* It would take Fran 6 hr alone and Sally 12 hr alone. ◀

DO EXERCISE 2.

▶ **EXAMPLE 3** A salesperson leaves Homeville and travels the 600 mi to Fartown at a certain speed. The return trip is made at a speed that is 10 mph slower. Total time for the round trip was 22 hr. How fast did the salesperson travel on each part of the trip?

1. *Familiarize.* We first make a drawing and label it with the known information.

Homeville 600 miles Fartown
r mph t_1 hours

600 miles
$r - 10$ mph t_2 hours
22 hours for round trip

We let $r =$ the speed on the first part of the trip. Then $r - 10 =$ the speed on the return trip. We can organize these facts in a table.

	Distance	Speed	Time
Trip out	600	r	t_1
Return trip	600	$r - 10$	t_2

2. *Translate.* Since we do not know either time, but only the total, we have called the times t_1 and t_2. Since the total was 22 hr, we can write one equation at once:

$$t_1 + t_2 = 22.$$

Recalling the motion formula $d = rt$ and solving for t, we get $t = d/r$. Now we go back to the table. For each line, we write $t = d/r$:

$$t_1 = \frac{600}{r} \quad \text{and} \quad t_2 = \frac{600}{r - 10}.$$

Substituting into our first equation, we get

$$\frac{600}{r} + \frac{600}{r - 10} = 22.$$

3. *Solve.* The LCM is $r(r - 10)$.

$$r(r - 10) \cdot \left[\frac{600}{r} + \frac{600}{r - 10}\right] = r(r - 10) \cdot 22 \quad \textbf{Multiplying by the LCM}$$

$$r(r - 10) \cdot \frac{600}{r} + r(r - 10) \cdot \frac{600}{r - 10} = r(r - 10) \cdot 22$$

$$600(r - 10) + 600r = 22r^2 - 220r$$

$$600r - 6000 + 600r = 22r^2 - 220r$$

$$300r - 3000 + 300r = 11r^2 - 110r \quad \textbf{Multiplying by } \tfrac{1}{2}\textbf{, or dividing by 2}$$

$$0 = 11r^2 - 710r + 3000 \quad \textbf{Standard form}$$

$$0 = (11r - 50)(r - 60) \quad \textbf{Factoring}$$

$$r = 60 \quad or \quad r = \tfrac{50}{11}, \text{ or } 4\tfrac{6}{11} \quad \textbf{Using the principle of zero products}$$

4. *Check.* $4\frac{6}{11}$ is not a solution because it would make the return speed negative, but 60 checks.

5. *State.* The speed out was 60 mph, and the speed back was 50 mph. ◀

DO EXERCISE 3.

3. A stream flows at 2 mph. A boat travels 24 mi upstream and returns in a total time of 5 hr. What is the speed of the boat in still water? (*Hint:* The speed upstream $= r - 2$ and the speed downstream $= r + 2$, where r is the speed of the boat in still water.)

Complete this table to help with the familiarization.

	Distance	Speed	Time
Up-stream		$r - 2$	
Down-stream		$r + 2$	

ANSWER ON PAGE A-8

4. Two ships make the same voyage, a distance of 3000 nautical miles. The faster ship travels 10 knots faster than the slower one (a *knot* is 1 nautical mile per hour). The faster ship makes the voyage in 50 hr less time than the slower one. Find the speeds of the two ships.

Complete this table to help with the familiarization.

	Distance	Speed	Time
Faster ship	3000		$t - 50$
Slower ship	3000		t

ANSWER ON PAGE A-8

▶ **EXAMPLE 4** A car travels 300 mi at a certain speed. If the speed had been 10 mph faster, the trip would have been made in 1 hr less time. Find the speed.

1. *Familiarize.* We make a drawing, labeling it with the known information. We can also organize the information in a table.

300 miles

r mph t hours

300 miles

$r + 10$ mph $t - 1$ hours

Distance	Speed	Time
300	r	t
300	$r + 10$	$t - 1$

Recalling the motion formula $d = rt$ and solving for r, we get $r = d/t$. From the first line of the table, we obtain

$$r = \frac{300}{t}.$$

From the second line, we get

$$r + 10 = \frac{300}{t-1}.$$

2. *Translate.* We substitute for r from the first equation into the second and get a translation:

$$\frac{300}{t} + 10 = \frac{300}{t-1}.$$

3. *Solve.*

$$t(t-1)\left[\frac{300}{t} + 10\right] = t(t-1) \cdot \frac{300}{t-1} \quad \text{Multiplying by the LCM}$$

$$t(t-1) \cdot \frac{300}{t} + t(t-1) \cdot 10 = t(t-1) \cdot \frac{300}{t-1}$$

$$300(t-1) + 10(t^2 - t) = 300t$$

$$10t^2 - 10t - 300 = 0 \quad \text{Standard form}$$

$$t^2 - t - 30 = 0 \quad \text{Multiplying by } \tfrac{1}{10}\text{, or dividing by 10}$$

$$(t-6)(t+5) = 0 \quad \text{Factoring}$$

$$t = 6 \quad or \quad t = -5 \quad \text{Using the principle of zero products}$$

4. *Check.* Since negative time has no meaning in this problem, we try 6 hr. Remembering that $r = d/t$, we get $r = 300/6 = 50$ mph.

To check, we take the speed 10 mph faster, which is 60 mph, and see how long the trip would have taken at that speed:

$$t = \frac{d}{r} = \frac{300}{60} = 5 \text{ hr}.$$

This is 1 hr less than the trip actually took, so we have an answer to the problem.

5. *State.* The speed was 50 mph. ◀

DO EXERCISE 4.

NAME SECTION DATE

EXERCISE SET 8.6

a Solve.

1. $\dfrac{1}{x} = \dfrac{x-2}{24}$

2. $\dfrac{x+3}{14} = \dfrac{2}{x}$

3. $\dfrac{1}{2x-1} - \dfrac{1}{2x+1} = \dfrac{1}{4}$

4. $\dfrac{1}{4-x} - \dfrac{1}{2+x} = \dfrac{1}{4}$

5. $\dfrac{50}{x} - \dfrac{50}{x-5} = -\dfrac{1}{2}$

6. $3x + \dfrac{10}{x-1} = \dfrac{16}{x-1}$

7. $\dfrac{x+2}{x} = \dfrac{x-1}{2}$

8. $\dfrac{x}{3} - \dfrac{6}{x} = 1$

9. $x - 6 = \dfrac{1}{x+6}$

10. $x + 7 = \dfrac{1}{x-7}$

ANSWERS

1. ______

2. ______

3. ______

4. ______

5. ______

6. ______

7. ______

8. ______

9. ______

10. ______

ANSWERS

11. ______

12. ______

13. ______

14. ______

15. ______

16. ______

17. ______

18. ______

19. ______

20. ______

11. $\frac{2}{x} = \frac{x+3}{5}$

12. $\frac{x+3}{x} = \frac{x-4}{3}$

13. $\frac{40}{x} - \frac{20}{x-3} = \frac{8}{7}$

14. $\frac{6}{y^2-4} + \frac{3}{y(2-y)} = \frac{1}{y}$

15. $\frac{x+1}{3x+2} = \frac{2x-3}{3x-2} - 1 - \frac{36}{4-9x^2}$

16. $\frac{x^2-5x}{x+3} = x - 3 + \frac{1}{x}$

17. $\frac{x-2}{x+2} + \frac{x+2}{x-2} = \frac{10x-8}{4-x^2}$

18. $1 = \frac{2}{x+1} - \frac{4x}{(x+1)^2}$

19. $\frac{13}{7x-5} + \frac{11x}{3} = \frac{39}{21x-15}$

20. $\frac{4}{3x+7} - \frac{10x}{9} = \frac{36}{27x+63}$

21. $\frac{12}{x^2 - 9} = 1 + \frac{3}{x - 3}$

22. $\frac{x^2}{x - 2} - \frac{x + 4}{2} + \frac{2 - 4x}{x - 2} + 1 = 0$

b Solve.

23. During the first part of a trip, a canoeist travels 80 mi at a certain speed. The canoeist travels 35 mi on the second part of the trip at a speed 5 mph slower. The total time for the trip is 3 hr. What was the speed on each part of the trip?

24. During the first part of a trip, a car travels 120 mi at a certain speed. It travels 100 mi on the second part of the trip at a speed 10 mph slower. The total time for the trip is 4 hr. What was the speed on each part of the trip?

25. A car travels 280 mi at a certain speed. If the speed had been 5 mph faster, the trip would have been made in 1 hr less time. Find the speed.

26. A car travels 200 mi at a certain speed. If the speed had been 10 mph faster, the trip would have been made in 1 hr less time. Find the speed.

27. Airplane A travels 2800 km at a certain speed. Plane B travels 2000 km at a speed 50 km/h faster than plane A in 3 hr less time. Find the speed of each plane.

28. Airplane A travels 600 mi at a certain speed. Plane B travels 1000 mi at a speed 50 mph faster, but takes 1 hr longer. Find the speed of each plane.

29. Two pipes are connected to the same tank. When working together, they can fill the tank in 2 hr. The larger pipe, working alone, can fill the tank in 3 hr less time than the smaller one. How long would the smaller one take, working alone, to fill the tank?

30. Two hoses are connected to a swimming pool. When working together, they can fill the pool in 4 hr. The larger hose, working alone, can fill the pool in 6 hr less time than the smaller one. How long would the smaller one take, working alone, to fill the pool?

ANSWERS

21. ______

22. ______

23. ______

24. ______

25. ______

26. ______

27. ______

28. ______

29. ______

30. ______

ANSWERS

31. ______

32. ______

33. ______

34. ______

35. ______

36. ______

37. ______

38. ______

39. ______

40. ______

41. ______

31. A boat travels 1 km upstream and 1 km back. The time for the roundtrip is 1 hour. The speed of the stream is 2 km/h. What is the speed of the boat in still water?

32. A boat travels 10 mi upstream and 10 mi back. The time required for the roundtrip is 3 hr. The speed of the stream is 5 mph. What is the speed of the boat in still water?

33. During the first part of a trip, a canoeist travels 80 mi at a certain speed. The canoeist travels 25 mi on the second part of the trip at a speed 5 mph slower. The total time for the trip is 3 hr. What was the speed on each part of the trip?

34. Two pipes are connected to the same tank. When working together, they can fill the tank in 2 hr. The larger pipe, working alone, can fill the tank in 2 hr less time than the smaller one. How long would the smaller one take, working alone, to fill the tank?

SKILL MAINTENANCE

35. Solve: $\sqrt{3x+1} = \sqrt{2x-1} + 1$.

36. Add: $\dfrac{1}{x-1} + \dfrac{1}{x^2 - 3x + 2}$.

37. Multiply and simplify $\sqrt[3]{18y^3}\,\sqrt[3]{4x^2}$. Assume that all expressions under radicals represent nonnegative numbers.

SYNTHESIS

38. ▦ At a factory, smokestack A pollutes the air 2.13 times as fast as smokestack B. Together they yield a certain amount of pollution in 16.3 hr. How long would it take each to pollute this much alone?

39. ▦ At Pizza Perfect, Ron can make 100 large pizza crusts in 1.2 hr less time than Chester. Together they can do the job in 1.8 hr. How long does it take each to do the job alone?

40. Solve: $\dfrac{4}{2x+i} - \dfrac{1}{x-i} = \dfrac{2}{x+i}$.

41. Find a when the reciprocal of $a - 1$ is $a + 1$.

8.7 Graphs of Quadratic Equations of the Type $y = a(x - h)^2 + k$

The bar graph shown below shows the rise and fall of the rate of unemployment during a recent year. A curve drawn along the graph would approximate the graph of a **quadratic equation.**

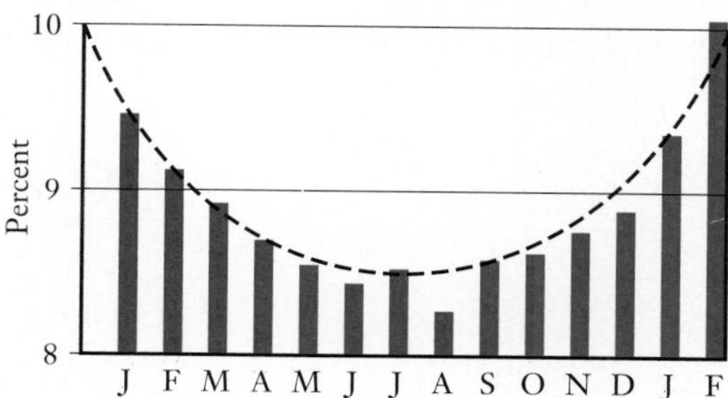

In this section you will learn to graph quadratic equations of the form

$$y = a(x - h)^2 + k, \quad a \neq 0.$$

The polynomial on the right side of the equation is of second degree, or **quadratic.**

a Graphing Quadratic Equations of the Type $y = ax^2$

Graphs of quadratic equations, $y = ax^2 + bx + c$ (where $a \neq 0$), are always cup-shaped. They all have a **line of symmetry** like the dashed lines shown in the figures below. If you fold on this line, the two halves will match exactly. The curve goes on forever. The top or bottom point at which the curve changes is called the **vertex.** The second coordinate of the vertex is either the largest value of y or the smallest value of y. The vertex is also thought of as a turning point.

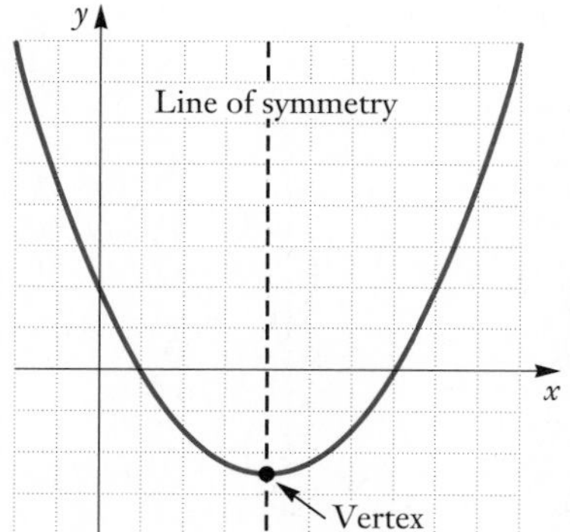

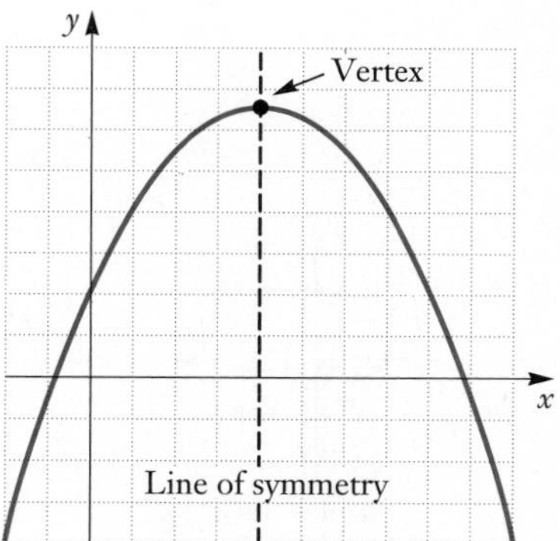

These curves are called **parabolas.** Some parabolas are thin and others are wide, but they all have the same general shape. We will be studying graphs of parabolas in this section, as well as Sections 8.8 and 9.1.

To graph a quadratic equation, we begin by choosing some numbers for x and computing the corresponding values of y.

OBJECTIVES

After finishing Section 8.7, you should be able to:

a Graph equations of the type $y = ax^2$, finding the vertex and the line of symmetry.

b Graph equations of the type $y = a(x - h)^2$, finding the vertex and the line of symmetry.

c Graph equations of the type $y = a(x - h)^2 + k$, finding the vertex, the line of symmetry, and the maximum or minimum y-value.

FOR EXTRA HELP

Tape 16B

Tape 15A

MAC: 8
IBM: 8

1. Graph: $y = \frac{1}{2}x^2$.

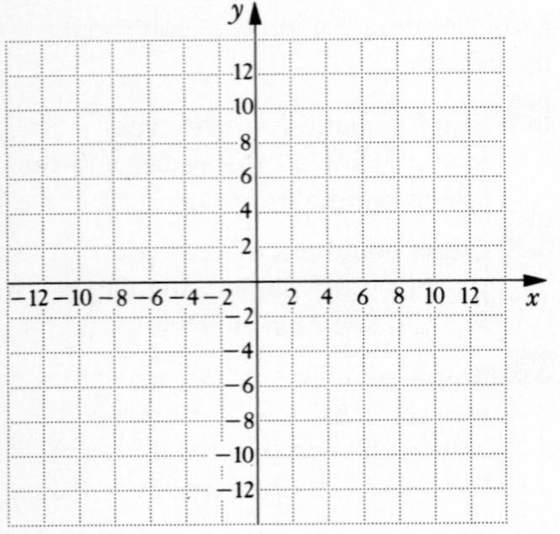

ANSWER ON PAGE A-9

► **EXAMPLE 1** Graph: $y = x^2$.

We choose numbers for x and find the corresponding values for y. Then we plot the ordered pairs (x, y) resulting from the computations and connect them with a smooth curve.

For $x = 3$, $y = x^2 = 3^2 = 9$.
For $x = 2$, $y = x^2 = 2^2 = 4$.
For $x = 1$, $y = x^2 = 1^2 = 1$.
For $x = 0$, $y = x^2 = 0^2 = 0$.
For $x = -1$, $y = x^2 = (-1)^2 = 1$.
For $x = -2$, $y = x^2 = (-2)^2 = 4$.
For $x = -3$, $y = x^2 = (-3)^2 = 9$.

x	y
3	9
2	4
1	1
0	0
−1	1
−2	4
−3	9

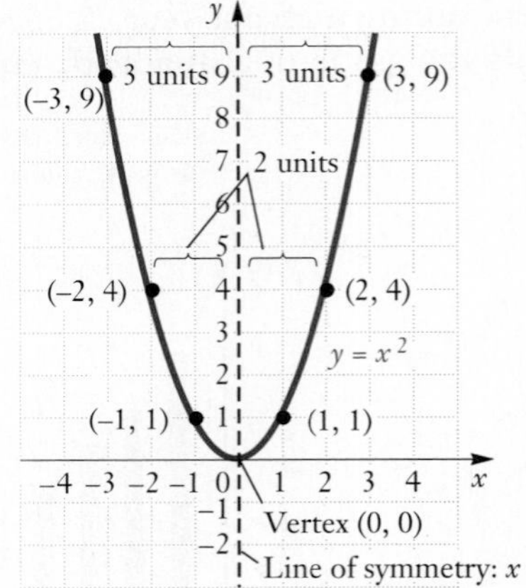

◄

In Example 1, the vertex is the point (0, 0). The second coordinate of the vertex, 0, is the smallest y-value. The y-axis is the axis of symmetry. Note that the symmetry would save us some computations; the points $(-1, 1)$, $(-2, 4)$, and $(-3, 9)$ can be found from the mirror images of (1, 1), (2, 4), and (3, 9) across the line $x = 0$. Parabolas whose equations are $y = ax^2$ always have the origin (0, 0) as the vertex and the y-axis as the line of symmetry.

Some common errors in drawing graphs of quadratic equations are shown in the following figure.

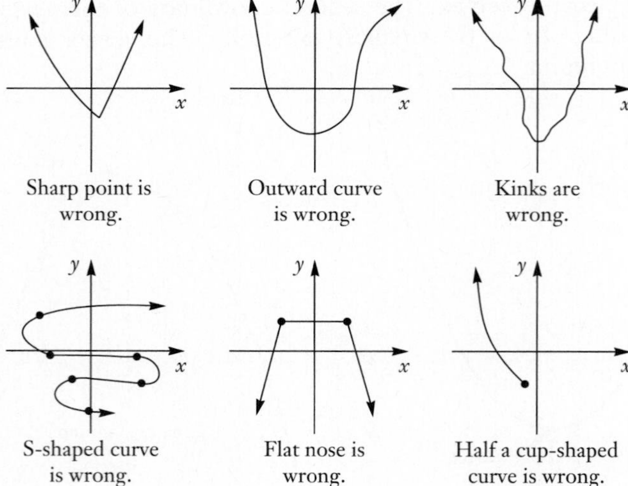

DO EXERCISE 1.

From Margin Exercise 1, you can see that the graph of $y = \frac{1}{2}x^2$ is a wider parabola than the graph of $y = x^2$. The graph of $y = 2x^2$ is thinner, as shown here, but the vertex and the line of symmetry have not changed.

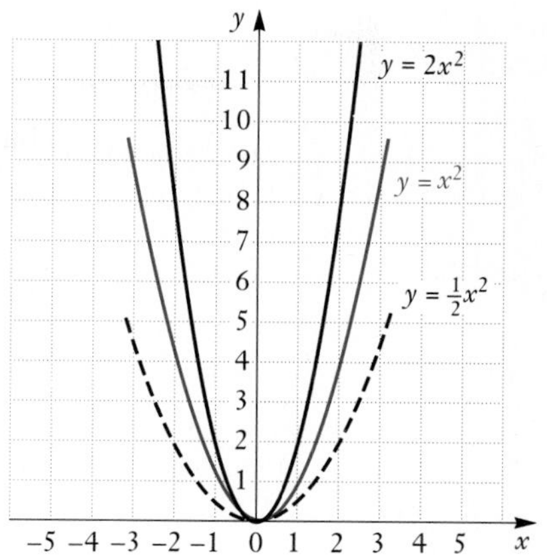

The graph of $y = ax^2$ is a parabola with the vertical axis (the y-axis, or line $x = 0$) as its line of symmetry and the origin as its vertex.

If $|a| > 1$, then the parabola is thinner than $y = x^2$.

If $0 < |a| < 1$, then the parabola is wider than $y = x^2$.

If $a > 0$, then the parabola opens upward; if $a < 0$, the parabola opens downward.

DO EXERCISES 2 AND 3.

b Graphing Quadratic Equations of the Type $y = a(x - h)^2$

► **EXAMPLE 2** Graph: $y = (x - 3)^2$.

We choose some values of x and compute y using $y = (x - 3)^2$.

x	y	
3	0	← Vertex
2	1	
1	4	
0	9	← y-intercept
−1	16	
4	1	
5	4	

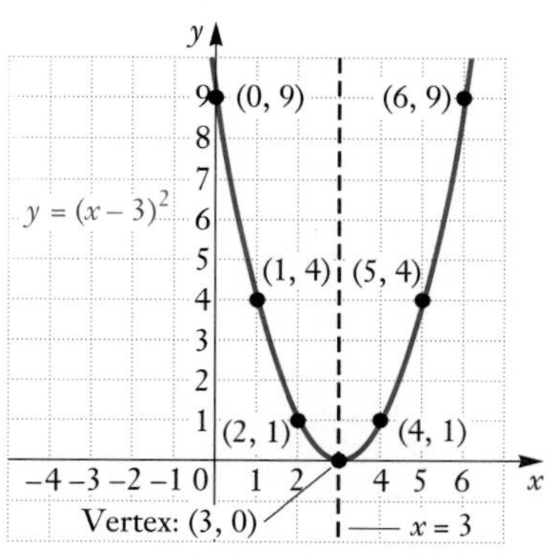

We note that $y = 16$ when $x = -1$ and y-values get larger and larger as x gets more and more negative. Thus we go back to positive values to fill out the table. Note that the line $x = 3$ is now the line of symmetry and the point (3, 0) is the vertex. If we had known that $x = 3$ is the line of symmetry at the outset, then we could have computed some values on one side, such as (4, 1), (5, 4), and (6, 9), and then used these and the symmetry to get their mirror images (2, 1), (1, 4), and (0, 9) without further computation. ◄

Graph.

2. $y = 3x^2$

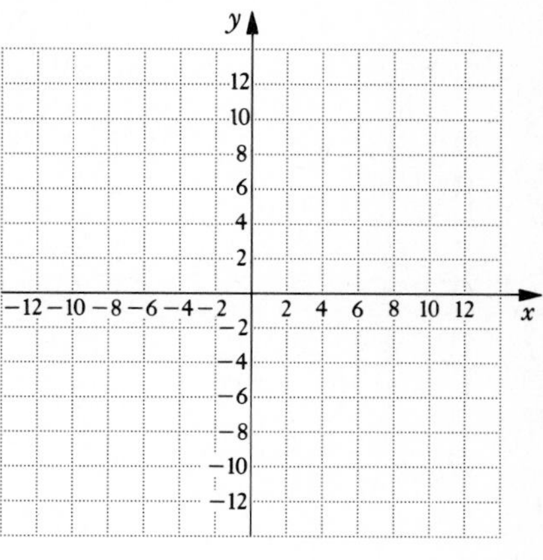

3. $y = -2x^2$

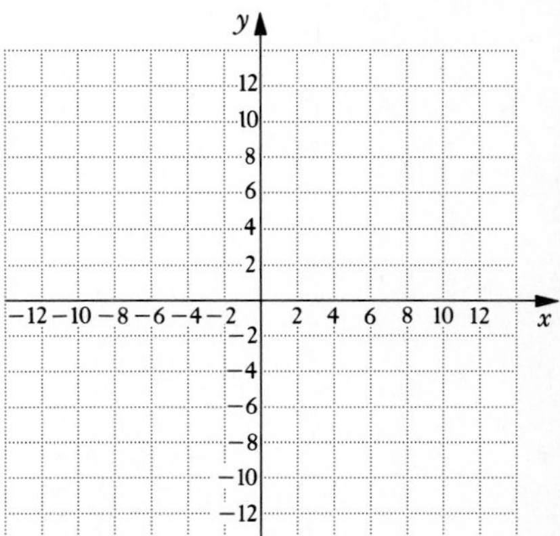

ANSWERS ON PAGE A-9

Graph the equation. Find and label the vertex and the line of symmetry.

4. $y = \frac{1}{2}(x - 4)^2$

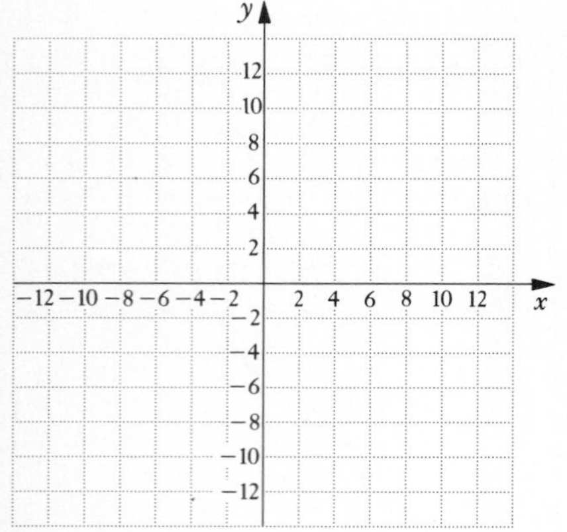

5. $y = -\frac{1}{2}(x - 4)^2$

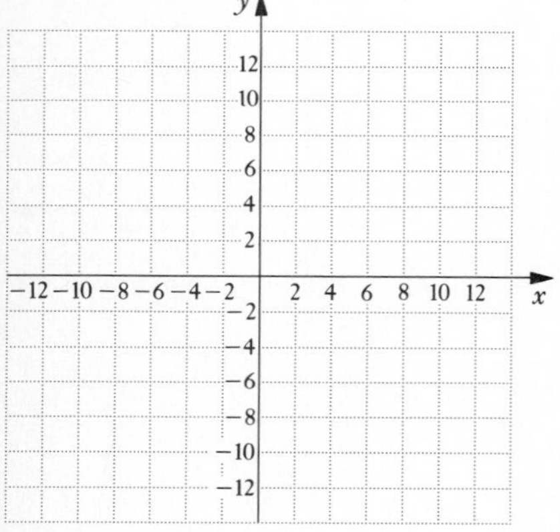

ANSWERS ON PAGE A-9

The graph of $y = (x - 3)^2$ looks just like the graph of $y = x^2$, except that it is moved, or translated, three units to the right.

> **The graph of $y = a(x - h)^2$ looks just like the graph of $y = ax^2$, except that it is translated to the left or right. If $h > 0$, then the parabola is translated to the right. If $h < 0$, then the parabola is translated to the left. The vertex is $(h, 0)$ and the line of symmetry is $x = h$.**

▶ **EXAMPLE 3** Graph: $y = -2(x + 3)^2$.

We can express the equation in the equivalent form $y = -2[x - (-3)]^2$. Then we know that the graph looks like that of $y = 2x^2$ translated three units to the left, and it will also open downward since $-2 < 0$. The vertex is $(-3, 0)$. The line of symmetry is $x = -3$. Plotting points as needed, we obtain the graph shown here.

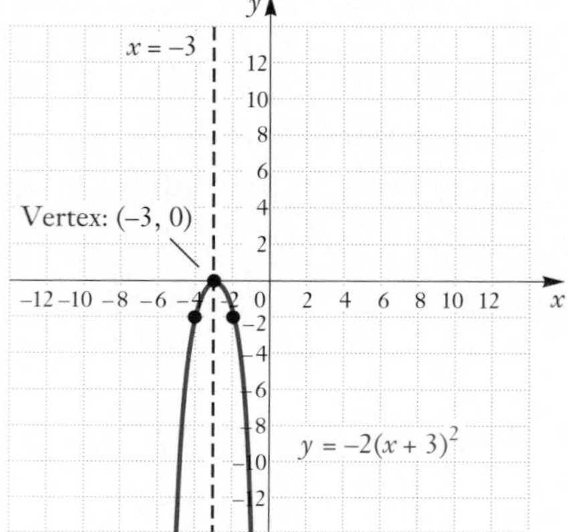

◀

DO EXERCISES 4 AND 5.

C Graphing Quadratic Equations of the Type $y = a(x - h)^2 + k$

What happens to the graph of $y = a(x - h)^2$ if we add a constant k? Suppose $k > 0$. This increases each y-value k units, so the graph is moved, or translated, k units up. Suppose $k < 0$. This decreases each y-value $|k|$ units, so the graph is moved, or translated, $|k|$ units down. The vertex will be (h, k) and the line of symmetry will still be $x = h$.

Note that if a parabola opens up ($a > 0$), the y-value at the vertex is the smallest, or **minimum** value. That is, it is less than the y-value at any other point. If the parabola opens down ($a < 0$), the y-value at the vertex will be the largest, or **maximum** value.

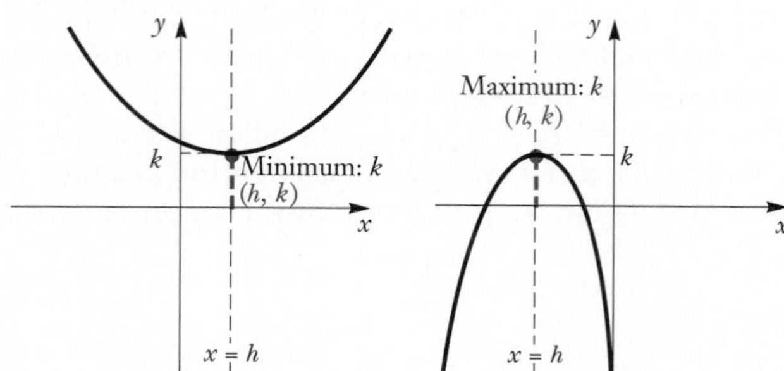

The graph of $y = a(x - h)^2 + k$ looks just like the graph of $y = a(x - h)^2$, except that it is translated up or down. If $k > 0$, the parabola is translated k units up. If $k < 0$, the parabola is translated k units down. The vertex is (h, k), and the line of symmetry is $x = h$. If $a > 0$, then k is the minimum y-value. If $a < 0$, then k is the maximum y-value.

▶ **EXAMPLE 4** Graph $y = 2(x + 3)^2 - 5$. Find the vertex, the line of symmetry, and the maximum or minimum value.

We first express the equation in the equivalent form

$$y = 2[x - (-3)]^2 - 5.$$

The graph looks like that of $y = 2[x - (-3)]^2$ translated 5 units down. The vertex is now $(-3, -5)$, and the line of symmetry is $x = -3$. Since $2 > 0$, we know that -5 is the minimum y-value.

If we wish, we can compute points as needed to complete the graph shown here.

x	y	
-3	-5	← Vertex
0	13	← y-intercept
-1	3	
1	27	
-2	-3	

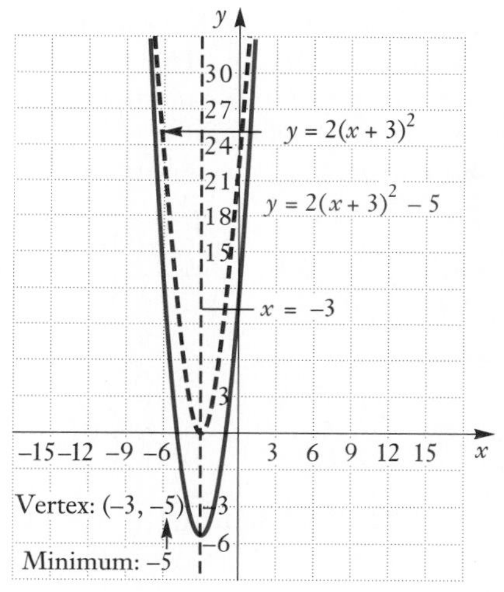

◀

▶ **EXAMPLE 5** Graph $y = \frac{1}{2}(x - 3)^2 + 5$. Find the vertex, the line of symmetry, and the maximum or minimum value.

We know that the graph looks like that of $y = \frac{1}{2}x^2$ translated 3 units to the right and 5 units up. The vertex is $(3, 5)$, and the line of symmetry is $x = 3$. Since $\frac{1}{2} > 0$, we know that 5, the second coordinate of the vertex, is the minimum y-value.

We compute a few points as needed. The graph is shown on the following page.

x	y	
3	5	← Vertex
0	$9\frac{1}{2}$	← y-intercept
1	7	
2	$5\frac{1}{2}$	
6	$9\frac{1}{2}$	

Graph the equation. Find the vertex, the line of symmetry, and the maximum or minimum y-value.

6. $y = \frac{1}{2}(x + 2)^2 - 4$

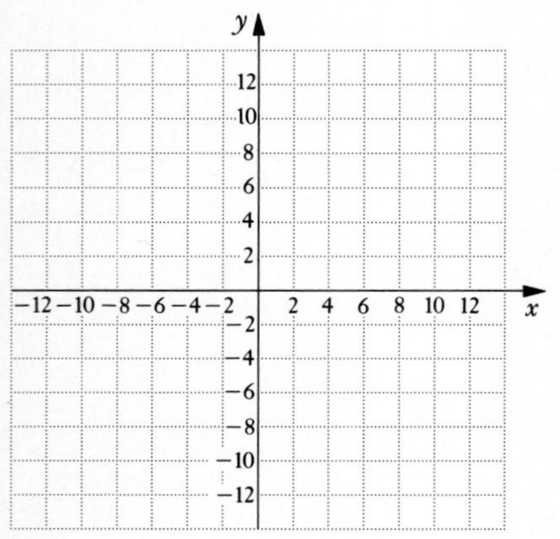

7. $y = -2(x - 5)^2 + 3$

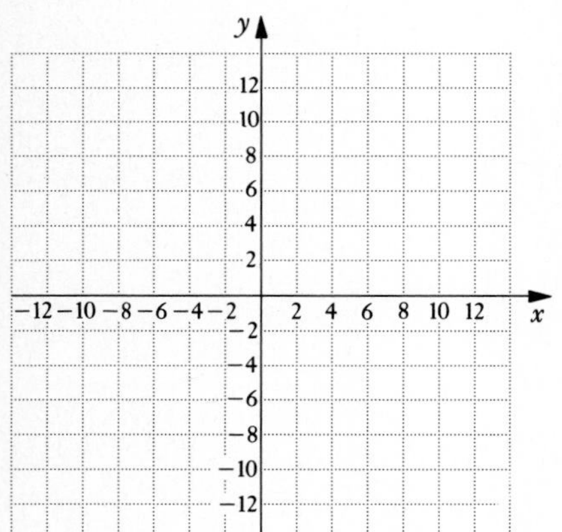

ANSWERS ON PAGE A-9

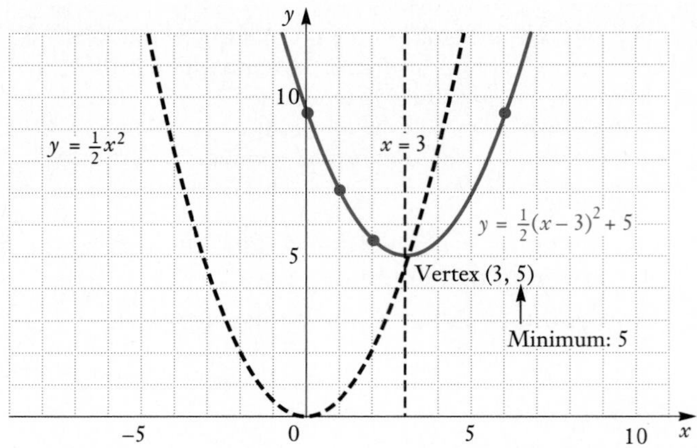

▶ **EXAMPLE 6** Graph $y = -2(x + 3)^2 + 5$. Find the vertex, the line of symmetry, and the maximum or minimum value.

We first express the equation in the equivalent form

$$y = -2[x - (-3)]^2 + 5.$$

We know that the graph looks like that of $y = -2x^2$ translated 3 units to the left and 5 units up. The vertex is $(-3, 5)$, and the line of symmetry is $x = -3$. Since $-2 < 0$, we know that 5, the second coordinate of the vertex, is the maximum y-value.

We compute a few points as needed. The graph is shown here.

x	y	
-3	5	Vertex
0	-13	y-intercept
-6	-13	
-2	3	
-4	3	

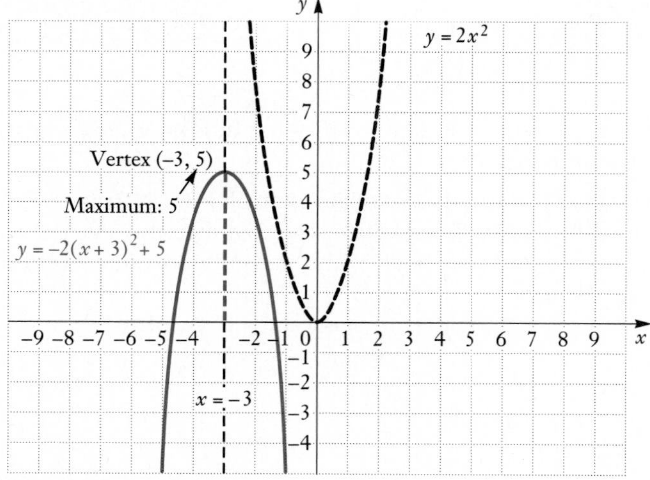

DO EXERCISES 6 AND 7.

NAME SECTION DATE

EXERCISE SET 8.7

a Graph the equation. Find and label the vertex and the line of symmetry.

1. $y = 5x^2$

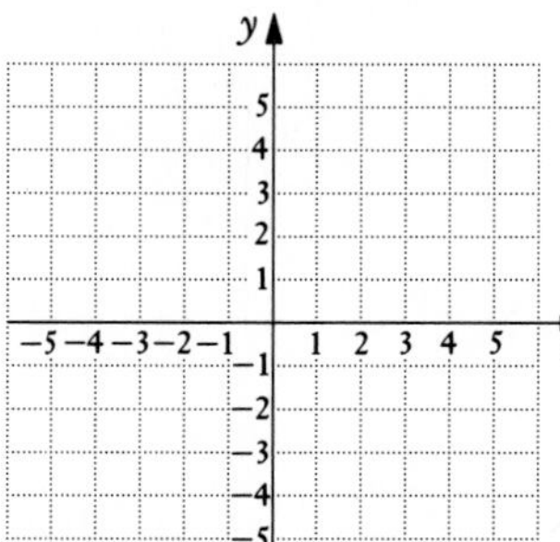

2. $y = 4x^2$

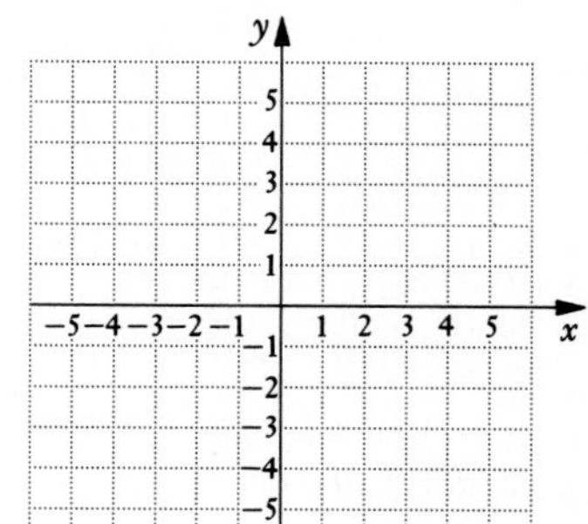

3. $y = \frac{1}{4}x^2$

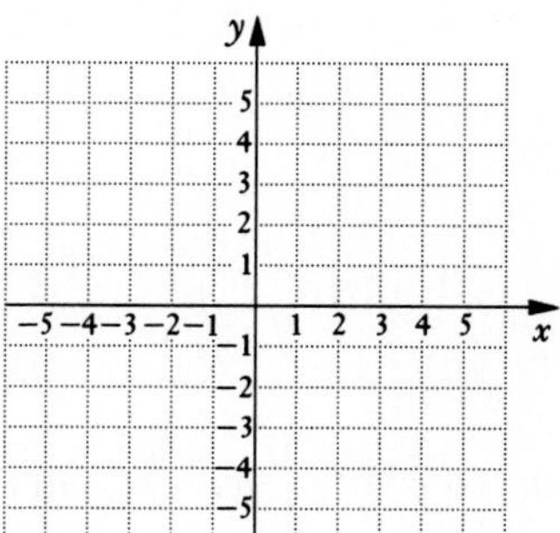

4. $y = \frac{1}{3}x^2$

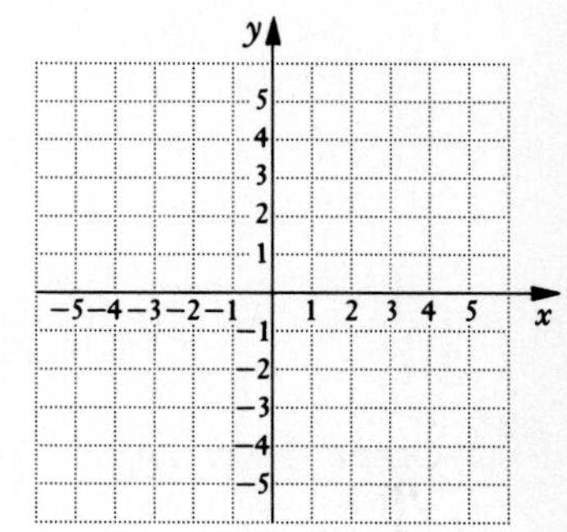

5. $y = -\frac{1}{2}x^2$

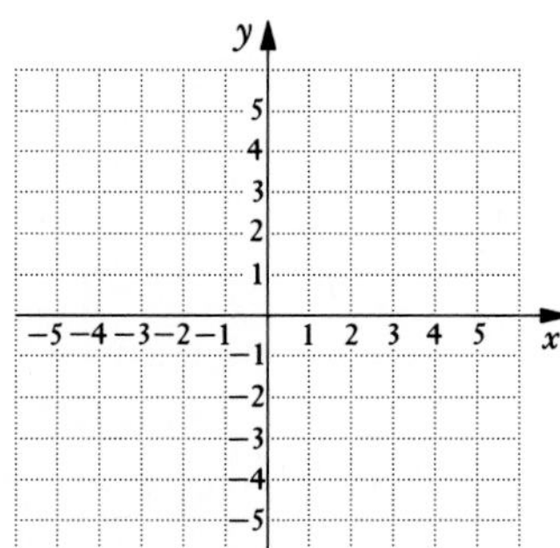

6. $y = -\frac{1}{4}x^2$

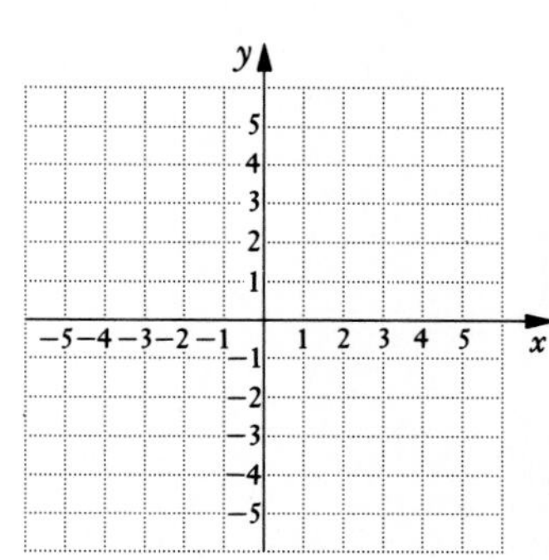

7. $y = -4x^2$

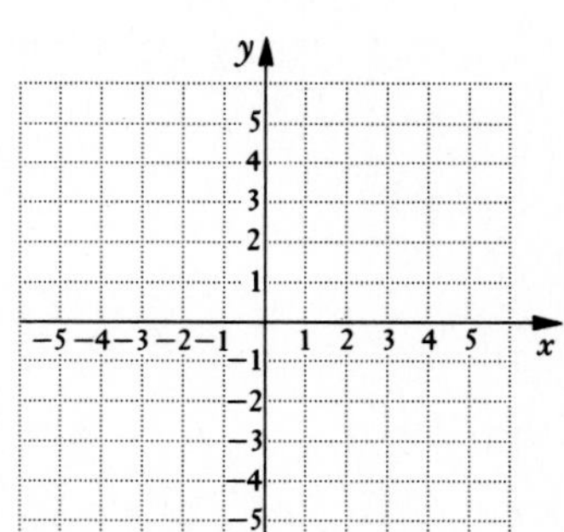

8. $y = -3x^2$

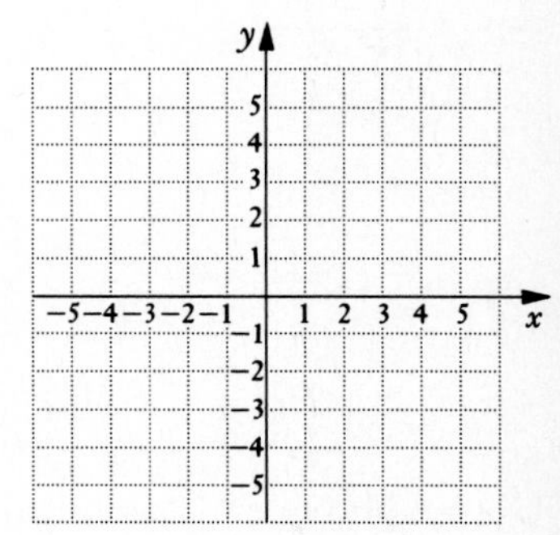

b Graph the equation. Find and label the vertex and the line of symmetry.

9. $y = (x - 3)^2$

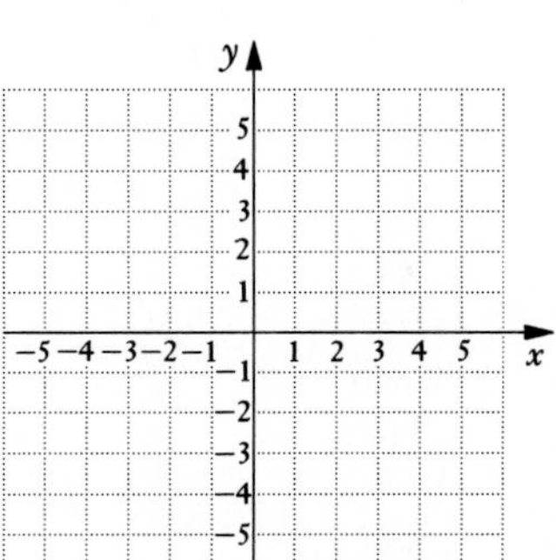

10. $y = (x + 1)^2$

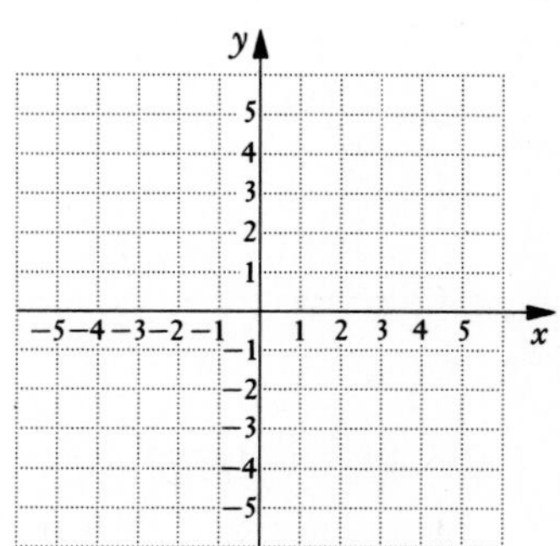

11. $y = 2(x - 4)^2$

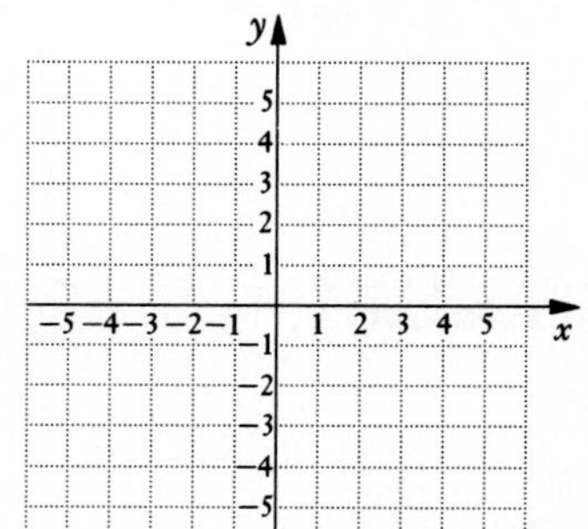

12. $y = -4(x - 1)^2$

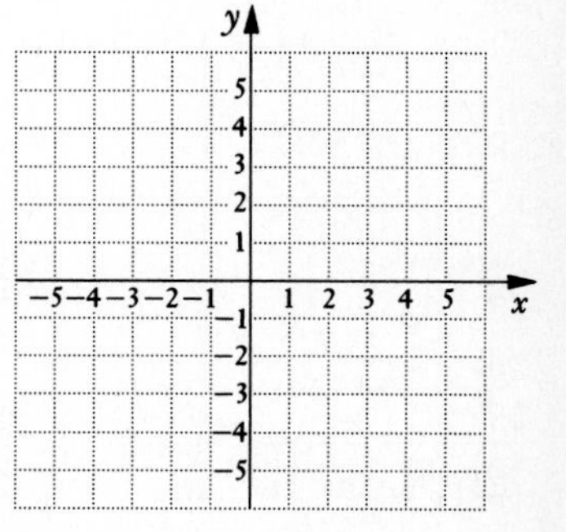

13. $y = -2(x + 2)^2$

14. $y = 2(x + 4)^2$

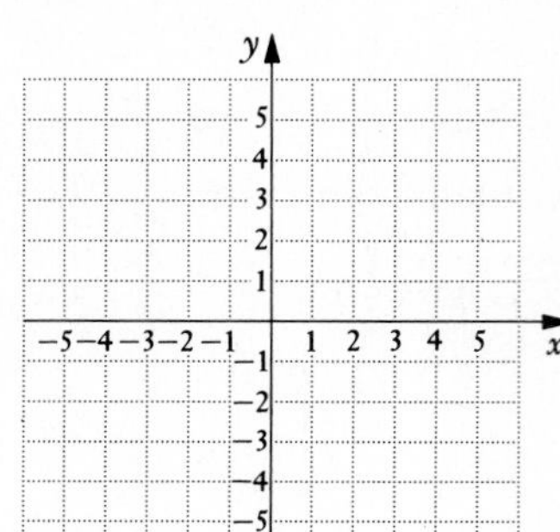

15. $y = 3(x - 1)^2$

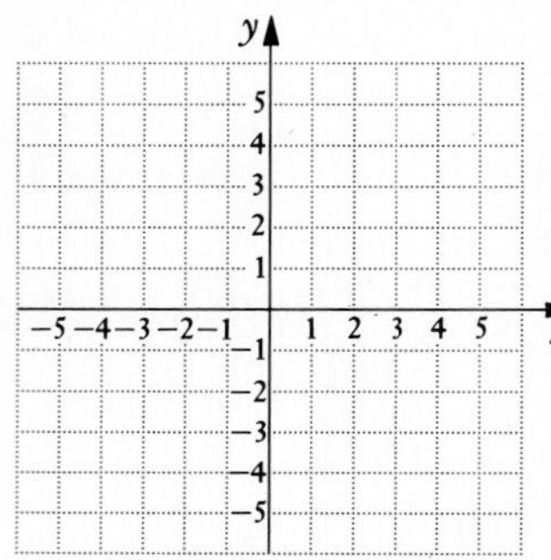

16. $y = -4(x - 2)^2$

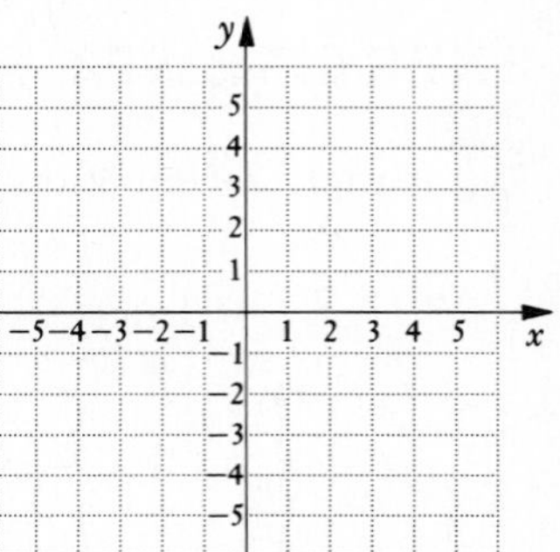

C Graph the equation. Find and label the vertex and the line of symmetry. Find the maximum or minimum value.

17. $y = (x - 3)^2 + 1$

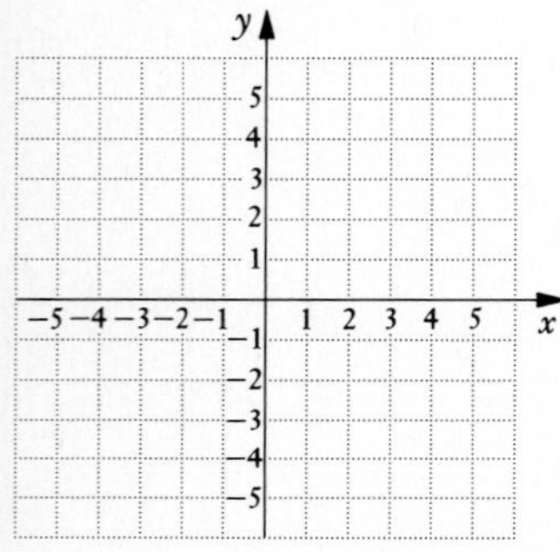

18. $y = (x + 2)^2 - 3$

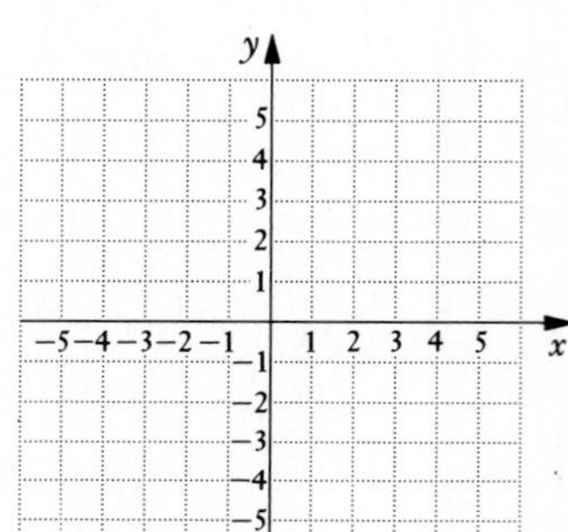

19. $y = -3(x + 4)^2 + 1$

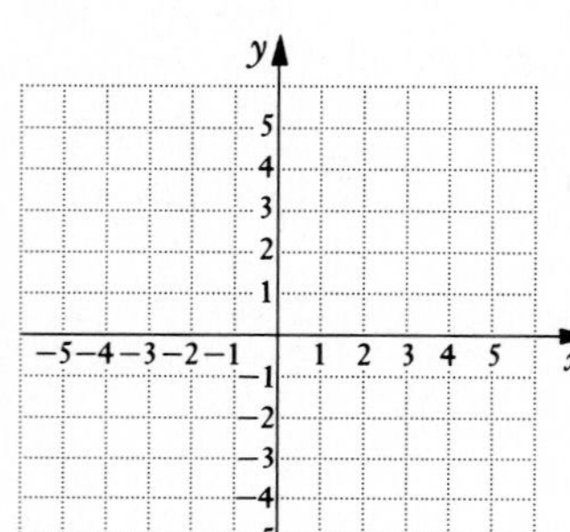

20. $y = \frac{1}{2}(x - 1)^2 - 3$

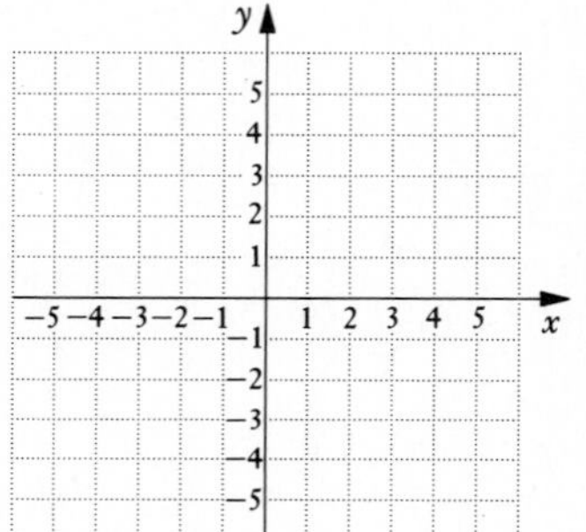

Use graph paper. Graph the function and find the maximum or minimum value.

21. $y = \frac{1}{2}(x + 1)^2 + 4$

22. $y = -2(x - 5)^2 - 3$

23. $y = -2(x + 2)^2 - 3$

24. $y = 3(x - 4)^2 + 2$

25. $y = -(x + 1)^2 - 2$

26. $y = -(x + 1)^2 + 1$

ANSWERS

27. ________

28. ________

29. ________

30. ________

SKILL MAINTENANCE

Solve.

27. $\sqrt{x} = -7$

28. $x = 7 + 2\sqrt{x + 1}$

29. $x - 5 = \sqrt{x + 7}$

30. $\sqrt{x + 2} - \sqrt{2x + 2} + 1 = 0$

8.8 Graphs of Quadratic Equations of the Type $y = ax^2 + bx + c$

a Finding the Vertex

The procedures discussed in Section 8.7 enable us to graph any quadratic equation of the type $y = a(x - h)^2 + k$. Suppose we have a general equation in the form $y = ax^2 + bx + c$. By *completing the square,* we can always rewrite the polynomial $ax^2 + bx + c$ in the form $a(x - h)^2 + k$.

How do we graph a general equation? A key is knowing the vertex. By plotting the point and then choosing x-values on both sides of the vertex, we can compute more points and complete the graph.

▶ **EXAMPLE 1** Graph: $y = x^2 - 6x + 4$.

We first find the vertex. We do this by completing the square, as follows:

$$y = x^2 - 6x + 4 = (x^2 - 6x) + 4.$$

We complete the square inside the parentheses. We take half the x-coefficient,

$$\frac{-6}{2} = -3, \qquad \text{and square it:} \quad (-3)^2 = 9.$$

Then we add $9 - 9$ inside the parentheses.

$$y = (x^2 - 6x + 9 - 9) + 4$$
$$y = (x^2 - 6x + 9) + (-9 + 4)$$
$$y = (x - 3)^2 - 5.$$

The vertex is $(3, -5)$. The line of symmetry is $x = 3$. The coefficient of x^2 is 1, which is positive, so the graph opens up. We choose some x-values on both sides of the vertex and graph the parabola.

For $x = 2$, $y = x^2 - 6x + 4 = 2^2 - 6(2) + 4 = 4 - 12 + 4 = -4$.
For $x = 1$, $y = x^2 - 6x + 4 = 1^2 - 6(1) + 4 = 1 - 6 + 4 = -1$.
For $x = 0$, $y = x^2 - 6x + 4 = 0^2 - 6(0) + 4 = 0 - 0 + 4 = 4$.
For $x = 4$, $y = x^2 - 6x + 4 = 4^2 - 6(4) + 4 = 16 - 24 + 4 = -4$.
For $x = 5$, $y = x^2 - 6x + 4 = 5^2 - 6(5) + 4 = 25 - 30 + 4 = -1$.
For $x = 6$, $y = x^2 - 6x + 4 = 6^2 - 6(6) + 4 = 36 - 36 + 4 = 4$.

The last three points can be found by symmetry, without calculating.

x	y	
3	−5	← Vertex
2	−4	
1	−1	
0	4	← y-intercept
4	−4	
5	−1	
6	4	

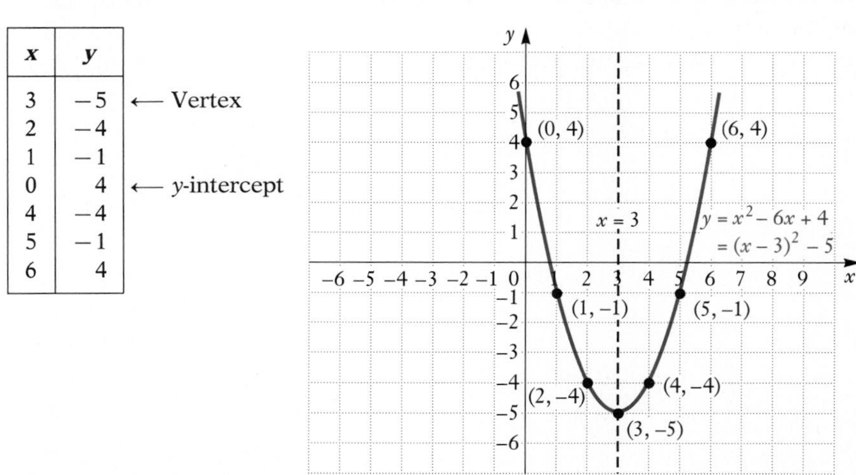

◀

OBJECTIVES

After finishing Section 8.8, you should be able to:

- **a** Graph a quadratic equation.
- **b** Find the x-intercepts of a quadratic equation.
- **c** Solve maximum and minimum problems involving quadratic equations.
- **d** Solve problems by fitting a quadratic equation to three data points.

FOR EXTRA HELP

Tape 16C | Tape 15A | MAC: 8 IBM: 8

▶ **EXAMPLE 2** Graph: $y = -2x^2 + 10x - 7$.

We first find the vertex by completing the square. We factor out -2 from the first two terms of the expression. This makes the coefficient of x^2 inside the parentheses 1:

$$\begin{aligned} y &= -2x^2 + 10x - 7 \\ &= -2(x^2 - 5x) - 7. \end{aligned}$$

Now we complete the square as before. We take half of the x-coefficient and square it to get $\frac{25}{4}$. Then we add $\frac{25}{4} - \frac{25}{4}$ inside the parentheses:

$$y = -2\left(x^2 - 5x + \frac{25}{4} - \frac{25}{4}\right) - 7$$

$$y = -2\left(x^2 - 5x + \frac{25}{4}\right) - 2\left(-\frac{25}{4}\right) - 7$$

Multiplying by -2, using a distributive law, and rearranging terms

$$y = -2\left(x - \frac{5}{2}\right)^2 + \frac{11}{2}.$$

The vertex is $(\frac{5}{2}, \frac{11}{2})$. The line of symmetry is $x = \frac{5}{2}$. The coefficient of x^2, -2, is negative, so the graph opens down. We choose some x-values on both sides of the vertex, compute y-values, and graph the parabola. Be sure to find y when $x = 0$. This gives us the y-intercept.

x	y	
$\frac{5}{2}$	$\frac{11}{2}$	← Vertex
0	-7	← y-intercept
1	1	
2	5	
3	5	
4	1	
5	-7	

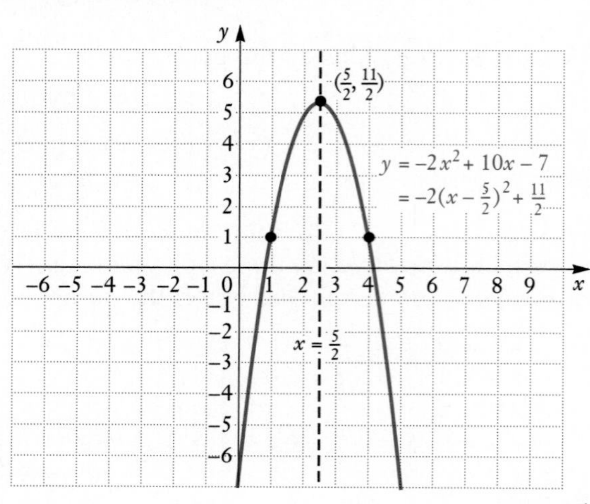

◀

We can find a formula for computing the vertex. We do this by completing the square, in a manner like Examples 1 and 2. This proof is similar to the proof of the quadratic formula. We first factor a out of the first two terms:

$$\begin{aligned} y &= ax^2 + bx + c \\ &= a\left(x^2 + \frac{b}{a}x\right) + c. \end{aligned}$$

Factoring a out of the first two terms. Check by multiplying.

Half of the x-coefficient, b/a, is $b/2a$. We square it to get $b^2/4a^2$ and add $b^2/4a^2 - b^2/4a^2$ inside the parentheses. Then we multiply a back through, as follows, and factor:

$$y = a\left(x^2 + \frac{b}{a}x + \frac{b^2}{4a^2} - \frac{b^2}{4a^2}\right) + c$$

$$y = a\left(x^2 + \frac{b}{a}x + \frac{b^2}{4a^2}\right) + a\left(-\frac{b^2}{4a^2}\right) + c$$

$$y = a\left(x + \frac{b}{2a}\right)^2 + \frac{-b^2}{4a} + \frac{4ac}{4a}$$

$$y = a\left[x - \left(-\frac{b}{2a}\right)\right]^2 + \frac{4ac - b^2}{4a}.$$

Thus we have the following.

For a parabola given by the quadratic equation $y = ax^2 + bx + c$, the vertex of the parabola is

$$\left(-\frac{b}{2a}, \frac{4ac - b^2}{4a}\right).$$

The x-coordinate of the vertex is $-b/2a$. The line of symmetry is $x = -b/2a$. The second coordinate of the vertex is found by substituting into the formula above, but can usually be found easier by substituting $x = -b/2a$ directly into the quadratic equation and computing y.

Let us look back at Examples 1 and 2 to see how we can find the vertex directly. In Example 1,

$$\text{the } x\text{-coordinate of the vertex is } -\frac{b}{2a} = -\frac{-6}{2(1)} = 3.$$

Substituting 3 for x in the equation, we find the second coordinate of the vertex:

$$y = x^2 - 6x + 4 = (3)^2 - 6(3) + 4 = 9 - 18 + 4 = -5.$$

The vertex is $(3, -5)$. The line of symmetry is $x = 3$.

In Example 2,

$$\text{the } x\text{-coordinate of the vertex is } -\frac{b}{2a} = -\frac{10}{2(-2)} = \frac{5}{2}.$$

Substituting $\frac{5}{2}$ into the equation, we find the second coordinate of the vertex:

$$y = -2x^2 + 10x - 7 = -2\left(\frac{5}{2}\right)^2 + 10\left(\frac{5}{2}\right) - 7 = -2\left(\frac{25}{4}\right) + 25 - 7 = \frac{11}{2}.$$

The vertex is $(\frac{5}{2}, \frac{11}{2})$. The line of symmetry is $x = \frac{5}{2}$.

We have actually developed two methods for finding the vertex. One is by completing the square and the other is by using a formula. You should consult with your instructor about which method to use.

DO EXERCISES 1 AND 2.

1. For $y = x^2 - 4x + 7$:

   a) Find the vertex and the line of symmetry.

   b) Graph the equation.

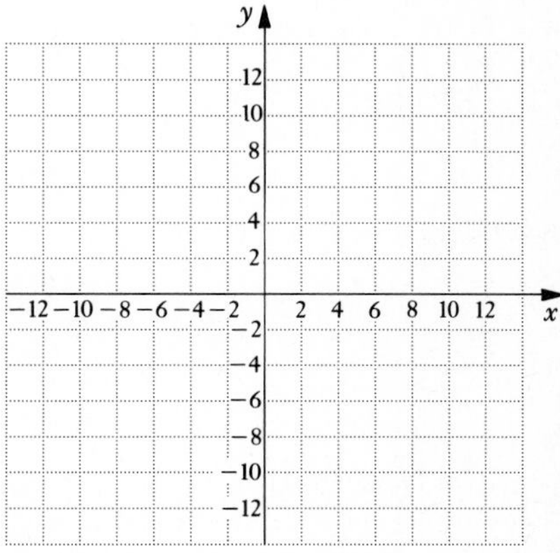

2. For $y = -4x^2 + 12x - 5$:

   a) Find the vertex and the line of symmetry.

   b) Graph the equation.

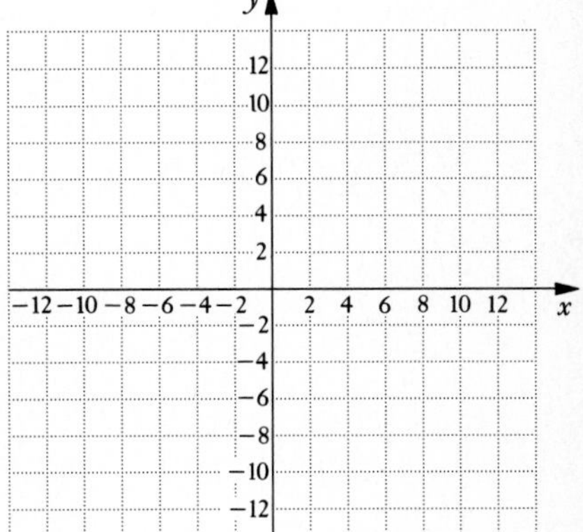

ANSWERS ON PAGE A-9

3. Find the x-intercepts:

$$y = x^2 - 2x - 5.$$

Find the x-intercepts if they exist.

4. $y = x^2 - 2x - 3$

5. $y = x^2 + 8x + 16$

6. $y = -2x^2 - 4x - 3$

ANSWERS ON PAGE A-9

b Finding the x-Intercepts of a Quadratic Equation

The points at which a graph crosses the x-axis are called its **x-intercepts.** These are, of course, the points at which $y = 0$.

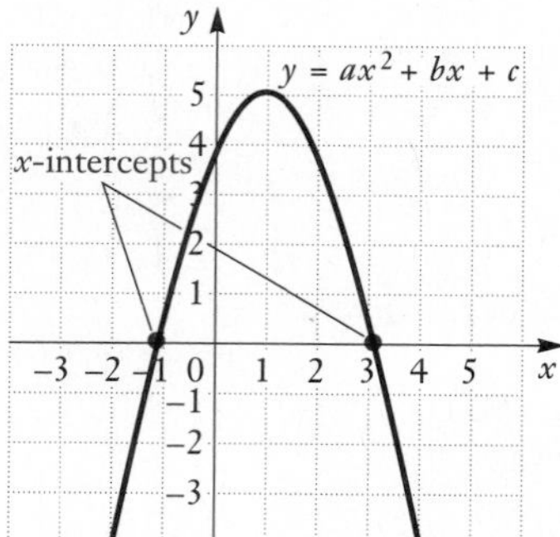

To find the x-intercepts of a quadratic equation $y = ax^2 + bx + c$, we solve the equation

$$0 = ax^2 + bx + c.$$

▶ **EXAMPLE 3** Find the x-intercepts of the graph of $y = x^2 - 2x - 2$. We solve the equation

$$x^2 - 2x - 2 = 0.$$

The equation is difficult to factor, so we use the quadratic formula and get $x = 1 \pm \sqrt{3}$. The x-intercepts are $(1 - \sqrt{3}, 0)$ and $(1 + \sqrt{3}, 0)$. For plotting, we approximate, to get $(-0.7, 0)$ and $(2.7, 0)$. We sometimes refer to the x-coordinates as intercepts. ◀

It can be useful to have the x-intercepts when graphing an equation.

DO EXERCISE 3.

The discriminant, $b^2 - 4ac$, tells us how many real-number solutions the equation $0 = ax^2 + bx + c$ has, so it also indicates how many intercepts there are. Compare the following.

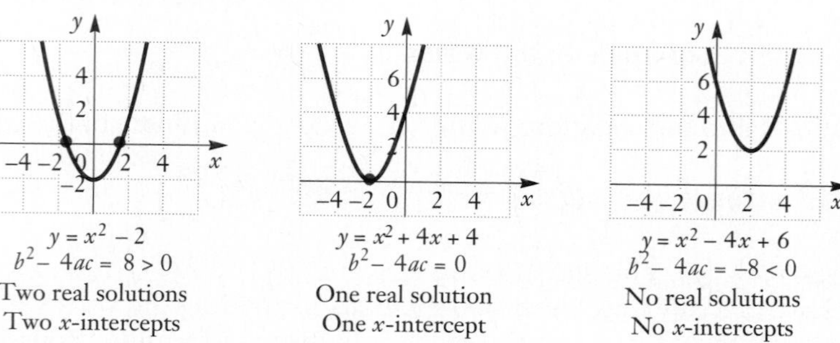

$y = x^2 - 2$	$y = x^2 + 4x + 4$	$y = x^2 - 4x + 6$
$b^2 - 4ac = 8 > 0$	$b^2 - 4ac = 0$	$b^2 - 4ac = -8 < 0$
Two real solutions	One real solution	No real solutions
Two x-intercepts	One x-intercept	No x-intercepts

DO EXERCISES 4–6.

c Maximum and Minimum Problems

In a quadratic equation, the y-value at the vertex will be either a minimum (when $a > 0$) or a maximum (when $a < 0$). In certain problems, we want to find a minimum or a maximum. If the situation can be translated to a quadratic equation, we can solve by finding the y-value at the vertex.

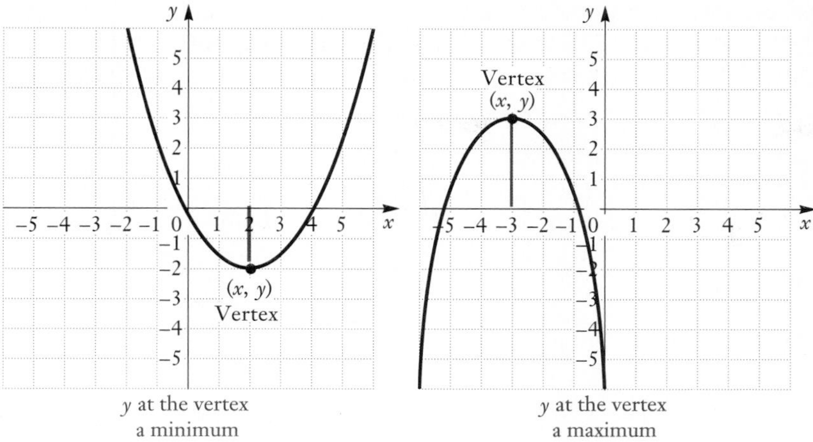

y at the vertex a minimum

y at the vertex a maximum

Consider a quadratic equation of the type $y = ax^2 + bx + c$. If a is positive, the graph opens up and y has a minimum value. If a is negative, the graph opens down and y has a maximum value.

► **EXAMPLE 4** What are the dimensions of the largest rectangular pen that can be enclosed with 64 m of fence?

We make a drawing and label it. The perimeter must be 64 m, so we have

$$2w + 2l = 64. \quad \textbf{(1)}$$

We wish to find the maximum area, so we try to find a quadratic equation for the area defined as an expression in one variable. We know that

$$A = lw. \quad \textbf{(2)}$$

Solving (1) for l, we get $l = 32 - w$. Substituting in (2), we get a quadratic equation:

$$A = (32 - w)w = -w^2 + 32w.$$

The w-coordinate of the vertex is

$$w = -\frac{b}{2a} = -\frac{32}{2(-1)} = 16.$$

Substituting into the equation, we find the A-coordinate of the vertex:

$$A = -w^2 + 32w = -(16)^2 + 32(16) = 256.$$

The coefficient of w^2 is negative, so we know that 256 is a maximum. The maximum value (area) is 256. It occurs when $w = 16$. Thus the dimensions are 16 m by 16 m. ◄

DO EXERCISES 7 AND 8.

d Fitting Quadratic Equations to Data

In many problems, a quadratic equation can be used to describe the situation. We can find a quadratic equation if we know three inputs and their outputs. Each such ordered pair is called a **data point.**

It might be helpful to review the solution of systems of three equations in three variables in Section 4.4 before doing this example.

7. What is the maximum product of two numbers whose sum is 30?

8. What are the dimensions of the largest rectangular pen that can be enclosed with 100 m of fence?

ANSWERS ON PAGE A-9

9. Find the quadratic equation that fits the data points (1, 0), (−1, 4), and (2, 1).

10. The following table shows the accident records in a city. It has values that a quadratic equation will fit.

Age of driver	Number of accidents (in a year)
20	400
40	150
60	400

a) Assuming that a quadratic function will describe the situation, find the number of accidents in terms of age.

b) Use the equation to calculate the number of accidents in which a typical 16-year-old is involved.

ANSWERS ON PAGE A-9

▶ **EXAMPLE 5** The instruction booklet for a video cassette recorder (VCR) includes a table relating the counter readings and the time the tape has run.

Counter readings, N	Time, t, tape has run (in hours)
000	0
300	1
500	2

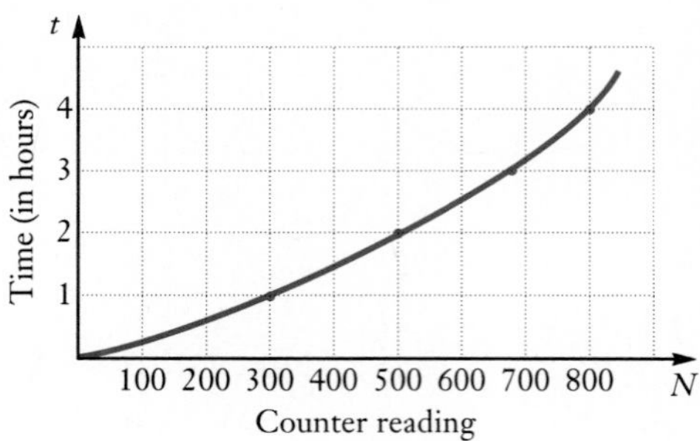

The counter readings do not give times for programs on the half-hour.

a) Find a quadratic equation that fits the data.

b) Use the equation to find the counter reading after the tape has run for $1\frac{1}{2}$ hr.

a) We look for an equation

$$t = aN^2 + bN + c,$$

where N is the number of revolutions shown on the counter and t is the time in hours. We use the three data points (000, 0), (300, 1), and (500, 2) to find a, b, and c:

$$0 = a(0)^2 + b(0) + c, \text{ or } 0 = c;$$
$$1 = a(300)^2 + b(300) + c, \text{ or } 1 = 90{,}000a + 300b + c;$$
$$2 = a(500)^2 + b(500) + c, \text{ or } 2 = 250{,}000a + 500b + c.$$

Since $c = 0$, the system reduces to a system of two equations in two variables:

$$1 = 90{,}000a + 300b,$$
$$2 = 250{,}000a + 500b.$$

Solving, we get $a = 0.00000333$ and $b = 0.00233$. Thus the equation we are looking for is

$$t = 0.00000333N^2 + 0.00233N.$$

b) To find the counter reading after $1\frac{1}{2}$ hr, we substitute 1.5 for t and solve for N:

$$1.5 = 0.00000333N^2 + 0.00233N$$
$$0 = 0.00000333N^2 + 0.00233N - 1.5.$$

Using the quadratic formula, we get

$$N = \frac{-b \pm \sqrt{b^2 - 4ac}}{2a}$$
$$= \frac{-0.00233 \pm \sqrt{(0.00233)^2 - 4(0.00000333)(-1.5)}}{2(0.00000333)}$$
$$\approx 407. \quad \text{Considering only the positive solution}$$

Thus after $1\frac{1}{2}$ hr the counter reading should be about 407. ◀

DO EXERCISES 9 AND 10.

NAME SECTION DATE

EXERCISE SET 8.8

a For each quadratic equation, **(a)** find the vertex and the line of symmetry and **(b)** graph the equation.

1. $y = x^2 - 2x - 3$

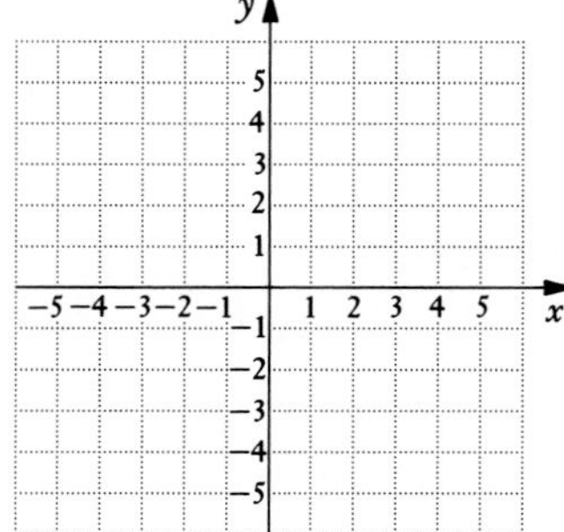

2. $y = x^2 + 2x - 5$

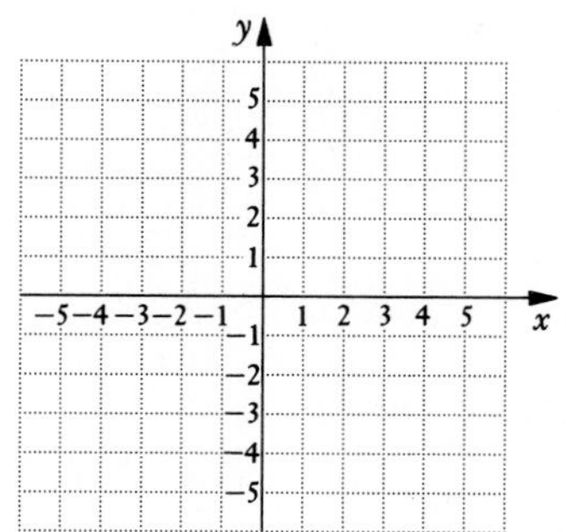

3. $y = -x^2 + 4x + 1$

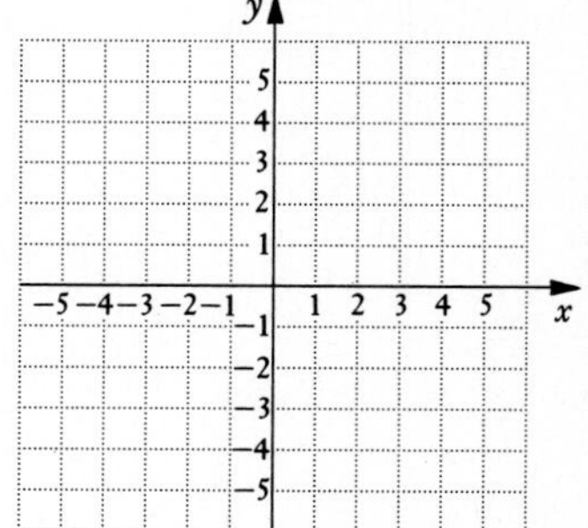

4. $y = -x^2 - 4x + 3$

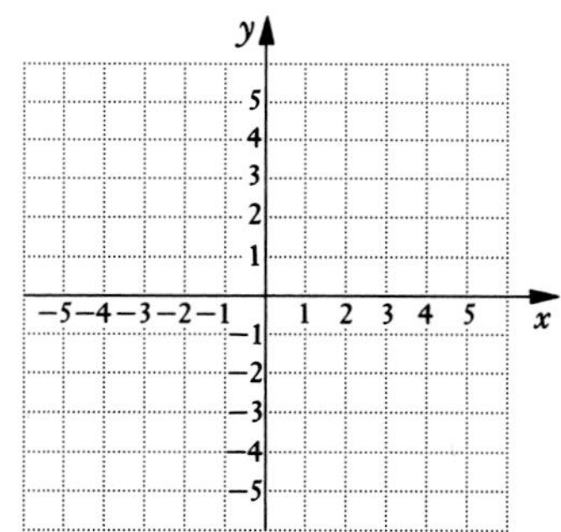

5. $y = 3x^2 - 24x + 50$

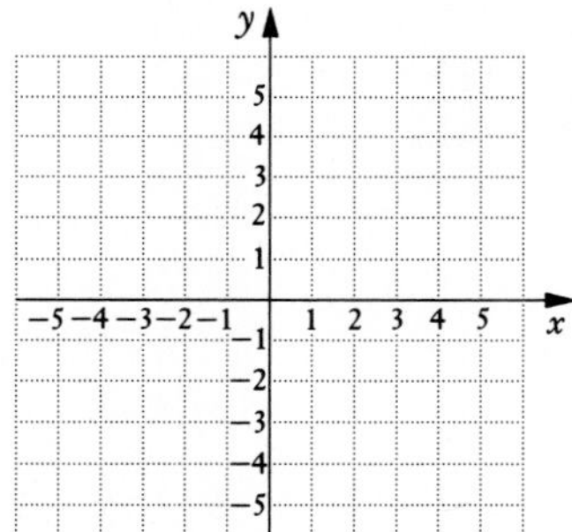

6. $y = 4x^2 + 8x - 3$

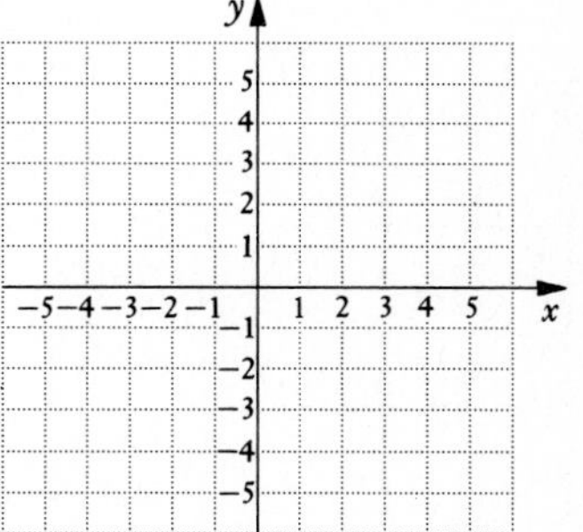

7. $y = -2x^2 + 2x + 1$

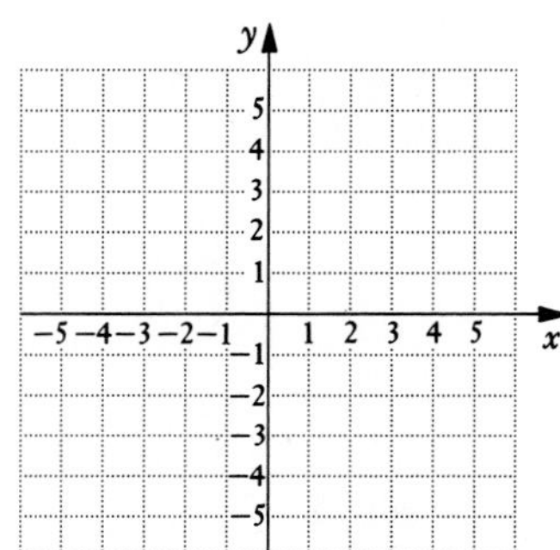

8. $y = -2x^2 - 2x + 3$

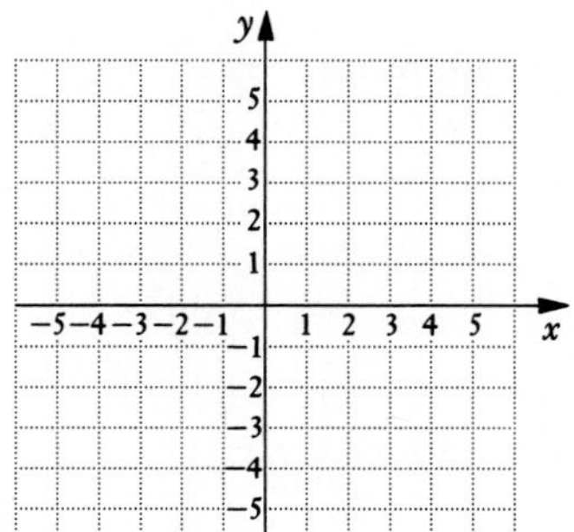

9. $y = 5 - x^2$

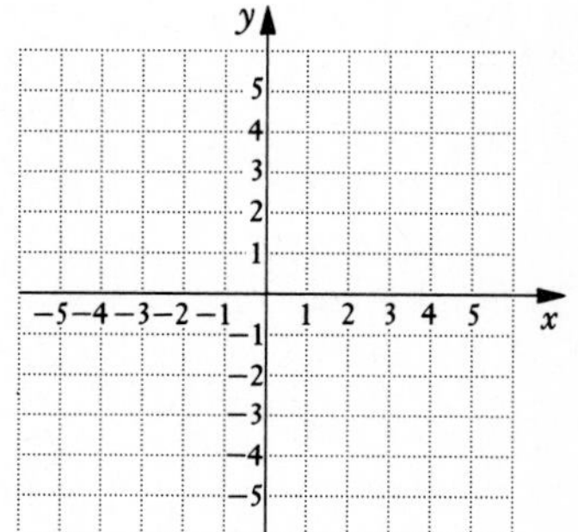

ANSWERS

10. ______
11. ______
12. ______
13. ______
14. ______
15. ______
16. ______
17. ______
18. ______
19. ______
20. ______
21. ______
22. ______
23. ______
24. ______
25. ______

b Find the x-intercepts.

10. $y = x^2 - 4x + 1$

11. $y = x^2 + 6x + 10$

12. $y = -x^2 + 2x + 3$

13. $y = -x^2 + 3x + 4$

14. $y = 4x^2 + 12x + 9$

15. $y = 3x^2 - 6x + 1$

16. $y = 2x^2 - 4x + 6$

17. $y = 2x^2 + 4x - 1$

c Solve.

18. A rancher is fencing off a rectangular area with a fixed perimeter of 76 ft. What dimensions would yield the maximum area? What is the maximum area?

19. A carpenter is building a rectangular room with a fixed perimeter of 68 ft. What dimensions would yield the maximum area? What is the maximum area?

20. What is the maximum product of two numbers whose sum is 22? What numbers yield this product?

21. What is the maximum product of two numbers whose sum is 45? What numbers yield this product?

22. What is the minimum product of two numbers whose difference is 4? What are the numbers?

23. What is the minimum product of two numbers whose difference is 6? What are the numbers?

24. What is the minimum product of two numbers whose difference is 5? What are the numbers?

25. What is the minimum product of two numbers whose difference is 7? What are the numbers?

26. *Projectile height.* A ball is thrown into the air. Its height h, in feet, t seconds after it is thrown is given by

$$h = -16t^2 + 80t + 96.$$

Find **(a)** its maximum height and when it attains it and **(b)** when it reaches the ground.

27. *Projectile height.* A rocket is fired into the air. Its height h, in feet, t seconds after blastoff is given by

$$h = -16t^2 + 64t + 2240.$$

Find **(a)** its maximum height and when it attains it and **(b)** when it reaches the ground.

28. *Cost.* The total daily costs C for a manufacturer of doors are given by

$$C = 2x^2 - 40x + 2000.$$

How many doors should be manufactured daily in order to minimize costs?

29. *Profit.* The total profit P on the production and sale of x units of a product is given by

$$P = -x^2 + 980x - 3000.$$

How many units should be produced and sold in order to maximize profit?

d Find the quadratic equation that fits the set of data points.

30. $(1, 4), (-1, -2), (2, 13)$

31. $(1, 4), (-1, 6), (-2, 16)$

32. $(1, 5), (2, 9), (3, 7)$

33. $(1, -4), (2, -6), (3, -6)$

34. *Predicting earnings.* A business earns \$38 in the first week, \$66 in the second week, and \$86 in the third week. The manager graphs the points (1, 38), (2, 66), and (3, 86) and uses a quadratic equation to describe the situation.

a) Find a quadratic equation that fits the data.

b) Using the equation, predict the earnings for the fourth week.

ANSWERS

26. a) ______

b) ______

27. a) ______

b) ______

28. ______

29. ______

30. ______

31. ______

32. ______

33. ______

34. a) ______

b) ______

ANSWERS

35. a) ______

b) ______

36. a) ______

b) ______

37. a) ______

b) ______

38. ______

39. ______

40. ______

41. ______

42. See graph.

43. See graph.

44. ______

45. ______

46. ______

35. *Predicting earnings.* A business earns \$1000 in its first month, \$2000 in the second month, and \$8000 in the third month. The manager plots the points (1, 1000), (2, 2000), and (3, 8000) and uses a quadratic equation to describe the situation.

a) Find a quadratic equation that fits the data.

b) Using the equation, predict the earnings for the fourth month.

36. a) Find a quadratic equation that fits the following data.

Travel speed (in km/h)	60	80	100
Number of daytime accidents (for every 200 million km)	200	130	100

b) Use the equation to calculate the number of daytime accidents that occur at 50 km/h.

37. Pizza Unlimited has the following prices for pizzas.

Diameter	*Price*
8 in.	\$3.00
12 in.	\$4.25
16 in.	\$5.75

Is price related to diameter by a quadratic equation? It probably should be because the price should be proportional to the area, and the area is a quadratic equation of the diameter. (The area of a circular region is given by $A = \pi r^2$ or $(\pi/4) \cdot d^2$.)

a) Fit a quadratic equation to the data points (8, 3), (12, 4.25), and (16, 5.75).

b) Use the equation to find the price of a 14-in. pizza.

SKILL MAINTENANCE

Multiply and simplify.

38. $\sqrt[4]{5x^3y^5}\,\sqrt[4]{125x^2y^3}$

39. $\sqrt{9a^3}\,\sqrt{16ab^4}$

Solve.

40. $\sqrt{4x - 4} = \sqrt{x + 4} + 1$

41. $\sqrt{5x - 4} + \sqrt{13 - x} = 7$

SYNTHESIS

42. Graph: $y = |x^2 - 1|$.

43. Graph: $y = |x^2 + 6x + 4|$.

44. The sum of the base and the height of a triangle is 38 cm. Find the dimensions for which the area is a maximum, and find the maximum area.

45. The perimeter of rectangle *RSTV* is 44 ft. Find the least possible length of the diagonal *RT*.

46. A horticulturist has 180 ft of fencing with which to form a rectangular garden. A greenhouse will provide one side of the garden, and the fencing will be used for the other three sides. What is the area of the largest region that can be enclosed?

8.9 Nonlinear Inequalities

a Quadratic and Other Polynomial Inequalities

Inequalities like the following are called **quadratic inequalities:**

$$x^2 + 3x - 10 < 0, \qquad 5x^2 - 3x + 2 \geq 0.$$

In each case, we have a polynomial of degree 2 on the left. We will consider solving such inequalities in three ways. The first two provide understanding and the last yields the fastest method.

The first method for solving quadratic inequalities is by considering the graph of a related equation.

▶ **EXAMPLE 1** Solve: $x^2 + 3x - 10 > 0$.

Consider the equation $y = x^2 + 3x - 10$ and its graph. Its graph opens up since the leading coefficient ($a = 1$) is positive.

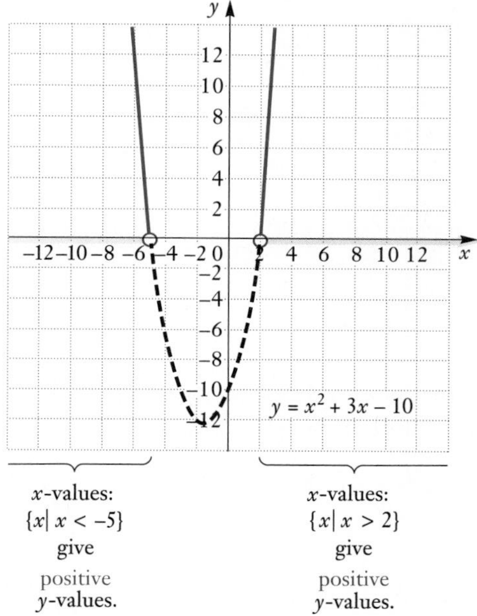

Values of y will be positive to the left and right of the intercepts, as shown. We find the intercepts by setting the polynomial equal to 0 and solving:

$$x^2 + 3x - 10 = 0$$
$$(x + 5)(x - 2) = 0$$
$$x + 5 = 0 \quad or \quad x - 2 = 0$$
$$x = -5 \quad or \quad x = 2.$$

Then the solution set of the inequality is

$$\{x | x < -5 \text{ or } x > 2\}.$$ ◀

DO EXERCISE 1.

We can solve any inequality by considering the graph of a related equation and finding intercepts as in Example 1. In some cases, we may need to use the quadratic formula to find the intercepts.

OBJECTIVES

After finishing Section 8.9, you should be able to:

a Solve quadratic and other polynomial inequalities.

b Solve rational inequalities.

FOR EXTRA HELP

Tape 17A

Tape 15B

MAC: 8
IBM: 8

1. Solve by graphing:

$$x^2 + 2x - 3 > 0.$$

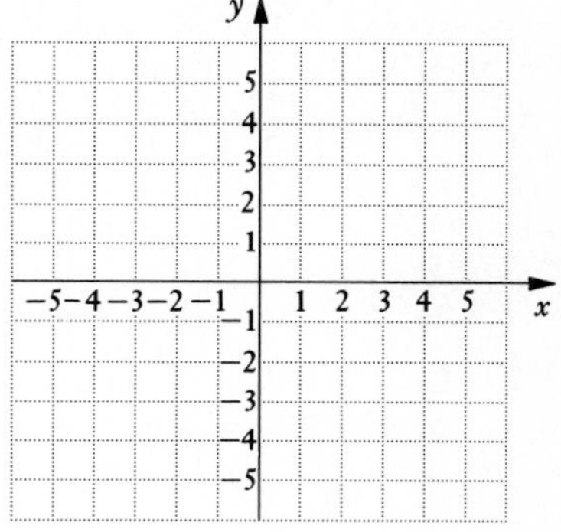

ANSWER ON PAGE A-9

2. Solve by graphing:

$x^2 + 2x - 3 < 0.$

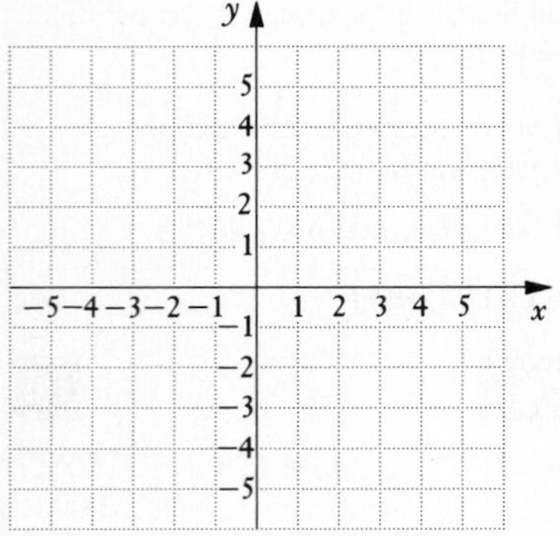

3. Solve by graphing:

$x^2 + 2x - 3 \leq 0.$

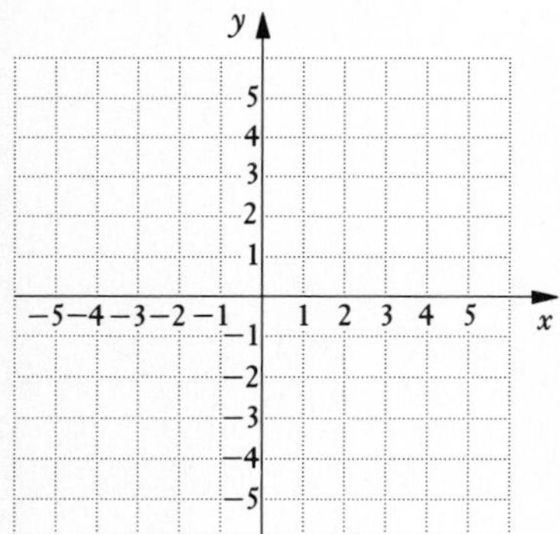

▶ **EXAMPLE 2** Solve: $x^2 + 3x - 10 < 0$.

Looking again at the graph of $y = x^2 + 3x - 10$ or at least visualizing it tells us that the function is negative for those inputs x such that x is between -5 and 2.

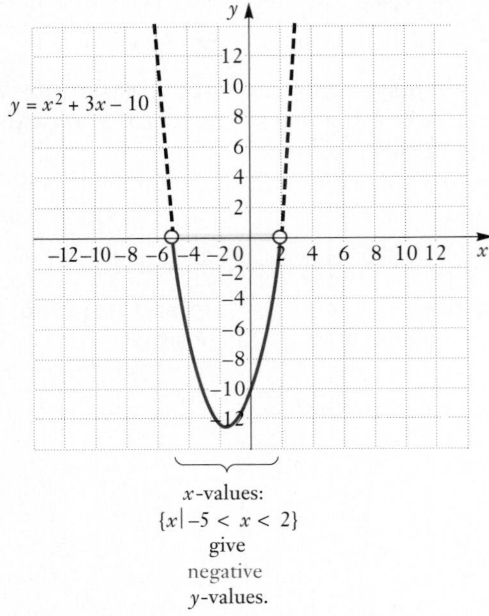

That is, the solution set is $\{x | -5 < x < 2\}$. ◀

DO EXERCISE 2.

When an inequality contains $\leq$ or $\geq$, the x-values of the intercepts must be included. Thus the solution set of the inequality $x^2 + 3x - 10 \geq 0$ is

$$\{x | x \leq -5 \text{ or } x \geq 2\}.$$

DO EXERCISE 3.

Let us now consider another method of solving inequalities that works for any polynomial that is factored into a product of first-degree polynomials.

▶ **EXAMPLE 3** Solve: $x^2 + 3x - 10 < 0$.

We factor the inequality, obtaining $(x + 5)(x - 2) < 0$. The solutions of $(x + 5)(x - 2) = 0$ are -5 and 2. They are not solutions of the inequality, but they divide the real-number line in a natural way, pictured as follows. When we graphed the equation $y = x^2 + 3x - 10$ as in Examples 1 and 2, we saw that the intercepts $(-5, 0)$ and $(2, 0)$ provide a division of the x-axis, which is a number line. The product $(x + 5)(x - 2)$ is positive or negative for values other than -5 and 2, depending on the signs of the factors $x + 5$ and $x - 2$. We can determine this efficiently with a diagram, as follows.

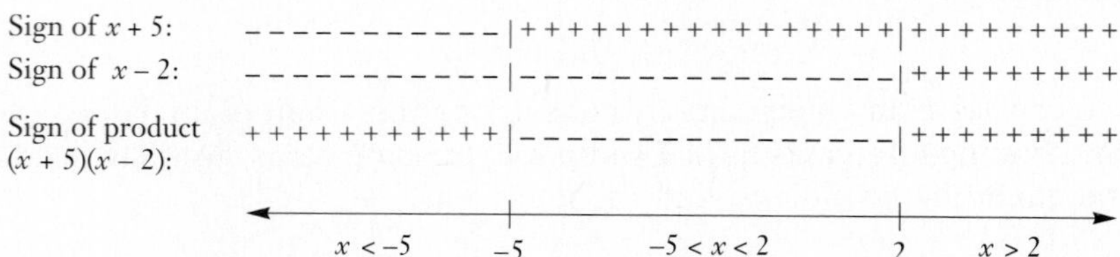

To set up the diagram, we first solve $x + 5 > 0$. We get $x > -5$. Thus, $x + 5$ is positive for all numbers to the right of -5. We indicate that with the + signs. Accordingly, $x + 5 < 0$ for all numbers to the left of -5. We indicate that with the − signs.

Similarly, we solve $x - 2 > 0$, and get $x > 2$. Thus, $x - 2$ is positive for all numbers to the right of 2. Accordingly, $x - 2 < 0$ for all numbers to the left of 2. We indicate this with the + and − signs.

Finally, we determine the signs of the product by using the rules for multiplication. In order for the product $(x + 5)(x - 2)$ to be negative, one factor must be positive and the other negative. In the table, this situation occurs only when $-5 < x < 2$. The solution set of the inequality is $\{x \mid -5 < x < 2\}$. ◀

Note that setting up a diagram as in Example 3 also tells us the solutions of $(x + 5)(x - 2) > 0$. That solution set is $\{x \mid x < -5 \text{ or } x > 2\}$.

DO EXERCISES 4 AND 5.

We now arrive at the fastest method for solving quadratic inequalities. The preceding discussion provides the understanding for this method. In Example 4, we see that the intercepts divide the number line up into intervals. If a particular equation has a positive output for one number in an interval, it will be positive for all the numbers in the interval. Thus we can merely make a test substitution in each interval in order to solve the inequality. This is very similar to our method of using test points to graph a linear inequality in a plane.

▶ **EXAMPLE 4** Solve: $x^2 + 3x - 10 < 0$.

We set the polynomial equal to 0 and solve. The solutions of $x^2 + 3x - 10 = 0$, or $(x + 5)(x - 2) = 0$, are -5 and 2. We then locate them on a number line as follows. Note that the numbers divide the number line into three intervals A, B, and C.

A B C

−5 2

We choose a test number in interval A, say -7, and substitute -7 for x in the equation $y = x^2 + 3x - 10$:

$$y = (-7)^2 + 3(-7) - 10$$
$$= 49 - 21 - 10 = 18.$$

Note that $18 > 0$, so the y-values will be positive for any number in interval A.

Next we try a test number in interval B, say 1, and find the corresponding y-value:

$$y = 1^2 + 3(1) - 10$$
$$= 1 + 3 - 10 = -6.$$

Note that $-6 < 0$, so the y-values will be negative for any number in interval B.

Solve using the method of Example 3.

4. $x^2 + 3x > 4$

5. $x^2 + 3x \leq 4$

ANSWERS ON PAGE A-9

Solve using the method of Example 4.

6. $x^2 + 3x > 4$

7. $x^2 + 3x \leq 4$

8. Solve: $6x(x + 1)(x - 1) < 0$.

ANSWERS ON PAGE A-9

Next we try a test number in interval C, say 4, and find the corresponding y-value:

$$\begin{aligned} y &= 4^2 + 3(4) - 10 \\ &= 16 + 12 - 10 = 18. \end{aligned}$$

Note that $18 > 0$, so the y-values will be positive for any number in interval C. We are looking for numbers x for which $x^2 + 3x - 10 < 0$. Thus any number x in interval B is a solution. If the inequality had been $\leq$ or $\geq$, we would also need to include the intercepts -5 or 2 in the solution set. The solution set is $\{x \mid -5 < x < 2\}$. ◀

To solve a quadratic inequality:

1. **Get 0 on one side, set the expression on the other side equal to 0, and solve to find the intercepts.**
2. **Use the numbers found in step (1) to divide the number line into intervals.**
3. **Substitute a number from each interval into the related equation. If the y-value is positive, then the expression will be positive for all numbers in the interval. If the y-value is negative, then the expression will be negative for all numbers in the interval.**
4. **Select the intervals for which the inequality is satisfied and write set-builder notation for the solution set.**

DO EXERCISES 6 AND 7.

▶ **EXAMPLE 5** Solve: $5x(x + 3)(x - 2) \geq 0$.

The solutions of $5x(x + 3)(x - 2) = 0$ are -3, 0, and 2. They divide the real-number line into four intervals as follows.

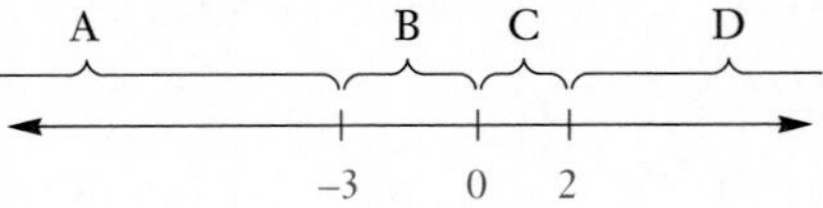

We try test numbers in each interval:

A: Test -5, $y = 5(-5)(-5 + 3)(-5 - 2) = -350$.

B: Test -2, $y = 5(-2)(-2 + 3)(-2 - 2) = 40$.

C: Test 1, $y = 5(1)(1 + 3)(1 - 2) = -20$.

D: Test 3, $y = 5(3)(3 + 3)(3 - 2) = 90$.

The expression is positive for values of x in intervals B and D. Since the inequality symbol is $\geq$, we will need to include the intercepts. The solution set of the inequality is

$$\{x \mid -3 \leq x \leq 0 \textit{ or } 2 \leq x\}. \quad ◀$$

DO EXERCISE 8.

b Rational Inequalities

We adapt the preceding method when an inequality involves rational expressions. We call these **rational inequalities.**

▶ **EXAMPLE 6** Solve: $\frac{x-3}{x+4} \geq 2$.

We write the related equation by changing the $\geq$ symbol to $=$:

$$\frac{x-3}{x+4} = 2.$$

Then we solve this related equation. We multiply on both sides of the equation by the LCM, which is $x + 4$:

$$(x+4) \cdot \frac{x-3}{x+4} = (x+4) \cdot 2$$
$$x - 3 = 2x + 8$$
$$-11 = x.$$

In the case of rational inequalities, we also need to determine those replacements that are not meaningful. These are those that make the denominator 0. We set the denominator equal to 0 and solve:

$$x + 4 = 0$$
$$x = -4.$$

Now we use the numbers -11 and -4 to divide the number line into intervals, as follows:

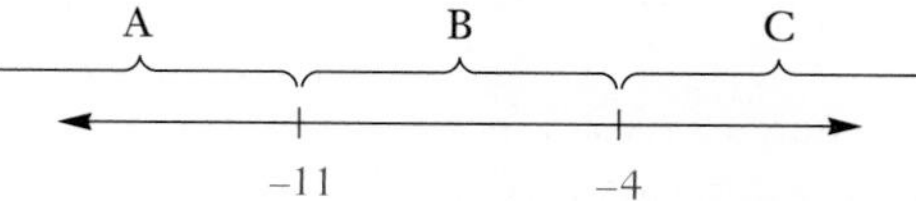

We try test numbers in each interval to see if each satisfies the original inequality.

A: Test -15,

$\frac{x-3}{x+4} \geq 2$	
$\frac{-15-3}{-15+4}$	2
$\frac{18}{11}$	FALSE

Since the inequality is false for $x = -15$, the number -15 is not a solution of the inequality. Interval A is *not* part of the solution set.

Solve.

9. $\dfrac{x+1}{x-2} \geq 3$

10. $\dfrac{x}{x-5} < 2$

ANSWERS ON PAGE A-9

B: Test -8,

$$\begin{array}{c|c} \dfrac{x-3}{x+4} & \geq 2 \\ \hline \dfrac{-8-3}{-8+4} & 2 \\ \dfrac{11}{4} & \quad \text{TRUE} \end{array}$$

Since the inequality is true for $x = -8$, the number -8 is a solution of the inequality. Interval B is part of the solution set.

C: Test 1,

$$\begin{array}{c|c} \dfrac{x-3}{x+4} & \geq 2 \\ \hline \dfrac{1-3}{1+4} & 2 \\ -\dfrac{2}{5} & \quad \text{FALSE} \end{array}$$

Since the inequality is false for $x = 1$, the number 1 is not a solution of the inequality. Interval C is not part of the solution set.

The solution set includes the interval B. The number -11 is also included since the inequality symbol is $\geq$ and -11 is a solution of the related equation. The number -4 is not included since it is not a meaningful replacement. Thus the solution set of the original inequality is

$$\{x \mid -11 \leq x < -4\}.$$ ◀

To solve a rational inequality:

1. **Change the inequality symbol to an equals sign and solve the related equation.**
2. **Find the replacements that are not meaningful.**
3. **Use the numbers found in steps (1) and (2) to divide the number line into intervals.**
4. **Substitute a number from each interval into the inequality. If the number is a solution, then the interval to which it belongs is part of the solution set.**
5. **Select the intervals for which the inequality is satisfied and write set-builder notation for the solution set.**

DO EXERCISES 9 AND 10.

NAME SECTION DATE

EXERCISE SET 8.9

a Solve.

1. $(x-5)(x+3)>0$
2. $(x-4)(x+1)>0$
3. $(x+1)(x-2)\leq 0$
4. $(x-5)(x+3)\leq 0$
5. $x^2-x-2<0$
6. $x^2+x-2<0$
7. $9-x^2\leq 0$
8. $4-x^2\geq 0$
9. $x^2-2x+1\geq 0$
10. $x^2+6x+9<0$
11. $x^2+8<6x$
12. $x^2-12>4x$
13. $3x(x+2)(x-2)<0$
14. $5x(x+1)(x-1)>0$
15. $(x+3)(x-2)(x+1)>0$
16. $(x-1)(x+2)(x-4)<0$
17. $(x+3)(x+2)(x-1)<0$
18. $(x-2)(x-3)(x+1)<0$

b Solve.

19. $\dfrac{1}{x-4}<0$
20. $\dfrac{1}{x+5}>0$
21. $\dfrac{x+1}{x-3}>0$
22. $\dfrac{x-2}{x+5}<0$
23. $\dfrac{3x+2}{x-3}\leq 0$
24. $\dfrac{5-2x}{4x+3}\leq 0$

ANSWERS

1. ______
2. ______
3. ______
4. ______
5. ______
6. ______
7. ______
8. ______
9. ______
10. ______
11. ______
12. ______
13. ______
14. ______
15. ______
16. ______
17. ______
18. ______
19. ______
20. ______
21. ______
22. ______
23. ______
24. ______

ANSWERS

25. ______
26. ______
27. ______
28. ______
29. ______
30. ______
31. ______
32. ______
33. ______
34. ______
35. ______
36. ______
37. ______
38. ______
39. ______
40. ______
41. ______
42. ______
43. ______
44. ______
45. ______
46. a) ______
b) ______
47. a) ______
b) ______

25. $\dfrac{x-1}{x-2} > 3$

26. $\dfrac{x+1}{2x-3} < 1$

27. $\dfrac{(x-2)(x+1)}{x-5} < 0$

28. $\dfrac{(x+4)(x-1)}{x+3} > 0$

29. $\dfrac{x}{x-2} \geq 0$

30. $\dfrac{x+3}{x} \leq 0$

31. $\dfrac{x-5}{x} < 1$

32. $\dfrac{x}{x-1} > 2$

33. $\dfrac{x-1}{(x-3)(x+4)} < 0$

34. $\dfrac{x+2}{(x-2)(x+7)} > 0$

35. $2 < \dfrac{1}{x}$

36. $\dfrac{1}{x} \leq 3$

37. $\dfrac{(x-1)(x+2)}{(x+3)(x-4)} > 0$

38. $\dfrac{(x+3)(x+2)}{(x-1)(x+1)} < 0$

39. $\dfrac{x^2+3x-10}{x^2-x-56} \leq 0$

40. $\dfrac{x^2-11x+30}{x^2-8x-9} \geq 0$

SYNTHESIS

Solve.

41. $x^2 - 2x \leq 2$

42. $x^2 + 2x > 4$

43. $x^4 + 2x^2 > 0$

44. $x^4 + 3x^2 \leq 0$

45. $\left|\dfrac{x+2}{x-1}\right| < 3$

46. *Total profit.* A company determines that its total profit on the production and sale of x units of a product is given by

$$P = -x^2 + 812x - 9600.$$

a) A company makes a profit for those nonnegative values of x for which $P > 0$. Find the values of x for which the company makes a profit.

b) A company loses money for those nonnegative values of x for which $P < 0$. Find the values of x for which the company loses money.

47. *Height of a thrown object.* The equation

$$S = -16t^2 + 32t + 1920$$

gives the height S of an object thrown from a cliff 1920 ft high, after time t seconds.

a) For what times is the height greater than 1920 ft?

b) For what times is the height less than 640 ft?

SUMMARY AND REVIEW: CHAPTER 8

IMPORTANT PROPERTIES AND FORMULAS

Principle of Square Roots: $x^2 = k$ has solutions $x = \sqrt{k}$ and $-\sqrt{k}$.

Compound-Interest Formula: $A = P(1 + r)^t$

Quadratic Formula: $x = \dfrac{-b \pm \sqrt{b^2 - 4ac}}{2a}$; Discriminant: $b^2 - 4ac$

The vertex of the graph of $y = ax^2 + bx + c$ is $\left(-\dfrac{b}{2a}, \dfrac{4ac - b^2}{4a}\right)$.

The line of symmetry is $x = -\dfrac{b}{2a}$.

REVIEW EXERCISES

The review sections and objectives to be tested in addition to the material in this chapter are [4.3a], [6.2a, b], [7.2b], and [7.7a, b].

Solve.

1. $2x^2 - 7 = 0$

2. $14x^2 + 5x = 0$

3. $x^2 - 12x + 27 = 0$

4. $4x^2 + 3x + 1 = 0$

5. $x^2 - 7x + 13 = 0$

6. $4x(x - 1) + 15 = x(3x + 4)$

7. $x^2 + 4x + 1 = 0$. Give exact solutions and approximate solutions to the nearest tenth.

Solve.

8. The width of a rectangle is 5 cm less than the length. The area is 126 cm^2. Find the length and the width.

9. A picture frame measures 16 in. by 12 in., and 140 in^2 of picture shows. Find the width of the frame.

10. When \$1000 is invested at interest rate r, compounded annually, it will grow to \$1690 in 2 years. What is the interest rate?

Solve.

11. $\dfrac{x}{x-2} + \dfrac{4}{x-6} = 0$

12. $\dfrac{x}{5} = \dfrac{x+3}{x+7}$

13. $\dfrac{x}{4} - \dfrac{4}{x} = 2$

14. $15 + \dfrac{6}{x-2} = \dfrac{8}{x+2}$

15. During the first part of a trip, a car travels 50 mi at a certain speed. It travels 80 mi on the second part of the trip at a speed 10 mph slower. The total time for the trip is 3 hr. What is the speed on each part of the trip?

16. Working together, two people can do a job in 4 hr. Person A requires 6 hr longer, working alone, than person B. How long would it take person B to do this job alone?

Determine the nature of the solutions of the equation.

17. $x^2 + 3x - 6 = 0$

18. $x^2 + 2x + 5 = 0$

19. Write a quadratic equation having the solutions $\frac{1}{5}, -\frac{3}{5}$.

20. Write a quadratic equation having -4 as its only solution.

21. Solve $N = 3\pi\sqrt{\frac{1}{p}}$ for p.

22. Solve $2A = \frac{3B}{T^2}$ for T.

Solve.

23. $x^4 - 13x^2 + 36 = 0$

24. $15x^{-2} - 2x^{-1} - 1 = 0$

25. $(x^2 - 4)^2 - (x^2 - 4) - 6 = 0$

26. For $y = -3x^2 - 12x - 8$:

a) Find the vertex.
b) Find the line of symmetry.
c) Graph the equation.

27. For $y = x^2 - x + 6$:

a) Find the vertex.
b) Find the line of symmetry.
c) Graph the equation.

28. Find the x-intercepts: $y = x^2 - 9x + 14$.

29. What is the minimum product of two numbers whose difference is 22? What numbers yield this product?

30. Find the quadratic equation that fits the data points $(0, -2)$, $(1, 3)$, and $(3, 7)$.

Solve.

31. $x^2 < 6x + 7$

32. $\frac{x - 5}{x + 3} < 0$

SKILL MAINTENANCE

33. Metal alloy A is 75% silver. Metal alloy B is 25% silver. How much of each should be mixed in order to produce 300 kg of an alloy that is 60% silver?

34. Solve: $\sqrt{5x - 1} + \sqrt{2x} = 5$.

35. Add and simplify: $\frac{x}{x^2 - 3x + 2} + \frac{2}{x^2 - 5x + 6}$.

36. Multiply and simplify: $\sqrt[3]{9t^6}\,\sqrt[3]{3s^4t^9}$.

SYNTHESIS

37. Solve: $\frac{26}{x + 13} - \frac{14}{x + 7} = \frac{12x}{x^2 + 20x + 91}$.

38. Find h and k, where $3x^2 - hx + 4k = 0$, the sum of the solutions is 20, and the product is 80.

39. The average of two positive integers is 171. One of the numbers is the square root of the other. Find the integers.

❖ THINKING IT THROUGH

1. Explain as many characteristics as you can of the graph of a parabola $y = ax^2 + bx + c$.

2. Explain a general strategy for solving a quadratic equation.

NAME SECTION DATE

TEST: CHAPTER 8

Solve.

1. $3x^2 - 4 = 0$

2. $4x(x-2) - 3x(x+1) = -18$

3. $x^2 + x + 1 = 0$

4. $x^2 + 4x = 2$. Give exact solutions and approximate the solutions to the nearest tenth.

5. $\frac{1}{4-x} + \frac{1}{2+x} = \frac{3}{4}$

6. $x^4 - 5x^2 + 5 = 0$

Solve.

7. A rectangle is 4.5 in. longer than it is wide. Its area is 34 in^2. Find the length and the width.

8. Two pipes can fill a tank in $1\frac{1}{2}$ hr. One pipe requires 4 hr longer running alone to fill the tank than the other. How long would it take for the faster pipe, working alone, to fill the tank?

9. Determine the nature of the solutions of the equation $x^2 + 5x + 17 = 0$.

10. Write a quadratic equation having solutions $\sqrt{3}$ and $3\sqrt{3}$.

11. Solve $V = \frac{1}{3}\pi(R^2 + r^2)$ for r.

12. When \$2000 is invested at interest rate r, compounded annually, it will grow to \$2880 in 2 years. What is the interest rate?

ANSWERS

1. ______

2. ______

3. ______

4. ______

5. ______

6. ______

7. ______

8. ______

9. ______

10. ______

11. ______

12. ______

ANSWERS

13. a) ______

b) ______

c) ______

14. a) ______

b) ______

c) ______

15. ______

16. ______

17. ______

18. ______

19. ______

20. ______

21. ______

22. ______

23. ______

24. ______

25. ______

26. ______

13. For $y = -x^2 - 2x$:

a) Find the vertex.
b) Find the line of symmetry.
c) Graph the equation.

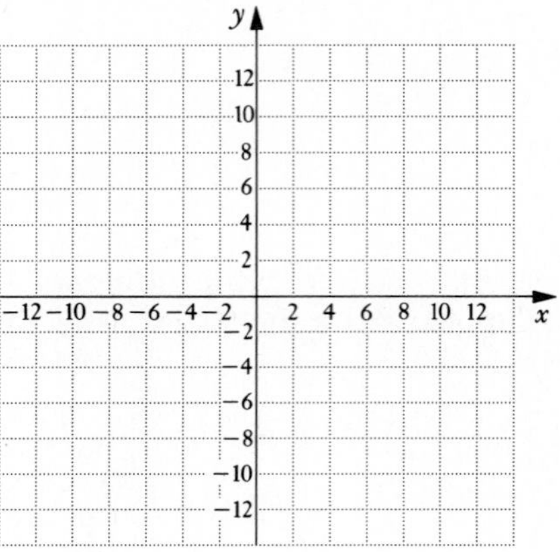

14. For $y = 4(x - 3)^2 + 5$:

a) Find the vertex.
b) Find the line of symmetry.
c) Graph the equation.

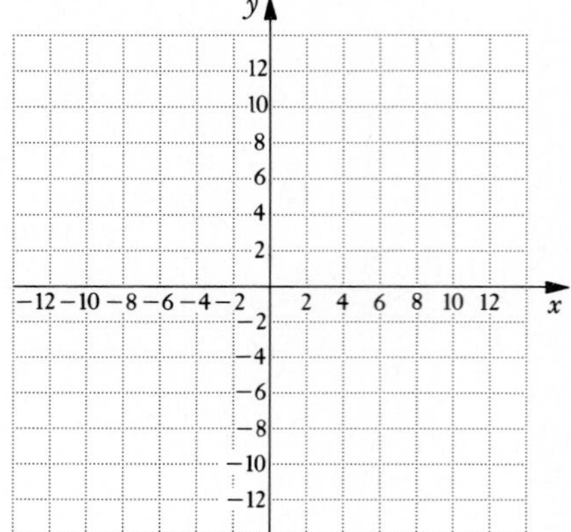

15. Find the x-intercepts:

$$y = -x^2 + 4x - 1.$$

16. Find the x-intercepts:

$$y = x^2 - x - 6.$$

17. What is the minimum product of two numbers having a difference of 8?

18. Find the quadratic equation that fits the data points (0, 0), (3, 0), and (5, 2).

Solve.

19. $(x + 2)(x - 1)(x - 2) > 0$

20. $\dfrac{(x + 4)(x - 1)}{(x + 2)} < 0$

SKILL MAINTENANCE

21. Solve: $\sqrt{x + 3} = x - 3$.

22. Multiply and simplify:

$$\sqrt[4]{2a^2b^3}\,\sqrt[4]{a^4b}.$$

23. Subtract and simplify:

$$\frac{x}{x^2 + 15x + 56} - \frac{7}{x^2 + 13x + 42}.$$

24. The perimeter of a hexagon with all six sides congruent is the same as the perimeter of a square. One side of the hexagon is three less than the side of the square. Find the perimeter of each polygon.

SYNTHESIS

25. One solution of $kx^2 + 3x - k = 0$ is -2. Find the other solution.

26. Solve:

$$\frac{88}{x - 11} - \frac{56}{x - 7} = \frac{32x}{x^2 - 18x + 77}.$$

CUMULATIVE REVIEW: CHAPTERS 1–8

Simplify.

1. $\dfrac{-9.1}{-13}$

2. $-3(x-1)+4(2x+5)$

3. $(-4a^2b^{-3})^{-3}$

4. $\dfrac{3.2 \times 10^{-7}}{8.0 \times 10^{8}}$

5. $(4+8x^2-5x)-(-2x^2+3x-2)$

6. $(2x^2-x+3)(x-4)$

7. $\dfrac{a^2-16}{5a-15} \cdot \dfrac{2a-6}{a+4}$

8. $\dfrac{y}{y^2-y-42} \div \dfrac{y^2}{y-7}$

9. $\dfrac{2}{m+1}+\dfrac{3}{m-5}-\dfrac{m^2-1}{m^2-4m-5}$

10. $\dfrac{\frac{1}{x}-\frac{1}{y}}{x+y}$

11. $(9x^3+5x^2+2) \div (x+2)$

12. $\sqrt{0.36}$

13. $\sqrt{9x^2-36x+36}$

14. $6\sqrt{45}-3\sqrt{20}$

15. $\dfrac{2\sqrt{3}-4\sqrt{2}}{\sqrt{2}-3\sqrt{6}}$

16. $(8^{2/3})^4$

17. $(3+2i)(5-i)$

18. $\dfrac{6-2i}{3i}$

Solve.

19. $3(4x-5)+6=3-(x+1)$

20. $F=\dfrac{mv^2}{r}$ for r

21. $5-3(2x+1) \leq 8x-3$

22. $3x-2<-6$ *or* $x+3>9$

23. $|4x-1| \leq 14$

24. $5x+10y=-10,$ $-2x-3y=5$

25. $2x+y-z=9,$ $4x-2y+z=-9,$ $2x-y+2z=-12$

26. $10x^2+28x-6=0$

27. $\dfrac{2}{n}-\dfrac{7}{n}=3$

28. $\dfrac{1}{2x-1}=\dfrac{3}{5x}$

29. $A=\dfrac{mh}{m+a}$ for m

30. $\sqrt{2x-1}=6$

31. $\sqrt{x-2}+1=\sqrt{2x-6}$

32. $16(t-1)=t(t+8)$

33. $x^2-3x+16=0$

34. $\dfrac{18}{x+1}-\dfrac{12}{x}=\dfrac{1}{3}$

35. $P=\sqrt{a^2-b^2}$ for a

Factor.

36. $2t^2 - 7t - 30$

37. $a^2 + 3a - 54$

38. $24a^3 + 18a^2 - 20a - 15$

39. $-3a^3 + 12a^2$

40. $64a^2 - 9b^2$

41. $3a^2 - 36a + 108$

42. $\frac{1}{27}a^3 - 1$

43. $(x + 1)(x - 1) + (x + 1)(x + 2)$

Graph.

44. $2x - y = 11$

45. $x + y = 2$

46. $y \geq 6x - 5$

47. $x < -3$

48. $3x - y > 6,$
$4x + y \leq 3$

49. $y = x^2 - 1$

50. $y = -2x^2 + 3$

51. Solve: $4x^2 - 25 > 0$.

52. Find an equation of the line with slope $\frac{1}{2}$ through $(-4, 2)$.

53. Find an equation of the line parallel to the line $3x + y = 4$ through $(0, 1)$.

54. The total number of refugees coming into the United States recently from Asia, Eastern Europe, and the USSR was 63,254. There were 12,603 more from the USSR than from Eastern Europe. There were 14,594 more from Asia than from the USSR. How many refugees came from each location?

55. The cost of mailing a letter in Japan is 3 cents more than the cost in Norway. The cost of mailing 2 letters in Norway is 39 cents more than the cost of mailing 1 letter in Japan. What does it cost to mail a letter in each country?

SYNTHESIS

56. Solve $ax^2 - bx = 0$ for x.

57. Simplify: $\left[\frac{1}{(-3)^{-2}} - (-3)^1\right] \cdot [(-3)^2 + (-3)^{-2}]$.

58. Solve: $\frac{2x + 1}{x} = 3 + 7\sqrt{\frac{2x + 1}{x}}$.

59. Factor: $\frac{a^3}{8} + \frac{8b^3}{729}$.

60. Pam can do a certain job in a hours working alone. Elaine can do the same job in b hours working alone. Working together, it takes them t hours to do the job.

a) Find a formula for t.
b) Solve the formula for a.
c) Solve the formula for b.

INTRODUCTION We now consider graphs of second-degree equations in two variables in which one or both variables are raised to the second power. It is a curious fact that we can obtain these graphs by cutting a cone with a plane. Thus they are called *conic sections.*

We will also consider a certain kind of relationship between sets called a *function.* Functions are very important in applications and in mathematics beyond this text.

The review sections to be tested in addition to the material in this chapter are 7.2, 7.5, 8.2, and 8.6. ❖

Conic Sections, Relations, and Functions

AN APPLICATION

A size 7 dress in the United States is a size 39 dress in France. Similarly, a size 10 dress in the United States corresponds to a size 42 dress in France. Find a function that describes this situation.

THE MATHEMATICS

A function that describes the situation is

$$f(x) = x + 32.$$

❖ POINTS TO REMEMBER: CHAPTER 9

Pythagorean Theorem:	$a^2 + b^2 = c^2$
Skills of completing the square:	Sections 8.1, 8.2, and 8.8
Graphing skills:	Sections 3.1–3.3, 8.7, and 8.8
Skills of solving systems of equations:	Chapter 4

PRETEST: CHAPTER 9

1. Find the distance between the points $(-6, 2)$ and $(2, -3)$. Give an exact answer and an approximation to three decimal places.

2. Find the midpoint of the segment with endpoints $(-6, 2)$ and $(2, -3)$.

3. Find the center and the radius of the circle
$$(x - 2)^2 + (y + 7)^2 = 16.$$

4. Find the center and the radius of the circle
$$x^2 + y^2 - 8x + 6y - 11 = 0.$$

5. Find an equation of the circle having center $(2, -9)$ and radius 3.

6. Given the function f described by $f(x) = 5 - x^2$, find $f(-1)$.

Graph.

7. $x^2 - y^2 = 4$

8. $x^2 + y^2 = 4$

9. $\dfrac{x^2}{4} - \dfrac{y^2}{25} = 1$

10. $\dfrac{x^2}{9} + \dfrac{y^2}{16} = 1$

11. $y = x^2 - 4x + 2$

12. $x = y^2 + 4y$

Solve.

13. $2x^2 - y^2 = 2,$
$3x + 2y = -1$

14. $2x^2 - y^2 = 7,$
$xy = 20$

15. The area of a rectangle is 12 m^2 and the length of a diagonal is 5 m. Find the dimensions of the rectangle.

16. The perimeter of a square is 12 ft more than the perimeter of another square. The area of the larger square exceeds the area of the smaller square by 81 ft^2. Find the perimeter of each square.

17. Graph: $f(x) = x^2 - 1$.

18. Given $f(x) = x^2 + 3$ and $g(x) = x - 5$, find $(f + g)(x)$, $(f - g)(x)$, $fg(x)$, $(f/g)(x)$, $ff(x)$, $f \circ g(x)$, and $g \circ f(x)$.

Find a formula for the inverse.

19. $f(x) = 2x - 7$

20. $f(x) = \dfrac{x - 4}{x + 1}$

9.1 Conic Sections: Parabolas and Circles

In Sections 9.1 and 9.2, we study curves formed by cross-sections of cones. The curves formed are called **conic sections.** These curves are graphs of second-degree equations in two variables. Some are shown below:

CONIC SECTIONS IN THREE DIMENSIONS

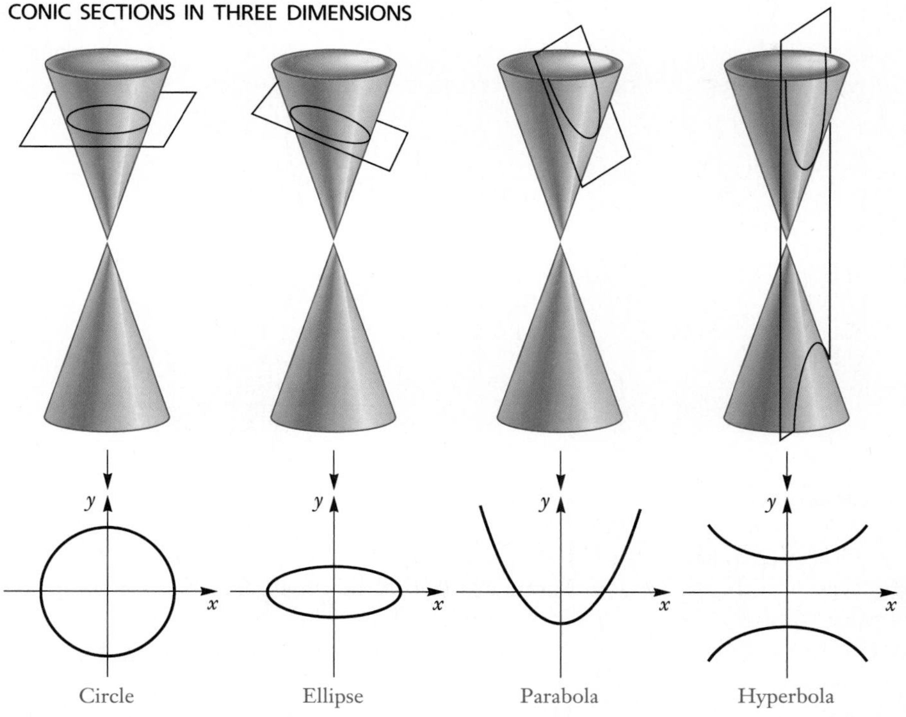

CONIC SECTIONS GRAPHED IN A PLANE

a Parabolas

When a cone is cut as shown in the third figure above, the conic section formed is a **parabola.** As we discussed in Chapter 8, graphs of quadratic equations are parabolas. General equations of parabolas are quadratic.

> ***Parabolas* have equations as follows:**
>
> $\boldsymbol{y = ax^2 + bx + c}$ **(Line of symmetry parallel to the y-axis);**
>
> $\boldsymbol{x = ay^2 + by + c}$ **(Line of symmetry parallel to the x-axis).**

In Chapter 8, we discussed quite extensively graphs of parabolas of the type $y = ax^2 + bx + c$. Our goal here is to review those graphs and extend those ideas to equations of the type $x = ay^2 + by + c$. These parabolas have lines of symmetry parallel to the x-axis and open to the right or to the left.

▶ **EXAMPLE 1** Graph: $y = x^2 - 4x + 8$.

We first find the vertex and the line of symmetry. We can do so in either of two ways. The first way is by completing the square:

$$\begin{aligned} y &= (x^2 - 4x) + 8 \\ &= (x^2 - 4x + 4 - 4) + 8 \qquad [\tfrac{-4}{2}]^2 = 4 \\ &= (x^2 - 4x + 4) + (-4 + 8) \\ &= (x - 2)^2 + 4. \end{aligned}$$

OBJECTIVES

After finishing Section 9.1, you should be able to:

- **a** Graph parabolas.
- **b** Use the distance formula to find the distance between two points whose coordinates are known.
- **c** Use the midpoint formula to find the midpoint of a segment when the coordinates of its endpoints are known.
- **d** Given an equation of a circle in standard form, find its center and radius and graph it; and given the center and radius of a circle, write an equation of the circle.
- **e** Given an equation of a circle not in standard form, complete the square to find standard form and to find the center and radius.

FOR EXTRA HELP

Tape 17B | Tape 15B | MAC: 9 IBM: 9

1. Graph: $y = x^2 + 4x + 7$.

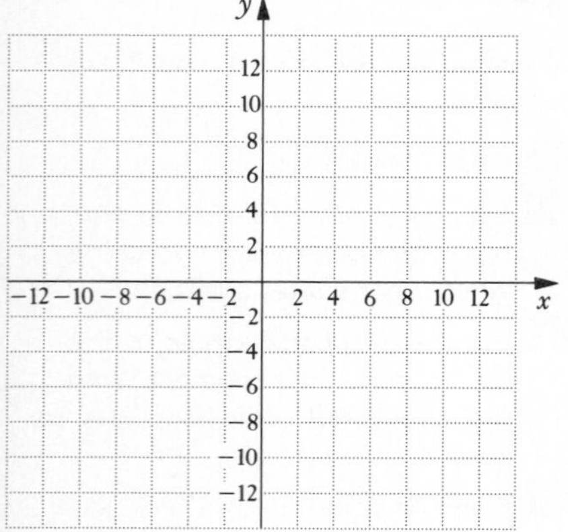

Parabolas have many applications in electricity, mechanics, and optics. The cross-section of a radar reflector is a parabola. Cables that support bridges are shaped like parabolas. (Free-hanging cables have a different shape, called a *catenary*.)

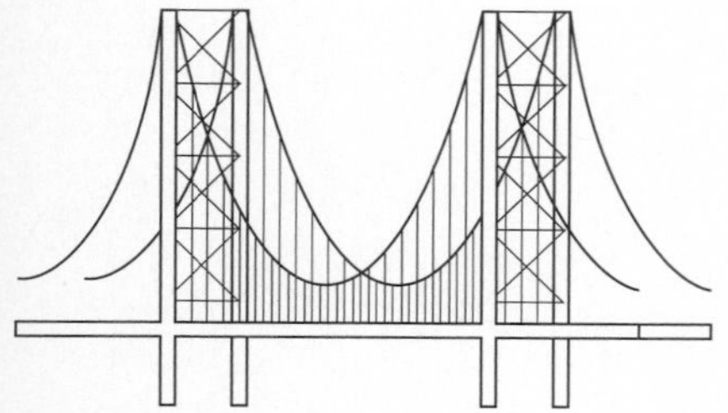

ANSWER ON PAGE A-9

The second way is to use the formula for the first coordinate of the vertex:

$$x = -\frac{b}{2a} = -\frac{-4}{2(1)} = 2; \qquad y = x^2 - 4x + 8 = 2^2 - 4(2) + 8 = 4.$$

Either way we know that the vertex is (2, 4) and the line of symmetry is $x = 4$. We choose some x-values on both sides of the vertex and compute the corresponding y-values. Then we plot the points and graph the parabola. Since the coefficient of x^2, 1, is positive, we know that the graph opens upward. It is easy to find the y-intercept by finding y when $x = 0$.

x	y	
2	4	← Vertex
0	8	← y-intercept
1	5	
3	5	
4	8	
5	13	
−1	13	

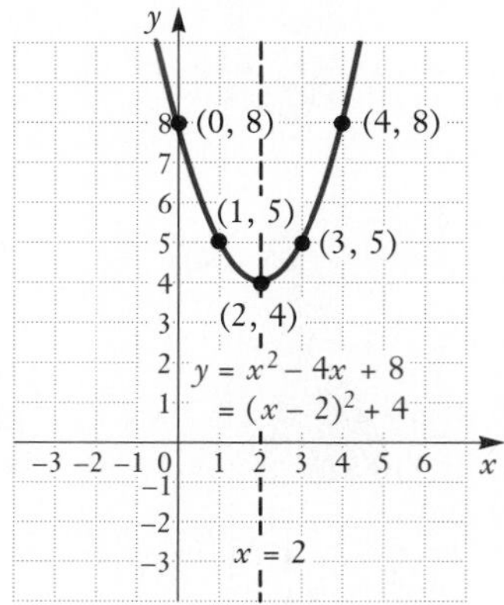

DO EXERCISE 1. ◀

Graphing parabolas of the type $x = ay^2 + by + c$ is similar to those of the type $y = ax^2 + bx + c$. There are two differences. The first is that we choose values for y and then compute x-values. The other is that the graph opens either to the right or the left and has an axis of symmetry parallel to the x-axis.

The graph of the parabola $x = ay^2 + by + c$ opens to the right if $a > 0$. It opens to the left if $a < 0$.

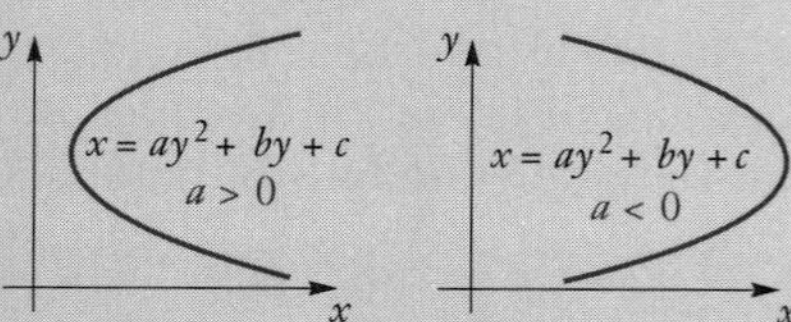

▶ **EXAMPLE 2** Graph: $x = y^2 - 4y + 8$.

The equation $x = y^2 - 4y + 8$ is like the one in Example 1 except that x and y are interchanged. The graphs have the same shape, but this graph is horizontal. We find the vertex and the line of symmetry using either of the two ways shown in Example 1. The difference is that the line of symmetry comes from the y-coordinate:

$$x = y^2 - 4y + 8 = (y - 2)^2 + 4;$$

$$y = -\frac{b}{2a} = 2 \quad \text{and} \quad x = 2^2 - 4(2) + 8 = 4.$$

The vertex is (4, 2) instead of (2, 4) and the line of symmetry is $y = 2$. We choose some y-values on both sides of the vertex and compute the corresponding x-values. Then we plot the points and graph the parabola. Since the coefficient of y^2, 1, is positive, we know that the graph opens to the right. It is easy to find the x-intercept by finding y when $x = 0$.

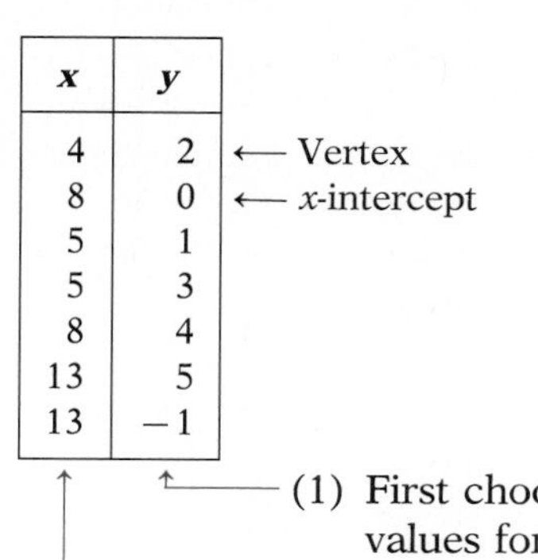

x	y	
4	2	← Vertex
8	0	← x-intercept
5	1	
5	3	
8	4	
13	5	
13	−1	

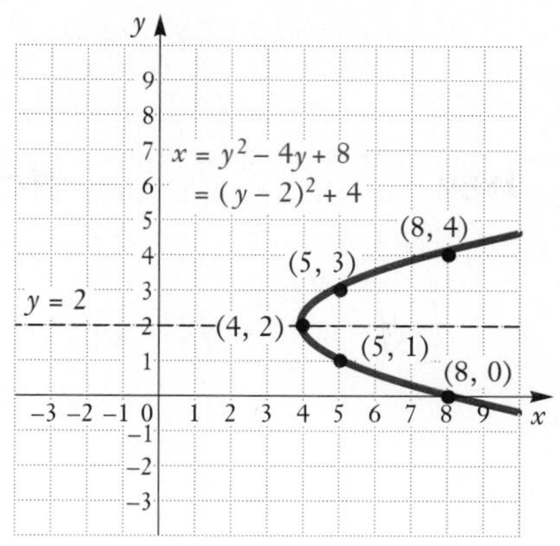

DO EXERCISE 2.

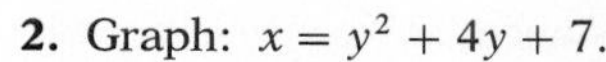

2. Graph: $x = y^2 + 4y + 7$.

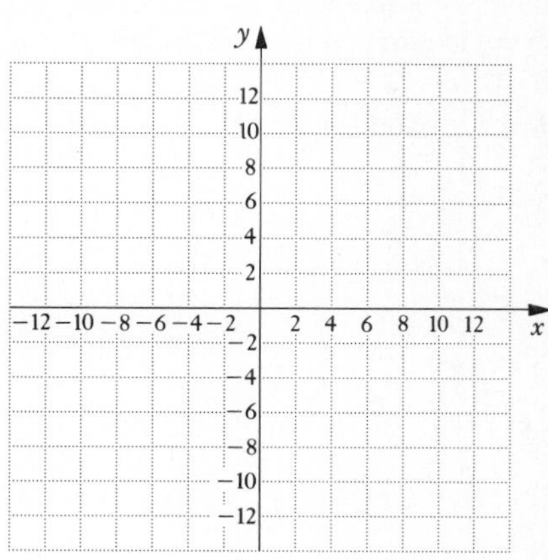

b The Distance Formula

We now develop a formula for finding the distance between any two points on a graph when we know their coordinates. First, we consider points on a vertical or a horizontal line.

If points are on a vertical line, they have the same first coordinate.

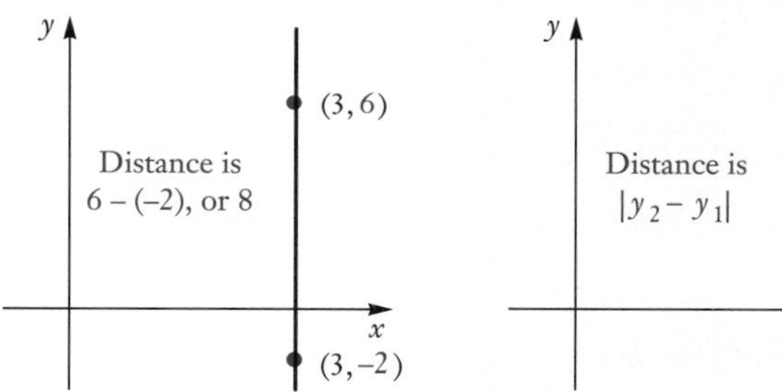

On the left above, we can find the distance between the points by taking the absolute value of the difference of their second coordinates:

$$|6-(-2)| = |8| = 8. \qquad \text{The distance is 8.}$$

If we subtract the opposite way, we get

$$|-2-6| = |-8| = 8.$$

If points are on a horizontal line, we take the absolute value of the difference of their first coordinates.

▶ **EXAMPLES** Find the distance between these points.

3. $(-5, 13)$ and $(-5, 2)$

We take the absolute value of the difference of the second coordinates, since the first coordinates are the same. The distance is

$$|13-2| = |11| = 11.$$

4. $(7, -3)$ and $(-5, -3)$

Since the second coordinates are the same, we take the absolute value of the difference of the first coordinates. The distance is

$$|-5-7| = |-12| = 12. \quad ◀$$

DO EXERCISES 3 AND 4.

Find the distance between the pair of points.

3. $(7, 12)$ and $(7, -2)$

4. $(6, 2)$ and $(-5, 2)$

ANSWERS ON PAGE A-9

Find the distance between the pair of points. Where appropriate, find an approximation to three decimal places.

5. (2, 6) and (−4, −2)

6. (−2, 1) and (4, 2)

ANSWERS ON PAGE A-9

Next we consider two points that are not on either a vertical or a horizontal line, such as (x_1, y_1) and (x_2, y_2) in the figure below. By drawing horizontal and vertical lines through these points, we form a right triangle. The vertex of the right angle of this triangle has coordinates (x_2, y_1). The legs have lengths $|x_2 - x_1|$ and $|y_2 - y_1|$. The distance we want is the length of the hypotenuse. By the Pythagorean theorem (Section 7.8),

$$d^2 = |x_2 - x_1|^2 + |y_2 - y_1|^2.$$

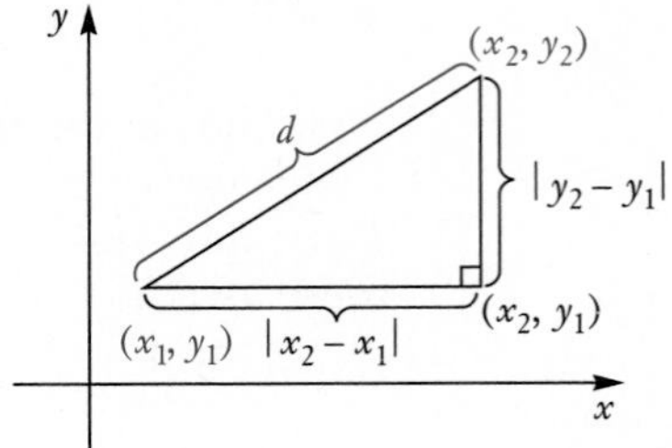

Since squares of numbers are never negative, we don't really need the absolute-value signs. Thus we have

$$d^2 = (x_2 - x_1)^2 + (y_2 - y_1)^2.$$

Taking the principal square root, we get the distance formula.

> The Distance Formula
>
> **The distance between any two points (x_1, y_1) and (x_2, y_2) is given by**
> $$d = \sqrt{(x_2 - x_1)^2 + (y_2 - y_1)^2}.$$

This formula holds even when the two points are on a vertical or a horizontal line.

▶ **EXAMPLE 5** Find the distance between (4, −3) and (−5, 4). Find an exact answer and an approximation to three decimal places.

We substitute into the distance formula:

$$d = \sqrt{(-5 - 4)^2 + [4 - (-3)]^2}$$
$$d = \sqrt{(-9)^2 + 7^2} = \sqrt{130} \approx 11.402.$$

Note that the distance formula has a radical sign in it. Do not make the mistake of writing $d = (x_2 - x_1)^2 + (y_2 - y_1)^2$. ◀

DO EXERCISES 5 AND 6.

C Midpoints of Segments

The distance formula can be used to verify or derive a formula for finding the coordinates of the midpoint of a segment when the coordinates of the endpoints are known. We will not derive the formula but simply state it.

> The Midpoint Formula
>
> **If the endpoints of a segment are (x_1, y_1) and (x_2, y_2), then the coordinates of the midpoint are**
> $$\left(\frac{x_1 + x_2}{2}, \frac{y_1 + y_2}{2}\right).$$
> **(We obtain the coordinates of the midpoint by averaging the coordinates of the endpoints.)**

▶ **EXAMPLE 6** Find the midpoint of the segment with endpoints $(-2, 3)$ and $(4, -6)$.

Using the midpoint formula, we obtain

$$\left(\frac{-2+4}{2}, \frac{3+(-6)}{2}\right), \quad \text{or} \quad \left(\frac{2}{2}, \frac{-3}{2}\right), \quad \text{or} \quad \left(1, -\frac{3}{2}\right). \qquad ◀$$

DO EXERCISES 7 AND 8.

Find the midpoint of the segment with the given endpoints.

7. $(-3, 1)$ and $(6, -7)$

8. $(10, -7)$ and $(8, -3)$

d Circles

Another conic section, or curve, shown in the figure at the beginning of this section is a **circle.** We can describe a circle in a plane as follows.

> **A *circle* is defined as the set of all points in a plane that are a fixed distance from a point in that plane.**

Let us find an equation for a circle. We call the center (h, k) and let the radius have length r. Suppose that (x, y) is any point on the circle. By the distance formula, we have

$$\sqrt{(x-h)^2 + (y-k)^2} = r.$$

Squaring both sides gives an equation of the circle in standard form: $(x-h)^2 + (y-k)^2 = r^2$. When $h = 0$ and $k = 0$, the circle is centered at the origin. Otherwise, we can think of that circle being translated $|h|$ units horizontally and $|k|$ units vertically.

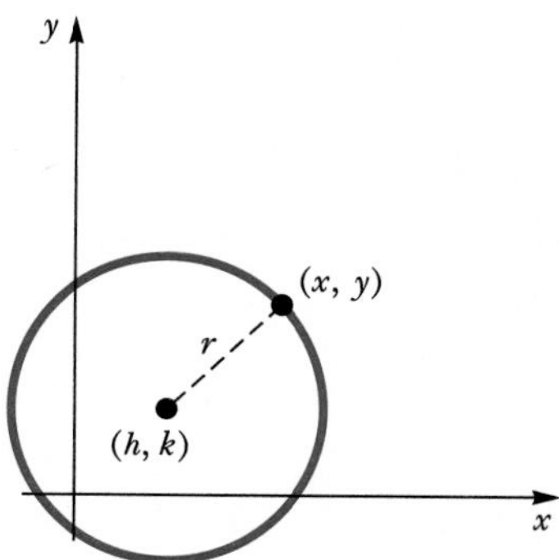

> Equation of a Circle
>
> **A circle with center (h, k) and radius r has equation**
>
> $$(x-h)^2 + (y-k)^2 = r^2. \qquad \textbf{(Standard form)}$$

▶ **EXAMPLE 7** Find the center and the radius and graph this circle:

$$(x+2)^2 + (y-3)^2 = 16.$$

First, we find an equivalent equation in standard form:

$$[x-(-2)]^2 + (y-3)^2 = 4^2.$$

Thus the center is $(-2, 3)$, and the radius is 4. We draw the graph, shown at the top of the following page, by locating the center and then using a compass, setting its radius at 4, to draw the circle.

ANSWERS ON PAGE A-9

9. Find the center and the radius of the circle $(x-5)^2 + (y+\frac{1}{2})^2 = 9$. Then graph the circle.

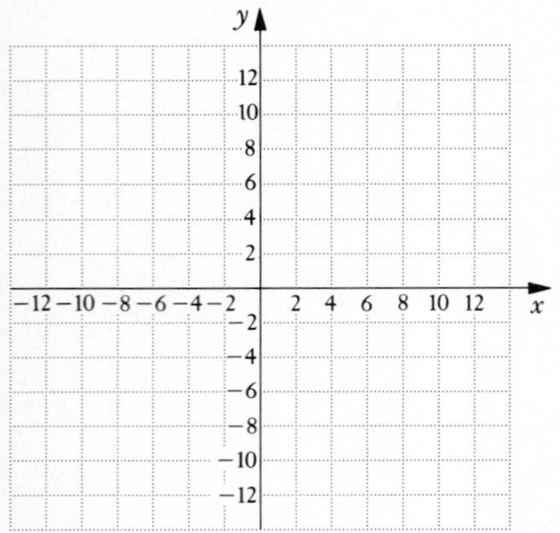

10. Find an equation of a circle with center $(-3, 1)$ and radius 6.

11. Find the center and the radius of the circle $x^2 + y^2 = 64$.

12. Show that this is an equation of a circle. Find the center and the radius.

$$x^2 + y^2 + 8x - 10y - 8 = 0$$

ANSWERS ON PAGE A-9

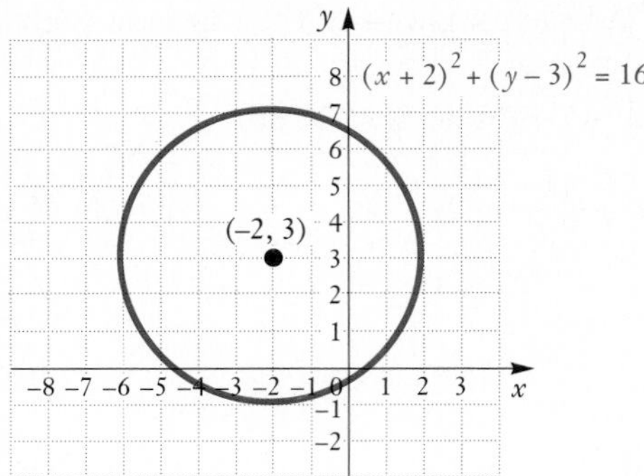

DO EXERCISE 9.

► **EXAMPLE 8** Write an equation of a circle with center $(9, -5)$ and radius $\sqrt{2}$.

We use standard form $(x-h)^2 + (y-k)^2 = r^2$ and substitute:

$$(x-9)^2 + [y-(-5)]^2 = (\sqrt{2})^2 \quad \textbf{Substituting}$$
$$(x-9)^2 + (y+5)^2 = 2. \quad \textbf{Simplifying}$$

DO EXERCISE 10.

If a circle is centered at the origin, then $(h, k) = (0, 0)$.

A circle centered at the origin with radius *r* has equation
$$x^2 + y^2 = r^2.$$

DO EXERCISE 11.

e Equations of Circles Not in Standard Form

By completing the square, we can show that certain equations have graphs that are circles.

► **EXAMPLE 9** Show that this is an equation of a circle. Find the center and the radius.

$$x^2 + y^2 + 6x - 4y + 9 = 0$$

We first regroup the terms and then complete the square twice, once with $x^2 + 6x$ and once with $y^2 - 4y$:

$$x^2 + y^2 + 6x - 4y + 9 = 0$$
$$(x^2 + 6x) + (y^2 - 4y) + 9 = 0$$
$$(x^2 + 6x + 9 - 9) + (y^2 - 4y + 4 - 4) + 9 = 0$$
$$(x^2 + 6x + 9) + (y^2 - 4y + 4) - 9 - 4 + 9 = 0$$
$$(x+3)^2 + (y-2)^2 = 4$$
$$[x-(-3)]^2 + (y-2)^2 = 2^2.$$

This shows that the graph of the equation is a circle. The center is $(-3, 2)$ and the radius is 2.

Remember that the standard form is $(x-h)^2 + (y-k)^2 = r^2$.

DO EXERCISE 12.

NAME SECTION DATE

EXERCISE SET 9.1

a Graph the equation.

1. $y = x^2$

2. $x = y^2$

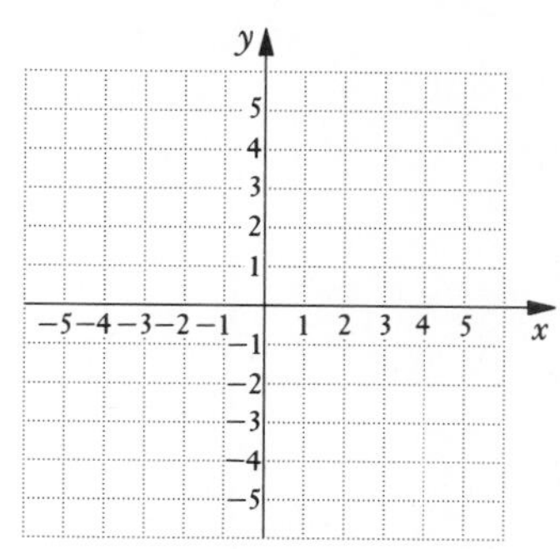

3. $x = y^2 + 4y + 1$

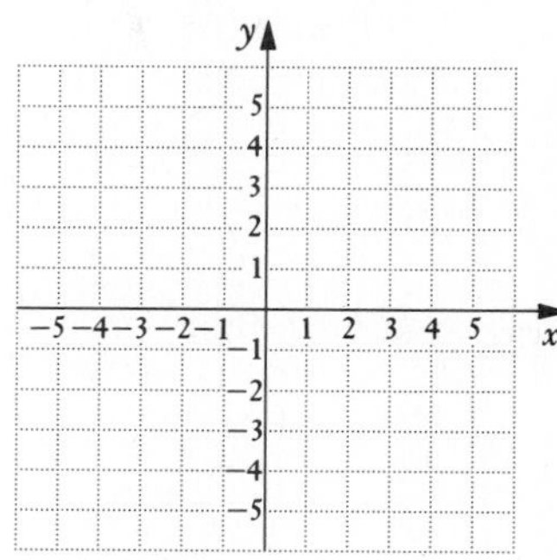

4. $y = x^2 - 2x + 3$

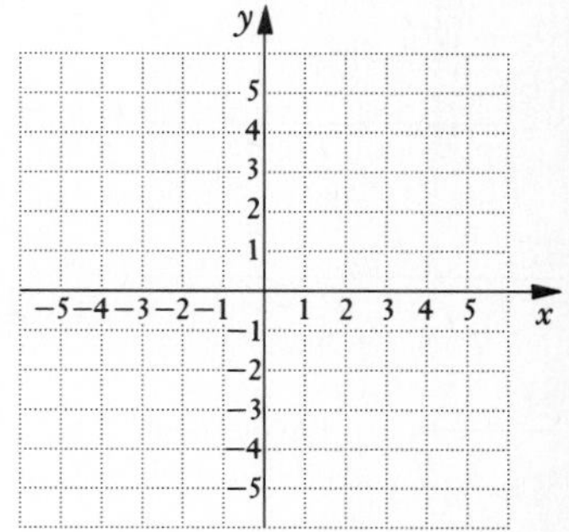

5. $y = -x^2 + 4x - 5$

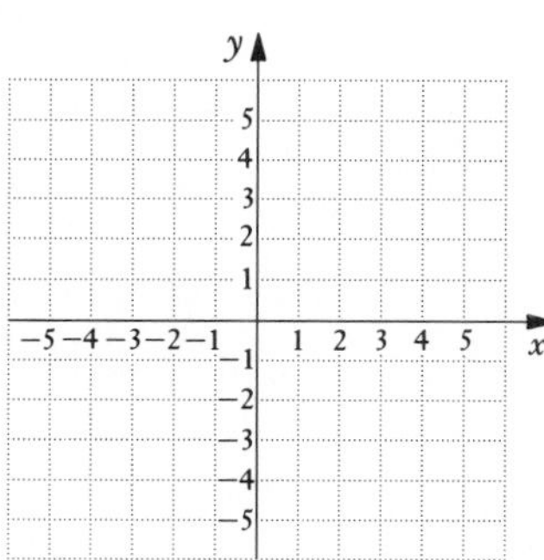

6. $x = 4 - 3y - y^2$

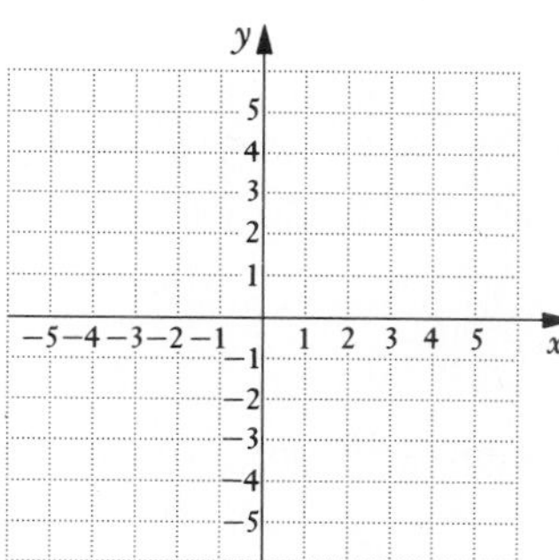

7. $x = -3y^2 - 6y - 1$

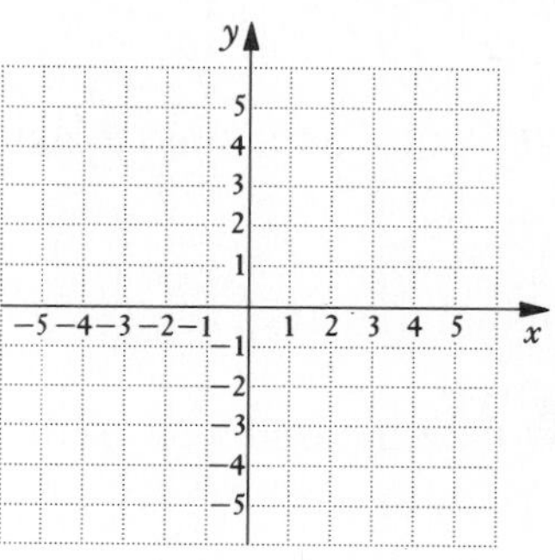

8. $y = -5 - 8x - 2x^2$

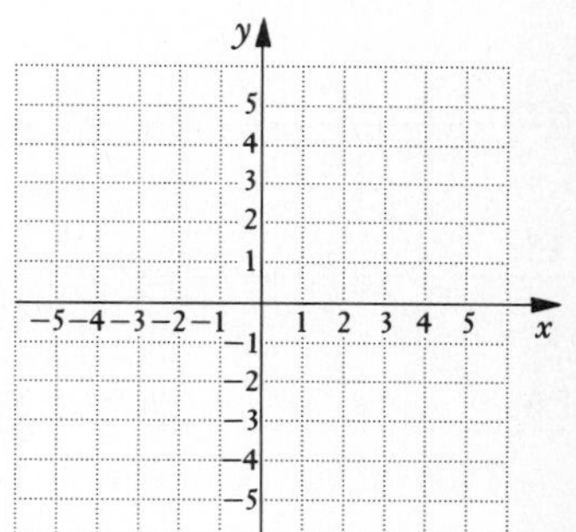

b Find the distance between the pair of points. Where appropriate, find an approximation to three decimal places.

9. (9, 5) and (6, 1)

10. (1, 10) and (7, 2)

11. (0, −7) and (3, −4)

12. (6, 2) and (6, −8)

13. (2, 2) and (−2, −2)

14. (5, 21) and (−3, 1)

15. (8.6, −3.4) and (−9.2, −3.4)

16. (5.9, 2) and (3.7, −7.7)

ANSWERS

9. ______

10. ______

11. ______

12. ______

13. ______

14. ______

15. ______

16. ______

ANSWERS

17. ______

18. ______

19. ______

20. ______

21. ______

22. ______

23. ______

24. ______

25. ______

26. ______

27. ______

28. ______

29. ______

30. ______

31. ______

32. ______

33. ______

34. ______

17. $\left(\frac{5}{7}, \frac{1}{14}\right)$ and $\left(\frac{1}{7}, \frac{11}{14}\right)$

18. $(0, \sqrt{7})$ and $(\sqrt{6}, 0)$

19. $(-23, 10)$ and $(56, -17)$

20. $(34, -18)$ and $(-46, -38)$

21. (a, b) and $(0, 0)$

22. $(0, 0)$ and (p, q)

23. $(\sqrt{2}, -\sqrt{3}$ and $(-\sqrt{7}, \sqrt{5})$

24. $(\sqrt{8}, \sqrt{3})$ and $(-\sqrt{5}, -\sqrt{6})$

25. $(1000, -240)$ and $(-2000, 580)$

26. $(-3000, 560)$ and $(-430, -640)$

C Find the midpoint of the segment with the given endpoints.

27. $(-3, 6)$ and $(2, -8)$

28. $(6, 7)$ and $(7, -9)$

29. $(8, 5)$ and $(-1, 2)$

30. $(-1, 2)$ and $(1, -3)$

31. $(-8, -5)$ and $(6, -1)$

32. $(8, -2)$ and $(-3, 4)$

33. $(-3.4, 8.1)$ and $(2.9, -8.7)$

34. $(4.1, 6.9)$ and $(5.2, -6.9)$

35. $\left(\frac{1}{6}, -\frac{3}{4}\right)$ and $\left(-\frac{1}{3}, \frac{5}{6}\right)$

36. $\left(-\frac{4}{5}, -\frac{2}{3}\right)$ and $\left(\frac{1}{8}, \frac{3}{4}\right)$

37. $(\sqrt{2}, -1)$ and $(\sqrt{3}, 4)$

38. $(9, 2\sqrt{3})$ and $(-4, 5\sqrt{3})$

d Find the center and the radius of the circle. Then graph the circle.

39. $(x + 1)^2 + (y + 3)^2 = 4$

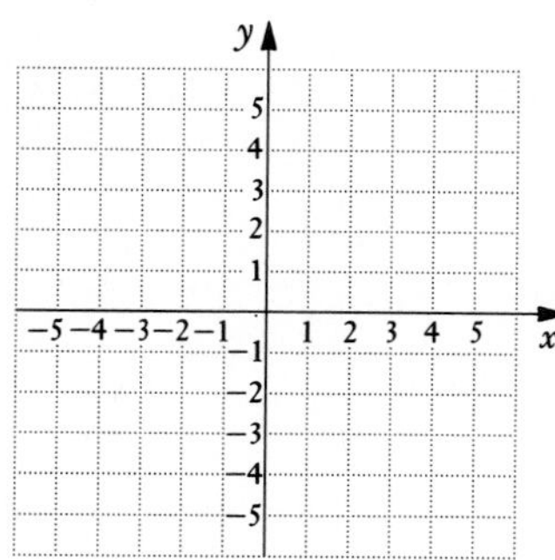

40. $(x - 2)^2 + (y + 3)^2 = 1$

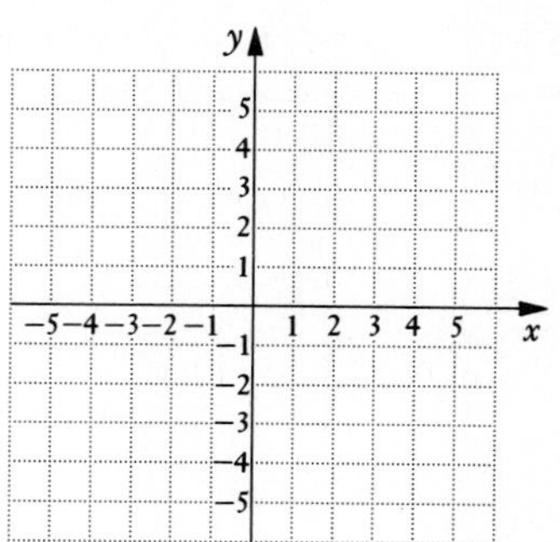

41. $(x - 3)^2 + y^2 = 2$

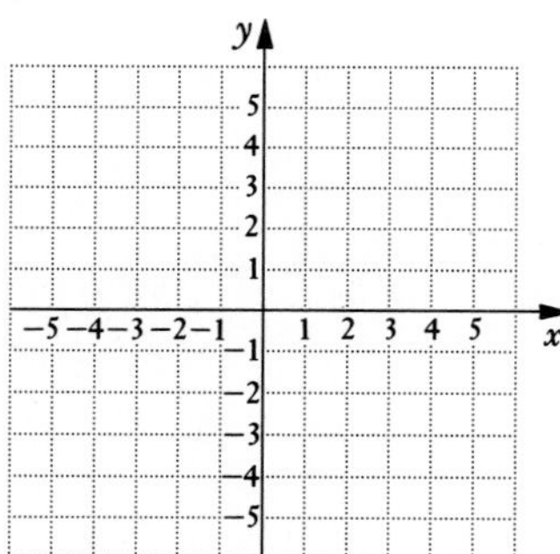

42. $x^2 + (y - 1)^2 = 3$

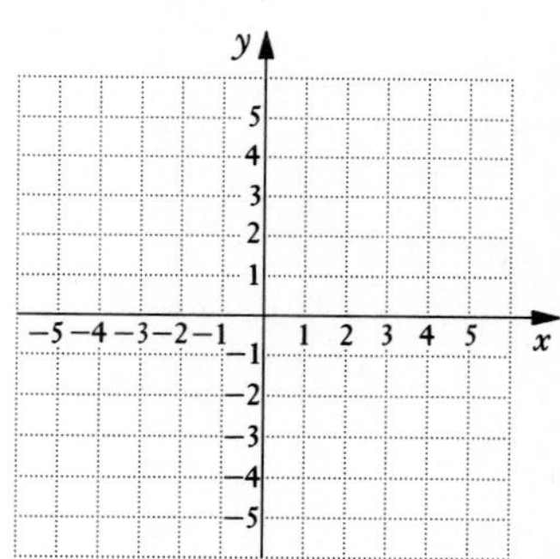

43. $x^2 + y^2 = 25$

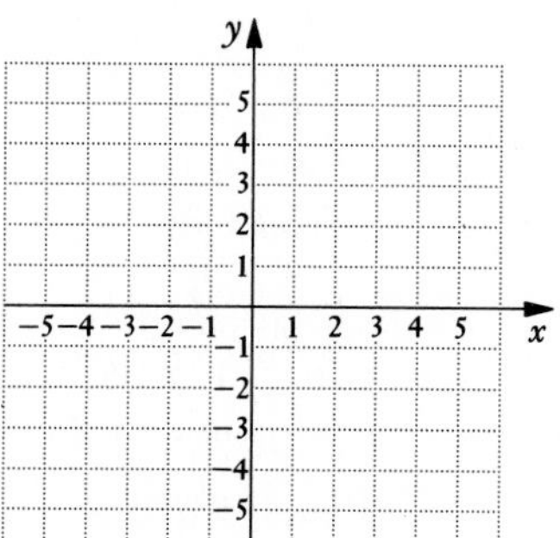

44. $x^2 + y^2 = 9$

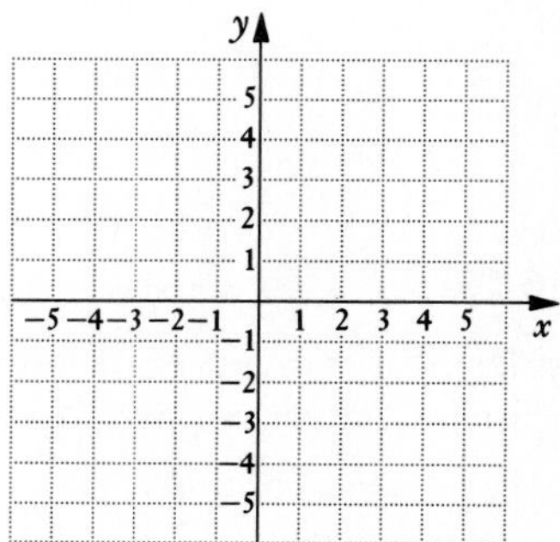

ANSWERS

35. ______

36. ______

37. ______

38. ______

39. ______

40. ______

41. ______

42. ______

43. ______

44. ______

ANSWERS

45. ______
46. ______
47. ______
48. ______
49. ______
50. ______
51. ______
52. ______
53. ______
54. ______
55. ______
56. ______
57. ______
58. ______
59. ______
60. ______
61. ______
62. ______
63. ______
64. ______
65. ______
66. ______
67. ______
68. ______
69. ______

Find an equation of the circle having the given center and radius.

45. Center (0, 0), radius 7

46. Center (0, 0), radius 4

47. Center $(-2, 7)$, radius $\sqrt{5}$

48. Center (5, 6), radius $2\sqrt{3}$

e Find the center and the radius of the circle.

49. $x^2 + y^2 + 8x - 6y - 15 = 0$

50. $x^2 + y^2 + 6x - 4y - 15 = 0$

51. $x^2 + y^2 - 8x + 2y + 13 = 0$

52. $x^2 + y^2 + 6x + 4y + 12 = 0$

53. $x^2 + y^2 - 4x = 0$

54. $x^2 + y^2 + 10y - 75 = 0$

SKILL MAINTENANCE

Solve.

55. $2x + 3y = 8,$
$x - 2y = -3$

56. $x^4 - 20x^2 + 64 = 0$

SYNTHESIS

Find an equation of a circle satisfying the given conditions.

57. Center (0, 0), passing through $(1/4, \sqrt{31}/4)$

58. Center $(-4, 1)$, passing through $(2, -5)$

59. Center $(-3, -2)$ and tangent to the y-axis

60. The endpoints of a diameter are $(7, 3)$ and $(-1, -3)$.

Find the distance between the given points.

61. $(-1, 3k)$ and $(6, 2k)$

62. (a, b) and $(-a, -b)$

63. $(6m, -7n)$ and $(-2m, n)$

64. $(\sqrt{d}, -\sqrt{3c})$ and $(\sqrt{d}, \sqrt{3c})$

65. $(-3\sqrt{3}, 1 - \sqrt{6})$ and $(\sqrt{3}, 1 + \sqrt{6})$

If the sides of a triangle have lengths a, b, and c and $a^2 + b^2 = c^2$, then the triangle is a right triangle. Determine whether the given points are vertices of a right triangle.

66. $(9, 6)$, $(-1, 2)$, and $(1, -3)$

67. $(-8, -5)$, $(6, 1)$, and $(-4, 5)$

68. Find the point on the y-axis that is equidistant from (2, 10) and (6, 2).

69. Find the midpoint of the segments with the endpoints $(2 - \sqrt{3}, 5\sqrt{2})$ and $(2 + \sqrt{3}, 3\sqrt{2})$.

9.2 Conic Sections: Ellipses and Hyperbolas

OBJECTIVES

After finishing Section 9.2, you should be able to:

a Given the standard equation of an ellipse, graph the ellipse.

b Given an equation of a hyperbola in standard form, graph it.

FOR EXTRA HELP

Tape 18A

Tape 16A

MAC: 9
IBM: 9

a Ellipses

When a cone is cut at an angle, as shown below, the conic section formed is an *ellipse.*

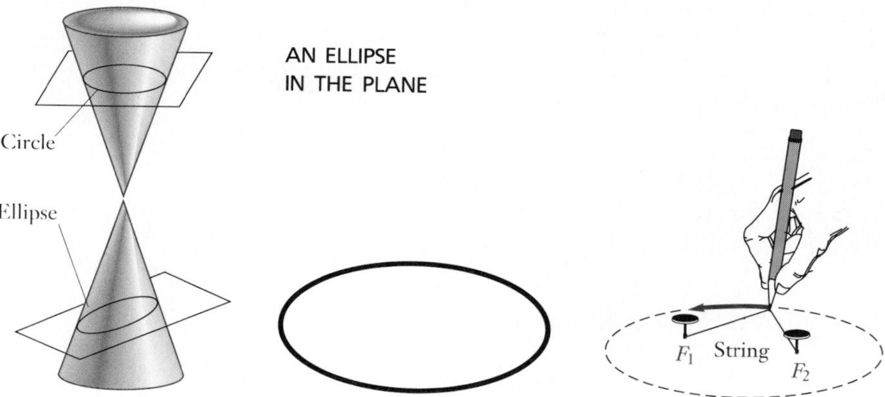

You can draw an ellipse by sticking two tacks in a piece of cardboard. Then tie a string to the tacks, place a pencil as shown, and draw. The formal mathematical definition is related to this method of drawing.

An **ellipse** is defined as the set of all points in a plane such that the *sum* of the distances from two fixed points F_1 and F_2 (called **foci**) is constant. In the preceding drawing, the tacks are at the foci. Ellipses have equations as follows.

> **Equation of an Ellipse**
>
> **An ellipse with its center at the origin has equation**
>
> $$\frac{x^2}{a^2} + \frac{y^2}{b^2} = 1, \ a, b > 0, \quad a \neq b. \qquad \textbf{(Standard form)}$$

> We can think of a circle as a special kind of ellipse. A circle is formed when $a = b$ and the angle at which the cutting plane cuts the cone is 90°. It is also formed when the foci, F_1 and F_2, are the same point. An ellipse with its foci close together is very nearly a circle.

When graphing ellipses, it helps to first find the intercepts. If we replace x by 0, we can find the y-intercepts:

$$\frac{0^2}{a^2} + \frac{y^2}{b^2} = 1$$

$$\frac{y^2}{b^2} = 1$$

$$y^2 = b^2$$

$$y = \pm b.$$

Thus the y-intercepts are $(0, b)$ and $(0, -b)$. Similarly, the x-intercepts are $(a, 0)$ and $(-a, 0)$.

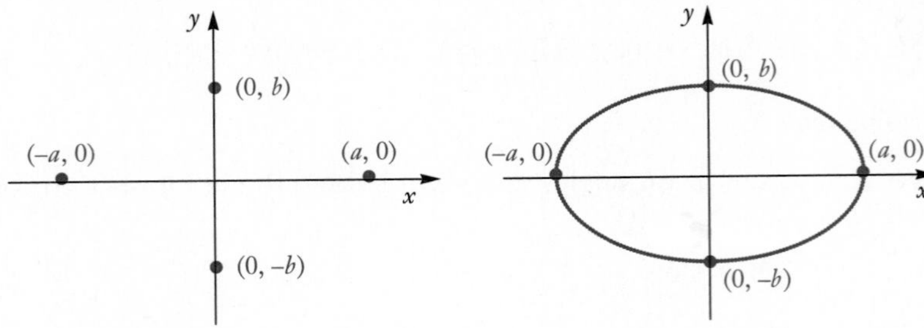

Plotting these points and filling in an oval-shaped curve, we get a graph of the ellipse. If a more precise graph is desired, we can plot more points.

> **For the ellipse**
>
> $$\frac{x^2}{a^2} + \frac{y^2}{b^2} = 1,$$
>
> **the x-intercepts are $(-a, 0)$ and $(a, 0)$, and the y-intercepts are $(0, -b)$ and $(0, b)$.**

▶ **EXAMPLE 1** Graph the ellipse

$$\frac{x^2}{4} + \frac{y^2}{9} = 1.$$

Note that

$$\frac{x^2}{4} + \frac{y^2}{9} = \frac{x^2}{2^2} + \frac{y^2}{3^2}.$$

Thus the x-intercepts are $(-2, 0)$ and $(2, 0)$, and the y-intercepts are $(0, -3)$ and $(0, 3)$. We plot these points and connect them with an oval-shaped curve. To be accurate, we might find some other points on the curve. We let $x = 1$ and solve for y:

$$\begin{aligned}
\frac{1^2}{4} + \frac{y^2}{9} &= 1 \\
36\left(\frac{1}{4} + \frac{y^2}{9}\right) &= 36 \cdot 1 \\
36 \cdot \frac{1}{4} + 36 \cdot \frac{y^2}{9} &= 36 \\
9 + 4y^2 &= 36 \\
4y^2 &= 27 \\
y^2 &= \frac{27}{4} \\
y = \pm\sqrt{\frac{27}{4}} &= \pm\frac{3\sqrt{3}}{2} \approx \pm 2.6.
\end{aligned}$$

Thus, $(1, 2.6)$ and $(1, -2.6)$ can also be plotted and used to draw the graph. Similarly, the points $(-1, -2.6)$ and $(-1, 2.6)$ can also be computed and plotted.

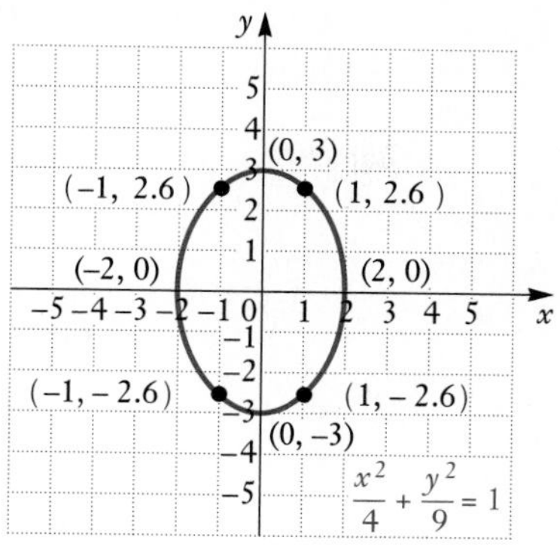

DO EXERCISES 1 AND 2.

b Hyperbolas

A **hyperbola** looks somewhat like a pair of parabolas, but the actual shapes are different. A hyperbola has two vertices, and the line through the vertices is known as an *axis*. The point halfway between the vertices is called the *center*. We consider only hyperbolas centered at the origin.

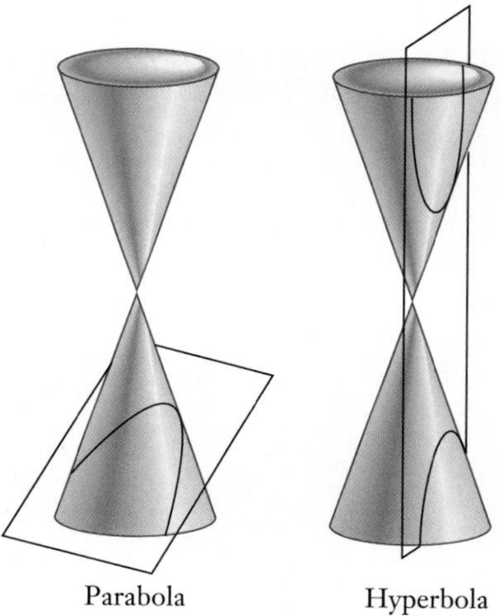

Hyperbolas **with their centers at the origin have equations as follows:**

$$\frac{x^2}{a^2} - \frac{y^2}{b^2} = 1 \quad \textbf{(Axis horizontal);}$$

$$\frac{y^2}{b^2} - \frac{x^2}{a^2} = 1 \quad \textbf{(Axis vertical).}$$

Note carefully that these equations have a 1 on the right and a minus sign between the terms.

Graph the ellipse.

1. $\frac{x^2}{9} + \frac{y^2}{4} = 1$

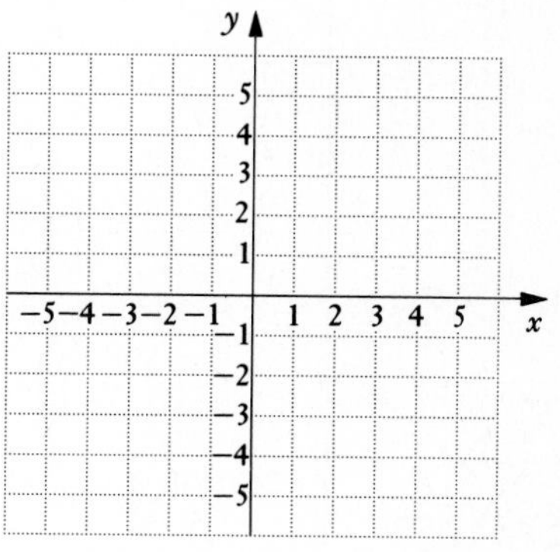

2. $\frac{x^2}{9} + \frac{y^2}{25} = 1$

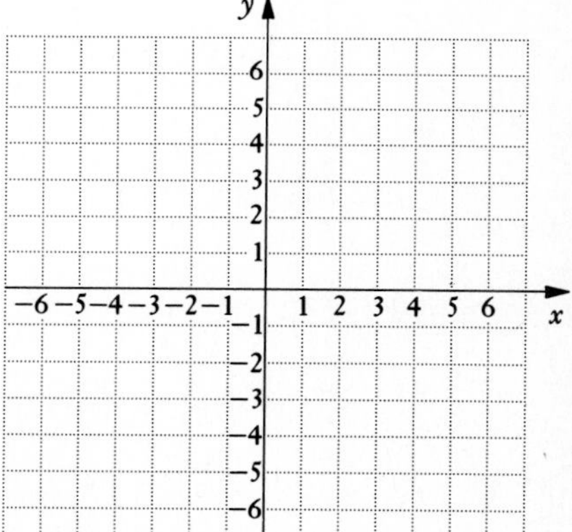

ANSWERS ON PAGE A-10

3. Graph: $\frac{x^2}{16} - \frac{y^2}{25} = 1.$

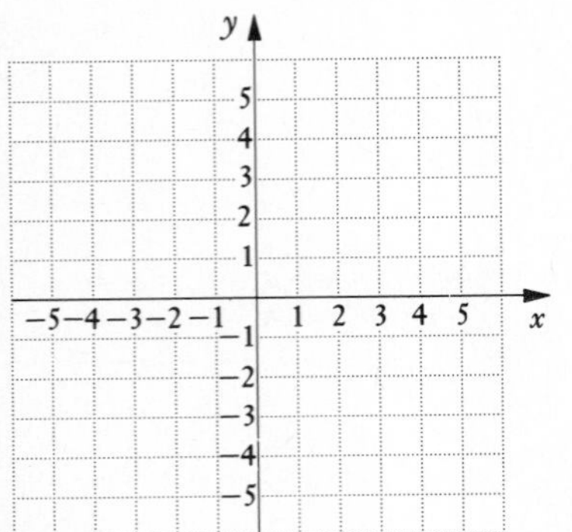

ANSWER ON PAGE A-10

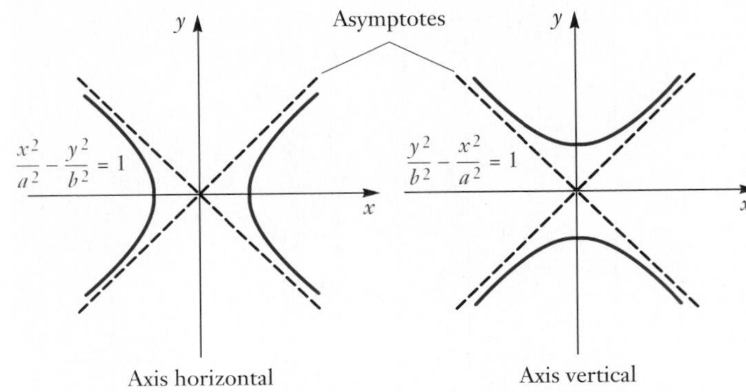

To graph a hyperbola, it helps to begin by graphing the lines called *asymptotes*.

> Asymptotes of a Hyperbola
>
> **For hyperbolas with equations as given above, the *asymptotes* are the lines**
>
> $$y = \frac{b}{a}x \quad \text{and} \quad y = -\frac{b}{a}x.$$

As a hyperbola gets farther away from the origin, it gets closer and closer to its asymptotes. The larger $|x|$ gets, the closer the graph gets to an asymptote. The asymptotes act to "constrain" the graph of a hyperbola. Parabolas do *not* have asymptotes. There are no lines to constrain parabolas.

The next thing to do is to find the vertices. Then it is easy to sketch the curve.

▶ **EXAMPLE 2** Graph: $\frac{x^2}{4} - \frac{y^2}{9} = 1.$

a) Note that

$$\frac{x^2}{4} - \frac{y^2}{9} = \frac{x^2}{2^2} - \frac{y^2}{3^2},$$

so $a = 2$ and $b = 3$. The asymptotes are thus

$$y = \frac{3}{2}x \quad \text{and} \quad y = -\frac{3}{2}x.$$

We sketch them.

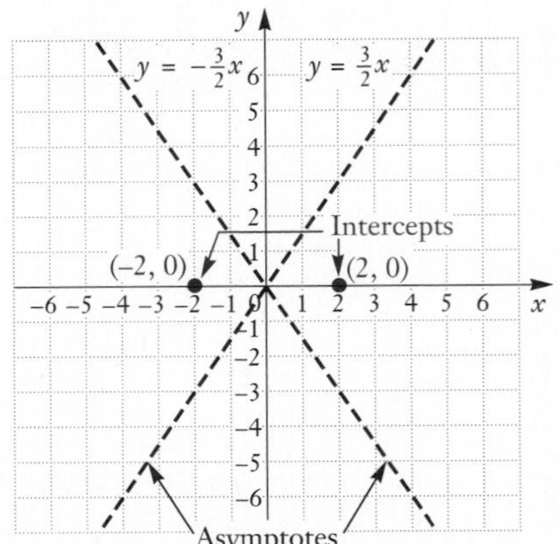

b) Next we find the intercepts, or *vertices*. If we let $y = 0$ (simply cover up the term containing y), we see that $x^2/2^2 = 1$, so $x = \pm 2$. The intercepts are $(2, 0)$ and $(-2, 0)$.

There are intercepts on only one axis. If we let $x = 0$ (simply cover up the term containing x), we see that $y^2/9 = -1$, and this equation has no real-number solutions.

c) We plot the intercepts. Then through each intercept, we draw a smooth curve that approaches the asymptotes closely.

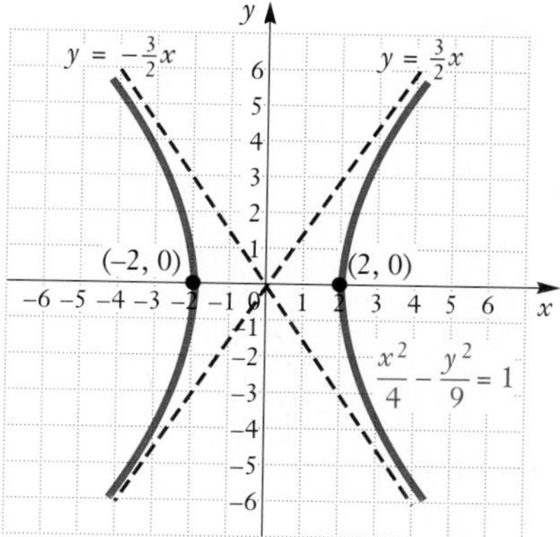

DO EXERCISE 3 ON THE PRECEDING PAGE.

▶ **EXAMPLE 3** Graph: $\dfrac{y^2}{36} - \dfrac{x^2}{4} = 1.$

a) Note that

$$\frac{y^2}{36} - \frac{x^2}{4} = \frac{y^2}{6^2} - \frac{x^2}{2^2} = 1.$$

The intercept distance is this number in the term without the minus sign. There is a y in this term, so the intercepts are on the y-axis.

The asymptotes are thus $y = \frac{6}{2}x$ and $y = -\frac{6}{2}x$, or $y = 3x$ and $y = -3x$. Using the numbers 6 and 2, we can quickly sketch a rectangle to use as a guide. To find the corners, we go 6 units (up or down) in the y-direction and 2 units (right or left) in the x-direction. The asymptotes go through the corners.

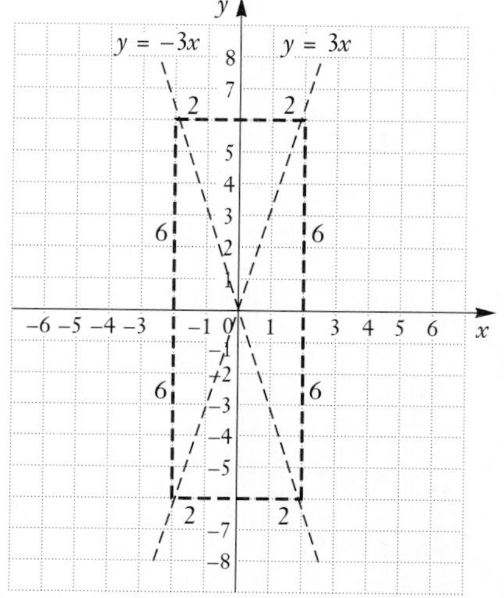

4. Graph.

a) $\dfrac{y^2}{9} - \dfrac{x^2}{49} = 1$

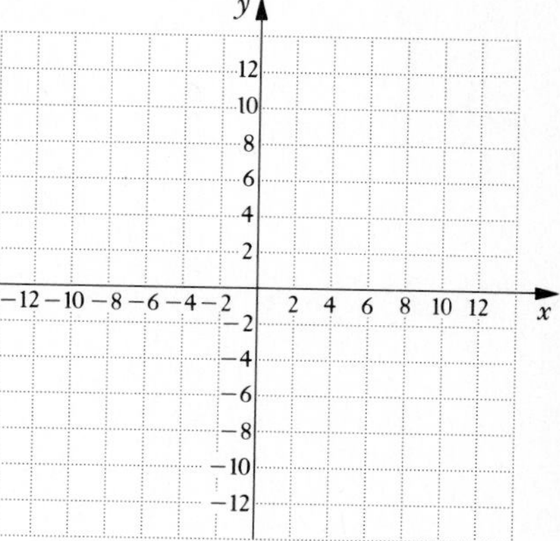

b) $\dfrac{x^2}{49} - \dfrac{y^2}{9} = 1$

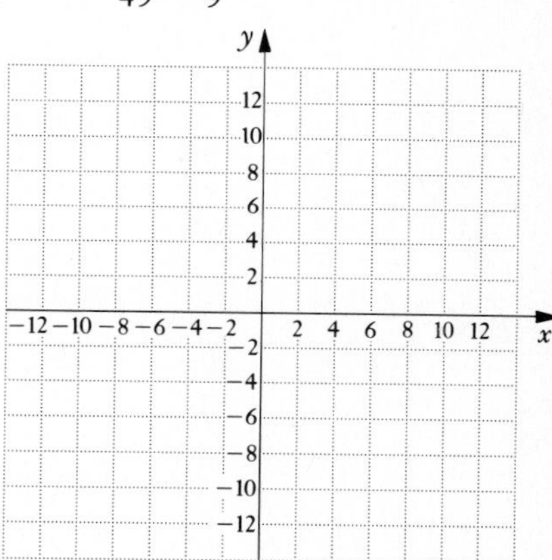

The sound waves from an airplane breaking the sound barrier form a hyperbola across the ground.

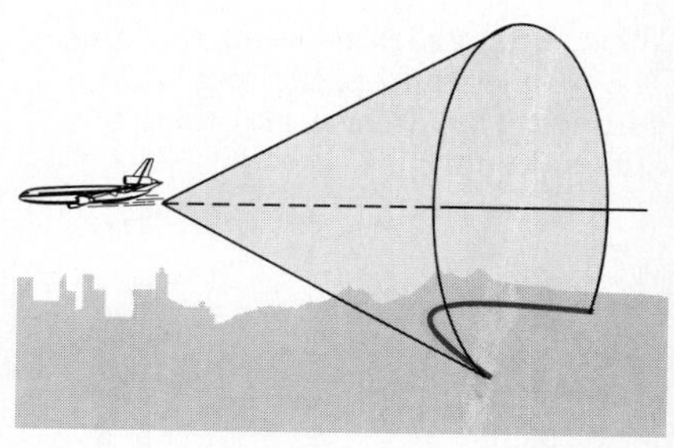

ANSWERS ON PAGE A-10

b) Now we draw curves through the intercepts, $(0, -6)$ and $(0, 6)$, toward the asymptotes as shown.

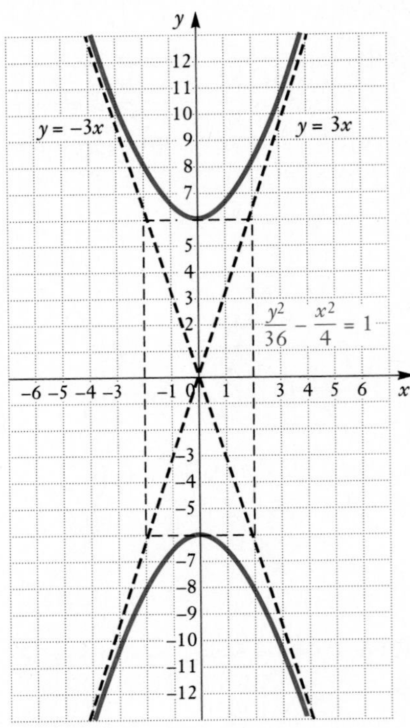

◀

DO EXERCISE 4 ON THE PRECEDING PAGE.

Planets and comets have orbits around the sun that are ellipses. The sun is located at one focus. (The plural is "foci.")

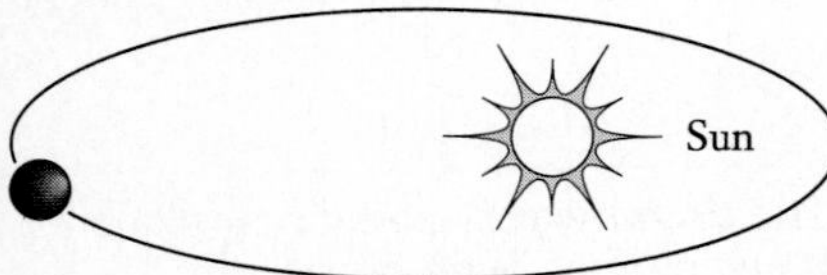

Planetary orbit

Whispering galleries are also ellipses. The people stand at the foci and whisper. They hear each other, but others do not.

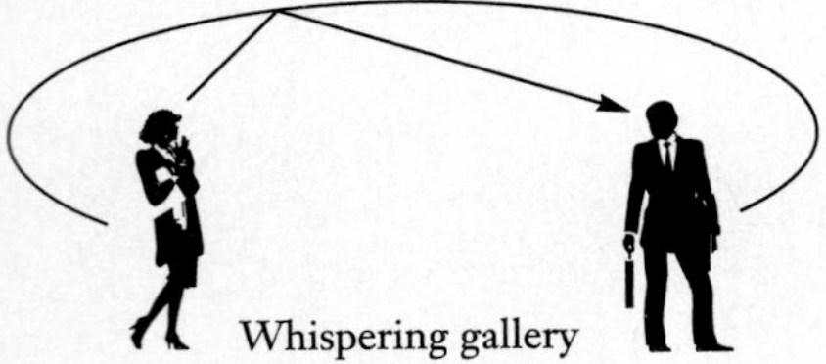

Whispering gallery

NAME SECTION DATE

EXERCISE SET 9.2

a Graph the ellipse.

1. $\frac{x^2}{4} + \frac{y^2}{1} = 1$

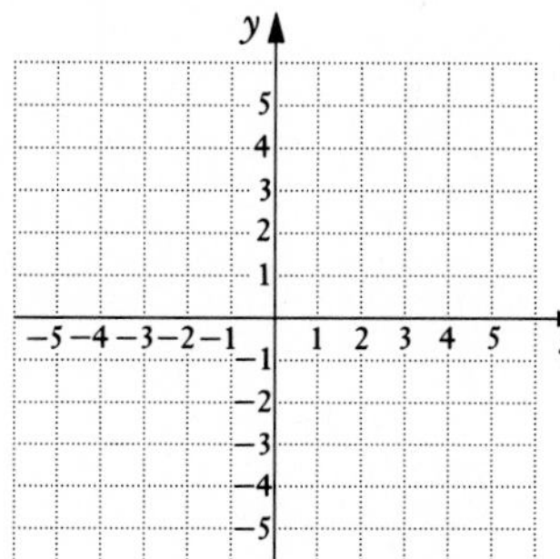

2. $\frac{x^2}{1} + \frac{y^2}{4} = 1$

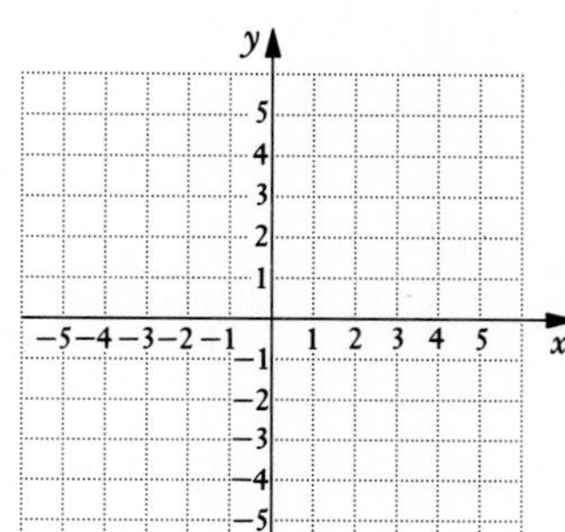

3. $\frac{x^2}{16} + \frac{y^2}{25} = 1$

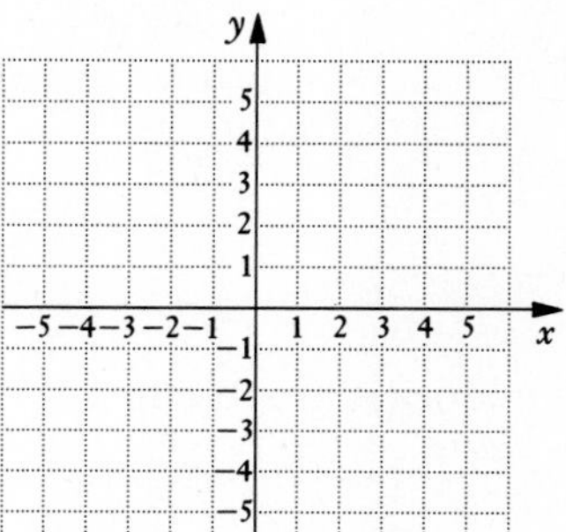

4. $\frac{x^2}{9} + \frac{y^2}{25} = 1$

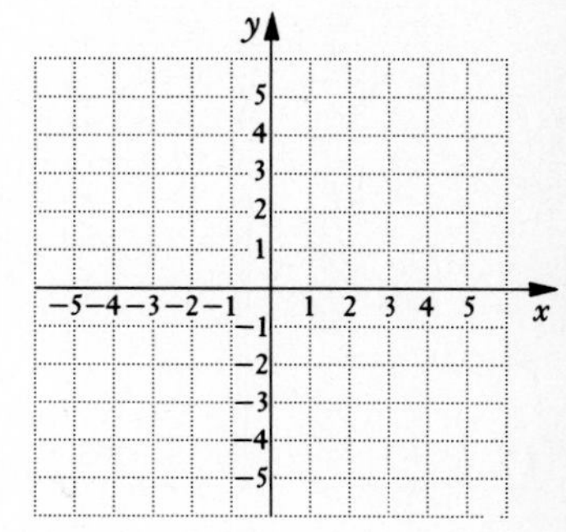

5. $4x^2 + 9y^2 = 36$
(*Hint:* Divide by 36.)

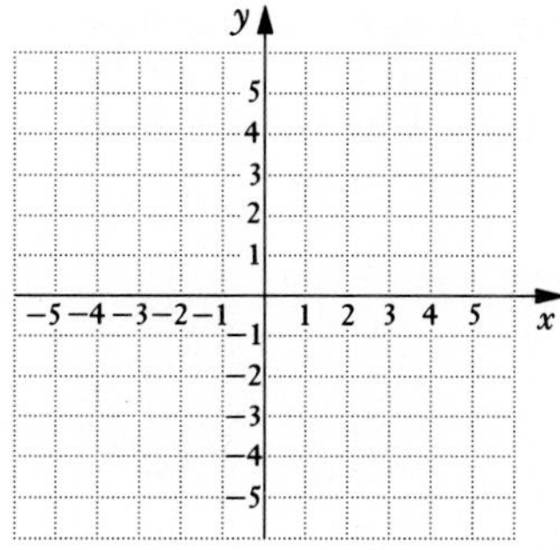

6. $9x^2 + 4y^2 = 36$

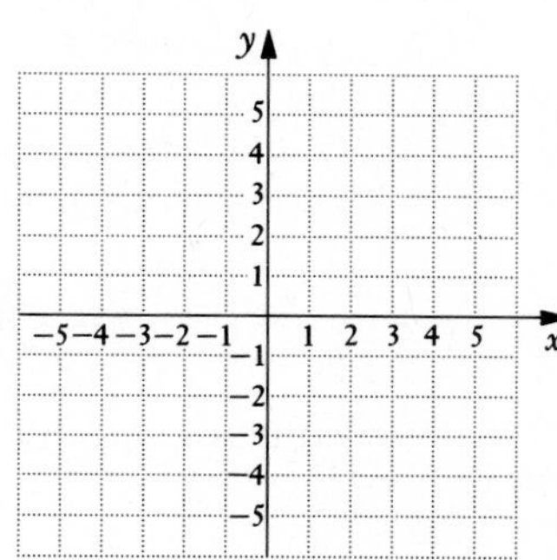

7. $x^2 + 4y^2 = 4$

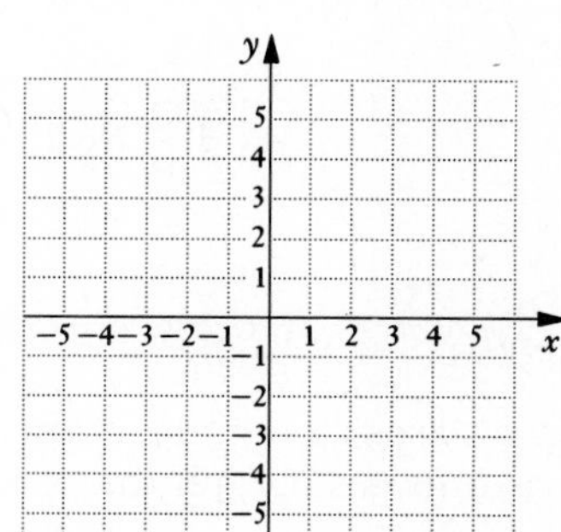

8. $9x^2 + 16y^2 = 144$

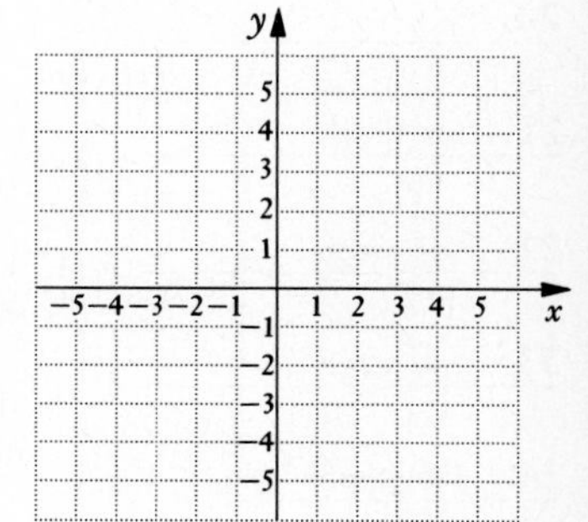

b Graph the hyperbola.

9. $\frac{x^2}{16} - \frac{y^2}{16} = 1$

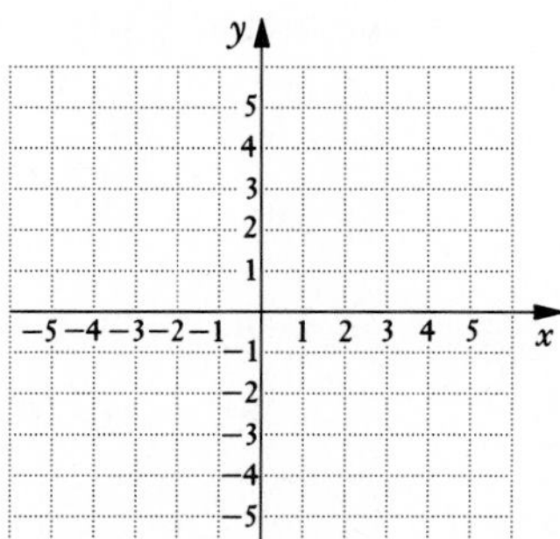

10. $\frac{y^2}{9} - \frac{x^2}{9} = 1$

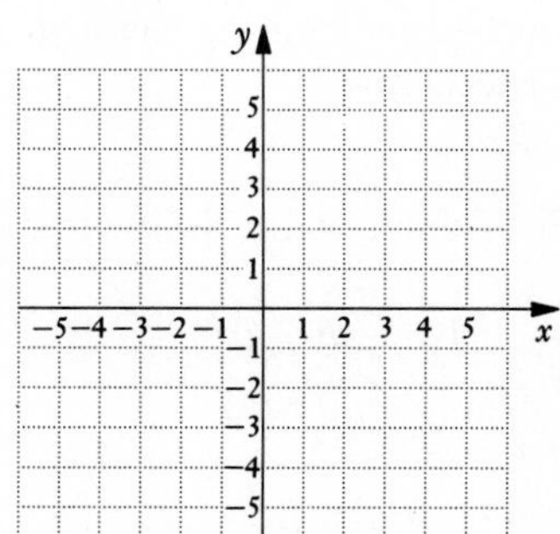

11. $\frac{y^2}{16} - \frac{x^2}{9} = 1$

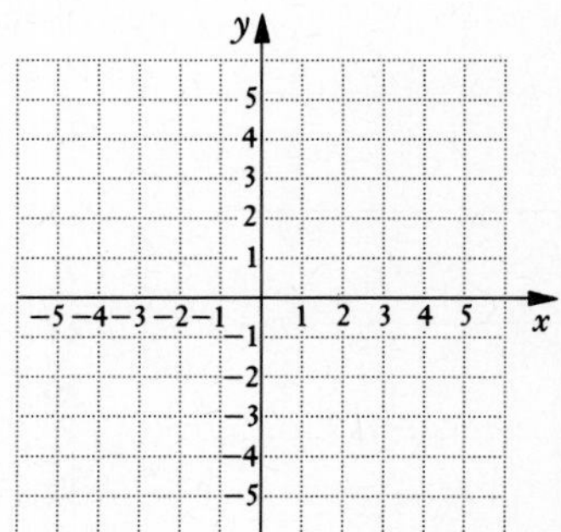

12. $\frac{x^2}{9} - \frac{y^2}{4} = 1$

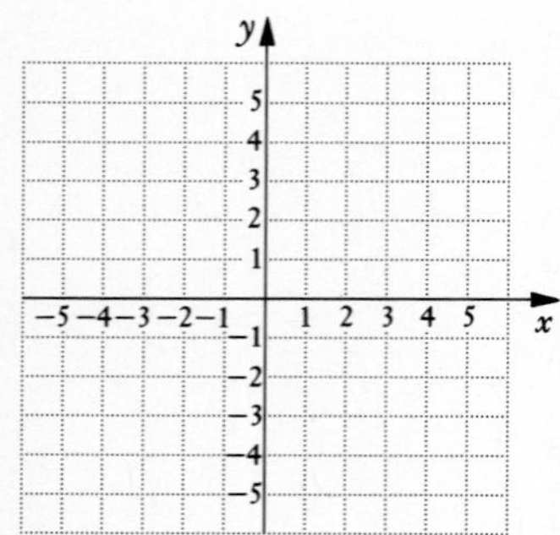

13. $\frac{x^2}{25} - \frac{y^2}{36} = 1$

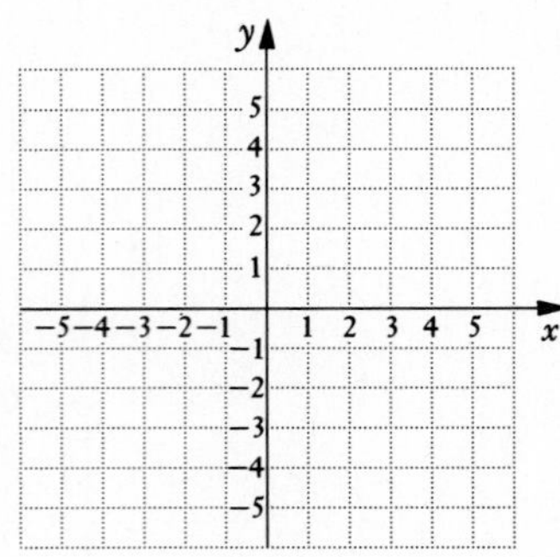

14. $\frac{y^2}{9} - \frac{x^2}{25} = 1$

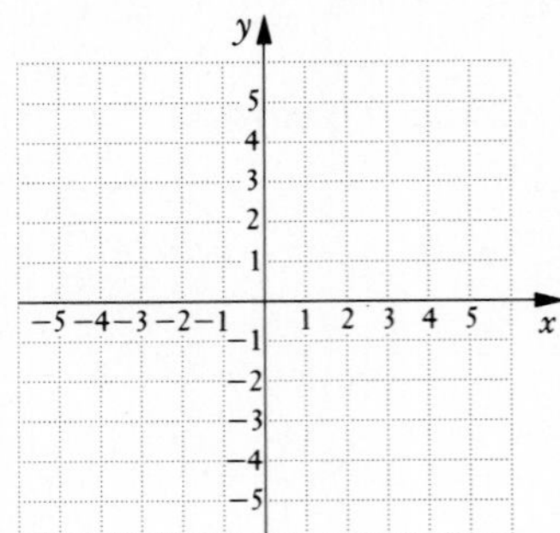

ANSWERS

17. ____________

18. ____________

19. ____________

20. ____________

21. See graph.

22. See graph.

23. See graph.

24. See graph.

25. See graph.

26. See graph.

27. ____________

28. ____________

29. ____________

30. ____________

31. ____________

32. ____________

33. ____________

Graph using graph paper.

15. $x^2 - y^2 = 4$

16. $y^2 - x^2 = 25$

SKILL MAINTENANCE

17. Simplify: $\sqrt[3]{125t^{15}}$.

18. Solve: $2x^2 + 10 = 0$.

19. Rationalize the denominator:

$$\frac{4\sqrt{2} - 5\sqrt{3}}{6\sqrt{3} - 8\sqrt{2}}.$$

20. An airplane travels 500 mi at a certain speed. A larger plane travels 1620 mi at a speed that is 320 mph faster, but takes 1 hr longer. Find the speed of each plane.

SYNTHESIS

In Exercises 21–24, the graph of the given equation is a hyperbola with the *x*- and *y*-axes as asymptotes. Graph the hyperbola.

21. $y = \frac{6}{x}$

22. $y = \frac{1}{x}$

23. $y = -\frac{1}{x}$

24. $y = -\frac{4}{x}$

The standard form of an ellipse with center (h, k) is

$$\frac{(x-h)^2}{a^2} + \frac{(y-k)^2}{b^2} = 1.$$

The *vertices* are $(a + h, k)$, $(-a + h, k)$, $(h, b + k)$, and $(h, -b + k)$. For each equation of an ellipse, find an equivalent equation in standard form. Find the center and the vertices. Then graph the ellipse.

25. $16x^2 + y^2 + 96x - 8y + 144 = 0$

26. $4x^2 + 25y^2 - 8x + 50y = 71$

Classify the graph of each of the following equations as a circle, an ellipse, a parabola, or a hyperbola.

27. $x^2 + y^2 - 10x + 8y - 40 = 0$

28. $y + 1 = 2x^2$

29. $1 - 3y = 2y^2 - x$

30. $9x^2 - 4y^2 - 36x + 24y - 36 = 0$

31. $4x^2 + 25y^2 - 8x - 100y + 4 = 0$

32. $\frac{x^2}{7} + \frac{y^2}{7} = 1$

33. Find an equation of a parabola satisfying the following conditions: line of symmetry parallel to the *y*-axis and passing through the points $(0, 3)$, $(-1, 6)$, $(2, 9)$.

9.3 Nonlinear Systems of Equations

All the systems of equations that we have studied so far have been linear. We now consider systems of two equations in two variables in which at least one equation is not linear.

OBJECTIVES

After finishing Section 9.3, you should be able to:

a Solve systems of two equations in two variables in which at least one equation is nonlinear.

b Solve problems involving such systems of equations.

FOR EXTRA HELP

Tape 18B

Tape 16A

MAC: 9
IBM: 9

a Algebraic Solutions

We consider systems of one first-degree and one second-degree equation. For example, the graphs may be a circle and a line. If so, there are three possibilities.

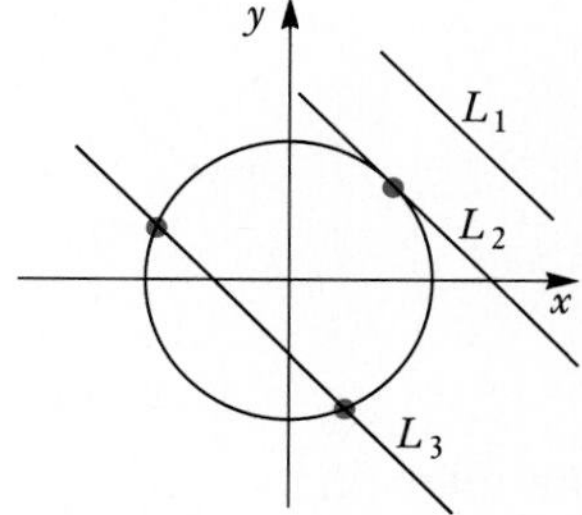

For L_1 there is no point of intersection, hence no solution in the set of real numbers. For L_2 there is one point of intersection, hence one real-number solution. For L_3 there are two points of intersection, hence two real-number solutions.

These systems can be solved graphically by finding the points of intersection. In solving algebraically, we use the substitution method.

► **EXAMPLE 1** Solve the following system:

$x^2 + y^2 = 25,$ (1) (The graph is a circle.)

$3x - 4y = 0.$ (2) (The graph is a line.)

We first solve the linear equation (2) for x:

$x = \frac{4}{3}y.$ (3)

We then substitute $\frac{4}{3}y$ for x in Equation (1) and solve for y:

$$\begin{aligned}(\tfrac{4}{3}y)^2 + y^2 &= 25\\ \tfrac{16}{9}y^2 + y^2 &= 25\\ \tfrac{25}{9}y^2 &= 25\\ y^2 &= 9\\ y &= \pm 3.\end{aligned}$$

Now we substitute these numbers for y in Equation (3) and solve for x:

$x = \frac{4}{3}(3) = 4; \qquad x = \frac{4}{3}(-3) = -4.$

Check: For (4, 3):

$x^2 + y^2 = 25$	
$4^2 + 3^2$	25
$16 + 9$	
25	

$3x - 4y = 0$	
$3(4) - 4(3)$	0
$12 - 12$	
0	

For $(-4, -3)$:

$$\begin{array}{c|c} x^2 + y^2 & = 25 \\ \hline (-4)^2 + (-3)^2 & 25 \\ 16 + 9 & \\ 25 & \end{array}$$

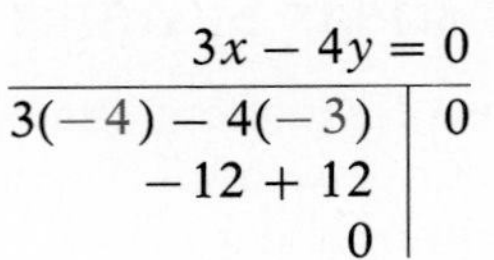

$$\begin{array}{c|c} 3x - 4y & = 0 \\ \hline 3(-4) - 4(-3) & 0 \\ -12 + 12 & \\ 0 & \end{array}$$

The pairs $(4, 3)$ and $(-4, -3)$ check, so they are solutions. We can see the solutions in the graph. The graph of Equation (1) is a circle, and the graph of Equation (2) is a line. The graphs intersect at the points $(4, 3)$ and $(-4, -3)$. ◄

DO EXERCISES 1 AND 2.

► **EXAMPLE 2** Solve the following system:

$$y + 3 = 2x, \qquad (1)$$
$$x^2 + 2xy = -1. \qquad (2)$$

We first solve the linear equation (1) for y:

$$y = 2x - 3. \qquad (3)$$

We then substitute $2x - 3$ for y in Equation (2) and solve for x:

$$x^2 + 2x(2x - 3) = -1$$
$$x^2 + 4x^2 - 6x = -1$$
$$5x^2 - 6x + 1 = 0$$
$$(5x - 1)(x - 1) = 0 \quad \text{Factoring}$$
$$5x - 1 = 0 \quad \text{or} \quad x - 1 = 0 \quad \text{Using the principle of zero products}$$
$$x = \tfrac{1}{5} \quad \text{or} \quad x = 1.$$

Now we substitute these numbers for x in Equation (3) and solve for y:

$$y = 2(\tfrac{1}{5}) - 3 = -\tfrac{13}{5},$$
$$y = 2(1) - 3 = -1.$$

The pairs $(\frac{1}{5}, -\frac{13}{5})$ and $(1, -1)$ check, so they are solutions. ◄

DO EXERCISE 3.

► **EXAMPLE 3** Solve:

$$x + y = 5, \qquad \text{(The graph is a line.)}$$
$$y = 3 - x^2. \qquad \text{(The graph is a parabola.)}$$

We substitute $3 - x^2$ for y in the first equation:

$$x + 3 - x^2 = 5$$
$$-x^2 + x - 2 = 0$$
$$x^2 - x + 2 = 0.$$

Solve. Sketch the graphs to confirm the solutions.

1. $x^2 + y^2 = 25,$
   $y - x = -1$

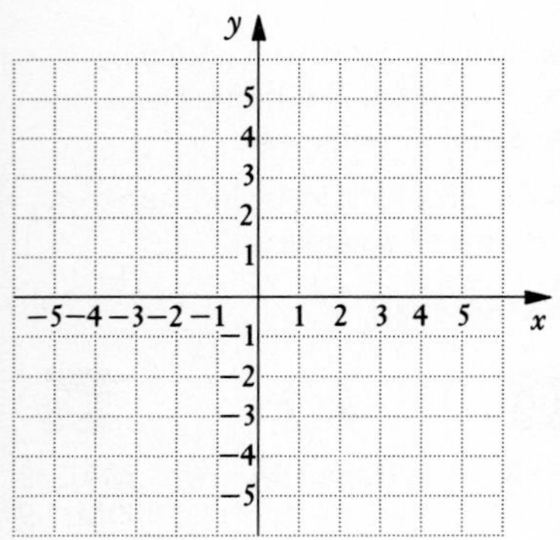

2. $y = x^2 - 2x - 1,$
   $y = x + 3$

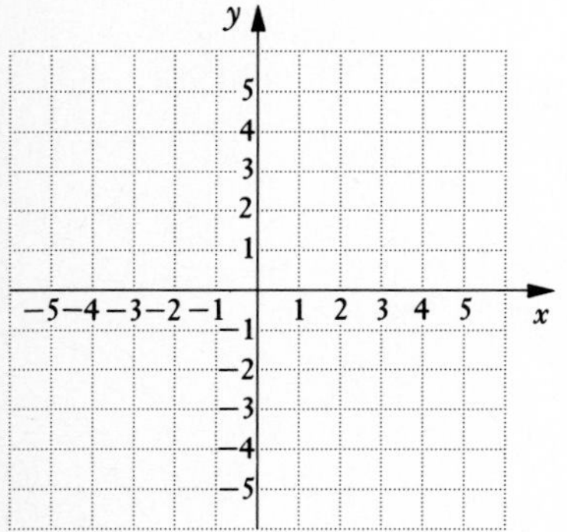

3. Solve:

$$y + 3x = 1,$$
$$x^2 - 2xy = 5.$$

ANSWERS ON PAGE A-10

To solve this equation, we need the quadratic formula:

$$x=\frac{-b\pm\sqrt{b^2-4ac}}{2a}=\frac{-(-1)\pm\sqrt{(-1)^2-4(1)(2)}}{2(1)}$$
$$=\frac{1\pm\sqrt{1-8}}{2}=\frac{1\pm\sqrt{-7}}{2}=\frac{1}{2}\pm\frac{\sqrt{7}}{2}i.$$

Then solving the first equation for y, we obtain $y = 5 - x$. Substituting values for x gives us

$$y=5-\left(\frac{1}{2}+\frac{\sqrt{7}}{2}i\right)=\frac{9}{2}-\frac{\sqrt{7}}{2}i \quad\text{and}\quad y=5-\left(\frac{1}{2}-\frac{\sqrt{7}}{2}i\right)=\frac{9}{2}+\frac{\sqrt{7}}{2}i.$$

The solutions are

$$\left(\frac{1}{2}+\frac{\sqrt{7}}{2}i, \frac{9}{2}-\frac{\sqrt{7}}{2}i\right)$$

and

$$\left(\frac{1}{2}-\frac{\sqrt{7}}{2}i, \frac{9}{2}+\frac{\sqrt{7}}{2}i\right).$$

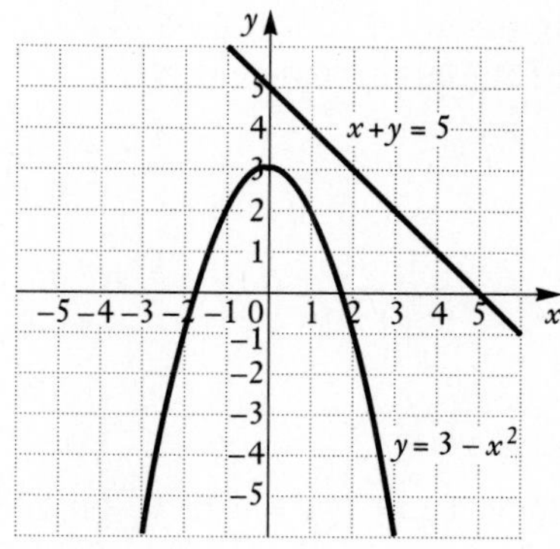

There are no real-number solutions. Note in the figure at right that the graphs do not intersect. Getting only complex-number solutions tells us that the graphs do not intersect. ◀

DO EXERCISE 4.

4. Solve:

$$9x^2 - 4y^2 = 36,$$
$$5x + 2y = 0.$$

Two second-degree equations can have common solutions in various ways. If the graphs happen to be a circle and a hyperbola, for example, there are five possibilities, as shown below.

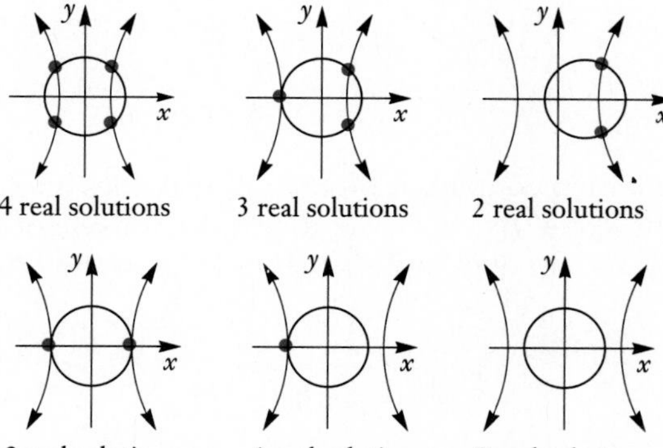

To solve systems of two second-degree equations, we can use either the substitution method or the elimination method. The elimination method is generally used when each equation is of the form $Ax^2 + By^2 = C$. Then we can eliminate an x^2- or a y^2-term in a manner similar to the procedure we used for systems of linear equations in Chapter 4.

ANSWER ON PAGE A-10

5. Solve:

$$2y^2 - 3x^2 = 6,$$
$$5y^2 + 2x^2 = 53.$$

ANSWER ON PAGE A-10

► **EXAMPLE 4** Solve:

$$2x^2 + 5y^2 = 22, \quad (1)$$
$$3x^2 - y^2 = -1. \quad (2)$$

Here we use the elimination method:

$$\begin{aligned} 2x^2 + 5y^2 &= 22 \\ 15x^2 - 5y^2 &= -5 \quad \textbf{Multiplying by 5 on both sides of Equation (2)} \\ \hline 17x^2 \qquad &= 17 \quad \textbf{Adding} \\ x^2 &= 1 \\ x &= \pm 1. \end{aligned}$$

If $x = 1$, $x^2 = 1$, and if $x = -1$, $x^2 = 1$, so substituting either 1 or -1 for x in Equation (2) gives us

$$\begin{aligned} 3x^2 - y^2 &= -1 \\ 3 \cdot 1 - y^2 &= -1 \quad \textbf{Substituting 1 for } x^2 \\ 3 - y^2 &= -1 \\ -y^2 &= -4 \\ y^2 &= 4 \\ y &= \pm 2. \end{aligned}$$

Thus if $x = 1$, $y = 2$ or $y = -2$, and if $x = -1$, $y = 2$ or $y = -2$. The possible solutions are $(1, 2)$, $(1, -2)$, $(-1, 2)$, and $(-1, -2)$.

Check: Since $(2)^2 = 4$, $(-2)^2 = 4$, $(1)^2 = 1$, and $(-1)^2 = 1$, we can check all four pairs at one time.

$2x^2 + 5y^2 = 22$	
$2(\pm 1)^2 + 5(\pm 2)^2$	22
$2 + 20$	
22	

$3x^2 - y^2 = -1$	
$3(\pm 1)^2 - (\pm 2)^2$	-1
$3 - 4$	
-1	

The solutions are $(1, 2)$, $(1, -2)$, $(-1, 2)$, and $(-1, -2)$. ◄

DO EXERCISE 5.

When one equation contains a product of variables and the other equation is of the form $Ax^2 + By^2 = C$, we often solve for one of the variables in the equation with the product and then substitute in the other.

► **EXAMPLE 5** Solve:

$$x^2 + 4y^2 = 20, \quad (1)$$
$$xy = 4. \quad (2)$$

Here we use the substitution method. First, we solve Equation (2) for y:

$$y = \frac{4}{x}.$$

Then we substitute $4/x$ for y in Equation (1) and solve for x:

$$x^2 + 4\left(\frac{4}{x}\right)^2 = 20$$

$$x^2 + \frac{64}{x^2} = 20$$

$$x^4 + 64 = 20x^2$$ **Multiplying by x^2**

$$x^4 - 20x^2 + 64 = 0$$ **Obtaining standard form. This equation is reducible to quadratic.**

$$u^2 - 20u + 64 = 0$$ **Letting $u = x^2$**

$$(u - 16)(u - 4) = 0$$ **Factoring**

$$u = 16 \quad or \quad u = 4.$$ **Using the principle of zero products**

Now we substitute x^2 for u and solve these equations:

$$x^2 = 16 \quad or \quad x^2 = 4$$

$$x = \pm 4 \quad or \quad x = \pm 2.$$

Then $x = 4$ or $x = -4$ or $x = 2$ or $x = -2$. Since $y = 4/x$, if $x = 4$, $y = 1$; if $x = -4$, $y = -1$; if $x = 2$, $y = 2$; and if $x = -2$, $y = -2$. The ordered pairs $(4, 1)$, $(-4, -1)$, $(2, 2)$, and $(-2, -2)$ check. They are the solutions. ◀

DO EXERCISE 6.

6. Solve:

$$x^2 + xy + y^2 = 19,$$
$$xy = 6.$$

b Solving Problems

We now consider solving problems in which the translation is to a system of equations in which at least one is not linear.

▶ **EXAMPLE 6** For a building at a community college, an architect wants to lay out a rectangular piece of ground that has a perimeter of 204 m and an area of 2565 m^2. Find the dimensions of the piece of ground.

1. *Familiarize.* We draw a picture of the field, labeling the drawing. We let l = the length and w = the width.

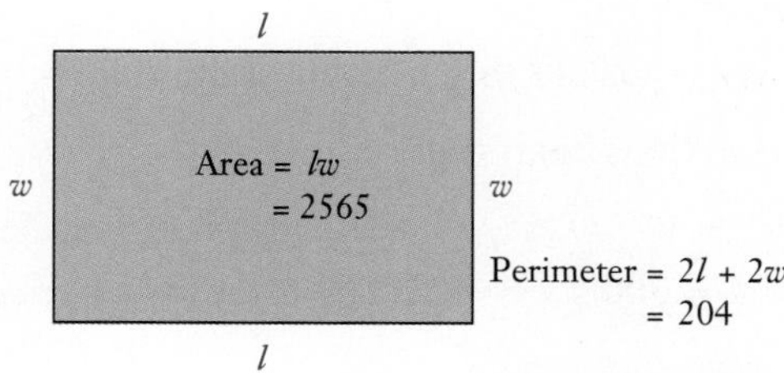

2. *Translate.* We then have the following translation:

Perimeter: $2w + 2l = 204$;
Area: $lw = 2565$.

3. *Solve.* We solve the system

$$2w + 2l = 204, \quad \text{(The graph is a line.)}$$
$$lw = 2565. \quad \text{(The graph is a hyperbola.)}$$

We solve the second equation for l and get $l = 2565/w$. Then we substi-

ANSWER ON PAGE A-10

7. The perimeter of a rectangular field is 34 m, and the length of a diagonal is 13 m. Find the dimensions of the field.

tute $2565/w$ for l in the first equation and solve for w:

$$2w + 2\left(\frac{2565}{w}\right) = 204$$

$$2w^2 + 2(2565) = 204w \quad \textbf{Multiplying by } w$$

$$2w^2 - 204w + 2(2565) = 0 \quad \textbf{Standard form}$$

$$w^2 - 102w + 2565 = 0 \quad \textbf{Multiplying by } \tfrac{1}{2}$$

$$w = \frac{-(-102) \pm \sqrt{(-102)^2 - 4 \cdot 1 \cdot 2565}}{2 \cdot 1}$$

Quadratic formula. Factoring could also be used, but the numbers are quite large.

$$w = \frac{102 \pm \sqrt{144}}{2} = \frac{102 \pm 12}{2}$$

$$w = 57 \quad \text{or} \quad w = 45.$$

If $w = 57$, then $l = 2565/w = 2565/57 = 45$. If $w = 45$, then $l = 2565/w = 2565/45 = 57$. Since length is usually considered to be longer than width, we have the solution $l = 57$ and $w = 45$, or $(57, 45)$.

4. *Check.* If $l = 57$ and $w = 45$, the perimeter is $2 \cdot 57 + 2 \cdot 45$, or 204. The area is $57 \cdot 45$, or 2565. The numbers check.
5. *State.* The answer is that the length is 57 m and the width is 45 m. ◀

DO EXERCISE 7.

8. The area of a rectangle is 2 in^2, and the length of a diagonal is $\sqrt{5}$ in. Find the dimensions of the rectangle.

▶ **EXAMPLE 7** The area of a rectangular Oriental rug is 300 yd^2, and the length of a diagonal is 25 yd. Find the dimensions of the rug.

1. *Translate.* We draw a picture and label it. Note that there is a right triangle in the figure. We let l = the length and w = the width.

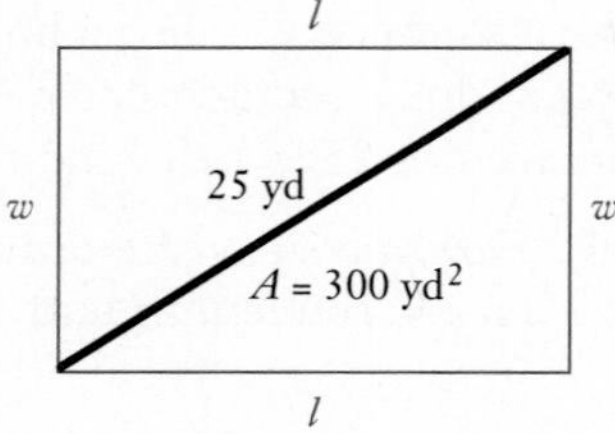

2. *Translate.* We translate to a system of equations.

From the Pythagorean theorem:

$$l^2 + w^2 = 25^2, \quad \text{(The graph is a circle.)}$$
$$lw = 300. \quad \text{(The graph is a hyperbola.)}$$

3. *Solve.* We solve the system

$$l^2 + w^2 = 625, \quad (1)$$
$$lw = 300. \quad (2)$$

We get $(20, 15)$ and $(-20, -15)$.

4. *Check.* Lengths of sides cannot be negative, so we need check only $(20, 15)$. In the right triangle, $15^2 + 20^2 = 225 + 400 = 625$, which is 25^2. The area is $15 \cdot 20$, or 300, so we have a solution.
5. *State.* The answer is that the length is 20 yd and the width is 15 yd. ◀

ANSWERS ON PAGE A-10

DO EXERCISE 8.

NAME SECTION DATE

EXERCISE SET 9.3

a Solve.

1. $x^2 + y^2 = 25,$
   $y - x = 1$

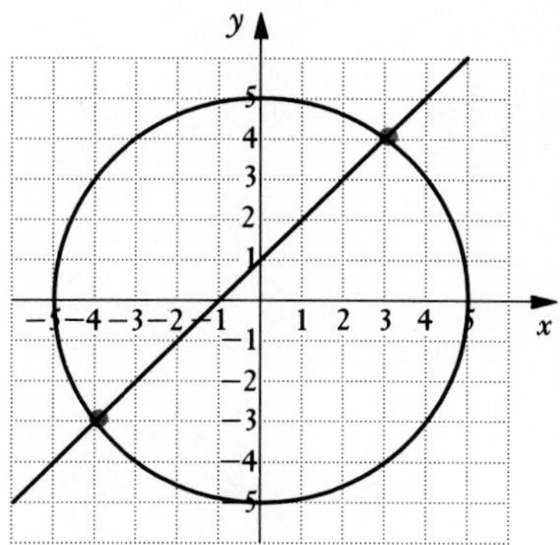

2. $x^2 + y^2 = 100,$
   $y - x = 2$

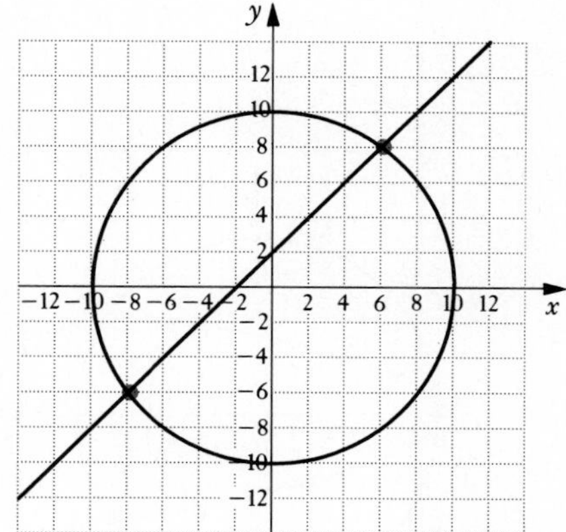

3. $4x^2 + 9y^2 = 36,$
   $3y + 2x = 6$

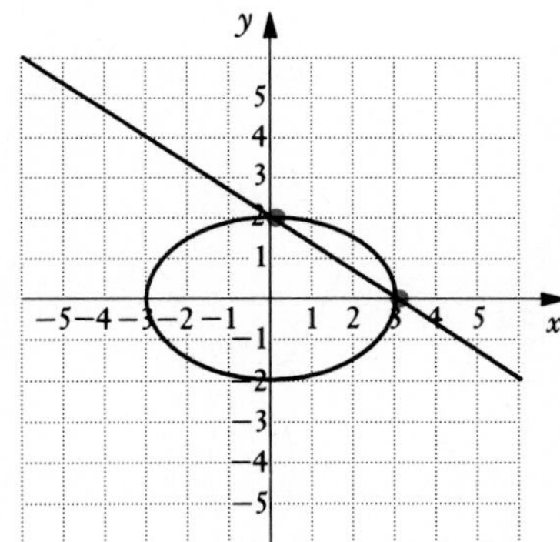

4. $9x^2 + 4y^2 = 36,$
   $3x + 2y = 6$

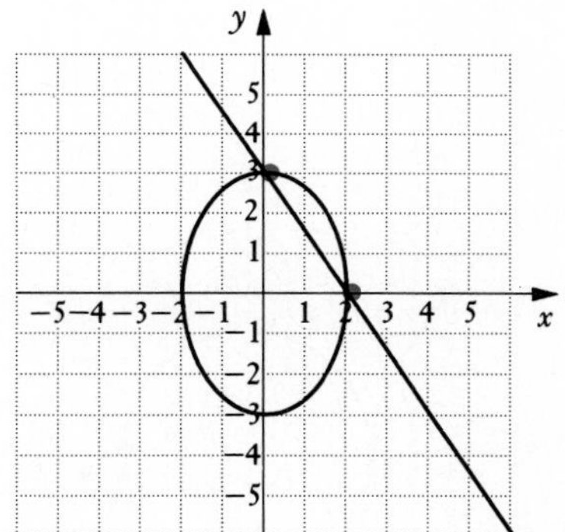

5. $y^2 = x + 3,$
   $2y = x + 4$

6. $y = x^2,$
   $3x = y + 2$

7. $x^2 - xy + 3y^2 = 27,$
   $x - y = 2$

8. $2y^2 + xy + x^2 = 7,$
   $x - 2y = 5$

9. $x^2 - xy + 3y^2 = 5,$
   $x - y = 2$

10. $m^2 + 3n^2 = 10,$
    $m - n = 2$

11. $2y^2 + xy = 5,$
    $4y + x = 7$

12. $a + b = -6,$
    $ab = -7$

ANSWERS

1. ______
2. ______
3. ______
4. ______
5. ______
6. ______
7. ______
8. ______
9. ______
10. ______
11. ______
12. ______

ANSWERS

13. ____________

14. ____________

15. ____________

16. ____________

17. ____________

18. ____________

19. ____________

20. ____________

21. ____________

22. ____________

23. ____________

24. ____________

13. $2a + b = 1,$
$b = 4 - a^2$

14. $4x^2 + 9y^2 = 36,$
$x + 3y = 3$

15. $x^2 + y^2 = 5,$
$x - y = 8$

16. $4x^2 + 9y^2 = 36,$
$y - x = 8$

17. $x^2 + y^2 = 25,$
$y^2 = x + 5$

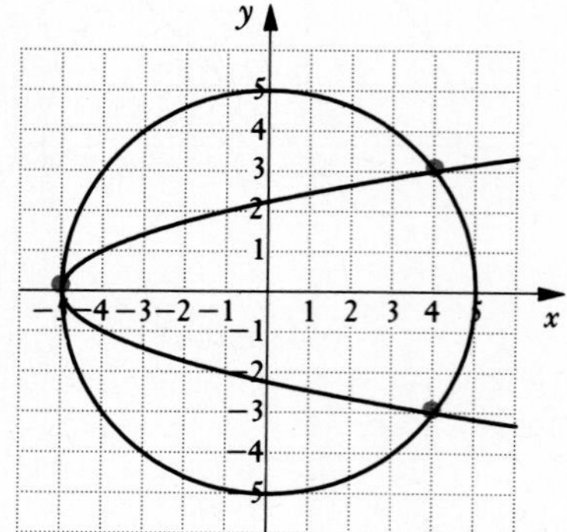

18. $y = x^2,$
$x = y^2$

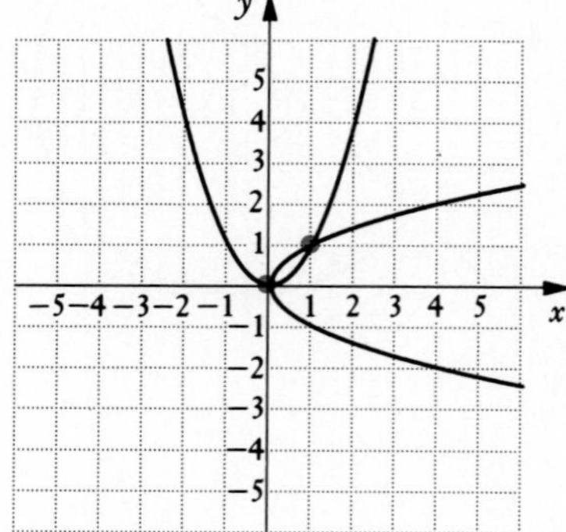

19. $x^2 + y^2 = 9,$
$x^2 - y^2 = 9$

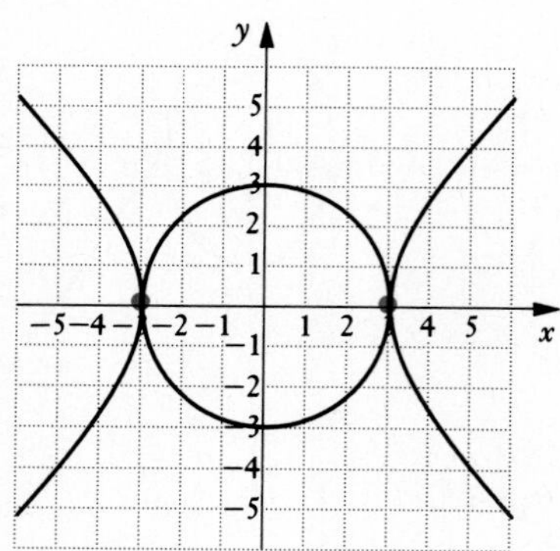

20. $y^2 - 4x^2 = 4,$
$4x^2 + y^2 = 4$

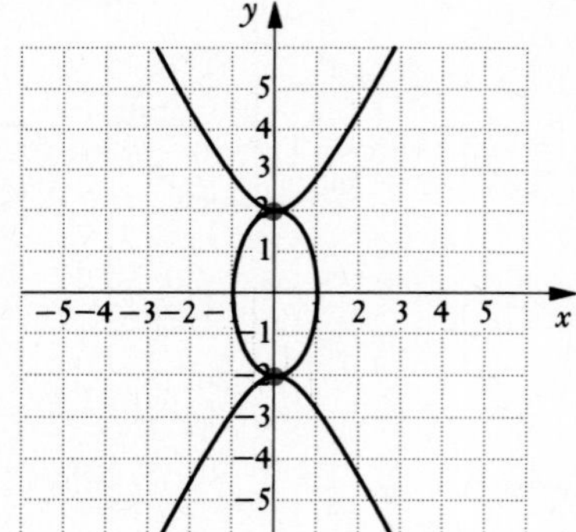

21. $x^2 + y^2 = 5,$
$xy = 2$

22. $x^2 + y^2 = 20,$
$xy = 8$

23. $x^2 + y^2 = 13,$
$xy = 6$

24. $x^2 + y^2 + 6y + 5 = 0,$
$x^2 + y^2 - 2x - 8 = 0$

25. $2xy + 3y^2 = 7,$
$3xy - 2y^2 = 4$

26. $xy - y^2 = 2,$
$2xy - 3y^2 = 0$

27. $4a^2 - 25b^2 = 0,$
$2a^2 - 10b^2 = 3b + 4$

28. $m^2 - 3mn + n^2 + 1 = 0,$
$3m^2 - mn + 3n^2 = 13$

29. $ab - b^2 = -4,$
$ab - 2b^2 = -6$

30. $a^2 + b^2 = 14,$
$ab = 3\sqrt{5}$

31. $x^2 + y^2 = 25,$
$9x^2 + 4y^2 = 36$

32. $x^2 + y^2 = 1,$
$9x^2 - 16y^2 = 144$

b Solve.

33. A computer parts company wants to make a rectangular memory board that has a perimeter of 28 cm and a diagonal of length 10 cm. What are the dimensions of the board?

34. A bathroom tile company wants to make a new rectangular tile that has a perimeter of 6 m and a diagonal of length $\sqrt{5}$ m. What are the dimensions of the tile?

35. A rectangle has an area of 20 in^2 and a perimeter of 18 in. Find its dimensions.

36. A rectangle has an area of 2 yd^2 and a perimeter of 6 yd. Find its dimensions.

37. It will take 210 yd of fencing to enclose a rectangular field. The area of the field is 2250 yd^2. What are the dimensions of the field?

38. The diagonal of a rectangle is 1 ft longer than the length of the rectangle and 3 ft longer than twice the width. Find the dimensions of the rectangle.

39. The product of the lengths of the legs of a right triangle is 156. The hypotenuse has length $\sqrt{313}$. Find the lengths of the legs.

40. The product of two numbers is 60. The sum of their squares is 136. Find the numbers.

41. A garden contains two square peanut beds. Find the length of each bed if the sum of their areas is 832 ft^2 and the difference of their areas is 320 ft^2.

42. A certain amount of money saved for 1 yr at a certain interest rate yielded \$7.50. If the principal had been \$25 more and the interest rate 1% less, the interest would have been the same. Find the principal and the rate.

ANSWERS

25. ____

26. ____

27. ____

28. ____

29. ____

30. ____

31. ____

32. ____

33. ____

34. ____

35. ____

36. ____

37. ____

38. ____

39. ____

40. ____

41. ____

42. ____

ANSWERS

43. ______

44. ______

45. ______

46. ______

47. ______

48. ______

49. ______

50. ______

51. ______

52. ______

53. ______

54. ______

55. ______

56. ______

57. ______

58. ______

59. ______

43. The area of a rectangle is $\sqrt{3}$ m^2, and the length of a diagonal is 2 m. Find the dimensions.

44. The area of a rectangle is $\sqrt{2}$ m^2, and the length of a diagonal is $\sqrt{3}$ m. Find the dimensions.

SKILL MAINTENANCE

Solve.

45. $3x^2 + 6 = 5x$

46. $3x^2 = 5x + 6$

Simplify.

47. $\sqrt{48}$

48. $\sqrt[4]{32a^{24}d^9}$

Given $\dfrac{\sqrt{x} - \sqrt{h}}{\sqrt{x} + \sqrt{h}}$:

49. Rationalize the numerator.

50. Rationalize the denominator.

51. A boat travels 4 mi upstream and 4 mi back downstream. The total time for the trip is 3 hr. The speed of the stream is 2 mph. Find the speed of the boat in still water.

SYNTHESIS

52. Find the equation of a circle that passes through the points (4, 6), (−6, 2), and (1, −3).

53. Find the equation of a circle that passes through (−2, 3) and (−4, 1) and whose center is on the line $5x + 8y = -2$.

54. A piece of wire 100 cm long is to be cut into two pieces and those pieces are each to be bent to make a square. The area of one square is to be 144 cm^2 greater than that of the other. How should the wire be cut?

55. Find the equation of an ellipse centered at the origin that passes through the points (2, −3) and $(1, \sqrt{13})$.

56. Four squares with sides 5 in. long are cut from the corners of a rectangular metal sheet that has an area of 340 in^2. The edges are bent up to form an open box with a volume of 350 in^3. Find the dimensions of the box.

57. *Business: Break-even points.* A company keeps records of the total revenue (money taken in) from the sale of x units of a product. It determines that total revenue R is given by

$$R = 100x - x^2.$$

It also keeps records of the total cost of producing x units of the same product. It determines that the total cost C is given by

$$C = 20x + 1500.$$

A **break-even point** is a value of x for which total revenue is the same as total cost; that is, $R = C$. Find the break-even points.

Solve.

58. $p^2 + q^2 = 13,$

$\dfrac{1}{pq} = -\dfrac{1}{6}$

59. $a + b = \dfrac{5}{6},$

$\dfrac{a}{b} + \dfrac{b}{a} = \dfrac{13}{6}$

9.4 Correspondences, Relations, and Functions

a Correspondences, Relations, and Functions

Consider a set of years. To each year, there corresponds the cost of a first-class postage stamp at the end of that year.

Year	Cost of first-class postage at year's end
Domain (Set of inputs)	**Range (Set of outputs)**
1948 →	3
1958 →	4
1964 →	5
1974 →	10
1978 →	15
1983 →	20
1984 →	20
1985 →	22
1988 →	25

Rather than draw arrows each time to show a correspondence, we can use ordered pairs. For example, instead of the correspondence

1948 ⟶ 3

we might write the ordered pair

(1948, 3).

▶ **EXAMPLE 1** Write the correspondence as a set of ordered pairs.

a) *Domain* / *Range*

f: $a \to 4$, $b \to 0$, $c \to 0$

b) *Domain* / *Range*

g: $3 \to 5$, $4 \to 9$, $4 \to -7$, $5 \to -7$, $6 \to -7$

c) *U.S. Senators (Domain)* / *State (Range)*

h: Cranston → California, Wilson → California, Bentsen → Texas, Gramm → Texas, Danforth → Missouri, Bond → Missouri

d) *State (Domain)* / *U.S. Senator (Range)*

p: California → Cranston, California → Wilson, Texas → Bentsen, Texas → Gramm, Missouri → Danforth, Missouri → Bond

Each arrow gives an ordered pair. Thus we have the following.

a) $f = \{(a, 4), (b, 0), (c, 0)\}$

b) $g = \{(3, 5), (4, 9), (4, -7), (5, -7), (6, -7)\}$

c) $h =$ {(Cranston, California), (Wilson, California), (Bentsen, Texas), (Gramm, Texas), (Danforth, Missouri), (Bond, Missouri)}

d) $p =$ {(California, Cranston), (California, Wilson), (Texas, Bentsen), (Texas, Gramm), (Missouri, Danforth), (Missouri, Bond)}

DO EXERCISES 1–3. (EXERCISE 3 IS ON THE FOLLOWING PAGE.)

OBJECTIVES

After finishing Section 9.4, you should be able to:

- **a** Write a correspondence as a relation, or set of ordered pairs. Find the domain and the range of a relation. Determine whether a relation is a function.
- **b** Given a function described by a formula, find function values (outputs) for specified values (inputs).
- **c** Graph simple functions, and given a graph, determine whether it is a graph of a function.
- **d** Find the domain of a function given by a formula.

FOR EXTRA HELP

Tape 18C Tape 16B MAC: 9 IBM: 9

Write the correspondence as a relation, or set of ordered pairs.

1. WOMEN'S DRESS SIZES

d:

Domain (United States)	Range (France)
6 →	38
8 →	40
10 →	42
12 →	44
14 →	46
16 →	48
18 →	50

2.

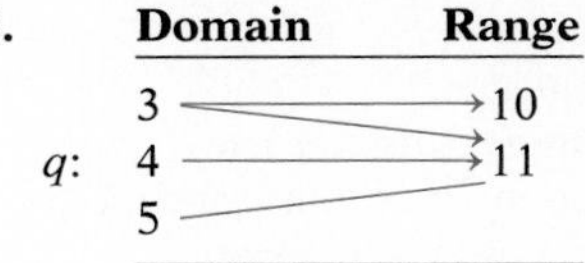

Domain	Range
3 →	10
4 →	11
5 →	11

q

ANSWERS ON PAGE A-10

3.

	Domain		Range
	6	→	3
h:	7	→	4
	−9	→	6

(Arrows: 6 → 3, 7 → 3, −9 → 4, −9 → 6)

4. Find the domain and the range of the relation d in the solution to Margin Exercise 1.

5. Find the domain and the range of the relation q in the solution to Margin Exercise 2.

6. Find the domain and the range of the relation h in the solution to Margin Exercise 3.

7. Determine whether the correspondence d in Margin Exercise 1 is a function.

8. Determine whether the correspondence q in Margin Exercise 2 is a function.

9. Determine whether the correspondence h in Margin Exercise 3 is a function.

ANSWERS ON PAGE A-10

A *relation* is a set of ordered pairs.

Each of the sets of ordered pairs in the solution to Example 1 is an example of a relation.

The set of all first members of ordered pairs in a relation is called the *domain*.

The set of all second members of ordered pairs in a relation is called the *range*.

▶ **EXAMPLE 2** Find the domain and the range of each of the relations in the solution of Example 1.

a) Domain of $f = \{a, b, c\}$; range of $f = \{4, 0\}$. Note that we need list the element 0 only once in the range.

b) Domain of $g = \{3, 4, 5, 6\}$; range of $g = \{5, 9, -7\}$

c) Domain of $h =$ {Cranston, Wilson, Bentsen, Gramm, Danforth, Bond}; range of $h =$ {California, Texas, Missouri}

d) Domain of $p =$ {California, Texas, Missouri}; range of $p =$ {Cranston, Wilson, Bentsen, Gramm, Danforth, Bond} ◀

DO EXERCISES 4–6.

Look again at the postage correspondence at the beginning of this section. Note that to each year there corresponds *exactly* one cost. Such a correspondence is called a **function.**

A *function* is a correspondence (or rule) that assigns to each member of some set (called the *domain*) exactly one member of a set (called the *range*).

Sometimes the members of the domain are called **inputs** and the members of the range are called **outputs.** Note that each input has exactly one output, although in the case of 1983 and 1984, those outputs are the same.

▶ **EXAMPLE 3** Determine whether each of the correspondences in Example 1 is a function.

a) The correspondence f is a function because each member of the domain is matched to only one member of the range.

b) The correspondence g is not a function because the member 4 of the domain is matched to more than one member of the range.

c) The correspondence h is a function.

d) The correspondence p is not a function because to each state there correspond two U.S. Senators. In a function, a member of the domain can be matched with only one member of the range. ◀

DO EXERCISES 7–9.

A function can also be described as a special kind of relation.

A *function* is a relation in which *no* two ordered pairs can have the same first coordinate and different second coordinates.

▶ **EXAMPLE 4** Determine whether each relation in the solution to Example 1 is a function.

a) The relation f is a function because no two ordered pairs have the same first coordinates and different second coordinates.

b) The relation g is *not* a function because the ordered pairs $(4, 9)$ and $(4, -7)$ have the same first coordinate and different second coordinates.

c) The relation h is a function because no two ordered pairs have the same first coordinates and different second coordinates.

d) The relation p is *not* a function because the ordered pairs (California, Cranston) and (California, Wilson) have the same first coordinate and different second coordinates. There are other pairs that also show that p is not a function, but we need only two, as given. ◀

DO EXERCISES 10–12.

10. Determine whether the relation d in the solution of Margin Exercise 1 is a function.

11. Determine whether the relation q in the solution of Margin Exercise 2 is a function.

12. Determine whether the relation h in the solution of Margin Exercise 3 is a function.

ANSWERS ON PAGE A-10

b Formulas of Functions

Usually a function involves ordered pairs of numbers. A formula often expresses the correspondence between the numbers.

For example, you see a flash of lightning and count the number of seconds before you hear the thunder. If you multiply the number of seconds by $\frac{1}{5}$, you can determine how far away the lightning was. If it takes 10 sec to hear the thunder, then the lightning was 2 mi away. This sets up a correspondence

$$10 \longrightarrow 2$$

and an ordered pair

$$(10, 2).$$

Suppose that $x =$ the number of seconds until you hear the thunder and $y =$ the distance; then you have $y = \frac{1}{5}x$. The value of y depends on the value of x.

In the case of an ordered pair (x, y), a formula is a "recipe" for determining y, given x. Since the value of y depends on the number x, we call y the **dependent variable** and x the **independent variable.** We often use the notation $f(x)$ for the output corresponding to an input x from a function f.

To understand function notation, it helps to think of a *function machine.* Think of putting a member of the domain (an *input*) into the machine. The machine knows the correspondence and gives us a member of the range (the *output*).

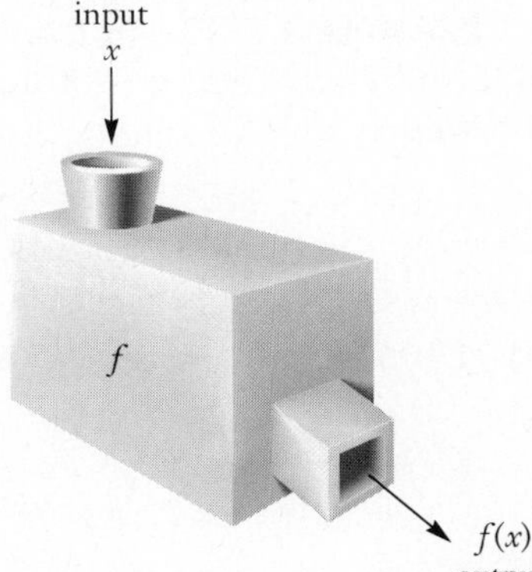

The function has been named f. We call the input x, and its output is named $f(x)$. This is read "f of x," or "f at x," or "the value of f at x." Note that $f(x)$ does *not* mean "f times x."

Some functions can be described by formulas or equations. For example, $f(x) = 2x + 3$ describes the function that takes an input x, multiplies it by 2, and then adds 3.

Find the function values.

13. $f(x) = 2x + 5$; find $f(6)$, $f(-2)$, and $f(0)$.

14. $P(x) = 2x^3 - x + 1$; find $P(2)$, $P(-2)$, and $P(0)$.

15. $g(x) = |x| - 3$; find $g(-1)$, $g(2)$, and $g(3)$.

16. The function

$$g(s) = 2s + 12$$

gives a correspondence between men's shirt sizes s in the United States and Britain and those $g(s)$ in Continental Europe. What sizes in Continental Europe correspond to sizes $14\frac{1}{2}$ and 16 in the United States and Britain?

17. For the function f given by $f(x) = 5x - 3$, find the following.

a) $f(6)$

b) $f(-4)$

c) $f(h)$

d) $f(a + h)$

18. For the function g given by $g(x) = x^2 + 5$, find the following.

a) $g(2.1)$

b) $g(-7)$

c) $g(2y)$

d) $g(h)$

e) $g(a + h)$

19. For the function F given by $F(x) = 3x^2 + 1$, find the following.

a) $F(-2)$

b) $F(h)$

c) $F(a + h)$

d) $F(a + h) - F(a)$

ANSWERS ON PAGE A-10

To find the output $f(4)$, we take the input, 4, double it, and add 3 to get 11. That is, we substitute 4 into the formula for $f(x)$:

$$f(4) = 2 \cdot 4 + 3 = 11.$$

► **EXAMPLES** Find the function value.

5. $f(x) = 3x + 2$; find $f(5)$.
$f(5) = 3 \cdot 5 + 2 = 17$

Remember: $f(x)$ does *not* mean f times x. Rather, $f(x)$ is the output when the input is x.

6. $g(x) = 5x^2 - 4$; find $g(3)$.
$g(3) = 5(3)^2 - 4 = 41$

7. $A(x) = 3x^2 + 2x$; find $A(-2)$.
$A(-2) = 3(-2)^2 + 2(-2) = 8$ ◄

DO EXERCISES 13–15.

Outputs are also called **function values.** In Example 5, $f(5) = 17$. We say that "the function value at 5 is 17," or "when x is 5, the value of the function is 17."

► **EXAMPLE 8** *Dress sizes.* Margin Exercise 1 showed a table of the correspondence between women's dress sizes in the United States with those in France. There is a formula for this function:

$$f(x) = x + 32.$$

What sizes in France correspond to sizes 20 and 4 in the United States?

We find $f(20)$ and $f(4)$:

$$f(20) = 20 + 32 = 52,$$
$$f(4) = 4 + 32 = 36.$$ ◄

DO EXERCISE 16.

When a function is given by a formula or equation, we find function values by substituting into the formula. That is true even if the inputs are not specific numbers. We may use such symbols as $2m$ or $a + h$ for inputs, but we still proceed by substituting.

► **EXAMPLE 9** The function g, given by $g(x) = x^2 - x$, takes an input x and squares it. Then it subtracts x from the result. Find the following function values: $g(-3)$, $g(3m)$, $g(a + h)$, and $g(a) + g(a + h)$.

a) $g(-3) = (-3)^2 - (-3)$ **Substituting the input, -3**
$= 9 + 3$ **Squaring -3 and subtracting -3**
$= 12$ **Simplifying**

b) $g(3m) = (3m)^2 - 3m$ **Substituting the input, $3m$**
$= 9m^2 - 3m$

c) $g(a + h) = (a + h)^2 - (a + h)$ **Substituting the input, $a + h$**
$= a^2 + 2ah + h^2 - a - h$ **Squaring $a + h$ and subtracting $a + h$**

d) $g(a) + g(a + h)$
$= [a^2 - a] + [(a + h)^2 - (a + h)]$ **Substituting the inputs, a and $a + h$**
$= [a^2 - a] + [a^2 + 2ah + h^2 - a - h]$ **Squaring $a + h$ and subtracting $a + h$**
$= a^2 - a + a^2 + 2ah + h^2 - a - h$ **Removing parentheses**
$= 2a^2 + 2ah + h^2 - 2a - h$ **Collecting like terms** ◄

DO EXERCISES 17–19.

C Graphs of Functions

To graph a function such as $f(x) = 3x + 2$, we do just what we do to graph $y = 3x + 2$. We find ordered pairs (x, y) or $(x, f(x))$ and plot them.

▶ **EXAMPLE 10** Graph: $f(x) = 3x + 2$.

A list of function values is shown in the table.

x	$f(x) = 3x + 2$
0	2
1	5
−1	−1
−2	−4
2	8

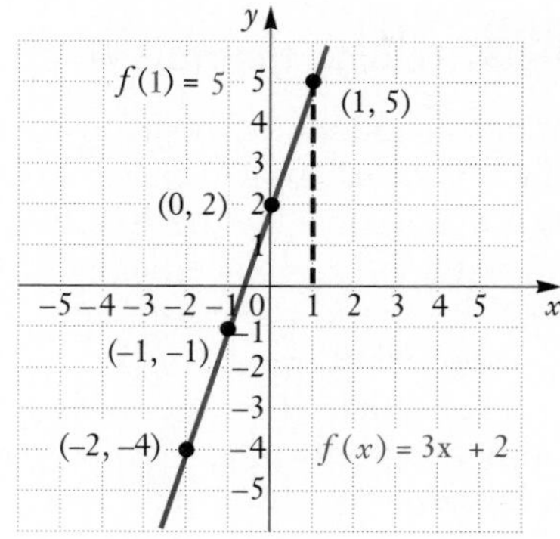

We plot the points and connect them. The graph is a straight line. ◀

DO EXERCISE 20.

▶ **EXAMPLE 11** Graph: $g(x) = |x| - 3$.

A list of function values is shown in the table.

x	$g(x) = \|x\| - 3$
0	−3
1	−2
2	−1
3	0
−1	−2
−2	−1
−3	0

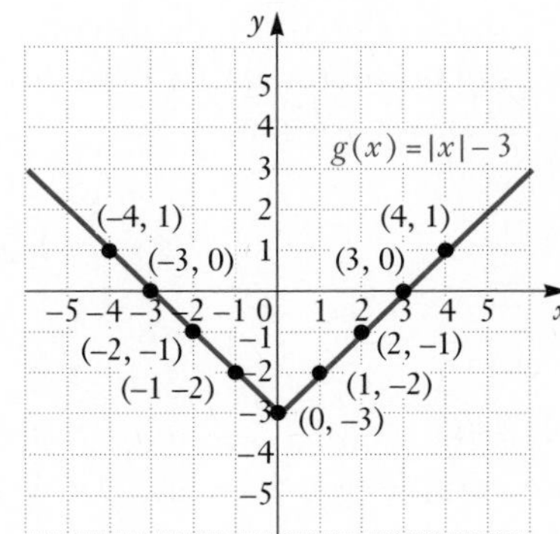

We plot these points and connect them. The graph is V-shaped. Note that as x increases through negative values, $|x| - 3$ decreases. As x increases through positive values, $|x| - 3$ increases. ◀

DO EXERCISE 21.

Some graphs are graphs of functions and some are not. If a vertical line drawn anywhere on the graph intersects the curve in more than one point, then there will be more than one number in the range corresponding to a member of the domain. The graph is then not the graph of a function.

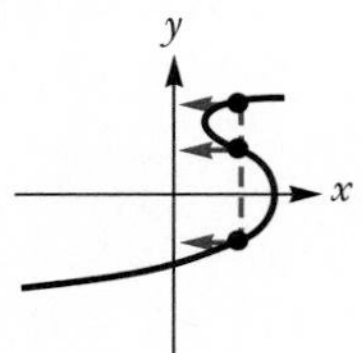

Not a function. Three y-values correspond to one x-value.

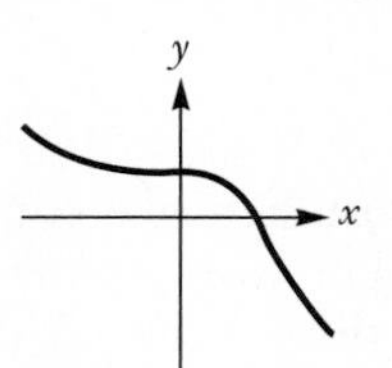

A function

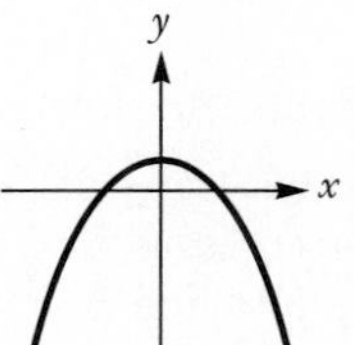

A function

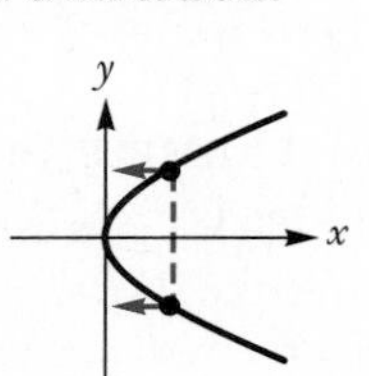

Not a function. Two y-values correspond to one x-value.

20. Graph: $f(x) = 2x - 3$.

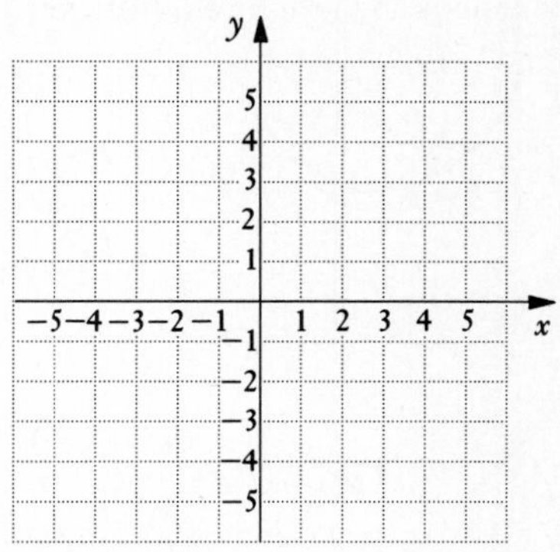

21. Graph: $g(x) = 4 - |x|$.

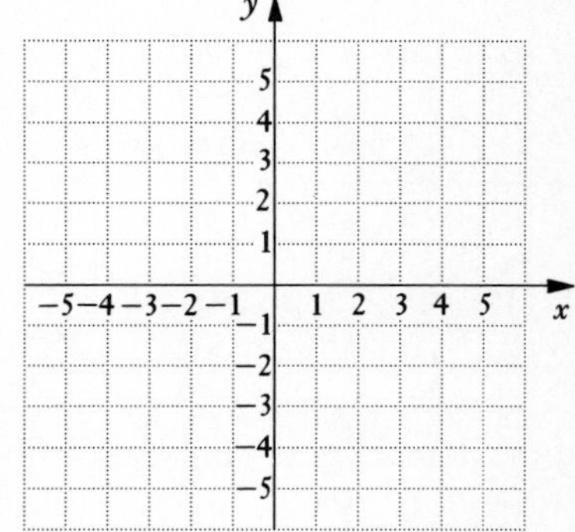

ANSWERS ON PAGE A-10

Determine whether each of the following is the graph of a function.

22. 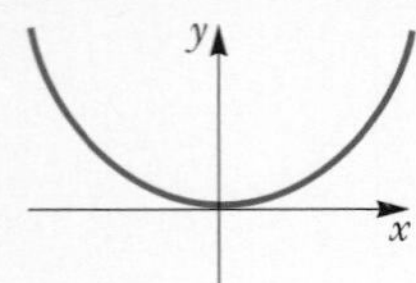

23. 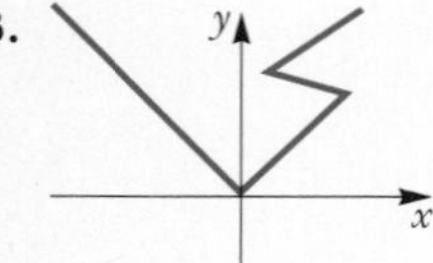

24. 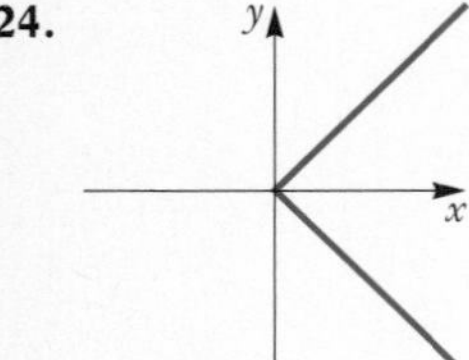

25. 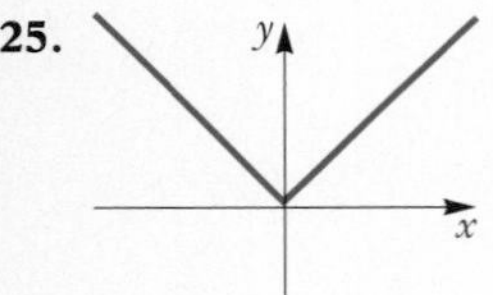

Find the domain of the function.

26. $f(x) = \dfrac{5}{x^2 + x - 12}$

27. $g(x) = \sqrt{4 - 5x}$

28. $p(x) = x^3 - 7x^2 + 2x - 4$

ANSWERS ON PAGE A-10

> **The Vertical-Line Test**
>
> **If it is possible for a vertical line to intersect a graph more than once, the graph is not the graph of a function.**

DO EXERCISES 22–25.

d Finding the Domain of a Function Given by a Formula

When a function f is given by a formula, the domain is understood to be the set of all real numbers that are meaningful replacements for x. For example, consider the function f given by $f(x) = 1/x$. The number 0 is not a meaningful replacement because division by 0 is not possible. All other real numbers are meaningful replacements, so the domain consists of all non-zero real numbers, $\{x | x \neq 0\}$.

Let's take the function g given by $g(x) = \sqrt{x}$. If we consider only real-number inputs and outputs, the negative numbers are not meaningful replacements. Thus the domain of g consists of all nonnegative numbers, $\{x | x \geq 0\}$.

► **EXAMPLE 12** Find the domain of the function f given by

$$f(x) = \frac{4}{x^2 - x - 2}.$$

The formula is meaningful so long as a replacement for x does not make the denominator 0. To find those replacements that do make the denominator 0, we solve $x^2 - x - 2 = 0$:

$$\begin{aligned} x^2 - x - 2 &= 0 \\ (x + 1)(x - 2) &= 0 \\ x + 1 = 0 \quad &or \quad x - 2 = 0 \\ x = -1 \quad &or \quad x = 2. \end{aligned}$$

Thus the domain consists of the set of all real numbers except -1 and 2. We can name this set $\{x | x \neq -1 \text{ and } x \neq 2\}$. ◄

► **EXAMPLE 13** Find the domain of the function g given by $g(x) = \sqrt{2x + 3}$.

Again considering only real-number inputs and outputs, we see that the formula is meaningful for all real numbers x for which $2x + 3 \geq 0$. To find those numbers, we solve the inequality $2x + 3 \geq 0$:

$$\begin{aligned} 2x + 3 &\geq 0 \\ 2x &\geq -3 \\ x &\geq -\tfrac{3}{2}. \end{aligned}$$

The domain is $\{x | x \geq -\frac{3}{2}\}$. ◄

► **EXAMPLE 14** Find the domain of $p(x) = x^2 + |x|$.

There are no restrictions on the numbers we can substitute into this formula. We can square any real number, we can take the absolute value of any real number, and we can add the results. Thus the domain is the entire set of real numbers. ◄

DO EXERCISES 26–28.

NAME SECTION DATE

EXERCISE SET 9.4

ANSWERS

1. ______
2. ______
3. ______
4. ______
5. ______
6. ______
7. ______
8. ______
9. ______
10. ______
11. ______
12. ______

a Write the correspondence as a relation, or set of ordered pairs.

1. d:

Domain (City)	**Range (Team)**
New York	→ Mets
New York	→ Yankees
Los Angeles	→ Rams
Los Angeles	→ Raiders
Houston	→ Astros

2. e:

Domain (City)	**Range (Population)**
Cleveland	→ 2,788,400
Dallas	→ 3,348,000
Boston	→ 4,026,000
Miami	→ 2,799,000

3. f:

Domain	**Range**
−4	→ 5
−5	→ 5
−6	→ 5
−7	→ 2

4. g:

Domain	**Range**
23	→ −47
34	→ −57
43	→ −57
54	→ −68
54	→ −79

Determine whether the relation is a function.

5. $c = \{(9, 1), (-5, -2), (2, -1), (3, -9)\}$

6. $s = \{(6, 8), (8, 9), (6, 6), (-2, 10)\}$

7. $t = \{(20, 7), (21, -5), (22, 7), (20, 0), (23, 0)\}$

8. $m = \{(2, 3), (4, 3), (7, 3), (0, 3), (17, 3), (-3, 3)\}$

9. Find the domain and the range of the relation in Exercise 5.

10. Find the domain and the range of the relation in Exercise 6.

11. Find the domain and the range of the relation in Exercise 7.

12. Find the domain and the range of the relation in Exercise 8.

ANSWERS

13.

14.

15.

16.

17.

18.

19.

20.

21.

22.

23.

24.

25.

26.

27.

28.

29.

30.

31.

32.

b Given the function *a* described by $g(x) = -2x - 4$, find each of the following.

13. $g(-3)$ **14.** $g(-\frac{1}{2})$ **15.** $g(\frac{3}{2})$ **16.** $g(1.2)$

Given the function *h* described by $h(x) = 3x^2$, find each of the following.

17. $h(-1)$ **18.** $h(1)$ **19.** $h(2)$ **20.** $h(-2)$

Given the function *P* described by $P(x) = x^3 - x$, find each of the following.

21. $P(2)$ **22.** $P(-2)$ **23.** $P(-1)$ **24.** $P(0)$

Given the function *Q* described by $Q(x) = x^4 - 2x^3 + x^2 - x + 2$, find each of the following.

25. $Q(-2)$ **26.** $Q(2)$ **27.** $Q(0)$ **28.** $Q(-1)$

29. *Pressure at sea depth.* The function $P(d) = 1 + (d/33)$ gives the pressure, in *atmospheres* (atm), at a depth d in the sea (d is in feet). Note that $P(0) = 1$ atm, $P(33) = 2$ atm, and so on. Find the pressure at 20 ft, 30 ft, and 100 ft.

30. *Record revolutions.* The function $R(t) = 33\frac{1}{3}t$ gives the number of revolutions of a $33\frac{1}{3}$ rpm record as a function of the time t (in minutes) that it turns. Find the number of revolutions at 5 min, 20 min, and 25 min.

31. *Consumer's demand function.* As the price of a product increases, consumer's purchase, or *demand,* for the product decreases. Suppose under certain conditions in our economy that the demand for sugar is related to price by the demand function

$$D(p) = -2.7p + 16.3,$$

where p is the price of a 5-lb bag of sugar and $D(p)$ is the quantity of 5-lb bags, in millions, purchased at price p. Find the quantity purchased when the price is \$1 per 5-lb bag, \$2 per 5-lb bag, \$3 per 5-lb bag, \$4 per 5-lb bag, and \$5 per 5-lb bag.

32. *Seller's supply function.* As the price of a product increases, the seller is willing to sell, or *supply,* more of the product. Suppose under certain conditions in our economy that the supply of sugar is related to price by the supply function

$$S(p) = 7p - 5.9,$$

where p is the price of a 5-lb bag of sugar and $S(p)$ is the quantity of 5-lb bags, in millions, that a seller allows to be purchased at price p. Find the quantity supplied for sale when the price is \$1 per 5-lb bag, \$2 per 5-lb bag, \$3 per 5-lb bag, \$4 per 5-lb bag, and \$5 per 5-lb bag.

c Graph the function.

33. $f(x) = -2x - 3$

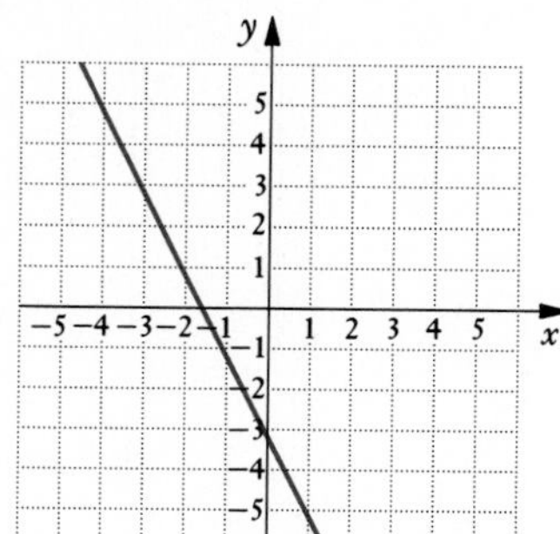

34. $g(x) = x + 4$

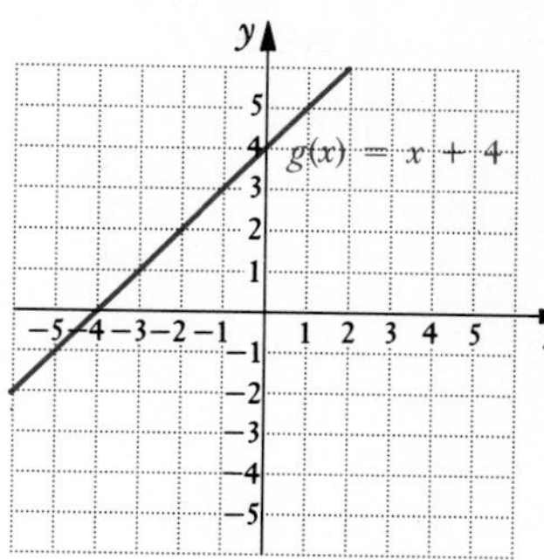

35. $h(x) = |x|$

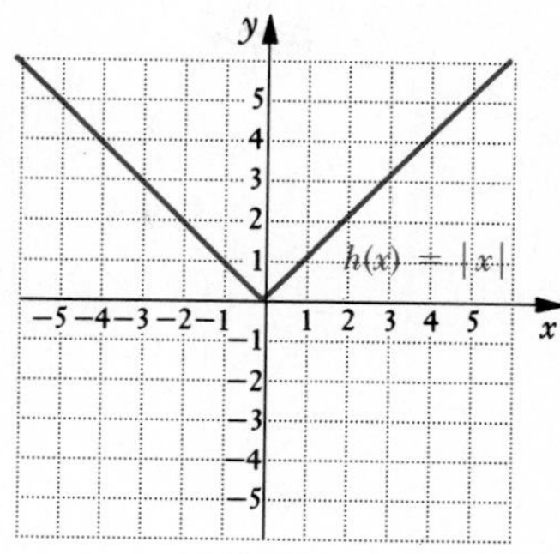

36. $f(x) = |x| + 2$

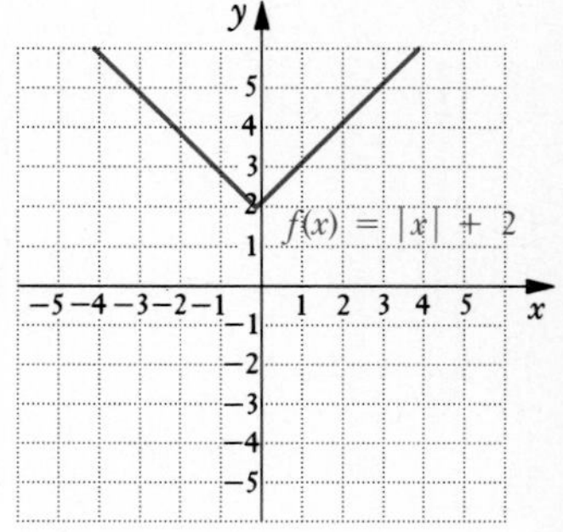

37. $f(x) = -5$

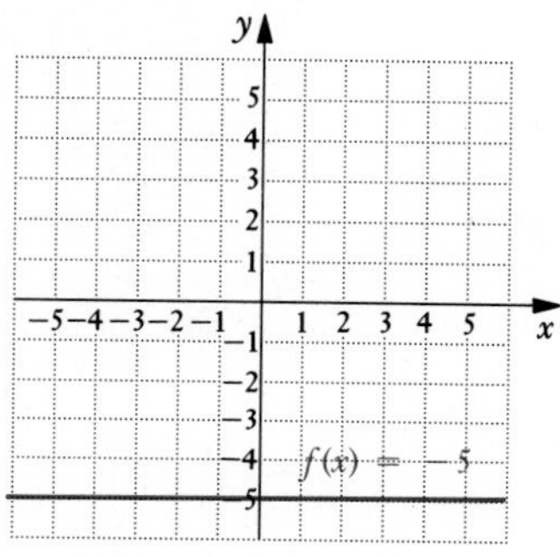

38. $g(x) = x^2$

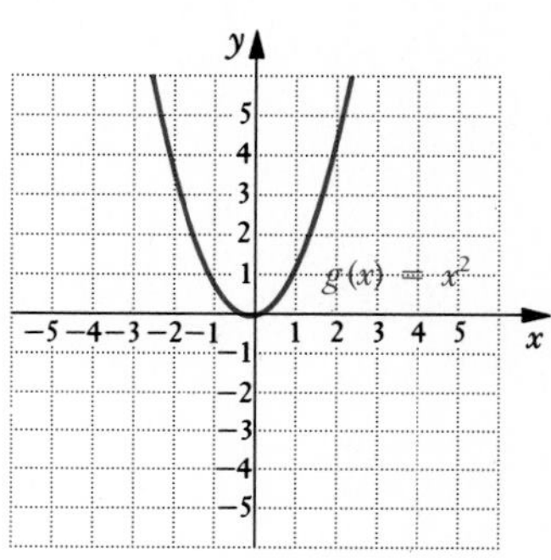

39. $f(x) = x^2 + 2$

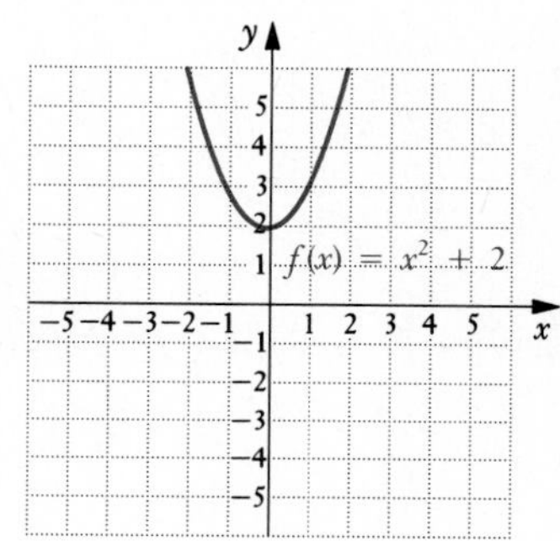

40. $f(x) = x^2 - 4$

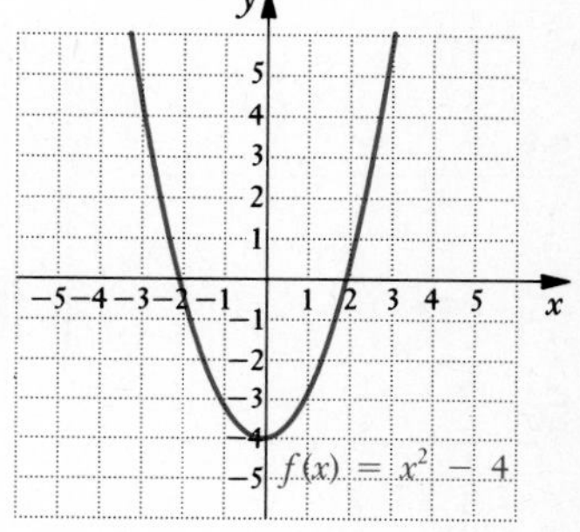

41. $f(x) = 3 - x^2$

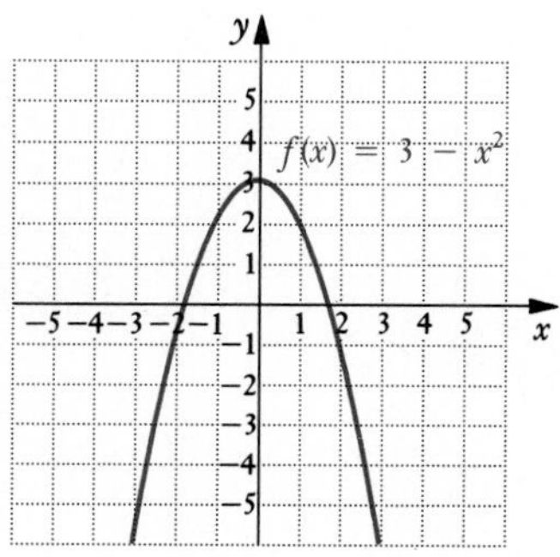

42. $h(x) = 1 - x^2$

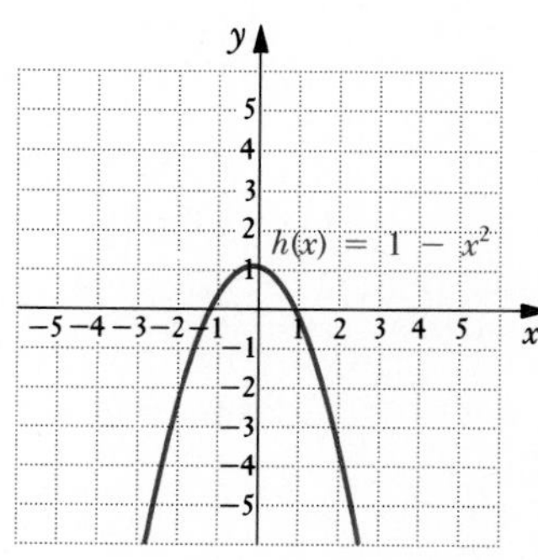

43. $g(x) = -\dfrac{4}{x}$

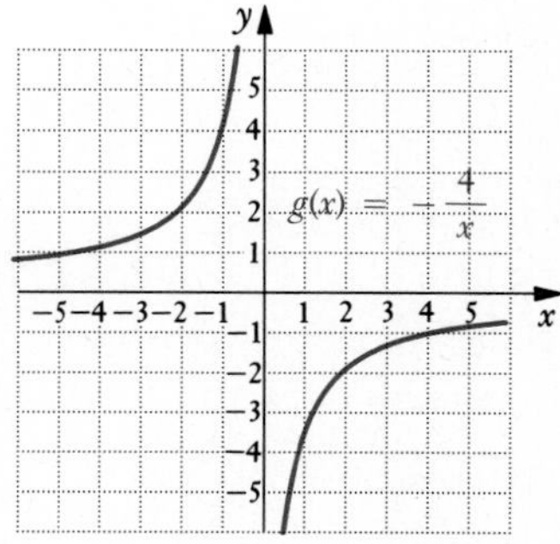

44. $f(x) = \dfrac{2}{x}$

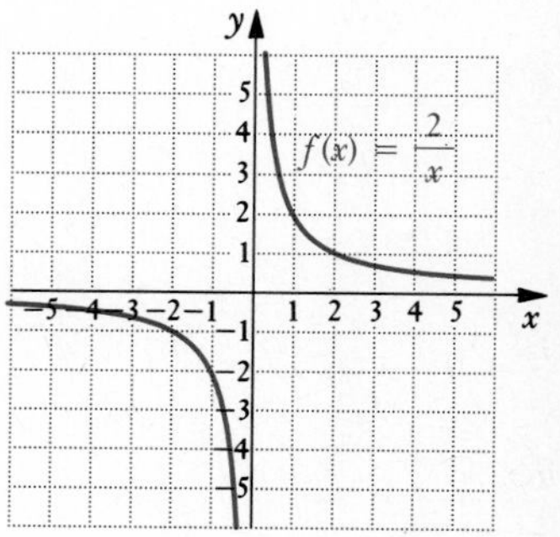

45. $f(x) = 3 - |x|$

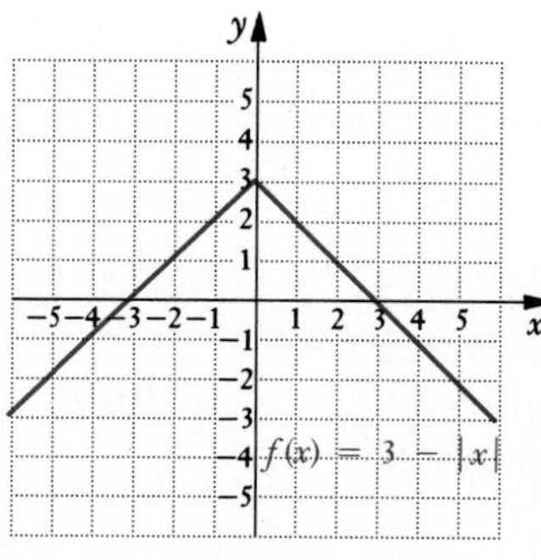

46. $g(x) = |3 - x|$

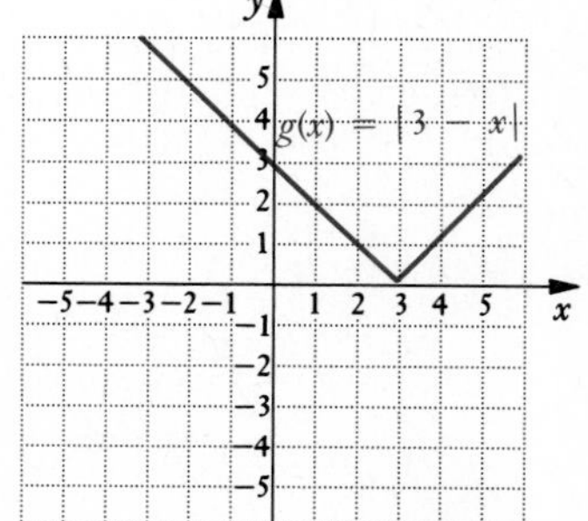

ANSWERS

47. ______

48. ______

49. ______

50. ______

51. ______

52. ______

53. ______

54. ______

55. ______

56. ______

57. ______

58. ______

59. ______

60. ______

61. ______

62. ______

63. See graph.

64. See graph.

65. See graph.

66. ______

67. ______

Determine whether each of the following is the graph of a function.

47.

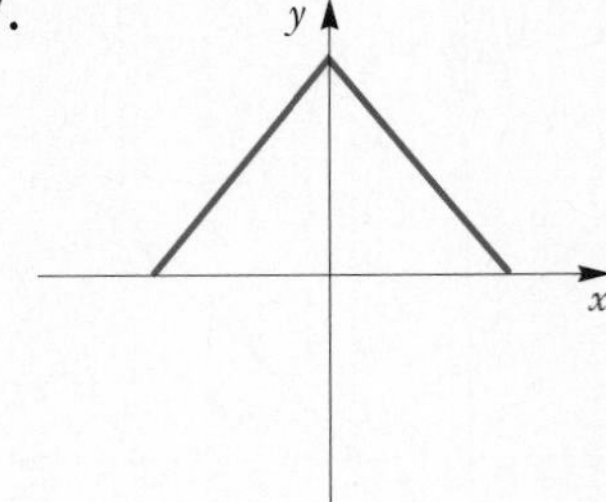

48.

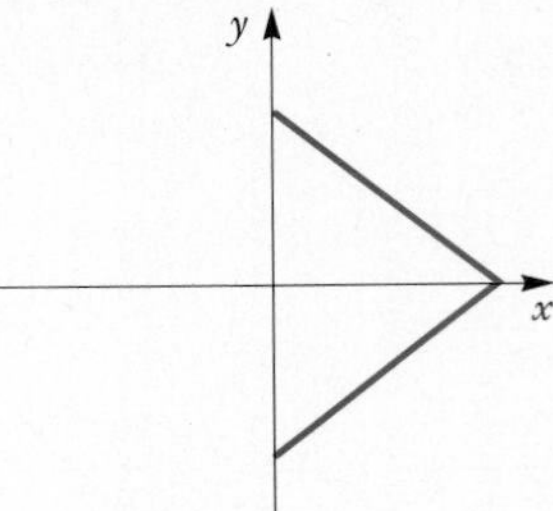

49.

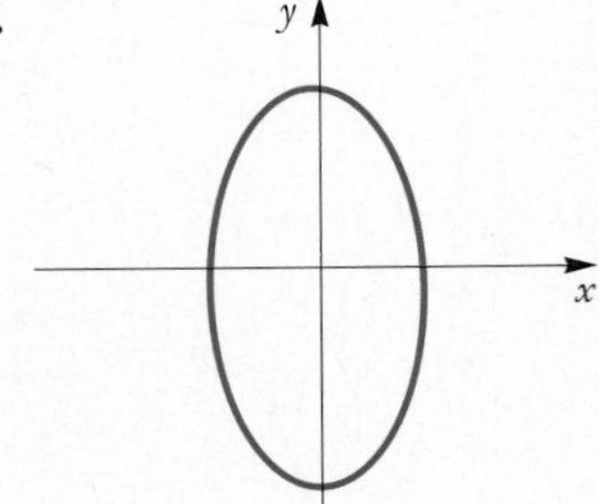

50.

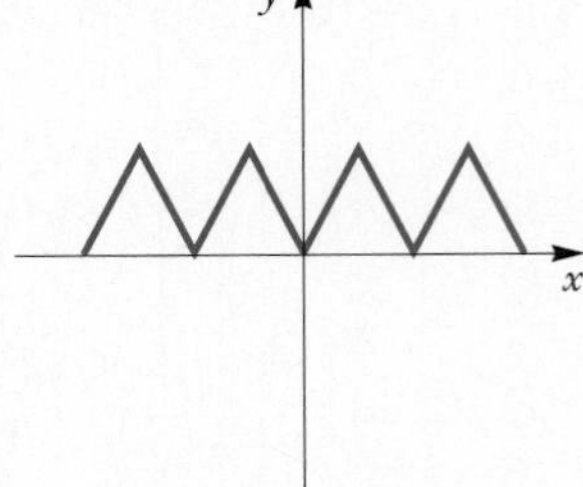

d Find the domain of the function.

51. $p(x) = x^3 - x^2 + x - 7$

52. $p(x) = |x| - 3x$

53. $f(x) = \dfrac{7}{x^2 - 25}$

54. $f(x) = \dfrac{8}{x^2 - 9}$

55. $g(x) = \dfrac{x}{x^2 + 8x + 15}$

56. $g(x) = \dfrac{4x}{x^2 - 9x + 20}$

57. $f(x) = \sqrt{6 + 8x}$

58. $f(x) = \sqrt{3x - 9}$

59. $g(x) = \dfrac{4}{5 - 2x}$

60. $h(x) = \dfrac{6}{7 + 3x}$

SYNTHESIS

Find the range of the function for the given domain.

61. $f(x) = 3x + 5$, when the domain is the set of whole numbers less than 4

62. $g(t) = t^2 - 5$, when the domain is the set of integers between -4 and 2

Graph.

63. $f(x) = \dfrac{|x|}{x}$

64. $g(x) = |x| + x$

65. $h(x) = |x| - x$

66. Sketch a graph that is not a function.

67. Draw the graph of $|y| = x$. Is this the graph of a function?

9.5 The Algebra of Functions

a The Sum, Difference, Product, and Quotient of Functions

In this section, we consider ways of combining given functions to form new functions.

Any function that can be described by a polynomial, such as

$$f(x) = 5x^7 + 3x^5 - 4x^2 - 5,$$

is called a **polynomial function.** Here are some special kinds of polynomial functions.

CONSTANT FUNCTIONS: $f(x) = c$, for any x

For example, for the constant function $f(x) = 4$, the output for any input is 4. Thus, $f(0) = 4$, $f(-1) = 4$, $f(0.37) = 4$, and so on.

LINEAR FUNCTIONS: $f(x) = mx + b$

For example, $f(x) = 3x + 2$, $g(x) = x - 7$, and $h(x) = 5 - 3x$ are all linear functions. Each is described by a first-degree polynomial. The graph of each function is a straight line.

QUADRATIC FUNCTIONS: $f(x) = ax^2 + bx + c$

For example, $f(x) = x^2 - 3x$, $g(x) = -3x^2 + 4x - 2$, and $h(x) = 13 - 4x^2$ are all quadratic functions. Each is described by a second-degree polynomial.

Consider the following two polynomial functions:

$$f(x) = x^2 - 5 \quad \text{and} \quad g(x) = x + 7.$$

Then $f(2) = 2^2 - 5 = -1$ and $g(2) = 2 + 7 = 9$. Suppose we add these function values: $f(2) + g(2) = -1 + 9 = 8$. Doing this for any input x creates a new function $\mathbf{f + g}$ called the **sum** of f and g, described by $\mathbf{f(x) + g(x)}$.

Suppose we subtract these function values: $f(2) - g(2) = -1 - 9 = -10$. Doing this for any input x creates a new function $\mathbf{f - g}$ called the **difference** of f and g, described by $\mathbf{f(x) - g(x)}$.

Suppose we multiply these function values: $f(2) \cdot g(2) = (-1)(9) = -9$. Doing this for any input x creates a new function $\mathbf{fg}$ called the **product** of f and g, described by $\mathbf{f(x) \cdot g(x)}$.

Suppose we divide these function values: $f(2)/g(2) = (-1)/9 = -1/9$. Doing this for any input x for which $g(x)$ is not 0 creates a new function $\mathbf{f/g}$ called the **quotient** of f and g, described by $\mathbf{f(x)/g(x)}$.

From any functions f and g, we can form new functions defined as:

1. **The *sum* $f + g$:** $\quad (f + g)(x) = f(x) + g(x)$;
2. **The *difference* $f - g$:** $\quad (f - g)(x) = f(x) - g(x)$;
3. **The *product* fg:** $\quad fg(x) = f(x) \cdot g(x)$;
4. **The *quotient* f/g:** $\quad (f/g)(x) = f(x)/g(x)$, **where** $g(x) \neq 0$.

OBJECTIVES

After finishing Section 9.5, you should be able to:

a Given two functions f and g, find their sum, difference, product, and quotient.

b Find the composition of functions and express certain functions as a composition of functions.

FOR EXTRA HELP

Tape 18C

Tape 16B

MAC: 9
IBM: 9

1. Given $f(x) = x^2 + 3$ and $g(x) = x^2 - 3$, find each of the following.

a) $(f + g)(x)$

b) $(f - g)(x)$

c) $fg(x)$

d) $(f/g)(x)$

e) $ff(x)$

ANSWERS ON PAGE A-10

► **EXAMPLE 1** Given f and g described by $f(x) = x^2 - 5$ and $g(x) = x + 7$, find $(f + g)(x)$, $(f - g)(x)$, $fg(x)$, $(f/g)(x)$, and $gg(x)$.

$$(f + g)(x) = f(x) + g(x) = (x^2 - 5) + (x + 7) = x^2 + x + 2;$$
$$(f - g)(x) = f(x) - g(x) = (x^2 - 5) - (x + 7) = x^2 - x - 12;$$
$$fg(x) = f(x) \cdot g(x) = (x^2 - 5)(x + 7) = x^3 + 7x^2 - 5x - 35;$$
$$(f/g)(x) = f(x)/g(x) = \frac{x^2 - 5}{x + 7};$$
$$gg(x) = g(x) \cdot g(x) = (x + 7)(x + 7) = x^2 + 14x + 49$$ ◄

Note that the sum, difference, and product of polynomials are also polynomial functions, but the quotient may not be.

DO EXERCISE 1.

b The Composition of Functions

In the real world, functions frequently occur in which some variable depends on the choice of a third variable. For instance, the number of employees hired by a firm may depend on the firm's profits, which may in turn depend on the number of items the firm produces. Such functions are called **composite functions.**

We can visualize the composition of functions as follows. Suppose

x = the number of items the firm manufactures,
g = the function describing the profit, and
f = the function describing the number of employees hired.

The machine in the following figure shows how to combine these functions.

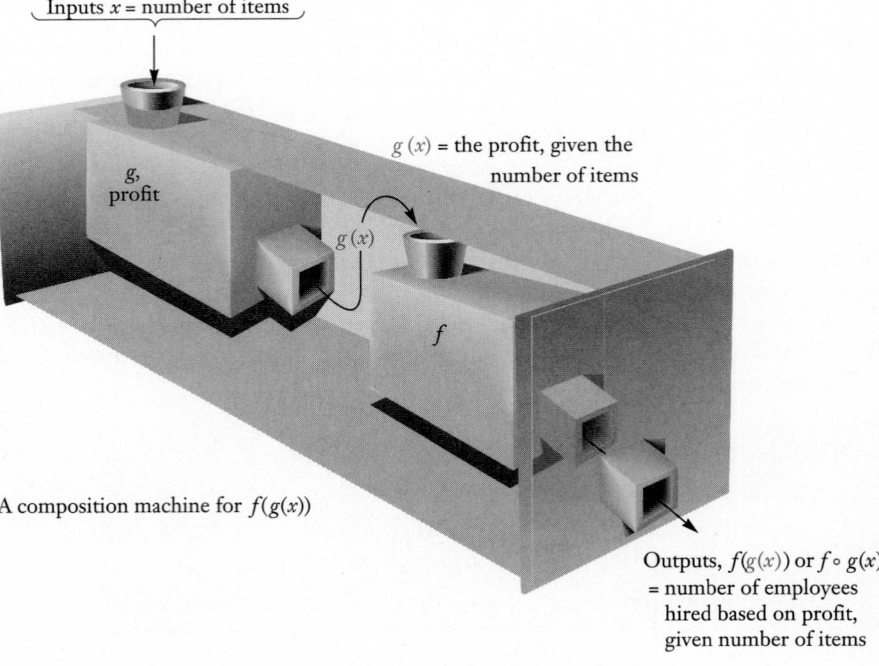

A composition machine for $f(g(x))$

The resulting new function is $f(g(x))$, or $f \circ g(x)$. It relates the number of items the firm manufactures, x, directly to the number of employees hired.

> **The *composite function* $f \circ g$, the *composition* of f and g, is defined as**
> $$f \circ g(x) = f(g(x)).$$

In the following example, we see how to find formulas for composite functions.

There is a function g that gives a correspondence between women's shoe sizes in the United States and Italy. The function is given by $g(x) = 2(x + 12)$, where x is a shoe size in the United States and $g(x)$ is a shoe size in Italy. For example, a shoe size of 4 in the United States corresponds to a shoe size of $g(4) = 2(4 + 12)$, or 32 in Italy.

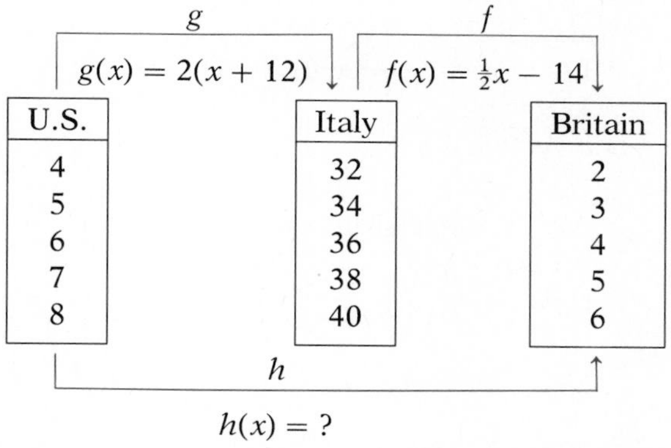

g: $g(x) = 2(x + 12)$ (U.S. → Italy) f: $f(x) = \frac{1}{2}x - 14$ (Italy → Britain)

U.S.	Italy	Britain
4	32	2
5	34	3
6	36	4
7	38	5
8	40	6

h (U.S. → Britain)

$h(x) = ?$

There is also a function that gives a correspondence between women's shoe sizes in Italy and those in Britain. The function is given by $f(x) = \frac{1}{2}x - 14$, where x is a shoe size in Italy and $f(x)$ is the corresponding shoe size in Britain. For example, a shoe size of 32 in Italy corresponds to a shoe size of $f(32) = \frac{1}{2}(32) - 14$, or 2 in Britain.

It seems reasonable to assume that a shoe size of 4 in the United States corresponds to a shoe size of 2 in Britain and that there is a function h that describes this correspondence. Can we find a formula for h? If we look at the tables, we might guess that such a formula is $h(x) = x - 2$, and that is indeed correct. For more complicated formulas, however, we need to do some algebra.

A shoe size x in the United States corresponds to a shoe size $g(x)$ in Italy, where

$$g(x) = 2(x + 12).$$

Now $2(x + 12)$ is a shoe size in Italy. If we replace x in $f(x)$ by $2(x + 12)$, we can find the corresponding shoe size in Britain:

$$\begin{aligned} f(g(x)) &= \tfrac{1}{2}[2(x + 12)] - 14 = \tfrac{1}{2}[2x + 24] - 14 \\ &= x + 12 - 14 = x - 2. \end{aligned}$$

This gives a formula for h: $h(x) = x - 2$. Thus a shoe size of 4 in the United States corresponds to a shoe size of $h(4) = 4 - 2$, or 2 in Britain. The function h is the **composition** of f and g, symbolized by $f \circ g$.

To find $f \circ g(x)$, we substitute $g(x)$ for x in $f(x)$.

2. Given $f(x) = x + 5$ and $g(x) = x^2 - 1$, find $f \circ g(x)$ and $g \circ f(x)$.

3. Given $f(x) = 4x + 5$ and $g(x) = \sqrt[3]{x}$, find $f \circ g(x)$ and $g \circ f(x)$.

4. Find $f(x)$ and $g(x)$ such that $h(x) = f \circ g(x)$. Answers may vary.

a) $h(x) = \sqrt[3]{x^2 + 1}$

b) $h(x) = \dfrac{1}{(x + 5)^4}$

ANSWERS ON PAGE A-10

▶ **EXAMPLE 2** Given $f(x) = 3x$ and $g(x) = 1 + x^2$:

a) Find $f \circ g(5)$ and $g \circ f(5)$.

b) Find $f \circ g(x)$ and $g \circ f(x)$.

We consider each function separately:

$$f(x) = 3x \qquad \textbf{This function multiplies each input by 3.}$$

and

$$g(x) = 1 + x^2. \qquad \textbf{This function adds 1 to the square of each input.}$$

a) $f \circ g(5) = f(g(5)) = f(1 + 5^2) = f(26) = 3(26) = 78;$
$g \circ f(5) = g(f(5)) = g(3 \cdot 5) = g(15) = 1 + 15^2 = 1 + 225 = 226$

b)
$$\begin{aligned} f \circ g(x) &= f(g(x)) \\ &= f(1 + x^2) \qquad \textbf{Substituting } 1 + x^2 \textbf{ for } x \\ &= 3(1 + x^2) \\ &= 3 + 3x^2; \end{aligned}$$

$$\begin{aligned} g \circ f(x) &= g(f(x)) \\ &= g(3x) \qquad \textbf{Substituting } 3x \textbf{ for } x \\ &= 1 + (3x)^2 \\ &= 1 + 9x^2 \end{aligned}$$ ◀

DO EXERCISE 2.

Note in Example 2 that $f \circ g(5) \neq g \circ f(5)$ and, in general, $f \circ g(x) \neq g \circ f(x)$.

▶ **EXAMPLE 3** Given $f(x) = \sqrt{x}$ and $g(x) = x - 1$, find $f \circ g(x)$ and $g \circ f(x)$.

$$f \circ g(x) = f(g(x)) = f(x - 1) = \sqrt{x - 1};$$
$$g \circ f(x) = g(f(x)) = g(\sqrt{x}) = \sqrt{x} - 1$$ ◀

DO EXERCISE 3.

It is important to be able to recognize how a function can be expressed as a composition. Such a situation can occur in a study of calculus.

▶ **EXAMPLE 4** Find $f(x)$ and $g(x)$ such that $h(x) = f \circ g(x)$:

$$h(x) = (7x + 3)^2.$$

This is $7x + 3$ to the 2nd power. Two functions that can be used for the composition are $f(x) = x^2$ and $g(x) = 7x + 3$. We can check by forming the composition:

$$h(x) = f \circ g(x) = f(g(x)) = f(7x + 3) = (7x + 3)^2.$$

This is the most "obvious" answer to the question. There can be other less obvious answers. For example, if

$$f(x) = (x - 1)^2 \quad \text{and} \quad g(x) = 7x + 4,$$

then

$$h(x) = f \circ g(x) = f(g(x)) = f(7x + 4) = (7x + 4 - 1)^2 = (7x + 3)^2.$$ ◀

DO EXERCISE 4.

NAME SECTION DATE

EXERCISE SET 9.5

ANSWERS

a Given the functions f and g as follows, find $(f+g)(x)$, $(f-g)(x)$, $fg(x)$, $(f/g)(x)$, and $ff(x)$.

1. $f(x) = x + 3,\ g(x) = x - 3$

2. $f(x) = x^2 - 2,\ g(x) = x + 1$

3. $f(x) = 2x^2 - 3x + 1,\ g(x) = x^3$

4. $f(x) = x - 1,\ g(x) = x^2 + x + 1$

5. $f(x) = -5x^2,\ g(x) = 4x^3$

6. $f(x) = 28x^5,\ g(x) = 7x^2$

7. $f(x) = 20,\ g(x) = -5$

8. $f(x) = 25,\ g(x) = 16$

b Find $f \circ g(x)$ and $g \circ f(x)$.

9. $f(x) = 5x - 8,\ g(x) = 7 - 2x$

10. $f(x) = 9 - 6x,\ g(x) = 0.37x + 4$

11. $f(x) = 3x^2 + 2,\ g(x) = 2x - 1$

12. $f(x) = 4x + 3,\ g(x) = 2x^2 - 5$

13. $f(x) = 4x^2 - 1,\ g(x) = \frac{2}{x}$

14. $f(x) = \frac{3}{x},\ g(x) = 2x^2 + 3$

15. $f(x) = x^2 + 1,\ g(x) = x^2 - 1$

16. $f(x) = \frac{1}{x^2},\ g(x) = x + 2$

1. ______
2. ______
3. ______
4. ______
5. ______
6. ______
7. ______
8. ______
9. ______
10. ______
11. ______
12. ______
13. ______
14. ______
15. ______
16. ______

ANSWERS

17. ______

18. ______

19. ______

20. ______

21. ______

22. ______

23. ______

24. ______

25. ______

26. ______

27. ______

28. ______

29. ______

30. ______

31. ______

32. ______

33. ______

34. ______

35. ______

36. ______

37. ______

38. ______

39. ______

40. ______

41. ______

Find $f(x)$ and $g(x)$ such that $h(x) = f \circ g(x)$. Answers may vary.

17. $h(x) = (5 - 3x)^2$

18. $h(x) = 4(3x - 1)^2 + 9$

19. $h(x) = (3x^2 - 7)^5$

20. $h(x) = \sqrt{5x + 2}$

21. $h(x) = \dfrac{1}{x - 1}$

22. $h(x) = \dfrac{3}{x} + 4$

23. $h(x) = \dfrac{1}{\sqrt{7x + 2}}$

24. $h(x) = \sqrt{x - 7} - 3$

25. $h(x) = \dfrac{x^3 + 1}{x^3 - 1}$

26. $h(x) = (\sqrt{x} + 5)^4$

SKILL MAINTENANCE

Simplify.

27. $\sqrt[3]{27x^6y^{12}}$

28. $\sqrt[5]{32p^{10}q^{17}}$

Solve.

29. $2x^2 - 7x + 4 = 0$

30. $5x^2 - 4x = 2x^2 + 8x - 3$

SYNTHESIS

For each function, find and simplify

$$\frac{f(x + h) - f(x)}{h}.$$

31. $f(x) = 3x + 7$

32. $f(x) = 6 - 4x$

33. $f(x) = -7$

34. $f(x) = 18$

35. $f(x) = x^2$

36. $f(x) = -4x^2$

37. $f(x) = \dfrac{1}{x}$

38. $f(x) = \dfrac{5}{x}$

39. $f(x) = \sqrt{x}$

40. $f(x) = x^3$

41. An organization determines that the cost per person of chartering a bus is given by the function

$$C(x) = \frac{100 + 5x}{x},$$

where x = the number of people in the group and $C(x)$ is in dollars. Find the cost of chartering a bus for 40 people; for 80 people.

9.6 Inverses of Relations and Functions

a Inverses of Relations

Consider the relation r given as follows:

$$r = \{(-4, 3), (-1, 2), (2, 5)\}.$$

Suppose we *interchange* the first and second coordinates. The relation we obtain is called the **inverse** of the relation r. It is given as follows:

$$\text{Inverse of } r = \{(3, -4), (2, -1), (5, 2)\}.$$

The *inverse* of a relation is found by interchanging the first and second coordinates in the original relation.

DO EXERCISES 1 AND 2.

When a relation is defined by an equation, interchanging x and y produces an equation of the *inverse* relation.

The solutions of the inverse relation are found from the solutions of the original relation by interchanging the first and second coordinates.

► **EXAMPLE 1** Find an equation of the inverse of the relation $y = 3x - 4$.

We interchange x and y and obtain $x = 3y - 4$. ◄

► **EXAMPLE 2** Find an equation of the inverse of the relation $y = x^2 + x - 2$.

We interchange x and y and obtain $x = y^2 + y - 2$. ◄

DO EXERCISES 3 AND 4.

b Inverses and One-to-One Functions

Let us consider the following two functions. We think of them as correspondences.

COST OF A 60-SECOND COMMERCIAL DURING THE SUPER BOWL

Domain (Set of inputs)	Range (Set of outputs)
1967 →	$80,000
1977 →	$324,000
1981 →	$550,000
1983 →	$800,000
1988 →	$1,350,000

U.S. SENATORS

Domain (Set of inputs)	Range (Set of outputs)
Cranston →	California
Wilson →	California
Chiles →	Florida
Graham →	Florida
Simon →	Illinois
Dixon →	Illinois

OBJECTIVES

After finishing Section 9.6, you should be able to:

- **a** Find the inverse of a relation if the relation is given as a set of ordered pairs, and find an equation of the inverse of a relation if it is given as an equation.
- **b** Given a function, determine whether it is one-to-one and has an inverse that is a function.
- **c** Find a formula for the inverse of a function, if it exists.
- **d** Graph inverse relations and functions.

FOR EXTRA HELP

Tape NC

Tape 17A

MAC: 9
IBM: 9

Find the inverse.

1. $r = \{(8, -6), (1, -2), (8, 6), (-4, 3)\}$

2. $h = \{(-1, 3), (-2, 4), (-3, 5), (-4, 6), (-5, 7)\}$

Find an equation of the inverse.

3. $y = 2x + 7$

4. $y = x^3 - 2x + |x|$

ANSWERS ON PAGE A-10

For each function, find its inverse correspondence and determine whether that inverse is a function.

5. WOMEN'S DRESS SIZES

Domain (United States)	Range (France)
6 →	38
8 →	40
10 →	42
12 →	44
14 →	46
16 →	48
18 →	50

6. SPORTS TEAMS

Domain	Range
Lakers →	Los Angeles
Dodgers →	Los Angeles
Rams →	Los Angeles
Knickerbockers →	New York
Yankees →	New York
Giants →	New York

ANSWERS ON PAGE A-10

Suppose we reverse the arrows. We obtain what is called the **inverse correspondence** or **relation.** Are these new correspondences functions?

COST OF A 60-SECOND COMMERCIAL DURING THE SUPER BOWL

Range (Set of outputs)	Domain (Set of inputs)
1967 ←	$80,000
1977 ←	$324,000
1981 ←	$550,000
1983 ←	$800,000
1988 ←	$1,350,000

U.S. SENATORS

Range (Set of outputs)	Domain (Set of inputs)
Cranston ←	California
Wilson ←	California
Chiles ←	Florida
Graham ←	Florida
Simon ←	Illinois
Dixon ←	Illinois

We see that the inverse of the first correspondence is a function, but that the inverse of the second correspondence is not a function.

Recall that for each input, a function provides exactly one output. However, nothing in our definition of function prevents having the same output for two or more different inputs. Thus it is possible for different inputs to correspond to the same output in the range. Only when this possibility is excluded is the inverse also a function.

In the Super Bowl function, different inputs have different outputs. It is what is called a **one-to-one** function. In the U.S. Senator function, the input *Simon* has the output *Illinois,* and the input *Dixon* also has the output *Illinois.* Thus it is not a one-to-one function.

A function f is *one-to-one* if different inputs have different outputs. That is,

if $a \neq b$, then $f(a) \neq f(b)$.

The inverse of a function is a function only when it is one-to-one.

DO EXERCISES 5 AND 6.

► **EXAMPLE 3** Consider the function f given by

$$f = \{(2, 4), (3, 5), (7, 4), (6, 6)\}.$$

a) Determine whether the function is one-to-one.

b) Determine whether the function has an inverse that is a function.

We solve as follows:

a) The function is *not* one-to-one. There are two ordered pairs, namely (2, 4) and (7, 4), that have different first coordinates but the same second coordinates. Thus, $2 \neq 7$, but $f(2) = f(7)$.

b) The inverse is not a function since it is not one-to-one.

We can answer the second question directly by looking at the inverse. The inverse of the function is $\{(4, 2), (5, 3), (4, 7), (6, 6)\}$. The inverse is *not* a function because there are two ordered pairs, namely (4, 2) and (4, 7), that have the same first coordinates but different second coordinates. Thus, f is not one-to-one, so the inverse is not a function. ◄

DO EXERCISES 7 AND 8 ON THE FOLLOWING PAGE.

How can we tell graphically whether a function is one-to-one and thus has an inverse that is a function?

▶ **EXAMPLE 4** Shown here is the graph of a function. Determine whether the function is one-to-one and thus has an inverse that is a function.

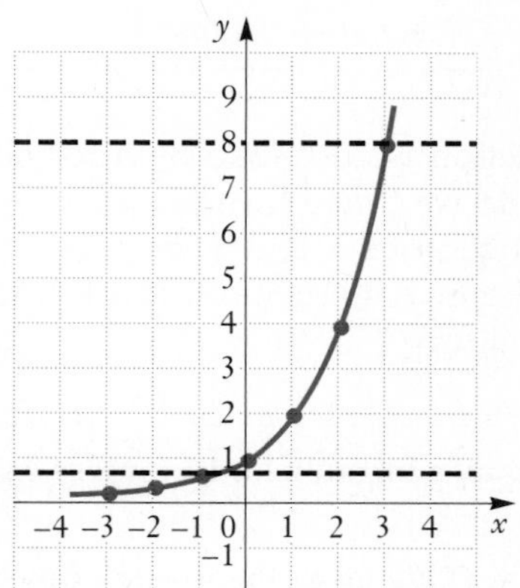

A function is one-to-one if different inputs have different outputs. In other words, no two x-values will have the same y-value. For this function, we cannot find two x-values that have the same y-value. Note also that no horizontal line can be drawn that will cross the graph more than once. The function is one-to-one and its inverse is a function. ◀

The Horizontal-Line Test

A function is one-to-one and has an inverse that is a function if there is no horizontal line that crosses the graph more than once.

A graph is that of a function if no vertical line crosses the graph more than once. A function has an inverse that is also a function if no horizontal line crosses the graph more than once.

▶ **EXAMPLE 5** Determine whether the function $f(x) = x^2$ is one-to-one and has an inverse that is also a function.

The graph is shown below. There are many horizontal lines that cross the graph more than once, so this function is not one-to-one and does not have an inverse that is a function.

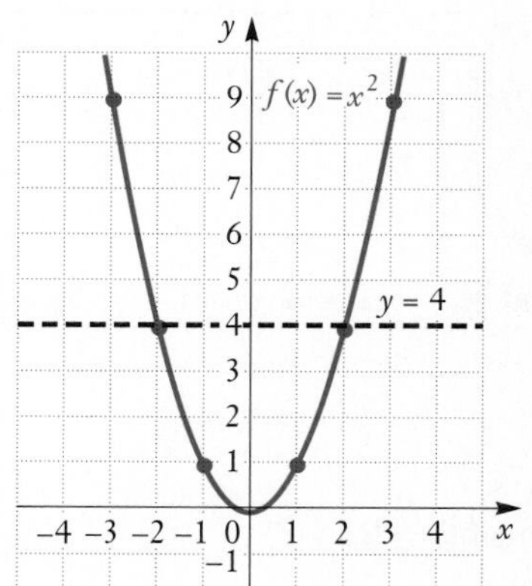

DO EXERCISES 9–12. (EXERCISES 11 AND 12 ARE ON THE FOLLOWING PAGE.)

7. Consider the function f given by

$$f = \{(2, 4), (3, 5), (-7, -5), (9, 11).\}$$

a) Determine whether the function is one-to-one.

b) Determine whether the function has an inverse that is a function.

8. Consider the function g given by

$$g = \{(2, 4), (3, 5), (7, 5), (9, 11)\}.$$

a) Determine whether the function is one-to-one.

b) Determine whether the function has an inverse that is a function.

Determine whether the function is one-to-one and thus has an inverse that is also a function.

9. $f(x) = 4 - x$

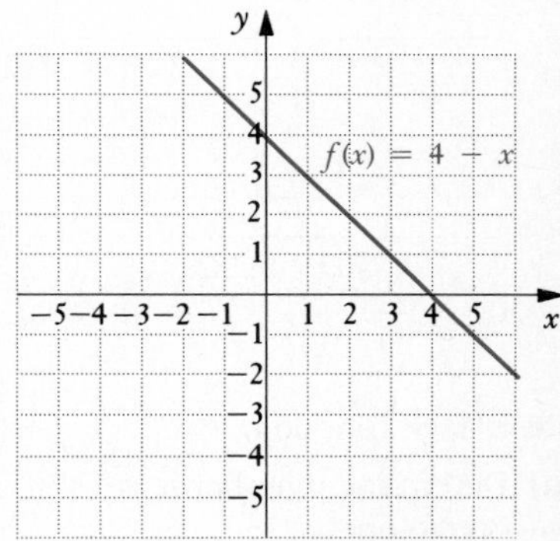

10. $f(x) = x^2 - 1$

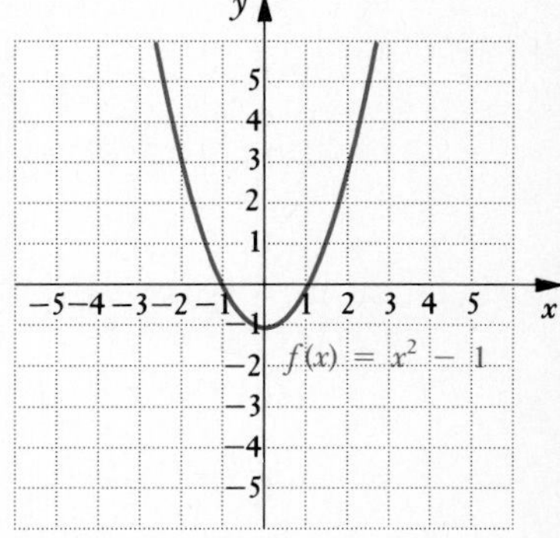

ANSWERS ON PAGE A-11

11. $f(x) = 5x + 8$ (Sketch this graph yourself.)

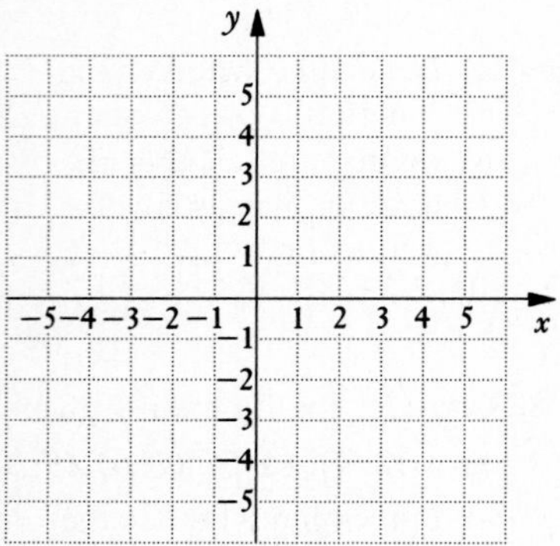

12. $f(x) = |x| - 3$ (Sketch this graph yourself.)

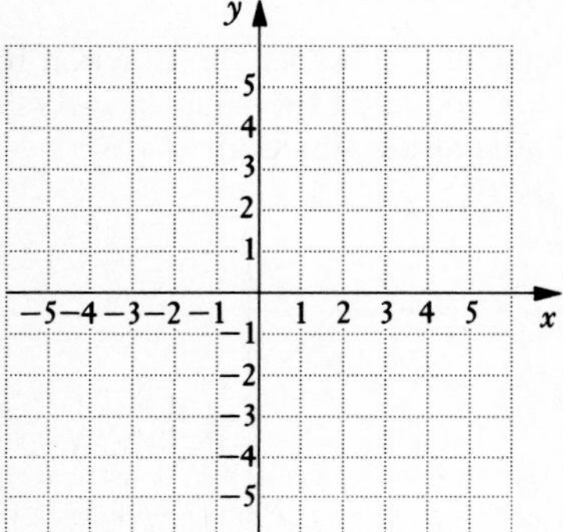

Given each function:

a) Determine whether it is one-to-one.

b) If it is one-to-one, find a formula for the inverse.

13. $f(x) = 3 - x$

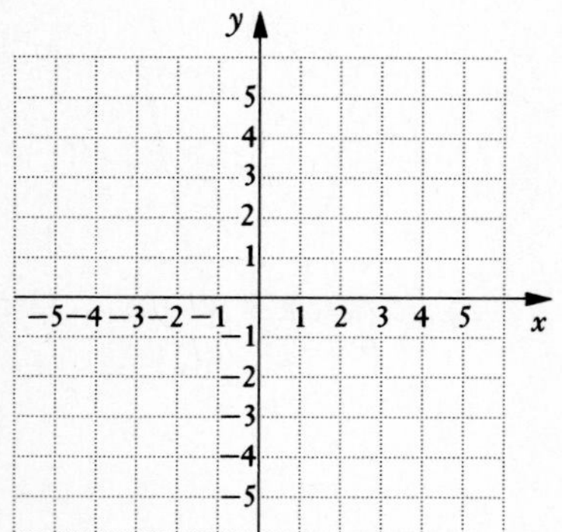

ANSWERS ON PAGE A-11

C Finding a Formula for an Inverse

If the inverse of a function f is also a function, it can be named f^{-1}, read "f-inverse."

> CAUTION! The -1 in f^{-1} is *not* an exponent!

Suppose that a function is described by a formula. If it has an inverse that is a function, how do we find a formula for its inverse? If for any equation with two variables such as x and y we interchange the variables, we obtain an equation of the inverse relation. If it is a function, we proceed as follows to find a formula for f^{-1}.

> **If a function is one-to-one, a formula for its inverse can be found as follows:**
> 1. **Replace $f(x)$ by y.**
> 2. **Interchange x and y. (This gives the inverse function.)**
> 3. **Solve for y.**
> 4. **Replace y by $f^{-1}(x)$.**

▶ **EXAMPLE 6** Given $f(x) = x + 1$:

a) Determine whether the function is one-to-one.

b) If it is one-to-one, find a formula for $f^{-1}(x)$.

We solve as follows.

a) The graph of $f(x) = x + 1$ is shown below. It passes the horizontal-line test, so it is one-to-one. Thus its inverse is a function.

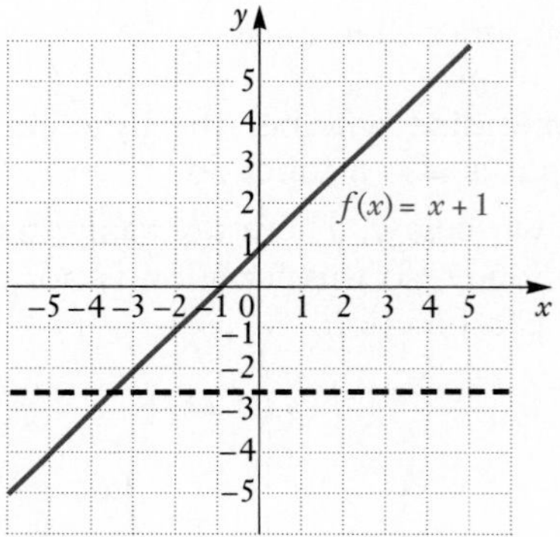

b) **1.** Replace $f(x)$ by y: $y = x + 1$.

2. Interchange x and y: $x = y + 1$. **This gives the inverse function.**

3. Solve for y: $x - 1 = y$.

4. Replace y by $f^{-1}(x)$: $f^{-1}(x) = x - 1$. ◀

▶ **EXAMPLE 7** Given $g(x) = 2x - 3$:

a) Determine whether the function is one-to-one.

b) If it is one-to-one, find a formula for $g^{-1}(x)$.

We solve as follows.

a) The graph of $g(x) = 2x - 3$ is shown below. It passes the horizontal-line test and is one-to-one.

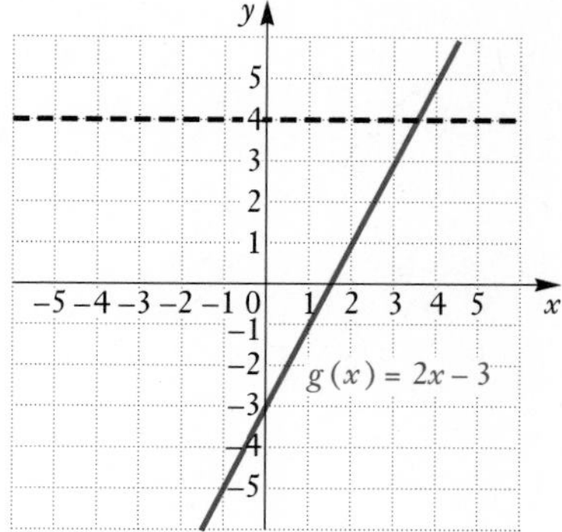

b) 1. Replace $g(x)$ by y: $y = 2x - 3$.

2. Interchange x and y: $x = 2y - 3$.

3. Solve for y: $x + 3 = 2y$

$$\frac{x+3}{2} = y.$$

4. Replace y by $g^{-1}(x)$: $g^{-1}(x) = \dfrac{x+3}{2}$. ◀

DO EXERCISES 13 AND 14. (EXERCISE 13 IS ON THE PRECEDING PAGE.)

d Graphs of Inverse Relations and Functions

▶ **EXAMPLE 8** Consider the relation r given by

$$r = \{(-4, 3), (-1, 2), (2, 5)\}.$$

Graph the relation. Then find the inverse relation and draw its graph in color.

The relation r is shown in black. The inverse of the relation is

$$\{(3, -4), (2, -1), (5, 2)\}$$

and is shown in color.

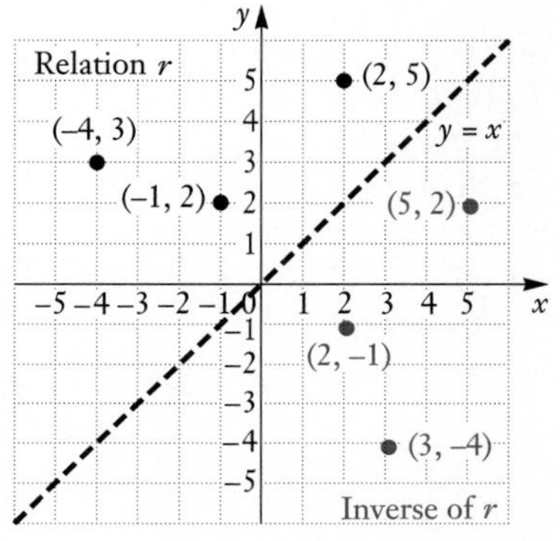

DO EXERCISE 15.

14. $g(x) = 3x - 2$

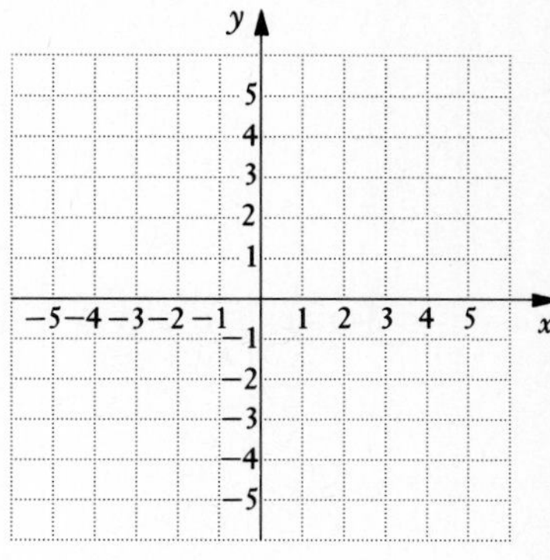

15. Consider the relation r given by

$$r = \{(4, 2), (3, -1), (0, -2)\}.$$

Graph the relation. Then find the inverse relation and draw its graph in color.

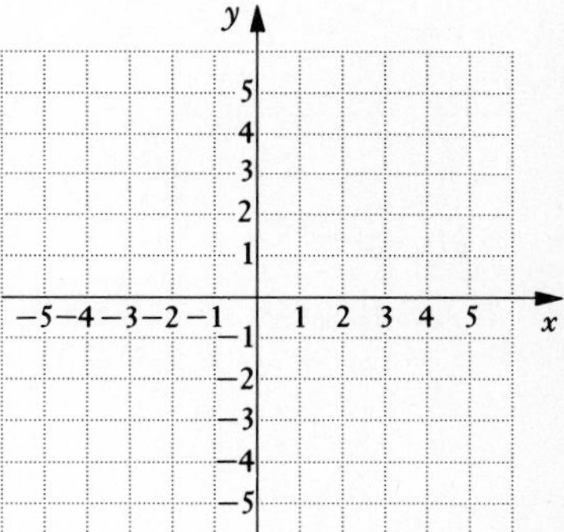

Graph the relation in black. Then find the inverse and graph it in color.

16. $y = 2x - 3$

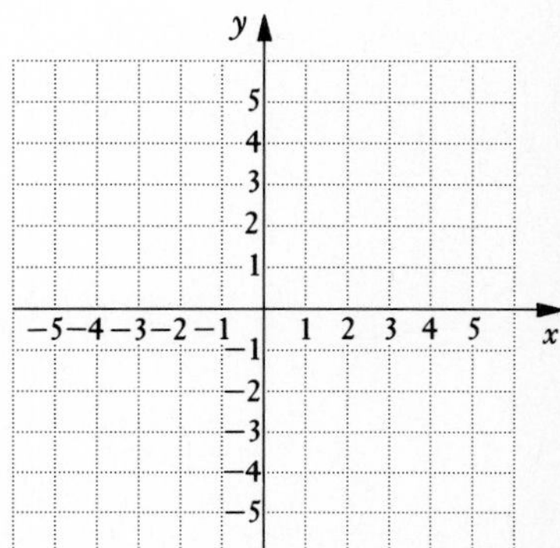

ANSWERS ON PAGE A-11

17. $y = 3 - x^2$

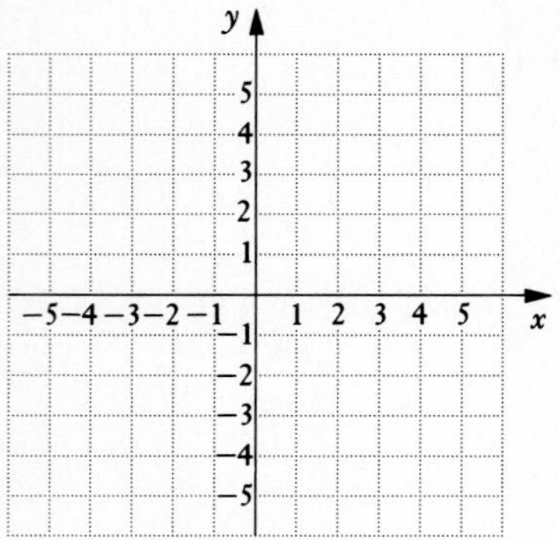

18. Graph the inverse of this relation.

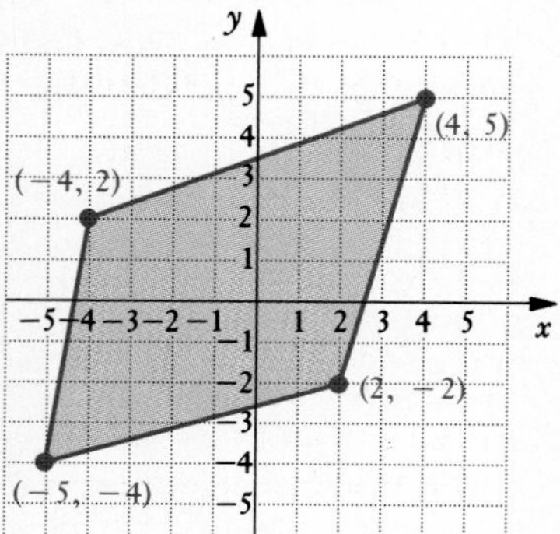

Note in Example 3 that we can obtain the inverse by reflecting each ordered pair across the line $y = x$. Interchanging x and y to find the inverse has the effect of reflecting each ordered pair in the original relation across the line $y = x$.

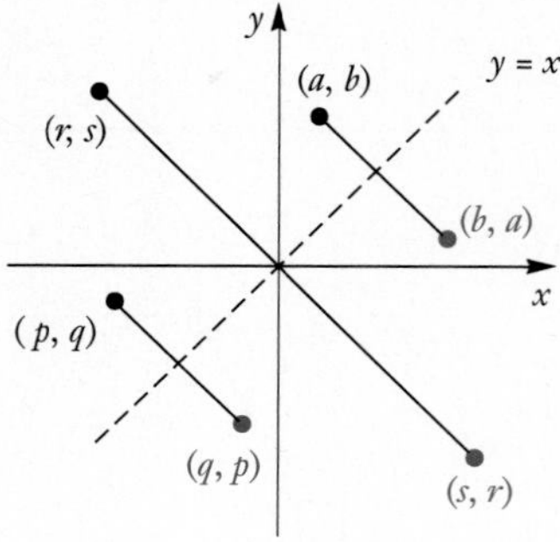

▶ **EXAMPLES** In each case, graph the relation in black. Then find the inverse and graph it in color.

9. Relation: $y = 4 - 2x$
Inverse: $x = 4 - 2y$

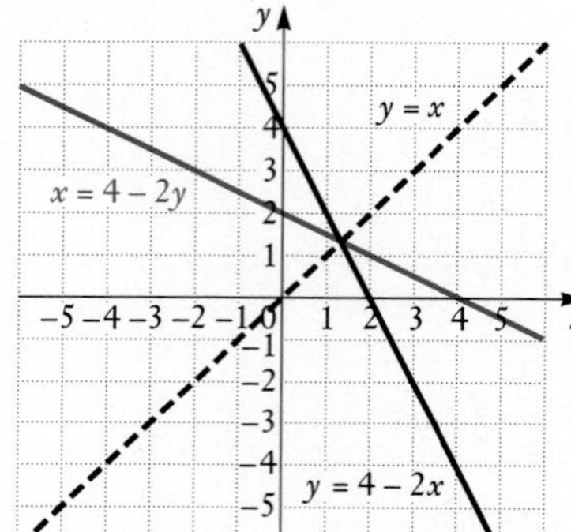

10. Relation: $y = x^2$
Inverse: $x = y^2$

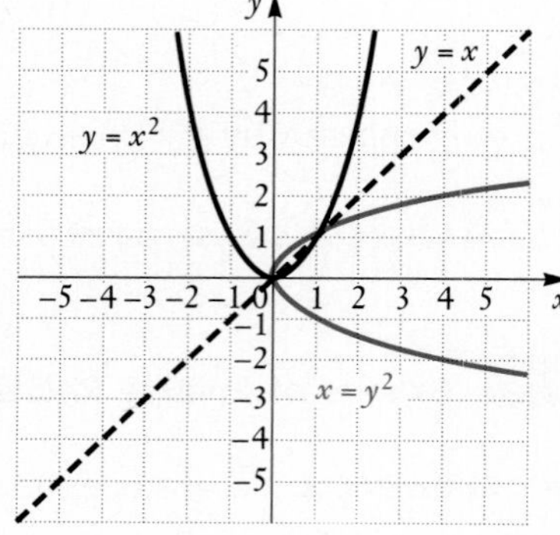

◀

Look back at the graphs of parabolas that we considered in Examples 1 and 2 of Section 9.1. The equations $y = x^2 - 4x + 8$ and $x = y^2 - 4y + 8$ are inverses.

DO EXERCISES 16–18. (EXERCISE 16 IS ON THE PRECEDING PAGE.)

▶ **EXAMPLE 11** Graph $g(x) = 2x - 3$ and $g^{-1}(x) = (x + 3)/2$ using the same set of axes. Then compare.

The graph of each function is shown here. Note that the graph of g^{-1} can be obtained by reflecting the graph of g across the line $y = x$. That is, if we graph $g(x) = 2x - 3$ and $y = x$ and fold the paper along the line $y = x$, the graph of $g^{-1}(x) = (x + 3)/2$ will be the result of "flipping" the graph of $g(x) = 2x - 3$ across the line.

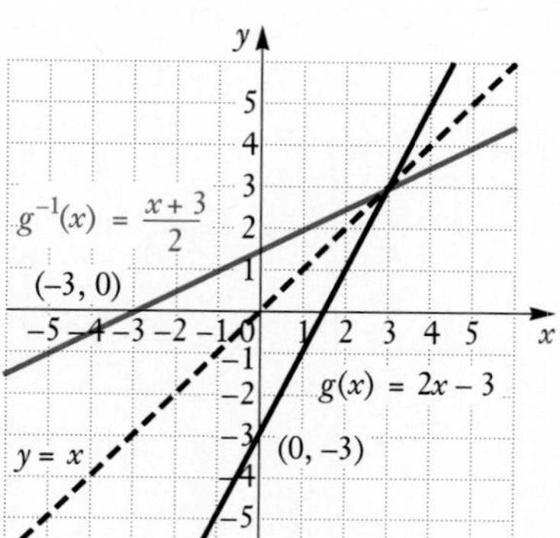

◀

The graph of f^{-1} is a reflection of the graph of f across the line $y = x$.

DO EXERCISE 19.

▶ **EXAMPLE 12** Given $f(x) = x^3 + 2$:

a) Determine whether the function is one-to-one.

b) If it is one-to-one, find a formula for its inverse.

c) Graph the inverse.

We solve as follows.

a) The graph of $f(x) = x^3 + 2$ is shown below. It passes the horizontal-line test and thus has an inverse.

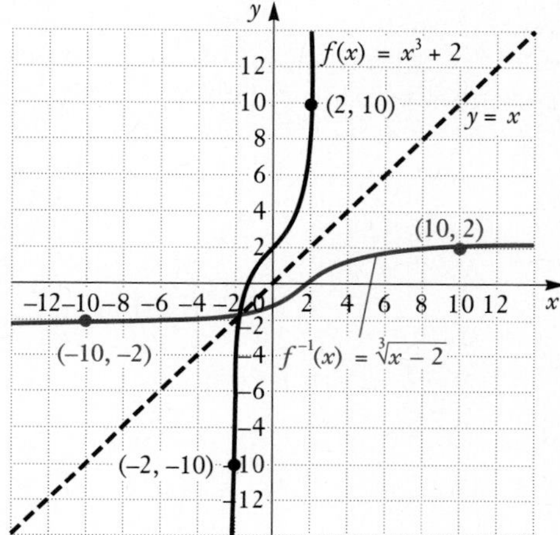

b) **1.** Replace $f(x)$ by y: $\quad y = x^3 + 2.$

2. Interchange x and y: $\quad x = y^3 + 2.$

3. Solve for y: $\quad x - 2 = y^3$

$$\sqrt[3]{x-2} = y.$$

Since a number has only one cube root, we can solve for y.

4. Replace y by $f^{-1}(x)$: $\quad f^{-1}(x) = \sqrt[3]{x-2}.$

c) We find the graph by flipping the graph of $f(x) = x^3 + 2$ over the line $y = x$, interchanging x- and y-coordinates. The graph can also be found by substituting into $f^{-1}(x) = \sqrt[3]{x-2}$ to find function values. The graph is shown above using the same set of axes. ◀

DO EXERCISE 20.

19. Graph $g(x) = 3x - 2$ and $g^{-1}(x) = (x + 2)/3$ using the same set of axes.

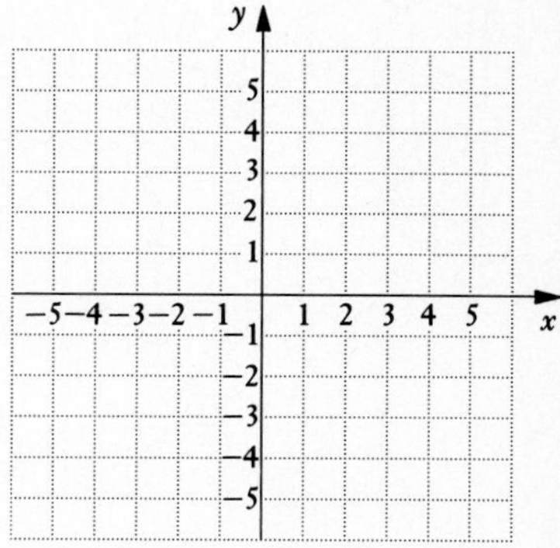

20. Given $f(x) = x^3 + 1$:

a) Determine whether the function is one-to-one.

b) If it is one-to-one, find a formula for its inverse.

c) Graph the function and its inverse using the same set of axes.

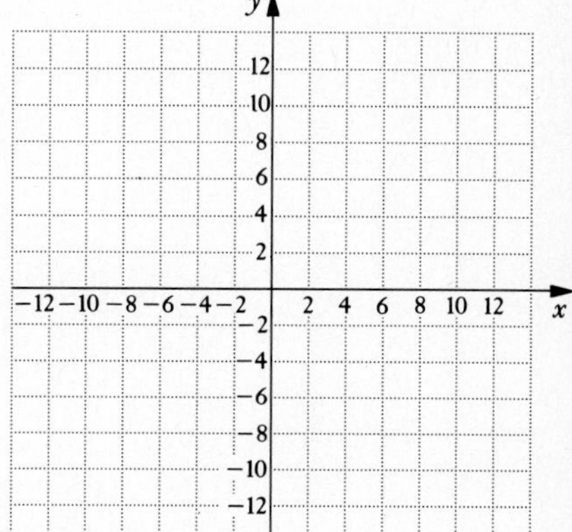

ANSWERS ON PAGE A-11

❖ SIDELIGHTS

Computer–Calculator Application: Monthly Mortgage Payments

You are about to buy a house. The total cost is \$150,000 and you have a down payment of \$10,000. A financial institution is going to lend you \$140,000 at 10% over a period of 20 years. What will your monthly payment be? You can use the following formula:

$$M = P\left[\frac{\frac{r}{12}\left(1+\frac{r}{12}\right)^n}{\left(1+\frac{r}{12}\right)^n - 1}\right],$$

where $M =$ the monthly payment, $n =$ the total number of payments, $r =$ the interest rate, and $P =$ the principal (the amount of the loan).

We will find your monthly payment in the above situation. You will need a calculator with a power key $\boxed{a^b}$. The total number of monthly payments is $20 \cdot 12$, or 240. The interest rate is 0.10. The principal is \$140,000. We substitute:

$$M = P\left[\frac{\frac{r}{12}\left(1+\frac{r}{12}\right)^n}{\left(1+\frac{r}{12}\right)^n - 1}\right]$$

$$= \$140{,}000\left[\frac{\frac{0.10}{12}\left(1+\frac{0.10}{12}\right)^{240}}{\left(1+\frac{0.10}{12}\right)^{240} - 1}\right]$$

$$= \$1351.03$$

Your monthly payment is \$1351.03.

EXERCISES

1. In order to reduce the cost of the monthly payment in the loan considered above, the time of the loan is changed to 30 years. What is the monthly payment?
2. Interest rates are often changing. Suppose the loan of \$140,000 for 20 years has the interest rate changed to 11.5%. What is the monthly payment?
3. A loan of \$80,000 is made for 30 years at an interest rate of 10.2%. What is the monthly payment?
4. A rule of thumb to use in deciding whether you can afford a house is to spend no more than 25% of your monthly salary for a mortgage payment. If you make \$22,200 per year, can you afford the house in Exercise 1?
5. If you are making \$68,000 per year, can you afford the house in Exercise 1?

NAME SECTION DATE

EXERCISE SET 9.6

a, **d** Find the inverse of the relation. Graph the original relation. Then graph the inverse in color.

1. $\{(1, 2), (6, -3), (-3, -5)\}$

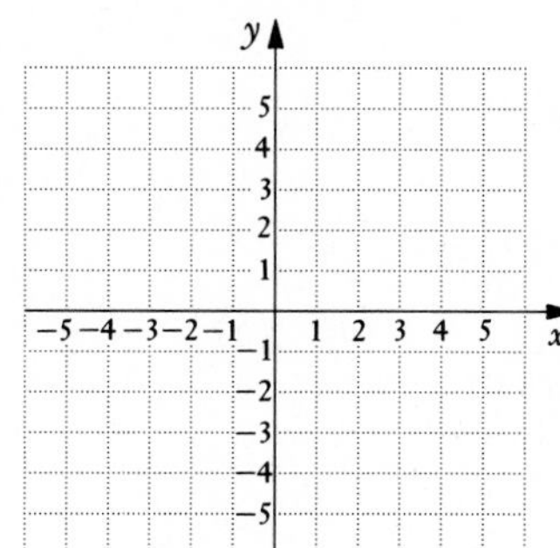

2. $\{(3, -1), (5, 2), (5, -3), (2, 0)\}$

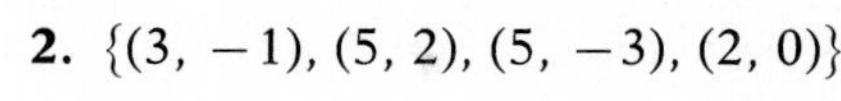

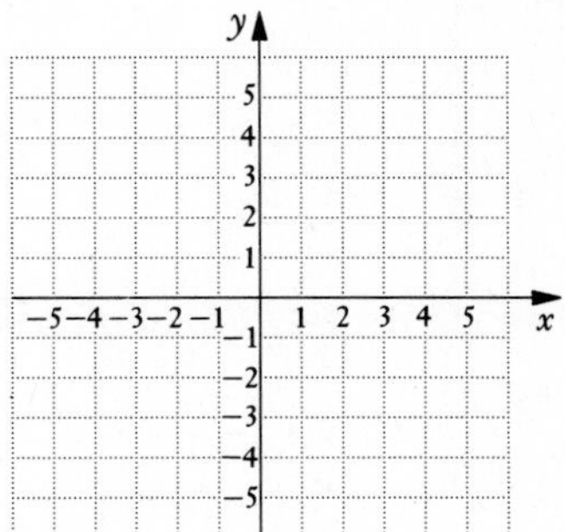

3. $\{(6, 4), (-2, 4), (3, -5), (4, -4)\}$

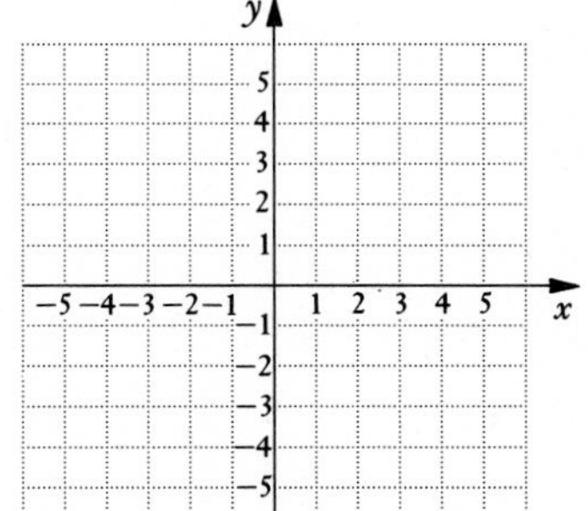

4. $\{(-2, -2), (-2, 5)\}$

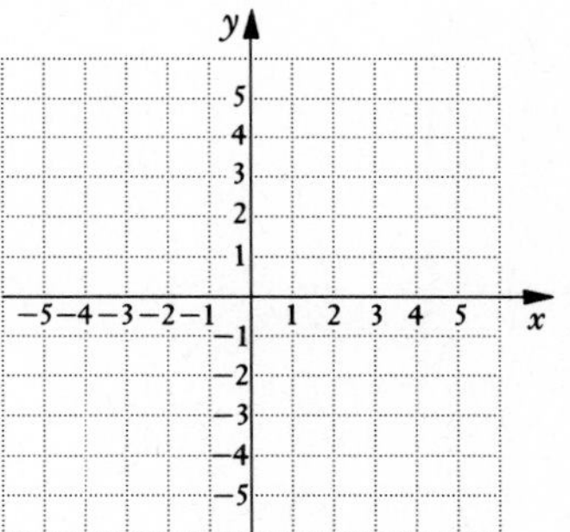

b, **c** Determine whether the function is one-to-one. If it is, find a formula for its inverse.

5. $f(x) = 3x - 4$

6. $f(x) = 5 - 2x$

7. $f(x) = x^2 - 3$

8. $f(x) = 1 - x^2$

9. $g(x) = |x|$

10. $h(x) = |x| - 1$

11. $f(x) = |x + 3|$

12. $f(x) = |x - 2|$

13. $g(x) = \dfrac{-2}{x}$

14. $h(x) = \dfrac{1}{x}$

ANSWERS

1. See graph.

2. See graph.

3. See graph.

4. See graph.

5. ______

6. ______

7. ______

8. ______

9. ______

10. ______

11. ______

12. ______

13. ______

14. ______

ANSWERS

15. ______
16. ______
17. ______
18. ______
19. ______
20. ______
21. ______
22. ______
23. ______
24. ______
25. ______
26. ______
27. ______
28. ______
29. ______
30. ______
31. ______
32. ______
33. ______
34. ______
35. ______
36. ______

C Find a formula for the inverse.

15. $f(x) = x + 2$

16. $f(x) = x + 7$

17. $f(x) = 4 - x$

18. $f(x) = 9 - x$

19. $g(x) = x - 5$

20. $g(x) = x - 8$

21. $f(x) = 3x$

22. $f(x) = 4x$

23. $g(x) = 3x + 2$

24. $g(x) = 4x + 7$

25. $h(x) = \dfrac{2}{x + 5}$

26. $h(x) = \dfrac{1}{x - 8}$

27. $f(x) = \dfrac{2x + 1}{5x + 3}$

28. $f(x) = \dfrac{3x + 2}{5}$

29. $g(x) = \dfrac{x - 3}{x + 4}$

30. $g(x) = \dfrac{2x - 1}{5x + 3}$

31. $f(x) = x^3 - 1$

32. $f(x) = x^3 + 5$

33. $G(x) = (x - 2)^3$

34. $G(x) = (x + 7)^3$

35. $f(x) = \sqrt[3]{x}$

36. $f(x) = \sqrt[3]{x - 4}$

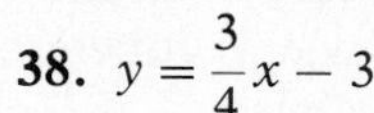

d Graph the relation. Then find an equation of the inverse and draw its graph.

37. $y = -\frac{1}{2}x + 2$

38. $y = \frac{3}{4}x - 3$

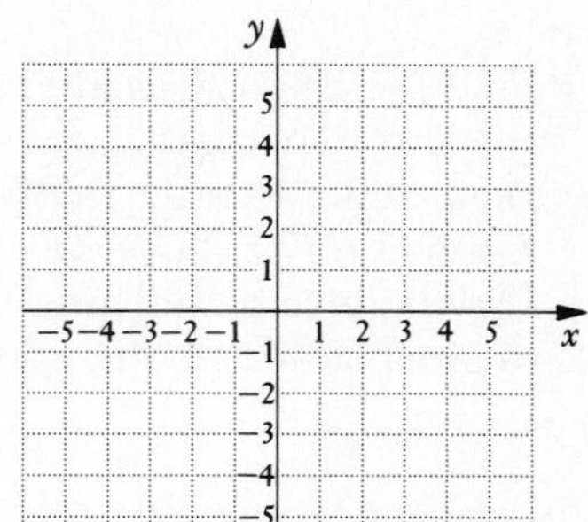

39. $y = x^2 - 3$

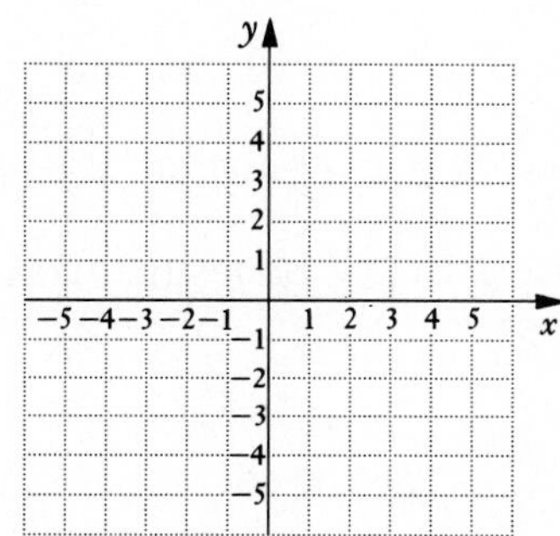

40. $y = |x|$

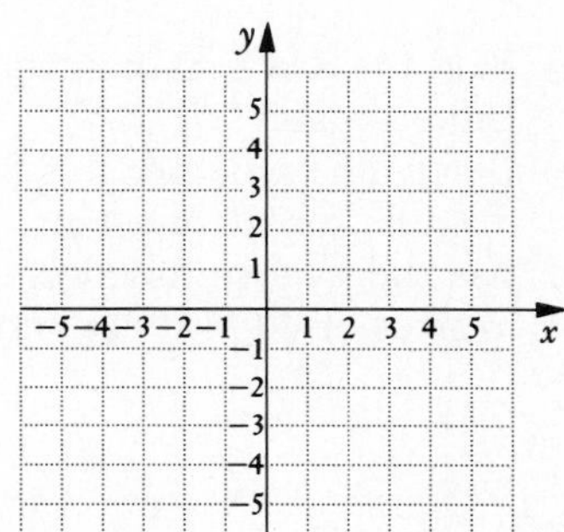

Graph the function and its inverse using the same set of axes.

41. $f(x) = \frac{1}{2}x - 3$

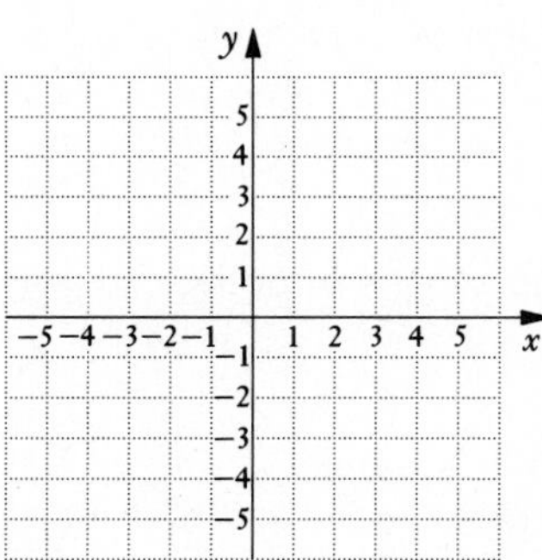

42. $g(x) = x + 4$

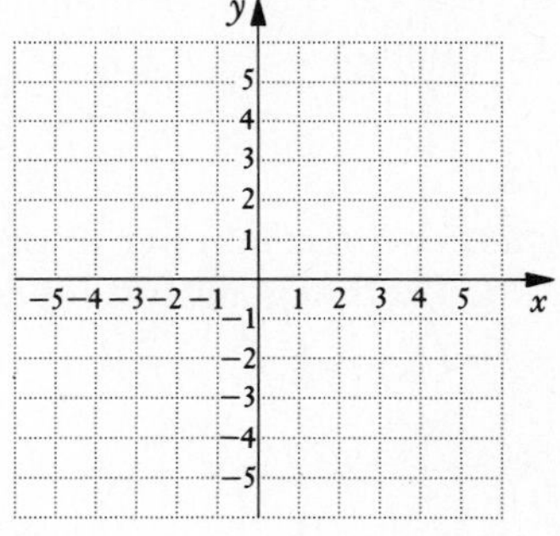

43. $f(x) = x^3$

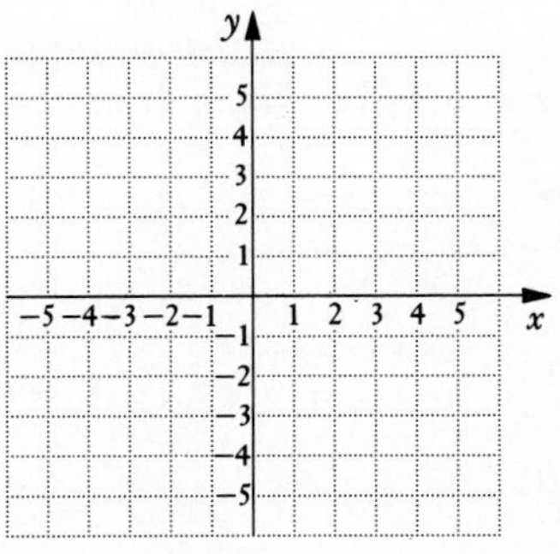

44. $f(x) = x^3 - 1$

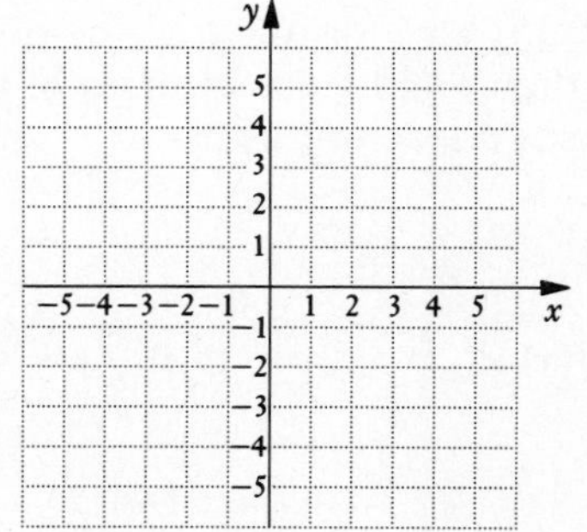

ANSWERS

37. See graph.

38. See graph.

39. See graph.

40. See graph.

41. See graph.

42. See graph.

43. See graph.

44. See graph.

ANSWERS

45. a) ______

b) ______

c) ______

46. a) ______

b) ______

c) ______

47. ______

48. ______

49. ______

50. ______

51. ______

52. ______

53. ______

54. ______

55. ______

56. ______

57. ______

45. *Women's dress sizes in the United States and France.* Sizes of clothing and shoes are not the same in different countries. For example, a size-6 dress in the United States is size 38 in France. A function that will convert dress sizes in the United States to those in France is

$$f(x) = x + 32.$$

a) Find the dress sizes in France that correspond to sizes of 8, 10, 14, and 18 in the United States.
b) Determine whether this function has an inverse that is a function. If so, find a formula for the inverse.
c) Use the inverse function to find dress sizes in the United States that correspond to sizes of 40, 42, 46, and 50 in France.

46. *Women's dress sizes in the United States and Italy.* A size-6 dress in the United States is size 36 in Italy. A function that will convert dress sizes in the United States to those in Italy is

$$f(x) = 2(x + 12).$$

a) Find the dress sizes in Italy that correspond to sizes of 8, 10, 14, and 18 in the United States.
b) Determine whether this function has an inverse that is a function. If so, find a formula for the inverse.
c) Use the inverse function to find dress sizes in the United States that correspond to sizes of 40, 44, 52, and 60 in Italy.

SYNTHESIS

47. Does the constant function $f(x) = 4$ have an inverse that is a function? If so, find a formula. If not, explain why.

48. Compare the graphs of the hyperbolas $x^2 - y^2 = 1$ and $y^2 - x^2 = 1$.

49. An organization determines that the cost per person of chartering a bus is given by the function

$$C(x) = \frac{100 + 5x}{x},$$

where $x =$ the number of people in the group and $C(x)$ is in dollars. Determine $C^{-1}(x)$ and explain this inverse function.

Find $f \circ f^{-1}(x)$ and $f^{-1} \circ f(x)$.

50. $f(x) = \frac{4}{5}x,\ f^{-1}(x) = \frac{5}{4}x$

51. $f(x) = \frac{x + 7}{3},\ f^{-1}(x) = 3x - 7$

52. $f(x) = \frac{1 - x}{x},\ f^{-1}(x) = \frac{1}{x + 1}$

53. $f(x) = x^3 - 5,\ f^{-1}(x) = \sqrt[3]{x + 5}$

If $f \circ g(x) = x$, for any x in the domain of g, and $g \circ f(x) = x$, for any x in the domain of f, then f and g are inverses of each other. Determine whether these functions are inverses of each other.

54. $f(x) = \frac{1}{2},\ g(x) = 2$

55. $f(x) = \sqrt[5]{x},\ g(x) = x^5$

56. $f(x) = \sqrt{x},\ x \geq 0,\ g(x) = x^2$

57. $f(x) = \frac{2x - 3}{4x + 7},\ g(x) = \frac{7x - 4}{3x + 2}$

SUMMARY AND REVIEW: CHAPTER 9

IMPORTANT PROPERTIES AND FORMULAS

Distance Formula: $d = \sqrt{(x_1 - x_2)^2 + (y_1 - y_2)^2}$

Circle: $(x - h)^2 + (y - k)^2 = r^2$

Ellipse: $\dfrac{x^2}{a^2} + \dfrac{y^2}{b^2} = 1$

Parabola

$y = ax^2 + bx + c,\ a > 0$
$= a(x - h)^2 + k$, opens upward;
$x = ay^2 + by + c,\ a > 0$
$= a(y - k)^2 + h$, opens to the right;

$y = ax^2 + bx + c,\ a < 0$
$= a(x - h)^2 + k$, opens downward;
$x = ay^2 + by + c,\ a < 0$
$= a(y - k)^2 + h$, opens to the left

Hyperbola

$\dfrac{x^2}{a^2} - \dfrac{y^2}{b^2} = 1$, axis is horizontal;

$\dfrac{y^2}{b^2} - \dfrac{x^2}{a^2} = 1$, axis is vertical

REVIEW EXERCISES

The review sections and objectives to be tested in addition to the material in this chapter are [7.2a], [7.5a, b], [8.2a], and [8.6a].

1. Find the distance between the pair of points. Where appropriate, find an approximation to three decimal places.

$(-2, -5)$ and $(-6, 4)$

2. Find the midpoint of the segment with the given endpoints.

$(-2, -5)$ and $(-6, 4)$

Find the center and the radius.

3. $(x + 2)^2 + (y - 3)^2 = 2$

4. $(x - 5)^2 + y^2 = 49$

5. $x^2 + y^2 + 8x - 6y - 10 = 0$

6. Find an equation of the circle with center $(-4, 3)$ and radius $4\sqrt{3}$.

7. Find an equation of the circle with center $(7, -2)$ and radius $2\sqrt{5}$.

Graph.

8. $\dfrac{x^2}{16} + \dfrac{y^2}{4} = 1$

9. $\dfrac{y^2}{9} - \dfrac{x^2}{4} = 1$

10. $x^2 + y^2 = 16$

11. $x = y^2 + 2y - 2$

12. $y = -2x^2 - 2x + 3$

13. $x^2 + y^2 + 6x - 8y - 39 = 0$

Solve.

14. $x^2 - y^2 = 33,$
$x + y = 11$

15. $x^2 - 2x + 2y^2 = 8,$
$2x + y = 6$

16. $x^2 - y = 3,$
$2x - y = 3$

17. $x^2 + y^2 = 25,$
$x^2 - y^2 = 7$

18. $x^2 - y^2 = 3,$
$y = x^2 - 3$

19. $x^2 + y^2 = 18,$
$2x + y = 3$

20. $x^2 + y^2 = 100,$
$2x^2 - 3y^2 = -120$

21. $x^2 + 2y^2 = 12,$
$xy = 4$

Solve.

22. A rectangle has a perimeter of 38 m and an area of 84 m^2. What are its dimensions?

23. Find two positive integers whose sum is 12 and the sum of whose reciprocals is $\frac{3}{8}$.

24. The perimeter of a square is 12 cm more than the perimeter of another square. Its area exceeds the area of the other by 39 cm^2. Find the perimeter of each square.

25. The sum of the areas of two circles is $130\pi\ \text{ft}^2$. The difference of the areas is $112\pi\ \text{ft}^2$. Find the radius of each circle.

Determine whether the relation is a function.

26. $\{(-1, 2), (-3, 4), (-4, 5), (-4, 6)\}$

27. $\{(9, -2), (8, -3), (7, -4), (6, -5)\}$

Determine whether each of the following is the graph of a function.

28.

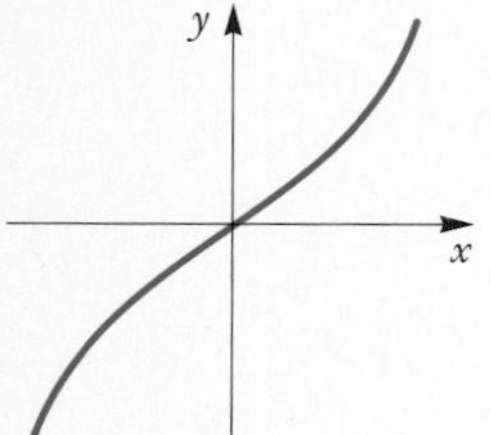

29.

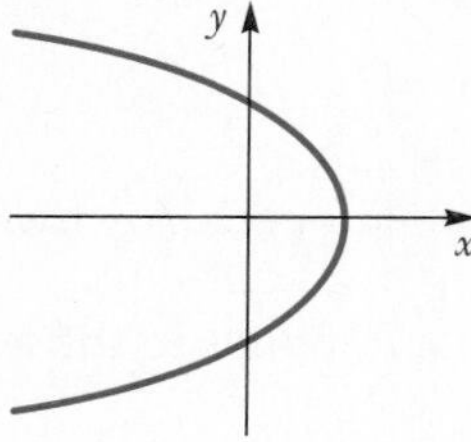

30. Given the function f described by $f(x) = 3x^2 - 2x + 7$, find $f(-1)$ and $f(0)$.

31. Given the function g described by $g(x) = 5 - x^3$, find $g(0)$ and $g(-2)$.

32. Given $f(x) = x^2 + 7$ and $g(x) = x^2 + 4$, find $(f + g)(x)$, $(f - g)(x)$, $fg(x)$, $(f/g)(x)$, $gg(x)$, $f \circ g(x)$, and $g \circ f(x)$.

33. Graph $f(x) = 2 - |x|$.

34. Find $f(x)$ and $g(x)$ such that $h(x) = f \circ g(x)$, where $h(x) = \sqrt{2x - 7}$.

35. Determine whether $f(x) = 4 - x^2$ is one-to-one.

36. Find the inverse of the relation $\{(3, 4), (3, 6), (-3, 7)\}$. Is the inverse a function?

Find a formula for the inverse.

37. $f(x) = x - 2$

38. $g(x) = \dfrac{2x - 3}{7}$

39. $f(x) = \dfrac{2}{x + 5}$

40. $g(x) = 8x^3$

SKILL MAINTENANCE

41. Simplify: $\sqrt[3]{81a^8b^{10}}$.

42. Solve: $x^2 + 2x + 5 = 0$.

43. Rationalize the numerator: $\dfrac{4 - \sqrt{a}}{2 + \sqrt{a}}$.

44. A boat travels 20 mi upstream and 20 mi downstream. The time required for the trip is 6 hr. The speed of the stream is 4 mph. Find the speed of the boat in still water.

SYNTHESIS

45. Find an equation of the circle that passes through $(4, -20)$, $(10, -2)$, and $(-4, -4)$.

46. Find an equation of the ellipse with vertices $(-8, 0)$, $(8, 0)$, $(0, -2)$, and $(0, 2)$.

❖ THINKING IT THROUGH

1. We have studied techniques for solving systems of equations in this chapter. How do the equations differ from those we studied with systems earlier in the text?
2. Explain the idea of a function in as many ways as possible.
3. Explain the uses of the vertical- and horizontal-line tests.

NAME SECTION DATE

TEST: CHAPTER 9

1. Find the distance between $(-6, 2)$ and $(6, 8)$. Find an exact answer and an approximation to three decimal places.

2. Find the midpoint of the segment with endpoints $(-6, 2)$ and $(6, 8)$.

Find the center and the radius of the circle.

3. $(x + 2)^2 + (y - 3)^2 = 64$

4. $x^2 + y^2 + 4x - 6y + 4 = 0$

5. Write an equation of the circle with center $(2, -7)$ and radius $3\sqrt{2}$.

Graph.

6. $y = x^2 - 4x - 1$

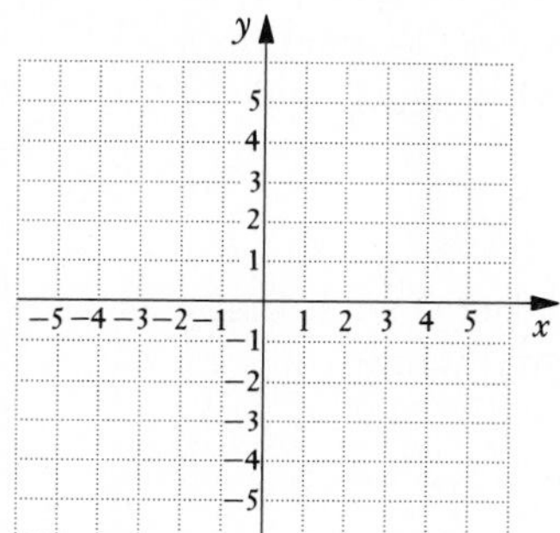

7. $x^2 + y^2 = 36$

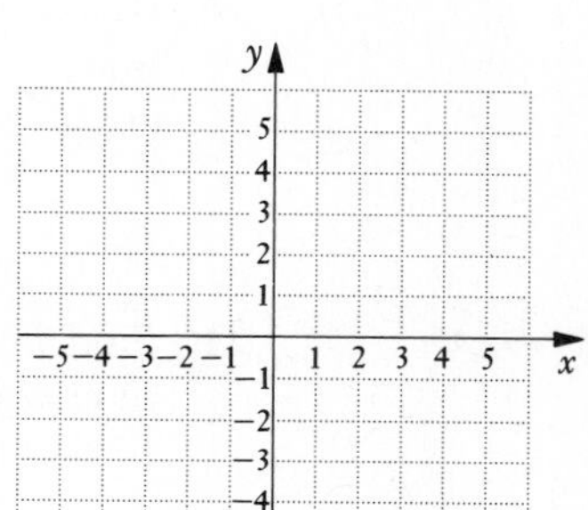

8. $\dfrac{x^2}{9} - \dfrac{y^2}{4} = 1$

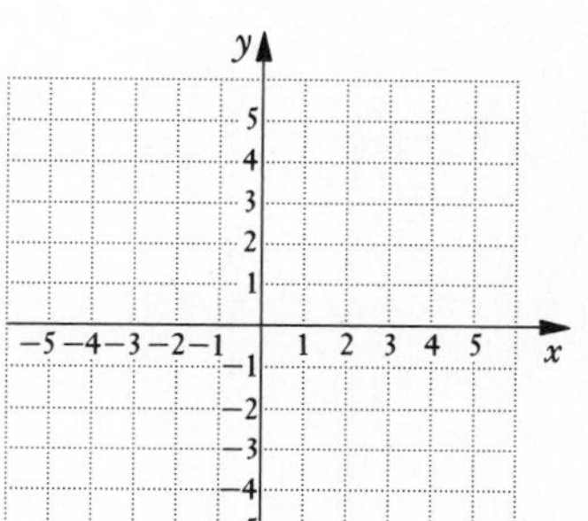

9. $\dfrac{x^2}{4} + \dfrac{y^2}{36} = 1$

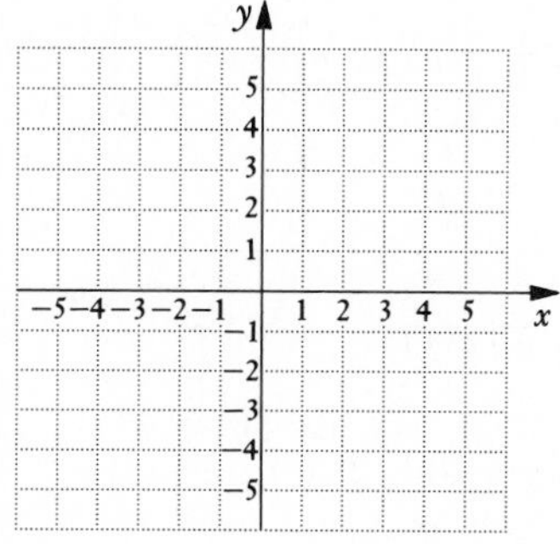

10. $y = 2x^2 - 10x + 7$

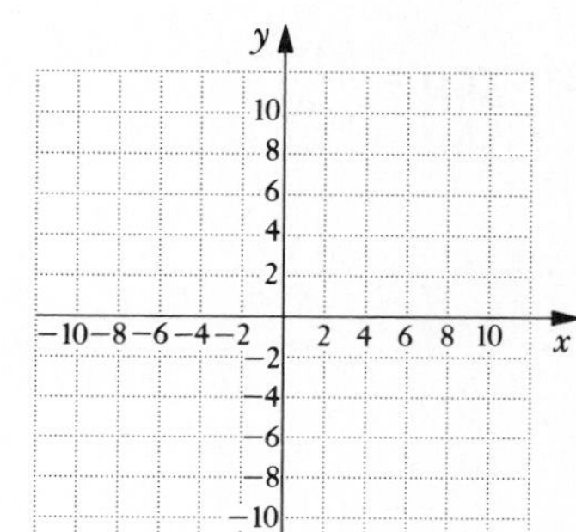

11. $x = -y^2 + 4y$

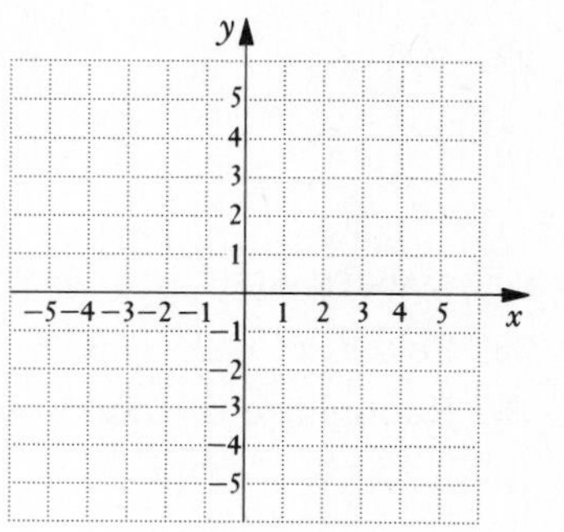

Solve.

12. $\dfrac{x^2}{16} + \dfrac{y^2}{9} = 1,$
$3x + 4y = 12$

13. $x^2 + y^2 = 16,$
$\dfrac{x^2}{16} - \dfrac{y^2}{9} = 1$

ANSWERS

1. ______

2. ______

3. ______

4. ______

5. ______

6. ______

7. ______

8. ______

9. ______

10. ______

11. ______

12. ______

13. ______

ANSWERS

14. ____________

15. ____________

16. ____________

17. ____________

18. ____________

19. ____________

20. ____________

21. ____________

22. ____________

23. ____________

24. ____________

25. ____________

26. ____________

27. ____________

28. ____________

29. ____________

30. ____________

31. ____________

14. In a fractional expression, the sum of the values of the numerator and the denominator is 23. The product of their values is 13. Find the values of the numerator and the denominator.

15. A rectangle with diagonal of length $5\sqrt{5}$ has an area of 22. Find the dimensions of the rectangle.

16. Two squares are such that the sum of their areas is 8 m^2 and the difference of their areas is 2 m^2. Find the length of a side of each square.

17. A rectangle has a diagonal of length 20 ft and a perimeter of 56 ft. Find the dimensions of the rectangle.

18. Graph: $g(x) = |x| - 5$.

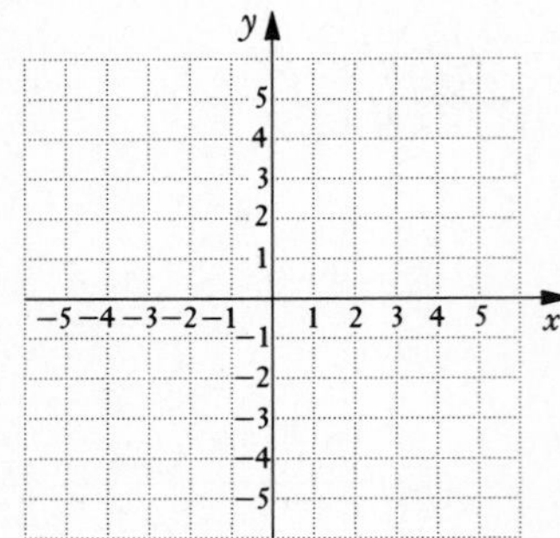

19. Given the function f described by $f(x) = x^2 - 3x + 5$, find $f(-2)$.

20. Given $f(x) = x - 4$ and $g(x) = x^2 + 5$, find $(f + g)(x)$, $(f - g)(x)$, $fg(x)$, $(f/g)(x)$, $ff(x)$, $f \circ g(x)$, and $g \circ f(x)$.

21. Find $f(x)$ and $g(x)$ such that $h(x) = f \circ g(x)$, where $h(x) = (5x + 8)^3$.

22. Determine whether $g(x) = |x| - 5$ is one-to-one.

Find a formula for the inverse.

23. $f(x) = 4x - 3$

24. $g(x) = \dfrac{x - 2}{4}$

25. $f(x) = \dfrac{x + 1}{x - 2}$

SKILL MAINTENANCE

26. Solve: $x^2 + 2x = 5$.

27. Simplify: $\sqrt[3]{48a^5b^{18}}$.

28. Rationalize the denominator:

$$\frac{4 - \sqrt{a}}{2 + \sqrt{a}}.$$

29. A boat travels 30 mi upstream and 10 mi downstream. The total time for the trip is 5 hr. The speed of the stream is 2 mph. Find the speed of the boat in still water.

SYNTHESIS

30. Find an equation of the ellipse with vertices (0, 2), (5, −1), (10, 2), and (5, 5).

31. The sum of two numbers is 36 and the product is 4. Find the sum of the reciprocals of the numbers.

CUMULATIVE REVIEW: CHAPTERS 1–9

Solve.

1. $\frac{1}{3}x - \frac{1}{5} \geq \frac{1}{5}x - \frac{1}{3}$

2. $|x| > 6.4$

3. $3 \leq 4x + 7 < 31$

4. $3x + y = 4,$
$-6x - y = -3$

5. $x - y + 2z = 3,$
$-x + z = 4,$
$2x + y - z = -3$

6. $2x^2 = x + 3$

7. $3x - \frac{6}{x} = 7$

8. $\sqrt{x+5} = x - 1$

9. $x(x + 10) = -21$

10. $2x^2 + x + 1 = 0$

11. $x^4 - 13x^2 + 36 = 0$

12. $\frac{3}{x-3} - \frac{x+2}{x^2+2x-15} = \frac{1}{x+5}$

13. $-x^2 + 2y^2 = 7,$
$x^2 + y^2 = 5$

14. $P = \frac{3}{4}(M + 2N)$, for N

15. $\frac{1}{p} + \frac{1}{q} = \frac{1}{f}$, for p

16. $x^2 - 1 \geq 0$

Simplify.

17. $\left|-\frac{8}{3} - \frac{10}{3}\right|$

18. $2^4 - (25 \div 5 - 49 \div 7)$

19. $\frac{5x^2y^5z^{-4}}{-15x^2yz}$

20. $(2x + 3)(x^2 - 2x - 1)$

21. $(3x^2 + x^3 - 1) -$
$(2x^3 + x + 5)$

22. $\frac{2m^2 + 11m - 6}{m^3 + 1} \cdot \frac{m^2 - m + 1}{m + 6}$

23. $\frac{x}{x-1} + \frac{2}{x+1} - \frac{2x}{x^2-1}$

24. $\frac{1 - \frac{5}{x}}{x - 4 - \frac{5}{x}}$

25. $(x^4 + 3x^3 - x + 4) \div (x + 1)$

26. $\frac{\sqrt{75x^5y^2}}{\sqrt{3xy}}$

27. $4\sqrt{50} - 3\sqrt{18}$

28. $(16^{3/2})^{1/2}$

29. $(2 - i\sqrt{2})(5 + 3i\sqrt{2})$

30. $\frac{5+i}{2-4i}$

Graph.

31. $4y - 3x = 12$

32. $y < -2$

33. $x + y \leq 0,$
$x \geq -4,$
$y \geq -1$

34. $y = 2x^2 - 8x + 9$

35. $(x - 1)^2 + (y + 1)^2 = 9$

36. $x = y^2 + 1$

Factor.

37. $2x^4 - 12x^3 + x - 6$

38. $3a^2 - 12ab - 135b^2$

39. $x^2 - 17x + 72$

40. $81m^4 - n^4$

41. $16x^2 - 16x + 4$

42. $81a^3 - 24$

43. $10x^2 + 66x - 28$

44. $6x^3 + 27x^2 - 15x$

45. Find an equation of the line containing the points $(1, 4)$ and $(-1, 0)$.

46. Find an equation of the line containing the point $(1, 2)$ and perpendicular to the line whose equation is $2x - y = 3$.

47. The sum of the squares of three consecutive even integers is equal to eight more than three times the square of the second number. Find the integers.

48. Find the center and the radius of the circle $x^2 - 16x + y^2 + 6y + 68 = 0$.

49. z varies directly as x and inversely as the cube of y, and $z = 5$ when $x = 4$ and $y = 2$. What is z when $x = 10$ and $y = 5$?

50. Given the function f described by $f(x) = x^3 - 2$, find $f(-2)$.

51. Find the distance between the points $(2, 1)$ and $(8, 9)$.

52. Find the midpoint of the segment with endpoints $(-1, -3)$ and $(3, 0)$.

53. Rationalize the denominator:

$$\frac{5 + \sqrt{a}}{3 - \sqrt{a}}.$$

54. The residents of Honolulu annually spend the most money per person on fast food in the United States. In one year, the total amount spent per person in Honolulu plus the total amount spent per person in New York City was \$555. The amount spent per person in Honolulu was \$99 more than twice that spent per person in New York City. How much was spent that year per person on fast food in each city?

55. Jane can finish a cross-stitch pattern in 3 hr. Laura can finish the same pattern in $1\frac{1}{2}$ hr. If they could work together, how long would it take to finish the pattern?

56. Find $(f + g)(x)$ and $(f - g)(x)$ when $f(x) = 2x - 5$ and $g(x) = x^2 - 3x + 1$.

57. Find $f^{-1}(x)$ when $f(x) = 2x - 3$.

SYNTHESIS

58. Solve:

$$\frac{x^3 + 8}{x + 2} = x^2 - 2x + 4.$$

59. Describe the graph of

$$\frac{x^2}{a^2} + \frac{y^2}{b^2} = 1$$

when $a^2 = b^2$.

60. Find an equation of a circle that passes through $(-2, 3)$ and $(-4, 1)$ and whose center is on the line $5x + 8y = -2$.

61. The square of a certain number exceeds twice the square of another number by $\frac{1}{8}$. The sum of their squares is $\frac{5}{16}$. Find the numbers.

INTRODUCTION The functions that we consider here are important for their rich applications to many fields. We will look at such applications as compound interest and population growth, but there are many others.

The basis of the theory concerns exponents. We define some functions having variable exponents, exponential functions; the rest follows from those functions and properties.

The review sections to be tested in addition to the material in this chapter are 7.9, 8.4, 8.5, and 9.4. ❖

Exponential and Logarithmic Functions

10

AN APPLICATION

The population of the world passed the 5.0 billion mark in 1987. The exponential growth rate was 2.8% per year. Find an exponential growth function for world population growth.

THE MATHEMATICS

The exponential growth function that fits the data is given by

$$P(t) = 5e^{0.028t},$$

where P is the population, in billions, t years after 1987.

❖ POINTS TO REMEMBER: CHAPTER 10

Product Rule of Exponents: $a^n a^m = a^{m+n}$
Power Rule of Exponents: $(a^n)^m = a^{mn}$

Quotient Rule of Exponents: $\dfrac{a^m}{a^n} = a^{m-n}$

PRETEST: CHAPTER 10

Graph.

1. $y = 2^x$

2. $y = \log_2 x$

3. Convert to a logarithmic equation:

$$a^3 = 1000.$$

4. Convert to an exponential equation:

$$\log_2 32 = t.$$

5. Express as a single logarithm:

$$2 \log_a M + \log_a N - \frac{1}{3} \log_a Q.$$

6. Express in terms of logarithms of x, y, and z:

$$\log_a \sqrt[4]{\frac{x^3 y}{z^2}}.$$

Solve.

7. $\log_x \frac{1}{4} = -2$

8. $\log_5 x = 2$

9. $3^x = 8.6$

10. $16^{x-3} = 4$

11. $\log (x^2 - 4) - \log (x - 2) = 1$

12. $\ln x = 0$

Find each of the following using a calculator.

13. $\log 714$

14. $\ln 0.0008464$

15. antilog (-3.4679)

16. $\text{antilog}_e\ (4.2024)$

17. $\text{antilog}_e\ (-2.34)$

18. $\log (0.0234)$

19. The consumer price index compares the cost of goods and services over various years using \$100 worth of goods and services in 1967 as a base. It is given by

$$P = \$100e^{0.06t},$$

where $t =$ the number of years since 1967. Goods and services that cost \$100 in 1967 will cost how much in 1998?

20. How old is an animal bone that has lost 76% of its carbon-14?

10.1 Exponential Functions

The following graph shows years in which the cost of first-class postage increased. A curve drawn along the graph approximates the graph of an *exponential function.* We now consider such graphs.

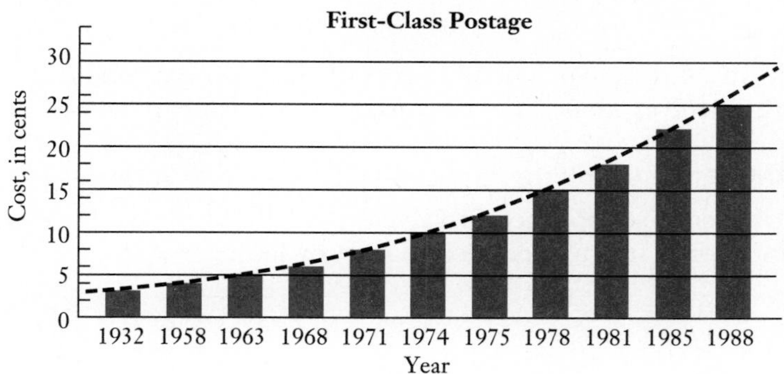

a Graphing Exponential Functions

We now develop the meaning of exponential expressions with irrational exponents. In Chapter 7, we gave meaning to exponential expressions with rational-number exponents such as

$$8^{1/4}, \quad 3^{-3/4}, \quad 7^{2.34}, \quad 5^{1.73}.$$

For example, $5^{1.73}$, or $5^{173/100}$, means to raise 5 to the 173rd power and then take the 100th root. Examples of expressions with irrational exponents are

$$5^{\sqrt{3}}, \quad 7^{\pi}, \quad 9^{-\sqrt{2}}.$$

Since we can approximate irrational numbers with decimal approximations, we can also approximate expressions with irrational exponents. For example, consider $5^{\sqrt{3}}$. As rational values of r get close to $\sqrt{3}$, 5^r gets close to some real number. Note the following:

r closes in on $\sqrt{3}$.	5^r closes in on some real number p.
r	5^r
$1 < \sqrt{3} < 2$	$5 = 5^1 < p < 5^2 = 25$
$1.7 < \sqrt{3} < 1.8$	$15.426 = 5^{1.7} < p < 5^{1.8} = 18.119$
$1.73 < \sqrt{3} < 1.74$	$16.189 = 5^{1.73} < p < 5^{1.74} = 16.452$
$1.732 < \sqrt{3} < 1.733$	$16.241 = 5^{1.732} < p < 5^{1.733} = 16.267$

As r closes in on $\sqrt{3}$, 5^r closes in on some real number p. We define $5^{\sqrt{3}}$ to be that number p. To seven decimal places, we have

$$5^{\sqrt{3}} \approx 16.2424508.$$

Any positive irrational exponent can be defined in a similar way. Negative irrational exponents are then defined in the same way as negative integer exponents. Then the expression a^x has meaning for any real number x. The general laws of exponents still hold, but we will not prove that here. We now define exponential functions.

The function $f(x) = a^x$, where a is a positive constant different from 1, is called the *exponential function,* base a.

OBJECTIVES

After finishing Section 10.1, you should be able to:

a Graph exponential equations and functions.

b Graph exponential equations when x and y have been interchanged.

c Solve problems involving applications of exponential functions and their graphs.

FOR EXTRA HELP

Tape 19A

Tape 17A

MAC: 10
IBM: 10

We restrict the base a to be positive to avoid the possibility of taking even roots of negative numbers such as the square root of -1, $(-1)^{1/2}$, which is not a real number. We restrict the base from being 1 because $f(x) = 1^x = 1$ and this function does not have an inverse. We will see that all other exponential functions with the base restriction do have inverses.

The following are examples of exponential functions:

$$f(x) = 2^x, \qquad f(x) = (\tfrac{1}{2})^x, \qquad f(x) = (0.4)^x.$$

Note that, in contrast to polynomial functions like $f(x) = x^2$ and $f(x) = x^3$, the variable is *in the exponent.* Let us consider graphs of exponential functions.

► **EXAMPLE 1** Graph the exponential function $y = f(x) = 2^x$.

We compute some function values, thinking of y as $f(x)$, and list the results in a table. It is a good idea to start by letting $x = 0$.

$$f(0) = 2^0 = 1;$$
$$f(1) = 2^1 = 2;$$
$$f(2) = 2^2 = 4;$$
$$f(3) = 2^3 = 8;$$
$$f(-1) = 2^{-1} = \frac{1}{2^1} = \frac{1}{2};$$
$$f(-2) = 2^{-2} = \frac{1}{2^2} = \frac{1}{4};$$
$$f(-3) = 2^{-3} = \frac{1}{2^3} = \frac{1}{8}.$$

x	y, or $f(x)$
0	1
1	2
2	4
3	8
-1	$\frac{1}{2}$
-2	$\frac{1}{4}$
-3	$\frac{1}{8}$

Next, we plot these points and connect them with a smooth curve.

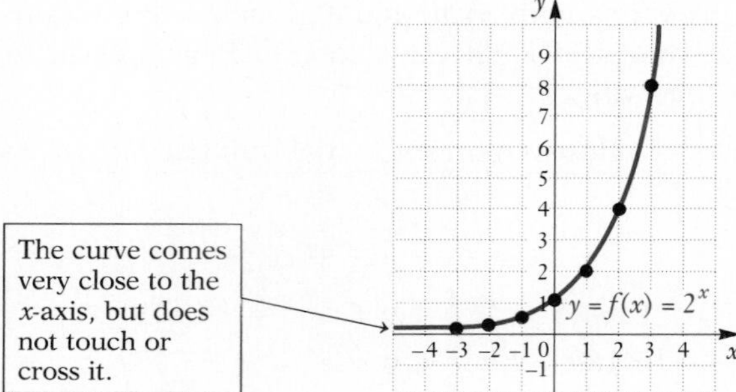

In graphing, be sure to plot enough points to determine how steeply the curve rises.

Note that as x increases, the function values increase indefinitely. As x decreases, the function values decrease, getting very close to 0. The x-axis, or the line $y = 0$, is an *asymptote,* meaning that the curve comes very close to but never touches the axis. We studied asymptotes in Chapter 9. ◄

DO EXERCISE 1.

1. Graph: $y = f(x) = 3^x$.

   a) Complete this table of solutions.

x	y, or $f(x)$
0	
1	
2	
3	
-1	
-2	
-3	

   b) Plot the points from the table and connect them with a smooth curve.

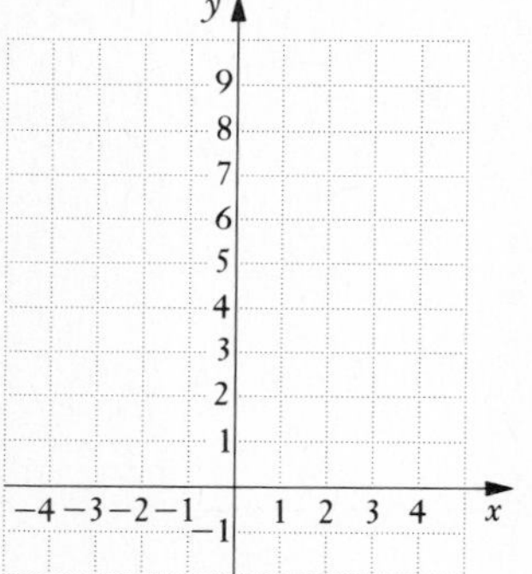

ANSWERS ON PAGE A-11

▶ **EXAMPLE 2** Graph the exponential function $y = f(x) = (\frac{1}{2})^x$.

We compute some function values, thinking of y as $f(x)$, and list the results in a table. Before we do this, note that

$$y = f(x) = (\tfrac{1}{2})^x = (2^{-1})^x = 2^{-x}.$$

Then we have

$$f(0) = 2^{-0} = 1;$$
$$f(1) = 2^{-1} = \frac{1}{2^1} = \frac{1}{2};$$
$$f(2) = 2^{-2} = \frac{1}{2^2} = \frac{1}{4};$$
$$f(3) = 2^{-3} = \frac{1}{2^3} = \frac{1}{8};$$
$$f(-1) = 2^{-(-1)} = 2^1 = 2;$$
$$f(-2) = 2^{-(-2)} = 2^2 = 4;$$
$$f(-3) = 2^{-(-3)} = 2^3 = 8.$$

x	y, or $f(x)$
0	1
1	$\frac{1}{2}$
2	$\frac{1}{4}$
3	$\frac{1}{8}$
−1	2
−2	4
−3	8

We plot these points and draw the curve. Note that this graph is a reflection across the y-axis of the graph in Example 1. The line $y = 0$ is again an asymptote.

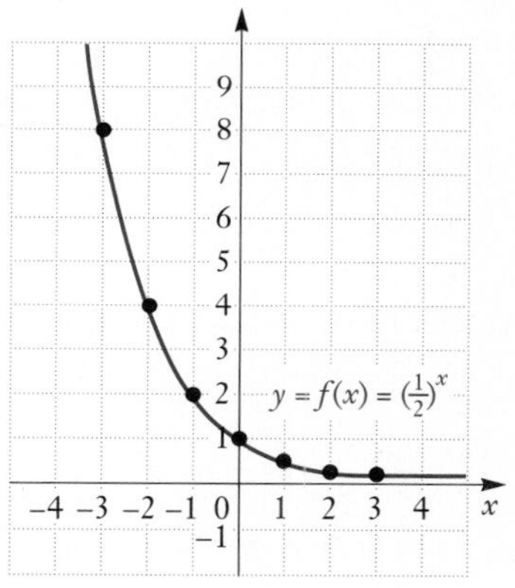

◀

DO EXERCISE 2.

The preceding examples illustrate exponential functions with various bases. Let us list some of these characteristics. Keep in mind that the definition of an exponential function, $f(x) = a^x$, requires that the base be positive and different from 1.

A. When $a = 1$, then

$$f(x) = a^x = 1^x = 1.$$

Thus the graph is of the horizontal line $y = 1$. The graph is not one-to-one, which is why we restrict 1 from being the base of an exponential function.

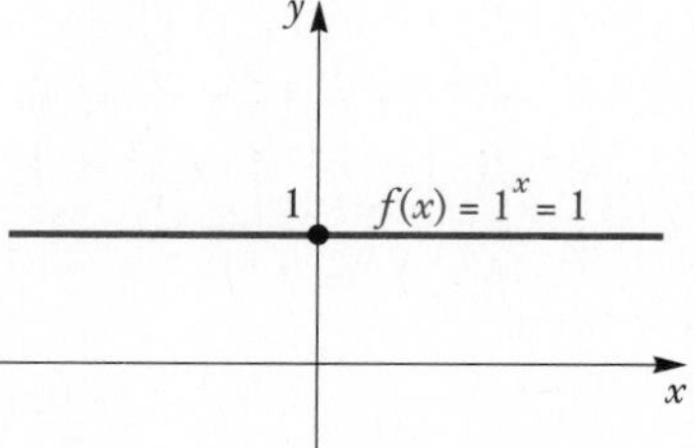

2. Graph: $y = \left(\frac{1}{3}\right)^x$.

a) Complete this table of solutions.

x	y, or $f(x)$
0	
1	
2	
3	
−1	
−2	
−3	

b) Plot the points from the table and connect them with a smooth curve.

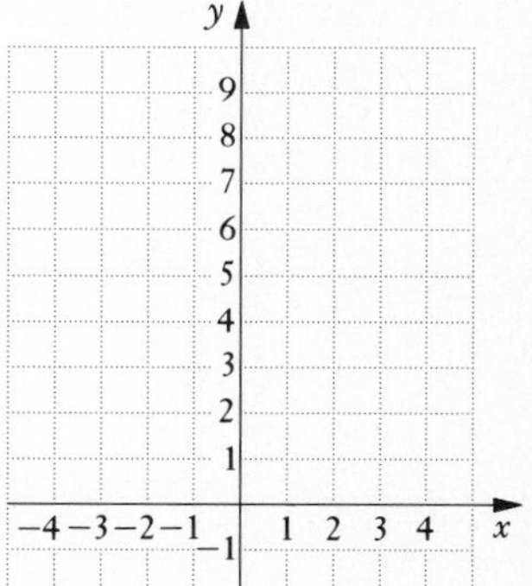

ANSWERS ON PAGE A-11

Graph.

3. $f(x) = 4^x$

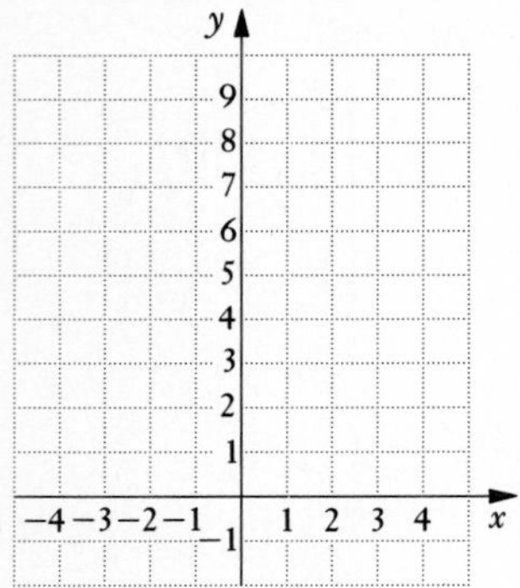

4. $f(x) = \left(\frac{1}{4}\right)^x$

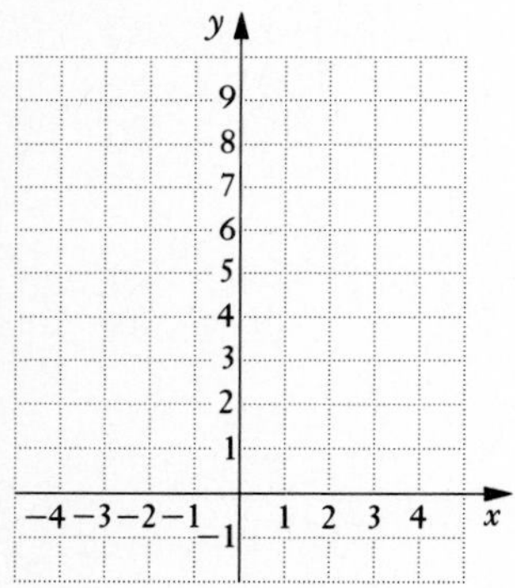

ANSWERS ON PAGE A-11

B. When $a > 1$, the function $f(x) = a^x$ increases from left to right. The greater the value of a, the steeper the curve. As x gets smaller and smaller, the curve gets closer to the line $y = 0$: It is an asymptote.

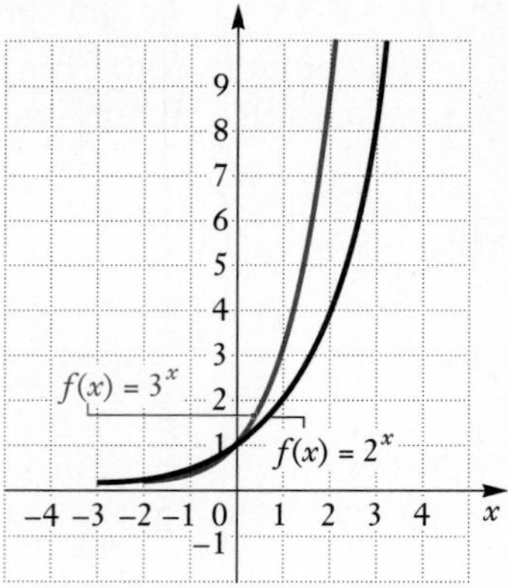

C. When $0 < a < 1$, the function $f(x) = a^x$ decreases from left to right. As a approaches 1, the curve becomes less steep. As x gets larger and larger, the curve gets closer and closer to the line $y = 0$: It is an asymptote.

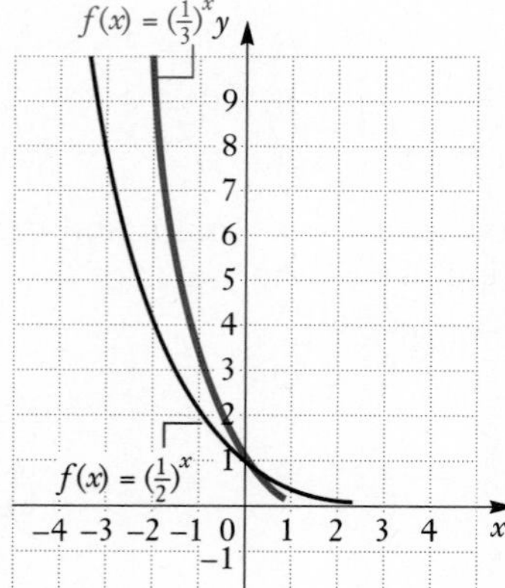

Note that all such functions $f(x) = a^x$ go through the point (0, 1). That is, the y-intercept is (0, 1).

DO EXERCISES 3 AND 4.

▶ **EXAMPLE 3** Graph: $y = f(x) = 2^{x-2}$.

We construct a table of values. Then we plot the points and connect them with a smooth curve. Be sure to note that $x - 2$ is the *exponent.*

$$f(0) = 2^{0-2} = 2^{-2} = \frac{1}{2^2} = \frac{1}{4};$$

$$f(1) = 2^{1-2} = 2^{-1} = \frac{1}{2^1} = \frac{1}{2};$$

$$f(2) = 2^{2-2} = 2^0 = 1;$$

$$f(3) = 2^{3-2} = 2^1 = 2;$$

$$f(4) = 2^{4-2} = 2^2 = 4;$$

$$f(-1) = 2^{-1-2} = 2^{-3} = \frac{1}{2^3} = \frac{1}{8};$$

$$f(-2) = 2^{-2-2} = 2^{-4} = \frac{1}{2^4} = \frac{1}{16}.$$

x	y, or $f(x)$
0	$\frac{1}{4}$
1	$\frac{1}{2}$
2	1
3	2
4	4
−1	$\frac{1}{8}$
−2	$\frac{1}{16}$

The graph looks just like the graph of $y = 2^x$, but it is translated 2 units to the right. The y-intercept of $y = 2^x$ is (0, 1). The y-intercept of $y = 2^{x-2}$ is $(0, \frac{1}{4})$. The line $y = 0$ is still an asymptote.

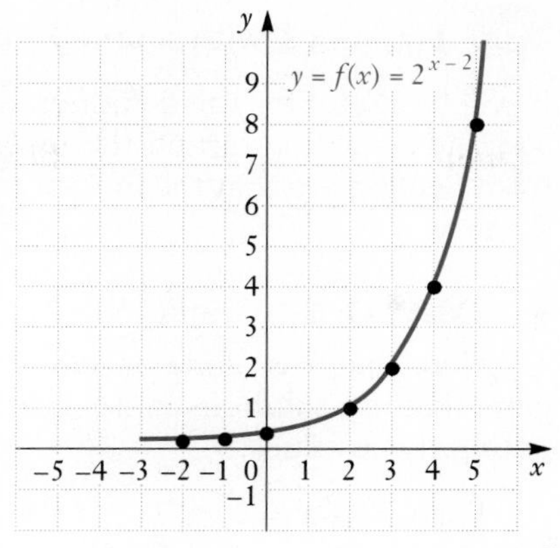

DO EXERCISE 5.

5. Graph: $y = 2^{x+2}$.

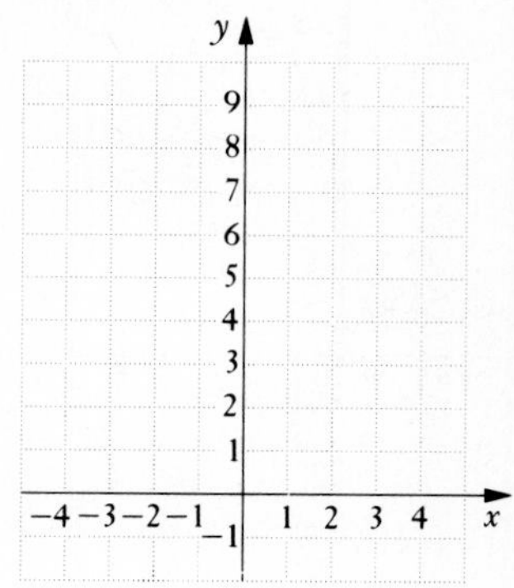

► **EXAMPLE 4** Graph: $y = f(x) = 2^x - 3$.

We construct a table of values. Then we plot the points and connect them with a smooth curve. Note that the only expression in the exponent is x.

$$f(0) = 2^0 - 3 = 1 - 3 = -2;$$
$$f(1) = 2^1 - 3 = 2 - 3 = -1;$$
$$f(2) = 2^2 - 3 = 4 - 3 = 1;$$
$$f(3) = 2^3 - 3 = 8 - 3 = 5;$$
$$f(4) = 2^4 - 3 = 16 - 3 = 13;$$
$$f(-1) = 2^{-1} - 3 = \frac{1}{2} - 3 = -\frac{5}{2};$$
$$f(-2) = 2^{-2} - 3 = \frac{1}{4} - 3 = -\frac{11}{4}.$$

x	y, or $f(x)$
0	-2
1	-1
2	1
3	5
4	13
-1	$-\frac{5}{2}$
-2	$-\frac{11}{4}$

The graph looks like the graph of $y = 2^x$, but it is translated down 3 units. The y-intercept is (0, 2). The line $y = -3$ is an asymptote. The curve gets closer to this line as x gets smaller and smaller.

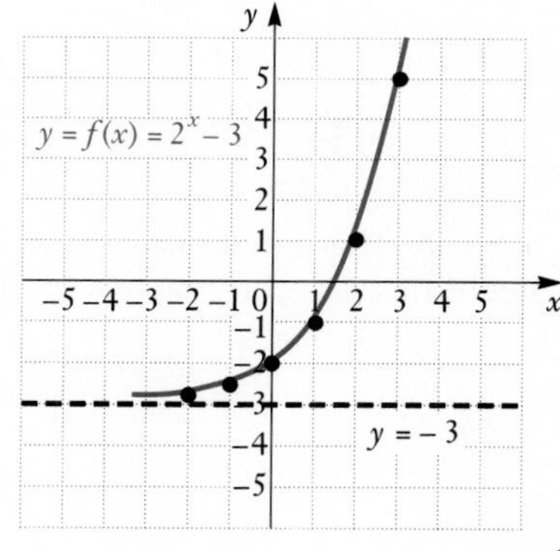

DO EXERCISE 6.

6. Graph: $y = 2^x - 4$.

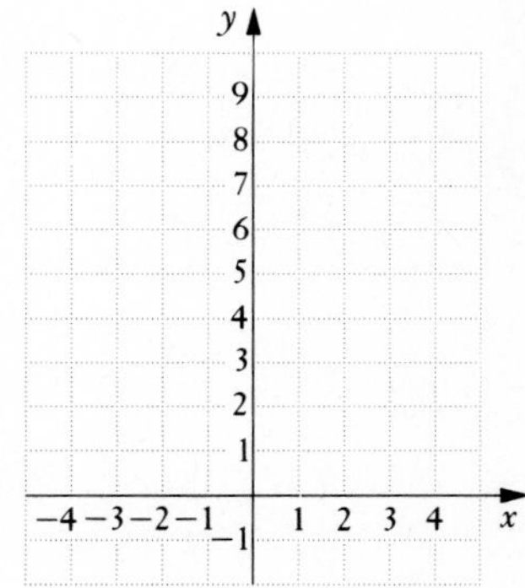

ANSWERS ON PAGE A-11

7. Graph: $x = 3^y$.

b Equations with x and y Interchanged: Inverses

It will be useful for the definition of a new function, which we will consider later, to be able to graph the inverse of an equation like $y = 2^x$. We know from Section 9.6 that the inverse of $y = 2^x$ is $x = 2^y$, found by interchanging x and y.

► **EXAMPLE 5** Graph: $x = 2^y$.

Note that x is alone on one side of the equation. We can find ordered pairs that are solutions more easily by choosing values for y and then computing the x-values.

For $y = 0$, $x = 2^0 = 1$.
For $y = 1$, $x = 2^1 = 2$.
For $y = 2$, $x = 2^2 = 4$.
For $y = 3$, $x = 2^3 = 8$.

For $y = -1$, $x = 2^{-1} = \dfrac{1}{2^1} = \dfrac{1}{2}$.

For $y = -2$, $x = 2^{-2} = \dfrac{1}{2^2} = \dfrac{1}{4}$.

For $y = -3$, $x = 2^{-3} = \dfrac{1}{2^3} = \dfrac{1}{8}$.

x	y
1	0
2	1
4	2
8	3
$\frac{1}{2}$	-1
$\frac{1}{4}$	-2
$\frac{1}{8}$	-3

(1) Choose values for y.
(2) Compute values for x.

We plot the points and connect them with a smooth curve.

This curve does not touch or cross the y-axis.

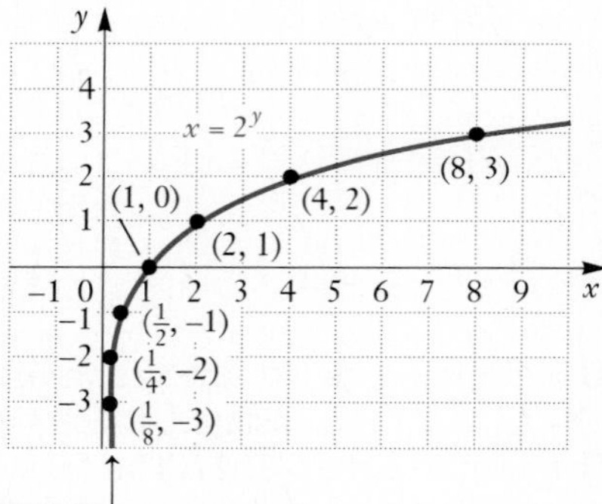

Note that this curve looks just like the graph of $y = 2^x$, except that it is reflected across the line $y = x$, as shown below.

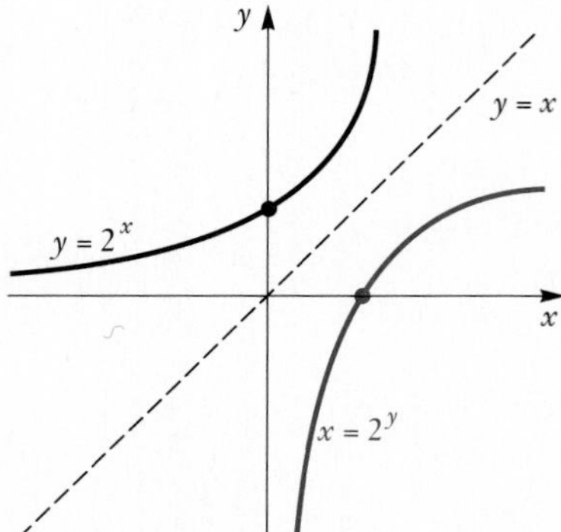

◄

ANSWER ON PAGE A-12

DO EXERCISE 7.

C Applications of Exponential Functions

► **EXAMPLE 6** *Interest compounded annually.* The amount of money A that a principal P will be worth after t years at interest rate r, compounded annually, is given by the formula

$$A = P(1 + r)^t.$$

Suppose that \$100,000 is invested at 8% interest, compounded annually.

a) If $P = \$100,000$ and $r = 8\% = 0.08$, we can substitute these values and

b) Find the amount of money in the account at $t = 0$, $t = 4$, $t = 8$, and $t = 10$.

c) Graph the function.

We solve as follows:

a) If $P = \$100,000$ and $r = 8\% = 0.08$, we can substitute these values and form the following function:

$$A(t) = \$100,000(1 + 0.08)^t = \$100,000(1.08)^t.$$

b) To find the function values, you might find a calculator with a power key helpful.

$$A(0) = \$100,000(1.08)^0 = \$100,000(1) = \$100,000;$$
$$A(4) = \$100,000(1.08)^4 = \$100,000(1.36048896) \approx \$136,048.90;$$
$$A(8) = \$100,000(1.08)^8 = \$100,000(1.85093021) \approx \$185,093.02;$$
$$A(10) = \$100,000(1.08)^{10} = \$100,000(2.158924997) \approx \$215,892.50$$

c) We use the function values computed in (b) with others, if we wish, to draw the graph as follows. Note that the axes are scaled differently because of the large numbers.

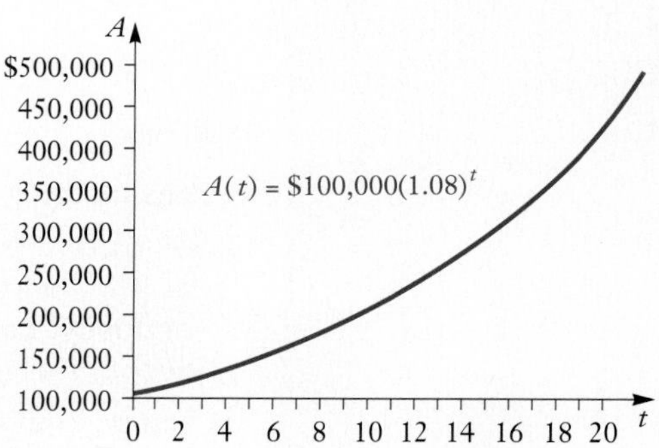

◄

DO EXERCISE 8.

8. Suppose that \$40,000 is invested at 7% interest, compounded annually.

a) Find a function for the amount in the account after t years.

b) Find the amount of money in the account at $t = 0$, $t = 4$, $t = 8$, and $t = 10$.

c) Graph the function.

ANSWERS ON PAGE A-12

❖ SIDELIGHTS

The Compound-Interest Formula

When interest is compounded quarterly, we can find a formula like the one considered in this section as follows:

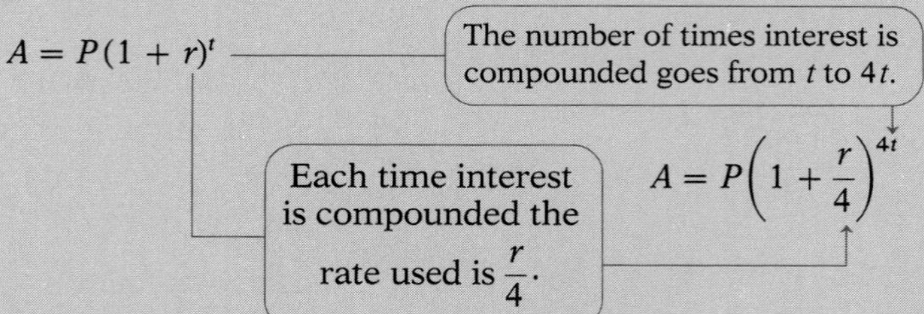

In general, the following formula for compound interest can be used. The amount of money A in an account is given by

$$A = P\left(1 + \frac{r}{n}\right)^{nt},$$

where $P =$ the principal, $r =$ the annual interest rate, $n =$ the number of times per year that interest is compounded, and $t =$ the time, in years.

EXAMPLE Suppose that $1000 is invested at 8%, compounded quarterly. How much is in the account at the end of 2 years?

We use the above equation, substituting 1000 for P, 0.08 for r, 4 for n (compounding quarterly), and 2 for t. Then we get

$$A = P\left(1 + \frac{r}{n}\right)^{nt} = 1000\left(1 + \frac{0.08}{4}\right)^{4 \cdot 2}$$
$$= 1000(1.02)^8 \approx \$1171.66.$$ ◀

EXERCISES

A calculator with an x^y key will be needed for these exercises.

1. Suppose that $1000 is invested at 8%, compounded semiannually. How much is in the account at the end of 2 years?

2. Suppose that $1000 is invested at 8%, compounded monthly. How much is in the account at the end of 2 years?

3. Suppose that $20,000 is invested at 9.2%, compounded quarterly. How much is in the account at the end of 10 years?

4. Suppose that $420,000 is invested at 9.6%, compounded daily. How much is in the account at the end of 30 years?

5. Suppose that $10,000 is invested at 8.4%. How much is in the account at the end of 1 year, if interest is compounded:
   a) annually?
   b) semiannually?
   c) quarterly?
   d) daily?
   e) hourly?

6. Suppose that $10,000 is invested at 10%. How much is in the account at the end of 1 year, if interest is compounded:
   a) annually?
   b) semiannually?
   c) quarterly?
   d) daily?
   e) hourly?

NAME SECTION DATE

EXERCISE SET 10.1

a Graph.

1. $y = f(x) = 2^x$

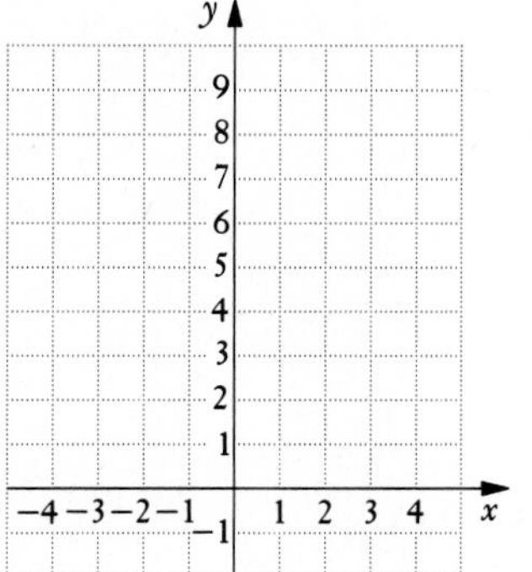

2. $y = f(x) = 3^x$

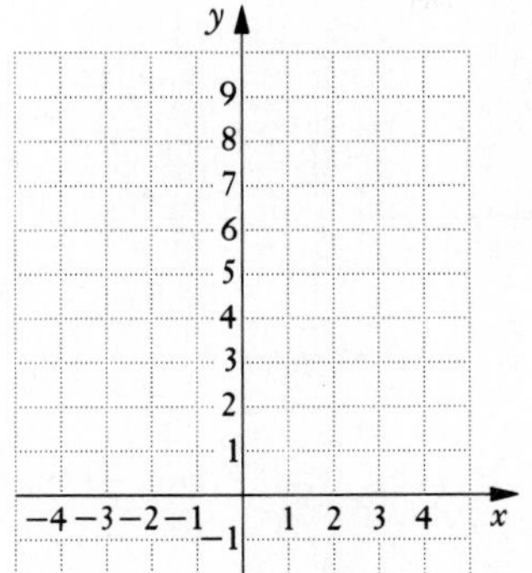

3. $y = 5^x$

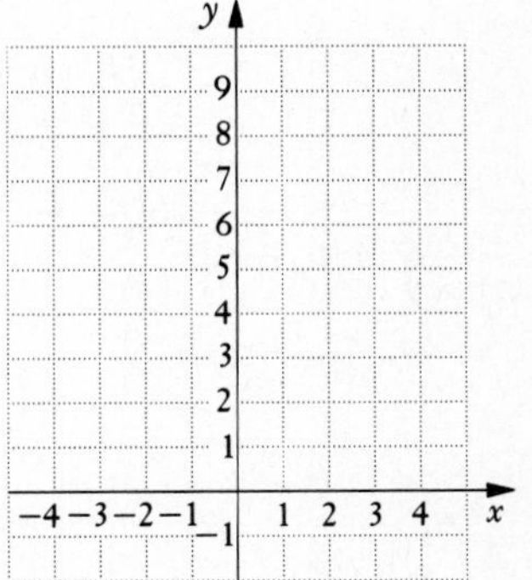

4. $y = 6^x$

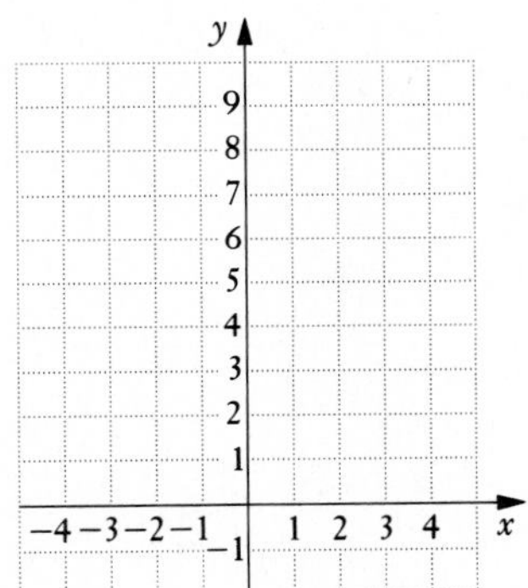

5. $y = 2^{x+1}$

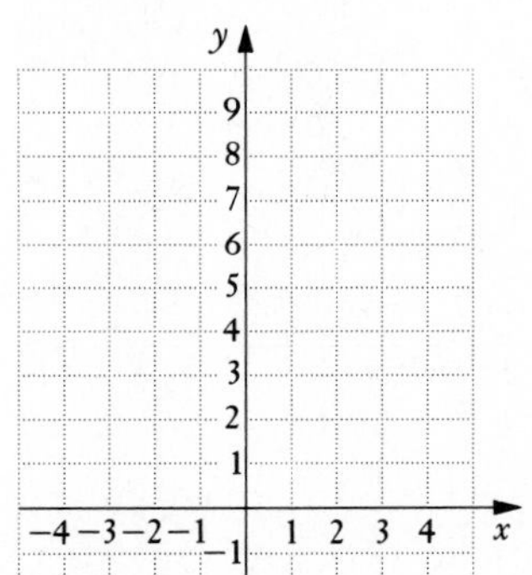

6. $y = 2^{x-1}$

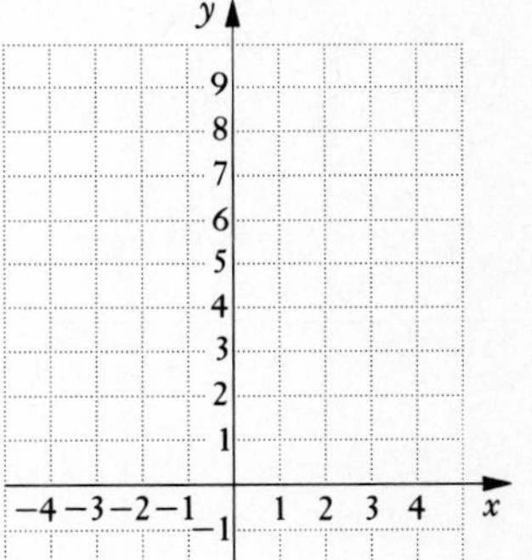

7. $y = 3^{x-2}$

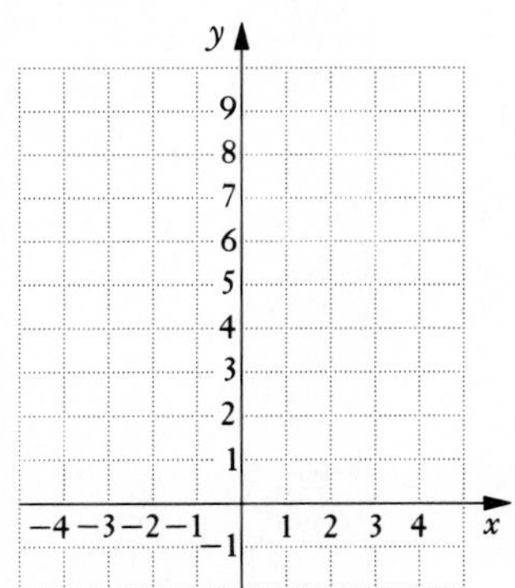

8. $y = 3^{x+2}$

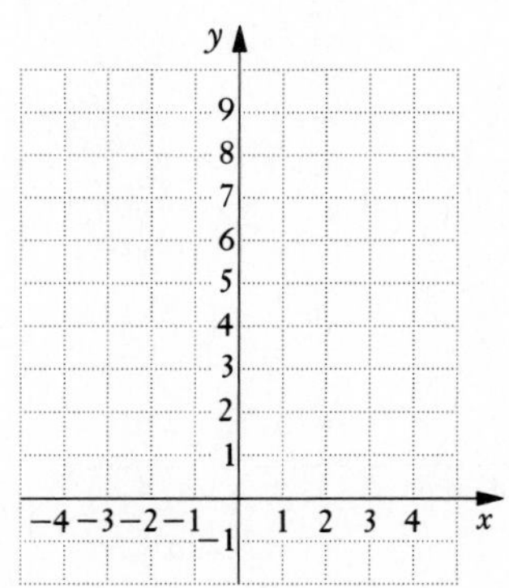

9. $y = 2^x - 3$

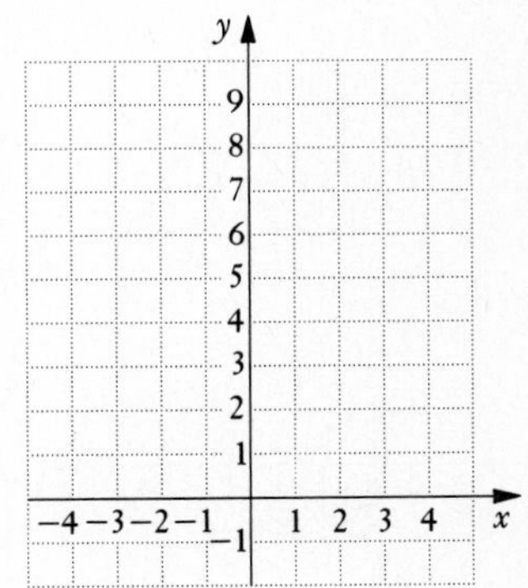

10. $y = 2^x + 1$

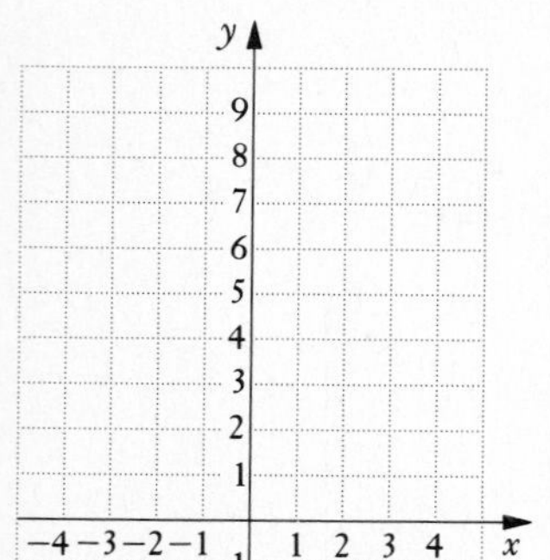

11. $y = 5^{x+3}$

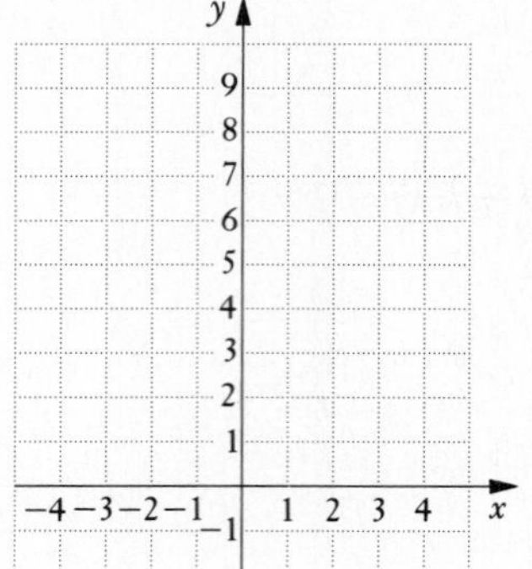

12. $y = 6^{x-4}$

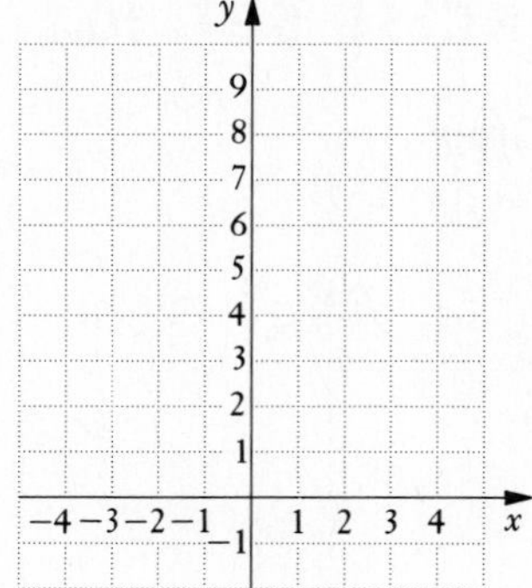

13. $y = \left(\frac{1}{2}\right)^x$

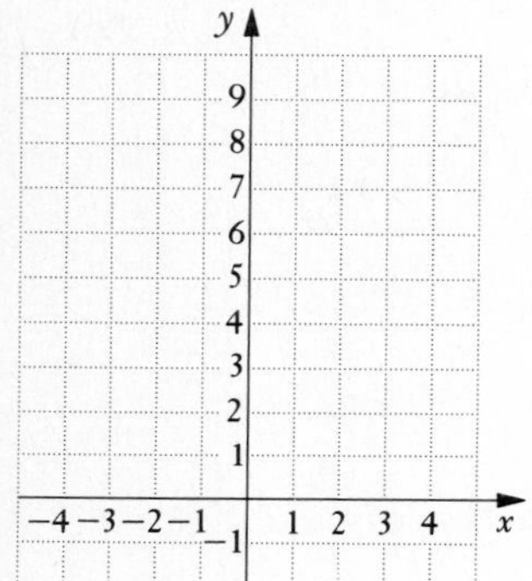

14. $y = \left(\frac{1}{3}\right)^x$

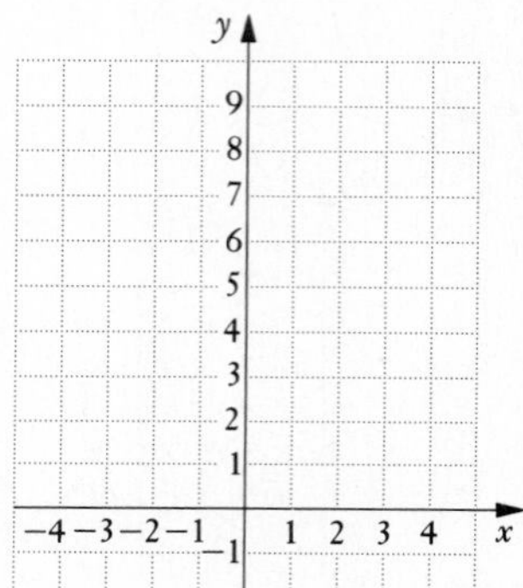

15. $y = \left(\frac{1}{5}\right)^x$

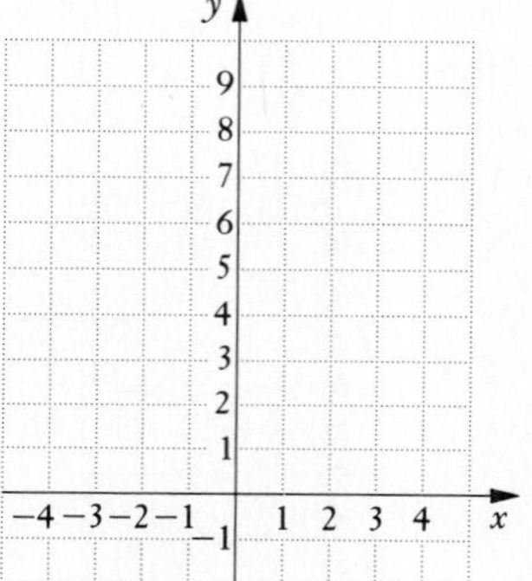

16. $y = \left(\frac{1}{4}\right)^x$

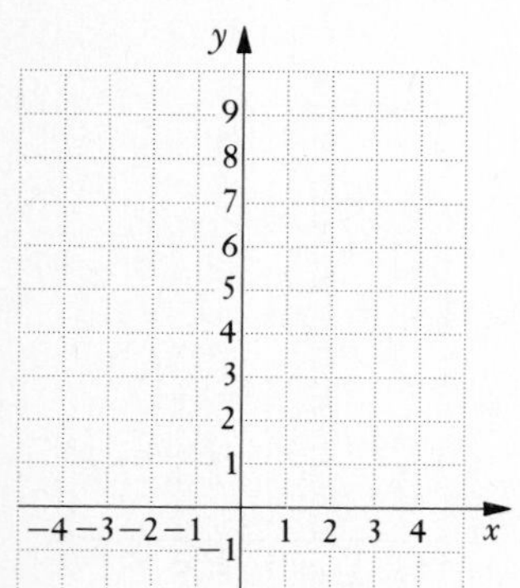

17. $y = 2^{2x-1}$

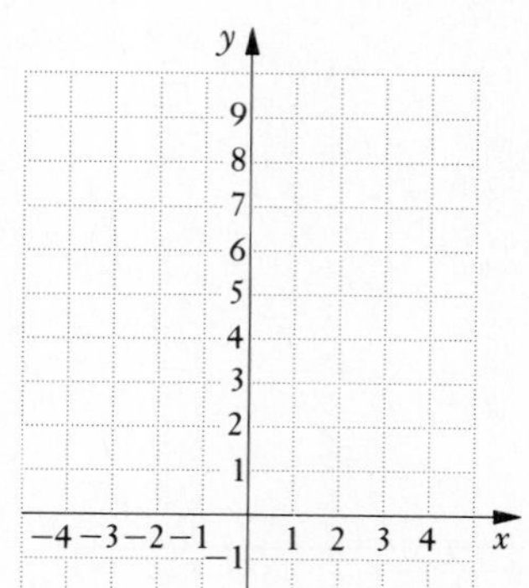

18. $y = 3^{3-x}$

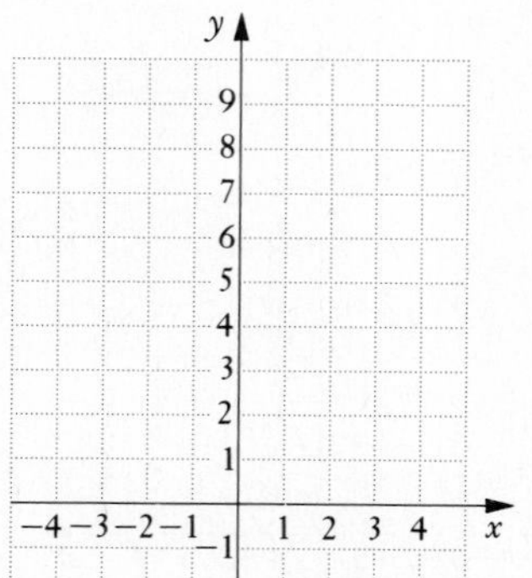

b Graph.

19. $x = 2^y$

20. $x = 6^y$

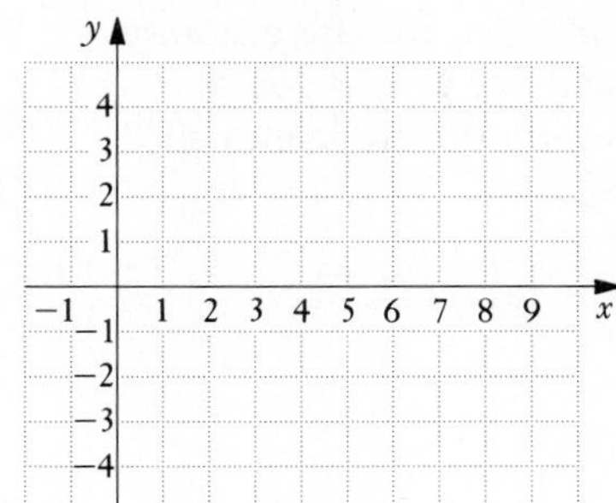

21. $x = \left(\frac{1}{2}\right)^y$

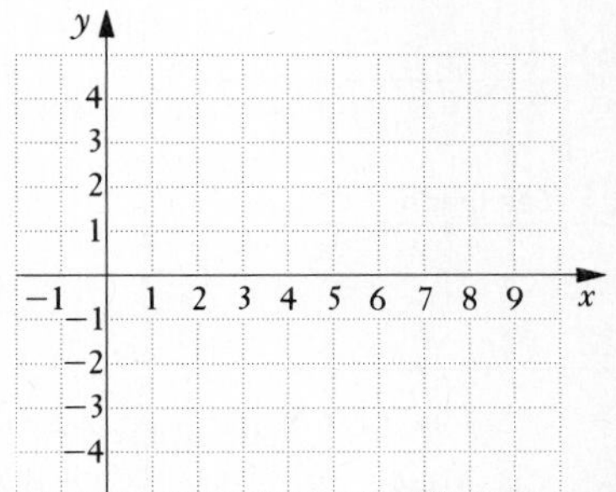

22. $x = \left(\frac{1}{3}\right)^y$

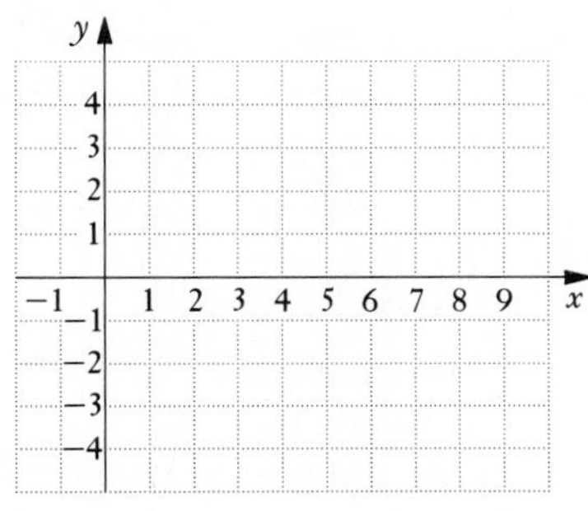

23. $x = 5^y$

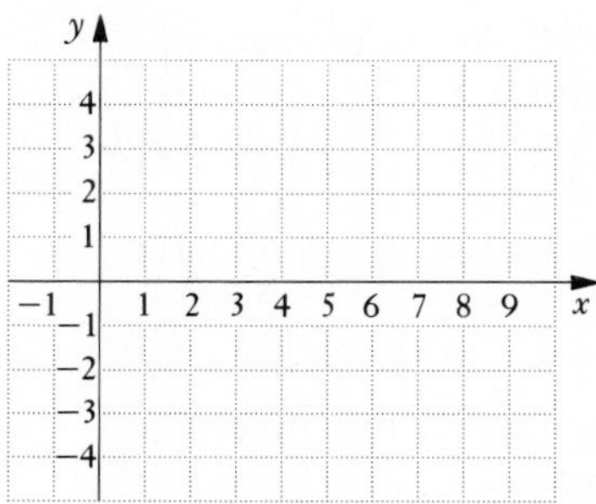

24. $x = \left(\frac{2}{3}\right)^y$

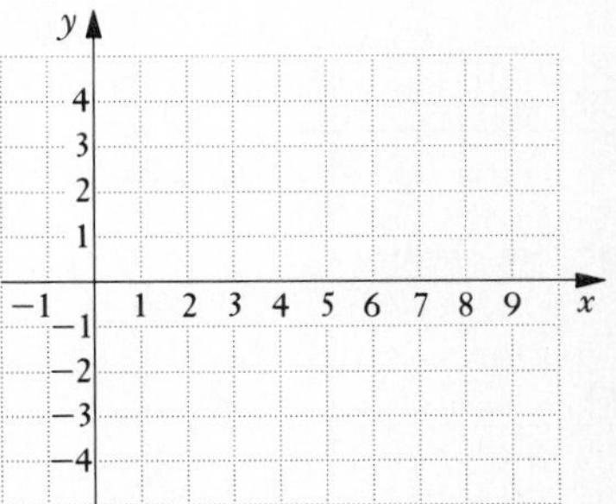

Graph both equations using the same set of axes.

25. $y = 2^x, \quad x = 2^y$

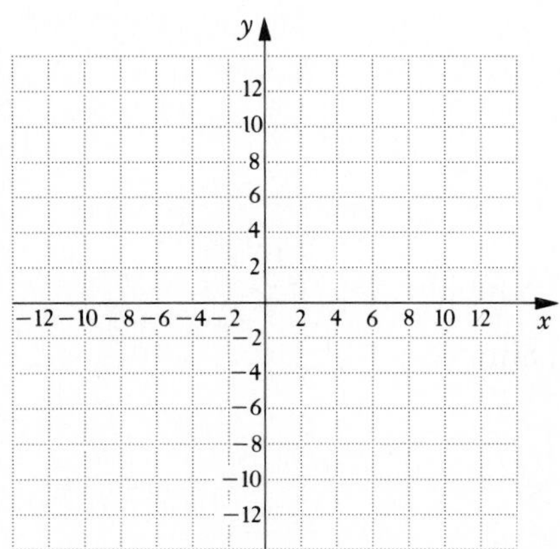

26. $y = \left(\frac{1}{2}\right)^x, \quad x = \left(\frac{1}{2}\right)^y$

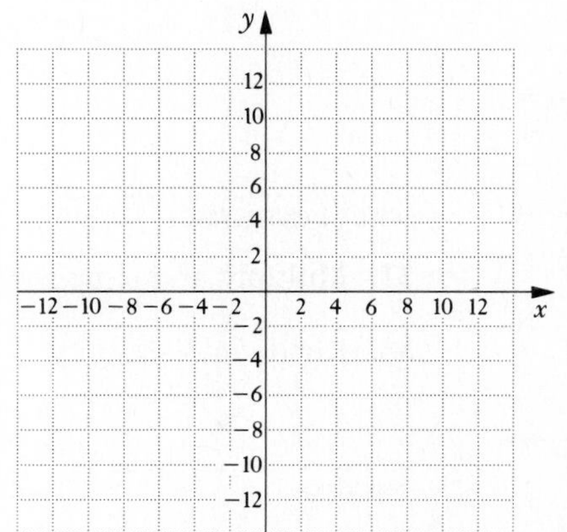

ANSWERS

27. a) ______

b) ______

c) See graph.

28. a) ______

b) See graph.

29. a) ______

b) See graph.

30. a) ______

b) See graph.

31. ______

32. ______

33. ______

34. ______

35. See graph.

36. See graph.

37. See graph.

38. See graph.

39. See graph.

40. See graph.

C

27. *Interest compounded annually.* Suppose that \$50,000 is invested at 8% interest, compounded annually.

a) Find a function for the amount in the account after t years.

b) Find the amount of money in the account at $t = 0$, $t = 4$, $t = 8$, and $t = 10$.

c) Graph the function.

28. *Recycling aluminum cans.* It is known that one fourth of all aluminum cans distributed will be recycled each year. A beverage company distributes 250,000 cans. The number still in use after time t, in years, is given by the function

$$N(t) = 250{,}000(\tfrac{1}{4})^t.$$

a) How many cans are still in use after 0 years? 1 year? 4 years? 10 years?

b) Graph the function.

29. *Salvage value.* An office machine is purchased for \$5200. Its value each year is about 80% of the value the preceding year. Its value after t years is given by the exponential function

$$V(t) = \$5200(0.8)^t.$$

a) Find the value of the machine after 0 years, 1 year, 2 years, 5 years, and 10 years.

b) Graph the function.

30. *Compact discs.* The number of compact discs purchased each year is increasing exponentially. The number N, in millions, purchased is given by

$$N(t) = 7.5(6)^{0.5t},$$

where $t = 0$ corresponds to 1985, $t = 1$ corresponds to 1986, and so on, t being the number of years after 1985.

a) Find the number of compact discs sold in 1986, 1987, 1988, 1990, 1995, and 2000.

b) Graph the function.

SKILL MAINTENANCE

31. Multiply and simplify: $x^{-5} \cdot x^3$.

32. Simplify: $(x^{-3})^4$.

33. Divide and simplify: $\dfrac{x^{-3}}{x^4}$.

34. Simplify: 5^0.

SYNTHESIS

Graph.

35. $y = 2^x + 2^{-x}$

36. $y = (\frac{1}{2})^x - 1$

37. $y = 3^x + 3^{-x}$

38. $y = 2^{-(x-1)^2}$

39. $y = |2^{x^2} - 1|$

40. $y = |2^x - 2|$

10.2 Logarithmic Functions

We now consider a kind of function called a *logarithm function,* or *logarithmic function.*

a Graphs of Logarithmic Functions

Consider the exponential function $f(x) = 2^x$, or $y = 2^x$. Does this function have an inverse that is a function? We can see from the graph that this function is one-to-one and that it does have an inverse f^{-1} that is a function, because it passes the horizontal-line test. For now the only equation we have describing the inverse is $x = 2^y$. If we could solve it for y and replace y by $f^{-1}(x)$, we would have a formula for the inverse. But, as yet we do not have a way to solve for y.

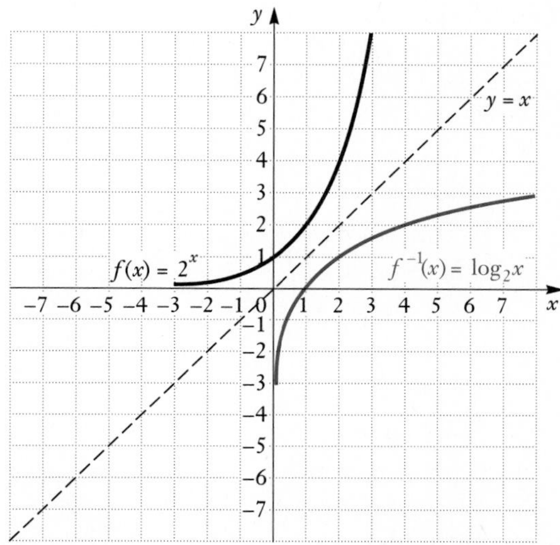

Consider the input 3 and the original function

$$f(x) = 2^x.$$

Then

$$f(3) = 2^3 = 8.$$

Suppose we know only the output 8. The inverse function $f^{-1}(x)$ of $f(x) = 2^x$, if it exists, would tell us what original input gave us an output of 8. To find the original input, we need to find "the power to which 2 was raised to get 8." Mathematicians have invented a shorter name for this thought process. It is called the **logarithmic function, base 2,** denoted

$$f^{-1}(x) = \log_2 x.$$

We read "$\log_2 x$" as "the logarithm, base 2, of x." So $\log_2 8$ is the power to which we raise 2 to get 8. Thus, $\log_2 8 = 3$.

Suppose $x = 5$:

$$f(5) = 2^5 = 32,$$
$$f^{-1}(32) = \log_2 32 = 5$$

($\log_2 32$ is the power to which we raise 2 to get 32).

OBJECTIVES

After finishing Section 10.2, you should be able to:

- a Graph logarithmic functions.
- b Convert from exponential equations to logarithmic equations.
- c Convert from logarithmic equations to exponential equations.
- d Solve certain logarithmic equations.

FOR EXTRA HELP

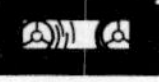
Tape 19B

Tape 17B

MAC: 10
IBM: 10

1. Write the meaning of $\log_2 64$. Then find $\log_2 64$.

Suppose $x = 4$:

$$f(4) = 2^4 = 16,$$
$$f^{-1}(16) = \log_2 16 = 4$$

($\log_2 16$ is the power to which we raise 2 to get 16).

Suppose we want to find $f^{-1}(13) = \log_2 13$, or the number to which we raise 2 to get 13. As yet, we have no simpler way to write this.

DO EXERCISE 1.

For any exponential function $f(x) = a^x$, the inverse is called a **logarithmic function, base *a*.** The graph of the inverse can, of course, be obtained by reflecting the graph of $y = a^x$ across the line $y = x$. Interchanging x and y in $y = a^x$, we obtain $x = a^y$, which is equivalent to $y = \log_a x$.

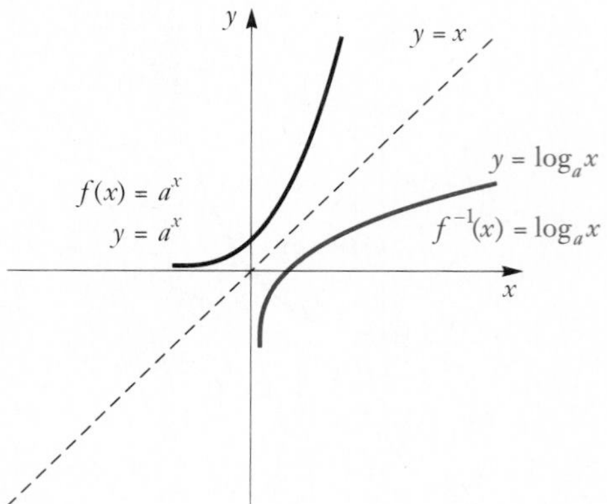

Definition of Logarithms

The inverse of $f(x) = a^x$ is given by

$$f^{-1}(x) = \log_a x.$$

We read "$\log_a x$" as the "logarithm, base a, of x." We define $y = \log_a x$ as that number y such that $x = a^y$, where $x > 0$ and a is a positive constant other than 1.

It is helpful in dealing with logarithmic functions to remember that the logarithm of a number is an **exponent.** It is the exponent y in $x = a^y$. You might also think to yourself, "The logarithm, base a, of a number x is the power to which a must be raised in order to get x."

A logarithm is an exponent.

The following is a comparison of exponential and logarithmic functions.

Exponential Function	*Logarithmic Function*
$y = a^x$	$x = a^y$
$f(x) = a^x$	$f^{-1}(x) = \log_a x$
$a > 0,\ a \neq 1$	$a > 0,\ a \neq 1$
Domain = The set of real numbers	Range = The set of real numbers
Range = The set of positive numbers	Domain = The set of positive numbers

ANSWER ON PAGE A-12

Why do we exclude 1 from being a logarithmic base? If we did include it, we would be considering $x = 1^y = 1$. The graph of this equation is a vertical line, which is not a function. It does not pass the vertical-line test.

► **EXAMPLE 1** Graph: $y = \log_5 x$.

The equation $y = \log_5 x$ is equivalent to $5^y = x$. We can find ordered pairs that are solutions by choosing values for y and computing the x-values.

For $y = 0$, $x = 5^0 = 1$.
For $y = 1$, $x = 5^1 = 5$.
For $y = 2$, $x = 5^2 = 25$.
For $y = 3$, $x = 5^3 = 125$.
For $y = -1$, $x = 5^{-1} = \frac{1}{5}$.
For $y = -2$, $x = 5^{-2} = \frac{1}{25}$.

x, or 5^y	y
1	0
5	1
25	2
125	3
$\frac{1}{5}$	-1
$\frac{1}{25}$	-2

(1) Select y.
(2) Compute x.

We plot the ordered pairs and connect them with a smooth curve. The graph of $y = 5^x$ has been shown only for reference.

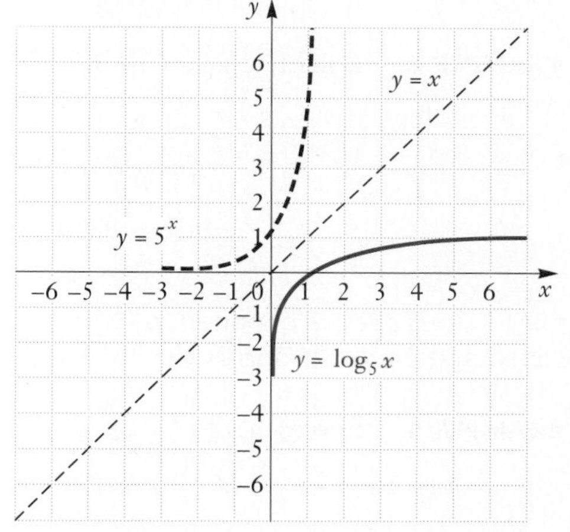

◄

DO EXERCISE 2.

2. Graph: $y = \log_3 x$.

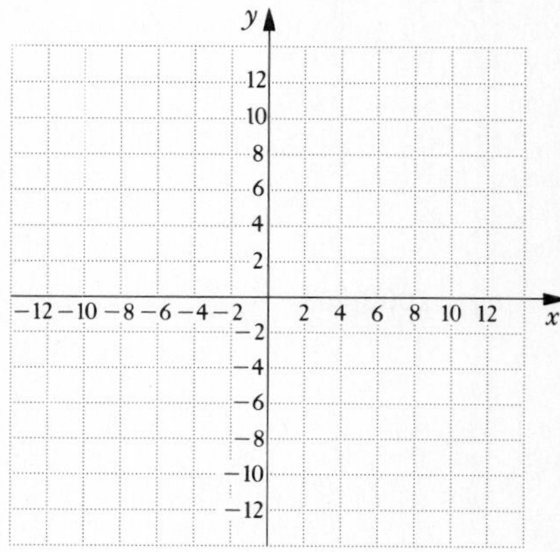

b From Exponential Equations to Logarithmic Equations

We use the definition of logarithms to convert from exponential equations to logarithmic equations.

> $y = \log_a x$ **is equivalent to** $a^y = x$.

CAUTION! **Be sure to memorize this relationship!** It is probably the most important definition in the chapter. Many times this definition will be a justification for a proof or a procedure that we are considering.

► **EXAMPLES** Convert to logarithmic equations.

2. $8 = 2^x \rightarrow x = \log_2 8$ — **The exponent is the logarithm. The base remains the base.**
3. $y^{-1} = 4 \rightarrow -1 = \log_y 4$
4. $a^b = c \rightarrow b = \log_a c$ ◄

DO EXERCISES 3–6.

Convert to a logarithmic equation.

3. $6^0 = 1$

4. $10^{-3} = 0.001$

5. $16^{0.25} = 2$

6. $m^T = P$

ANSWERS ON PAGE A-12

Convert to an exponential equation.

7. $\log_2 32 = 5$

8. $\log_{10} 1000 = 3$

9. $\log_a Q = 7$

10. $\log_t M = x$

Solve.

11. $\log_{10} x = 4$

12. $\log_x 81 = 4$

13. $\log_2 x = -2$

ANSWERS ON PAGE A-12

c From Logarithmic Equations to Exponential Equations

We also use the definition of logarithms to convert from logarithmic equations to exponential equations.

▶ **EXAMPLES** Convert to exponential equations.

5. $y = \log_3 5 \rightarrow 3^y = 5$ — The logarithm is the exponent. The base does not change.

6. $-2 = \log_a 7 \rightarrow a^{-2} = 7$

7. $a = \log_b d \rightarrow b^a = d$ ◀

DO EXERCISES 7–10.

d Solving Certain Logarithmic Equations

Certain equations involving logarithms can be solved by first converting to exponential equations. We will solve more complicated equations later.

▶ **EXAMPLE 8** Solve: $\log_2 x = -3$.

$$\log_2 x = -3$$
$$2^{-3} = x \quad \text{Converting to an exponential equation}$$
$$\frac{1}{8} = x \quad \text{Computing } 2^{-3}$$

Check: $\log_2 \frac{1}{8}$ is the exponent to which we raise 2 to get $\frac{1}{8}$. Since $2^{-3} = \frac{1}{8}$, we know that $\frac{1}{8}$ checks and is the solution. ◀

▶ **EXAMPLE 9** Solve: $\log_x 16 = 2$.

$$\log_x 16 = 2$$
$$x^2 = 16 \quad \text{Converting to an exponential equation}$$
$$x = 4 \quad \text{or} \quad x = -4 \quad \text{Using the principle of zero products}$$

Check: $\log_4 16 = 2$ because $4^2 = 16$. Thus, 4 is a solution. Since all logarithm bases must be positive, $\log_{-4} 16$ is not defined. Therefore, -4 is not a solution. ◀

DO EXERCISES 11–13.

Solving an equation like $\log_b a = x$ amounts to finding the logarithm, base b, of the number a. You have done this before in graphing logarithmic functions. To think of finding logarithms as solving equations may help in some cases.

▶ **EXAMPLE 10** Find $\log_{10} 1000$.

Method 1. Let $\log_{10} 1000 = x$. Then,

$$10^x = 1000 \quad \text{Converting to an exponential equation}$$
$$10^x = 10^3$$
$$x = 3. \quad \text{The exponents are the same.}$$

Therefore, $\log_{10} 1000 = 3$.

Method 2. Think of the meaning of $\log_{10} 1000$. It is the exponent to which we raise 10 to get 1000. That exponent is 3. Therefore, $\log_{10} 1000 = 3$. ◀

▶ **EXAMPLE 11** Find $\log_{10} 0.01$.

Method 1. Let $\log_{10} 0.01 = x$. Then

$$10^x = 0.01$$
$$10^x = \frac{1}{100}$$
$$10^x = 10^{-2}$$
$$x = -2.$$

Therefore, $\log_{10} 0.01 = -2$.

Method 2. $\log_{10} 0.01$ is the exponent to which we raise 10 to get 0.01. Noting that

$$0.01 = \frac{1}{100} = \frac{1}{10^2} = 10^{-2},$$

we see that the exponent is -2. Therefore, $\log_{10} 0.01 = -2$. ◀

▶ **EXAMPLE 12** Find $\log_5 1$.

Method 1. Let $\log_5 1 = x$. Then

$5^x = 1$ Converting to an exponential equation

$5^x = 5^0$

$x = 0.$

Therefore, $\log_5 1 = 0$.

Method 2. $\log_5 1$ is the exponent to which we raise 5 to get 1. That exponent is 0. Therefore, $\log_5 1 = 0$. ◀

DO EXERCISES 14–16.

Example 12 illustrates an important property of logarithms.

> **For any base a,**
>
> $$\log_a 1 = 0.$$
>
> **The logarithm, base a, of 1 is always 0.**

The proof follows from the fact that $a^0 = 1$. This is equivalent to the logarithmic equation $\log_a 1 = 0$.

Another property follows similarly. We know that $a^1 = a$ for any real number a. In particular, it holds for any positive number a. This is equivalent to the logarithmic equation $\log_a a = 1$.

> **For any base a,**
>
> $$\log_a a = 1.$$

DO EXERCISES 17–20.

Find each of the following.

14. $\log_{10} 10{,}000$

15. $\log_{10} 0.0001$

16. $\log_7 1$

Simplify.

17. $\log_3 1$

18. $\log_3 3$

19. $\log_c c$

20. $\log_c 1$

ANSWERS ON PAGE A-12

❖ SIDELIGHTS

Handling Dimension Symbols (Part II)*

We can treat dimension symbols much like numerals or variables, because we obtain correct results that way. We can change units by substituting or by multiplying by 1, as shown below.

EXAMPLE 1 Convert 3 ft to inches.

Method 1. We have 3 ft. We know that 1 ft = 12 in., so we substitute 12 in. for ft:

$$3 \text{ ft} = 3 \cdot 12 \text{ in.} = 36 \text{ in.}$$

Method 2. We want to convert from "ft" to "in." We multiply by 1 using a symbol for 1 with "ft" on the bottom since we are converting from "ft," and with "in." on the top since we are converting to "in."

$$\begin{aligned} 3 \text{ ft} &= 3 \text{ ft} \cdot \frac{12 \text{ in.}}{1 \text{ ft}} \\ &= \frac{3 \cdot 12}{1} \cdot \frac{\text{ft}}{\text{ft}} \cdot \text{in.} = 36 \text{ in.} \end{aligned}$$ ◀

We can multiply by 1 several times to make successive conversions. In the following example, we first convert mi/hr to ft/hr and then to ft/sec.

EXAMPLE 2 Convert 60 mi/hr to ft/sec.

$$\begin{aligned} 60 \frac{\text{mi}}{\text{hr}} &= 60 \frac{\text{mi}}{\text{hr}} \cdot 60 \frac{5280 \text{ ft}}{1 \text{ mi}} \\ &= 60 \frac{\text{mi}}{\text{hr}} \cdot \frac{5280 \text{ ft}}{1 \text{ mi}} \cdot \frac{1 \text{ hr}}{60 \text{ min}} \cdot \frac{1 \text{ min}}{60 \text{ sec}} \\ &= \frac{60 \cdot 5280}{60 \cdot 60} \cdot \frac{\text{mi}}{\text{mi}} \cdot \frac{\text{ft}}{\text{sec}} = 88 \frac{\text{ft}}{\text{sec}}. \end{aligned}$$

EXERCISES

Make these unit changes.

1. Change 3.2 lb to oz.
2. Change 6.2 km to m.
3. Change 35 mi/hr to ft/min.
4. Change $375 per day to dollars per minute.
5. Change 8 ft to inches.
6. Change 25 yd to ft.
7. How many years ago is 1 million seconds ago?
8. How many years ago is 1 billion seconds ago?
9. How many years ago is 1 trillion seconds ago?
10. Change 20 lb to oz (16 oz = 1 lb).
11. Change $60 \frac{\text{lb}}{\text{ft}}$ to $\frac{\text{oz}}{\text{in.}}$.
12. Change $44 \frac{\text{ft}}{\text{sec}}$ to $\frac{\text{mi}}{\text{hr}}$.
13. Change 2 days to seconds.
14. Change 128 hr to days.
15. Change 216 in^2 to ft^2.
16. Change 1440 man-hours to man-days.
17. Change $80 \frac{\text{lb}}{\text{ft}^3}$ to $\frac{\text{ton}}{\text{yd}^3}$.
18. Change the speed of light, 186,000 mi/sec, to mi/yr. Let 365 days = 1 yr.

* For more on Handling Dimension Symbols, see p. 300.

NAME SECTION DATE

EXERCISE SET 10.2

a Graph.

1. $y = \log_2 x$

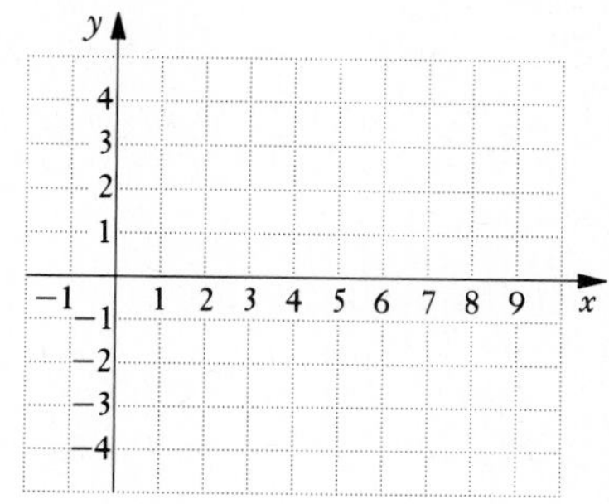

2. $y = \log_{10} x$

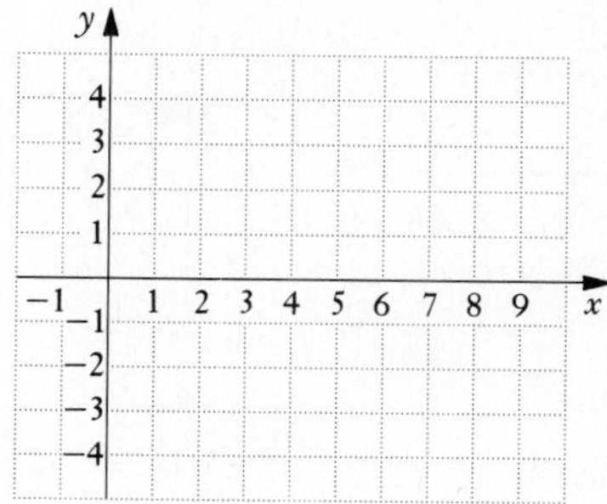

3. $y = \log_6 x$

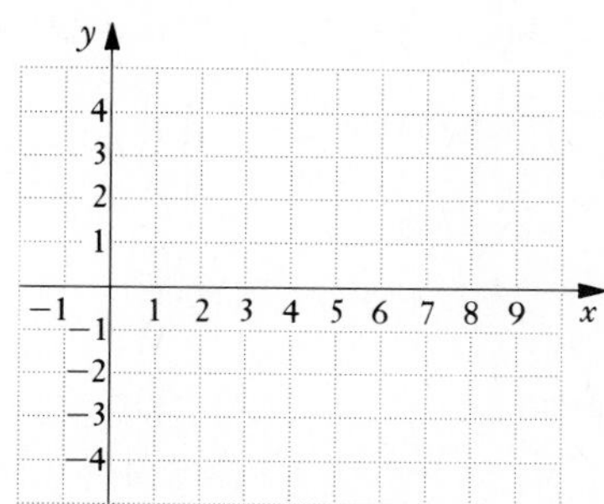

4. $y = \log_{1/2} x$

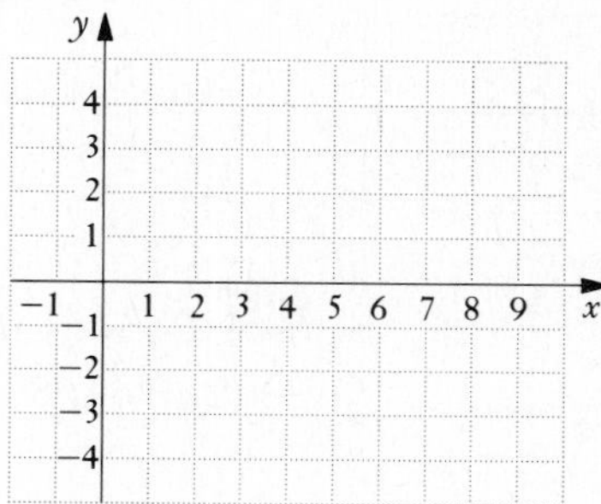

Graph both functions using the same set of axes.

5. $f(x) = 3^x, \quad f^{-1}(x) = \log_3 x$

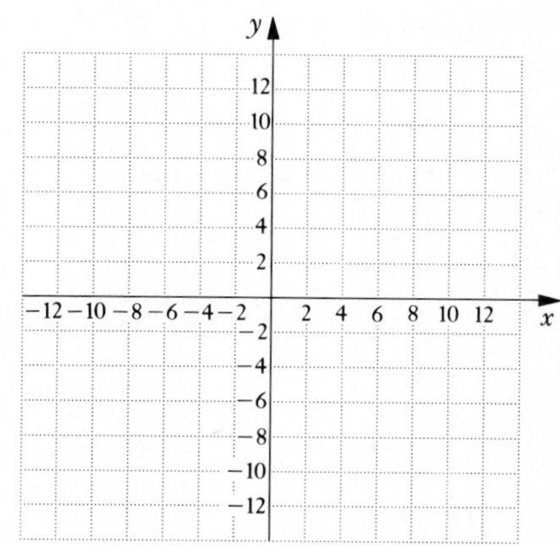

6. $f(x) = 4^x, \quad f^{-1}(x) = \log_4 x$

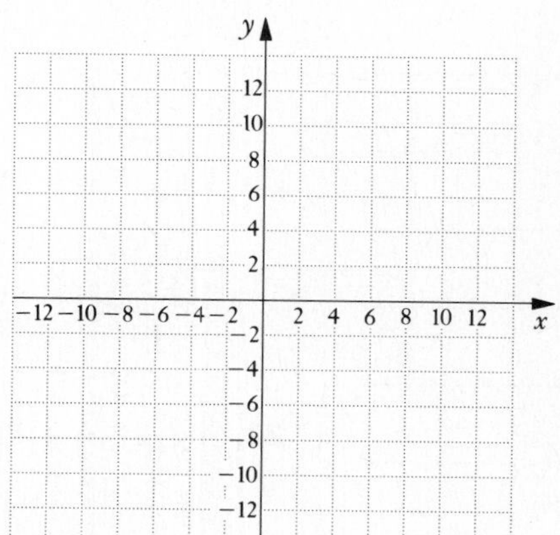

ANSWERS

7. ______

8. ______

9. ______

10. ______

11. ______

12. ______

13. ______

14. ______

15. ______

16. ______

17. ______

18. ______

19. ______

20. ______

21. ______

22. ______

23. ______

24. ______

25. ______

26. ______

27. ______

28. ______

29. ______

30. ______

31. ______

32. ______

33. ______

34. ______

b Convert to a logarithmic equation.

7. $10^3 = 1000$ **8.** $10^2 = 100$ **9.** $5^{-3} = \frac{1}{125}$ **10.** $4^{-5} = \frac{1}{1024}$

11. $8^{1/3} = 2$ **12.** $16^{1/4} = 2$ **13.** $10^{0.3010} = 2$ **14.** $10^{0.4771} = 3$

15. $e^2 = t$ **16.** $p^k = 3$ **17.** $Q^t = x$ **18.** $P^m = V$

19. $e^2 = 7.3891$ **20.** $e^3 = 20.0855$ **21.** $e^{-2} = 0.1353$ **22.** $e^{-4} = 0.0183$

c Convert to an exponential equation.

23. $t = \log_3 8$ **24.** $h = \log_7 10$ **25.** $\log_5 25 = 2$

26. $\log_6 6 = 1$ **27.** $\log_{10} 0.1 = -1$ **28.** $\log_{10} 0.01 = -2$

29. $\log_{10} 7 = 0.845$ **30.** $\log_{10} 3 = 0.4771$ **31.** $\log_e 20 = 2.9957$

32. $\log_e 10 = 2.3036$ **33.** $\log_t Q = k$ **34.** $\log_m P = a$

d Solve.

35. $\log_3 x = 2$

36. $\log_4 x = 3$

37. $\log_x 16 = 2$

38. $\log_x 64 = 3$

39. $\log_2 16 = x$

40. $\log_5 25 = x$

41. $\log_3 27 = x$

42. $\log_4 16 = x$

43. $\log_x 13 = 1$

44. $\log_x 23 = 1$

45. $\log_6 x = 0$

46. $\log_9 x = 1$

47. $\log_2 x = -1$

48. $\log_3 x = -2$

49. $\log_8 x = \frac{1}{3}$

50. $\log_{32} x = \frac{1}{5}$

Find each of the following.

51. $\log_{10} 100$

52. $\log_{10} 100{,}000$

53. $\log_{10} 0.1$

54. $\log_{10} 0.001$

ANSWERS

35. ______

36. ______

37. ______

38. ______

39. ______

40. ______

41. ______

42. ______

43. ______

44. ______

45. ______

46. ______

47. ______

48. ______

49. ______

50. ______

51. ______

52. ______

53. ______

54. ______

ANSWERS

55.

56.

57.

58.

59.

60.

61.

62.

63.

64.

65.

66.

67.

68.

69.

70.

71. See graph.

72. See graph.

73. See graph.

74.

75.

76.

77.

78.

79.

80.

81.

82.

83.

55. $\log_{10} 1$ **56.** $\log_{10} 10$ **57.** $\log_5 625$ **58.** $\log_2 64$

59. $\log_7 49$ **60.** $\log_5 125$ **61.** $\log_2 8$ **62.** $\log_8 64$

63. $\log_5 \dfrac{1}{25}$ **64.** $\log_2 \dfrac{1}{16}$ **65.** $\log_3 1$ **66.** $\log_4 4$

67. $\log_e e$ **68.** $\log_e 1$ **69.** $\log_{27} 9$ **70.** $\log_8 2$

SYNTHESIS

71. Graph both equations using the same set of axes.

$$y = (\tfrac{3}{2})^x, \qquad y = \log_{3/2} x$$

Graph.

72. $y = \log_2 (x - 1)$ **73.** $y = \log_3 |x + 1|$

Solve.

74. $|\log_3 x| = 3$ **75.** $\log_{125} x = \frac{2}{3}$ **76.** $\log_4 (3x - 2) = 2$

77. $\log_8 (2x + 1) = -1$ **78.** $\log_{10} (x^2 + 21x) = 2$

Simplify.

79. $\log_{1/4} \frac{1}{64}$ **80.** $\log_{81} 3 \cdot \log_3 81$

81. $\log_{10} (\log_4 (\log_3 81))$ **82.** $\log_2 (\log_2 (\log_4 256))$

83. $\log_{1/5} 25$

10.3 Properties of Logarithmic Functions

The ability to manipulate logarithmic expressions is important in many applications and in more advanced mathematics. We now establish some basic properties that are useful in manipulating logarithmic expressions.

a Logarithms of Products

Property 1: The Product Rule

For any positive numbers M and N,

$$\log_a M \cdot N = \log_a M + \log_a N.$$

(The logarithm of a product is the sum of the logarithms of the factors. The number a can be any logarithmic base.)

▶ **EXAMPLE 1** Express as a sum of logarithms: $\log_2 (4 \cdot 16)$.

$\log_2 (4 \cdot 16) = \log_2 4 + \log_2 16$ **By Property 1** ◀

▶ **EXAMPLE 2** Express as a single logarithm: $\log_{10} 0.01 + \log_{10} 1000$.

$\log_{10} 0.01 + \log_{10} 1000 = \log_{10} (0.01 \times 1000)$ **By Property 1**

$= \log_{10} 10$ ◀

DO EXERCISES 1–4.

A Proof of Property 1 (*Optional*). Let $\log_a M = x$ and $\log_a N = y$. Converting to exponential equations, we have $a^x = M$ and $a^y = N$. Then we multiply the latter two equations to obtain

$$M \cdot N = a^x \cdot a^y = a^{x+y}.$$

Converting back to a logarithmic equation, we get

$$\log_a M \cdot N = x + y.$$

Remembering what x and y represent, we get

$$\log_a M \cdot N = \log_a M + \log_a N.$$

b Logarithms of Powers

The second basic property is as follows.

Property 2: The Power Rule

For any positive number M and any real number k,

$$\log_a M^k = k \cdot \log_a M.$$

(The logarithm of a power of M is the exponent times the logarithm of M. The number a can be any logarithmic base.)

▶ **EXAMPLES** Express as a product.

3. $\log_a 9^{-5} = -5 \log_a 9$ **By Property 2**
4. $\log_a \sqrt[4]{5} = \log_a 5^{1/4}$ **Writing exponential notation**

$= \frac{1}{4} \log_a 5$ **By Property 2** ◀

DO EXERCISES 5 AND 6.

OBJECTIVES

After finishing Section 10.3, you should be able to:

- **a** Express the logarithm of a product as a sum of logarithms, and conversely.
- **b** Express the logarithm of a power as a product.
- **c** Express the logarithm of a quotient as a difference of logarithms, and conversely.
- **d** Convert from logarithms of products, quotients, and powers to expressions in terms of individual logarithms, and conversely.
- **e** Simplify expressions of the type $\log_a a^k$.

FOR EXTRA HELP

Tape 19C

Tape 17B

MAC: 10
IBM: 10

Express as a sum of logarithms.

1. $\log_5 (25 \cdot 5)$
2. $\log_b PQ$

Express as a single logarithm.

3. $\log_3 7 + \log_3 5$
4. $\log_a C + \log_a A + \log_a B + \log_a I + \log_a N$

Express as a product.

5. $\log_7 4^5$
6. $\log_a \sqrt{5}$

ANSWERS ON PAGE A-12

7. Express as a difference of logarithms:

$$\log_b \frac{P}{Q}.$$

8. Express as a single logarithm: $\log_2 125 - \log_2 25$.

ANSWERS ON PAGE A-12

A Proof of Property 2 (*Optional*). Let $x = \log_a M$. Then we convert to an exponential equation to get $a^x = M$. Raising both sides to the kth power, we obtain

$$(a^x)^k = M^k, \quad \text{or} \quad a^{xk} = M^k.$$

Converting back to a logarithmic equation, we get $\log_a M^k = xk$. But $x = \log_a M$, so

$$\log_a M^k = (\log_a M)k = k \cdot \log_a M.$$

c Logarithms of Quotients

> Property 3: The Quotient Rule
>
> **For any positive numbers *M* and *N*,**
>
> $$\log_a \frac{M}{N} = \log_a M - \log_a N.$$
>
> **(The logarithm of a quotient is the logarithm of the dividend minus the logarithm of the divisor. The number *a* can be any logarithmic base.)**

► **EXAMPLE 5** Express as a difference of logarithms: $\log_t \frac{6}{U}$.

$$\log_t \frac{6}{U} = \log_t 6 - \log_t U \qquad \text{By Property 3}$$ ◄

► **EXAMPLE 6** Express as a single logarithm: $\log_b 17 - \log_b 27$.

$$\log_b 17 - \log_b 27 = \log_b \frac{17}{27} \qquad \text{By Property 3}$$ ◄

► **EXAMPLE 7** Express as a single logarithm: $\log_{10} 10{,}000 - \log_{10} 100$.

$$\log_{10} 10{,}000 - \log_{10} 100 = \log_{10} \frac{10{,}000}{100} = \log_{10} 100$$ ◄

DO EXERCISES 7 AND 8.

A Proof of Property 3 (*Optional*). The proof makes use of Property 1 and Property 2.

$$\begin{aligned}
\log_a \frac{M}{N} &= \log_a MN^{-1} \\
&= \log_a M + \log_a N^{-1} && \text{By Property 1} \\
&= \log_a M + (-1)\log_a N && \text{By Property 2} \\
&= \log_a M - \log_a N
\end{aligned}$$

d Using the Properties Together

► **EXAMPLES** Express in terms of logarithms of x, y, and z.

8. $$\begin{aligned}
\log_a \frac{x^2y^3}{z^4} &= \log_a (x^2y^3) - \log_a z^4 && \text{Using Property 3} \\
&= \log_a x^2 + \log_a y^3 - \log_a z^4 && \text{Using Property 1} \\
&= 2\log_a x + 3\log_a y - 4\log_a z && \text{Using Property 2}
\end{aligned}$$

9. $\log_a \sqrt[4]{\frac{xy}{z^3}} = \log_a \left(\frac{xy}{z^3}\right)^{1/4}$ **Writing exponential notation**

$= \frac{1}{4} \log_a \frac{xy}{z^3}$ **Using Property 2**

$= \frac{1}{4}(\log_a xy - \log_a z^3)$ **Using Property 3 (note the parentheses)**

$= \frac{1}{4}(\log_a x + \log_a y - 3 \log_a z)$ **Using Properties 1 and 2**

$= \frac{1}{4} \log_a x + \frac{1}{4} \log_a y - \frac{3}{4} \log_a z$ **Distributive law**

10. $\log_b \frac{xy}{m^3 n^4} = \log_b xy - \log_b m^3 n^4$ **Using Property 3**

$= (\log_b x + \log_b y) - (\log_b m^3 + \log_b n^4)$ **Using Property 1**

$= \log_b x + \log_b y - \log_b m^3 - \log_b n^4$ **Removing parentheses**

$= \log_b x + \log_b y - 3 \log_b m - 4 \log_b n$ **Using Property 2** ◀

DO EXERCISES 9–11.

► **EXAMPLES** Express as a single logarithm.

11. $\frac{1}{2} \log_a x - 7 \log_a y + \log_a z$

$= \log_a x^{1/2} - \log_a y^7 + \log_a z$ **Using Property 2**

$= \log_a \frac{\sqrt{x}}{y^7} + \log_a z$ **Using Property 3**

$= \log_a \frac{z\sqrt{x}}{y^7}$ **Using Property 1**

12. $\log_a \frac{b}{\sqrt{x}} + \log_a \sqrt{bx}$

$= \log_a b - \log_a \sqrt{x} + \log_a \sqrt{bx}$ **Using Property 3**

$= \log_a b - \frac{1}{2} \log_a x + \frac{1}{2} \log_a (bx)$ **Using Property 2**

$= \log_a b - \frac{1}{2} \log_a x + \frac{1}{2}(\log_a b + \log_a x)$ **Using Property 1**

$= \log_a b - \frac{1}{2} \log_a x + \frac{1}{2} \log_a b + \frac{1}{2} \log_a x$

$= \frac{3}{2} \log_a b$ **Collecting like terms**

$= \log_a b^{3/2}$ **Using Property 2**

Example 12 could also be done as follows:

$$\log_a \frac{b}{\sqrt{x}} + \log_a \sqrt{bx} = \log_a \frac{b}{\sqrt{x}} \sqrt{bx} \quad \textbf{Using Property 1}$$

$$= \log_a b\sqrt{b}, \text{ or } \log_a b^{3/2}.$$ ◀

DO EXERCISES 12 AND 13.

Express in terms of logarithms of x, y, z, and w.

9. $\log_a \sqrt{\frac{z^3}{xy}}$

10. $\log_a \frac{x^2}{y^3 z}$

11. $\log_a \frac{x^3 y^4}{z^5 w^9}$

Express as a single logarithm.

12. $5 \log_a x - \log_a y + \frac{1}{4} \log_a z$

13. $\log_a \frac{\sqrt{x}}{b} - \log_a \sqrt{bx}$

ANSWERS ON PAGE A-12

Given

$\log_a 2 = 0.301,$

$\log_a 5 = 0.699,$

find each of the following.

14. $\log_a 4$

15. $\log_a 10$

16. $\log_a \frac{2}{5}$

17. $\log_a \frac{5}{2}$

18. $\log_a \frac{1}{5}$

19. $\log_a \sqrt{a^3}$

20. $\log_a 5a$

21. $\log_a 16$

Simplify.

22. $\log_2 2^6$

23. $\log_{10} 10^{3.2}$

24. $\log_e e^{12}$

ANSWERS ON PAGE A-12

► **EXAMPLES** Given $\log_a 2 = 0.301$ and $\log_a 3 = 0.477$, find each of the following.

13. $\log_a 6$ $\quad \log_a 6 = \log_a (2 \cdot 3) = \log_a 2 + \log_a 3$ **Property 1**

$= 0.301 + 0.477 = 0.778$

14. $\log_a \frac{2}{3}$ $\quad \log_a \frac{2}{3} = \log_a 2 - \log_a 3$ **Property 3**

$= 0.301 - 0.477 = -0.176$

15. $\log_a 81$ $\quad \log_a 81 = \log_a 3^4 = 4 \log_a 3$ **Property 2**

$= 4(0.477) = 1.908$

16. $\log_a \frac{1}{3}$ $\quad \log_a \frac{1}{3} = \log_a 1 - \log_a 3$ **Property 3**

$= 0 - 0.477 = -0.477$

17. $\log_a \sqrt{a}$ $\quad \log_a \sqrt{a} = \log_a a^{1/2} = \frac{1}{2} \log_a a = \frac{1}{2} \cdot 1 = \frac{1}{2}$ **Property 2**

18. $\log_a 2a$ $\quad \log_a 2a = \log_a 2 + \log_a a$ **Property 1**

$= 0.301 + 1 = 1.301$

19. $\log_a 5$ *No way to find using these properties.*

$(\log_a 5 \neq \log_a 2 + \log_a 3)$

20. $\frac{\log_a 3}{\log_a 2} = \frac{\log_a 3}{\log_a 2} = \frac{0.477}{0.301} = 1.58$ *We simply divided, not using any property.* ◄

DO EXERCISES 14–21.

e The Logarithm of the Base to a Power

The final property that we will consider is as follows.

> Property 4
>
> **For any base a,**
>
> $$\log_a a^k = k.$$
>
> **The logarithm, base a, of a to a power is the power.**

A Proof of Property 4 (*Optional*). The proof involves Property 2 and the fact that $\log_a a = 1$:

$$\log_a a^k = k(\log_a a) \quad \textbf{Using Property 2}$$
$$= k \cdot 1 \quad \textbf{Using } \log_a a = 1$$
$$= k.$$

► **EXAMPLES** Simplify.

21. $\log_3 3^7 = 7$

22. $\log_{10} 10^{5.6} = 5.6$

23. $\log_e e^{-t} = -t$ ◄

DO EXERCISES 22–24.

> CAUTION! Keep in mind that, in general,
>
> $\log_a (M + N) \neq \log_a M + \log_a N,$
>
> $\log_a (M - N) \neq \log_a M - \log_a N,$
>
> $\log_a MN \neq (\log_a M)(\log_a N)$, and
>
> $\log_a (M/N) \neq (\log_a M) \div (\log_a N).$

NAME SECTION DATE

EXERCISE SET 10.3

a Express as a sum of logarithms.

1. $\log_2 (32 \cdot 8)$
2. $\log_3 (27 \cdot 81)$
3. $\log_4 (64 \cdot 16)$
4. $\log_5 (25 \cdot 125)$
5. $\log_c Bx$
6. $\log_t 5Y$

Express as a single logarithm.

7. $\log_a 6 + \log_a 70$
8. $\log_b 65 + \log_b 2$
9. $\log_c K + \log_c y$
10. $\log_t H + \log_t M$

b Express as a product.

11. $\log_a x^3$
12. $\log_b t^5$
13. $\log_c y^6$
14. $\log_{10} y^7$
15. $\log_b C^{-3}$
16. $\log_c M^{-5}$

c Express as a difference of logarithms.

17. $\log_a \frac{67}{5}$
18. $\log_t \frac{T}{7}$
19. $\log_b \frac{3}{4}$
20. $\log_a \frac{y}{x}$

Express as a single logarithm.

21. $\log_a 15 - \log_a 7$
22. $\log_b 42 - \log_b 7$

d Express in terms of logarithms.

23. $\log_a x^2y^3z$
24. $\log_a 5xy^4z^3$
25. $\log_b \frac{xy^2}{z^3}$
26. $\log_b \frac{p^2q^5}{m^4n^7}$
27. $\log_c \sqrt[3]{\frac{x^4}{y^3z^2}}$
28. $\log_a \sqrt{\frac{x^6}{p^5q^8}}$
29. $\log_a \sqrt[4]{\frac{m^8n^{12}}{a^3b^5}}$
30. $\log_a \sqrt{\frac{a^6b^8}{a^2b^5}}$

Express as a single logarithm and simplify if possible.

31. $\frac{2}{3} \log_a x - \frac{1}{2} \log_a y$
32. $\frac{1}{2} \log_a x + 3 \log_a y - 2 \log_a x$
33. $\log_a 2x + 3(\log_a x - \log_a y)$
34. $\log_a x^2 - 2 \log_a \sqrt{x}$

ANSWERS

1. ______
2. ______
3. ______
4. ______
5. ______
6. ______
7. ______
8. ______
9. ______
10. ______
11. ______
12. ______
13. ______
14. ______
15. ______
16. ______
17. ______
18. ______
19. ______
20. ______
21. ______
22. ______
23. ______
24. ______
25. ______
26. ______
27. ______
28. ______
29. ______
30. ______
31. ______
32. ______
33. ______
34. ______

ANSWERS

35.
36.
37.
38.
39.
40.
41.
42.
43.
44.
45.
46.
47.
48.
49.
50.
51.
52.
53.
54.
55.
56.
57.
58.
59.
60.
61.
62.
63.
64.
65.
66.
67.
68.

35. $\log_a \dfrac{a}{\sqrt{x}} - \log_a \sqrt{ax}$

36. $\log_a (x^2 - 4) - \log_a (x - 2)$

Given $\log_b 3 = 1.099$ and $\log_b 5 = 1.609$, find each of the following.

37. $\log_b 15$

38. $\log_b \dfrac{3}{5}$

39. $\log_b \dfrac{5}{3}$

40. $\log_b \dfrac{1}{3}$

41. $\log_b \dfrac{1}{5}$

42. $\log_b \sqrt{b}$

43. $\log_b \sqrt{b^3}$

44. $\log_b 3b$

45. $\log_b 5b$

46. $\log_b 9$

e Simplify.

47. $\log_t t^9$

48. $\log_p p^4$

49. $\log_e e^m$

50. $\log_Q Q^{-2}$

Solve for x.

51. $\log_3 3^4 = x$

52. $\log_5 5^7 = x$

53. $\log_e e^x = -7$

54. $\log_a a^x = 2.7$

SKILL MAINTENANCE

Compute and simplify. Express answers in the form $a + bi$, where $i^2 = -1$.

55. i^{29}

56. $(2 + i)(2 - i)$

57. $\dfrac{2 + i}{2 - i}$

58. $(7 - 8i) - (-16 + 10i)$

SYNTHESIS

Express as a single logarithm and simplify if possible.

59. $\log_a (x^8 - y^8) - \log_a (x^2 + y^2)$

60. $\log_a (x + y) + \log_a (x^2 - xy + y^2)$

Express as a sum or difference of logarithms.

61. $\log_a \sqrt{1 - s^2}$

62. $\log_a \dfrac{c - d}{\sqrt{c^2 - d^2}}$

Determine whether each is true or false.

63. $\dfrac{\log_a P}{\log_a Q} = \log_a \dfrac{P}{Q}$

64. $\dfrac{\log_a P}{\log_a Q} = \log_a P - \log_a Q$

65. $\log_a 3x = \log_a 3 + \log_a x$

66. $\log_a 3x = 3 \log_a x$

67. $\log_a (P + Q) = \log_a P + \log_a Q$

68. $\log_a x^2 = 2 \log_a x$

10.4 Finding Logarithmic Function Values on a Calculator

Before calculators became so widely available, common logarithms were extensively used in calculations. In fact, that is why logarithms were invented. Any positive number different from 1 can be used as the base of a logarithmic function. However, some numbers are easier to use than others, and there are logarithmic bases that fit into certain applications more naturally than others. Base-10 logarithms are called **common logarithms.** They are useful because they have the same base as our *commonly* used decimal system for naming numbers.

Another logarithmic base that is used a great deal today is an irrational number named e. The number e is about 2.7182818. Logarithms, base e, are called **natural logarithms.** We will consider e and natural logarithms later in this section. We first consider common logarithms.

a Common Logarithms on a Calculator

Before the invention of calculators, tables were developed to list common logarithms. Today we find common logarithms using calculators.

The abbreviation **log,** with no base written, is understood to mean logarithm base 10, or a common logarithm. Thus

$$\log 17 \quad \text{means} \quad \log_{10} 17.$$

On scientific calculators, the key for common logarithms is usually marked [LOG]. To find the common logarithm of a number, we enter that number and press the [LOG] key.

► **EXAMPLE 1** Find log 53,128.

We enter 53,128 and then press the [LOG] key. We find that

$\log 53{,}128 \approx 4.7253.$ **Rounded to four decimal places** ◄

Keep in mind that 4.7253 is the exponent used with a base of 10 to get 53,128. That is, $10^{4.7253} \approx 53{,}128$.

► **EXAMPLE 2** Find log 378,000.

$$\log 378{,}000 \approx 5.5775$$ ◄

► **EXAMPLE 3** Find log 0.000128.

We enter 0.000128 and then press the [LOG] key. We find that

$\log 0.000128 \approx -3.8928.$ **Rounded to four decimal places** ◄

► **EXAMPLE 4** Find log 0.052763.

$$\log 0.052763 \approx -1.2777$$ ◄

DO EXERCISES 1–4.

The inverse of a logarithmic function is, of course, an exponential function. The inverse of finding a logarithm is also called finding an **antilogarithm.** To find an antilogarithm, we evaluate an exponent:

$$f(x) = \log x, \qquad f^{-1}(x) = \text{antilog } x = 10^x.$$

OBJECTIVES

After finishing Section 10.4, you should be able to:

a Find logarithms and antilogarithms, base 10, using a calculator.

b Find logarithms and antilogarithms, base e, using a calculator.

c Use the change-of-base formula to find logarithms to bases other than e or 10.

FOR EXTRA HELP

Tape 19D | Tape 18A | MAC: 10 IBM: 10

Using a calculator, find each of the following.

1. log 8,021,544

2. log 0.000895

3. log 44.2

4. log 0.35357

ANSWERS ON PAGE A-12

Using a calculator, find each of the following.

5. antilog 5.1053

6. antilog 0.001832

7. antilog (-3.42)

8. antilog (-0.344567)

ANSWERS ON PAGE A-12

Generally, there is no key on a calculator marked "antilog." It is up to you to know that to find the inverse, or antilogarithm, you must use the $\boxed{10^x}$ key, if there is one. If there is no such key, then you must raise 10 to the x power using an exponential key.

▶ **EXAMPLE 5** Find antilog 2.1792.

a) *Using the* $\boxed{10^x}$ *key.* We enter 2.1792 and then press the $\boxed{10^x}$ key. We find that

$$\text{antilog } 2.1792 = 10^{2.1792} \approx 151.078.$$

b) *Using an exponential key.* We enter both 10 and 2.1792. Pressing the keys in the order appropriate for your particular calculator (you must read the instructions), you will find that

$$\text{antilog } 2.1792 = 10^{2.1792} \approx 151.078.$$ ◀

▶ **EXAMPLES**

6. antilog $(-4.678834) = 10^{-4.678834} \approx 0.00002095$
7. antilog $(-1.3) = 10^{-1.3} \approx 0.050119$
8. antilog $5.000403 = 10^{5.000403} \approx 100{,}092.84$ ◀

DO EXERCISES 5–8.

b The Base *e* and Natural Logarithms on a Calculator

The compound-interest formula, which we considered partially in Chapter 8, is

$$A = P\left(1 + \frac{r}{n}\right)^{nt},$$

where A is the amount that an initial investment P will be worth after t years at interest rate r, compounded n times per year. Suppose that \$1 is an initial investment at 100% interest for 1 year (no bank would pay this!). The formula above becomes a function A defined in terms of the number of compounding periods n:

$$A(n) = \left(1 + \frac{1}{n}\right)^n.$$

Let us find some function values. We round to six decimal places and use a calculator with a power key $\boxed{y^x}$.

n	$A(n) = \left(1 + \frac{1}{n}\right)^n$
1 (compounded annually)	\$2.00
2 (compounded semiannually)	\$2.25
3	\$2.370370
4 (compounded quarterly)	\$2.441406
5	\$2.488320
100	\$2.704814
365 (compounded daily)	\$2.714567
8760 (compounded hourly)	\$2.718121

The numbers in this table get closer and closer to a very famous number in mathematics, called e. The number e occurs in a great many applications. It may seem like a strange one to use as a logarithmic base since it is an irrational number. Its decimal representation does not terminate or repeat:

$$e \approx 2.7182818284\ldots.$$

Logarithms to the base e are called *natural logarithms.* The abbreviation **ln** is generally used with natural logarithms. Thus the symbol

$$\ln 23 \quad \text{means} \quad \log_e 23.$$

On scientific calculators, the key for the natural logarithmic function is marked [LN].

▶ **EXAMPLE 9** Find ln 3568.

We enter 3568 and then press the [LN] key. We find that

$$\ln 3568 \approx 8.1798. \quad \textbf{Rounded to four decimal places}$$ ◀

▶ **EXAMPLE 10** Find ln 2.

$$\ln 2 \approx 0.6931$$ ◀

▶ **EXAMPLE 11** Find ln 0.0007659.

We enter 0.0007659 and then press the [LN] key. We find that

$$\ln 0.0007659 \approx -7.1745.$$ ◀

▶ **EXAMPLE 12** Find ln 0.1223476.

$$\ln 0.1223476 \approx -2.1009$$ ◀

DO EXERCISES 9–12.

To find the antilogarithm base e, we use the [e^x] key, if there is one. If not, we use a power key [y^x] and an approximation for e, say 2.71828.

▶ **EXAMPLE 13** Find antilog$_e$ 2.4837.

a) *Using the* [e^x] *key.* We enter 2.4837 and then press the [e^x] key. We find that

$$\text{antilog}_e\, 2.4837 = e^{2.4837} \approx 11.9855.$$

b) *Using an exponential key.* We enter both 2.4837 and an approximate value of e, say, 2.71828. Pressing keys in the order appropriate for your calculator (you must read the instructions), you will find that

$$\text{antilog}_e\, 2.4837 = e^{2.4837} \approx 11.9855.$$ ◀

▶ **EXAMPLES**

14. $\text{antilog}_e\,(-5.6734) = e^{-5.6734} \approx 0.003436$
15. $\text{antilog}_e\,(-2) = e^{-2} \approx 0.1353$
16. $\text{antilog}_e\, 0.931 = e^{0.931} \approx 2.5370$ ◀

DO EXERCISES 13–16.

Using a calculator, find each of the following.

9. ln 65,030

10. ln 0.001042

11. ln 0.28

12. ln 1233.923

Using a calculator, find each of the following.

13. antilog$_e$ 5.1485

14. antilog$_e$ 0.05051

15. antilog$_e$ (-1.2)

16. antilog$_e$ (-4.566678)

ANSWERS ON PAGE A-12

17. Find $\log_6 7$ using common logarithms.

18. Find $\log_2 46$ using natural logarithms.

ANSWERS ON PAGE A-12

C Changing Logarithmic Bases

Most calculators give the values of both common logarithms and natural logarithms. To find a logarithm with some other base, we can use the following conversion formula.

> The Change-of-Base Formula
>
> **For any logarithmic bases a and b and any positive number M,**
>
> $$\log_b M = \frac{\log_a M}{\log_a b}.$$

A Proof of the Change-of-Base Formula. Let $x = \log_b M$. Then, writing an equivalent exponential equation, we have $b^x = M$. Next we take the logarithmic base a on both sides. This gives us

$$\log_a b^x = \log_a M.$$

By Property 2,

$$x \log_a b = \log_a M,$$

and solving for x, we obtain

$$x = \frac{\log_a M}{\log_a b}.$$

But $x = \log_b M$, so we have

$$\log_b M = \frac{\log_a M}{\log_a b},$$

which is the change-of-base formula.

▶ **EXAMPLE 17** Find $\log_4 7$ using common logarithms.

Let $a = 10$, $b = 4$, and $M = 7$. Then we substitute into the change-of-base formula:

$$\log_4 7 = \frac{\log_{10} 7}{\log_{10} 4} \qquad \textbf{Substituting}$$

$$\approx \frac{0.8451}{0.6021} \approx 1.4036.$$

To check, we use a calculator with a power key $\boxed{y^x}$ to verify that

$$4^{1.4036} \approx 7.$$ ◀

DO EXERCISE 17.

We can also use base e for a conversion.

▶ **EXAMPLE 18** Find $\log_5 29$ using natural logarithms.

Substituting e for a, 5 for b, and 29 for M, we have

$$\log_5 29 = \frac{\log_e 29}{\log_e 5} \qquad \textbf{Using the change-of-base formula}$$

$$= \frac{\ln 29}{\ln 5} \approx \frac{3.3673}{1.6094} \approx 2.0923.$$ ◀

DO EXERCISE 18.

NAME SECTION DATE

EXERCISE SET 10.4

a Use a calculator to find each of the following common logarithms and antilogarithms.

1. log 2 **2.** log 5 **3.** log 8 **4.** log 11

5. log 6.34 **6.** log 5.02 **7.** log 62.4 **8.** log 11.4

9. log 437 **10.** log 295 **11.** log 13,400 **12.** log 93,100

13. log 0.57 **14.** log 0.69 **15.** log 0.052 **16.** log 0.387

17. log 0.00621 **18.** log 0.00483 **19.** antilog 3 **20.** antilog 5

21. antilog 2.7 **22.** antilog (14.8) **23.** antilog 0.477133

24. antilog (0.06532) **25.** antilog (−0.5465) **26.** antilog (−0.3404)

27. $10^{-2.9523}$ **28.** $10^{4.8982}$

b Find each of the following logarithms and antilogarithms, base *e*, using a calculator.

29. ln 2 **30.** ln 3 **31.** ln 8

32. ln 12 **33.** ln 62 **34.** ln 30

35. ln 0.0062 **36.** ln 0.00073 **37.** $\text{antilog}_e\ 3.6052$

ANSWERS

1.
2.
3.
4.
5.
6.
7.
8.
9.
10.
11.
12.
13.
14.
15.
16.
17.
18.
19.
20.
21.
22.
23.
24.
25.
26.
27.
28.
29.
30.
31.
32.
33.
34.
35.
36.
37.

ANSWERS

38. ______
39. ______
40. ______
41. ______
42. ______
43. ______
44. ______
45. ______
46. ______
47. ______
48. ______
49. ______
50. ______
51. ______
52. ______
53. ______
54. ______
55. ______
56. ______
57. ______
58. ______
59. ______
60. ______
61. ______
62. ______
63. ______
64. ______
65. ______
66. ______
67. ______
68. ______
69. ______
70. See graph.
71. See graph.
72. ______
73. ______
74. ______
75. ______

38. antilog$_e$ 4.9312 **39.** antilog$_e$ (-6.0751) **40.** antilog$_e$ (-2.3001)

41. antilog$_e$ 0.00567 **42.** antilog$_e$ 0.01111 **43.** antilog$_e$ 23.2

44. antilog$_e$ 57 **45.** ln 3460 **46.** ln 9030

47. ln 0.0351 **48.** ln 0.00468 **49.** antilog$_e$ 7.4012

50. antilog$_e$ 6.3058 **51.** $e^{2.0325}$ **52.** $e^{-1.3783}$

C Find each of the following logarithms using the change-of-base formula.

53. $\log_6 100$ **54.** $\log_3 18$ **55.** $\log_2 10$ **56.** $\log_7 50$

57. $\log_{200} 30$ **58.** $\log_{100} 30$ **59.** $\log_{0.5} 5$ **60.** $\log_{0.1} 3$

61. $\log_2 0.2$ **62.** $\log_2 0.08$ **63.** $\log_\pi 58$ **64.** $\log_\pi 200$

SKILL MAINTENANCE

Solve for x.

65. $ax^2 - b = 0$ **66.** $ax^2 - bx = 0$

Solve.

67. $x^{1/2} - 6x^{1/4} + 8 = 0$ **62.** $2y - 7\sqrt{y} + 3 = 0$

SYNTHESIS

69. Find a formula for converting natural logarithms to common logarithms.

70. Using function values obtained on a calculator, plot points and draw a precise graph of $y = g(x) = e^x$.

71. Using function values obtained on a calculator and values obtained in Exercise 70, plot points and draw a precise graph of $y = g^{-1}(x) = \ln x$.

Simplify.

72. $\dfrac{\log_3 8}{\log_3 5}$ **73.** $\dfrac{\log_2 47}{\log_2 16}$

Solve for x.

74. $\log 95x^2 = 3.0177$ **75.** $\dfrac{4.31}{\ln x} = \dfrac{28}{3.01}$

10.5 Solving Exponential and Logarithmic Equations

a Solving Exponential Equations

Equations with variables in exponents, such as $5^x = 12$ and $2^{7x} = 64$, are called **exponential equations.** Sometimes, as is the case with $2^{7x} = 64$, we can write each side as a power of the same number:

$$2^{7x} = 2^6.$$

Since the base is the same, 2, the exponents are the same. We can set them equal and solve:

$$7x = 6$$
$$x = \tfrac{6}{7}.$$

We use the following property.

> **For any $a > 0$, $a \neq 1$,**
>
> $a^x = a^y$ **is equivalent to** $x = y$.

▶ **EXAMPLE 1** Solve: $2^{3x-5} = 16$.

Note that $16 = 2^4$. Thus we can write each side as a power of the same number:

$$2^{3x-5} = 2^4.$$

Since the base is the same, 2, the exponents must be the same. Thus,

$$3x - 5 = 4$$
$$3x = 9$$
$$x = 3.$$

Check:

2^{3x-5}	$= 16$
$2^{3\cdot 3-5}$	16
2^{9-5}	
2^4	
16	

The solution is 3. ◀

DO EXERCISES 1 AND 2.

When it does not seem possible to write each side as a power of the same base, we can take the common or natural logarithm on each side and then use Property 2.

▶ **EXAMPLE 2** Solve: $5^x = 12$.

$$5^x = 12$$
$$\log 5^x = 12 \qquad \text{Taking the common logarithm on both sides}$$
$$x \log 5 = \log 12 \qquad \text{Property 2}$$
$$x = \frac{\log 12}{\log 5}$$

← CAUTION! This is not $\log 12 - \log 5$!

OBJECTIVES

After finishing Section 10.5, you should be able to:

a Solve exponential equations.

b Solve logarithmic equations.

FOR EXTRA HELP

Tape 19E　Tape 18A　MAC: 10 IBM: 10

Solve.

1. $3^{2x} = 9$

2. $4^{2x-3} = 64$

ANSWERS ON PAGE A-12

3. Solve: $7^x = 20$.

4. Solve: $e^{0.3t} = 80$.

5. Solve: $\log_5 x = 2$.

ANSWERS ON PAGE A-12

This is an exact answer. We cannot simplify further, but we can approximate using a calculator:

$$x = \frac{\log 12}{\log 5} \approx \frac{1.0792}{0.6990} \approx 1.5439.$$

You can check this answer by finding $5^{1.5439}$ using a $\boxed{y^x}$ key on a calculator. You get an answer close to 12. ◀

DO EXERCISE 3.

If the base is e, we can make our work easier by taking the logarithm base e on both sides.

▶ **EXAMPLE 3** Solve: $e^{0.06t} = 1500$.

We take the natural logarithm on both sides:

$$\begin{aligned} e^{0.06t} &= 1500 \\ \ln e^{0.06t} &= \ln 1500 && \text{Taking ln on both sides} \\ 0.06t &= \ln 1500 && \text{Here we use Property 4: } \log_a a^k = k. \\ t &= \frac{\ln 1500}{0.06}. \end{aligned}$$

We can approximate using a calculator:

$$t = \frac{\ln 1500}{0.06} \approx \frac{7.3132}{0.06} \approx 121.89.$$ ◀

DO EXERCISE 4.

b Solving Logarithmic Equations

Equations containing logarithmic expressions are called **logarithmic equations.** We solved some logarithmic equations in Section 10.2 by converting to equivalent exponential equations.

▶ **EXAMPLE 4** Solve: $\log_2 x = 3$.

We obtain an equivalent exponential expression:

$$x = 2^3$$
$$x = 8.$$

The solution is 8. ◀

DO EXERCISE 5.

To solve a logarithmic equation, first try to obtain a single logarithmic expression on one side and then write an equivalent exponential equation.

▶ **EXAMPLE 5** Solve: $\log_4 (8x - 6) = 3$.

We already have a single logarithmic expression, so we write an equivalent exponential equation:

$$8x - 6 = 4^3 \quad \text{Writing an equivalent exponential equation}$$
$$8x - 6 = 64$$
$$8x = 70$$
$$x = \tfrac{70}{8}, \text{ or } \tfrac{35}{4}$$

Check:

$\log_4 (8x - 6) = 3$	
$\log_4 (8 \cdot \frac{35}{4} - 6)$	3
$\log_4 (70 - 6)$	
$\log_4 64$	
3	

The solution is $\frac{35}{4}$. ◀

DO EXERCISE 6.

6. Solve: $\log_3 (5x + 7) = 2$.

▶ **EXAMPLE 6** Solve: $\log x + \log (x - 3) = 1$.

Here we have common logarithms. It will help us follow the solution to write in the 10's.

$$\log_{10} x + \log_{10} (x - 3) = 1$$
$$\log_{10} [x(x - 3)] = 1 \quad \text{Using Property 1 to obtain a single logarithm}$$
$$x(x - 3) = 10^1 \quad \text{Writing an equivalent exponential expression}$$
$$x^2 - 3x = 10$$
$$x^2 - 3x - 10 = 0$$
$$(x + 2)(x - 5) = 0 \quad \text{Factoring}$$
$$x + 2 = 0 \quad \textit{or} \quad x - 5 = 0 \quad \text{Using the principle of zero products}$$
$$x = -2 \quad \textit{or} \quad x = 5$$

Check: For -2:

$\log x + \log (x - 3) = 1$	
$\log (-2) + \log (-2 - 3)$	1

The number -2 does *not* check because negative numbers do not have logarithms.

For 5:

$\log x + \log (x - 3) = 1$	
$\log 5 + \log (5 - 3)$	1
$\log 5 + \log 2$	
$\log (5 \cdot 2)$	
$\log 10$	
1	

The solution is 5. ◀

DO EXERCISE 7.

7. Solve: $\log x + \log (x + 3) = 1$.

ANSWERS ON PAGE A-12

8. Solve:

$\log_3 (2x - 1) - \log_3 (x - 4) = 2.$

▶ **EXAMPLE 7** Solve: $\log_2 (x + 7) - \log_2 (x - 7) = 3.$

$$\log_2 (x + 7) - \log_2 (x - 7) = 3$$

$$\log_2 \frac{x+7}{x-7} = 3 \quad \textbf{Using Property 3 to obtain a single logarithm}$$

$$\frac{x+7}{x-7} = 2^3 \quad \textbf{Writing an equivalent exponential expression}$$

$$\frac{x+7}{x-7} = 8$$

$$x + 7 = 8(x - 7) \quad \textbf{Multiplying by the LCM, } x - 7$$

$$x + 7 = 8x - 56 \quad \textbf{Using a distributive law}$$

$$63 = 7x$$

$$\frac{63}{7} = x$$

$$9 = x$$

Check:

$$\begin{array}{r|l} \log_2 (x + 7) - \log_2 (x - 7) & = 3 \\ \hline \log_2 (9 + 7) - \log_2 (9 - 7) & 3 \\ \log_2 16 - \log_2 2 & \\ \log_2 \frac{16}{2} & \\ \log_2 8 & \\ 3 & \end{array}$$

The solution is 9. ◀

DO EXERCISE 8.

ANSWER ON PAGE A-12

NAME SECTION DATE

EXERCISE SET 10.5

a Solve.

1. $2^x = 8$ **2.** $3^x = 81$ **3.** $4^x = 256$ **4.** $5^x = 125$

5. $2^{2x} = 32$ **6.** $4^{3x} = 64$ **7.** $3^{5x} = 27$ **8.** $5^{7x} = 625$

9. $2^x = 9$ **10.** $2^x = 30$ **11.** $2^x = 10$ **12.** $2^x = 33$

13. $5^{4x-7} = 125$ **14.** $4^{3x+5} = 16$ **15.** $3^{x^2} \cdot 3^{4x} = \frac{1}{27}$ **16.** $3^{5x} \cdot 9^{x^2} = 27$

17. $4^x = 7$ **18.** $8^x = 10$ **19.** $e^t = 100$ **20.** $e^t = 1000$

21. $e^{-t} = 0.1$ **22.** $e^{-t} = 0.01$ **23.** $e^{-0.02t} = 0.06$ **24.** $e^{0.07t} = 2$

25. $2^x = 3^{x-1}$ **26.** $3^{x+2} = 5^{x-1}$ **27.** $(2.8)^x = 41$ **28.** $(3.4)^x = 80$

b Solve.

29. $\log_3 x = 3$ **30.** $\log_5 x = 4$ **31.** $\log_2 x = -3$ **32.** $\log_4 x = \frac{1}{2}$

33. $\log x = 1$ **34.** $\log x = 3$ **35.** $\log x = -2$ **36.** $\log x = -3$

37. $\ln x = 2$ **38.** $\ln x = 1$ **39.** $\ln x = -1$ **40.** $\ln x = -3$

ANSWERS

1. ______
2. ______
3. ______
4. ______
5. ______
6. ______
7. ______
8. ______
9. ______
10. ______
11. ______
12. ______
13. ______
14. ______
15. ______
16. ______
17. ______
18. ______
19. ______
20. ______
21. ______
22. ______
23. ______
24. ______
25. ______
26. ______
27. ______
28. ______
29. ______
30. ______
31. ______
32. ______
33. ______
34. ______
35. ______
36. ______
37. ______
38. ______
39. ______
40. ______

ANSWERS

41. ______
42. ______
43. ______
44. ______
45. ______
46. ______
47. ______
48. ______
49. ______
50. ______
51. ______
52. ______
53. ______
54. ______
55. ______
56. ______
57. ______
58. ______
59. ______
60. ______
61. ______
62. ______
63. ______
64. ______
65. ______
66. ______
67. ______
68. ______
69. ______
70. ______
71. ______
72. ______
73. ______

41. $\log_5 (2x - 7) = 3$

42. $\log_2 (7 - 6x) = 5$

43. $\log x + \log (x - 9) = 1$

44. $\log x + \log (x + 9) = 1$

45. $\log x - \log (x + 3) = -1$

46. $\log (x + 9) - \log x = 1$

47. $\log_2 (x + 1) + \log_2 (x - 1) = 3$

48. $\log_2 x + \log_2 (x - 2) = 3$

49. $\log_4 (x + 6) - \log_4 x = 2$

50. $\log_4 (x + 3) - \log_4 (x - 5) = 2$

51. $\log_4 (x + 3) + \log_4 (x - 3) = 2$

52. $\log_5 (x + 4) + \log_5 (x - 4) = 2$

53. $\log_3 (2x - 6) - \log_3 (x + 4) = 2$

54. $\log_4 (2 + x) - \log_4 (3 - 5x) = 3$

SKILL MAINTENANCE

Solve.

55. $x^2 + y^2 = 25,$
$y - x = 1$

56. $2x + 3y = 15,$
$xy = 9$

57. $2x^2 + 1 = y^2,$
$2y^2 + x^2 = 22$

58. $3x^2 + 2y^2 = 77,$
$y^2 + 2x^2 = 43$

SYNTHESIS

Solve.

59. $8^x = 16^{3x+9}$

60. $27^x = 81^{2x-3}$

61. $\log_6 (\log_2 x) = 0$

62. $\log_x (\log_3 27) = 3$

63. $\log_5 \sqrt{x^2 - 9} = 1$

64. $x \log \frac{1}{8} = \log 8$

65. $\log (\log x) = 5$

66. $2^{x^2+4x} = \frac{1}{8}$

67. $\log x^2 = (\log x)^2$

68. $\log_5 |x| = 4$

69. $\log_a a^{x^2+4x} = 21$

70. $\log_3 |5x - 7| = 2$

71. $3^{2x} - 8 \cdot 3^x + 15 = 0$

72. $\sqrt{x} \cdot \sqrt[3]{x} \cdot \sqrt[4]{x} \cdot \sqrt[5]{x} = 146$

73. If $x = (\log_{125} 5)^{\log_5 125}$, what is the value of $\log_3 x$?

10.6 Applications of Exponential and Logarithmic Functions

a We now consider applications of exponential and logarithmic functions. A calculator with logarithmic and power keys would be most helpful.

► **EXAMPLE 1** *Interest compounded annually.* Suppose that \$100,000 is invested at 8% interest, compounded annually. In t years, it will grow to the amount A given by the function

$A(t) = \$100{,}000(1.08)^t$. (See Example 6 in Section 10.1.)

a) After what amount of time will there be \$500,000 in the account?

b) Let T = the amount of time it takes for the \$100,000 to double itself; T is called the **doubling time.** Find the doubling time.

We solve as follows.

a) We set $A(t) = \$500{,}000$ and solve for t:

$$500{,}000 = 100{,}000(1.08)^t$$

$$\frac{500{,}000}{100{,}000} = (1.08)^t$$

$$5 = (1.08)^t$$

$$\log 5 = \log (1.08)^t \quad \text{Taking the common logarithm on both sides}$$

$$\log 5 = t \log 1.08 \quad \text{Property 2}$$

$$\frac{\log 5}{\log 1.08} = t.$$

We cannot simplify further, but we can approximate using a calculator:

$$t = \frac{\log 5}{\log 1.08} \approx \frac{0.69897}{0.03342} \approx 20.9.$$

It will take about 20.9 years for the \$100,000 to grow to \$500,000.

> *Calculator note.* When doing a calculation like this on your calculator, it is not necessary to stop and round the approximate values of the logarithms. Just find the numbers and divide. Answers to the exercises will be found that way. You may note some slight variation in the last one or two decimal places if you round as you go. You might check with your instructor regarding such variation.

b) To find the doubling time, we set $A(t) = \$200{,}000$ and $t = T$ and then solve for T:

$$200{,}000 = 100{,}000(1.08)^T$$

$$2 = (1.08)^T$$

$$\log 2 = \log (1.08)^T \quad \text{Taking the common logarithm on both sides}$$

$$\log 2 = T \log 1.08 \quad \text{Property 2}$$

$$T = \frac{\log 2}{\log 1.08} \approx \frac{0.30103}{0.03342} \approx 9.0.$$

The doubling time is about 9 years. ◄

DO EXERCISE 1.

OBJECTIVE

After finishing Section 10.6, you should be able to:

a Solve problems involving applications of exponential and logarithmic functions.

FOR EXTRA HELP

Tape 19F

Tape 18B

MAC: 10
IBM: 10

1. Suppose that \$40,000 is invested at 7% interest, compounded annually.

   a) After what amount of time will there be \$250,000 in the account?

   b) Find the doubling time.

ANSWERS ON PAGE A-12

2. The population of certain cities is given below. Find the walking speed of people in each city.

a) New York: population 7,900,000

b) Key West, Florida: 29,600

c) Albuquerque, New Mexico: 290,000

3. What will the population of the United States be in 1998? in 2005?

ANSWERS ON PAGE A-12

▶ **EXAMPLE 2** *Psychology: Walking speed.* The psychologists Bornstein and Bornstein found in a study that the average walking speed R of a person living in a city of population P, in thousands, is given by the function

$$R(P) = 0.37 \ln P + 0.05,$$

where R is in feet per second. The population of Seattle, Washington, is 531,000. Find the average walking speed of people living in Seattle.

We substitute 531 for P, since P is in thousands:

$$R(531) = 0.37 \ln (531) + 0.05 \qquad \textbf{Substituting}$$
$$\approx 0.37(6.2748) + 0.05 \qquad \textbf{Finding the natural logarithm on a calculator}$$
$$\approx 2.4 \text{ ft/sec.}$$

The average walking speed of people living in Seattle is 2.4 ft/sec. ◀

DO EXERCISE 2.

Growth

The equation

$$P(t) = P_0 e^{kt}$$

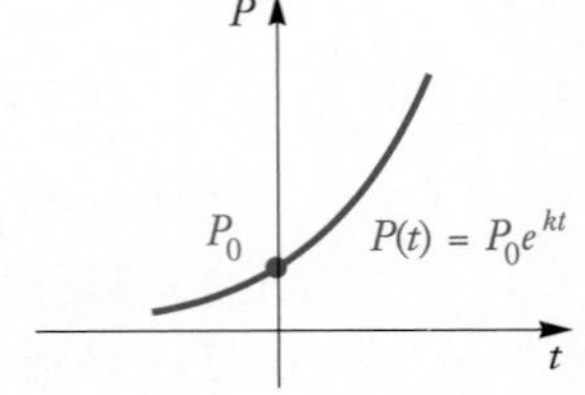

is an effective model of many kinds of population growth, whether it be a population of people or a population of money. In this equation, P_0 is the number of people at time 0, P is the population after time t, and k is a positive constant that depends on the situation. The constant k is often called the **exponential growth rate.**

▶ **EXAMPLE 3** *Growth of the United States.* In 1985 the population of the United States was 234 million and the exponential growth rate was 0.8% per year.

a) Find the exponential growth function.
b) What will the population be in 2000?

We solve as follows.

a) At $t = 0$ (1985), the population was 234 million. We substitute 234 for P_0 and 0.8%, or 0.008, for k to obtain the exponential growth function

$$P(t) = 234e^{0.008t}.$$

b) In 2000, we have $t = 15$. That is, 15 years have passed since 1985. To find the population in 2000, we substitute 15 for t:

$$P(15) = 234e^{0.008(15)} \qquad \textbf{Substituting 15 for } t$$
$$= 234e^{0.12}$$
$$\approx 234(1.1275) \qquad \textbf{Finding } e^{0.12} \textbf{ using a calculator}$$
$$\approx 264 \text{ million.}$$

The population of the United States in 2000 will be about 264 million. ◀

DO EXERCISE 3.

► **EXAMPLE 4** *The cost of a first-class postage stamp.* The cost of a first-class postage stamp became 3¢ in 1932 and the exponential growth rate was 3.8% per year. The exponential growth function is found by substituting 3 for P_0 and 3.8%, or 0.038, for k to obtain the function

$$P(t) = 3e^{0.038t}$$

a) The cost of first-class postage increased to 25¢ in 1988. Use the given function to see what the predicted cost is for 1988 and compare with the actual cost.

b) What will the cost of a first-class postage stamp be in 2000?

c) When will the cost of a first-class postage stamp be $1.00?

We solve as follows.

a) At $t = 0$ (1932), the cost of a stamp was 3¢. The year 1988 is 56 years from 1932; thus we have $t = 56$. To find the cost in 1988, according to the function, we substitute 56 for t:

$$\begin{aligned} P(56) &= 3e^{0.038(56)} && \textbf{Substituting 56 for } t \\ &= 3e^{2.128} \\ &\approx 3(8.3981) && \textbf{Finding } 3^{2.128} \textbf{ using a calculator} \\ &\approx 25.2¢. \end{aligned}$$

The function seems to be a fairly accurate predictor of first-class postage. The actual amount is 25¢.

b) In the year 2000, we have $t = 68$. That is, 68 yr have passed. To find the cost in 2000, we substitute 68 for t:

$$\begin{aligned} P(68) &= 3e^{0.038(68)} && \textbf{Substituting 68 for } t \\ &= 3e^{2.584} \\ &\approx 3(13.2500) && \textbf{Finding } e^{2.584} \textbf{ using a calculator} \\ &\approx 40. \end{aligned}$$

The cost of a first-class stamp in the year 2000 will be about 40¢.

c) To find when the cost will be $1.00, we substitute 100¢ for $P(t)$ and solve for t:

$$\begin{aligned} 100 &= 3e^{0.038t} \\ \tfrac{100}{3} &= e^{0.038t} \\ \ln \tfrac{100}{3} &= \ln e^{0.038t} && \textbf{Taking the natural logarithm on both sides} \\ \ln 33.3333 &\approx 0.038t && \textbf{Using Property 3 and Property 4, recalling that } \ln e^{0.038t} = \log_e e^{0.038t} \\ 3.5066 &\approx 0.038t \\ 92 &\approx t. \end{aligned}$$

In 92 years from 1932, or in 2024, we predict that the cost of first-class postage will be $1. ◄

DO EXERCISE 4.

4. **a)** What will the cost of a first-class stamp be in 1998?

b) When will a first-class stamp cost $2?

ANSWERS ON PAGE A-12

5. Predict the number of heart transplants in 2000.

In order to fit an exponential growth function to a situation, we must determine P_0 and k from given data. Then we can make predictions.

▶ **EXAMPLE 5** *Medicine: Heart transplants.* In 1967 Dr. Christian Barnard of South Africa stunned the world by performing the first heart transplant. There was 1 transplant in 1967. In 1987, there were 1418 such transplants.

a) Find an exponential growth function that fits the data.

b) Use the function to predict the number of heart transplants in 1995.

We solve as follows.

a) The exponential growth function is

$$N(t) = N_0 e^{kt}.$$

We assume that $t = 0$ corresponds to 1967 and that $N(0) = N_0 = N_0 e^{k(0)} = 1$. Then the growth function is

$$N(t) = e^{kt}.$$

To find k, we can use the fact that at $t = 20$ (1987), the number of transplants was 1418. Then we substitute and solve for k as follows:

$$1418 = e^{k(20)}$$

$$1418 = e^{20k}$$

$$\ln 1418 = \ln e^{20k}$$ **Taking the natural logarithm on both sides**

$$\ln 1418 = 20k$$ ***Remember:*** **$\log_a a^k = k$, so $\ln e^{20k} = 20k$, since $\ln e^{20k} = \log_e e^{20k}$.**

$$\frac{\ln 1418}{20} = k$$

$$\frac{7.2570}{20} \approx k$$ **Finding ln 1418 using a calculator**

$$0.363 \approx k.$$

Thus the exponential growth function is

$$N(t) = e^{0.363t}.$$

b) The year 1995 is 28 years from 1967. We let $t = 28$ and find $N(28)$:

$$N(28) = e^{0.363(28)} \approx 25{,}952.$$

Thus, according to the exponential growth function, there will be about 25,952 heart transplants in 1995. ◀

DO EXERCISE 5.

ANSWER ON PAGE A-12

Decay

The function

$$P(t) = P_0 e^{-kt}$$

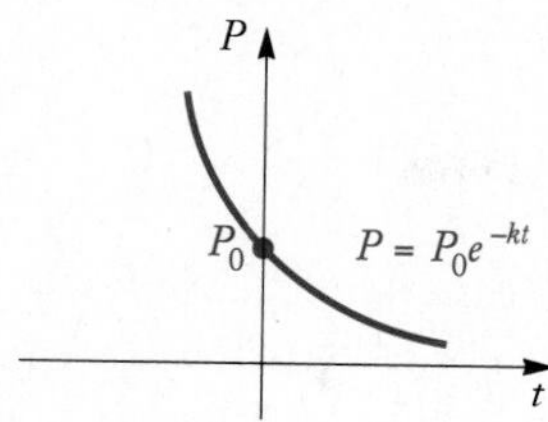

is an effective model of the decline, or decay, of a population. An example is the decay of a radioactive substance. Here P_0 is the amount of the substance at time $t = 0$, P is the amount of the substance left after time t, and k is a positive constant that depends on the situation. The constant k is called the **decay rate.** The **half-life** of bismuth is 5 days. This means that half of an amount of bismuth will cease to become radioactive in 5 days.

▶ **EXAMPLE 6** *Carbon dating.* The radioactive element carbon-14 has a half-life of 5750 years. The percentage of carbon-14 present in the remains of animal bones can be used to determine age. How old is an animal bone that has lost 40% of its carbon-14?

We first find k. To do so, we use the concept of half-life. When $t = 5750$ (half-life), P will be half of P_0. Then

$$\frac{1}{2}P_0 = P_0 e^{-k(5750)}, \quad \text{or} \quad \frac{1}{2} = e^{-5750k}.$$

We take the natural logarithm on both sides:

$$\ln \frac{1}{2} = \ln e^{-5750k} = -5750k.$$

Then

$$k = \frac{\ln 0.5}{-5750} \approx \frac{-0.6931}{-5750} \approx 0.00012.$$

Now we have the function

$$P(t) = P_0 e^{-0.00012t}.$$

(*Note:* This equation can be used for any subsequent carbon-dating problem.) If an animal bone has lost 40% of its carbon-14 from an initial amount P_0, then $60\%(P_0)$ is the amount present. To find the age t of the bone, we solve the following equation for t:

$$\begin{aligned} 60\%P_0 &= P_0 e^{-0.00012t} \\ 0.6 &= e^{-0.00012t} \\ \ln 0.6 &= \ln e^{-0.00012t} \\ -0.5108 &= -0.00012t \\ t &= \frac{-0.5108}{-0.00012} \approx 4257. \end{aligned}$$

The animal bone is about 4257 years old.

DO EXERCISE 6.

6. An animal bone has lost 30% of its carbon-14. How old is the bone?

ANSWER ON PAGE A-12

❖ SIDELIGHTS

Careers Involving Mathematics

You are about to finish this course in intermediate algebra. If you have done well, you might be considering a career in mathematics or one that involves mathematics. If either is the case, the following information may be valuable to you.

CAREERS INVOLVING MATHEMATICS The following is the result of a survey conducted by *The Jobs Related Almanac*, published by the American References Inc. of Chicago. It used the criteria of salary, stress, work environment, outlook, security, and physical demands to rate the desirability of 250 jobs. The top 10 of the 250 jobs listed were:

1. Actuary
2. Computer programmer
3. Computer systems analyst
4. Mathematician
5. Statistician

(The top 5 involve mathematics.)

6. Hospital administrator
7. Industrial engineer
8. Physicist
9. Astrologer
10. Paralegal.

Two things are interesting to note. First, the top five rated professions involve a heavy use of mathematics. The top, actuary, involves the application of mathematics to insurance. The second point of interest is that choices like doctor, lawyer, and astronaut are *not* in the top ten.

Perhaps you might be interested in a career in teaching mathematics. This profession will be expanding increasingly in the next ten years. The field of mathematics will need well-qualified mathematics teachers in all areas from elementary to junior high to secondary to two-year college to college instruction. Some questions you might ask yourself in making a decision about a career in mathematics teaching are the following.

1. Do you find yourself carefully observing the strengths and weaknesses of your teachers?
2. Are you deeply interested in mathematics?
3. Are you interested in the ways of learning? If a student is struggling with a topic, would it be challenging to you to discover two or three other ways to present the material so that the student might understand?
4. Are you able to put yourself in the place of the students in order to help them be successful in learning mathematics?

If you are interested in a career involving mathematics, the next courses you would take are *precalculus algebra and trigonometry* and *calculus*. You might want to seek out a counselor in the mathematics department at your college for further assistance.

WHAT KIND OF SALARIES ARE THERE IN VARIOUS FIELDS? The College Placement Council published the following comparisons of the average salaries of graduating students with bachelors degrees who were taking the following jobs:

Subject area	Annual salary
All engineering	\$27,800
Computer science	\$26,400
Mathematics	\$25,900
Sciences other than math and computer science	\$22,200
Humanities and social science	\$21,800
Accounting	\$21,700
All business	\$21,300

Many people choose to go on to earn a masters degree. Here are salaries in the same fields for students just graduating with a masters degree:

Subject area	Annual salary
All engineering	\$34,000
Computer science	\$33,800
Mathematics	\$27,900
Sciences other than math and computer science	\$27,400
Humanities and social science	\$22,300
Accounting	\$26,000
Business administration	\$30,700

NAME SECTION DATE

EXERCISE SET 10.6

ANSWERS

a Solve.

1. *Interest compounded annually.* Suppose that \$50,000 is invested at 6% interest, compounded annually. After time t, in years, it grows to the amount A given by the function

$$A(t) = \$50{,}000(1.06)^t.$$

   a) After what amount of time will there be \$450,000 in the account?
   b) Find the doubling time.

2. *Turkey consumption.* The amount of turkey consumed by each person in this country is increasing exponentially. Assume that $t = 0$ corresponds to 1937. The amount of turkey, in pounds per person, consumed t years after 1937 is given by the function

$$N(t) = 2.3(3)^{0.033t}.$$

   a) After what amount of time will each person consume 20 lb of turkey?
   b) What is the doubling time on the consumption of turkey?

3. *Recycling aluminum cans.* It is known that one fourth of all aluminum cans distributed will be recycled each year. A beverage company distributes 250,000 cans. The number still in use after time t, in years, is given by the function

$$N(t) = 250{,}000(\tfrac{1}{4})^t.$$

   a) After what amount of time will 60,000 cans still be in use?
   b) After what amount of time will only 10 cans still be in use?

4. *Salvage value.* An office machine is purchased for \$5200. Its value each year is about 80% of the value the preceding year. Its value after t years is given by the exponential function

$$V(t) = \$5200(0.8)^t.$$

   a) After what amount of time will the salvage value be \$1200?
   b) After what amount of time will the salvage value be half its original value? This is known as *half-life*.

1. a) ______

b) ______

2. a) ______

b) ______

3. a) ______

b) ______

4. a) ______

b) ______

ANSWERS

5. a) ______

b) ______

6. a) ______

b) ______

c) ______

7. ______

8. ______

9. ______

10. ______

11. a) ______

b) ______

c) ______

12. a) ______

b) ______

5. *Compact discs.* The number of compact discs purchased each year is increasing exponentially. The number N purchased, in millions, is given by

$$N(t) = 7.5(6)^{0.5t},$$

where $t = 0$ corresponds to 1985, $t = 1$ corresponds to 1986, and so on, t being the number of years after 1985.

a) After what amount of time will 1 billion compact discs be sold in a year?

b) What is the doubling time on the sale of compact discs?

6. *Business: Advertising.* A model for advertising response is given by

$$N(a) = 1000 + 200 \log a, \quad a \geq 1,$$

where $N(a)$ = the number of units sold, and a = the amount spent on advertising, in thousands of dollars.

a) How many units were sold after spending \$1000 ($a = 1$) on advertising?

b) How many units were sold after spending \$5000?

c) How much must be spent in order to sell 1276 units?

Various cities and their populations are given below. Find the walking speed of people in each city. (See Example 2.)

7. Lincoln, Nebraska: 175,000

8. Pierre, South Dakota: 10,300

9. Rome, New York: 50,400

10. Lansing, Michigan: 134,000

11. *World population growth.* The population of the world passed the 5.0 billion mark in 1987. The exponential growth rate was 2.8% per year.

a) Find the exponential growth function.

b) Predict the population of the world in 1996 and in 2000.

c) When will the world population be 6.0 billion?

12. *Consumer price index.* The consumer price index is often in the news. It compares the cost of goods and services over various years, where \$100 worth of goods and services in 1967 is used as a base (P_0). Although rates vary for different years, an average exponential growth rate is 6%.

a) Find the exponential function for the consumer price index.

b) Goods and services that cost \$100 in 1967 will cost how much in 2000?

13. *Cost of a Hershey bar.* The cost of a Hershey chocolate bar in 1962 was 5¢ and is increasing at an exponential growth rate of 9.7%.

a) Find an exponential function describing the growth of the cost of a Hershey bar.

b) What will a Hershey bar cost in the year 2010?

c) When will a Hershey bar cost $5?

d) What is the doubling time of the cost of a Hershey bar?

14. *Cost of a 60-second commercial during the Super Bowl.* The cost of a 60-second commercial during the 1967 Super Bowl was $80,000. In 1988 it was $1,350,000.

a) Find an exponential growth function for the cost of a Super Bowl commercial. Assume that $C_0 = \$80$ thousand. That is, at $t = 0$ (1967), $C = \$80$ (in thousands).

b) What will the cost of a 60-second commercial be in 1995?

c) When will the cost of a commercial be $3,000,000?

d) What is the doubling time for the cost of a Super Bowl commercial?

15. *Cost of a double-dip ice cream cone.* In 1970 the cost of a double-dip ice cream cone was 52¢. In 1978 it was 66¢. Assuming the exponential model:

a) Find the value k ($P_0 = 52$). Write the exponential growth function.

b) Estimate the cost of a cone in 1994.

c) After what period of time will the cost of a cone be twice that of 1970?

d) When will the cost of a cone be $3?

16. *Interest compounded continuously.* Suppose that P_0 is invested in a savings account in which interest is compounded continuously at 9% per year. That is, the balance $P(t)$ after time t, in years, is

$$P(t) = P_0 e^{kt}.$$

a) Find the exponential function for $P(t)$ in terms of P_0 and 0.09.

b) Suppose that $1000 is invested. What is the balance after 1 year? after 2 years?

c) When will an investment of $1000 double itself?

ANSWERS

13. a) ______

b) ______

c) ______

d) ______

14. a) ______

b) ______

c) ______

d) ______

15. a) ______

b) ______

c) ______

d) ______

16. a) ______

b) ______

c) ______

ANSWERS

17. a) ______

b) ______

c) ______

d) ______

18. a) ______

b) ______

c) ______

d) ______

19. ______

20. ______

21. a) ______

b) ______

17. *Value of a Van Gogh painting.* The Van Gogh painting "Irises," shown below, sold for \$84,000 in 1947, but was sold again for \$53,900,000 in 1987. Assume that the growth in the value V of the painting was exponential.

a) Find the value k and determine the exponential growth function, assuming $P_0 = 84{,}000$.
b) Estimate the value of the painting after 50 years.
c) What is the doubling time for the value of the painting?
d) After what amount of time will the value of the painting be \$1 billion?

Van Gogh's "Irises," a 28-by-32-inch oil on canvas.

18. *Exponential growth of the value of a baseball card.* The collecting of baseball cards and other memorabilia has become a profitable hobby. The card shown below contains a photograph of Dale Murphy in his rookie season of 1977. The value of that card in 1983 was \$17.50. Its value in 1987 was \$48.00. Assume that the value of the card has grown exponentially.

a) Find the value k and determine the exponential growth function, assuming $V_0 = \$17.50$.
b) Estimate the value of the card in 1995 and in 2000.
c) What is the doubling time for the value of the card?
d) After what amount of time will the value of the card be \$2000?

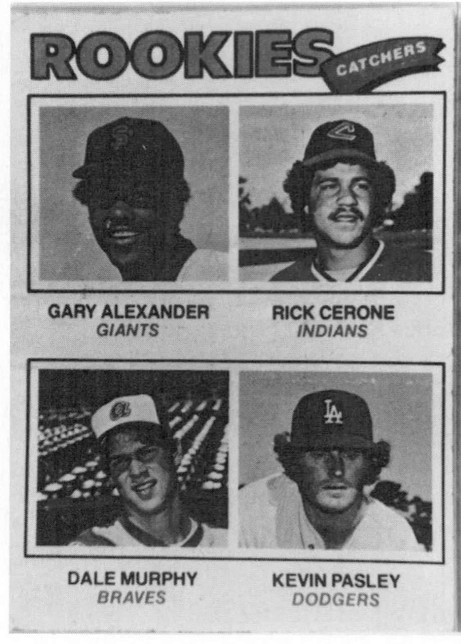

19. An ivory tusk has lost 20% of its carbon-14. How old is the tusk?

20. A piece of wood has lost 10% of its carbon-14. How old is the piece of wood?

21. *Salvage value.* A business estimates that the salvage value V of a piece of machinery after t years is given by

$$V(t) = \$28{,}000e^{-t}.$$

a) What did the machinery cost initially?
b) What is the salvage value after 2 years?

SUMMARY AND REVIEW: CHAPTER 10

IMPORTANT PROPERTIES AND FORMULAS

Exponential Function: $f(x) = a^x$

Interest Compounded Annually: $A = P(1 + r)^t$

Interest Compounded n Times per Year: $A = P\left(1 + \frac{r}{n}\right)^{nt}$

Definition of Logarithms: $y = \log_a x$ is that number y such that $x = a^y$, where $x > 0$ and a is a positive constant other than 1.

Properties of Logarithms:

$\log M = \log_{10} M$, $\quad \log_a 1 = 0$, $\quad \ln M = \log_e M$, $\quad \log_a a = 1$,

$\log_a MN = \log_a M + \log_a N$, $\quad \log_a a^k = k$, $\quad \log_a M^k = k \cdot \log_a M$, $\quad \log_b M = \dfrac{\log_a M}{\log_a b}$,

$\log_a \dfrac{M}{N} = \log_a M - \log_a N$, $\quad e \approx 2.1782818284\ldots$

Growth: $P(t) = P_0 e^{kt}$

Decay: $P(t) = P_0 e^{-kt}$

Carbon Dating: $P(t) = P_0 e^{-0.00012t}$

REVIEW EXERCISES

The review sections and objectives to be tested in addition to the material in this chapter are [7.9b, c, d, e], [8.4a], [8.5a], and [9.4a].

Graph.

1. $f(x) = 3^{x-2}$ **2.** $x = 3^{y-2}$ **3.** $y = \log_3 x$ **4.** $f(x) = \log_{1/2} x$

Convert to an exponential equation.

5. $\log_4 16 = x$ **6.** $\log_{10} 2 = 0.3010$ **7.** $\log_{1/2} 8 = -3$ **8.** $\log_{16} 8 = \frac{3}{4}$

Convert to a logarithmic equation.

9. $10^4 = 10{,}000$ **10.** $25^{1/2} = 5$ **11.** $7^{-2} = \frac{1}{49}$ **12.** $(2.718)^3 = 20.1$

Express in terms of logarithms of x, y, and z.

13. $\log_a x^4y^2z^3$ **14.** $\log_a \dfrac{xy}{z^2}$ **15.** $\log \sqrt[4]{\dfrac{z^2}{x^3y}}$ **16.** $\log_q \left(\dfrac{x^2y^{1/3}}{z^4}\right)$

Express as a single logarithm.

17. $\log_a 8 + \log_a 15$ **18.** $\log_a 72 - \log_a 12$ **19.** $\frac{1}{2}\log a - \log b - 2\log c$ **20.** $\frac{1}{3}[\log_a x - 2\log_a y]$

Simplify.

21. $\log_m m$ **22.** $\log_m 1$ **23.** $\log_m m^{17}$

Given $\log_a 2 = 1.8301$ and $\log_a 7 = 5.0999$, find each of the following.

24. $\log_a 14$ **25.** $\log_a \frac{2}{7}$ **26.** $\log_a 28$ **27.** $\log_a 3.5$ **28.** $\log_a \sqrt{7}$

Find each of the following using a calculator.

29. log 0.00627 **30.** log 72,800,000 **31.** antilog 4.4742 **32.** antilog (−1.4425)

33. antilog 2.3294 **34.** log 0.004937 **35.** log 394,900 **36.** antilog (−6.7889)

37. ln 23,912.2 **38.** ln 0.06774 **39.** $\text{antilog}_e\,(-10.56)$ **40.** $\text{antilog}_e\,45$

Find each of the following logarithms using the change-of-base formula.

41. $\log_5 2$

42. $\log_{12} 70$

Solve.

43. $\log_3 x = -2$

44. $\log_x 32 = 5$

45. $\log x = -4$

46. $\ln x = 2$

47. $4^{2x-5} = 16$

48. $4^x = 8.3$

49. $\log_4 16 = x$

50. $\log(x^2 - 9) - \log(x - 3) = 1$

51. $\log_4 x + \log_4 (x - 6) = 2$

52. $\log x + \log(x - 15) = 2$

53. $\log_3 (x - 4) = 3 - \log_3 (x + 4)$

54. *Forgetting.* In a business class, students were tested at the end of the course on a final examination. They were tested again after 6 months. The forgetting formula was determined to be

$$S(t) = 62 - 18 \log(t + 1),$$

where t is the time, in months, after taking the first test.

a) The average score when they first took the test is when $t = 0$. Find their average score on the final exam.

b) What was the average score after 6 months?

c) After what time was the average score 34?

55. *Cost of a prime-rib dinner.* The average cost C of a prime rib dinner was \$4.65 in 1962. In 1986 it was \$15.81. Assuming the exponential growth function:

a) Find k and write the exponential growth function.

b) How much will a prime-rib dinner cost in 2002?

c) When will the average cost of a prime-rib dinner be \$20?

d) What is the doubling time?

56. How old is a skeleton that has lost 34% of its carbon-14?

SKILL MAINTENANCE

57. Solve $aT^2 + bT = Q$ for T.

58. Solve: $x^4 - 80 = 11x^2$.

Multiply.

59. $(2 - 3i)(4 + i)$

60. i^{20}

61. Divide: $\dfrac{4 - 5i}{1 + 3i}$.

62. Add: $(4 - 5i) + (1 + 3i)$.

63. Solve: $3x - 2y = 0$,
$4 - x = xy$.

SYNTHESIS

Solve.

64. $\log(\log x) = 3$

65. $3^{x^2 - 3x} = 81$

66. $5^{x+y} = 25$,
$2^{2x-y} = 64$

❖ THINKING IT THROUGH

1. Compare the following properties of exponents to the properties of logarithms:

$$a^n a^m = a^{n+m}, \qquad \frac{a^n}{a^m} = a^{n-m}, \qquad (a^n)^m = a^{nm}.$$

2. Describe the types of equations we are able to solve after studying this chapter that are not like those we studied earlier.

3. Find the logarithms, base 2, of 2, 4, 8, 64, $\frac{1}{2}$, $\frac{1}{16}$, and 1. Discuss your results.

NAME SECTION DATE

TEST: CHAPTER 10

Graph.

1. $f(x) = 2^{x+3}$

2. $f(x) = \log_3 x$

Convert to a logarithmic equation.

3. $4^{-3} = x$

4. $256^{1/2} = 16$

Convert to an exponential equation.

5. $\log_4 16 = 2$

6. $m = \log_7 49$

7. Express in terms of logarithms of a, b, and c:

$$\log \frac{a^3 b^{1/2}}{c^2}.$$

8. Express as a single logarithm:

$$\frac{1}{3} \log_a x - 3 \log_a y + 2 \log_a z$$

Simplify.

9. $\log_t t^{23}$

10. $\log_p p$

11. $\log_c 1$

Given that $\log_a 2 = 0.301$, $\log_a 6 = 0.778$, and $\log_a 7 = 0.845$, find each of the following.

12. $\log_a \frac{2}{7}$

13. $\log_a \sqrt{24}$

14. $\log_a 21$

Find each of the following using a calculator.

15. log 0.0123

16. antilog 5.6484

17. antilog (−7.2614)

18. log 12,340

ANSWERS

1. ______
2. ______
3. ______
4. ______
5. ______
6. ______
7. ______
8. ______
9. ______
10. ______
11. ______
12. ______
13. ______
14. ______
15. ______
16. ______
17. ______
18. ______

ANSWERS

19. ______

20. ______

21. ______

22. ______

23. ______

24. ______

25. ______

26. ______

27. ______

28. ______

29. a) ______

b) ______

30. a) ______

b) ______

c) ______

d) ______

31. ______

32. ______

33. ______

34. ______

35. ______

36. ______

37. ______

19. $\ln 0.01234$

20. $\text{antilog}_e (5.6774)$

21. Find $\log_{18} 31$ using the change-of-base formula.

Solve.

22. $\log_x 25 = 2$

23. $\log_4 x = \frac{1}{2}$

24. $\log x = 4$

25. $5^{4-3x} = 125$

26. $7^x = 1.2$

27. $\ln x = \frac{1}{4}$

28. $\log (x^2 - 1) - \log (x - 1) = 1$

29. *Walking speed.* The average walking speed R of people living in a city of population P, in thousands, is given by

$$R = 1.98 \log P + 0.05,$$

where R is in feet per second.

a) The population of Akron, Ohio, is 660,000. Find the average walking speed.

b) A city has an average walking speed of 2.6 ft/sec. Find the population.

30. *Population of the USSR.* The population of the USSR was 209 million in 1959 and the exponential growth rate was 1% per year.

a) Write an exponential function describing the growth of the population of the USSR.

b) What will the population be in 1996? in 2010?

c) When will the population be 250 million?

d) What is the doubling time?

31. How old is an animal bone that has lost 43% of its carbon-14?

SKILL MAINTENANCE

32. Solve: $y - 9\sqrt{y} + 8 = 0.$

33. Solve $S = at^2 - bt$ for t.

34. Multiply: $(2 + 5i)(2 - 5i)$.

35. Solve: $x^2 + y^2 = 5,$
$y - 2x = 0.$

SYNTHESIS

36. Solve: $\log_5 |2x - 7| = 4.$

37. If $\log_a x = 2$, $\log_a y = 3$, and $\log_a z = 4$, find

$$\log_a \frac{\sqrt[3]{x^2 z}}{\sqrt[3]{y^2 z^{-1}}}.$$

CUMULATIVE REVIEW: CHAPTERS 1-10

1. Evaluate $\dfrac{x^0 + y}{-z}$ when $x = 6$, $y = 9$, and $z = -5$.

Simplify.

2. $\left|-\dfrac{5}{2} + \left(-\dfrac{7}{2}\right)\right|$

3. $(-2x^2y^{-3})^{-4}$

4. $(-5x^4y^{-3}z^2)(-4x^2y^2)$

5. $\dfrac{3x^4y^6z^{-2}}{-9x^4y^2z^3}$

6. $2x - 3 - 2[5 - 3(2 - x)]$

7. $3^3 + 2^2 - (32 \div 4 - 16 \div 8)$

Solve.

8. $8(2x - 3) = 6 - 4(2 - 3x)$

9. $(5x - 2)(4x + 20) = 0$

10. $4x - 3y = 15,$
    $3x + 5y = 4$

11. $x + y - 3z = -1,$
    $2x - y + z = 4,$
    $-x - y + z = 1$

12. $7 = x^2 - 6x$

13. $x(x - 3) = 10$

14. $\dfrac{7}{x^2 - 5x} - \dfrac{2}{x - 5} = \dfrac{4}{x}$

15. $\dfrac{8}{x + 1} + \dfrac{11}{x^2 - x + 1} = \dfrac{24}{x^3 + 1}$

16. $\sqrt{x - 1} = \sqrt{x + 4} - 1$

17. $\sqrt[3]{2x} = 1$

18. $3x^2 + 75 = 0$

19. $x - 8\sqrt{x} + 15 = 0$

20. $x^4 - 13x^2 + 36 = 0$

21. $y = x^2 - 6x + 8,$
    $x - y = 2$

22. $x^2 + y^2 = 16,$
    $x^2 - y^2 = 16$

23. $\log_8 x = 1$

24. $\log_x 49 = 2$

25. $6^x = 9$

26. $\log x - \log (x - 8) = 1$

27. $|2x + 5| < 3$

28. $|2x - 3| \geq 9$

Solve.

29. $D = \dfrac{ab}{b + a}$, for a

30. $\dfrac{1}{p} + \dfrac{1}{q} = \dfrac{1}{f}$, for q

31. $M = \dfrac{2}{3}(A + B)$, for B

Evaluate.

32. $\begin{vmatrix} 6 & -5 \\ 4 & -3 \end{vmatrix}$

33. $\begin{vmatrix} 7 & -6 & 0 \\ -2 & 1 & 2 \\ -1 & 1 & -1 \end{vmatrix}$

Solve.

34. Twenty-four plus five times a number is eight times the number. Find the number.

35. The perimeter of a rectangular garden is 112 m. The length is 16 m more than the width. Find the length and the width.

36. A rectangular field has a perimeter of 74 m. The length is 5 m longer than the width. Find the area.

37. The sum of the areas of two squares is 34 ft^2. The difference of the two areas is 2 ft^2. Find the lengths of the sides of the squares.

38. Phil can build a shed from a lumber kit in 10 hr. Jenny can build the same shed in 12 hr. How long would it take them, working together, to build the shed?

39. Solution A is 30% sulfuric acid. Solution B is 80% sulfuric acid. How many liters of each should be mixed together in order to get 100 L of a solution that is 50% sulfuric acid?

40. A boat can move at a speed of 5 km/h in still water. The boat travels 42 km downstream in the same time that it takes to travel 12 km upstream. What is the speed of the stream?

41. What is the minimum product of two numbers whose difference is 14? What are the numbers that yield this product?

42. The speed of a passenger train is 13 mph faster than the speed of a freight train. The passenger train travels 160 mi in the same time that it takes the freight train to travel 108 mi. Find the speed of each train.

Forgetting. Students in a biology class took a final examination. A forgetting formula for the exam was determined to be

$$S(t) = 78 - 15 \log (t + 1),$$

where t is the number of months after the final was taken.

43. The average score when the students first took the test is when $t = 0$. Find the students' average score on the final exam.

44. What would the average score be on a retest after 4 months?

Population growth. The population of Europe west of Russia was 430 million in 1961, and the exponential growth rate was 1% per year.

45. Write an exponential equation describing the growth of the population of Europe.

46. What will the population be in 1991? in 2001?

47. y varies directly as the square of x and inversely as z, and $y = 2$ when $x = 5$ and $z = 100$. What is y when $x = 3$ and $z = 4$?

Perform the indicated operations and simplify.

48. $(5p^2q^3 - 4p^3q + 6pq - p^2 + 3) + (2p^2q^3 + 2p^3q + p^2 - 5pq - 9)$

49. $(11x^2 - 6x - 3) - (3x^2 + 5x - 2)$

50. $(3x^2 - 2y)^2$

51. $(5a + 3b)(2a - 3b)$

52. $\dfrac{x^2 + 8x + 16}{2x + 6} \div \dfrac{x^2 + 3x - 4}{x^2 - 9}$

53. $\dfrac{1 + \dfrac{3}{x}}{x - 1 - \dfrac{12}{x}}$

54. $\dfrac{a^2 - a - 6}{a^3 - 27} \cdot \dfrac{a^2 + 3a + 9}{6}$

55. $\dfrac{3}{x + 6} - \dfrac{2}{x^2 - 36} + \dfrac{4}{x - 6}$

Factor.

56. $xy - 2xz + xw$

57. $1 - 125x^3$

58. $6x^2 + 8xy - 8y^2$

59. $x^4 - 4x^3 + 7x - 28$

60. $a^2 - 10a + 25 - 81b^2$

61. $2m^2 + 12mn + 18n^2$

62. $x^4 - 16y^4$

63. For the function h described by $h(x) = -3x^2 + 4x + 8$, find $h(-2)$.

64. Divide: $(x^4 - 5x^3 + 2x^2 - 6) \div (x - 3)$.

65. Multiply $(5.2 \times 10^4)(9.3 \times 10^{-6})$. Write scientific notation for the answer.

66. Divide $\dfrac{3.2 \times 10^5}{6.8 \times 10^{-9}}$. Write scientific notation for the answer.

For the radical expressions that follow, assume all variables represent nonnegative numbers.

67. Divide and simplify: $\dfrac{\sqrt[3]{40xy^8}}{\sqrt[3]{5xy}}$.

68. Multiply and simplify: $\sqrt{7xy^3} \cdot \sqrt{28x^2y}$.

69. Rationalize the numerator: $\dfrac{\sqrt{3} - \sqrt{5}}{\sqrt{2}}$.

70. Rationalize the numerator: $\dfrac{3 - \sqrt{y}}{2 - \sqrt{y}}$.

71. Write as a single radical expression: $\dfrac{\sqrt[5]{x + 5}}{\sqrt{x + 5}}$.

72. Multiply these complex numbers: $(1 + i\sqrt{3})(6 - 2i\sqrt{3})$.

73. Divide these complex numbers: $\dfrac{3 - 2i}{4 - 3i}$.

74. Write a quadratic equation whose solutions are $\frac{2}{5}$ and $-\frac{3}{5}$.

75. Find an equation of the line containing the points $(0, -3)$ and $(-1, 2)$.

76. Find an equation of the line containing the point $(-3, 5)$ and perpendicular to the line whose equation is $2x + y = 6$.

Graph.

77. $5x = 15 + 3y$

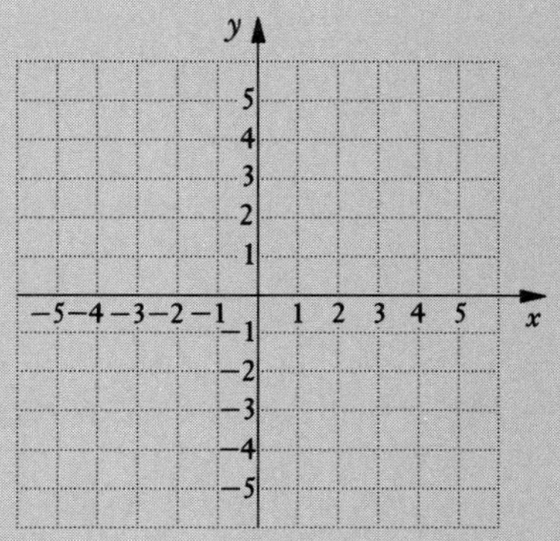

78. $x^2 - y^2 = 16$

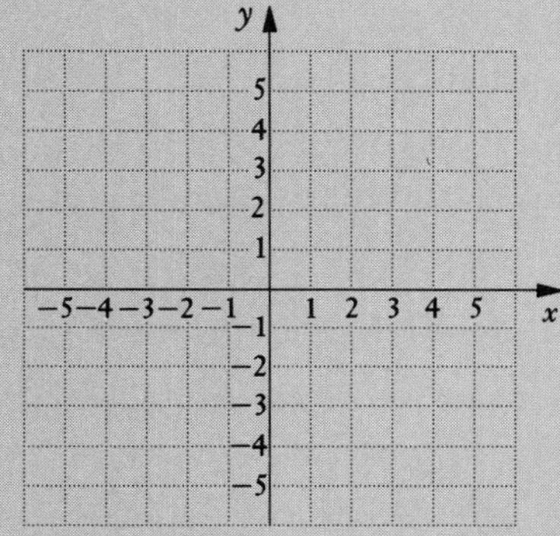

79. $y = 2x^2 - 4x - 1$

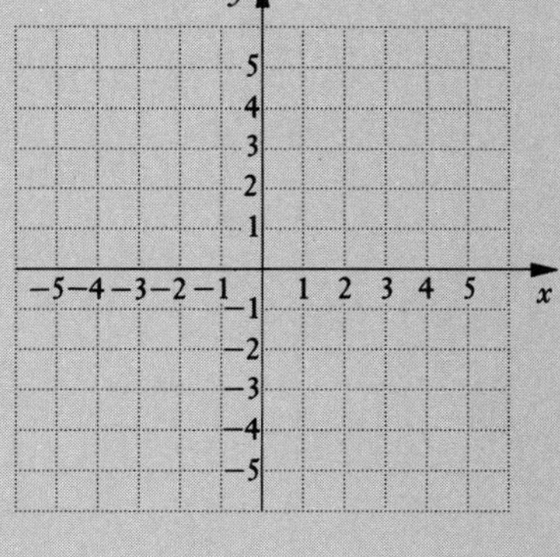

80. $y = \log_3 x$

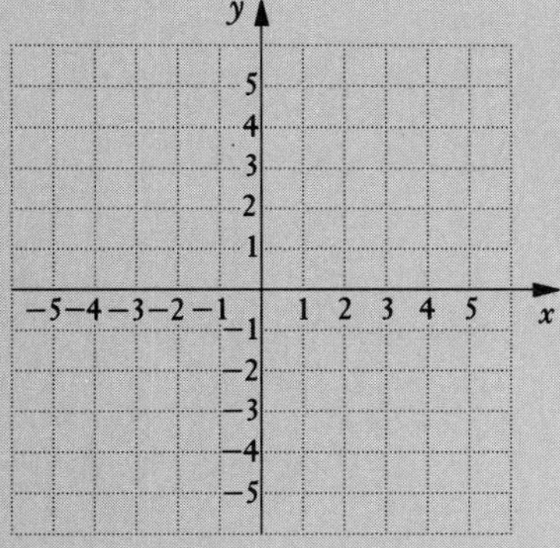

81. $y = 3^x$

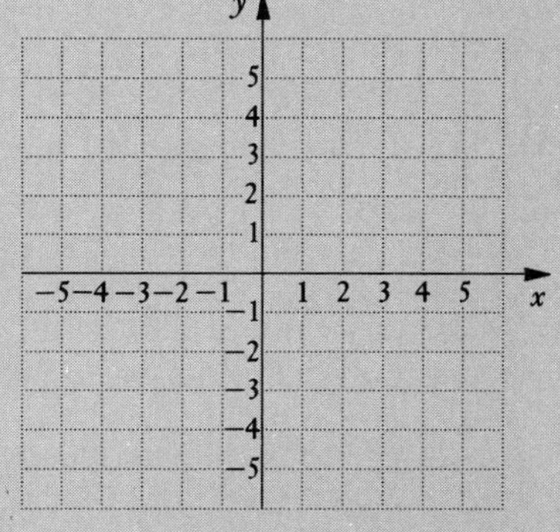

82. $-2x - 3y \leq 6$

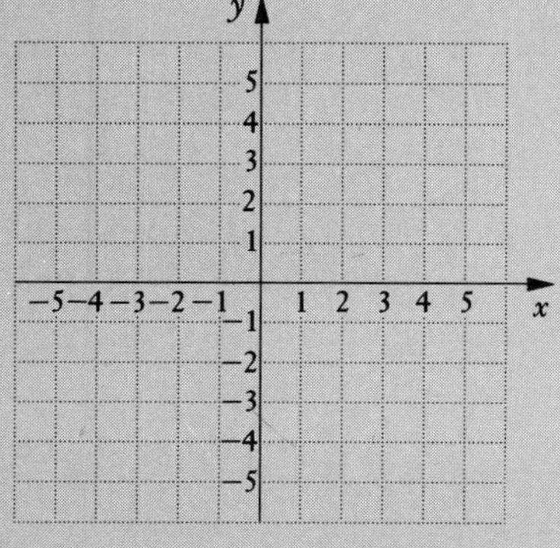

83. Graph $y = 2(x + 3)^2 + 1$.

a) Label the vertex.
b) Draw the line of symmetry.
c) Find the maximum or minimum value.

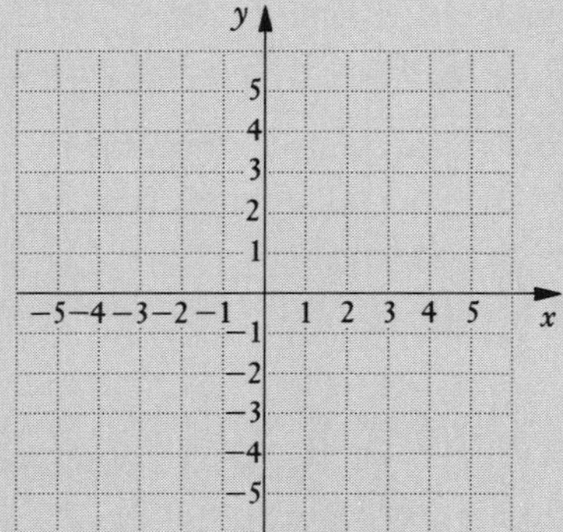

84. Graph. Find the vertices.

$$x - 2y \le 0,$$
$$y - 2x \le 2,$$
$$x + y \le 6$$

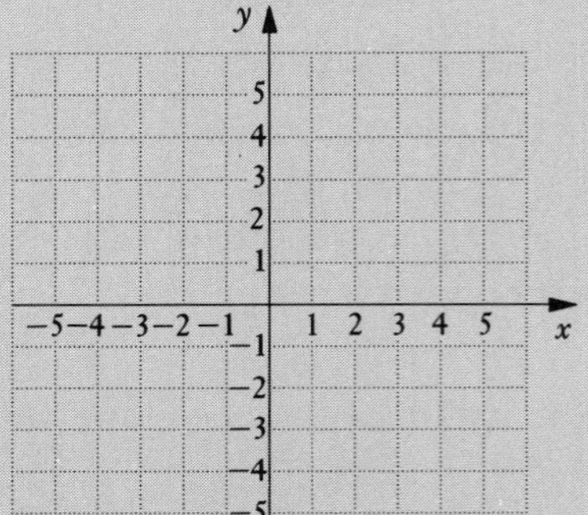

85. Find the center and radius of the circle
$$x^2 + y^2 + 6x + 4y - 7 = 0.$$

86. Write an equation of the circle whose center is $(4, -8)$ and whose radius is $2\sqrt{7}$.

87. Express in terms of logarithms of a, b, and c:
$$\log\left(\frac{a^2c^3}{b}\right).$$

88. Express as a single logarithm:
$$3 \log x - \frac{1}{2} \log y - 2 \log z.$$

89. Convert to an exponential equation: $\log_a 5 = x$.

90. Convert to a logarithmic equation: $x^3 = t$.

91. Given $f(x) = x^2 + 7$ and $g(x) = 2x - 5$, find $(f + g)(x)$, $fg(x)$, and $f \circ g(x)$.

92. Find $f(x)$ and $g(x)$ such that $h(x) = f \circ g(x)$, where
$$h(x) = \frac{1}{x^2 + 3}.$$

Find each of the following using a calculator.

93. $\log 0.05566$

94. antilog 5.4453

95. $\ln 12.78$

96. $\text{antilog}_e\,(-3.6762)$

Find a formula for the inverse.

97. $f(x) = 5x - 7$

98. $f(x) = \dfrac{x - 7}{x + 4}$

99. Solve: $(x + 5)(x + 4)(x - 1) > 0$.

SYNTHESIS

100. Find a parabola such that its line of symmetry is parallel to the x-axis, the vertex is $(2, 1)$, and it passes through the point $(18, -1)$.

101. Solve: $\dfrac{5}{3x - 3} + \dfrac{10}{3x + 6} = \dfrac{5x}{x^2 + x - 2}$.

102. Solve: $\log \sqrt{2x} = \sqrt{\log 2x}$.

103. A bicyclist travels 280 mi at a certain speed. If the speed had been increased 5 mph, the trip could have been made in 1 hr less time. Find the actual speed.

NAME SECTION DATE

FINAL EXAMINATION

Simplify.

1. $(-9x^2y^3)(5x^4y^{-7})$
2. $|-3.5 + 9.8|$
3. $2y - [3 - 4(5 - 2y) - 3y]$
4. $(10 \cdot 8 - 9 \cdot 7)^2 - 54 \div 9 - 3$
5. Evaluate $\dfrac{ab - ac}{bc}$ when $a = -2$, $b = 3$, and $c = -4$.

Perform the indicated operations and simplify.

6. $(5a^2 - 3ab - 7b^2) - (2a^2 + 5ab + 8b^2)$
7. $(-3x^2 + 4x^3 - 5x - 1) + (9x^3 - 4x^2 + 7 - x)$
8. $(2a - 1)(3a + 5)$
9. $(3a^2 - 5y)^2$
10. $\dfrac{1}{x-2} - \dfrac{4}{x^2-4} + \dfrac{3}{x+2}$
11. $\dfrac{x^2 - 6x + 8}{3x + 9} \cdot \dfrac{x+3}{x^2 - 4}$
12. $\dfrac{3x + 3y}{5x - 5y} \div \dfrac{3x^2 + 3y^2}{5x^3 - 5y^3}$
13. $\dfrac{x - \dfrac{a^2}{x}}{1 + \dfrac{a}{x}}$

Factor.

14. $4x^2 - 12x + 9$
15. $27a^3 - 8$
16. $a^3 + 3a^2 - ab - 3b$
17. $15y^4 + 33y^2 - 36$
18. For the function described by $f(x) = 3x^2 - 4x$, find $f(-2)$.
19. Divide: $(7x^4 - 5x^3 + x^2 - 4) \div (x - 2)$.

Solve.

20. $9(x - 1) - 3(x - 2) = 1$
21. $x^2 - 2x = 48$
22. $\dfrac{6}{x} + \dfrac{6}{x+2} = \dfrac{5}{2}$
23. $\dfrac{7x}{x-3} - \dfrac{21}{x} + 11 = \dfrac{63}{x^2 - 3x}$
24. $5x + 3y = 2,$
    $3x + 5y = -2$
25. $x + y - z = 0,$
    $3x + y + z = 6,$
    $x - y + 2z = 5$
26. $\sqrt{x - 5} = 5 - \sqrt{x}$
27. $x^4 - 29x^2 + 100 = 0$
28. $x^2 + y^2 = 8,$
    $x^2 - y^2 = 2$
29. $5^x = 8$

ANSWERS

1. ____
2. ____
3. ____
4. ____
5. ____
6. ____
7. ____
8. ____
9. ____
10. ____
11. ____
12. ____
13. ____
14. ____
15. ____
16. ____
17. ____
18. ____
19. ____
20. ____
21. ____
22. ____
23. ____
24. ____
25. ____
26. ____
27. ____
28. ____
29. ____

ANSWERS

30. ____

31. ____

32. ____

33. ____

34. ____

35. ____

36. ____

37. ____

38. ____

39. ____

40. ____

41. ____

42. ____

43. ____

44. ____

45. ____

30. $\log (x^2 - 25) - \log (x + 5) = 3$

31. $\log_4 x = -2$

32. $7^{2x+3} = 49$

33. $|2x - 1| \leq 5$

34. $7x^2 + 14 = 0$

35. $x^2 + 4x = 3$

36. $|2y + 3| > 7$

Solve.

37. The perimeter of a rectangle is 34 ft. The length of a diagonal is 13 ft. Find the dimensions of the rectangle.

38. A telephone company charges \$0.40 for the first minute and \$0.25 for every other minute of a long-distance called placed before 5 P.M. The rates after 5 P.M. drop to \$0.30 for the first minute and \$0.20 for every other minute of the call. A certain call placed before 5 P.M. costs \$4.20. How much would a call of the same duration placed after 5 P.M. cost?

39. Find three consecutive integers whose sum is 198.

40. A pentagon with all five sides congruent has a perimeter equal to that of an octagon with all eight sides congruent. One side of the pentagon is two less than three times one side of the octagon. What is the perimeter of each figure?

41. A chemist has two solutions of ammonia and water. Solution A is 6% ammonia. Solution B is 2% ammonia. How many liters of each solution are needed in order to obtain 80 L of a solution that is 3.2% ammonia?

42. An airplane can fly 190 mi with the wind in the same time that it takes to fly 160 mi against the wind. The speed of the wind is 30 mph. How fast can the plane fly in still air?

43. A can do a certain job in 21 min. B can do the same job in 14 min. How long would it take to do the job if the two worked together?

44. The centripetal force F of an object moving in a circle varies directly as the square of the velocity v and inversely as the radius r of the circle. If $F = 8$ when $v = 1$ and $r = 10$, what is F when $v = 2$ and $r = 16$?

45. A farmer wants to fence in a rectangular area next to a river. (Note that no fence will be needed along the river.) What is the area of the largest region that can be fenced in with 100 ft of fencing?

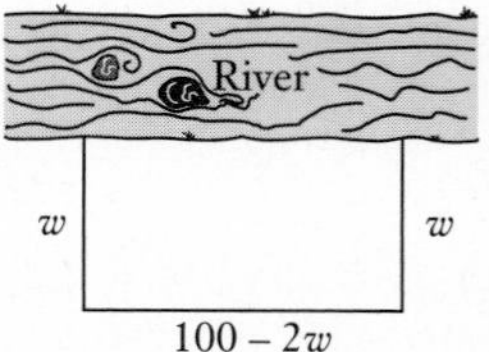

Graph.

46. $3x - y = 6$

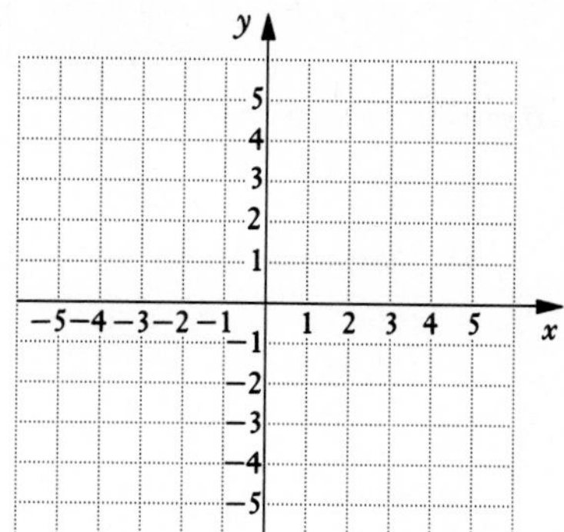

47. $\dfrac{x^2}{25} + \dfrac{y^2}{4} = 1$

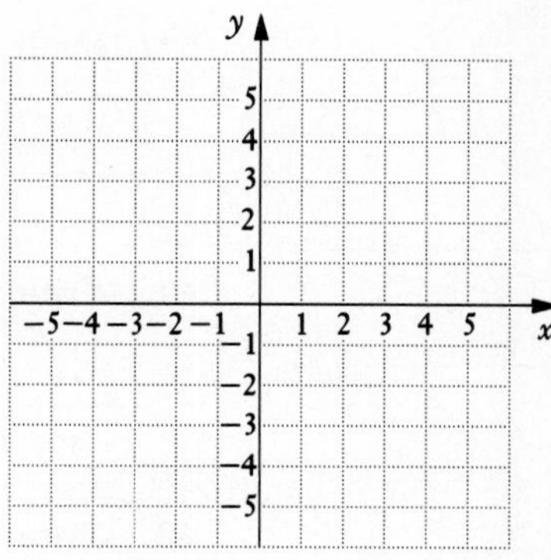

48. $y = \log_2 x$

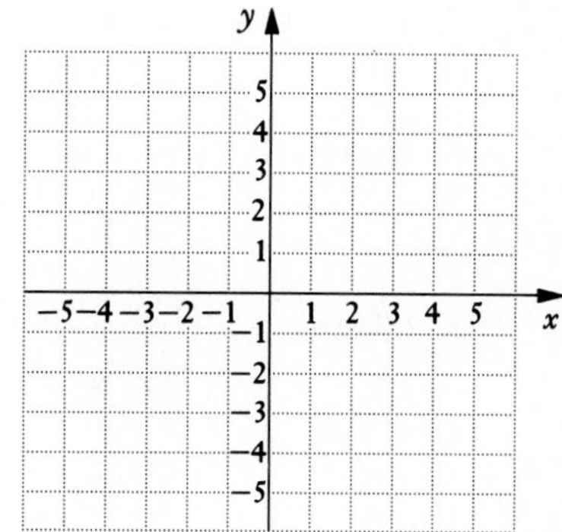

49. $2x - 3y < -6$

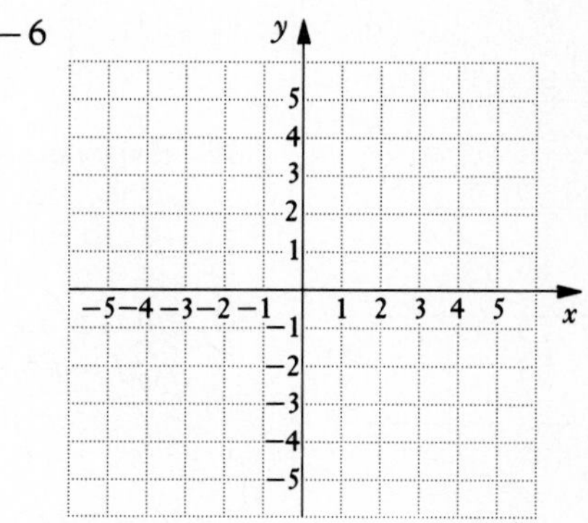

50. Graph: $y = -2(x - 3)^2 + 1$.

a) Label the vertex.

b) Draw the line of symmetry.

c) Find the maximum or minimum value.

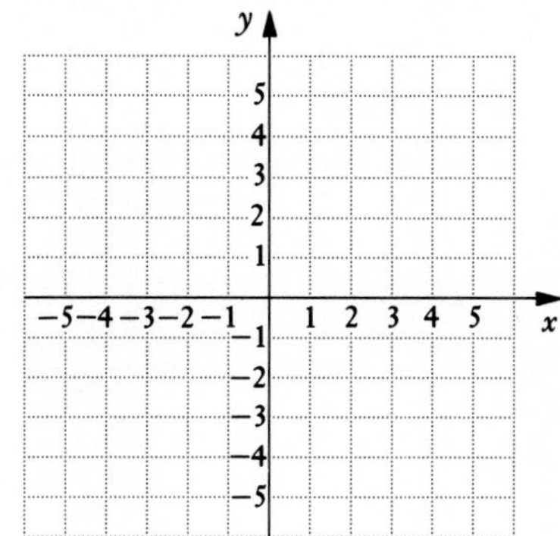

51. Graph. Find the vertices.

$$y - x \geq 1,$$
$$y - x \leq 3,$$
$$2 \leq x \leq 5$$

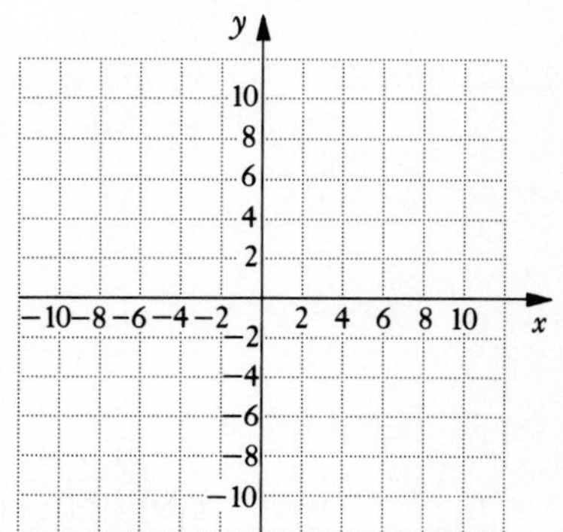

52. Solve $A = P + Prt$ for r.

53. Solve $I = \dfrac{R}{R + r}$ for R.

54. Find an equation of the line containing the point $(-1, 4)$ and perpendicular to the line whose equation is $x - y = 6$.

Evaluate.

55. $\begin{vmatrix} -5 & -7 \\ 4 & 6 \end{vmatrix}$

56. $\begin{vmatrix} 2 & -1 & 1 \\ 1 & 2 & 0 \\ 3 & -1 & 1 \end{vmatrix}$

57. Multiply $(8.9 \times 10^{-17})(7.6 \times 10^4)$. Write scientific notation for the answer.

58. Multiply and simplify: $\sqrt{8x}\sqrt{8x^3y}$.

ANSWERS

52. ______________

53. ______________

54. ______________

55. ______________

56. ______________

57. ______________

58. ______________

ANSWERS

59. ____

60. ____

61. ____

62. ____

63. ____

64. ____

65. ____

66. ____

67. ____

68. ____

69. ____

70. ____

71. ____

72. ____

73. ____

74. ____

75. ____

76. ____

77. ____

78. ____

79. ____

80. ____

59. Divide and simplify: $\dfrac{\sqrt[3]{15x}}{\sqrt[3]{3y^2}}$.

60. Rationalize the denominator:
$$\frac{1-\sqrt{x}}{1+\sqrt{x}}.$$

61. Write a single radical expression:
$$\frac{\sqrt[3]{(x+1)^5}}{\sqrt{(x+1)^3}}.$$

62. Multiply these complex numbers:
$$(3+2i)(4-7i).$$

63. Write a quadratic equation whose solutions are $5\sqrt{2}$ and $-5\sqrt{2}$.

64. Find the center and radius of the circle
$$x^2+y^2-4x+6y-23=0.$$

65. Express as a single logarithm:
$$\tfrac{2}{3}\log_a x-\tfrac{1}{2}\log_a y+5\log_a z.$$

66. Convert to an exponential equation:
$$\log_a c=5.$$

67. Find a formula for the inverse of $f(x)=x^3-8$.

68. Given $f(x)=x^2$ and $g(x)=x+3$, find $(f+g)(x)$, $ff(x)$, $fg(x)$, $f\circ g(x)$, and $g\circ f(x)$.

Find each of the following using a calculator.

69. log 5677.2

70. antilog (-4.8904)

71. ln 5677.2

72. $\text{antilog}_e\,(-4.8904)$

Population growth. The Virgin Islands has one of the highest exponential growth rates in the world, 9.6%. In 1970 the population was 75,150.

73. Write an exponential equation describing the growth of the population of the Virgin Islands.

74. What will the population be in 1990?

75. For the function f described by $f(x)=x^2-3x$, find $f(-2)$, $f(0)$, and $f(a+b)$.

SYNTHESIS

Solve.

76. $\dfrac{9}{x}-\dfrac{9}{x+12}=\dfrac{108}{x^2+12x}$

77. $\log_2(\log_3 x)=2$

78. y varies directly as the cube of x and x is multiplied by 0.5. What is the effect on y?

79. Divide these complex numbers:
$$\frac{2\sqrt{6}+4\sqrt{5}i}{2\sqrt{6}-4\sqrt{5}i}.$$

80. Diaphantos, a famous mathematician, spent $\frac{1}{6}$ of his life as a child, $\frac{1}{12}$ as a young man, and $\frac{1}{7}$ as a bachelor. Five years after he was married, he had a son who died 4 years before his father at half his father's final age. How long did Diaphantos live?

Tables

TABLE 1

POWERS, ROOTS, AND RECIPROCALS

n	n^2	n^3	$\sqrt{n}$	$\sqrt[3]{n}$	$\frac{1}{n}$
1	1	1	1.000	1.000	1.0000
2	4	8	1.414	1.260	.5000
3	9	27	1.732	1.442	.3333
4	16	64	2.000	1.587	.2500
5	25	125	2.236	1.710	.2000
6	36	216	2.449	1.817	.1667
7	49	343	2.646	1.913	.1429
8	64	512	2.828	2.000	.1250
9	81	729	3.000	2.080	.1111
10	100	1,000	3.162	2.154	.1000
11	121	1,331	3.317	2.224	.0909
12	144	1,728	3.464	2.289	.0833
13	169	2,197	3.606	2.351	.0769
14	196	2,744	3.742	2.410	.0714
15	225	3,375	3.873	2.466	.0667
16	256	4,096	4.000	2.520	.0625
17	289	4,913	4.123	2.571	.0588
18	324	5,832	4.243	2.621	.0556
19	361	6,859	4.359	2.668	.0526
20	400	8,000	4.472	2.714	.0500
21	441	9,261	4.583	2.759	.0476
22	484	10,648	4.690	2.802	.0455
23	529	12,167	4.796	2.844	.0435
24	576	13,824	4.899	2.884	.0417
25	625	15,625	5.000	2.924	.0400
26	676	17,576	5.099	2.962	.0385
27	729	19,683	5.196	3.000	.0370
28	784	21,952	5.292	3.037	.0357
29	841	24,389	5.385	3.072	.0345
30	900	27,000	5.477	3.107	.0333
31	961	29,791	5.568	3.141	.0323
32	1,024	32,768	5.657	3.175	.0312
33	1,089	35,937	5.745	3.208	.0303
34	1,156	39,304	5.831	3.240	.0294
35	1,225	42,875	5.916	3.271	.0286
36	1,296	46,656	6.000	3.302	.0278
37	1,369	50,653	6.083	3.332	.0270
38	1,444	54,872	6.164	3.362	.0263
39	1,521	59,319	6.245	3.391	.0256
40	1,600	64,000	6.325	3.420	.0250

n	n^2	n^3	$\sqrt{n}$	$\sqrt[3]{n}$	$\frac{1}{n}$
41	1,681	68,921	6.403	3.448	.0244
42	1,764	74,088	6.481	3.476	.0238
43	1,849	79,507	6.557	3.503	.0233
44	1,936	85,184	6.633	3.530	.0227
45	2,025	91,125	6.708	3.557	.0222
46	2,116	97,336	6.782	3.583	.0217
47	2,209	103,823	6.856	3.609	.0213
48	2,304	110,592	6.928	3.634	.0208
49	2,401	117,649	7.000	3.659	.0204
50	2,500	125,000	7.071	3.684	.0200
51	2,601	132,651	7.141	3.708	.0196
52	2,704	140,608	7.211	3.733	.0192
53	2,809	148,877	7.280	3.756	.0189
54	2,916	157,464	7.348	3.780	.0185
55	3,025	166,375	7.416	3.803	.0182
56	3,136	175,616	7.483	3.826	.0179
57	3,249	185,193	7.550	3.849	.0175
58	3,364	195,112	7.616	3.871	.0172
59	3,481	205,379	7.681	3.893	.0169
60	3,600	216,000	7.746	3.915	.0167
61	3,721	226,981	7.810	3.936	.0164
62	3,844	238,328	7.874	3.958	.0161
63	3,969	250,047	7.937	3.979	.0159
64	4,096	262,144	8.000	4.000	.0156
65	4,225	274,625	8.062	4.021	.0154
66	4,356	287,496	8.124	4.041	.0152
67	4,489	300,763	8.185	4.062	.0149
68	4,624	314,432	8.246	4.082	.0147
69	4,761	328,509	8.307	4.102	.0145
70	4,900	343,000	8.367	4.121	.0143

n	n^2	n^3	$\sqrt{n}$	$\sqrt[3]{n}$	$\frac{1}{n}$
71	5,041	357,911	8.426	4.141	.0141
72	5,184	373,248	8.485	4.160	.0139
73	5,329	389,017	8.544	4.179	.0137
74	5,476	405,224	8.602	4.198	.0135
75	5,625	421,875	8.660	4.217	.0133
76	5,776	438,976	8.718	4.236	.0132
77	5,929	456,533	8.775	4.254	.0130
78	6,084	474,552	8.832	4.273	.0128
79	6,241	493,039	8.888	4.291	.0127
80	6,400	512,000	8.944	4.309	.0125
81	6,561	531,441	9.000	4.327	.0123
82	6,724	551,368	9.055	4.344	.0122
83	6,889	571,787	9.110	4.362	.0120
84	7,056	592,704	9.165	4.380	.0119
85	7,225	614,125	9.220	4.397	.0118
86	7,396	636,056	9.274	4.414	.0116
87	7,569	658,503	9.327	4.431	.0115
88	7,744	681,472	9.381	4.448	.0114
89	7,921	704,969	9.434	4.465	.0112
90	8,100	729,000	9.487	4.481	.0111
91	8,281	753,571	9.539	4.498	.0110
92	8,464	778,688	9.592	4.514	.0109
93	8,649	804,357	9.644	4.531	.0108
94	8,836	830,584	9.695	4.547	.0106
95	9,025	857,375	9.747	4.563	.0105
96	9,216	884,736	9.798	4.579	.0104
97	9,409	912,673	9.849	4.595	.0103
98	9,604	941,192	9.899	4.610	.0102
99	9,801	970,299	9.950	4.626	.0101
100	10,000	1,000,000	10.000	4.642	.0100

TABLE 2

COMMON LOGARITHMS

x	0	1	2	3	4	5	6	7	8	9
1.0	.0000	.0043	.0086	.0128	.0170	.0212	.0253	.0294	.0334	.0374
1.1	.0414	.0453	.0492	.0531	.0569	.0607	.0645	.0682	.0719	.0755
1.2	.0792	.0828	.0864	.0899	.0934	.0969	.1004	.1038	.1072	.1106
1.3	.1139	.1173	.1206	.1239	.1271	.1303	.1335	.1367	.1399	.1430
1.4	.1461	.1492	.1523	.1553	.1584	.1614	.1644	.1673	.1703	.1732
1.5	.1761	.1790	.1818	.1847	.1875	.1903	.1931	.1959	.1987	.2014
1.6	.2041	.2068	.2095	.2122	.2148	.2175	.2201	.2227	.2253	.2279
1.7	.2304	.2330	.2355	.2380	.2405	.2430	.2455	.2480	.2504	.2529
1.8	.2553	.2577	.2601	.2625	.2648	.2672	.2695	.2718	.2742	.2765
1.9	.2788	.2810	.2833	.2856	.2878	.2900	.2923	.2945	.2967	.2989
2.0	.3010	.3032	.3054	.3075	.3096	.3118	.3139	.3160	.3181	.3201
2.1	.3222	.3243	.3263	.3284	.3304	.3324	.3345	.3365	.3385	.3404
2.2	.3424	.3444	.3464	.3483	.3502	.3522	.3541	.3560	.3579	.3598
2.3	.3617	.3636	.3655	.3674	.3692	.3711	.3729	.3747	.3766	.3784
2.4	.3802	.3820	.3838	.3856	.3874	.3892	.3909	.3927	.3945	.3962
2.5	.3979	.3997	.4014	.4031	.4048	.4065	.4082	.4099	.4116	.4133
2.6	.4150	.4166	.4183	.4200	.4216	.4232	.4249	.4265	.4281	.4298
2.7	.4314	.4330	.4346	.4362	.4378	.4393	.4409	.4425	.4440	.4456
2.8	.4472	.4487	.4502	.4518	.4533	.4548	.4564	.4579	.4594	.4609
2.9	.4624	.4639	.4654	.4669	.4683	.4698	.4713	.4728	.4742	.4757
3.0	.4771	.4786	.4800	.4814	.4829	.4843	.4857	.4871	.4886	.4900
3.1	.4914	.4928	.4942	.4955	.4969	.4983	.4997	.5011	.5024	.5038
3.2	.5051	.5065	.5079	.5092	.5105	.5119	.5132	.5145	.5159	.5172
3.3	.5185	.5198	.5211	.5224	.5237	.5250	.5263	.5276	.5289	.5302
3.4	.5315	.5328	.5340	.5353	.5366	.5378	.5391	.5403	.5416	.5428
3.5	.5441	.5453	.5465	.5478	.5490	.5502	.5514	.5527	.5539	.5551
3.6	.5563	.5575	.5587	.5599	.5611	.5623	.5635	.5647	.5658	.5670
3.7	.5682	.5694	.5705	.5717	.5729	.5740	.5752	.5763	.5775	.5786
3.8	.5798	.5809	.5821	.5832	.5843	.5855	.5866	.5877	.5888	.5899
3.9	.5911	.5922	.5933	.5944	.5955	.5966	.5977	.5988	.5999	.6010
4.0	.6021	.6031	.6042	.6053	.6064	.6075	.6085	.6096	.6107	.6117
4.1	.6128	.6138	.6149	.6160	.6170	.6180	.6191	.6201	.6212	.6222
4.2	.6232	.6243	.6253	.6263	.6274	.6284	.6294	.6304	.6314	.6325
4.3	.6335	.6345	.6355	.6365	.6375	.6385	.6395	.6405	.6415	.6425
4.4	.6435	.6444	.6454	.6464	.6474	.6484	.6493	.6503	.6513	.6522
4.5	.6532	.6542	.6551	.6561	.6571	.6580	.6590	.6599	.6609	.6618
4.6	.6628	.6637	.6646	.6656	.6665	.6675	.6684	.6693	.6702	.6712
4.7	.6721	.6730	.6739	.6749	.6758	.6767	.6776	.6785	.6794	.6803
4.8	.6812	.6821	.6830	.6839	.6848	.6857	.6866	.6875	.6884	.6893
4.9	.6902	.6911	.6920	.6928	.6937	.6946	.6955	.6964	.6972	.6981
5.0	.6990	.6998	.7007	.7016	.7024	.7033	.7042	.7050	.7059	.7067
5.1	.7076	.7084	.7093	.7101	.7110	.7118	.7126	.7135	.7143	.7152
5.2	.7160	.7168	.7177	.7185	.7193	.7202	.7210	.7218	.7226	.7235
5.3	.7243	.7251	.7259	.7267	.7275	.7284	.7292	.7300	.7308	.7316
5.4	.7324	.7332	.7340	.7348	.7356	.7364	.7372	.7380	.7388	.7396
x	0	1	2	3	4	5	6	7	8	9

TABLE 2 (*continued*)

x	0	1	2	3	4	5	6	7	8	9
5.5	.7404	.7412	.7419	.7427	.7435	.7443	.7451	.7459	.7466	.7474
5.6	.7482	.7490	.7497	.7505	.7513	.7520	.7528	.7536	.7543	.7551
5.7	.7559	.7566	.7574	.7582	.7589	.7597	.7604	.7612	.7619	.7627
5.8	.7634	.7642	.7649	.7657	.7664	.7672	.7679	.7686	.7694	.7701
5.9	.7709	.7716	.7723	.7731	.7738	.7745	.7752	.7760	.7767	.7774
6.0	.7782	.7789	.7796	.7803	.7810	.7818	.7825	.7832	.7839	.7846
6.1	.7853	.7860	.7868	.7875	.7882	.7889	.7896	.7903	.7910	.7917
6.2	.7924	.7931	.7938	.7945	.7952	.7959	.7966	.7973	.7980	.7987
6.3	.7993	.8000	.8007	.8014	.8021	.8028	.8035	.8041	.8048	.8055
6.4	.8062	.8069	.8075	.8082	.8089	.8096	.8102	.8109	.8116	.8122
6.5	.8129	.8136	.8142	.8149	.8156	.8162	.8169	.8176	.8182	.8189
6.6	.8195	.8202	.8209	.8215	.8222	.8228	.8235	.8241	.8248	.8254
6.7	.8261	.8267	.8274	.8280	.8287	.8293	.8299	.8306	.8312	.8319
6.8	.8325	.8331	.8338	.8344	.8351	.8357	.8363	.8370	.8376	.8382
6.9	.8388	.8395	.8401	.8407	.8414	.8420	.8426	.8432	.8439	.8445
7.0	.8451	.8457	.8463	.8470	.8476	.8482	.8488	.8494	.8500	.8506
7.1	.8513	.8519	.8525	.8531	.8537	.8543	.8549	.8555	.8561	.8567
7.2	.8573	.8579	.8585	.8591	.8597	.8603	.8609	.8615	.8621	.8627
7.3	.8633	.8639	.8645	.8651	.8657	.8663	.8669	.8675	.8681	.8686
7.4	.8692	.8698	.8704	.8710	.8716	.8722	.8727	.8733	.8739	.8745
7.5	.8751	.8756	.8762	.8768	.8774	.8779	.8785	.8791	.8797	.8802
7.6	.8808	.8814	.8820	.8825	.8831	.8837	.8842	.8848	.8854	.8859
7.7	.8865	.8871	.8876	.8882	.8887	.8893	.8899	.8904	.8910	.8915
7.8	.8921	.8927	.8932	.8938	.8943	.8949	.8954	.8960	.8965	.8971
7.9	.8976	.8982	.8987	.8993	.8998	.9004	.9009	.9015	.9020	.9025
8.0	.9031	.9036	.9042	.9047	.9053	.9058	.9063	.9069	.9074	.9079
8.1	.9085	.9090	.9096	.9101	.9106	.9112	.9117	.9122	.9128	.9133
8.2	.9138	.9143	.9149	.9154	.9159	.9165	.9170	.9175	.9180	.9186
8.3	.9191	.9196	.9201	.9206	.9212	.9217	.9222	.9227	.9232	.9238
8.4	.9243	.9248	.9253	.9258	.9263	.9269	.9274	.9279	.9284	.9289
8.5	.9294	.9299	.9304	.9309	.9315	.9320	.9325	.9330	.9335	.9340
8.6	.9345	.9350	.9355	.9360	.9365	.9370	.9375	.9380	.9385	.9390
8.7	.9395	.9400	.9405	.9410	.9415	.9420	.9425	.9430	.9435	.9440
8.8	.9445	.9450	.9455	.9460	.9465	.9469	.9474	.9479	.9484	.9489
8.9	.9494	.9499	.9504	.9509	.9513	.9518	.9523	.9528	.9533	.9538
9.0	.9542	.9547	.9552	.9557	.9562	.9566	.9571	.9576	.9581	.9586
9.1	.9590	.9595	.9600	.9605	.9609	.9614	.9619	.9624	.9628	.9633
9.2	.9638	.9643	.9647	.9652	.9657	.9661	.9666	.9671	.9675	.9680
9.3	.9685	.9689	.9694	.9699	.9703	.9708	.9713	.9717	.9722	.9727
9.4	.9731	.9736	.9741	.9745	.9750	.9754	.9759	.9763	.9768	.9773
9.5	.9777	.9782	.9786	.9791	.9795	.9800	.9805	.9809	.9814	.9818
9.6	.9823	.9827	.9832	.9836	.9841	.9845	.9850	.9854	.9859	.9863
9.7	.9868	.9872	.9877	.9881	.9886	.9890	.9894	.9899	.9903	.9908
9.8	.9912	.9917	.9921	.9926	.9930	.9934	.9939	.9943	.9948	.9952
9.9	.9956	.9961	.9965	.9969	.9974	.9978	.9983	.9987	.9991	.9996
x	**0**	**1**	**2**	**3**	**4**	**5**	**6**	**7**	**8**	**9**

Answers

MARGIN EXERCISE ANSWERS

CHAPTER 1

Margin Exercises, Section 1.1, pp. 3-6

1. $64 + x = 71$; 7 yr **2.** 84 **3.** 68 **4.** 50 **5.** 120 **6.** 96 sq ft **7.** 24 **8.** 2.81 **9.** $47 + y$, or $y + 47$ **10.** $y - 16$ **11.** $16 - x$ **12.** $\frac{1}{4}t$ **13 .** $8x + 6$, or $6 + 8x$ **14.** $m - n$ **15.** $69\%z$, or $0.69z$ **16.** $xy - 300$ **17.** $a + b$

Margin Exercises, Section 1.2, pp. 9-14

1. -9 **2.** 6 **3.** 0 **4.** $\{1, 3, 5, 7, 9, 11, 13\}$; $\{x \mid x$ is an odd whole number between 0 and 14$\}$ **5.** **(a)** 20; **(b)** 20, 0; **(c)** 20, -10, 0; **(d)** $\sqrt{7}, -\sqrt{2}$, 9.34334333433334...; **(e)** 20, -10, -5.34, 18.999, $\frac{11}{45}$, 0, $-\frac{2}{3}$; **(f)** 20, -10, -5.34, 18.999, $\frac{11}{45}$, $\sqrt{7}$, $-\sqrt{2}$, 0, $-\frac{2}{3}$, 9.34334333433334... **6.** $<$ **7.** $>$ **8.** $>$ **9.** $<$ **10.** $>$ **11.** $>$ **12.** $<$ **13.** $>$ **14.** $>$ **15.** $6 < x$ **16.** $7 \geq -4$

17. True **18.** True **19.** False

20. [number line: -1 0]

21. [number line: 0 5]

22. $\frac{1}{4}$ **23.** 2 **24.** $\frac{3}{2}$ **25.** 2.3

Margin Exercises, Section 1.3, pp. 17-22

1. 2 **2.** 4 **3.** -4 **4.** 0 **5.** -18 **6.** -18.6 **7.** $-\frac{29}{5}$ **8.** $-\frac{7}{10}$ **9.** 0 **10.** -7.4 **11.** -3 **12.** -3.3 **13.** $-\frac{1}{4}$ **14.** $\frac{1}{10}$ **15.** 14 **16.** $-\frac{2}{3}$ **17.** 0 **18.** -9 **19.** $\frac{3}{5}$ **20.** -5.9 **21.** $\frac{2}{3}$ **22.** **(a)** -11; **(b)** 17; **(c)** 0; **(d)** $-x$; **(e)** x **23.** 17 **24.** -16 **25.** -3 **26.** -29.6 **27.** 3 **28.** 1 **29.** $\frac{3}{2}$ **30.** **(a)** -6; **(b)** -40; **(c)** 6 **31.** 10, 5, 0, -5, -10, -15, -20, -25, -30 **32.** -24 **33.** -28.35 **34.** -8 **35.** -10, -5, 0, 5, 10, 15, 20, 25 **36.** 72 **37.** $\frac{8}{15}$ **38.** 42.77 **39.** 2 **40.** -25

41. −3 **42.** 0.2 **43.** Undefined **44.** 0 **45.** Undefined **46.** Undefined **47.** $\frac{8}{3}$ **48.** $-\frac{5}{4}$ **49.** $\frac{1}{18}$ **50.** $-\frac{1}{4.3}$, or $-\frac{10}{43}$

51. $\frac{1}{0.5}$, or 2

52.

	Opposite (additive inverse)	Reciprocal (multiplicative inverse)
$\frac{2}{3}$	$-\frac{2}{3}$	$\frac{3}{2}$
$\frac{4}{5}$	$-\frac{4}{5}$	$\frac{5}{4}$
$-\frac{3}{4}$	$\frac{3}{4}$	$-\frac{4}{3}$
0.25	−0.25	4
8	−8	$\frac{1}{8}$
−5	5	$-\frac{1}{5}$
0	0	Does not exist

53. $-\frac{6}{7}$ **54.** $\frac{36}{7}$

Margin Exercises, Section 1.4, pp. 27-34

1.

	$x+2x$	$3x$
$x=5$	15	15
$x=-6$	−18	−18
$x=4.8$	14.4	14.4

2.

	$6x-x$	$5x$
$x=3$	15	15
$x=-6$	−30	−30
$x=4.8$	24	24

3. $\frac{2y}{3y}$ **4.** $\frac{2}{3}$ **5.** $-\frac{5}{3}$ **6.** 2; 2 **7.** −14; −14 **8.** 21; 21 **9.** 440; 440 **10.** $5+y$; ba; $mn+8$, or $nm+8$, or $8+nm$ **11.** The following are three possibilities: $2\cdot(x\cdot y)$; $(2\cdot y)\cdot x$; $(y\cdot 2)\cdot x$ **12.** 40; 40 **13.** −48; −48 **14.** (a) −5; (b) −3; not equivalent **15.** (a) 50; (b) 50 **16.** (a) 5841; (b) 5841 **17.** $8y-80$ **18.** $ax+ay-az$ **19.** $40x-60y+5z$ **20.** $-5x, -7y, 67t, -\frac{4}{5}$ **21.** $9(x+y)$ **22.** $a(c-y)$ **23.** $6(x-2)$ **24.** $5(7x-5y+3w+1)$ **25.** $b(s+t-w)$ **26.** $20x$ **27.** $-7x$ **28.** $6x$ **29.** $-6x$ **30.** $23.4x+3.9y$ **31.** $\frac{22}{15}x-\frac{19}{12}y+23$ **32.** −24 **33.** 0 **34.** 10 **35.** $-9x$ **36.** $24t$ **37.** $-7+y$ **38.** $-x+y$ **39.** $-9x-6y-11$ **40.** $-23x+7y+2$ **41.** $3x+2y+1$ **42.** $2x+5z-24$ **43.** $-3x+2y$ **44.** $-\frac{1}{4}t-41w+5d+23$ **45.** $3x-8$ **46.** $4y+1$ **47.** $-2x-9y-6$ **48.** $15x+9y-6$ **49.** $-x-2y$ **50.** $23x-10y$ **51.** $3a-3b+\frac{23}{4}$

Margin Exercises, Section 1.5, pp. 39-42

1. 8^4 **2.** m^3 **3.** $(4y)^5$ **4.** $3\cdot3\cdot3\cdot3$, or 81 **5.** $\frac{1}{4}\cdot\frac{1}{4}$, or $\frac{1}{16}$ **6.** $y\cdot y$ **7.** (0.2)(0.2)(0.2), or 0.008 **8.** $5x\cdot5x\cdot5x\cdot5x$, or $625xxxx$ **9.** $25y^2$ **10.** −8 **11.** 8 **12.** −31 **13.** 1 **14.** 1 **15.** 1 **16.** $\frac{1}{m^4}$ **17.** $\frac{1}{(-4)^3}$ **18.** x^3 **19.** q^{-3} **20.** $(-5)^{-4}$ **21.** −4117 **22.** 551 **23.** 9; 33 **24.** −25 **25.** 6 **26.** $\frac{54}{241}$ **27.** 34 **28.** $23x+52$ **29.** $12a+12$

Margin Exercises, Section 1.6, pp. 47-52

1. 8^4 **2.** y^5 **3.** $-18x^3$ **4.** $-\frac{75}{x^{14}}$ **5.** $-42x^{8n}$ **6.** $-\frac{10y^2}{x^{12}}$ **7.** $60y$ **8.** 4^3 **9.** 5^6 **10.** $\frac{1}{10^6}$ **11.** $-5x^{2n}$ **12.** $-\frac{2y^{10}}{x^4}$ **13.** $\frac{3}{2}a^3b^2$ **14.** 3^{42} **15.** z^{20} **16.** $\frac{1}{t^{14m}}$ **17.** $8x^3y^3$ **18.** $\frac{16y^{14}}{x^4}$ **19.** $-32x^{20}y^{10}$ **20.** $\frac{1000y^{21}}{x^{12}z^6}$ **21.** x^9y^{12} **22.** $\frac{9x^4}{y^{16}}$ **23.** $-\frac{8a^9}{27b^{21}}$ **24.** 5.88×10^{12} mi **25.** 2.57×10^{-10} **26.** 0.0000000000004567 **27.** 93,000,000 mi **28.** 7.462×10^{-13} **29.** 5.6×10^{-15} **30.** 2.0×10^3 **31.** 5.5×10^2 **32.** 0.00001224; 0.001224%; 1.224×10^{-5} **33.** 1.90164×10^{27} kg

CHAPTER 2

Margin Exercises, Section 2.1, pp. 63-70

1. a, f **2.** b, c **3.** d, e **4.** $89+2$ and $3+4$ represent the same number **5.** $6-7$ and $4-5$ represent the same number **6.** 5, 8, 0, 2 **7.** 3 **8.** (a) Yes; (b) no **9.** −7 **10.** $-\frac{17}{20}$ **11.** 38 **12.** 140.3 **13.** $\frac{5}{4}$ **14.** −49 **15.** $\frac{3}{14}$ **16.** −3 **17.** $\frac{4}{3}$ **18.** $4-5y=2$; $\frac{2}{5}$ **19.** $63x-91=30$; $\frac{121}{63}$ **20.** 3 **21.** $-\frac{29}{5}$ **22.** $\frac{17}{2}$ **23.** No solution **24.** All real numbers are solutions. **25.** −2 **26.** $-\frac{19}{8}$

Margin Exercises, Section 2.2, pp. 73-78

1. 8 ft, 24 ft **2.** $\frac{1}{2}$ **3.** 14% **4.** 80°, 40°, 60° **5.** 10.385 sec **6.** $124 **7.** 17, 19 **8.** $l=13$ m; $w=6\frac{1}{2}$ m

Margin Exercises, Section 2.3, pp. 83-84

1. $b=\frac{2A}{h}$ **2.** $c=\frac{5}{3}P-10$, or $\frac{5P-30}{3}$ **3.** $m=\frac{H-2r}{3}$
4. $Q=\frac{T}{1+iy}$ **5.** $b=\frac{2A-ha}{h}$, or $\frac{2A}{h}-a$

Margin Exercises, Section 2.4, pp. 87-94

1. Yes **2.** No **3.** Yes
4. $\{x \mid x>3\}$;
5. $\{x \mid x \le 3\}$;
6. $\{x \mid x \le -2\}$;
7. $\left\{x \,\middle|\, x \ge \frac{19}{12}\right\}$ **8.** $\{y \mid y<-2\}$
9. $\left\{y \,\middle|\, y \le \frac{3}{10}\right\}$;
10. $\left\{y \,\middle|\, y < -\frac{5}{12}\right\}$;
11. $\{x \mid x \ge 12\}$;
12. $\left\{y \,\middle|\, y \le -\frac{1}{5}\right\}$ **13.** $\left\{x \,\middle|\, x < \frac{1}{2}\right\}$ **14.** $\left\{y \,\middle|\, y \ge \frac{17}{9}\right\}$
15. $x<-5$ **16.** $n>4.85$ **17.** $C \le \$6700$
18. $y \ge \$27{,}500$ **19.** $13+2x<78.2$ **20.** $4(a+b)>6\frac{1}{2}$
21. $\{t \mid t>64\}$, or 64 years after 1984
22. $\{n \mid n<100\}$

Margin Exercises, Section 2.5, pp. 105-106

1. $\{0,3\}$ **2.**

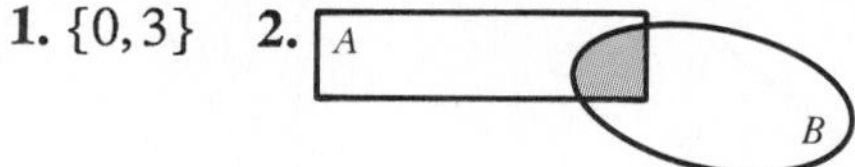

3.
4. $\{x \mid x<-3\}$; **5.** $\varnothing$
6. $\{x \mid -2<x<1\}$ **7.** $\{x \mid 2 \le x \le 6\}$
8. $\{0, 1, 3, 4, 7, 9\}$ **9.**
10.

11. $\{x \mid x<1 \text{ or } x \ge 6\}$ **12.** $\left\{x \,\middle|\, x \ge \frac{7}{2} \text{ or } x<-2\right\}$
13. All real numbers are solutions. **14.** $\{x \mid x \le 2\}$

Margin Exercises, Section 2.6, pp. 111-116

1. $7|x|$ **2.** x^8 **3.** $5a^2|b|$ **4.** $\frac{7|a|}{b^2}$ **5.** $9|x|$ **6.** 29 **7.** 5
8. p **9.** $\{6,-6\}$;
10. $\varnothing$ **11.** $\{0\}$ **12.** $\{2,-2\}$ **13.** $\left\{\frac{17}{4}, -\frac{17}{4}\right\}$ **14.** $\{4,-4\}$
15. $\{3,5\}$ **16.** $\left\{-\frac{13}{3}, 7\right\}$ **17.** $\varnothing$ **18.** $\left\{\frac{7}{4}, -\frac{1}{6}\right\}$ **19.** $\left\{-\frac{7}{2}\right\}$
20. $\{5,-5\}$;
21. $\{x \mid -5<x<5\}$;
22. $\{x \mid x \le -5 \text{ or } x \ge 5\}$;
23. $\left\{2, -\frac{10}{3}\right\}$;
24. $\{x \mid -2<x<5\}$;
25. $\left\{x \,\middle|\, 1 \le x \le \frac{11}{3}\right\}$
26. $\left\{x \,\middle|\, x \le -\frac{7}{3} \text{ or } x \ge 1\right\}$;

CHAPTER 3

Margin Exercises, Section 3.1, pp. 129-134

1.- 9.

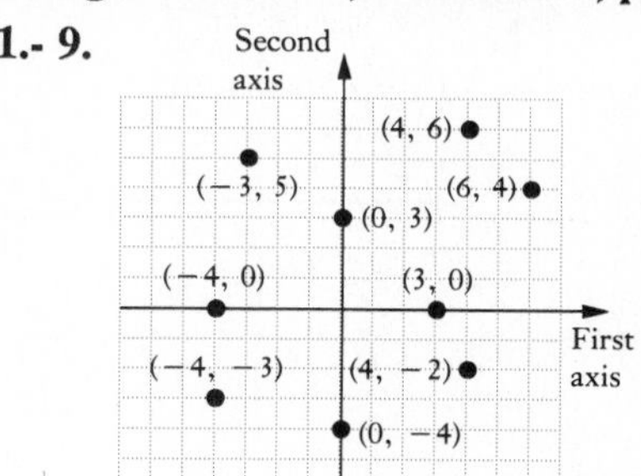

10. Both negative **11.** First positive, second negative
12. No **13.** No **14.** Yes
15. **16.**

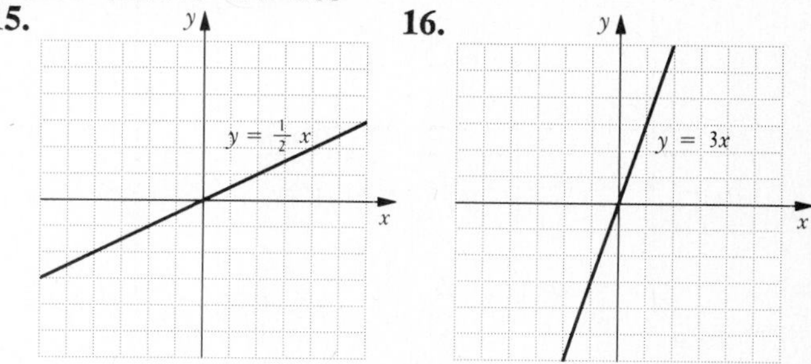

17. **18.**

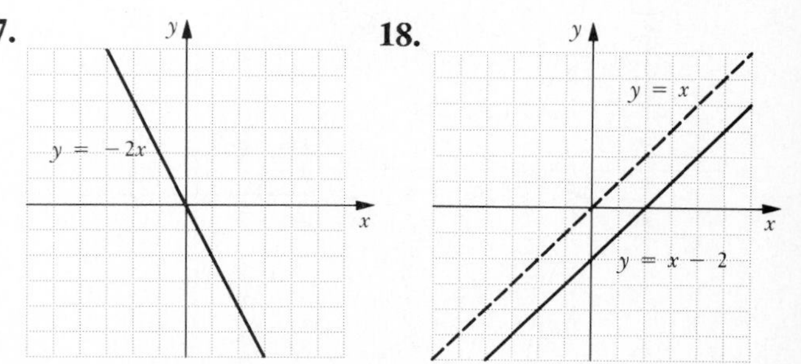

19.

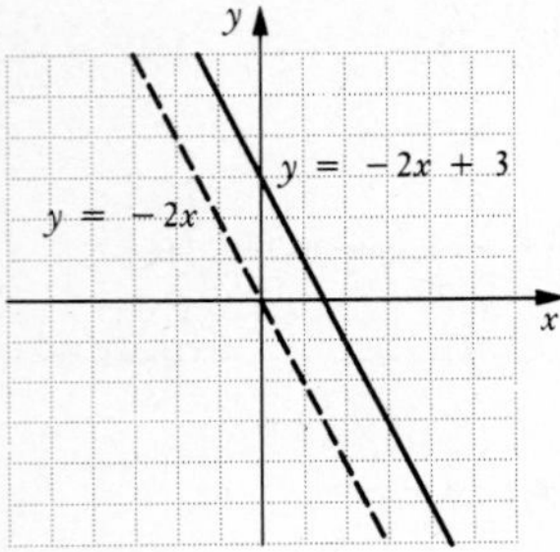

20. **(a)** 25.44, 26.8, 27.82; **(b)**

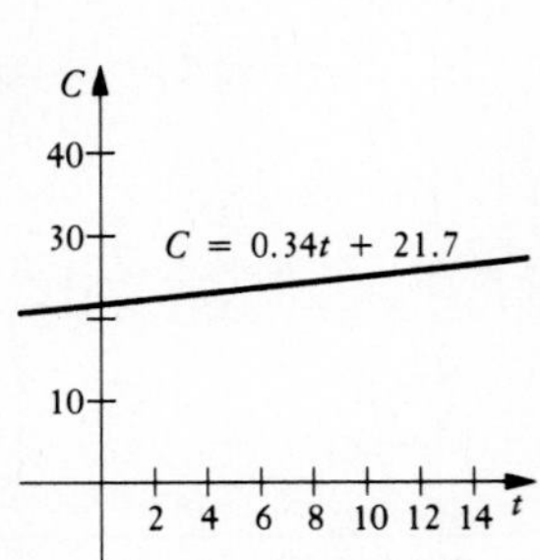

Margin Exercises, Section 3.2, pp. 137-138

1.

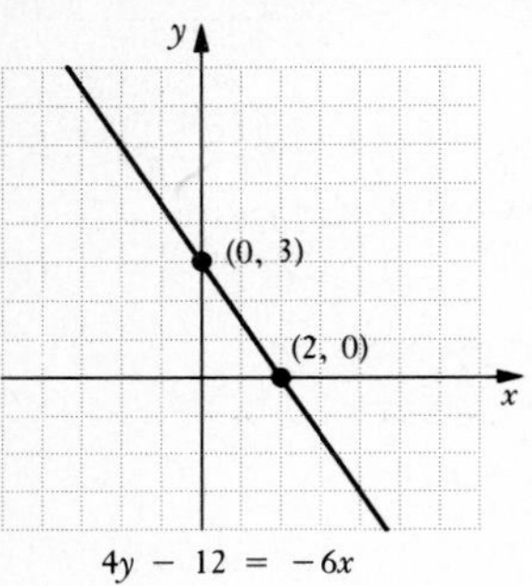

2.

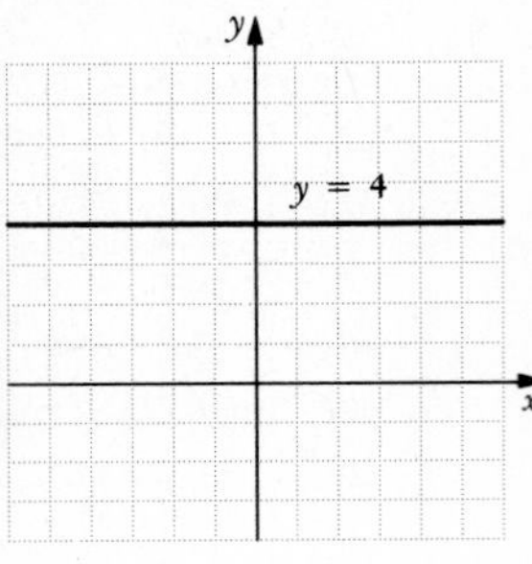

3.

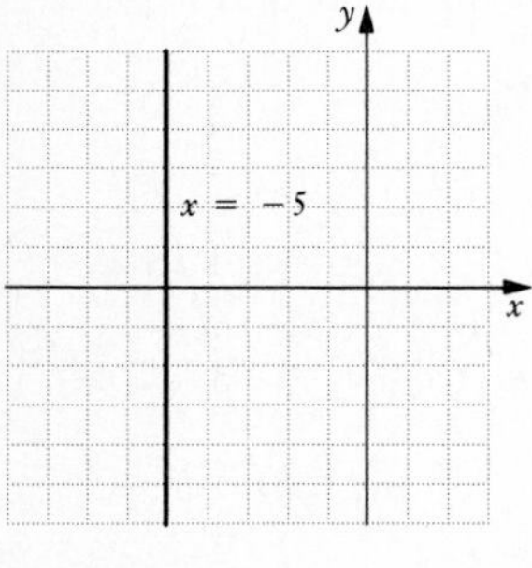

4.

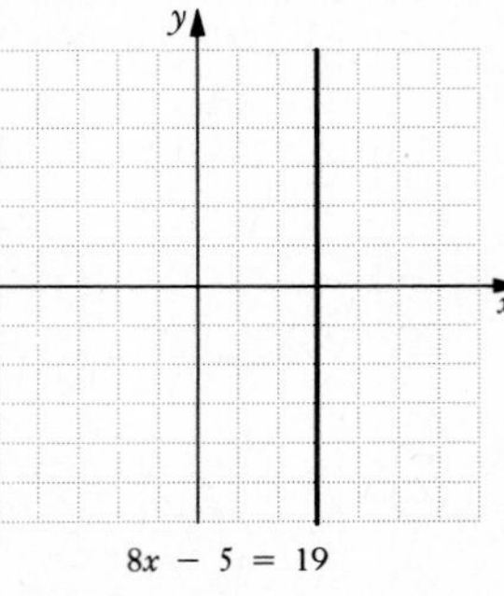

5.

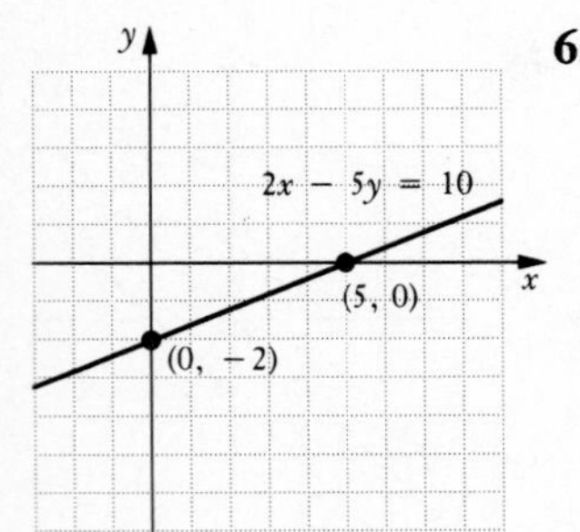

6.

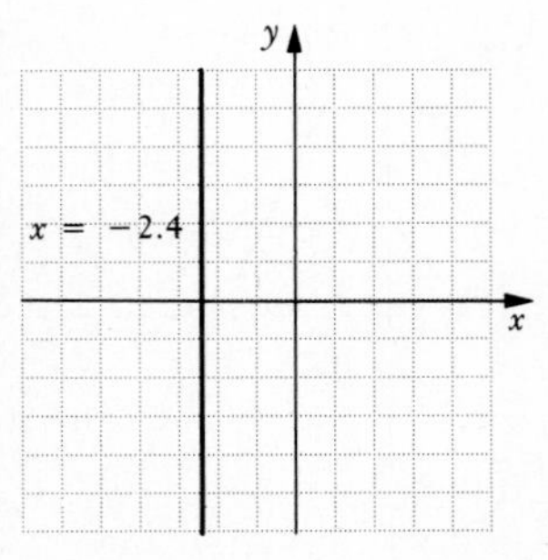

Margin Exercises, Section 3.3, pp. 143-148

1. $m = -1$

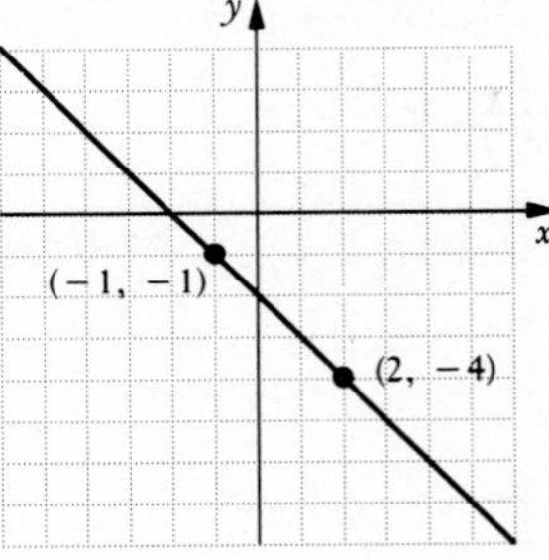

2. $m = -\frac{1}{3}$

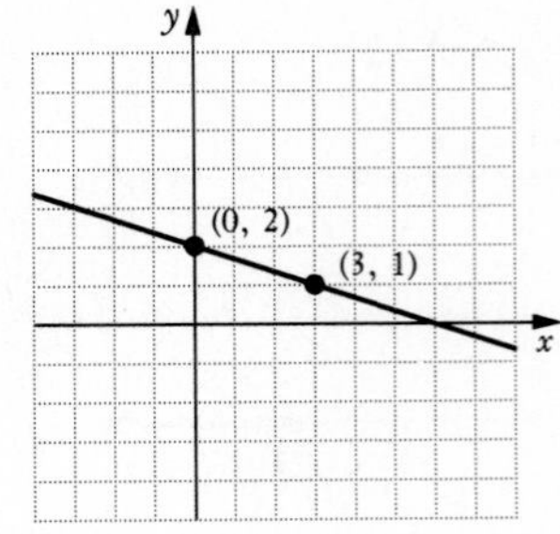

3. Undefined **4.** 0 **5.** -2.3 **6.** $\frac{2}{3}$ **7.** Slope is -8; y-intercept is $(0, 23)$. **8.** Slope is $\frac{1}{2}$; y-intercept is $\left(0, -\frac{5}{2}\right)$. **9.** $y = 3.4x - 8$

10.

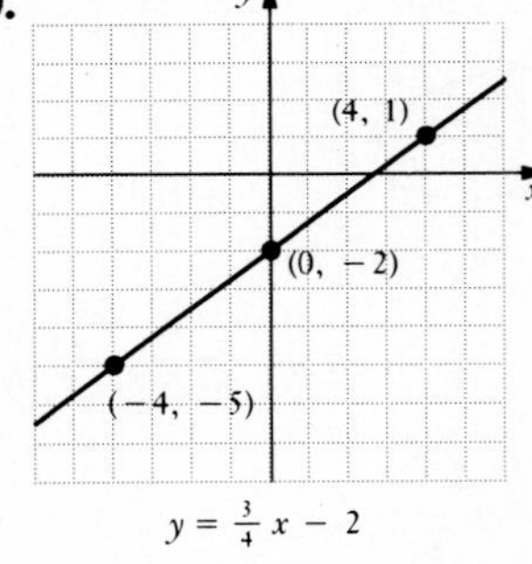

11.

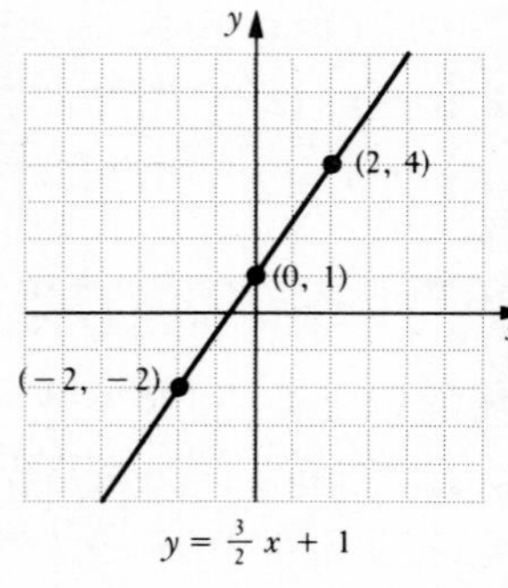

12.

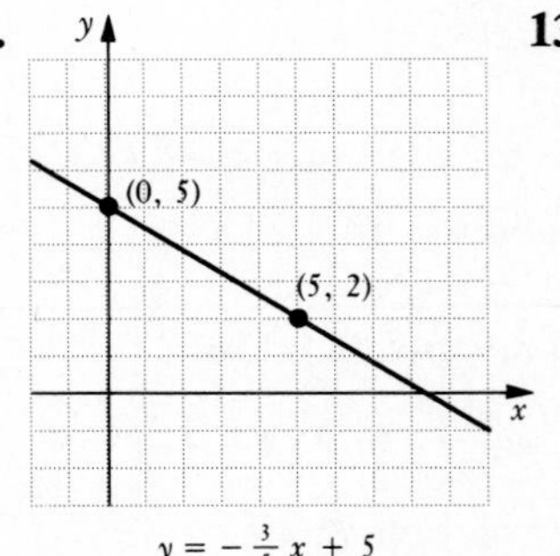

13.

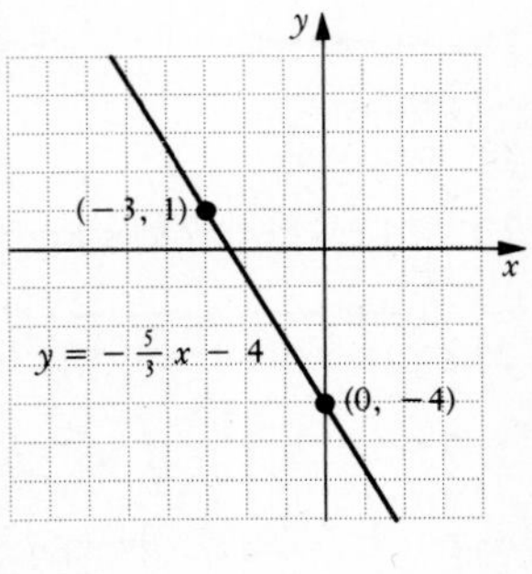

14. 12%

Margin Exercises, Section 3.4, pp. 153-156

1. $y = 3x - 5$ **2.** $y = -\frac{4}{3}x - \frac{71}{12}$ **3.** $y = -\frac{5}{3}x + \frac{11}{3}$
4. $y = -17x - 56$ **5.** Yes **6.** No **7.** No **8.** $y = \frac{8}{7}x - \frac{23}{7}$
9. Yes **10.** No **11.** $y = -2x + 14$

Margin Exercises, Section 3.5, pp. 159-160

1. (a) $R = -0.01t + 10.43$; **(b)** 9.69 sec, 9.65 sec; **(c)** 2063

Margin Exercises, Section 3.6, pp. 163-166

1. No **2.** Yes

3.

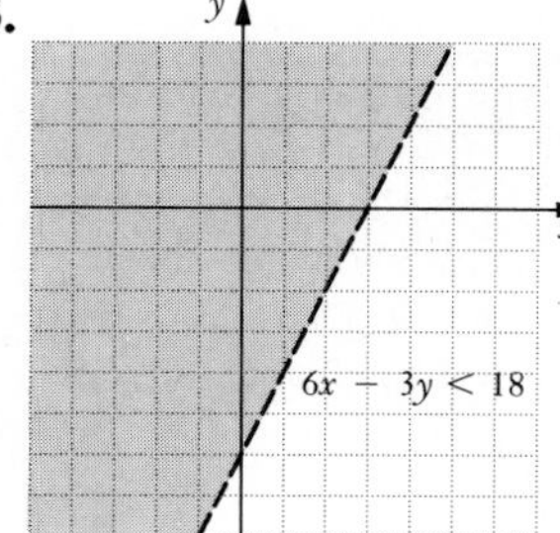

4.

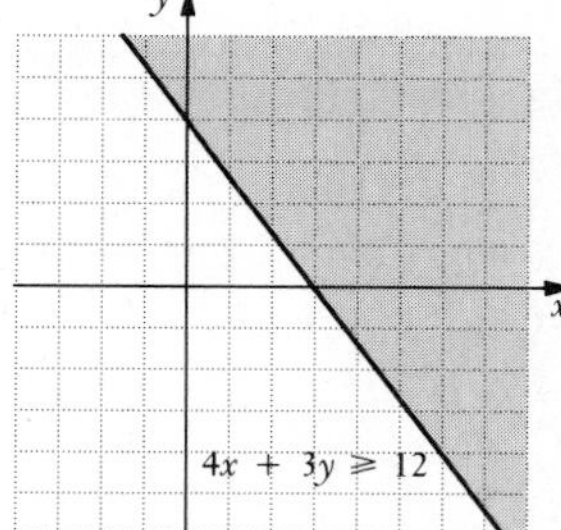

5.

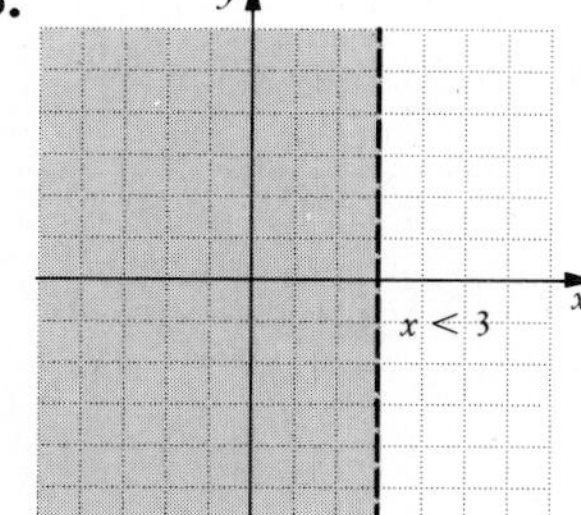

6.

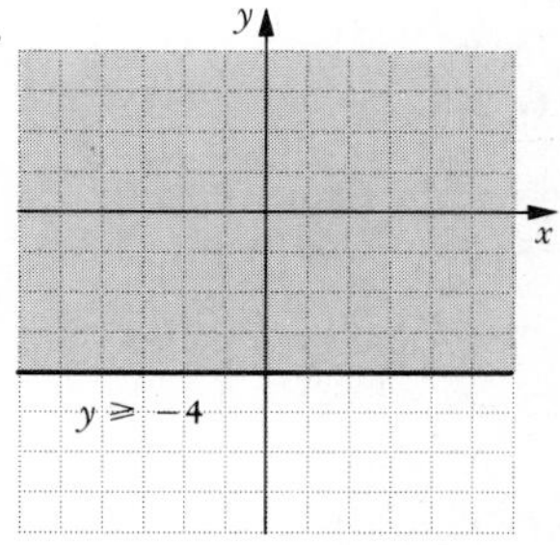

CHAPTER 4

Margin Exercises, Section 4.1, pp. 177-180

1. Yes **2.** No **3.** $(0,1)$ **4.** $(2,1)$ **5.** No Solution
6. Consistent: 3, 4; inconsistent: 5 **7.** Infinitely many solutions. **8.** Independent: 3, 4, 5; dependent: 7

Margin Exercises, Section 4.2, pp. 183-188

1. $(2,4)$ **2.** $(5,2)$ **3.** $(-3,2)$ **4.** $(13,16)$ **5.** $(1,4)$
6. $\left(\frac{1}{3},\frac{1}{2}\right)$ **7.** $(2,3)$ **8.** $(-127,100)$ **9.** Infinite number of solutions **10.** Infinite number of solutions
11. $2x + 3y = 1$, $3x - y = 7$; $(2,-1)$ **12.** $9x + 10y = 5$, $9x - 4y = 3$; $\left(\frac{25}{63},\frac{1}{7}\right)$

Margin Exercises, Section 4.3, pp. 191-198

1. $48\frac{1}{3}°$, $131\frac{2}{3}°$ **2.** 22 white, 8 red
3. 30 L of 5%, 70 L of 15%

5% weedkiller	5% weedkiller	Mixture
x liters	y liters	100 liters
5%	15%	12%
$0.05x$	$0.15y$	12 liters

4. \$1800 at 7%, \$1900 at 9%

First Investment	Second Investment	Total
x	y	\$3700
7%	9%	
1 yr	1 yr	
$0.07x$	$0.09y$	\$297

5. w: 100 ft; l: 160 ft
6. 280 km

Distance	Rate	Time	
d	35 km/h	t	⟵ $d = 35t$
d	40 km/h	$t - 1$	⟵ $d = 40(t-1)$

7. 180 mph

Distance	Rate	Time	
d	$r + 20$	4 hr	⟵ $d = 4(r+20)$
d	$r - 20$	5 hr	⟵ $d = 5(r-20)$

Margin Exercises, Section 4.4, pp. 203-206

1. (a) No; **(b)** *yes* **2.** $(2,1,-1)$ **3.** $\left(2,-2,\frac{1}{2}\right)$
4. $(20,30,50)$

Margin Exercises, Section 4.5, pp. 209-212

1. 2 −3, 4 **2.** 64°, 32°, 84° **3.** \$400 at 7%, \$500 at 8%, \$1600 at 9%

Margin Exercises, Section 4.6, pp. 217-220

1. −5 **2.** −7 **3.** −6 **4.** 93 **5.** $(3,-1)$ **6.** $(1,3,-2)$

Margin Exercises, Section 4.7, pp. 223-226

1. 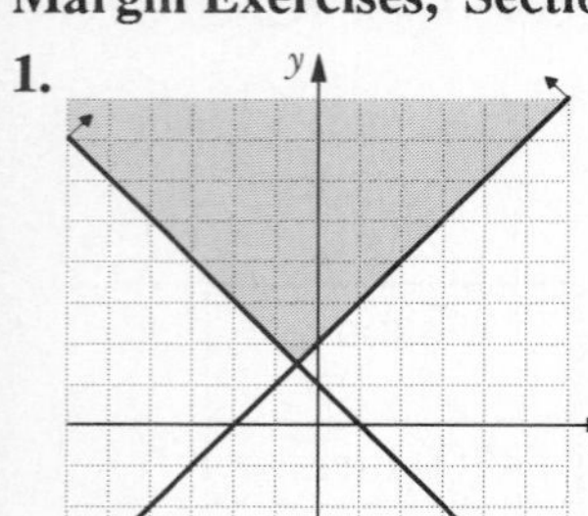

2. 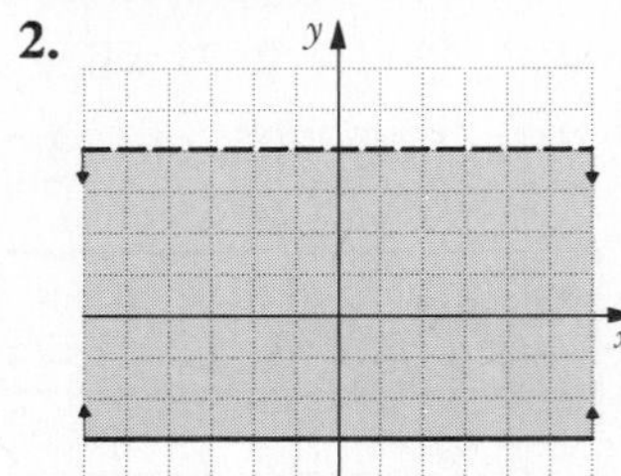

3. 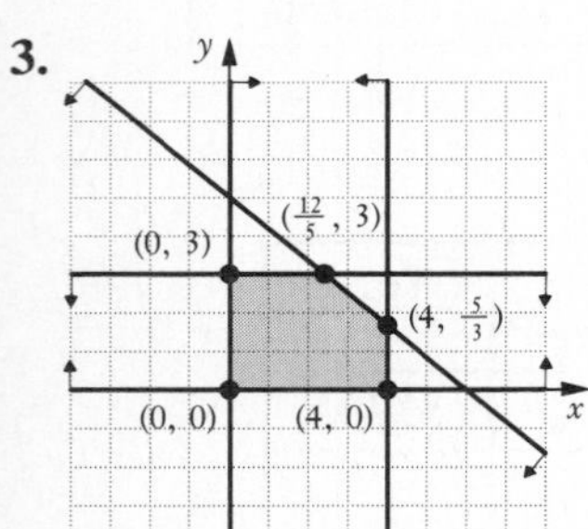

4. 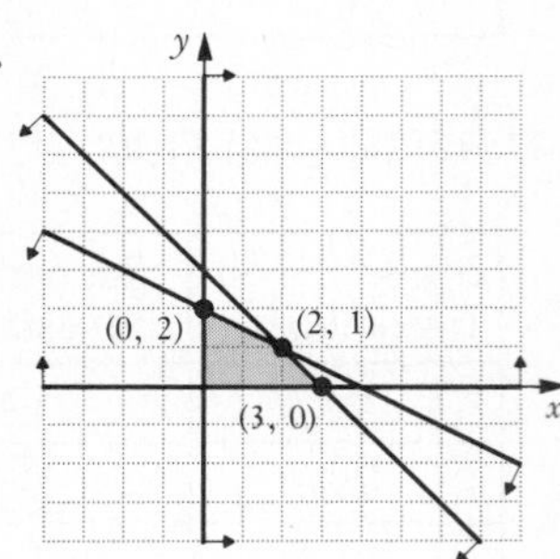

CHAPTER 5

Margin Exercises, Section 5.1, pp. 237-242

1. $-92x^5, -8x^4, x^2, 5; -92x^5$ **2.** 5, –4, –2, 1, –1, –5; 5 **3. (a)** 2, 3, 1, 0; 3; $-5x^3$; –5 **(b)** 1, 0, 1, 7, 3; 7; $7x^2y^3z^2$; 7 **4.** Monomials: c, d, e; binomials: a, f, h; trinomials: b, g **5. (a)** $5+10x-6x^2+7x^3-x^4$; **(b)** $-x^4+7x^3-6x^2+10x+5$ **6. (a)** $-5+x^3+5x^4y-3y^2+3x^2y^3$; **(b)** $3x^2y^3-3y^2+5x^4y+x^3-5$ **7. (a)** 5; **(b)** 13; **(c)** 13 **8.** 11 cents/km **9.** $3y-4x+4xy^2$ **10.** $8xy^3+2x^3y-9x^2y$ **11.** $-4x^3+2x^2-4x+2$ **12.** $10y^5-4y^2+5$ **13.** $5p^2q^4-8p^2q^2+5$ **14.** $-\left(4x^3-5x^2+\frac{1}{4}x-10\right); -4x^3+5x^2-\frac{1}{4}x+10$ **15.** $-\left(8xy^2-4x^3y^2-9x-\frac{1}{5}\right); -8xy^2+4x^3y^2+9x+\frac{1}{5}$ **16.** $-\left(-9y^5-8y^4+\frac{1}{2}y^3-y^2+y-1\right); 9y^5+8y^4-\frac{1}{2}y^3+y^2-y+1$ **17.** $3x^2+5$ **18.** $14y^3-2y+4$ **19.** p^2-6p-2 **20.** $3y^5-3y^4+5y^3-2y^2-3$ **21.** $9p^4q-10p^3q^2+4p^2q^3+9q^4$ **22.** $y^3-y^2+\frac{4}{3}y+0.1$

Margin Exercises, Section 5.2, pp. 247-250

1. $-18y^3$ **2.** $24x^8y^3$ **3.** $-90x^4y^9z^{12}$ **4.** $-6y^2-18y$ **5.** $8xy^3-10xy$ **6.** $5x^3+15x^2-4x-12$ **7.** $6y^2+y-12$ **8.** $p^4+p^3-12p^2-5p+15$ **9.** $2x^4-8x^3+4x^2-21x+20$ **10.** $8x^5+12x^4-20x^3+4x^2-15x+6$ **11.** $a^5-2a^4b+3a^3b+a^3b^2-7a^2b^2+5ab^3-b^4$ **12.** $y^2+6y-40$ **13.** $2p^2+7pq-15q^2$ **14.** $x^3y^3+x^2y^3+2x^2y^2+2xy^2$ **15.** $a^2-2ab+b^2$ **16.** $x^2+16x+64$ **17.** $9x^2-42x+49$ **18.** $m^6+\frac{1}{2}m^3n+\frac{1}{16}n^2$ **19.** x^2-64 **20.** $16y^2-49$ **21.** $7.84a^2-16.81b^2$ **22.** $9w^2-\frac{9}{25}q^4$ **23.** $49x^4y^2-4y^2$ **24.** $4x^2+12x+9-25y^2$ **25.** $81x^4-16y^4$

Margin Exercises, Section 5.3, pp. 255-256

1. $3(x^2-2)$ **2.** $4x^3(x^2-2)$ **3.** $3y^2(3y^2-5y+1)$ **4.** $3x^2y(2-7xy+y^2)$ **5.** $-8(x-4)$ **6.** $-3(x^2+5x-3)$ **7.** $(p+q)(2x+y+2)$ **8.** $(y+3)(2y-11)$ **9.** Cannot be factored by grouping **10.** $(y^2-2)(5y+2)$

Margin Exercises, Section 5.4, pp. 259-266

1. $(x+2)(x+3)$ **2.** $(y+2)(y+5)$ **3.** $(m-2)(m-6)$ **4.** $(t-3)(t-8)$, or $(3-t)(8-t)$ **5.** $x(x+6)(x-2)$ **6.** $(y-6)(y+2)$ **7.** $(x+10)(x-11)$ **8.** Not factorable **9.** $(x-2y)(x-3y)$ **10.** $(p-8q)(p+2q)$ **11.** $(x^2-2)(x^2-9)$ **12.** $(p^3+3)(p^3-2)$ **13.** $(x-7)(3x+8)$ **14.** $(3x+2)(x+1)$ **15.** $2(4y-1)(3y-5)$ **16.** $2x^3(2x-3)(5x-4)$ **17.** $(3x+4)(x+5)$ **18.** $4(2x-1)(2x+3)$ **19.** $(2x+3)(2x-1)$ **20.** $(4x+1)(x+9)$ **21.** $y^2(5y+4)(2y-3)$ **22.** $a(36a^2+21a+1)$

Margin Exercises, Section 5.5, pp. 271-274

1. (a), (b), (d) **2.** $(x+7)^2$ **3.** $(3y-5)^2$ **4.** $(4x+9y)^2$ **5.** $(4x^2-5y^3)^2$ **6.** $-2(2a-3b)^2$ **7.** $3(a-5b)^2$ **8.** $(y+2)(y-2)$ **9.** $(7x^2+5y^5)(7x^2-5y^5)$ **10.** $(5xy+2a)(5xy-2a)$ **11.** $(3x+4y)(3x-4y)$ **12.** $5(2x+y)(2x-y)$ **13.** $y^2(9x^2+4)(3x+2)(3x-2)$ **14.** $(a+4)(a-4)(a+1)$ **15.** $(x+1+p)(x+1-p)$ **16.** $(y-4+3m)(y-4-3m)$ **17.** $(x+4-10t)(x+4+10t)$ **18.** $[8p+(x+4)][8p-(x+4)]$, or $(8p+x+4)(8p-x-4)$

Margin Exercises, Section 5.6, pp. 277-278

1. $(x-2)(x^2+2x+4)$ **2.** $(4-y)(16+4y+y^2)$ **3.** $(3x+y)(9x^2-3xy+y^2)$ **4.** $(2y+z)(4y^2-2yz+z^2)$ **5.** $(m+n)(m^2-mn+n^2)(m-n)(m^2+mn+n^2)$ **6.** $(2xy)(2x^2+3y^2)(4x^4-6x^2y^2+9y^4)$ **7.** $(3x+2y)(9x^2-6xy+4y^2)(3x-2y)(9x^2+6xy+4y^2)$ **8.** $(x-0.3)(x^2+0.3x+0.09)$

Margin Exercises, Section 5.7, pp. 281-282

1. $3y(y+2x)(y-2x)$ **2.** $7(a-1)(a^2+a+1)$

3. $(2x+3y)(4x^2-6xy+9y^2)(2x-3y)(4x^2+6xy+9y^2)$
4. $(x-2)(3-bx)$ **5.** $5(y^4+4x^6)$ **6.** $3(x-2)(2x+3)$
7. $(a-b)^2(a+b)$ **8.** $3(x+3a)^2$
9. $2(x-5+3b)(x-5-3b)$

Margin Exercises, Section 5.8, pp. 285-290

1. 4, 2 **2.** $\frac{1}{2}$, –3 **3.** 0, 2 **4.** –5 **5.** 0, 2, –3 **6.** 8, –6
7. $l = 8$ cm, $w = 3$ cm **8.** 13 m, 12 m

Margin Exercises, Section 5.9, pp. 295-300

1. $\frac{x^2}{2}+8x+3$ **2.** $5y^3-2y^2+6y$ **3.** $\frac{x^2y^2}{2}+5xy+8$
4. $x+5$ **5.** $3y^3-2y^2+6y-4$ **6.** $y^2-8y-24$, R -66; or $y^2-8y-24+\frac{-66}{y-3}$ **7.** $x^2+10x+10$, R 5; or $x^2+10x+10+\frac{5}{x-1}$ **8.** $y-11$, R $3y-27$; or $y-11+\frac{3y-27}{y^2-3}$ **9.** $2x^2+2x+14$, R 34; or $2x^2+2x+14+\frac{34}{x-3}$ **10.** $x^2-4x+13$, R -30; or $x^2-4x+13+\frac{-30}{x+2}$ **11.** y^2-y+1

CHAPTER 6

Margin Exercises, Section 6.1, pp. 311-316

1. All real numbers except $-\frac{5}{2}$ **2.** All real numbers except 2 and 5 **3.** $\frac{(3x+2y)x}{(5x+4y)x}$ **4.** $\frac{(2x^2-y)(3x+2)}{(3x+4)(3x+2)}$ **5.** $\frac{-1(2a-5)}{-1(a-b)}$ **6.** $\frac{4}{5}$
7. $7x$ **8.** $2a+3$ **9.** $\frac{3x+2}{2(x+2)}$ **10.** $\frac{2(y+2)}{y-1}$ **11.** $\frac{3(x-y)(x-y)}{x+y}$
12. $a-b$ **13.** $\frac{x-5}{x+3}$ **14.** $\frac{1}{x+7}$ **15.** y^3-9 **16.** $\frac{x+5}{2(x-5)}$
17. $\frac{2ab(a+b)}{a-b}$ **18.** a^2-4

Margin Exercises, Section 6.2, pp. 321-326

1. 90 **2.** 72 **3.** $\frac{47}{60}$ **4.** $\frac{97}{72}$ **5.** $5a^3b^2$
6. $(y+3)(y+4)(y+4)$ **7.** $2x^2(x-3)(x+3)(x+2)$
8. $2(a+b)(a-b)$, or $2(a+b)(b-a)$ **9.** $\frac{12+y}{y}$
10. $3x+1$ **11.** $\frac{a-b}{b+2}$ **12.** $\frac{y+12}{x^2+y^2}$ **13.** $\frac{1-b^2}{3}$ **14.** $\frac{2x^2+11}{x-5}$
15. $\frac{3+7x}{4y}$ **16.** $\frac{11x^2}{2x-y}$ **17.** $\frac{9x^2+28y}{21x}$ **18.** $\frac{3}{x+y}$ **19.** $\frac{a+12}{a(a+3)}$
20. $\frac{3y^2+12y+3}{(y-4)(y-3)(y+5)}$ **21.** $\frac{2}{x-1}$

Margin Exercises, Section 6.3, pp. 331-334

1. $\frac{2}{3}$ **2.** –31 **3.** No solution **4.** –3 **5.** 4, –3 **6.** 7 **7.** –13

Margin Exercises, Section 6.4, pp. 337-342

1. $2\frac{2}{5}$ hr **2.** Pipe A, 32 hr; pipe B, 96 hr **3.** $2\frac{23}{95}$, or 2.24
4. 1620 **5.** 280 miles **6.** 35.5 mph

Margin Exercises, Section 6.5, pp. 347-348

1. $T=\frac{PV}{k}$ **2.** $t=\frac{uv}{v+u}$ **3.** $T=\frac{A+Vnt}{Vn}$ **4.** $M=\frac{Fd^2}{km}$ **5.** $t=\frac{5L}{r-w}$

Margin Exercises, Section 6.6, pp. 351-354

1. $\frac{7(2y+1)}{2(7y-1)}$ **2.** $\frac{x}{x+1}$ **3.** $\frac{b+a}{b-a}$ **4.** $\frac{a^2b^2}{b^2+ab+a^2}$ **5.** $\frac{7(2y+1)}{2(7y-1)}$ **6.** $\frac{x}{x+1}$
7. $\frac{b+a}{b-a}$ **8.** $\frac{a^2b^2}{b^2+ab+a^2}$

Margin Exercises, Section 6.7, pp. 357-362

1. $\frac{2}{5}$; $y=\frac{2}{5}x$ **2.** 0.7; $y=0.7x$ **3.** 50 volts
4. 176,250 tons **5.** 0.6; $y=\frac{0.6}{x}$ **6.** $7\frac{1}{2}$ hr **7.** $y=7x^2$
8. $y=\frac{9}{x^2}$ **9.** $y=\frac{1}{2}xz$ **10.** $y=\frac{5xz^2}{w}$ **11.** 490 m **12.** 1 ohm

CHAPTER 7

Margin Exercises, Section 7.1, pp. 375-378

1. 3, –3 **2.** 6, –6 **3.** 11, –11 **4.** 1 **5.** 6 **6.** $\frac{9}{10}$ **7.** 0.08
8. **(a)** 4; **(b)** –4; **(c)** Does not exist **9.** **(a)** 7; **(b)** –7; **(c)** Does not exist **10.** **(a)** 12; **(b)** –12; **(c)** Does not exist **11.** $28+x$ **12.** $\frac{y}{y+3}$ **13.** $|y|$ **14.** 24 **15.** $|5y|$, or $5|y|$ **16.** $|4y|$, or $4|y|$ **17.** $|x+7|$ **18.** $|2(x-2)|$, or $2|x-2|$ **19.** $|7(y+5)|$, or $7|y+5|$ **20.** $|x-3|$
21. –4 **22.** $3y$ **23.** $2(x+2)$ **24.** $-\frac{7}{4}$ **25.** $-2x$
26. $3x+2$ **27.** **(a)** 3; **(b)** –3; **(c)** Does not exist
28. $|2(x-2)|$, or $2|x-2|$ **29.** $|x|$ **30.** $|x+3|$

Margin Exercises, Section 7.2, pp. 381-384

1. $\sqrt{133}$ **2.** $\sqrt{x^2-4y^2}$ **3.** $\sqrt[4]{2821}$ **4.** $\sqrt[3]{8x^5+40x}$ **5.** $4\sqrt{2}$
6. $2\sqrt[3]{10}$ **7.** $10\sqrt{3}$ **8.** $6y$ **9.** $(x+2)\sqrt{3}$ **10.** $2bc\sqrt{3ab}$
11. $2\sqrt[3]{2}$ **12.** $3xy^2\sqrt[3]{3xy^2}$ **13.** $(a+b)\sqrt[3]{a+b}$ **14.** $3\sqrt{2}$
15. $6y\sqrt{7}$ **16.** $3x\sqrt[3]{4y}$ **17.** $7\sqrt{3ab}$
18. $2(y+5)\sqrt[3]{(y+5)^2}$ **19.** 4.796 **20.** 9.849 **21.** 12.649
22. 18.466 **23.** –3.487 **24.** 9.071

Margin Exercises, Section 7.3, pp. 387-388

1. 5 **2.** $56\sqrt{xy}$ **3.** $5a$ **4.** $\frac{20}{7}$ **5.** $\frac{2ab}{3}$ **6.** $\frac{5}{6}$ **7.** $\frac{x}{10}$ **8.** $\frac{3x\sqrt[3]{2x^2}}{5}$

9. 25 **10.** $6y\sqrt{6y}$ **11.** $2a^4b\sqrt{2ab}$

Margin Exercises, Section 7.4, pp. 391-392

1. $13\sqrt{2}$ **2.** $10\sqrt[4]{5x}-\sqrt{7}$ **3.** $19\sqrt{5}$ **4.** $(3y+4)\sqrt[3]{y^2}+2y^2$ **5.** $2\sqrt{x-1}$ **6.** $5\sqrt{6}+3\sqrt{14}$ **7.** $a\sqrt[3]{3}-\sqrt[3]{2a^2}$ **8.** $-4-9\sqrt{6}$ **9.** $3\sqrt{ab}-4\sqrt{3a}+6\sqrt{3b}-24$ **10.** -3 **11.** $p-q$ **12.** $20-4y\sqrt{5}+y^2$ **13.** $58+12\sqrt{6}$

Margin Exercises, Section 7.5, pp. 397-400

1. $\frac{\sqrt{10}}{5}$ **2.** $\frac{3\sqrt{10}}{2}$ **3.** $\frac{\sqrt{10}}{5}$ **4.** $\frac{2\sqrt{3ab}}{3b}$ **5.** $\frac{\sqrt[4]{56}}{2}$ **6.** $\frac{x\sqrt[3]{12x^2y^2}}{2y}$ **7.** $\frac{7\sqrt[3]{2x^2y}}{2y^2}$ **8.** $\frac{11}{\sqrt{66}}$ **9.** $\frac{3a^2}{\sqrt[3]{18a^2c^2}}$ **10.** c^2-b **11.** $a-b$ **12.** $\frac{\sqrt{3}+y}{\sqrt{3}+y}$ **13.** $\frac{\sqrt{2}-\sqrt{3}}{\sqrt{2}-\sqrt{3}}$ **14.** $-7-6\sqrt{2}$ **15.** $6-2\sqrt{2}$ **16.** $\frac{5+\sqrt{7}}{5+\sqrt{7}}$ **17.** $\frac{\sqrt{a}-\sqrt{b}}{\sqrt{a}-\sqrt{b}}$ **18.** $\frac{4}{3\sqrt{2}-\sqrt{10}-3\sqrt{6}+\sqrt{30}}$

Margin Exercises, Section 7.6, pp. 405-408

1. $\sqrt[4]{y}$ **2.** $\sqrt{3a}$ **3.** $\sqrt[4]{16}$, or 2 **4.** $\sqrt[3]{125}$, or 5 **5.** $\sqrt[5]{a^3b^2c}$ **6.** $19^{1/3}$ **7.** $(abc)^{1/2}$ **8.** $\left(\frac{x^2y}{16}\right)^{1/5}$ **9.** $\sqrt{x^3}$, or $(\sqrt{x})^3$, which simplifies to $x\sqrt{x}$ **10.** $\sqrt[3]{8^2}$, or $(\sqrt[3]{8})^2$, which simplifies to 4 **11.** $\sqrt{4^5}$, or $(\sqrt{4})^5$, which simplifies to 32 **12.** $(7abc)^{4/3}$ **13.** $6^{7/5}$ **14.** $\frac{1}{\sqrt[4]{16}}$, or $\frac{1}{2}$ **15.** $\frac{1}{(3xy)^{7/8}}$ **16.** $7^{14/15}$ **17.** $5^{2/6}$, or $5^{1/3}$ **18.** $9^{2/5}$ **19.** $\sqrt{a}$ **20.** x **21.** $\sqrt{2}$ **22.** ab^2 **23.** xy^3 **24.** $\sqrt[4]{xy^2}$ **25.** $\sqrt[4]{63}$ **26.** $\sqrt[4]{a-b}$ **27.** $\sqrt[6]{x^4y^3z^5}$ **28.** $\sqrt[4]{ab}$

Margin Exercises, Section 7.7, pp. 411-414

1. 100 **2.** No solution **3.** 1 **4.** 2, 5 **5.** 4 **6.** 17 **7.** 9 **8.** 27 **9.** 5

Margin Exercises, Section 7.8, pp. 417-418

1. $\sqrt{41}$; 6.403 **2.** $\sqrt{6}$; 2.449 **3.** $\sqrt{200}$; 14.142 **4.** $\sqrt{14,500}$; 120.416 ft **5.** 7.4 ft

Margin Exercises, Section 7.9, pp. 421-426

1. $i\sqrt{5}$, or $\sqrt{5}i$ **2.** $5i$ **3.** $-i\sqrt{11}$, or $-\sqrt{11}i$ **4.** $-6i$ **5.** $3i\sqrt{6}$, or $3\sqrt{6}i$ **6.** $15-3i$ **7.** $-12+6i$ **8.** $3-5i$ **9.** $12-7i$ **10.** -10 **11.** $-\sqrt{34}$ **12.** 42 **13.** $-9-12i$ **14.** $-35-25i$ **15.** $-14+8i$ **16.** $11+10i$ **17.** $5+12i$ **18.** $-i$ **19.** 1 **20.** i **21.** -1 **22.** $8-i$ **23.** 3 **24.** $-7-6i$ **25.** $-1+i$ **26.** $6-3i$ **27.** $-9+5i$ **28.** $\pi+\frac{1}{4}i$ **29.** 53 **30.** 10 **31.** $2i$ **32.** $\frac{10}{17}-\frac{11}{17}i$ **33.** Yes **34.** Yes

CHAPTER 8

Margin Exercises, Section 8.1, pp. 439-446

1. $\frac{3}{5}$, 1 **2.** 4, -4; or ±4 **3.** $\sqrt{3}$, $-\sqrt{3}$; or $\pm\sqrt{3}$ **4.** $\frac{2\sqrt{6}}{3}$, $-\frac{2\sqrt{6}}{3}$; or $\pm\frac{2\sqrt{6}}{3}$ **5.** $\frac{\sqrt{2}}{2}i$, $-\frac{\sqrt{2}}{2}i$; or $\pm\frac{\sqrt{2}}{2}i$ **6.** $1\pm\sqrt{5}$ **7.** $-8\pm\sqrt{11}$ **8.** -2, -4 **9.** 10, -2 **10.** $-3\pm\sqrt{10}$ **11.** $\frac{1\pm\sqrt{22}}{3}$ **12.** \$3603.65 **13.** 20% **14.** About 9.5 sec

Margin Exercises, Section 8.2, pp. 449-452

1. (a) $-\frac{1}{2}$, 4; (b) $-\frac{1}{2}$, 4 **2.** $\frac{-1\pm\sqrt{22}}{3}$ **3.** $\frac{1\pm i\sqrt{7}}{2}$ **4.** $-\frac{1}{3}$, 2 **5.** 2.1, 0.2 **6.** 1.2, -1.9

Margin Exercises, Section 8.3, pp. 455-456

1. Two real **2.** One real **3.** Two nonreal **4.** $3x^2+7x-20=0$ **5.** $x^2+25=0$ **6.** $x^2+\sqrt{2}x-4=0$

Margin Exercises, Section 8.4, pp. 459-460

1. ±3, ±1 **2.** 4 **3.** 4, 2, -1, -3 **4.** $\frac{1}{2}$, $-\frac{1}{3}$

Margin Exercises, Section 8.5, pp. 463-466

1. $w_2=\frac{w_1}{A^2}$ **2.** $r=\sqrt{\frac{V}{\pi h}}$ **3.** $s=\frac{-R+\sqrt{R^2-\frac{4L}{c}}}{2L}$ **4.** $b=\frac{pa}{\sqrt{1+p^2}}$ **5.** 2 ft **6.** A to B is 6 mi; A to C is 8 mi **7.** A to B is $\sqrt{31}-1$, or 4.57 mi; A to C is $1+\sqrt{31}$, or 6.57 mi

Margin Exercises, Section 8.6, pp. 471-474

1. $-\frac{2}{5}$, 10 **2.** 36 hr **3.** 10 mph

Distance	Speed	Time
24	$r-2$	t_1
24	$r+2$	t_2

4. 20 knots, 30 knots

Distance	Speed	Time
3000	$r+10$	$t-50$
3000	r	t

Margin Exercises, Section 8.7, pp. 479-484

1.

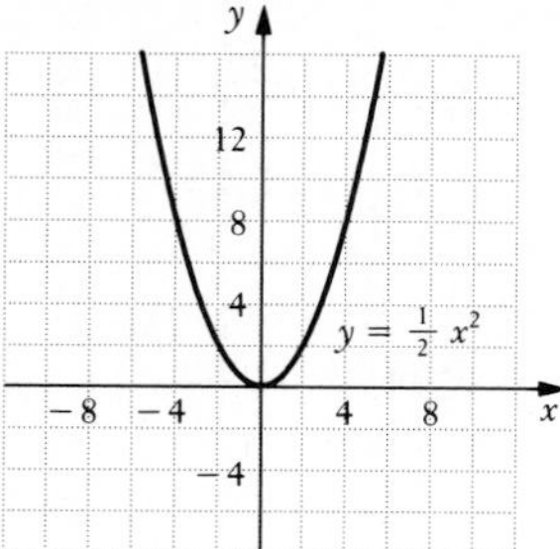

2.

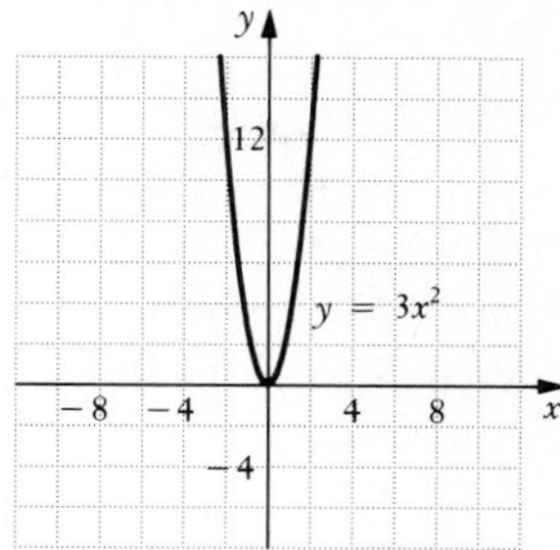

3.

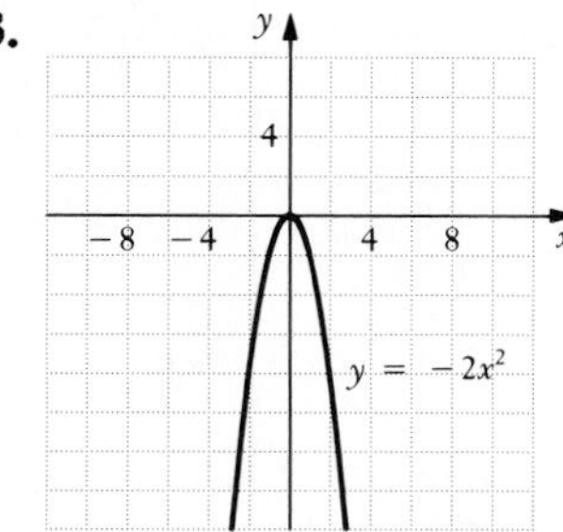

4.

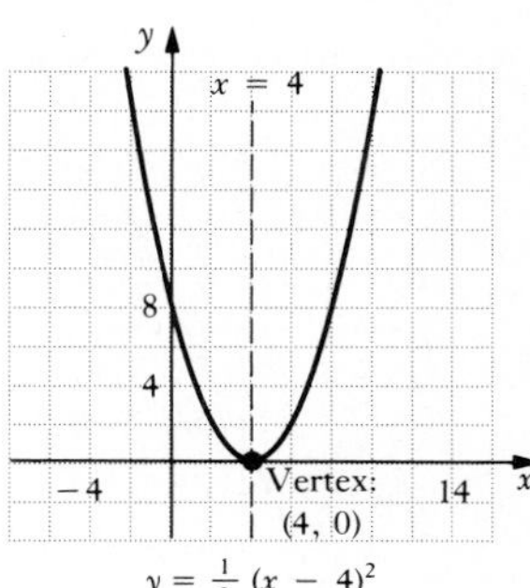

$y = \frac{1}{2}(x-4)^2$

5.

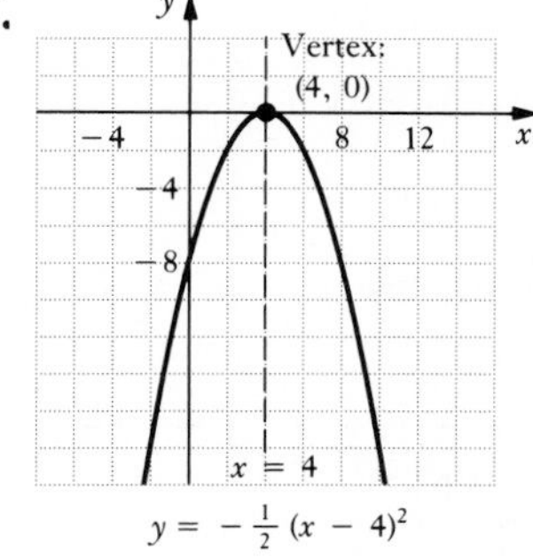

$y = -\frac{1}{2}(x-4)^2$

6.

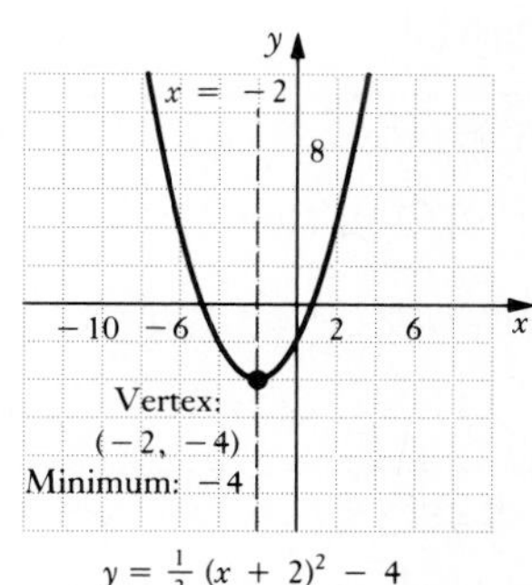

$y = \frac{1}{2}(x+2)^2 - 4$

7.

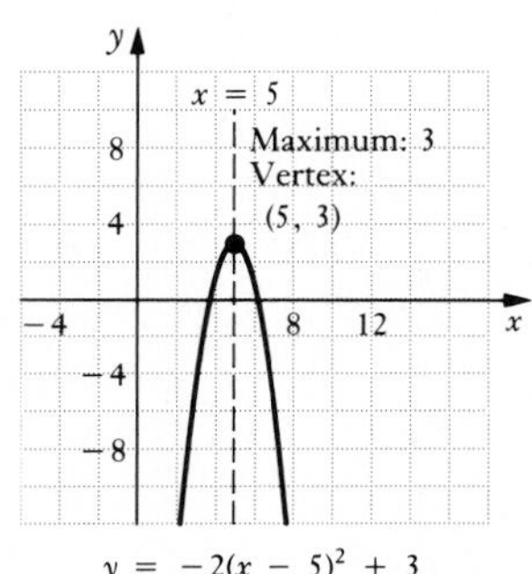

$y = -2(x-5)^2 + 3$

Margin Exercises, Section 8.8, pp. 487-492

1. (a) $(2, 3)$, $x = 2$; **(b)**

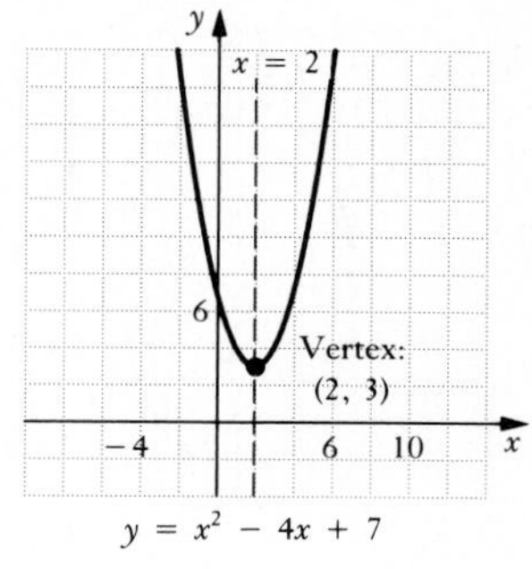

$y = x^2 - 4x + 7$

2. (a) 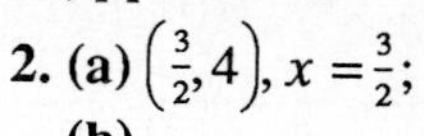 $\left(\frac{3}{2}, 4\right)$, $x = \frac{3}{2}$; **(b)**

$x = \frac{3}{2}$

Vertex: $\left(\frac{3}{2}, 4\right)$

$y = -4x^2 + 12x - 5$

3. $(1-\sqrt{6}, 0)$, $(1+\sqrt{6}, 0)$ **4.** $(-1, 0)$, $(3, 0)$ **5.** $(-4, 0)$
6. None **7.** 225 **8.** 25 m by 25 m **9.** $y = x^2 - 2x + 1$
10. (a) $y = 0.625x^2 - 50x + 1150$; **(b)** 510

Margin Exercises, Section 8.9, pp. 497-502

1. $\{x \mid x < -3 \text{ or } x > 1\}$ **2.** $\{x \mid -3 < x < 1\}$
3. $\{x \mid -3 \le x \le 1\}$ **4.** $\{x \mid x < -4 \text{ or } x > 1\}$
5. $\{x \mid -4 \le x \le 1\}$ **6.** $\{x \mid x < -4 \text{ or } x > 1\}$
7. $\{x \mid -4 \le x \le 1\}$ **8.** $\{x \mid x < -1 \text{ or } 0 < x < 1\}$
9. $\left\{x \,\middle|\, 2 < x \le \frac{7}{2}\right\}$ **10.** $\{x \mid x < 5 \text{ or } x > 10\}$

CHAPTER 9

Margin Exercises, Section 9.1, pp. 513-518

1.

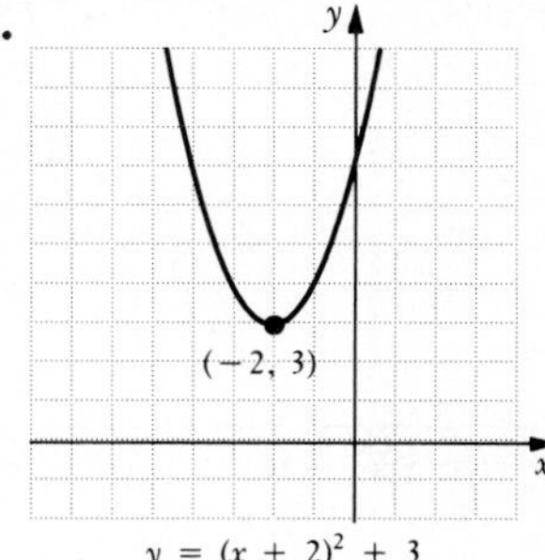

$y = (x+2)^2 + 3$

2.

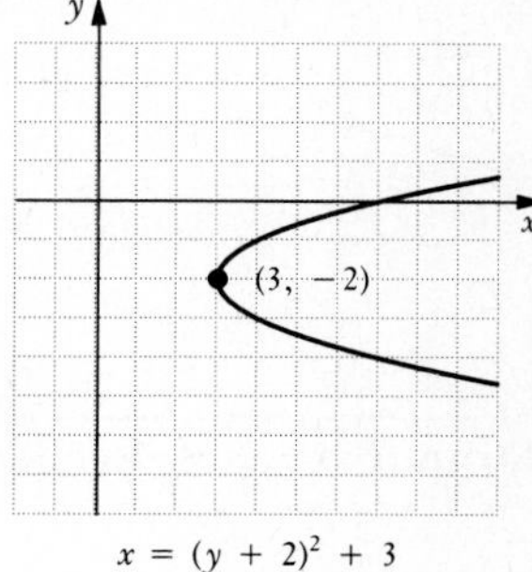

$x = (y+2)^2 + 3$

3. 14 **4.** 11 **5.** 10 **6.** $\sqrt{37} \approx 6.083$ **7.** $\left(\frac{3}{2}; -3\right)$ **8.** $(9, -5)$

9.

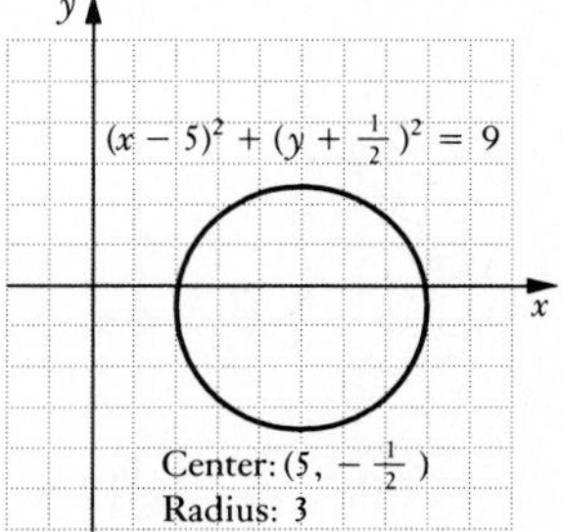

10. $(x+3)^2+(y-1)^2=36$ **11.** $(0,0)$, $r=8$
12. $(x+4)^2+(y-5)^2=49$; center: $(-4,5)$; radius: 7

Margin Exercises, Section 9.2, pp. 523-528

1.

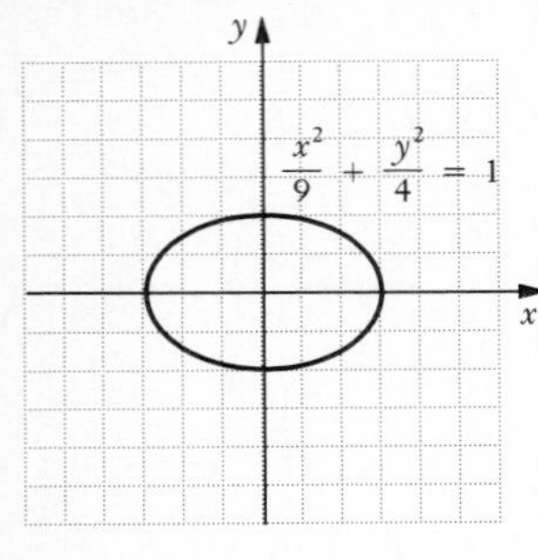

2.

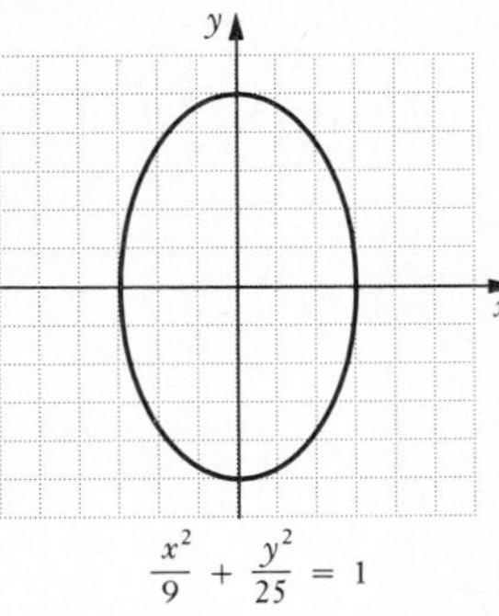

3.

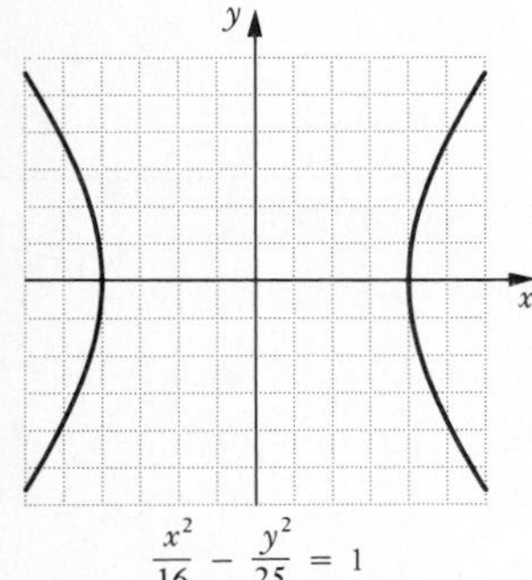

4. (a)

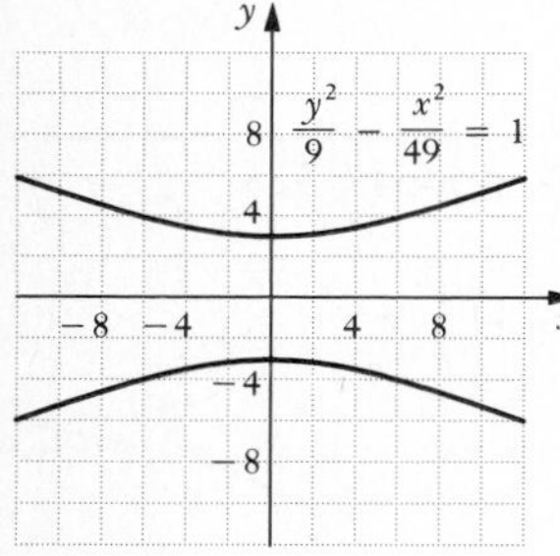

(b)

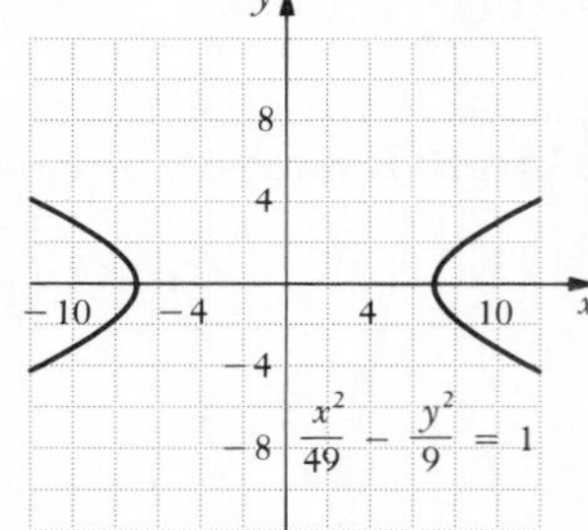

Margin Exercises, Section 9.3, pp. 531-536

1.

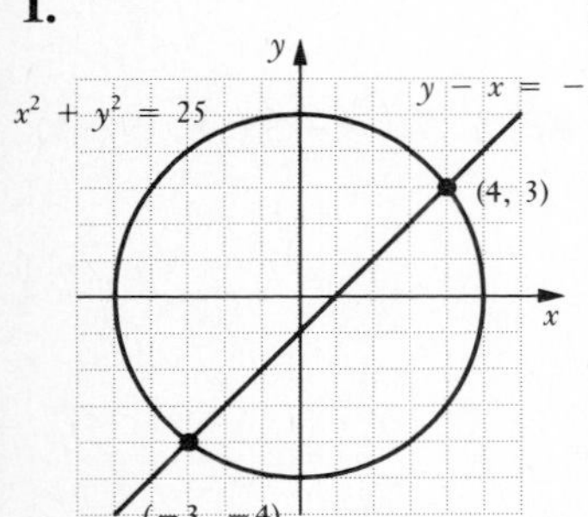

2.

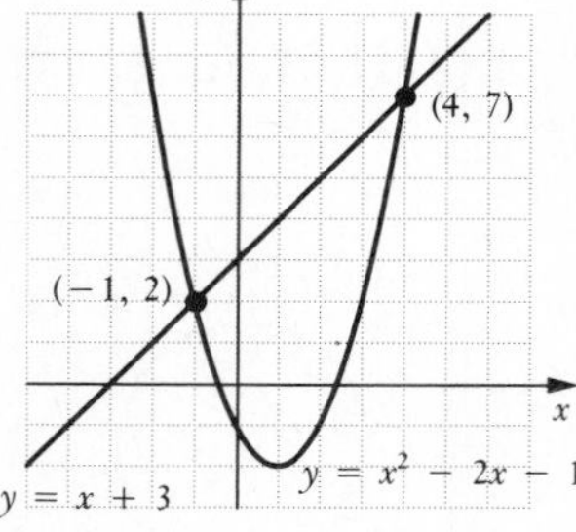

3. $\left(-\frac{5}{7},\frac{22}{7}\right)$, $(1,-2)$ **4.** $\left(-\frac{3}{2}i,\frac{15}{4}i\right)$, $\left(\frac{3}{2}i,-\frac{15}{4}i\right)$ **5.** $(2,3)$
$(2,-3)$ $(-2,3)$, $(-2,-3)$ **6.** $(3,2)$, $(-3,-2)$, $(2,3)$, $(-2,-3)$
7. 12 m by 5 m **8.** Length: 2 in.; width: 1 in.

Margin Exercises, Section 9.4, pp. 541-546

1. $d=\{(6,38),(8,40),(10,42),(12,44),(14,46),(16,48),(18,50)\}$ **2.** $q=\{(3,10),(3,11),(4,11),(5,11)\}$ **3.** $h=\{(6,3),(7,3),(-9,4),(-9,6)\}$ **4.** Domain of $d=\{6, 8, 10, 12, 14, 16, 18\}$; range of $d=\{38, 40, 42, 44, 46, 48, 50\}$ **5.** Domain of $q=\{3, 4, 5\}$; range of $q=\{10, 11\}$ **6.** Domain of $h=\{6, 7, -9\}$; range of $h=\{3, 4, 6\}$ **7.** Yes **8.** No **9.** No **10.** Yes **11.** No **12.** No **13.** 17, 1, 5 **14.** 15, −13, 1 **15.** −2, −1, 0 **16.** 41, 44 **17. (a)** 27; **(b)** −23; **(c)** $5h-3$; **(d)** $5a+5h-3$ **18. (a)** 9.41; **(b)** 54; **(c)** $4y^2+5$; **(d)** h^2+5; **(e)** $a^2+2ah+h^2+5$ **19. (a)** 13; **(b)** $3h^2+1$; **(c)** $3a^2+6ah+3h^2+1$; **(d)** $6ah+3h^2$

20.

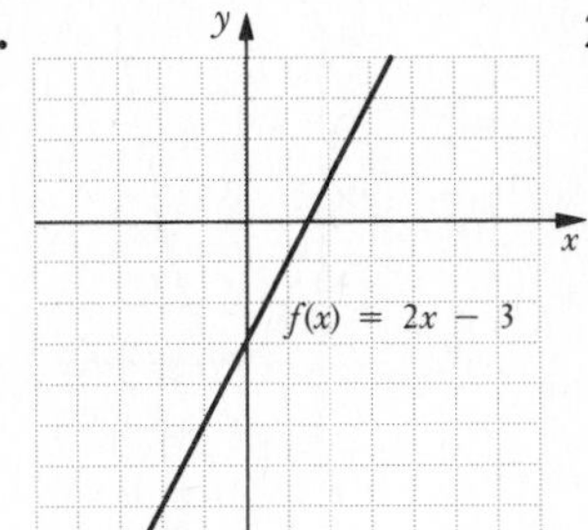

21.

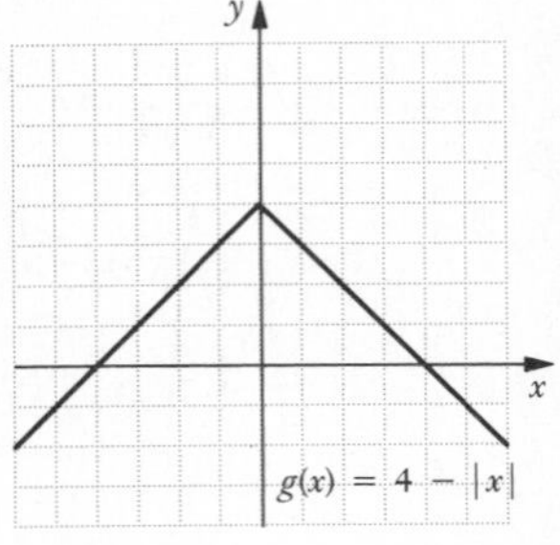

22. Yes **23.** No **24.** No **25.** Yes
26. $\{x \mid x \neq 3 \text{ and } x \neq -4\}$ **27.** $\left\{x \,\middle|\, x \leq \frac{4}{5}\right\}$ **28.** All real numbers

Margin Exercises, Section 9.5, pp. 551-554

1. (a) $2x^2$; **(b)** 6; **(c)** x^4-9; **(d)** $\frac{x^2+3}{x^2-3}$; **(e)** x^4+6x^2+9
2. x^2+4; $x^2+10x+24$ **3.** $4\sqrt[3]{x}+5$; $\sqrt[3]{4x+5}$ **4. (a)** $f(x)=\sqrt[3]{x}$, $g(x)=x^2+1$; **(b)** $f(x)=\frac{1}{x^4}$, $g(x)=x+5$

Margin Exercises, Section 9.6, pp. 557-564

1. $\{(-6,8),(-2,1),(6,8),(3,-4)\}$
2. $\{(3,-1),(4,-2),(5,-3),(6,-4),(7,-5)\}$ **3.** $x=2y+7$
4. $x=y^3-2y+|y|$

5.

Women's Dress Sizes

Range (USA)		Domain (France)
6	⟵	38
8	⟵	40
10	⟵	42
12	⟵	44
14	⟵	46
16	⟵	48
18	⟵	50

Yes, a function.

6.

Sports Teams		
Range	Domain	
Lakers ← Dodgers ← Rams ←	Los Angeles	Not a function.
Knickerbockers ← Yankees ← Giants ←	New York	

7. (a) Yes; (b) Yes **8.** (a) No; (b) No **9.** Yes **10.** No **11.** Yes **12.** No **13.** (a) Yes; (b) $f^{-1}(x) = 3 - x$ **14.** (a) Yes; (b) $g^{-1}(x) = \frac{x+2}{3}$

15.

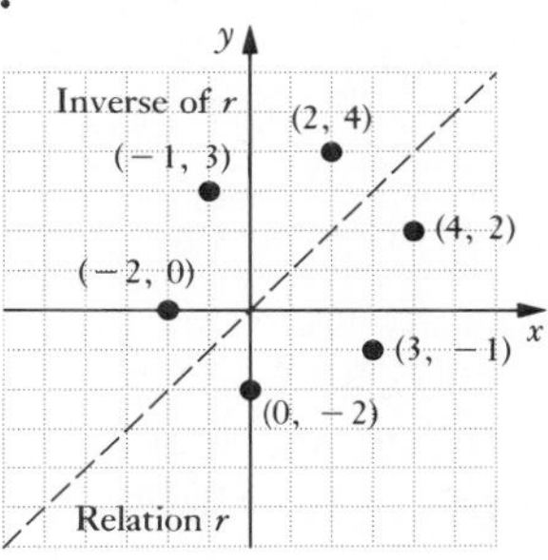

16.

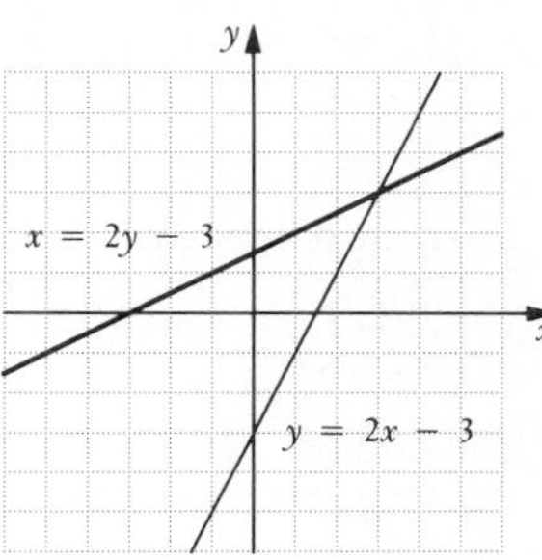

17.

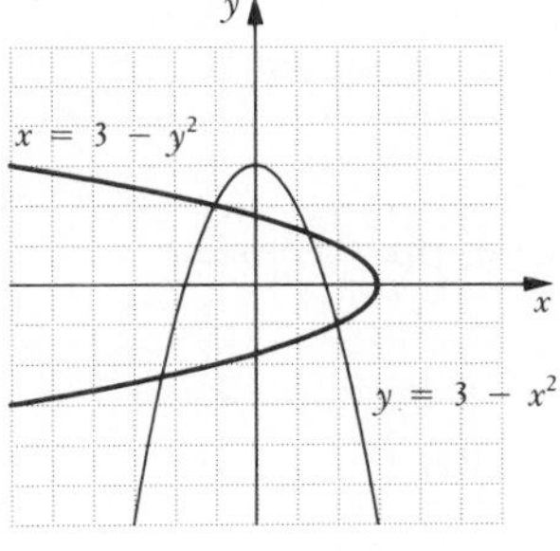

18.

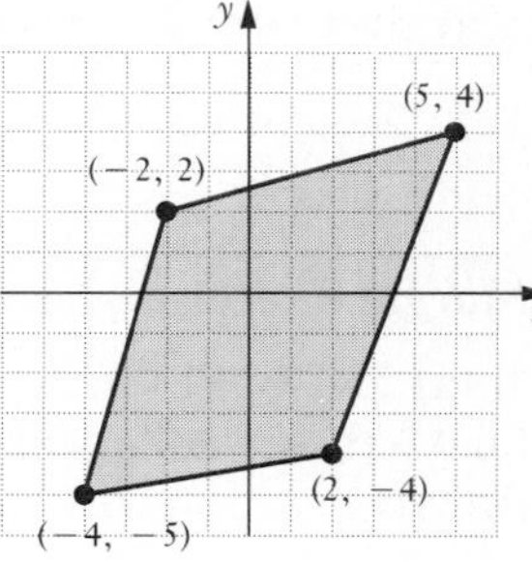

19.

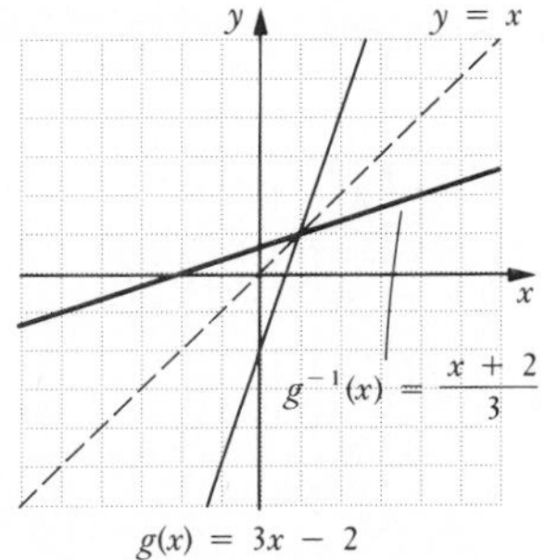

20. (a) Yes; (b) $f^{-1}(x) = \sqrt[3]{x - 1}$;

(c)

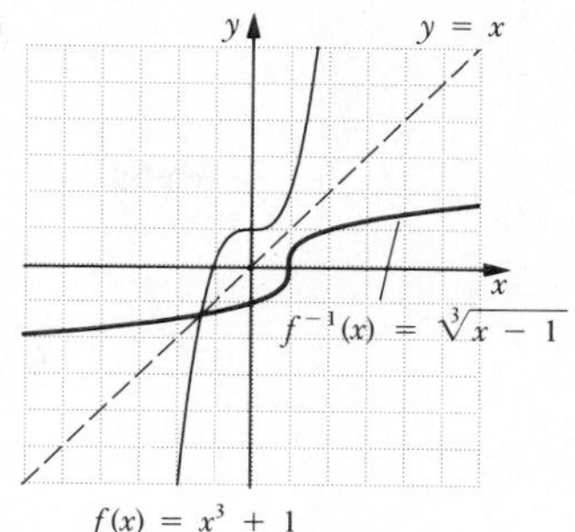

CHAPTER 10

Margin Exercises, Section 10.1, pp. 577-584

1. (a)

x	y, or $f(x)$
0	1
1	3
2	9
3	27
−1	$\frac{1}{3}$
−2	$\frac{1}{9}$
−3	$\frac{1}{27}$

(b)

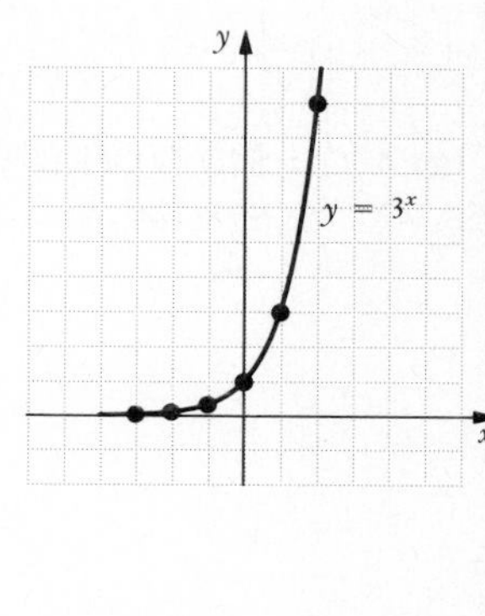

2. (a)

x	y, or $f(x)$
0	1
1	$\frac{1}{3}$
2	$\frac{1}{9}$
3	$\frac{1}{27}$
−1	3
−2	9
−3	27

(b)

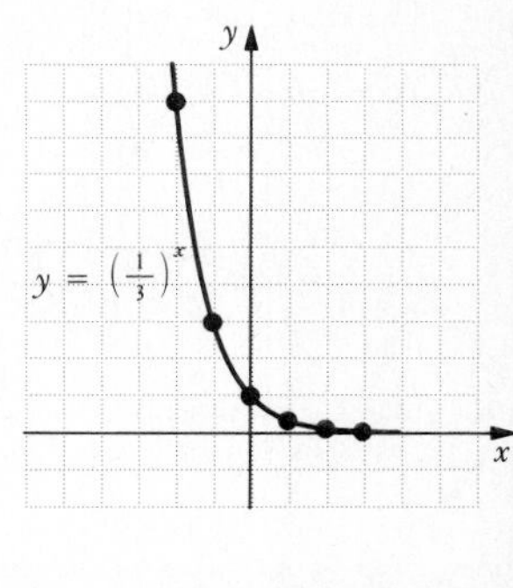

3.

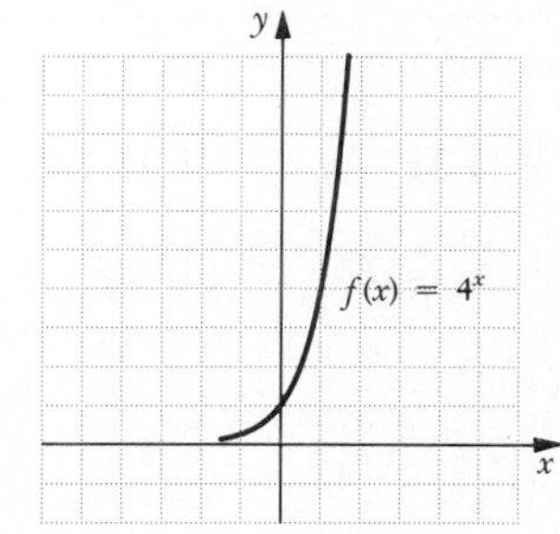

4.

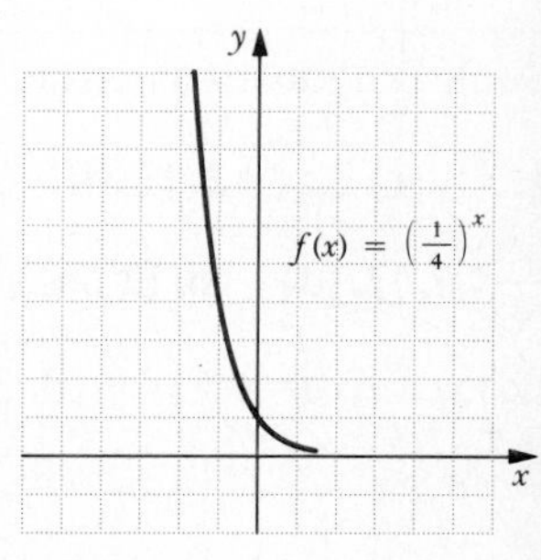

5.

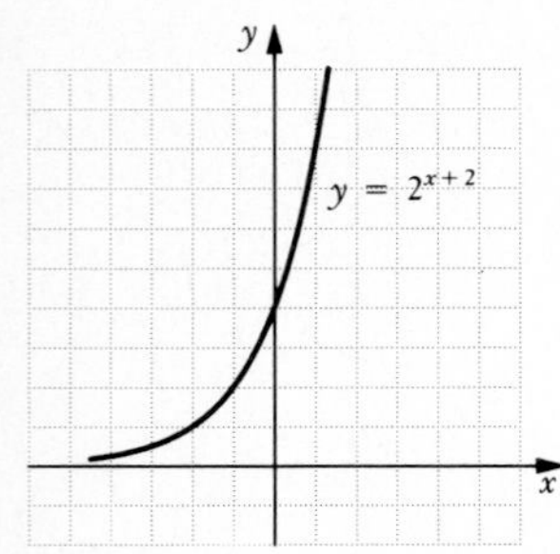

6.

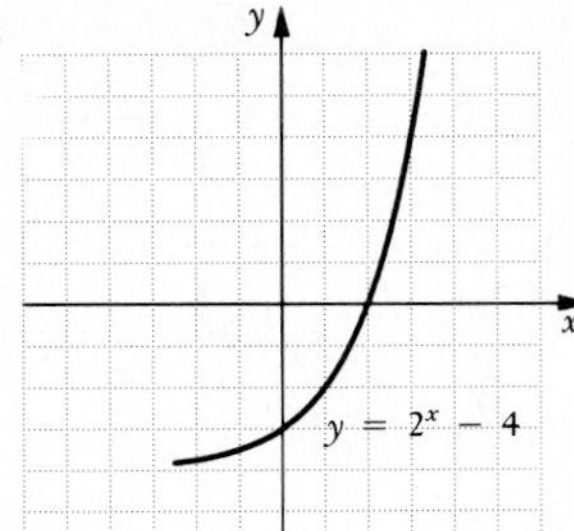

7.

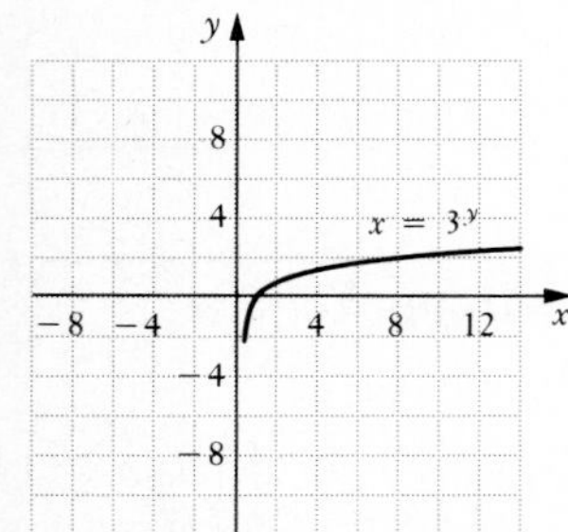

8. **(a)** $A(t) = \$40{,}000(1.07)^t$; **(b)** \$40,000, \$52,431.84; \$68,727.45; \$78,686.05; **(c)**

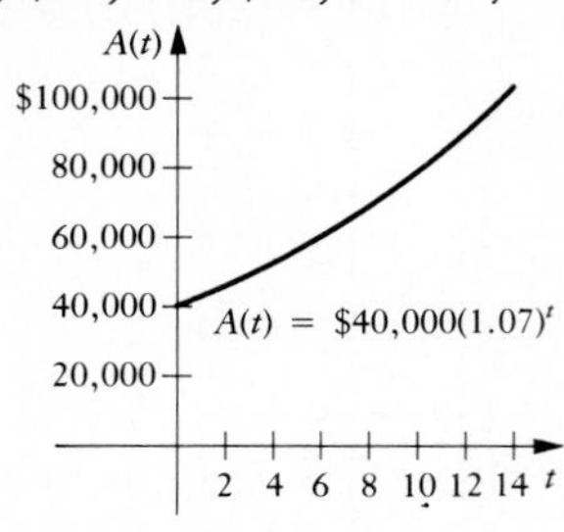

Margin Exercises, Section 10.2, pp. 589-594

1. $\log_2 64$ is the power to which we raise 2 to get 64; 6
2.

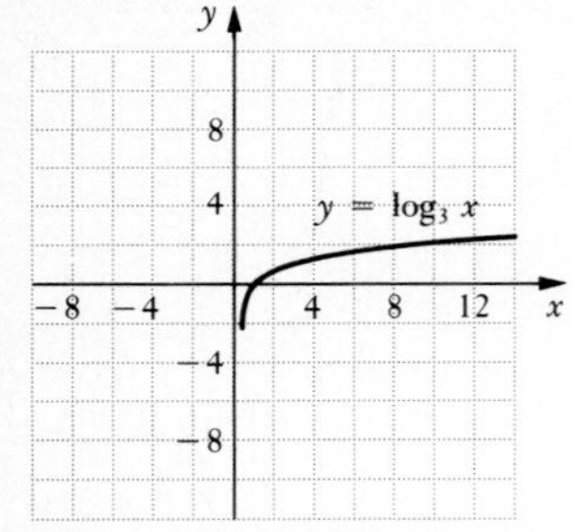

3. $0 = \log_6 1$ **4.** $-3 = \log_{10} 0.001$ **5.** $0.25 = \log_{16} 2$ **6.** $T = \log_m P$ **7.** $2^5 = 32$ **8.** $10^3 = 1000$ **9.** $a^7 = Q$ **10.** $t^x = M$ **11.** 10,000 **12.** 3 **13.** $\frac{1}{4}$ **14.** 4 **15.** −4 **16.** 0 **17.** 0 **18.** 1 **19.** 1 **20.** 0

Margin Exercises, Section 10.3, pp. 599-602

1. $\log_5 25 + \log_5 5$ **2.** $\log_b P + \log_b Q$ **3.** $\log_3 35$ **4.** $\log_a (CABIN)$ **5.** $5 \log_7 4$ **6.** $\frac{1}{2} \log_a 5$ **7.** $\log_b P - \log_b Q$ **8.** $\log_2 5$ **9.** $\frac{3}{2} \log_a z - \frac{1}{2} \log_a x - \frac{1}{2} \log_a y$ **10.** $2 \log_a x - 3 \log_a y - \log_a z$ **11.** $3 \log_a x + 4 \log_a y - 5 \log_a z - 9 \log_a w$ **12.** $\log_a \frac{x^5 z^{1/4}}{y}$ **13.** $\log_a \frac{1}{b\sqrt{b}}$, or $\log_a b^{-3/2}$ **14.** 0.602 **15.** 1 **16.** −0.398 **17.** 0.398 **18.** −0.699 **19.** $\frac{3}{2}$ **20.** 1.699 **21.** 1.204 **22.** 6 **23.** 3.2 **24.** 12

Margin Exercises, Section 10.4, pp. 605-608

1. 6.9043 **2.** −3.0482 **3.** 1.6454 **4.** −0.4515 **5.** 127,438.309 **6.** 1.0042 **7.** 0.00038 **8.** 0.4523 **9.** 11.0826 **10.** −6.8666 **11.** −1.2730 **12.** 7.1180 **13.** 172.173 **14.** 1.0518 **15.** 0.3012 **16.** 0.0104 **17.** 1.086 **18.** 5.5236

Margin Exercises, Section 10.5, pp. 611-614

1. 1 **2.** 3 **3.** 1.5395 **4.** 14.6068 **5.** 25 **6.** $\frac{2}{5}$ **7.** 2 **8.** 5

Margin Exercises, Section 10.6, pp. 617-622

1. **(a)** 27.1 yr; **(b)** 10.2 yr **2.** **(a)** 3.4 ft/sec; **(b)** 1.3 ft/sec; **(c)** 2.1 ft/sec; **3.** 260 million; 275 million **4.** **(a)** 37¢; **(b)** 2043 **5.** 159,373 **6.** 2972 yr

EXERCISE SET AND TEST ANSWERS

Book Diagnostic Pretest, p. xxiii

1. **[1.3c]** 3.57 **2.** **[1.3d]** $-\frac{1}{9}$ **3.** **[1.5c]** $6x + 8$ **4.** **[1.6a,b]** $\frac{4y^{14}}{9x^6}$ **5.** **[2.1d]** $\frac{1}{4}$ **6.** **[2.5a]** $\{x \mid -3 \le x \le 2\}$ **7.** **[2.6e]** $\{x \mid x < -2 \text{ or } x > -1\}$ **8.** **[2.2a]** $\frac{5}{2}$

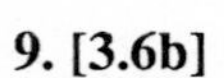

9. **[3.6b]**

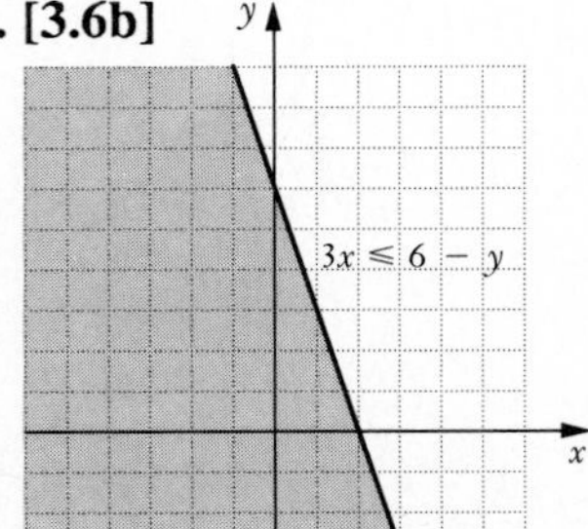

10. **[3.2b]**

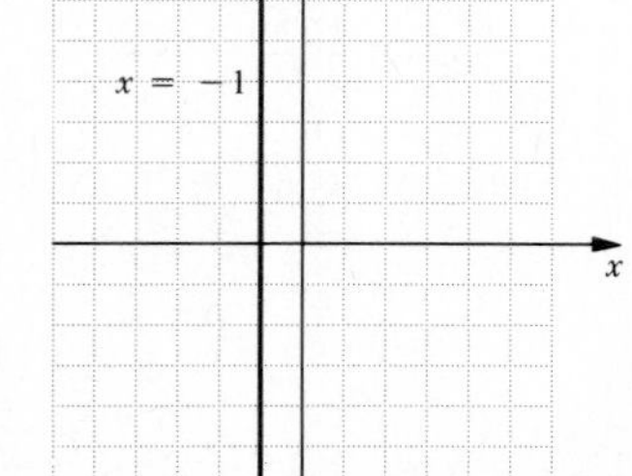

11. **[3.4b]** $y = -x + 4$ **12.** **[3.4c]** $y = -\frac{1}{3}x + 3$
13. **[4.2b]** $\left(\frac{13}{7}, \frac{4}{7}\right)$ **14.** **[4.4b]** (3, 0, −2) **15.** **[4.3b]** 21 mph
16. **[4.7a]**

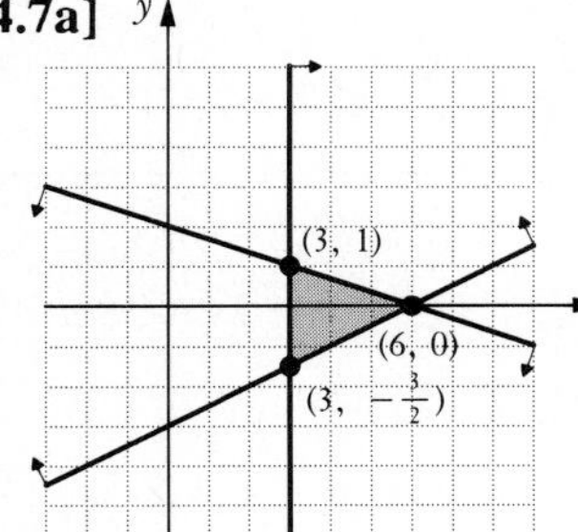

17. **[5.2c]** $9x^2 - 30xy + 25y^2$ **18.** **[5.5b]** $(x^2 + 1)(x + 1)(x - 1)$ **19.** **[5.8a]** −2, 9 **20.** **[5.8b]** $-\frac{1}{3}, 2$ **21.** **[6.1e]** $\frac{2x+6}{x^2-2x-3}$

22. **[6.6a]** $x - 1$ **23.** **[6.3a]** 1 **24.** **[6.4a]** $1\frac{5}{7}$ hr **25.** **[7.4b]** $6 - 4y\sqrt{3} + 4x\sqrt{3x} - 8xy\sqrt{x}$ **26.** **[7.4b]** $24\sqrt{3}$ **27.** **[7.5c]** $\frac{2x+\sqrt{xy}-y}{x-y}$ **28.** **[7.7a]** −3 **29.** **[8.2a]** $-1 \pm 2i$ **30.** **[8.4a]** −1, 6 **31.** **[8.6a]** −2, 1
32. **[8.8a]**

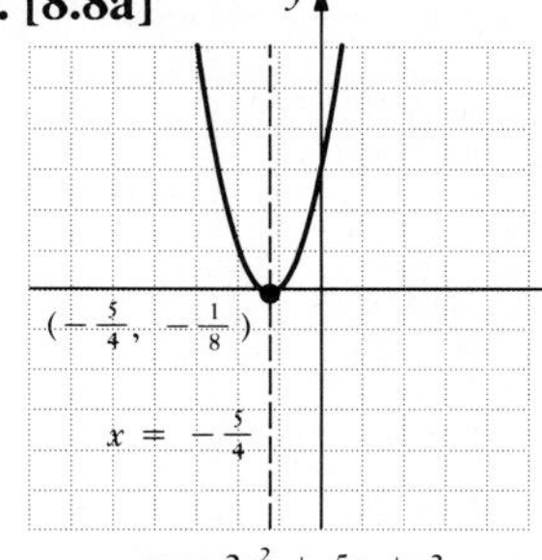

33. **[9.2a]**

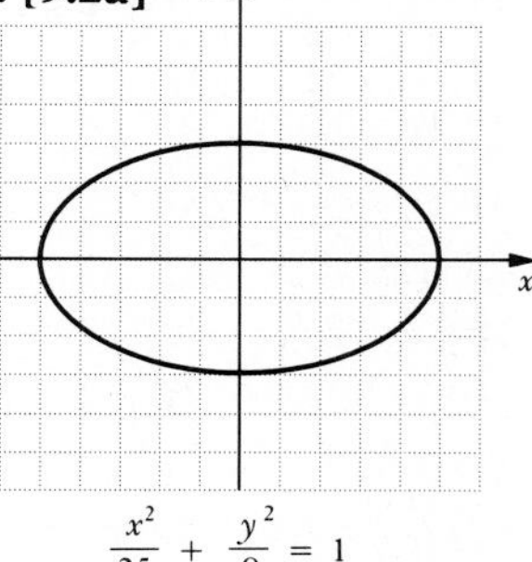

34. **[9.1d,e]**

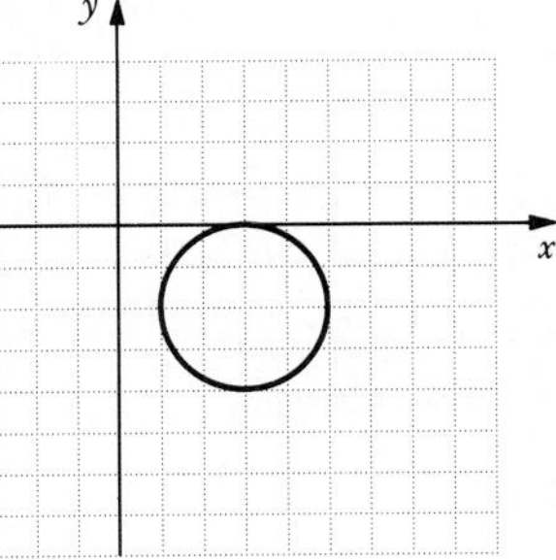

$x^2 + y^2 - 6x + 4y + 9 = 0$

35. **[9.3a]** $(2\sqrt{2}, \sqrt{2}), (-2\sqrt{2}, -\sqrt{2}), (\sqrt{2}i, -2\sqrt{2}i), (-\sqrt{2}i, 2\sqrt{2}i)$
36. **[9.6c]** $f^{-1}(x) = \frac{x-5}{2}$ **37.** **[10.2a]**

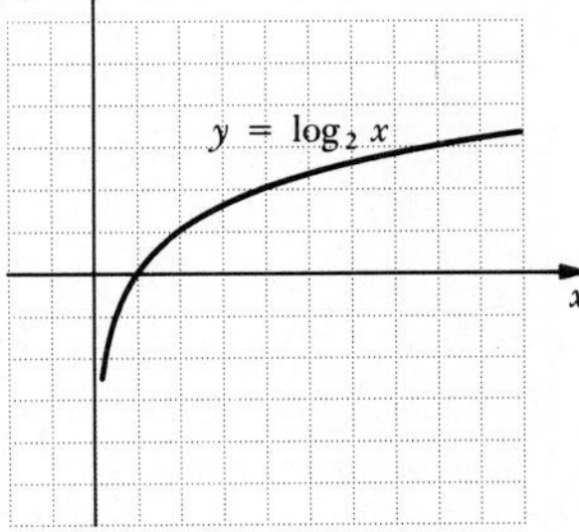

38. **[10.3d]** $\log\frac{\sqrt{x}}{y}$ **39.** **[10.2d]** 0 **40.** **[10.5b]** $\sqrt{3}$

CHAPTER 1

Pretest: Chapter 1, p. 2

1. **[1.1a]** 17 **2.** **[1.1b]** $0.47x$ **3.** **[1.2b]** True **4.** **[1.2d]** 3.6 **5.** **[1.3a]** 16 **6.** **[1.3c]** −11.6 **7.** **[1.3d]** $-\frac{6}{7}$ **8.** **[1.3e]** 8 **9.** **[1.4c]** $-4x + 12y$ **10.** **[1.4c]** $6(x - 3xy + 4)$ **11.** **[1.4d]** $-x + 6y + 11$ **12.** **[1.4e]** $-4x - 56$ **13.** **[1.4e]** $2x - 5y + 24$ **14.** **[1.6a]** $-\frac{20}{x^3}$ **15.** **[1.6a]** $-\frac{4}{x^2y^2}$ **16.** **[1.6b]** $\frac{x^{12}}{25y^{10}}$ **17.** **[1.6c]** 7.86×10^{-8} **18.** **[1.6c]** 4.5789×10^{11} **19.** **[1.6c]** 78,900,000,000,000 **20.** **[1.6c]** 0.00000789

Exercise Set 1.1, p. 7

1. 13 yd; 22 yd; 60 yd **3.** 144 sq cm **5.** 291 **7.** 5 **9.** 4 **11.** $22\frac{1}{2}$ **13.** $400 **15.** $7 + m$, or $m + 7$ **17.** $c - 11$ **19.** $26 + q$, or $q + 26$ **21.** $a + b$, or $b + a$ **23.** $y - x$ **25.** $28\%x$, or $0.28x$ **27.** $a + b$ **29.** $2x$ **31.** $7t$ **33.** $17 - b$ **35.** $y + 8$ **37.** $x - 54$ **39.** $54 - x$ **41.** $x + 3y$ **43.** $d = rt$ **45.** 6

Exercise Set 1.2, p. 15

1. 2, 14, $\sqrt{16}$ **3.** −5, 0, 2, $-\frac{1}{2}$, −3, $\frac{7}{8}$, 14, $-\frac{8}{3}$, 2.43, $7\frac{1}{2}$, $\sqrt{16}$, $-\frac{15}{3}$ **5.** −5, 0, 2, $-\frac{1}{2}$, −3, $\frac{7}{8}$, 14, $-\frac{8}{3}$, 2.43, $7\frac{1}{2}$, $\sqrt{3}$, $\sqrt{16}$, $-\frac{15}{3}$, 0.121221222122221... **7.** 12, 0 **9.** −11, 12, 0 **11.** $-\sqrt{5}$, π, −3.565665666566665... **13.** $\{m, a, t, h\}$ **15.** {0, 2, 4, 6, 8, 10, 12, 14} **17.** {2, 4, 6, 8, ...} **19.** $\{x \mid x$ is a whole number less than or equal to 5$\}$ **21.** $\left\{\frac{a}{b} \mid a \text{ and } b \text{ are integers and } b \neq 0\right\}$ **23.** $\{x \mid x > -3\}$ **25.** > **27.** < **29.** < **31.** < **33.** > **35.** < **37.** > **39.** < **41.** $x < -8$ **43.** $y \geq -12$ **45.** False **47.** True

49. −2 0

51. −2 0

53. −4 0

55. 0 2

57. 5 **59.** 18 **61.** 46 **63.** $\frac{2}{3}$ **65.** 0 **67.** 1280 **69.** 0, 10, −6, $\frac{1}{2}$, −0.7, $-\frac{5}{8}$, 4.7, −0.001, 1000, −987.6; answers may vary **71.** ≤ **73.** ≤ **75.** $\frac{1}{8}\%$, 0.3%, 0.009, 1%, 1.1%, $\frac{9}{100}$, $\frac{1}{11}$, $\frac{99}{1000}$, 0.11, $\frac{1}{8}$, $\frac{2}{7}$, 0.286

Exercise Set 1.3, p. 23

1. −28 **3.** 5 **5.** −16 **7.** −4 **9.** −7 **11.** −24 **13.** 1.2 **15.** −8.86 **17.** $\frac{1}{7}$ **19.** $-\frac{4}{3}$ **21.** $\frac{1}{10}$ **23.** $\frac{7}{20}$ **25.** −7 **27.** 2.8 **29.** −10 **31.** 0 **33.** −2 **35.** −12 **37.** 0 **39.** −46 **41.** 5 **43.** 15 **45.** −11.6 **47.** −29.25 **49.** $-\frac{7}{2}$ **51.** $-\frac{5}{12}$ **53.** $-\frac{41}{24}$

55. $-\frac{7}{15}$ **57.** -21 **59.** -8 **61.** 16 **63.** -126 **65.** 34.2 **67.** $-\frac{12}{35}$ **69.** 2 **71.** 60 **73.** 26.46 **75.** 1 **77.** $-\frac{8}{27}$ **79.** -2 **81.** -7 **83.** 7 **85.** 0.3 **87.** Undefined **89.** 0 **91.** Undefined **93.** $\frac{4}{3}$ **95.** $-\frac{8}{7}$ **97.** $\frac{1}{15}$ **99.** $\frac{1}{4.4}$, or $\frac{10}{44}$, or $\frac{5}{22}$ **101.** $-\frac{6}{77}$ **103.** 25 **105.** -6 **107.** 5 **109.** -120 **111.** $-\frac{9}{8}$ **113.** $\frac{5}{3}$ **115.** $\frac{3}{2}$ **117.** $\frac{9}{64}$ **119.** -2 **121.** $0.\overline{923076}$ **123.** -1.62 **125.** Undefined **127.** *First row:* $\frac{2}{3}, -\frac{2}{3}, \frac{3}{2}$; *Second row:* $-\frac{5}{4}, \frac{5}{4}, -\frac{4}{5}$; *Third row:* 0, 0, Does not exist; *Fourth row:* 1, -1, 1; *Fifth row:* -4.5, 4.5, $-\frac{1}{4.5}$ or $-\frac{10}{45}$ or $-\frac{2}{9}$ **129.** 8 **131.** $\frac{3}{10}$

Exercise Set 1.4, p. 35

1. $\frac{6x}{10}$ **3.** $\frac{3x}{4x}$ **5.** $\frac{5}{3}$ **7.** $\frac{4}{x}$ **9.** $9+m$ **11.** qp **13.** $ab+8$, $ba+8$, or $8+ba$ **15.** $yx+z$, $z+yx$, or $z+xy$ **17.** $(m+n)+2$ **19.** $7\cdot(x\cdot y)$ **21.** $a+(8+b)$, $(a+8)+b$, $b+(a+8)$; others are possible. **23.** $(7\cdot b)\cdot a$, $b\cdot(a\cdot 7)$, $(b\cdot a)\cdot 7$; others are possible. **25.** -8 **27.** -4 **29.** -14 **31.** 1 **33.** \$127.20 **35.** $3a+3$ **37.** $4x-4y$ **39.** $-10a-15b$ **41.** $2ab-2ac+2ad$ **43.** $2\pi rh+2\pi r$ **45.** $\frac{1}{2}ha+\frac{1}{2}hb$ **47.** $4a$, $-5b$, 6 **49.** $2x$, $-3y$, $-2z$ **51.** $18(x+y)$ **53.** $9(p-1)$ **55.** $7(x-3)$ **57.** $x(y+1)$ **59.** $2(x-y+z)$ **61.** $3(x+2y-1)$ **63.** $a(b+c-d)$ **65.** $\frac{1}{4}\pi r(r+s)$ **67.** $9a$ **69.** $-3b$ **71.** $15y$ **73.** $11a$ **75.** $-8t$ **77.** $10x$ **79.** $8x-8y$ **81.** $2c+10d$ **83.** $22x+18$ **85.** $1.19x+0.93y$ **87.** $-\frac{2}{15}a-\frac{1}{3}b-27$ **89.** $2(l+w)$ **91.** $4b$ **93.** $-a-2$ **95.** $-b+3$, or $3-b$ **97.** $-t+y$, or $y-t$ **99.** $-a-b-c$ **101.** $-8x+6y-13$ **103.** $2c-5d+3e-4f$ **105.** $1.2x-56.7y+34z+\frac{1}{4}$ **107.** $3a+5$ **109.** $m+1$ **111.** $5d-12$ **113.** $-7x+14$ **115.** $-9x+17$ **117.** $17x+3y-18$ **119.** $10x-19$ **121.** 7 **123.** $-\frac{1}{2}$ **125.** -900 **127.** 4050 **129.** Yes **131.** No

Exercise Set 1.5, p. 43

1. 4^5 **3.** 5^6 **5.** m^4 **7.** $(3a)^4$ **9.** $5^2c^3d^4$ **11.** $2\cdot 2\cdot 2\cdot 2\cdot 2$, or 32 **13.** $(-3)\cdot(-3)\cdot(-3)\cdot(-3)$, or 81 **15.** $x\cdot x\cdot x\cdot x$ **17.** $(-4b)(-4b)(-4b)$, or $-64bbb$ **19.** $ab\cdot ab\cdot ab\cdot ab$ **21.** 5 **23.** 1 **25.** 1 **27.** $\frac{7}{8}$ **29.** $\frac{1}{x^3}$ **31.** a^2 **33.** $\frac{1}{(-11)^1}$ **35.** 3^{-4} **37.** b^{-3} **39.** $(-16)^{-2}$ **41.** -5 **43.** -117 **45.** 2 **47.** 8 **49.** -358 **51.** 144; 74 **53.** -576 **55.** 2599 **57.** 36 **59.** 5619.712 **61.** About $-12{,}182{,}840{,}000$ **63.** 3 **65.** 3 **67.** 16 **69.** -310 **71.** 2 **73.** 1875 **75.** 7804.48 **77.** 12 **79.** 8 **81.** 16 **83.** -86 **85.** 37 **87.** -1 **89.** 22 **91.** -55 **93.** 12 **95.** -549 **97.** -144 **99.** 2 **101.** $-\frac{31}{76}$ **103.** $\frac{61}{13}$ **105.** $44a-22$ **107.** -190 **109.** $12x+30$ **111.** $3x+30$ **113.** $9x-18$ **115.** $-4x+808$ **117.** $-14y-186$ **119.** $17x+14y+129$ **121.** $-42x-360y-276$ **123.** 228 ft; 2808 ft^2 **125.** True; $(-x)^2=(-x)(-x)=(-1)x(-1)x=(-1)(-1)x\cdot x=x^2$ **127.** $-9z+5x$ **129.** $-x+19$

Exercise Set 1.6, p. 53

1. 5^9 **3.** $\frac{1}{8^4}$ **5.** $\frac{1}{8^6}$ **7.** $\frac{1}{b^3}$ **9.** a^3 **11.** $72x^5$ **13.** $-28m^5n^5$ **15.** $-\frac{14}{x^{11}}$ **17.** $\frac{105}{x^{2t}}$ **19.** 6^5 **21.** 4^5 **23.** $\frac{1}{10^9}$ **25.** 9^2 **27.** $\frac{1}{x^{10n}}$ **29.** $\frac{1}{w^{5q}}$ **31.** a^5 **33.** 1 **35.** $-\frac{4x^9}{3y^2}$ **37.** $\frac{3x^3}{2y^2}$ **39.** 4^6 **41.** $\frac{1}{8^{12}}$ **43.** 6^{12} **45.** $27x^6y^6$ **47.** $\frac{y^8}{4x^6}$ **49.** $\frac{a^4}{36b^6c^2}$ **51.** $\frac{1}{4^9\cdot 3^{12}}$ **53.** $\frac{8}{27}x^9y^3$ **55.** $\frac{a^{10}b^5}{5^{10}}$ **57.** $\frac{6^{30}2^{12}y^{36}}{z^{48}}$ **59.** $\frac{64}{x^{24}y^{12}}$ **61.** $\frac{5^7b^{28}}{3^7a^{35}}$ **63.** 10^a **65.** $3a^{-x-4}$ **67.** $-5x^{a+1}y^{-1}$ **69.** 8^{4xy} **71.** 12^{6b-2ab} **73.** $5^{2c}x^{2ac-2c}y^{2bc+2c}$, or $25^cx^{2ac-2c}y^{2bc+2c}$ **75.** $2x^{a+2}y^{b-2}$ **77.** 4.7×10^{10} **79.** \$$9.32\times 10^{11}$ **81.** 1.6×10^{-8} **83.** 7×10^{-11} **85.** 673,000,000 **87.** 0.000066 cm **89.** 0.00000000048 **91.** 0.0000000008923 **93.** 9.66×10^{-5} **95.** 1.3338×10^{-11} **97.** 2.5×10^3 **99.** 5.0×10^{-4} **101.** 9.125×10^6; 3.74×10^{-2} **103.** 1.35×10^{12} km **105.** 2.3725×10^9 gal **107.** 1.512×10^{10} ft^3; 1.324512×10^{14} ft^3 **109.** $19x+4y-20$ **111.** 2^{21} **113.** $\frac{1}{a^{14}b^{27}}$ **115.** $4x^{2a}y^{2b}$

Summary and Review: Chapter 1, p. 57

1. [1.1a] 314 cm^2 **2.** [1.1a] 3 **3.** [1.1a] 8 **4.** [1.1b] $5x$ **5.** [1.1b] $t-9$ **6.** [1.1b] $28\%y$, or $0.28y$ **7.** [1.2b] $<$ **8.** [1.2b] $x<18$ **9.** [1.2b] False **10.** [1.2b] True

11. [1.2c]
−4 0

12. [1.2c]
0 1

13. [1.2d] 7.23 **14.** [1.2d] 0 **15.** [1.2d] $\frac{21}{8}$ **16.** [1.3a] -2 **17.** [1.3a] -7.9 **18.** [1.3a] $-\frac{5}{2}$ **19.** [1.3c] -5 **20.** [1.3c] -26.6 **21.** [1.3c] $\frac{19}{4}$ **22.** [1.3d] 10.26 **23.** [1.3d] $-\frac{3}{7}$ **24.** [1.3d] 168 **25.** [1.3e] -4 **26.** [1.3e] 21 **27.** [1.3e] -7 **28.** [1.3e] $-\frac{35}{60}$, or $-\frac{7}{12}$ **29.** [1.3e] $-\frac{77}{15}$ **30.** [1.3e] $\frac{8}{3}$ **31.** [1.3e] -24 **32.** [1.4b] $a+11$ **33.** [1.4b] $y\cdot 8$ **34.** [1.4b] $9+(a+b)$ **35.** [1.4b] $(8x)y$ **36.** [1.3b] 7 **37.** [1.3b] -2.3 **38.** [1.3b] 0 **39.** [1.4c] $-6x+3y$ **40.** [1.4c] $8abc+4ab$ **41.** [1.4c] $5(x+2y-z)$ **42.** [1.4c] $pt(r+s)$ **43.** [1.4d] $-3x+5y$ **44.** [1.4d] $16c-4$ **45.** [1.4e] $9c-4d+3$ **46.** [1.4e] $x+3$ **47.** [1.4e] $6x+15$ **48.** [1.5c] $22x-14$ **49.** [1.5c] $-17m+36$ **50.** [1.5c] 59 **51.** [1.5c] -116 **52.** [1.6a]

$-\frac{10x^7}{y^5}$ **53.** [1.6a] $-\frac{3y^3}{2x^4}$ **54.** [1.6b] $\frac{a^8}{9b^2c^6}$ **55.** [1.6b] $\frac{81y^{40}}{16x^{24}}$ **56.** [1.6c] 6.875×10^9 **57.** [1.6c] 1.312×10^{-1} **58.** [1.2a] $\sqrt{4}, -\frac{2}{3}, 0.45\overline{45}, -23.788$ **59.** [1.4c; 1.6a] a, i; d, f; h, j **60.** [1.6a, b] x^{12y} **61.** [1.6a, b] 32 **62.** [1.5c] $-31a$ **63.** [1.5c] $-0.99703x + 0.99699y$

Test: Chapter 1, p. 59

1. [1.1a] 18 **2.** [1.1b] 36%m, or $0.36m$

3. [1.2c] (number line: −1, 0)

4. [1.2c] (number line: −2, 0)

5. [1.2b] True **6.** [1.2b] True **7.** [1.2b] $a < 5$ **8.** [1.2b] $>$ **9.** [1.6c] 4.37×10^{-5} **10.** [1.2d] $\frac{7}{8}$ **11.** [1.2d] 13.4 **12.** [1.2d] 0 **13.** [1.3b] −8 **14.** [1.3b] 13 **15.** [1.3b] $-\frac{1}{4}$ **16.** [1.3a] −2 **17.** [1.3a] −13.1 **18.** [1.3a] $-\frac{17}{4}$ **19.** [1.3c] −1 **20.** [1.3c] −29.7 **21.** [1.3c] $\frac{25}{4}$ **22.** [1.3d] −33.62 **23.** [1.3d] $\frac{3}{4}$ **24.** [1.3d] −528 **25.** [1.3e] −5 **26.** [1.3e] 15 **27.** [1.3e] $-\frac{7}{5}$ **28.** [1.3e] $-\frac{28}{15}$ **29.** [1.3e] $\frac{22}{7}$ **30.** [1.3e] −82 **31.** [1.4b] qp **32.** [1.4b] $4 + t$ **33.** [1.4b] $(3 + t) + w$ **34.** [1.4b] $4(ab)$ **35.** [1.4c] $-6a + 8b$ **36.** [1.4c] $3\pi rs + 3\pi r$ **37.** [1.4c] $a(b - c + 2d)$ **38.** [1.4c] $h(2a + 1)$ **39.** [1.4d] $10y - 5x$ **40.** [1.4d] $21a + 14$ **41.** [1.4e] $9x - 7y + 22$ **42.** [1.4e] $-7x + 14$ **43.** [1.4e] $-11y + 30$ **44.** [1.5c] $10x - 21$ **45.** [1.5c] $-a + 68$ **46.** [1.5c] 756 **47.** [1.5c] 25; −105 **48.** [1.6a] $-\frac{6a^9}{b^5}$ **49.** [1.6a] $-\frac{3y^2}{2x^4}$ **50.** [1.6a] $-50a^{9n}$ **51.** [1.6a] $-\frac{5}{x^{4t}}$ **52.** [1.6b] $\frac{a^{12}}{81b^8c^4}$ **53.** [1.6a, b] $-\frac{125y^{24}}{27x^{15}}$ **54.** [1.6b] $\frac{a^{30}b^{18}}{(200)^6c^6}$ **55.** [1.6a, b] $\frac{16a^{48}}{b^{48}}$ **56.** [1.6c] 1.196×10^{22} kg **57.** [1.2a] $\sqrt{7}, \pi$ **58.** [1.4b, c; 1.6b] b, e; d, f, h; i, j **59.** [1.6a, b] m^{8b} **60.** [1.5a] No, $-x^2 = -1 \cdot x^2$, and $(-x)^2 = (-x)(-x) = (-1 \cdot x)(-1 \cdot x) = 1 \cdot x \cdot x = x^2$.

CHAPTER 2

Pretest: Chapter 2, p. 62

1. [2.1c] −7 **2. [2.1b]** 21.3 **3. [2.1d]** −1 **4. [2.1d]** $\frac{14}{5}$ **5. [2.4b]** $\{x \mid x > -3\}$ **6. [2.4c]** $\{x \mid x \geq -3\}$ **7. [2.5a]** $\left\{x \,\middle|\, -1 \leq x \leq \frac{2}{3}\right\}$ **8. [2.6c]** −1, 8 **9. [2.5b]** $\left\{a \,\middle|\, a < -2 \text{ or } a > -\frac{1}{2}\right\}$ **10. [2.6e]** $\left\{a \,\middle|\, a < -2 \text{ or } a > -\frac{4}{3}\right\}$ **11. [2.6d]** $-\frac{11}{2}, \frac{3}{4}$ **12. [2.4c]** $\left\{x \,\middle|\, x \leq \frac{19}{14}\right\}$ **13. [2.3a]** $r = \frac{P}{3q}$ **14. [2.6b]** 2.4 **15. [2.2a]** Length: 45 m; width: 40 m **16. [2.2a]** \$750 **17. [2.4d]** Over $111\frac{1}{9}$ miles **18. [2.5a]** $\{-3, 4\}$ **19. [2.5b]** $\{-1, 0, 1\}$ **20. [2.6a]** $\frac{5}{2}\left|\frac{y}{x}\right|$

Sidelight: Repeating Decimals and Fractional Notation, p. 70

1. $\frac{4}{9}$ **2.** $\frac{1}{6}$ **3.** $\frac{25}{3}$ **4.** $\frac{4}{33}$ **5.** $\frac{41}{333}$ **6.** $\frac{113}{900}$ **7.** $\frac{127}{300}$ **8.** $\frac{6121}{990}$

Exercise Set 2.1, p. 71

1. 9 **3.** −4 **5.** 23 **7.** −39 **9.** 86.86 **11.** $\frac{1}{6}$ **13.** 4 **15.** −22 **17.** $\frac{1}{6}$ **19.** 32 **21.** −6 **23.** $-\frac{1}{6}$ **25.** 18 **27.** 11 **29.** −12 **31.** 8 **33.** 2 **35.** 21 **37.** −12 **39.** 2 **41.** −1 **43.** $\frac{18}{5}$ **45.** 0 **47.** 1 **49.** All real numbers **51.** No solution **53.** 2 **55.** 2 **57.** 7 **59.** 5 **61.** $-\frac{3}{2}$ **63.** 5 **65.** $\frac{23}{66}$ **67.** $\frac{5}{32}$ **69.** $-\frac{18x^2}{y^{11}}$ **71.** $-12x + 8y - 4z$ **73.** $\frac{3}{2}$ **75.** −6

Exercise Set 2.2, p. 79

1. 8 in.; 4 in. **3.** $1\frac{3}{5}$ m; $2\frac{2}{5}$ m **5.** McDonalds: \$1,500,000; Burger King: \$1,025,000 **7.** \$24,000,000 **9.** 215 mi **11.** 45 min **13.** 9 **15.** $\frac{25}{3}$ **17.** \$103.50 **19.** \$650 **21.** 32°, 96°, 52° **23.** $l = 31$ m; $w = 17$ m **25.** 12, 13, 14 **27.** \$8000 **29.** \$1644 **31.** 16 **33.** No solution **35.** 98% **37.** 9, 11, 13 **39.** 46 cm; 54 cm **41.** 120 **43.** 84 in^2 **45.** 130 **47.** $51\frac{1}{49}$ in^2

Exercise Set 2.3, p. 85

1. $l = \frac{A}{w}$ **3.** $I = \frac{W}{E}$ **5.** $t = \frac{d}{r}$ **7.** $t = \frac{I}{Pr}$ **9.** $m = \frac{E}{c^2}$ **11.** $l = \frac{P - 2w}{2}$, or $\frac{P}{2} - w$ **13.** $a^2 = c^2 - b^2$ **15.** $x = \frac{C - By}{A}$ **17.** $r^2 = \frac{A}{\pi}$ **19.** $h = \frac{2}{11}W + 40$ **21.** $r^3 = \frac{3V}{4\pi}$ **23.** $c = \frac{2A + hd}{h}$, or $\frac{2A}{h} + d$ **25.** $m = \frac{Fr}{v^2}$ **27.** $a = \frac{P}{5 - 3b}$; $b = \frac{P - 5a}{-3a}$, or $\frac{5}{3} - \frac{P}{3a}$ **29.** −5 **31.** −2 **33.** $a = \frac{2(s - v_1 t)}{t^2}$; $v_1 = \frac{s - \frac{1}{2}at^2}{t}$ **35.** $V_1 = \frac{P_2 V_2 T_1}{P_1 T_2}$; $P_2 = \frac{P_1 V_1 T_2}{T_1 V_2}$ **37.** 72 cm^2

Exercise Set 2.4, p. 95

1. No, no, no, yes **3.** No, yes, yes, no

5. $\{x \mid x > -5\}$; (number line: −5, 0)

7. $\{y \mid y < 6\}$; (number line: 0, 6)

9. $\{a \mid a \leq -21\}$; (number line: −21, 0)

11. $\{t \mid t \geq -5\}$; −5 0

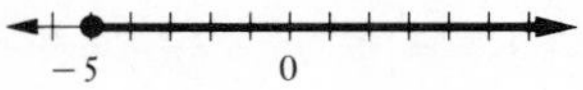

13. $\{y \mid y > -6\}$; −6 0

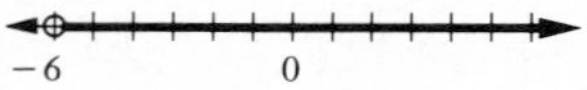

15. $\{x \mid x \leq 9\}$; 0 9

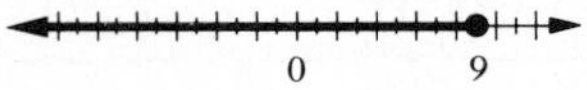

17. $\{x \mid x \geq 3\}$; 0 3

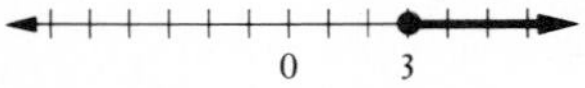

19. $\{x \mid x < -60\}$; −60 0

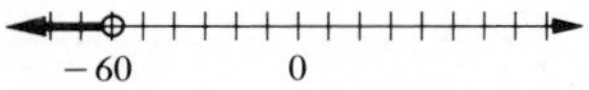

21. $\{x \mid x \leq 0.9\}$ **23.** $\left\{x \,\middle|\, x \leq \frac{5}{6}\right\}$

25. $\{x \mid x < 6\}$; 0 6

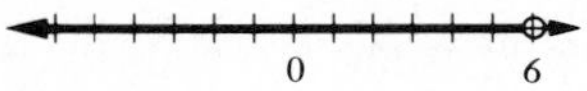

27. $\{y \mid y \leq -3\}$; −3 0

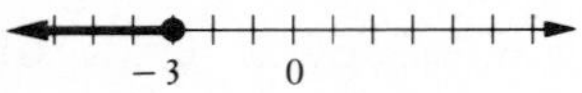

29. $\left\{y \,\middle|\, y > \frac{2}{3}\right\}$ **31.** $\left\{x \,\middle|\, x \geq \frac{45}{4}\right\}$ **33.** $\left\{x \,\middle|\, x \leq \frac{1}{2}\right\}$ **35.** $\left\{y \,\middle|\, y \leq -\frac{53}{6}\right\}$ **37.** $\left\{x \,\middle|\, x > -\frac{2}{17}\right\}$ **39.** $\left\{m \,\middle|\, m > \frac{7}{3}\right\}$ **41.** $\{r \mid r < -3\}$ **43.** $\{x \mid x \geq 2\}$ **45.** $\{y \mid y < 5\}$ **47.** $\left\{x \,\middle|\, x \leq \frac{4}{7}\right\}$ **49.** $\{x \mid x > 8\}$ **51.** $\left\{x \,\middle|\, x \geq \frac{13}{2}\right\}$ **53.** $\left\{x \,\middle|\, x < \frac{11}{18}\right\}$ **55.** $\left\{x \,\middle|\, x \geq -\frac{51}{31}\right\}$ **57.** $\{a \mid a \leq 2\}$ **59.** $x < 8$ **61.** $p \geq \$6$ **63.** $p \leq \$17.95$ **65.** $24 - 3y < 16 + y$ **67.** $15(x + y) \geq 78$ **69.** $\{m \mid m \leq 330\}$ **71.** $\{s \mid s \geq 84\}$ **73.** $1.99 **75.** $\{x \mid x > \$1450\}$ **77.** Gross sales greater than $15,000 **79.** $\left\{n \,\middle|\, n < 166\frac{2}{3}\right\}$ **81.** $\{s \mid s > 8\}$ **83.** **(a)** 44,000; 165,978; 275,758.2; **(b)** 1988 and after **85.** $6(-2a + 5bl)$ **87.** $-8a + 22b$ **89.** **(a)** $\{p \mid p < 10\}$; **(b)** $\{p \mid p > 10\}$ **91.** False; $-3 < -2$, but $(-3)^2 > (-2)^2$. **93.** All real numbers **95.** All real numbers except 0

Exercise Set 2.5, p. 107

1. $\{6, 8\}$ **3.** $\varnothing$ **5.** $\{1, 2, 3, 4\}$

7. 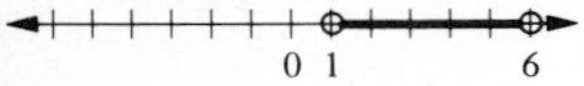 0 1 6

9.  −6 0 2

11. $\{x \mid -4 \leq x < 5\}$ **13.** $\{x \mid x \geq 2\}$ **15.** $\varnothing$ **17.** $\{x \mid -4 < x < 6\}$ **19.** $\{y \mid -2 < y \leq 2\}$ **21.** $\left\{x \,\middle|\, -\frac{5}{3} \leq x \leq \frac{4}{3}\right\}$ **23.** $\{x \mid -1 < x \leq 6\}$ **25.** $\left\{x \,\middle|\, -\frac{3}{2} \leq x < \frac{9}{2}\right\}$ **27.** $\{x \mid 10 < x \leq 14\}$ **29.** $\left\{x \,\middle|\, -\frac{7}{2} < x < \frac{37}{2}\right\}$ **31.** $\{1, 4, 5, 6, 7, 8, 11\}$ **33.** $\{1, 2, 3, 4, 5, 6, 8\}$ **35.** $\{4, 8, 11\}$

37. 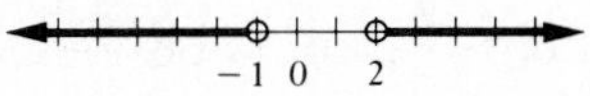 −1 0 2

39. 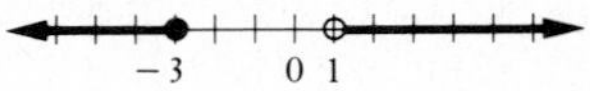 −3 0 1

41. $\{x \mid x < -9 \text{ or } x > -5\}$ **43.** $\left\{x \,\middle|\, x \leq \frac{5}{2} \text{ or } x \geq 11\right\}$ **45.** $\{x \mid x \geq -3\}$ **47.** $\left\{x \,\middle|\, x \leq -\frac{5}{4} \text{ or } x > -\frac{1}{2}\right\}$ **49.** All real numbers **51.** $\{x \mid x > 2 \text{ or } x < -4\}$ **53.** $\left\{x \,\middle|\, x < \frac{79}{4} \text{ or } x > \frac{89}{4}\right\}$ **55.** $\left\{x \,\middle|\, x \leq -\frac{13}{2} \text{ or } x \geq \frac{29}{2}\right\}$ **57.** $-\frac{32y^{30}}{x^{20}}$ **59.** $-\frac{4a^{17}}{5b^{15}}$ **61.** **(a)** $1945.4° \leq F < 4820°$; **(b)** $1761.44° \leq F < 3956°$ **63.** $\{x \mid -5 < x < 11\}$ **65.** $\{x \mid -4 < x \leq 1\}$ **67.** $\left\{x \,\middle|\, \frac{2}{5} \leq x \leq \frac{4}{5}\right\}$ **69.** $\left\{x \,\middle|\, -\frac{1}{8} < x < \frac{1}{2}\right\}$ **71.** $\{x \mid 10 < x \leq 18\}$

Exercise Set 2.6, p. 117

1. $3|x|$ **3.** $9x^2$ **5.** $4x^2$ **7.** $8|y|$ **9.** $\frac{4}{|x|}$ **11.** $\frac{x^2}{|y|}$ **13.** $4|x|$ **15.** 34 **17.** 11 **19.** 6.3 **21.** 5 **23.** $\{3, -3\}$ **25.** $\varnothing$ **27.** $\{0\}$ **29.** $\{15, -9\}$ **31.** $\left\{\frac{7}{2}, -\frac{1}{2}\right\}$ **33.** $\left\{\frac{23}{4}, -\frac{5}{4}\right\}$ **35.** $\{11, -11\}$ **37.** $\{389, -389\}$ **39.** $\{8, -8\}$ **41.** $\{7, -7\}$ **43.** $\left\{\frac{11}{5}, -\frac{11}{5}\right\}$ **45.** $\{8, -7\}$ **47.** $\{2, -12\}$ **49.** $\left\{\frac{7}{2}, -\frac{5}{2}\right\}$ **51.** $\varnothing$ **53.** $\left\{-\frac{13}{54}, -\frac{7}{54}\right\}$ **55.** $\left\{\frac{3}{4}, -\frac{11}{2}\right\}$ **57.** $\left\{-\frac{3}{2}\right\}$ **59.** $\left\{5, -\frac{3}{5}\right\}$ **61.** All real numbers **63.** $\left\{-\frac{3}{2}\right\}$ **65.** $\left\{\frac{24}{23}, 0\right\}$ **67.** $\left\{32, \frac{8}{3}\right\}$ **69.** $\{x \mid -3 < x < 3\}$ **71.** $\{x \mid x \leq -2 \text{ or } x \geq 2\}$ **73.** $\{x \mid 2 < x < 4\}$ **75.** $\{x \mid -5 \leq x \leq -3\}$ **77.** $\left\{x \,\middle|\, -\frac{1}{2} \leq x \leq \frac{7}{2}\right\}$ **79.** $\left\{y \,\middle|\, y < -\frac{3}{2} \text{ or } y > \frac{17}{2}\right\}$ **81.** $\left\{x \,\middle|\, x \leq -\frac{5}{4} \text{ or } x \geq \frac{23}{4}\right\}$ **83.** $\{y \mid -9 < y < 15\}$ **85.** $\left\{x \,\middle|\, -\frac{7}{2} \leq x \leq \frac{1}{2}\right\}$ **87.** $\left\{y \,\middle|\, y < -\frac{4}{3} \text{ or } y > 4\right\}$ **89.** $\left\{x \,\middle|\, x \leq -\frac{5}{4} \text{ or } x \geq \frac{23}{4}\right\}$ **91.** $\left\{x \,\middle|\, -\frac{9}{2} < x < 6\right\}$ **93.** $\left\{x \,\middle|\, x \leq -\frac{25}{6} \text{ or } x \geq \frac{23}{6}\right\}$ **95.** $\{x \mid -5 < x < 19\}$ **97.** $\left\{x \,\middle|\, x \leq -\frac{2}{15} \text{ or } x \geq \frac{14}{15}\right\}$ **99.** $\{m \mid -12 \leq m \leq 2\}$ **101.** $\left\{x \,\middle|\, \frac{1}{2} \leq x \leq \frac{5}{2}\right\}$ **103.** $\{x \mid 0.49705 \leq x \leq 0.50295\}$ **105.** −49.3 **107.** 252.3 **109.** $\left\{x \,\middle|\, x \geq \frac{5}{2}\right\}$ **111.** $\{x \mid x \geq -5\}$ **113.** $\left\{1, -\frac{1}{4}\right\}$ **115.** $\varnothing$ **117.** All real numbers **119.** $|x| < 3$ **121.** $|x| \geq 6$ **123.** $|x + 3| > 5$

Summary and Review: Chapter 2, p. 121

1. [2.1b] 8 **2.** [2.1c] $\frac{3}{7}$ **3.** [2.1d] $\frac{22}{5}$ **4.** [2.1d] $-\frac{1}{13}$ **5.** [2.1d] −0.2 **6.** [2.1d] 5 **7.** [2.3a] $d = \frac{11}{4}(C - 3)$

8. [2.3a] $b = \frac{A-2a}{-3}$, or $\frac{2a-A}{3}$ **9.** [2.2a] 15 m, 12 m **10.** [2.2a] 13, 14, 15 **11.** [2.2a] 14, 16 **12.** [2.2a] $2600 **13.** [2.4a] $\{x \mid x \leq -4\}$ **14.** [2.4a] $\{x \mid x > 1\}$ **15.** [2.4a] $\{a \mid a \leq -21\}$ **16.** [2.4a] $\{y \mid y \geq -7\}$ **17.** [2.4b] $\left\{y \middle| y > -\frac{15}{4}\right\}$ **18.** [2.4b] $\{y \mid y > -30\}$ **19.** [2.4c] $\{x \mid x > -3\}$ **20.** [2.4c] $\left\{y \middle| y < -\frac{6}{5}\right\}$ **21.** [2.4c] $\{x \mid x < -3\}$ **22.** [2.4c] $\{y \mid y > -10\}$ **23.** [2.4c] $\left\{x \middle| x \leq -\frac{5}{2}\right\}$ **24.** [2.4d] 92 or greater **25.** [2.4d] $10,000 **26.** [2.5a] $\varnothing$ **27.** [2.5a] $\{x \mid -7 < x \leq 2\}$ **28.** [2.5a] $\left\{x \middle| -\frac{5}{4} < x < \frac{5}{2}\right\}$ **29.** [2.5b] $\{x \mid x < -3 \text{ or } x > 1\}$ **30.** [2.5b] $\{x \mid x < -11 \text{ or } x \geq -6\}$ **31.** [2.5b] $\{x \mid x \leq -6 \text{ or } x \geq 8\}$ **32.** [2.6a] $\frac{3}{|x|}$ **33.** [2.6a] $\frac{2|x|}{y^2}$ **34.** [2.6a] $\frac{4}{|y|}$ **35.** [2.6b] 62 **36.** [2.6c] $\{6, -6\}$ **37.** [2.6e] $\varnothing$ **38.** [2.6e] $\{x \mid x \leq -3.5 \text{ or } x \geq 3.5\}$ **39.** [2.6c] $\{9, -5\}$ **40.** [2.6e] $\left\{x \middle| -\frac{17}{2} < x < \frac{7}{2}\right\}$ **41.** [2.6e] $\left\{x \middle| x \leq -\frac{11}{3} \text{ or } x \geq \frac{19}{3}\right\}$ **42.** [2.6d] $\left\{-14, \frac{4}{3}\right\}$ **43.** [2.6c] $\varnothing$ **44.** [2.5a] $\{1, 5, 9\}$ **45.** [2.5b] $\{1, 2, 3, 5, 6, 9\}$ **46.** [1.3a] 33 **47.** [1.3c] $\frac{1}{6}$ **48.** [1.3d] 2349 **49.** [1.3e] $\frac{4}{5}$ **50.** [1.4c] $20x - 30y + 70$ **51.** [1.4c] $8(5x - y + 2)$ **52.** [1.4d] $-2x + 5y + 12$ **53.** [1.6a] $-\frac{12b^3}{a}$ **54.** [1.6b] $\frac{b^{32}}{256a^{20}}$ **55.** [1.6a] $-4a^7b^3$ **56.** [2.6e] $\left\{x \middle| -\frac{8}{3} \leq x \leq -2\right\}$ **57.** [2.4b] False; $-4 < 3$, but $(-4)^2 > 9$.

Test: Chapter 2, p. 123

1. [2.1b] 22 **2.** [2.1c] $\frac{2}{3}$ **3.** [2.1d] 2.2 **4.** [2.1d] -2 **5.** [2.3a] $W = \frac{11}{2}(h - 40)$ **6.** [2.6c] $\{-6, 12\}$ **7.** [2.2a] 10 **8.** [2.2a] $12.50 **9.** [2.4a] $\{x \mid x < 14\}$ **10.** [2.4b] $\{y \mid y > -50\}$ **11.** [2.4c] $\{y \mid y \leq -2\}$ **12.** [2.4c] $\left\{a \middle| a \leq \frac{11}{5}\right\}$ **13.** [2.4c] $\{y \mid y > 1\}$ **14.** [2.4c] $\left\{x \middle| x > \frac{5}{2}\right\}$ **15.** [2.4c] $\left\{x \middle| x \leq \frac{7}{4}\right\}$ **16.** [2.4d] Over $66\frac{2}{3}$ mi **17.** [2.4d] 94 and greater **18.** [2.5a] $\{x \mid x \geq 4\}$ **19.** [2.5a] $\{x \mid -1 < x < 6\}$ **20.** [2.5a] $\left\{x \middle| -\frac{2}{5} < x \leq \frac{9}{5}\right\}$ **21.** [2.5b] $\left\{x \middle| x < -4 \text{ or } x > -\frac{5}{2}\right\}$ **22.** [2.5b] All real numbers are solutions. **23.** [2.5b] $\{x \mid x < 3\} \text{ or } x > 6\}$ **24.** [2.6a] $\frac{7}{|x|}$ **25.** [2.6a] $2|x|$ **26.** [2.6b] 8.4 **27.** [2.6c] $\{9, -9\}$ **28.** [2.6e] $\{x \mid x < -3 \text{ or } x > 3\}$ **29.** [2.6e] $\left\{x \middle| -\frac{7}{8} < x < \frac{11}{8}\right\}$ **30.** [2.6e] $\left\{x \middle| x \leq -\frac{13}{5} \text{ or } x \geq \frac{7}{5}\right\}$ **31.** [2.6d] $\{1\}$ **32.** [2.6c] $\varnothing$ **33.** [2.6e] $\{x \mid -99 \leq x \leq 111\}$ **34.** [2.5a] $\{3, 5\}$ **35.** [2.5b] $\{1, 3, 5, 7, 9, 11, 13\}$ **36.** [1.3a] $\frac{1}{8}$ **37.** [1.3c] $\frac{11}{8}$ **38.** [1.3d] 120 **39.** [1.3e] -250 **40.** [1.4c] $-16a + 24b$ **41.** [1.4c] $2(3a - 5b + 6)$ **42.** [1.4d] $4.7x - 6.4y$ **43.** [1.6a] $-\frac{20a}{b^{22}}$ **44.** [1.6b] $\frac{b^{48}}{625a^{24}}$ **45.** [2.6e] $\varnothing$ **46.** [2.5a] $\left\{x \middle| \frac{1}{5} < x < \frac{4}{5}\right\}$

Cumulative Review: Chapters 1-2, p. 125

1. [1.1a] 29 **2.** [1.1a] 2 **3.** [1.3a] -15 **4.** [1.2d] $\frac{4}{7}$ **5.** [1.3c] 17 **6.** [1.4e] $x + 5$ **7.** [1.5a] 1 **8.** [1.5b] $\frac{1}{625}$ **9.** [1.5c] 40 **10.** [1.5c] -64 **11.** [1.5c] $-x - 23$ **12.** [1.6a] $\frac{1}{a^3}$ **13.** [1.6a] $6x^2$ **14.** [1.6a] $\frac{1}{9}$ **15.** [1.6b] $\frac{9a^8}{4b^{14}}$ **16.** [2.6a] $3|x|$ **17.** [2.1b] 18 **18.** [2.1c] $-\frac{8}{15}$ **19.** [2.1d] 4 **20.** [2.1d] 4 **21.** [2.1d] 5 **22.** [2.3a] $m = \frac{y-b}{x}$ **23.** [2.3a] $y^2 = r^2 - x^2$ **24.** [2.3a] $p = \frac{y^2}{4x}$ **25.** [2.3a] $m = \frac{3I}{t^2}$ **26.** [2.4a] $\{x \mid x \leq -3\}$ **27.** [2.4c] $\{x \mid x > -2\}$ **28.** [2.4c] $\left\{x \middle| x \geq \frac{3}{4}\right\}$ **29.** [2.4c] $\left\{x \middle| x < -\frac{1}{3}\right\}$ **30.** [2.5a] $\{x \mid 2 < x \leq 7\}$ **31.** [2.5b] All real numbers **32.** [2.6c] $\varnothing$ **33.** [2.6c] $\{-3, 7\}$ **34.** [2.6c] $\{-12, 12\}$ **35.** [2.6e] $\left\{x \middle| -\frac{13}{6} \leq x \leq \frac{19}{6}\right\}$ **36.** [2.6d] $\left\{\frac{1}{3}, 9\right\}$ **37.** [1.3d] 10.5 **38.** [1.3d] $-\frac{5}{24}$ **39.** [1.4c] $-24a + 8b$ **40.** [1.4c] $3xy - 6xz$ **41.** [1.4c] $5(x - 9)$ **42.** [1.4c] $6(a - 3b + 9)$ **43.** [1.1b] $x + 3$, or $3 + x$ **44.** [1.1b] $8 - m$ **45.** [1.2a] $-\sqrt{3}, \sqrt{67}$ **46.** [1.2a] $-7.2, -5, 0, \frac{3}{11}, 2, 8.\overline{12}$

47. [2.5b] [number line: 0 1 2]

48. [2.6b] 26 **49.** [1.2a] $\{x \mid x < -2\}$ **50.** [1.2b] $>$ **51.** [1.6c] $\$4.58 \times 10^7$ **52.** [2.2a] 47 ft, 39 ft **53.** [2.2a] $760 **54.** [2.2a] 415, 417 **55.** [2.2a] $140 **56.** [1.3b] 6.4 **57.** [2.5a] $\{1, 7\}$ **58.** [2.5b] $\{0, 1, 2, 3, 4\}$ **59.** [1.4c] $\frac{2}{3}$ **60.** [1.5c] $14x + 18$ **61.** [1.6c] $\frac{4096x^{24}}{81y^{14}}$

CHAPTER 3

Pretest: Chapter 3, p. 128

1. [3.2a]

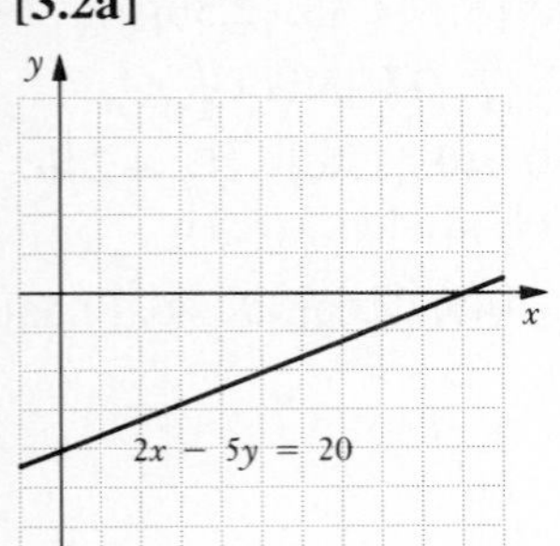

2. [3.2b]

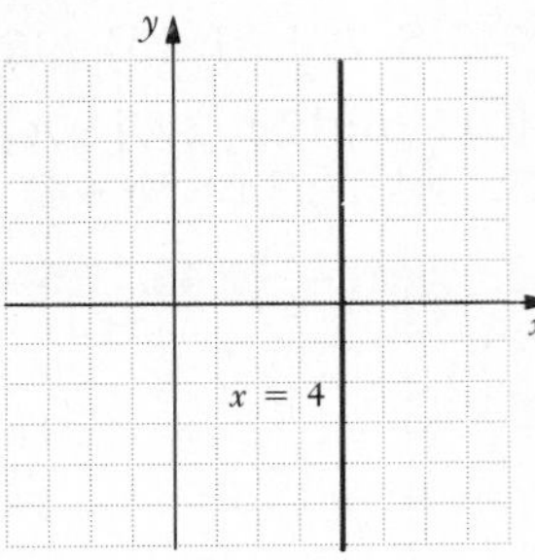

3. [3.3d]

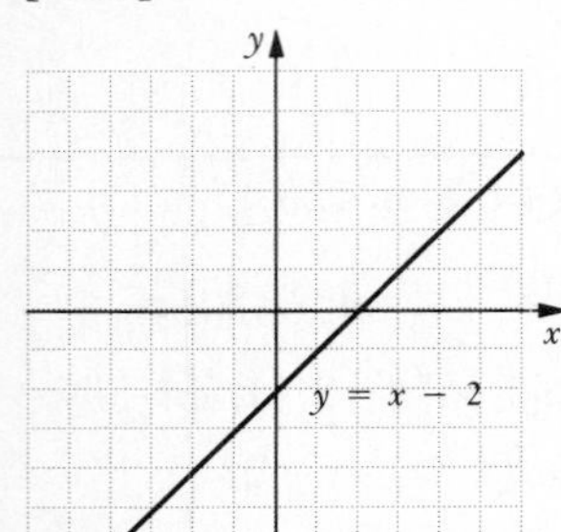

4. [3.2b]

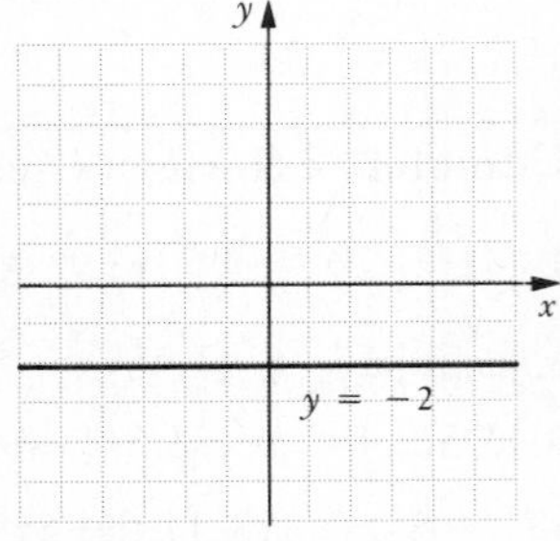

5. [3.3c] 5; (0,–3) **6. [3.3a]** 0 **7. [3.4b]** $y = x + 10$ **8. [3.4a]** $y = 2x + 7$ **9. [3.4c]** $y = \frac{2}{7}x + \frac{31}{7}$ **10. [3.4c]** $y = -\frac{7}{2}x + 12$ **11. [3.4c]** Perpendicular **12. [3.4c]** Parallel **13. [3.2a]** (6, 0) **14. [3.2a]** (0, –4)

Exercise Set 3.1, p. 135

1.

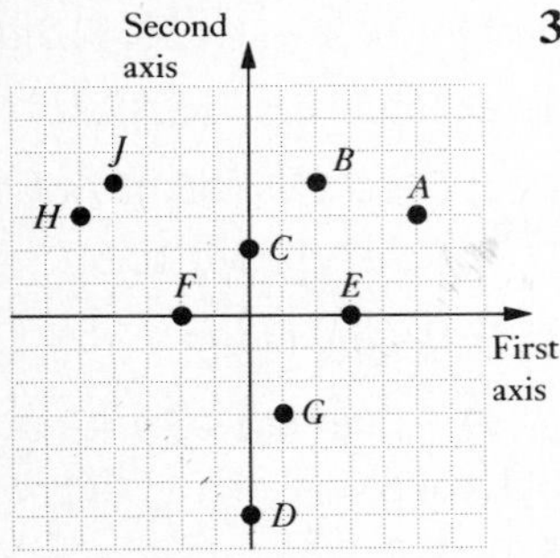

3.

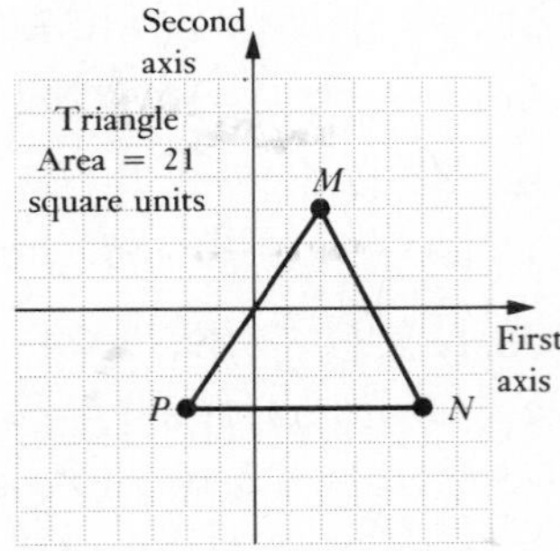

5. Yes **7.** No **9.** No

11.

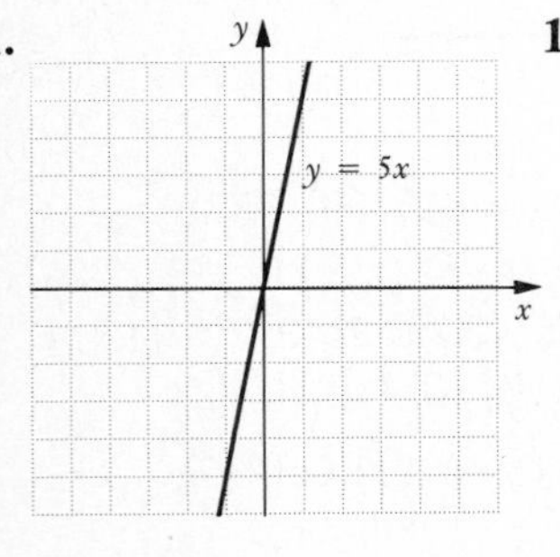

13.

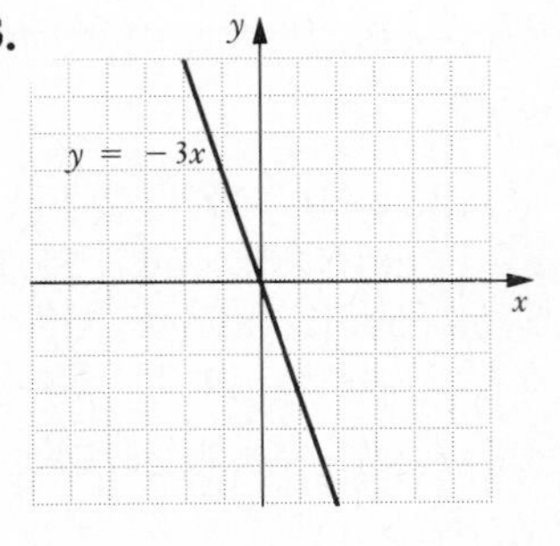

15.

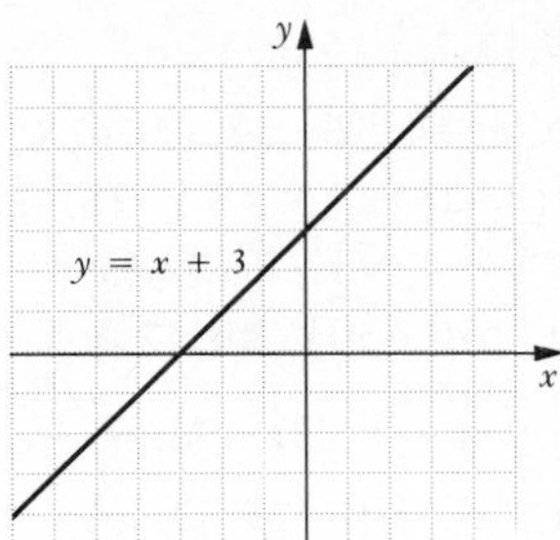

17.

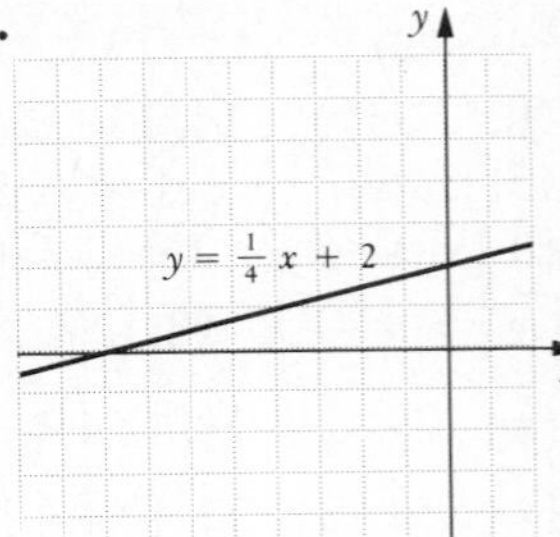

19.

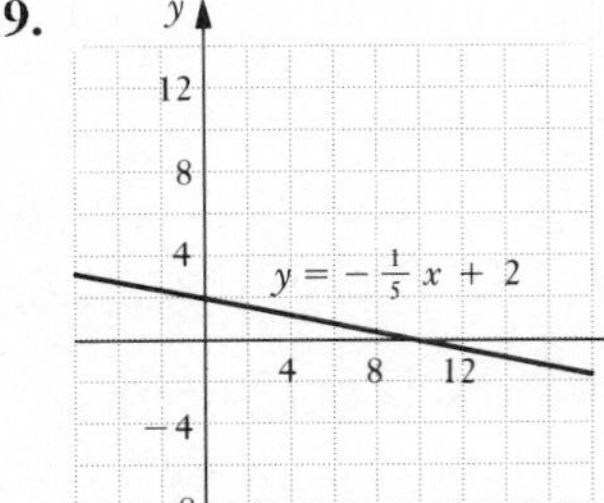

21.

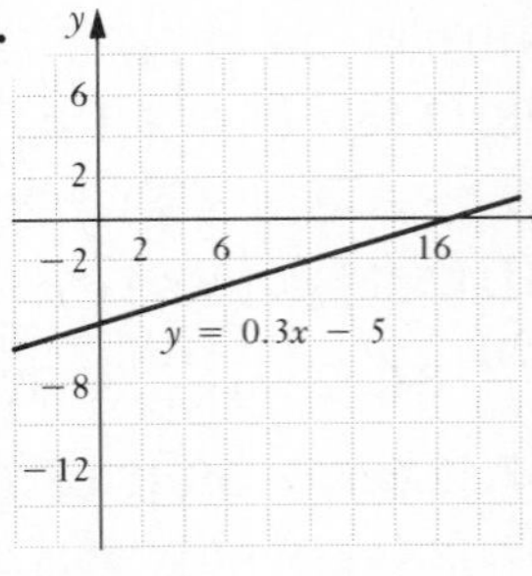

23. (a) $-17\frac{7}{9}°$, 0°, 5°, 100°; **25.** $\{x \mid x > -24\}$ **27.** 4

(b)

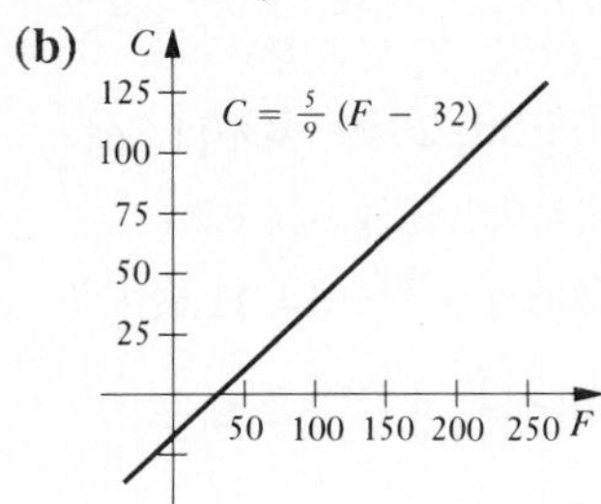

Exercise Set 3.2, p. 139

1.

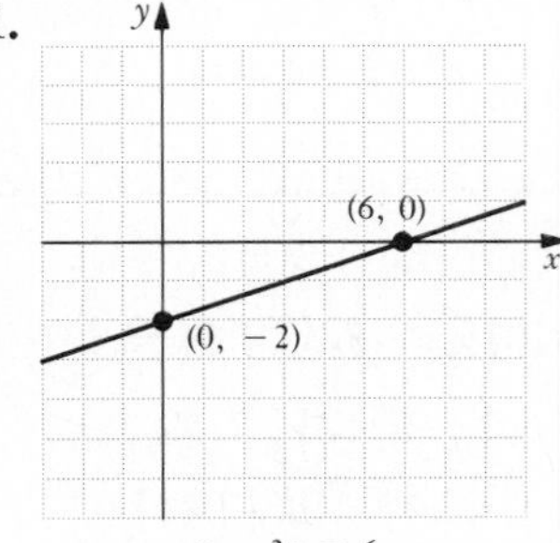

3.

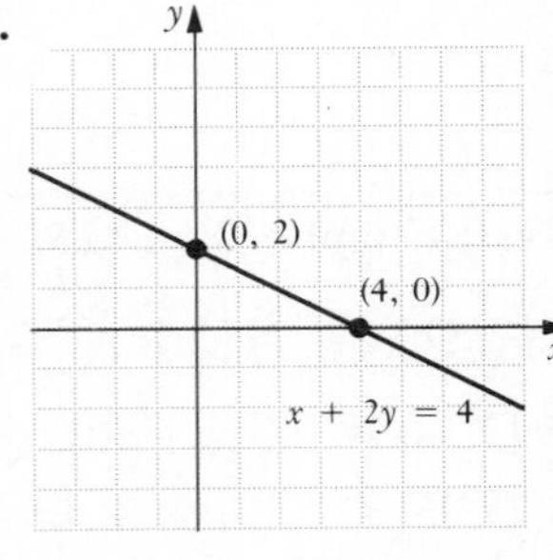

5.

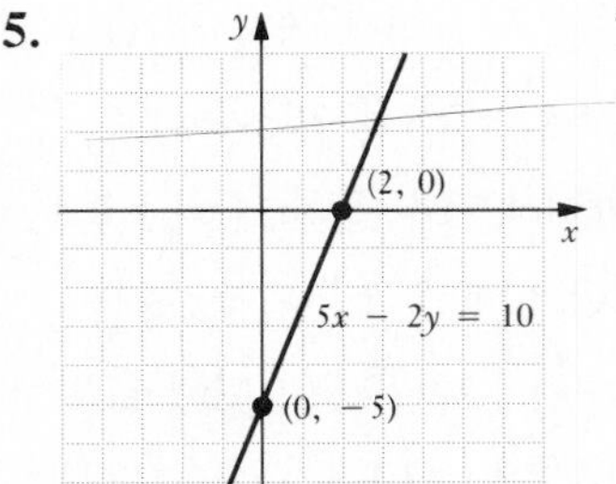

7.

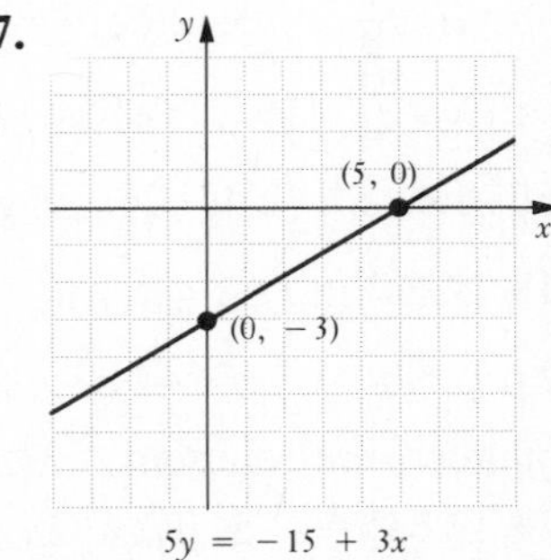

9.

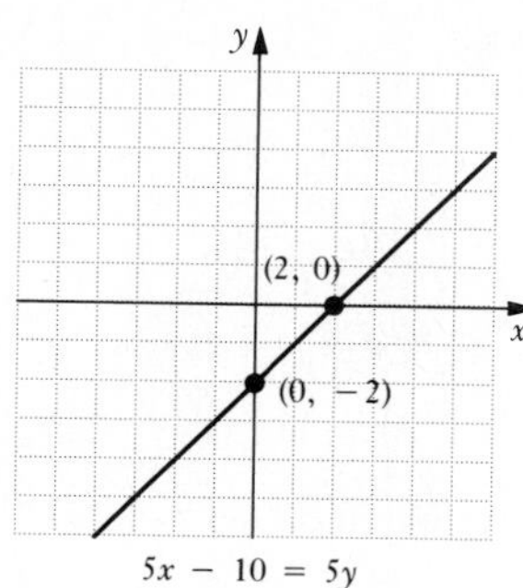

11.

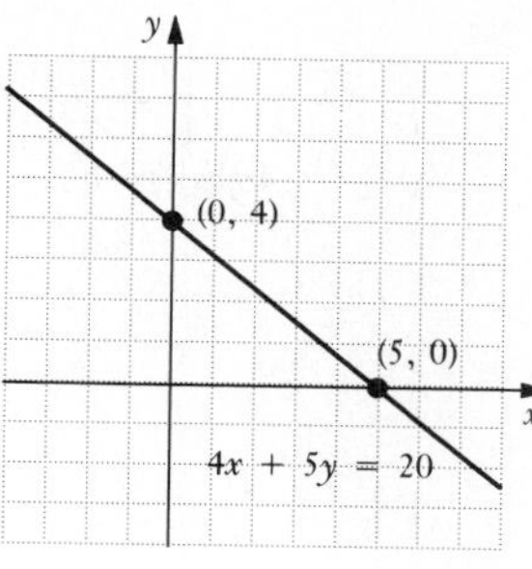

13.

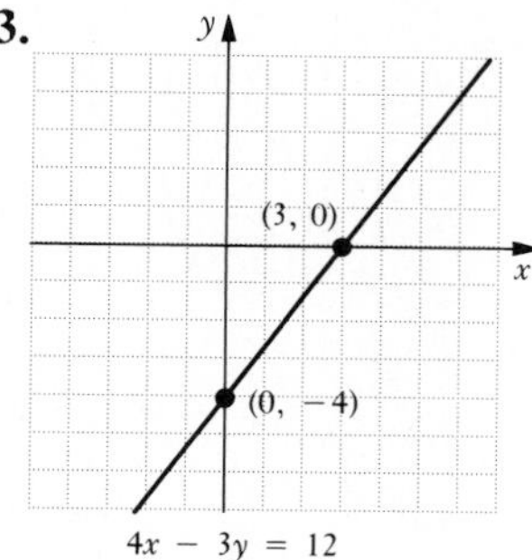

15.

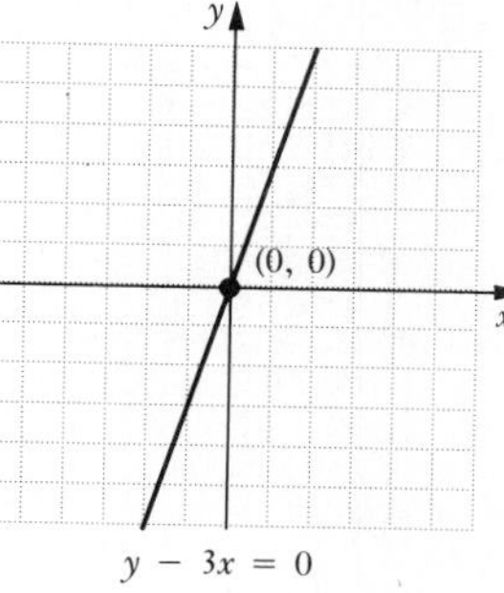

17.

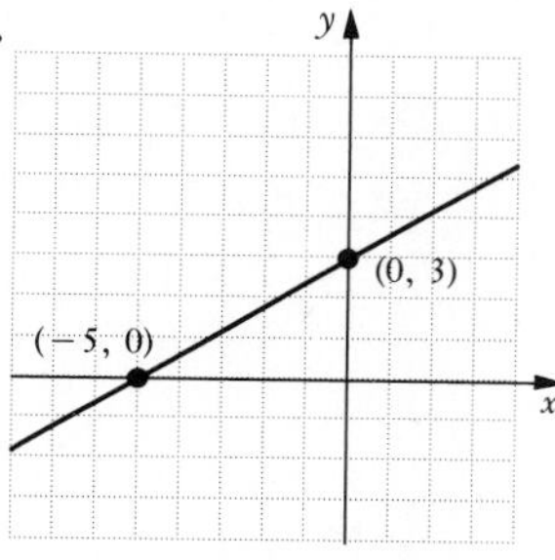

19.

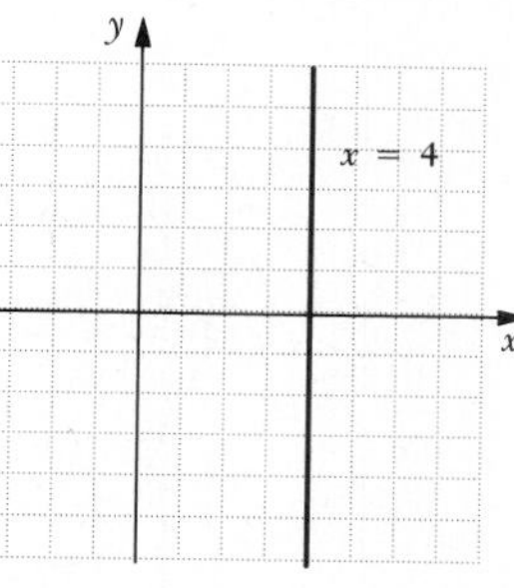

21.

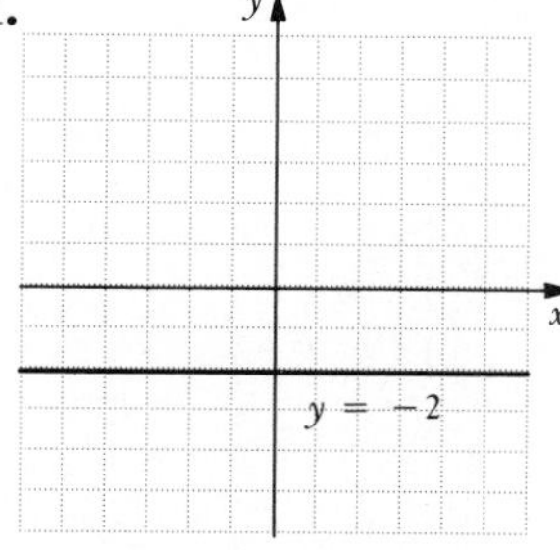

23.

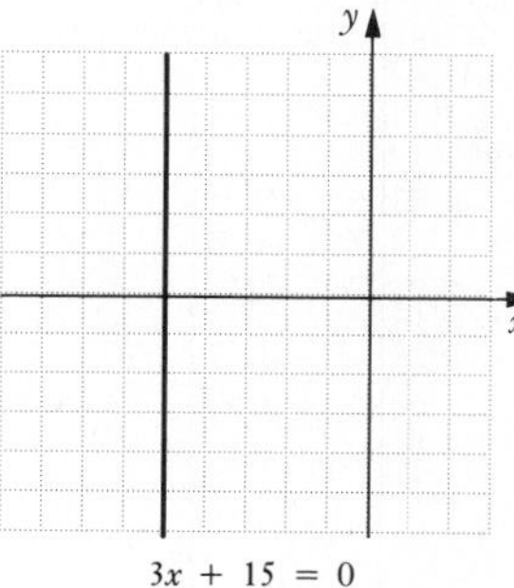

25.

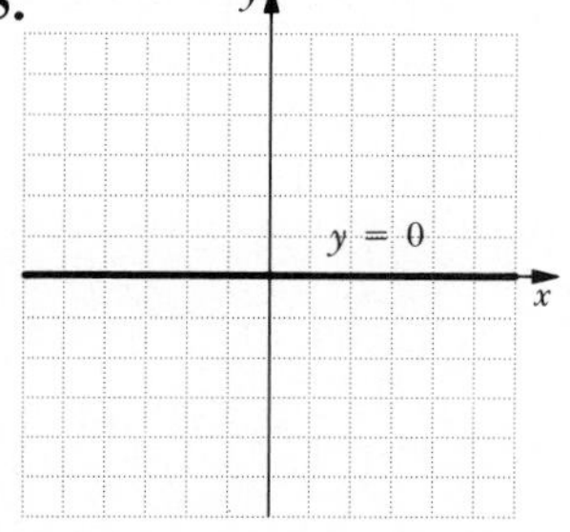

27.

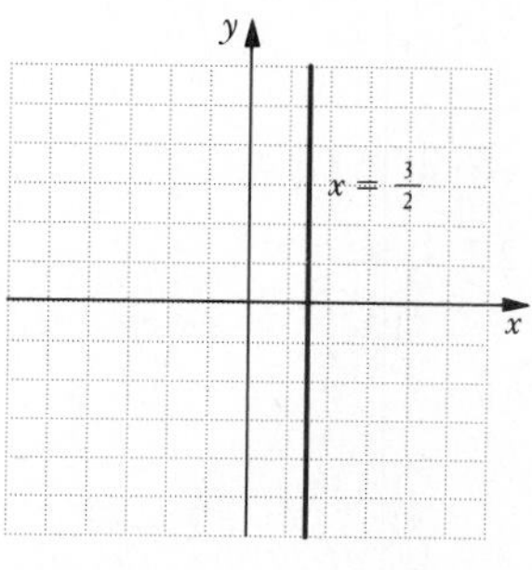

29. 11% **31.** $x = 0$ **33.** $(-4, 5)$ **35.** $x = 12$ **37.** $-\frac{3}{4}$

Exercise Set 3.3, p. 149

1. -2 **3.** 5 **5.** $-\frac{1}{3}$ **7.** Undefined **9.** 3 **11.** Undefined **13.** 0 **15.** 0 **17.** Undefined **19.** $\frac{1}{2}$ **21.** Undefined **23.** 0 **25.** 2 **27.** Slope is -8; y-intercept is $(0, -9)$. **29.** Slope is 3.8; y-intercept is $(0, 0)$. **31.** Slope is $-\frac{2}{3}$; y-intercept is $\left(0, \frac{8}{3}\right)$. **33.** Slope is $-\frac{8}{7}$; y-intercept is $\left(0, -\frac{24}{7}\right)$. **35.** Slope is 3; y-intercept is $(0, -2)$. **37.** Slope is $-\frac{3}{2}$; y- intercept is $\left(0, -\frac{3}{4}\right)$. **39.** $y = 5x + 8$ **41.** $y = 5.8x - 1$ **43.** $y = -\frac{7}{3}x - 5$

45.

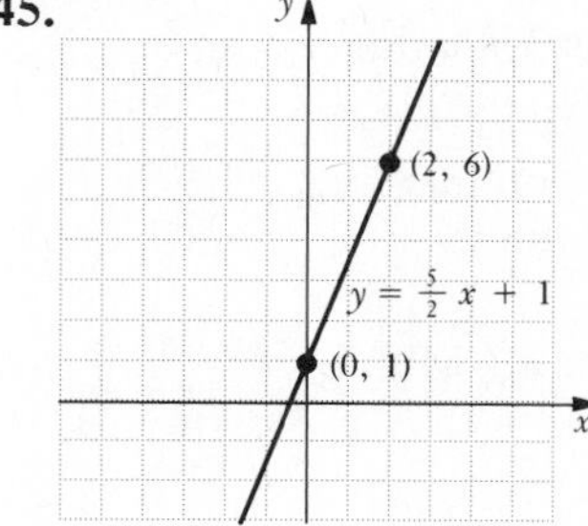

47.

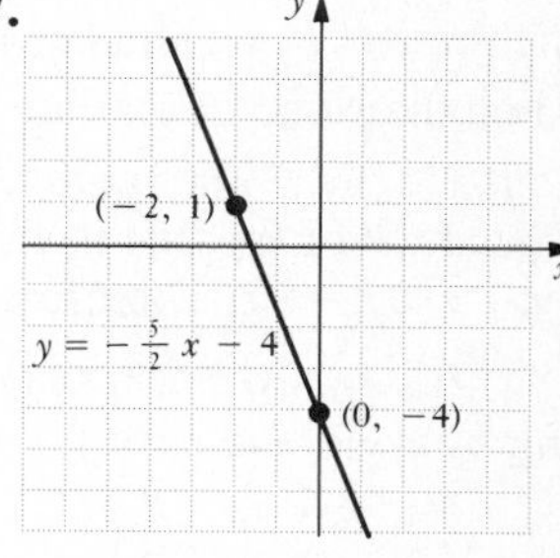

49.

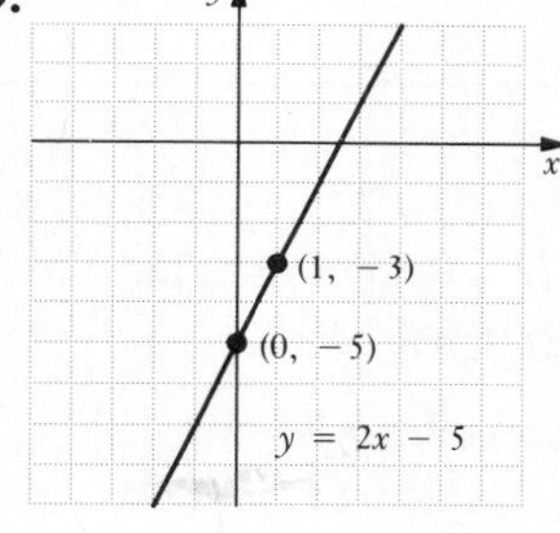

51.

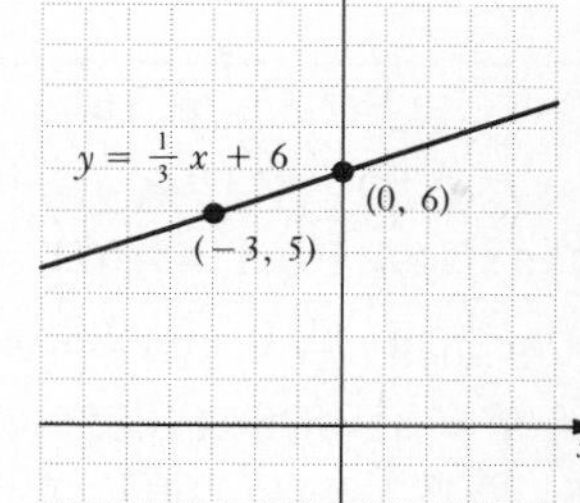

53.

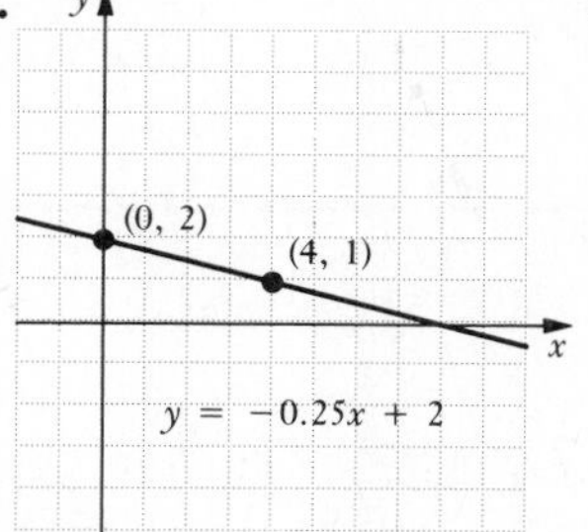

55.

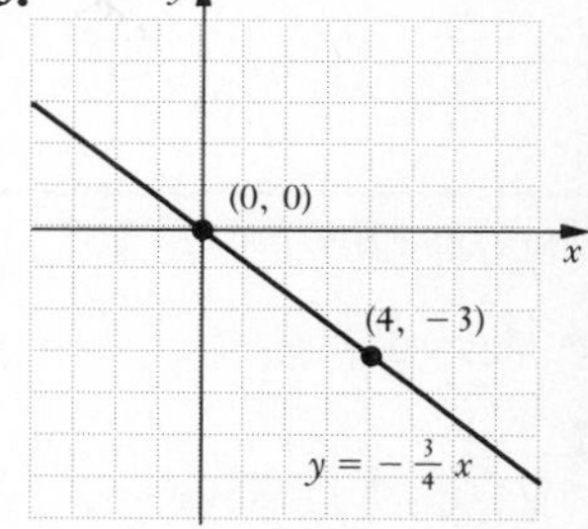

57. 4% **59.** 0.78 ft **61.** 15; 20 **63.** $\left\{x \middle| x \le -\frac{24}{5} \text{ or } x \ge 8\right\}$ **65.** $m = -8$; y-intercept $\left(0, -\frac{7}{2}\right)$ **67.** $m = -\frac{3}{7}$; y-intercept $\left(0, -\frac{6}{77}\right)$ **69. (a)** $-\frac{5c}{4b}$; **(b)** Undefined; **(c)** $\frac{a+d}{f}$

Exercise Set 3.4, p. 157

1. $y = 4x - 10$ **3.** $y = -2x + 15$ **5.** $y = 3x + 2$ **7.** $y = -2x + 16$ **9.** $y = -7$ **11.** $y = \frac{2}{3}x - \frac{8}{3}$ **13.** $y = \frac{1}{2}x + \frac{7}{2}$ **15.** $y = x$ **17.** $y = \frac{5}{2}x + 5$ **19.** $y = \frac{3}{2}x$ **21.** $y = \frac{2}{5}x$ **23.** $y = 13x - \frac{15}{4}$ **25.** Yes **27.** No **29.** Yes **31.** Yes **33.** No **35.** No **37.** $y = -\frac{1}{2}x + \frac{17}{2}$ **39.** $y = \frac{5}{7}x - \frac{17}{7}$ **41.** $y = \frac{1}{3}x + 4$ **43.** $y = \frac{1}{2}x + 4$ **45.** $y = \frac{4}{3}x - 6$ **47.** $y = \frac{5}{2}x + 9$ **49.** $\{x \mid x \geq \frac{7}{3}\}$ **51.** $\{-7, \frac{1}{3}\}$ **53.** $y = \frac{5}{12}x + \frac{41}{12}$ **55.** $k = 7$ **57.** Suppose that the points (x_1, y_1) and (x_2, y_2) are on a line. Then the slope of that line is $(y_2 - y_1)/(x_2 - x_1)$. Consider any other point (x, y) on that line. The slope of the line is also given by $(y - y_1)/(x - x_1)$. Therefore, $(y - y_1)/(x - x_1) = (y_2 - y_1)/(x_2 - x_1)$. Multiplying by $x - x_1$ we have

$$y - y_1 = \frac{y_2 - y_1}{x_2 - x_1}(x - x_1).$$

Exercise Set 3.5, p. 161

1. (a) (0, 72), (20, 75); **(b)** $E = 0.15t + 72$; **(c)** 78.9; 79.5 **3. (a)** (165, 70), (145, 67); **(b)** $H = 0.15W + 45.25$; **(c)** $64\frac{3}{4}$ in. **5. (a)** (0, 3.85), (20, 3.70); **(b)** $R = -0.0075t + 3.85$; **(c)** 3.34 min; 3.31 min; **(d)** $2003.\overline{3}$, or in 2004 **7. (a)** (2, 1.75), (3, 2.00); **(b)** $C = 0.25M + 1.25$; **(c)** \$3.00 **9.** \$1300 **11.** 100.03916 cm; 99.96796 cm

Exercise Set 3.6, p. 167

1. Yes **3.** No

5.

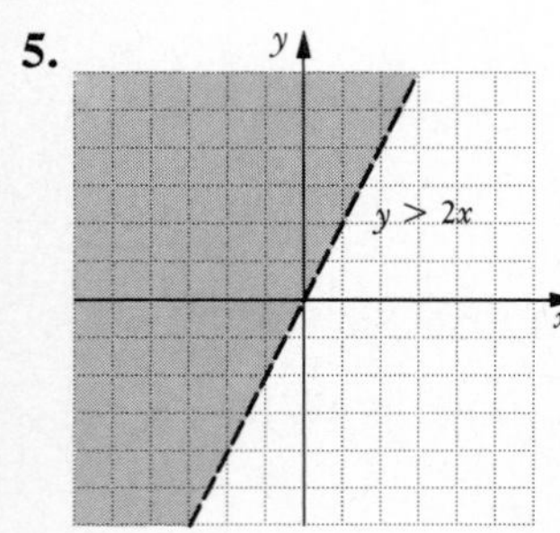

7.

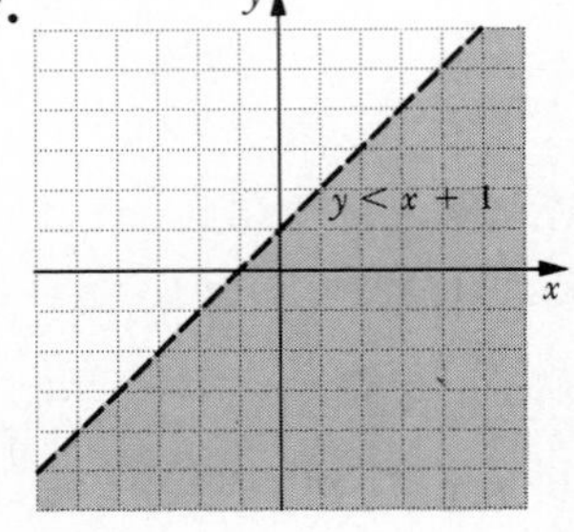

9.

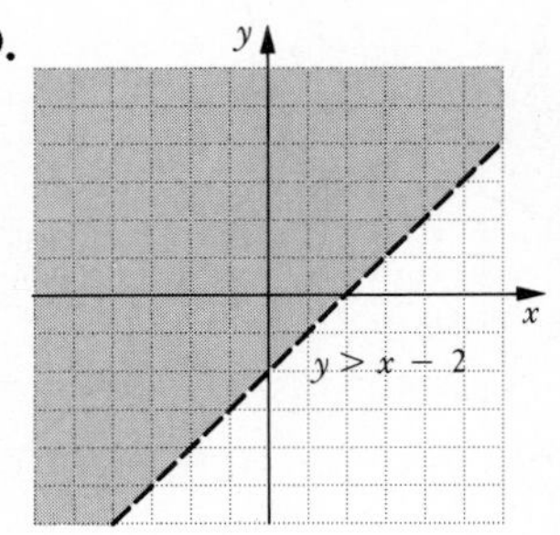

11.

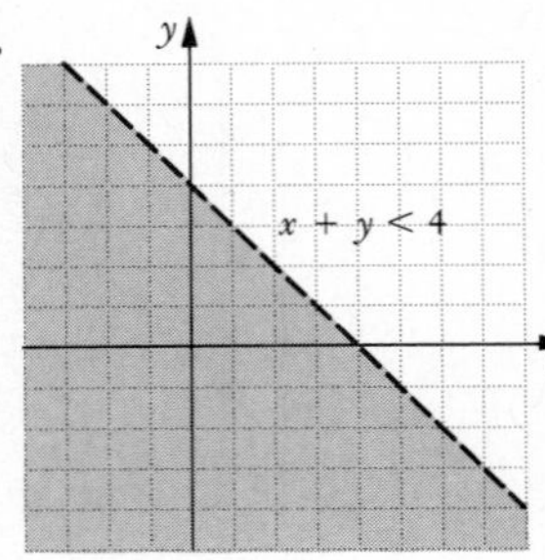

13.

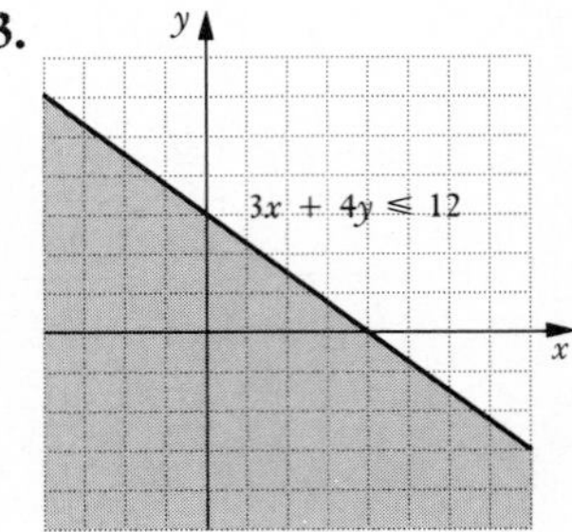

15.

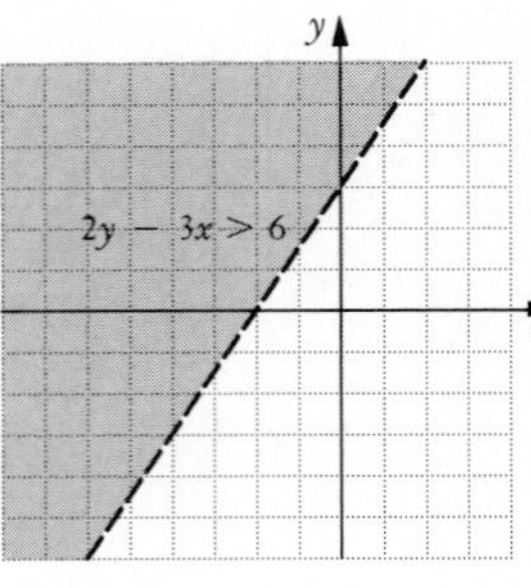

17.

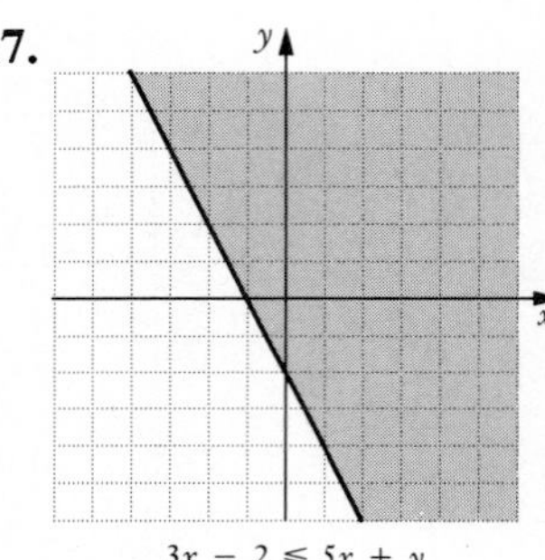

19.

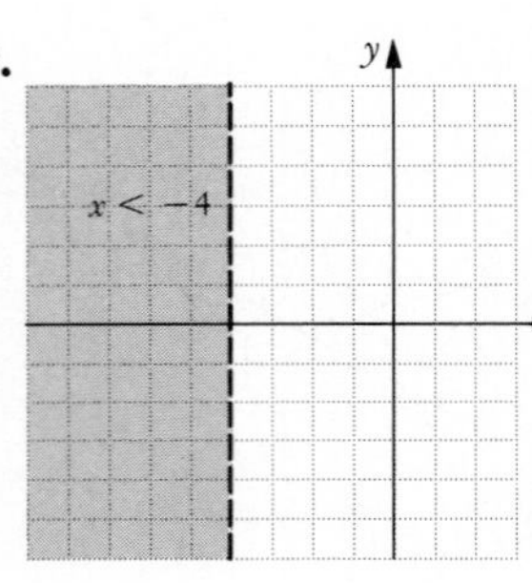

21.

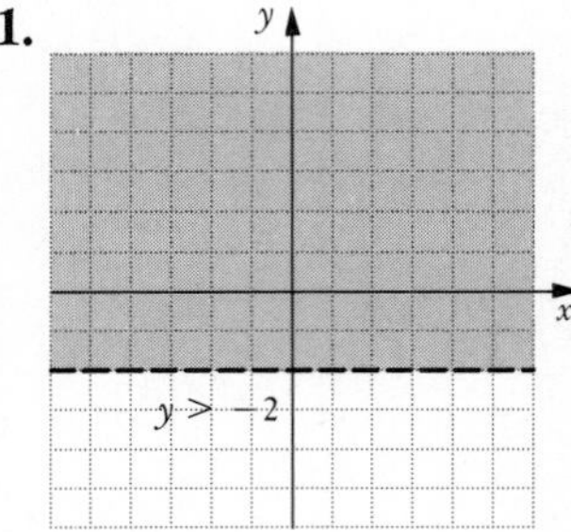

23.

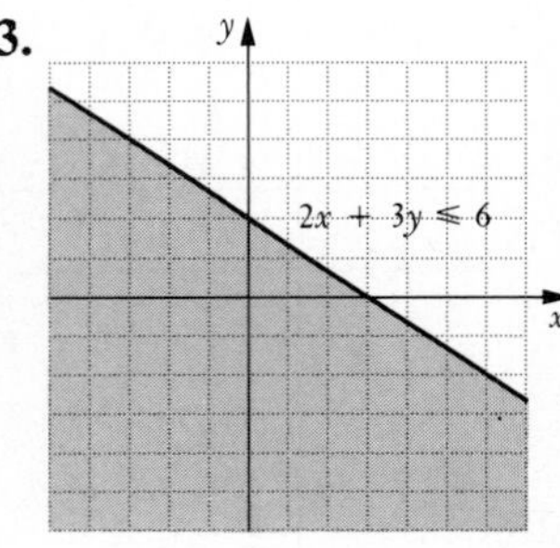

25. 36.5 ft; 41.5 ft

27.

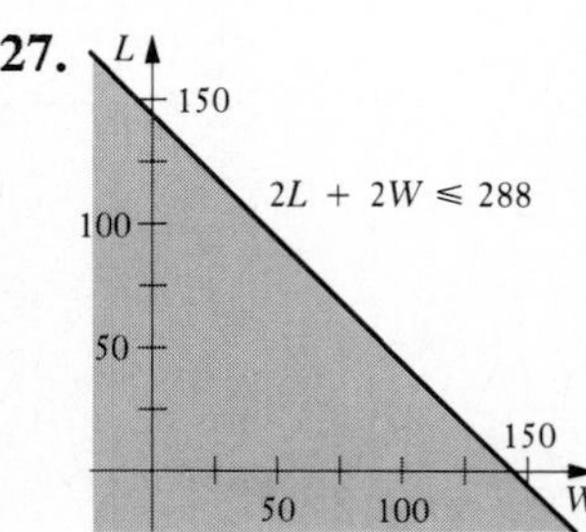

29.

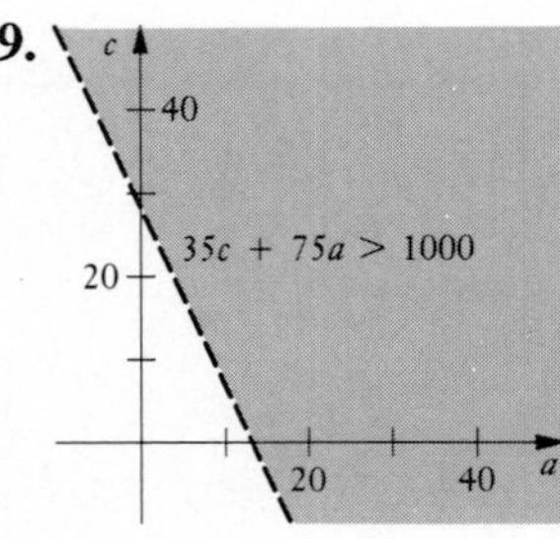

Summary and Review: Chapter 3, p. 169

1. [3.3d]

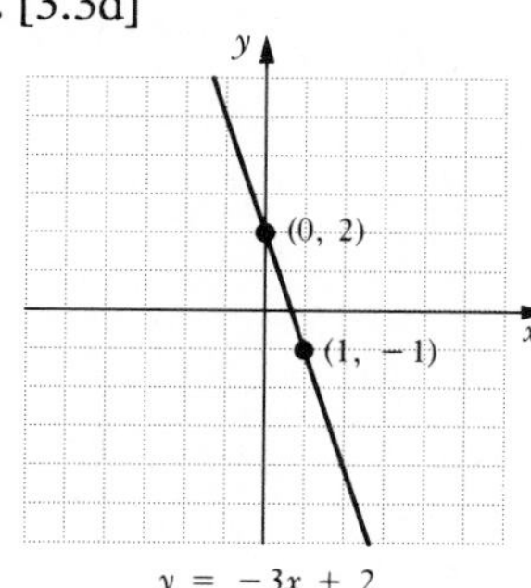

2. [3.3d]

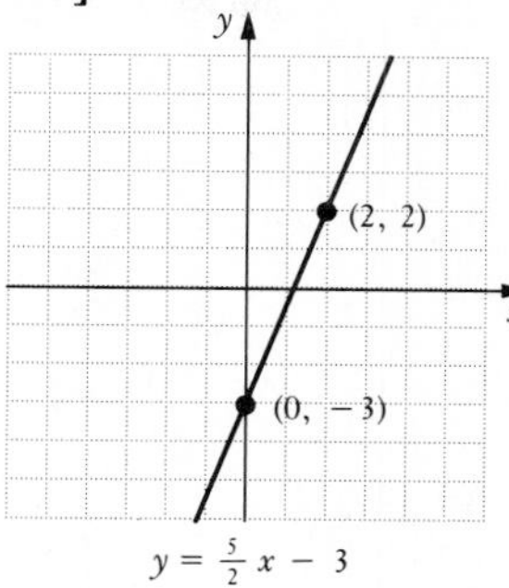

3. [3.2b]

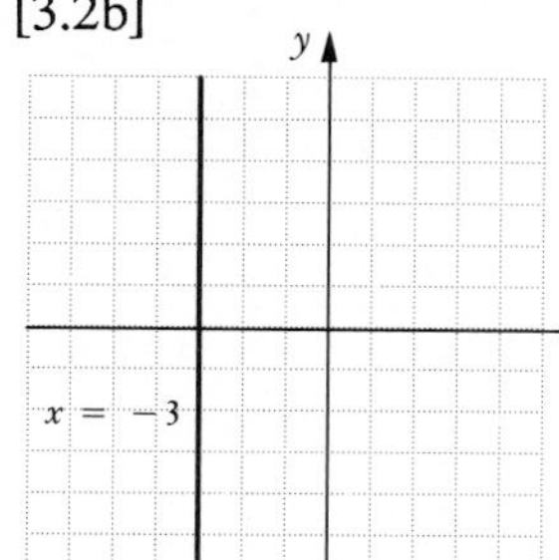

4. [3.2b]

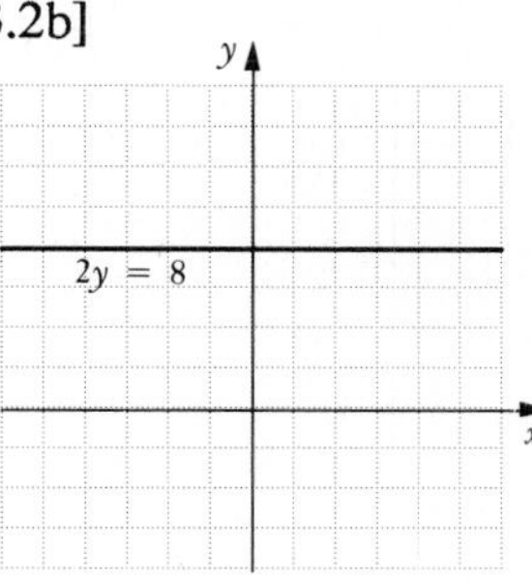

5. [3.2a]

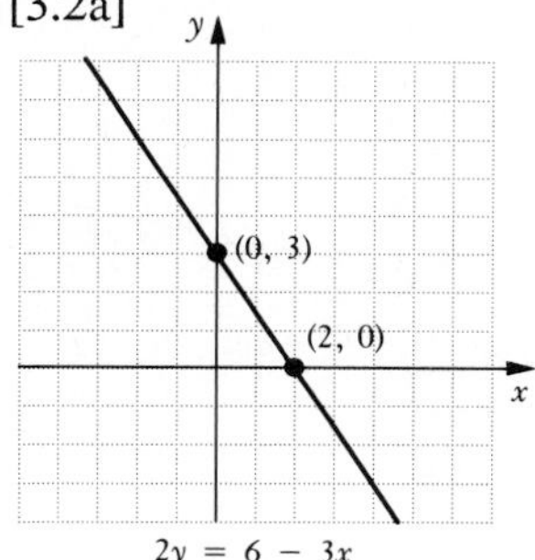

6. [3.2a]

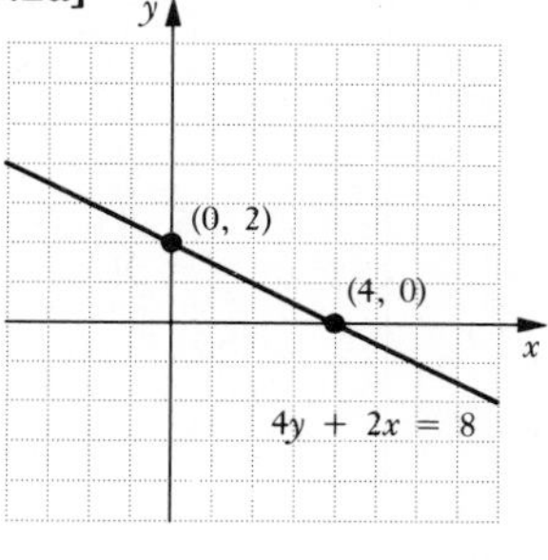

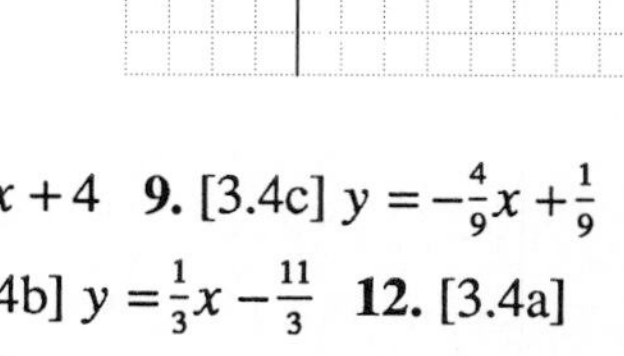

7. [3.3a] $\frac{11}{3}$ **8.** [3.4a] $y=-3x+4$ **9.** [3.4c] $y=-\frac{4}{9}x+\frac{1}{9}$ **10.** [3.4c] $y=\frac{2}{3}x-\frac{7}{3}$ **11.** [3.4b] $y=\frac{1}{3}x-\frac{11}{3}$ **12.** [3.4a] $y=2x+1$ **13.** [3.4b] $y=-\frac{3}{2}x$ **14.** [3.4c] $y=-\frac{5}{7}x+9$ **15.** [3.4c] $y=\frac{1}{3}x+\frac{1}{3}$ **16.** [3.4c] Perpendicular **17.** [3.4c] Parallel **18.** [3.4c] Parallel **19.** [3.4c] Perpendicular **20.** [3.5a] **(a)** $(2.5, 2.75)$, $(3.0, 2.90)$; **(b)** $C=0.3M+2$; **(c)** \$3.50

21. [3.6b]

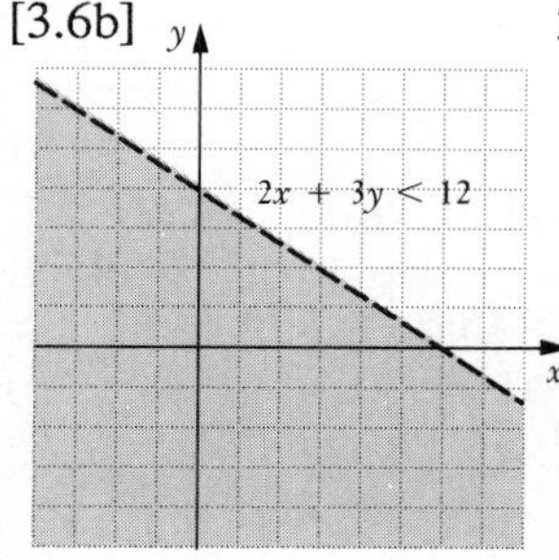

22. [3.6b]

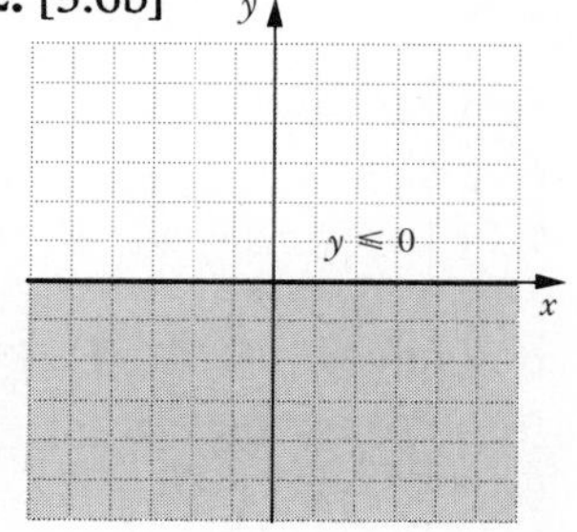

23. [3.6b]

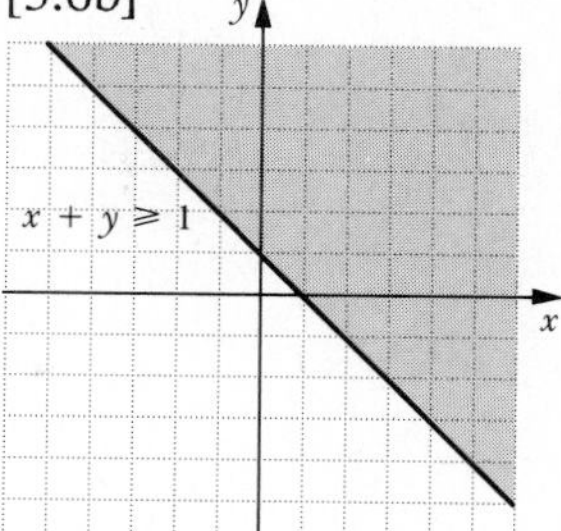

24. [2.4c] $\left\{x \mid x<-\frac{77}{4}\right\}$ **25.** [2.6c] $\left\{-3, \frac{13}{2}\right\}$ **26.** [2.6e] $\left\{x \mid -3 \le x \le \frac{13}{2}\right\}$ **27.** [2.6e] $\left\{x \mid x<-3 \text{ or } x>\frac{13}{2}\right\}$ **28.** [2.5b] $\left\{x \mid x \neq \frac{21}{2}\right\}$ **29.** [2.2a] 69, 76, 83 **30.** [3.4c] $y=2x+\frac{5}{7}$ **31.** [3.5a] \$31,000

Test: Chapter 3, p. 171

1. [3.3d]

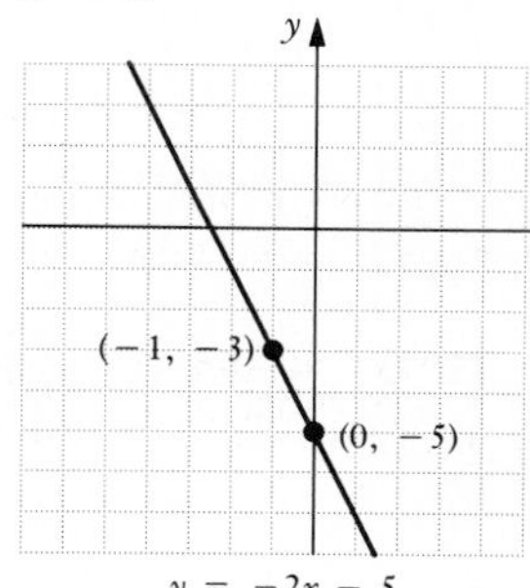

2. [3.3d]

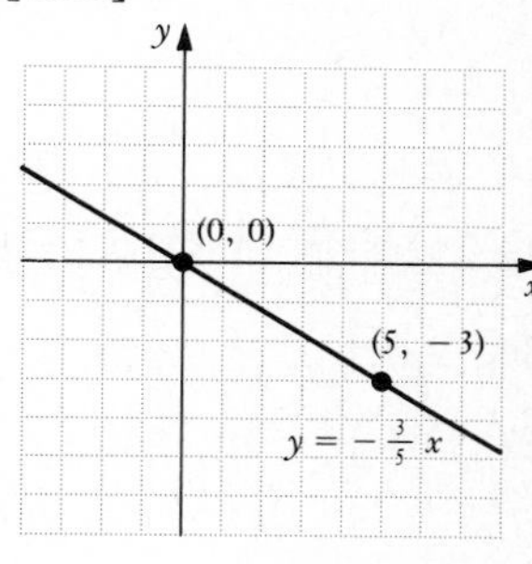

3. [3.2a]

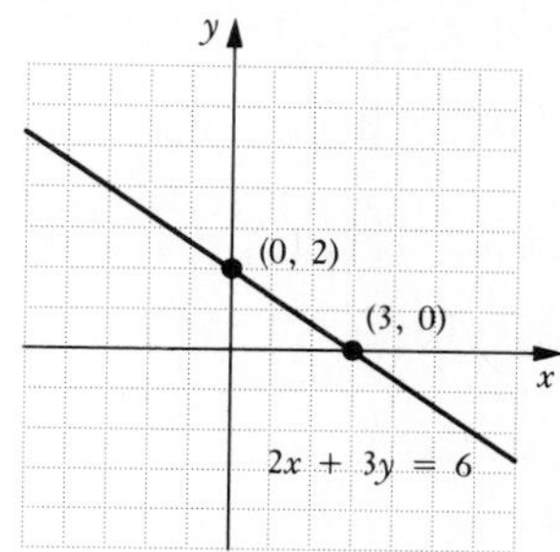

4. [3.2b]

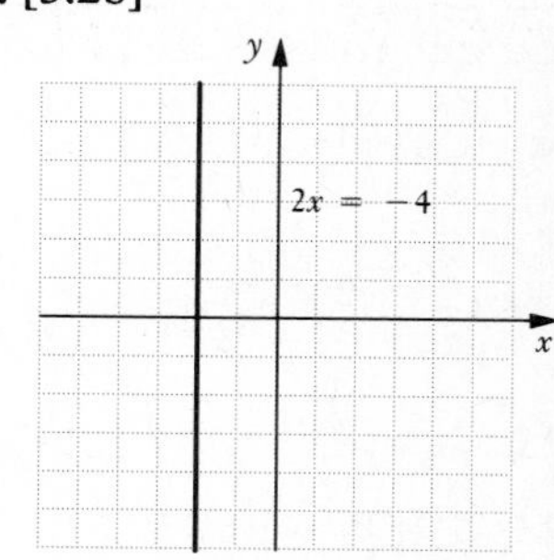

5. [3.3a] $-\frac{9}{7}$ **6.** [3.4a] $y=-3x+1$ **7.** [3.4b] $y=-\frac{3}{2}x$ **8.** [3.4c] $y=\frac{1}{2}x-3$ **9.** [3.4c] $y=3x-1$ **10.** [3.4c] Parallel **11.** [3.4c] Perpendicular **12.** [3.5a] **(a)** $(250, 100)$, $(300, 110)$; **(b)** $C=0.2M+50$; **(c)** \$150

13. [3.6b] **14.** [3.6b]

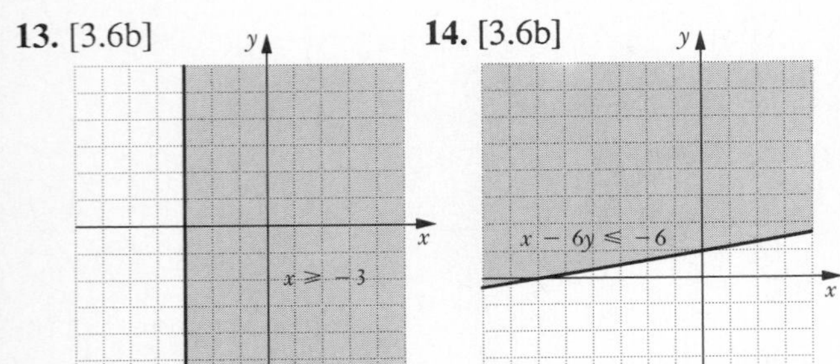

15. [3.6b]

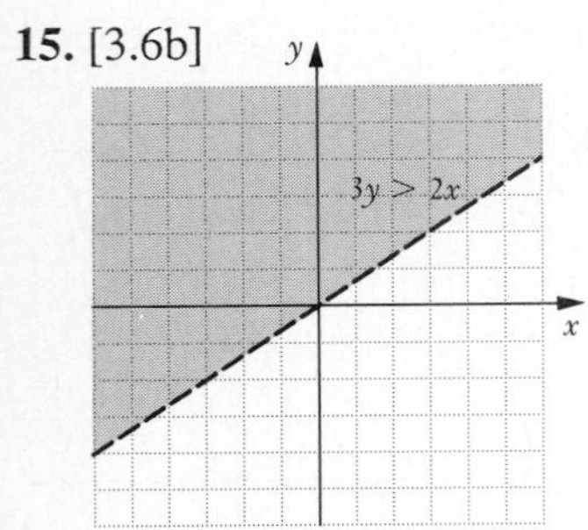

16. [2.4c] $\left\{x \middle| x \geq \frac{13}{9}\right\}$ **17.** [2.5b] $\left\{x \middle| x < -2 \text{ or } x > \frac{32}{7}\right\}$
18. [2.6e] $\left\{x \middle| -2 < x < \frac{32}{7}\right\}$ **19.** [2.6c] $\left\{-2, \frac{32}{7}\right\}$
20. [2.6e] $\left\{x \middle| x \leq -2 \text{ or } x \geq \frac{32}{7}\right\}$ **21.** [2.2a] \$35
22. [3.4c] $\frac{24}{5}$ **23.** [3.4c] $y = -\frac{1}{2}x - \frac{3}{5}$

Cumulative Review: Chapters 1-3, p. 173

1. [1.5a] 31 **2.** [1.2d] 2.2 **3.** [1.2d] $\frac{2}{3}$ **4.** [1.3c] $\frac{1}{2}$
5. [1.3d] 36.48 **6.** [1.4e] $-10x + 32$ **7.** [1.6a] $\frac{32x^8}{y}$
8. [1.6a] $-\frac{9}{x^2y^2}$ **9.** [1.5c] $23x + 31$ **10.** [1.5c] -328
11. [1.5c] -1 **12.** [2.1b] -22 **13.** [2.1d] $\frac{15}{88}$ **14.** [2.1c] 20
15. [2.1d] $-\frac{21}{4}$ **16.** [2.1d] -5 **17.** [2.3a] $x = \frac{W - By}{A}$
18. [2.3a] $A = \frac{M}{1+4B}$ **19.** [2.4a] $\{y \mid y \leq 7\}$ **20.** [2.4c] $\left\{x \middle| x < -\frac{3}{2}\right\}$ **21.** [2.4c] $\left\{x \middle| x > -\frac{1}{11}\right\}$ **22.** [2.5b] All real numbers **23.** [2.5a] $\{x \mid -7 < x \leq 4\}$ **24.** [2.5a] $\left\{x \middle| -2 \leq x \leq \frac{3}{2}\right\}$ **25.** [2.6c] $\{-8, 8\}$ **26.** [2.6e] $\{y \mid y < -4 \text{ or } y > 4\}$ **27.** [2.6e] $\left\{x \middle| -\frac{3}{2} \leq x \leq 2\right\}$

28. [3.1c]

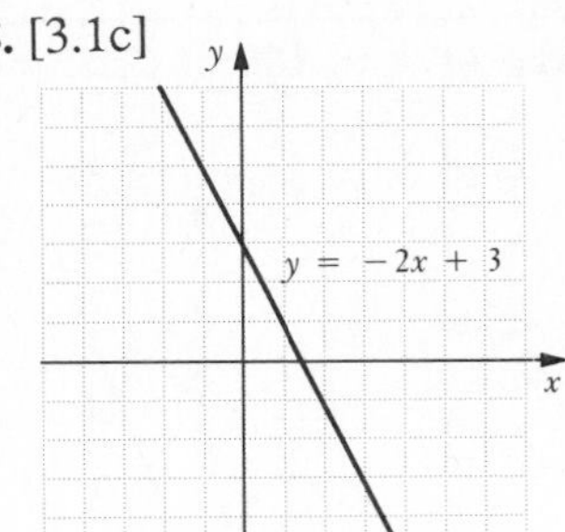

29. [3.2a]

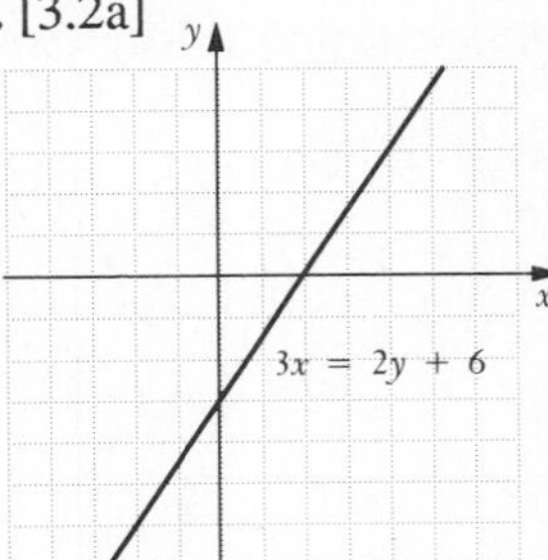

30. [3.2b]

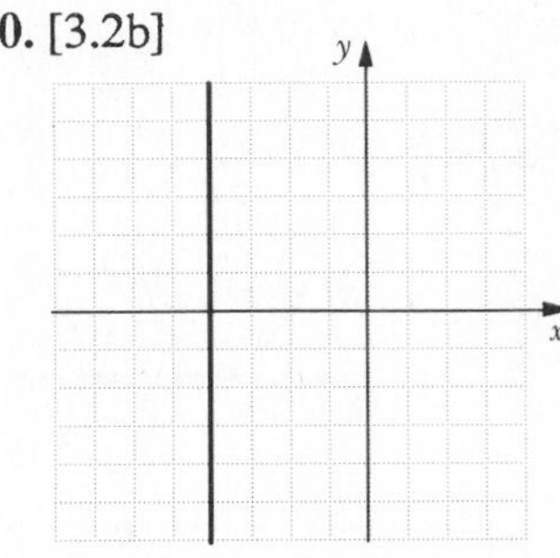

31. [3.2b]

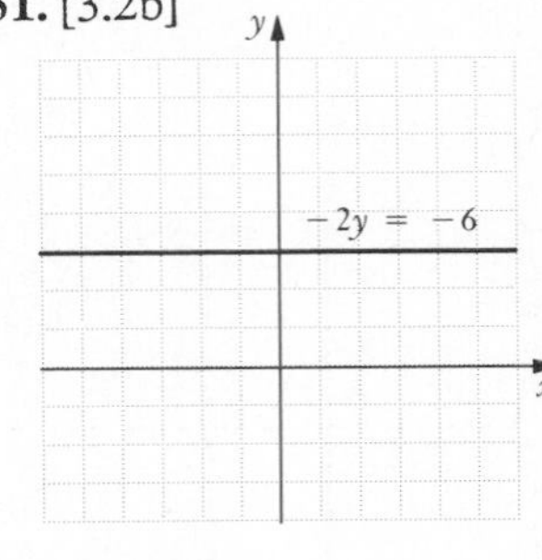

32. [3:6b]

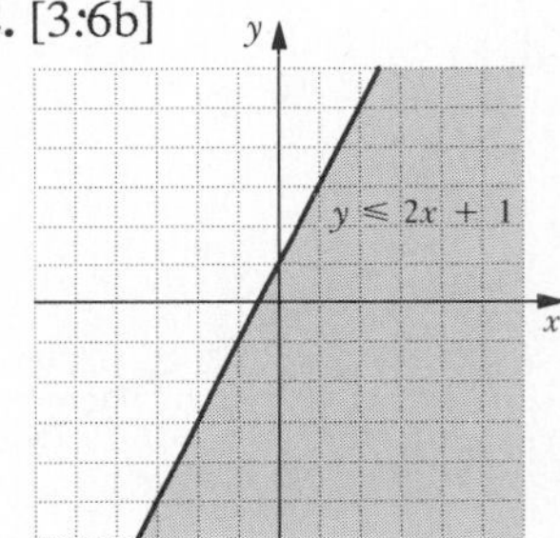

33. [3.6b]

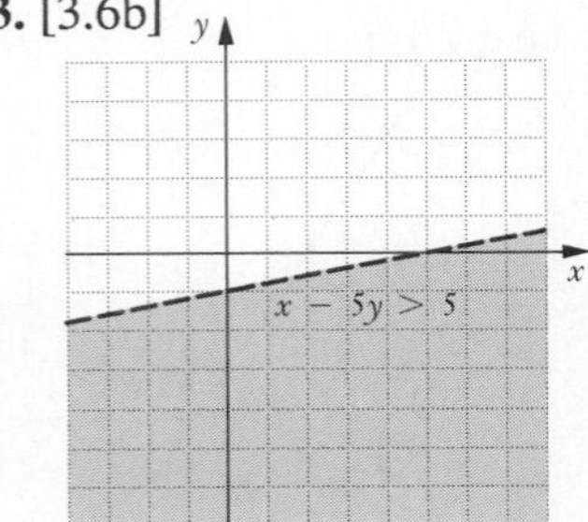

34. [3.4c] $y = -4x - 22$ **35.** [3.4c] $y = \frac{1}{4}x - 5$ **36.** [3.3c] $m = \frac{9}{4}$, y-intercept $(0, -3)$ **37.** [3.3a] $m = \frac{4}{3}$ **38.** [3.4a] $y = -3x - 5$ **39.** [3.4b] $y = -\frac{1}{10}x + \frac{12}{5}$ **40.** [2.2a] 32.88
41. [2.2a] \$67,000 **42.** [2.2a] 4 **43.** [2.2a] \$9000
44. [2.2a] $w = 17$ m, $l = 23$ m **45.** [3.5a] \$151,000
46. [1.5a] $-12x^{2a}y^{b+y+3}$ **47.** [2.5a] $\{x \mid 6 < x \leq 10\}$
48. [3.4c] (1), (4)

CHAPTER 4

Pretest: Chapter 4, p. 176

1. [4.1a] $(1, 2)$ **2. [4.2a]** $(1, 2)$ **3. [4.2b]** $(-3, 2)$
4. [4.1b] Consistent **5. [4.1b]** Independent **6. [4.6c]** $(-3, 2)$ **7. [4.4b]** $(1, 2, 3)$ **8. [4.6a]** -17 **9. [4.6b]** -5
10. [4.6c] $(3, -1, 2)$ **11. [4.3a]** \$3000 at 8%; \$4500 at 10% **12. [4.5a]** 29, 46, 134 **13. [4.3a]** *A*: $64\frac{8}{13}$ lb; *B*: $55\frac{5}{13}$ lb **14. [4.3b]** 54 mph

15. [4.7a]

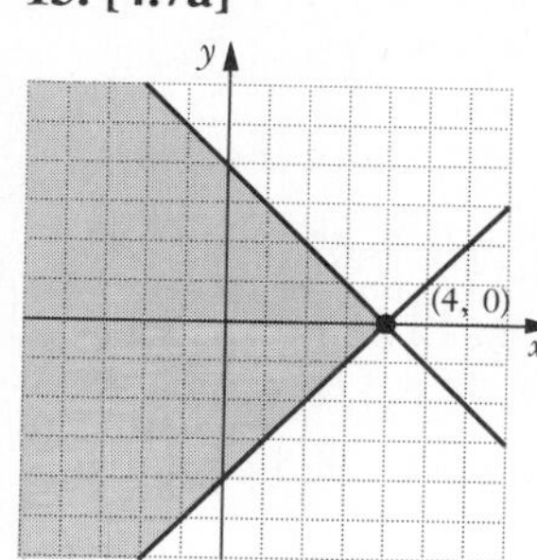

16. [4.7a]

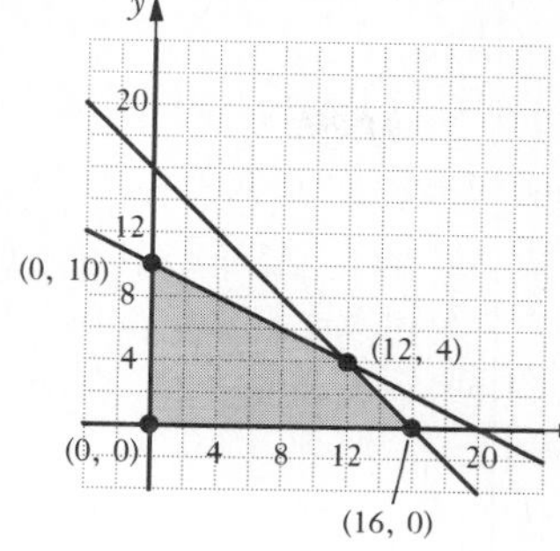

Exercise Set 4.1, p. 181

1. Yes **3.** No **5.** Yes **7.** (3, 1); consistent, independent **9.** (1, −2); consistent, independent **11.** (4, −2); consistent, independent **13.** (2, 1); consistent, independent **15.** $\left(\frac{5}{2}, -2\right)$; consistent, independent **17.** (3, −2); consistent, independent **19.** No solution; inconsistent, independent **21.** (4, −5); consistent, independent **23.** $\left(\frac{15}{7}, -\frac{22}{7}\right)$; consistent, independent **25.** Infinitely many solutions; consistent, dependent **27.** $A: -\frac{17}{4}, B: -\frac{12}{5}$ **29.** $2x = y$; $x = y - 3$ **31.** $x + y = 1$; $3x + 3y = 3$

Exercise Set 4.2, p. 189

1. $(-4, 3)$ **3.** $(-3, -15)$ **5.** $(2, -2)$ **7.** $(-2, 1)$ **9.** $(1, 2)$ **11.** $(-1, 2)$ **13.** $\left(\frac{128}{31}, -\frac{17}{31}\right)$ **15.** $(6, 2)$ **17.** $\left(\frac{140}{13}, -\frac{50}{13}\right)$ **19.** $(4, 6)$ **21.** $\left(\frac{110}{19}, -\frac{12}{19}\right)$ **23.** Infinite number of solutions **25.** No solution **27.** $\left(\frac{1}{2}, -\frac{1}{2}\right)$ **29.** $\left(-\frac{4}{3}, -\frac{19}{3}\right)$ **31.** $\left(\frac{1000}{11}, -\frac{1000}{11}\right)$ **33.** $(140, 60)$ **35.** 1.3 **37.** $(23.118879, -12.039964)$ **39.** $\left(\frac{a+2b}{7}, \frac{a-5b}{7}\right)$ **41.** $p = 2$; $q = -\frac{1}{3}$

Exercise Set 4.3, p. 199

1. 5 and −47 **3.** 8 white, 22 yellow **5.** l: 94 ft, w: 50 ft **7.** 119°, 61° **9.** 12 field goals, 6 free throws **11.** 23 wins, 14 ties **13.** 4 30-sec, 8 60-sec **15.** 24 and 8 **17.** 150 lb of soybean meal, 200 lb of corn meal **19.** 5 liters of each **21.** \$4100 at 14%, \$4700 at 16% **23.** \$725 at 12%, \$425 at 11% **25.** Hot dog: \$1.75; hamburger: \$2.65 **27.** Carlos 28, Maria 20 **29.** 17 quarters, 13 fifty-cent pieces **31.** 375 km;

Distance	Rate	Time	
d	75	$t+2$	⟶ $d = 75(t+2)$
d	125	t	⟶ $d = 125t$

33. $1\frac{3}{4}$ hours **35.** 24 mph **37.** Headwind: 40 mph; plane: 620 mph **39.** $1\frac{1}{3}$ hr **41.** l: $57\frac{3}{5}$ in.; w: $20\frac{2}{5}$ in. **43.** 180 **45.** $4\frac{4}{7}$ liters **47.** 261 city mi; 204 highway mi **49.** 143 gal

Exercise Set 4.4, p. 207

1. Yes **3.** $(1, 2, 3)$ **5.** $(-1, 5, -2)$ **7.** $(3, 1, 2)$ **9.** $(-3, -4, 2)$ **11.** $(2, 4, 1)$ **13.** $(-3, 0, 4)$ **15.** $(2, 2, 4)$ **17.** $\left(\frac{1}{2}, 4, -6\right)$ **19.** $\left(\frac{1}{2}, \frac{1}{3}, \frac{1}{6}\right)$ **21.** $\left(\frac{1}{2}, \frac{2}{3}, -\frac{5}{6}\right)$ **23.** $(15, 33, 9)$ **25.** $(3, 4, -1)$ **27.** $c = \frac{2F + td}{t}$, or $c = \frac{2F}{t} + d$ **29.** $(1, -2, 4, -1)$

Exercise Set 4.5, p. 213

1. 4, 2, −1 **3.** 34°, 104°, 42° **5.** \$41.1 billion on newspaper, \$36 billion on television, \$7.7 billion on radio **7.** 25°, 50°, 105° **9.** Egg: 274 mg; cupcake: 19 mg; pizza: 9 mg **11.** Steak: 1; baked potatoes: 3; asparagus: 4 **13.** Oriental: 385; black: 200; white: 154 **15.** A: 900; B: 1300; C: 1500 **17.** \$45,000 at 8%; \$15,000 at 6%; \$20,000 at 9% **19.** 6 par-3 holes, 8 par-4 holes, 4 par-5 holes **21.** 180° **23.** 35

Exercise Set 4.6, p. 221

1. 3 **3.** 36 **5.** 0 **7.** 0 **9.** −10 **11.** −3 **13.** 5 **15.** 0 **17.** $(2, 0)$ **19.** $\left(-\frac{25}{2}, -\frac{11}{22}\right)$ **21.** $\left(\frac{3}{2}, \frac{5}{2}\right)$ **23.** $(-4, 3)$ **25.** $(2, -1, 4)$ **27.** $(1, 2, 3)$ **29.** $\left(\frac{3}{2}, -4, 3\right)$ **31.** $(2, -2, 1)$ **33.** $\frac{333}{245}$ **35.** $\frac{3}{5}$ **37.** 12 **39.** 10

Exercise Set 4.7, p. 227

1.

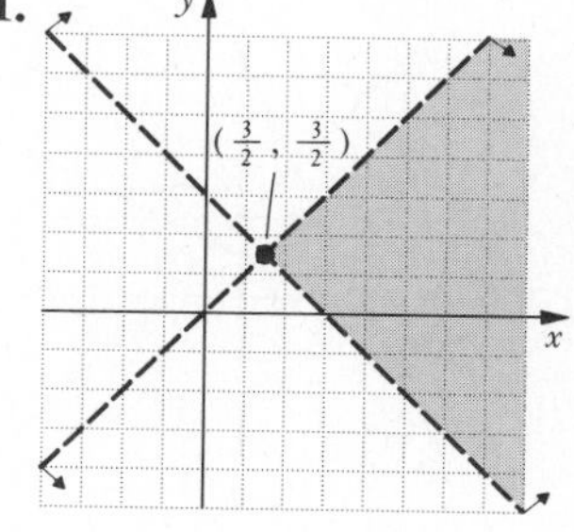

3.

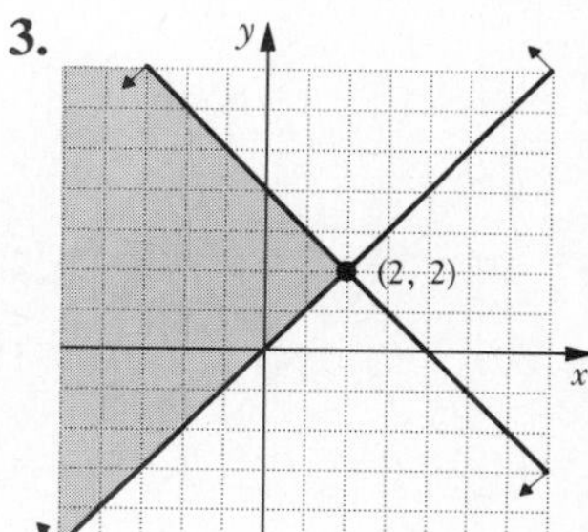

5.

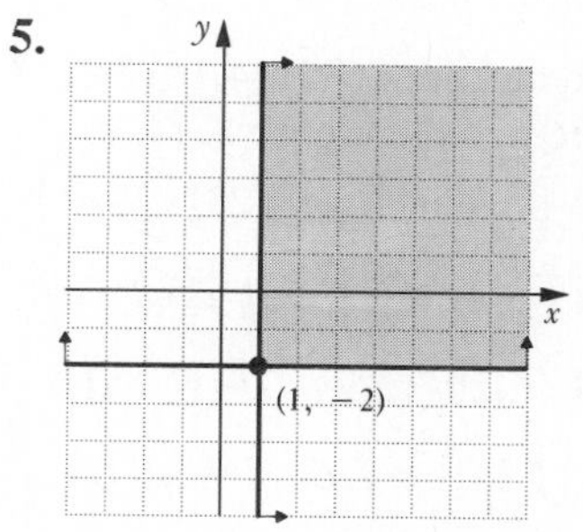

7.

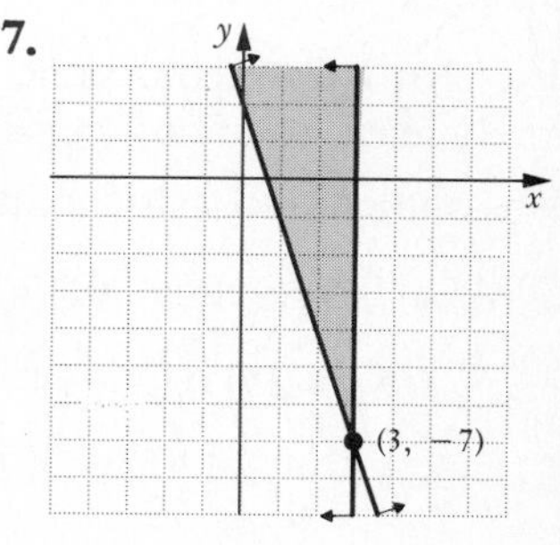

9.

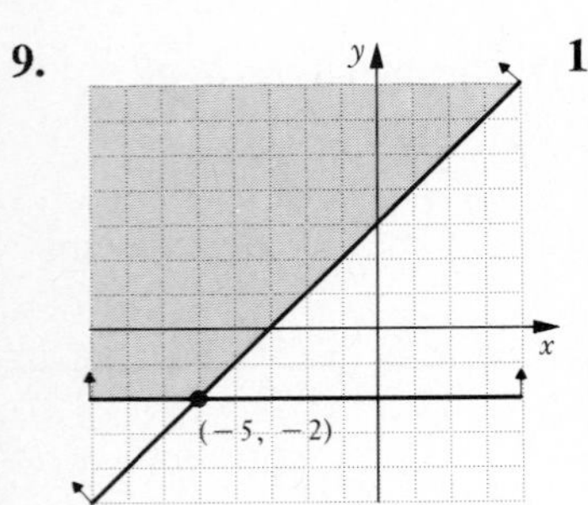

11.

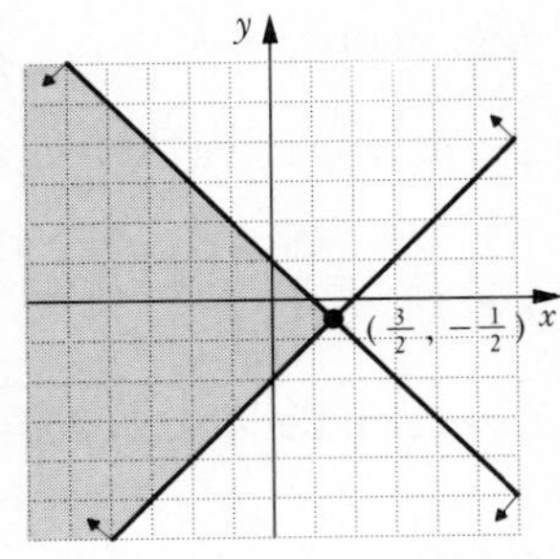

13.

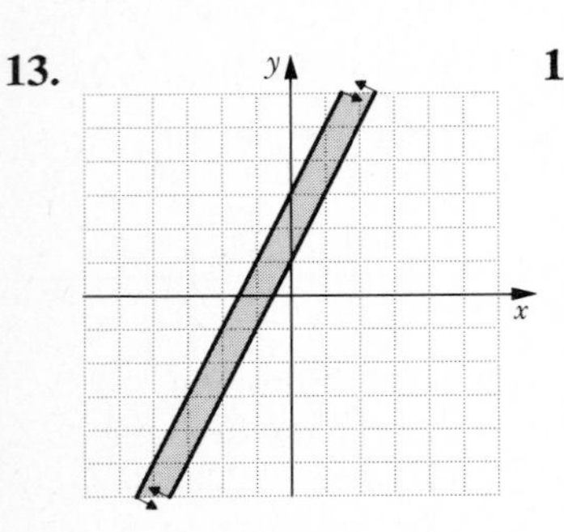

15.

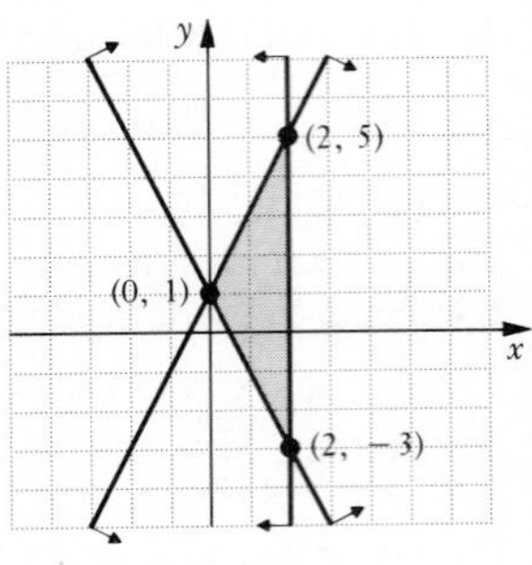

17.

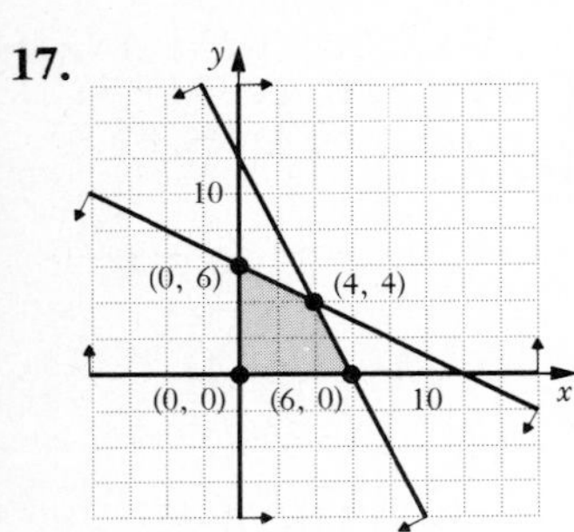

19.

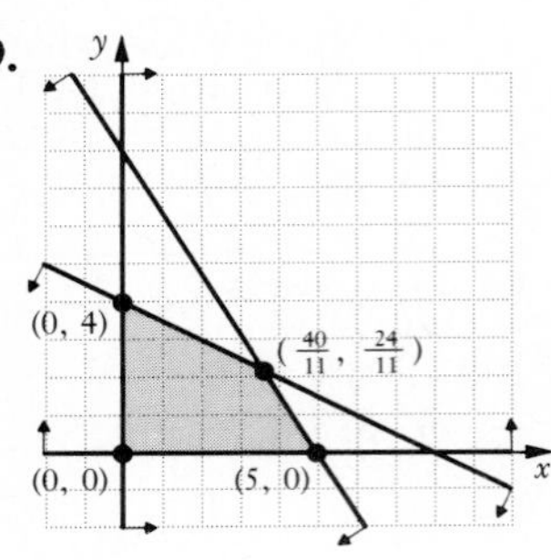

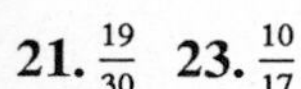

21. $\frac{19}{30}$ 23. $\frac{10}{17}$

25.

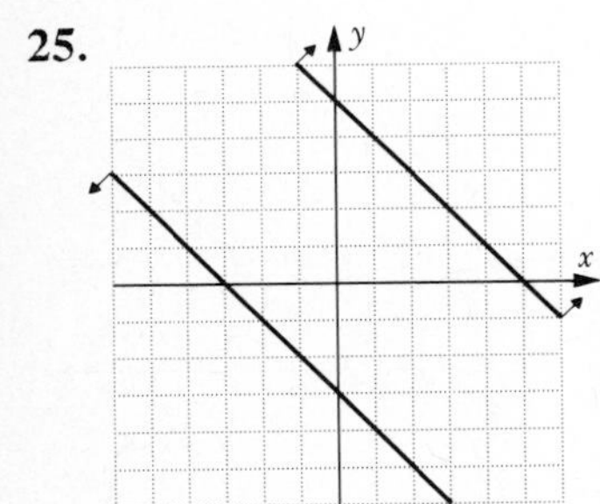

27.

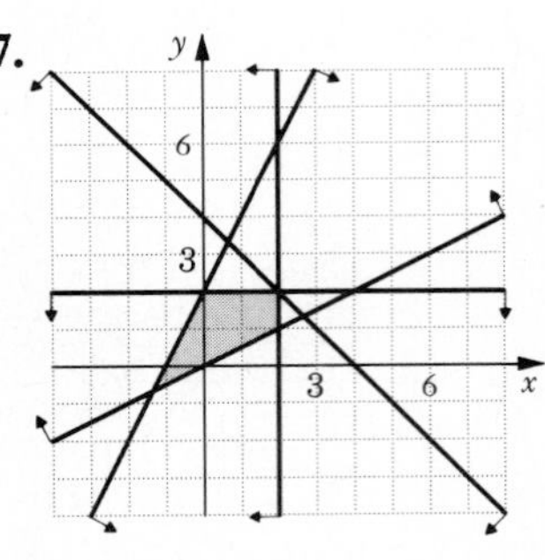

Summary and Review: Chapter 4, p. 229

1. [4.1b] (−2, 1);consistent, independent 2. [4.1b] Infinitely many solutions; consistent, dependent 3. [4.1b] No solution; inconsistent, independent 4. [4.2a] $\left(\frac{2}{5}, -\frac{4}{5}\right)$ 5. [4.2a] No solution 6. [4.2a] $\left(-\frac{11}{15}, -\frac{43}{30}\right)$ 7. [4.2b] $\left(\frac{37}{19}, \frac{53}{19}\right)$ 8. [4.2b] $\left(\frac{76}{17}, -\frac{2}{119}\right)$ 9. [4.2b] (2,2) 10. [4.3a] \$7, \$6 11. [4.3b] $5\frac{1}{2}$ hr 12. [4.3a] 10 L of 30% alcohol, 30 L of 50% alcohol 13. [4.4b] (10, 4, −8) 14. [4.4b] $\left(-\frac{7}{3}, \frac{125}{27}, \frac{20}{27}\right)$ 15. [4.4b] (2, 0, 4) 16. [4.2b] No solution 17. [4.4b] $\left(\frac{8}{9}, -\frac{2}{3}, \frac{10}{9}\right)$ 18. [4.2b] Infinite number of solutions 19. [4.4b] $\left(2, \frac{1}{3}, -\frac{2}{3}\right)$ 20. [4.5a] 90°, $67\frac{1}{2}$°, $22\frac{1}{2}$° 21. [4.5a] 641 22. [4.5a] 5 \$20 bills, 15 \$5 bills, 19 \$1 bills 23. [4.6a] 2 24. [4.6b] 9 25. [4.6c] (6,−2) 26. [4.6c] (−3, 0, 4) 27. [4.6c] $\left(55, -\frac{89}{2}\right)$ 28. [4.6c] $\left(2, -\frac{1}{2}, 5\right)$

29. [4.7a]

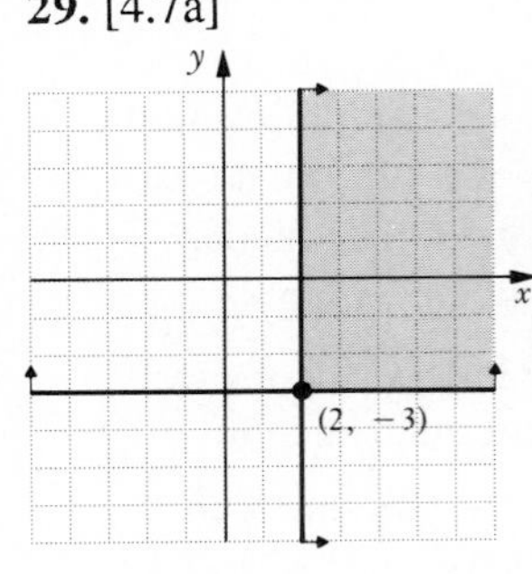

30. [4.7a]

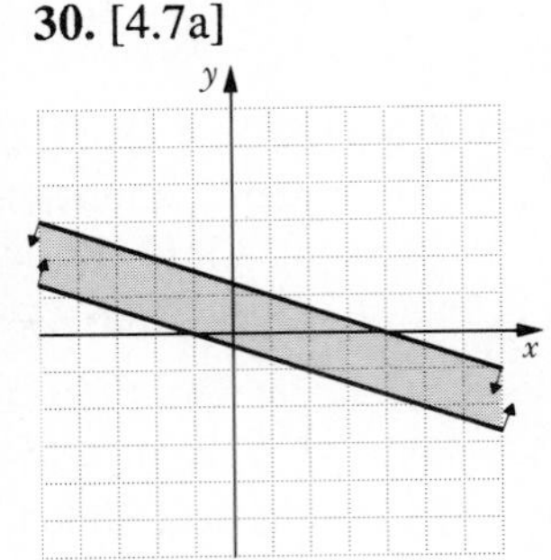

31. [4.7a]

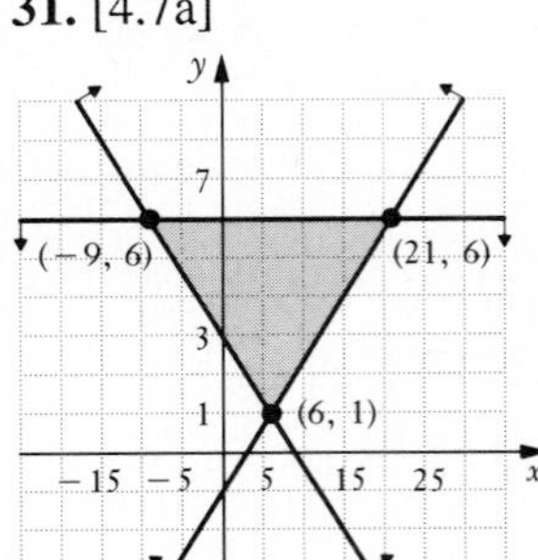

32. [2.1d] $-\frac{4}{3}$ 33. [2.3a] $t = \frac{Q}{a-4}$ 34. [1.5c] $-22x + 72$ 35. [3.3c] $\frac{5}{8}$ 36. [2.1d], [4.6b] $-10\frac{1}{2}$

Test: Chapter 4, p. 231

1. [4.1b](−2, 1);consistent, independent 2. [4.1b] No solution; inconsistent, independent 3. [4.1b] Infinite number of solutions; consistent, dependent 4. [4.2a] $\left(3, -\frac{11}{3}\right)$ 5. [4.2a] $\left(\frac{15}{7}, -\frac{18}{7}\right)$ 6. [4.2b] $\left(-\frac{3}{2}, -\frac{3}{2}\right)$ 7. [4.2b] No solution 8. [4.3a] l: 30, w: 18 9. [4.3a] 70 two-piece dinners, 62 three-piece dinners 10. [4.4b] Infinite number of solutions 11. [4.4b] $\left(2, -\frac{1}{2}, -1\right)$ 12. [4.4b] No solution 13. [4.4b] (0, 1, 0) 14. [4.6c] $\left(\frac{34}{107}, -\frac{104}{107}\right)$ 15. [4.6c] (3, 1, −2) 16. [4.3b] 120 km/h 17. [4.3a] $48\frac{8}{9}$ lb of A; $71\frac{1}{9}$ lb of B

18. [4.7a]

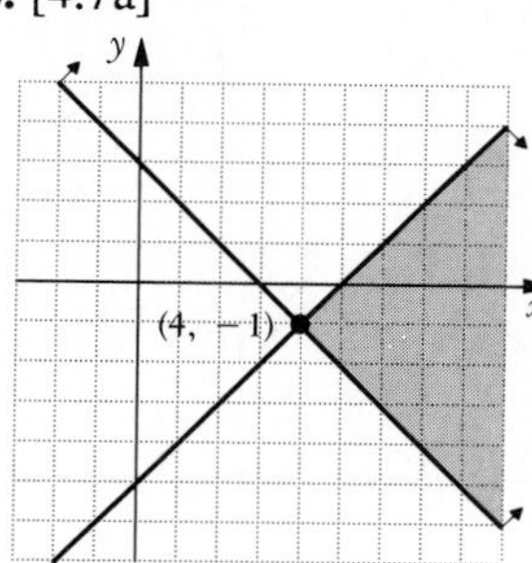

19. [4.7a]

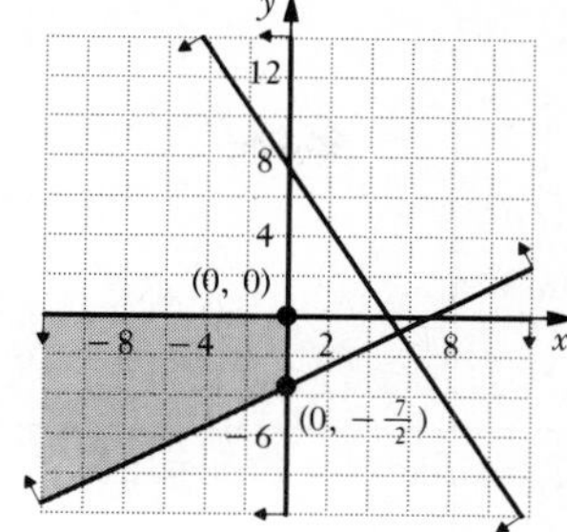

20. [3.3b] $-\frac{11}{5}$ **21.** [2.3a] $a = \frac{P+3b}{4}$ **22.** [2.1d] $\frac{9}{5}$ **23.** [1.5c] $-20t - 16$ **24.** [4.3a] $m = 7, b = 10$

Cumulative Review: Chapters 1-4, p. 233

1. [1.1a] -1 **2.** [1.1a] -14 **3.** [1.2d] 13 **4.** [1.2d] 0 **5.** [1.3d] 87.78 **6.** [1.3e] $-\frac{5}{21}$ **7.** [1.4e] $3x+33$ **8.** [1.5c] 10 **9.** [1.6a] $-\frac{2a^{11}}{5b^{33}}$ **10.** [1.5c] $16b+1$ **11.** [1.6a] y^{10} **12.** [2.1d] $\frac{10}{9}$ **13.** [2.1d] 6 **14.** [2.3a] $h = \frac{A}{\pi r^2}$ **15.** [2.3a] $P = \frac{3L}{m} - k$, or $\frac{3L-km}{m}$ **16.** [2.4c] $\{x \mid x > -1\}$ **17.** [2.5b] $\{x \mid x \leq 3 \text{ or } x \geq 7\}$ **18.** [2.5a] $\left\{x \,\middle|\, \frac{1}{3} < x \leq \frac{13}{3}\right\}$ **19.** [2.6e] $\left\{y \,\middle|\, y \leq -\frac{3}{2} \text{ or } y \geq \frac{9}{4}\right\}$ **20.** [2.6c] $\{-5, 3\}$

21. [3.2b]

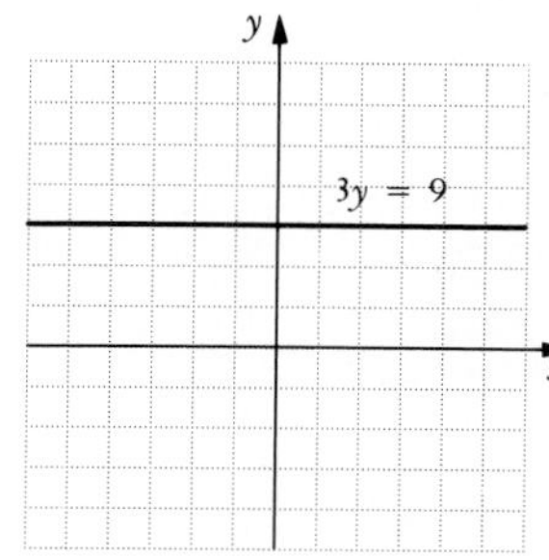

22. [3.1c]

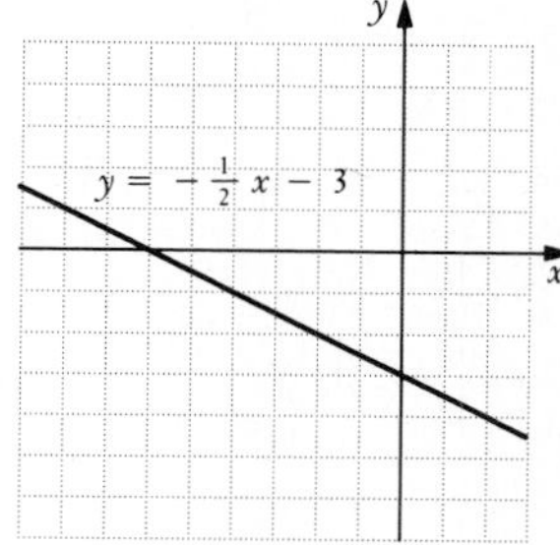

23. [3.1c]

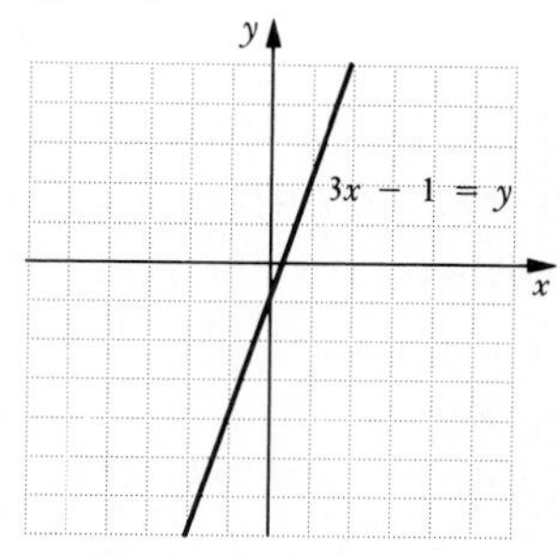

24. [3.6b]

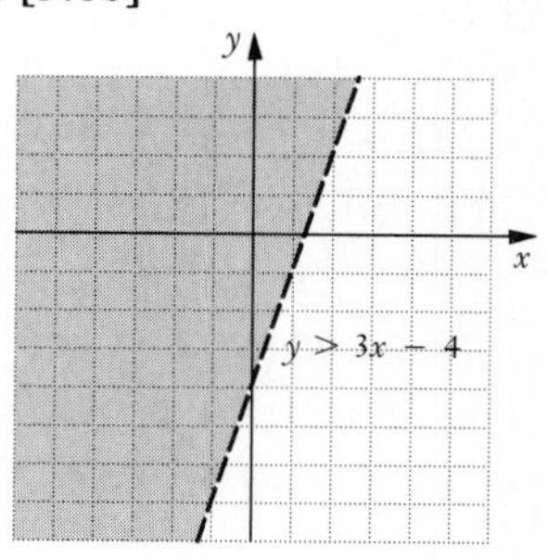

25. [3.2a]

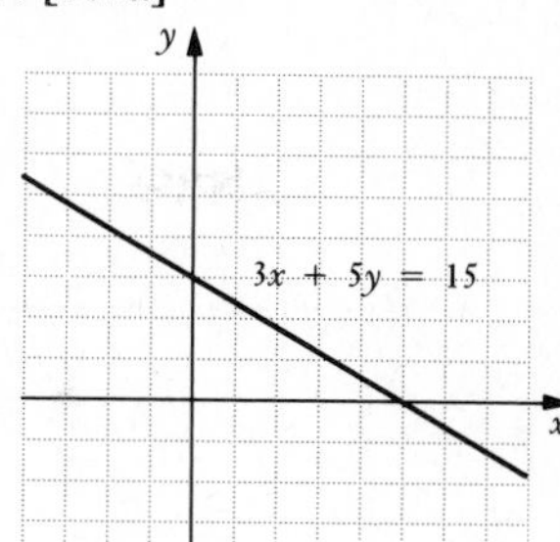

26. [3.6b]

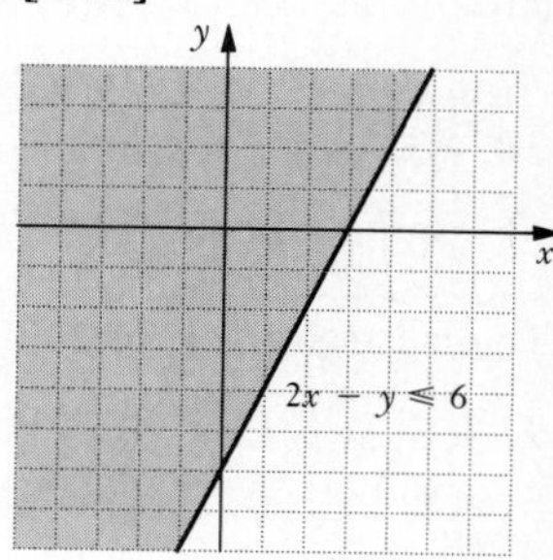

27. [4.1b] $(3, -1)$ **28.** [4.2b] $\left(\frac{8}{5}, -\frac{1}{5}\right)$ **29.** [4.2b] $(1, 1)$ **30.** [4.2b] $\left(-\frac{23}{13}, \frac{22}{13}\right)$ **31.** [4.4b] $(0, -1, 2)$ **32.** [4.4b] $(2, 0, -1)$ **33.** [4.6c] $\left(-\frac{2}{3}, \frac{1}{6}\right)$ **34.** [4.6c] $\left(\frac{1}{4}, -\frac{1}{2}, \frac{3}{4}\right)$ **35.** [4.6a] -11 **36.** [4.6b] 7

37. [4.7a]

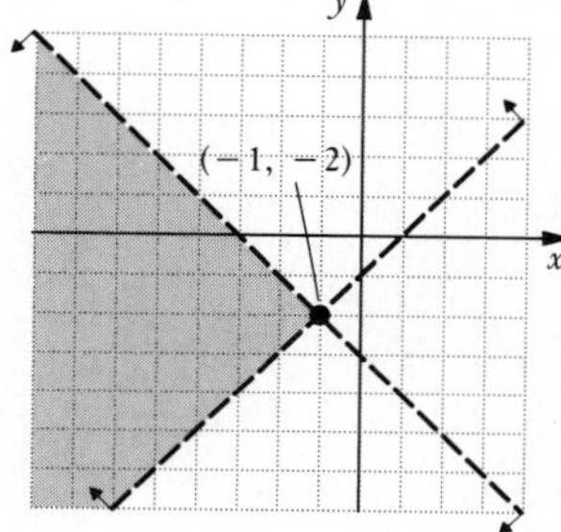

38. [4.7a]

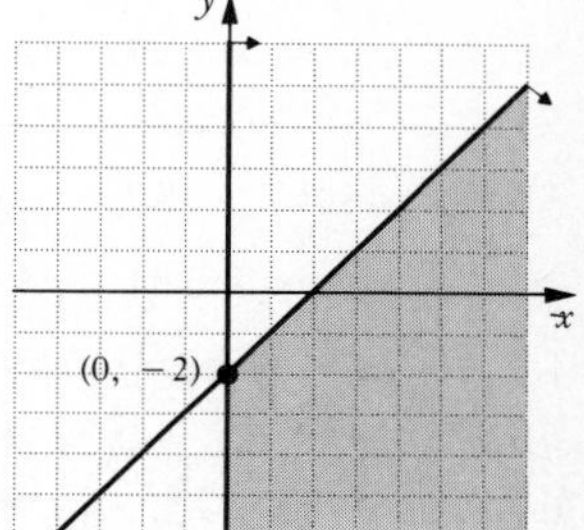

39. [3.3c] $m = \frac{4}{5}$, y-intercept $(0, 4)$ **40.** [3.3c] $m = -\frac{1}{2}$, y-intercept $\left(0, \frac{7}{4}\right)$ **41.** [3.4a] $y = -3x + 17$ **42.** [3.4c] $y = \frac{1}{3}x + 4$ **43.** [2.2a] 4 m, 6 m **44.** [2.2a] United States: \$5.25; Japan: \$11.34 **45.** [4.3a] 19 nickels, 15 dimes **46.** [4.5a] \$60 **47.** [3.5a] In 1993 **48.** [4.5a] First: 74.5; second: 68.5; third: 82 **49.** [4.5a] 20 **50.** [4.2b] $m = -\frac{5}{9}$, $b = -\frac{2}{9}$

CHAPTER 5

Pretest: Chapter 5, p. 236

1. [5.1b] 6 **2. [5.1d]** $6x^2y + xy$ **3. [5.1f]** $5m^3 - 6m^2 + 12$ **4. [5.1a]** $4xy^5 + x^2y^3 - 3x^6y^2 - 2y$ **5. [5.2a]** $x^4 - 2x^3 + 2x - 1$ **6. [5.2b]** $8y^2 + 18yz - 5z^2$ **7. [5.2d]** $a^2 - 9b^2$ **8. [5.2c]** $25t^2 - 30tm^2 + 9m^4$ **9. [5.4b]** $(2x-1)(2x+3)$ **10. [5.5a]** $2(5m+2)^2$ **11. [5.6a]** $4t^3(t+1)(t^2-t+1)$ **12. [5.4a]** $(a+2)(a+4)$ **13. [5.5b]** $(x+7y)(x-7y)$ **14. [5.3b]** $(y^2+4)(y+3)$ **15. [5.8a]** $-\frac{2}{3}, \frac{4}{3}$ **16. [5.8b]** $-5, 7$ **17. [5.9b]** $x + 17$, R 23; or $x + 17 + \frac{23}{x-3}$ **18. [5.9c]** $x^3 - 3x^2 + 7x - 14$, R 12; or $x^3 - 3x^2 + 7x - 14 + \frac{12}{x+2}$

Sidelight: Factors and Sums, p. 242

First row: 90, −432, −63; *second row:* 7, −18, −36, −14, 12, −6, −21, −11; *third row:* 9, −2, −2, 10, −8, −8, −8, −10, 21; *fourth row:* −19, −6

Exercise Set 5.1, p. 243

1. 4, 3, 2, 1, 0; 4; $-11x^4$; −11 **3.** 3, 7, 6, 0; 7; $2y^7$; 2 **5.** 5, 6, 2, 1, 0; 6; $4a^2b^4$; 4 **7.** $-4y^3-6y^2+7y+23$ **9.** $-xy^3+x^2y^2+x^3y+1$ **11.** $-5a^5y^5-4a^2y^3+3ay$ **13.** $12+4x-5x^2+3x^4$ **15.** $3xy^3+x^2y^2-9x^3y+2x^4$ **17.** $-7ab+4ax-7ax^2+4x^6$ **19.** 54; 2 **21.** $-45; -\frac{235}{27}$ **23.** 28; 190 **25.** About 44.5 in^2 **27.** \$40,000 **29.** \$31,000 **31.** $P=-x^2+280x-7000$ **33.** $2x^2$ **35.** $3x+y$ **37.** $a+6$ **39.** $-6a^2b-2b^2$ **41.** $9x^2+2xy+15y^2$ **43.** $-x^2y+4y+9xy^2$ **45.** $5x^2+2y^2+5$ **47.** $6a+b+c$ **49.** $-4a^2-b^2+3c^2$ **51.** $-2x^2+x-xy-1$ **53.** $5x^2y-4xy^2+5xy$ **55.** $9r^2+9r-9$ **57.** $-\frac{2}{15}xy+\frac{19}{12}xy^2+1.7x^2y$ **59.** $-(5x^3-7x^2+3x-6)$; $-5x^3+7x^2-3x+6$ **61.** $-(-12y^5+4ay^4-7by^2)$; $12y^5-4ay^4+7by^2$ **63.** $13x-6$ **65.** $-4x^2-3x+13$ **67.** $2a-4b+3c$ **69.** $-2x^2+6x$ **71.** $-4a^2+8ab-5b^2$ **73.** $9-9y$ **75.** $16ab+8a^2b+3ab^2$ **77.** $0.06y^4+0.032y^3-0.94y^2+0.93$ **79.** x^4-x^2-1 **81.** $\{x \mid -7<x<11\}$ **83.** $5x^2-8x$ **85.** $47x^{4a}+40x^{3a}+30x^{2a}+x^a+4$ **87.** $x^{5b}+4x^{4b}+x^{3b}-6x^{2b}-9x^b$

Exercise Set 5.2, p. 251

1. $10y^3$ **3.** $-20x^3y$ **5.** $-10x^5y^6$ **7.** $6x-2x^2$ **9.** $3a^2b+3ab^2$ **11.** $15c^3d^2-25c^2d^3$ **13.** $15x^2+x-2$ **15.** s^2-9t^2 **17.** $x^2-2xy+y^2$ **19.** $2y^2+9xy-56x^2$ **21.** $a^4-5a^2b^2+6b^4$ **23.** x^3-64 **25.** x^3+y^3 **27.** $a^4+5a^3-2a^2-9a+5$ **29.** $4a^3b^2+4a^3b-10a^2b^2-2a^2b+3ab^3+7ab^2-6b^3$ **31.** $x^2+\frac{1}{2}x+\frac{1}{16}$ **33.** $3.25x^2-0.9xy-28y^2$ **35.** a^2+5a+6 **37.** y^2+y-6 **39.** $4a^2+\frac{4}{3}a+\frac{1}{9}$ **41.** $x^2-4xy+4y^2$ **43.** $b^2-\frac{5}{6}b+\frac{1}{6}$ **45.** $2x^2+13x+18$ **47.** $400a^2-6.4ab+0.0256b^2$ **49.** $4x^2-4xy-3y^2$ **51.** x^2+6x+9 **53.** $4x^4-12x^2y^2+9y^4$ **55.** $a^4b^4+2a^2b^2+1$ **57.** $0.01a^4-a^2b+25b^2$ **59.** c^2-4 **61.** $4a^2-1$ **63.** $9m^2-4n^2$ **65.** $x^4-y^2z^2$ **67.** $m^4-m^2n^2$ **69.** $\frac{1}{4}p^2-\frac{4}{9}q^2$ **71.** x^4-1 **73.** $a^4-2a^2b^2+b^4$ **75.** $a^2+2ab+b^2-1$ **77.** $4x^2+12xy+9y^2-16$ **79.** $A=P+2Pi+Pi^2$ **81.** *A*: 75; *B*: 84; *C*: 63 **83.** $-\frac{4}{3}x^6y^{11}$ **85.** $-\frac{1}{4}r^{22}s^{10}$ **87.** $\frac{1}{4}a^{7x}b^{2y+2}$ **89.** $y^{3n+3}z^{n+3}-4y^4z^{3n}$ **91.** $y^{12}-6y^{10}+15y^8-20y^6+15y^4-6y^2+1$ **93.** $9x^{10}-\frac{30}{11}x^5+\frac{25}{121}$ **95.** $x^3-\frac{1}{343}$ **97.** x^4-25x^2+144 **99.** $a^3-b^3+3b^2-3b+1$ **101.** $x^{a^2-b^2}$

Exercise Set 5.3, p. 257

1. $2a(2a+1)$ **3.** $y^2(y+9)$ **5.** $3x^2(2-x^2)$ **7.** $4xy(x-3y)$ **9.** $3(y^2-y-3)$ **11.** $2a(2b-3c+6d)$ **13.** $5(2a^4+3a^2-5a-6)$ **15.** $-3(x-4)$ **17.** $-6(y+12)$ **19.** $-2(x^2-2x+6)$ **21.** $-3y(y-8)$ **23.** $-3y(y^2-4y+5)$ **25.** $-1(x^2-3x+7)$ **27.** $-a(a^3-2a^2+13)$ **29.** $(a+c)(b-2)$ **31.** $(x-2)(2x+13)$ **33.** $2a^2(x-y)$ **35.** $(a+b)(c+d)$ **37.** $(b^2+2)(b-1)$ **39.** $(y^2+1)(y-8)$ **41.** $12(x^2+3)(2x-3)$ **43.** $x^2(x^4+x^3-x+1)$ **45.** $(2y^2+5)(y^2+3)$ **47.** \$1200 at $5\frac{1}{2}$%; \$600 at 7%; \$1700 at 8% **49.** $2(2y^{2a}+5)(y^{2a}+3)$ **51.** $R=0.4x(700-x)$ **53.** 20

Exercise Set 5.4, p. 267

1. $(x+5)(x+4)$ **3.** $(t-5)(t-3)$ **5.** $(x-9)(x+3)$ **7.** $2(y-4)(y-4)$ **9.** $(p+9)(p-6)$ **11.** $(x+5)(x+9)$ **13.** $(y+9)(y-7)$ **15.** $(t-7)(t-4)$ **17.** $(x+5)(x-2)$ **19.** $(x+2)(x+3)$ **21.** $(8-x)(7+x)$ **23.** $y(8-y)(4+y)$ **25.** $(x^2+16)(x^2-5)$ **27.** Not factorable **29.** $(x+9y)(x+3y)$ **31.** $(x-7)(x-7)$ **33.** $(x^2+49)(x^2+1)$ **35.** $(x^3+6)(x^3-7)$ **37.** $(3x+2)(x-6)$ **39.** $x(3x-5)(2x+3)$ **41.** $(3a-4)(a-2)$ **43.** $(5y+2)(7y+4)$ **45.** $2(5t-3)(t+1)$ **47.** $4(2x+1)(x-4)$ **49.** $x(3x-4)(4x-5)$ **51.** $x^2(7x+1)(2x-3)$ **53.** $(3a-4)(a+1)$ **55.** $(3x+1)(3x+4)$ **57.** $(3-z)(1+12z)$ **59.** $(-2t+3)(2t+5)$ **61.** $x(3x+1)(x-2)$ **63.** $(24x+1)(x-2)$ **65.** $(7x+3)(3x+4)$ **67.** $4(10x^4+4x^2-3)$ **69.** $(4a-3b)(3a-2b)$ **71.** $(2x-3y)(x+2y)$ **73.** $(3x-4y)(2x-7y)$ **75.** $(3x-5y)(3x-5y)$ **77.** $(3x^3-2)(x^3+1)$ **79.** **(a)** 224 ft; 288 ft; 320 ft; 288 ft; 128 ft; **(b)** $h=-16(t-7)(t+2)$ **81.** $(pq+4)(pq+3)$ **83.** $\left(x+\frac{4}{5}\right)\left(x-\frac{1}{5}\right)$ **85.** $(y+0.5)(y-0.1)$ **87.** $(7ab+6)(ab+1)$ **89.** $3(x+15)(x-11)$ **91.** $6x(x+9)(x+4)$ **93.** $(x+a)(x+b)$ **95.** $(bx+a)(dx+c)$ **97.** $(x-4)(x+8)$ **99.** ±76, ±28, ±20

Exercise Set 5.5, p. 275

1. $(y-3)^2$ **3.** $(x+7)^2$ **5.** $(x+1)^2$ **7.** $(y-6)^2$ **9.** $y(y-9)^2$ **11.** $3(2a+3)^2$ **13.** $2(x-10)^2$ **15.** $(1-4d)^2$, or $(4d-1)^2$ **17.** $(y+2)^2(y-2)^2$ **19.** $(0.5x+0.3)^2$ **21.** $(p-q)^2$ **23.** $(a+2b)^2$ **25.** $(5a-3b)^2$ **27.** $(y^3+13)^2$ **29.** $(4x^5-1)^2$ **31.** $(x^2+y^2)^2$ **33.** $(x+4)(x-4)$ **35.** $(p+7)(p-7)$ **37.** $(pq+5)(pq-5)$ **39.** $6(x+y)(x-y)$ **41.** $4x(y^2+z^2)(y+z)(y-z)$ **43.** $a(2a+7)(2a-7)$ **45.** $3(x^4+y^4)(x^2+y^2)(x+y)(x-y)$ **47.** $a^2(3a+5b^2)(3a-5b^2)$ **49.** $\left(\frac{1}{5}+x\right)\left(\frac{1}{5}-x\right)$ **51.** $(0.2x+0.3y)(0.2x-0.3y)$ **53.** $(m+2)(m-2)(m-7)$ **55.** $(a+b)(a-b)(a-2)$ **57.** $(a+b+10)(a+b-10)$ **59.** $(a+b+3)(a+b-3)$ **61.** $(r-1+2s)(r-1-2s)$ **63.** $2(m+n+5b)(m+n-5b)$ **65.** $(3+a+b)(3-a-b)$ **67.** $\left(\frac{1}{2}p-\frac{2}{5}\right)^2$ **69.** $(x^{2a}-y^b)(x^{2a}+y^b)$ **71.** $(3x^n-1)^2$ **73.** $(3a+4+7b)(3a+4-7b)$

Exercise Set 5.6, p. 279

1. $(x+2)(x^2-2x+4)$ **3.** $(y-4)(y^2+4y+16)$ **5.** $(w+1)(w^2-w+1)$ **7.** $(2a+1)(4a^2-2a+1)$ **9.** $(y-2)(y^2+2y+4)$ **11.** $(2-3b)(4+6b+9b^2)$ **13.** $(4y+1)(16y^2-4y+1)$ **15.** $(2x+3)(4x^2-6x+9)$ **17.** $(a-b)(a^2+ab+b^2)$ **19.** $\left(a+\frac{1}{2}\right)\left(a^2-\frac{1}{2}a+\frac{1}{4}\right)$ **21.** $2(y-4)(y^2+4y+16)$ **23.** $3(2a+1)(4a^2-2a+1)$ **25.** $r(s+4)(s^2-4s+16)$ **27.** $5(x-2z)(x^2+2xz+4z^2)$ **29.** $(x+0.1)(x^2-0.1x+0.01)$ **31.** $8(2x^2-t^2)(4x^4+2x^2t^2+t^4)$ **33.** $2y(y-4)(y^2+4y+16)$ **35.** $(z-1)(z^2+z+1)(z+1)(z^2-z+1)$ **37.** $(t^2+4y^2)(t^4-4t^2y^2+16y^4)$ **39.**

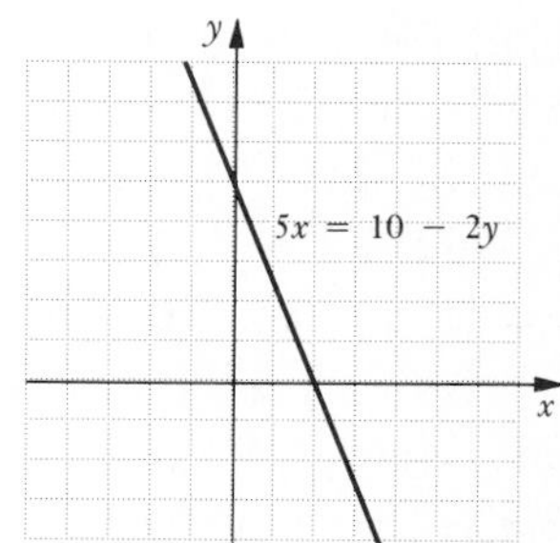

41. $\left\{x \mid -\frac{33}{5} \le x \le 9\right\}$ **43.** 1; 19; 19; 7; 25 **45.** $(x^{2a}+y^b)(x^{4a}-x^{2a}y^b+y^{2b})$ **47.** $3(x^a+2y^b)(x^{2a}-2x^ay^b+4y^{2b})$

49. $\frac{1}{3}\left(\frac{1}{2}xy+z\right)\left(\frac{1}{4}x^2y^2-\frac{1}{2}xyz+z^2\right)$ **51.** $7\left(x-\frac{1}{2}\right)\left(x^2+\frac{1}{2}x+\frac{1}{4}\right)$ **53.** $y(3x^2+3xy+y^2)$ **55.** $4(3a^2+4)$

Exercise Set 5.7, p. 283

1. $(x+12)(x-12)$ **3.** $(2x+3)(x+4)$ **5.** $3(x^2+2)(x^2-2)$ **7.** $(a+5)^2$ **9.** $2(x-11)(x+6)$ **11.** $(3x+5y)(3x-5y)$ **13.** $(m+1)(m^2-m+1)(m-1)(m^2+m+1)$ **15.** $(x+3+y)(x+3-y)$ **17.** $2(5x-4y)(25x^2+20xy+16y^2)$ **19.** $(m^3+10)(m^3-2)$ **21.** $(c-b)(a+d)$ **23.** $(2c-d)^2$ **25.** $(x^2+2)(2x-7)$ **27.** $2(x+2)(x-2)(x+3)$ **29.** $2(2x+3y)(4x^2-6xy+9y^2)$ **31.** $(6y-5)(6y+7)$ **33.** $(a^4+b^4)(a^2+b^2)(a+b)(a-b)$ **35.** $ab(a+4b)(a-4b)$ **37.** $(a+c)(b-2)$ **39.** $5(x-y)^2(x+y)$ **41.** $(9ab+2)(3ab+4)$ **43.** $y(2y-5)(4y^2+10y+25)$ **45.** $\left(\frac{17}{6},-4,-\frac{23}{12}\right)$ **47.** $(6y^2-5x)(5y^2-12x)$ **49.** $5\left(x-\frac{1}{3}\right)\left(x^2+\frac{1}{3}x+\frac{1}{9}\right)$ **51.** $x(x-2p)$ **53.** $y(y-1)^2(y-2)$ **55.** $(x+1)(x^2+1)(x-1)^3$ **57.** $(2x+y-r+3s)(2x+y+r-3s)$ **59.** $c(c^w+1)^2$ **61.** $y(y^4+1)(y^2+1)(y+1)(y-1)$ **63.** $3x(x+5)$

Sidelight: Computer-Calculator Exercises: Graphing and Solving Polynomial Equations, p. 290

1. −2, 1 **2.** −1.41, 0, 1.41 **3.** −2.08, 0.46, 3.12 **4.** 0, 700 **5.** −3.10, −0.65, 0.65, 3.10 **6.** −1, 1 **7.** −2, −1.41, 1 **8.** −3, −1, 2, 3

Exercise Set 5.8, p. 291

1. −7, 4 **3.** 4 **5.** 6 **7.** −5, −4 **9.** 0, −8 **11.** −3, 3 **13.** −6, 6 **15.** 7, −9 **17.** −4, 8 **19.** $-2, -\frac{2}{3}$ **21.** $\frac{1}{2}, \frac{3}{4}$ **23.** 0, 6 **25.** $\frac{2}{3}, -\frac{3}{4}$ **27.** −2, 2 **29.** $\frac{2}{3}, -\frac{5}{7}$ **31.** $0, \frac{1}{5}$ **33.** 7, −2 **35.** 0, −2, 3 **37.** 0, −8, 8 **39.** $-\frac{5}{2}, \frac{9}{2}$ **41.** $l = 12$ cm, $w = 8$ cm **43.** 9 and 11 **45.** 3 cm **47.** $h = 7$ cm, $b = 16$ cm **49.** 2 **51.** −2, −1, 0; 11, 12, 13 **53.** 40 m, 41 m **55.** 7 sec **57.** $\left(\frac{1306}{5}, \frac{208}{5}, 16\right)$ **59.** 4 **61.** −1, 11 **63.** $-\frac{5}{3}$, 4, 5 **65.** −3, −2, 2 **67.** 3 and 14 **69.** 54 cm^2

Sidelight: Handling Dimension Symbols (Part I), p. 300

1. 68 ft **2.** 82 km/hr **3.** 45 g **4.** 8.6 lb **5.** 15 mi/hr **6.** 32 km/hr **7.** 3.3 m/sec **8.** 19 ft/min **9.** $4\,\frac{\text{in.-lb}}{\text{sec}}$

10. $24 \frac{\text{man-hr}}{\text{da}}$ **11.** 12 yd **12.** 220 mi **13.** 16ft^3 **14.** $\frac{1}{4}\frac{\text{lb}^2}{\text{ft}^2}$ **15.** $\frac{\$970}{\text{day}}$ **16.** $\frac{\$3.20}{\text{hr}}$

Exercise Set 5.9, p. 301

1. $6x^4 - 3x^2 + 8$ **3.** $x^2y^3 - 2xy^2 + 3y$ **5.** $x + 7$ **7.** $a - 12$, R 32; or $a - 12 + \frac{32}{a+4}$ **9.** $x + 2$, R 4; or $x + 2 + \frac{4}{x+5}$ **11.** $2y^2 - y + 2$, R 6; or $2y^2 - y + 2 + \frac{6}{2y+4}$ **13.** $2y^2 + 2y - 1$, R 8; or $2y^2 + 2y - 1 + \frac{8}{5y-2}$ **15.** $2x^2 - x - 9$, R $(3x + 12)$; or $2x^2 - x - 9 + \frac{3x+12}{x^2+2}$ **17.** $2x^3 + 5x^2 + 17x + 51$, R $152x$; or $2x^3 + 5x^2 + 17x + 51 + \frac{152x}{x^2-3x}$ **19.** $x^2 - x + 1$, R -4; or $x^2 - x + 1 + \frac{-4}{x-1}$ **21.** $a + 7$, R -47; or $a + 7 + \frac{-47}{a+4}$ **23.** $x^2 - 5x - 23$, R -43; or $x^2 - 5x - 23 + \frac{-43}{x-2}$ **25.** $3x^2 - 2x + 2$, R -3; or $3x^2 - 2x + 2 + \frac{-3}{x+3}$ **27.** $y^2 + 2y + 1$, R 12; or $y^2 + 2y + 1 + \frac{12}{y-2}$ **29.** $3x^3 + 9x^2 + 2x + 6$ **31.** $x^2 + 3x + 9$ **33.** $y^3 + 2y^2 + 4y + 8$ **35.**

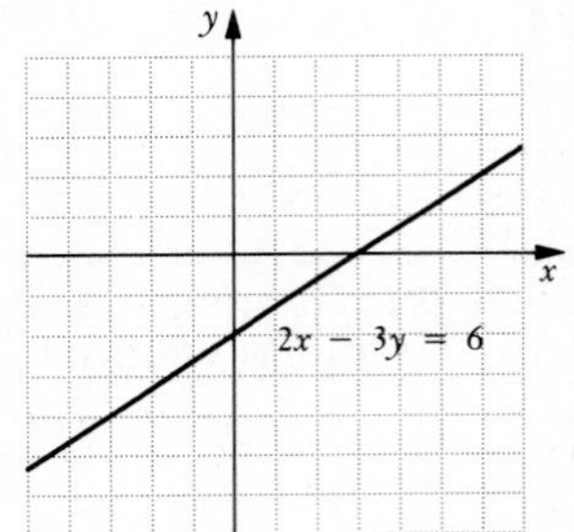

37. $\frac{2}{7}, -\frac{4}{5}$ **39.** $2x^4 - 2x^3 + 5x^2 - 4x - 1$, R $(5x + 2)$; or $2x^4 - 2x^3 + 5x^2 - 4x - 1 + \frac{5x+2}{x^2+x+1}$ **41.** $x^2 + 2y$ **43.** $a^6 - a^5b + a^4b^2 - a^3b^3 + a^2b^4 - ab^5 + b^6$

Summary and Review: Chapter 5, p. 303

1. [5.1b] 0; −6 **2.** [5.1b] 4; −31 **3.** [5.1a] (a) 7, 11, 3, 2; 11; (b) $-7x^8y^3$; −7; (c) $-3x^2 + 2x^3 + 3x^6y - 7x^8y^3$; (d) $-7x^8y^3 + 3x^6y + 2x^3 - 3x^2$ **4.** [5.1c] $-x^2y - 2xy^2$ **5.** [5.1c] $ab + 12ab^2 + 4$ **6.** [5.1d] $-x^3 + 2x^2 + 5x + 2$ **7.** [5.1d] $x^3 + 6x^2 - x - 4$ **8.** [5.1d] $13x^2y - 8xy^2 + 4xy$ **9.** [5.1f] $9x - 7$ **10.** [5.1f] $-2a + 6b + 7c$ **11.** [5.1f] $16p^2 - 8p$ **12.** [5.1f] $6x^2 - 7xy + y^2$ **13.** [5.2a] $-18x^3y^4$ **14.** [5.2a] $x^8 - x^6 + 5x^2 - 3$ **15.** [5.2b] $8a^2b^2 + 2abc - 3c^2$ **16.** [5.2d] $4x^2 - 25y^2$ **17.** [5.2c] $4x^2 - 20xy + 25y^2$ **18.** [5.2a] $20x^4 - 18x^3 - 47x^2 + 69x - 27$ **19.** [5.2c] $x^4 + 8x^2y^3 + 16y^6$ **20.** [5.2a] $x^3 - 125$ **21.** [5.2b] $x^2 - \frac{1}{2}x + \frac{1}{18}$ **22.** [5.3a] $3y^2(3y^2 - 1)$ **23.** [5.3a] $3x(5x^3 - 6x^2 + 7x - 3)$ **24.** [5.4a] $(a - 9)(a - 3)$ **25.** [5.4b] $(3m + 2)(m + 4)$ **26.** [5.5a] $(5x + 2)^2$ **27.** [5.7a] $4(y + 2)(y - 2)$ **28.** [5.3b] $(x - y)(a + 2b)$ **29.** [5.3a] $4(x^4 + x^2 + 5)$ **30.** [5.6a] $(3x - 2)(9x^2 + 6x + 4)$ **31.** [5.6a] $(0.4b - 0.5c)(0.16b^2 + 0.2bc + 0.25c^2)$ **32.** [5.7a] $y(y^2 + 1)(y + 1)(y - 1)$ **33.** [5.3a] $2z^6(z^2 - 8)$ **34.** [5.7a] $2y(3x^2 - 1)(9x^4 + 3x^2 + 1)$ **35.** [5.6a] $(1 + a)(1 - a + a^2)$ **36.** [5.7a] $4(3x - 5)^2$ **37.** [5.4b] $(3t + p)(2t + 5p)$ **38.** [5.5c] $(x + 3)(x - 3)(x + 2)$ **39.** [5.5c] $(a - b + 2t)(a - b - 2t)$ **40.** [5.8a] 10 **41.** [5.8a] $\frac{2}{3}, \frac{3}{2}$ **42.** [5.8a] $\frac{1}{2}, \frac{5}{4}$ **43.** [5.8b] 5 **44.** [5.8b] −7, −5, −3; 3, 5, 7 **45.** [5.8b] *l*: 8 in.; *w*: 5 in. **46.** [5.9a] $4s^2 - 3s - 2rs^2$ **47.** [5.9b] $y^2 + 4y + 16$ **48.** [5.9b] $4x + 3$, R $(-9x - 5)$; or $4x + 3 + \frac{-9x-5}{x^2+1}$ **49.** [5.9c] $x^2 + 6x + 20$, R 54; or $x^2 + 6x + 20 + \frac{54}{x-3}$ **50.** [5.9c] $4x^2 - 6x + 18$, R -59; or $4x^2 - 6x + 18 + \frac{-59}{x+3}$ **51.** [4.4b] $(0, -2, 7)$ **52.** [3.2a]

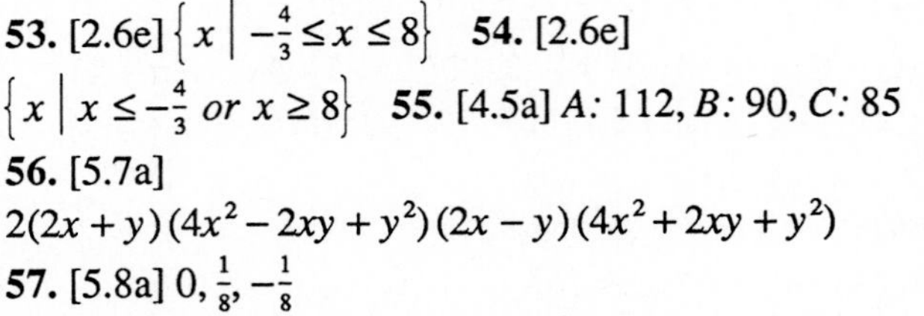

53. [2.6e] $\left\{x \,\middle|\, -\frac{4}{3} \le x \le 8\right\}$ **54.** [2.6e] $\left\{x \,\middle|\, x \le -\frac{4}{3} \text{ or } x \ge 8\right\}$ **55.** [4.5a] *A*: 112, *B*: 90, *C*: 85 **56.** [5.7a] $2(2x + y)(4x^2 - 2xy + y^2)(2x - y)(4x^2 + 2xy + y^2)$ **57.** [5.8a] $0, \frac{1}{8}, -\frac{1}{8}$

Test: Chapter 5, p. 305

1. [5.1b] 4; 2 **2.** [5.1a] 9 **3.** [5.1a] $5x^5y^4$ **4.** [5.1a] $5x^5y^4 - 2x^4y - 4x^2y + 3xy^3$ **5.** [5.1c] $3xy + 3xy^2$ **6.** [5.1d] $-3x^3 + 3x^2 - 6y - 7y^2$ **7.** [5.1d] $7a^3 - 6a^2 + 3a - 3$ **8.** [5.1d] $7m^3 + 2m^2n + 3mn^2 - 7n^3$ **9.** [5.1f] $6a - 8b$ **10.** [5.1f] $7x^2 - 7x + 13$ **11.** [5.1f] $2y^2 + 5y + y^3$ **12.** [5.2a] $64x^3y^3$ **13.** [5.2b] $12a^2 - 4ab - 5b^2$ **14.** [5.2a] $x^3 - 2x^2y + y^3$ **15.** [5.2a] $-3m^4 - 13m^3 + 5m^2 + 26m - 10$ **16.** [5.2c] $16y^2 - 72y + 81$ **17.** [5.2d] $x^2 - 4y^2$ **18.** [5.3a] $x(9x + 7)$ **19.** [5.3a] $8y^2(3y + 2)$ **20.** [5.5c] $(y + 2)(y - 2)(y + 5)$ **21.** [5.4a] $(p - 14)(p + 2)$ **22.** [5.4b] $(6m + 1)(2m + 3)$ **23.** [5.5b] $(3y + 5)(3y - 5)$

24. [5.7a] $3(r-1)(r^2+r+1)$ **25.** [5.5a] $(3x-5)^2$ **26.** [5.5b] $(z+1+b)(z+1-b)$ **27.** [5.5b] $(x^4+y^4)(x^2+y^2)(x+y)(x-y)$ **28.** [5.5c] $(y+4+10t)(y+4-10t)$ **29.** [5.7a] $5(2a+b)(2a-b)$ **30.** [5.7a] $2(4x-1)(3x-5)$ **31.** [5.7a] $2ab(2a^2+3b^2)(4a^4-6a^2b^2+9b^4)$ **32.** [5.8a] $-3, 6$ **33.** [5.8a] $-5, 5$ **34.** [5.8a] $-\frac{3}{2}, -7$ **35.** [5.8b] l: 8 cm; w: 5 cm **36.** [5.9b] $x-7$, R -46; or $x-7+\frac{-46}{x-3}$ **37.** [5.9c] $6y^3+3y^2-6y-16$, R 38; or $6y^3+3y^2-6y-16+\frac{38}{y+2}$ **38.** [2.6e] $\left\{x \mid -6 < x < \frac{2}{3}\right\}$ **39.** [2.6e] $\left\{x \mid x < -6 \text{ or } x > \frac{2}{3}\right\}$ **40.** [3.2a]

41. [4.4b] $(5, -1, -2)$ **42.** [4.5a] 31 multiple choice, 26 true-false, 13 fill-in **43.** (a)[5.2a] x^5+x+1; (b) [5.7a] $(x^2+x+1)(x^3-x^2+1)$ **44.** [5.4b] $(3x^n+4)(2x^n-5)$

Cumulative Review: Chapters 1-5, p. 307

1. [1.1a] 1 **2.** [1.2d] 0 **3.** [1.3c] -4 **4.** [1.4e] $-a-5$ **5.** [1.5c] $5x+20$ **6.** [1.6b] $\frac{a^{60}}{(-3)^{10}b^{20}}$, or $\frac{a^{60}}{59{,}049b^{20}}$ **7.** [5.1d] $-2x^2+x-xy-1$ **8.** [5.1f] $-2x^2+6x$ **9.** [5.2a] $a^4+a^3-8a^2-3a+9$ **10.** [5.2b] $x^2+13x+36$ **11.** [2.1d] 2 **12.** [2.1d] 13 **13.** [2.3a] $b=\frac{2A-ha}{h}$, or $\frac{2A}{h}-a$ **14.** [2.4c] $\left\{x \mid x \ge -\frac{7}{9}\right\}$ **15.** [2.5b] $\left\{x \mid x < \frac{5}{4} \text{ or } x > 4\right\}$ **16.** [2.6e] $\{x \mid -2 < x < 5\}$ **17.** [4.4b] $(1, 3, -9)$ **18.** [4.2b] $(4, -2)$ **19.** [4.2b] $\left(\frac{19}{8}, \frac{1}{8}\right)$ **20.** [4.4b] $(-1, 0, -1)$ **21.** [4.6c] $\left(\frac{3}{2}, -\frac{1}{3}\right)$ **22.** [4.6c] $\left(\frac{1}{3}, \frac{1}{2}\right)$ **23.** [4.6c] $\left(\frac{1}{2}, \frac{1}{2}, \frac{1}{4}\right)$ **24.** [5.8a] $-3, -8$ **25.** [5.8a] $\frac{1}{2}, 7$ **26.** [4.6b] 59 **27.** [5.3a] $3x^2(x-4)$ **28.** [5.3b] $(2x+1)(x+1)(x^2-x+1)$ **29.** [5.4a] $(x-2)(x+7)$ **30.** [5.4b] $(4a-3)(5a-2)$ **31.** [5.5b] $(2x+5)(2x-5)$ **32.** [5.5a] $2(x-7)^2$ **33.** [5.6a] $(a+4)(a^2-4a+16)$ **34.** [5.6a] $(2x-1)(4x^2+2x+1)$ **35.** [5.7a] $(a^3+6)(a^3-2)$ **36.** [5.7a] $x^2y^2(2x+y)(2x-y)$

37. [3.1c] **38.** [3.1c] **39.** [3.2b] **40.** [3.2b] **41.** [3.6b] **42.** [3.6b] **43.** [4.7a] **44.** [4.7a]

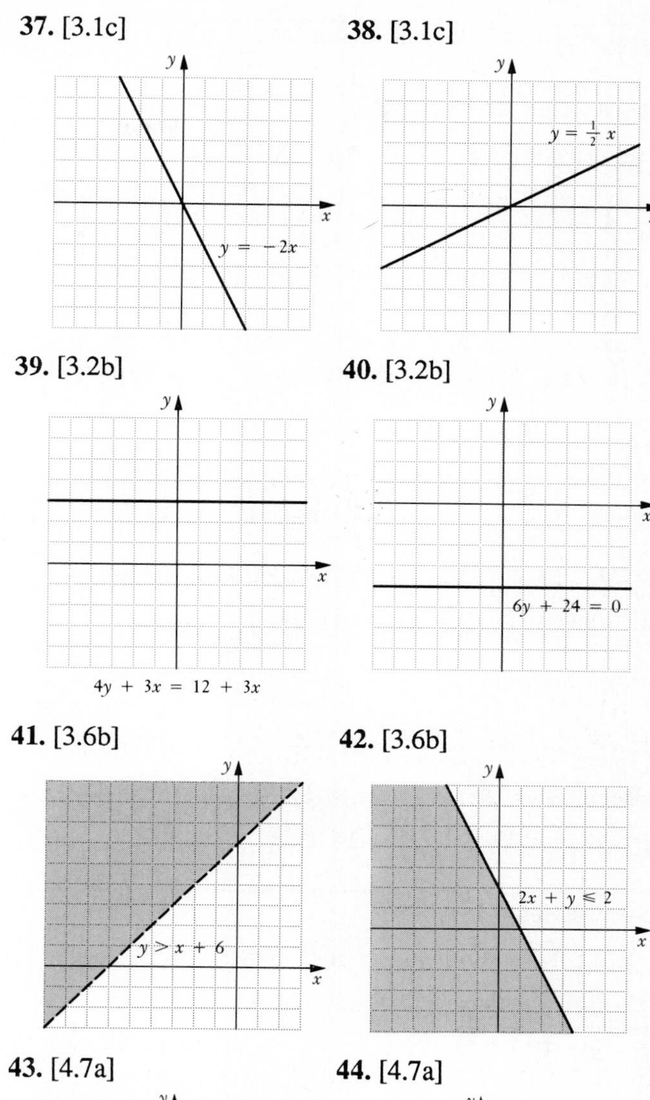

45. [3.4c] $y=-\frac{1}{2}x+\frac{17}{2}$ **46.** [3.4c] $y=\frac{4}{3}x-6$ **47.** [3.4b] $x=-1$ **48.** [3.4a] $y=-3x+7$ **49.** [2.2a] \$1170 for service, \$1716 for disposables **50.** [2.2a] Yes **51.** [2.2a] Caracas: \$0.14/gal; Lagos: \$0.30/gal **52.** [4.5a] 1500 on A, 1900 on B, 2300 on C **53.** [2.6e] $\{x \mid x \le 1\}$ **54.** [3.5a] 100.03916 cm, 99.96796 cm **55.** [4.3a] Bert: 32 yr; Sally: 14 yr **56.** [4.5a] First: \$4.50; second: \$10.00; third: \$18.00

CHAPTER 6

Pretest: Chapter 6, p. 310

1. **[6.2a]** $2x^2(x-1)(x-2)$ **2.** **[6.2b]** $\frac{y+9}{(y+4)^2(y+5)}$ **3.** **[6.2b]** $\frac{4-a}{a-1}$ **4.** **[6.2c]** $\frac{1}{2(y+8)}$ **5.** **[6.1e]** $\frac{(y-11)^2}{(y-2)^2}$ **6.** **[6.1d]** $\frac{2}{3(y+1)}$ **7.** **[6.6a]** $\frac{y-1}{y+1}$ **8.** **[6.3a]** $\frac{132}{17}$ **9.** **[6.3a]** -25 **10.** **[6.5a]** $b=\frac{c}{aM-a}$ **11.** **[6.7e]** $y=\frac{1}{4}\cdot\frac{xz^2}{w}$ **12.** **[6.7b]** 5.625 yd^3 **13.** **[6.4c]** A: 250 km/h; B: 200 km/h **14.** **[6.4a]** $\frac{12}{7}$ hr

Exercise Set, 6.1, p. 317

1. All real numbers except 4 **3.** All real numbers except 6 and 9 **5.** $\frac{3x(x+1)}{3x(x+3)}$ **7.** $\frac{(t-3)(t+3)}{(t+2)(t+3)}$ **9.** $\frac{3y}{5}$ **11.** $\frac{2}{3p^4}$ **13.** $a-3$ **15.** $\frac{2x+3}{8}$ **17.** $\frac{y-3}{y+3}$ **19.** $\frac{t+4}{t-4}$ **21.** $\frac{x-8}{x+4}$ **23.** $\frac{a^2+ab+b^2}{a+b}$ **25.** $\frac{1}{3x^3}$ **27.** $\frac{(x-4)(x+4)}{x(x+3)}$ **29.** $\frac{y+4}{2}$ **31.** $\frac{(2x+3)(x+5)}{7x}$ **33.** $c-2$ **35.** $\frac{1}{x+y}$ **37.** $\frac{3x^5}{2y^3}$ **39.** 3 **41.** $\frac{(y-3)(y+2)}{y}$ **43.** $\frac{2a+1}{a+2}$ **45.** $\frac{(x+4)(x+2)}{3(x-5)}$ **47.** $\frac{y(y^2+3)}{(y+3)(y-2)}$ **49.** $\frac{x^2+4x+16}{(x+4)(x+4)}$ **51.** $\frac{2s}{r+2s}$ **53.** 12 field goals; 3 free throws **55.** 24,640 m^2 **57.** $\frac{x-3}{(x+1)(x+3)}$ **59.** $\frac{m-t}{m+t+1}$ **61.** $\frac{x^2+xy+y^2+x+y}{x-y}$

Exercise Set, 6.2, p. 327

1. 36 **3.** 144 **5.** 72 **7.** 45 **9.** $\frac{11}{10}$ **11.** $\frac{43}{36}$ **13.** $\frac{251}{240}$ **15.** $12x^2y$ **17.** $3(y-3)(y+3)$ **19.** $30a^3b^2$ **21.** $5(y-3)(y-3)$ **23.** $(x+2)(x-2)$, or $(x+2)(2-x)$ **25.** $(2r+3)(r-4)(3r-1)(r+4)$ **27.** $x^3(x-2)(x-2)(x^2+4)$ **29.** $10x^3(x-1)(x-1)(x+1)(x^2+1)$ **31.** 2 **33.** $\frac{3y+5}{y-2}$ **35.** $a+b$ **37.** $\frac{11}{x}$ **39.** $\frac{1}{x+5}$ **41.** $\frac{2y^2+22}{y^2-y-20}$ **43.** $\frac{x+y}{x-y}$ **45.** $\frac{3x-4}{x^2-3x+2}$ **47.** $\frac{8x+1}{x^2-1}$ **49.** $\frac{2x-14}{15(x+5)}$ **51.** $\frac{-a^2+7ab-b^2}{a^2-b^2}$ **53.** $\frac{y}{y^2-5y+6}$ **55.** $\frac{3y-10}{y^2-y-20}$ **57.** $\frac{3y^2-3y-29}{(y+8)(y-3)(y-4)}$ **59.** $\frac{2x^2-13x+7}{(x+3)(x-1)(x-3)}$ **61.** 0 **63.** $\frac{3}{x+2}$ **65.** $\frac{-3x^2-3x-4}{x^2-1}$ **67.** $\frac{-7x-11}{2x^2-x-3}$ **69.** $\frac{c+12}{2c^2+3c-2}$ **71.** $\frac{-2}{x-y}$, or $\frac{2}{y-x}$ **73.**

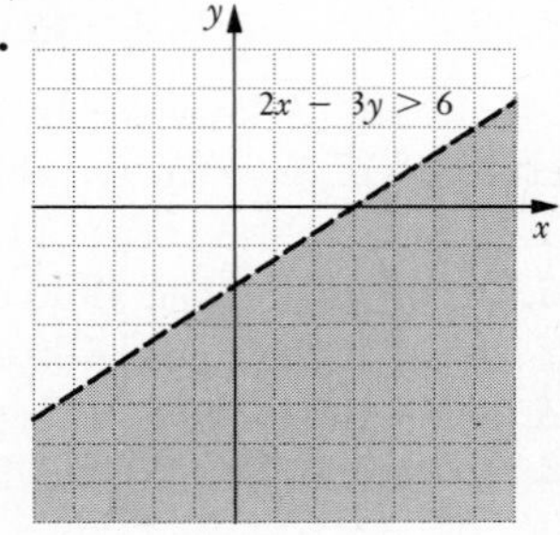

75. $3y^2(4y-1)(y-1)$ **77.** $(x-7)^2$ **79.** $x^4(x+1)(x-1)(x^2+1)(x^2+x+1)(x^2-x+1)$ **81.** $\frac{2b-c-a}{a-b+c}$ **83.** $\frac{x}{3x+1}$

Exercise Set, 6.3, p. 335

1. $\frac{51}{2}$ **3.** -2 **5.** 144 **7.** $-5, -1$ **9.** 2 **11.** $\frac{17}{4}$ **13.** 11 **15.** No solution **17.** 2 **19.** $\frac{3}{5}$ **21.** 5 **23.** -145 **25.** $-\frac{10}{3}$ **27.** -3 **29.** $\frac{31}{5}$ **31.** $\frac{85}{12}$ **33.** $-6, 5$ **35.** No solution **37.** $\frac{17}{4}$ **39.** No solution **41.** 2 **43.** $(x+3)(4x-17)$ **45.** $(1-t)(1+t)(1+t+t^2)(1-t+t^2)$ **47.** $-\frac{7}{2}$ **49.** All real numbers except $-4, -3$

Exercise Set, 6.4, p. 343

1. $-3, -7$ **3.** $\frac{30}{11}$ **5.** $3\frac{3}{14}$ hr **7.** $8\frac{4}{7}$ hr **9.** $3\frac{9}{52}$ hr **11.** A, 10 days; B, 40 days **13.** 11,160,000 mi **15.** 200 **17.** 936 km **19.** 954 **21.** (a) 4.8 T; (b) 48 lb **23.** 1 **25.** 7 mph **27.** 375 km **29.** Train A, 46 mph; train B, 58 mph **31.** 9 km/h **33.** 261 mi, city; 204 mi, highway **35.** $21\frac{9}{11}$ min after 4:00 **37.** Boat, 14 km/h; stream, 10 km/h **39.** $3\frac{3}{4}$ km/h **41.** 48 km/h **43.** $t=\frac{2}{3}$ hr **45.** 48,780,000

Exercise Set, 6.5, p. 349

1. $d_1=\frac{d_2W_1}{W_2}$ **3.** $t=\frac{2s}{v_1+v_2}$ **5.** $r_2=\frac{Rr_1}{r_1-R}$ **7.** $s=\frac{Rg}{g-R}$ **9.** $p=\frac{qf}{q-f}$ **11.** $E=\frac{Inr}{n-I}$ **13.** $H=Sm(t_1-t_2)$ **15.** $e=\frac{rE}{R+r}$ **17.** $r=\frac{A-P}{Pt}$ **19.**

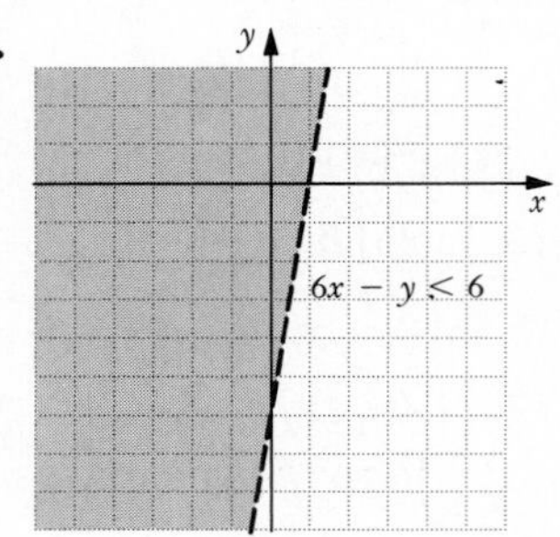

21. $(t+2b)(t^2-2bt+4b^2)$ **23.** $\frac{t}{a}+\frac{t}{b}=1$ **25.** $a=\frac{bt}{b-t}$

Exercise Set, 6.6, p. 355

1. $\frac{1+4x}{1-3x}$ **3.** $\frac{x^2-1}{x^2+1}$ **5.** $\frac{3y+4x}{4y-3x}$ **7.** $\frac{x+y}{x}$ **9.** $\frac{a^2(b-3)}{b^2(a-1)}$ **11.** $\frac{1}{a-b}$ **13.** $\frac{-1}{x(x+h)}$ **15.** $\frac{y-3}{y+5}$ **17.** $\frac{x+1}{5-x}$ **19.** $\frac{5x-16}{4x+1}$ **21.** $\frac{ab(b-a)}{a^2-ab+b^2}$ **23.** $\frac{5}{6}$ **25.** $\frac{11x+8}{4x+3}$ **27.** $\frac{b-a}{ab}$ **29.** a

Exercise Set, 6.7, p. 363

1. 8; $y = 8x$ **3.** 3.6; $y = 3.6x$ **5.** $\frac{8}{5}$; $y = \frac{8}{5}x$ **7.** 6 amperes **9.** \$4.29 **11.** 50 kg **13.** 60; $y = \frac{60}{x}$ **15.** 36; $y = \frac{36}{x}$ **17.** 0.32; $y = \frac{0.32}{x}$ **19.** $\frac{2}{9}$ ampere **21.** $685\frac{5}{7}$ kg **23.** $y = 15x^2$ **25.** $y = \frac{0.0015}{x^2}$ **27.** $y = xz$ **29.** $y = \frac{3}{10}xz^2$ **31.** $y = \frac{xz}{5wp}$ **33.** $355\frac{5}{9}$ ft **35.** 199.4 lb **37.** 97 **39.** 729 gal **41.** 55 correct, 20 wrong **43.** $I = 0.185P$ **45.** $\frac{\pi}{4}$ **47.** Q varies directly as the square of p and inversely as the cube of q

Summary and Review: Chapter 6, p. 367

1. [6.2a] $48x^3$ **2.** [6.2a] $(x-7)(x+7)(3x+1)$ **3.** [6.2a] $(x+5)(x-4)(x-2)$ **4.** [6.1d] $\frac{y-8}{2}$ **5.** [6.1d] $\frac{(x-2)(x+5)}{x-5}$ **6.** [6.1e] $\frac{3a-1}{a-3}$ **7.** [6.1e] $\frac{(x^2+4x+16)(x-6)}{(x+4)(x+2)}$ **8.** [6.2b] $\frac{x-3}{(x+1)(x+3)}$ **9.** [6.2b] $\frac{x^2+11xy+y^2}{(x-y)(x+y)}$ **10.** [6.2b] $\frac{2x^3+2x^2y+2xy^2-2y^3}{(x-y)(x+y)}$ **11.** [6.2c] $\frac{-y}{(y+4)(y-1)}$ **12.** [6.3a] 2 **13.** [6.3a] $\frac{28}{11}$ **14.** [6.3a] 6 **15.** [6.3a] No solution **16.** [6.3a] 3 **17.** [6.3a] $-\frac{11}{3}$ **18.** [6.4a] $5\frac{1}{7}$ hr **19.** [6.4c] 742.5 mi **20.** [6.4c] 24 mph **21.** [6.5a] $s = \frac{Rg}{g-R}$; $g = \frac{Rs}{s-R}$ **22.** [6.5a] $m = \frac{H}{S(t_1-t_2)}$; $t_1 = \frac{H+Smt_2}{Sm}$ **23.** [6.7a] $y = 4x$ **24.** [6.7c] $y = \frac{2500}{x}$ **25.** [6.7d] 1.92 ft **26.** [6.7b] About 77.7 **27.** [6.7f] 500 watts **28.** [6.7f] $\frac{15}{2}$ m **29.** [6.6a] $\frac{3}{4}$ **30.** [6.6a] $\frac{a^2b^2}{2(a^2-ab+b^2)}$ **31.** [6.6a] $\frac{(y+11)(y+5)}{(y-5)(y+2)}$

32. [3.6b]

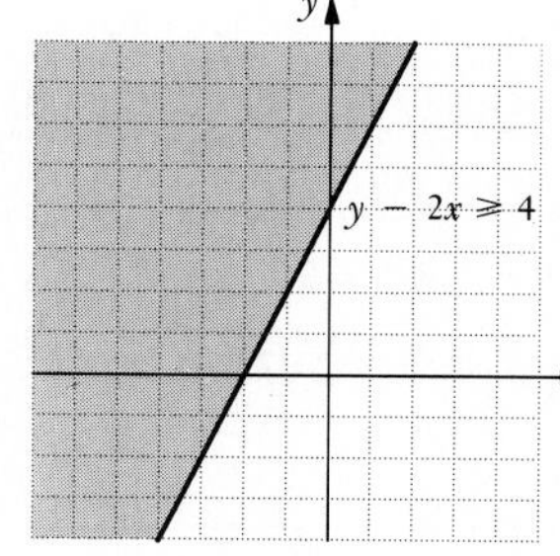

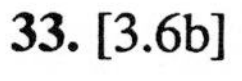
33. [3.6b]

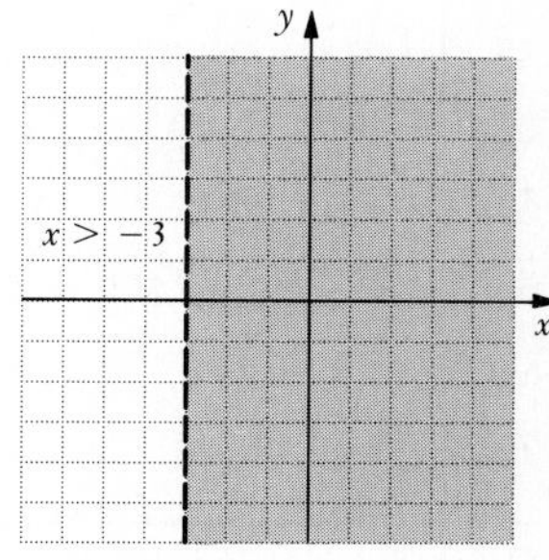

34. [5.6a] $(5x-2y)(25x^2+10xy+4y^2)$ **35.** [5.4b] $(x+6)(6x-7)$ **36.** [4.3a] 15, 20 **37.** [6.3a] All real numbers except 0 and 13 **38.** [6.1c, e] a^2+ab+b^2

Test: Chapter 6, p. 369

1. [6.2a] $(x+3)(x-2)(x+5)$ **2.** [6.1d] $\frac{2(x+5)}{x-2}$ **3.** [6.2b] $\frac{x-6}{(x+4)(x+6)}$ **4.** [6.1e] $\frac{y+4}{2}$ **5.** [6.2b] $x+y$ **6.** [6.2c] $\frac{3x}{(x-1)(x+1)}$ **7.** [6.2c] $\frac{a^3+a^2b+ab^2+ab-b^2-2}{a^3-b^3}$ **8.** [6.3a] 9 **9.** [6.3a] No solution **10.** [6.3a] No solution **11.** [6.3a] $\frac{17}{8}$ **12.** [6.4a] $2\frac{2}{5}$ hr **13.** [6.4c] 11 km/h **14.** [6.4c] 8 km/h **15.** [6.4a] $3\frac{1}{3}$ hr **16.** [6.5a] $t = \frac{mr}{2-m}$ **17.** [6.7e] $Q = \frac{5}{2}xy$ **18.** [6.7b] \$495 **19.** [6.7d] $2\frac{5}{8}$ hr **20.** [6.7f] 6.664 cm² **21.** [6.6a] $\frac{x+1}{x}$ **22.** [6.6a] $\frac{b^2-ab+a^2}{a^2b^2}$ **23.** [5.6a] $(4a+1)(16a^2-4a+1)$ **24.** [5.4b] $8(2t+3)(t-3)$ **25.** [5.4a] $(p-3q)(p+5q)$

26. [3.6b]

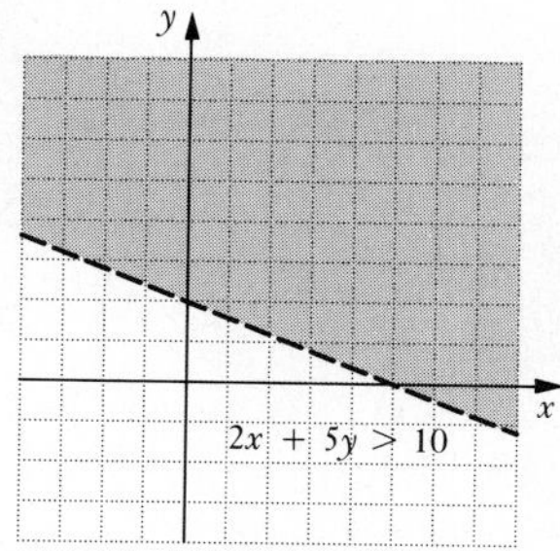

27. [4.3a] 6 30 – sec; 6 60 – sec **28.** [6.3a] All real numbers except 0 and 15 **29.** [6.2a] $(1-t^6)(1+t^6)$

Cumulative Review: Chapters 1-6, p. 371

1. [1.1a] $\frac{4}{3}$ **2.** [1.2d] 12 **3.** [1.3e] $-\frac{8}{9}$ **4.** [1.5c] $47b - 51$ **5.** [1.5c] 224 **6.** [1.6b] $\frac{x^6}{4y^8}$ **7.** [5.1f] $16p^2 - 8p$ **8.** [5.2c] $36m^2 - 12mn + n^2$ **9.** [5.2b] $15a^2 - 14ab - 8b^2$ **10.** [6.1d] $\frac{y-2}{3}$ **11.** [6.1e] $\frac{3x-5}{x+4}$ **12.** [6.2b] $\frac{6x+13}{20(x-3)}$ **13.** [6.2c] $\frac{4x+1}{(x+2)(x-2)}$ **14.** [6.6a] $\frac{y+2x}{3y-x}$ **15.** [6.6a] $\frac{y^3-2y}{y^3-1}$ **16.** [5.9b] $2x^2 - 11x + 23 + \frac{-49}{x+2}$ **17.** [2.1d] $\frac{15}{2}$ **18.** [2.3a] $C = \frac{5}{9}(F-32)$ **19.** [2.4c] $\left\{x \mid -3 < x < -\frac{3}{2}\right\}$ **20.** [2.6e] $\{x \mid x \leq -2.1 \text{ or } x \geq 2.1\}$ **21.** [4.2b] Infinite number of solutions **22.** [4.4b] (3, 2, –1) **23.** [5.8a] $\frac{1}{4}$ **24.** [5.8a] $-2, \frac{7}{2}$ **25.** [6.3a] No solution **26.** [6.3a] –1 **27.** [6.5a] $a = -\frac{bP}{P-3}$, or $\frac{bP}{3-P}$ **28.** [4.7a] (–2, 1) **29.** [4.7a] $\left(\frac{5}{8}, \frac{1}{16}, -\frac{3}{4}\right)$ **30.** [5.3a] $2x^2(2x+9)$ **31.** [5.3b] $(2a-1)(4a^2-3)$ **32.** [5.4a] $(x-6)(x+14)$ **33.** [5.4b] $(2x+5)(3x-2)$ **34.** [5.5b] $(4y+9)(4y-9)$ **35.** [5.5a] $(t-8)^2$ **36.** [5.6a] $8(2x+1)(4x^2-2x+1)$ **37.** [5.6a] $(0.3b-0.2c)(0.09b^2+0.06bc+0.04c^2)$ **38.** [5.7a] $x^2(x^2+1)(x+1)(x-1)$ **39.** [5.7a] $(4x-1)(5x+3)$

40. [3.1c] **41.** [3.2b]

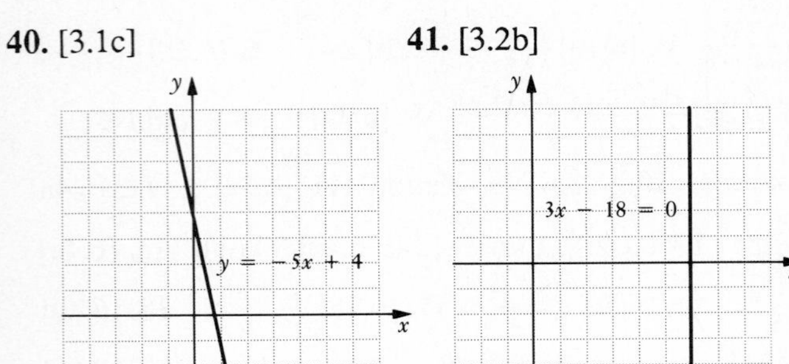

42. [3.6b] **43.** [4.7a]

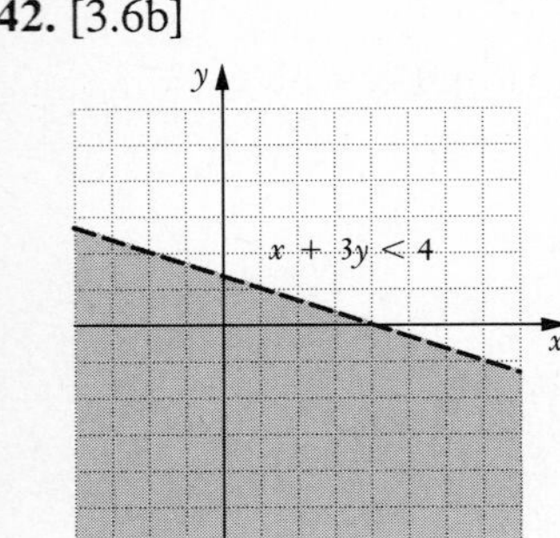

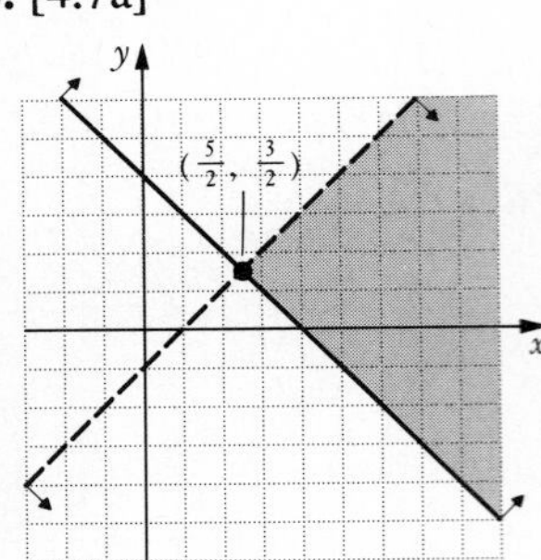

44. [3.4a] $y = -\frac{1}{2}x - 1$ **45.** [3.4c] $y = \frac{1}{2}x - \frac{5}{2}$ **46.** [2.4d] $I < \$45{,}333.3\overline{3}$ **47.** [2.2a] 20.9% **48.** [4.3a] Detroit: \$4.09; Chicago: \$1.66 **49.** [4.5a] Win: 38; lose: 30; tie: 13 **50.** [6.4a] $5\frac{1}{7}$ hr **51.** [6.7f] 60 years old **52.** [6.2a] 7410 **53.** [4.5a] $a = 1$, $b = -5$, $c = 6$ **54.** [6.3a] All real numbers except 9 and -5 **55.** [5.8a] $0, \frac{1}{4}, -\frac{1}{4}$

CHAPTER 7

Pretest: Chapter 7, p. 374

1. [7.1b] $|t|$ **2. [7.1c]** $3x$ **3. [7.1d]** $|y|$ **4. [7.3c]** $3a\sqrt[3]{3ab^2}$ **5. [7.4a]** $4\sqrt{5}$ **6. [7.2b]** $12x^2\sqrt{x}$ **7. [7.4b]** $25 - 4\sqrt{6}$ **8. [7.3a]** $2\sqrt{a}$ **9. [7.5c]** $\frac{3x^2 - 4x\sqrt{z} + z}{9x^2 - z}$ **10. [7.7a]** 13 **11. [7.7a]** $-\frac{7}{6}$ **12. [7.7b]** 0, 8 **13. [7.8a]** $\sqrt{39}$; 6.245 **14. [7.8a]** $4\sqrt{2}$ ft; 5.657 ft **15. [7.6d]** $\sqrt[6]{x^3(x-4)^2}$ **16. [7.9f]** No **17. [7.9b]** $-5 + 13i$ **18. [7.9c]** $24 - 7i$ **19. [7.9e]** $-\frac{14}{13} - \frac{5}{13}i$ **20. [7.9d]** $-i$

Sidelight: Wind Chill Temperature, p. 378

1. 0°F **2.** −29° F **3.** −10° F **4.** −22° F **5.** −64° F **6.** −94° F

Exercise Set 7.1, p. 379

1. 4, −4 **3.** 12, −12 **5.** 20, −20 **7.** $-\frac{7}{6}$ **9.** 14 **11.** $-\frac{4}{9}$ **13.** 0.3 **15.** −0.07 **17.** $p^2 + 4$ **19.** $\frac{x}{y+4}$ **21.** $|4x|$, or $4|x|$ **23.** $|7c|$, or $7|c|$ **25.** $|a+1|$ **27.** $|x-2|$ **29.** $|2x+7|$ **31.** 3 **33.** $-4x$ **35.** −6 **37.** $0.7(x+1)$ **39.** 5 **41.** −1 **43.** $-\frac{2}{3}$ **45.** $|x|$ **47.** $|5a|$, or $5|a|$ **49.** 6 **51.** $|a+b|$ **53.** y **55.** $x - 2$ **57.** −2, 1 **59.** $-\frac{7}{2}, \frac{7}{2}$ **61.** (a) 13; (b) 15; (c) 18; (d) 20

Exercise Set 7.2, p. 385

1. $2\sqrt{2}$ **3.** $2\sqrt{6}$ **5.** $2\sqrt{10}$ **7.** $6x^2\sqrt{5}$ **9.** $3x^2\sqrt[3]{2x^2}$ **11.** $2t^2\sqrt[3]{10t^2}$ **13.** $2\sqrt[4]{5}$ **15.** $3x^2y^2\sqrt[4]{3y^2}$ **17.** $(x+y)\sqrt[3]{x+y}$ **19.** $-2xy\sqrt[3]{3xy^2}$ **21.** $2xy^3\sqrt[5]{3x^2}$ **23.** $3\sqrt{10}$ **25.** $3\sqrt[3]{2}$ **27.** $30\sqrt{3}$ **29.** $5bc^2\sqrt{2b}$ **31.** $2y^3\sqrt[3]{2}$ **33.** $(b+3)^2$ **35.** $4a^3b\sqrt{6ab}$ **37.** $4\sqrt[4]{4}$ **39.** $2a\sqrt[4]{5a}$ **41.** $6x^2y^4\sqrt{15xy}$ **43.** $a^2(b-c)^3\sqrt[5]{a(b-c)^3}$ **45.** 13.416 **47.** 7.477 **49.** 0.454 **51.** 4.414 **53.** (a) 20 mph; (b) 37.4 mph; (c) 42.4 mph **55.** $0, \frac{2}{3}$ **57.** 2.4×10^{-2} **59.** 84

Exercise Set 7.3, p. 389

1. $\sqrt{7}$ **3.** 3 **5.** $y\sqrt{5y}$ **7.** $2\sqrt[3]{a^2b}$ **9.** $3\sqrt{2xy}$ **11.** $2x^2y^2$ **13.** $\sqrt{x^2 + xy + y^2}$ **15.** $\frac{4}{5}$ **17.** $\frac{4}{3}$ **19.** $\frac{7}{y}$ **21.** $\frac{5y\sqrt{y}}{x^2}$ **23.** $\frac{2x\sqrt[3]{x^2}}{3y}$ **25.** $\frac{3x}{2}$ **27.** $\frac{pq^2\sqrt[4]{p}}{r^3}$ **29.** $\frac{2x\sqrt[5]{x^3}}{y^2}$ **31.** $\frac{x^2\sqrt[6]{x}}{yz^2}$ **33.** $6a\sqrt{6a}$ **35.** $4b\sqrt[3]{4b}$ **37.** $54a^3b\sqrt{2b}$ **39.** $2c\sqrt[3]{18cd^2}$ **41.** $x\sqrt[3]{49xy^2}$ **43.** $9y^2\sqrt[4]{x^2y^2}$ **45.** 8 **47.** Length, 20; width, 5 **49.** a^3bxy^2 **51.** $2yz\sqrt{2z}$

Exercise Set 7.4, p. 393

1. $8\sqrt{3}$ **3.** $3\sqrt[3]{5}$ **5.** $13\sqrt[3]{y}$ **7.** $7\sqrt{2}$ **9.** $6\sqrt[3]{3}$ **11.** $21\sqrt{3}$ **13.** $38\sqrt{5}$ **15.** $122\sqrt{2}$ **17.** $9\sqrt[3]{2}$ **19.** $29\sqrt{2}$ **21.** $(1+6a)\sqrt{5a}$ **23.** $(2-x)\sqrt[3]{3x}$ **25.** $3\sqrt{2y-2}$ **27.** $(x+3)\sqrt{x-1}$ **29.** $15\sqrt[3]{4}$ **31.** $2\sqrt{6} - 18$ **33.** $\sqrt{6} - \sqrt{10}$ **35.** $2\sqrt{15} - 6\sqrt{3}$ **37.** −6 **39.** $3a\sqrt[3]{2}$ **41.** 1 **43.** −12 **45.** 44 **47.** 1 **49.** 3 **51.** −19 **53.** $a - b$ **55.** $1 + \sqrt{5}$ **57.** $7 + 3\sqrt{3}$ **59.** −6 **61.** $a + \sqrt{3a} + \sqrt{2a} + \sqrt{6}$ **63.** $2\sqrt[3]{9} - 3\sqrt[3]{6} - 2\sqrt[3]{4}$ **65.** $7 + 4\sqrt{3}$ **67.** $3 - \sqrt[5]{24} - \sqrt[5]{81} + \sqrt[5]{72}$ **69.** 6 **71.** $14 + 2\sqrt{15} - 6\sqrt{2} - 2\sqrt{30}$ **73.** $3\sqrt[3]{3} + 2\sqrt[3]{9} - 8$

Exercise Set 7.5, p. 401

1. $\frac{\sqrt{30}}{5}$ **3.** $\frac{\sqrt{30}}{3}$ **5.** $\frac{2\sqrt{15}}{5}$ **7.** $\frac{2\sqrt[3]{6}}{3}$ **9.** $\frac{\sqrt[3]{75ac^2}}{5c}$ **11.** $\frac{y\sqrt[3]{9yx^2}}{3x^2}$ **13.** $\frac{\sqrt[3]{x^2y^2}}{xy}$ **15.** $\frac{\sqrt{10x}}{6}$ **17.** $\frac{\sqrt[3]{100xy}}{5x^2y}$ **19.** $\frac{\sqrt[4]{2xy}}{2x^2y}$ **21.** $\frac{\sqrt[5]{432x^2y^4}}{3xy^2}$ **23.** $\frac{7}{\sqrt{21x}}$ **25.** $\frac{2}{\sqrt{6}}$ **27.** $\frac{52}{3\sqrt{91}}$ **29.** $\frac{7}{\sqrt[3]{98}}$ **31.** $\frac{7x}{\sqrt{21xy}}$ **33.** $\frac{5y^2}{x\sqrt[3]{150x^2y^2}}$ **35.** $\frac{ab}{3\sqrt{ab}}$

37. $\frac{40+5\sqrt{6}}{58}$ **39.** $-2\sqrt{35}-2\sqrt{21}$ **41.** $\frac{\sqrt{15}+20-6\sqrt{2}-8\sqrt{30}}{-77}$
43. $\frac{x-2\sqrt{xy}+y}{x-y}$ **45.** $\frac{3\sqrt{6}+4}{2}$ **47.** $\frac{2x-5\sqrt{xy}+2y}{4x-y}$ **49.** $-\frac{11}{4\sqrt{3}-20}$
51. $-\frac{22}{\sqrt{6}+5\sqrt{2}+5\sqrt{3}+25}$ **53.** $\frac{x-y}{x+2\sqrt{xy}+y}$ **55.** $\frac{7}{43\sqrt{2}+66}$ **57.** $\frac{2-9x}{2-4\sqrt{2x}+3x}$
59. $\frac{a^2b-c^2}{ab-c\sqrt{b}+ac\sqrt{b}-c^2}$ **61.** 6 **63.** 1 **65.** $\frac{\sqrt{10}+2\sqrt{5}-2\sqrt{3}}{10}$ **67.** $\frac{b^2+\sqrt{b}}{b^2+b+1}$
69. $6a\sqrt[3]{36ab^2}$ **71.** $\frac{x-19}{x+31+10\sqrt{x+6}}$ **73.** $-\frac{3\sqrt{a^2-3}}{a^2-3}$ **75.** $1-\sqrt{w}$

Exercise Set 7.6, p. 409

1. $\sqrt[4]{x}$ **3.** 2 **5.** $\sqrt[5]{a^2b^2}$ **7.** 8 **9.** $20^{1/3}$ **11.** $(xy^2z)^{1/5}$
13. $(3mn)^{3/2}$ **15.** $(8x^2y)^{5/7}$ **17.** $\frac{1}{x^{1/3}}$ **19.** $x^{2/3}$ **21.** $5^{7/8}$
23. $7^{1/4}$ **25.** $8.3^{7/20}$ **27.** $10^{3/20}$ **29.** $a^{3/2}$ **31.** $a^{8/3}b^{5/2}$
33. $\sqrt[3]{a^2}$ **35.** $2y^2$ **37.** $2\sqrt[4]{2}$ **39.** $\sqrt[3]{2x}$ **41.** $2c^2d^3$ **43.** $\frac{m^2n^4}{2}$
45. $\sqrt[4]{r^2s}$ **47.** $3ab^3$ **49.** $2x^3y^8$ **51.** $2\sqrt{2}p^3q^8$ **53.** $\sqrt[6]{243}$
55. $\sqrt[6]{4x^5}$ **57.** $\sqrt[6]{x^5-4x^4+4x^3}$ **59.** $\sqrt[6]{a+b}$ **61.** $\sqrt[12]{a^8b^9}$
63. $\sqrt[12]{s^3t^{16}}$, or $t\sqrt[12]{s^3t^4}$ **65.** $xy\sqrt[4]{xy^3}$ **67.** $a^2b^2c^2\sqrt[6]{a^2bc^2}$
69. $\sqrt[12]{\frac{q^2}{p^3}}$ **71.** $x^3\cdot\sqrt[6]{x}$ **73.** $\sqrt[4]{2xy^2}$ **75.** $\sqrt[6]{p+q}$ **77.** x^3

Exercise Set 7.7, p. 415

1. $\frac{3}{2}$ **3.** $\frac{49}{5}$ **5.** 168 **7.** $\frac{80}{3}$ **9.** −27 **11.** No solution **13.** 3
15. 19 **17.** −6 **19.** $\frac{1}{64}$ **21.** 5 **23.** 9 **25.** 7 **27.** $\frac{80}{9}$ **29.** 6, 2 **31.** −1 **33.** No solution **35.** 3 **37.** 0, −2.8 **39.** 6912
41. 0 **43.** −3, −6 **45.** 2 **47.** 0, $\frac{125}{4}$ **49.** 2 **51.** $\frac{1}{2}$ **53.** 3

Exercise Set 7.8, p. 419

1. $\sqrt{34}$; 5.831 **3.** $\sqrt{288}$; 16.971 **5.** 5 **7.** $\sqrt{31}$; 5.568
9. $\sqrt{12}$; 3.464 **11.** $\sqrt{n-1}$ **13.** $\sqrt{325}$; 18.028 ft
15. Neither; both have the same area, 300 ft^2 **17.** $s\sqrt{2}$
19. 7.1 ft **21.** 8 ft **23.** 12 in. **25.** (0,−4), (0,4) **27.** 3, 8
29. 50 ft^2

Exercise Set 7.9, p. 427

1. $i\sqrt{15}$, or $\sqrt{15}i$ **3.** $4i$ **5.** $-2i\sqrt{3}$, or $-2\sqrt{3}i$ **7.** $i\sqrt{3}$, or $\sqrt{3}i$ **9.** $9i$ **11.** $7i\sqrt{2}$, or $7\sqrt{2}i$ **13.** $-7i$ **15.** $4-2\sqrt{15}i$
17. $(2+2\sqrt{3})i$ **19.** $8+i$ **21.** $9-5i$ **23.** $7+4i$
25. $-2-3i$ **27.** $-1+i$ **29.** $11+6i$ **31.** −18 **33.** $-\sqrt{10}$
35. 21 **37.** $35+20i$ **39.** $1+5i$ **41.** $18+14i$
43. $38+9i$ **45.** $2-46i$ **47.** $5-12i$ **49.** $-5+12i$
51. $-5-12i$ **53.** $-i$ **55.** 1 **57.** −1 **59.** i **61.** −1
63. $-125i$ **65.** 8 **67.** $1-26i$ **69.** 0 **71.** 0 **73.** 1
75. $5-8i$ **77.** $2-\frac{\sqrt{6}}{2}i$ **79.** $\frac{8}{5}+\frac{1}{5}i$ **81.** $-i$ **83.** $-\frac{3}{7}-\frac{8}{7}i$
85. $\frac{6}{5}-\frac{2}{5}i$ **87.** $-\frac{8}{41}+\frac{10}{41}i$ **89.** $-\frac{4}{3}i$ **91.** $-\frac{1}{2}-\frac{1}{4}i$ **93.** $\frac{3}{5}+\frac{4}{5}i$
95. Yes **97.** No **99.** 7 **101.** $-4-8i$ **103.** $-3-4i$
105. $-88i$ **107.** 8 **109.** $\frac{3}{5}+\frac{9}{5}i$ **111.** 1

Summary and Review: Chapter 7, p. 431

1. [7.1b] $9|a|$ **2.** [7.1b] $|c-3|$ **3.** [7.1b] $|x-3|$
4. [7.1b] $|2x+1|$ **5.** [7.1d] −2 **6.** [7.1c] $-\frac{1}{3}$ **7.** [7.1d] $|x|$ **8.** [7.1d] 3 **9.** [7.2c] 15.652 **10.** [7.2c] 10.583
11. [7.2c] −0.236 **12.** [7.2b] $3xy\sqrt{2y}$ **13.** [7.2b] $3a\sqrt[3]{a^2b^2}$ **14.** [7.3a] $y\sqrt[3]{6}$ **15.** [7.3a] $\frac{5\sqrt{x}}{2}$ **16.** [7.3c] $8xy^2$
17. [7.3c] $2a\sqrt[3]{2ab^2}$ **18.** [7.4a] $30\sqrt[3]{5}$ **19.** [7.4a] $15\sqrt{2}$
20. [7.4b] $9-\sqrt[3]{4}$ **21.** [7.4b] $-43-2\sqrt{10}$ **22.** [7.4b] $8-2\sqrt{7}$ **23.** [7.5c] $\frac{2a-\sqrt{ab}-b}{a-b}$ **24.** [7.5c] $\frac{4a-b}{2a+\sqrt{ab}-b}$
25. [7.6a] $(8x^6y^2)^{4/5}$ **26.** [7.6a] $\sqrt[4]{(5a)^3}$ **27.** [7.6d] $\sqrt[12]{x^4y^3}$
28. [7.6d] $\sqrt[12]{x^3(x-3)^4}$ **29.** [7.7b] 4 **30.** [7.7a] 13
31. [7.8a] 25 **32.** [7.8a] $\sqrt{46}$; 6.782 **33.** [7.8a] $\sqrt{481}$; 21.932 ft **34.** [7.9a] $-2\sqrt{2}i$ **35.** [7.9b] $-2-9i$
36. [7.9b] $1+i$ **37.** [7.9c] 29 **38.** [7.9d] i **39.** [7.9c] $9-12i$ **40.** [7.9e] $\frac{2}{5}+\frac{3}{5}i$ **41.** [7.9e] 3 **42.** [7.9f] No
43. [5.8b] 12, 13, 14 **44.** [6.3a] $\frac{23}{7}$ **45.** [5.8b] 15 ft by 12 ft **46.** [6.1d] $\frac{x(x+2)}{(x+y)(x-3)}$ **47.** [7.7b] 3 **48.** [7.4a] $-28\sqrt{3}-22\sqrt{5}$

Test: Chapter 7, p. 433

1. [7.1b] $6|y|$ **2.** [7.1b] $|x+5|$ **3.** [7.1c] −2 **4.** [7.1d] 4 **5.** [7.2c] 12.166 **6.** [7.2b] $2x^3$ **7.** [7.3b] $x\sqrt{5x}$
8. [7.3c] $4a\sqrt[3]{4ab^2}$ **9.** [7.4a] $38\sqrt{2}$ **10.** [7.4b] −20
11. [7.5c] $\frac{a+2\sqrt{ab}+b}{a-b}$ **12.** [7.6a] $(5xy^2)^{5/2}$ **13.** [7.6d] $\sqrt[4]{2x^2y-12xy+18y}$ **14.** [7.7b] 7 **15.** [7.9a] $3i\sqrt{2}$, or $3\sqrt{2}i$ **16.** [7.9b] $7+5i$ **17.** [7.9c] $-2i$ **18.** [7.9d] $-i$
19. [7.9d] $-1-512i$ **20.** [7.9e] $-\frac{77}{50}+\frac{7}{25}i$ **21.** [7.9f] No
22. [7.8a] $\sqrt{98}$; 9.899 **23.** [7.8a] 2 **24.** [7.8a] $\sqrt{16,200}$; 127.279 ft **25.** [5.8a] 5 **26.** [6.1e] $\frac{x-3}{x-4}$
27. [6.3a] No solution **28.** [5.8b] 16, 18; −18, −16
29. [7.7b] 3 **30.** [7.9c, e] $-\frac{17}{4}i$ **31.** [7.4a] $(25x-30y-9xy)\sqrt{2xy}$

Cumulative Review: Chapters 1-7, p. 435

1. [1.3d] $\frac{21}{50}$ **2.** [1.3c] −37.2 **3.** [1.5c] $25c-31$ **4.** [1.5c] 174 **5.** [1.6b] $\frac{64x^{21}}{27y^{15}}$ **6.** [5.1d] $-3x^3+9x^2+3x-3$

7. [5.2c] $4x^4-4x^2y+y^2$ **8.** [5.2a] $15x^4-x^3-9x^2+5x-2$ **9.** [6.1d] $\frac{(x+4)(x-7)}{x+7}$ **10.** [6.2c] $\frac{-2x+4}{(x+2)(x-3)}$ **11.** [6.6a] $\frac{y-6}{y-9}$ **12.** [5.9b] $y^2+y-2+\frac{-1}{y+2}$ **13.** [7.1c] $-2x$ **14.** [7.2a] $4\,|\,x-1\,|$ **15.** [7.4a] $57\sqrt{3}$ **16.** [7.2b] $4xy^2\sqrt{y}$ **17.** [7.5c] $\sqrt{30}+\sqrt{15}$ **18.** [7.1d] $\frac{m^2n^4}{2}$ **19.** [7.6c] $6^{8/9}$ **20.** [7.9b] $3+5i$ **21.** [7.9e] $\frac{7}{61}-\frac{16}{61}i$ **22.** [2.1d] 2 **23.** [2.3a] $c=8M+3$ **24.** [2.4c] $\{a\mid a>-7\}$ **25.** [2.5a] $\{x\mid -10<x<13\}$ **26.** [2.6c] $\frac{4}{3},\frac{8}{3}$ **27.** [4.2b] $(5,3)$ **28.** [4.4b] $(-1,0,4)$ **29.** [5.8a] $\pm\frac{25}{7}$ **30.** [6.3a] -5 **31.** [6.3a] $\frac{1}{3}$ **32.** [6.5a] $R=\frac{nE-nrI}{I}$ **33.** [7.7a] 6 **34.** [7.7b] $-\frac{1}{4}$ **35.** [4.6c] $\left(\frac{1}{2},3\right)$ **36.** [4.6c] $\left(-3,\frac{1}{3},-\frac{1}{4}\right)$

37. [3.1c]

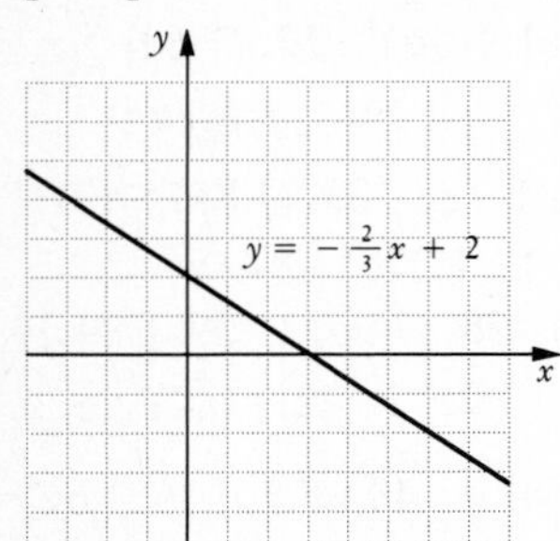

38. [3.2a]

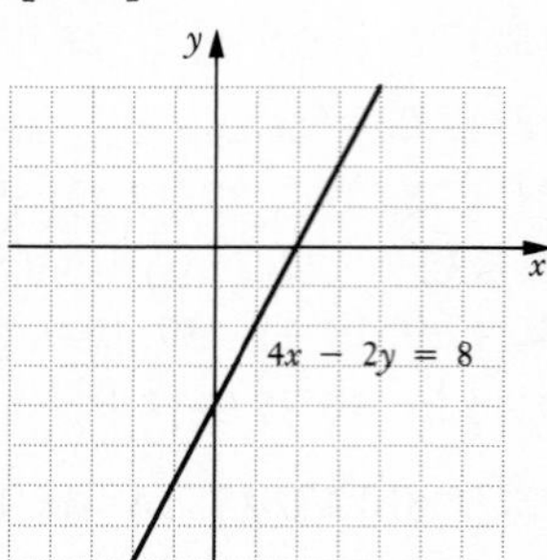

39. [3.6b]

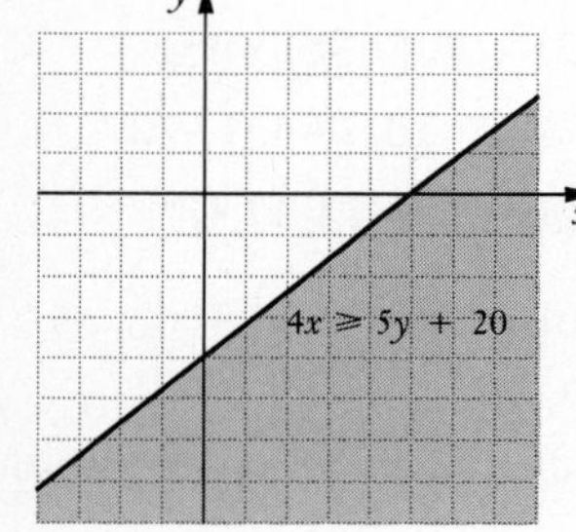

40. [4.7a]

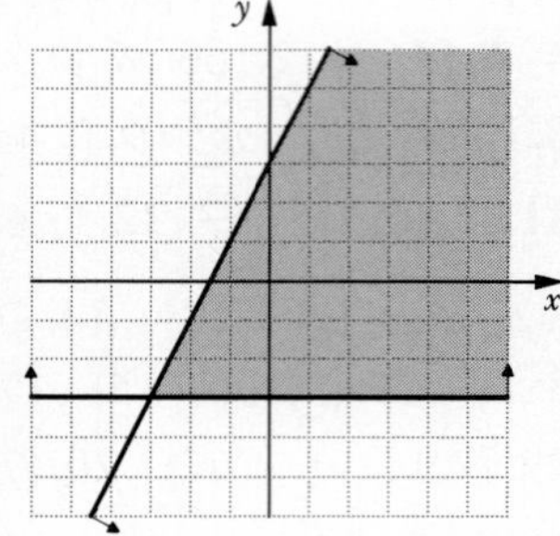

41. [5.3a] $6xy^2(2x-5y)$ **42.** [5.4b] $(3x+4)(x-7)$ **43.** [5.4a] $(y+11)(y-12)$ **44.** [5.6a] $(3y+2)(9y^2-6y+4)$ **45.** [5.5b] $(2x+25)(2x-25)$ **46.** [4.3a] Finland: \$7.99, Sweden: \$6.45 **47.** [3.3c] $m=\frac{3}{2}$; y-intercept $(0,-4)$ **48.** [3.4c] $y=-\frac{1}{3}x+\frac{13}{3}$ **49.** [4.3a] U of P: 457, MCV: 212 **50.** [6.4a] 1 hr **51.** [6.7d] $h=125$ ft, $A=1000$ ft^2 **52.** [7.8a] $d=3\sqrt{5}$ ft

53. [2.6e]

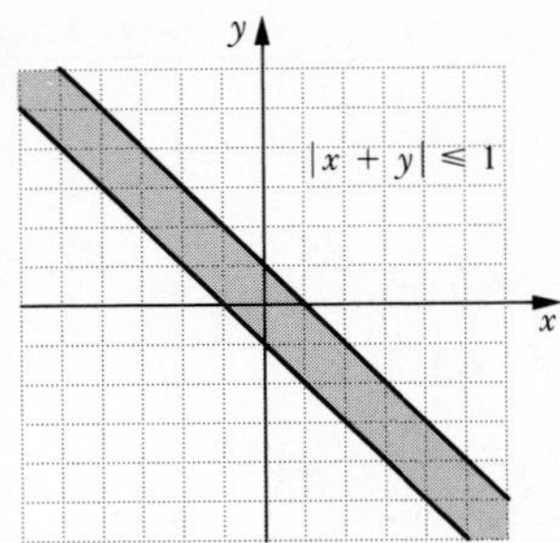

54. [7.7b] $-\frac{8}{9}$ **55.** [5.7a] $\left(x-\frac{5}{2}\right)\left(x-\frac{1}{2}\right)$

CHAPTER 8

Pretest: Chapter 8, p. 438

1. [8.1b] $0,-3$ **2. [8.2a]** $-2+2i,-2-2i$ **3. [8.1a]** 5 **4. [8.4a]** $\frac{\sqrt{3}}{3},-\frac{\sqrt{3}}{3},\sqrt{2},-\sqrt{2}$ **5. [8.6a]** $-1,\frac{5}{6}$ **6. [8.2b]** $\frac{-2\pm\sqrt{6}}{2}$; $-2.2, 0.2$ **7. [8.5a]** $T=\frac{1}{RW^2}$ **8. [8.3b]** $3x^2+4x-4=0$ **9. [8.7c]** **(a)** $(3,-2)$; **(b)** $x=3$; **(c)**

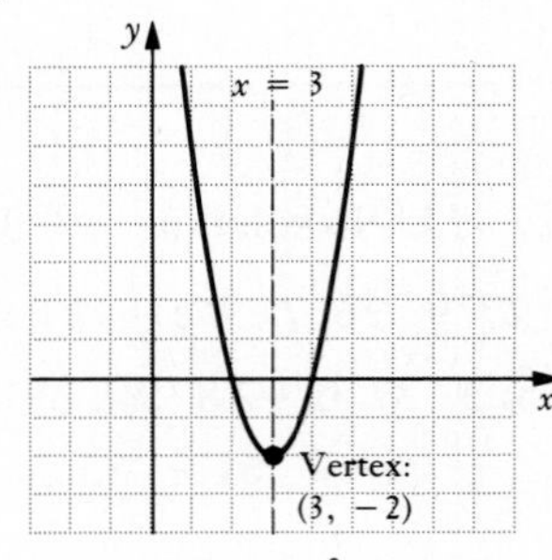

10. [8.8b] $(3+\sqrt{5},0),(3-\sqrt{5},0)$ **11. [8.8d]** $y=4x^2-5x+1$ **12. [8.5b]** 12, 14, 16 **13. [8.6b]** 40 mph; 48 mph **14. [8.8c]** 5625 ft^2 **15. [8.9a]** $\{x\mid x<-5 \text{ or } 0<x<3\}$ **16. [8.9b]** $\{x\mid x<-3 \text{ or } x\geq 2\}$

Exercise Set 8.1, p. 447

1. $\pm\sqrt{5}$ **3.** $\pm\frac{2}{5}i$ **5.** $\pm\frac{\sqrt{6}}{2}$ **7.** $5,-9$ **9.** $3\pm\sqrt{21}$ **11.** $13\pm2\sqrt{2}$ **13.** $7\pm2i$ **15.** $9\pm\sqrt{34}$ **17.** $\frac{3\pm\sqrt{14}}{2}$ **19.** $5,-11$ **21.** $9,5$ **23.** $-1\pm\sqrt{6}$ **25.** $9\pm\sqrt{91}$ **27.** $\frac{-1\pm\sqrt{5}}{2}$ **29.** $\frac{5\pm\sqrt{53}}{2}$ **31.** $\frac{-3\pm\sqrt{57}}{4}$ **33.** $\frac{9\pm\sqrt{105}}{4}$ **35.** $2,-8$ **37.** $-11\pm\sqrt{19}$ **39.** $5\pm\sqrt{29}$ **41.** $\frac{-7\pm\sqrt{57}}{2}$ **43.** $7,-4$ **45.** $\frac{3\pm\sqrt{17}}{4}$ **47.** $\frac{3\pm\sqrt{145}}{4}$ **49.** $\frac{2\pm\sqrt{7}}{3}$ **51.** $-\frac{1}{2}\pm\frac{\sqrt{7}}{2}i$ **53.** $2\pm3i$ **55.** $6\frac{1}{4}\%$ **57.** 20% **59.** 4% **61.** About 9.3 sec

63.

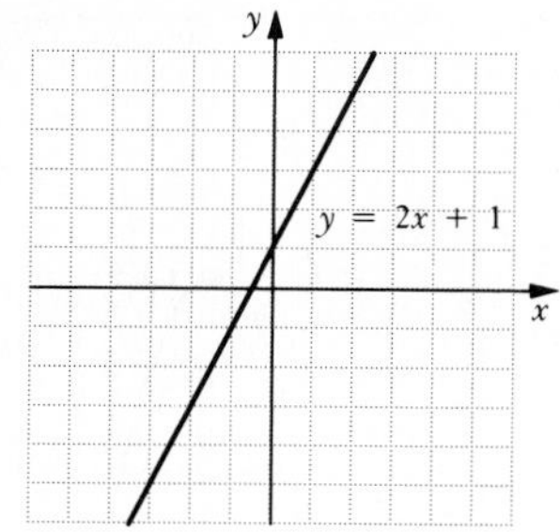

65. 4.6 **67.** $b = 16, -16$ **69.** $0, \frac{7}{2}, -8, -\frac{10}{3}$ **71.** *A:* 15 km/h; *B:* 8 km/h

Exercise Set 8.2, p. 453

1. $-3 \pm \sqrt{5}$ **3.** $-1, -\frac{5}{3}$ **5.** $\frac{1 \pm i\sqrt{3}}{2}$ **7.** $2 \pm 3i$ **9.** $\frac{-3 \pm \sqrt{41}}{2}$ **11.** $-1 \pm 2i$ **13.** $0, -1$ **15.** $0, -\frac{9}{14}$ **17.** $\frac{2}{5}$ **19.** $-1, -2$ **21.** 5, 10 **23.** $\frac{13 \pm \sqrt{509}}{10}$ **25.** $1 \pm 2i$ **27.** $\frac{2}{3}, \frac{3}{2}$ **29.** $-2, 3$ **31.** $\frac{3}{4}, -2$ **33.** $\frac{1 \pm 3i}{2}$ **35.** $2, -1 \pm i\sqrt{3}$ **37.** 1.3, –5.3 **39.** 5.2, 0.8 **41.** 2.8, –1.3 **43.** 1.9, –0.3 **45.** 30 lb of *A*, 20 lb of *B* **47.** 1.8692840, –0.3251940 **49.** $\frac{1 \pm \sqrt{1 + 8\sqrt{5}}}{4}$ **51.** $\frac{-i \pm i\sqrt{1 + 4i}}{2}$ **53.** $\frac{-1 \pm 3\sqrt{5}}{6}$

Exercise Set 8.3, p. 457

1. One real **3.** Two nonreal **5.** Two real **7.** One real **9.** Two nonreal **11.** Two real **13.** Two real **15.** One real **17.** $x^2 + 2x - 99 = 0$ **19.** $x^2 - 14x + 49 = 0$ **21.** $25x^2 - 20x - 12 = 0$ **23.** $4x^2 - 2(c + d)x + cd = 0$ **25.** $x^2 - 4\sqrt{2}x + 6 = 0$ **27.** 6 30-sec; 6 60-sec **29.** The product of the solutions is $\left(\frac{-b + \sqrt{b^2 - 4ac}}{2a}\right)\left(\frac{-b - \sqrt{b^2 - 4ac}}{2a}\right) = \frac{(-b)^2 - \left(\sqrt{b^2 - 4ac}\right)^2}{4a^2} = \frac{b^2 - b^2 + 4ac}{4a^2} = \frac{4ac}{4a^2} = \frac{c}{a}$. **31.** (a) $k = -\frac{3}{5}$; (b) $-\frac{1}{3}$ **33.** $x^2 - \sqrt{3}x + 8 = 0$

Exercise Set 8.4, p. 461

1. $\pm\sqrt{5}$ **3.** 81, 1 **5.** 7, –1, 5, 1 **7.** $\frac{1}{3}, -\frac{1}{2}$ **9.** 1 **11.** 4, 1, 6, –1 **13.** $\pm\sqrt{2 + \sqrt{6}}, \pm\sqrt{2 - \sqrt{6}}$ **15.** 2, –1 **17.** $\frac{\pm\sqrt{15}}{3}, \frac{\pm\sqrt{6}}{2}$ **19.** 125, –1 **21.** 6, 1 **23.** $-\frac{3}{2}$ **25.** $3 \pm \sqrt{10}, -1 \pm \sqrt{2}$ **27.** $3x^2\sqrt{x}$ **29.** $\frac{x^3 + x^2 + 2x + 2}{x^3 - 1}$ **31.** 28.5 **33.** $\frac{100}{99}$ **35.** 259 **37.** 1, 3

Exercise Set 8.5, p. 467

1. $s = \sqrt{\frac{A}{6}}$ **3.** $r = \sqrt{\frac{Gm_1m_2}{F}}$ **5.** $c = \sqrt{\frac{E}{m}}$ **7.** $b = \sqrt{c^2 - a^2}$ **9.** $k = \frac{3 + \sqrt{9 + 8N}}{2}$ **11.** $r = \frac{-\pi h + \sqrt{\pi^2h^2 + 2\pi A}}{2\pi}$ **13.** $g = \frac{4\pi^2 L}{T^2}$ **15.** $D = \sqrt{\frac{32LV}{g(P_1 - P_2)}}$ **17.** $v = \frac{c\sqrt{m^2 - m_0^2}}{m}$ **19.** Length: 6 ft; width: 2 ft **21.** Length: 24 yd; width: 12 yd **23.** 2 in. **25.** 10 ft, 24 ft **27.** 4 ft, 3 ft **29.** 52, 53 **31.** Length: $2 + \sqrt{14} \approx 5.7$ ft; width: $\sqrt{14} - 2 \approx 1.7$ ft **33.** Length: $16\sqrt{2} \approx 22.6$ yd; width: $8\sqrt{2} \approx 11.3$ yd **35.** $\frac{17 - \sqrt{109}}{2} \approx 3$ in. **37.** $7 + \sqrt{239} \approx 22.5$ ft; $-7 + \sqrt{239} \approx 8.5$ ft **39.** $2\sqrt{5}i$ **41.** No solution

Exercise Set 8.6, p. 475

1. 6, –4 **3.** $\pm\frac{3}{2}$ **5.** 25, –20 **7.** 4, –1 **9.** $\pm\sqrt{37}$ **11.** 2, –5 **13.** $10, \frac{21}{2}$ **15.** –3, 2 **17.** 0, –5 **19.** 0 **21.** $\frac{-3 \pm \sqrt{57}}{2}$ **23.** First part: 40 mph; second part: 35 mph **25.** 35 mph **27.** *A:* 350 km/h; *B:* 400 km/h **29.** 6 hr **31.** 3.24 km/h **33.** First part: 36.33 mph; second part: 31.33 mph **35.** 5, 1 **37.** $2y\sqrt[3]{9x^2}$ **39.** Chester: 4.3 hr; Ron: 3.1 hr **41.** $\pm\sqrt{2}$

Exercise Set 8.7, p. 485

1.

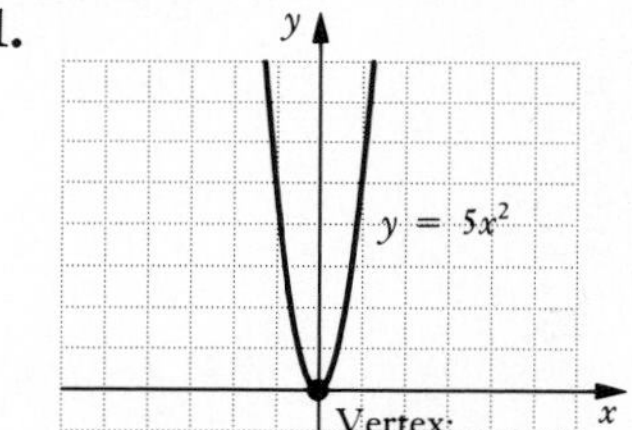

3.

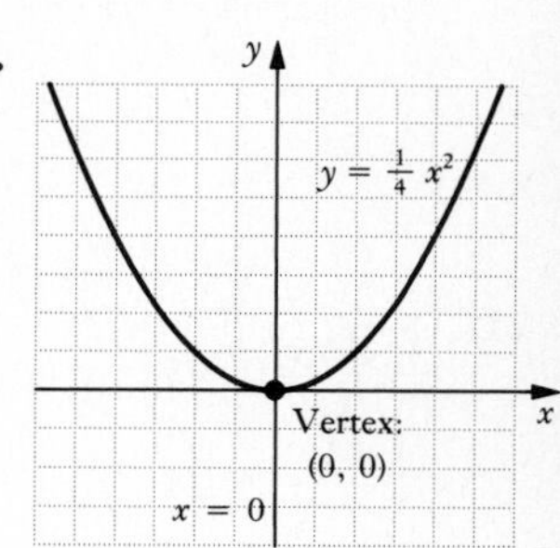

5.

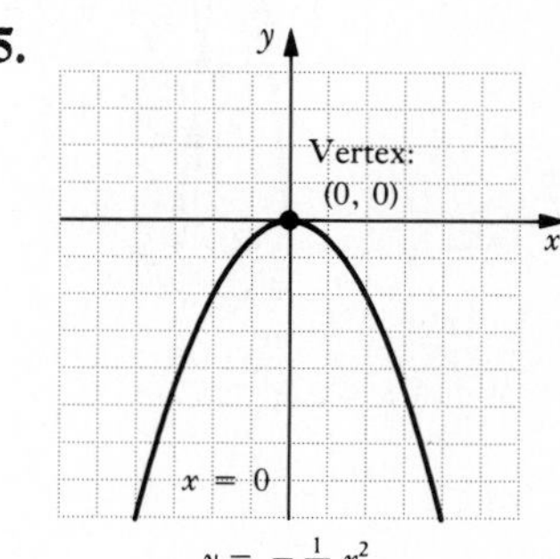

7.

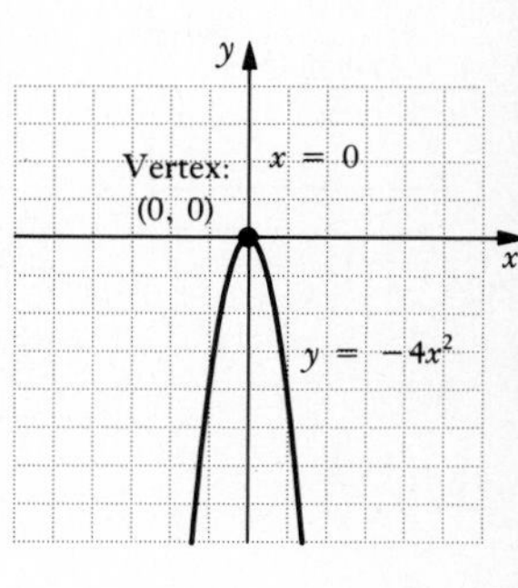

9.

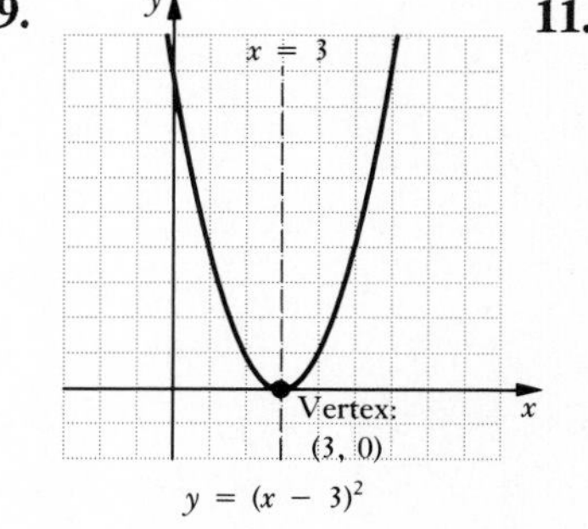

11.

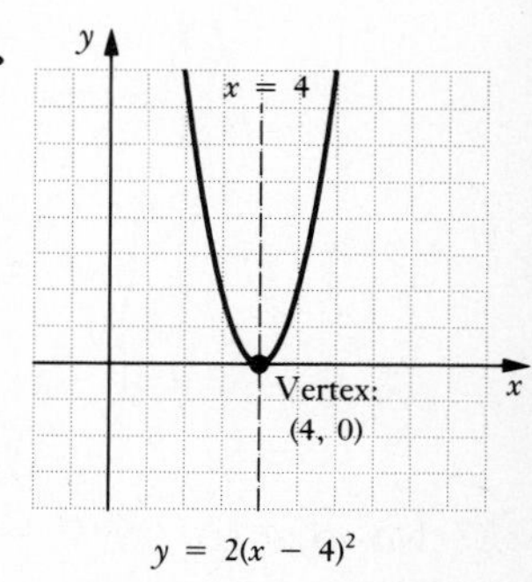

13.

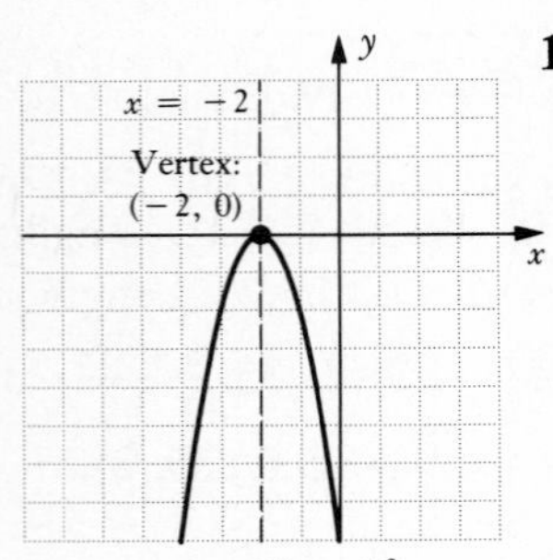

15.

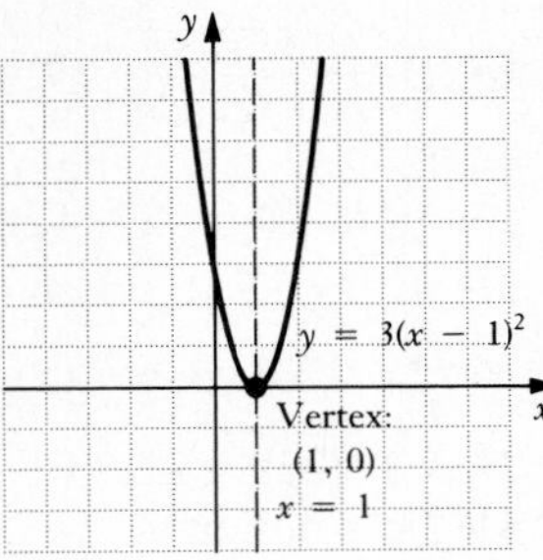

17.

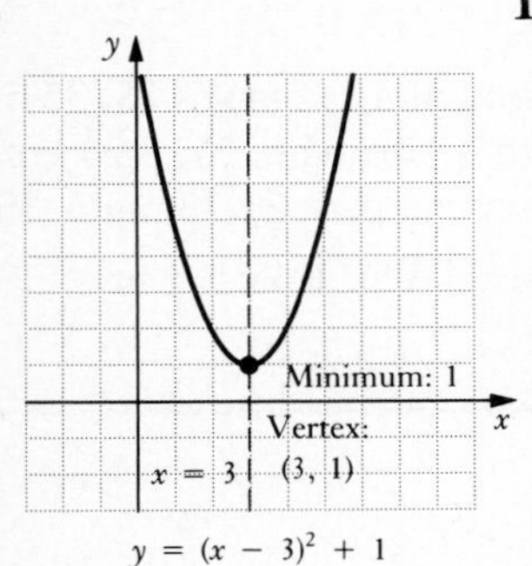

19.

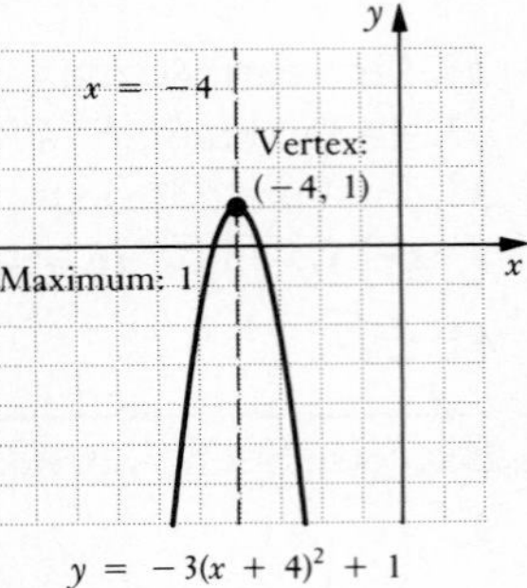

21.

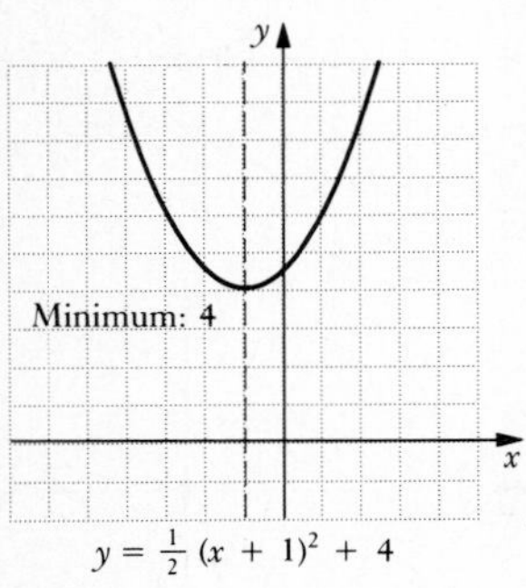

23.

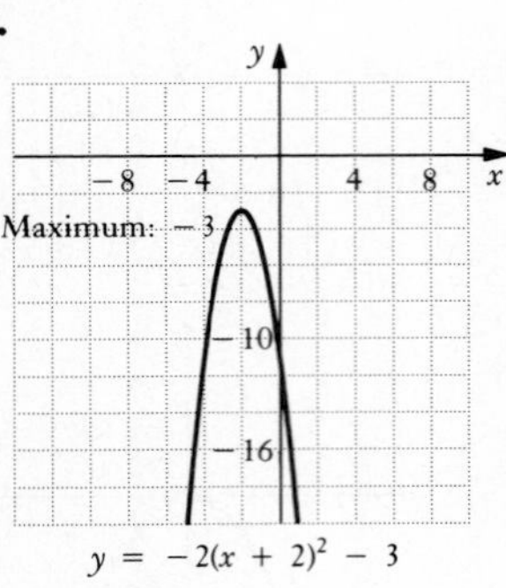

25.

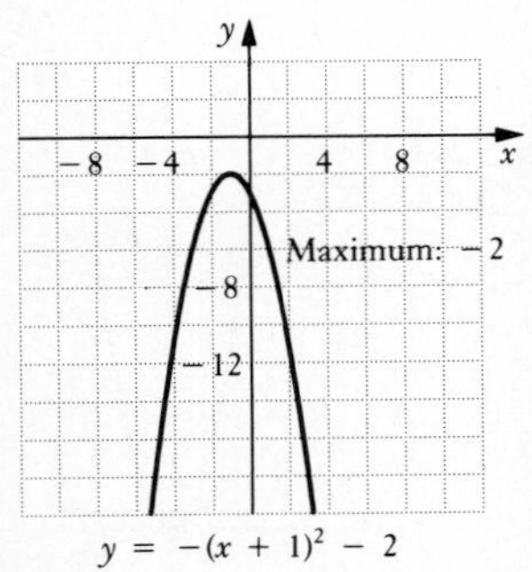

27. No solution **29.** 9

Exercise Set 8.8, p. 493

1.

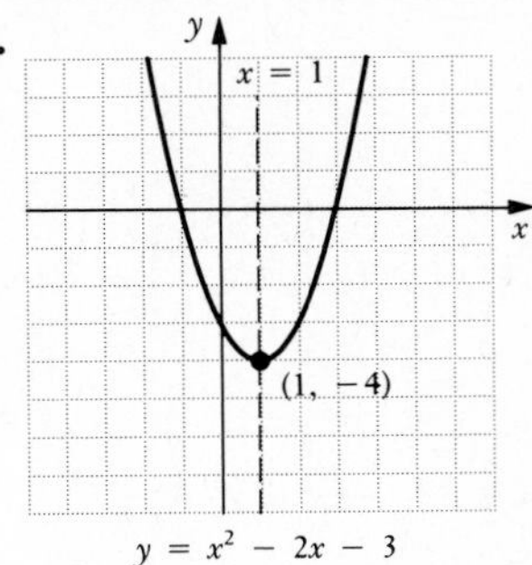

3.

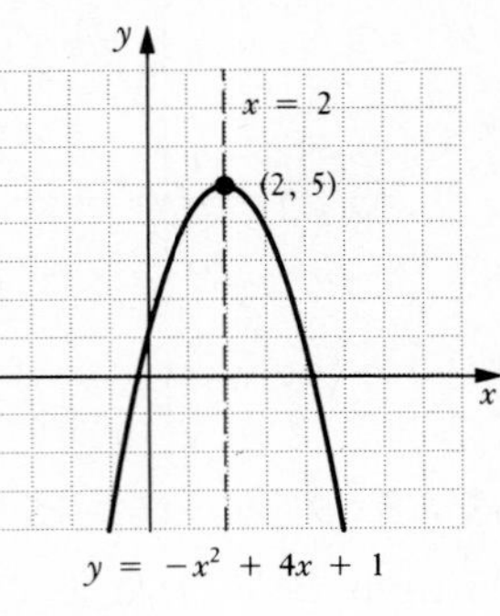

5.

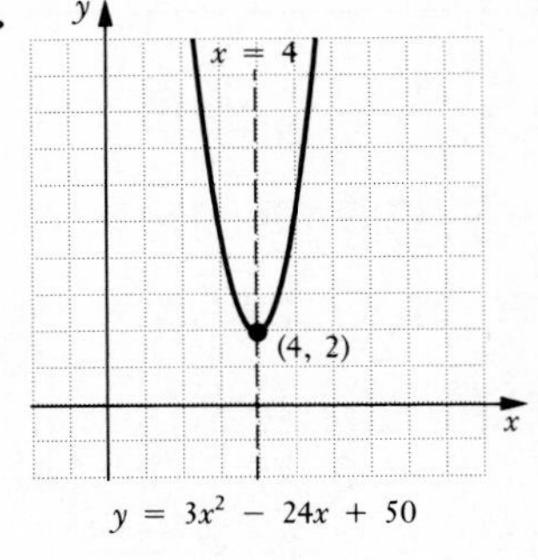

7.

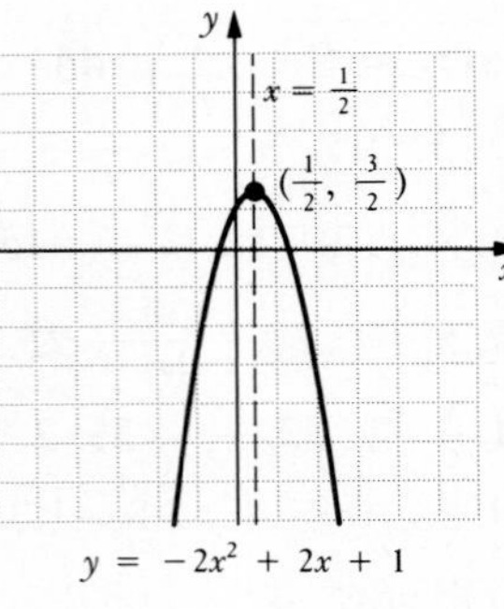

9.

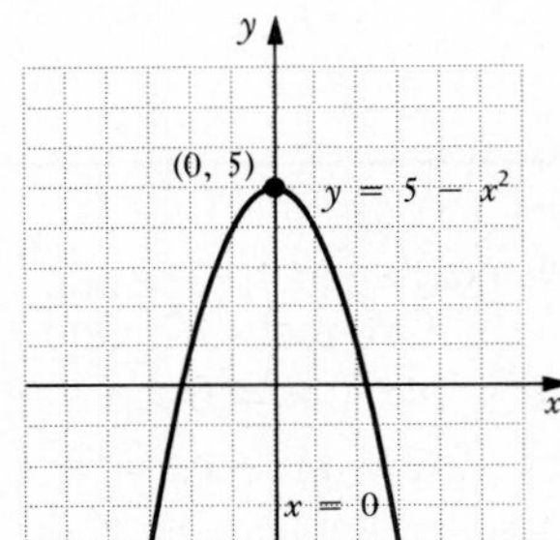

11. None **13.** $(-1,0)$, $(4,0)$ **15.** $\left(\frac{3-\sqrt{6}}{3},0\right),\left(\frac{3+\sqrt{6}}{3},0\right)$ **17.** $\left(\frac{-2-\sqrt{6}}{2},0\right),\left(\frac{-2+\sqrt{6}}{2},0\right)$ **19.** 17 ft by 17 ft; 289 ft^2 **21.** 506.25; 22.5 and 22.5 **23.** –9; 3 and –3 **25.** $-\frac{49}{4}$; $\frac{7}{2}$ and $-\frac{7}{2}$ **27. (a)** 2304 ft; 2 sec; **(b)** 14 sec **29.** 490 **31.** $y=3x^2-x+2$ **33.** $y=x^2-5x$ **35. (a)** $y=2500x^2-6500x+5000$; **(b)** \$19,000 **37. (a)** $y=0.0078125x^2+0.15625x+1.25$; **(b)** \$4.97 **39.** $12a^2b^2$ **41.** 4 **43.**

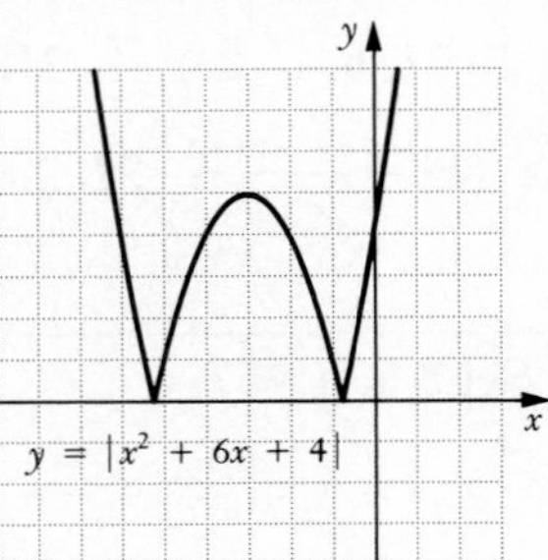

45. $11\sqrt{2}$ ft

Exercise Set 8.9, p. 503

1. $\{x \mid x < -3 \text{ or } x > 5\}$ **3.** $\{x \mid -1 \le x \le 2\}$ **5.** $\{x \mid -1 < x < 2\}$ **7.** $\{x \mid x \le -3 \text{ or } x \ge 3\}$ **9.** All real numbers **11.** $\{x \mid 2 < x < 4\}$ **13.** $\{x \mid x < -2 \text{ or } 0 < x < 2\}$ **15.** $\{x \mid -3 < x < -1 \text{ or } x > 2\}$ **17.** $\{x \mid x < -3 \text{ or } -2 < x < 1\}$ **19.** $\{x \mid x < 4\}$ **21.** $\{x \mid x < -1 \text{ or } x > 3\}$ **23.** $\left\{x \,\middle|\, -\frac{2}{3} \le x < 3\right\}$ **25.** $\left\{x \,\middle|\, 2 < x < \frac{5}{2}\right\}$ **27.** $\{x \mid x < -1 \text{ or } 2 < x < 5\}$ **29.** $\{x \mid x \le 0 \text{ or } x > 2\}$ **31.** $\{x \mid x > 0\}$ **33.** $\{x \mid x < -4 \text{ or } 1 < x < 3\}$ **35.** $\left\{x \,\middle|\, 0 < x < \frac{1}{2}\right\}$ **37.** $\{x \mid x < -3 \text{ or } -2 < x < 1 \text{ or } x > 4\}$ **39.** $\{x \mid -7 < x \le -5 \text{ or } 2 \le x < 8\}$ **41.** $\{x \mid 1 - \sqrt{3} \le x \le 1 + \sqrt{3}\}$ **43.** All real numbers except 0 **45.** $\left\{x \,\middle|\, x < \frac{1}{4} \text{ or } x > \frac{5}{2}\right\}$ **47. (a)** $\{t \mid 0 < t < 2\}$; **(b)** $\{t \mid t > 10\}$

Summary and Reveiew: Chapter 8, p. 505

1. [8.1a] $\pm\frac{\sqrt{14}}{2}$ **2.** [8.2a] $0, -\frac{5}{14}$ **3.** [8.2a] 3, 9 **4.** [8.2a] $\frac{-3 \pm i\sqrt{7}}{8}$ **5.** [8.2a] $\frac{7 \pm i\sqrt{3}}{2}$ **6.** [8.2a] 3, 5 **7.** [8.2a, b] $-2 \pm \sqrt{3}$; $-0.3, -3.7$ **8.** [8.5b] 9 cm, 14 cm **9.** [8.5b] 1 in. **10.** [8.1c] 30% **11.** [8.6a] 4, −2 **12.** [8.6a] −5, 3 **13.** [8.6a] $4 \pm 4\sqrt{2}$ **14.** [8.6a] $\frac{1 \pm \sqrt{481}}{15}$ **15.** [8.6b] First part: 50 mph; second part: 40 mph **16.** [8.6b] 6 hr **17.** [8.3a] Two real **18.** [8.3a] Two nonreal **19.** [8.3b] $25x^2 + 10x - 3 = 0$ **20.** [8.3b] $x^2 + 8x + 16 = 0$ **21.** [8.5a] $p = \frac{9\pi^2}{N^2}$ **22.** [8.5a] $T = \sqrt{\frac{3B}{2A}}$ **23.** [8.4a] 2, −2, 3, −3 **24.** [8.4a] 3, −5 **25.** [8.4a] $\pm\sqrt{7}, \pm\sqrt{2}$

26. [8.7a] **(a)** $(-2, 4)$; **(b)** $x = -2$; **(c)**

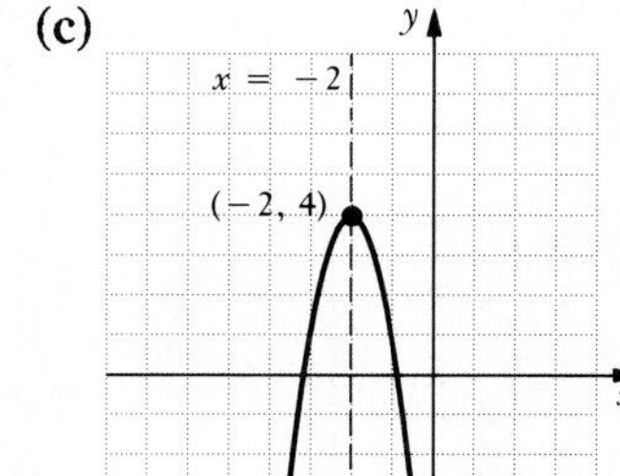

$y = -3x^2 - 12x - 8$

27. [8.7a] **(a)** $\left(\frac{1}{2}, \frac{23}{4}\right)$; **(b)** $x = \frac{1}{2}$; **(c)**

$y = x^2 - x + 6$

28. [8.8b] (7, 0), (2, 0) **29.** [8.8c] −121; 11 and −11 **30.** [8.8d] $y = -x^2 + 6x - 2$ **31.** [8.9a] $\{x \mid -1 < x < 7\}$ **32.** [8.9b] $\{x \mid -3 < x < 5\}$ **33.** [4.3a] 210 kg of *A*; 90 kg of *B* **34.** [7.7b] 2 **35.** [6.2b] $\frac{x+1}{(x-3)(x-1)}$ **36.** [7.2b] $3t^5 s \sqrt[3]{s}$ **37.** [8.6a] All real numbers except −13 and −7 **38.** [8.3b] $h = 60, k = 60$ **39.** [4.3a] 18 and 324

Test: Chapter 8, p. 507

1. [8.1a] $\pm\frac{2\sqrt{3}}{3}$ **2.** [8.2a] 9, 2 **3.** [8.2a] $\frac{-1 \pm i\sqrt{3}}{2}$ **4.** [8.2a, b] $-2 \pm \sqrt{6}$; −4.4, 0.4 **5.** [8.6a] 0, 2 **6.** [8.4a] $\pm\sqrt{\frac{5+\sqrt{5}}{2}}$, $\pm\sqrt{\frac{5-\sqrt{5}}{2}}$ **7.** [8.5b] Length: 8.5 in., width: 4 in. **8.** [8.6b] 2 hr **9.** [8.3a] Two nonreal **10.** [8.3b] $x^2 - 4\sqrt{3}x + 9 = 0$ **11.** [8.5a] $r = \sqrt{\frac{3V}{\pi} - R^2}$ **12.** [8.1c] 20%

13. [8.7a] **(a)** $(-1, 1)$ **(b)** $x = -1$ **(c)**

$y = -x^2 - 2x$

14. [8.7a] **(a)** $(3, 5)$ **(b)** $x = 3$ **(c)**

$y = 4(x - 3)^2 + 5$

15. [8.8b] $(2 + \sqrt{3}, 0), (2 - \sqrt{3}, 0)$ **16.** [8.8b] (3, 0) and (−2, 0) **17.** [8.8c] −16; 4 and −4 **18.** [8.8d] $y = \frac{1}{5}x^2 - \frac{3}{5}x$ **19.** [8.9a] $\{x \mid -2 < x < 1 \text{ or } x > 2\}$ **20.** [8.9b] $\{x \mid x < -4 \text{ or } -2 < x < 1\}$ **21.** [7.7a] 6 **22.** [7.2b] $ab\sqrt[4]{2a^2}$ **23.** [6.2b] $\frac{x-8}{(x+6)(x+8)}$ **24.** [4.3a] 36 **25.** [8.3b] $\frac{1}{2}$ **26.** [8.6a] All real numbers except 7 and 11

Cumulative Review: Chapters 1-8, p. 509

1. [1.3e] 0.7 **2.** [1.4e] $5x + 23$ **3.** [1.6b] $-\frac{b^9}{64a^6}$ **4.** [1.6c] 4.0×10^{-16} **5.** [5.1f] $10x^2 - 8x + 6$ **6.** [5.2a] $2x^3 - 9x^2 + 7x - 12$ **7.** [6.1d] $\frac{2a-8}{5}$ **8.** [6.1e] $\frac{1}{y^2+6y}$ **9.** [6.2c] $\frac{-m^2+5m-6}{(m+1)(m-5)}$ **10.** [6.6a] $\frac{y-x}{xy(x+y)}$ **11.** [5.9b] $9x^2 - 13x + 26 + \frac{-50}{x+2}$ **12.** [7.1b] 0.6 **13.** [7.1b] $3(x-2)$ **14.** [7.4a] $12\sqrt{5}$ **15.** [7.5c] $\frac{2\sqrt{6}+18\sqrt{2}-24\sqrt{3}-8}{-52}$ **16.** [7.6c] $2^8 = 256$ **17.** [7.9c] $17 + 7i$ **18.** [7.9e] $-\frac{2}{3} - 2i$ **19.** [2.1d] $\frac{11}{3}$ **20.** [2.3a] $r = \frac{mv^2}{F}$ **21.** [2.4c] $\left\{x \,\middle|\, x \ge \frac{5}{14}\right\}$ **22.** [2.5b] $\left\{x \,\middle|\, x < -\frac{4}{3} \text{ or } x > 6\right\}$ **23.** [2.6e] $\left\{x \,\middle|\, -\frac{13}{4} \le x \le \frac{15}{4}\right\}$ **24.** [4.2b] $(-4, 1)$ **25.** [4.4b] $\left(\frac{1}{2}, 3, -5\right)$ **26.** [5.8a] $\frac{1}{5}, -3$ **27.** [6.3a] $-\frac{5}{3}$ **28.** [6.3a] 3 **29.** [6.5a] $m = \frac{aA}{h-A}$ **30.** [7.7a] $\frac{37}{2}$ **31.** [7.7b] 11 **32.** [5.8a] 4 **33.** [8.2a] $\frac{3 \pm i\sqrt{55}}{2}$ **34.** [8.6a] $\frac{17 \pm \sqrt{145}}{2}$ **35.** [8.5a]

$a = \sqrt{P^2 + b^2}$ **36.** [5.4b] $(2t+5)(t-6)$ **37.** [5.4a] $(a+9)(a-6)$ **38.** [5.3b] $(6a^2-5)(4a+3)$ **39.** [5.3a] $-3a^2(a-4)$ **40.** [5.5b] $(8a+3b)(8a-3b)$ **41.** [5.5b] $3(a-6)^2$ **42.** [5.6a] $\left(\frac{1}{3}a-1\right)\left(\frac{1}{9}a^2+\frac{1}{3}a+1\right)$ **43.** [5.3a] $(x+1)(2x+1)$

44. [3.1c]

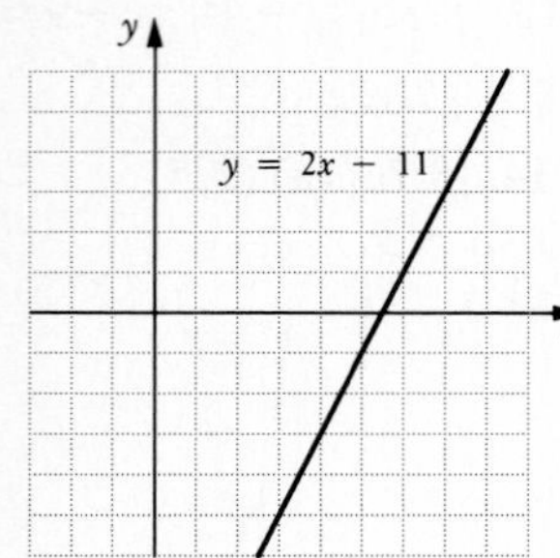

45. [3.2a]

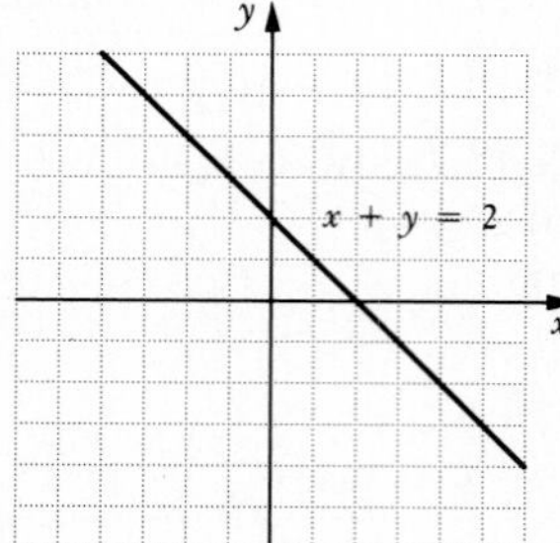

46. [3.6b]

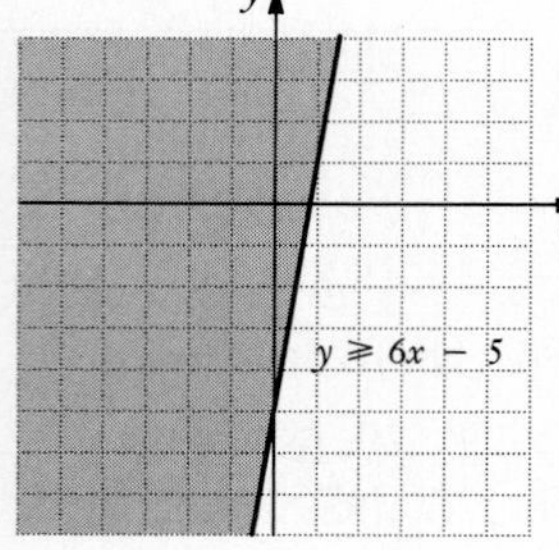

47. [3.6b]

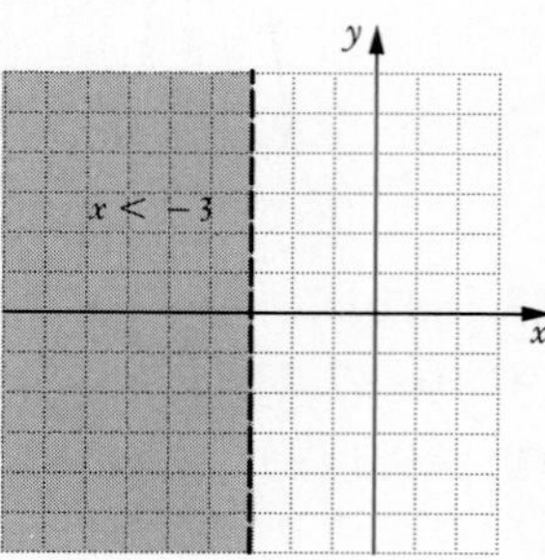

48. [4.7a]

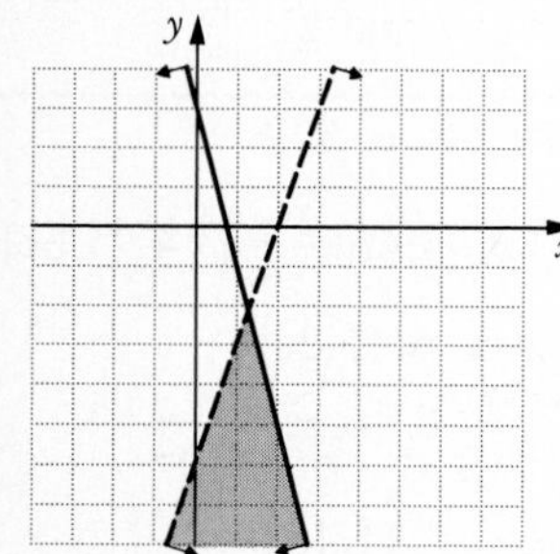

49. [8.8a]

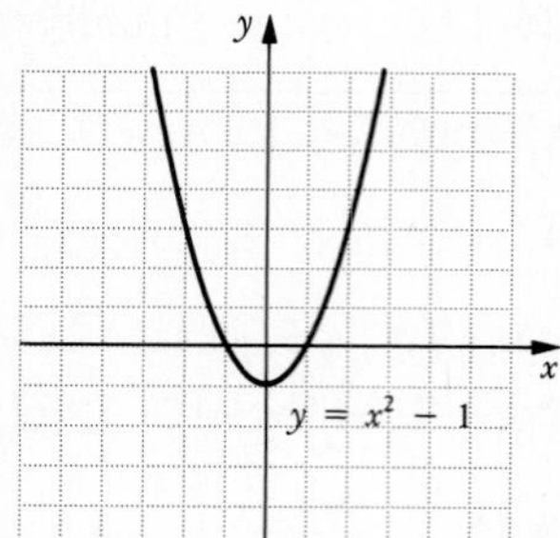

50. [8.8a]

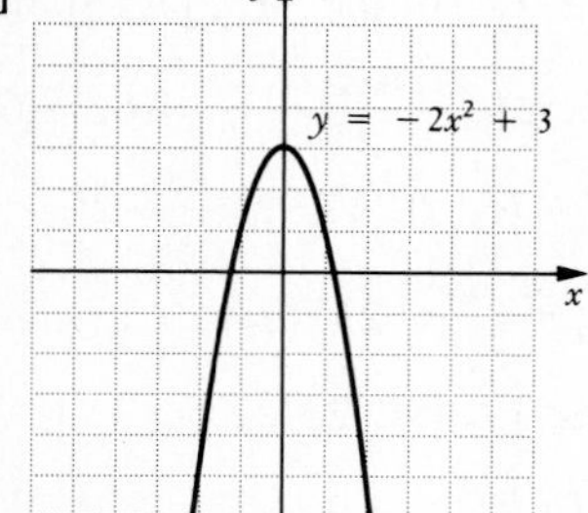

51. [8.9a] $\left\{x \,\middle|\, x < -\frac{5}{2} \text{ or } x > \frac{5}{2}\right\}$ **52.** [3.4a] $y = \frac{1}{2}x + 4$ **53.** [3.4c] $y = -3x + 1$ **54.** [4.3a] Asia: 35,015; Eastern Europe: 7818; USSR: 20,421 **55.** [4.3a] Japan: 45 cents; Norway: 42 cents **56.** [5.8a] $0, \frac{b}{a}$ **57.** [1.6a] $\frac{328}{3}$ **58.** [7.7a] $\frac{2}{51 \pm 7\sqrt{61}}$, *or* $\frac{-51 \pm 7\sqrt{61}}{194}$ **59.** [5.6a] $\left(\frac{a}{2}+\frac{2b}{9}\right)\left(\frac{a^2}{4}-\frac{ab}{9}+\frac{4b^2}{81}\right)$ **60.** [6.4a] **(a)** $t = \frac{ab}{a+b}$; **(b)** $a = \frac{bt}{b-t}$; **(c)** $b = \frac{at}{a-t}$

CHAPTER 9

Pretest: Chapter 9, p. 512

1. [9.1b] $\sqrt{89}$; 9.434 **2. [9.1c]** $\left(-2, -\frac{1}{2}\right)$ **3. [9.1d]** Center: $(2, -7)$; radius: 4 **4. [9.1e]** Center: $(4, -3)$; radius: 6 **5. [9.1d]** $(x-2)^2 + (y+9)^2 = 9$ **6. [9.4b]** 4

7. [9.2b]

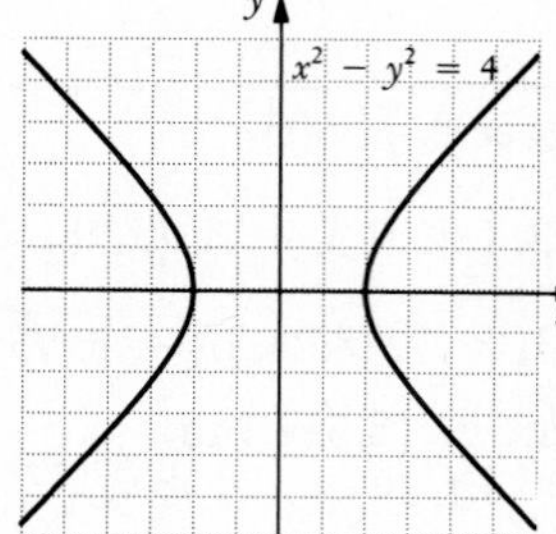

8. [9.1d]

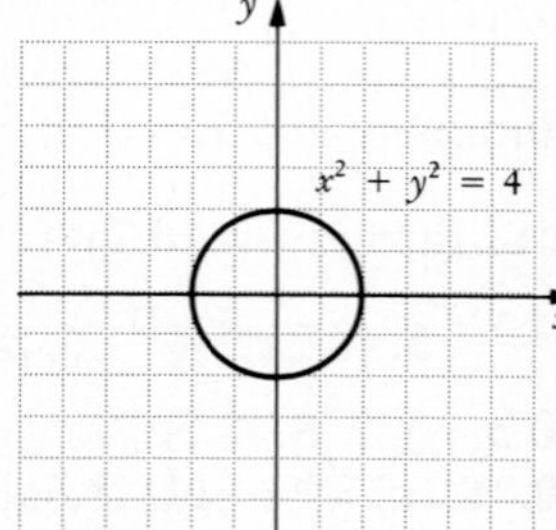

9. [9.2b]

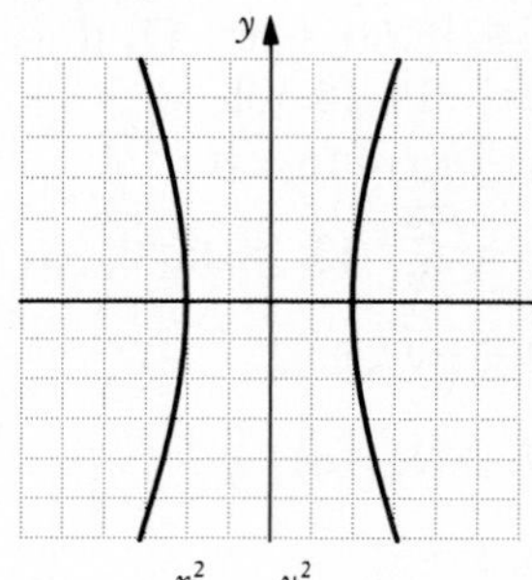

10. [9.2a]

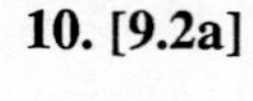

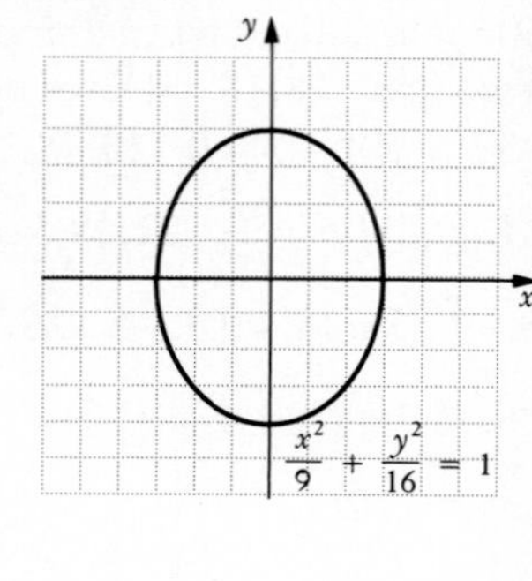

11. [9.1a]

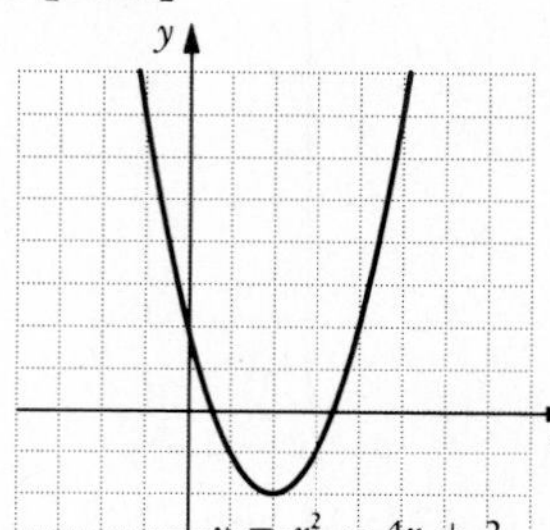

12. [9.1a]

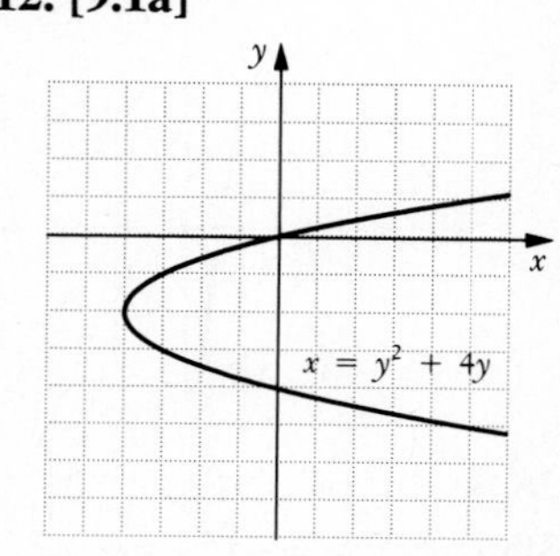

13. [9.3a] $(-3, 4)$ **14. [9.3a]** $(4, 5), (-4, -5)$ $\left(-\frac{5i\sqrt{2}}{2}, 4i\sqrt{2}\right)$, $\left(\frac{5i\sqrt{2}}{2}, -4i\sqrt{2}\right)$ **15. [9.3b]** 3 m by 4 m **16. [9.3b]** 60 ft; 48 ft

17. [9.4c]

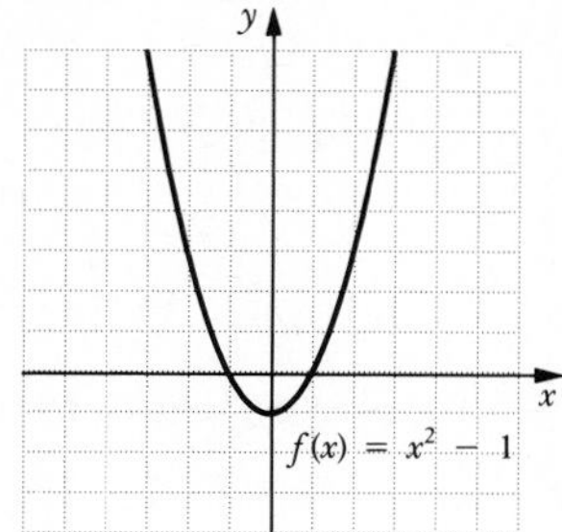

18. [9.5a,b] $(f+g)(x) = x^2 + x - 2$; $(f-g)(x) = x^2 - x + 8$; $fg(x) = x^3 - 5x^2 + 3x - 15$; $(f/g)(x) = \frac{x^2+3}{x-5}$; $ff(x) = x^4 + 6x^2 + 9$; $f \circ g(x) = x^2 - 10x + 28$; $g \circ f(x) = x^2 - 2$ **19.** $f^{-1}(x) = \frac{x+7}{2}$
20. [9.6c] $f^{-1}(x) = \frac{-x-4}{x-1}$

Exercise Set 9.1, p. 519

1.

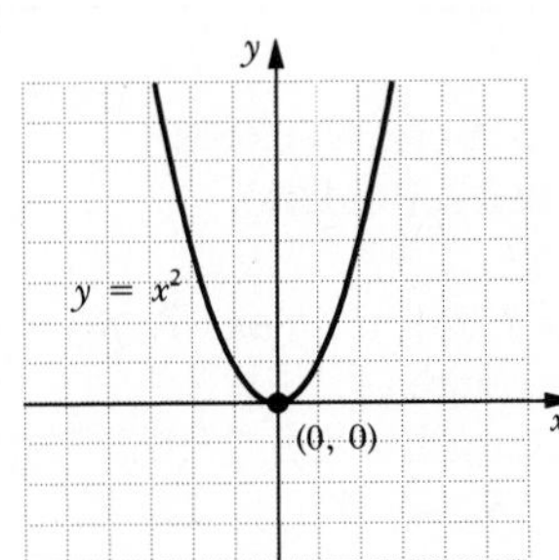

3.

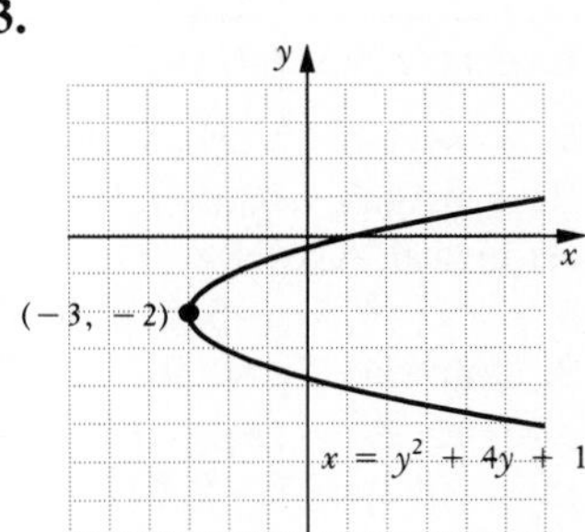

5.

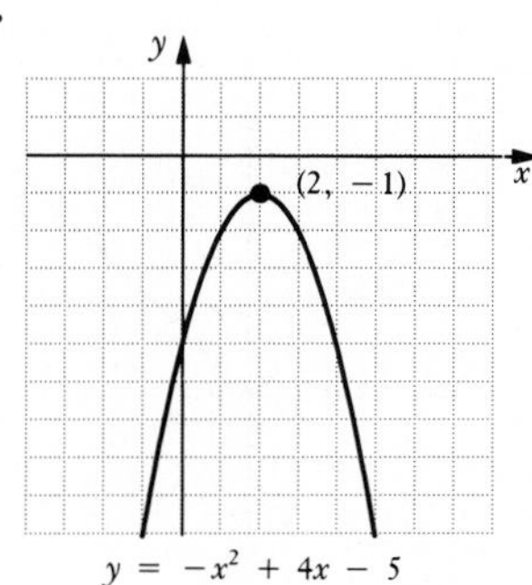

7.

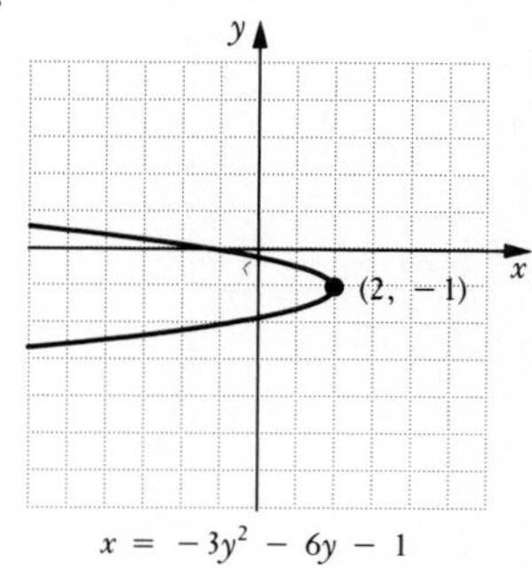

9. 5 **11.** $\sqrt{18} \approx 4.243$ **13.** $\sqrt{32} \approx 5.657$ **15.** 17.8 **17.** $\frac{\sqrt{41}}{7} \approx 0.915$ **19.** $\sqrt{6970} \approx 83.487$ **21.** $\sqrt{a^2+b^2}$ **23.** $\sqrt{17+2\sqrt{14}+2\sqrt{15}}$ **25.** $20\sqrt{24,181} \approx 3110.048$ **27.** $\left(-\frac{1}{2}, -1\right)$ **29.** $\left(\frac{7}{2}, \frac{7}{2}\right)$ **31.** $(-1, -3)$ **33.** $(-0.25, -0.3)$ **35.** $\left(-\frac{1}{12}, \frac{1}{24}\right)$ **37.** $\left(\frac{\sqrt{2}+\sqrt{3}}{2}, \frac{3}{2}\right)$

39.

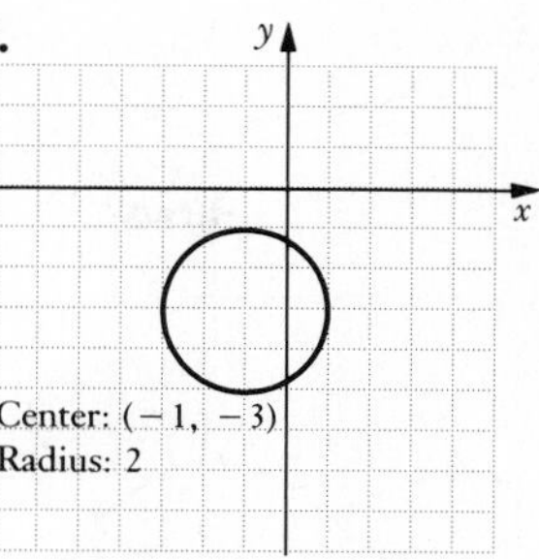

41.

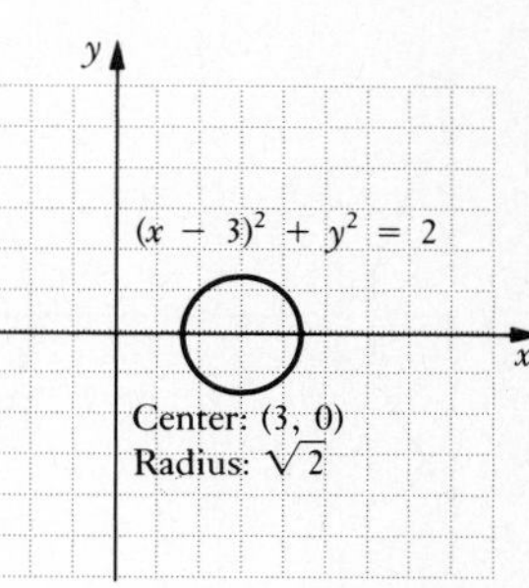

43.

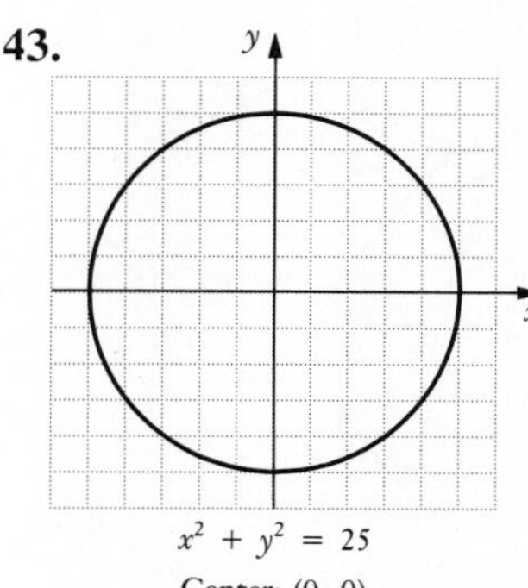

45. $x^2 + y^2 = 49$ **47.** $(x+2)^2 + (y-7)^2 = 5$ **49.** $(-4, 3)$, $r = 2\sqrt{10}$ **51.** $(4, -1)$, $r = 2$ **53.** $(2, 0)$ $r = 2$ **55.** $(1, 2)$ **57.** $x^2 + y^2 = 2$ **59.** $(x+3)^2 + (y+2)^2 = 9$ **61.** $\sqrt{49+k^2}$ **63.** $8\sqrt{m^2+n^2}$ **65.** $6\sqrt{2}$ **67.** Yes **69.** $(2, 4\sqrt{2})$

Exercise Set 9.2, p. 529

1.

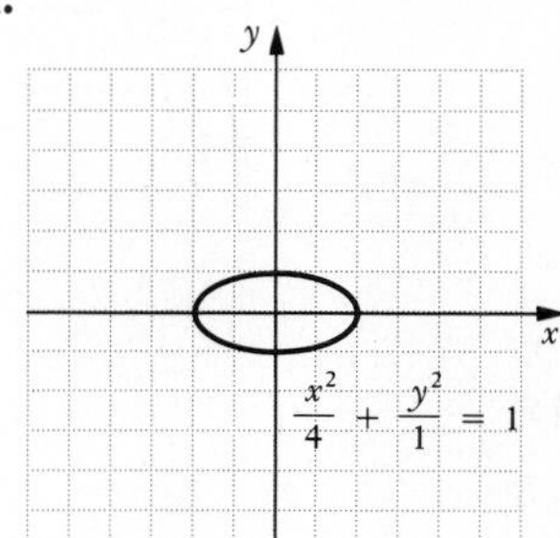

3.

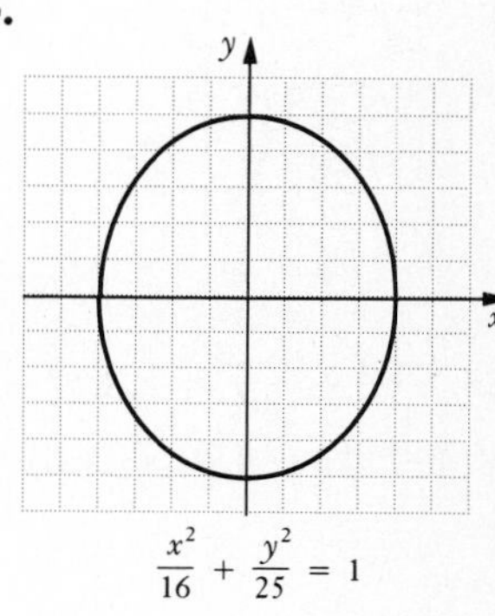

5.

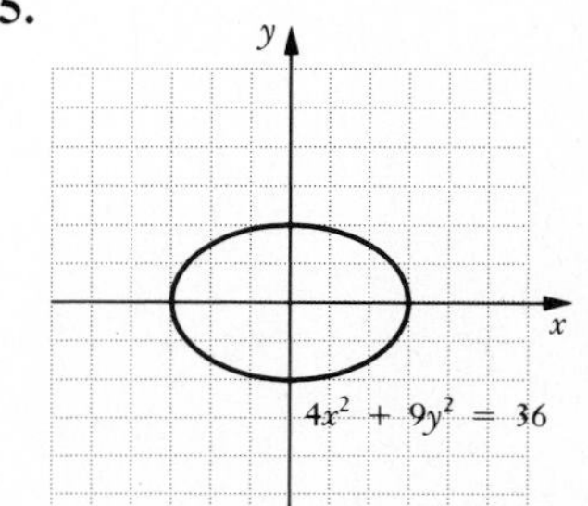

7.

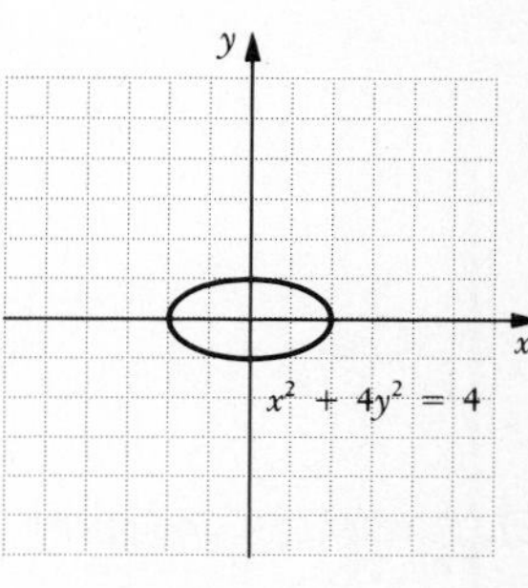

9.

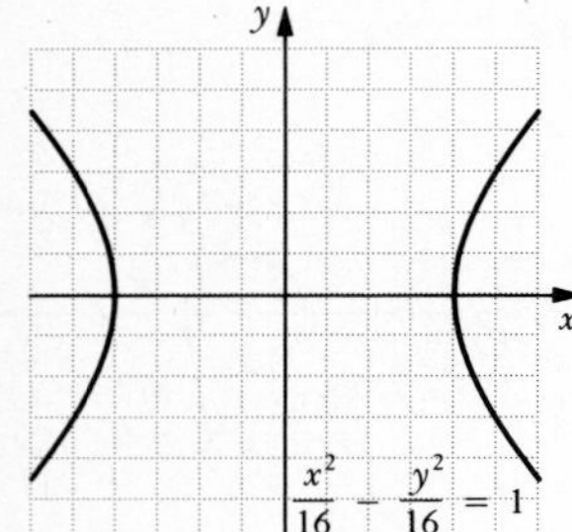

11.

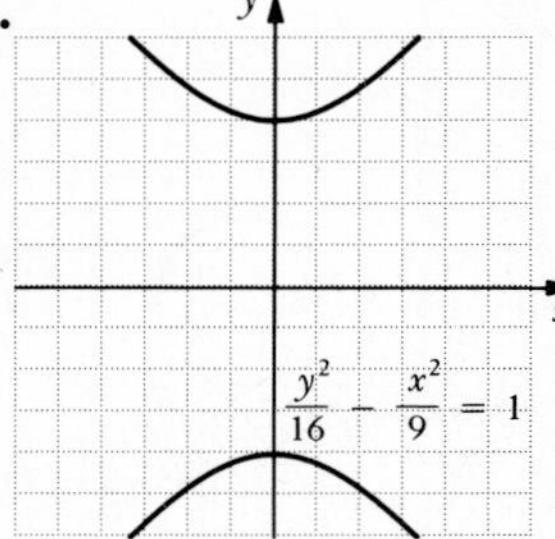

13.

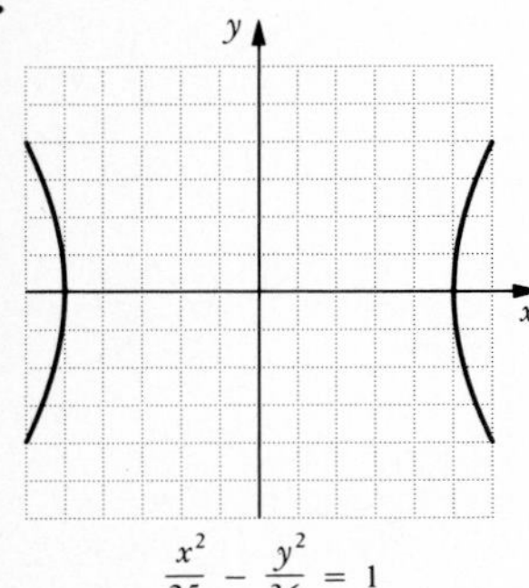

$\frac{x^2}{25} - \frac{y^2}{36} = 1$

15.

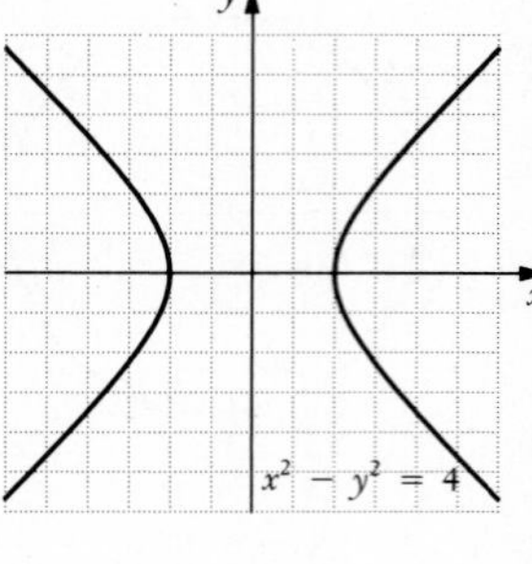

17. $5t^5$ **19.** $\frac{13+8\sqrt{6}}{10}$

21.

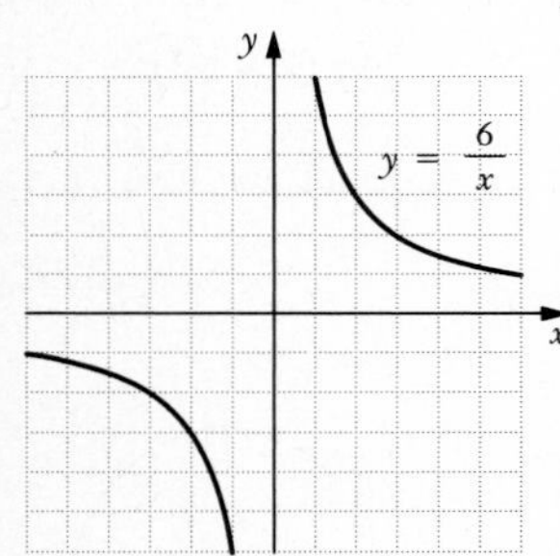

23.

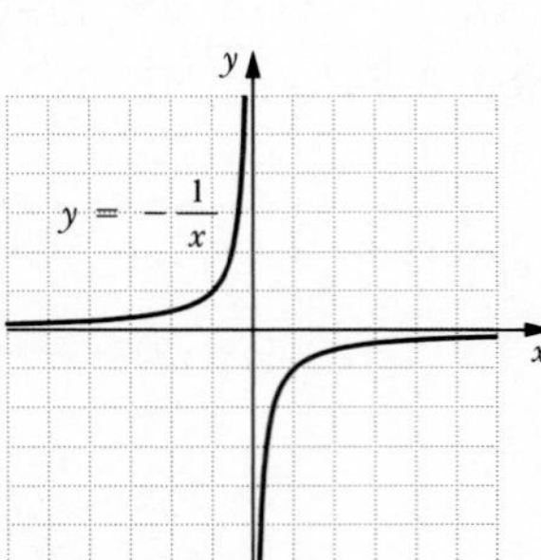

25. $\frac{(x+3)^2}{1} + \frac{(y-4)^2}{16} = 1$; C: $(-3, 4)$;
V: $(-2,4), (-4,4), (-3,8), (-3,0)$

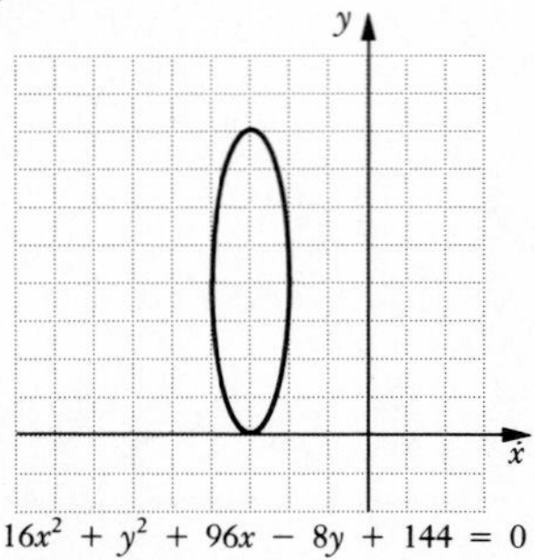

27. Circle **29.** Parabola **31.** Ellipse **33.** $y = 2x^2 - x + 3$

Exercise Set 9.3, p. 537

1. $(-4,-3), (3,4)$ **3.** $(0,2), (3,0)$ **5.** $(-2,1)$
7. $\left(\frac{5+\sqrt{70}}{3}, \frac{-1+\sqrt{70}}{3}\right), \left(\frac{5-\sqrt{70}}{3}, \frac{-1-\sqrt{70}}{3}\right)$ **9.** $\left(\frac{7}{3}, \frac{1}{3}\right), (1,-1)$
11. $\left(-3, \frac{5}{2}\right), (3,1)$ **13.** $(3,-5), (-1,3)$ **15.** $\left(\frac{8+3i\sqrt{6}}{2}, \frac{-8+3i\sqrt{6}}{2}\right)$, $\left(\frac{8-3i\sqrt{6}}{2}, \frac{-8-3i\sqrt{6}}{2}\right)$ **17.** $(-5,0), (4,3), (4,-3)$ **19.** $(3,0)$, $(-3,0)$ **21.** $(1,2)\ (-1,-2), (2,1), (-2,-1)$ **23.** $(2,3)$, $(-2,-3), (3,2), (-3,-2)$ **25.** $(2,1), (-2,-1)$
27. $\left(2,-\frac{4}{5}\right), (5,2), (-5,2)\ \left(-2,-\frac{4}{5}\right)$ **29.** $(\sqrt{2},-\sqrt{2})$, $(-\sqrt{2},\sqrt{2})$ **31.** $\left(\frac{8i\sqrt{5}}{5}, \frac{3\sqrt{105}}{5}\right), \left(-\frac{8i\sqrt{5}}{5}, \frac{3\sqrt{105}}{5}\right), \left(\frac{8i\sqrt{5}}{5}, -\frac{3\sqrt{105}}{5}\right)$, $\left(-\frac{8i\sqrt{5}}{5}, -\frac{3\sqrt{105}}{5}\right)$ **33.** Length: 8 cm; width: 6 cm
35. Length: 5 in.; width: 4 in. **37.** Length: 75 yd; width: 30 yd **39.** 13 and 12 **41.** 24 ft, 16ft **43.** Length: $\sqrt{3}$ m; width: 1 m **45.** $\frac{5 \pm i\sqrt{47}}{6}$ **47.** $4\sqrt{3}$ **49.** $\frac{x-h}{x+2\sqrt{xh}+h}$
51. 3.7 mph **53.** $(x+2)^2 + (y-1)^2 = 4$
55. $4x^2 + 3y^2 = 43$ **57.** 30, 50 **59.** $\left(\frac{1}{2}, \frac{1}{3}\right), \left(\frac{1}{3}, \frac{1}{2}\right)$

Exercise Set 9.4, p. 547

1. d = {(New York, Mets), (New York, Yankees), (Los Angeles, Rams), (Los Angeles, Raiders), (Houston, Astros)}
3. $f = \{(-4,5), (-5,5), (-6,5), (-7,2)\}$ **5.** Yes **7.** No
9. Domain of $c = \{9, -5, 2, 3\}$; range of $c = \{1, -2, -1, -9\}$ **11.** Domain of $t = \{20, 21, 22, 23\}$; range of $t = \{7, -5, 0\}$ **13.** 2 **15.** −7 **17.** 3 **19.** 12
21. 6 **23.** 0 **25.** 40 **27.** 2 **29.** $\frac{53}{33}$ atm, $\frac{21}{11}$ atm, $\frac{133}{33}$ atm
31. 13.6 million, 10.9 million, 8.2 million, 5.5 million, 2.8 million

33.

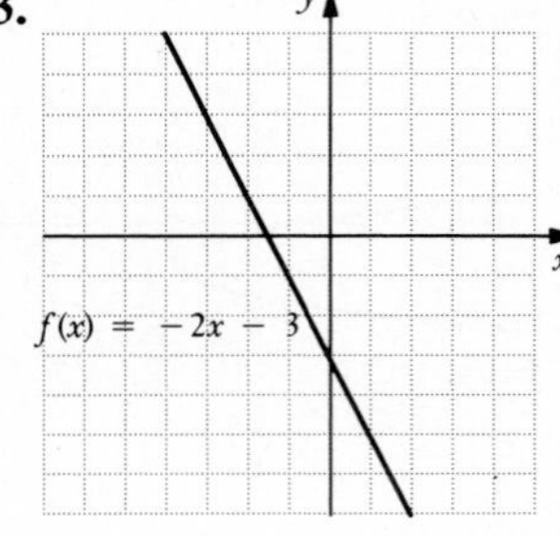

35.

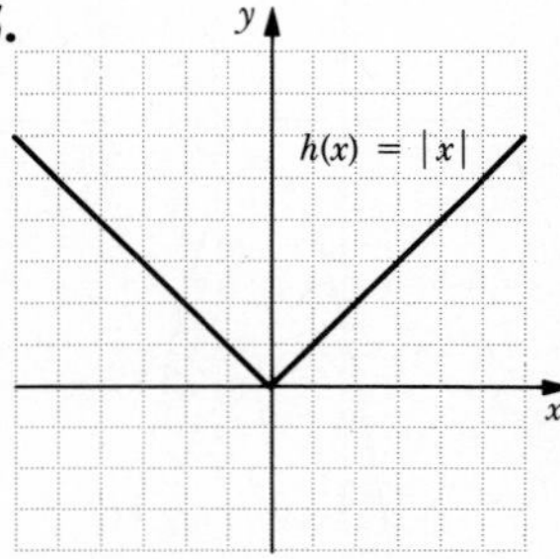

37.

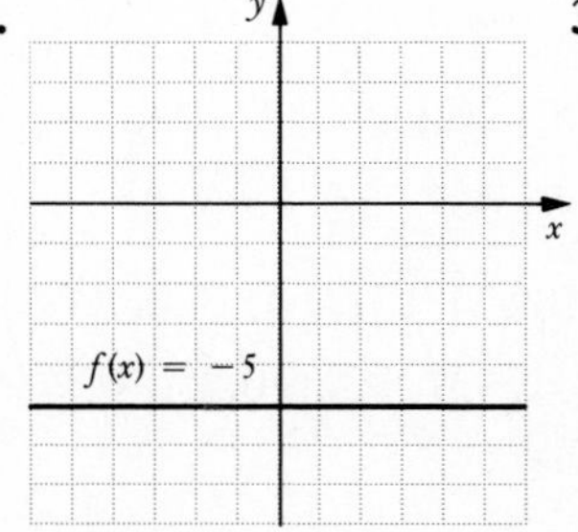

39.

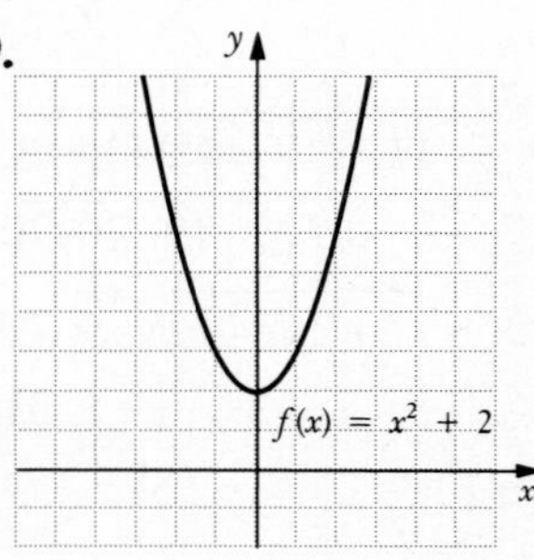

41.

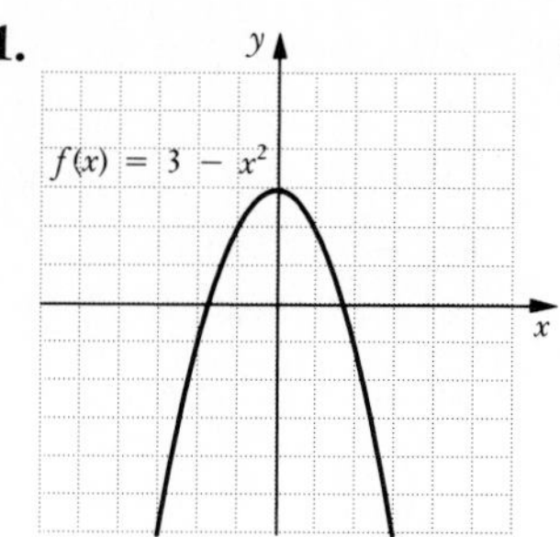

43.

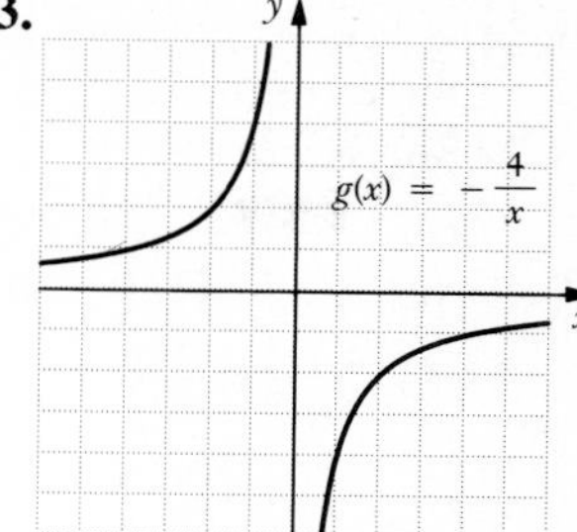

45.

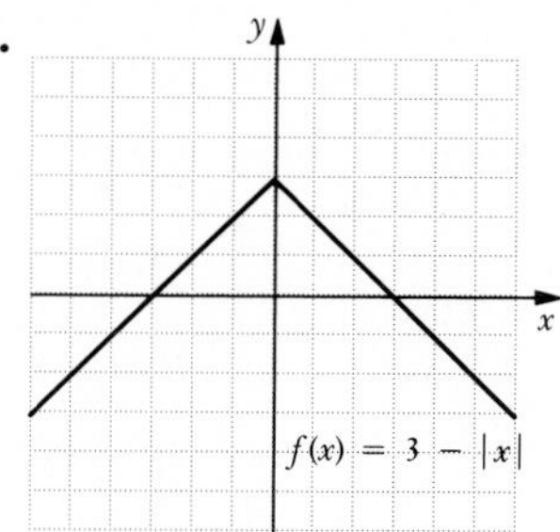

47. Yes **49.** No **51.** All real numbers **53.** $\{x \mid x \neq 5 \text{ and } x \neq -5\}$ **55.** $\{x \mid x \neq -3 \text{ and } x \neq -5\}$ **57.** $\left\{x \,\middle|\, x \geq -\frac{3}{4}\right\}$ **59.** $\left\{x \,\middle|\, x \neq \frac{5}{2}\right\}$ **61.** Range of $f = \{5, 8, 11, 14\}$

63.

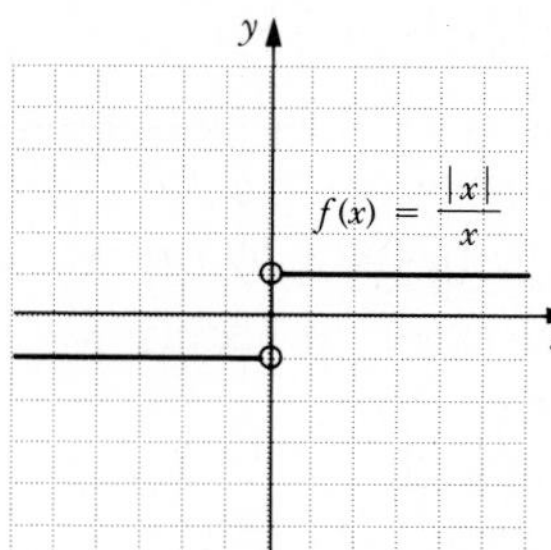

65.

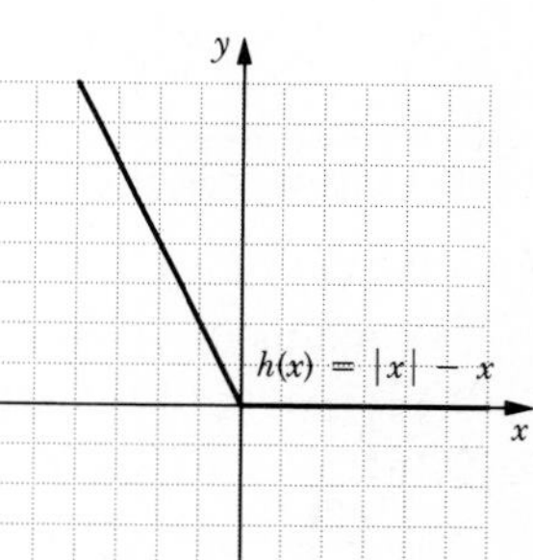

67.

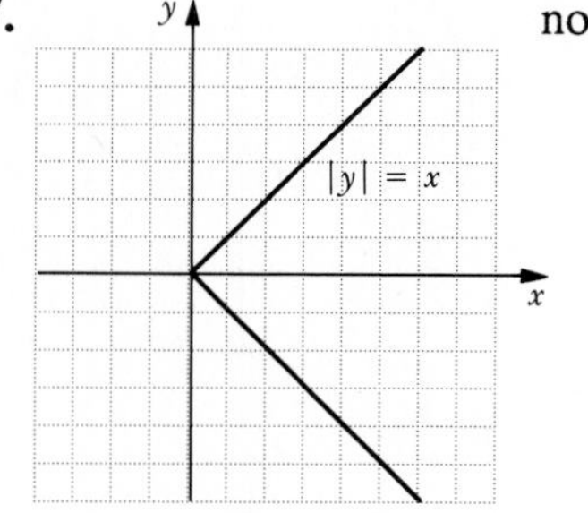

no

Exercise Set 9.5, p. 555

1. $2x$; 6; x^2-9; $\frac{x+3}{x-3}$; x^2+6x+9 **3.** x^3+2x^2-3x+1; $-x^3+2x^2-3x+1$; $2x^5-3x^4+x^3$; $\frac{2x^2-3x+1}{x^3}$; $4x^4-12x^3+13x^2-6x+1$ **5.** $4x^3-5x^2$; $-4x^3-5x^2$; $-20x^5$; $\frac{-5}{4x}$; $25x^4$ **7.** 15; 25; −100; −4; 400 **9.** $-10x+27$; $-10x+23$ **11.** $12x^2-12x+5$; $6x^2+3$ **13.** $\frac{16}{x^2}-1$; $\frac{2}{4x^2-1}$ **15.** x^4-2x^2+2; x^4+2x^2 **17.** $f(x)=x^2$; $g(x)=5-3x$ **19.** $f(x)=x^5$; $g(x)=3x^2-7$ **21.** $f(x)=\frac{1}{x}$; $g(x)=x-1$ **23.** $f(x)=\frac{1}{\sqrt{x}}$; $g(x)=7x+2$ **25.** $f(x)=\frac{x+1}{x-1}$; $g(x)=x^3$ **27.** $3x^2y^4$ **29.** $\frac{7\pm\sqrt{17}}{4}$ **31.** 3 **33.** 0 **35.** $2x+h$ **37.** $\frac{-1}{x(x+h)}$ **39.** $\frac{1}{\sqrt{x+h}+\sqrt{x}}$ **41.** \$300; \$500

Sidelight: Computer-Calculator Application: Monthly Mortgage Payments, p. 564

1. \$1228.60 **2.** \$1493.00 **3.** \$713.91 **4.** No **5.** Yes

Exercise Set 9.6, p. 565

1.

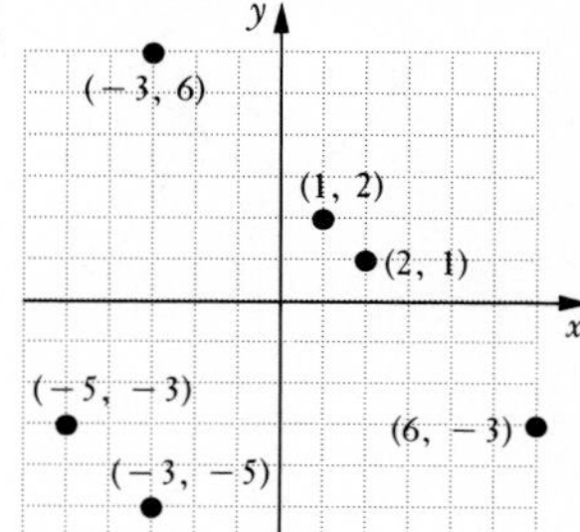

3.

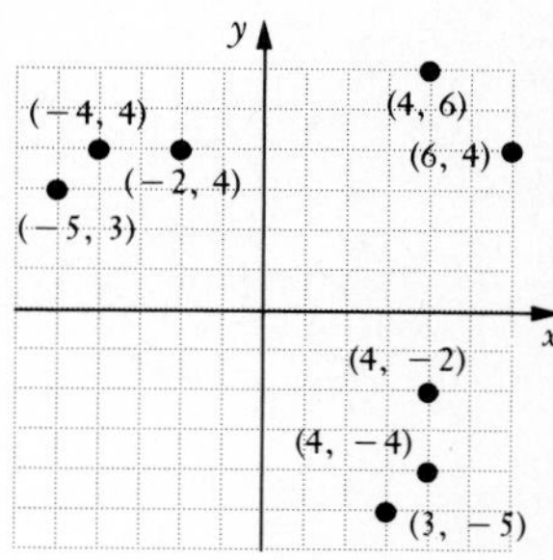

5. $f^{-1}(x)=\frac{x+4}{3}$ **7.** Not one-to-one **9.** Not one-to-one **11.** Not one-to-one **13.** $g^{-1}(x)=\frac{-2}{x}$ **15.** $f^{-1}(x)=x-2$ **17.** $f^{-1}(x)=4-x$ **19.** $g^{-1}(x)=x+5$ **21.** $f^{-1}(x)=\frac{1}{3}x$ **23.** $g^{-1}(x)=\frac{x-2}{3}$ **25.** $h^{-1}(x)=\frac{2}{x}-5$ **27.** $f^{-1}(x)=\frac{1-3x}{5x-2}$ **29.** $g^{-1}(x)=\frac{4x+3}{1-x}$ **31.** $f^{-1}(x)=\sqrt[3]{x+1}$ **33.** $G^{-1}(x)=\sqrt[3]{x}+2$ **35.** $f^{-1}(x)=x^3$

37.

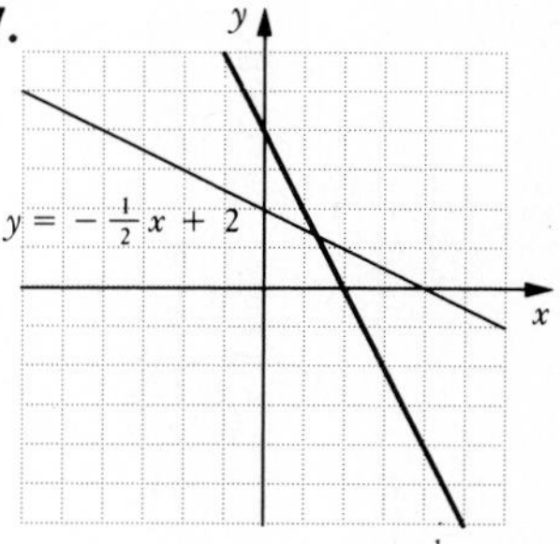

39.

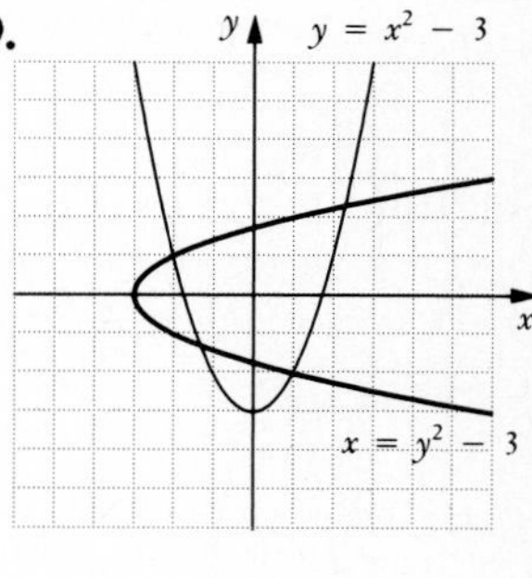

41.

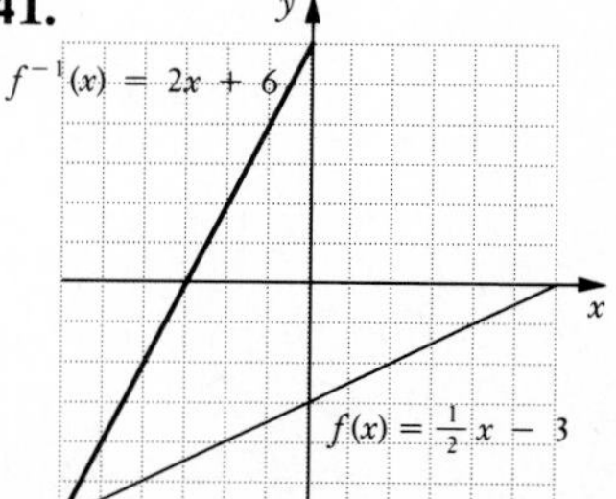

43.

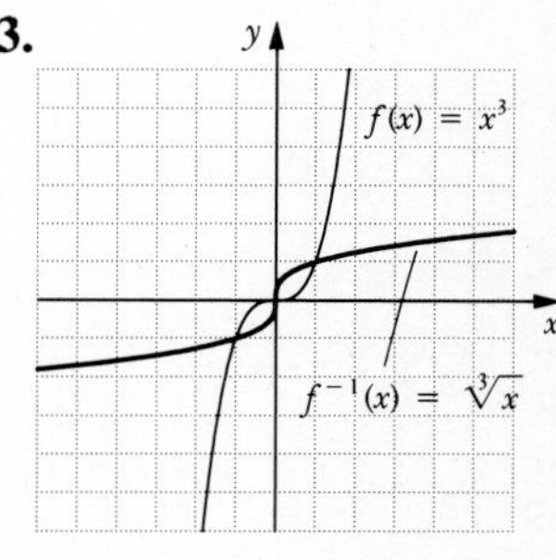

45. (a) 40, 42, 46, 50; (b) Yes, $f^{-1}(x) = x - 32$; (c) 8, 10, 14, 18 **47.** No; it is not one-to one; it does not pass the horizontal line test. **49.** $C^{-1}(x) = \frac{100}{x-5}$; the inverse function gives the number of people in a group where x is the cost per person of chartering a bus. **51.** x; x **53.** x; x **55.** Yes **57.** No

Summary and Review: Chapter 9, p. 569

1. [9.1b] $\sqrt{97} \approx 9.849$ **2.** [9.1c] $\left(-4, -\frac{1}{2}\right)$ **3.** [9.1d] Center: $(-2, 3)$; radius: $\sqrt{2}$ **4.** [9.1d] Center: $(5, 0)$; radius: 7 **5.** [9.1e] Center: $(-4, 3)$; radius: $\sqrt{35}$ **6.** [9.1d] $(x+4)^2 + (y-3)^2 = 48$ **7.** [9.1d] $(x-7)^2 + (y+2)^2 = 20$

8. [9.2a]

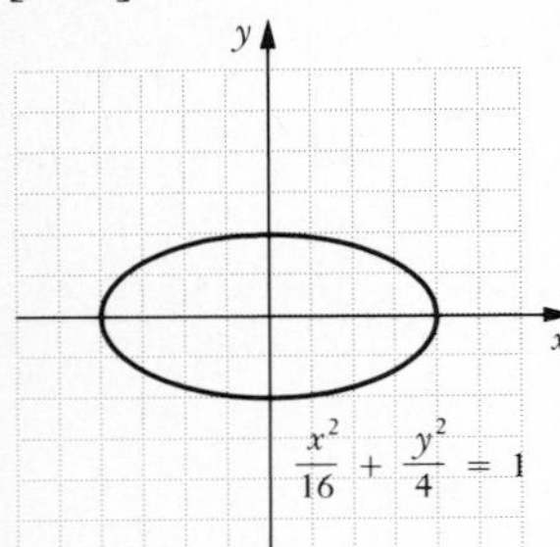

9. [9.2b]

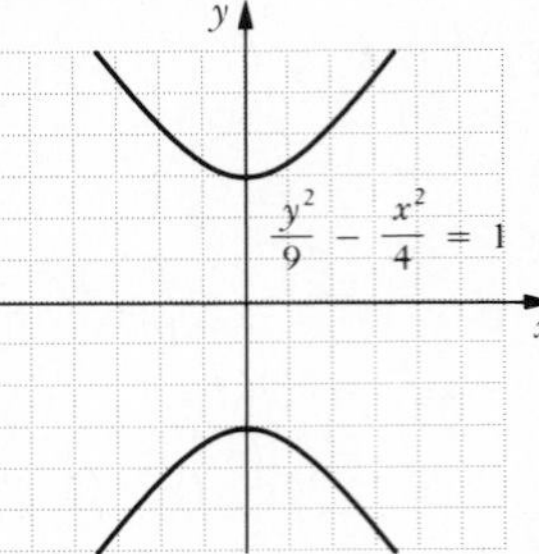

10. [9.1d]

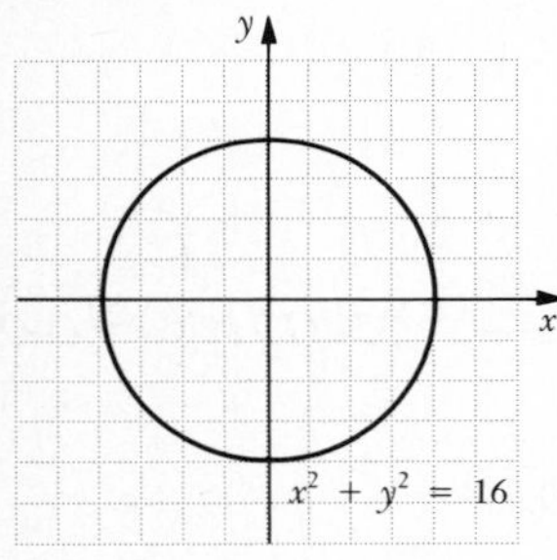

11. [9.1a]

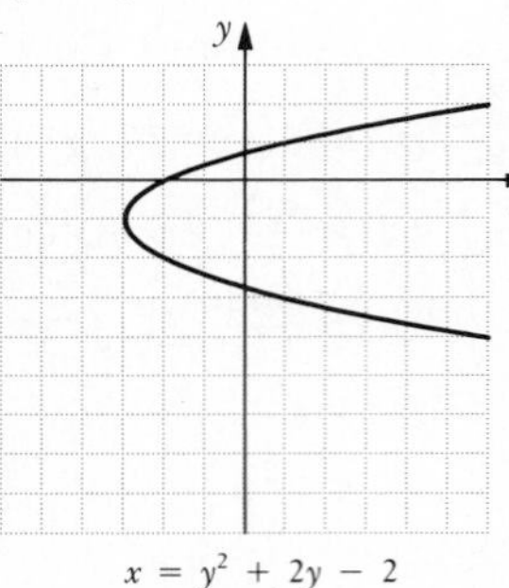

12. [9.1a]

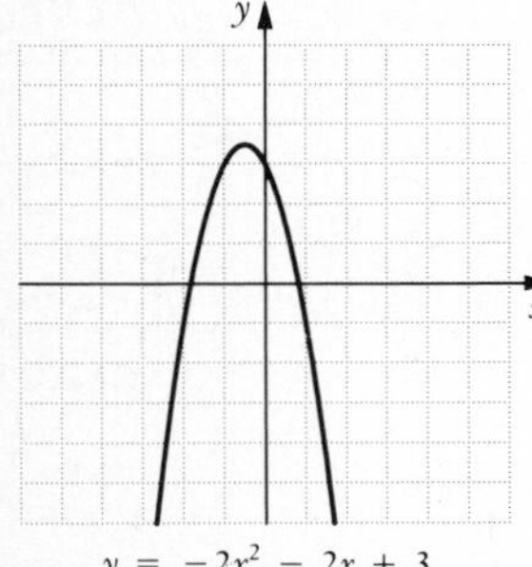

13. [9.1d, e]

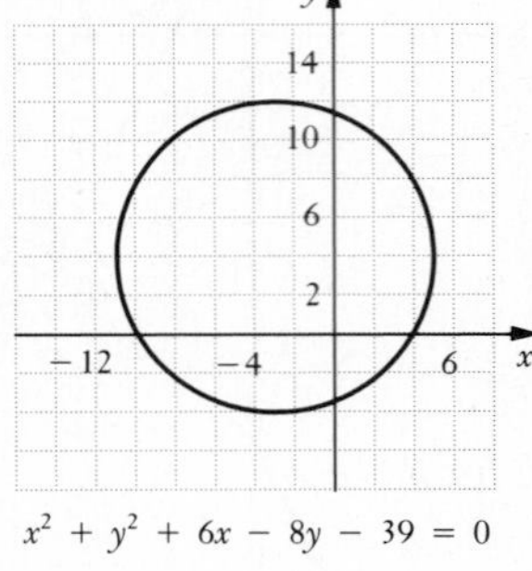

14. [9.3a] $(7, 4)$ **15.** [9.3a] $(2, 2)$, $\left(\frac{32}{9}, -\frac{10}{9}\right)$ **16.** [9.3a] $(0, -3)$, $(2, 1)$ **17.** [9.3a] $(4, 3)$, $(4, -3)$, $(-4, 3)$, $(-4, -3)$ **18.** [9.3a] $(2, 1)$, $(\sqrt{3}, 0)$, $(-2, 1)$, $(-\sqrt{3}, 0)$ **19.** [9.3a] $(3, -3)$, $\left(-\frac{3}{5}, \frac{21}{5}\right)$ **20.** [9.3a] $(6, 8)$, $(6, -8)$, $(-6, 8)$, $(-6, -8)$

21. [9.3a] $(2, 2)$, $(-2, -2)$, $(2\sqrt{2}, \sqrt{2})$, $(-2\sqrt{2}, -\sqrt{2})$ **22.** [9.3b] Length: 12 m; width: 7 m **23.** [9.3b] 4 and 8 **24.** [9.3b] 32 cm, 20 cm **25.** [9.3b] 11 ft, 3 ft **26.** [9.4a] No **27.** [9.4a] Yes **28.** [9.4c] Yes **29.** [9.4c] No **30.** [9.4b] 12, 7 **31.** [9.4b] 5, 13 **32.** [9.5a, b] $2x^2 + 11$; 3; $x^4 + 11x^2 + 28$; $\frac{x^2+7}{x^2+4}$; $x^4 + 8x^2 + 16$; $x^4 + 8x^2 + 23$; $x^4 + 14x^2 + 53$ **33.** [9.4c]

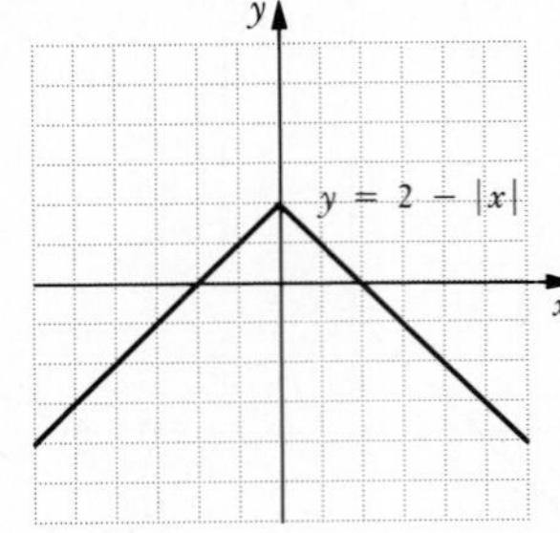

34. [9.5b] $f(x) = \sqrt{x}$; $g(x) = 2x - 7$; answers may vary **35.** [9.6b] No **36.** [9.6a] $\{(4, 3), (6, 3), (7, -3)\}$; yes **37.** [9.6c] $f^{-1}(x) = x + 2$ **38.** [9.6c] $g^{-1}(x) = \frac{7x+3}{2}$ **39.** [9.6c] $f^{-1}(x) = \frac{2}{x} - 5$ **40.** [9.6c] $g^{-1}(x) = \frac{\sqrt[3]{x}}{2}$ **41.** [7.2a] $3a^2b^3\sqrt[3]{3a^2b}$ **42.** [8.2a] $-1 \pm 2i$ **43.** [7.5a] $\frac{16-a}{8+6\sqrt{a}+a}$ **44.** [8.6b] About 8.5 mph **45.** [9.3b] $(x-4)^2 + (y+10)^2 = 100$ **46.** [9.3b] $\frac{x^2}{64} + \frac{y^2}{4} = 1$

Test: Chapter 9, p. 571

1. [9.1b] $6\sqrt{5} \approx 13.416$ **2.** [9.1c] $(0, 5)$ **3.** [9.1d] Center: $(-2, 3)$; radius: 8 **4.** [9.1e] Center: $(-2, 3)$; radius: 3 **5.** [9.1d] $(x-2)^2 + (y+7)^2 = 18$

6. [9.1a]

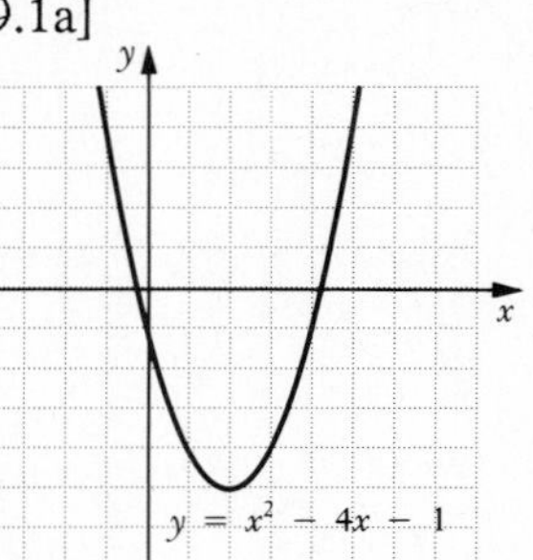

7. [9.1d]

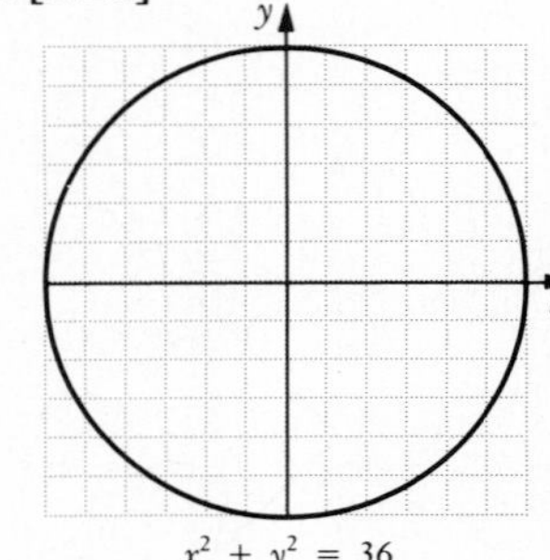

8. [9.2b]

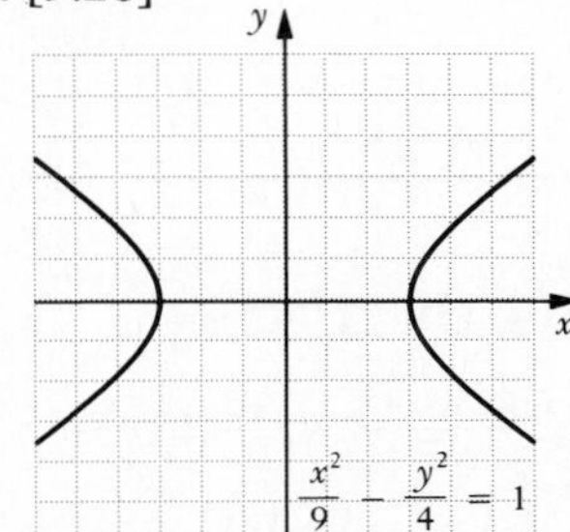

9. [9.2a]

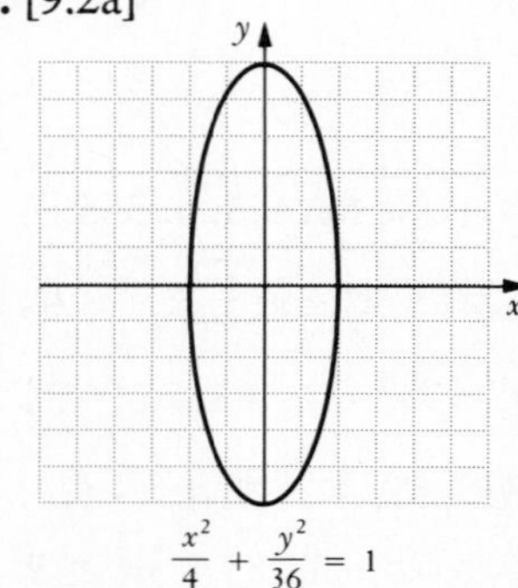

10. [9.1a]

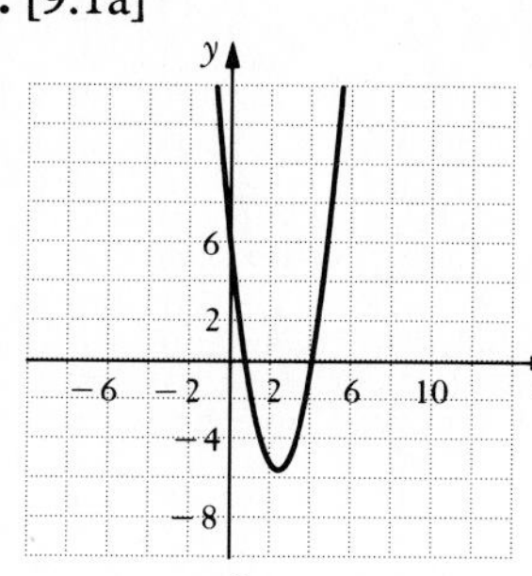

11. [9.1a]

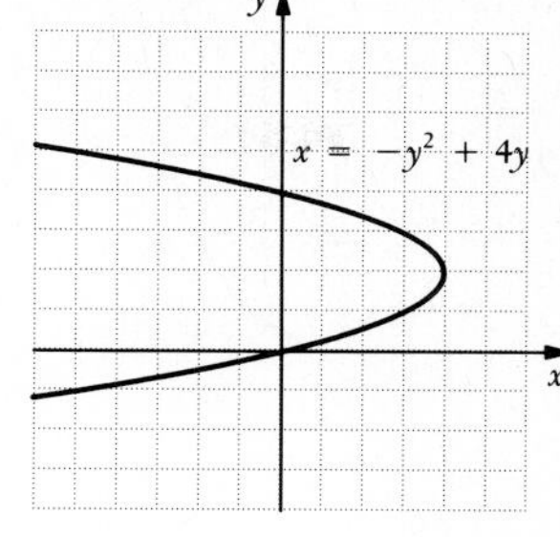

12. [9.3a] (0, 3), (4, 0) **13.** [9.3a] (4, 0), (−4, 0) **14.** [9.3b] $\frac{23+3\sqrt{53}}{2}, \frac{23-3\sqrt{53}}{2}$ **15.** [9.3b] 11 by 2 **16.** [9.3b] $\sqrt{5}$ m, $\sqrt{3}$ m **17.** [9.3b] 16 ft, 12 ft

18. [9.4c]

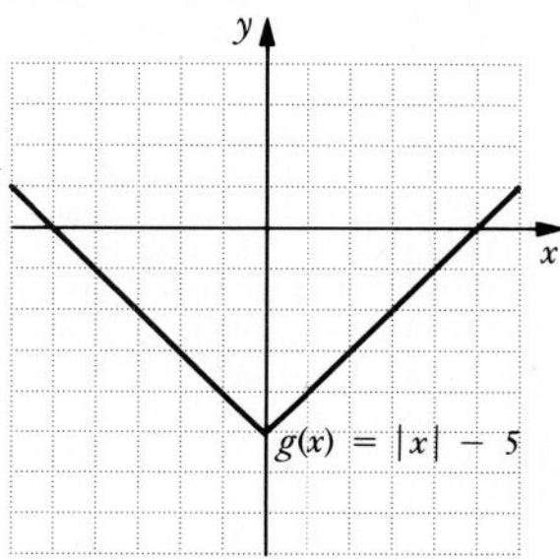

19. [9.4b] 15

20. [9.5a, b] x^2+x+1; $-x^2+x-9$; $x^3-4x^2+5x-20$; $\frac{x-4}{x^2+5}$; $x^2-8x+16$; x^2+1; $x^2-8x+21$ **21.** [9.5b] $f(x)=x^3$; $g(x)=5x+8$; answers may vary **22.** [9.6b] No **23.** [9.6c] $f^{-1}(x)=\frac{x+3}{4}$ **24.** [9.6c] $g^{-1}(x)=4x+2$ **25.** [9.6c] $f^{-1}(x)=\frac{1+2x}{x-1}$ **26.** [8.2a] $-1\pm\sqrt{6}$ **27.** [7.2a] $2ab^6\sqrt[3]{6a^2}$ **28.** [7.5a] $\frac{8-6\sqrt{a}+a}{4-a}$ **29.** [8.6b] About 9.3 mph **30.** [9.3b] $\frac{(x-5)^2}{25}+\frac{(y-2)^2}{9}=1$ **31.** [9.3b] 9

Cumulative Review: Chapters 1-9, p. 572

1. [2.4c] $\{x \mid x \geq 1\}$ **2.** [2.6e] $\{x \mid x < -6.4 \text{ or } x > 6.4\}$ **3.** [2.5a] $\{x \mid -1 \leq x < 6\}$ **4.** [4.2b] $\left(-\frac{1}{3}, 5\right)$ **5.** [4.4b] (−1, 2, 3) **6.** [5.8a] $-1, \frac{3}{2}$ **7.** [6.3a] $-\frac{2}{3}, 3$ **8.** [7.7a] 4 **9.** [8.1a] −7, −3 **10.** [8.2a] $\frac{-1\pm i\sqrt{7}}{4}$ **11.** [5.8a] −3, −2, 2, 3 **12.** [6.3a] −16 **13.** [9.3a] (1, 2), (−1, 2), (1, −2), (−1, −2) **14.** [2.3a] $N=\frac{4P-3M}{6}$ **15.** [6.5a] $p=\frac{qf}{q-f}$ **16.** [8.9a] $\{x \mid x \leq -1 \text{ or } x \geq 1\}$ **17.** [1.2d] 6 **18.** [1.5c] 18 **19.** [1.6a] $-\frac{y^4}{3z^5}$ **20.** [5.2a] $2x^3-x^2-8x-3$ **21.** [5.1f] $-x^3+3x^2-x-6$ **22.** [6.1d] $\frac{2m-1}{m+1}$ **23.** [6.2c] $\frac{x+2}{x+1}$

24. [6.6a] $\frac{1}{x+1}$ **25.** [5.9b] $x^3+2x^2-2x+1+\frac{3}{x+1}$ **26.** [7.3a] $5x^2\sqrt{y}$ **27.** [7.4a] $11\sqrt{2}$ **28.** [7.6c] 8 **29.** [7.9c] $16+i\sqrt{2}$ **30.** [7.9e] $\frac{3}{10}+\frac{11}{10}i$

31. [3.2a]

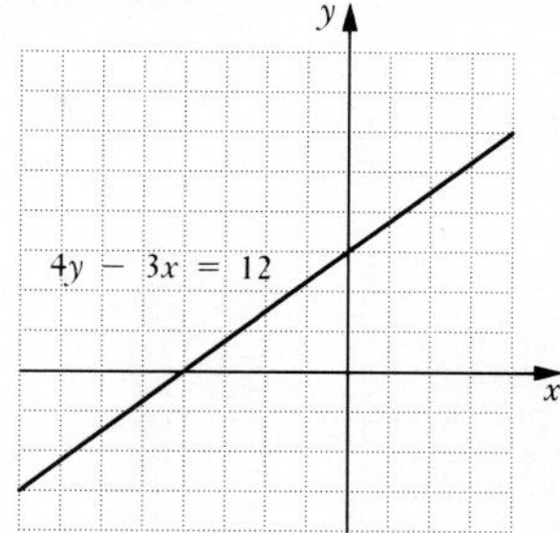

32. [3.6b]

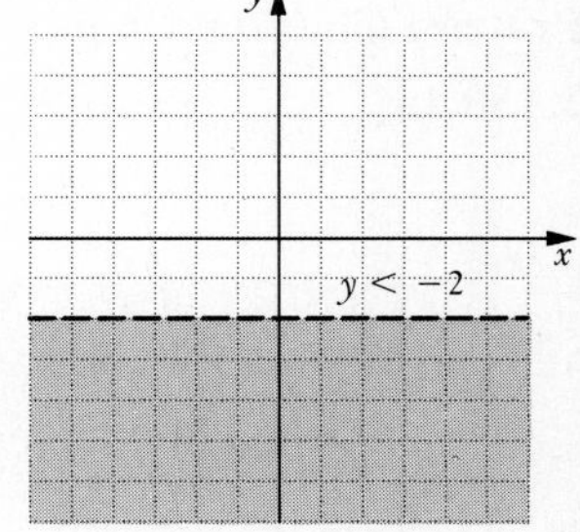

33. [4.7a]

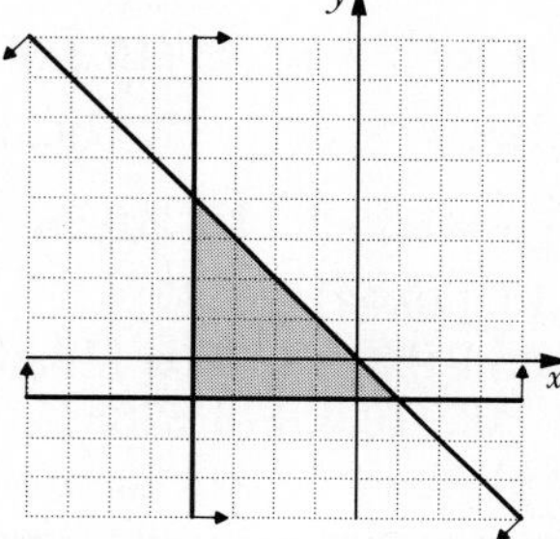

34. [8.8a]

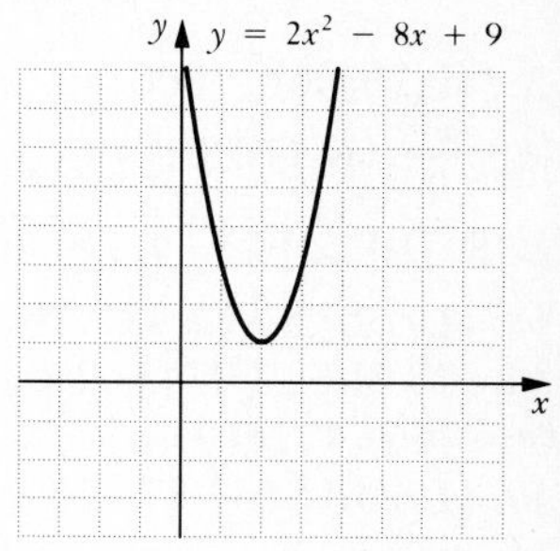

35. [9.1d]

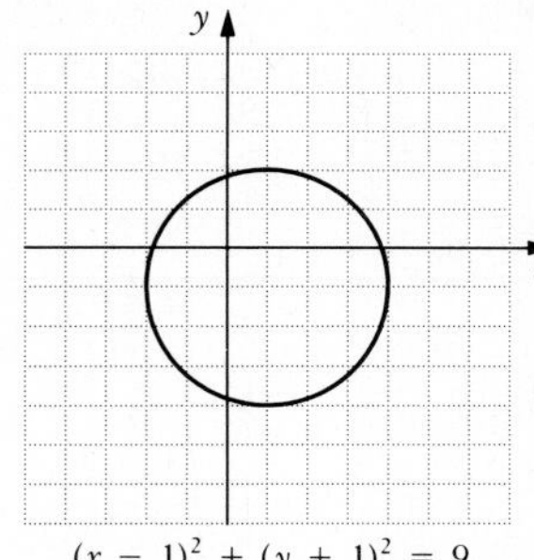

36. [8.8a]

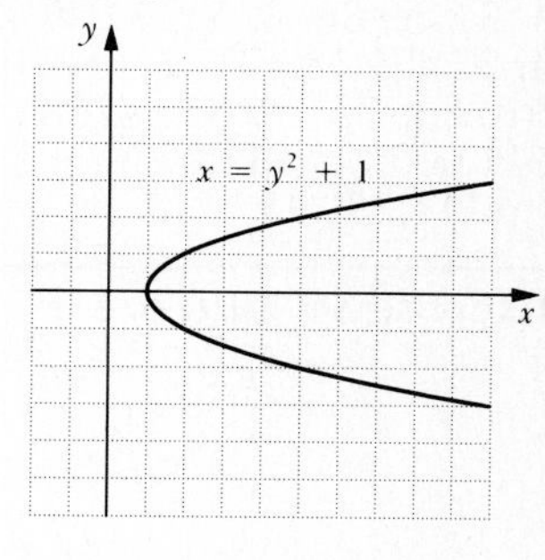

37. [5.3b] $(2x^3+1)(x-6)$ **38.** [5.4a] $3(a-9b)(a+5b)$ **39.** [5.4a] $(x-8)(x-9)$ **40.** [5.5b] $(9m^2+n^2)(3m+n)(3m-n)$ **41.** [5.5a] $4(2x-1)^2$ **42.** [5.6a] $3(3a-2)(9a^2+6a+4)$ **43.** [5.4b] $2(5x-2)(x+7)$ **44.** [5.7a] $3x(2x-1)(x+5)$ **45.** [3.4b] $y=2x+2$ **46.** [3.4c] $y=-\frac{1}{2}x+\frac{5}{2}$ **47.** [8.5b] All sets of three consecutive even integers satisfy this condition. **48.** [9.1e] Center: (8, −3); radius: $\sqrt{5}$ **49.** [6.7e] $z=\frac{4}{5}$ **50.** [9.4b] $f(-2)=-10$ **51.** [9.1b] $d=10$ **52.** [9.1c] $\left(1, -\frac{3}{2}\right)$ **53.** [9.5c] $\frac{15+8\sqrt{a}+a}{9-a}$ **54.** [4.3a] Honolulu: $403 per person; New York City: $152 per person **55.** [6.4a] 1 hr **56.** [9.5a] $(f+g)(x)=x^2-x-4$, $(f-g)(x)=-x^2+5x-6$ **57.** [9.6c] $f^{-1}(x)=\frac{1}{2}(x+3)$

58. [6.3a] All real numbers except −2 **59.** [9.1d] Circle centered at the origin with radius a. **60.** [9.1d] $(x+2)^2+(y-1)^2=4$ **61.** [9.3a] $\frac{1}{2}, \frac{1}{4}$; $\frac{1}{2}, -\frac{1}{4}$; $-\frac{1}{2}, \frac{1}{4}$; or $-\frac{1}{2}, -\frac{1}{4}$

CHAPTER 10

Pretest: Chapter 10, p. 576

1. **[10.1a]**

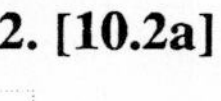

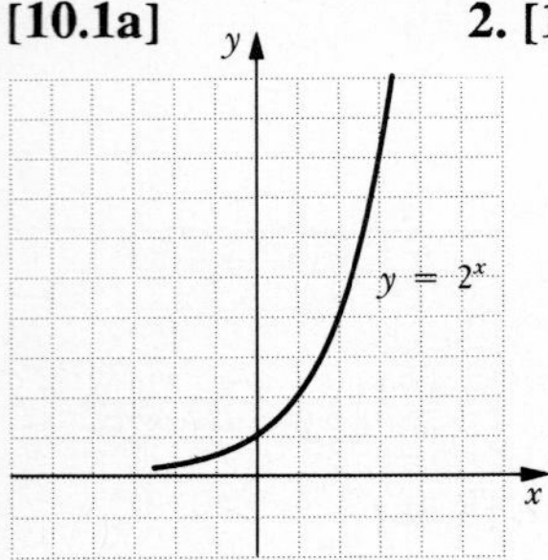

2. **[10.2a]**

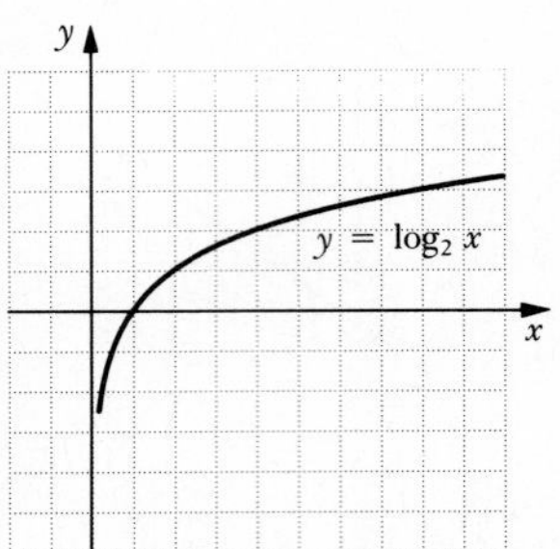

3. **[10.2b]** $\log_a 1000 = 3$ **4.** **[10.2c]** $2^t = 32$ **5.** **[10.3d]** $\log_a \frac{M^2N}{\sqrt[3]{Q}}$ **6.** **[10.3d]** $\frac{3}{4}\log_a x + \frac{1}{4}\log_a y - \frac{1}{2}\log_a z$ **7.** **[10.2d]** 2 **8.** **[10.2d]** 25 **9.** **[10.5a]** $\frac{\log 8.6}{\log 3} \approx 1.9586$ **10.** **[10.5a]** $\frac{7}{2}$ **11.** **[10.5b]** 8 **12.** **[10.5b]** 1 **13.** **[10.4a]** 2.8537 **14.** **[10.4b]** −7.0745 **15.** **[10.4a]** 0.0003405 **16.** **[10.4b]** 66.8466 **17.** **[10.4b]** 0.09633 **18.** **[10.4a]** −1.6308 **19.** **[10.6a]** \$642.37 **20.** **[10.6a]** About 11,893 years old

Sidelight: The Compound-Interest Formula, p. 584

1. \$1169.86 **2.** \$1172.89 **3.** \$49,665.56 **4.** \$7,479,186.72 **5.** (a) \$10,840; (b) \$10,857.64; (c) \$10,866.83; (d) \$10,876.18; (e) \$10,876.28 **6.** (a) \$11,000; (b) \$11,025; (c) \$11,038.13; (d) \$11,051.56; (e) \$11,051.75

Exercise Set 10.1, p. 575

1.

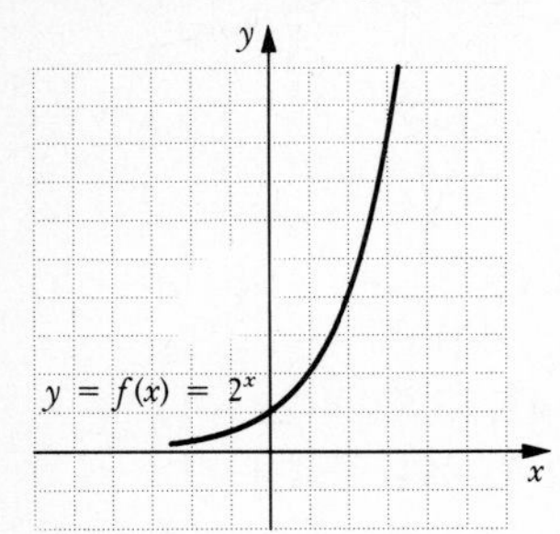

3.

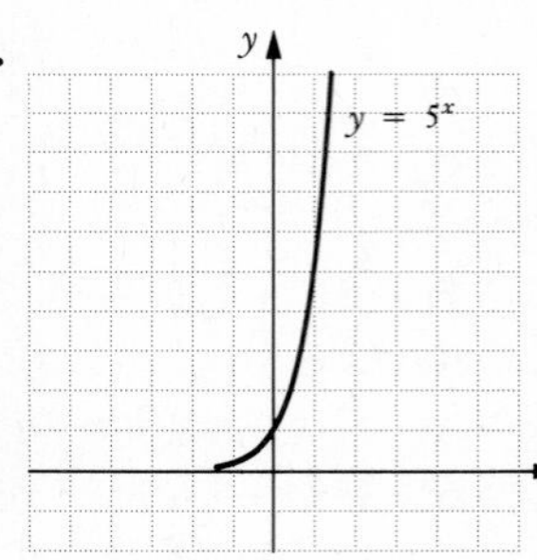

5.

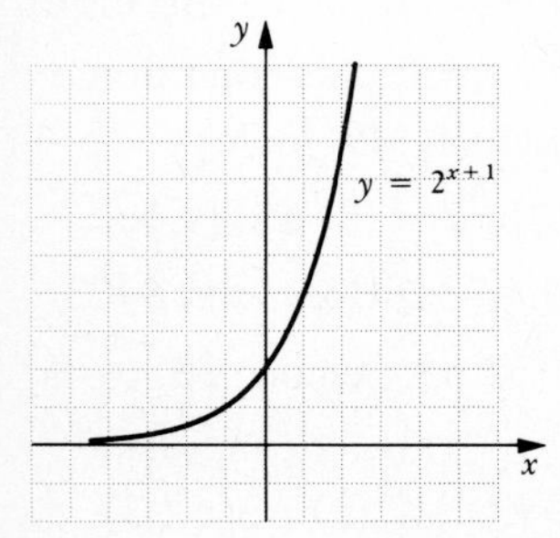

7.

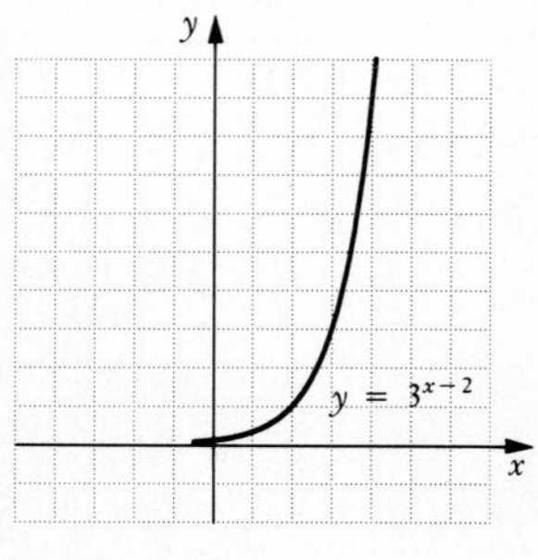

9.

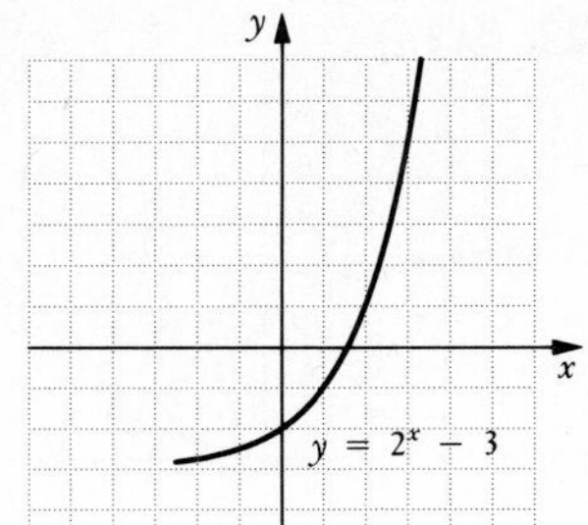

11.

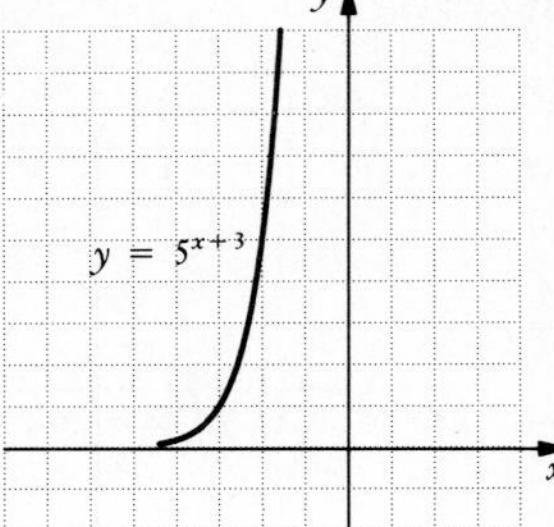

13.

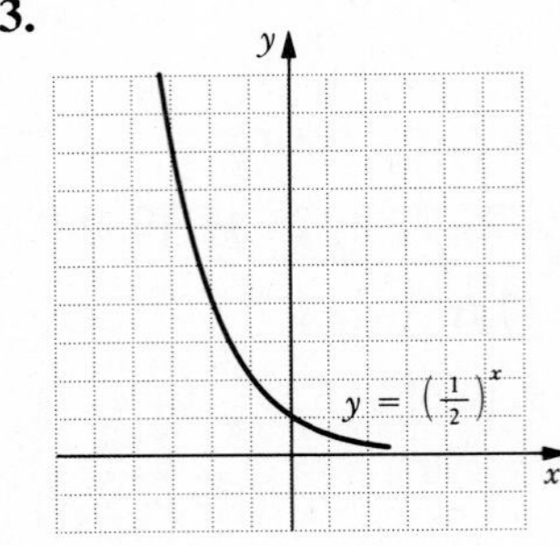

15.

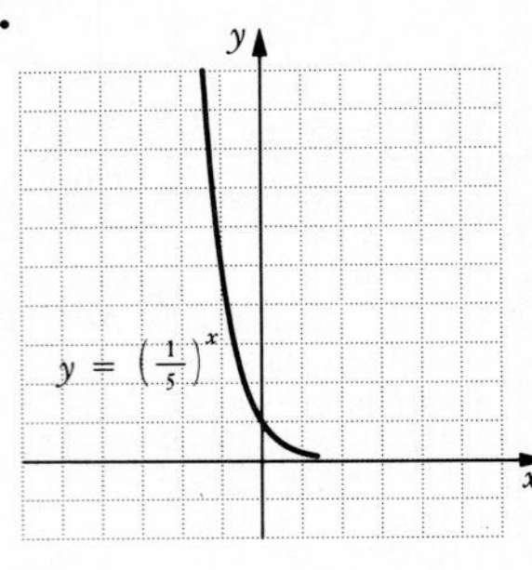

17.

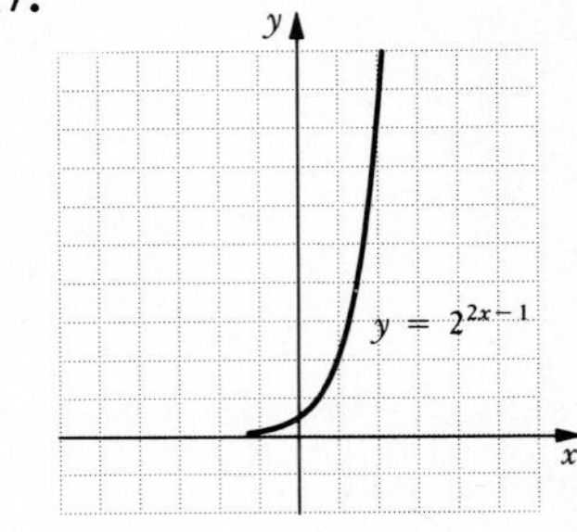

19.

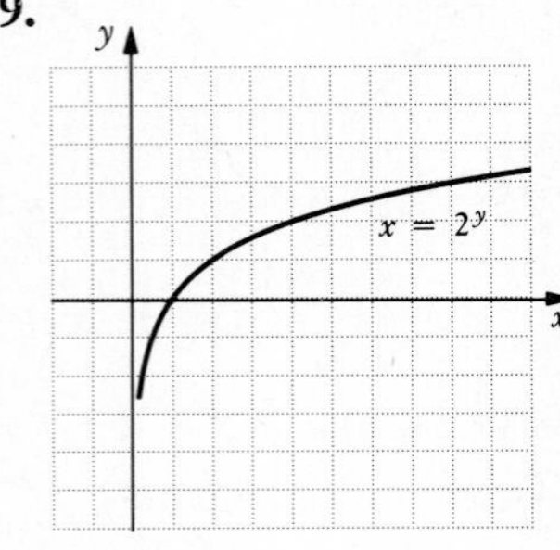

21.

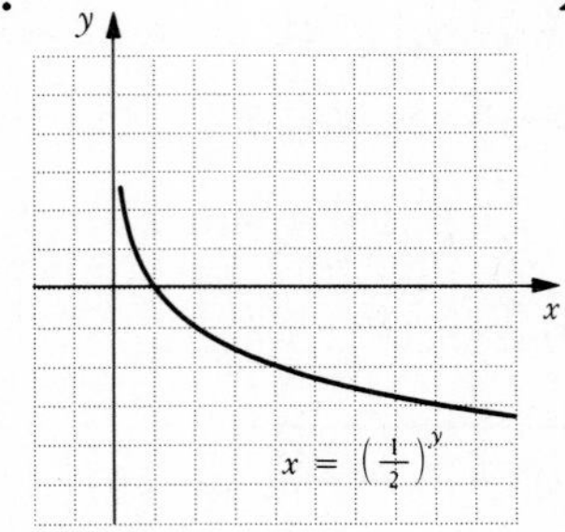

23.

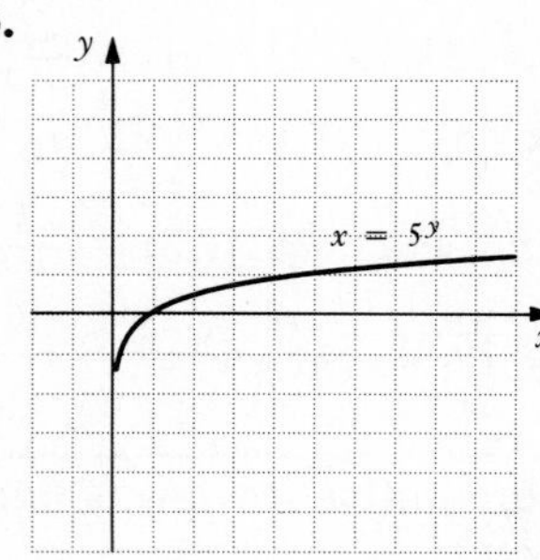

25.

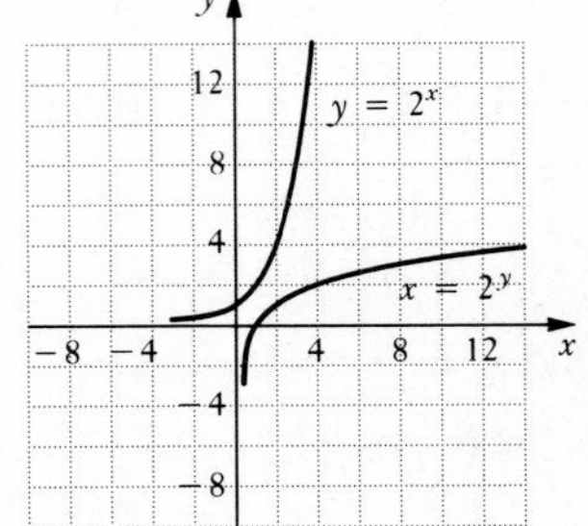

27. **(a)** $A(t) = \$50,000(1.08)^t$; **(b)** \$50,000; \$68,024.45;

\$92,546.51; \$107,946.25; **(c)**

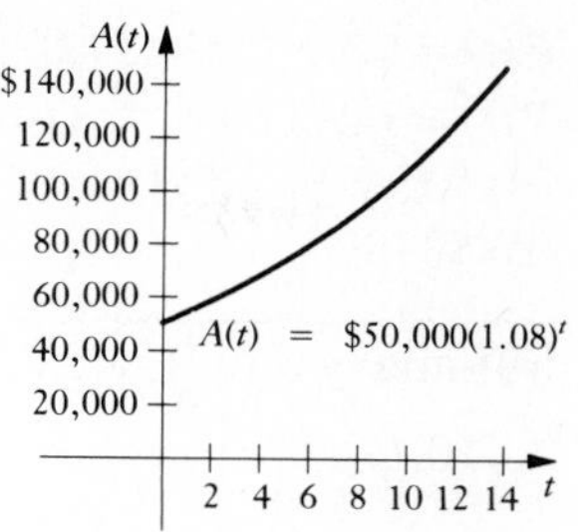

29. (a) \$5200, \$4160, \$3328, \$1703.94; \$558.35; **(b)**

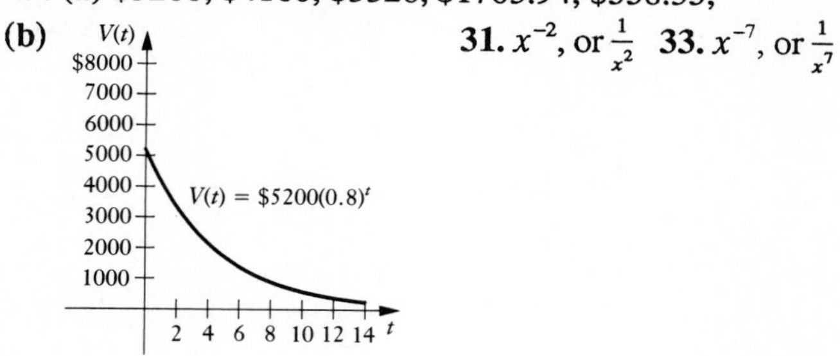

31. x^{-2}, or $\frac{1}{x^2}$ **33.** x^{-7}, or $\frac{1}{x^7}$

35. **37.**

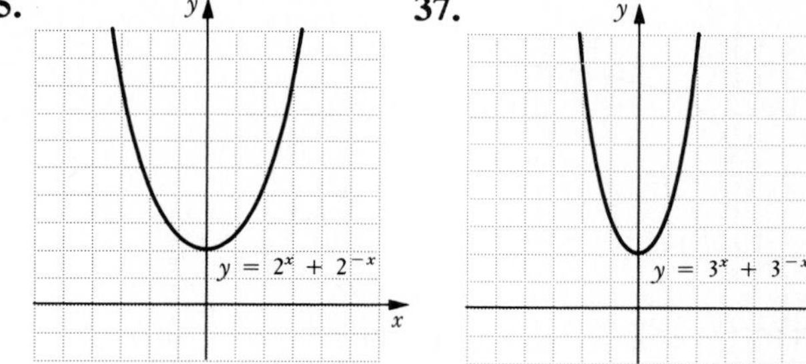

39.

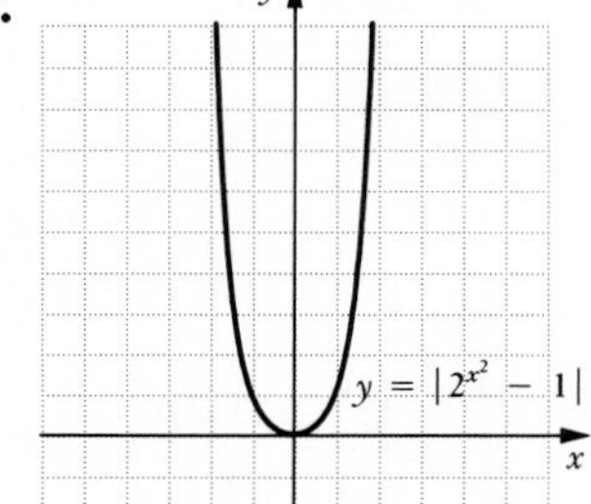

Sidelight: Handling Dimension Symbols (Part II), p. 594

1. 51.2 oz **2.** 6200 m **3.** 3080 $\frac{\text{ft}}{\text{min}}$ **4.** Approx. $\frac{\$0.26}{\text{min}}$ **5.** 96 in. **6.** 75 ft **7.** Approx. 0.03 yr **8.** Approx. 31.7 yr **9.** Approx. 31,710 yr **10.** 320 oz **11.** 80 $\frac{\text{oz}}{\text{in.}}$ **12.** 30 $\frac{\text{mi}}{\text{hr}}$ **13.** 172,800 sec **14.** 16/3 days **15.** 3/2 ft^2 **16.** 60 man-days **17.** 1.08 $\frac{\text{ton}}{\text{yd}^3}$ **18.** 5,865,696,000,000 $\frac{\text{mi}}{\text{yr}}$

Exercise Set 10.2, p. 595

1. **3.**

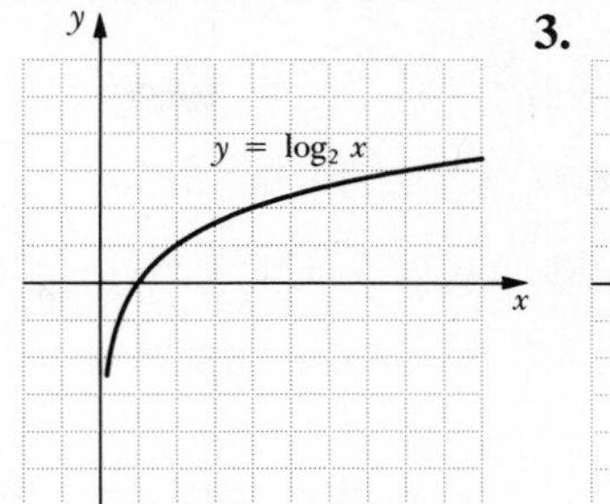

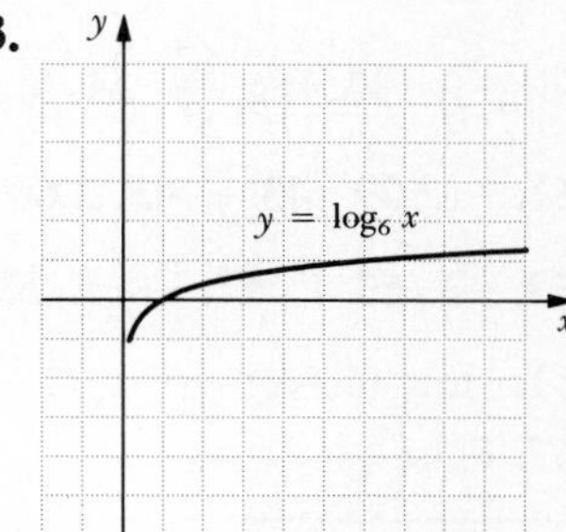

5.

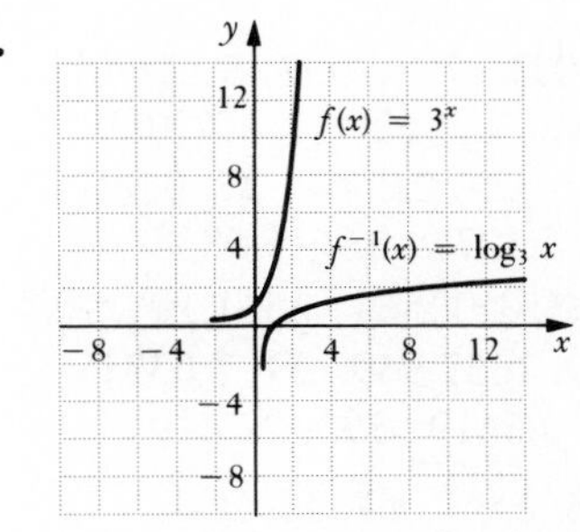

7. $3 = \log_{10} 1000$ **9.** $-3 = \log_5 \frac{1}{125}$ **11.** $\frac{1}{3} = \log_8 2$ **13.** $0.3010 = \log_{10} 2$ **15.** $2 = \log_e t$ **17.** $t = \log_Q x$ **19.** $2 = \log_e 7.3891$ **21.** $-2 = \log_e 0.1353$ **23.** $3^t = 8$ **25.** $5^2 = 25$ **27.** $10^{-1} = 0.1$ **29.** $10^{0.845} = 7$ **31.** $e^{2.9957} = 20$ **33.** $t^k = Q$ **35.** 9 **37.** 4 **39.** 4 **41.** 3 **43.** 13 **45.** 1 **47.** $\frac{1}{2}$ **49.** 2 **51.** 2 **53.** −1 **55.** 0 **57.** 4 **59.** 2 **61.** 3 **63.** −2 **65.** 0 **67.** 1 **69.** $\frac{2}{3}$

71. **73.**

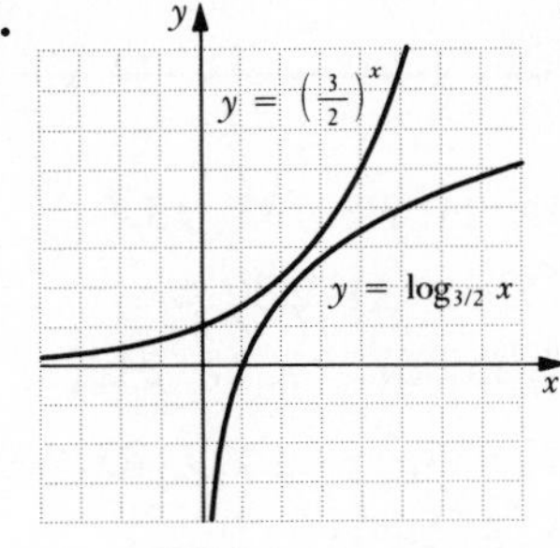

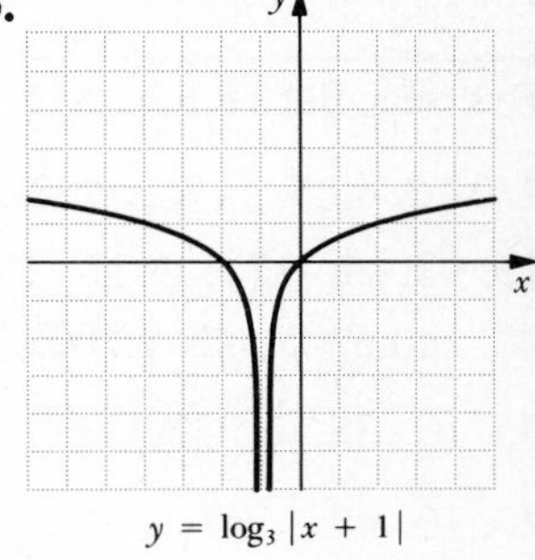

75. 25 **77.** $-\frac{7}{16}$ **79.** 3 **81.** 0 **83.** −2

Exercise Set 10.3, p. 603

1. $\log_2 32 + \log_2 8$ **3.** $\log_4 64 + \log_4 16$ **5.** $\log_c B + \log_c x$ **7.** $\log_a 420$ **9.** $\log_c Ky$ **11.** $3\log_a x$ **13.** $6\log_c y$ **15.** $-3\log_b C$ **17.** $\log_a 67 - \log_a 5$ **19.** $\log_b 3 - \log_b 4$ **21.** $\log_a \frac{15}{7}$ **23.** $2\log_a x + 3\log_a y + \log_a z$ **25.** $\log_b x + 2\log_b y - 3\log_b z$ **27.** $\frac{4}{3}\log_c x - \log_c y - \frac{2}{3}\log_c z$

29. $2\log_a m + 3\log_a n - \frac{3}{4} - \frac{5}{4}\log_a b$ **31.** $\log_a \frac{x^{2/3}}{y^{1/2}}$, or $\log_a \frac{\sqrt[3]{x^2}}{\sqrt{y}}$ **33.** $\log_a \frac{2x^4}{y^3}$ **35.** $\log_a \frac{\sqrt{a}}{x}$ **37.** 2.708 **39.** 0.51 **41.** −1.609 **43.** $\frac{3}{2}$ **45.** 2.609 **47.** 9 **49.** m **51.** 4 **53.** −7 **55.** i **57.** $\frac{3}{5}+\frac{4}{5}i$ **59.** $\log_a (x^6 - x^4y^2 + x^2y^4 - y^6)$ **61.** $\frac{1}{2}\log_a (1-s) + \frac{1}{2}\log_a (1+s)$ **63.** False **65.** True **67.** False

Exercise Set 10.4, p. 609

1. 0.3010 **3.** 0.9031 **5.** 0.8021 **7.** 1.7952 **9.** 2.6405 **11.** 4.1271 **13.** −0.2441 **15.** −1.2840 **17.** −2.2069 **19.** 1000 **21.** 501.187 **23.** 3 **25.** 0.2841 **27.** 0.00112 **29.** 0.6931 **31.** 2.0794 **33.** 4.1271 **35.** −5.0832 **37.** 36.789 **39.** 0.002299 **41.** 1.00569 **43.** 1.19×10^{10} **45.** 8.1490 **47.** −3.3496 **49.** 1637.949 **51.** 7.6331 **53.** 2.5702 **55.** 3.3219 **57.** 0.6419 **59.** −2.3219 **61.** −2.3219 **63.** 3.5471 **65.** $\pm\sqrt{\frac{b}{a}}$ **67.** 16, 256 **69.** $\log x = 0.4343 \ln x$ **71.**

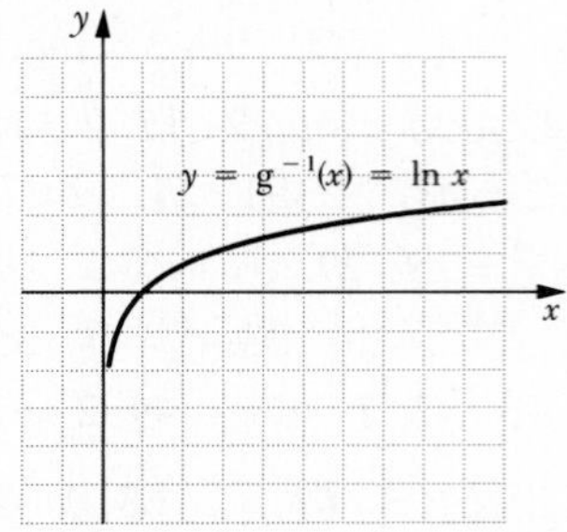

73. 1.3886 **75.** 1.5893

Exercise Set 10.5, p. 615

1. 3 **3.** 4 **5.** $\frac{5}{2}$ **7.** $\frac{3}{5}$ **9.** 3.1699 **11.** 3.32193 **13.** $\frac{5}{2}$ **15.** −3, −1 **17.** 1.4037 **19.** 4.6052 **21.** 2.3026 **23.** 140.6705 **25.** 2.7095 **27.** 3.6067 **29.** 27 **31.** $\frac{1}{8}$ **33.** 10 **35.** $\frac{1}{100}$ **37.** $e^2 \approx 7.3891$ **39.** $\frac{1}{e} \approx 0.3679$ **41.** 66 **43.** 10 **45.** $\frac{1}{3}$ **47.** 3 **49.** $\frac{2}{5}$ **51.** 5 **53.** No solution **55.** (3, 4), (−4, −3) **57.** (2, 3), (2, −3), (−2, 3), (−2, −3) **59.** −4 **61.** 2 **63.** $\pm\sqrt{34}$ **65.** $10^{100,000}$ **67.** 1, 100 **69.** 3, −7 **71.** 1, $\log_3 5 \approx 1.465$ **73.** −3

Exercise Set 10.6, p. 623

1. (a) 37.7 yr; **(b)** 11.9 yr **3. (a)** After 1 yr; **(b)** 7.3 yr **5. (a)** After 5.5 yr; in 1990; **(b)** 0.8 yr **7.** 2.0 ft/sec **9.** 1.5 ft/sec **11. (a)** $P = 5e^{0.028t}$, where t is the number of years after 1987; **(b)** 6.4 billion, 7.2 billion; **(c)** 1994 **13. (a)** $C = 5e^{0.097t}$ where t is the number of years after 1962; **(b)** \$5.26; **(c)** 2009; **(d)** 7.1 yr **15. (a)** $P(t) = 52e^{0.03t}$; **(b)** \$1.07; **(c)**23.1 yr; **(d)** 2028 **17. (a)** $P(t) = 84{,}000e^{0.1616t}$; **(b)** \$271,000,000; **(c)** 4.3 yr; **(d)** After 58 yr **19.** 1860 yr **21. (a)** \$28,000; **(b)** \$3789

Summary and Review: Chapter 10, p. 627

1. [10.1a]

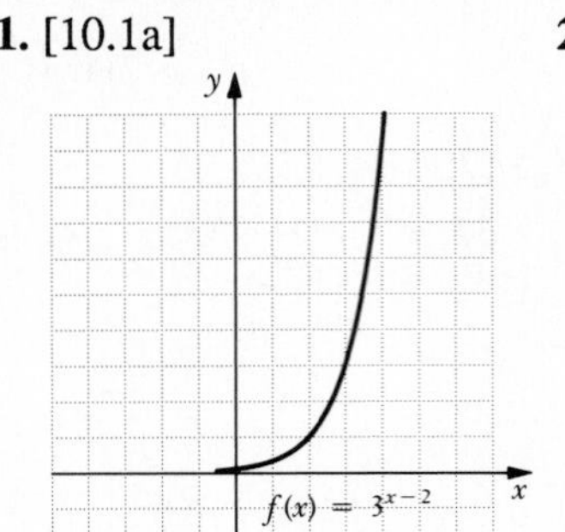

2. [10.1b]

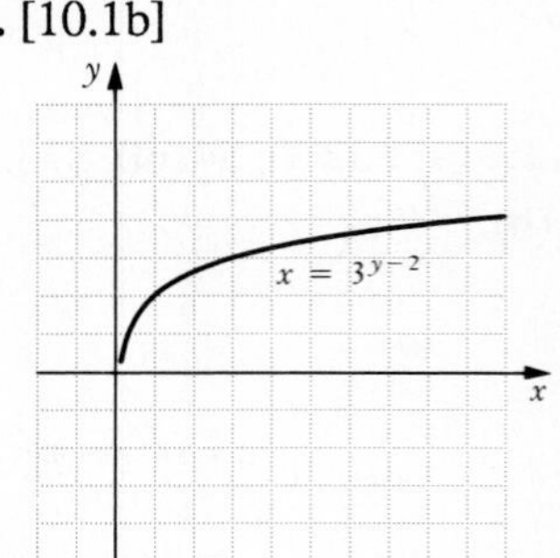

3. [10.2a]

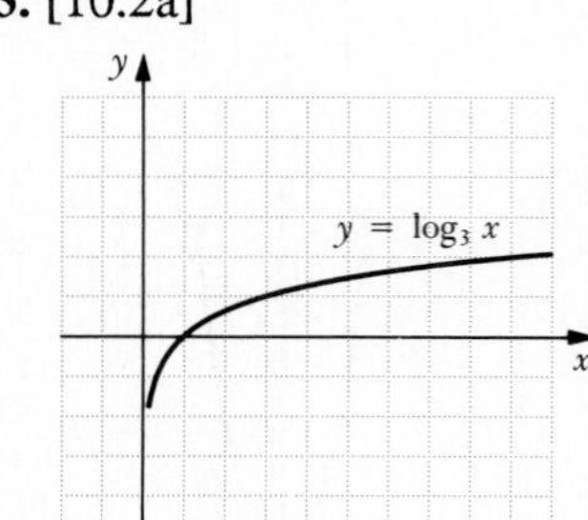

4. [10.2a]

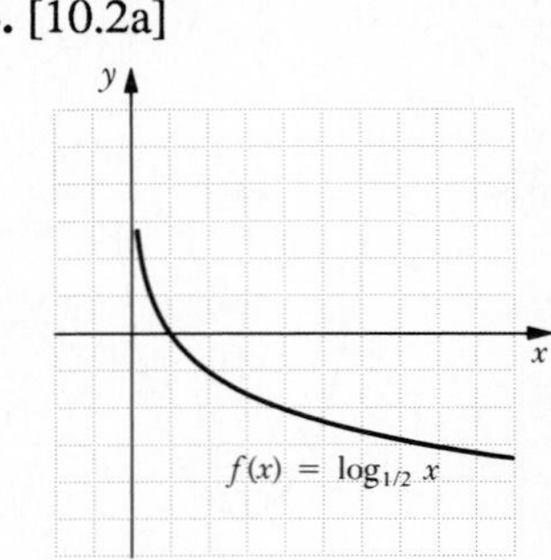

5. [10.2c] $4^x = 16$ **6.** [10.2c] $10^{0.3010} = 2$ **7.** [10.2c] $\left(\frac{1}{2}\right)^{-3} = 8$ **8.** [10.2c] $16^{3/4} = 8$ **9.** [10.2b] $4 = \log 10{,}000$ **10.** [10.2b] $\frac{1}{2} = \log_{25} 5$ **11.** [10.2b] $-2 = \log_7 \frac{1}{49}$ **12.** [10.2b] $3 = \log_{2.718} 20.1$ **13.** [10.3d] $4\log_a x + 2\log_a y + 3\log_a z$ **14.** [10.3d] $\log_a x + \log_a y - 2\log_a z$ **15.** [10.3d] $\frac{1}{2}\log z - \frac{3}{4}\log x - \frac{1}{4}\log y$ **16.** [10.3d] $2\log_q x + \frac{1}{3}\log_q y - 4\log_q z$ **17.** [10.3a] $\log_a 120$ **18.** [10.3c] $\log_a 6$ **19.** [10.3d] $\log \frac{\sqrt{a}}{bc^2}$ **20.** [10.3d] $\log_a \sqrt[3]{\frac{x}{y^2}}$ **21.** [10.2d] 1 **22.** [10.2d] 0 **23.** [10.3e] 17 **24.** [10.3d] 6.93 **25.** [10.3d] −3.2698 **26.** [10.3d] 8.7601 **27.** [10.3d] 3.2698 **28.** [10.3d] 2.54995 **29.** [10.4a] −2.2027 **30.** [10.4a] 7.8621 **31.** [10.4a] 29,798.88 **32.** [10.4a] 0.0361 **33.** [10.4a] 213.501 **34.** [10.4a] −2.3065 **35.** [10.4a] 5.5965 **36.** [10.4a] 0.000000163 **37.** [10.4b] 10.0821 **38.** [10.4b] −2.6921 **39.** [10.4b] 0.0000259 **40.** [10.4b] 3.4934×10^{19} **41.** [10.4c] 0.4307 **42.** [10.4c] 1.7097 **43.** [10.2d] $\frac{1}{9}$ **44.** [10.2d] 2 **45.** [10.2d] $\frac{1}{10{,}000}$ **46.** [10.5b] $e^2 \approx 7.3891$

47. [10.5a] $\frac{7}{2}$ **48.** [10.5a] 1.5266 **49.** [10.2d] 2 **50.** [10.5b] 7 **51.** [10.5b] 8 **52.** [10.5b] 20 **53.** [10.5b] $\sqrt{43}$ **54.** [10.6a] **(a)** 62; **(b)** 46.8; **(c)** 35 months **55.** [10.6a] **(a)** $C(t) = 4.65e^{0.051t}$; **(b)** \$35.76; **(c)** 1991; **(d)** 13.6 yr **56.** [10.6a] 3463 yr **57.** [8.5a] $T = \frac{-b \pm \sqrt{b^2+4aQ}}{2a}$ **58.** [8.4a] $\pm 4, \pm i\sqrt{5}$ **59.** [7.9c] $11-10i$ **60.** [7.9d] 1 **61.** [7.9e] $-\frac{11}{10}-\frac{17}{10}i$ **62.** [7.9b] $5-2i$ **63.** [9.3a] $\left(\frac{4}{3}, 2\right), (-2,-3)$ **64.** [10.5b] 10^{1000} **65.** [10.5a] 4, −1 **66.** [10.5a] $\left(\frac{8}{3}, -\frac{2}{3}\right)$

Test: Chapter 10, p. 629

1. [10.1a]

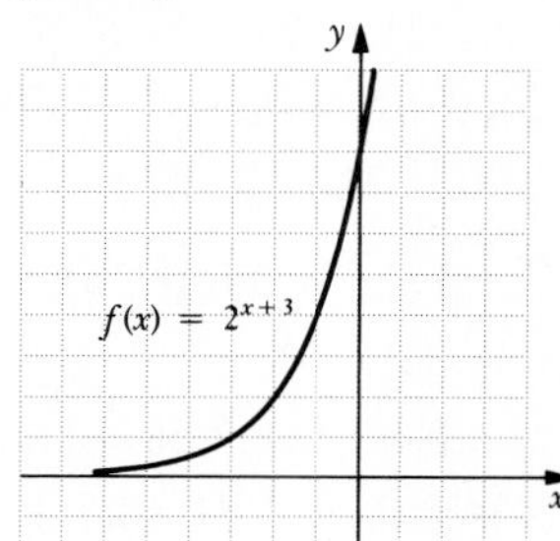

2. [10.2a]

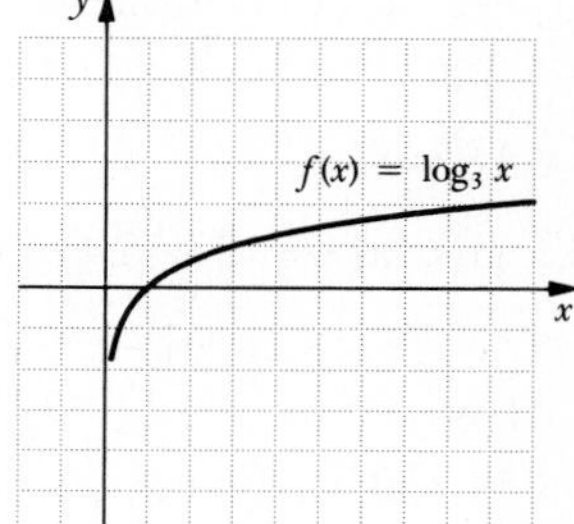

3. [10.2b] $-3 = \log_4 x$ **4.** [10.2b] $\frac{1}{2} = \log_{256} 16$ **5.** [10.2c] $4^2 = 16$ **6.** [10.2c] $7^m = 49$ **7.** [10.3d] $3\log a + \frac{1}{2}\log b - 2\log c$ **8.** [10.3d] $\log_a \frac{\sqrt[3]{xz^2}}{y^3}$ **9.** [10.3e] 23 **10.** [10.2d] 1 **11.** [10.2d] 0 **12.** [10.3d] −0.544 **13.** [10.3d] 0.69 **14.** [10.3d] 1.322 **15.** [10.4a] −1.9101 **16.** [10.4a] 445,040.98 **17.** [10.4a] 0.0000000548 **18.** [10.4a] 4.0913 **19.** [10.4b] −4.3949 **20.** [10.4b] 292.189 **21.** [10.4c] 1.1881 **22.** [10.2d] 5 **23.** [10.2d] 2 **24.** [10.2d] 10,000 **25.** [10.5a] $\frac{1}{3}$ **26.** [10.5a] 0.0937 **27.** [10.5b] $e^{1/4} \approx 1.284$ **28.** [10.5b] 9 **29.** [10.6a] **(a)** 5.6 ft/sec; **(b)** 19,400 **30.** [10.6a] **(a)** $P(t) = 209e^{0.01t}$; **(b)** 303 million, 348 million; **(c)** 1977; **(d)** 69.3 yr **31.** [10.6a] 4684 yr **32.** [8.4a] 1, 64 **33.** [8.5a] $t = \frac{b+\sqrt{b^2+4aS}}{2a}$ **34.** [7.9c] 29 **35.** [9.3a] (1,2), (−1,−2) **36.** [10.5b] 316, −309 **37.** [10.3d] 2

Cumulative Review: Chapters 1-10, p. 631

1. [1.1a; 1.5a] −2 **2.** [1.3a; 1.2d] 6 **3.** [1.6b] $\frac{y^{12}}{16x^8}$ **4.** [1.6a] $\frac{20x^6z^2}{y}$ **5.** [1.6a] $-\frac{y^4}{3z^5}$ **6.** [1.5c] $-4x-1$ **7.** [1.5c] 25 **8.** [2.1d] $\frac{11}{2}$ **9.** [5.8a] $\frac{2}{5}, -5$ **10.** [4.2b] (3,−1) **11.** [4.4b] (1,−2,0) **12.** [5.8a] 7, −1 **13.** [5.8a] 5, −2 **14.** [6.3a] $\frac{9}{2}$ **15.** [6.3a] $\frac{5}{8}$ **16.** [7.7b] 5 **17.** [7.7a] $\frac{1}{2}$ **18.** [8.1a] $\pm 5i$ **19.** [8.4a] 9, 25 **20.** [8.4a] ±3, ±2 **21.** [9.3a] (5,3), (2,0) **22.** [9.3a] (4,0), (−4,0) **23.** [10.2d] 8 **24.** [10.2d] 7 **25.** [10.5a] 1.2263 **26.** [10.5b] $\frac{80}{9}$ **27.** [2.6e] $\{x \mid -4 < x < -1\}$ **28.** [2.6e] $\{x \mid x \le -3 \text{ or } x \ge 6\}$ **29.** [6.5a] $a = \frac{Db}{b-D}$ **30.** [6.5a] $q = \frac{pf}{p-f}$ **31.** [6.5a] $B = \frac{3M-2A}{2}$, or $B = \frac{3}{2}M - A$ **32.** [4.6a] 2 **33.** [4.6b] 3 **34.** [2.2a] 8 **35.** [4.3a] 36 m, 20 m **36.** [4.3a] 336 m^2 **37.** [9.3b] $3\sqrt{2}$ ft, 4 ft **38.** [6.4a] $5\frac{5}{11}$ hr **39.** [4.3a] 60 L of *A*; 40 L of *B* **40.** [6.4c] $2\frac{7}{9}$ km/h **41.** [8.8c] −49; −7 and 7 **42.** [6.4c] Freight: 27 mph; passenger: 40 mph **43.** [10.6a] 78 **44.** [10.6a] 67.5 **45.** [10.6a] $P = 430e^{0.01t}$ **46.** [10.6a] 580 million; 641 million **47.** [6.7e] 18 **48.** [5.1d] $7p^2q^3 - 2p^3q + pq - 6$ **49.** [5.1f] $8x^2 - 11x - 1$ **50.** [5.2c] $9x^4 - 12x^2y + 4y^2$ **51.** [5.2b] $10a^2 - 9ab - 9b^2$ **52.** [6.1e] $\frac{(x+4)(x-3)}{2(x-1)}$ **53.** [6.6a] $\frac{1}{x-4}$ **54.** [6.1d] $\frac{a+2}{6}$ **55.** [6.2c] $\frac{7x+4}{(x-6)(x+6)}$ **56.** [5.3a] $x(y-2z+w)$ **57.** [5.6a] $(1-5x)(1+5x+25x^2)$ **58.** [5.7a] $2(3x-2y)(x+2y)$ **59.** [5.3b] $(x^3+7)(x-4)$ **60.** [5.5c] $(a-5+9b)(a-5-9b)$ **61.** [5.7a] $2(m+3n)^2$ **62.** [5.7a] $(x-2y)(x+2y)(x^2+4y^2)$ **63.** [9.4b] −12 **64.** [5.9b, c] $x^3 - 2x^2 - 4x - 12$, R −42; or $x^3 - 2x^2 - 4x - 12 + \frac{-42}{x-3}$ **65.** [1.6c] 4.836×10^{-1} **66.** [1.6c] 4.7×10^{13} **67.** [7.3a] $2y^2\sqrt[3]{y}$ **68.** [7.2b] $14xy^2\sqrt{x}$ **69.** [7.5b] $\frac{-2}{\sqrt{6}+\sqrt{10}}$ **70.** [7.5a] $\frac{6+\sqrt{y}-y}{4-y}$ **71.** [7.6d]

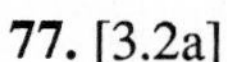

$\sqrt[10]{\frac{1}{(x+5)^3}}$ **72.** [7.9c] $12+4i\sqrt{3}$ **73.** [7.9e] $\frac{18}{25}+\frac{1}{25}i$ **74.** [8.3b] $25x^2+5x-6=0$ **75.** [3.4b] $y=-5x-3$ **76.** [3.4c] $y = \frac{1}{2}x + \frac{13}{2}$

77. [3.2a]

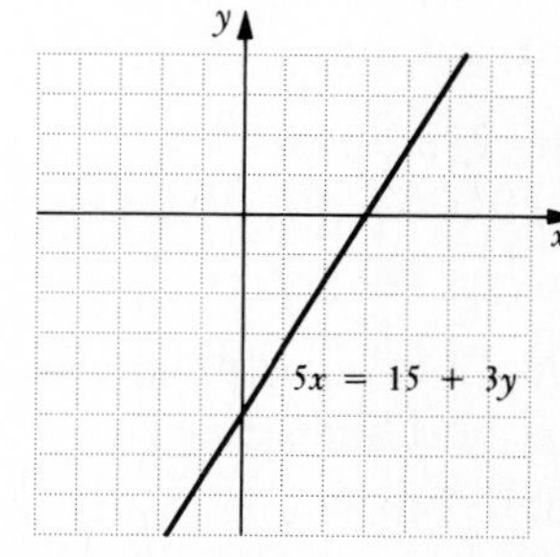

78. [9.2b]

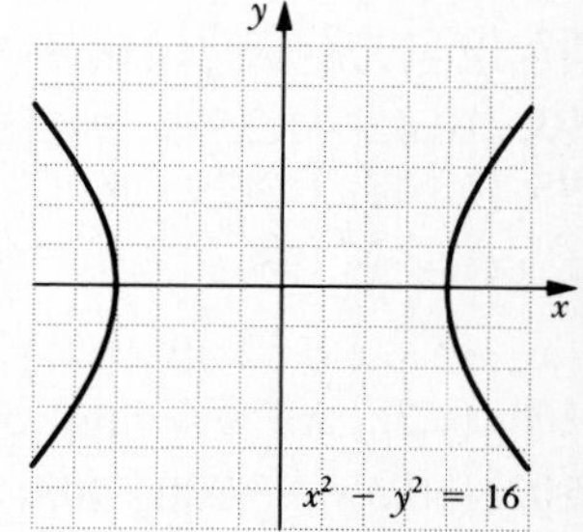

79. [9.1a]

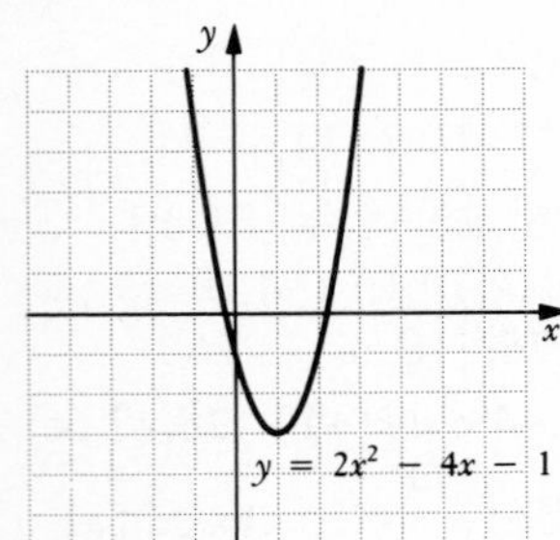

80. [10.2a]

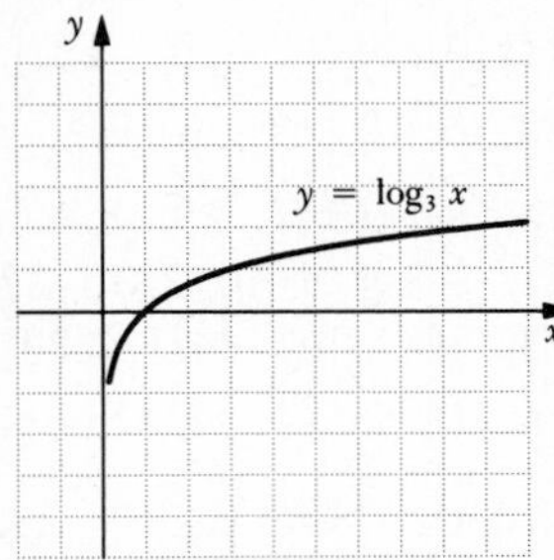

81. [10.1a]

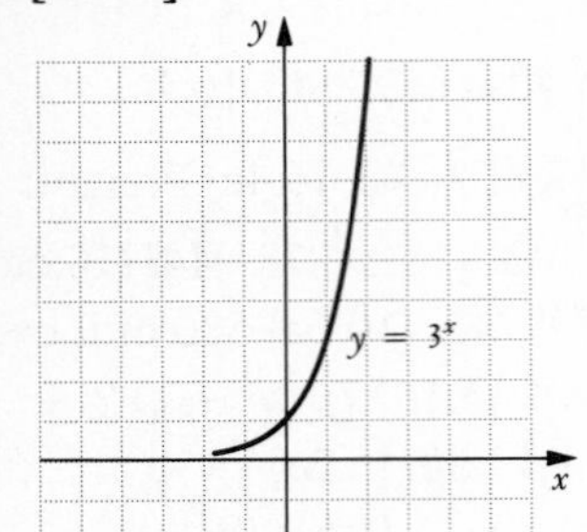

82. [3.6b]

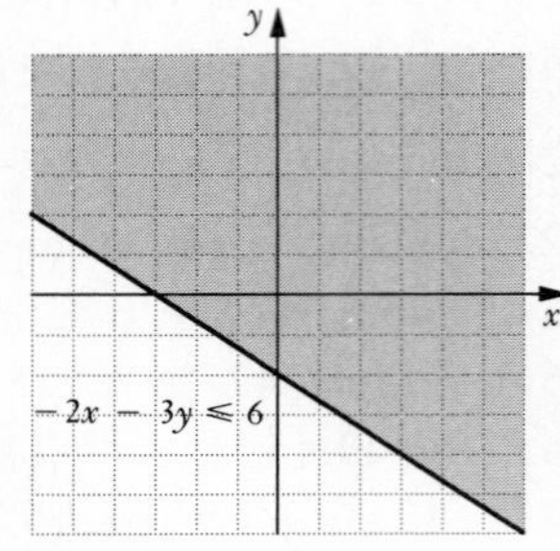

83. [8.7c]

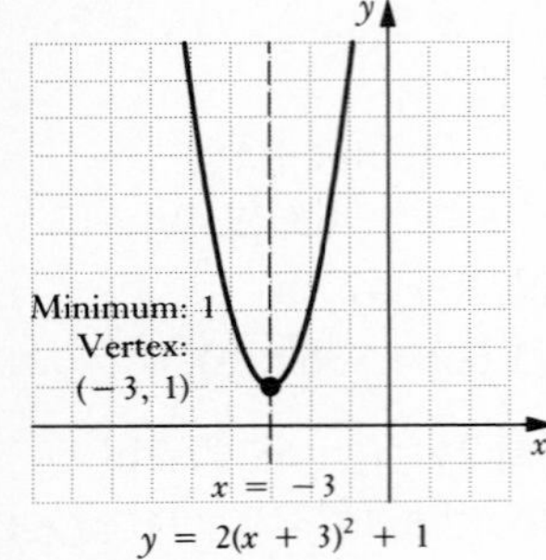

84. [4.7a]

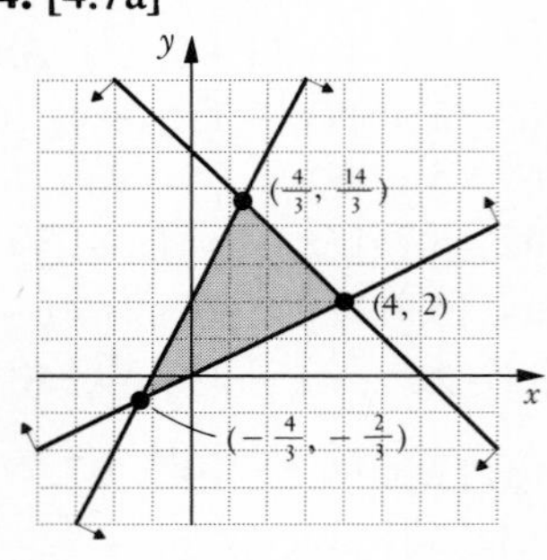

85. [9.1e] Center: (–3, –2); radius: $2\sqrt{5}$

86. [9.1d] $(x-4)^2 + (y+8)^2 = 28$ **87.** [10.3d] $2\log a + 3\log c - \log b$ **88.** [10.3d] $\log\left(\frac{x^3}{y^{1/2}z^2}\right)$

89. [10.2c] $a^x = 5$ **90.** [10.2b] $\log_x t = 3$ **91.** [9.5a, b] $x^2 + 2x + 2$; $2x^3 - 5x^2 + 14x - 35$; $4x^2 - 20x + 32$

92. [9.5b] $f(x) = \frac{1}{x}$; $g(x) = x^2 + 3$; answers may vary

93. [10.4a] –1.2545 **94.** [10.4a] 278,804.64

95. [10.4b] 2.5479 **96.** [10.4b] 0.0253 **97.** [9.6c] $f^{-1}(x) = \frac{x+7}{5}$ **98.** [9.6c] $f^{-1}(x) = \frac{4x+7}{1-x}$ **99.** [8.9a] $\{x \mid -5 < x < -4 \text{ or } x > 1\}$ **100.** [9.1a] $x = 4y^2 - 8y + 6$

101. [6.3a] All real numbers except 1 and –2

102. [10.5b] $\frac{1}{2}$, 5000 **103.** [8.6b] 35 mph

Final Examination, p. 635

1. [1.6a] $\frac{-45x^6}{y^4}$ **2.** [1.3a; 1.2d] 6.3 **3.** [1.5c] $-3y + 17$

4. [1.5c] 280 **5.** [1.1a] $\frac{7}{6}$ **6.** [5.1f] $3a^2 - 8ab - 15b^2$

7. [5.1d] $13x^3 - 7x^2 - 6x + 6$ **8.** [5.2b] $6a^2 + 7a - 5$

9. [5.2c] $9a^4 - 30a^2y + 25y^2$ **10.** [6.2c] $\frac{4}{x+2}$ **11.** [6.1d] $\frac{x-4}{3(x+2)}$ **12.** [6.1e] $\frac{(x+y)(x^2+xy+y^2)}{x^2+y^2}$ **13.** [6.6a] $x - a$

14. [5.5a] $(2x-3)^2$ **15.** [5.6a] $(3a-2)(9a^2+6a+4)$

16. [5.3b] $(a^2 - b)(a+3)$ **17.** [5.7a] $3(y^2+3)(5y^2-4)$

18. [9.4b] 20 **19.** [5.9b, c] $7x^3 + 9x^2 + 19x + 38$, R 72; or $7x^3 + 9x^2 + 19x + 38 + \frac{72}{x-2}$ **20.** [2.1d] $\frac{2}{3}$ **21.** [5.8a] 8, –6

22. [6.3a] $-\frac{6}{5}$, 4 **23.** [6.3a] No solution **24.** [4.2b] (1, –1) **25.** [4.4b] (2, –1, 1) **26.** [7.7b] 9 **27.** [8.4a] ±5, ±2 **28.** [9.3a] $(\sqrt{5}, \sqrt{3})$, $(\sqrt{5}, -\sqrt{3})$, $(-\sqrt{5}, \sqrt{3})$, $(-\sqrt{5}, -\sqrt{3})$

29. [10.5a] 1.2920 **30.** [10.5b] 1005 **31.** [10.2d] $\frac{1}{16}$

32. [10.5a] $-\frac{1}{2}$ **33.** [2.6e] $\{x \mid -2 \leq x \leq 3\}$ **34.** [8.1a] $\pm\sqrt{2}\,i$ **35.** [8.2a, c] $-2 \pm \sqrt{7}$ **36.** [2.6e] $\{y \mid y < -5 \text{ or } y > 2\}$ **37.** [9.3b] 5 ft by 12 ft **38.** [2.2a] $3.34 **39.** [2.2a] 65, 66, 67 **40.** [4.3a] $11\frac{3}{7}$ **41.** [4.3a] 24 L of *A*; 56 L of *B* **42.** [6.4c] 350 mph **43.** [6.4a] $8\frac{2}{5}$ min **44.** [6.7e] 20 **45.** [8.8c] 1250 ft²

46. [3.2a]

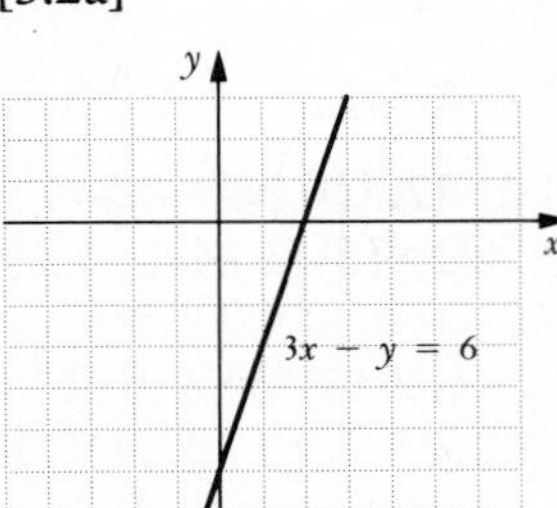

47. [9.2a]

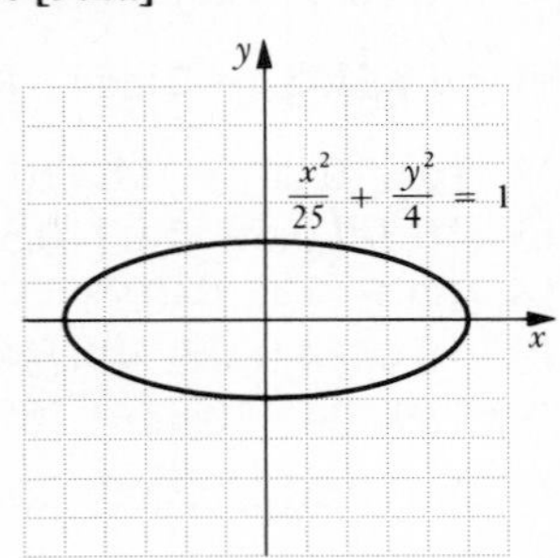

48. [10.2a]

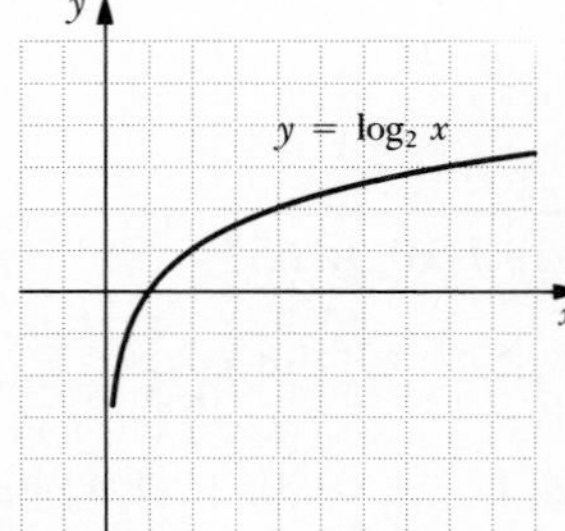

49. [3.6b]

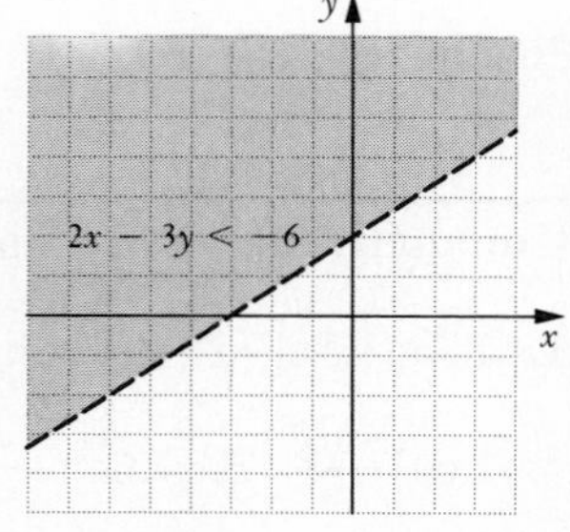

50. [8.7c]

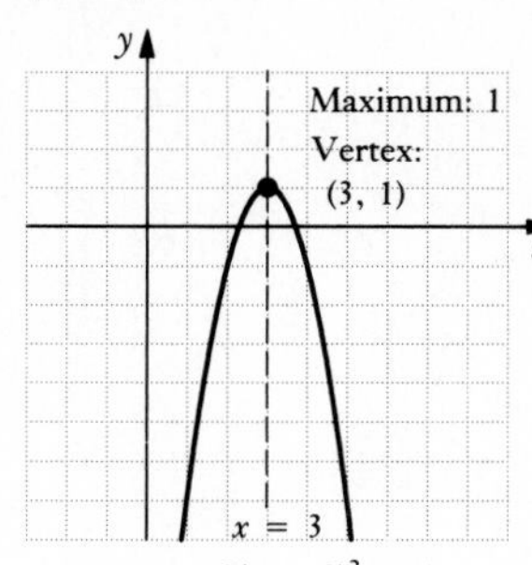

51. [4.7a]

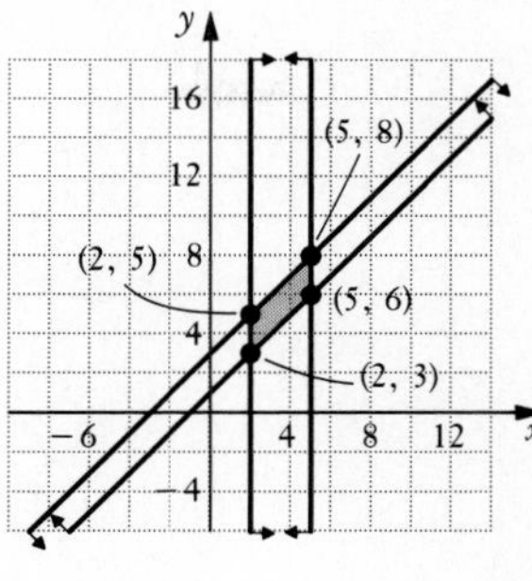

52. [2.3a] $r = \frac{A-P}{Pt}$ **53.** [6.5a] $R = \frac{Ir}{1-I}$ **54.** [3.4c] $y = -x + 3$ **55.** [4.6a] -2 **56.** [4.6b] -2 **57.** [1.6c] 6.764×10^{-12} **58.** [7.2b] $8x^2\sqrt{y}$ **59.** [7.3a] $\frac{\sqrt[3]{5xy}}{y}$ **60.** [7.5a] $\frac{1-2\sqrt{x}+x}{1-x}$ **61.** [7.6d] $\sqrt[6]{x+1}$ **62.** [7.9c] $26-13i$ **63.** [8.3b] $x^2-50=0$ **64.** [9.1e] Center: $(2, -3)$; radius: 6 **65.** [10.3d] $\log_a \frac{\sqrt[3]{x^2 \cdot z^5}}{\sqrt{y}}$ **66.** [10.2c] $a^5 = c$ **67.** [9.6c] $f^{-1}(x) = \sqrt[3]{x+8}$ **68.** [9.5a, b] x^2+x+3; x^4; x^3+3x^2; x^2+6x+9; x^2+3 **69.** [10.4a] 3.7541 **70.** [10.4a] 0.0000129 **71.** [10.4b] 8.6442 **72.** [10.4b] 0.0075 **73.** [10.6a] $P = 75{,}150e^{0.096t}$ **74.** [10.6a] 512,595 **75.** [9.4b] 10; 0; $a^2+2ab+b^2-3a-3b$ **76.** [6.3a] All real numbers except 0 and -12 **77.** [10.5b] 81 **78.** [6.7e] y gets divided by 8 **79.** [7.9e] $-\frac{7}{13}+\frac{2\sqrt{30}}{13}i$ **80.** [2.2a] 84 yr

Index

O

P